The Industrial Electronics Handbook
SECOND EDITION

POWER ELECTRONICS AND MOTOR DRIVES

The Industrial Electronics Handbook
SECOND EDITION

FUNDAMENTALS OF INDUSTRIAL ELECTRONICS

POWER ELECTRONICS AND MOTOR DRIVES

CONTROL AND MECHATRONICS

INDUSTRIAL COMMUNICATION SYSTEMS

INTELLIGENT SYSTEMS

The Electrical Engineering Handbook Series

Series Editor
Richard C. Dorf
University of California, Davis

Titles Included in the Series

The Industrial Electronics Handbook
SECOND EDITION

POWER ELECTRONICS AND MOTOR DRIVES

Edited by

Bogdan M. Wilamowski
J. David Irwin

CRC Press
Taylor & Francis Group
Boca Raton London New York

CRC Press is an imprint of the
Taylor & Francis Group, an **informa** business

CRC Press
Taylor & Francis Group
6000 Broken Sound Parkway NW, Suite 300
Boca Raton, FL 33487-2742

First issued in paperback 2017

© 2011 by Taylor and Francis Group, LLC
CRC Press is an imprint of Taylor & Francis Group, an Informa business

No claim to original U.S. Government works

ISBN-13: 978-1-4398-0285-4 (hbk)
ISBN-13: 978-1-138-07747-8 (pbk)

Library of Congress Cataloging-in-Publication Data

Power electronics and motor drives / editors, Bogdan M. Wikamowski and J. David Irwin.
    p. cm.
"A CRC title."
Includes bibliographical references and index.
ISBN 978-1-4398-0285-4 (alk. paper)
1. Power electronics. 2. Electric motors--Power supply. 3. Electric power supplies to apparatus--Design and construction. I. Wikamowski, Bogdan M. II. Irwin, J. David. III. Title.

TK7881.15.P665 2010
621.46--dc22

2010020061

**Visit the Taylor & Francis Web site at
http://www.taylorandfrancis.com**

**and the CRC Press Web site at
http://www.crcpress.com**

Contents

PART I   Semiconductor Devices

PART II   Electrical Machines

PART III Conversion

PART IV Motor Drives

PART V Power Electronic Applications

PART VI Power Systems

Preface

The field of industrial electronics covers a plethora of problems that must be solved in industrial practice. Electronic systems control many processes that begin with the control of relatively simple devices like electric motors, through more complicated devices such as robots, to the control of entire fabrication processes. An industrial electronics engineer deals with many physical phenomena as well as the sensors that are used to measure them. Thus, the knowledge required by this type of engineer is not only traditional electronics but also specialized electronics, for example, that required for high-power applications. The importance of electronic circuits extends well beyond their use as a final product in that they are also important building blocks in large systems, and thus the industrial electronics engineer must also possess a knowledge of the areas of control and mechatronics. Since most fabrication processes are relatively complex, there is an inherent requirement for the use of communication systems that not only link the various elements of the industrial process but are tailor-made for the specific industrial environment. Finally, the efficient control and supervision of factories requires the application of intelligent systems in a hierarchical structure to address the needs of all components employed in the production process. This need is accomplished through the use of intelligent systems such as neural networks, fuzzy systems, and evolutionary methods. The Industrial Electronics Handbook addresses all these issues and does so in five books outlined as follows:

1. *Fundamentals of Industrial Electronics*
2. *Power Electronics and Motor Drives*
3. *Control and Mechatronics*
4. *Industrial Communication Systems*
5. *Intelligent Systems*

The editors have gone to great lengths to ensure that this handbook is as current and up to date as possible. Thus, this book closely follows the current research and trends in applications that can be found in *IEEE Transactions on Industrial Electronics*. This journal is not only one of the largest engineering publications of its type in the world, but also one of the most respected. In all technical categories in which this journal is evaluated, its worldwide ranking is either number 1 or number 2. As a result, we believe that this handbook, which is written by the world's leading researchers in the field, presents the global trends in the ubiquitous area commonly known as industrial electronics.

Universities throughout the world typically provide an excellent education on the various aspects of electronics; however, they normally focus on traditional low-power electronics. In contrast, in the industrial environment there is a need for high-power electronics that is used to control electromechanical systems in addition to the low-power electronics typically employed for analog and digital systems. In order to address this need, Part I focuses on special high-power semiconductor devices. The most common interface between an electronic system and a moving mechanical system is an electric motor.

Motors come in many types and sizes and, therefore, in order to efficiently drive them, engineers must have a comprehensive understanding of the object to be controlled. Therefore, Part II not only describes the various types of electric motors and their principles of operation, but covers their limitations as well. Since electrical power can be delivered in either ac or dc, there is a need for high-efficiency devices that perform the necessary conversion between these different types of powers. These aspects are covered in Part III. It is believed that electric motors represent the soul of the industry and as such play a fundamental role in our daily lives. This preeminent position they occupy is a direct result of the fact that the majority of electric energy is consumed by electric motors. Therefore, it is important that these motors be efficient converters of electrical power into mechanical power, and the drive mechanisms be efficient as well. Part IV is dedicated to a presentation of very specialized electronic circuits for the efficient control of electric motors. In addition to its use in electric motors, power electronics has many other applications, such as lighting, renewable energy conversion, and automotive electronics, and these topics are covered in Part V. The last part, Part VI, deals with the power electronics that is employed in very-high-power electrical systems for the transmission of energy.

For MATLAB® and Simulink® product information, please contact

The MathWorks, Inc.
3 Apple Hill Drive
Natick, MA, 01760-2098 USA
Tel: 508-647-7000
Fax: 508-647-7001
E-mail: info@mathworks.com
Web: www.mathworks.com

Acknowledgments

The editors wish to express their heartfelt thanks to their wives Barbara Wilamowski and Edie Irwin for their help and support during the execution of this project.

Editorial Board

Editors

 **Bogdan M. Wilamowski** received his MS in computer engineering in 1966, his PhD in neural computing in 1970, and Dr. habil. in integrated circuit design in 1977. He received the title of full professor from the president of Poland in 1987. He was the director of the Institute of Electronics (1979–1981) and the chair of the solid state electronics department (1987–1989) at the Technical University of Gdansk, Poland. He was a professor at the University of Wyoming, Laramie, from 1989 to 2000. From 2000 to 2003, he served as an associate director at the Microelectronics Research and Telecommunication Institute, University of Idaho, Moscow, and as a professor in the electrical and computer engineering department and in the computer science department at the same university. Currently, he is the director of ANMSTC—Alabama Nano/Micro Science and Technology Center, Auburn, and an alumna professor in the electrical and computer engineering department at Auburn University, Alabama. Dr. Wilamowski was with the Communication Institute at Tohoku University, Japan (1968–1970), and spent one year at the Semiconductor Research Institute, Sendai, Japan, as a JSPS fellow (1975–1976). He was also a visiting scholar at Auburn University (1981–1982 and 1995–1996) and a visiting professor at the University of Arizona, Tucson (1982–1984). He is the author of 4 textbooks, more than 300 refereed publications, and has 27 patents. He was the principal professor for about 130 graduate students. His main areas of interest include semiconductor devices and sensors, mixed signal and analog signal processing, and computational intelligence.

Dr. Wilamowski was the vice president of the IEEE Computational Intelligence Society (2000–2004) and the president of the IEEE Industrial Electronics Society (2004–2005). He served as an associate editor of *IEEE Transactions on Neural Networks*, *IEEE Transactions on Education*, *IEEE Transactions on Industrial Electronics*, the *Journal of Intelligent and Fuzzy Systems*, the *Journal of Computing*, and the *International Journal of Circuit Systems and IES Newsletter*. He is currently serving as the editor in chief of *IEEE Transactions on Industrial Electronics*.

Professor Wilamowski is an IEEE fellow and an honorary member of the Hungarian Academy of Science. In 2008, he was awarded the Commander Cross of the Order of Merit of the Republic of Poland for outstanding service in the proliferation of international scientific collaborations and for achievements in the areas of microelectronics and computer science by the president of Poland.

J. David Irwin received his BEE from Auburn University, Alabama, in 1961, and his MS and PhD from the University of Tennessee, Knoxville, in 1962 and 1967, respectively.

In 1967, he joined Bell Telephone Laboratories, Inc., Holmdel, New Jersey, as a member of the technical staff and was made a supervisor in 1968. He then joined Auburn University in 1969 as an assistant professor of electrical engineering. He was made an associate professor in 1972, associate professor and head of department in 1973, and professor and head in 1976. He served as head of the Department of Electrical and Computer Engineering from 1973 to 2009. In 1993, he was named Earle C. Williams Eminent Scholar and Head. From 1982 to 1984, he was also head of the Department of Computer Science and Engineering. He is currently the Earle C. Williams Eminent Scholar in Electrical and Computer Engineering at Auburn.

Dr. Irwin has served the Institute of Electrical and Electronic Engineers, Inc. (IEEE) Computer Society as a member of the Education Committee and as education editor of *Computer*. He has served as chairman of the Southeastern Association of Electrical Engineering Department Heads and the National Association of Electrical Engineering Department Heads and is past president of both the IEEE Industrial Electronics Society and the IEEE Education Society. He is a life member of the IEEE Industrial Electronics Society AdCom and has served as a member of the Oceanic Engineering Society AdCom. He served for two years as editor of *IEEE Transactions on Industrial Electronics*. He has served on the Executive Committee of the Southeastern Center for Electrical Engineering Education, Inc., and was president of the organization in 1983–1984. He has served as an IEEE Adhoc Visitor for ABET Accreditation teams. He has also served as a member of the IEEE Educational Activities Board, and was the accreditation coordinator for IEEE in 1989. He has served as a member of numerous IEEE committees, including the Lamme Medal Award Committee, the Fellow Committee, the Nominations and Appointments Committee, and the Admission and Advancement Committee. He has served as a member of the board of directors of IEEE Press. He has also served as a member of the Secretary of the Army's Advisory Panel for ROTC Affairs, as a nominations chairman for the National Electrical Engineering Department Heads Association, and as a member of the IEEE Education Society's McGraw-Hill/Jacob Millman Award Committee. He has also served as chair of the IEEE Undergraduate and Graduate Teaching Award Committee. He is a member of the board of governors and past president of Eta Kappa Nu, the ECE Honor Society. He has been and continues to be involved in the management of several international conferences sponsored by the IEEE Industrial Electronics Society, and served as general cochair for IECON'05.

Dr. Irwin is the author and coauthor of numerous publications, papers, patent applications, and presentations, including *Basic Engineering Circuit Analysis*, 9th edition, published by John Wiley & Sons, which is one among his 16 textbooks. His textbooks, which span a wide spectrum of engineering subjects, have been published by Macmillan Publishing Company, Prentice Hall Book Company, John Wiley & Sons Book Company, and IEEE Press. He is also the editor in chief of a large handbook published by CRC Press, and is the series editor for Industrial Electronics Handbook for CRC Press.

Dr. Irwin is a fellow of the American Association for the Advancement of Science, the American Society for Engineering Education, and the Institute of Electrical and Electronic Engineers. He received an IEEE Centennial Medal in 1984, and was awarded the Bliss Medal by the Society of American Military Engineers in 1985. He received the IEEE Industrial Electronics Society's Anthony J. Hornfeck Outstanding Service Award in 1986, and was named IEEE Region III (U.S. Southeastern Region) Outstanding Engineering Educator in 1989. In 1991, he received a Meritorious Service Citation from the IEEE Educational Activities Board, the 1991 Eugene Mittelmann Achievement Award from the IEEE Industrial Electronics Society, and the 1991 Achievement Award from the IEEE Education Society. In 1992, he was named a Distinguished Auburn Engineer. In 1993, he received the IEEE Education Society's McGraw-Hill/Jacob Millman Award, and in 1998 he was the recipient of the

IEEE Undergraduate Teaching Award. In 2000, he received an IEEE Third Millennium Medal and the IEEE Richard M. Emberson Award. In 2001, he received the American Society for Engineering Education's (ASEE) ECE Distinguished Educator Award. Dr. Irwin was made an honorary professor, Institute for Semiconductors, Chinese Academy of Science, Beijing, China, in 2004. In 2005, he received the IEEE Education Society's Meritorious Service Award, and in 2006, he received the IEEE Educational Activities Board Vice President's Recognition Award. He received the Diplome of Honor from the University of Patras, Greece, in 2007, and in 2008 he was awarded the IEEE IES Technical Committee on Factory Automation's Lifetime Achievement Award. In 2010, he was awarded the electrical and computer engineering department head's Robert M. Janowiak Outstanding Leadership and Service Award. In addition, he is a member of the following honor societies: Sigma Xi, Phi Kappa Phi, Tau Beta Pi, Eta Kappa Nu, Pi Mu Epsilon, and Omicron Delta Kappa.

Contributors

Ayman A. Alabduljabbar
King Abdul Aziz City for Science
 and Technology
Riyadh, Saudi Arabia

Kamal Al-Haddad
École de Technologie Supérieure
Montreal, Quebec, Canada

Francisco Javier Azcondo
Electronics Technology System and Automation
 Engineering Department
School of Industrial and Telecommunications
 Engineering
University of Cantabria
Santander, Spain

Pavol Bauer
Department of Electrical Sustainable Energy
Delft University of Technology
Delft, the Netherlands

Nicola Bianchi
Department of Electrical Engineering
Universita of Padova
Padova, Italy

Elżbieta Bogalecka
Faculty of Electrical and Control Engineering
Gdańsk University of Technology
Gdańsk, Poland

Aldo Boglietti
Dipartimento di Ingegneria Elettrica
Politecnico di Torino
Torino, Italy

Jean-François Brudny
Faculté des Sciences Appliquées
Laboratoire Systèmes Electrotechniques
 et Environnement
Univ Lille Nord de France
UArtois, Béthune, France

Jian Cao
Electrical and Computer Engineering Department
Illinois Institute of Technology
Chicago, Illinois

Bertrand Cassoret
Laboratoire Systèmes Electrotechniques
 et Environnement
Université d'Artois
Bethune, France

Andrea Cavagnino
Dipartimento di Ingegneria Elettrica
Politecnico di Torino
Torino, Italy

Henry Chung
Department of Electronic Engineering
City University of Hong Kong
Kowloon, Hong Kong

Jorge Duarte
Electromechanics and Power Electronics Group
Eindhoven University of Technology
Eindhoven, Netherlands

Ali Emadi
Electrical and Computer Engineering Department
Illinois Institute of Technology
Chicago, Illinois

Babak Fahimi
Department of Electrical Engineering
University of Texas at Arlington
Arlington, Texas

Leopoldo Garcia Franquelo
Electronics Engineering Department
University of Sevilla
Sevilla, Spain

K. Gopakumar
Centre for Electronics Design and Technology
Indian Institute of Science
Bangalore, India

Charles A. Gross
Department of Electrical and Computer
    Engineering
Auburn University
Auburn, Alabama

Josep M. Guerrero
Department of Automatic Control Systems
    and Computer Engineering
Technical University Catalonia
Barcelona, Spain

Shu-Yuen (Ron) Hui
Department of Electronic Engineering
City University of Hong Kong
Kowloon, Hong Kong

and

Department of Electrical and Electronic
    Engineering
Imperial College London
London, United Kingdom

Grzegorz Iwański
Institute of Control and Industrial Electronics
Warsaw University of Technology
Warsaw, Poland

Marek Jasiński
Institute of Control and Industrial Electronics
Warsaw University of Technology
Warsaw, Poland

Hadi Y. Kanaan
Department of Electrical Engineering
St. Joseph University
Mar Roukoz, Lebanon

Marian P. Kazmierkowski
Institute of Control and Industrial Electronics
Warsaw University of Technology
Warsaw, Poland

Włodzimierz Koczara
Institute of Control and Industrial Electronics
Warsaw University of Technology
Warsaw, Poland

Samir Kouro
Department of Electrical and Computer
    Engineering
Ryerson University
Toronto, Ontario, Canada

Mahesh Krishnamurthy
Electrical and Computer Engineering Department
Illinois Institute of Technology
Chicago, Illinois

Zbigniew Krzemiński
Faculty of Electrical and Control Engineering
Gdańsk University of Technology
Gdańsk, Poland

Friederich Kupzog
Institute of Computer Technology
Vienna University of Technology
Vienna, Austria

Mario Lazzari
Dipartimento di Ingegneria Elettrica
Politecnico di Torino
Torino, Italy

Jean-Philippe Lecointe
Laboratoire Systèmes Electrotechniques
    et Environnement
Université d'Artois
Bethune, France

José I. León
Electronics Engineering Department
University of Sevilla
Sevilla, Spain

Emil Levi
School of Engineering
Liverpool John Moores University
Liverpool, United Kingdom

Xin Li
P. D. Ziogas Power Electronics Laboratory
Department of Electrical and Computer
   Engineering
Concordia University
Montreal, Quebec, Canada

Elena Lomonowa
Electromechanics and Power Electronics Group
Eindhoven University of Technology
Eindhoven, Netherlands

Leo Lorenz
Infineon Technologies
Neubiberg, Germany

Mariusz Malinowski
Institute of Control and Industrial Electronics
Warsaw University of Technology
Warsaw, Poland

Anton Mauder
Infineon Technologies
Neubiberg, Germany

Jovica V. Milanović
School of Electrical and Electronic Engineering
The University of Manchester
Manchester, United Kingdom

Artur Moradewicz
Electrotechnical Institute
Warsaw, Poland

István Nagy
Department of Automation and Applied
   Informatics
Budapest University of Technology
   and Economics
Budapest, Hungary

Franz Josef Niedernostheide
Infineon Technologies
Neubiberg, Germany

Teresa Orłowska-Kowalska
Institute of Electrical Machines, Drives
   and Measurements
Wroclaw University of Technology
Wroclaw, Poland

Peter Palensky
Austrian Institute of Technology
Vienna, Austria

Igor Papič
Faculty of Electrical Engineering
University of Ljubljana
Ljubljana, Slovenia

Giovanni Petrone
Dipartimento di Ingegneria dell'Informazione
   ed Ingegneria Elettrica
Università di Salerno
Fisciano, Italy

M.A. Rahman
Faculty of Engineering and Applied Science
Memorial University of Newfoundland
St. John's, Newfoundland and Labrador, Canada

Salem Rahmani
High Institute of Medical Technologies
École de Technologie Supérieure
Montreal, Quebec, Canada

José Rodríguez
Electronics Engineering Department
Universidad Tecnica Federico Santa Maria
Valparaiso, Chile

Raphael Romary
Faculté des Sciences Appliquées
Laboratoire Systèmes Electrotechniques
   et Environnement
Univ Lille Nord de France
UArtois, Béthune, France

Roland Rupp
Infineon Technologies
Neubiberg, Germany

Hans Joachim Schulze
Infineon Technologies
Neubiberg, Germany

Christoph Sonntag
Electromechanics and Power Electronics Group
Eindhoven University of Technology
Eindhoven, Nertherlands

Giovanni Spagnuolo
Dipartimento di Ingegneria dell'Informazione
    ed Ingegneria Elettrica
Università di Salerno
Fisciano, Italy

Zoltán Sütö
Department of Automation and Applied
    Informatics
Budapest University of Technology
    and Economics
Budapest, Hungary

Krzysztof Szabat
Institute of Electrical Machines, Drives
    and Measurements
Wroclaw University of Technology
Wroclaw, Poland

Juan C. Vasquez
Department of Automatic Control Systems
    and Computer Engineering
Technical University Catalonia
Barcelona, Spain

Patrick Wheeler
Department of Electrical and Electronic
    Engineering
University of Nottingham
Nottingham, United Kingdom

Sheldon S. Williamson
P. D. Ziogas Power Electronics Laboratory
Department of Electrical and Computer
    Engineering
Concordia University
Montreal, Quebec, Canada

Bin Wu
Department of Electrical and Computer
    Engineering
Ryerson University
Toronto, Ontario, Canada

Yan Zhang
ABB Corporate Research
Baden, Switzerland

Semiconductor Devices

I

1

Electronic Devices for Power Switching: The Enabling Technology for Power Electronic System Development

Leo Lorenz
Infineon Technologies

Hans Joachim Schulze
Infineon Technologies

Franz Josef Niedernostheide
Infineon Technologies

Anton Mauder
Infineon Technologies

Roland Rupp
Infineon Technologies

1.1 Introduction

Power semiconductor switches are primarily used to control the flow of electrical energy between the energy source and the load, and to do so with great precision, with extremely fast control times, and with low dissipated power. The application of IC technologies on state-of-the-art power semiconductor devices has resulted in advanced components with low power dissipation, simple drive characteristics, good control dynamics, and switching power extending into the megawatt range.

Power semiconductor devices and control ICs are the key elements of power electronic systems—despite the fact that their costs are minimal in many applications, relative to the overall system costs. Improving

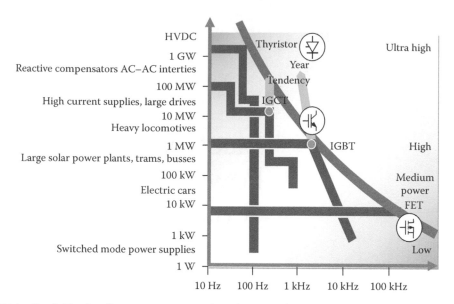

FIGURE 1.1 Key fields of application versions switching frequency for power semiconductor devices.

their characteristics along with an increasing functionality reduces the system cost and opens opportunities for new fields of applications. New system trends are moving toward high switching frequency, reducing or eliminating bulky ferrites and electrolytes, as well as soft switching topologies for higher efficiency and low harmonies.

In electrical energy transfer, electronic devices are generally required to operate in "switch mode." This means they should have ideal switch-like characteristics: they appear like a short-circuit passing current with minimal voltage drop across it in the on state; in the other side, they block the flow of current by supporting full supply voltage across it appearing like an open circuit in the off state. They operate in a different mode from power amplifying devices, which allow power transfer according to a linear relationship with an input signal, such as audio amplification. In switch mode operation, an electronic control signal is applied to turn the switch ON, and removed to turn the device OFF. For present devices, the control signal is typically in the 5–12 V range while the power supply voltage can be in the 20 V–8 kV range.

Solid state switch mode devices have been used for controlling power transfer for over 50 years. Demands for the rational use of energy, miniaturization of electronic systems, and electronic power management systems have been the driving force behind the revolutionary development of power semiconductor devices over the last five decades [1].

As shown in Figure 1.1, the power semiconductor switches cover all applications in the power range from 1 W needed for charging the battery of a mobile phone, up to the GW range needed for energy transmission lines (HVDC lines). As pointed out in this diagram, the bipolar devices (e.g., thyristor, integrated gate-commutated thyristor [IGCT]) are a key technology for ultrahigh power systems while the MOS-controlled devices (e.g., insulated gate bipolar transistor [IGBT], power MOSFET including SMART power systems) are the driving components for medium and low power electronic conversion systems. In the top power end, the switching frequency is below several 100 Hz, the medium power is dominated in the range of 10 kHz, but the system development for lower power is driven by several 100 kHz.

Advances in power electronic systems over the last three to four decades have been marked by five major inventions. Light-triggered thyristors and IGCTs in the top-end power range, IGBTs in the mid- and high-end power range, power MOSFET in the low-end power range, and SMART power systems for monolithic system integration, are mainly applied in automotive power. The bipolar transistor and the gate turn-off (GTO) thyristor do not play a significant role in present development. For this reason, these device types are not focused on in this chapter.

1.2 Brief History and Basics of Key Power Semiconductor Devices

1.2.1 Bipolar Device: Thyristor

The first device developed 40 years ago, with many significant development steps, was the Si thyristor, a four-layer *p-n-p-n* structure allowing for very low resistance when turned on, and the ability to block voltage of up to 10 kV in the off state. It has a positive feedback mechanism for the buildup of current, once one of the *p-n* junctions in the structure is turned on. This is usually achieved by injecting a control current. The major drawback with the thyristor is that it cannot be turned off by applying a control signal. The same positive feedback mechanism that governs the current flow in a thyristor can only be stopped through "natural commutation," that is, when the conditions in the circuit to which the thyristor is connected lead to current reversal through the device. Controlled turn-off mechanisms based on current transfer to ancillary circuits for short periods have been developed for thyristors. However, they were unsuited for rapid ON/OFF switch mode operations. Nevertheless, they were widely used in low-frequency switching applications due to their excellent on-state characteristics. They remain in use as rectifiers and inverters used in HVDC power transmission and as solid state control elements in static VAR compensators used for power factor optimization in the power network. The required voltage rating, up to 1000 kV for HVDC transmission systems, is obtained through serial connection of individual devices rated at 8–10 kV. Similarly, the current rating is obtained by parallel connection of device stacks, with each device typically rated for up to 6 kA [2,3].

1.2.2 Unipolar Device: Power MOSFET

A kind of revolution in switch mode control of power transfer was brought about by the advent of fully voltage controllable solid state devices capable of sustaining high off-scale voltages in the mid-1970s. This was the power MOSFET.

Current flow is vertical from drain, through an inversion channel placed on the top surface at right angles to the main current flow path, and into the source. The ability to control the current flow by application of a gate voltage to turn the device on and removal of the gate voltage to turn the device OFF are its main control features. This control principle of applying a gate voltage to a metal-oxide semiconductor (MOS) structure to create a conducting channel was, of course, well established for the low voltage MOSFET, and reliable gate fabrication technology was developed for integrated circuits by the mid-1970s. The advance of the power MOSFET was the double-diffused channel structure, with the channel being created in a diffused-body region rather than in the substrate, which allowed the device to have a *p-n* junction blocking region to support a large voltage in the off state. A power switch, however, with high current conduction in the on state is required. In the power MOSFET, this was achieved by replicating millions of cells like those shown in Figure 1.2.

Since the power MOSFET is a unipolar device and its current is carried only by charge carriers of one polarity (electrons for an *n*-channel device and holes for a *p*-channel device), it can be switched very fast (like resistors). This makes the power MOSFET ideally suited for high frequency switching.

Its major limitation, however, also arises from the unipolar nature of current flow, especially for high length of the lowly doped drift region, that also has to be increased together with a reduction in the doping concentration. Both these changes in design parameters tend to increase the on-state resistance of a power MOSFET switch according to the relationship $R_{on} \sim V_{max}^{2.5}$. However, if the on-state voltage is high, the static loss in the switch will be unacceptable. Because of this reason, the DMOSFET device shown in Figure 1.2 is not practical for use as a power switch at voltage ratings in excess of 800. It can, however, be switched at frequencies as high as 5 MHz.

1.2.3 MOS-Controlled Bipolar Mode Power Device IGBT

The insulated gate bipolar transistor (IGBT) has a MOS gate control structure identical to that of a power MOSFET. The only difference is that the n^+ drain contact of the power MOSFET is replaced by a p^+ minority carrier injector in the IGBT (Figure 1.3).

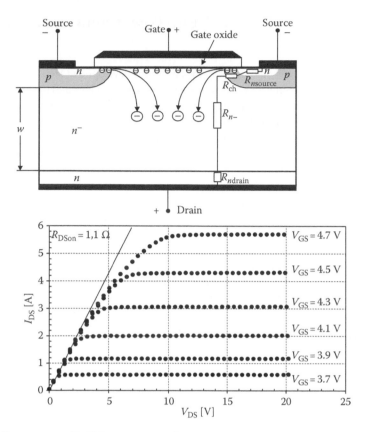

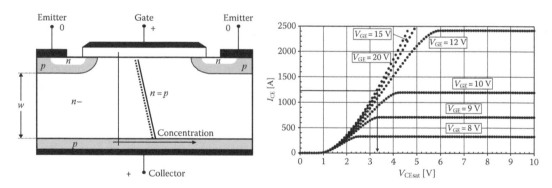

FIGURE 1.2   Cell structure and *I–V* characteristics of a power MOSFET.

FIGURE 1.3   Cell structure and *I–V* characteristics of IGBT.

Using this simple and elegant adaptation, a whole new class of hybrid MOS-bipolar solid state devices, being particularly aimed at power switching was demonstrated in the early 1980s. When the MOS channel is turned on, the *p-n* diode at the high-voltage terminal (anode) is turned on, and minority carriers (holes) are injected into the *n*-drift region. This is the classical conductivity modulation effect that can be achieved in a semiconductor by having charge carriers of two polarities carrying the current flow. Hence the on-state resistance in the IGBT drift region is much lower than that in a MOSFET. In principle, the IGBT has all the advantages afforded by voltage control, inherent in a MOSFET, together with the low on-state voltage enabled by bipolar conduction. However, the large stored charge in the *n*-drift also severely reduces its high frequency and hard switching capability.

Over the last two decades, major efforts have been directed at optimizing the trade-off between low on-resistance and high turn-off losses in the IGBT. There efforts have led to the point where the IGBT is the device of choice for all power control applications at voltages from 600 up to 6500 V.

1.2.4 Key Power Device Development and Their Major Characteristics

Originating from these basic structures, huge development steps have advanced the power semiconductor switches to the enabling technology for all energy efficiency power electronic system developments. Based on these principles, many new device families have become available, for example, light-triggered thyristor (LTT), power diodes, non-punch-through IGBTs (NPT-IGBTs), super junction power MOSFET (SJ-MOSFET), SiC devices (silicon carbide–based devices), and SMART power systems. In the following sections, these device concepts will be shown and their characteristics will be discussed.

1.3 Bipolar Devices

1.3.1 Thyristor and LTT

The thyristor is a four-layer p^+-n-p-n^+ device. Since three p-n junctions are connected in series, the thyristor is able to block a negative (reverse blocking mode) as well as a positive voltage (forward blocking mode) applied between the anode (p^+-layer) and the cathode (n^+-layer). For positive anode-to-cathode voltages, switching of voltages up to more than 10 kV and currents up to several kA is possible by feeding a short current pulse in the inner p-layer. Such a trigger current can be provided either by a third electrical gate terminal or by using a light pulse (Figure 1.4). In the latter case, the light impinging into the device creates electron-hole pairs that are separated in the space–charge region of the reverse-biased inner p-n junction. The hole current flowing toward the cathode layer is used to trigger the thyristor. Utilization of light-triggered thyristors is of particular benefit in applications with thyristors connected in series, since optoelectronic coupling and galvanic isolation is an inherent feature of light-triggered thyristor systems [4].

In order to minimize the turn-on current that is required to trigger the thyristor, several auxiliary thyristors, the so-called amplifying gate structures, are usually connected between the central trigger area (gate terminal or light-sensitive area) and the main cathode area of the thyristor. Figure 1.4 shows two and four of such amplifying gate (AG) structures for the electrically-triggered and the light-triggered thyristor, respectively. The trigger sensitivity of each amplifying gate can be adjusted easily, for example, by the width of its n^+-emitter and/or the sheet resistivity of the p-base below the same. As a rule of thumb, the minimum trigger current of two successive amplifying gates differs by a factor between 3 and 10.

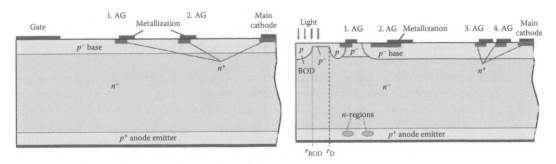

FIGURE 1.4 Electrical-triggered (left) and light-triggered thyristor (right).

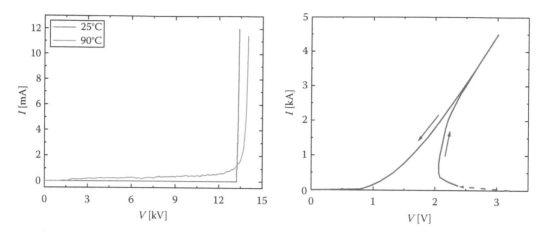

FIGURE 1.5   Forward blocking current voltage characteristic of a 13 kV thyristor (left). Typical on-state characteristic of a high-voltage thyristor (right). (Data from Niedernostheide, F.-J. et al., 13-kV rectifiers: Studies on diodes and asymmetric thyristors, *Proceedings of the ISPSD'03*, Cambridge, U.K., pp. 122–125, 2003.)

   Typical forward blocking and on-state current–voltage characteristics of high-power thyristors are depicted in Figure 1.5. The hysteresis in the on-state characteristic results from the fact that the current has to distribute across the extended main cathode area after turn-on. In the 5 in. thyristor considered here, the current distributes over the entire cathode area, not until the current exceeds approximately 3 kA. Current spreading during the turn-on process and the final on-state voltage V_T can be controlled by several measures: For large-area thyristors, the outermost AG is typically designed in such a way that the main cathode area is triggered along a preferably extended section, resulting in an AG structure that is distributed over the thyristor area (Figure 1.6). In addition, current spreading is influenced by the emitter shorts and the charge-carrier lifetime in the thyristor. Emitter shorts are local resistive connections distributed over the main cathode area and provide a bypass of the emitter junctions. Such emitter shorts are necessary to reduce the dV/dt sensitivity of the main cathode. However, extended emitter shorts distributed with a high density over the active area reduce the current-spreading velocity and lead to higher on-state voltages. These trade-off relationships have to be carefully accounted for when designing the emitter shorts. The same is valid for decreasing the charge-carrier lifetime, improving the dV/dt capability, and reducing the circuit-commutated turn-off period t_q (the minimum time delay that is necessary, after a thyristor having been switched off by forced commutation, before the thyristor can withstand a positively biased voltage pulse), so as to decrease the current-spreading velocity and increase the on-state voltage V_T. The charge-carrier lifetime can be adjusted very accurately by creating recombination centers. This can be achieved either by diffusion of heavy metals such as gold or platinum, or by creating irradiation defects by means of electron or light-ion irradiation. Since gold-related trap centers usually cause high leakage currents, in particular at elevated operating temperatures, and the recombination rate of platinum-related trap centers decreases significantly under low-injection conditions, the most used technique recently to adjust the charge-carrier lifetime is based on irradiation-induced defects.

   Finding the optimum charge-carrier lifetime is also of particular importance for optimizing the turn-off behavior (Figure 1.7). Reducing the reverse-recovery charge Q_{rr} and, consequently, the turn-off losses E_{off} is essential, since a standard thyristor cannot be actively turned off by a control signal. Instead, turn-off is usually achieved by commutating the anode-to-cathode voltage. As soon as the thyristor has reached the applied reverse voltage, the remaining charge carriers can disappear only by recombination. Thus, to accelerate the turn-off process a short charge-carrier lifetime is advantageous. Figure 1.8 illustrates typical t_q–V_T and Q_{rr}–V_T trade-off relationships.

FIGURE 1.6 Top view on a light-triggered thyristor, the line pattern in the blank covering the main cathode area represents the shape of the distributed outermost AG.

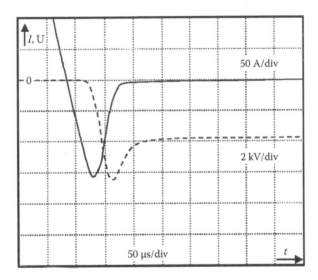

FIGURE 1.7 Typical turn-off characteristics of a high-voltage thyristor switched off by forced commutation.

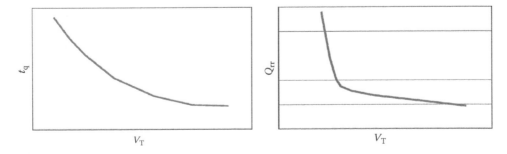

FIGURE 1.8 Schematic $t_q - V_T$ trade-off relationship (left) and $Q_{rr} - V_T$ trade-off relationship (right) of a high-voltage thyristor.

Thyristors with high blocking voltages can be used not only for high-voltage direct-current (HVDC) transmission applications requiring a total blocking voltage capability up to 1 MV, but also in miscellaneous pulse-power applications, such as accelerators, cable analysis systems, crowbar applications (e.g., klystron protection), discharge of capacitive and inductive storages (e.g., series-capacitors protection), electromagnetic forming, spare of ignitrons, sterilization of foods and medical instruments, or switch gears. Today's commercial thyristors have maximum current ratings up to several kiloamperes, surge current capabilities of a few tens of kiloamperes, blocking voltage capability higher than 8 kV, and device areas up to 6 in.

For many applications, thyristors require protection against a variety of failure modes. For example, the thyristor must be protected against destruction caused by overvoltage pulses or voltages with a voltage rise rate exceeding the maximum rated rise rate. In addition, for HVDC transmission applications, it is necessary to avoid premature device turn-on when a forward voltage pulse is applied during the circuit-commutated turn-off period, because a thyristor is not able to withstand a forward voltage pulse with the rated blocking voltage or the rated maximum dV/dt value until the charge-carrier plasma is completely removed from the n-base. Such protection requirements can be achieved by the implementation of extensive monitoring and electrical protection circuitry. However, recent developments in thyristor switches are aimed at reducing external electrical protection circuits by integrating the corresponding protection functions directly into the thyristor pellet [5,6] as given in the following:

- Integration of an overvoltage protection function can be achieved by implementing a break over diode (BOD) in the light-sensitive area of a light-triggered thyristor (Figure 1.4). The voltage level V_{BOD}, at which the overvoltage protection function is activated, can be adjusted by the distance between the central p region with radius r_{BOD} and the concentric p ring with an inner radius r_p. For large distances, the breakdown voltage is essentially determined by the curvature of the central p region. A reduction of the distance results in a reduction of the electric-field strength at the center of the BOD for a given voltage. For sufficiently small distances, the breakdown voltage approaches the value of the uniform p-n^- junction [7].

- By designing the innermost AG such that its dV/dt sensitivity is higher than that of the other AGs and the main cathode, a safe turn-on of the device starting from the innermost AG is ensured, when the voltage rises at a rate higher than the threshold dV/dt rate of the innermost AG. By this means, a dV/dt protection function is integrated into the device in addition to the overvoltage protection function. Apart from the geometrical dimensions of the AGs, the sheet resistivity of the p-base is an important parameter to adjust the dV/dt sensitivity of the AGs.

- In order to protect the thyristor from being destroyed during the circuit-commuted turn-off period, the thyristor should be turned on in a controlled way by the AG region when the thyristor is loaded by a forward voltage pulse during the circuit-commutated turn-off time. However, since the AGs usually turn off earlier than the main cathode area, there are typically fewer free charge carriers below the AG structure compared to the main cathode area. Two measures can be used to overcome this problem: First, the radial distribution of the charge-carrier lifetime should be modified such that it is reduced in the main cathode area of the device compared to the AG region. Secondly, phosphorus islands implemented into the p-emitter in the inner AG structure (Figure 1.4, right) form the emitter of local n-p-n transistors when a reverse voltage is applied to the device and therefore provide further support for re-triggering in the AG region when a forward voltage pulse is applied to the device. The carrier injection of these islands can be controlled by their sizes and their doping profile.

Integrating these three protection functions provides a completely self-protected, directly light-triggered thyristor, ensuring a reliable operation with a drastically reduced monitoring and protection circuitry.

1.3.2 Gate Turn-Off Thyristor and Integrated Gate-Commutated Thyristor

1.3.2.1 The GTO Thyristor

A gate turn-off (GTO) thyristor is a special type of thyristor. GTO thyristors, as opposed to normal thyristors, are fully controllable switches that can be turned on and off by their third lead, the gate lead. Thyristors can only be turned off by reducing the on-state current below the holding current. Therefore, thyristors are not suitable for applications with DC power sources. The GTO thyristor can be turned on by a gate signal, and can also be turned off by a gate signal of negative polarity.

Turn-on is accomplished by a positive voltage pulse between the gate and cathode terminals. The typical gate voltage is in the range of 15 V. The turn-on phenomenon in GTO thyristors is, however, not as reliable as in a thyristor and a small positive gate current must be maintained even after turn-on to improve reliability. Amplifying gate structures, which are very helpful for the turn-on of the thyristor, are not implemented in GTO thyristors.

Turn-off is induced by a negative voltage pulse between the gate and cathode terminals. Some of the forward current (about one-third to one-fifth) is used to induce a cathode-gate voltage, which in turn results in a decrease of the forward current, and the GTO thyristor will switch off. Usually, the carrier lifetime in the base region has to be reduced by a well-defined creation of recombination centers to shorten the tail phase and to keep the turn-off losses low. These recombination centers can be generated by electron or helium irradiation, resulting in crystal defects effecting deep levels in the band gap.

The cross section and the top view of a GTO thyristor are illustrated in Figure 1.9. There are many small emitter mesa structures distributed along the device, which are identical in width and length, to guarantee a relatively homogeneous flow of the turn-off current. The homogeneity of the current flow during the turn-off period is a very critical point because such inhomogeneities result in current filamentation [9] and with it in dynamic avalanche. The resulting local self-heating effects can be so strong that the device burns out. Therefore, the maximum current, which can be turned off without destroying the device, can be significantly reduced by inhomogeneities of the turn-off current induced, for example, by an inhomogeneous distribution of the carrier lifetime in the n-base, of the p-base resistance, of the penetration depth of the n-emitter/p-base junction, or of the contact resistance between metallization and semiconductor. Also, mechanical stress effects can play an important role. Therefore, it is extremely important to guarantee clean processing [10] and homogeneous doping processes.

To keep the electrical field strength induced by dynamic avalanche as low as possible, the transistor gain α_{pnp} has to be chosen very carefully. For that purpose, the hole injection by the p-emitter has to be limited, for example, by a vertically inhomogeneous carrier lifetime reduction with a high recombination rate below the p-emitter or by a limitation of the emitter efficiency by a relatively small doping concentration of the p-emitter.

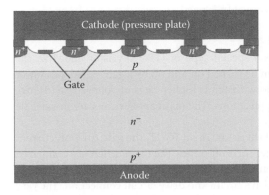

FIGURE 1.9 Cross-section (left) and top view of a GTO thyristor with mesa cathode structure (right).

GTO thyristors suffer from long switch-off times, whereby after the forward current falls, there is a long tail time where residual current continues to flow until all remaining charge from the device is taken away. This long current tail restricts the maximum switching frequency to approximately 1 kHz. It may be noted, however, that the turn-off time of comparable symmetrical controlled rectifiers (SCRs) is about 10 times that of a GTO thyristor. Thus, switching frequency of GTO thyristors is much better than that of SCRs. The main applications of such GTO thyristors are in variable speed motor drives, high-power inverters, and traction.

GTO thyristors are available either with or without reverse blocking capability. Reverse blocking capability enhances the forward voltage drop and the dynamic losses because of the need to have a thick, low doped base region. GTO thyristors capable of blocking reverse voltage are known as symmetrical GTO thyristors. Usually, the reverse blocking voltage rating and forward blocking voltage rating are about the same. The typical application for symmetrical GTO thyristors is in current source inverters.

GTO thyristors incapable of blocking reverse voltage are known as asymmetrical GTO thyristors. They typically have a reverse breakdown rating in tens of volts or less. By the use of the anode shorts, the forward blocking capability of the device is enhanced due to the reduced transistor current gain α_{pnp}, especially for high temperature operation. Asymmetrical GTO thyristors are used, where either a reverse conducting diode is applied in parallel (for example, in voltage source inverters), or where reverse voltage would never occur (for example, in switching power supplies or DC traction choppers). Asymmetrical GTO thyristors can be fabricated with a reverse-conducting diode in the same package. These are known as reverse conducting (RC) GTO thyristors.

Unlike the IGBT, the GTO thyristor requires external devices to shape the turn-on and turn-off currents to prevent device destruction. During turn-on, the device has a maximum dI/dt rating limiting the rise of current. This is to allow the entire bulk of the device to reach turn-on before full current is reached. If this rating is exceeded, the area of the device nearest the gate contacts will overheat and melt from overcurrent. The rate of dI/dt is usually controlled by adding a saturable reactor. Reset of the saturable reactor usually places a minimum off-time requirement on GTO thyristor-based circuits.

During turn-off, the forward voltage of the device must be limited until the current becomes small. The limit is usually around 20% of the forward blocking voltage rating. If the voltage rises too fast during turn-off, not all of the device will turn off, and current filamentation occurs so that the GTO thyristor will be destroyed due to self-heating effects induced by the high voltage and current focused on a small portion of the device. Substantial snubber circuits have to be added around the device to limit the rise of voltage at turn-off. Resetting the snubber circuit usually places a minimum on-time requirement on GTO thyristor based circuits.

The minimum on and off time is handled in DC motor chopper circuits by using a variable switching frequency at the lowest and highest duty cycle. This is observable in traction applications, where the frequency will ramp up as the motor starts, then the frequency stays constant over most of the speed ranges, and finally the frequency drops back down to zero at full speed.

1.3.2.2 The IGCT

The integrated gate-commutated thyristor (IGCT) is a special type of GTO thyristor and, like the GTO thyristor, a fully controllable power switch. It can be turned on and off by a gate signal, has lower conduction losses as compared to GTO thyristors, and withstands higher rates of voltage rise (dV/dt), such that no snubber circuits are required for most applications. The main applications are in variable frequency inverters, drives, and traction.

The structure of an IGCT is very similar to a GTO thyristor. In an IGCT, the gate turn-off current is greater than the anode current. This results in shorter turn-off times. The main difference compared with a GTO thyristor is a reduction in cell size, combined with a much more substantial gate connection, resulting in a much lower inductance in the gate drive circuit and drive circuit connection. The very high gate currents and the fast dI/dt rise of the gate current means that regular wires cannot be used to connect the gate drive to the IGCT. The drive circuit printed circuit board (PCB) is integrated into the package of the device. The drive circuit surrounds the device and a large circular conductor attaching to the edge of the IGCT die is used. The large contact area and short distance reduces both the inductance and resistance of the connection.

The IGCT's much shorter turn-off times compared with GTO thyristors allows it to operate at higher frequencies. Up to several kilohertz for very short periods of time are possible. However, because of high switching losses, typical operating frequencies are up to 500 Hz.

IGCTs are also available either with or without reverse blocking capability. IGCTs capable of blocking reverse voltage are known as symmetrical IGCTs. The typical application for symmetrical IGCTs is in current source inverters. IGCTs incapable of blocking reverse voltage are known as asymmetrical IGCTs. They typically have a reverse breakdown rating in tens of volts or less. Such IGCTs are used where either a reverse conducting diode is applied in parallel or where reverse voltage would never occur. Asymmetrical IGCT can be fabricated with a reverse-conducting diode in the same package. These are known as reverse conducting (RC) IGCTs.

1.3.3 Power Diodes

There are three major uses of power diodes in power electronic systems—line rectifiers, snubber diodes, and freewheeling diodes—which have different requirements on the electrical characteristics of the diode.

A line rectifier allows a current flow during one half wave of the applied sinusoidal voltage and has to block the current flow during the next (e.g., negative) half wave of the voltage. The basic requirement is a low forward voltage drop that leads to low forward losses and the capability to carry large surge currents, which may occur especially during turning on of the system. On the other hand, these line rectifiers have to block the peak voltage of the line and some voltage peaks, for example, those caused by transients of other loads. The transition from forward to blocking operation is rather slow, depending on the line frequency (typically 50 or 60 Hz) and the peak voltage. The voltage slope is in the range of a few V μs^{-1} or below, even at high peak voltages in the range of a few kV.

The requirements of high blocking voltage and high current capability for the same device are supported by a *p-i-n*-structure. Technically, these devices frequently use a slightly *n*-doped material as the example in Figure 1.10 shows. The voltage-sustaining layer has a width and doping concentration adjusted to the required blocking capability. As a rule of thumb, the thickness of the voltage-sustaining layer is 10 μm per

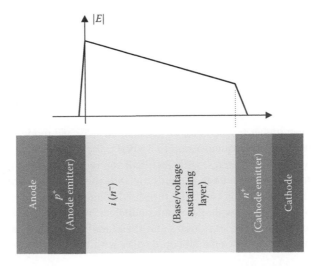

FIGURE 1.10 Cross-section of a *p-i-n* diode and distribution of the electric field in blocking operation.

100 V blocking voltage, for example 100 μm for a 1000 V device. The maximum doping concentration of the voltage-sustaining layer is below 10^{17} cm^{-3}, approximately, divided by the blocking voltage in V.

During forward operation, the voltage-sustaining layer is flooded by electrons and holes coming from the anode and cathode emitters and resulting in a charge plasma with a much higher carrier concentration compared to the background doping and thus in a lowered series resistance of the line rectifier. Line rectifiers require strong emitter structures at anode and cathode to build up much excess charge for low series resistance and low conduction losses of the device.

Before the line rectifier can be turned from forward into blocking operation, the excess charge stored in the voltage-sustaining layer must be removed. Thus, high excess charge leads to high turn-off energy losses of the diode, but since the operating frequency is low, the total losses are still dominated by the conduction losses in forward operation.

The threshold of the *p-n* junction leads to the lower limit for the forward voltage drop. For Si-based diodes, the minimum is around 0.7 V.

In contrast to line rectifiers, snubber diodes and freewheeling diodes are operated at higher frequencies (some 100 Hz up to 20 kHz) and with higher voltage slopes during commutation as the switching of the diode from forward to blocking operation is called. The turn-off losses of these diodes cannot be neglected, thus an optimum operating point must be found depending on the operating frequency.

Snubber diodes are used in the connection from a power switch (e.g., GTO) to a capacitor of a snubber network, which reduces inductive peak voltages when turning off the power switch. A snubber diode should have a high current capability when turned on and low excess charge before a reverse voltage is applied to the diode.

To reduce the excess charge during static operation, recombination centers are introduced into the base of the diode. When reducing the carrier lifetime, the carrier concentration during the forward pulses is reduced. On the other hand, strong anode and cathode emitters lead to the required high surge current capability of the snubber diode. The forward current drops automatically when the peak voltage at the power switch ends, thus the turn-off behavior of a snubber diode is of minor importance.

In contrast to other diodes, for freewheeling diodes the turn-off characteristic is of high importance. The switching characteristic is dominated by the carrier distribution during forward operation of the chip [11] and the doping profile in the voltage-sustaining layer. To reduce the switching losses, the electronic designer strives for decreasing the switching time of the diode. Freewheeling diodes are used, for example in converters in conjunction with GTO and IGBT switches.

Faster switching, however, leads to more critical conditions for a hard cut off of the reverse current, which is not desired. Second, the stress on the diode during commutation is critical. At the time when the high reverse current is extracting the excess charge of the diode, already considerable reverse voltage lies at the diode terminals. Of course, the stress must not exceed the capability of the freewheeling diode.

Softer switching and higher robustness at commutation are the enablers for reduced dynamic losses of the diode. In recent years, considerable softer switching of freewheeling diodes was achieved [12]. Also, the understanding of the robustness led to significantly improved robustness [13–15] also in the area of higher blocking voltages up to 6.5 kV.

For applications at even higher frequencies, for example in switched mode power supplies (SMPS) where diodes are commutated at frequencies up to 300 kHz, it can be technically and economically advantageous to use two diodes in series, with each half the required blocking capability since the turn-off losses of diodes grow approximately quadratic with their blocking voltage. As a drawback, two diodes connected in series exhibit twice the threshold voltage. At the high end of switching frequencies, Schottky diodes based on wide-gap semiconductors, which behave like a small capacitors when they are commutated, provide least losses and therefore least system cost despite their being more expensive compared to conventional silicon devices with the same static forward current and blocking capability.

1.4 MOS-Controlled Bipolar Mode Device

Similar to unipolar MOS-controlled devices, the blocking capability of MOS-controlled bipolar mode devices increases with the thickness of the region along the space–charge region developing when a blocking voltage is applied. However, while the charge-carrier concentration in the on state for unipolar devices is mainly determined by the doping concentration of this region, it can be increased toward much higher values in bipolar devices. Consequently, switching losses and the switching behavior can be optimized to a large extent independent from the doping concentration of the drift region in MOS-controlled bipolar devices. The most successful MOS-controlled bipolar switch is the IGBT, which is employed in miscellaneous applications in the voltage range from 300 V up to 6.5 kV.

1.4.1 IGBT

1.4.1.1 Basic Concepts

Figure 1.11 shows three vertical IGBT designs with the aid of a planar DMOS cell. Similar to the MOSFET, the blocking voltage is sustained by the *p-n* junction formed by the *p*-body and the weakly doped *n*-base. The distinctive difference between the MOSFET and the IGBT is that the *n*-doped drain is replaced by a *p*-doped backside collector that is able to inject holes into the *n*-base. When the gate voltage exceeds the threshold voltage, the *n*-base will be flooded by electrons injected from the *n*-doped source layer through the *n*-channel and by holes from the *p*-doped backside layer. As a consequence, a charge-carrier plasma evolves in the *n*-base. The charge-carrier concentration in this plasma ($>10^{16}$ cm^{-3}) is typically several orders of magnitude higher than that of the doping concentration ($<10^{14}$ cm^{-3}) of the weakly doped *n*-base. Thus, despite the low doping concentration of the *n*-base that is required to sustain high blocking voltages in the off state, the voltage drop in the on-state voltage of the IGBT for a given current can be kept much lower than that of a MOSFET with the same blocking voltage capability due to the conductivity modulation in the *n*-base.

Each of the three IGBTs structures shown in Figure 1.11 has specific advantages: the non-punch-through (NPT) IGBT is characterized by the thick weakly *n*-doped drift region. Its width is chosen so long that the electric-field strength drops to very small values inside this drift region under any operating condition—even when the maximum rated voltage is applied between the emitter and collector contact. The desired trade-off between the saturation voltage and the turn-off losses can be adjusted easily by the implantation fluence of the backside emitter, without the need of an additional charge-carrier lifetime reduction. Moreover, the switching losses of an NPT IGBT depend only weakly on the operating temperature.

The drift region in the punch-through (PT) IGBT is much shorter compared to an NPT IGBT, resulting in a lower on-state voltage. To ensure the same blocking capability of the PT IGBT, however, an additional *n*-doped buffer layer between the drift region and the thick *p*-substrate is required. A major disadvantage of the PT IGBT concerns the alignment of the backside emitter efficiency by means of the buffer layer or an additional charge-carrier lifetime reduction.

The field-stop concept [16,17], or similar approaches like light-punch-through [18], soft-punch-through [19], or controlled-punch-through [20], combines the advantages of NPT and PT IGBTs. The design parameters of the field-stop layer mainly determine the blocking voltage capability and the turn-off behavior. A major challenge was, and still is, the handling of large-area and thin wafers to make the independent adjustment of the backside emitter efficiency by standard implantation processes possible. In recent years, sophisticated technology processes have been developed so that, for example, 600 V IGBTs with a thickness well below 70 μm can be fabricated from 8 in. wafers.

Another important step to reduce on-state and switching losses was achieved by modifying the cell structure and the development of the trench cell (Figure 1.11). The horizontal position of the *n*-channel along the front side changes to vertical position and provides potential for chip-area shrinking due to the transition from the planar design to the trench structure. However, just as important is the effect of

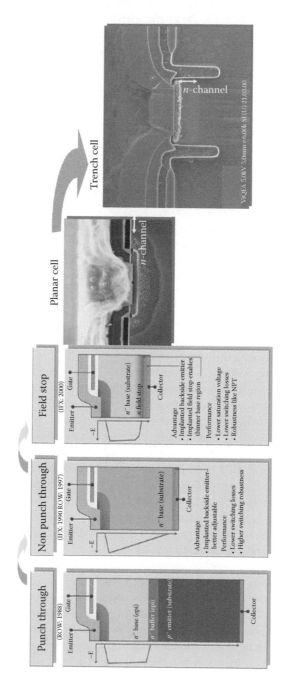

FIGURE 1.11 Evolution of the vertical structure (left) and the cell structure (right) in IGBT development. (Data from Laska, T. et al., Review of power semiconductor switches for hybrid and fuel cell automotive applications, *Proceedings of the APE'2006*, Berlin, Germany, CD-ROM, 2006.)

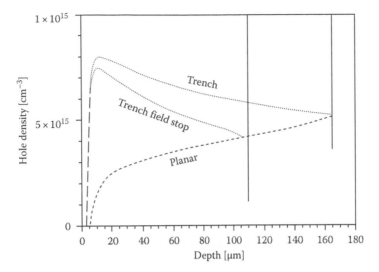

FIGURE 1.12 Vertical cross-sections of the hole distribution from the emitter to the cathode contact in IGBTs with different cell structures. The backside emitter is located at a depth of 165 μm for the trench and the planar cell and at about 110 μm for the trench field stop design. (Data from Laska, T. et al., Review of power semiconductor switches for hybrid and fuel cell automotive applications, *Proceedings of the APE'2006*, Berlin, Germany, CD-ROM, 2006.)

the trench structure on the vertical charge-carrier distribution between the cathode and the anode contact (Figure 1.12). The trench acts as a kind of bottleneck for the holes flowing from the backside anode contact toward the cathode, resulting in a drastic increase of the concentration near the cathode. For a properly designed trench structure, the increase in the hole concentration near the trench region can become so large that the hole concentration along the entire drift region exceeds that of a comparable planar cell. Because of charge neutrality, the electron distribution changes similarly. Since approximately three-fourths of the total load current is carried by the electron current, the on-state losses can be significantly reduced in the trench cell with relatively little influence on the switching losses, resulting in an improved E_{off}–V_{CEsat} trade-off relationship compared to the planar cell structure [16].

Thus, both decreasing the chip thickness and improving the cell design are key factors to reduce the active chip area. The evolution of the chip thickness and the chip area of IGBTs during the last years are illustrated in Figure 1.13. Another important aspect for improving the E_{off}–V_{CEsat} trade-off relationship concerns the cell density. Since the electron current through the channel acts as a base drive for the *p-n-p* transistor of the IGBT, decreasing the channel resistance results in a stronger hole injection

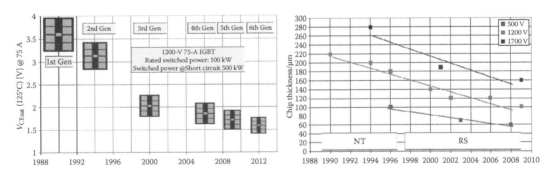

FIGURE 1.13 Evolution of the chip area and V_{CEsat} (left) and chip thickness (right) in IGBT development. (Data from Laska, T. et al., Review of power semiconductor switches for hybrid and fuel cell automotive applications, *Proceedings of the APE'2006*, Berlin, Germany, CD-ROM, 2006.)

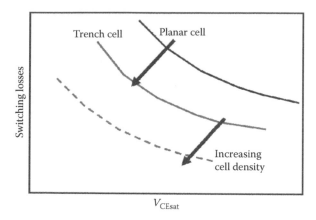

FIGURE 1.14 Schematic E_{off}–V_{CEsat} trade-off relationship: Comparison of different cell designs and influence of the cell density.

from the anode and, consequently, in lower on-state voltages. As the channel resistance decreases with the channel width, the increase in the cell density results in a significant improvement of the E_{off}–V_{CEsat} trade-off relationship (Figure 1.14). Typical output characteristic of a 75 A 1200 V IGBT are shown in Figure 1.15.

The turn-off behavior of a trench field-stop IGBT is shown in Figure 1.16. At the beginning of the turn-off period, the IGBT is in the conductive state, since the gate-emitter voltage is significantly higher than the threshold voltage. Consequently, the collector-emitter voltage drop is very

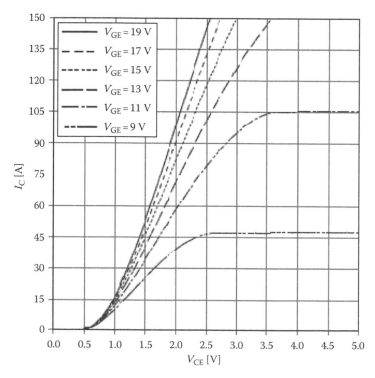

FIGURE 1.15 Typical output characteristic of a 75 A 1200 V IGBT at 125°C. (Data from Infineon Technologies AG, Neubiberg, Germany, data sheet.)

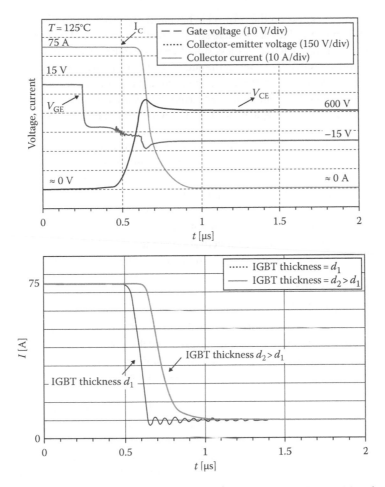

FIGURE 1.16 Turn-off characteristics $V_{GE}(t)$, $V_{CE}(t)$, and $I_C(t)$ of a 75 A 1200 V IGBT at 125°C under nominal conditions (top) and turn-off current $I_C(t)$ for two IGBTs with different device thickness and an additional stray inductance of 400 nH at 25°C (bottom). The measurements were performed with a module so that the measured gate signal represents not the gate potential of the IGBT but is shifted by the potential drop across an ohmic resistance inside the module.

low (<2 V). The collector current is limited by the load. Once the gate potential V_{GE} changes from 15 V to −15 V, a gate current rises to discharge the input capacitance that is essentially formed by the parallel connection of the gate-emitter and the gate-collector capacitance. The approximately exponential decay of the gate-emitter voltage continues until the threshold voltage is reached. Due to the inductive load, the collector current cannot drop immediately, but is maintained by the extraction of charge carriers. In this initial phase of the turn-off period, the turn-off characteristics of the IGBT are similar to that of a MOSFET. Before the current can start dropping, the collector-emitter voltage must rise. Once the threshold voltage is reached, the gate-emitter voltage is initially constant (Miller plateau), since nearly the entire gate current is needed to discharge the gate-collector capacitance. As this capacitance decreases with increasing collector voltage, the gate current can start to discharge the gate-emitter capacitance again (end of the Miller plateau) so that the gate-emitter potential drops further. The ensuing voltage overshoot is caused by the voltage drop induced by the stray inductance in the load circuit due to the decreasing collector current. A distinct difference compared to the turn-off behavior of the MOSFET is the appearance of the so-called tail current at the end of the turn-off phase. This tail current is caused by excess carriers in the IGBT that do not appear in the unipolar MOSFET.

The influence of the thickness on the turn-off behavior of an IGBT is also illustrated in Figure 1.16. Two IGBTs with different chip thicknesses were stressed with an additional stray inductance of 400 nH in the load circuit. The thinner IGBT has less stored excess carriers. Consequently, turn-off is faster compared with the thicker IGBT. However, at the end of the turn-off phase, the excess carrier density in the thinner IGBT is too low to support the load current, so that the current abruptly decreases, resulting in the excitation of voltage and current oscillations from the resonant *LC* resonant circuit that is formed by the stray inductance *L* and the capacitance *C* of the IGBT. The higher excess charge in the thicker IGBT, however, results in soft turn-off without any oscillations.

An important feature of the IGBT is its ability to withstand a short circuit for a certain time interval. Today's IGBTs are typically able to resist a short circuit for a period of 10 μs. This period provides enough time to detect the fault and turn off the IGBT by an external monitoring circuit. The short-circuit current is usually considerably higher than the rated current. Thus, if the nominal voltage is applied to the device during the short-circuit period, there is a huge energy dissipation in the IGBT, resulting in a strong self heating of the device. If the device is not turned off fast enough, the current will increase in a way that is no longer controllable due to the activation of the parasitic thyristor formed by the source, the *p*-body, the *n*-drift region, and the *p*-emitter, so that the device will eventually be destroyed.

1.4.1.2 Advanced Concepts

1.4.1.2.1 Reverse Conducting IGBT

In an RC-IGBT, a diode is monolithically integrated into the IGBT chip. For volume production, this concept was first realized with an optimization for soft-switching applications such as lamp ballast or inductive heating applications in the 600 V [22,23] and 1200 V class [24]. Meanwhile, also RC-IGBTs for hard switching applications, such as industrial inverters or drive applications have been developed on the basis of the NPT-technology [25–27].

Figure 1.17 shows the cross section of an RC-IGBT, based on a trench field-stop IGBT. The *n*-doped regions at the backside act as a cathode emitter, while the *p*-body of the IGBT and the highly *p*-doped anti latch-up region near the frontside act as an anode emitter of the integrated diode. Thus, the IGBT is able to conduct a current even when the polarity of the collector-emitter-voltage is reversely biased. Major challenges for RC-IGBT production are, particularly for thin wafers, the necessity of a backside photolithographic process, and particularly for higher load currents, the robustness of the diode. Moreover, integration of the diode and the IGBT into the same chip makes the independent adjustment of the charge-carrier distribution in the diode and the IGBT difficult. However, it has been shown that the Q_{rr}–V_f trade-off relationship of the diode can be significantly improved by lifetime control techniques sustaining a good IGBT performance.

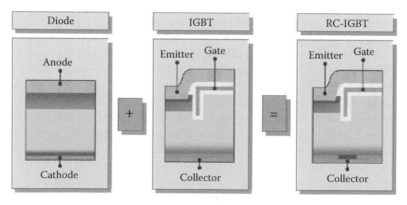

FIGURE 1.17 Integration of a diode and an IGBT resulting in a reverse conducting IGBT. (Data from Laska, T. et al., Review of power semiconductor switches for hybrid and fuel cell automotive applications, *Proceedings of the APE'2006*, Berlin, Germany, CD-ROM, 2006.)

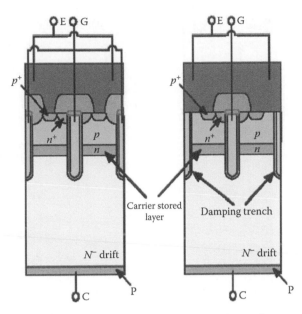

FIGURE 1.18 CSTBT (left) and CSTBT with inactive trenches (right). (Data from Nakamura, S. et al., Advanced wide cell pitch CSTBTs having light punch-through (LPT) structure, *Proceedings of the ISPSD'02*, Santa Fe, NM, 2002, pp. 277–280.)

1.4.1.2.2 Carrier Stored Trench Bipolar Transistor IGBT

As illustrated above in the light of the trench IGBT, the on-state and switching losses can be optimized by tailoring the charge-carrier distribution in the IGBT. The increase in the hole concentration in the trench IGBT, for example, results in a drastic decrease of the on-state voltage. In the carrier stored trench bipolar transistor (CSTBT), this increase in the hole concentration is further strengthened by the implementation of an additional *n*-doped layer below the channel region (Figure 1.18). The *n*-doped layer forms a barrier for holes moving from the anode to the cathode, resulting in an increase in the carrier concentration. If the doping concentration is properly designed, the blocking capability of the IGBT will not significantly be reduced.

A stripe-shaped trench design as indicated in the schematic of the CSTBT is typically characterized by a big gate capacity and a high short-circuit current. These disadvantages can be avoided by deactivating a part of the trenches. Such a trench deactivation can be easily achieved, for example, by connecting the respective trenches not to the gate contact but to the emitter contact (Figure 1.18).

1.4.1.2.3 Clustered IGBT

In order not to deteriorate the blocking capability of a CSTBT, the maximum doping concentration of the *n*-doped layer, and consequently, the increase in the carrier concentration is limited. The clustered-IGBT (CIGBT) shifts this limitation to higher values of the doping concentration by the implementation of an additional *p*-well directly below the *n*-doped layer (Figure 1.19, [28]). The CIGBT can be built as planar or trench IGBT. The floating *p*-well is part of an internal thyristor formed by the *p*-anode, the *n*-drift, the *p*-well, and the *n*-well. In Figure 1.19, the single gate contact of the CIGBT is divided into two parts in order to elucidate the function of the device: Turn-on of the CIGBT is essentially controlled by gate-2. If the gate voltage surpasses the threshold voltage, the *n*-drift and *n*-well region are connected to source potential and the potential of the floating *p*-well rises with increasing positive anode voltage. Once the *p*-well potential exceeds the built-in voltage of the *p-n* junction formed by the *p*-well and the *n*-well, the internal thyristor turns on without snap-back. In this operation mode, the load current of the CIGBT is controlled by the potential of gate-1. If the anode voltage is increased under this condition, the main part of voltage drops across the

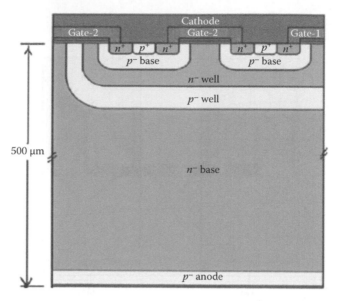

FIGURE 1.19 3.3 kV planar CIGBT. (Data from Sweet, M. et al., Experimental Demostration of 3.3 kV Planar CIGBT in NPT technology, *Proceedings of the ISPSD'08*, Orlando, FL, 2008, pp. 48–51.)

p-n junction formed by the *p*-base and the *n*-well. Turn-off of the CIGBT is achieved by reducing the gate voltage to zero so that the electron current feeding the internal thyristor is interrupted.

IGBTs with dynamic clamping capability for overvoltage self-protection

Recently, much work has been done to provide the IGBT with a so-called dynamic clamping capability that enables the device to regulate its voltage during turn-off to a level that is close to the rated voltage. The integration of such a dynamic overvoltage self-protection function renders superfluous external control and protection circuits for voltage clamping and gives the user more freedom concerning the selection of the optimum gate resistance for his application. However, ensuring that dynamic clamping works properly under all possible operating conditions is a challenging task. Important design parameters that can be used to enhance the dynamic clamping capability of a field-stop IGBT are the doping dose of the field-stop layer and the backside *p*-emitter. Both lowering the field-stop layer dose and increasing the backside *p*-emitter dose result in a better dynamic clamping capability. However, these doses cannot be adjusted arbitrarily without deteriorating other characteristics of the device. Reduction of the field-stop dose, for example, is limited by the required breakdown voltage. A higher backside emitter dose is in particular critical for fast-switching IGBTs, because more charge carriers have to be removed during the turn-off period.

Furthermore, shifting of the initial breakdown from the junction termination into the trench-cell area by appropriate cell and junction termination design is important to avoid destruction in the cell area close to the junction termination area. One possibility to achieve this shifting is to use different trench geometries for the trenches near the junction termination and that in the cell area. Another possibility is to lower the breakdown voltage in the cell area by increasing the doping concentration of the *n*-doped drift region locally, near the *p-n* junction formed by the *p*-body and the *n*-drift region. It is also possible to expose the junction termination area to light-ion irradiation, to shift the breakdown voltage of the junction termination area and its nearby trench cells to higher values.

As an example, Figure 1.20 illustrates the successful dynamic clamping for a turn-off current of twice the rated current of a 1200 A module with 16 parallel-connected 75 A 1200 V trench IGBT chips that are stressed by a 400 nH stray inductance. Even under this hard condition, the integrated dynamic clamping function operates reliably and the clamping voltage does not significantly exceed the rated voltage.

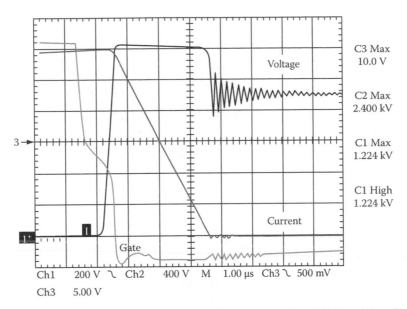

FIGURE 1.20 Overcurrent (twice the rated current) turn-off behavior of a 1200 A 1200 V module with a DC link voltage of 900 V, an estimated worst-case stray inductance of 400 nH at an ambient temperature of 125°C. (Data from Laska, T. et al., Field stop IGBTS with dynamic clamping capability—A new degree of freedom for future inverter designs, EPE 2005, *11th European Conference on Power and Electronics Applications*, Dresden, Germany, CD-ROM, 2005.)

Meanwhile, several investigations on IGBTs with dynamic clamping capability have been reported for voltage classes ranging from 1.2 up to 6.5 kV (e.g., [30–33]). It is worth noting that a CIGBT can also be provided with a self-clamping function: if the *p-n* junction formed by the *p*-base and the *n*-well region is properly designed, punch-through at a certain anode voltage results in a voltage clamp.

1.5 Unipolar Devices

1.5.1 High-Voltage Power MOSFET

In certain applications, minimizing the volume, weight, and of course the cost of transformers and other inductive devices is aimed at, which is done by increasing the operation or switching frequency of power devices. SMPS, as used in many consumer or information technology appliances, are one example for these applications. Here, switching frequencies between 30 and 300 kHz are common today with a blocking capability of the switches between 200 and 1000 V. Operation of IGBTs, for example, at these frequencies is possible, but result in rather high dynamic losses. Power MOSFETs have higher on-state resistance compared to IGBTs with the same chip area, but they also have no charge plasma resulting in much lower turn-off losses. On the other hand, the on-state resistance per chip area increases with the blocking capability as sketched in Figure 1.21.

The reason behind this is that higher blocking voltages require thicker voltage-sustaining layers with lower doping. Since the load current flows directly through this voltage-sustaining layer, it forms a series resistance that—at least for high-voltage power MOSFETs above 200 V—dominates by far the overall resistance of the device.

Selecting a 500 V MOSFET, for example, with low on-state resistance for a given application, would result in a device with a correspondingly large and expensive chip area. A large chip area exhibits large stray capacities, which have to be charged and discharged during each turn-on and turn-off of the device, resulting in comparative slow switching transients. These slow switching transients help to reduce electromagnetic interference (EMI) and ringing, mainly in disadvantageous layouts. But on

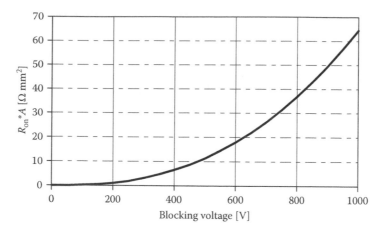

FIGURE 1.21 Dependence of the series resistance of a MOSFET of the blocking voltage.

the other hand, slow switching transients waste switching energy and, due to the rather high switching frequency, impair the efficiency of a design. Worse efficiency directly leads to oversize the power switches and heatsinks to solve thermal issues.

A professional system designer will strive for an EMI conform layout and using fast-switching devices to get an efficient and compact solution for the application at least cost.

The major tasks for the device manufacturer are to provide high-voltage power transistors with less switching losses, ergo smaller parasitic capacitances and lower on-state resistance per chip area. Both tasks can be accomplished by reducing the area-specific on-state resistance, since smaller chips with the same nominal on-state resistance also have smaller parasitic capacitances.

Practically a higher doping of the voltage-sustaining layer will lead to more carriers available for current transport. On the other hand, such a device would lead to a lower blocking voltage capability. The way out of this dilemma was the introduction of a doping of the opposite type close to the doping of the current bearing path [34,35] to have local high conductivity and global low doping due to the compensation. It took several years until the first devices were commercially available that successfully used this approach [36,37]. Since then, the doping compensation devices were the development path for improved high-voltage MOSFETs using different manufacturing approaches and optimization goals.

Beyond all these developments lies the same basic structure as in the right part of Figure 1.22, which is compared to a conventional power MOSFET (left part). Both devices basically have the same structures on the chip front side with a gate controlling an inversion channel.

Here, the voltage-sustaining layer consists of donor (n^-) and acceptor (p^-) doping situated in two individual regions. For blocking operation, the difference of donor and acceptor doping determines the blocking voltage. This net doping is comparable to the very low doping of the voltage-sustaining layer for a conventional power MOSFET. The donor doping of the most modern devices can be increased by a factor of 15 or more compared to standard MOSFETs and thus the on-state resistance is reduced by a factor of 7.5 or more.

Since the blocking characteristic is determined by the difference of a comparatively high donor and acceptor doping the control of this net doping becomes the most challenging task.

When building up a blocking voltage at closed channel, a space–charge region starts extending from the folded *pn*-junction into the *p* compensation columns and into the *n* current path. The width of this insulating region grows with rising blocking voltage as sketched in Figure 1.23. Already, at rather low blocking voltages applied between drain and source compared to the blocking capability, almost the whole area of current path and compensation column is depleted.

One major advantage is the reduction of the parasitic capacitances of compensation devices, especially at higher drain-source voltages as depicted in Figure 1.24. This leads to lower control power needs

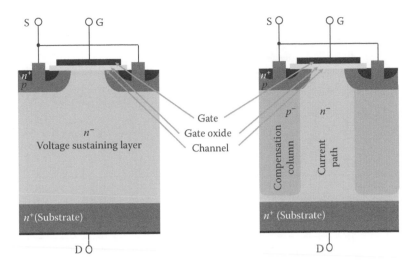

FIGURE 1.22 Left: Cross section of a conventional vertical *n*-channel power MOSFET: The load current is controlled by the gate and flows from the *n* source on the front surface of the chip towards the drain on the rear surface through the low doped voltage sustaining layer. Right: Cross section of a super junction vertical power MOSFET. During on-state the load current flows through the *n*-doped current path while during off-state the doping is compensated by an adjacent *p* column leading to a low net doping serving as voltage sustaining layer.

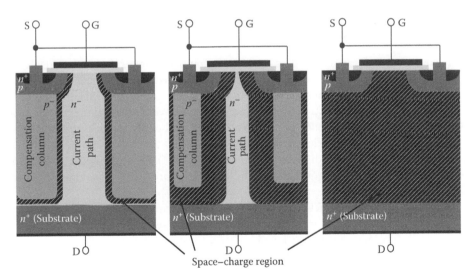

FIGURE 1.23 Left to right: Increasing blocking voltage and growth of the space charge–region. For a device with 600 V blocking capability the virtually fully depleted case at the right is already reached below 100 V.

and to much faster switching, thus lower overall dynamic losses. The nonuniform curves for the drain-source-capacitance originates from the building up of the insulating space–charge region with increasing blocking voltage between drain and source as depicted in Figure 1.23.

It is expected that power MOSFETs based on the compensation principle will continue to replace standard MOSFETs in applications, setting a new "standard." Further energy efficiency will be of interest also for consumer and low-cost information technology appliances; the pressure to use new generations of compensation devices with faster and therefore less-loss switching will also increase. This will lead to use professional layouts, also for low-cost solutions as they are already common for high-end power supplies today.

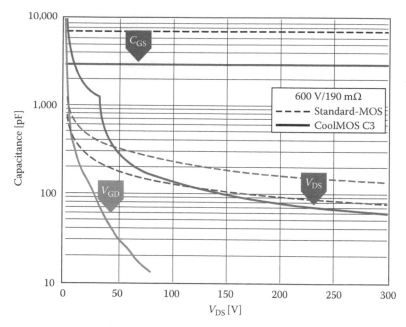

FIGURE 1.24 Comparison of standard MOSFETs and Infineon CoolMOS both with 190 mΩ on-state resistance.

1.5.2 Low-Voltage Power MOSFET

Low voltage power MOSFETs are widely used as switching transistors, for example, in AC/DC converters or DC/DC converters. Especially in the latter case, they are operated at high frequencies above 0.5 MHz making their parasitic capacitances more important as for high-voltage power MOSFETs. Also, the control losses are of higher importance since the output voltage is only a few volts higher as the control voltage and thus the relation between switched output power to needed control power is much smaller. The parasitic capacitances therefore are crucial for low voltage power MOSFETs.

To compare the performance of different low voltage power MOSFETs, the figure-of-merit on-state resistance multiplied by gate charge or multiplied by total charge are used—depending on the focus on control losses or switching losses.

Stray inductances as well as inductances on the application board and the transistor package are of high importance, because of the rather low voltage used. The fast switching leads to significant voltage drops across small stray inductances, which will influence the device behavior.

Compared to high-voltage devices where the losses are dominated by the conduction losses in the voltage-sustaining layer, low voltage power MOSFETs have more leveled distribution to the on-state losses. The impedance level in total must be lower in total, which leads to a different approach for the cell design.

The on-state resistances for the conducting inversion channel and the voltage drop over the voltage-sustaining layer are in the same order of magnitude. Additionally, the stray resistance of the package and interconnections play an important role leading to new package concepts with less parasitics; also stray inductances are reduced.

Low voltage power MOSFETs use trench gates most frequently (see Figure 1.25) compared to planar gate structures dominating high-voltage MOSFETs. Trench cells allow a denser packaging of the cells, higher channel widths and thus lower channel resistances. The area of the gate electrode opposite to the drain electrode is smaller, leading to a smaller gate-drain capacitance, thus less feedback (Miller effect) and faster switching. The area of the source electrode opposite to the drain electrode also is smaller, leading to smaller output capacitances and less switching losses.

Future developments for low power MOSFETs will focus on using finer structures to improve the figure of merit $R_{on} \times Q_{total}$.

1.6 Wide Bandgap Devices

SiC is well known as an ideal semiconductor for power electronic applications since several decades [38]; however it took until 2001 to introduce the first commercial devices based on SiC into the market [39]. The reason for this long pre-development time is the difficult substrate wafer manufacturing process. In fact, the first commercial SiC devices have been manufactured on 2 inch diameter wafer, meanwhile, only 7 years later, the wafer diameter used in production is already increased to 4. This makes the formerly exotic and expensive SiC technology much more affordable. The unique feature of both SiC diodes and switches in the 600 V and above range is that virtually lossless switching is enabled in combination with attractive conduction behavior, allowing benchmark efficiency and reduced complexity in modern power conversion systems.

1.6.1 SiC Schottky Diodes

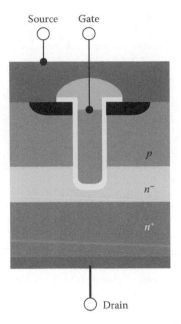

FIGURE 1.25 Cross section through a modern low-voltage transistor.

Other than *pn*-diodes, Schottky diodes do not show any dynamic changes in the charge-carrier density, when being forward biased, as shown in Figure 1.26, therefore no "reverse recovery" is also necessary when the bias changes sign. Figure 1.26 demonstrates the dynamic characteristic of a SiC Schottky diode in comparison to fast-switching Si-diodes.

Based on this principle, the equivalent model of the SiC Schottky diode is very simple. It consists of an ideal diode, possessing a temperature-dependent junction potential and temperature-dependent differential resistance with no switching losses and a depletion capacitor in parallel. When switching the unipolar diode off, only the displacement current of the capacitor can be observed instead of a

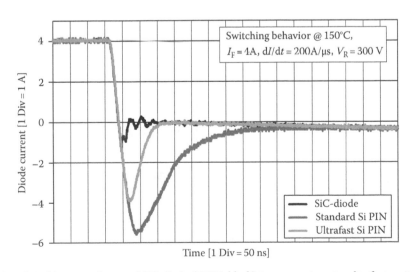

FIGURE 1.26 Switching waveforms of SiC diode (600 V, black) in comparison to ultrafast switching silicon diodes (gray for two 300 V diodes in series connection, dark gray for one 600 V diode).

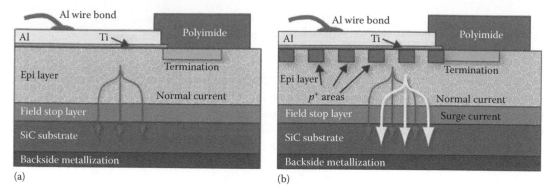

FIGURE 1.27 Schematic drawing of a plain Schottky diode (a) and a Merged-PN-Schottky concept (b). Carrier injection for forward voltages >3 V allow surge current capability increase. The epi layer is responsible for the blocking capability of the device.

typical bipolar reverse-recovery waveform. As expected, there is also no dependence of this capacitive "recovery" charge (Q_c) from temperature, forward current, or di/dt [40,41]. Of course, such Schottky diodes can also be realized in silicon, but at a voltage rating >150 V, they suffer significantly from both very high on-resistance and leakage current. Compared to ultrafast silicon diodes, the losses depend strongly on di/dt, current level, and temperature; the SiC diodes are independent on these boundaries.

The structure of a plain Schottky diode is simple, as indicated in Figure 1.27a. One of the drawbacks of this simple device is the very limited surge current capability. As the ohmic slope of its forward characteristic is purely governed by the mobility of the charged carriers (which depend via $1/T^2$ on temperature T), there exists a strong positive feedback mechanism between increasing current → increasing power dissipation → increasing R → increasing V_f → increasing power dissipation ..., what finally leads to a thermal destruction of the devices at surge currents only ~3 times higher than rated current within 10 ms.

How can this issue be circumvented without penalty on the switching behavior? The solution is shown in Figure 1.27b—it is the so-called merged pn-Schottky diode [42]. This concept takes advantage of the wide bandgap material properties of SiC. The forward characteristic of this merged SiC Schottky diode and SiC pn diode is shown in Figure 1.28. Under normal operating conditions, the high pn junction

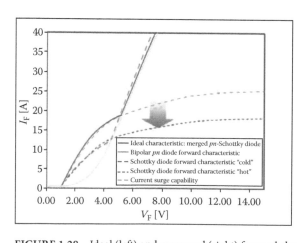

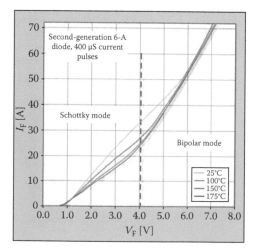

FIGURE 1.28 Ideal (left) and measured (right) forward characteristic of Infineon's ThinQ® 2G diodes.

potential (~3 V) of SiC precludes conduction of the *pn* structure. Only in surge current conditions, this forward voltage will be reached, and the *pn* structure will provide additional carrier injection for conductivity modulation of the drift region.

The *p* regions with low ohmic contact to the Schottky barrier have further benefits in this structure. They will concentrate the maximum electrical field away from the Schottky barrier surface. This allows the usage of a higher maximum field potential in the blocking mode, without degrading the barrier and compensates for the area used by the p^- wells. This also provides a true and consistent avalanche breakdown characteristic—which is not achieved by competitors with plain Schottky barrier structure. As demonstrated in Figure 1.28 during the normal operation (no overload), the SiC Schottky diode has a forward voltage drop of <1.5 V just as a Schottky diode and in overload (e.g., $I_L > 5 \times I_N$), the diode's forward characteristic is following the SiC *pn* diode structure. According to this characteristic, the overload performance is just as known in the surge current operation of any *pn*-diodes.

Even though the *p*-areas shown in Figure 1.27b do consume a certain area, there is no increase in the ohmic slope of the forward characteristic, as this effect is taken care for by an improved conductivity of the cell structure. The significant improvement in surge current capability comes therefore without any penalty.

Due to the very high breakdown field strength of SiC, the thickness of the required blocking layer is very small (<5 μm for 600 V SiC devices in comparison to 40–60 μm in 600 V Si-diodes). This allows even under surge current mode to ensure purely capacitive switching. Figure 1.29 shows a diode commutation at a current of 10 times the rated nominal current.

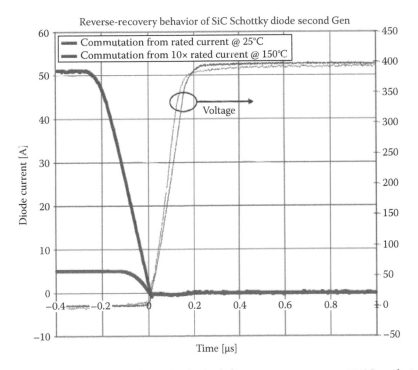

FIGURE 1.29   Commutation of a merged *pn* Schottky diode from 10× surge current at 150°C results in similar low switching losses as switching from 1x rated current at room temperature. Further this is completely independent from d*V*/d*t* and d*I*/d*t*.

The very small remaining capacitive switching losses are directly linked to the active area of the diode, which means that over dimensioning the SiC diode will increase those dynamic losses—this is the opposite, as what most designers are used to see for Si-diodes, where switching losses strongly depend on current density and T-rise due to self heating.

What limits the power that can be drawn from a SiC diode? With rising current, the conduction losses at full load will increase due to the ohmic behavior of the diode. The dissipated power must not drive the device into thermal runaway nor violate max junction temperature ratings. With the merged *pn*/Schottky concept, we get already a nearly temperature-independent forward characteristic (as can be seen from Figure 1.29, right-hand graph beyond at V_f of 5.5 V), which practically eliminates the problem of thermal runaway. For the reduction of junction temperature, an appropriate mounting technology is required, which takes away the thermal barrier from the 60 to 80 μm thick solder layer being conventionally used for mounting power devices into discrete packages.

In effect, the extremely good thermal conductivity of SiC is now directly coupled to the large and also good thermal conducting copper lead frame of the TO package. The conventional solder layer is replaced by an extremely thin diffusion zone that is only 2 μm wide. This results in a significant improvement of both steady-state thermal resistance R_{th} and transient thermal impedance Z_{th}.

Of course, SiC Schottky diodes are not limited to 600 V. Due to the comparatively low resistivity of the necessary blocking layers ("drift layers"), 1200 and 1700 V Schottky diodes have very attractive performance values in comparison with their Si counterparts. As for the 600 V level, the switching losses are only minimal and are due to capacitive displacement current. In fact, the capacity of 1200 V with a certain active area is even smaller than for a 600 V device, caused by the lower doping concentration in the drift layer. Thus, those diodes are ideal companions as freewheeling diodes for modern ultrafast Si-IGBTs.

1.6.2 SiC Power Switches

Even after 9 years of commercial availability of SiC diodes, there still is no SiC power switch in the market. The reason is surely not that such a device would not have plenty of application benefits. This is especially true in the voltage range of 1000 V and above, where unipolar switching devices like Si-MOSFETs are already very rare and of insufficient performance (best in class discrete devices have typically several Ohm on resistance). The main competition in this voltage range comes from IGBT-like devices, with their well-known restrictions with respect to switching losses and maximum frequency. Thus, many application engineers are looking for SiC-based alternatives for applications like solar converters, UPS, HEV, and high-precision drives.

There have been plenty of announcements of achievements made on SiC switches in the recent years, but no products have materialized. There are various reasons for this, but for sure one dominating issue is the insufficient quality of the SiC oxide interface. A SiC MOSFET suffers not only from a low channel mobility, which compensates a big part of the advantages of the physical properties of SiC, but also the reliability of the SiC MOS system with respect to the so-called extrinsic (early) failures is still questionable.

However, for the time being there is one device in favor, a so-called junction field effect transistor (JFET) power switch concept in SiC, which does not require a gate oxide and offers superior ruggedness in many application aspects (e.g., ESD, electrostatic discharge–ESD, avalanche, short-circuit conditions). However, this device is normally on (conducting without gate voltage), when best cost/performance is the target. This feature can be addressed with the help of a cascode configuration employing a low voltage MOSFET to generate the necessary voltage drop along the MOSFET source-drain path to pinch the JFET off. This principle is shown in Figure 1.30. For the SiC-JFET, very attractive area-specific on resistances have already been achieved: <6 M Ohm × cm² for devices with 1200 V blocking.

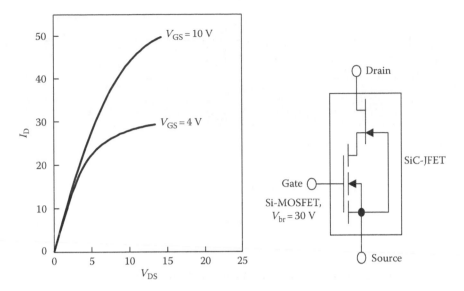

FIGURE 1.30 Schematic of a high voltage SiC-JFET/low voltage Si-MOSFET configuration (right). The output characteristic of this cascode is equivalent to the low voltage Si-MOSFET, whereas the blocking properties are determined by the SiC-JFET.

1.7 SMART Power Systems

Considerable challenges that system engineers are frequently faced with are the device selection, control functions, and the optimized operation along the SOA-diagram (Save Operating Area), together with the implementation of protective and diagnostic features in power semiconductor components. This problem has been solved with IC-compatible power switches, the so-called SMART-Power systems. These new generation of power semiconductor switches have integrated all the controlling protection and diagnostic-functions together with a communication interface on one chip. Depending on the power rating (voltage capability and current scaling), circuit complexity, and requirements toward safety isolation, there are several ways of realization, for example, monolithic integration in SMART power technology (SPT) or silicon on isolator (SOI), chip-on-chip (CoC), chip-by-chip (CbC) or multi-die assembly on substrate carriers [43].

In this new generation of devices, microelectronics and power electronics are combined for both systems and manufacturing. This has triggered a new area in system integration of power and microelectronics, a significant step toward system miniaturization, higher reliability, reduced dissipated power, and fully protected and communicable electrical systems. A broad spectrum of various semiconductor technologies for "SMART" solutions has been developed since the mid-1980s [44].

1.7.1 High-Voltage System Integration

In typical power conversion systems, for example motor control, the input voltage (supplied by the mains) has to be rectified and the output would be controlled to optimize power transfer for varying motor load conditions. There is a great demand expected for such electronic motor drive systems for domestic appliances that are required to meet energy efficiency guidelines. In such mass market consumer applications, economy of scale can be achieved by reducing electronic component count in the overall electronic systems.

Since these system solutions are entire power electronic circuits comprising a full bridge diode rectifier, IGBT converter and gate drive circuit can be integrated. For such integration, lateral power

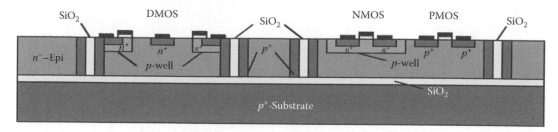

FIGURE 1.31 A PIC cell in SOI technology showing a power MOSFET and standard CMOS MOSFET.

switching devices such as lateral IGBTs or power MOSFETs (LDMOS, Figure 1.31) with all there terminals on the surface have to be used. The technology that is most promising for such integration is SOI,
which allows total static isolation between power devices. Figure 1.31 shows the integration of various
power devices using trench isolation.

1.7.2 SMART Power Technology for Low-Voltage Integration

Until the mid-1990s, bipolar technology was the dominant process for power IC's. Depending on their
breakdown voltage, these processes could only offer a limited element density. The maximum power
dissipation is restricted by the base current and the saturation voltage of the bipolar power transistors.
SMART power processes overcome these limitations. Analog functions can be realized with bipolar
transistors, CMOS logic allows complex logic functions, and DMOS power transistors result in negligible power dissipation. Technical requirements of the application and general cost considerations

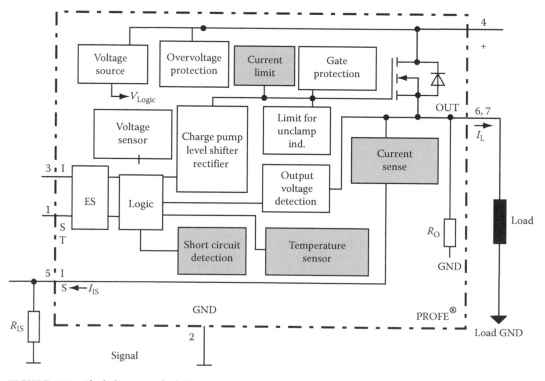

FIGURE 1.32 Block diagram of a fully protected high side switch.

restrict the choice of the optimum process, for example in selecting self-isolation or junction isolation, SMART-power technology.

The number of power output channels is a relevant orientation for the structure of the power transistor and this decides on the technology. In case of single output, the backside of the chip can act as the drain contact, which stands for self-isolation technology. Thus, the current flows vertically through the devices, which is optimal for the power losses generated. For multiple outputs, all contacts are normally placed on top of the chip and as a consequence, the current flows laterally. In this case, the junction isolation process is the favorite.

SMART power technology in self-isolation technologies is preferred for high current devices—the voltage rating is typically <100 V. The backside of the power MOSFET chip is used as the common drain of a single or multiple power DMOS transistor(s).

Various protection functions are included in the element—monolithically or in CoC-technology—and are indicated by the status output, for example, over-temperature, short circuit, open load, over- and under-voltages shutdown, reverse polarity, load dump protection, communication enforce, etc. (Figure 1.32).

For highly complex logic functions, CMOS technology is the best choice. Analog circuits can be fabricated in bipolar, and any number of separate DMOS power transistors can be integrated. If the application requires high logic density, multiple channels (as high-side or low-side switches), and moderate output power, the junction isolations technologies is the most suitable way for system integration.

1.8 Summary

For all power electronic systems to be realized, it is essential to be able to actively control the dynamic power transfer in all optimum manners. This can only be achieved through the use of the optimized electronic switching devices. For extremely high-power applications, the bipolar devices, for example, thyristors, LTTs, and IGCTs, including their future development are the driving technology today and in the next decades.

The high-power and medium-power system development will be dominated by the IGBT. A huge potential for further development is given. All applications requiring high and ultrahigh switching frequencies in order to minimize the power electronics converter at high energy efficiencies are driven by Power MOSFETs. For lower voltage (<100 V) as well as high voltage (>500 V), many advances have been made; new generations will follow.

All innovations in the car will be driven by SMART-Power systems. This technology covers power switches with simple logic functions integrated up to highly sophisticated system solutions with low power output stages.

References

1. L. Lorenz, Key milestone in the development of power semiconductor, *EPE Proceedings*, Dubrovnik, Croatia, 2002.
2. L. Lorenz and G.A. Amaratunga, Electronic devices for power switching and power integration circuits: The enabling technology for clean environment, *VDI/Proceeding*, 2000.
3. L. Lorenz and H. Mitlehner, Key power semiconductor devices concept for the next decade, *IAS/IEEE Proceeding*, 2002.
4. P.D. Taylor, *Thyristor Design and Realization*, John Wiley & Sons, Chichester, U.K., 1987.
5. H.-J. Schulze, F.-J. Niedernostheide, and U. Kellner-Werdehausen, Thyristor with integrated forward recovery protection, *Proceedings of the 2001 International Symposium on Power Semiconductor Devices and IC's*, Osaka, Japan, May 2001, pp. 199–202.
6. F.-J. Niedernostheide, H.-J. Schulze, and U. Kellner-Werdehausen, Self-protected high-power thyristors, *Proceedings of the Power Conversion and Intelligent Motion (PCIM 2001)*, Nürnberg, Germany, June 2001, pp. 51–56.
7. F.-J. Niedernostheide, H.-J. Schulze, H.-P. Felsl, T. Laska, U. Kellner-Werdehausen, and J. Lutz, Thyristors and IGBTs with integrated self-protection functions, *IET Circuits, Devices & Systems*, 1, 315–320, 2007.

8. F.-J. Niedernostheide, H.-J. Schulze, U. Kellner-Werdehausen, R. Barthelmeß, H. Schoof, J. Przybilla, R. Keller, and D. Pikorz, 13-kV rectifiers: Studies on diodes and asymmetric thyristors, *Proceedings of the ISPSD'03*, Cambridge, U.K., 2003, pp. 122–125.

9. H. Güldner, A. Thiede, L. Göhler, H.-J. Schulze, J. Sigg, J. Otto, and D. Metzner, *Proceedings of the ICPE'98*, Duisburg, Germany, 1998, p. 246.

10. H.-J. Schulze and B.O. Kolbesen, *Solid-State Electronics*, 42, 2187, 1998.

11. A. Porst, F. Auerbach, H. Brunner, G. Deboy, and F. Hille, Improvement of the diode characteristics using emitter-controlled principles (EMCON-diode), *Proceedings of the ISPSD*, Weimar, Germany, 1997, pp. 213–216.

12. A. Mauder, T. Laska, and L. Lorenz, Dynamic behaviour and ruggedness of advanced fast switching IGBTs and diodes, *Proceedings of the IEEE IAS 2003*, Salt Lake City, UT, pp. 995–999.

13. J. Biermann, K.-H. Hoppe, O. Schilling, J.G. Bauer, A. Mauder, E. Falck, H.-J. Schulze, H. Rüthing, and G. Achatz, New 3300V high power Emcon-HDR diode with high dynamic robustness, *Proceedings of the PCIM 2003*, Nuremberg, Germany, 2003, pp. 315–320.

14. L. Lorenz, A. Mauder, and J.G. Bauer, Rated overload characteristics of IGBTs for low voltage and high voltage devices, *IEEE Transactions on Industry Applications*, 40(5), 1273–1280, 2004.

15. M. Domeij, J. Lutz, and D. Silber, Stable and unstable dynamic avalanche in fast silicon power diodes, *Proceedings of the 31th European Solid-State Device Research Conference*, Nuremberg, Germany, September 2001, p. 263.

16. T. Laska, M. Münzer, F. Pfirsch, C. Schaeffer, and T. Schmidt, The field stop IGBT (FS IGBT)—A new power device concept with great improvement potential, *Proceedings of the ISPSD 2000*, Toulouse, France, 2000, pp. 335–358.

17. L. Lorenz, A. Mauder, and J.G. Bauer, Rated overload characteristics of IGBT for low voltage and high voltage devices, *IEEE Transactions on Power Electronics*, 2004.

18. K. Nakamura, S. Kusunoki, H. Nakamura, Y. Ishimura, Y. Tomomatsu, and M. Harada, Advanced wide cell pitch CSTBTs having light punch-through (LPT) structure, *Proceedings of the 14th ISPSD*, Santa Fe, NM, 2002, pp. 277–280.

19. S. Dewar, S. Linder, C. von Arx, A. Mukhitnov, and G. Debled, Soft punch through (SPT)—Setting new standards in 1200V IGBT, *Proceedings of the PCIM Europe*, Nuremberg, Germany, 2000.

20. J. Vobecky, M. Rahimo, A. Kopta, and S. Linder, Exploring the silicon design limits of thin wafer IGBT technology: The controlled punch-through (CPT) IGBT, *Proceedings of the ISPSD'08*, Orlando, FL, 2008, pp. 76–79.

21. T. Laska, M. Münzer, R. Rupp, and H. Rüthing, Review of power semiconductor switches for hybrid and fuel cell automotive applications, *Proceedings of the APE'2006*, Berlin, Germany, CD-ROM, 2006.

22. E. Griebl, O. Hellmund, M. Herfurth, H. Hüsken, and M. Pürschel, *LightMOS - IGBT with Integrated Diode for Lamp Ballast Applications*, Conference on Power Electronics and Intelligent Motion PCIM 2003, 79, 2003.

23. E. Griebl, L. Lorenz, and M. Pürschel, *LightMOS a new power semiconductor concept dedicated for lamp ballast application*, Conference Record of the 2003 IEEE Industry Applications Conference, pp. 768–772, 2003.

24. O. Hellmund, L. Lorenz, and H. Rüthing, *1200V Reverse Conducting IGBTs for Soft-Switching Applications*, China Power Electronics Journal, Edition 5/2005, pp. 20–22, 2005.

25. H. Takahashi, A. Yamamoto, S. Aono, and T. Minato, *1200V Reverse Conducting IGBT*, *Proceedings of the 16th ISPSD*, pp. 133–136, 2004.

26. K. Satoh, T. Iwagami, H. Kawafuji, S. Shirakawa, M. Honsberg, and E. Thal, *A new 3A/600V transfer mold IPM with RC (Reverse Conducting)-IGBT*, Conference for Power Conversion Intelligent Motion PCIM 2006, pp. 73–78, 2006.

27. H. Rüthing, F. Hille, F.-J. Niedernostheide, H.-J. Schulze, and B. Brunner, *600 V Reverse Conducting (RC-)IGBT for Drives Applications in Ultra-Thin Wafer Technology*, *Proceedings of the ISPSD'07*, pp. 89–92, 2007.

28. N. Nakamura, S. Kusunoki, H. Nakamura, Y. Ishimura, Y. Tomomatsu, and M. Harada, *Advanced wide cell pitch CSTBTs having light punch-through (LPT) structure*, Proceedings of the ISPSD'02, pp. 277–280, 2002.

29. M. Sweet, N. Luther-King, S.T. Kong, and E.M. Sankara Narayanan, Experimental demonstration of 3.3 kV planar CIGBT in NPT technology, *Proceedings of the ISPSD'08*, Orlando, FL, 2008, pp. 48-51.

30. M. Otsuki, Y. Onozawa, S. Yoshiwatari, and Y. Seki, 1200 V FS-IGBT module with enhanced dynamic clamping capability, *Proceedings of the ISPSD'04*, Kitakyushu, Japan, 2004, pp. 339–342.

31. T. Laska, M. Bässler, G. Miller, C. Schäffer, and F. Umbach, Field stop IGBTS with dynamic clamping capability—A new degree of freedom for future inverter designs, *EPE 2005, 11th European Conference on Power and Electronics Applications*, Dresden, Germany, CD-ROM, 2005.

32. M. Rahimo, A. Kopta, S. Eicher, U. Schlapbach, and S. Linder, Switching-Self-Clamping-Mode "SSCM", a breakthrough in SOA performance for high voltage IGBTs and diode, *Proceedings of the ISPSD'04*, Kitakyushu, Japan, 2004, pp. 437–440.

33. M. Rahimo, A. Kopta, and S. Linder, Novel enhanced-planar IGBT technology rated up to 6.5 kV for lower losses and higher SOA capability, *Proceedings of the ISPSD'06*, Napoli, Italy, 2006, pp. 33–36.

34. G. Deboy, L. Lorenz, M. Marz, and A. Knapp, CoolMOS—A new milestone in high voltage power MOS, *IEEE-ISPSD-Record 99*, Toronto, Canada.

35. J. David Coe, High voltage semiconductor devices, U.K. Patent Application GB 2 089 119 A, filed December 10, 1980.

36. L. Lorenz, G. Deboy, and I. Zverev, Matched pair of CoolMOS with SiC Schottky diode —Advantages in applications, *IEEE Transactions on Power Electronics*, 2004.

37. G. Deboy, M. März, J.-P. Stengl, H. Strack, J. Tihanyi, and H. Weber, A new generation of high voltage MOSFETs breaks the limit line of silicon, *Proceedings of the IEDM 1998*, San Francisco, CA, 1998, pp. 683–685.

38. B.J. Baliga, *Journal of Applied Physics*, 53, 1759–1764, 1982.

39. H. Kapels, R. Rupp, L. Lorenz, and I. Zverev, SiC Schottky diodes: A milestone in hard switching applications, *Proceedings of the PCIM 2001*, Nuremberg, Germany.

40. J. Hancock and L. Lorenz, Comparison of circuit design approaches in high frequency PFC converters for SiC Schottky diode and high performance silicon diodes, *Proceedings of the PCIM 2001*, Nuremberg, Germany, pp. 192–200.

41. I. Zverev, M. Treu, H. Kapels, O. Hellmund, R. Rupp, and J. Weiss, SiC Schottky rectifiers: performance, reliability and key application, *Proceedings of EPE*, Graz, Austria, 2001.

42. R. Rupp, M. Treu, S. Voss, F. Bjoerk, and T. Reimann, 2nd generation SiC Schottky diodes: A new benchmark in SiC device ruggedness, *Proceedings of the ISPSD*, Naples, Italy, 2006.

43. L. Lorenz, T. Reimann, U. Franke, J. Petzoldt, and R. Krummer, System integration-thermal aspects of chip utilization of power devices and control, *ISPSD Proceedings*, Cambridge, U.K., April 2003.

44. M. Stecher, M. Jensen, M. Denison, R. Rudolf, B. Strzalkowski, M. Muener, and L. Lorenz, Key technologies for system-integration in the automotive and industrial application, *IEEE Transactions on Power Electronics*, 20(3), 537–549, 2005.

II

Electrical Machines

2

AC Machine Windings

Andrea Cavagnino
Politecnico di Torino

Mario Lazzari
Politecnico di Torino

2.1 Introduction

Along the airgap of AC rotating electric machines, there are several *distributed windings*. These windings are positioned in the stator and/or rotor magnetic structure (typically inside slots), and they have to produce suitable magnetomotive force (mmf) waveforms in the airgap. The windings are usually distributed over a more or less wide airgap circumference arc. There are a wide range of AC windings, and it is not possible to provide a complete description of any winding type [1–8]. For this reason, the attention is focused on *symmetrical windings*, with particular reference to three-phase windings (induction and synchronous machine stator windings). In general, a distributed winding is defined as "symmetrical" if its distribution is characterized by two orthogonal symmetry axes. The aim of this chapter is to provide to the reader the basic elements of the winding's theory together with some practical aspects concerning the winding realization.

2.2 MMF and Magnetic Field Waveforms in the Airgap

2.2.1 Introduction

In a distributed winding, it is possible to distinguish the so-called *active lengths*, and the *head connections* or *endwindings*. The active lengths are constituted by conductors facing the airgap and they are positioned, in common machines, inside slots parallel with the machine axle. These conductors are active in the electromagnetic energy conversion by means of the interaction with the airgap magnetic field. The endwindings have the function to close the turns only, allowing the current to pass from one active length to another. In general, the endwindings do not directly influence the electromagnetic energy conversion.

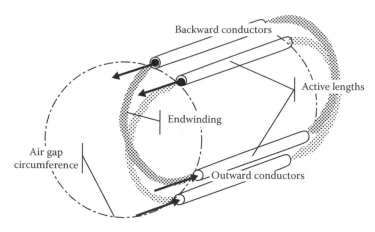

FIGURE 2.1   Layout of a distributed single-phase winding.

Considering a cross section perpendicular to the machine, an ideal single-phase winding is positioned inside two belts constituted by contiguous slots, as shown in Figure 2.1. In the former belt, the current flows in the conductors entering the section plane (outward conductors); in the latter, the current direction is reversed (backward conductors). From a functional point of view, it is not important how and which conductor of a belt is connected with the conductor of the other belt. On the contrary, the distribution of the active lengths along the airgap circumference is very important. In fact, the waveform of the mmf in the airgap and the induced electromotive force (emf) in the winding both depend on the arrangement of the active lengths, due to the variations of the linked flux.

In the following, the methodology to determine the airgap mmf waveform produced by distributed windings when the geometrical positions of the active lengths are known is presented. The proposed method is based on the following hypotheses:

- The radial thickness of the airgap is assumed constant with the angle.
- The lamination permeability is assumed infinite: the mmf produced by the winding drops in the airgap only.
- The slot-opening width is assumed infinitesimal: all the conductors inside the slot can be represented as a single dot-like conductor positioned in the slot center, close to the airgap.
- Three-phase "regular" windings with an integer number of slots per pole per phase are considered.

The meaning of the main symbols used in the following text is

$\hat{}$	Apex for the maximum value of a sinusoidal quantity
$\sim$	Apex for the rms value of a sinusoidal quantity
α	Angular coordinate of the airgap
$A(\alpha)$	Airgap mmf waveform
P	Pole pair number of the actual winding
p	Pole pair number of a generic distribution
q	Number of slots per pole per phase
Z_f	Number of conductors in series per phase
r_t	Airgap average radius
l_a	Active length of the conductors (equal to the slot axial length)

2.2.2 MMF Waveform Produced by a Single Full-Pitch Bobbin

Let us consider the electromagnetic structure shown in Figure 2.2. The $Z_f/2$ outward conductors and the $Z_f/2$ backward conductors of the bobbin are diametrically positioned (full-pitch bobbin), and they carry

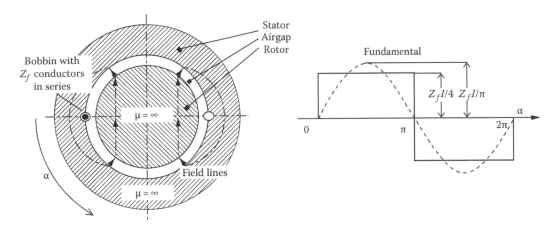

FIGURE 2.2 MMF waveform produced by a full-pitch bobbin with Z_f conductors.

current I. Due to the structure symmetry, it is possible to draw the magnetic field lines, as shown in the figure. The mmf absolute value linked with each field line is equal to $0.5 \cdot Z_f I$.

For all the field lines, the imposed mmf counterbalances the magnetic voltage drop in the two airgap crossings. As a consequence, taking into account the direction of the field lines, the airgap mmf waveform is a square wave, sqw(α), shown in (2.1) and at the right side of Figure 2.2. The mmf is assumed positive when the field lines cross the airgap from the stator to the rotor.

$$A(\alpha) = \frac{Z_f}{4} I \cdot \text{sqw}(\alpha) \tag{2.1}$$

The distribution $A(\alpha)$ can be decomposed into the sum of spatial harmonics (2.2), using the Fourier series:

$$A(\alpha) = \sum_{h=1,3,5,7,\ldots} \frac{Z_f}{\pi \cdot h} \cdot I \sin(h \cdot \alpha) \tag{2.2}$$

To study the rotating-field electric machines, the fundamental harmonic of the mmf distribution is particularly important. The fundamental airgap mmf waveform produced by a full-pitch bobbin with Z_f active lengths is as follows:

$$A_{\text{fundamental}}(\alpha) = A_1 \sin(\alpha); \quad A_1 = \frac{Z_f}{\pi} I \tag{2.3}$$

2.2.3 MMF Waveform Produced by a Single-Phase Distributed Winding

Typically, in rotating-field electric machines, the windings are subdivided in several bobbins positioned inside slots regularly spaced along the airgap. For example, Figure 2.3 shows a single-phase winding realized with three identical full-pitch bobbins connected in series. The bobbins are positioned inside three contiguous slots.

In general, let us define β as the angle between two contiguous slots, q as the slot pair number used by the winding, and Z_f as the total number of active conductors.

For this configuration, the airgap mmf spatial distribution, due to the current I flowing in the winding, can be obtained by summing up q square waveforms shifted by the angle β, as shown in Figure 2.3 for the case $q = 3$. In general, the resultant mmf can be written as follows:

$$A_{\text{bobbin}} = \frac{Z_f}{4q} I; \quad A_{\text{resultant}}(\alpha) = A_{\text{bobbin}} \sum_{i=1}^{q} \text{sqw}(\alpha + i \cdot \beta) \tag{2.4}$$

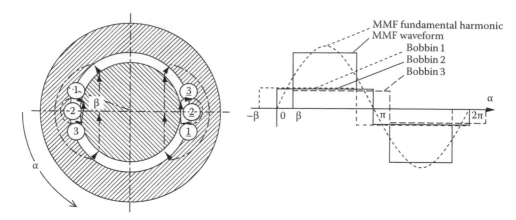

FIGURE 2.3 Airgap mmf waveform produced by a single-phase distributed winding with three identical full-pitch bobbins connected in series.

Also, in this case, the maximum value of the mmf distribution is equal to

$$A_{\max} = \frac{Z_f}{4} I \tag{2.4bis}$$

In order to evaluate the amplitude of the fundamental component of the resultant mmf distribution produced by this winding structure, it is possible to sum point-to-point the fundamental component due to each bobbin. These fundamental components can be represented as vectors with amplitude A_{bobbin} calculated from (2.4) and the positive direction parallel to the magnetic axis of the bobbin.

The resultant amplitude of the mmf can be obtained by a vectorial sum, represented by the polygonal line shown in Figure 2.4. The amplitude of the resultant fundamental mmf is given as follows:

$$A_{\text{fundamental}} = 2r \sin\left(q \cdot \frac{\beta}{2}\right) = \frac{Z_f I}{\pi} \cdot \frac{\sin(q \cdot \beta/2)}{q \cdot \sin(\beta/2)} \tag{2.5}$$

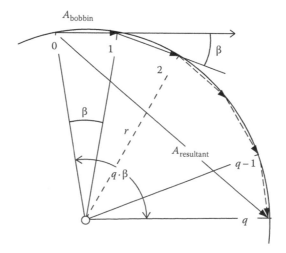

FIGURE 2.4 Vector diagram of the resultant mmf for a distributed winding.

TABLE 2.1   Distribution Coefficient

	Two-Phase Motor	Three-Phase Motor
q	K_d	K_d
1	1.0000	1.0000
2	0.9239	0.9659
3	0.9107	0.9598
4	0.9061	0.9577
5	0.9040	0.9567
6	0.9029	0.9561
8	0.9018	0.9556
∞	0.9003	0.9549

Distribution Coefficient

The general formula of the amplitude of the fundamental mmf produced in the airgap by a winding with Z_f conductors arranged in q pairs of diametrical slots shifted by an angle β is given as follows:

$$A_{\text{fundamental}} = K_d \cdot \frac{Z_f I}{\pi} \tag{2.6}$$

where K_d is the so-called distribution coefficient of the winding. This coefficient is

$$K_d = \frac{\sin(q \cdot \beta/2)}{q \cdot \sin(\beta/2)} \tag{2.7}$$

In Table 2.1, the phase winding distribution coefficients, for two-phase and three-phase motors, are given. These coefficients are evaluated considering that the slots are uniformly distributed. In this case, the angle β depends on the number of slots per pole per phase (q) and on the phase number (m), in accordance with the relation $\beta = 2\pi/N_{\text{slot}} = 2\pi/(2P \cdot m \cdot q)$, considering, in this case, $P = 1$ (see Section 2.2.5).

2.2.4 MMF Waveform Produced by a Shortened-Pitch Winding

In electric machines, the actual winding structure is often more complex than the full-pitch one considered previously in this chapter. Without considering very irregular winding structures, the attention is focused on shortened-pitch windings. In this winding type, the backward conductors of a turn are positioned at an angle less than 180°, with respect to the outward conductors (non-diametrical bobbin). This solution allows to obtain a lower length of the endwinding and a lower harmonic content in the airgap spatial mmf distribution (lower distortion of the mmf wave).

In order to analyze the shortened-pitch windings, let us consider a classical full-pitch phase winding (diametrical pitch) positioned inside $2q$ slots with $Z_c = Z_f/2q$ conductors per slot. Suppose that we divide this winding in two twin parts (layers) still positioned in $2q$ slots, but each one realized with $Z_c/2$ conductors, as shown in Figure 2.5a. Now, let us impose a relative rotation between the two layers, of an integer number of slots (n_r), obtaining the phase configuration shown in Figure 2.5b. The obtained winding is positioned inside a number of slots greater than those of the original full-pitch winding; in particular, the number of occupied slots by the conductors is $2(q + n_r)$, but these slots are not uniformly filled. This winding structure is called shortened-pitch winding, and n_r is the pitch shortening, expressed in number of slots.

The analysis of the mmf distribution produced by this different layer positioning can be done through the study of the two half-windings. In fact, if the two layers rotate each other by n_r slots (corresponding

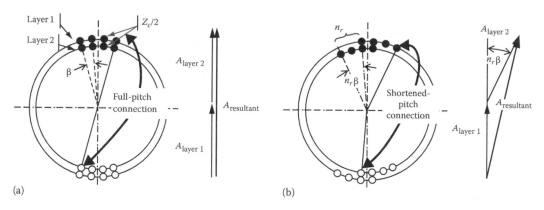

FIGURE 2.5 Double-layer winding with four slots per pole per phase: (a) full-pitch structure and (b) shortened-pitch structure with $n_r = 2$.

to angle $n_r\beta$), the vectors of the mmf fundamental components created by the two layers will also be shifted by the same angle.

Let us define A_{layer} as the amplitude of the mmf fundamental component produced by each layer and K_d as the distribution coefficient of the single layer. The amplitude of the resultant mmf can be calculated as

$$A_{\text{fundamental}} = 2A_{\text{layer}} \cdot \cos\frac{n_r\beta}{2}; \quad \text{where } A_{\text{layer}} = \frac{1}{2}K_d\frac{Z_f}{\pi}I$$

$$A_{\text{fundamental}} = K_d \cdot \cos\frac{n_r\beta}{2} \cdot \frac{Z_f I}{\pi}$$

$$K_r = \cos\frac{n_r\beta}{2} \tag{2.8}$$

The coefficient K_r is called shortening coefficient of the winding. This coefficient is equal to 1 when the two layers are superposed in the same slots ($n_r = 0$). Obviously, in this case, the winding is defined as full-pitch winding.

Winding Coefficient

The product of the distribution coefficient and the shortening one is commonly called the winding coefficient, K_a. Then, for a generic winding, the mmf fundamental distribution can be written as follows:

$$A_{\text{fundamental}} = K_a \cdot \frac{Z_f I}{\pi}; \quad \text{where } K_a = \cos\left(n_r \cdot \frac{\beta}{2}\right) \cdot \frac{\sin(q \cdot \beta/2)}{q \cdot \sin(\beta/2)} \tag{2.9}$$

Equation 2.9 is particularly important in order to describe the magnetization effect due to the distributed winding. In fact, through (2.9), any winding can be identified by an equivalent turn number, N', from the production of the fundamental mmf component point of view, defined as follows:

$$N' = K_a \cdot \frac{Z_f}{\pi} \tag{2.9bis}$$

If the number N' and the value of current I in the winding are known, it is possible to calculate the amplitude of the fundamental harmonic of the airgap mmf directly by using

$$A_{\text{fundamental}}(\alpha) = N' I \sin(\alpha) \tag{2.10}$$

It is important to state that two different windings with the same equivalent turn number, N', can be considered identical from the point of view of the phenomena linked to the fundamental mmf distribution in the airgap only.

Spatial Harmonics

Besides the fundamental component (2.10), in the airgap mmf waveform produced by a distributed winding positioned in slots, a great number of spatial harmonics are present. By means of the Fourier series, the actual spatial harmonics can be related to the odd harmonics of a square wave. In fact, as shown in (2.2), the mmf square wave due to a single diametrical bobbin can be considered as the basic component to study more complex winding structures.

The spatial harmonics can be conventionally considered as a secondary effect in the magnetization produced by the winding. In general, the presence of these harmonics is undesired and they are often considered as a disturbance. From this point of view, the analysis of the spatial harmonics is very important.

For each harmonic order, h, of the resultant mmf distribution, it is possible to define the harmonic winding coefficient, $K_{a,h}$, in accordance with

$$A_{\text{harmonic},h} = K_{a,h} \cdot \frac{Z_f I}{\pi \cdot h}; \quad K_{a,h} = \cos\left(n_r \cdot h \cdot \frac{\beta}{2}\right) \frac{\sin(q \cdot h \cdot \beta/2)}{q \cdot \sin(h \cdot \beta/2)} \tag{2.11}$$

Then, the resultant mmf distribution of the winding can be written as

$$A(\alpha) = \frac{Z_f I}{\pi} \sum_{h=1,3,5,7,\ldots} \frac{K_{a,h}}{h} \sin(h \cdot \alpha) \tag{2.12}$$

Let us define, on the analogy of (2.9bis), the following equivalent turn number for the hth harmonic:

$$N'_h = \frac{K_{a,h}}{h} \cdot \frac{Z_f}{\pi}$$

Then, (2.12) can be rewritten as

$$A(\alpha) = I \sum_{h=1,3,5,7,\ldots} N'_h \sin(h \cdot \alpha) \tag{2.13}$$

If the phase winding is symmetrical, in the mmf distribution spectrum there are odd spatial harmonics only.

As an example, Figure 2.6 shows the mmf spatial distribution for a symmetrical, shortened-pitch, double-layer phase winding with $q = 3$, $n_r = 1$, and $\beta = 30°$.

In Figure 2.7, a comparison of the harmonic winding coefficients for a full-pitch winding and a one-slot shortened-pitch winding (both windings with $q = 3$ and $\beta = 30°$) is shown.

Figure 2.7 highlights that the pitch shortening affects, in a modest way, the fundamental mmf harmonic ($h = 1$), and in a more sensible way, the amplitude of some spatial harmonics. In particular, the

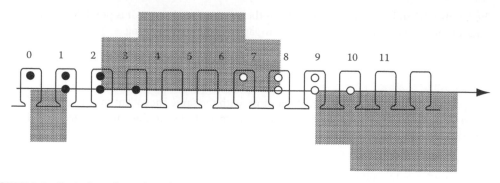

FIGURE 2.6   Typical mmf waveform for shortened-pitch winding.

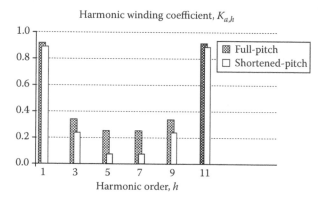

FIGURE 2.7   Harmonic winding coefficient comparison between a full-pitch winding with $q = 3$ and $\beta = 30°$, and a shortened-pitch winding with $q = 3$, $\beta = 30°$, and $n_r = 1$.

amplitudes of the 5th and 7th harmonics are considerably reduced. As a consequence, the pitch shortening can be considered as a simple method to reduce the amplitude of some spatial harmonics in the mmf distribution.

Equation 2.11 highlights that the winding coefficient value for a generic hth harmonic order does not depend on the sign of h. The harmonic winding coefficient values have a periodic trend with respect to the harmonic order. In particular, two spatial harmonics of order h' and h that verify the relation (2.14) are characterized with the same value of the winding factor.

$$h' = \pm h + kN_{\text{slots}} \tag{2.14}$$

As a consequence, the harmonics of an order of h' that respects

$$h' = kN_{\text{slots}} \pm 1 \tag{2.14a}$$

will have the same winding coefficient as of the fundamental harmonic. These spatial harmonics are conventionally called "toothing harmonics."*

---

* The name "toothing harmonics" does not have any relation with the magnetic anisotropy phenomena due to the presence of slots and teeth along the airgap. In fact, these harmonics are already present in the mmf distribution produced by the winding, and they depend on the number of slots only.

The toothing harmonics have typically a great impact on the spectrum of the spatial harmonics. In addition, these harmonics cannot be attenuated using a suitable positioning of the phase conductors, as is the case with the other harmonic orders, because this solution should involve a reduction of the fundamental component too. A mitigation of the effects of the toothing harmonics can be obtained using other methods, such as axial skewing of the slots.

2.2.5 Definition of the Winding Polarity (Pole Pair Concept)

In the analyses reported so far, windings with diametrical or quasi-diametrical turns have been considered. For this type of structures, the airgap mmf waveform has a unique sign alternation along the airgap circumference (see Figure 2.6). In other words, the windings create two magnetic polarities or poles (north and south). Usually, these windings are called two-pole windings or windings with one pole pair ($P = 1$).

In rotating-field electric machines, phase windings with the number of pole pairs greater than one ($P > 1$) are often adopted. In these cases, the airgap mmf distribution has more sign alternations along the whole circumference, and more magnetic polarities are produced in the airgap. The easiest method to produce a pole pair number greater than one is to repeat the disposition of the active winding lengths of an elementary two-pole winding, in a cyclic way, along the airgap circumference, as shown graphically in Figure 2.8.

Let us suppose that the P elementary windings are connected in series, in order to have the same current I in all the active lengths, and let us define Z_f as the total number of active conductors used in the winding with P pole pairs; then, the fundamental component of the mmf waveform produced by the winding with P pole pairs can be evaluated using (2.15). An example of a four-pole mmf distribution is shown in Figure 2.9.

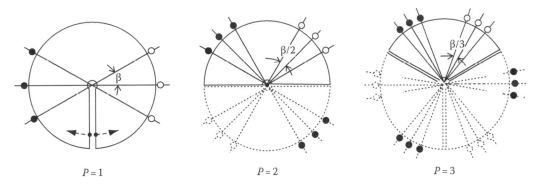

$P = 1$ $P = 2$ $P = 3$

FIGURE 2.8 Ideal procedure to realize a winding with different pole pair numbers deforming a two-pole winding.

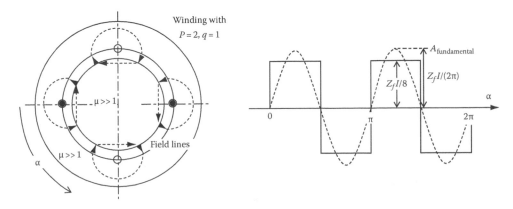

FIGURE 2.9 MMF waveform and field lines for a four-pole ($P = 2$) winding.

$$\hat{A}_{\text{fundamental}} = \frac{K_a \cdot Z_f I}{\pi \cdot P} \tag{2.15}$$

With respect to the calculation of the winding coefficient, K_a, it is important to observe that in this case, the angle β has to be evaluated from the electric-phase-rotation point of view. It is therefore useful to introduce the concept of electric angle, β_e, as the product of the slot pitch geometrical angle and the pole pair number of the winding:

$$\beta_e = P \cdot \beta \tag{2.16}$$

In this way, the winding coefficient for a winding with $2P$ magnetic poles can be calculated in the same formal way used for a two-pole winding ($P = 1$), using the electric angle, β_e, instead of the geometrical angle, β. Also, the equations that describe the fundamental mmf distribution for a two-pole winding (see (2.9), (2.9bis), and (2.10)) can be rewritten in a more general form thanks to the electric angle, as follows:

$$K_a = \cos\left(n_r \cdot \frac{\beta_e}{2} \right) \cdot \frac{\sin(q \cdot \beta_e/2)}{q \cdot \sin(\beta_e/2)}; \quad \text{winding coefficient} \tag{2.17}$$

$$N' = K_a \cdot \frac{Z_f}{\pi P}; \quad \text{equivalent turn number} \tag{2.18}$$

$$A_{\text{fundamental}}(\alpha) = N' I \sin(\alpha_e); \quad \text{fundamental mmf contribution} \tag{2.19}$$

where $\alpha_e = P\alpha$.

With reference to the spatial harmonics of the resultant mmf waveform, for a generic $2P$-pole winding, the following equations can be derived, on the analogy of (2.11) and (2.12):

$$K_{a,h} = \cos\left(n_r \cdot h \cdot \frac{\beta_e}{2} \right) \frac{\sin(q \cdot h \cdot \beta_e/2)}{q \cdot \sin(h \cdot \beta_e/2)} \tag{2.17bis}$$

$$N'_h = K_{a,h} \cdot \frac{Z_f}{\pi \cdot P \cdot h} \tag{2.18bis}$$

$$A(\alpha) = \hat{I} \sum_{h=1,3,5,7,\dots} N'_h \sin(h \cdot \alpha_e) \tag{2.19bis}$$

The use of the electric angle is very important because it allows to study a $2P$-pole winding as a simple two-pole winding. In fact, all the relations involving an angular airgap coordinate (α, β, etc.) written for a two-pole winding are still valid for a $2P$-pole winding if the electric angle is used instead of the geometrical angle.

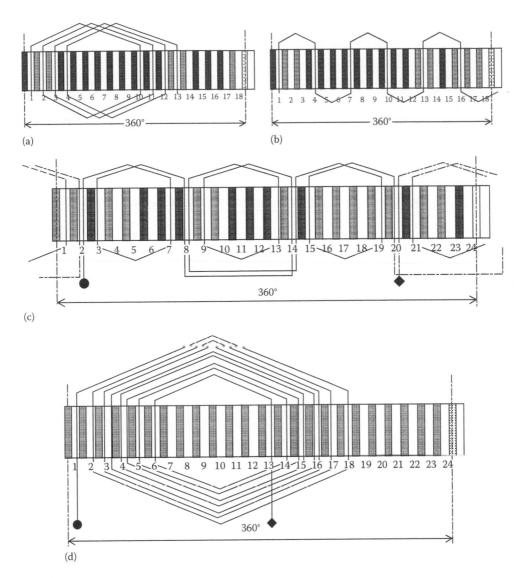

FIGURE E.2.1

Example 2.1

Let us consider the winding layouts reported in Figure E.2.1.
The winding polarity and the winding coefficient of these windings are

Winding A	$P = 1$	$q = 4$	$N_r = 0$	$\beta_e = 20°$	$K_a = 0.925$
Winding B	$P = 3$	$q = 1$	$N_r = 0$	$\beta_e = 20°$	$K_a = 1.000$
Winding C	$P = 2$	$q = 2$	$N_r = 1$	$\beta_e = 30°$	$K_a = 0.933$
Winding D	$P = 1$	$q = 4$	$N_r = 2$	$\beta_e = 15°$	$K_a = 0.925$

2.2.6 Airgap MMF Waveform Produced by a Single Conductor

In this section, the mmf distribution produced by a single conductor is analyzed. This particular winding structure can be considered as a theoretical case and it can be used as a starting point to develop

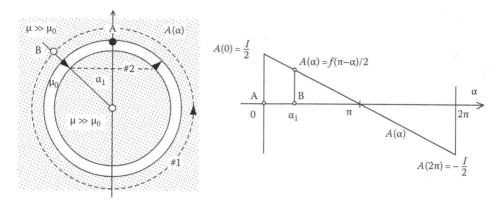

FIGURE 2.10 Magnetic potential distribution produced by an indefinite straight conductor positioned in a cylindrical airgap with constant thickness (point A).

a general theory of a nonconventional winding structure, such as the squirrel cage winding, typically used in induction machines.

Let us consider the geometrical situation shown in Figure 2.10, where two coaxial cylindrical magnetic structures are shown. The single conductor is positioned in the airgap (point A), and it carries current I, entering the drawing plane. Due to this current, two different paths for the magnetic field lines are possible:

- Path #1: magnetic field lines that are in the outer magnetic structure only
- Path #2: magnetic field lines that cross the airgap and are both in the outer and inner cylinders

Nevertheless, if the magnetic material permeability is high, most part of the field lines and, consequently, most part of the linked flux with the conductor will be in path #1, while the field lines with path #2 will be weaker because they have to cross the airgap. The linked flux associated with field lines in path #2 becomes negligible with respect to the total linked flux. As a consequence, it is possible to think that the magnetic voltage drop along the airgap circumference of the external structure is due to the presence of the field lines in path #1 only. For the same reasons, it can be assumed that in any point of the inner structure, the magnetic potential is about zero.

On the basis of previous remarks, it is possible to consider that the magnetic potential difference between the two coaxial structures is proportional to the angular coordinate, α, of the considered point along the airgap, as shown in Figure 2.10 and expressed as follows:

$$A(\alpha) = \frac{I}{2}\text{saw}(\alpha) \tag{2.20}$$

where $A(\alpha)$ is the airgap mmf distribution produced by the conductor and the conventional positive current, I, and the sawtooth function, saw(α), has unitary amplitude and period equal to 2π.

If Z_f conductors, each carrying the same current I, are concentrated at point A of Figure 2.10, (2.20) can be rewritten as

$$A(\alpha) = \frac{Z_f \cdot I}{2}\text{saw}(\alpha) \tag{2.20bis}$$

The mmf waveform is periodic with period 2π, and, using the Fourier series, it can be expressed as follows and as shown in Figure 2.11:

$$A(\alpha) = \sum_{h=1,2,3,\dots} \frac{Z_f}{\pi \cdot h} \cdot I \sin(h \cdot \alpha) \tag{2.21}$$

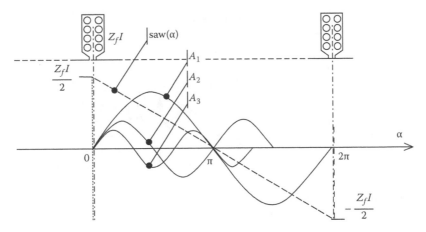

FIGURE 2.11 Harmonic decomposition of the airgap mmf waveform produced by Z_f conductors positioned in a single slot.

Contrary to the case of the diametrical turn, in the distribution spectrum, both odd and even spatial harmonics are present. The fundamental component of the mmf can be calculated using

$$A_{\text{fundamental}}(\alpha) = \frac{Z_f}{\pi} I \sin(\alpha) \tag{2.22}$$

Comparing (2.22) with (2.3), valid for a diametrical bobbin, the following equivalence considerations between a single conductor and a diametrical bobbin can be regarded:

- With respect to the fundamental component: a single conductor can be substituted by a fictitious diametrical bobbin with one active length only ($Z_f = 1$).
- With respect to the spatial harmonics: the amplitude of the spatial harmonics is proportional to $1/h$ in both cases, but for the single conductor, both the odd and even harmonic orders are present.

2.2.7 Airgap Magnetic Flux Density Waveform

When the airgap mmf distribution produced by a system of active windings is known, it is possible to estimate the magnetic field waveform, $H(\alpha)$, or the magnetic flux density waveform, $B(\alpha) = \mu_0 H(\alpha)$. This can be easily done if the following simplifications are adopted:

- No saturation phenomena are present: in this case, all the mmf distribution counterbalances the magnetic voltage drop in the airgap.
- The magnetic structure is assumed to be isotropic. In other words, the airgap thickness is considered constant in each direction.

Thanks to these hypotheses, the flux density distribution along the airgap circumference can be calculated using (2.23), and the two waveforms, $B(\alpha)$ and $A(\alpha)$, are similar in shape:

$$B_t(\alpha) = \mu_0 \frac{A(\alpha)}{l_t} \tag{2.23}$$

In reality, the slots that contain the windings do not have a negligible opening width, and, as a consequence, the airgap thickness, l_t, cannot be assumed constant with respect to the angular coordinate α.

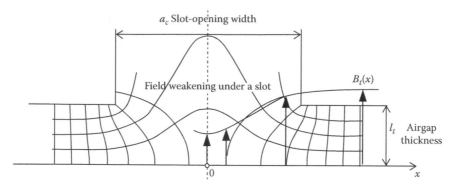

FIGURE 2.12 Slot-opening effect on airgap flux density.

From this point of view, (2.23) has to be considered inadequate for a point-to-point description of the airgap flux density distribution.

Corresponding to the slot opening, the magnetic field is weakened with respect to the field value under the tooth, as shown in Figure 2.12.

Supposing that just one of the airgap surfaces has the slots and the other one is smooth, it is possible to quantify in an analytical way the flux weakening near to the slot opening, if the following assumptions are made:

- Infinite permeability of the magnetic laminations
- Slots with indefinite deep and parallel borders
- Constant magnetic potential difference between faced surfaces

In this case, it is possible to determine an analytical expression for the normal component at the smooth surface of the airgap magnetic field. With reference to Figure 2.12, let us define the airgap linear coordinate x having its origin at the center line of the slot, and the following quantities:

τ_c Slot pitch
a_c Slot-opening width
l_t Airgap thickness
A Magnetic potential difference between the stator and the rotor
$B_{t,\max}$ Magnetic flux density under the center line of the tooth

Furthermore, let us define the parameter $\xi_a = a_c/2l_t$. The normal component of the airgap flux density, $B_{tn}(x)$, on a smooth surface can be evaluated by a Schwarz–Christoffel conformal transformation. The result is expressed in (2.24). In this equation, the intermediate variable w, related to the conformal transformation, is in the range of 0–1 when the coordinate x changes from 0 to ∞.

$$b(x) = \frac{B_{tn}(x)}{B_{t,\max}} = \frac{1}{\sqrt{1+\xi_a^2(1-w^2)}}; \quad \text{with } B_{t,\max} = \mu_0 \frac{A}{l_t}$$

$$\frac{2x}{a_c} = \frac{2}{\pi}\left[\arcsin\frac{\xi_a w}{\sqrt{1+\xi_a^2}} + \frac{1}{2\xi_a}\ln\frac{\sqrt{1+\xi_a^2(1-w^2)}+w}{\sqrt{1+\xi_a^2(1-w^2)}-w}\right]$$

(2.24)

Figure 2.13 shows the ratio $b(x)$ between the flux density value close to the slot, $B_{tn}(x)$, and the same value without the slot presence, $B_{t,\max}$. With reference to the results shown in Figure 2.13, it is possible to conclude that lower values of the flux density in correspondence to the slot opening reduce the magnetic

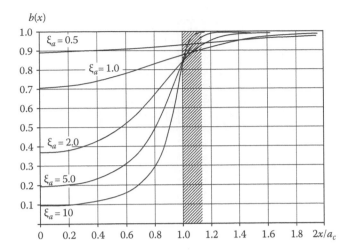

FIGURE 2.13 Airgap field-weakening function due to slot opening.

flux that crosses the airgap. Let us define $\Delta\Phi_c$, the magnetic flux per axial length unit of the machine, that is missing due to the slot opening effect. From (2.24), the following equation can be obtained for the $\Delta\Phi_c$ quantity:

$$\Delta\Phi_c \cong 2B_{t,\max} \cdot \frac{l_t}{\pi}\left(2\xi_a \cdot \arctan\xi_a - \ln\left(\xi_a^2 + 1\right)\right) \tag{2.25}$$

As a consequence, the magnetic flux in a slot pitch produced by the magnetic potential difference, A, can be written as

$$\Phi_d = B_{t,\max}\tau_c - \Delta\Phi_c = B_{t,\max}\left(\tau_c - \frac{2}{\pi}l_t\left(2\xi_a \cdot \arctan\xi_a - \ln\left(\xi_a^2 + 1\right)\right)\right) \quad \text{with } B_{t,\max} = \mu_0\frac{A}{l_t}$$

The same flux can be calculated considering the same magnetic voltage difference, A, on both the faced surfaces without slots, but using an increased value of the airgap thickness, as follows:

$$l_t' = K_C \cdot l_t$$

$$K_C = \frac{\tau_c}{\tau_c - (2/\pi)l_t\left(2\xi_a \arctan\xi_a - \ln\left(1 + \xi_a^2\right)\right)}; \quad \text{where } \xi_a = \frac{a_c}{2l_t} \tag{2.26}$$

The increase coefficient, K_C, is called Carter coefficient. For convenience, some values of this coefficient are given in Figure 2.14.

The Carter coefficient is a number grater than one, and it takes into account, in a global way, the airgap flux weakening due to slots. In the case of semi-closed slots, an approximated equation for the Carter coefficient estimation is

$$K_C \cong \frac{\tau_c}{\tau_c - l_t\left(4\xi_a^2/(5 + 2\xi_a)\right)} \tag{2.26bis}$$

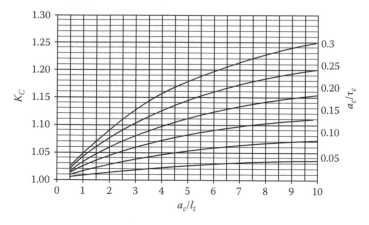

FIGURE 2.14 Values of the Carter coefficient.

Summarizing the considerations taken into account so far, the following conclusions can be drawn:

- If a point-to-point description of the flux density waveform is requested, (2.23) is unacceptable and it has to be at least substituted by the following relation:*

$$B_t(\alpha) = \mu_0 \frac{A(\alpha)}{l_t} \cdot b(\alpha) \tag{2.27}$$

where $b(\alpha)$ is the airgap flux density weakening function due to the slot opening stated in (2.24) and shown in Figure 2.13. The effect of the slots on the actual flux density waveform is shown in Figure 2.15.

- On the other hand, if just the amplitude of the fundamental components of the mmf and flux density waveforms have to be calculated, (2.23) can be rewritten as follows in order to approximately include the slot-opening effects:

$$\hat{B}_{t,\text{fundamental}} \cong \mu_0 \frac{\hat{A}_{\text{fundamental}}}{K_C \cdot l_t} \tag{2.28}$$

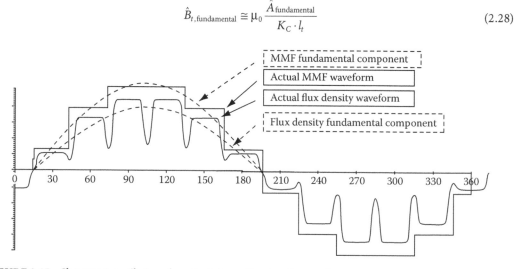

FIGURE 2.15 Slot-opening effect on the actual airgap flux density waveform.

---

* In actual fact, this formulation is not free of criticisms and it should be corrected taking into account the magnetic non-equipotential condition between the right and the left border of a slot when electric current in the conductors inside the considered slot is present.

FIGURE E.2.2

- In the not so infrequent case in which both the stator and rotor surfaces are slotted, the resultant Carter coefficient can be approximately evaluated as the product of the two Carter coefficients due to stator slots and rotor slots separately (considering one surface slotted and the other smooth). This approximation is generally acceptable for semi-closed slots.
- Equation 2.28 can be also used for the calculation of the flux density harmonics produced by the airgap mmf harmonics, but, in this case, the Carter coefficient for each harmonics has to be calculated with equations other than (2.26) and (2.26bis).

Example 2.2

Let us consider the double-layer stator winding of a three-phase, two-pole rotating-field machine (i.e., induction motor) with 18 slots, as shown in Figure E.2.2. The winding pitch is diametrical (full-pitch winding) and there are five conductors in series per slot per layer. Each phase winding structure uses three slots for the outgoing active conductors and three diametrical slots for the backward conductors, as shown in the figure.

Determine the maximum value of the mmf distribution and the amplitude of the mmf fundamental component when a phase current, I, equal to 8 A (instantaneous value) is supplied in the phase winding.

Number of slots per pole per phase	$q = 3$
Number of conductors in series per slot	$Z_c = 10$
MMF amplitude (maximum value)	$A_{max} = q \cdot Z_c \cdot I/2 = 3 \times 10 \times 8/2 = 120\,A$
Slot angular pitch	$\beta = 360°/18 = 20°$
Distribution coefficient (= winding coefficient)	$K_d = \mathrm{sen}(3 \times 10°)/(3 \cdot \mathrm{sen}(10°)) = 0.960$
Number of conductors in series per phase	$Z_f = 5 \cdot 12 = 60$
Amplitude of the mmf fundamental component	$A_{fund} = 0.960 \cdot 60 \cdot 8/3.14 = 146.6\,A$

Example 2.3

Determine the phase current value that produces the same amplitude of the mmf fundamental component as calculated in the previous example, when a pitch shortening of two slots is adopted. In this case, evaluate the new maximum value of the mmf distribution produced by the winding too.

Pitch shortening (in number of slots)	$n_r = 2$
Shortening coefficient	$K_r = \cos(2 \cdot 10°) = 0.940$
Phase current (to get $A_{fund} = 146.6\,A$)	$I' = 8.0/0.940 = 8.5\,A$
MMF amplitude (maximum value)	$A_{max} = q \cdot Z_c \cdot I'/2 = 3 \times 10 \times 8.5/2 = 127.5\,A$

Example 2.4

In Figure E.2.4, a single-layer stator winding of a three-phase rotating-field machine with 24 slots is shown. Using $Z_f = 96$ active conductors in series per phase, two winding structures, with different pole numbers, have to be realized: a two-pole winding ($P = 1$, layout (a) in the figure) and a four-pole winding ($P = 2$, layout (b) in the figure), respectively.

For the two structures, determine the amplitude of the mmf fundamental component if the phase current is $I = 7\,A$.

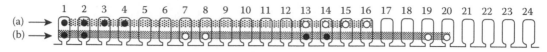

FIGURE E.2.4

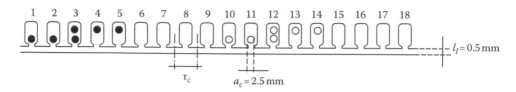

FIGURE E.2.5

	$P = 1$	$P = 2$
Number of pole pairs	$P = 1$	$P = 2$
Slot angular pitch	$\beta_e = 1 \cdot 360°/24 = 15°$	$\beta_e = 2 \cdot 360°/24 = 30°$
Number of slots per pole per phase	$q = 4$	$q = 2$
Winding coefficient ($n_r = 0$)	$K_a = 0.958$	$K_a = 0.966$
MMF fundamental component amplitude	$A_{fund} = 204.8\,A$	$A_{fund} = 103.3\,A$

Example 2.5

For a stator winding, the following data are known: 18 slots, $Z_f = 96$ conductors in series per phase, $q = 3$ slots/pole/phase, and $n_r = 2$ slots. As shown in the Figure E.2.5 the airgap radius is $R_t = 45\,mm$, the slot-opening width is $a_c = 2.5\,mm$, and the airgap thickness is $l_t = 0.5\,mm$.

Determine the phase current value that produces an airgap fundamental flux density amplitude equal to $B_{t,max} = 0.857\,T$.

Slot angular pitch	$\beta = 360°/18 = 20°$
Slot pitch (linear)	$\tau_c = 2\pi \cdot 45/18 = 15.7\,mm$
Half slot-opening width/airgap thickness ratio	$\xi_a = 2.5/(2 \cdot 0.5) = 2.5$
Carter coefficient	$K_c = 15.7/\{15.7 - 2 \cdot 0.5[2 \cdot 2.5 \cdot atn(2.5) - ln(1 + 2.5^2)]/\pi\} = 1.087$
Equivalent airgap thickness	$l_t' = 1.087 \cdot 0.5 = 0.543\,mm$
Distribution coefficient	$K_d = sin(3 \cdot 20°/2)/(3 \cdot sin(20°/2)) = 0.960$
Shortening coefficient	$K_r = cos(2 \cdot 20°/2) = 0.940$
Winding coefficient	$K_a = 0.960 \cdot 0.940 = 0.902$
Equivalent turn number	$N' = 0.902 \cdot 96/3.14 = 27.6$
Phase current[a]	$I = 0.857 \cdot 0.543 \cdot 10^{-3}/(1.256 \cdot 10^{-6} \cdot 27.6) = \mathbf{13.4\,A}$

[a]By the equation of the airgap fundamental flux density amplitude, it is possible to evaluate the phase current, as follows:

$$B_t = \frac{\mu_0}{l_t'} N'I \rightarrow I = B_t \frac{l_t'}{\mu_0 N'}$$

2.3 Rotating Magnetic Field

2.3.1 Rotating Magnetic Field in Three-Phase Windings

In AC electric motors, such as in AC generators, the winding positioned in the stator is, in the majority of cases, a three-phase winding. For this reason, the attention is focused on three-phase winding structures, considering them as a special case of more generic polyphase windings.

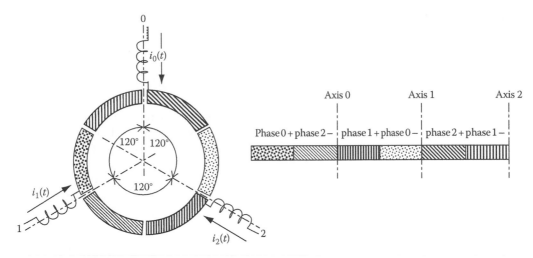

FIGURE 2.16   Typical layout of a two-pole three-phase winding.

In a three-phase winding, the stator slot number pole pair is typically a multiple of six ($N_S = m \cdot q \cdot 2P = 6 \cdot q \cdot P$) in order to place three identical single-phase windings, each one distributed in $2q$ diametrical or quasi-diametrical slots and with symmetry axes shifted by 120°, as shown in Figure 2.16.

If the three single-phase windings (called phases) are identical, they will have the same winding coefficient, K_a, and the same equivalent turn number, N' (see (2.9) and (2.9bis)).

Let us consider a symmetrical set of three sinusoidal currents (2.29) in the phases and the spatial distribution of the phase along the airgap shown in Figure 2.16.

$$i_k(t) = \hat{I} \cos\left(\omega \cdot t - k\frac{2\pi}{3}\right); \quad k = 0,1,2 \tag{2.29}$$

In this case, the presence of the spatial mmf harmonics is neglected and only the fundamental component is taken into account.

By opportunely choosing the origin of the angular coordinate α, for the kth phase, the following relation can be written:

$$A_k(\alpha,t) = N' \sin\left(\alpha - k\frac{2\pi}{3}\right) \cdot i_k(t); \quad N' = \frac{K_a \cdot Z_f}{\pi} \tag{2.30}$$

Equation 2.30 represents an mmf waveform fixed in the airgap (in the space) with an amplitude that changes with time, proportionally with the instantaneous value $i_k(t)$.

This spatial wave has the maximum in correspondence with the symmetry axis of the kth phase, and, if ω is the angular pulsation of the sinusoidal current, the waveform amplitude changes with time in a sinusoidal manner with pulsation ω. In this way, each single-phase winding produces its own pulsating mmf waveform in the airgap, related to the phase winding position. The resultant action in the airgap, from the mmf waveform point of view, can be obtained by summing up the single actions of each phase. The airgap flux density distribution is given by the following equation:

$$B_{t,3}(\alpha,t) = \frac{\mu_0}{l'_t} N' \hat{I} \sum_{k=0,1,2} \sin\left(\alpha - k\frac{2\pi}{3}\right) \cdot \cos\left(\omega \cdot t - k\frac{2\pi}{3}\right)$$

Obtaining the sum and after simple calculations, this relation can be reformulated as follows:

$$B_{t,3}(\alpha,t) = \frac{3}{2}\frac{\mu_0}{l_t'}N'\hat{I}\sin(\alpha - \omega \cdot t) \qquad (2.31)$$

Equation 2.31 shows the airgap flux density waveform as a function of the spatial coordinate α and time t. Equation 2.31 is still a sinusoidal wave along the airgap, but it is not fixed in the space and its spatial phase changes with time with law ωt. Equation 2.31 describes the concept of a rotating magnetic field.

On the basis of the previous comments and with the help of Figure 2.17, it is possible to conclude that the flux density waveform produced by a three-phase winding, supplied with a symmetrical set of three sinusoidal currents of pulsation ω, rotates along the airgap with an angular speed equal to the current pulsation.

In the case that the three-phase winding has P pole pairs, (2.29), (2.30), and (2.31) are still valid if the electric angles are used instead of the mechanical ones and the equivalent turn number is modified as in Section 2.2.2. The mmf distributions of each phase winding and the resultant flux density waveform are, respectively,

$$A_k(\alpha,t) = N'\sin\left(P\cdot\alpha - k\frac{2\pi}{3}\right)\cdot i_k(t); \quad N' = \frac{K_a \cdot Z_f}{\pi P} \qquad (2.30\text{bis})$$

$$B_{t,3}(\alpha,t) = \frac{3}{2}\frac{\mu_0}{l_t'}N'\hat{I}\sin(P\cdot\alpha - \omega \cdot t) \qquad (2.31\text{bis})$$

In particular, (2.30bis) and (2.31bis) highlight the following aspects:

- For a fixed number of total phase conductors, the mmf waveform amplitude produced by the winding is in inversely related to the pole pair number.
- A three-phase winding, with P pole pairs, produces in the airgap a magnetic field with the same number of magnetic polarity of the winding.
- The resultant field wave rotates along the airgap at an angular speed equal to ω/P.

As a consequence, it is possible to state that the winding polarity defines, even if through a discrete series, the speed of the rotating magnetic field. From a technical point of view, this aspect is very important in rotating-field electric machines.

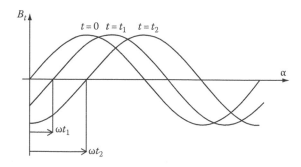

FIGURE 2.17 Graphical representation of a rotating magnetic field.

2.3.2 Rotating Magnetic Field in Squirrel Cage Windings

The squirrel cage can be considered as an atypical case of poly-phase windings, and it is frequently used as a rotor winding in induction machines. In fact, in each conductor (or bar) of the cage, the current is different from the currents in the other bars, and as a consequence, each bar can be considered as a phase winding.

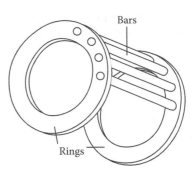

FIGURE 2.18   Squirrel cage winding.

From this point of view, the squirrel cage is a polyphase winding with a phase number m equal to the number of bars, N_R, and each phase is constituted by a unique conductor ($Z_f = 1$) (Figure 2.18).

In addition, the cage winding does not have its own magnetic pole number, as is the case in the traditional distributed winding. The current system in the cage is induced by the airgap rotating field produced by another distributed winding with P pole pairs. This induced current system, flowing in the cage, automatically generates an mmf distribution with the same pole pair number, P.

In this section, the magnetic effects due to this winding structure will be initially analyzed and discussed for a bar current system with two magnetic poles ($P = 1$).

In order to evaluate the fundamental component of the resultant mmf distribution in the airgap produced by the squirrel cage, let us consider a symmetrical set of sinusoidal currents in the N_R bars:

$$i_k(t) = \hat{I}\cos\left(\omega \cdot t - k\cdot\frac{2\pi}{N_R}\right); \quad k = 0,1,2,3,\ldots,N_R-1 \tag{2.32}$$

As stated in Section 2.2.6, the mmf fundamental waveform due to each bar can be calculated as follows, where N' is the equivalent turn number of one bar, from the fundamental mmf distribution production point of view:

$$A_k(\alpha,t) = N'\sin\left(\alpha - k\frac{2\pi}{N_R}\right)\cdot i_k(t); \quad N' = \frac{1}{\pi} \tag{2.33}$$

On the analogy of the three-phase winding case, the fundamental distribution of the airgap flux density for the whole cage can be determined using the following equation:

$$B_{t,N_R}(\alpha,t) = \frac{\mu_0}{l'_t}\frac{1}{\pi}\hat{I}\sum_{k=0,1,2}\sin\left(\alpha - k\frac{2\pi}{N_R}\right)\cdot\cos\left(\omega\cdot t - k\frac{2\pi}{N_R}\right)$$

and, after some calculations, we obtain

$$B_{t,N_R}(\alpha,t) = \frac{N_R}{2}\frac{\mu_0}{l'_t}\frac{1}{\pi}\cdot\hat{I}\sin(\alpha - \omega\cdot t) \tag{2.34}$$

Equation (2.34) is similar to (2.31), except for the coefficient $N_R/2$, instead of the coefficient $3/2$ of the three-phase winding. This is reasonable because the cage can be considered as a polyphase winding with N_R phases.

If the pole pair number of the inducing rotating field, through which the bar current system originates, is equal to P, then the bar current system is given by

$$i_k(t) = \hat{I}\cos\left(\omega \cdot t - kP \cdot \frac{2\pi}{N_R}\right); \quad k = 0,1,2,3,\ldots,N_R - 1 \tag{2.32bis}$$

In this case, the rotating flux density produced by the cage is given by

$$B_{t,N_R}(\alpha,t) = \frac{N_R}{2}\frac{\mu_0}{l_t'}\frac{1}{\pi \cdot P} \cdot \hat{I}\sin(P\alpha - \omega \cdot t) \tag{2.34bis}$$

2.3.3 Equivalence between Different Windings

The expressions of the fundamental distribution of the airgap rotating magnetic field for three-phase winding (2.31bis) and for the polyphase cage winding (2.34bis) are quite similar. For convenience, these equations are stated here again:

$$B_{t,3}(\alpha,t) = \frac{3}{2}\frac{\mu_0}{l_t'}N'\hat{I}\sin(P\alpha - \omega \cdot t); \quad N' = \frac{K_a \cdot Z_f}{\pi P} \tag{2.31bis}$$

$$B_{t,N_R}(\alpha,t) = \frac{N_R}{2}\frac{\mu_0}{l_t'}N'\hat{I}\sin(P\alpha - \omega \cdot t); \quad N' = \frac{1}{\pi P} \tag{2.34bis}$$

In both the cases, the fundamental field distribution is a wave with sinusoidal spatial distribution that rotates along the airgap with an angular speed equal to ω/P, where ω is the electric pulsation of the current system in the windings.

Equations 2.31bis and 2.34bis suggest the possible generalization of the rotating-field expression for polyphase windings with a generic phase number, m.

If Z_f is the number of conductors in series per phase, $\hat{I}$ is the amplitude of the symmetrical set of sinusoidal currents in m phases and ω is the electric pulsation, the rotating-field waveform for the m-phase winding can be evaluated as follows:

$$B_{t,m}(\alpha,t) = \frac{m}{2}\frac{\mu_0}{l_t'}N'\hat{I}\sin(P\alpha - \omega \cdot t); \quad N' = \frac{K_a \cdot Z_f}{\pi P} \tag{2.35}$$

On the basis of (2.35), the following conclusions can be drawn:

- For a three-phase winding, the fundamental field distribution can be calculated from (2.35) using $m = 3$.
- For a cage winding, the correspondent distribution can be obtained from (2.35) using $Z_f = 1$ (just a unique conductor in series per phase), $K_a = 1$ (winding factor), and $m = N_R$ (number of phases).
- The same value of fundamental flux density distribution in the airgap can be equivalently produced by
  - A winding (S) with $m^{(S)}$ phases, $Z_f^{(S)}$ conductors in series per phase, and a symmetrical set of sinusoidal currents of amplitude $I^{(S)}$;
  - A winding (R) with $m^{(R)}$ phases, $Z_f^{(R)}$ conductors in series per phase, and a symmetrical set of sinusoidal currents of amplitude $I^{(R)}$

$$B_t(\alpha,t) = \frac{m^{(S)}}{2} \frac{\mu_0}{l_t'} N'^{(S)} \hat{I}^{(S)} \sin(P\alpha - \omega \cdot t); \quad N'^{(S)} = \frac{K_a^{(S)} \cdot Z_f^{(S)}}{\pi P}$$

$$B_t(\alpha,t) = \frac{m^{(R)}}{2} \frac{\mu_0}{l_t'} N'^{(R)} \hat{I}^{(R)} \sin(P\alpha - \omega \cdot t); \quad N'^{(R)} = \frac{K_a^{(R)} \cdot Z_f^{(R)}}{\pi P}$$

(2.36)

- The current values $I^{(S)}$ and $I^{(R)}$ that verify the identity (2.36) can be defined as **equivalent**, and the ratio between these values is shown in (2.37). The coefficient K_I can be considered as a coefficient to report the current of the winding (R) to the winding (S). In other words, if the ratio in (2.37) is verified, it is possible to conclude that the polyphase current set $I^{(R)}$ in the (R) winding is equivalent, from the fundamental field production point of view, to the polyphase current set $I^{(S)}$ of the winding (S).

$$K_I = \frac{\hat{I}^{(S)}}{\hat{I}^{(R)}} = \frac{m^{(R)} N'^{(R)}}{m^{(S)} N'^{(S)}}$$

(2.37)

Examples 2.6

For a two-pole, full-pitch, three-phase winding, the following data are known: 12 slots and $Z_f = 132$ conductors in series per phase. The average airgap radius is $R_t = 20$ mm, the slot-opening width is $a_c = 2.5$ mm, and the airgap thickness is $l_t = 0.5$ mm.

Determine the rms value of the symmetrical three-phase current set that produces an airgap fundamental flux density amplitude equal to $B_{t,max} = 1$ T.

Slot pitch (linear)	$\tau_c = 2\pi \cdot 20/12 = 10.5$ mm
Half slot-opening width/airgap thickness ratio	$\xi_a = 2.5/(2 \cdot 0.5) = 2.5$
Carter coefficient	$K_C = 10.5/\{10.5 - 2 \cdot 0.5[2 \cdot 2.5 \cdot \text{atn}(2.5) - \ln(1 + 2.5^2)]/\pi\} = 1.137$
Equivalent airgap thickness	$l_t' = 1.137 \cdot 0.5 = 0.568$ mm
Slot angular pitch	$\beta = 360°/12 = 30°$
Winding coefficient	$K_a = \sin(2 \cdot 15°)/(2 \cdot \sin 15°) = 0.966$
Equivalent turn number	$N' = 0.966 \cdot 132/3.14 = 40.6$
rms phase current[a]	$\tilde{I} = 1.0 \cdot 1.414 \cdot 0.568 \cdot 10^{-3}/(3 \cdot 1.256 \cdot 10^{-6} \cdot 40.6) = \mathbf{5.2\ A}$

[a]Defining $\tilde{I}$ as the rms value of the three-phase current system, from (2.31bis) it is possible to write the following relation:

$$\tilde{I} = \hat{B}_t \frac{\sqrt{2}}{3} \frac{l_t'}{\mu_0 N'}$$

Example 2.7

A rotating-field machine consists of the following windings:

(a) Three-phase winding with $P = 1$, $Z_f = 234$, $q = 3$, and $n_r = 0$
(b) Squirrel cage winding with 48 slots (bars)

If the bar current is equal to 150 A_{rms}, calculate the phase current rms value of the three-phase winding that should generate the same airgap fundamental flux density waveform produced by the cage winding.

Calculation of the winding coefficient for the three-phase winding (a)

Slot angular pitch	$\beta^{(a)} = 360°/(6 \cdot 3) = 20°$
Winding coefficient	$K_a^{(a)} = \text{sen}(3 \cdot 10°)/(3 \cdot \text{sen} 10°) = 0.960$
Equivalent current	$\tilde{I}^{(a)} = K_I \cdot \tilde{I}^{(b)} = 48 \cdot 150/(3 \cdot 324 \cdot 0.960) = \mathbf{7.7\ A}$

2.3.4 Vectorial Representation of Airgap Distributions

The fundamental flux density waveform produced by the polyphase winding, such as any sinusoidal distribution along the airgap, can be symbolically represented by means of a vector.

Let us define B_t, the vector associated with the sinusoidal distribution of the flux density. This vector has a magnitude equal to the amplitude of the spatial waveform, and it is oriented where the sinusoidal distribution is maximum, as shown in Figure 2.19.

With the same procedure, it is possible to define the vector A, describing the fundamental mmf distribution that produces the flux density wave. In magnetic linearity conditions, the following relation is valid:

$$\hat{B}_t = \frac{\mu_0}{l_t'} A \tag{2.38}$$

In an electric machine with constant airgap thickness (isotropic magnetic structure), the vectors A and B_t are parallel.

The vector A is determined, as magnitude and orientation, by the amplitude and position of the fundamental mmf distribution produced by the polyphase winding. As a consequence, the vector A depends on the amplitude and the instantaneous phase of the symmetrical polyphase current system in the winding. Starting from this consideration, it is possible to conventionally include in Figure 2.19 a vector I, in phase with the vector A, defined as

$$\hat{A} = \frac{m}{2} N' \hat{I} \tag{2.39}$$

The meaning of I is different from the meaning of the vectors B_t and A. In fact, while these two vectors describe sinusoidal spatial distributions along the airgap of the corresponding quantities, the vector I can be interpreted, from the geometric point of view, in a different way. In particular, the projections of this vector on the magnetic axes of each phase represent the instantaneous values

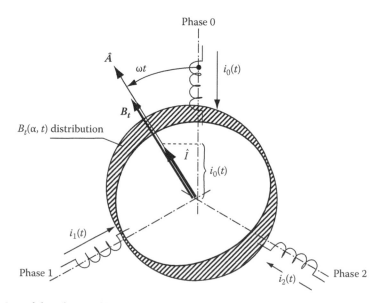

FIGURE 2.19   Spatial distribution of the rotating field and its vectorial distribution.

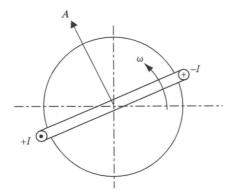

FIGURE 2.20 Rotating bobbin supplied with DC current (this structure is equivalent at a polyphase winding supplied with AC current).

of the respective phase currents (see Figure 2.19), which give rise to the magnetic effects represented by vectors A and B_t.

As the main advantage, the vectorial representation allows to represent the electromagnetic phenomena that occur in the airgap of a machine from a global and synthetic point of view, without the necessity to describe in detail all the local aspects concerning each winding. In other words, the actual polyphase winding previously used to approach the rotating-field theory can be substituted, from the point of view of the resultant effects, with an equivalent fictitious bobbin. In fact, (2.38) and (2.39) can correctly describe the fundamental mmf and flux density rotating waves produced by a diametrical concentrated bobbin, with an equivalent turn number N' equal to

$$N' = \frac{m}{2} \frac{K_a \cdot Z_f}{\pi}$$

This equivalent bobbin is supplied with a DC electric current and rotates at an angular speed ω, as shown in Figure 2.20.

2.3.5 Airgap Useful Flux

Let us define the *airgap useful flux* (or *pole flux*, or *machine flux*), the magnetic flux in the surface corresponding to a polar pitch (one pole) due to the fundamental airgap flux density waveform.

If $\hat{B}_t$ is the amplitude of this fundamental distribution, R_t the airgap radius, and L_a the axial length of the active conductors, the airgap useful flux for a machine with P pole pairs can be calculated as follows:

$$\hat{\Phi}_u = \int_0^{\pi/P} \hat{B}_t \sin P\alpha \cdot R_t \cdot L_a \, d\alpha; \qquad \hat{\Phi}_u = \hat{B}_t \cdot \frac{2R_t L_a}{P} \tag{2.40}$$

The airgap useful flux represents a very important quantity in the rotating-field electric machine study. In fact, the electromechanical conversion phenomenon in the machine can be analyzed thanks to this flux. Also the airgap useful flux can be represented with a spatial vector, Φ_u. This vector has the same direction and orientation of the spatial vector B_t, as shown in Figure 2.21.

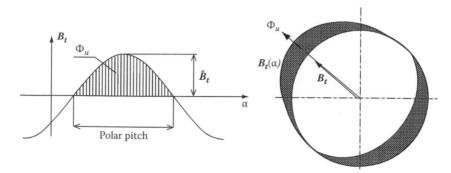

FIGURE 2.21   Airgap useful flux definition and its vectorial representation.

2.3.6 Harmonic Rotating Fields

In the rotating magnetic field theory previously analyzed, only the fundamental distributions of mmf and of flux density have been considered. In fact, in rotating AC electric machines, the "useful" energy electromechanical conversion depends, almost exclusively, on these fundamental distributions.

In the actual case, together with the fundamental field distribution, there are a lot of spatial field harmonics, as shown in (2.13). The main effect of these harmonics is the distortion of the ideal rotating-field wave, adding at the sinusoidal wave other waves rotating a different speed.

In this section, the effect of the spatial harmonics on the resultant rotating field, produced by a poly-phase winding, will be analyzed considering m symmetrical single-phase windings and two-pole structures (phase shift angle = $2\pi/m$).

The mmf distribution produced by the kth phase of the m-phase system is expressed as follows, where h is the harmonic order, Z_f is the conductor number in series per phase of each single-phase winding, and $K_{a,h}$ is the winding coefficient of the hth harmonic, as defined in (2.11):

$$A_k(\alpha) = I \sum_h N_h' \sin\left[h \cdot \left(\alpha - 2\pi \cdot \frac{k}{m} \right) \right] \quad \text{where } N_h' = \frac{K_{a,h} \cdot Z_f}{\pi \cdot h} \tag{2.41}$$

It is important to remember that, for a regular single-phase winding, the h value is in the set of odd positive numbers (1,3,5,7, …), while, for a squirrel cage winding, the h value can be a positive integer number (1,2,3, …). Let us suppose that the sinusoidal currents in the polyphase winding are the symmetrical set as follows, where k is the phase-numbering order:

$$i_k(t) = \hat{I}\cos\left(\omega \cdot t - k\frac{2\pi}{m} \right); \quad k = 0,1,2,\ldots,m-1 \tag{2.42}$$

The resultant mmf distribution due to the excited m-phase winding system can be calculated as

$$A_m(\alpha,t) = \hat{I} \sum_h \left\{ N_h' \cdot \sum_{k=0}^{m-1} \sin\left[h\left(\alpha - k \cdot \frac{2\pi}{m} \right) \right] \cos\left[\left(\omega \cdot t - k \cdot \frac{2\pi}{m} \right) \right] \right\}$$

The above relation can be rewritten in the following form:

$$A_m(\alpha,t) = \hat{I} \sum_h \frac{N_h'}{2} \sum_{k=0}^{m-1} \left\{ \sin\left[h \cdot \alpha - \omega \cdot t - (h-1)k \cdot \frac{2\pi}{m} \right] + \sin\left[h \cdot \alpha + \omega \cdot t - (h+1)k \cdot \frac{2\pi}{m} \right] \right\} \tag{2.43}$$

The corresponding flux density waveform in the airgap is equal to

$$B_{t,m}(\alpha,t) = \mu_0 \frac{A_m(\alpha,t)}{K_C \cdot l_t}$$

In (2.43), the second sum (with the index k from 0 to $m - 1$) is different for zero only for the hth orders of the spatial harmonic that are different from a unit with respect to a multiple of the phase number m.

It is possible to conclude that the flux density spatial harmonics, produced by the polyphase winding, can be grouped in two sets in accordance with the following conditions:

Case 1 $h = nm + 1$ (n integer ≥ 0)

$$B_h(\alpha,t) = \hat{B}_h \sin(h \cdot \alpha - \omega \cdot t) \tag{2.44}$$

Case 2 $h = nm - 1$ (n integer > 0)

$$B_h(\alpha,t) = \hat{B}_h \sin(h \cdot \alpha + \omega \cdot t) \tag{2.45}$$

In both the cases, the results obtained is $B_h = \dfrac{m}{2} \mu_0 \dfrac{N'_h \cdot I}{K_C \cdot l_t}$.

Remarks

- A polyphase winding, supplied in a symmetrical and equilibrated way, produces in the airgap a smaller number of flux density waves with respect to all the spatial harmonics created by each single-phase winding. The greater the phase number m, the lower the distorting harmonic content superimposed at the fundamental rotating field.
- The flux density spatial harmonics, corresponding to the function $\sin(h\alpha \pm \omega t)$, are sinusoidal waveform with a pole pair number equal to h ($2h$ magnetic poles).
- The absolute value of the rotational speed of a flux density wave with order h is $\omega_h = \omega/h$, and the greater the harmonic order h, the lower the speed. The rule that the rotational speed of an airgap field distribution is in inverse proportion to its pole pair number is still valid.
- The rotation direction of the field wave depends on the spatial harmonic order h of the winding. In particular,
  - The values h derived from case 1 ($h = nm + 1$) define harmonics that rotate likewise the fundamental one (direct rotation)
  - The values h derived from case 2 ($h = nm - 1$) define harmonics that rotate in the opposite direction to the fundamental one (inverse rotation)

In symmetrical three-phase windings, the phases produce harmonics with odd integer values for the order h. For these windings, (2.44) and (2.45) can be written as given in (2.44bis) and (2.45bis), respectively, as follows:

Case 1 $h = 6n + 1$ (n integer ≥ 0)

$$B_h(\alpha,t) = \hat{B}_h \sin(h \cdot \alpha - \omega \cdot t) \tag{2.44bis}$$

Case 2 $h = 6n - 1$ (n integer > 0)

$$B_h(\alpha,t) = \hat{B}_h \sin(h \cdot \alpha + \omega \cdot t) \tag{2.45bis}$$

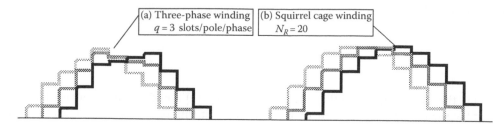

FIGURE 2.22 Rotating-field waveform at different time instants: (a) three-phase winding with 3 slots per pole per phase, (b) squirrel cage winding with 20 bars.

Since the harmonic waves rotate along the airgap at different speeds, the resultant waveform will change its shape during the rotation, as shown in Figure 2.22. Figure 2.22 highlights that the waveform distortion is bigger for the three-phase winding with respect to the 20-bars cage winding. In fact, as previously discussed, the cage winding can be considered as a 20-phase winding.

2.3.7 Windings for Linear AC Machines

Linear AC machines depict a special case of the traditional rotating ones. The distributed windings used in the linear machines can be analyzed as the traditional ones so far described.

If the ideal procedure of deformation of a two-pole machine, introduced in Section 2.2.5, is made until to rectify the machine, the winding result distributed over a straight line. In this case, the result is a **linear winding**, as shown in Figure 2.23.

In linear machines, the airgap field is not rotating, but is a linearly moving field with a linear speed, v. The span length, D, of the linear winding, shown in Figures 2.23 and 2.24, can be calculated using the airgap circumference radius, R_t, of the ideal starting winding. In particular, it is $D = 2\pi R_t$.

The linear speed of the field can be evaluated considering this new winding equivalent to the rotating-field winding. If ω is the electric pulsation of current I, flowing in the windings, a complete rotation of the field happens after a time period $T = 2\pi/\omega$. In the same time period, T, the fundamental airgap magnetic field, produced by the linear winding, covers the distance D. This means that the linear speed of the field is equal to

$$v = \frac{D}{T} = \frac{D}{2\pi}\omega \tag{2.46}$$

All the other aspects discussed for the conventional rotating-field polyphase windings (such as the spatial harmonic content, winding polarity, slot effects, and so on) are valid for the linear windings too.

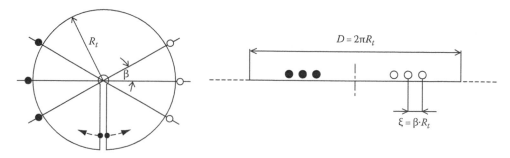

FIGURE 2.23 Ideal procedure to realize a linear winding.

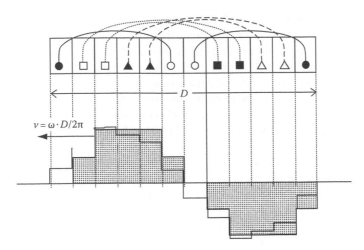

FIGURE 2.24 Two-pole, three-phase linear winding ($q = 2$ slots/pole/phase), and airgap field waveform at two different time instants.

2.3.8 Fractional-Slot Concentrated Windings

Nowadays, interest in AC windings with a non-integer number of slots per pole and per phase less than one ($q < 1$) is sensibly increased, in particular, in permanent magnet synchronous machines. In fact, this winding structure provides some technological advantages, such as the possibility to obtain very short, nonoverlapped endwindings. Although they have some disadvantages with respect to traditional distributed windings; for example, they create heavy mmf subharmonics (spatial harmonics with an order lower than the machine polarity) if no shrewdness is adopted to limit them. In this section, a short summary of the design rules of fractional-slot concentrated windings is given. A complete description of the theory for these winding types can be found in the literature [7]. As an example, in Figure 2.25, a three-phase, single-layer, fractional-slot winding with coils wound around the teeth is shown.

As is well known, the number of slots per pole and per phase of an electric machine is equal to $q = N_S/(2P \cdot m)$, where N_S represents the number of slots, m is the number of phases, and P is the number of pole pairs.

Fixing N_S, if the machine has a high number of pole pairs, then q decreases. In fact, the number of slots is determined once the diameter and the slot pitch of the machine are fixed. For an integer

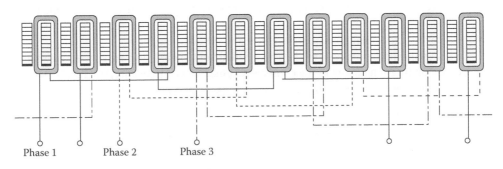

Phase 1   Phase 2   Phase 3

FIGURE 2.25 24-slot, 28-pole, three-phase fractional-slot concentrated winding ($q = 0.2857$ slots/pole/phase, $K_a = 0.9659$).

value of q, a large number of pole pairs, P, restricts the number of slots per pole and per phase, q, and this contributes to worsen the form of the induced emf. Winding arrangements with a number of slots per pole and per phase lower than unity become sometimes mandatory for the construction of AC machines with a large number of pole pairs. Fractional-slot windings with q less than unity may indeed yield a larger number of poles at a fixed number of slots by placing less than one slot per phase within each pole. In other terms, in each pole, conductors pertaining to one or more phases may be missing. In some cases, adopting a layout of this kind makes it possible to realize concentrated nonoverlapping windings that, at the same time, yield high values of the fundamental winding factor (Figure 2.25).

In fact, writing q as $b/2P$, and naming r the GCD $(b, 2P)$, it is possible to individuate r repetitions of an elementary winding. The elementary winding is composed of N_S/r slots and $P' = 2P/r$ pole pairs. The number of slots per phase of the elementary winding is therefore $q_r = qP'$. Generally, q_r is an integer number greater than unity.

The distribution factor, K_a, related to the working spatial mmf harmonic of the considered fractional-slot winding, is

$$K_a = \frac{1}{q_r} \frac{\sin((\pi \cdot P \cdot q_r)/N_S)}{\sin((\pi \cdot P)/N_S)} = \frac{1}{2q_r \sin(\pi/(6 \cdot q_r))} \tag{2.47}$$

It is possible to evaluate the combinations of number of poles, $2P$, and number of slots, N_S, that provide the maximum value of the coefficient K_a. Some comparative investigations on different fractional-slot winding arrangements together with a comprehensive analysis of winding factors for concentrated windings can be found in the technical literature.

2.3.9 Constructive Aspects of AC Distributed Windings

As reported in Section 2.2.1, in order to analyze the airgap mmf produced by a distributed winding, it is not important to know if the active conductors are interconnected. Anyway, the ways to connect the active conductors can be related to constructive opportunities, the spatial localization of the ends of the phase windings, and the necessity to avoid the presence of shaft currents. For these reasons, some aspects related to the winding realization are briefly described [8].

In general, the following classifications of the interconnection solutions are possible:

1. With respect to the endwinding layout:
   a. Concentric winding: in this solution, the endwindings are different from each other (Figure 2.26a).
   b. Crossed winding: in this case, the endwindings are all equal and overlapped (Figure 2.26b).
2. With respect to the connections between a pole and the adjacent poles:
   a. Type A winding: in this case, all the active conductors under a pole are connected with all the corresponding conductors in the consecutive pole. As a consequence, each phase winding is constituted by a number of coil sets equal to the pole pair number, P (Figure 2.27a).

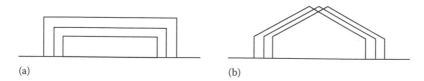

(a)   (b)

FIGURE 2.26   Endwinding layout: (a) concentric and (b) crossed type.

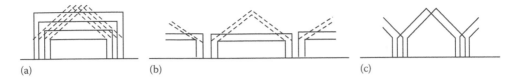

FIGURE 2.27 (a) Type A and (b and c) type B windings.

  b. Type B winding: the active conductors under a pole are connected both with conductors in the previous and in the consecutive pole. In this case, the number of coil sets is equal to the pole number, $2P$ (Figure 2.27b and c).

3. With respect to the winding realization:

  a. Winding realized with coils (wires or bars with a small cross section), as shown in Figure 2.28a.

  b. Undulating winding or winding realized with bars (single-layer winding with a single conductor in the slot): this type of winding is used in machines with high currents and the connections are progressive from one pole to the other, as shown in Figure 2.28b.

Each of the previous possibilities can be used in any classification. As a consequence, in principle, it is possible to have concentric or crossed windings realized with coils of both types A and B. In a similar manner, undulating concentric or undulating crossed windings, both of type A and B, are possible.

The concentric and the crossed undulating windings of type B, which have two directions in the progression of the connections along the circumference, require a "regression bar" in order to link the two directions (Figure 2.29).

In general, considering the whole winding structure, a double-layer winding can be considered as a type B winding. As shown in Figure 2.30, the shape of the endwindings, in an axial direction, is quite different for single- and double-layer windings.

In a single-phase winding, the type A winding structure must not be used, in order to avoid the presence of shaft voltage.

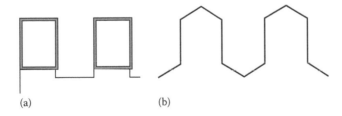

FIGURE 2.28 (a) Winding realized with coils and (b) undulating winding.

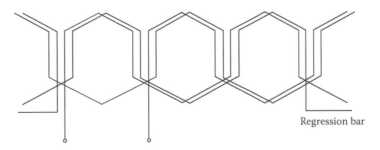

Regression bar

FIGURE 2.29 Type B, crossed undulating winding ($2P = 4$ poles, $q = 4$ slots/pole/phase).

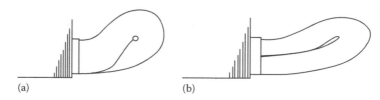

FIGURE 2.30   Endwinding shape in an axial direction: (a) single-layer winding and (b) double-layer winding.

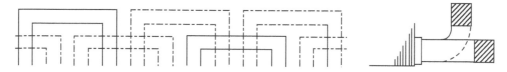

FIGURE 2.31   Endwindings positioned in two planes ($N_S = 24$, $2P = 4$, $q = 2$ slots/pole/phase).

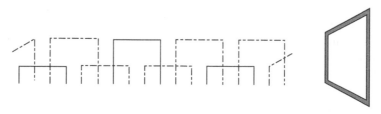

FIGURE 2.32   Endwindings positioned in two planes ($N_S = 18$, $2P = 6$, $q = 1$ slot/pole/phase). In this case a crooked coil is necessary.

FIGURE 2.33   Endwindings positioned in three planes ($N_S = 24$, $2P = 4$, $q = 2$ slots/pole/phase).

For concentric windings, the endwindings of each phase have to be positioned on different planes. With reference to the three-phase case, the following situations are possible:

1. Type A winding with P even (Figure 2.31): the endwindings are positioned onto two planes; each phase has $P/2$ straight coils and $P/2$ bent coils.
2. Type A winding with P odd (Figure 2.32): the endwindings are positioned onto two planes, but a crooked coil is requested in order to pass from one plane to the other.
3. The crooked coil suggests the so-called American winding type, where all the coils have the same crooked shape. The American winding structure can be realizable for any crossed winding type.
4. Type B winding (Figure 2.33): in this case, the endwindings are positioned on three different planes (one plane for each phase).

References

1. I. Boldea and S. A. Nasar, *The Induction Machine Handbook*, CRC Press, Boca Raton, FL, 2002, ISBN 0-8493-0004-5.
2. M.M. Liwschitz-Garik, *Winding Alternating Current Machines*, Van Nostrand Publications, New York, 1950.
3. W. Schuisky, *Berechnung Elektrischer Maschinen*, 1st edn., Springer-Verlag Publishers, Weinheim, Germany, 1960.
4. H. Sequez, The windings of electrical machines, *A.C. Machines*, vol. 3, Springer Verlag, Vienna, Austria, 1950 (In German).
5. E. Levi, *Polyphase Motors: A Direct Approach to Their Design*, John Wiley & Sons, New York, February 1984, ISBN-13: 978-0471898665.
6. P. L. Alger, *Induction Machines—Their Behavior and Uses*, Gordon and Breach Science Publishers SA, Basel, Switzerland, 1970, ISBN 2-88449-199-6.
7. N. Bianchi, M. Dai Prè, L. Alberti, and E. Fornasiero, Theory and design of fractional-slot PM machines, *IEEE IAS Tutorial Course Notes, Editorial CLEUP Editore*, Seattle, WA, September 2007, ISBN 978-88-6129-122-5.
8. G. Crisci, *Costruzione, schemi e calcolo degli avvolgimenti delle machine rotanti*, Editorial STEM Mucchi, Modena, Italy, 1977 (in Italian).

3

Multiphase AC Machines

Emil Levi
Liverpool John Moores University

3.1 Introduction

AC machines with three or more phases ($n \geq 3$) operate utilizing the principle of rotating field,* which is created by spatially shifting individual phases along the circumference of the machine by an angle that equals the phase shift in the multiphase system of voltages (currents), used to supply such a multiphase winding. Such machines are of either synchronous or induction type. All rotating fields in multiphase machines, caused by the fundamental harmonic of the supply, rotate at synchronous speed, governed with the stator winding frequency. When the rotor rotates at the same speed, the machine is of synchronous type. When the rotor rotates at a speed different from synchronous, the machine is called asynchronous or induction machine.

Principles of mathematical modeling of multiphase machines have been developed in the first half of the twentieth century [1–3]. These include a number of different mathematical transformations that replace original phase variables (voltages, currents, flux linkages) with some new fictitious variables, the principal aim being simplification of the system of dynamic equations that describes a multiphase ac machine. Matrices are customarily used in the process of the model transformation, typically in real form. A somewhat different and nowadays very popular approach, which utilizes space vectors and derives from Fortescue's symmetrical component (complex) transformation [1],

---

* What is customarily known as a two-phase winding is in essence a four-phase structure, since the spatial shift between magnetic axes of the phases, as well as the phase shift between phase currents, is equal to $\pi/2$.

was developed in [4]. Its principal advantage, when compared to the matrix method, is a more compact form of the resulting model (that is otherwise the same), which is also easier to relate to the physics of the machinery.

Following the extensive work, conducted in relation to multiphase machine modeling in the beginning of the last century, numerous textbooks have been published, which detail the model transformation procedures for induction and synchronous machines, as well as the applications of the models in analysis of ac machine transients [5–23]. The principles of multiphase machine modeling, model transformations, and resulting models for both induction and synchronous machine (including machines with an excitation winding, permanent magnet synchronous machines, and synchronous reluctance machines) are presented here in a compact and easy-to-follow manner. Although most of the industrial machines are with three phases, the general case of an *n*-phase machine is considered throughout, with subsequent discussion of the required particularization to different phase numbers.

Modeling of multiphase ac machines is customarily subject to a number of simplifying assumptions. In particular, it is assumed that all individual phase windings are identical and that the multiphase winding is symmetrical. This means that the spatial displacement between magnetic axes of any two consecutive phases is exactly equal to $\alpha = 2\pi/n$ electrical degrees.* Further, the winding is distributed across the circumference of the stator (rotor) and is designed in such a way that the magneto-motive force (mmf) and, consequently, flux have a distribution around the air-gap, which can be regarded as sinusoidal. This means that all the spatial harmonics of the mmf, except for the fundamental, are neglected. Next, the impact of slotting of stator (rotor) is neglected, so that the air-gap is regarded as uniform in machines with circular cross section of both stator and rotor (induction machines and certain types of synchronous machines). If there is a winding on the rotor, which is of a squirrel-cage type (as the case is in the most frequently used induction machines and in certain synchronous machines), bars of such a rotor winding are distributed in such a manner that the mmf of this winding has the same pole pair number as the stator winding and the complete winding can be regarded as equivalent to a winding with the same number of phases as the stator winding.

Some further assumptions relate to the parameters of the machines. In particular, resistances of stator (rotor) windings are assumed constant (temperature-related variation and frequency-related variation due to skin effect are thus neglected). Leakage inductances are also assumed constant, so that any leakage flux saturation and frequency-related leakage inductance variation are ignored. Nonlinearity of the ferromagnetic material is neglected, so that the magnetizing characteristic is regarded as linear. Consequently, magnetizing (mutual) inductances are constant. Finally, losses in the ferromagnetic material due to hysteresis and eddy currents are neglected, as are any parasitic capacitances.

The assumptions listed in the preceding two paragraphs enable formulation of the mathematical model of a multiphase machine in terms of phase variables. Of particular importance is the assumption on sinusoidal mmf distribution, which, combined with the assumed linearity of the ferromagnetic material, leads to constant inductance coefficients within a multiphase (stator or rotor) winding in all machines with uniform air-gap. In machines with nonuniform air-gap, however, inductance coefficients within a multiphase winding are governed by a sum of a constant term and the second harmonic, which imposes certain restrictions in the process of the model transformation. Hence, a machine with uniform air-gap is selected for the discussions of the modeling procedure and subsequent model derivation. The machine is a multiphase induction machine, since obtained dynamic models can easily be accommodated to various types of synchronous machines. Motoring convention for positive power flow is

---

* In certain multiphase ac machines this condition is not satisfied. The discussion of such machines is covered in Section 3.7.

utilized throughout, so that the positive direction for current is always from the supply source into the phase of the machine. The number of rotor bars (phases) is, for simplicity, taken as equal to the number of stator phases n.

3.2 Mathematical Model of a Multiphase Induction Machine in Original Phase-Variable Domain

Consider an n-phase induction machine. Let the phases of both stator and rotor be denoted with indices 1 to n, according to the spatial distribution of the windings, and let additional indices s and r identify the stator and the rotor, respectively. Schematic representation of the machine is shown in Figure 3.1, where magnetic axes of the stator winding are illustrated. The machine's phase windings are assumed to be connected in star, with a single isolated neutral point.

Since all the windings of the machine are of resistive-inductive nature, voltage equilibrium equation of any phase of either stator or rotor is of the same principal form, $v = Ri + d\psi/dt$. Here, v, i, and ψ stand for instantaneous values of the terminal phase to neutral voltage, phase current, and phase flux linkage, respectively, while R is the phase winding resistance. Since there are n phases on both stator and rotor, the voltage equilibrium equations can be written in a compact matrix form, separately for stator and rotor, as

$$[v_s] = [R_s][i_s] + \frac{d[\psi_s]}{dt}$$

$$[v_r] = [R_r][i_r] + \frac{d[\psi_r]}{dt}$$

(3.1)

where voltage, current, and flux linkage column vectors are defined as

$$[v_s] = \begin{bmatrix} v_{1s} & v_{2s} & v_{3s} & \cdots & v_{ns} \end{bmatrix}^t \qquad [v_r] = \begin{bmatrix} v_{1r} & v_{2r} & v_{3r} & \cdots & v_{nr} \end{bmatrix}^t$$

$$[i_s] = \begin{bmatrix} i_{1s} & i_{2s} & i_{3s} & \cdots & i_{ns} \end{bmatrix}^t \qquad [i_r] = \begin{bmatrix} i_{1r} & i_{2r} & i_{3r} & \cdots & i_{nr} \end{bmatrix}^t$$

$$[\psi_s] = \begin{bmatrix} \psi_{1s} & \psi_{2s} & \psi_{3s} & \cdots & \psi_{ns} \end{bmatrix}^t \qquad [\psi_r] = \begin{bmatrix} \psi_{1r} & \psi_{2r} & \psi_{3r} & \cdots & \psi_{nr} \end{bmatrix}^t$$

(3.2)

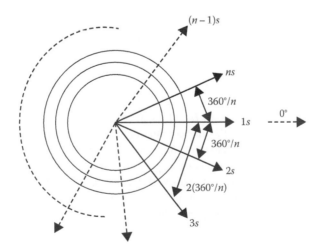

FIGURE 3.1 Schematic representation of an n-phase induction machine, showing magnetic axes of stator phases $((\alpha = 2\pi/n))$.

and $[R_s]$ and $[R_r]$ are diagonal $n \times n$ matrices, $[R_s] = \text{diag}(R_s)$, $[R_r] = \text{diag}[R_r]$. Since rotor winding in squirrel-cage induction machines and in synchronous machines (where it exists) is short-circuited, rotor voltages in (3.2) are zero. The exception is a slip-ring (wound rotor) induction machine, where rotor windings can be accessed from the stationary outside world and rotor voltages may thus be of nonzero value.

Connection between stator (rotor) phase flux linkages and stator/rotor currents can be given in a compact matrix form as

$$[\psi_s] = [L_s][i_s] + [L_{sr}][i_r]$$

$$[\psi_r] = [L_r][i_r] + [L_{sr}]^t[i_s]$$

(3.3)

where $[L_s]$, $[L_r]$, and $[L_{sr}]$ stand for inductance matrices of the stator winding, the rotor winding, and mutual stator-to-rotor inductances, respectively. Relationship $[L_{rs}] = [L_{sr}]^t$ holds true and it has been taken into account in (3.3). Due to the assumed perfectly cylindrical structure of both stator and rotor, and assumption of constant parameters, stator and rotor inductance matrices contain only constant coefficients:

$$[L_s] = \begin{bmatrix} L_{11s} & L_{12s} & L_{13s} & \cdots & L_{1ns} \\ L_{21s} & L_{22s} & L_{23s} & \cdots & L_{2ns} \\ L_{31s} & L_{32s} & L_{33s} & \cdots & L_{3ns} \\ \cdots & \cdots & \cdots & \cdots & \cdots \\ L_{n1s} & L_{n2s} & L_{n3s} & \cdots & L_{nns} \end{bmatrix}$$

(3.4a)

$$[L_r] = \begin{bmatrix} L_{11r} & L_{12r} & L_{13r} & \cdots & L_{1nr} \\ L_{21r} & L_{22r} & L_{23r} & \cdots & L_{2nr} \\ L_{31r} & L_{32r} & L_{33r} & \cdots & L_{3nr} \\ \cdots & \cdots & \cdots & \cdots & \cdots \\ L_{n1r} & L_{n2r} & L_{n3r} & \cdots & L_{nnr} \end{bmatrix}$$

(3.4b)

Here, for both stator and rotor, winding phase self-inductances are governed with $L_{11} = L_{22} = \ldots = L_{nn}$, while for mutual inductances within the stator (rotor) winding $L_{ij} = L_{ji}$ holds true, where $i \neq j$, $i, j = 1 \ldots n$. For example, in a three-phase winding $L_{12} = L_{13} = L_{21} = L_{31} = L_{23} = L_{32} = M \cos 2\pi/3$, since $\cos 2\pi/3 = \cos 4\pi/3$, so that there is a single value of all the mutual inductances within a winding. Also, $L_{ii} = L_l + M$, where L_l is the leakage inductance. However, taking as an example a five-phase winding, one has two different values of mutual inductances within a winding, $L_{12} = L_{21} = L_{15} = L_{51} = L_{23} = L_{32} = L_{34} = L_{43} = L_{45} = L_{54} = M \cos 2\pi/5$ and $L_{13} = L_{31} = L_{14} = L_{41} = L_{24} = L_{42} = L_{35} = L_{53} = L_{52} = L_{25} = M \cos 2(2\pi/5)$. In general, given an n-phase winding, there will be, due to symmetry, $(n-1)/2$ different mutual inductance values within the winding.

Stator-to-rotor mutual inductance matrix of (3.3) contains time-varying coefficients. Time dependence is indirect, through the instantaneous rotor position variation, since the position of any rotor phase winding magnetic axis constantly changes with respect to any stator phase winding magnetic axis, due to rotor rotation. Let the instantaneous position of the rotor phase 1 magnetic axis with respect to the stator phase 1 magnetic axis be θ degrees (electrical). Electrical rotor speed of rotation and the rotor position are related through

$$\theta = \int \omega \, dt$$

(3.5)

Due to the assumption of sinusoidal mmf distribution, mutual inductances between stator and rotor phase windings can be described with only the first harmonic terms, so that

$$[L_{sr}] = M \begin{bmatrix} \cos\theta & \cos(\theta-(n-1)\alpha) & \cos(\theta-(n-2)\alpha) & \dots & \cos(\theta-\alpha) \\ \cos(\theta-\alpha) & \cos\theta & \cos(\theta-(n-1)\alpha) & \dots & \cos(\theta-2\alpha) \\ \cos(\theta-2\alpha) & \cos(\theta-\alpha) & \cos\theta & \dots & \cos(\theta-3\alpha) \\ \dots & \dots & \dots & \dots & \dots \\ \cos(\theta-(n-1)\alpha) & \cos(\theta-(n-2)\alpha) & \cos(\theta-(n-3)\alpha) & \dots & \cos\theta \end{bmatrix} \quad (3.6)$$

Note that in (3.6) one has $\cos(\theta - (n - 1)\alpha) \equiv \cos(\theta + \alpha)$, $\cos(\theta - (n - 2)\alpha) \equiv \cos(\theta + 2\alpha)$, etc.

Model (3.1) through (3.6) completely describes the electrical part of a multiphase induction machine. Since there is only one degree of freedom for rotor movement, the equation of mechanical motion is

$$T_e - T_L = J\frac{d\omega_m}{dt} + k\omega_m \quad (3.7a)$$

where
  J is inertia of rotating masses
  k is the friction coefficient
  T_L is the load torque
  ω_m is the mechanical angular speed of rotation

The inductances of (3.6) are functions of electrical rotor position and, hence, according to (3.5), electrical rotor speed of rotation. The equation of mechanical motion (3.7) is, therefore, customarily given in terms of electrical speed of rotation ω, which is related to the mechanical angular speed of rotation through the number of magnetic pole pairs P, $\omega = P\omega_m$. Hence,

$$T_e - T_L = \frac{J}{P}\frac{d\omega}{dt} + \frac{1}{P}k\omega \quad (3.7b)$$

Equation of mechanical motion (3.7) is always of the same form, regardless of whether original variables or some new variables are used. Symbol T_e stands for the electromagnetic torque, developed by the machine. It in essence links the electromagnetic subsystem with the mechanical subsystem and is responsible for the electromechanical energy conversion. In general, electromagnetic torque is governed with

$$T_e = P\frac{1}{2}[i]^t \frac{d[L]}{d\theta}[i] \quad (3.8)$$

where

$$[L] = \begin{bmatrix} [L_s] & [L_{sr}] \\ [L_{rs}] & [L_r] \end{bmatrix} \quad (3.9a)$$

$$[i] = \begin{bmatrix} [i_s]^t & [i_r]^t \end{bmatrix}^t \quad (3.9b)$$

As stator and rotor winding inductance matrices, given with (3.4), do not contain rotor-position-dependent coefficients, Equation 3.8 reduces for smooth air-gap multiphase machines to

$$T_e = P[i_s]^t \frac{d[L_{sr}]}{d\theta}[i_r] \qquad (3.10)$$

This means that, in machines with uniform air-gap, electromagnetic torque is solely created due to the interaction of the stator and rotor windings.

Any multiphase induction machine is completely described, in terms of phase variables (or, as it is said, in the original phase domain) with the mathematical model given with (3.1) through (3.8) (or (3.10) instead of (3.8)). The model is composed of a total of $2n + 1$ first-order differential equations (3.1) and (3.7), where $2n$ differential equations are voltage equilibrium equations, while the $(2n + 1)$th differential equation is the mechanical equilibrium equation. In addition, there are $2n + 1$ algebraic equations (3.3) and (3.8). The first $2n$ algebraic equations provide correlation between flux linkages and currents of the machine, while the $(2n + 1)$th algebraic equation is the torque equation. Finally, the model is completed with an integral equation (3.5), which relates instantaneous rotor electrical position with the angular speed of rotation.

Substitution of flux linkages (3.3) into voltage equilibrium equations (3.1) and electromagnetic torque (3.10) into the equation of mechanical motion (3.7) eliminates algebraic equations, so that the machine model contains $2n + 1$ first-order differential equations in terms of winding currents, plus the integral equation (3.5). This is a system of nonlinear differential equations, with time-varying coefficients due to variable stator-to-rotor mutual inductances of (3.6). While solving this model directly, in terms of phase variables, is nowadays possible with the help of computers, this was not the case 100 years ago. Hence, a range of mathematical transformations of the basic phase-variable model has been developed, with the prime purpose of simplifying the model by the so-called change of variables. Model transformation is therefore considered next.

Before proceeding further, one important remark is due. Since stator and rotor variables and parameters in general apply to two different voltage levels, rotor winding is normally referred to the stator winding voltage level. This is in principle the same procedure that is customarily applied in conjunction with transformers, and it basically brings all the windings of the machine to the same voltage (and current) base. In all machines where the squirrel-cage rotor winding is used (induction machines and synchronous machines with damper winding), the actual values of rotor currents and rotor parameters cannot anyway be measured and, hence, this change of the rotor winding voltage level has no consequence on the subsequent model utilization since rotor voltages of (3.2) are by default equal to zero. However, if there is excitation at the rotor winding side, as the case may be with slip-ring induction machines (and as the case is with the field winding of the synchronous machines), in which case rotor winding voltages are not zero, it is important to have in mind that rotor voltages and currents (as well as parameters) will in what follows be values referred to the stator winding. No distinction is made here in terms of notation between original rotor winding variables and parameters, and corresponding values referred to the stator voltage level. As a matter of fact, it has already been implicitly assumed in the development of the model (3.1) through (3.10) that rotor winding has been referred to the stator winding.

3.3 Decoupling (Clarke's) Transformation and Decoupled Machine Model

Variables of an n-phase symmetrical induction machine can be viewed as belonging to an n-dimensional space. Since the stator winding is star connected and the neutral point is isolated, the effective number of the degrees of freedom is $(n-1)$; this applies to the rotor winding also. The machine model in the original phase-variable form can be transformed using decoupling (Clarke's) transformation matrix, which replaces the original sets of n variables with new sets of n variables. This transformation decomposes the original n-dimensional vector space into $n/2$ two-dimensional subspaces (planes) if the phase number is an even number. If the phase number is an odd number, the original space is decomposed into $(n-1)/2$ planes plus one single-dimensional quantity. The main property of the transformation is that new two-dimensional subspaces are mutually perpendicular, so that there is no coupling between them. Further, in each two-dimensional subspace, there is a pair of quantities, positioned along two mutually perpendicular axes. This leads to significant simplification of the model, compared to the original one in phase-variable form, as demonstrated next.

Let the correlation between any set of original phase variables and a new set of variables be defined as

$$[f]_{\alpha\beta} = [C][f_{1,2,\ldots n}] \tag{3.11}$$

where

$[f]_{\alpha\beta}$ stands for voltage, current, or flux linkage column matrix of either stator or rotor after transformation

$[f_{1,2,\ldots n}]$ is the corresponding column matrix in terms of phase variables

$[C]$ is the decoupling transformation matrix

It is the same for both stator and rotor multiphase windings and, for an arbitrary phase number n, it can be given as

$$
\underline{C} = \sqrt{\frac{2}{n}}
\begin{array}{c}
\alpha \\ \beta \\ x_1 \\ y_1 \\ x_2 \\ y_2 \\ \cdots \\ x_{\frac{n-4}{2}} \\ y_{\frac{n-4}{2}} \\ 0_+ \\ 0_-
\end{array}
\left[
\begin{array}{cccccccc}
1 & \cos\alpha & \cos 2\alpha & \cos 3\alpha & \cdots & \cos 3\alpha & \cos 2\alpha & \cos\alpha \\
0 & \sin\alpha & \sin 2\alpha & \sin 3\alpha & \cdots & -\sin 3\alpha & -\sin 2\alpha & -\sin\alpha \\
1 & \cos 2\alpha & \cos 4\alpha & \cos 6\alpha & \cdots & \cos 6\alpha & \cos 4\alpha & \cos 2\alpha \\
0 & \sin 2\alpha & \sin 4\alpha & \sin 6\alpha & \cdots & -\sin 6\alpha & -\sin 4\alpha & -\sin 2\alpha \\
1 & \cos 3\alpha & \cos 6\alpha & \cos 9\alpha & \cdots & \cos 9\alpha & \cos 6\alpha & \cos 3\alpha \\
0 & \sin 3\alpha & \sin 6\alpha & \sin 9\alpha & \cdots & -\sin 9\alpha & -\sin 6\alpha & -\sin 3\alpha \\
\cdots & \cdots & \cdots & \cdots & \cdots & \cdots & \cdots & \cdots \\
1 & \cos\left(\frac{n-2}{2}\right)\alpha & \cos 2\left(\frac{n-2}{2}\right)\alpha & \cos 3\left(\frac{n-2}{2}\right)\alpha & \cdots & \cos 3\left(\frac{n-2}{2}\right)\alpha & \cos 2\left(\frac{n-2}{2}\right)\alpha & \cos\left(\frac{n-2}{2}\right)\alpha \\
0 & \sin\left(\frac{n-2}{2}\right)\alpha & \sin 2\left(\frac{n-2}{2}\right)\alpha & \sin 3\left(\frac{n-2}{2}\right)\alpha & \cdots & -\sin 3\left(\frac{n-2}{2}\right)\alpha & -\sin 2\left(\frac{n-2}{2}\right)\alpha & -\sin\left(\frac{n-2}{2}\right)\alpha \\
\frac{1}{\sqrt{2}} & \frac{1}{\sqrt{2}} & \frac{1}{\sqrt{2}} & \frac{1}{\sqrt{2}} & \cdots & \frac{1}{\sqrt{2}} & \frac{1}{\sqrt{2}} & \frac{1}{\sqrt{2}} \\
\frac{1}{\sqrt{2}} & \frac{-1}{\sqrt{2}} & \frac{1}{\sqrt{2}} & \frac{-1}{\sqrt{2}} & \cdots & \frac{-1}{\sqrt{2}} & \frac{1}{\sqrt{2}} & \frac{-1}{\sqrt{2}}
\end{array}
\right]
\tag{3.12}
$$

Here once more $\alpha = 2\pi/n$. The coefficient in (3.12) in front of the matrix, $\sqrt{2/n}$, is associated with the powers of the original machine and the new machine, obtained after transformation. Selection as in (3.12) keeps the total powers invariant under the transformation.[*] Also, due to such a choice of the scaling factor, the transformation matrix satisfies the condition that $[C]^{-1} = [C]^t$, so that $[f_{1,2,\ldots n}] = [C]^t [f]_{\alpha\beta}$.

The first two rows in (3.12) define variables that will lead to fundamental flux and torque production (α–β components; stator-to-rotor coupling will appear only in the equations for α–β components). The last two rows define the two zero-sequence components and the last row of the transformation matrix (3.12) is omitted for all odd phase numbers n. In between, there are $(n-4)/2$ (or $(n-3)/2$ for n = odd) pairs of rows that define $(n-4)/2$ (or $(n-3)/2$ for n = odd) pairs of variables, termed further on x–y components. Upon application of (3.12) in conjunction with the phase-variable model (3.1) through (3.6) and (3.10),

---

[*] An alternative and frequently used form of the transformation (3.12) utilizes coefficient $2/n$ in front of the matrix. In such a case, powers per phase of the original and new machine are kept invariant in the transformation, but not the total powers. The transformation is then usually termed power-variant transformation and a scaling factor equal to $n/2$ appears in the torque equation after transformation.

assuming without any loss of generality that the phase number n is an odd number and that rotor n-phase winding is short-circuited, one gets the following new model equations:

$$v_{\alpha s} = R_s i_{\alpha s} + \frac{d\psi_{\alpha s}}{dt} = R_s i_{\alpha s} + (L_{ls} + L_m)\frac{di_{\alpha s}}{dt} + L_m \frac{d}{dt}(i_{\alpha r}\cos\theta - i_{\beta r}\sin\theta)$$

$$v_{\beta s} = R_s i_{\beta s} + \frac{d\psi_{\beta s}}{dt} = R_s i_{\beta s} + (L_{ls} + L_m)\frac{di_{\beta s}}{dt} + L_m \frac{d}{dt}(i_{\alpha r}\sin\theta + i_{\beta r}\cos\theta)$$

$$v_{x1s} = R_s i_{x1s} + \frac{d\psi_{x1s}}{dt} = R_s i_{x1s} + L_{ls}\frac{di_{x1s}}{dt}$$

$$v_{y1s} = R_s i_{y1s} + \frac{d\psi_{y1s}}{dt} = R_s i_{y1s} + L_{ls}\frac{di_{y1s}}{dt} \tag{3.13}$$

- -

$$v_{x((n-3)/2)s} = R_s i_{x((n-3)/2)s} + \frac{d\psi_{x((n-3)/2)s}}{dt} = R_s i_{x((n-3)/2)s} + L_{ls}\frac{di_{x((n-3)/2)s}}{dt}$$

$$v_{y((n-3)/2)s} = R_s i_{y((n-3)/2)s} + \frac{d\psi_{y((n-3)/2)s}}{dt} = R_s i_{y((n-3)/2)s} + L_{ls}\frac{di_{y((n-3)/2)s}}{dt}$$

$$v_{0s} = R_s i_{0s} + \frac{d\psi_{0s}}{dt} = R_s i_{0s} + L_{ls}\frac{di_{0s}}{dt}$$

$$v_{\alpha r} = 0 = R_r i_{\alpha r} + \frac{d\psi_{\alpha r}}{dt} = R_r i_{\alpha r} + (L_{lr} + L_m)\frac{di_{\alpha r}}{dt} + L_m \frac{d}{dt}(i_{\alpha s}\cos\theta + i_{\beta s}\sin\theta)$$

$$v_{\beta r} = 0 = R_r i_{\beta r} + \frac{d\psi_{\beta r}}{dt} = R_r i_{\beta r} + (L_{lr} + L_m)\frac{di_{\beta r}}{dt} + L_m \frac{d}{dt}(-i_{\alpha s}\sin\theta + i_{\beta s}\cos\theta)$$

$$v_{x1r} = 0 = R_r i_{x1r} + \frac{d\psi_{x1r}}{dt} = R_r i_{x1r} + L_{lr}\frac{di_{x1r}}{dt}$$

$$v_{y1r} = 0 = R_r i_{y1r} + \frac{d\psi_{y1r}}{dt} = R_r i_{y1r} + L_{lr}\frac{di_{y1r}}{dt} \tag{3.14}$$

- -

$$v_{x((n-3)/2)r} = 0 = R_r i_{x((n-3)/2)r} + \frac{d\psi_{x((n-3)/2)r}}{dt} = R_r i_{x((n-3)/2)r} + L_{lr}\frac{di_{x((n-3)/2)r}}{dt}$$

$$v_{y((n-3)/2)r} = 0 = R_r i_{y((n-3)/2)r} + \frac{d\psi_{y((n-3)/2)r}}{dt} = R_r i_{y((n-3)/2)r} + L_{lr}\frac{di_{y((n-3)/2)r}}{dt}$$

$$v_{0r} = 0 = R_r i_{0r} + \frac{d\psi_{0r}}{dt} = R_r i_{0r} + L_{lr}\frac{di_{0r}}{dt}$$

$$T_e = PL_m \left[\cos\theta(i_{\alpha r}i_{\beta s} - i_{\beta r}i_{\alpha s}) - \sin\theta(i_{\alpha r}i_{\alpha s} + i_{\beta r}i_{\beta s})\right] \tag{3.15}$$

Per-phase equivalent circuit magnetizing inductance is introduced in (3.13) through (3.15) as $L_m = (n/2)M$ and symbols L_{ls} and L_{lr} stand for leakage inductances of the stator and rotor windings, respectively. These are in essence the same parameters that appear in the well-known equivalent steady-state circuit of an induction machine and which can be obtained from standard no-load and locked rotor tests on the machine. Subscript + in designation of the zero-sequence component of (3.12) is omitted since there is a single such component when the phase number is an odd number.

Torque equation (3.15) shows that the torque is entirely developed due to the interaction of stator/rotor α–β current components and is independent of the value of x–y current components. This also follows from the α–β voltage equilibrium equations of both stator and rotor in (3.13) and (3.14), since these are the only axis component equations where coupling between stator and rotor remains to be present, through the rotor position angle θ. From rotor equations (3.14) it follows that, since the rotor winding is short-circuited and stator x–y components are decoupled from rotor x–y components, equations for rotor x–y components and the zero-sequence component equation can be omitted from further considerations.

The same applies to the stator zero-sequence component equation. Note that zero sequence is governed by the sum of all instantaneous phase quantities. Since winding is considered as star connected with isolated neutral, no zero-sequence current can flow in the stator winding (if the number of phases is even and such that $n \geq 6$, the second zero-sequence 0_- current component can flow if the supply is such that $v_{0_- s}$ is not zero). As far as the x–y stator current components are concerned, they will also be zero as long as the supply voltages upon application of the decoupling transformation do not yield nonzero stator voltage x–y components. Thus, under ideal symmetrical and balanced sinusoidal multiphase voltage supply, the total number of equations that has to be considered in the electromagnetic subsystem is only four differential equations (two pairs of α–β equations in (3.13) and (3.14)) instead of the $2n$ differential equations in the original phase-variable model.

As is obvious from (3.13) and (3.14), the basic form of the voltage equilibrium equations has not been changed by applying the decoupling transformation, and they are still governed with $v = Ri + d\psi/dt$. However, by comparing the phase-variable model of the previous section with the relevant equations obtained after application of decoupling transformation, it is obvious that a considerable simplification has been achieved. Regardless of the actual phase number, one only needs to consider further four voltage equilibrium equations, instead of $2n$, as long as the machine is supplied from a balanced symmetrical n-phase sinusoidal source. Torque equation (3.15) is also of a considerably simpler form than its counterpart in (3.10). Needless to say, Equations 3.5 and 3.7 do not change the form in the model transformation process. However, the problem of time-varying coefficients and nonlinearity of the system of differential equations has not been resolved.

3.4 Rotational Transformation

New fictitious α–β and x–y stator and rotor windings are still firmly attached to the corresponding machine's member, meaning that stator windings are stationary, while rotor windings rotate together with the rotor. In order to get rid of the time-varying inductance terms in (3.13) through (3.15), it is necessary to perform one more transformation, usually called rotational transformation. This means that the fictitious machine's windings, obtained after application of the decoupling transformation, are now transformed once more into yet another set of fictitious windings. This time, however, the transformation for stator and rotor variables is not the same any more.

As stator-to-rotor coupling takes place only in α–β equations, rotational transformation is applied only to these two pairs of equations. Its form for an n-phase machine is identical as for a three-phase machine, since x–y component equations do not need to be transformed. The transformation is defined in such a way that the resulting new sets of stator and rotor windings, which will replace α–β windings, rotate at the same angular speed, so-called speed of the common reference frame. Thus, relative motion between stator and rotor windings gets eliminated, leading to a set of differential equations with constant coefficients. Since in an induction machine air-gap is uniform and all inductances within both stator and rotor multiphase winding in (3.4) are constants, selection of the speed of the common reference frame is arbitrary. In other words, any convenient speed can be selected. Let us call such an angular speed arbitrary speed of the common reference frame, ω_a. This speed defines instantaneous position of the d-axis of the common reference frame with respect to the stationary stator phase 1 axis,

$$\theta_s = \int \omega_a \, dt \tag{3.16}$$

which will be used in the rotational transformation for stator quantities. Considering that rotor rotates, and therefore phase 1 of rotor has an instantaneous position θ with respect to stator phase 1, the angle between d-axis of the common reference frame and rotor phase 1 axis, which will be used in transformation of the rotor quantities, is determined with

$$\theta_r = \theta_s - \theta = \int (\omega_a - \omega)\, dt \tag{3.17}$$

The second axis of the common reference frame, which is perpendicular to the d-axis, is customarily labeled as q-axis. The correlation between variables obtained upon application of the decoupling transformation and new $d-q$ variables is defined similarly to (3.11):

$$[f_{dq}] = [D][f_{\alpha\beta}] \tag{3.18}$$

However, rotational transformation matrix $[D]$ is now different for the stator and rotor variables:

$$[D_s] = \begin{array}{c} ds \\ qs \\ x_{1s} \\ y_{1s} \\ \cdots \\ 0_s \end{array} \begin{bmatrix} \cos\theta_s & \sin\theta_s & 0 & 0 & \cdots & 0 \\ -\sin\theta_s & \cos\theta_s & 0 & 0 & \cdots & 0 \\ 0 & 0 & 1 & 0 & \cdots & 0 \\ 0 & 0 & 0 & 1 & \cdots & 0 \\ \cdots & \cdots & \cdots & \cdots & \cdots & \cdots \\ 0 & 0 & 0 & 0 & \cdots & 1 \end{bmatrix}$$

$$[D_r] = \begin{array}{c} dr \\ qr \\ x_{1r} \\ y_{1r} \\ \cdots \\ 0_r \end{array} \begin{bmatrix} \cos\theta_r & \sin\theta_r & 0 & 0 & \cdots & 0 \\ -\sin\theta_r & \cos\theta_r & 0 & 0 & \cdots & 0 \\ 0 & 0 & 1 & 0 & \cdots & 0 \\ 0 & 0 & 0 & 1 & \cdots & 0 \\ \cdots & \cdots & \cdots & \cdots & \cdots & 0 \\ 0 & 0 & 0 & 0 & \cdots & 1 \end{bmatrix} \tag{3.19}$$

As is evident from (3.19), rotational transformation is applied only to α–β equations, while x–y and zero-sequence equations do not change the form. The inverse relationship of (3.18), $[f_{\alpha\beta}] = [D]^{-1}\,[f_{dq}]$, is again a simple expression since once more $[D]^{-1} = [D]^t$. An illustration of the various spatial angles in the cross section of the machine is shown in Figure 3.2.

When the decoupled model (3.13) through (3.15) of an n-phase induction machine with sinusoidal winding distribution is transformed using (3.18) and (3.19), the set of voltage equilibrium and flux linkage equations in the common reference frame for a machine with an odd number of phases is obtained in the following form:

$$v_{ds} = R_s i_{ds} + \frac{d\psi_{ds}}{dt} - \omega_a \psi_{qs}$$

$$v_{qs} = R_s i_{qs} + \frac{d\psi_{qs}}{dt} + \omega_a \psi_{ds} \tag{3.20a}$$

$$v_{dr} = 0 = R_r i_{dr} + \frac{d\psi_{dr}}{dt} - (\omega_a - \omega)\psi_{qr}$$

$$v_{qr} = 0 = R_r i_{qr} + \frac{d\psi_{qr}}{dt} + (\omega_a - \omega)\psi_{dr}$$

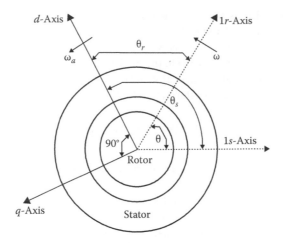

d-Axis 1r-Axis

θ_r

ω_a ω

θ_s

90° θ 1s-Axis

Rotor

q-Axis Stator

FIGURE 3.2 Illustration of various angles used in the rotational transformation of an induction machine's model (1s and 1r denote magnetic axes of the first stator and rotor phases).

$$v_{x1s} = R_s i_{x1s} + \frac{d\psi_{x1s}}{dt}$$

$$v_{y1s} = R_s i_{y1s} + \frac{d\psi_{y1s}}{dt}$$

$$v_{x2s} = R_s i_{x2s} + \frac{d\psi_{x2s}}{dt}$$ (3.20b)

$$v_{y2s} = R_s i_{y2s} + \frac{d\psi_{y2s}}{dt}$$

$$\dots\dots\dots\dots\dots$$

$$v_{0s} = R_s i_{0s} + \frac{d\psi_{0s}}{dt}$$

$$\psi_{ds} = (L_{ls} + L_m)i_{ds} + L_m i_{dr}$$

$$\psi_{qs} = (L_{ls} + L_m)i_{qs} + L_m i_{qr}$$

$$\psi_{dr} = (L_{lr} + L_m)i_{dr} + L_m i_{ds}$$ (3.21a)

$$\psi_{qr} = (L_{lr} + L_m)i_{qr} + L_m i_{qs}$$

$$\psi_{x1s} = L_{ls} i_{x1s}$$

$$\psi_{y1s} = L_{ls} i_{y1s}$$

$$\psi_{x2s} = L_{ls} i_{x2s}$$ (3.21b)

$$\psi_{y2s} = L_{ls} i_{y2s}$$

$$\dots\dots\dots\dots$$

$$\psi_{0s} = L_{ls} i_{0s}$$

Since rotor winding is regarded as short-circuited, zero-sequence and x–y component equations of the rotor have been omitted from (3.20) and (3.21). If there is a need to consider these equations (as the case may be if the rotor winding has more than three phases and is supplied from a power electronic converter in a slip-ring machine), one only needs to add to the model (3.20) and (3.21) rotor x–y equations of (3.14), which are of identical form as in (3.20b) and (3.21b) and only index s needs to be replaced with index r.

Upon application of the rotational transformation torque expression (3.15) becomes

$$T_e = PL_m \left[i_{dr}i_{qs} - i_{ds}i_{qr} \right] \tag{3.22}$$

Model (3.20) through (3.22) fully describes a general n-phase induction machine, of any odd phase number. If the number of phases is even, it is only necessary to add the equations for the second zero-sequence component, which are of the identical form as in (3.20) and (3.21) for the first zero-sequence component. However, the complete model needs to be considered only if the supply of the machine contains components that give rise to the stator voltage x–y components. If the machine is considered to be supplied with a set of symmetrical balanced sinusoidal n-phase voltages (of equal rms value and phase shift of exactly $2\pi/n$ between any two consecutive voltages), then stator voltage x–y components are all zero, regardless of the phase number. This means that analysis of an n-phase machine can be conducted under these conditions by using only stator and rotor d–q pairs of equations, in exactly the same manner as for a three-phase machine.

A closer inspection of the d–q voltage equilibrium equations in (3.20a) of the stator and the rotor shows that, upon application of the rotational transformation, these equations are not of the same form as in phase domain (i.e., the form is not any more $v = Ri + d\psi/dt$). The equations contain an additional term, a product of an angular speed and a corresponding flux linkage component. The reason for this is that, by means of rotational transformation, the speed of the windings has been changed. Instead of being zero and ω for the stator and the rotor, respectively, the speeds of new windings have been equalized and are now ω_a. The new additional terms account for this change and they represent rotational induced electromotive forces in fictitious d–q windings of the stator and the rotor.

A schematic representation of the fictitious machine that results upon application of the rotational transformation is shown in Figure 3.3. Assuming ideal symmetrical and balanced n-phase sinusoidal supply of the machine, the representation of the machine, regardless of the number of phases, is as in

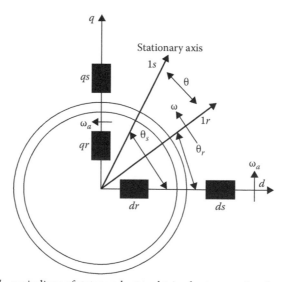

FIGURE 3.3 Fictitious d–q windings of stator and rotor obtained using rotational transformation.

Figure 3.3. What this means is that an *n*-phase machine can be replaced with an equivalent two-phase machine for modeling purposes. Zero sequence is along a line perpendicular to the $d - q$ plane (or, for even phase numbers, in a plane perpendicular to the $d - q$ plane). If the supply is such that x–y stator voltage components are not zero, the representation of a machine with five or more phases has to include also x–y voltage and flux linkage equations. However, since the equivalent x–y windings are situated in the planes perpendicular to the one of Figure 3.3, simultaneous graphical representation of all the new windings is not possible any more.

As can be seen from (3.21), time-varying inductance terms have been eliminated by means of rotational transformation. Hence, electromagnetic torque equation does not contain such time-varying terms either. The system of differential equations is now with constant coefficients. Further, if the speed of rotation is considered as constant, Equations 3.20 and 3.21 become linear differential equations, analysis of which can be done using, say, Laplace transform. This was just about the only technique available in the beginning of the last century, so that the model transformation has enabled initial analytical analyses of the transients to be conducted (albeit at a constant speed).

Electromagnetic torque equation (3.22) can be given in a number of alternative ways, by utilizing the correlations between $d - q$ axis stator/rotor currents and $d - q$ axis stator/rotor flux linkages of (3.21). Some alternative formulations of the electromagnetic torque are the following:

$$T_e = P(\psi_{ds} i_{qs} - \psi_{qs} i_{ds}) = P \frac{L_m}{L_r} (\psi_{dr} i_{qs} - \psi_{qr} i_{ds}) \tag{3.23}$$

As noted already, angular speed of the common reference frame can be selected freely in an induction machine. However, some selections are more favorable than the others.

For simulation of transients of a mains-fed squirrel-cage induction machine, the most opportune common reference frame is the stationary reference frame, such that $\omega_a = 0$, $\theta_s = 0$, since then the stator variables actually involve only decoupling transformation. It should be noted that rotor variables are practically never of interest in squirrel-cage induction machines since they are immeasurable anyway. The other frequently used reference frame is the synchronous reference frame, in which the common $d - q$ reference frame rotates at the angular speed equal to the angular frequency of the fundamental stator supply. Such a reference frame is very convenient for various analytical studies of, for example, inverter-supplied induction machines. The common reference frame fixed to the rotor ($\omega_a = \omega$) is only suitable if a slip-ring induction machine is under consideration, with a power electronic supply connected to the rotor winding.

A completely different selection of the angular speed of the common reference frame is utilized for the realization of high-performance induction motor drives with closed-loop control. Such control schemes are termed vector- or field-oriented control schemes, and the speed of the common reference frame is selected as speed of rotation of one of the rotating fields (stator, air-gap, or rotor) in the machine.

3.5 Complete Transformation Matrix

Since the relationship between original phase variables and variables obtained after decoupling transformation is governed by (3.11), while $d - q$ variables are related to variables obtained after decoupling transformation through (3.18), it is possible to express the two individual transformations as a single matrix transformation that will relate phase variables $1, 2, \ldots, n$ with $d - q$ variables. Let such a transformation matrix be denoted as $[T]$. From (3.11) and (3.18), one has $[f_{dq}] = [D][C][f_{1,2,\ldots,n}]$, so that $[T] = [D][C]$. Since the rotational transformation matrix is different for stator and rotor variables, the complete transformation matrix will also be different. Taking as an example a three-phase machine, the combined decoupling/rotational transformation matrix for stator variables will be

$$
[T_s] = \sqrt{\frac{2}{3}} \begin{array}{c} ds \\ qs \\ 0s \end{array} \begin{bmatrix} \cos\theta_s & \cos(\theta_s - \alpha) & \cos(\theta_s + \alpha) \\ -\sin\theta_s & -\sin(\theta_s - \alpha) & -\sin(\theta_s + \alpha) \\ \dfrac{1}{\sqrt{2}} & \dfrac{1}{\sqrt{2}} & \dfrac{1}{\sqrt{2}} \end{bmatrix}
\tag{3.24}
$$

In the general n-phase case one has, instead of (3.24), the following:

$[T_s] =$

$$
\sqrt{\frac{2}{n}}
\begin{array}{c} ds \\ qs \\ x_{1s} \\ y_{1s} \\ x_{2s} \\ y_{2s} \\ \cdots \\ x_{\frac{n-4}{2}s} \\ y_{\frac{n-4}{2}s} \\ 0_{+s} \\ 0_{-s} \end{array}
\begin{bmatrix}
\cos\theta_s & \cos(\theta_s-\alpha) & \cos(\theta_s-2\alpha) & \cos(\theta_s-3\alpha) & \cdots & \cos(\theta_s+3\alpha) & \cos(\theta_s+2\alpha) & \cos(\theta_s+\alpha) \\
-\sin\theta_s & -\sin(\theta_s-\alpha) & -\sin(\theta_s-2\alpha) & -\sin(\theta_s-3\alpha) & \cdots & -\sin(\theta_s+3\alpha) & -\sin(\theta_s+2\alpha) & -\sin(\theta_s+\alpha) \\
1 & \cos 2\alpha & \cos 4\alpha & \cos 6\alpha & \cdots & \cos 6\alpha & \cos 4\alpha & \cos 2\alpha \\
0 & \sin 2\alpha & \sin 4\alpha & \sin 6\alpha & \cdots & -\sin 6\alpha & -\sin 4\alpha & -\sin 2\alpha \\
1 & \cos 3\alpha & \cos 6\alpha & \cos 9\alpha & \cdots & \cos 9\alpha & \cos 6\alpha & \cos 3\alpha \\
0 & \sin 3\alpha & \sin 6\alpha & \sin 9\alpha & \cdots & -\sin 9\alpha & -\sin 6\alpha & -\sin 3\alpha \\
\cdots & \cdots & \cdots & \cdots & \cdots & \cdots & \cdots & \cdots \\
1 & \cos\left(\dfrac{n-2}{2}\right)\alpha & \cos 2\left(\dfrac{n-2}{2}\right)\alpha & \cos 3\left(\dfrac{n-2}{2}\right)\alpha & \cdots & \cos 3\left(\dfrac{n-2}{2}\right)\alpha & \cos 2\left(\dfrac{n-2}{2}\right)\alpha & \cos\left(\dfrac{n-2}{2}\right)\alpha \\
0 & \sin\left(\dfrac{n-2}{2}\right)\alpha & \sin 2\left(\dfrac{n-2}{2}\right)\alpha & \sin 3\left(\dfrac{n-2}{2}\right)\alpha & \cdots & -\sin 3\left(\dfrac{n-2}{2}\right)\alpha & -\sin 2\left(\dfrac{n-2}{2}\right)\alpha & -\sin\left(\dfrac{n-2}{2}\right)\alpha \\
\dfrac{1}{\sqrt{2}} & \dfrac{1}{\sqrt{2}} & \dfrac{1}{\sqrt{2}} & \dfrac{1}{\sqrt{2}} & \cdots & \dfrac{1}{\sqrt{2}} & \dfrac{1}{\sqrt{2}} & 1/\sqrt{2} \\
\dfrac{1}{\sqrt{2}} & \dfrac{-1}{\sqrt{2}} & \dfrac{1}{\sqrt{2}} & \dfrac{-1}{\sqrt{2}} & \cdots & \dfrac{-1}{\sqrt{2}} & \dfrac{1}{\sqrt{2}} & \dfrac{-1}{\sqrt{2}}
\end{bmatrix}
$$

$$
\tag{3.25}
$$

Transformation matrices for the rotor are, in form, identical to those for the stator (3.24) and (3.25), and it is only necessary to replace the angle of transformation θ_s with θ_r.

When the model of the machine is used for simulation purposes, it is typically necessary to apply the appropriate transformation matrix in both directions. For the sake of example, consider a three-phase induction machine, supplied from a three-phase voltage source. Hence, stator phase voltages are known. Corresponding $d-q$ axis voltage components are calculated using (3.24) for the selected reference frame:

$$
v_{ds} = \sqrt{\frac{2}{3}}\left(v_{1s}\cos\theta_s + v_{2s}\cos\left(\theta_s - \frac{2\pi}{3}\right) + v_{3s}\cos\left(\theta_s - \frac{4\pi}{3}\right) \right)
$$
$$
v_{qs} = -\sqrt{\frac{2}{3}}\left(v_{1s}\sin\theta_s + v_{2s}\sin\left(\theta_s - \frac{2\pi}{3}\right) + v_{3s}\sin\left(\theta_s - \frac{4\pi}{3}\right) \right)
\tag{3.26}
$$

These are the inputs of the $d-q$ axis model, together with the disturbance, load torque. The model is solved for the electromagnetic torque, rotor speed, and stator $d-q$ axis currents (rotor $d-q$ currents are usually not of interest; however, they are obtained too). Since actual stator phase currents are of interest, then $d-q$ axis stator current components have to be now transformed back into the phase domain, using inverse transformation:

$$i_{1s} = \sqrt{\frac{2}{3}}(i_{ds}\cos\theta_s - i_{qs}\sin\theta_s)$$

$$i_{2s} = \sqrt{\frac{2}{3}}\left(i_{ds}\cos\left(\theta_s - \frac{2\pi}{3}\right) - i_{qs}\sin\left(\theta_s - \frac{2\pi}{3}\right)\right) \qquad (3.27)$$

$$i_{3s} = \sqrt{\frac{2}{3}}\left(i_{ds}\cos\left(\theta_s - \frac{4\pi}{3}\right) - i_{qs}\sin\left(\theta_s - \frac{4\pi}{3}\right)\right)$$

Note that, due to assumed stator winding connection into star, with isolated neutral point, zero-sequence current cannot flow, and hence zero-sequence components are not considered.

Assuming that stator voltages are sinusoidal, balanced, and symmetrical, of rms value V, it is simple to show that the amplitude of $d-q$ axis voltage components in (3.26) is, regardless of the selected reference frame, equal to $\sqrt{3}V$. This is the consequence of the adopted power-invariant form of the transformation matrices. In general, for an n-phase machine, the amplitude is $\sqrt{n}V$. In contrast to this, if the transformation is power-variant and keeps transformed power per-phase equal (coefficient in (3.25) is $2/n$ rather than $\sqrt{2/n}$), amplitudes of $d-q$ axis components are equal to $\sqrt{2}V$ regardless of the phase number.

3.6 Space Vector Modeling

Since, upon application of the decoupling transformation, one gets pairs of axis components in mutually perpendicular planes and these pairs are in mutually perpendicular axes as well, it is possible to consider all the planes as complex and define one axis component as a real part and the other axis component as an imaginary part of a complex number. Such complex numbers are known as space vectors and they differ considerably from phasors (complex representatives of sinusoidal quantities). To start with, space vectors can be used for both sinusoidal and nonsinusoidal supply. Second, space vectors describe a machine in both transient and steady-state operating conditions. In what follows, space vectors are denoted with underlined symbols.

Consider decoupling transformation matrix (3.12). As can be seen, each pair of rows contains sine and cosine functions of the same angles. Let a complex operator $\underline{a}$ be introduced as $\underline{a} = \exp(j\alpha) = \cos\alpha + j\sin\alpha$, where once more $\alpha = 2\pi/n$. Each pair of rows in (3.12) then defines one space vector, with odd rows determining the real parts and the even rows imaginary parts of the corresponding complex numbers, that is, space vectors. Let f stand once more for voltage, current, or flux linkage of either the stator or the rotor. Space vectors are then governed with

$$\underline{f}_{\alpha-\beta} = f_\alpha + jf_\beta = \sqrt{\frac{2}{n}}\left(f_1 + \underline{a}f_2 + \underline{a}^2 f_3 + \cdots + \underline{a}^{(n-1)} f_n\right)$$

$$\underline{f}_{x1-y1} = f_{x1} + jf_{y1} = \sqrt{\frac{2}{n}}\left(f_1 + \underline{a}^2 f_2 + \underline{a}^4 f_3 + \cdots + \underline{a}^{2(n-1)} f_n\right)$$

$$\underline{f}_{x2-y2} = f_{x2} + jf_{y2} = \sqrt{\frac{2}{n}}\left(f_1 + \underline{a}^3 f_2 + \underline{a}^6 f_3 + \cdots + \underline{a}^{3(n-1)} f_n\right) \qquad (3.28)$$

- -

$$\underline{f}_{x\frac{n-3}{2}-y\frac{n-3}{2}} = f_{x\frac{n-3}{2}} + jf_{y\frac{n-3}{2}} = \sqrt{\frac{2}{n}}\left(f_1 + \underline{a}^{(n-1)/2} f_2 + \underline{a}^{2[(n-1)/2]} f_3 + \cdots + \underline{a}^{(n-1)^2/2} f_n\right)$$

It is again assumed that the phase number is an odd number and neutral point is isolated, so that zero sequence cannot be excited. It is therefore not included here, but it in general remains to be governed with the corresponding penultimate row of the decoupling transformation matrix (3.12).

Since rotational transformation is applied only to α–β components, then only the corresponding α–β space vector will undergo a further transformation, governed with (3.19) in real form. Of course, the transformation is once more different for stator and rotor quantities. The stator and rotor voltage, current, and flux linkage space vectors are obtained in the common reference frame by rotating corresponding α–β space vector by an angle, which is for stator θ_s and for rotor θ_r. This is done by means of the vector rotator, $\exp(-j\theta_s)$ for stator and $\exp(-j\theta_r)$ for rotor variables. Hence, space vectors that will describe the machine in an arbitrary common reference frame are governed with

$$\underline{f}_{d-q(s)} = f_{ds} + jf_{qs} = (f_{\alpha s} + jf_{\beta s})e^{-j\theta_s} = \sqrt{\frac{2}{n}}\left(f_{1s} + \underline{a}\,f_{2s} + \underline{a}^2 f_{3s} + \cdots + \underline{a}^{(n-1)}f_{ns}\right)e^{-j\theta_s}$$

$$\underline{f}_{d-q(r)} = f_{dr} + jf_{qr} = (f_{\alpha r} + jf_{\beta r})e^{-j\theta_r} = \sqrt{\frac{2}{n}}\left(f_{1r} + \underline{a}\,f_{2r} + \underline{a}^2 f_{3r} + \cdots + \underline{a}^{(n-1)}f_{nr}\right)e^{-j\theta_r}$$

(3.29)

To form the induction machine's model in terms of space vectors, it is only necessary to combine $d-q$ axis equations of the real model (3.20) and (3.21) as real and imaginary parts of the corresponding complex equations. Hence, the torque-producing part of the model is, regardless of the phase number, described with

$$\underline{v}_s = R_s\underline{i}_s + \frac{d\underline{\psi}_s}{dt} + j\omega_a\underline{\psi}_s$$

$$\underline{v}_r = 0 = R_r\underline{i}_r + \frac{d\underline{\psi}_r}{dt} + j(\omega_a - \omega)\underline{\psi}_r$$

(3.30)

$$\underline{\psi}_s = (L_{ls} + L_m)\underline{i}_s + L_m\underline{i}_r$$

$$\underline{\psi}_r = (L_{lr} + L_m)\underline{i}_r + L_m\underline{i}_s$$

(3.31)

Indices $d-q$, used in (3.29) to define space vectors, have been omitted in (3.30) and (3.31) for simplicity. In (3.30) and (3.31), space vectors are $\underline{v}_s = v_{ds} + jv_{qs}$, $\underline{i}_s = i_{ds} + ji_{qs}$, $\underline{\psi}_s = \psi_{ds} + j\psi_{qs}$ and $\underline{v}_r = v_{dr} + jv_{qr}$, $\underline{i}_r = i_{dr} + ji_{qr}$, $\underline{\psi}_r = \psi_{dr} + j\psi_{qr}$. Torque equation (3.22) can be given, using space vectors, as

$$T_e = PL_m\,\text{Im}\left(\underline{i}_s\underline{i}_r^*\right)$$

(3.32)

where
 * stands for complex conjugate
 Im denotes the imaginary part of the complex number

Equations 3.30 through 3.32 together with the equation of mechanical motion (3.7) fully describe a three-phase induction machine. If the machine has more than three phases and the supply is either not balanced or it contains additional time harmonics apart from the fundamental (so that $x-y$ stator voltage components are not zero), the model (3.30) through (3.32) needs to be complemented with additional space vector equations that describe x–y circuits of stator. Using again real model (3.20) and (3.21) and the definition of space vectors in (3.28), these additional equations are all of the same form

$$\underline{v}_{x-y(s)} = R_s \underline{i}_{x-y(s)} + \frac{d\underline{\psi}_{x-y(s)}}{dt}$$

$$\underline{\psi}_{x-y(s)} = L_{ls} \underline{i}_{x-y(s)}$$

(3.33)

and there are $(n-3)/2$ such voltage and flux linkage equations for x–y components 1 to $(n-3)/2$.

Model (3.30) and (3.31) is the dynamic model of an induction machine. Consider now steady-state operation with symmetrical balanced sinusoidal supply. Regardless of the selected common reference frame, model (3.30) and (3.31) under these conditions reduces to the well-known equivalent circuit of an induction machine, described with

$$\underline{v}_s = R_s \underline{i}_s + j\omega_s \left(L_s \underline{i}_s + L_m \underline{i}_r \right) = R_s \underline{i}_s + j\omega_s \left(L_{ls} \underline{i}_s + L_m (\underline{i}_s + \underline{i}_r) \right)$$

$$0 = R_r \underline{i}_r + j(\omega_s - \omega)\left(L_r \underline{i}_r + L_m \underline{i}_s \right) = R_r \underline{i}_r + j(\omega_s - \omega)(L_{lr} \underline{i}_r + L_m (\underline{i}_s + \underline{i}_r))$$

(3.34)

where ω_s stands for angular frequency of the stator supply. By defining slip s in the standard manner as $(\omega_s - \omega)/\omega_s$, introducing reactances as products of stator angular frequency and inductances, and defining magnetizing current space vector as $\underline{i}_m = \underline{i}_s + \underline{i}_r$, these equations reduce to the standard form

$$\underline{v}_s = R_s \underline{i}_s + jX_{ls} \underline{i}_s + jX_m (\underline{i}_s + \underline{i}_r)$$

$$0 = \left(\frac{R_r}{s} \right) \underline{i}_r + j\left[X_{lr} \underline{i}_r + X_m (\underline{i}_s + \underline{i}_r) \right]$$

(3.35)

which describes the equivalent circuit of Figure 3.4. The only (but important) differences, when compared to the phasor equivalent circuit, are that the quantities in the circuit of Figure 3.4 are now space vectors rather than phasors, and that there is no circuit of the form given in Figure 3.4 for each phase of the machine, there is a single circuit for the whole multiphase machine instead. The space vectors will also be of different time dependence, depending on the selected common reference frame. For example, in the stationary reference frame $\underline{v}_s(\omega_a = 0) = \sqrt{n}V \exp(j\omega_s t)$, while in the synchronous reference frame in which d-axis is aligned with the stator voltage space vector $\underline{v}_s(\omega_a = \omega_s) = \sqrt{n}V$.

Stator voltage space vector under symmetrical sinusoidal supply conditions is shown in Figure 3.5 for a three-phase machine. It travels around the circle of radius equal to $\sqrt{3}V$. Instantaneous projections of the space vector onto α- and β-axis represent space vector real and imaginary parts, in accordance with the definition in (3.28). Upon application of the vector rotator of (3.29) with $\theta_s = \int \omega_s dt = \omega_s t$ the

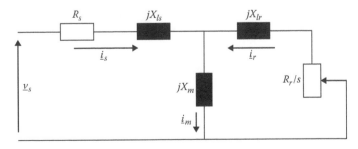

FIGURE 3.4 Equivalent circuit of an induction machine for steady-state operation with sinusoidal supply in terms of space vectors.

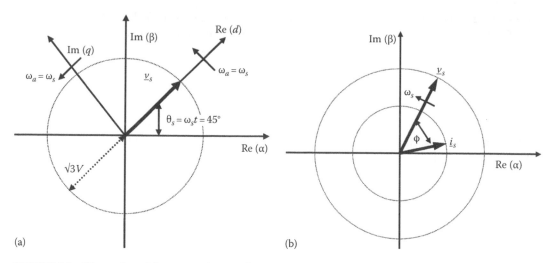

(a)   (b)

FIGURE 3.5   Illustration of the stator voltage and current space vectors for symmetrical sinusoidal supply conditions. (a) Stator voltage space vector and (b) stator voltage and current space vectors.

stator voltage space vector becomes aligned with the d-axis of the common rotating reference frame so that the q-component is zero. Since the $d-q$ system of axes rotates, its position continuously changes; thus, the illustration in Figure 3.5a applies to one specific instant in time, when the angle is 45°. Since the machine is in steady state, the stator current space vector is in essence determined with the ratio of the stator voltage space vector and impedance. The angle that appears between the stator voltage and stator current space vectors is the power factor angle ϕ (Figure 3.5b). Speed of rotation of the stator current space vector is of course equal to the speed of the voltage space vector, but the radius of the circle along which the stator current space vector travels is different.

If the machine has five or more phases and the stator supply is either not balanced/symmetrical, or it contains certain time harmonics that map into $x–y$ stator voltage components, then it becomes necessary to use additional equivalent circuits, one per each $x–y$ plane (i.e., only one for a five-phase machine, but two for a seven-phase machine, and so on). In principle, the form of equivalent circuits for $x–y$ components is governed with (3.33). However, since $x–y$ voltages may contain more than one frequency component, a separate equivalent circuit is needed for steady-state representation at each such frequency. Assuming, for the sake of illustration, that stator $x–y$ voltages contain a single-frequency component, the equivalent circuit is as given in Figure 3.6.

Whether or not the stator winding $x–y$ circuits are excited entirely depends on the properties of the stator winding supply. If the supply is a power electronic converter, which produces time harmonics in the

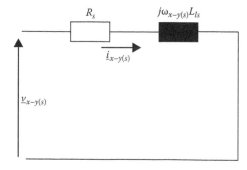

FIGURE 3.6   Equivalent circuit, applicable to each frequency component of every $x–y$ stator voltage space vector in machines with more than three phases.

TABLE 3.1 Harmonic Mapping into Different Planes for Five-Phase and Seven-Phase Systems ($j = 0,1,2,3...$)

Plane	Five-Phase System	Seven-Phase System
$\alpha - \beta$	$10j \pm 1$ (**1**, 9, 11...)	$14j \pm 1$ (**1**, 13, 15...)
$x_1 - y_1$	$10j \pm 3$ (**3**, 7, 13...)	$14j \pm 5$ (**5**, 9, 19...)
$x_2 - y_2$	n/a	$14k \pm 3$ (**3**, 11, 17...)
Zero-sequence	$5(2j + 1)$ (5, 15...)	$7(2j + 1)$ (7, 21...)

output phase voltage, then some of these harmonics will map into each x–y plane. As an example, Table 3.1 shows harmonic mapping, characteristic for five-phase and seven-phase stator windings [24]. As can be seen, one particular time harmonic in each x–y plane for each phase number is shown in bold font. These are the time harmonics of the supply that can be used, in addition to the fundamental, to produce an average torque. The idea is to increase the torque density available from the machine, and this applies equally to both generating operation [25] and motoring operation [26]. However, for this to be possible, it is necessary that the stator winding is of the concentrated type, so that, in addition to the fundamental space harmonic, there exist the corresponding low-order space harmonics of the mmf. In simple terms, this means that the spatial distribution of the mmf is not regarded as sinusoidal any more; it is quasi-rectangular instead. Modeling of such machines is beyond the scope of this article. It suffices to say that, while the decoupling transformation matrix remains the same, rotational transformation changes the form. Also, the starting phase-variable model in this case has to take into account the existence of the low-order spatial harmonics through appropriate harmonic inductance terms. In the final model, $d - q$ equations remain the same but electromagnetic torque equation and x–y circuit equations change.

3.7 Modeling of Multiphase Machines with Multiple Three-Phase Windings

In high-power applications, it is more and more common that, instead of using three-phase machines, machines with multiple three-phase windings are used. The most common case is a six-phase machine. The stator winding is composed of two three-phase windings, which are spatially shifted by 30°. The outlay is shown schematically in Figure 3.7 for an induction machine. Since there are now two three-phase windings, phases are labeled as a, b, c, and indices 1 and 2 apply to the two three-phase windings (index s is omitted). As can be seen from Figure 3.7, this spatial shift leads to asymmetrical positioning

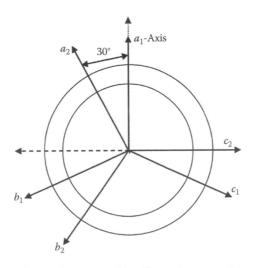

FIGURE 3.7 Asymmetrical six-phase induction machine, illustrating magnetic axes of the stator phases.

of the stator phase magnetic axes in the cross section of the machine. Such a type of the multiphase machine is, therefore, usually termed asymmetrical machine, since spatial shift between any two consecutive phases is not equal any more and it is not governed by $2\pi/n$. Instead, there is a shift between three-phase windings, equal to π/n. Furthermore, since the machine is based on three-phase windings and there are in general a of them, then the neutral points of each individual three-phase winding are kept isolated, so that there are a isolated neutral points.

Modeling principles, discussed so far, are valid for asymmetrical multiphase machines as well. As a matter of fact, final machine models in the common reference frame (3.20) through (3.22) and (3.30) through (3.33) remain to be valid, provided that decoupling transformation matrix (3.12) is adapted to the winding layout in Figure 3.7. In particular, [C] is, for an asymmetrical six-phase machine, given with [27]

$$
\underline{C} = \sqrt{\frac{2}{6}}
\begin{array}{c}
\alpha \\
\beta \\
x_1 \\
y_1 \\
0_+ \\
0_-
\end{array}
\overset{\displaystyle
\begin{array}{cccccc}
a_1 & b_1 & c_1 & a_2 & b_2 & c_2
\end{array}}
{\left[
\begin{array}{cccccc}
1 & \cos(2\pi/3) & \cos(4\pi/3) & \cos(\pi/6) & \cos(5\pi/6) & \cos(9\pi/6) \\
0 & \sin(2\pi/3) & \sin(4\pi/3) & \sin(\pi/6) & \sin(5\pi/6) & \sin(9\pi/6) \\
1 & \cos(4\pi/3) & \cos(8\pi/3) & \cos(5\pi/6) & \cos(\pi/6) & \cos(9\pi/6) \\
0 & \sin(4\pi/3) & \sin(8\pi/3) & \sin(5\pi/6) & \sin(\pi/6) & \sin(9\pi/6) \\
1 & 1 & 1 & 0 & 0 & 0 \\
0 & 0 & 0 & 1 & 1 & 1
\end{array}
\right]}
\tag{3.36}
$$

Here, the first three terms in each row relate to the first three-phase winding, while the second three terms relate to the second three-phase winding, as indicated in the row above the transformation matrix. The form of the last two rows in (3.36) takes into account that neutral points of the two windings are isolated.

Provided that the asymmetrical six-phase machine's phase-variable model is decoupled using (3.36), rotational transformation matrices (3.19) remain the same and identical equations are obtained in the $d-q$ common reference frame and in space vector form as for a symmetrical multiphase machine (of course, the complete transformation matrix of (3.25) has to be modified in accordance with (3.36)). One important note is however due in relation to the total number of $x-y$ equation pairs. Use of a individual and isolated neutral points means that, upon transformation, there will be

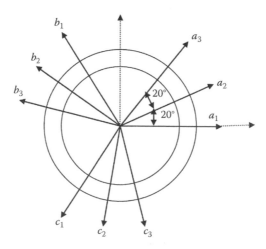

FIGURE 3.8 An asymmetrical nine-phase stator winding structure.

only $(n-a)$ voltage equilibrium equations to consider, since zero-sequence current cannot flow in any of the three-phase windings. Since $n = 3a$, then the total number of equations is $2a$. As the first pair is always for $d-q$ components, then the resulting number of $x-y$ voltage equation pairs is only $(a-1)$. This comes down to one $d-q$ pair and two $x-y$ pairs for an asymmetrical nine-phase machine with three isolated neutral points. Had the neutral points been connected, there would have been three pairs of $x-y$ equations.

As an example, consider an asymmetrical nine-phase machine, with disposition of stator phase magnetic axes shown in Figure 3.8. Stator phases of any of the three-phase windings are labeled again as a, b, c and additional index 1, 2, 3 denotes the particular three-phase winding. The angle between three-phase windings is $\alpha = \pi/n = 20°$. The winding may have a single neutral point or three isolated neutral points. Decoupling transformation matrix for the asymmetrical nine-phase winding with a single neutral point is determined with (the ordering of terms in the rows of the transformation matrix now corresponds to the spatial ordering of phases in Figure 3.8, as indicated in the row above the transformation matrix):

$$
[C] = \sqrt{\frac{2}{9}}
\begin{array}{c}
\alpha \\ \beta \\ x_1 \\ y_1 \\ x_2 \\ y_2 \\ x_3 \\ y_3 \\ 0
\end{array}
\begin{bmatrix}
1 & \cos(\alpha) & \cos(2\alpha) & \cos(6\alpha) & \cos(7\alpha) & \cos(8\alpha) & \cos(12\alpha) & \cos(13\alpha) & \cos(14\alpha) \\
0 & \sin(\alpha) & \sin(2\alpha) & \sin(6\alpha) & \sin(7\alpha) & \sin(8\alpha) & \sin(12\alpha) & \sin(13\alpha) & \sin(14\alpha) \\
1 & \cos(7\alpha) & \cos(14\alpha) & \cos(6\alpha) & \cos(13\alpha) & \cos(2\alpha) & \cos(12\alpha) & \cos(\alpha) & \cos(8\alpha) \\
0 & \sin(7\alpha) & \sin(14\alpha) & \sin(6\alpha) & \sin(13\alpha) & \sin(2\alpha) & \sin(12\alpha) & \sin(\alpha) & \sin(8\alpha) \\
1 & \cos(13\alpha) & \cos(8\alpha) & \cos(6\alpha) & \cos(\alpha) & \cos(14\alpha) & \cos(12\alpha) & \cos(7\alpha) & \cos(2\alpha) \\
0 & \sin(13\alpha) & \sin(8\alpha) & \sin(6\alpha) & \sin(\alpha) & \sin(14\alpha) & \sin(12\alpha) & \sin(7\alpha) & \sin(2\alpha) \\
1 & \cos(6\alpha) & \cos(12\alpha) & 1 & \cos(6\alpha) & \cos(12\alpha) & 1 & \cos(6\alpha) & \cos(12\alpha) \\
0 & \sin(6\alpha) & \sin(12\alpha) & 0 & \sin(6\alpha) & \sin(12\alpha) & 0 & \sin(6\alpha) & \sin(12\alpha) \\
\frac{1}{\sqrt{2}} & \frac{1}{\sqrt{2}} & \frac{1}{\sqrt{2}} & \frac{1}{\sqrt{2}} & \frac{1}{\sqrt{2}} & \frac{1}{\sqrt{2}} & \frac{1}{\sqrt{2}} & \frac{1}{\sqrt{2}} & \frac{1}{\sqrt{2}}
\end{bmatrix}
\begin{array}{c}
a_1 \; a_2 \; a_3 \; b_1 \; b_2 \; b_3 \; c_1 \; c_2 \; c_3
\end{array}
$$

(3.37)

and there are, in addition to the $\alpha-\beta$ components and zero-sequence component, three pairs of $x-y$ components. However, if the neutral points of three-phase windings are left isolated, the decoupling transformation matrix of (3.37) becomes

$$
[C] = \sqrt{\frac{2}{9}}
\begin{array}{c}
\alpha \\ \beta \\ x_1 \\ y_1 \\ x_2 \\ y_2 \\ 0_1 \\ 0_2 \\ 0_3
\end{array}
\begin{bmatrix}
1 & \cos(\alpha) & \cos(2\alpha) & \cos(6\alpha) & \cos(7\alpha) & \cos(8\alpha) & \cos(12\alpha) & \cos(13\alpha) & \cos(14\alpha) \\
0 & \sin(\alpha) & \sin(2\alpha) & \sin(6\alpha) & \sin(7\alpha) & \sin(8\alpha) & \sin(12\alpha) & \sin(13\alpha) & \sin(14\alpha) \\
1 & \cos(7\alpha) & \cos(14\alpha) & \cos(6\alpha) & \cos(13\alpha) & \cos(2\alpha) & \cos(12\alpha) & \cos(\alpha) & \cos(8\alpha) \\
0 & \sin(7\alpha) & \sin(14\alpha) & \sin(6\alpha) & \sin(13\alpha) & \sin(2\alpha) & \sin(12\alpha) & \sin(\alpha) & \sin(8\alpha) \\
1 & \cos(13\alpha) & \cos(8\alpha) & \cos(6\alpha) & \cos(\alpha) & \cos(14\alpha) & \cos(12\alpha) & \cos(7\alpha) & \cos(2\alpha) \\
0 & \sin(13\alpha) & \sin(8\alpha) & \sin(6\alpha) & \sin(\alpha) & \sin(14\alpha) & \sin(12\alpha) & \sin(7\alpha) & \sin(2\alpha) \\
1 & 0 & 0 & 1 & 0 & 0 & 1 & 0 & 0 \\
0 & 1 & 0 & 0 & 1 & 0 & 0 & 1 & 0 \\
0 & 0 & 1 & 0 & 0 & 1 & 0 & 0 & 1
\end{bmatrix}
\begin{array}{c}
a_1 \; a_2 \; a_3 \; b_1 \; b_2 \; b_3 \; c_1 \; c_2 \; c_3
\end{array}
$$

(3.38)

so that there are now only two pairs of $x-y$ components.

3.8 Modeling of Synchronous Machines

3.8.1 General Considerations

Modeling principles, detailed in preceding sections for multiphase induction machines, apply in general equally to synchronous machines, since the stator winding of all synchronous machines is identical as for an induction machine, regardless of the number of phases. However, the rotor of synchronous machines differs considerably from the induction machine's rotor, both in terms of the winding disposition used and in terms of its construction. Moreover, synchronous machines are much more versatile than induction machines and come in a variety of configurations.

Most of synchronous machines have excitation on rotor, which can be provided either by permanent magnets or by a dc-supplied excitation (or field) winding. The exception is synchronous reluctance machine, where rotor is not equipped with either magnets or the excitation winding. Further, rotor of a synchronous machine may or may not carry a squirrel-cage short-circuited winding, depending on whether the machine is designed to operate from mains or from a power electronic supply with closed-loop speed (position) control. Finally, rotor may be of circular cross section, but it may also have a so-called salient-pole structure.

Two principal geometries of the rotor are illustrated in Figure 3.9. Only one phase (1s) of the stator multiphase winding is shown and it is illustrated schematically with its magnetic axis. The rotor is shown as having an excitation winding, which is supplied from a dc source and which produces rotor field. This field is stationary with respect to rotor and acts along the d-axis. But, since rotor rotates at synchronous speed, the rotor field rotates at synchronous speed in the air-gap as well. In both types of synchronous machines, which are normally used for electric power generation and high-power motoring applications, rotor will either physically have a squirrel-cage winding (salient-pole rotor; not shown in Figure 3.9) or will behave as though there is a squirrel-cage winding (cylindrical rotor structure).

If permanent magnets are used instead of the excitation winding, then they may be either fixed along the circumference of a cylindrical rotor (surface-mounted permanent magnet synchronous machine, often abbreviated as SPMSM) or they may be embedded (or inset) into the rotor (interior PMSM or IPMSM). If the machine is designed for variable-speed operation with closed-loop control, the rotor will not have any windings. If the machine is aimed at line operation, then the rotor will have to have a squirrel-cage winding (recall that a synchronous motor develops torque at synchronous speed only; hence, if supplied from mains, it cannot start unless there is a squirrel-cage winding that will provide asynchronous torque at nonsynchronous speeds of rotation).

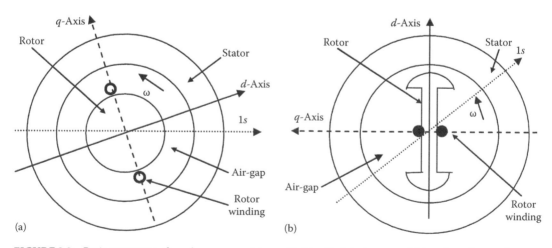

FIGURE 3.9 Basic structures of synchronous machines with (a) cylindrical rotor and (b) with a salient-pole rotor.

Transformations discussed in Sections 3.3 and 3.4 and given with (3.12) and (3.19) remain to be valid in exactly the same form for synchronous machine. However, what is different and therefore impacts considerably on the transformation procedure is the fact that the air-gap in a synchronous machine is not uniform any more. This is obvious for the salient-pole structure of Figure 3.9, but also applies to the cylindrical rotor structure, since the excitation winding occupies only a portion of the rotor circumference, so that the effective air-gap is the lowest in the d-axis and is the highest in the axis perpendicular to d-axis (i.e., q-axis). Nonuniform air-gap length means that the magnetic reluctance, seen by stator phase windings, continuously changes as rotor rotates. Note, however, that as far as the inductances of rotor windings are concerned, the situation is identical as for induction machines, since stator cross section is circular (and the same as in induction machines). Thus, rotor winding inductances will all be constant, as the case was in an induction machine.

As far as permanent magnet synchronous machines are concerned, in terms of magnetic behavior IMPSM corresponds to the salient-pole structure (since permeability of permanent magnets is very close to the permeability of the air, thus causing considerably higher magnetic reluctance in the rotor area where magnets are embedded, compared to the rotor area where there is only ferromagnetic material). On the other hand, SPMSMs behave similar to the machines with cylindrical rotor structure. Since magnets are effectively increasing the air-gap length and are placed uniformly on the rotor surface, the difference between the magnetic reluctance in SPMSMs along d- and q-axis is very small and is usually neglected.

Magnetic reluctance, seen by stator phase windings, varies continuously as the rotor rotates. It changes between two extreme values, the minimum one along d-axis and the maximum one along q-axis. Hence, one can define two corresponding extreme stator phase winding self-inductances, L_{sd} and L_{sq}. Assuming again that the spatial distribution of the mmf is sinusoidal, it can be shown that the stator phase 1 inductance is now governed with

$$L_{11s} = \frac{(L_{sd} + L_{sq})}{2} + \left[\frac{(L_{sd} - L_{sq})}{2}\right]\cos 2\theta \tag{3.39}$$

where angle θ is the instantaneous position of the rotor d-axis with respect to magnetic axis of stator phase 1 axis. Self-inductances of all the other phases are of the same form as in (3.39), with an appropriate shift that accounts for the spatial displacement of a particular phase with respect to phase 1. In (3.39) one has, using a three-phase machine as an example, $L_{sd} = L_{ls} + M_d$ and $L_{sq} = L_{ls} + M_q$, where M_d and M_q are mutual inductances within the stator winding along the two axes.

As can be seen from (3.39), self-inductance is a constant position-independent quantity if and only if the inductances along d- and q-axis are the same, which applies only if the air-gap is perfectly uniform. When there is a variation in the air-gap, the self-inductance contains the second harmonic of a continuously changing value as the rotor rotates. Self-inductance of (3.39) will during each revolution of the rotor take the maximum and minimum values (L_{sd} and L_{sq}) twice. Similar considerations also apply to mutual inductances within the multiphase stator winding, which will now also contain the second harmonic in addition to a constant value. Hence, in synchronous machines, all elements of the stator inductance matrix (3.4a) contain rotor-position-dependent terms, which is a very different situation when compared to an induction machine. Dependence of stator inductance matrix terms on rotor position also means that the electromagnetic torque of the machine (3.8) does not reduce any more to the form given in (3.10), since there is an additional term,

$$T_e = P[i_s]^t \frac{d[L_{sr}]}{d\theta}[i_r] + \left(\frac{P}{2}\right)[i_s]^t \frac{d[L_s]}{d\theta}[i_s] \tag{3.40}$$

The first torque component in (3.40) is again the consequence of the interaction of the stator and rotor windings (fundamental torque component) and it exists in all synchronous machines with excitation on rotor (using either permanent magnets or an excitation winding). The second component is, however, purely produced due to the variable air-gap and is called reluctance torque component. In synchronous reluctance machines, where there is no excitation on rotor, this torque component is the only one available if squirrel-cage rotor winding does not exist.

The consequence of the rotor-position-dependent inductances of the stator winding on modeling procedure is that any synchronous machine can be described with a set of differential equations with constant coefficients if and only if one selects the common reference frame as firmly fixed to the rotor. Hence, *d*-axis of the common reference frame is selected as the axis along which the rotor field winding (or permanent magnets) produces flux. Thus, in (3.19) one now has $\theta_s \equiv \theta$, which simultaneously means that $\theta_r \equiv 0$. Such transformation matrix is often called Park's transformation in literature. In simple terms, this means that rotational transformation is applied only to the stator fictitious windings, obtained after decoupling transformation. The machine is therefore modeled in the rotor reference frame. If the machine runs at synchronous speed, this coincides with the synchronous reference frame. However, in a more general case and especially in motoring applications, one needs to have in mind that fixing the reference frame to the rotor means that the transformation angle for stator variables has to be continuously recalculated using (3.5), where speed of rotation is a variable governed by (3.7).

As noted at the end of Section 3.2, it is important to observe that in synchronous machines with field winding there are separate voltage levels at the stator and field winding. It is assumed further on that the field winding (and the squirrel-cage winding, if it exists) has been referred already to the stator winding voltage level. Models are further given separately for synchronous machines with excitation winding and permanent magnet synchronous machines. Only the torque-producing part of the model is given, which is the same for all machines with three or more phases on the stator and in essence comes down to rearranging appropriately Equations 3.20a, 3.21a, and 3.22 of the induction machine model. If the machine has more than three phases, the models given further on need to be complemented with the *x*–*y* voltage and flux equations of the stator winding, (3.20b) and (3.21b). These remain to be given with identical expressions as for an induction machine and are therefore not repeated further on.

3.8.2 Synchronous Machines with Excitation Winding

Stator voltage equilibrium equations (3.20a) are in principle identical as for an induction machine, except that now $\omega_a = \omega$. Rotor short-circuited winding (damper winding) voltage equations are also the same as in (3.20a) with the last term set to zero, since $\omega_a = \omega$. Hence,

$$v_{ds} = R_s i_{ds} + \frac{d\psi_{ds}}{dt} - \omega\psi_{qs}$$

$$v_{qs} = R_s i_{qs} + \frac{d\psi_{qs}}{dt} + \omega\psi_{ds}$$

(3.41a)

$$0 = R_{rd} i_{dr} + \frac{d\psi_{dr}}{dt}$$

$$0 = R_{rq} i_{qr} + \frac{d\psi_{qr}}{dt}$$

(3.41b)

Resistances of the rotor damper winding along *d*- and *q*-axis are not necessarily the same, and this is taken into account in (3.41b). Zero-sequence voltage equation of the stator winding is the same as in (3.13) and is not repeated. Voltage equilibrium equation of the excitation winding, identified with index

f (which has not undergone any transformation, except for the voltage level referral to stator voltage level) is of the same form as for damper windings, except that the voltage is not zero:

$$v_f = R_f i_f + \frac{d\psi_f}{dt} \tag{3.41c}$$

Flux linkage equations of various windings, however, now involve two different values of the magnetizing inductance, L_{md} and L_{mq}, which is the consequence of the uneven air-gap. These inductances are related with the corresponding phase mutual inductance terms through $L_{md} = (n/2) M_d$ and $L_{mq} = (n/2) M_q$. Hence, flux linkages along d- and q-axis are

$$\psi_{ds} = (L_{ls} + L_{md}) i_{ds} + L_{md} i_{dr} + L_{md} i_f$$

$$\psi_{qs} = (L_{ls} + L_{mq}) i_{qs} + L_{mq} i_{qr}$$

$$\psi_{dr} = (L_{lrd} + L_{md}) i_{dr} + L_{md} i_{ds} + L_{md} i_f \tag{3.42}$$

$$\psi_{qr} = (L_{lrq} + L_{mq}) i_{qr} + L_{mq} i_{qs}$$

$$\psi_f = (L_{lf} + L_{md}) i_f + L_{md} i_{dr} + L_{md} i_{ds}$$

where $L_d = L_{ls} + L_{md}$ and $L_q = L_{ls} + L_{mq}$ are the self-inductances of the stator $d-q$ windings. The fact that the excitation winding produces flux along d-axis only has been accounted for in (3.42). In general, leakage inductances of the d- and q-axis damper windings may differ, and this is also taken into account in (3.42).

It should be noted that in certain cases damper winding of the rotor is modeled with one equivalent d-axis winding (as in (3.41b) and (3.42)) but with two equivalent q-axis windings. In such a case, one more voltage equilibrium equation and one more flux equation are needed for the q-axis. Their form is identical as for the q-axis damper winding in (3.41b) and (3.42), but the parameters (resistance and leakage inductance) are in general different.

Electromagnetic torque equation (3.40) upon transformation reduces in the rotor reference frame to a simple form,

$$T_e = P(\psi_{ds} i_{qs} - \psi_{qs} i_{ds}) \tag{3.43a}$$

which is exactly the same as for an induction machine (see (3.23)). However, if the stator flux $d-q$ axis flux linkage components are eliminated using (3.42), the resulting equation differs from the corresponding one for induction machines (3.22) due to the existence of the excitation winding and due to two different values of the magnetizing inductances along two axes:

$$T_e = P\left[L_{md}(i_{ds} + i_f + i_{dr}) i_{qs} - L_{mq}(i_{qs} + i_{qr}) i_{ds} \right] \tag{3.43b}$$

The form of (3.43b) can be re-arranged so that the fundamental torque component is separated from the reluctance torque component,

$$T_e = P\left[L_{md}(i_f + i_{dr}) i_{qs} - L_{mq} i_{qr} i_{ds} \right] + P(L_{md} - L_{mq}) i_{ds} i_{qs} \tag{3.43c}$$

which is convenient for subsequent discussions of permanent magnet and synchronous reluctance machine types.

Mechanical equation of motion of (3.7) is of course the same as for an induction machine. Relationship between original stator phase variables and transformed stator $d - q$ axis quantities is in the general case and in the three-phase case governed with (3.25) and (3.24), respectively, where $\theta_s \equiv \theta = \int \omega dt$.

3.8.3 Permanent Magnet Synchronous Machines

Since in permanent magnet synchronous machines field winding does not exist, the field winding equations ((3.41c) and the last of (3.42)) are omitted from the model. It is also observed that the permanent magnet flux ψ_m now replaces term $L_{md}i_f$ in the flux linkage equations of the d-axis. If the machine has a damper winding, it can again be represented with an equivalent dr–qr winding. Hence, voltage, flux, and torque equations of a permanent magnet machine can be given as

$$v_{ds} = R_s i_{ds} + \frac{d\psi_{ds}}{dt} - \omega\psi_{qs}$$

$$v_{qs} = R_s i_{qs} + \frac{d\psi_{qs}}{dt} + \omega\psi_{ds}$$

(3.44a)

$$0 = R_{rd}i_{dr} + \frac{d\psi_{dr}}{dt}$$

$$0 = R_{rq}i_{qr} + \frac{d\psi_{qr}}{dt}$$

(3.44b)

$$\psi_{ds} = (L_{ls} + L_{md})i_{ds} + L_{md}i_{dr} + \psi_m$$

$$\psi_{qs} = (L_{ls} + L_{mq})i_{qs} + L_{mq}i_{qr}$$

(3.45a)

$$\psi_{dr} = (L_{lrd} + L_{md})i_{dr} + L_{md}i_{ds} + \psi_m$$

$$\psi_{qr} = (L_{lrq} + L_{mq})i_{qr} + L_{mq}i_{qs}$$

(3.45b)

$$T_e = P\left[\psi_m i_{qs} + (L_{md}i_{dr}i_{qs} - L_{mq}i_{qr}i_{ds})\right] + P(L_{md} - L_{mq})i_{ds}i_{qs}$$

(3.46)

In torque equation (3.46), the first and the third component are the synchronous torques produced by the interaction of the stator and the rotor and due to uneven magnetic reluctance, respectively, while the second component is the asynchronous torque (the same conclusions apply to (3.43c), valid for a synchronous machine with a field winding). This component exists only when the speed is not synchronous, since at synchronous speed there is no electromagnetic induction in the short-circuited damper windings.

Model (3.44) through (3.46) describes an IPMSM. If the machine is not equipped with a damper winding, as the case will be in machines designed for variable-speed operation with power electronic supply, it is only necessary to remove from the model (3.44) through (3.46) all variables associated with the rotor winding. This comes down to omission of (3.44b) and (3.45b) and setting of rotor $d - q$ currents in (3.45a) and (3.46) to zero.

If the magnets are surface-mounted, it is usually assumed that the machine is with uniform air-gap, so that $L_{md} = L_{mq} = L_m$. This makes magnetizing inductances along the two axes equal in (3.45) and (3.46) and, consequently, eliminates the reluctance component in the torque equation (3.46).

Thus, for a SPMSM without damper winding, one gets an extremely simple model, which consist of the following equations:

$$v_{ds} = R_s i_{ds} + \frac{d\psi_{ds}}{dt} - \omega\psi_{qs}$$

$$v_{qs} = R_s i_{qs} + \frac{d\psi_{qs}}{dt} + \omega\psi_{ds} \tag{3.47}$$

$$\psi_{ds} = (L_{ls} + L_m)i_{ds} + \psi_m$$

$$\psi_{qs} = (L_{ls} + L_m)i_{qs} \tag{3.48}$$

$$T_e = P\psi_m i_{qs} \tag{3.49}$$

The electrical part of the model (3.47) and (3.48) is usually written with eliminated stator $d-q$ axis flux linkages, as

$$v_{ds} = R_s i_{ds} + L_s \frac{di_{ds}}{dt} - \omega L_s i_{qs}$$

$$v_{qs} = R_s i_{qs} + L_s \frac{di_{qs}}{dt} + \omega(\psi_m + L_s i_{ds}) \tag{3.50}$$

where $L_s = L_{ls} + L_m$ and the time derivative of permanent magnet flux is zero. The dynamic $d-q$ axis equivalent circuits for permanent magnet machines without damper winding are shown in Figure 3.10. These apply in general to IPMSMs; for SPMSM it is only necessary to set $L_d = L_q = L_s$. If the machine operates in steady state, with sinusoidal terminal phase voltages, speed of the reference frame coincides with synchronous speed and the di/dt terms in (3.47) (or (3.50)) become equal to zero. Hence, in steady-state operation with balanced symmetrical sinusoidal supply of the stator winding, one has for a SPMSM

$$v_{ds} = R_s i_{ds} - \omega L_s i_{qs}$$

$$v_{qs} = R_s i_{qs} + \omega(\psi_m + L_s i_{ds}) \tag{3.51}$$

$$T_e = P\psi_m i_{qs}$$

With regard to the correlation between stator phase and transformed variables, the same remarks apply as given in conjunction with a synchronous machine with excitation winding.

3.8.4 Synchronous Reluctance Machine

This type of synchronous machine does not have any excitation on rotor. Depending on whether the machine is designed for line operation or for power electronic supply, the rotor may or may not have the squirrel-cage winding. To get the model of this type of synchronous machine, it is only necessary to

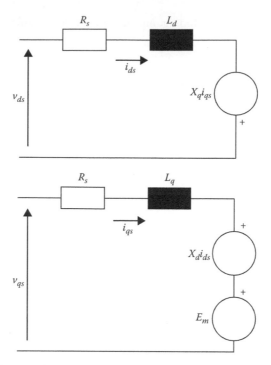

FIGURE 3.10 Equivalent dynamic $d-q$ circuits of permanent magnet synchronous machines ($X_d = \omega L_d$, $X_q = \omega L_q$, $E_m = \omega\psi_m$).

remove from the IPMSM model terms related to the permanent magnet flux linkage. Hence, from (3.44) through (3.46), one now gets

$$v_{ds} = R_s i_{ds} + \frac{d\psi_{ds}}{dt} - \omega\psi_{qs}$$

$$v_{qs} = R_s i_{qs} + \frac{d\psi_{qs}}{dt} + \omega\psi_{ds}$$

(3.52a)

$$0 = R_{rd} i_{dr} + \frac{d\psi_{dr}}{dt}$$

$$0 = R_{rq} i_{qr} + \frac{d\psi_{qr}}{dt}$$

(3.52b)

$$\psi_{ds} = (L_{ls} + L_{md})i_{ds} + L_{md}i_{dr}$$

$$\psi_{qs} = (L_{ls} + L_{mq})i_{qs} + L_{mq}i_{qr}$$

(3.53a)

$$\psi_{dr} = (L_{lrd} + L_{md})i_{dr} + L_{md}i_{ds}$$

$$\psi_{qr} = (L_{lrq} + L_{mq})i_{qr} + L_{mq}i_{qs}$$

(3.53b)

$$T_e = P\left[(L_{md}i_{dr}i_{qs} - L_{mq}i_{qr}i_{ds}) + (L_{md} - L_{mq})i_{ds}i_{qs}\right]$$

(3.54)

where the first component is the asynchronous torque, while the second component is the synchronous torque.

If the machine does not have squirrel-cage winding on rotor, rotor voltage equations (3.52b) and rotor flux linkage equations (3.53b) are omitted. Hence, the stator voltage equations and the electromagnetic torque in such a machine take an extremely simple form,

$$v_{ds} = R_s i_{ds} + L_d \frac{di_{ds}}{dt} - \omega L_q i_{qs}$$

$$v_{qs} = R_s i_{qs} + L_q \frac{di_{qs}}{dt} + \omega L_d i_{ds} \tag{3.55}$$

$$T_e = P(L_{md} - L_{mq}) i_{ds} i_{qs}$$

The form of the $d-q$ axis equivalent circuits is the same as in Figure 3.10, provided that the electromotive force term $\omega \psi_m$ is set to zero.

As noted already, permanent magnet machines and synchronous reluctance machines without rotor damper (squirrel-cage) windings are exclusively used in conjunction with power electronic supply and closed-loop control, which requires information on the instantaneous rotor position.

3.9 Concluding Remarks

A basic review of the modeling procedure, as applied in conjunction with multiphase ac machines with sinusoidal mmf distribution around the air-gap, has been provided. The material has been presented in a systematic way so that not only three-phase but also machines with any phase number are covered. All the types of ac machinery that operate on the basis of the rotating field have been encompassed. This includes both induction and synchronous machines of various designs. Given modeling procedure and the models in developed form are valid under the simplifying assumptions introduced in Section 3.2. In this context a couple of remarks seem appropriate.

In a number of cases the assumptions of constant machine parameters represent physically unjustifiable simplifications. This is sometimes due to the machine construction and sometimes due to the transient phenomenon under consideration. For example, frequency-dependent variation of parameters (resistance and leakage inductance) is of importance in rotor windings of squirrel-cage induction machines, which are often designed with deep-bar winding, or there may even exist physically two separate cage windings. In both cases the accuracy of the model is significantly improved if the rotor is represented as having two (rather than one) squirrel-cage windings. In terms of the final model, this comes down to expanding the Equations 3.20 through 3.22) (or 3.30 through 3.32) so that the representation contains voltage equilibrium and flux equations for two rotor windings (note that this also affects the torque equation (3.22)). For more detailed discussion the reader is referred to [22].

Assumption of constant stator leakage inductance is usually accurate enough. The exception are the investigations related to starting, reversing, re-closing, and similar transients of mains-fed induction machines, where the stator current may typically reach values of five to seven times the stator rated current. Means for accounting for stator leakage flux saturation in the $d-q$ axis models have been developed, and such modified models require knowledge of the stator leakage flux magnetizing curve, which can be obtained from locked rotor test.

The iron losses are of magnetic nature, and accounting for them in the $d-q$ axis models can only ever be approximate. The usual procedure is the same as in the steady-state equivalent circuit phasor representation. An equivalent iron loss resistance can be added in parallel to the magnetizing branch in the circuit of Figure 3.4. This of course requires expansion of the model with additional equations and an appropriate modification of the torque equation. It should be noted that such a representation of iron losses can only ever relatively accurately represent the phenomenon if the machine is supplied from a sinusoidal source.

By far the most frequently inadequate assumption is the one related to the linearity of the magnetizing characteristic, which has made the magnetizing (mutual) inductance (or inductances in synchronous machines) constant. This applies to both induction and synchronous machines. There are even situations where this assumption essentially means that a certain operating condition cannot be simulated at all; for example, self-excitation of a stand-alone squirrel-cage induction generator. It is for this reason that huge amount of work has been devoted during the last 30 years or so to the ways in which main flux saturation can be incorporated into the $d–q$ axis models of induction and synchronous machines. Numerous improved machine models, which account for magnetizing flux saturation (and therefore utilize the magnetizing characteristic of the machine), are nowadays available. Some methods are discussed in references [14,20,21]. In principle, the machine model always becomes considerably more complicated than the case is when saturation of the main flux is neglected.

Finally, resistances of all windings change with operating temperature. Since temperature does not exist as a variable in the $d-q$ models, this variation cannot be accounted for unless the $d-q$ model is coupled with an appropriate thermal model of the machine.

References

1. C.L. Fortescue, Method of symmetrical co-ordinates applied to the solution of polyphase networks, *AIEE Transactions, Part II*, 37, 1027–1140, 1918.
2. R.H. Park, Two-reaction theory of synchronous machines—I, *AIEE Transactions*, 48, 716–731, July 1929.
3. E. Clarke, *Circuit Analysis of A-C Power*, Vols. 1 and 2, John Wiley & Sons, New York, 1941 (Vol. 1) and 1950 (Vol. 2).
4. K.P. Kovács and I. Rácz, *Transiente Vorgänge in Wechselstrommaschinen*, Band I und Band II, Verlag der Ungarischen Akademie der Wissenschaften, Budapest, Hungary, 1959.
5. C. Concordia, *Synchronous Machines: Theory and Performance*, John Wiley & Sons, New York, 1951.
6. B. Adkins, *The General Theory of Electrical Machines*, Chapman & Hall, London, U.K., 1957.
7. D.C. White and H.H. Woodson, *Electromechanical Energy Conversion*, John Wiley & Sons, New York, 1959.
8. W.J. Gibbs, *Electric Machine Analysis Using Matrices*, Sir Isaac Pitman & Sons, London, U.K., 1962.
9. S. Seely, *Electromechanical Energy Conversion*, McGraw-Hill, New York, 1962.
10. K.P. Kovács, *Symmetrische Komponenten in Wechselstrommaschinen*, Birkhäuser Verlag, Basel, Switzerland, 1962.
11. M.G. Say, *Introduction to the Unified Theory of Electromagnetic Machines*, Pitman Publishing, London, U.K., 1971.
12. H. Späth, *Elektrische Maschinen*, Springer-Verlag, Berlin/Heidelberg, Germany, 1973.
13. N.N. Hancock, *Matrix Analysis of Electrical Machinery* (2nd edn.), Pergamon Press, Oxford, U.K., 1974.
14. P.M. Anderson and A.A. Fouad, *Power System Control and Stability*, The Iowa State University Press, Ames, IA, 1980.
15. J. Lesenne, F. Notelet, and G. Seguier, *Introduction à l'électrotechnique approfondie*, Technique et Documentation, Paris, France, 1981.
16. Ph. Barret, *Régimes transitoires des machines tournantes électriques*, Eyrolles, Paris, France, 1982.
17. J. Chatelain, *Machines électriques*, Dunod, Paris, France, 1983.
18. A. Ivanov-Smolensky, *Electrical Machines*, Part 3, Mir Publishers, Moscow, Russia, 1983.
19. I.P. Kopylov, *Mathematical Models of Electric Machines*, Mir Publishers, Moscow, Russia, 1984.
20. K.P. Kovács, *Transient Phenomena in Electrical Machines*, Akadémiai Kiadó, Budapest, Hungary, 1984.
21. P.C. Krause, *Analysis of Electric Machinery*, McGraw-Hill, New York, 1986.
22. I. Boldea and S.A. Nasar, *Electric Machine Dynamics*, Macmillan Publishing, New York, 1986.
23. G.J. Retter, *Matrix and Space-Phasor Theory of Electrical Machines*, Akadémiai Kiadó, Budapest, Hungary, 1987.

24. E. Levi, Multiphase electric machines for variable-speed applications, *IEEE Transactions on Industrial Electronics*, 55(5), 1893–1909, 2008.
25. T.A. Lipo and F.X. Wang, Design and performance of a converter optimized AC machine, *IEEE Transactions on Industry Applications,* 20, 834–844, 1984.
26. D.F. Gosden, An inverter-fed squirrel cage induction motor with quasi-rectangular flux distribution, *Proceedings of the Electric Energy Conference*, Adelaide, Australia, 1987, pp. 240–246.
27. E. Levi, R. Bojoi, F. Profumo, H.A. Toliyat, and S. Williamson, Multiphase induction motor drives—A technology status review, *IET—Electric Power Applications*, 1(4), 489–516, 2007.

4

Induction Motor

Aldo Boglietti
Politecnico di Torino

4.1 General Considerations and Constructive Characteristics

The operation of the induction motor is based on the rotating magnetic field discovered, from the theoretical point of view by Galileo Ferraris in 1885. Nicola Tesla later developed the first applications and the first induction motors as they are now known. The rotating magnetic field is produced by a polyphase winding, which is built-in in a fixed magnetic structure in the following named the stator. This stator magnetic field induces a system of electromotive force (e.m.f.) and current in a polyphase winding built-in in a rotating magnetic structure in the following named the rotor. Since the stator and the rotor are separated by an air gap with constant thickness, the induction motor magnetic structure is isotropic. The stator and rotor windings are positioned in slots punched in the stator and rotor laminations, as shown in Figure 4.1.

The rotor windings can have a number of phases that are different from those of the stator, but the stator and rotor pole number should be the same. For simplicity, the theoretical approach reported in the following, will use a two pole motor as reference. The pole number will be included in the equations when this value is requested to define equations having general validity. The motor supply is supposed to be a symmetric three-phase sinusoidal voltage. The electrical circuits hereafter adopted are considered as taking into account the phasor theory, but the related equations are reported considering the module amplitude of the phasor only.

With a three-phase sinusoidal supply, the stator windings will be able to produce a rotating field, which will interact with the stator and rotor windings. In a phase stator winding, an e.m.f. will be induced, with amplitude, E_s, equal to

$$E_s = K_s \Phi \omega_s \tag{4.1}$$

where
  K_s is the stator winding constant
  Φ is the machine flux
  ω_s is the rotating magnetic field angular speed

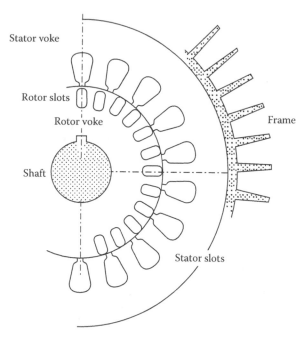

FIGURE 4.1 Induction motor cross section.

Let us consider that the rotor is still with the windings open. The rotor windings see the rotating magnetic field, which rotates at an angular speed equal to ω_s. As a consequence, the phase rotor–induced e.m.f. in the rotor phase winding can be written as

$$E_r = K_r \Phi \omega_s \tag{4.2}$$

where
    K_r is the rotor winding constant
    Φ is the machine flux
    ω_s is the rotating magnetic field angular speed

Both K_s and K_r coefficients take into account the characteristic of the stator and rotor winding (such as number of turns, winding topology, etc.).

In these conditions, the induction motor is equivalent to a transformer, in which the flux variation is due to a sinusoidal flux with constant amplitude rotating in space, while in a traditional transformer the flux is fixed in space but it changes its amplitude in time. As a consequence, in first approximation, for the induction motor, it is possible to define a voltage transformation ratio equal to $t = K_s/K_r$.

In the case of a rotor with a different phase number with respect to that of the stator, the induction motor allows a modification, both of the voltage and of the phase number between the primary winding (stator) and the secondary (rotor). Now, let us consider the rotor rotating at a mechanical angular speed, ω_m, and with the rotor windings still open. The rotor e.m.f. will be now equal to

$$E_r = K_r \Phi (\omega_s - \omega_m) \tag{4.3}$$

The difference between the two speeds $(\omega_s - \omega_m)$ is the relative rotor speed with respect to the stator magnetic field, and it is defined as absolute slip. The ratio between the absolute slip and the rotating magnetic field speed, ω_s, is defined as relative slip (usually referred to simply as slip), and it is a fundamental quantity in the study of the induction motor. The slip is defined by the following relation:

$$s = \frac{\omega_s - \omega_m}{\omega_s} \tag{4.4}$$

The percentage slip is obviously defined by

$$s_{\%} = \frac{\omega_s - \omega_m}{\omega_s} 100 \tag{4.5}$$

With the rotor still, the slip is one, while with the rotor rotating at the same speed of the stator magnetic field, the slip is zero. It is now possible to determine the frequency of the rotor electrical quantities when the rotor is rotating at the speed ω_m. The relation among the rotating magnetic field speed (expressed in rotation per minute), the stator voltage frequency, f_s, and the pole pair number p is

$$n_s = \frac{60 f_s}{p} \tag{4.6}$$

As a consequence, the stator frequency is

$$f_s = \frac{n_s p}{60} \tag{4.7}$$

In an analogous manner, the rotor frequency can be written as

$$f_r = \frac{(n_s - n_m) p}{60} \tag{4.8}$$

Multiplying and dividing the relation by ω_s, it is possible to get the following important relation:

$$f_r = \frac{(n_s - n_m) p}{60} \frac{n_s}{n_s} = \left(\frac{n_s - n_m}{n_s} \right) \frac{n_s p}{60} = s f_s \tag{4.9}$$

that shows that the rotor electrical quantities depend on the stator supply frequency and the slip. With the rotor still, the frequency of the rotor quantities is equal to that of the stator, while with the rotor rotating at the speed of the stator magnetic field, the frequency of the rotor quantities is zero. Now, let us consider the rotor still, but with the rotor windings closed in short circuit. In these conditions, the rotor e.m.f. is able to induce rotor currents, which together with the flux density in the air gap produce mechanical forces, according to the well-known electromagnetic equations. As a consequence, in the presence of short-circuited rotor windings, the rotor is under the action of a torque that leads the rotor to follow the rotating magnetic field of the stator.

As previously mentioned, the induction motor can be considered completely equivalent to a rotating field transformer, and as a consequence, the equivalent circuit topology of the induction motor is similar

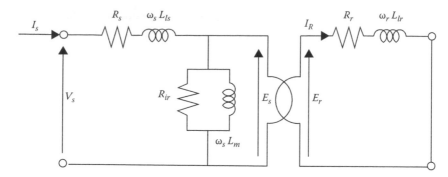

FIGURE 4.2   Induction motor equivalent circuit referred to a single phase.

to that of the transformer, with the secondary winding in short circuit. The equivalent circuit of single phase induction motor is presented in Figure 4.2, with a description of the following parameters:

V_s Stator phase voltage
R_s Phase stator winding resistance
L_{ls} Phase stator leakage inductance
ω_s Pulsation of the stator electrical quantities
R_{ir} Phase resistance equivalent to the iron losses
L_m Magnetizing inductance
E_s Stator e.m.f.
E_r Rotor e.m.f.
R_r Phase rotor winding resistance
L_{lr} Phase rotor leakage inductance
ω_r Pulsation of the rotor electrical quantities

   With respect to the well-known transformer equivalent circuit, it is important to highlight that the stator and rotor quantities are not at the same frequency. From a theoretical point of view, in addition to the classical voltage transformation, the ideal transformer shown in Figure 4.2 has to be able to interface two circuits at different frequencies. At the moment, the ideal transformer has to be considered as a special device able to modify the voltage amplitude and the frequencies. It is important to underline that the ideal transformer represents, from the physical point of view, the air gap between the stator and the rotor, where the electrical energy of the stator is transformed into the mechanical energy of the rotor. Since in system at different frequencies the average power is always zero, in the air gap, the quantities will have to act together at the same frequency. This is true because, with respect to a reference frame steady in the air gap, the stator quantities are seen at the pulsation, ω_s. The rotor is rotating at the angular speed ω_m, while the rotor quantities are at the pulsation ω_r. As a consequence, they are seen at a resulting pulsation equal to $\omega_r + \omega_m$. As previously shown, the sum $\omega_r + \omega_m$ is exactly the pulsation ω_s. As a consequence, with respect to the reference frame positioned in the air gap, the stator and rotor quantities have the same frequency and the energy transfer between stator and rotor is possible, together with the electromechanical conversion. In the circuit shown in Figure 4.2, the two networks at different frequencies can be led back to a single frequency circuit, thanks to some easy considerations on the rotor electrical equations. The rotor e.m.f. can be rewritten as

$$E_r = K_r \Phi 2\pi f_r \tag{4.10}$$

Since the rotor frequency is defined by $f_r = sf_s$ the rotor e.m.f. can be written as

$$E_r = K_r \Phi 2\pi s f_s \tag{4.11}$$

Now it is possible to introduce the concept of rotor e.m.f. at unitary slip $E_r(1)$, that is the rotor e.m.f. when the rotor is still and the rotor quantities are at the same frequency of the stator ones. As a consequence, the rotor e.m.f. can be written as

$$E_r = sE_r(1) \tag{4.12}$$

A similar approach can be used for the rotor leakage reactance; in fact it can be defined by

$$X_{lr} = \omega_r L_{lr} = 2\pi f_r L_{lr} = 2\pi s f_s L_{lr} \tag{4.13}$$

Introducing the concept of leakage reactance at unitary slip, $X_{lr}(1)$ (rotor leakage reactance with rotor still), it is possible to write

$$X_{lr} = sX_{lr}(1) \tag{4.14}$$

With the introduction of the slip in E_r and X_{lr} relations, all the rotor quantities are now referred to the stator frequencies, and in Figure 4.3 the new corresponding equivalent circuit is depicted.

Neglecting the index (1) for writing plainness, the equation of the rotor circuit can be rewritten as

$$sE_r = R_r I_r + jsX_{lr}I_r \tag{4.15}$$

With the hypothesis of slip always different by zero, it is possible to divide the previous relation by the slip, getting

$$E_r = \frac{R_r}{s}I_r + jX_{lr}I_r \tag{4.16}$$

and obtaining the equivalent circuit reported in Figure 4.4.

It is important to underline that in the equivalent circuit reported in Figure 4.4, the skin effect present in the rotor cage bars is neglected. It is now possible to write the power balance of the induction motor on the basis of the equivalent circuit reported in Figure 4.4. The active power absorbed by the stator results

$$P_s = 3V_s I_s \cos\varphi_s \tag{4.17}$$

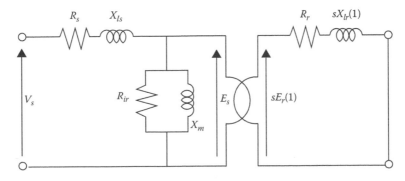

FIGURE 4.3 Induction motor equivalent circuit with the rotor quantities referred to the stator frequencies.

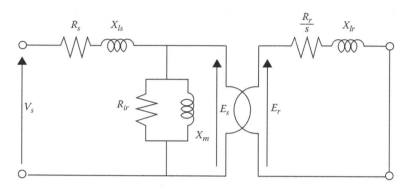

FIGURE 4.4 Induction motor equivalent circuit with the R_r/s rotor resistance.

In the stator, both the stator joule losses, $P_{js} = 3R_sI_s^2$, and the iron losses, $P_{fe} = 3(E_s^2/R_{ir})$, are active. The difference between the absorbed electrical power and the stator losses is the power transmitted by the stator to the rotor P_T

$$P_T = P_s - P_{js} - P_{fe} \qquad (4.18)$$

With reference to the single phase equivalent circuit of Figure 4.4, the transmitted power has to be attributed to the resistance R_r/s and the following relation can be written as

$$P_T = 3\frac{R_r}{s}I_r^2 \qquad (4.19)$$

Since, from the physical point of view, the rotor winding has an actual resistance, R_r, the rotor joule losses active in the rotor are

$$P_{jr} = 3R_rI_r^2 \qquad (4.20)$$

The power balance in the rotor demonstrates that the difference between the transmitted power and the rotor joule losses must be the converted mechanical power P_m

$$P_m = P_T - P_{jr} = 3\frac{R_r}{s}I_r^2 - 3R_rI_r^2 = 3\frac{1-s}{s}R_rI_r^2 \qquad (4.21)$$

where, the power involved in resistance $((l-s)/s)R_r$ represents the mechanical power. The new equivalent circuit is reported in Figure 4.5.

Using the previous equations, the following relation between the transmitted power and the rotor joule losses can be obtained:

$$P_{jr} = sP_T \qquad (4.22)$$

This relation shows that the slip can be considered as a power splitter of the transmitted power between the rotor joule losses and the mechanical power. In particular, at unitary slip, all the transmitted power is dissipated in the rotor as joule losses, while for a generic slip, "s," the mechanical power is defined by the ratio $(1-s)/s$.

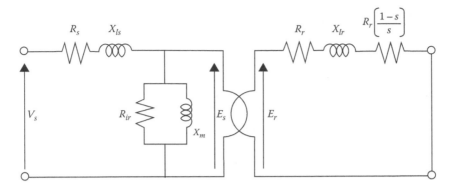

FIGURE 4.5 Induction motor equivalent circuit after the separation between the rotor resistance, R_r, and the equivalent resistance to the mechanical power $R_r((1-s)/s)$.

4.2 Torque Characteristic Determination

On the basis of the equivalent circuit previously defined, it is possible to obtain the mechanical torque produced by the machine and the torque–speed characteristic. The electromagnetic torque active in the air gap T_T is the ratio between the transmitted power and the angular speed of the rotating magnetic field, as shown in the following equation:

$$T_T = \frac{P}{\omega_s} = \frac{3(R_r/s)I_r^2}{\omega_s} \tag{4.23}$$

This equation is very important because it demonstrates the univocal proportionality between torque and transmitted power. This means that to get a torque it is necessary to transfer the right transmitted power to the rotor. The mechanical torque is defined by the ratio between the electric power of the resistance $((1-s)/s)R_r$ and the rotor angular speed as reported in the relation

$$T_m = \frac{P_m}{\omega_r} = \frac{3((1-s)/s)R_rI_r^2}{\omega_r} \tag{4.24}$$

Remembering that the rotor speed and the rotating magnetic field speed are linked by the slip through $\omega_r = (1-s)\omega_s$, it is possible to obtain the following relation:

$$T_m = \frac{P_m}{\omega_r} = \frac{3((1-s)/s)R_rI_r^2}{\omega_r} = \frac{3((1-s)/s)R_rI_r^2}{(1-s)\omega_s} = T_T \tag{4.25}$$

The previous equation shows the perfect equality between the electromagnetic torque and the mechanical torque. Obviously, the mechanical torque included all the friction and windage losses present in the rotor. The net torque can be obtained subtracting these losses from mechanical torque previously determined. In addition, it is interesting to underline that with the rotor still (slip = 1), the torque is different by zero and this torque is the machine starting torque.

It is now possible to move the rotor parameters from the rotor side to that of the stator; in this way the ideal transformer can be avoided because it is always in short circuit, obtaining the final equivalent circuit reported in Figure 4.6.

With respect to the typical equivalent circuit of the transformer, it is not possible to move the no load parameters, R_{ir} and X_m up to the stator parameters, because of the high value of the magnetizing current.

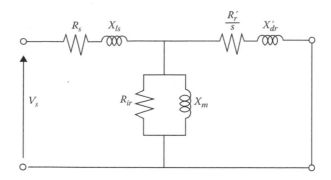

FIGURE 4.6   Induction motor equivalent circuit with the rotor quantities reported to the stator side.

In fact, in the induction motors the presence of the air gap requires a magnetizing current, which can be the 40%–60% of the rated current, depending on the motor size. Using the previous equivalent circuit, it is possible to define in an analytical way, the induction motor torque characteristic. In order to simplify the circuit under analysis, it is possible to determine the Thevenin-equivalent circuit at the rotor connections. In addition, to make the equation writing easier, all the apex will not be reported anymore, remembering that the rotor parameter value has been reported to the stator. The Thevenin rotor equivalent voltage can be written as

$$V_{eq} = \frac{V_s}{(R_s + jX_{ls}) + Z_{p0}} z_{p0} \tag{4.26}$$

Where Z_{p0} is the parallel between the resistance equivalent to the iron losses, R_{ir}, and the magnetizing reactance, X_m. The Thevenin-equivalent impedance results

$$Z_{eq} = \frac{(R_s + jX_{ls}) \times Z_{p0}}{(R_s + jX_{ls}) + Z_{p0}} = R_{eq} + X_{eq} \tag{4.27}$$

The new simplified circuit is reported in Figure 4.7.

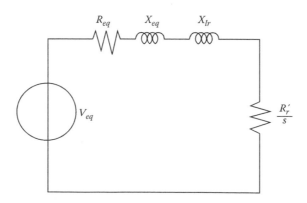

FIGURE 4.7   Thevenin-equivalent circuit of Figure 4.6.

The amplitude of the rotor current phasor can be easily computed by

$$|I_r| = \frac{V_{eq}}{\sqrt{\left(R_{eq} + (R_r/s)\right)^2 + (X_{eq} + X_{lr})^2}} \tag{4.28}$$

As a consequence, the transmitted power can be obtained by the following relation:

$$P_T = 3\frac{R_r}{s}I_r^2 = 3\frac{R_r}{s}\frac{V_{eq}^2}{(R_{eq} + (R_r/r))^2 + (X_{eq} + X_{lr})^2} \tag{4.29}$$

Taking into account the pole pair number p, the electromagnetic torque can be written as

$$T_m = 3\frac{p}{\omega_s}V_{eq}^2\frac{R_r/s}{(R_{eq} + (R_r/s))^2 + (X_{eq} + X_{lr})^2} \tag{4.30}$$

The quadratic relationship between the torque and the supply voltage is immediately evident. Consequently, this means a high sensitivity of the torque with the voltage variation. It is now possible to determine the characteristic between the torque and the slip from the graphical point of view using some considerations on the torque-slip function limit, for slip equal to zero and slip equal to infinity. With the slips leaning toward zero, the approximation $R_{eq} \ll R_r/s$ and $R_r^2/S^2 \gg (X_{eq} + X_{lr})^2$ can be assumed; the torque relation for small slips can be written as

$$T_T(s \rightarrow 0) \cong 3\frac{p}{\omega_s}V_{eq}^2\frac{s}{R_r} \tag{4.31}$$

This means that for small slips, the torque characteristic is linear with the slip. For slips leaning toward infinity, the inequality $R_{eq} \gg R_r/s$ can be assumed.

As a consequence, the torque relation can be written as

$$T_T(s \rightarrow \infty) \cong 3\frac{p}{\omega_s}V_{eq}^2\frac{R_r/s}{(R_{eq})^2 + (X_{eq} + X_{lr})^2} \tag{4.32}$$

This means that for infinite slip, the torque characteristic can be assumed as a hyperbolic function with respect to the slip. On the basis of these considerations, the torque vs. slip characteristic can be drawn as reported in Figure 4.8, where negative slip (brake or generator operations) and positive slip (motor operations) are considered in the slip range −1 to +1.

Remembering the relation between the slip and the mechanical rotor speed $\omega_m = \omega_s(1 - s)$, it is possible to get immediately the speed vs. mechanical rotor speed as reported in Figure 4.9. The two characteristics are mirrored, in fact at slip equal to one the rotor speed is zero, while with slip equal to zero, the rotor speed is equal to the rotating magnetic field.

In the mechanical characteristic, the starting torque (torque at speed equal to zero) and the peak torque are very evident.

The stable part of the torque–speed characteristic is delimited by the peak torque and the speed ω_s. On the basis of the torque characteristic, it is possible to demonstrate the rotor power balance as shown in Figure 4.10, where P_T is the rectangular area $T_L\omega_s$, P_{mech} is the rectangular area $T_L\omega_{mech}$ and P_{jr} is the rectangular area $P_T - P_{mech}$. As a consequence, it is very evident, as previously discussed, that the

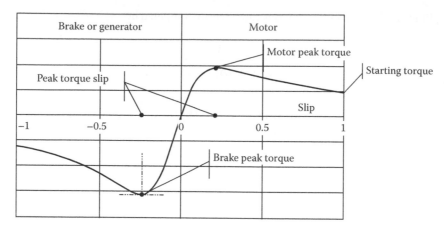

FIGURE 4.8   Induction motor torque vs. slip characteristic.

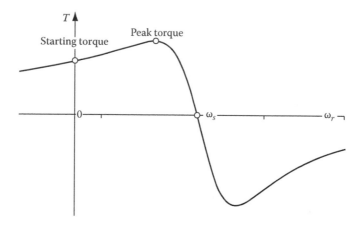

FIGURE 4.9   Induction motor torque vs. mechanical speed characteristic.

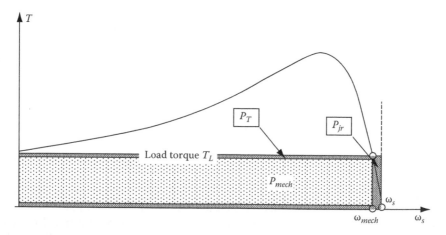

FIGURE 4.10   Induction motor rotor power balance.

function of the slip is as a power splitter of the transmitted power, P_T, between the rotor joule losses, P_{jr}, and the converted mechanical power, P_{mech}.

4.2.1 Starting Torque and Current

Imposing in the torque relation a slip equal to one, the starting torque value can be obtained as

$$T_{start} = 3\frac{p}{\omega_s}V_{eq}^2\frac{R_r}{(R_{eq} + R_r)^2 + (X_{eq} + X_{lr})^2} \tag{4.33}$$

In the same way, the starting current can be determined by

$$I_{start} = \frac{V_{eq}}{\sqrt{(R_{eq} + R_r)^2 + (X_{eq} + X_{lr})^2}} \tag{4.34}$$

The denominator of the previous relation is the short circuit impedance, often defined as locked rotor impedance. It is evident that the starting current corresponds to the locked rotor current, also called short circuit current. The starting current assumes a very high value with respect to the rated one, and it represents a serious problem for the motor itself and the motor supply source. Different techniques can be adopted for limiting the starting current. In particular, the following techniques are the most used:

- Connection of starting reactances between the supply source and the motor. These reactances have to be short-circuited after the motor is started.
- Start to delta connection modification during the starting transient. The motor is started with the motor windings connected in star and after a defined time interval (depending on the motor size and the motor and load inertia), the windings are switched to delta connection. Obviously, the procedure requires a delta connection motor during normal work conditions. During the starting condition, the star connection reduces the voltage applied to each phase of a factor equal to $\sqrt{3}$. Consequently, the line starting current is reduced by a factor equal to 3 such as the torque capability.
- Soft started devices based on solid state power electronic components. Presently, this technique is the most used for its high efficiency and its capability to control the starting current during the starting transient.

Obviously, all previous methods involve a reduction of the supply voltage and correspond to a quadratic reduction of the available torque. For this reason, in order to guarantee a correct starting of the motor, it is very important to check that the actual starting torque of the motor is still higher than the load starting torque.

4.2.2 Peak Torque

Using the analytical expression for torque, it is possible to determine the peak torque relationship. Due to the linear relation between torque and transmitted power, the peak torque condition corresponds to the peak of the transmitted power. In sinusoidal supply, the peak of the active power transfer is obtained when the equivalent impedance value of the supply source is equal to the load resistance. As a consequence, the peak transmitted power will be obtained when the following condition is verified:

$$\sqrt{R_{eq}^2 + (X_{eq} + X_{lr})^2} = \frac{R_r}{s} \tag{4.35}$$

The slip corresponding to the peak of the transmitted power, called peak torque slip, s_{Tx}, is defined by

$$s_{Tx} = \frac{R_r}{\sqrt{R_{eq}^2 + (X_{eq} + X_{lr})^2}} \tag{4.36}$$

Since in the induction motors the term $X_{eq} + X_{lr}$ is practically equal to the total motor leakage reactance X_{lt} and the total leakage reactance is greater than the equivalent resistance ($X_{lt} \gg R_{eq}$), the peak torque slip can be simplified as shown in the following relation:

$$s_{Tx} \cong \frac{R_r}{X_{lt}}$$

Including the simplified peak torque slip, s_x, in the torque equation, it is possible to obtain the value of the motor peak torque as follows:

$$T_x \cong 3 \frac{p}{\omega_s} V_{eq}^2 \frac{X_{lt}}{X_{lt}^2 + (R_{eq} + X_{lt})^2} \tag{4.37}$$

In addition, since ($X_{lt} \gg R_{eq}$), the final simplified equation of the peak torque can be written as

$$T_x \cong 3 \frac{p}{\omega_s} V_{eq}^2 \frac{1}{2X_{lt}} \tag{4.38}$$

It is very evident that the peak torque is inversely proportional to the total leakage reactance. In other words, the motor leakage reactance is a key parameter during the design of the induction motor, because it sets the motor capability to produce high peak torque. Induction motors for industrial applications have a ratio between peak torque and rated torque in the range 1.5–2.5. As a consequence, the induction motors have a good torque overload capability.

4.3 Induction Motor Name Plate Data

The main induction motor name plate data are the following:

Rated power P_R: It is the mechanical rated power at the motor shaft.
Rated speed n_R: It is the motor speed of the motor when it is working at the rated torque.
Rated torque T_R: It is obtainable by the ratio between the rated power and the rated angular speed.
Rated voltage V_R: It is the line to line voltage and it depends on the winding connection.
Rated current I_R: It is the line current when the motor works at the rated power.
Rated power factor $\cos \varphi_R$: It is the motor power factor in rated condition.

Using the previous name plate data, it is possible to define the absorbed electrical rated power using the following relation:

$$P_e = \sqrt{3} \cdot V_R \cdot I_R \cdot \cos \varphi_R$$

The efficiency is computed as the ratio between the rated power and the electrical rated power previously defined.

In addition, on the motor data sheets, the following three ratios, which are very important for a correct choice of motor with respect to the load demands, are always reported:

- Ratio between peak torque and rated torque
- Ratio between starting torque and rated torque
- Ratio between the starting current and the rated current

Taking into account the loss segregation, it is possible to define the following loss values, which can be computed by the equivalent circuit previously discussed:

- Stator joule losses, $3R_s I_s^2$
- Rotor joule losses, $3R_r I_r^2$
- Iron losses, $3(E^2/R_{ir})$

Mechanical, windage, and additional losses

All the equivalent circuit parameters, the mechanical, windage, and additional losses are obtainable by no-load and locked rotor test defined by international standards such as IEEE 112 method B.

4.4 Induction Motor Topologies

From the electrical point of view, the induction motor stator is constituted by a three-phase winding system, which has the duty to produce the rotating magnetic field. In first approximation, the stator characteristics do not influence the motor torque–speed characteristics. On the contrary, the motor torque–speed characteristics are largely dependent on the used rotor type. Wound rotor and squirrel cage rotor topologies can be realized.

4.4.1 Wound Rotor

The rotor windings are realized with copper wires in the same way as with the stator.

Stators and rotors can have different number of phases, but they must have the same pole number. The rotor windings are usually star connected and the three free terminals are connected to a system of three rings (as shown in Figure 4.11). Obviously, brushes to realize the short circuit connection or to connect possible external loads are necessary. A typical load is a three-phase resistance system used to limit the starting currents and to increase the starting torque. Due to the high cost of this type of machine, the production of wound rotor induction motors has been given up now. Wound rotor induction motors are still used in applications where very high power is required (typically MW machines at medium voltage), because substitution with squirrel cage machines is very onerous from the economic point of view.

4.4.2 Squirrel Cage Rotor

The squirrel cage rotor winding is realized with a system of aluminum or copper bars fit in the rotor slots. The bar ends are connected to two short circuit rings of the same material, as show in Figure 4.12.

In particular, the aluminum rotor cage is built using a die cast process, where the complete squirrel cage (bars plus rings) is realized at the same time. From the electrical point of view, the cage is able to

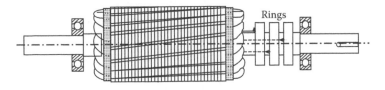

FIGURE 4.11 Wound rotor induction motor.

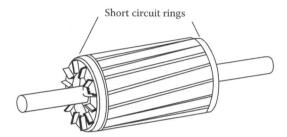

FIGURE 4.12   Squirrel cage induction motor rotor.

produce by itself an equivalent winding system with a pole number equal to that of the stator and a phase number equal to the bar number.

4.4.2.1 Double Cage Rotor

A double cage rotor can be built with inner and outer rotor cages. Due to its position in the rotor lamination, the inner cage has a higher leakage reactance with respect to the outer one. This is because the inner cage is completely surrounded by the rotor magnetic material, which increases the slot leakage flux. On the contrary, the outer cage is close to the machine air gap with a lower slot leakage flux. The different leakage reactance values of the two cages play an important role during the motor starting, when the rotor quantities have the same frequency as those of the stator. The current distribution in the two cages is practically imposed by the two leakage reactances and the current moves from the inner cage that has a higher reactance, to the outer one that has a lower leakage reactance. At the same time, due to the shift of the current from the inner cage to the outer one, the total available bar section will be reduced with an increase of the equivalent rotor resistance. This phenomenon, briefly described below, is well known as the skin effect. The skin effect produces a continuous variation of the equivalent rotor resistance during the starting transient, with a starting current reduction and a starting torque increase. In other words, the equivalent rotor resistance increases with the electrical rotor frequency ($f_r = s \cdot f_s$); as a consequence, the equivalent rotor resistance is the highest one with the rotor still (slip = 1), and it is the minimum one during the normal working condition (where percentage slip value of few percent is typical). It is evident the possibility of getting different torque characteristics curves using the more appropriate leakage reactance and resistance for the two cages during the design of the motor. In this way, it is possible to produce motors with torque vs. speed characteristics, fitting the load torque vs. speed.

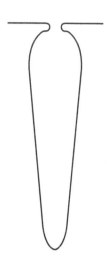

FIGURE 4.13   Example of deep rotor bar shape.

4.4.2.2 Deep Bar Rotor

In small and medium power motors, the skin effect can be obtained using a single cage where the bars have a high height–width ratio, as shown in Figure 4.13.

The skin effect is present because the lower parts of the bars have a higher leakage reactance value with respect to the outer ones. Most of the induction motors are built with deep bar squirrel cage rotors.

4.5 Induction Motor Speed Regulation

Since the stable part of the torque vs. speed characteristic has a high slope, the induction motor can be considered as an almost constant speed machine. As a consequence, a variation of the load torque leads to a small rotor speed variation. From the application point of view, the induction motor speed regulation has always been requested by the users, which requires often a wide range of speed variation.

Three methods can be used to modify or to regulate the induction motor speed, as listed in the following:

1. Pole number variation
2. Rotor resistance variation
3. Supply frequency regulation

4.5.1 Pole Number Variation

It is possible to change the machine pole number using opportune configurations of the stator windings. As a consequence, since the rotating magnetic field speed (in rpm) is linked to the pole number ($n_s = 60\ f/p$) the rotor speed can be changed in a discrete way. From the practical point of view, the rotor speed is switched between two values. For example, this technique was widely used in single phase induction motors for old style washing machine applications, where the lower speed was selected for the washing condition, and the higher speed was selected for the drying condition.

4.5.2 Speed Regulation Using Rotor Resistance

Due to the necessity of connecting the rotor winding to external resistances, this speed regulation can be used for wound rotor machines only, as shown in Figure 4.14. Starting from the peak torque slip relation, $s_{Tx} = R_r/X_{lr}$, and the peak torque equation, $T_x - 3\left(p V_{eq}^2\ /\ \omega_s \right)(1\ /\ 2X_{ll})$, it is evident that, in first approximation, a rotor resistance variation leads to a slip variation without a modification of the peak torque value. In particular, an increase of the rotor resistance leads to a torque–speed characteristic inclination, as shown in Figure 4.14.

As a consequence, at constant load torque, the rotor will rotate at lower speed depending on the rotor resistance increase. This behavior can be well understood on the basis of the motor power balance. As previously discussed, at constant torque, a constant transmitted power will be delivered to the rotor, $P_T = 3(R_r/s)I_r^2$. Since the relation between transmitted power and torque is $P_T - T_T\omega_s$, it is possible to write the torque as

$$T_T = \frac{3}{\omega_s} \frac{R_r}{s} I_r^2 \tag{4.39}$$

This equation shows that at constant torque, the current is constant too when the ratio between the rotor resistance and the slip is unchanged. As a consequence, with the increase of the rotor resistance, a greater amount of the transmitted power will be dissipated as rotor joule losses with a reduction of the converted mechanical power. At constant torque, this means a reduction of the rotor speed. This

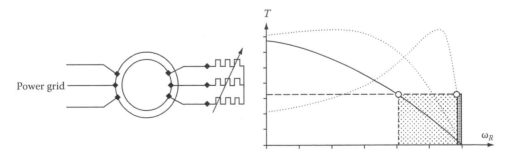

FIGURE 4.14 Effect of the rotor resistance on the torque–speed characteristic.

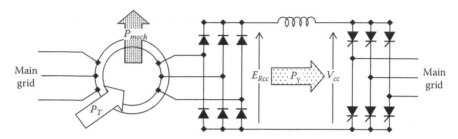

FIGURE 4.15   Speed regulation using power converter between rotor and main terminals.

speed regulation technique was used frequently in the past, but due to its low efficiency is totally given up now. Anyway, this speed regulation finds application in large induction motors, where thank the solid state power converter, the difference between the transmitted power and the converted mechanical power can be recycled in the main supply. The power converter is positioned between the rotor output and the main and allows interfacing of the rotor circuit at the rotor frequency with the main grid at the stator frequency as shown in Figure 4.15. As previously mentioned, this speed regulation can still be found in large machines where high dynamic speed regulation is not mandatory.

4.5.3 Supply Frequency Regulation

While in the previous solutions the speed regulation has been made at constant supply frequency, in this case the supply frequency is regulated for getting a continuous speed regulation in accordance to the equation $n_s = 60f/p$.

In order to have a frequency regulation, it is necessary to connect a frequency converter, usually called inverter, between the main and the induction motor. All the inverter topologies do not produce sinusoidal voltage, but in this analysis it is convenient to assume that the motor is fed by an ideal sinusoidal voltage supply regulated in frequency. In order to better understand the phenomena involved in the machine when the supply frequency is modified, it is convenient to consider the stator e.m.f. reported hereafter:

$$E_s = K_s \Phi \omega_s = K_s \Phi 2\pi f_s \tag{4.40}$$

This equation shows that in order to have a constant machine flux, stator voltage and frequency must change at the same time as evident in the following equation:

$$\Phi = \frac{E_s}{K_s 2\pi f_s} \tag{4.41}$$

As a consequence, the inverter has to be able to regulate the frequency and the voltage at the same time, following a linear law shown in Figure 4.16. Under this hypothesis, the flux is constant and the torque vs. speed characteristic is moved with the frequency as reported in Figure 4.17. This regulation is usually limited to a maximum frequency and voltage equal to the rated ones.

In order to boost the rotor speed over the rated one, it is necessary to increase the frequency at constant voltage accepting a machine flux reduction and its consequent reduction in torque production capability at rated current, as shown in Figure 4.18. As a consequence, in analogy with the speed regulation in separated excitation DC motor, it is possible to define a speed regulation at constant torque and at constant power.

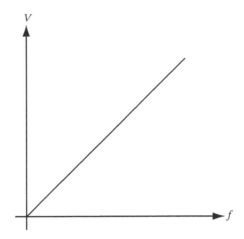

FIGURE 4.16 Induction motor voltage vs. frequency regulation.

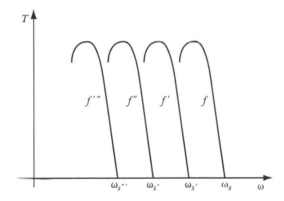

FIGURE 4.17 Variation of the torque vs. speed characteristic following the voltage vs. frequency regulation reported in Figure 4.16.

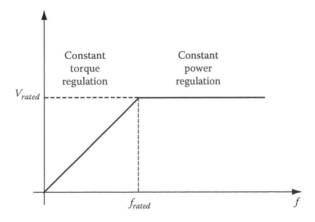

FIGURE 4.18 Constant torque and constant power regions for induction motor speed regulation.

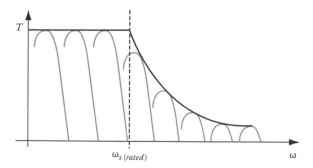

FIGURE 4.19 Peak torque limits for speed regulation in the constant torque and the constant power regions.

The modification of the torque vs. speed characteristic, in the whole frequency range, is shown in Figure 4.19. During the speed regulation at constant power, a reduction of the peak torque is unavoidable. In first approximation, this phenomenon can be analyzed considering the relation of the peak torque at constant voltage and variable frequency. Since the total leakage reactance is $X_{lt} = 2\pi f_s L_{lt}$, the peak torque relation can be rewritten as

$$T_x = 3p\frac{V_{eq}^2}{\omega_s}\frac{1}{2L_{lt}\omega_s} = 3p\frac{V_{eq}^2}{4\pi L_{lt}f_s^2} \tag{4.42}$$

where the peak torque at constant voltage and variable frequency is inversely proportional to the square of the frequency.

Considering a constant power load (load torque inversely proportional to the speed), the peak torque curve and the torque curve at constant power will have to cross in one point that defines the maximum frequency (and then, the maximum speed) for the speed regulation at constant power, as shown in Figure 4.20. Obviously, it is possible to increase the speed, leaving the constant power condition, and following a reduced power curve imposed by the peak torque trend. Nevertheless, this load condition does not find practical use.

Using the motor equivalent circuit, it is possible to demonstrate that, in first approximation, the ratio between the motor rated speed and the maximum speed at constant power is equal to the ratio between the peak torque and the rated torque. As a consequence, in order to have a wide range of speed regulation at constant power, a high ratio between the peak power and the rated power is requested. This behavior can be obtained with an ad hoc rotor design.

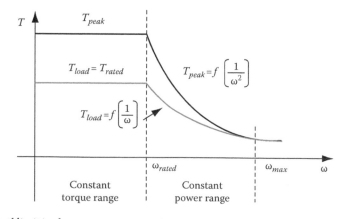

FIGURE 4.20 Speed limit in the constant power region.

4.6 Final Considerations

This chapter is not absolutely exhaustive, because in order to make the approach straightforward, several simplifications have been done. Obviously, a complete analysis of induction motors can be found in electrical machine books describing motor theory and in electrical drive books describing speed regulation [1–4]. In particular, it is important to highlight that the validity of equivalent circuit discussed in this chapter is limited to steady state conditions. Transient analysis requires a dynamic model, which is out of the scope of this chapter.

References

1. Amin, B., *Induction Motors*, Springer, Berlin/Heidelberg, Germany, 2001.
2. Alger, P. L., *Induction Machines Their Behavior and Uses*, Gordon & Breach Science Publishers, Basel, Switzerland, 1970.
3. Boldea, I. and Nasai, S. A., *The Induction Machine Handbook*, CRC Press, Boca Raton, FL, 2002.
4. Levi, E. E., *Polyphase Motors: A Direct Approach to Their Design*, John Wiley & Sons, New York, February 1984.

5

Permanent Magnet Machines

M.A. Rahman
*Memorial University
of Newfoundland*

Power electronics has emerged as the key enabling technology for all aspects of modern electric machines. Broadly speaking, rotating electric machines are of two types: dc and ac machines. In each category, permanent magnet (PM) materials are widely employed to achieve efficient performances.

The basic principle of electromagnetic energy conversion is well known. The electromotive force (EMF), simply called voltage (V) is generated in a rotating machine by the **B**l**v** principle involving the cross product of the magnetic field vector (**B**) and the velocity vector l**v**, where l is the rotor length and **v** is the velocity of the prime mover coupled with the rotor. Maximum voltage is generated when the **B** and l**v** vectors are orthogonally disposed. Similarly, the electromagnetic force, and hence torque, is developed in an electric motor by the principle of **B**l*i*, in which l*i* is the current vector. Maximum electromagnetic torque is developed when the **B** and l*i* vectors are orthogonally disposed in design, and **i** is the current flowing in the conductor of the rotor. In a typical motor, the field is radial and the current is axial in the rotor; thus the developed electromagnetic force or torque is circumferential in nature. It is to be noted that the magnetic field density, **B**, is common in both voltage generation and torque development. In most conventional electric machines, the magnetic field is provided by wire-wound dc in the pole structure of the rotating electric machine.

In PM machines, the dc field is replaced by modern hard PMs having good magnetic flux density and very large coercive force. The PM machines consist of both PM generators and PM motors. The PM dc generators are hardly used in recent times. The need is met by the large-scale availability of power electronics–based ac–dc rectifier converters. The applications of PM ac generators are also somewhat limited, as most of the large electric utility generators do not use PM excitation systems for various reasons. PMs are employed in diesel/gas-based PM ac generators for isolated standard ac power supply sources at construction sites, etc. In recent years, the PM ac generators are used in automobiles as starter/alternator in standard cars and recently in hybrid electric vehicles for charging the on-board battery modules. The PM ac generators are of specific application types. Recently, higher rating wind generators with surface-mounted PMs (SPM) are getting introduced in wind energy systems. Thus, with the scope of PM ac generators being limited, these will not be covered in this PM machines chapter.

Over the last 30 years, significant advances have taken place in PM motors technology [3]. There are various reasons for the emergence of the energy-efficient PM motors [1,2]. Broadly, the PM motors can be classified into three categories: PM dc motors, PM brushless dc motors, and PM ac synchronous motors. The PM dc motors are used in control and general purpose applications. In a way, it is a separately excited dc motor, where the dc excited electromagnetic pole (field) structure in the stationary part is replaced by PMs using PM materials ranging from low grade barium ferrite or Alnico to high grade neodymium boron iron [5]. The rotor (armature) is of the typical commutator type, with carbon brush-gear assembly. Unlike in conventional separately excited dc motor that requires two sources of dc power, the PM dc motor is a singly fed industrial drive. A multi-quadrant ac–dc converter provides the dc voltage to the armature for motor operation. The PM dc motors have been dominating the field of adjustable speed drives until the 1990s. This type of dc motors is traditionally used in motion control and industrial drive applications. However, certain limitations are associated with PM dc motors such as narrow range of speed operations, lack of robustness, wear of brush gears, and low load capability. In addition, the commutator bars and brushes of the dc motor need periodic maintenance that makes the motor less reliable and unsuitable to operate in harsh environments. These shortcomings have encouraged researchers to find alternatives to the PM dc motors for high performance variable speed operations, where high reliability and minimum maintenance are prime requirements.

The second category of PM motors is the PM brushless dc (BL dc) motors. It is basically an electronically commutated PM synchronous motor, which is sequentially switched-on by means of 3-phase voltage inverters with trapezoidal waveforms. It is obvious that the commutator bars and brush gear assembly are completely dispensed with in the case of BL dc motors. The PM brushless dc motors are extensively used as efficient industrial drives for machine tools, computer hard disk drives and control applications.

It is well known that the workhorse of modern ac industrial drives is the induction motor, because of its ruggedness, reliability, simplicity, good efficiency, and low cost. The standard squirrel-cage induction motors are cheap and widely available internationally in mass scale production. However, there are several limitations of induction motors, which discourage their use in high performance constant speed drive applications. An induction motor always operates with a slip at lagging power factor. It cannot develop any torque at constant synchronous speed. The cost and complexity of the control equipment for the induction motor drives are generally high. The performance of the induction motor drive system is less efficient due to the slip power loss. The modern inverter-fed induction motors are also subjected to non-sinusoidal voltage and current waveforms, resulting in two major detrimental effects: additional power losses and torque pulsations. The dynamic control of an induction motor drive system and its real-time implementation depends on the sophisticated modeling of motor and the estimation of motor parameters in addition to complicated control circuitry. For constant speed operation, wire-wound dc field excited synchronous motors are traditionally utilized for variable power factor operation with the inherent limitation of requiring both ac and dc sources of power.

Unlike in the wire-wound synchronous motors, the rotor excitation of the PM synchronous motors is provided by PMs. The PM ac synchronous motors do not need extra dc power supply or field windings in order to provide rotor excitation. So, the power losses related to the field windings are eliminated in the PM ac motors [1,2].

The limitations of both the ac induction motor and conventional synchronous motor drives are overcome by the singly fed PM ac motors. The control of PM ac motor drive is relatively quite simple. Furthermore, it meets all the attributes of modern high performance industrial drives. Considerable improvement in the dynamics of the PM ac synchronous motors can be achieved because of high air gap magnetic flux density, low rotor inertia, and decoupling control characteristics of speed and flux. These modern energy-efficient PM ac synchronous motors are getting widely accepted in applications requiring high performances in order to meet the competitive drive market place for quality products and improved services because of their advantageous features such as high-torque-to-current ratio, high-power-to-weight ratio, high efficiency, high power factor, low noise, and robustness.

The basic classification of the PM ac synchronous motors is shown in Figure 5.1. It is also often called PM synchronous motor, where the letters ac (alternating current) is dropped for the sake of brevity. Broadly, the PM synchronous motors can be classified into stator line-fed and stator inverter-fed types. The stator line-fed PM synchronous motors with rotor conduction bars use the cage winding to provide the starting torque of the motor at line voltage and frequency. The stator inverter-fed PM synchronous motors with rotor conduction cage are similar in construction of the stator line-fed PM synchronous motors. The stator inverter-fed PM synchronous motors with rotor conduction cage can be operated in both open

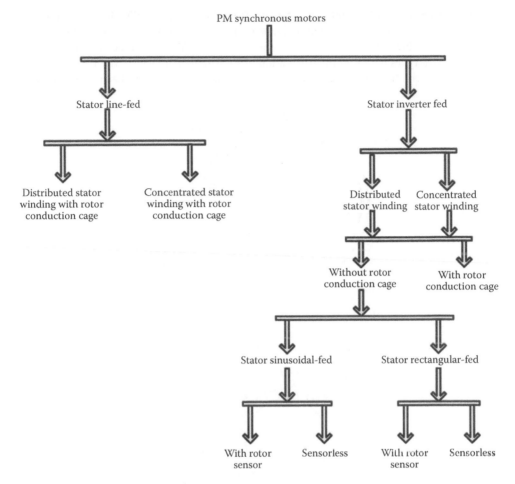

FIGURE 5.1 Classification of PM synchronous motors.

loop and closed loop conditions at variable voltage and/or variable frequency. The stator inverter-fed PM synchronous motors without rotor cage use the feedback of rotor positioning sensor(s) to start smoothly from standstill up to the steady state operating speed. The rotor position can be sensed using an absolute encoder or an incremental encoder, or can be estimated using the position sensorless approaches. The PM synchronous motors have been implemented in a variety of application fields, which include automobiles, air conditioner, aerospace, machine tools, servo drive, ship propulsion drive, etc. The PM synchronous motors of 3–10 horsepower (hp) ratings have been almost exclusively used as high efficient compressor drive motors of Japanese air conditioners. Recently, PM synchronous motors of higher than 1 MW ratings have been successfully designed and used for cycloconverter-fed propulsion drives for navy ships.

Based on the use of rotor position sensors, the PMSM drives can be again classified into two categories: (1) those with sensor PM synchronous motor drives and (2) sensorless PM synchronous motor drives. In sensorless drives, the rotor position is estimated from motor currents, voltages, and motor parameters using an observer or using a computational technique. The implementation of the sensorless scheme for the PM synchronous motor drive can be difficult, as it requires sophisticated algorithms to estimate the rotor position. The estimated rotor position is not accurate because of the variation of parameters for various operating conditions of PM synchronous motor drives.

Based on the orientation of magnets in the rotor, the PM synchronous motors can be further classified into three categories: (1) interior type, where the PMs are buried within the rotor core [4,6,36]; (2) surface-mounted type, where the PMs are mounted on the surface of the rotor [11]; and (3) inset type, where the PMs are fully or partially inset into the rotor core from the air gap end [7]. The cross sections of the interior PM (IPM) type, surface-mounted permanent magnet (SPM), and inset type PM synchronous motors are shown in Figures 5.2 through 5.4, respectively. The PM synchronous motors can be again classified into three types based on the orientation of rotor magnetic field of the PMs. These include radial type, circumferential type, and axial type of PM synchronous motors. Each type has its relative

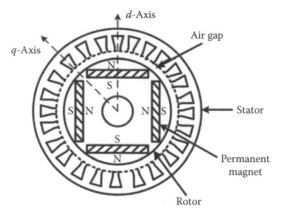

FIGURE 5.2 Cross section of the interior type PM (straight magnet) synchronous motor.

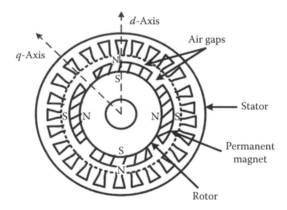

FIGURE 5.3 Cross section of the surface-mounted type PM synchronous motor.

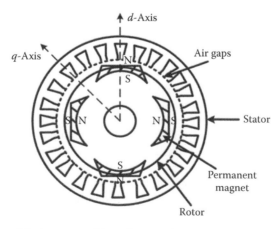

FIGURE 5.4 Cross section of the inset type PM synchronous motor.

advantages and limitations for specific applications. Generally, the axial and circumferential types of PM are less energy efficient. Hence, the interior permanent magnet (IPM) types are widely used in high efficiency category [8–10,12–40].

The direction of magnetic field of the interior type permanent magnet synchronous motors (IPMSM) and the inset type PM synchronous motors is predominantly radial. The direction of magnetic field is also radial for the case of surface-mounted type permanent magnet synchronous motors (PMSM). The directions of magnetic field of the PM synchronous motors of Figures 5.2 through 5.4 are radial. The majority of commercially available PM synchronous motors are constructed with PMs buried inside the rotor iron core. These types of motors are known as IPM synchronous motors. The arrangement of PMs inside the rotor core of IPM synchronous machine produces several significant effects on the operating characteristics of the motor. Burying the magnets inside the rotor of the IPMSM provides a mechanically robust rotor since the magnets are physically contained and protected. On the other hand, the magnets of the surface-mounted type PM synchronous motors (PMSM) are held protected against centrifugal forces by means of an adhesive or a high strength non-magnetic band (sleeve) during the high speed operation. Therefore, the rotor of the surface-mounted type (PMSM) is less robust than the interior type (IPMSM) for high speed applications. The relative permeability of PMs in surface-mounted PM motor being almost equal to that of the air, the PMSM behaves like a non-salient pole synchronous machine. The rotor of an IPMSM with radial magnetization is easy and economical to manufacture in mass volume. Moreover, as the PMs are buried within the rotor core of the IPMSM, it provides a smooth surface at rotor air gap, and it has uniform air gap length. The IPMSM is an inherently high efficiency powerful machine [18].

The control and operation of IPMSM drive system forms the core of high performance and efficient IPM motor technology for large-scale industrial applications in the first decade of the twenty-first century [29–40]. The reasons are due to the fact that the interior permanent magnet synchronous motor (IPMSM) is a hybrid machine in which the reluctance torque and the electrical torque are simultaneously developed [21]. It is fundamentally a salient pole type synchronous machine, and hence it produces more power. This leads to more output torque. Figure 5.5 shows the rotor and cross section of a four-pole, three-phase IPM motor with a typical stator slot and a rotor conduction cage, as well as a V-shaped neodymium boron iron (NdBFe) magnet [36].

The IPM machine shown in Figure 5.5 with V-shaped neodymium boron iron PMs is a focused high-flux PM motor. The magnet arrangements are designed within the rotor, such that it creates variations

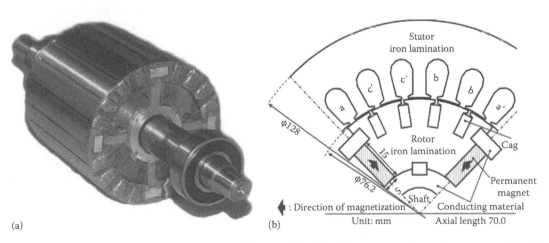

(a)   (b)

FIGURE 5.5 (a) Rotor and (b) its cross section of the 4-pole (V-shaped) IPM motor. (From Binn, K. J. et al., *IEE Proc. Part B*, 125(3), 203, 1978. With permission.)

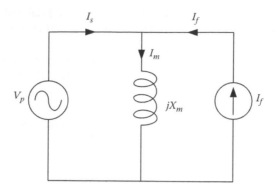

FIGURE 5.6   Norton's equivalent circuit of an IPM motor.

of machine inductance along the direct (d) and quadrature (q) axes. It behaves like a salient-pole synchronous motor having direct and quadrature $d–q$ axis machine inductances. The main challenges of designing an IPM synchronous motor are as follows:

- Create variations of $d–q$ axis inductances without varying air gap.
- Vary and control excitation of permanently excited rotor of IPM rotor.
- Optimum variation PM torque and reluctance torque for specific applications.
- Use of intelligent power converter and inverter modules for IPM drive.
- Reduction of weight, size, and cost of IPM motor.

The variation of a PM ac synchronous motor can be operated at variable power factor by using the Norton's equivalent circuit, where the Kirchchoff's current laws can be easily applied.

Figure 5.6 shows the Norton's equivalent circuit of an IPM motor in which $I_s + I_f = I_m$, where I_s = per phase stator input current, I_f = per phase field current due to PM excitation, and I_m = per phase stator magnetizing current, respectively. V_p is the per phase stator input voltage and X_m is the per phase magnetizing reactance.

It is to be noted that the equivalent field current I_f due to permanent magnet excitation is constant, and hence its magnitude cannot be altered, but the angle β, between current phasors I_f and I_m, can be controlled to operate the IPM motor at leading, unity, and lagging power factors as shown in Figure 5.7.

The power factors of a modern IPM motor can also be easily controlled by changing the direct axis current from positive to negative values, in which the stator input phasor current I_s of the IPM motor is resolved into $d–q$ axis components such that $I_s = I_q + jI_d$ where I_q = the q-axis or torque component of the stator current, and I_d = the d-axis or flux component of the stator current, respectively. Figure 5.8 shows the operation of the IPM motor again at leading, unity, and lagging power factors by injecting +ve or −ve direct axis component I_d of the stator, such that the magnitude of the resultant d-axis or flux component of the stator current is altered in variable modes from +ve, 0, and −ve for operating

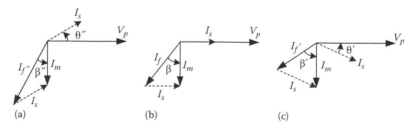

FIGURE 5.7   Current phasor diagrams of an IPM motor at (a) leading, (b) unity, and (c) lagging power factors.

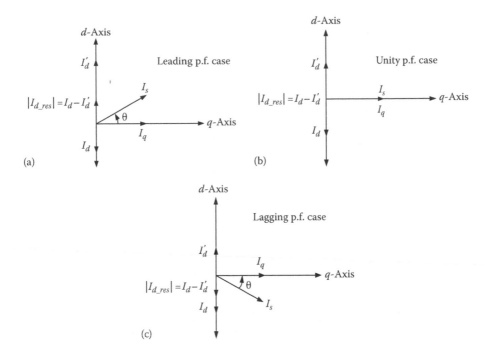

FIGURE 5.8 Direct-axis control of stator current for variable power factors operation of IPM motor: (a) leading power factor operation, (b) unity power factor operation, and (c) lagging power factor operation.

at leading, unity, and lagging power factors without changing the q-axis or the torque component of the stator current.

The developed torque of an IPM synchronous motor can expressed in the following form as

$$T_d = \frac{p}{2}\lambda_m i_q + \frac{p}{2}(L_q - L_d)i_d i_q$$

$$A = B + C$$

where

the term A is the total developed motor torque

the term B is the electrical torque like that of dc separately excited dc motor, except the fact that the flux is provided by the PM

the term C is the reluctance torque due to the difference of d–q axis inductances multiplied by the d–q axis current components of the stator current

λ_m is the flux linkage due to PM excitation

L_d and L_q are the d–q axis inductances, respectively

i_d and i_q are the d–q axis currents, respectively

p is the number of poles

It is obvious that the sum A is always greater than its parts B or C.

The IPMSM is a singly fed modern hybrid machine that combines both the electromagnet and reluctance torques in one compact unit. Figure 5.9 shows the one-quadrant structure of a three-dimensional finite element generated flux density contour of the partial V-shaped interior PM motor [23,24,36].

The performances of line-start IPM synchronous motor of Figure 5.5 are presented in Table 5.1. A comparison of the performances of the 600 W IPM motor and an identically rated standard squirrel-cage

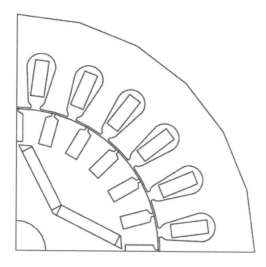

FIGURE 5.9 One-quadrant structure of a partial V-shaped IPM motor.

TABLE 5.1 Comparison of Performances for IPM and Induction Motors

Quantity	IPM Motor		Induction Motor
Input voltage: V_i (V)	130	140	200
Input current: I_i (A)	3.11	2.91	3.43
Input power: P_i (W)	687	696	818
Rotor speed: n (rpm)	1500	1500	1434
Torque: T (N m)	3.82	3.82	4.00
Efficiency: η (%)	87.3	86.2	73.3
Power factor: p.f. (%)	98.1	98.6	68.8
Output power: P_o (W)	600	600	600
Eff. × p.f. product (%)	85.6	85.0	50.4
Maximum output: P_{om} (W)	960	1115	1240

Source: Kurihara, K. and Rahman, M.A., *IEEE Trans. Ind. Appl.*, 40(3), 789, 2004. With permission.

induction motor in Table 5.1 clearly establishes that in each category the IPM motor yields superior results than those of the induction motor. It is worth noting that the efficiency as well as the efficiency-power factor product of the IPM motor is significantly higher by over 35% than those for the induction motor [36].

For modern line-start as well as soft-start inverter-fed ac motor drives, the IPM synchronous motors usher in good news for an energy-hungry world.

The power ratings of modern energy-efficient IPM motors have been dramatically expanded by three orders of magnitude during the past two decades. A close examination reveals that several knowledge-based technological advancements in new PM materials with very large coercive force, intelligent control, and industrial electronics system as well as fierce global market forces combined, sometimes in fortuitous ways, to accelerate the development of IPM technology that we find available today. The IPM motor technology includes not only more powerful hybrid IPM synchronous motors, but also the combination of intelligent power electronics module, variable power factor control, direct torque control, indirect vector control, maximum torque per Ampere control, minimization of torque ripples, field weakening control using new intelligent techniques, optimization of reluctance, and electrical torques with minimum losses over wide speed ranges for high performance ac synchronous motor drives. There are many specific applications that require advanced research and development in modern IPM motor drive systems. The following list of sample references may provide a state-of-the-art survey of significant as well as incremental but important contributions in chronological order over the past 55 years for further studies.

References

1. F.M. Merril, Permanent magnet excited synchronous motors, *AIEE Transactions*, 74, 1754–1760, 1955.
2. K.J. Binn, W.R. Barnard, and M.A. Jabbar, Hybrid permanent magnet synchronous motors, *IEE Proceedings, Part B*, 125(3), 203–208, 1978.
3. M.A. Rahman, Permanent magnet synchronous motors—A review of the state of design art, *Proceedings of International Conference on Electric Machines (ICEM)*, Vol. 1, Athens, Greece, September 15–17, 1980, pp. 312–319.
4. V.B. Honsinger, Field and parameters of interior type ac permanent magnet motors, *IEEE Transactions on Power Apparatus and System*, 101(4), 867–876, 1981.
5. M.A. Rahman and G.R. Slemon, Promising applications of neodymium boron iron magnets, *IEEE Transactions on Magnetics*, 21(5), 1712–1716, 1985.
6. M.A. Rahman, T.A. Little, and G.R. Slemon, Analytical models for permanent magnet synchronous motors, *IEEE Transactions on Magnetics*, 21(5), 1741–1743, 1985.
7. T. Sebastian, G.R. Slemon, and M.A. Rahman, Modelling of permanent magnet synchronous motors, *IEEE Transactions on Magnetics*, 22(5), 1069–1071, 1986.
8. T.M. Jahns, G.B. Kliman, and T.W. Neumann, Interior PM synchronous motors for adjustable speed drives, *IEEE Transactions on Industry Applications*, 22(4), 738–747, 1986.
9. T.M. Jahns, Flux-weakening regime operation of permanent magnet synchronous motor drive, *IEEE Transactions on Industry Applications*, 23(4), 681–689, 1987.
10. T. Sebastian and G.R. Slemon, Operating limits of inverter driven permanent magnet synchronous motor drives, *IEEE Transactions on Industry Applications*, 23(2), 327–333, 1987.
11. M.A. Rahman, Analytical model of exterior-type permanent magnet synchronous motors, *IEEE Transactions on Magnetics*, 23(5), 3625–3627, September 1987.
12. G.R. Slemon and T. Li, Reduction of cogging torques in permanent magnet synchronous motors, *Transactions on Magnetics*, 4(6), 2901–2903, 1988.
13. M.A. Rahman and A. M. Osheiba, Parameter sensitivity analysis of line start permanent magnet motors, *Electric Machines and Power Systems Journal*, 14(3–4), 195–212, 1988.
14. B.K. Bose and P.M. Szczesny, A microcontroller based control of an advanced IPM synchronous motor drive system for electric vehicle propulsion, *IEEE Transactions on Vehicular Technology*, 35(4), 547–559, 1988.
15. A.M. Osheiba, M.A. Rahman, A.D. Esmail, and M.A. Choudhury, Stability of interior permanent magnet synchronous motors, *Electric Machines and Power Systems Journal*, 16(6), 411–430, 1989.
16. S. Morimoto, Y. Takeda, T. Hirasa, and K. Taniguchi, Expansion of operating limits for permanent magnet motors by optimum flux-weakening, *IEEE Transactions on Industry Applications*, 26(5), 966–971, 1990.
17. M.A. Rahman and A.M. Osheiba, Performance of large line-start permanent magnet synchronous motors, *IEEE Transactions on Energy Conversion*, 5(1), 211–217, 1990.
18. R.F. Schiferl and T.A. Lipo, Power capability of salient pole permanent magnet synchronous motors in variable speed drive applications, *IEEE Transactions on Industry Applications*, 27(1), 115–123, 1991.
19. A.B. Kulkarni and M. Ehsani, A novel position sensor elimination technique for interior permanent magnet motor drive, *IEEE Transactions on Industry Applications*, 28(1), 141–150, 1992.
20. Z.Q. Zhu and D. Howe, Influence of design parameters on cogging torque in permanent magnet machines, *IEEE Transactions on Energy Conversion*, 15(2), 407–412, 1992.

21. M.A. Rahman, Combination hysteresis, reluctance, permanent magnet motor, U.S. Patent 5,187,401, issue date: February 16, 1993.

22. S. Morimoto, M. Sanada, and Y. Takeda, Effects of compensation of magnetic saturation in flux-weakening controlled permanent magnet synchronous motor drives, *IEEE Transactions on Industry Applications*, 30(6), 1632–1637, 1994.

23. Ping Zhou, M.A. Rahman, and M.A. Jabbar, Field and circuit analysis of permanent magnet synchronous machines, *IEEE Transactions on Magnetics*, 30(4), 1350–1359, 1994.

24. M.A. Rahman and P. Zhou, Field circuit analysis of brushless permanent magnet synchronous motors, *IEEE Transactions on Industrial Electronics*, 43(2), 256–267, April 1996.

25. M. Ooshima, A. Chiba, T. Fukao, and M.A. Rahman, Design and analysis of radial force in a permanent magnet type bearingless motor, *IEEE Transactions on Industrial Electronics*, 43(2), 292–299, 1996.

26. M.A. Rahman and M.A. Hoque, On-line adaptive artificial neural network based vector control of permanent magnet synchronous motors, *IEEE Transactions on Energy Conversion*, 13(4), 311–318, 1998.

27. Y. Honda, T. Higaki, S. Morimoto, and Y. Takeda, Rotor design optimization of a multi-layer interior permanent magnet synchronous motor, *IEE Proceedings, Electric Power Applications*, 135(2), 119–124, 1998.

28. S. Vaez, V.I. John, and M.A. Rahman, An on-line loss minimization controller for interior permanent magnet motor drives, *IEEE Transactions on Energy Conversion*, 14(4), 1435–1440, 1999.

29. L. Zhong, M.F. Rahman, W.Y. Hu, K.W. Lim, and M.A. Rahman, Direct torque controller for permanent magnet synchronous motor drive, *IEEE Transactions on Energy Conversion*, 14(3), 637–642, 1999.

30. M.N. Uddin and M.A. Rahman, Fuzzy logic based speed controller for IPM synchronous motor drive, *Journal of Advanced Computational Intelligence*, 4(3), 212–219, 2000.

31. A. Consoli, G. Scarcella, and A. Testa, Industry application of zero-speed sensorless control techniques for PM synchronous motors, *IEEE Transactions on Industry Applications*, 37(2), 513–521, 2001.

32. W.L. Soong and E. Ertugrul, Field weakening performance of interior permanent magnet motors, *IEEE Transactions on Industry Applications*, 38(5), 1251–1258, 2002.

33. M.A. Rahman, M. Vilathgamuwa, M.N. Uddin, and K.J. Tseng, Non-linear control of interior permanent magnet synchronous motor, *IEEE Transactions on Industry Applications*, 39(2), 408–416, 2003.

34. J. He, K. Ide, T. Sawa, and S.K. Sul, Sensorless rotor position estimation of interior permanent magnet motor from initial states, *IEEE Transactions on Industry Applications*, 39(3), 761–767, 2003.

35. M.F. Rahman, L. Zhong, M.E. Haque, and M.A. Rahman, Direct torque controlled interior permanent magnet synchronous motor drive, *IEEE Transactions on Energy Conversion*, 18(1), 17–22, 2003.

36. K. Kurihara and M.A. Rahman, High efficiency line-start interior permanent magnet synchronous motors, *IEEE Transactions on Industry Applications*, 40(3), 789–796, 2004.

37. M.N. Uddin, M.A. Abido, and M.A. Rahman, Development and implementation of a hybrid intelligent controller for interior permanent magnet synchronous motor drive, *IEEE Transaction on Industry Applications*, 40(1), 68–76, 2004.

38. Y. Jeong, R.D. Lorentz, T.M. Jahns, and S.K. Sul, Initial position estimation of an IPM synchronous machine using carrier frequency injection methods, *IEEE Transaction on Industry Applications*, 41(1), 38–45, 2005.

39. M.A. Rahman, T.S. Radwan, R.M. Milasi, C. Lucas, and B.N. Arrabi, Implementation of emotional controller for interior permanent magnet synchronous motor drive, *IEEE Transactions on Industry Applications*, 44(5), 1466–1476, 2008.

40. M.A.S.K. Khan and M.A. Rahman, Implementation of a new wavelet controller for interior permanent magnet motor drives, *IEEE Transactions on Industry Applications*, 44(6), 1957–1965, 2008.

6

Permanent Magnet Synchronous Motors

Nicola Bianchi
Universita of Padova

This chapter deals with the permanent magnet (PM) synchronous motors, supplied by current-controlled voltage source inverter. They are formed by a rotor containing PMs, a stator with a distributed multi-phase winding, typically a three-phase winding. The phase coils of such a winding are fed by sinewave currents synchronous with the corresponding flux linkages due to the PM flux.

There are two key advantages in using the PMs to create the main magnetic flux of the machine. First, the space required by the PMs for the magnetization is small, so that the motor design exhibits several degrees of freedom. Second, since there are no losses for magnetization, the PM motors feature high torque density and high efficiency.

The increasing interest toward PM motors is also due to the high energy density of the modern PMs, showing high residual flux density and high coercive force. In addition, the PM specific cost is decreasing, making the cost of the PM motor competitive with other motor types. As a consequence, the PM synchronous motors are more and more used in several applications. The power ratings of the PM synchronous motors are widening, and today they range from fractions of Watts to some million of Watts.

After a brief introduction on the PM characteristics, this chapter illustrates the key features of the PM synchronous motors. Different geometrical topologies are presented, including both integral-slot and fractional-slot winding PM motors. Finally, some control strategies are described, highlighting the relationship between the PM motor performance and its rotor geometry.

6.1 Rotor Configurations

The stator of the PM synchronous motor is the same of the induction motor. Conversely, the rotor can assume different topologies, according to how the PM is placed in the rotor. The motors are distinguished in three classes: surface-mounted PM (SPM) motors, inset PM motor, and interior PM (IPM) motor.

Figure 6.1a shows a cross section of a 4-pole 24-slot SPM motor. There are four PMs mounted with alternating polarity on the surface of the rotor. Since the PM permeability is close to the air permeability, the rotor is isotropic. Figure 6.1b shows a 4-pole inset PM motor. Its rotor is similar to the SPM rotor, the difference being the iron tooth between each couple of adjacent PMs. As in the SPM motor, the main flux is due to the PMs. The rotor teeth yield a moderate anisotropy. When the rotor is anisotropic, the motor

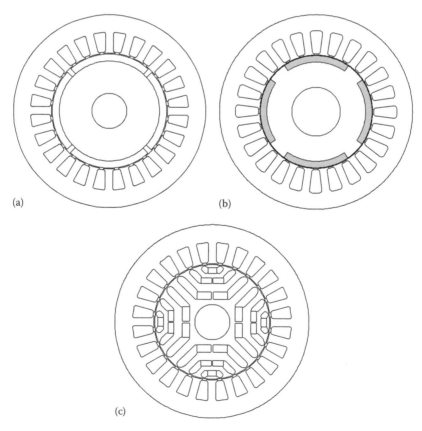

(a)   (b)

(c)

FIGURE 6.1 PM synchronous motors with (a) SPM, (b) inset PM, and (c) IPM rotor.

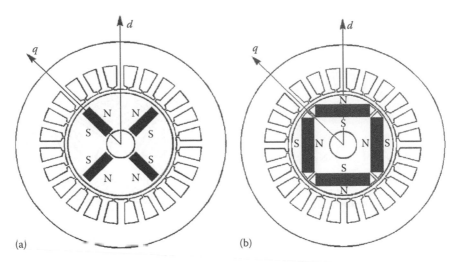

FIGURE 6.2 Four-pole IPM motor with (a) tangentially and (b) radially magnetized PMs. (Adapted from Bianchi, N., Analysis of the IPM motor, in *Design, Analysis and Control of Interior PM Synchronous Machines, Tutorial Course Notes*, Seattle, WA, October 3, 2004.)

exhibits two torque components: the PM torque and the reluctance torque [1]. Figure 6.1c shows a 4-pole IPM motor, whose rotor is characterized by three flux barriers per pole. The high number of flux barriers per pole yields a high rotor anisotropy [2]. Both torque components are high, hence the IPM motor exhibits a high torque density and it is well-suited for flux-weakening operations, up to very high speeds.

Hereafter, the positive rotor direction is in the counterclockwise direction. The rotor position is represented by the d- and the q-axis that are locked with the rotor. The d-axis is chosen as the PM flux axis, and the q-axis leads the d-axis of $\pi/2$ electrical radians. The d- and q-axis define the synchronous (rotating) reference frame.

As far as the IPM motor is concerned, it can be distinguished according to the direction of the magnetization of the PMs inside the rotor [3]. They can be

- Tangentially magnetized PMs, as in Figure 6.2a
- Radially magnetized PMs, as in Figure 6.2b

In the first configuration, the PMs have tangential magnetization and alternating polarity: then the flux in the air gap corresponds to the sum of the flux of two PMs. Rotor of this type is generally designed with a high number of poles, so that the sum of the surface of two PMs results higher than the pole surface, yielding a concentration of the flux in the air gap. A nonmagnetic shaft is required in order to avoid flux leakage.

In the second configuration, the PMs have radial magnetization and alternating polarity. The PM surface is lower than the pole surface, yielding a lower flux density in the air gap. This configuration can be designed with two or more flux barriers per pole. Figure 6.3a shows an IPM motor with two flux barriers per pole, and Figure 6.3b shows an IPM motor obtained with an axially laminated rotor. Both rotors yield a high rotor anisotropy. Such IPM motors, characterized by high anisotropy and moderate PM flux, are often called PM-assisted synchronous reluctance (PMASR) motors.

A photo of the laminations of an 8-pole tangentially magnetized PM rotor is reported in Figure 6.4a. An IPM motor is shown in Figure 6.4b, together with the rotor laminations with two flux barriers per pole. The laminations refer to two different figures of flux barriers.

All the IPM rotors exhibit magnetic paths with different permeance, from which the possibility of developing a reluctance torque. Since the differential permeability of the PM is close to the air permeability, the d-axis magnetic permeance results to be lower than the q-axis inductance, yielding $L_d < L_q$, contrarily to the common wound rotor synchronous machines. The saliency ratio (or anisotropy ratio) is defined as the ratio $\xi = L_q/L_d$.

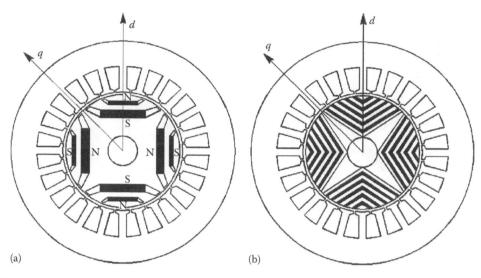

FIGURE 6.3 IPM motor with (a) two flux-barriers per pole and (b) axially laminated rotor. (Adapted from Bianchi, N., Analysis of the IPM motor, in *Design, Analysis and Control of Interior PM Synchronous Machines, Tutorial Course Notes*, Seattle, WA, October 3, 2004.)

FIGURE 6.4 IPM motor prototypes: (a) a rotor of an IPM motor with tangentially magnetized PMs and (b) an IPM motor and laminations with radially magnetized PMs.

6.2 Hard Magnetic Material (Permanent Magnet)

There are two main types of magnetic materials: the soft magnetic materials and the hard magnetic materials [4]. The soft magnetic materials are easily magnetized and demagnetized, and they are used to carry magnetic flux. Conversely, the hard magnetic materials are hardly magnetized and demagnetized, and they are generally referred to as PMs. They typically exhibit a very wide

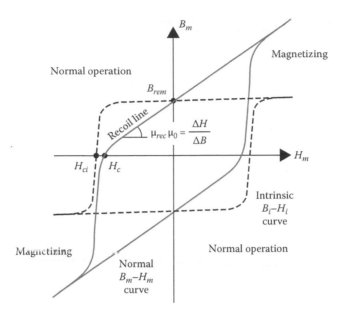

FIGURE 6.5 Hard magnetic material hysteresis loop and characteristic parameters.

magnetic hysteresis loop, as shown in Figure 6.5. The PMs are magnetized in quadrant I (or III) and operate in quadrant II (or IV). Properties and performance characteristics are described on quadrant II. Both the intrinsic (dashed line) and normal (solid line) hysteresis loops are drawn in Figure 6.5, as shown in most PM data sheets. The intrinsic curve represents the added magnetic flux that the PM material produces. The normal curve represents the total magnetic flux that is carried in combination by the air and by the PM [5]. Commonly, it is used to determine the actual flux density of the PM motor.

There are two important quantities associated with the demagnetization curve that are the residual flux density (or remanence) B_{rem} and the coercive force H_c. PMs are designed to operate on the linear part of the demagnetization curve, between B_{rem} and H_c. The differential relative magnetic permeability of the recoil line is labeled μ_{rec} and is slightly higher than unity. Such a recoil line is usually approximated by

$$B_m = B_{rem} + \mu_{rec}\mu_0 H_m \tag{6.1}$$

The product $B_m H_m$ is called the PM energy density. The maximum energy density product $\{B_m H_m\}_{max}$ is a relative measure of the strength of a PM and is always listed on the material data sheet.

The operating point moves from B_{rem} toward H_c, or from H_c toward B_{rem}, as the external system acts to demagnetize or to magnetize the PM, respectively. As long as the operating point remains on the linear slope of curve, the magnetizing and demagnetizing cycles are reversible. Conversely, if the demagnetizing field becomes large enough to move the operating point beyond the linear region, (i.e., beyond the knee of the demagnetizing curve, defined by the point $H_{knee}B_{knee}$), the subsequent magnetizing cycle follows the recoil line at lower flux density. This means that the PM is "irreversibly" demagnetized (in the sense that it requires a new process of magnetization).

The key properties of some common PM materials are listed in Table 6.1, and the normal PM demagnetization curves are shown in Figure 6.6. Figure 6.7 shows the effect of the temperature on PM demagnetization curve of a Neodymium Iron Boron (NdFeB) magnet.

TABLE 6.1 Main Properties of Hard Magnetic Material

	B_{rem} (T)	H_c (kA/m)	Curie T (°C)	Operating T_{max} (°C)	Density (kg/m³)	$\{B_mH_m\}_{max}$ (kJ/m³)
Ferrite	0.38	250	450	300	4800	30
Alnico	1.20	50	860	540	7300	45
SmCo	0.85	570	775	250	8300	140
NdFeB	1.15	880	310	180	7450	260

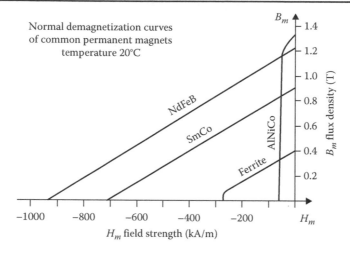

FIGURE 6.6 Demagnetization curves of common PM materials.

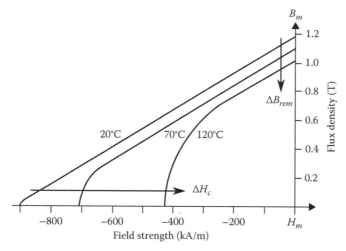

FIGURE 6.7 Effect of temperature on PM demagnetization curve.

6.2.1 Magnetic Device with PM

Figure 6.8a shows a magnetic device including an iron core, a PM, and an air gap. The Ampere law around the dashed line is

$$H_m t_m + H_g g = 0 \tag{6.2}$$

where
 H_m and H_g are the magnetic field strengths of the PM and air gap, respectively
 t_m and g are the PM and the air gap thickness, respectively

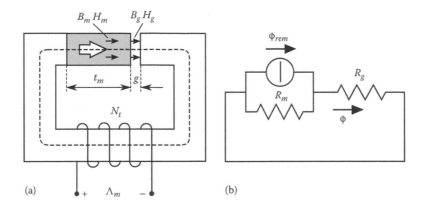

FIGURE 6.8 (a) Magnetic device and (b) equivalent magnetic network.

The magnetic drop in the iron is neglected because of $\mu_{Fe} \gg \mu_0$.

By neglecting any leakage flux, the Gauss law yields

$$\Phi = B_m A_m = B_g A_g \tag{6.3}$$

where

B_m and B_g are the flux density in the PM and in the air gap, respectively
A_m and A_g are the cross-sectional areas of the PM and air gap, respectively

Being the constitutive equation of air $B_g = \mu_0 H_g$, we can state:

$$H_m t_m + \frac{B_g}{\mu_0} g = 0 \tag{6.4}$$

or

$$H_m t_m + \frac{B_m A_m}{\mu_0 A_g} g = 0 \tag{6.5}$$

Then

$$B_m = -\left(\mu_0 \frac{t_m}{g} \frac{A_g}{A_m}\right) H_m \tag{6.6}$$

that is, the equation of a straight line on the H_m–B_m plane. The operating point of the PM is determined as the intersection of the PM demagnetization curve and the load line, defined by (6.6), as shown in Figure 6.9. From (6.2) and (6.6), it results in

$$B_m = B_{rem}\left(\frac{1}{1 + \left(\mu_{rec} g / t_m\right)\left(A_m / A_g\right)}\right) \tag{6.7}$$

The equivalent magnetic network is drawn in Figure 6.8b. The PM is represented by the residual flux generator $\Phi_{rem} = B_{rem} A_m$ in parallel with the reluctance $R_m = t_m / (\mu_{rec} \mu_0 A_m)$. Then, $R_g = g / (\mu_0 A_m)$ is the air gap reluctance. The flux Φ computed by the magnetic network corresponds to the flux of Equation 6.3.

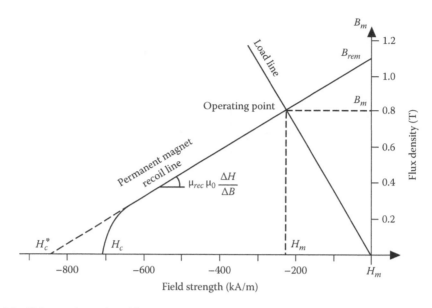

FIGURE 6.9   PM operating point without current.

A coil of N_t turns is included in the device. It links the magnetic flux Φ produced by the PM. Such a PM flux linkage is labeled as $\Lambda_m = N_t\Phi$.

6.2.2 Impact of the Current

Let us refer now to the magnetic device of Figure 6.10a, in which the coil formed by N_t turns carries a current I. The Gauss law remains as in (6.3), while the Ampere law is rewritten as

$$H_m t_m + H_g g = N_t I \tag{6.8}$$

Rearranging the equations, the load line results in

$$B_m = -\left(\mu_0 \frac{t_m}{g} \frac{A_g}{A_m}\right) H_m + \mu_0 \frac{A_g}{A_m} \frac{N_t I}{g} \tag{6.9}$$

It remains a straight line, but it is translated along the H_m axis of a quantity ΔH_i proportional to the current, as shown in Figure 6.11. According to the sign of the current, the translation is toward the positive magnetic fields (magnetizing current, i.e., positive according to the convention of Figure 6.10a) or toward the negative magnetic fields (demagnetizing current, i.e., negative according to the same convention). As above, the operating point of the PM is determined as the intersection between the PM demagnetization curve and the load line (6.9). With a magnetizing (positive) current, the flux density in the PM, and then in the other parts of the circuit, increases. In this case, it should be verified that no iron part is saturated. With a demagnetizing (negative) current, the flux density in the PM decreases. In this case, it should be verified that the minimum flux density is not below the knee of the PM demagnetizing curve, where an irreversible demagnetization of the PM occurs. The corresponding equivalent network is represented in Figure 6.10b.

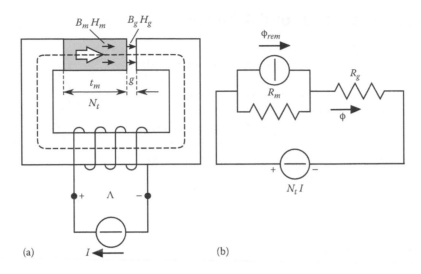

FIGURE 6.10 (a) Magnetic circuit and (b) equivalent magnetic network.

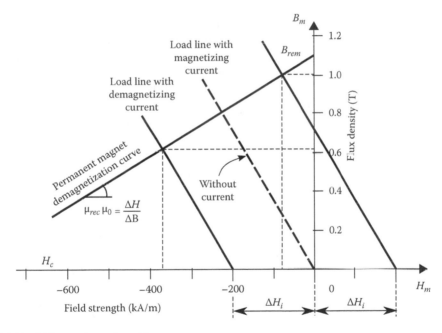

FIGURE 6.11 PM operating point with current.

6.2.3 Parameters

When the coil carries a current I, there is an additional flux in the magnetic circuit that is superimposed on the flux due to the PM. Therefore, the coil links the total flux, due to both PM and current. The total flux linkage is expressed as $\Lambda = \Lambda_m + \Lambda_i$, where $\Lambda_i = LI$ indicates the portion of flux linkage due to the current I only. The parameter L is the inductance of the circuit; it depends on the geometry of the magnetic circuit and the magnetic property of the materials. The inductance L is computed as $L = (\Lambda - \Lambda_m)/I$. Alternatively, L can be computed by assuming a demagnetized PM (i.e., assuming the residual flux density equal to zero, $B_{rem} = 0$), so that all the flux linked by the coil is due to the current, and then $L = \Lambda_i/I$.

6.3 Magnetic Analysis of PM Motor

The magnetic analysis presented in the previous section is easily extended to the study of the PM motors [6]. Two motors with $Q = 24$ slots and $p = 2$ pole pairs are considered, and with SPM and IPM rotor, respectively. Thanks to the motor symmetry, only one pole of the motor is analyzed.

Figure 6.12 shows the geometry of the SPM motor. For convenience, it is drawn with the a-phase axis placed parallel to the d-axis (i.e., the PM magnetization axis). The b-phase axis and c-phase axis lead the a-phase axis of $2\pi/3$ and $4\pi/3$ electrical radians, respectively. The stator winding is characterized by N conductors per phase, and a winding factor equal to k_w.

6.3.1 No-Load Operation (SPM Motor)

Figure 6.13a shows the flux lines at no load, that is, due to the PM only. The flux density distribution along the air gap is reported in Figure 6.13b. The figure highlights the decrease of the flux density corresponding to the slot openings of the stator. In order

FIGURE 6.12   Geometry of one pole of the SPM motor.

to avoid the irreversible demagnetization of the PM, it is imperative to verify that the minimum flux density in the PM remains higher than the flux density of the knee of the demagnetizing curve.

The flux due to the PM is linked by each stator winding. It varies according to the position of the rotor. According to Figure 6.12, phase a links the maximum flux since the d-axis is aligned to the a-axis. Such a maximum flux linkage is referred to as Λ_m.

The PM flux linkage can be computed integrating the vector magnetic potential A_z over the stator slot surfaces, that is

$$\Lambda_a = L_{stk} \sum_{q=1}^{Q_s} n_{aq} \frac{1}{S_{slot}} \int_{S_{slot}} A_z \, dS \tag{6.10}$$

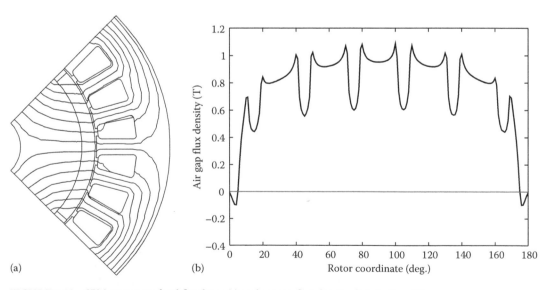

(a)   (b)

FIGURE 6.13   SPM motor: no-load flux lines (a) and air gap flux density distribution (b).

where

L_{stk} is the stack length of the motor

S_{slot} is the cross area of the slot

n_{aq} is the number of conductors of the phase a within the qth slot of the stator

The PM flux linkage can be also estimated analytically as

$$\Lambda_m = \frac{k_w N}{2} \Phi \tag{6.11}$$

where Φ is the magnetic flux per pole, given by

$$\Phi = B_g \frac{\pi D L_{stk}}{2p} \tag{6.12}$$

where

B_g is the average flux density in a pole

D is the stator inner diameter

6.3.2 Operation with *d*-Axis Stator Current

The d-axis current produces a flux along the d-axis. A positive d-axis current is magnetizing, increasing the flux produced by the PM. Conversely, a negative d-axis current is demagnetizing, since it weakens the PM flux.

Figure 6.14a shows the flux plot due to the d-axis current I_d only (i.e., without PMs). According to the Figure 6.12, the phase currents are $i_a = I_d$ and $i_b = i_c = -I_d/2$. The flux lines are similar to the flux lines due to PMs, shown in Figure 6.13a. The air gap flux density distribution is reported in Figure 6.14b, using solid line. For the sake of comparison, the flux density distribution at no load is reported using dashed line.

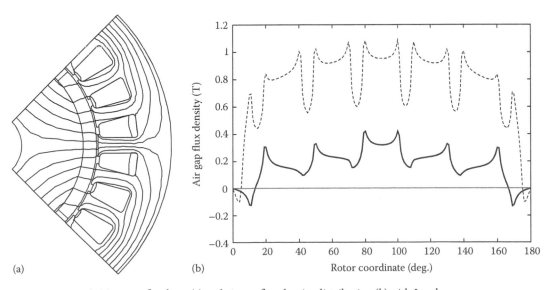

(a) (b)

FIGURE 6.14 SPM motor: flux lines (a) and air gap flux density distribution (b) with I_d only.

The synchronous inductance is computed by dividing the d-axis flux linkage (achieved by means of the abc-to-dq transformation) by the d-axis current. An analytical estimation of such an inductance is

$$L = \frac{3}{\pi}\mu_0 \left(\frac{k_w N}{2p} \right)^2 \frac{DL_{stk}}{g + t_m/\mu_{rec}} \qquad (6.13)$$

6.3.3 Operation with q-Axis Stator Current

The q-axis current induces a flux in quadrature to the flux due to the PM. Figure 6.15a shows the flux plot due to q-axis current only (i.e., without PMs). According to the Figure 6.12, the phase currents are $i_a = 0$, $i_b = -\sqrt{3}I_q/2$ and $i_c = \sqrt{3}I_q/2$.

Figure 6.15b shows the air gap flux density distribution due to the q-axis current only, using solid line. From the comparison with the no-load distribution (dashed line), it is noticing that the effect of the q-axis current is to increase the flux density in half a pole, and to decrease the flux density in the other half. PM permeability being similar to the air permeability μ_0, in the SPM motor, the q-axis inductance is practically the same as the d-axis inductance.

6.3.4 Inductance in an IPM Motor

A similar analysis can be carried out on an IPM motor, whose geometry is shown in Figure 6.16a. Figure 6.16b shows the flux lines at no load, that is, due to the PM buried within the flux barrier.

When the motor is supplied by only q-axis current, the flux lines go through the rotor without crossing the flux barrier, as shown in Figure 6.16c. This means that the flux barrier does not obstruct the q-axis flux so that the q-axis inductance L_q assumes a high value. Such an inductance can be estimated by (6.13), simply substituting g for $g + t_m/\mu_{rec}$.

Conversely, when the motor is supplied by d-axis current, the flux lines cross the flux barrier, as shown in Figure 6.16d. In this case, the flux barrier represents a magnetic reluctance and the d-axis inductance

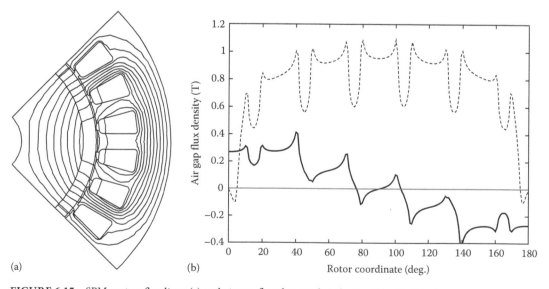

(a)   (b)

FIGURE 6.15 SPM motor: flux lines (a) and air gap flux density distribution (b) with I_q only.

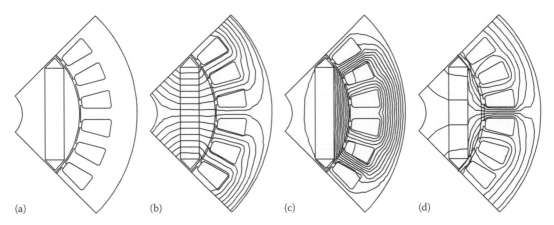

FIGURE 6.16 Flux lines in an IPM motor. (a) Geometry, (b) no-load, (c) only I_q, and (d) only I_d.

L_d results lower than L_q. The analytical estimation of L_d is tightly dependent on the geometry of the flux barriers [7]. The ratio between the two inductances, that is, $\xi = L_q/L_d$ is called saliency ratio [8].

6.3.5 Magnetic Model of the PM Synchronous Motor

In the synchronous d–q reference frame (which is rotating at the electrical angular speed ω), the d- and q-axis flux linkage components are given by

$$\lambda_d = \Lambda_m + L_d i_d$$
$$\lambda_q = L_q i_q$$
(6.14)

and the d- and q-axis voltage components result in

$$v_d = R i_d + \frac{d\lambda_d}{dt} - \omega\lambda_q$$
$$v_q = R i_q + \frac{d\lambda_q}{dt} + \omega\lambda_d$$
(6.15)

or, using (6.14):

$$v_d = R i_d + L_d \frac{d i_d}{dt} - \omega L_q i_q$$
$$v_q = R i_q + L_q \frac{d i_q}{dt} + \omega(\Lambda_m + L_d i_d)$$
(6.16)

Figure 6.17 shows the steady-state vector diagram of the PM synchronous motor in d–q reference frame [9].

The PM synchronous motor model can be represented by the equivalent circuit shown in Figure 6.18, where R_{fe} is introduced so as to take into account the iron losses. Such an equivalent iron loss resistance is not a constant but it depends on the operating frequency [10].

6.3.6 Effect of Saturation

When iron saturation occurs, the inductances in (6.14) vary with the currents, and they decrease when the currents increase. In addition, the saturation produces an interaction between the d- and the q-axis

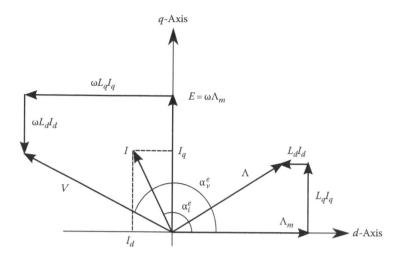

FIGURE 6.17   Steady-state vector diagram for PM synchronous motor in d–q reference frame.

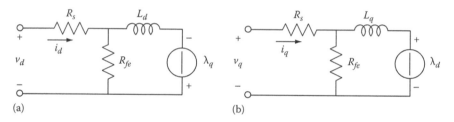

FIGURE 6.18   Equivalent circuits of PM synchronous motor in d–q reference frame including iron losses resistance: (a) d-axis circuit and (b) q-axis circuit.

quantities, which is indicated as cross-coupling effect. The magnetic model describing the relationship between the d- and the q-axis flux linkages and current is more complex, given by

$$\lambda_d = \lambda_d(i_d, i_q)$$
$$\lambda_q = \lambda_q(i_d, i_q) \tag{6.17}$$

Such a model is used for an accurate estimation of the motor performance, for instance, to precisely predict the average torque, the torque ripple, or the capability to sensorless detect the rotor position. In any case, the relations are restricted to be single-valued functions, because it is assumed that energy stored in the electromagnetic fields can be described by state functions [11].

6.4 Electromechanical Torque

Let us consider the rotor position θ_m, and d- and q-axis currents i_d and i_q as state variables. In the synchronous reference frame, the motor torque is given by

$$T = \frac{3}{2} p \left(\lambda_d i_q - \lambda_q i_d \right) + \frac{\partial W'_m}{\partial \theta_m} \tag{6.18}$$

where

p is the number of pole pairs

W'_m is the magnetic coenergy, which must be considered as a state function of the state variables θ_m, i_d, and i_q, that is, $W'_m = W'_m(\theta_m, i_d, i_q)$ [11]

The first term of the second member of (6.18) is labeled as T_{dq}, that is

$$T_{dq} = \frac{3}{2}p(\lambda_d i_q - \lambda_q i_d).$$ (6.19)

Adopting the space phasor notation, so as $\vec{\lambda} = \lambda_d + j\lambda_q$ and $\vec{i} = i_d + ji_q$, the torque (6.19) can be rearranged as

$$T_{dq} = \frac{3}{2}p(\vec{\lambda} \times \vec{i}).$$ (6.20)

where $\times$ means the cross vector product [12]. The relationship (6.20) is independent of the particular reference frame. Therefore, it is not limited to the synchronous reference frame, but it can be used in stationary and any other reference frame.

With sinusoidal current waveforms, d- and q-axis currents are constant with the rotor position. Then, the partial derivative of the magnetic coenergy with the rotor position in (6.18) results in

$$\frac{\partial W'_m}{\partial \theta_m} = \frac{3}{2}p\left(i_d \frac{\partial \lambda_d}{\partial \theta_m} + i_q \frac{\partial \lambda_q}{\partial \theta_m}\right) - \frac{\partial W_m}{\partial \theta_m}$$ (6.21)

where W_m is the magnetic energy that again has to be expressed as a function of the state variables, that is, $W_m = W_m(\theta_m, i_d, i_q)$.

Let us remark that both λ_d and λ_q vary with θ_m, so that they can be expressed by means of the Fourier series expansion. The rate of change of a flux linkage harmonic of vth order is proportional to the flux linkage harmonic amplitude times the order v. The variation of the flux linkages λ_d and λ_q is lower than the variation of their rates of change that appear in (6.21). Therefore, the torque ripple is mainly described in the torque term expressed by (6.21). On the contrary, the torque term T_{dq}, Equation 6.19, is slightly affected by the harmonics of the flux linkages and it results to be suitable for the computation of the average torque.

In an ideal system, besides the d- and q-axis currents, the d- and q-axis flux linkages are constant, as well as the magnetic energy. Therefore, the quantity given in (6.21) is equal to zero. The motor torque is constant and exactly equal to the term T_{dq}, given by (6.19).

6.4.1 Computation of Cogging Torque

Cogging torque is the ripple torque due to the interaction between the PM flux and the stator teeth. Since the stator currents are zero, it is $T_{dq} = 0$ and, from (6.18) and (6.21), the cogging torque results to be equal to

$$T_{cog} = \frac{\partial W'_m}{\partial \theta_m} = -\frac{\partial W_m}{\partial \theta_m}$$ (6.22)

Figure 6.19a shows the cogging torque versus rotor position of the SPM motor of Figure 6.1a. Solid line refers to the torque computation by means of the Maxwell stress tensor, directly computed from finite element field solution [13–15], while the circles refer to the torque computation (6.22). A further comparison between predictions and measurements of cogging torque of an SPM motor is reported in [16].

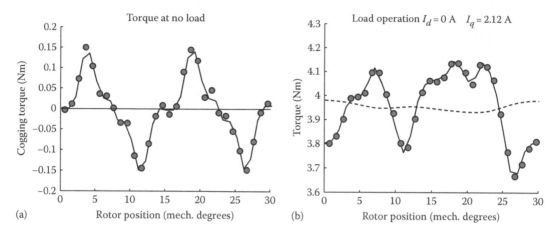

(a)   (b)

FIGURE 6.19 Torque behavior at no load (cogging torque) and under load of the SPM motor of Figure 6.1a. Solid line refers to the computation using Maxwell stress tensor, dashed line refer to (6.20), circles refer to (6.19).

6.4.2 Computation under Load (SPM Motor)

Figure 6.19b shows the torque behavior versus rotor position of the SPM motor fed by q-axis current only, while d-axis current is zero. Solid line refers to the Maxwell stress tensor computation. The circles refer to the torque computation (6.18). The dashed line refers to the torque computation T_{dq}, given by (6.19). As expected, the behavior of T_{dq} is smooth and close to the average torque.

6.4.3 Computation under Load (IPM Motor)

Similar results are found when an IPM motor is considered, as the IPM motor shown in Figure 6.1c. The nominal current is fixed to $\hat{I}_n = 4.3$ A (peak) with an electrical phase angle of $\alpha_i^e = 130$ electrical degrees. Figure 6.20a shows the behavior of the torque versus rotor position, under load. Solid line refers to the Maxwell stress tensor computation, highlighting a high torque ripple. Dashed line refers to the torque

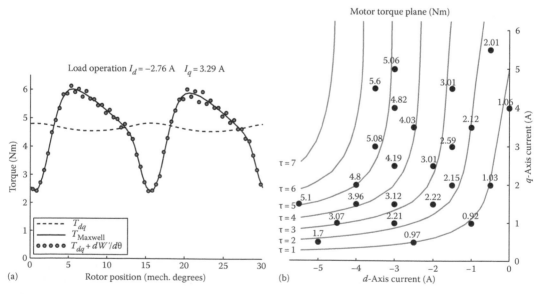

(a)   (b)

FIGURE 6.20 IPM motor of Figure 7.1c under load: (a) torque behavior at $\hat{I} = 4.3$ A and $\alpha_i^e = 130°$ and (b) predicted T_{dq} and measured (dots) torque in (i_d, i_q) plane.

computation (6.19): it is very close to the average torque, represented by the thin line in Figure 6.20a. Finally, the circles report the torques computed by (6.18).

Figure 6.20b compares the average motor torque predicted (solid lines) and measured (dots) corresponding to different *d*- and *q*-axis currents. The solid lines are the constant torque curves obtained from the finite element simulation. The fair agreement between measurements and simulations is evident.

6.5 Reduction of the Torque Ripple

Several applications require smooth motor running, in order to avoid vibration and acoustic noise. Different techniques are adopted in designing the PM motors in order to eliminate or minimize the torque ripple [16,17]. In particular, the cogging torque of SPM motors and torque ripple of IPM motors are particularly despised.

6.5.1 Reduction of the Cogging Torque in SPM Motors

It is convenient to consider the motor cogging torque T_{cog} as the sum of the interactions of each edge of the rotor PMs with the stator slot openings. Each of them is considered independent from the others. Figure 6.21a shows half a PM pole and a single slot opening moving with respect to the PM edge, where θ_m indicates the angular position between the slot axis and the PM edge. The magnetic energy W_m, sum of the air and the PM energy contributions, is a function of the angular position θ_m. The variation of W_m with θ_m is large if the slot opening is near the PM edge.

The elementary torque due to the interaction of the slot opening with the PM edge, that is, T_{edge}, corresponds to $-dW_m/d\theta_m$, see (6.22). Since W_m is monotonously decreasing with θ_m, T_{edge} is always positive (with respect to θ_m direction). It is zero when the slot is in the middle of the PM and when the slot is far from the PM. Then, T_{edge} exhibits a peak when the slot is near the PM edge, corresponding to the maximum rate of variation of magnetic energy with θ_m, as illustrated in Figure 6.21a. The peak of T_{edge} does not appear exactly at the PM edge. The same elementary torque behavior, but with opposite sign, occurs considering the interaction of the slot opening and the other PM edge. Finally, the total T_{cog} is obtained as the sum of all the elementary torques due to the interaction of the PM edges with the slot openings.

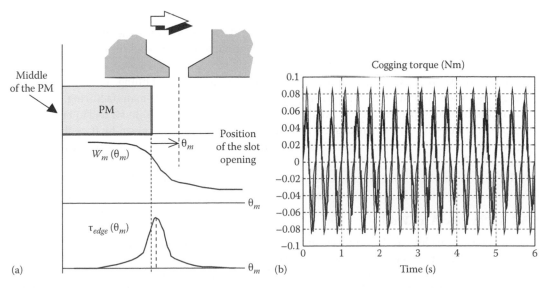

FIGURE 6.21 Simple model of the cogging torque mechanism, based on the superposition of PM edge torques T_{edge}, and a comparison between predicted (thin line) and measured (bold line) cogging torque (motor rated torque is 3 N m). (Modified from Bianchi, N. and Bolognani, S., *IEEE Trans. Ind. Appl.*, 38(5), 1259, 2002.)

TABLE 6.2   Number N_p of Cogging Torque Periods per Slot Pitch Rotation

$2p$	2	2	2	2	4	4	4	8	8	8
Q	3	6	9	12	6	9	12	6	9	15
GCD	1	2	1	2	2	1	4	2	1	1
N_p	2	1	2	1	2	4	1	4	8	8

Source: Bianchi, N. and Bolognani, S., *IEEE Trans. Ind. Appl.*, 38(5), 1259, 2002.

6.5.1.1 Number of T_{cog} Periods in a Slot Pitch Rotation

The number N_p of periods of the T_{cog} waveform during a rotation of a slot pitch depends on the number of stator slots Q and poles $2p$. For a rotor with identical PM poles, equally spaced around the rotor, the number of T_{cog} periods during a slot pitch rotation is given by

$$N_p = \frac{2p}{GCD\{Q, 2p\}} \tag{6.23}$$

where GCD means Greater Common Divisor. Thus, the mechanical angle corresponding to each period is $\alpha_{\tau_c} = 2\pi/(N_p Q)$. Table 6.2 reports the values of N_p for some common combinations of Q and $2p$.

The value of N_p is an index that shows if the elementary cogging torque waveforms are in phase or not. When N_p is low, positive (and negative) elementary torques T_{edge} occur at the same rotor position, so that they are superimposed, yielding a high T_{cog}. Conversely, when N_p is high, the elementary torques T_{edge} are distributed along the slot pitch, yielding a low T_{cog}.

An example of measured cogging torque is reported in Figure 6.21b, referring to a 9-slot 6-pole motor. During the test, the motor has been rotated at 10 r/min, so that a complete round is accomplished in 6 s. As expected, with $N_p = 2$, there are two periods of T_{cog} per each stator slot, that is, $2 \cdot 9 = 18$ periods per each complete round of the rotor.

6.5.1.2 Skewing

Skewing rotor PMs, or alternatively stator slots, is a classical method to reduce the cogging torque. The cogging torque is almost completely eliminated, with a continuous skewing angle θ_{sk} equal to the period α_{τ_c} of the cogging torque, that is, $\theta_{sk} = 2\pi/(N_p Q)$.

However, the stator skewing makes almost impossible the automatic slot filling, and the rotor skewing requires expensive PMs with complex shapes. To make easier the rotor manufacturing, the skewing is approximated by placing the PM axially skewed by N_s discrete steps, as illustrated in Figure 6.22. The optimal value of the mechanical skew angle between two modules is given by $\theta_{ss} = \theta_{sk}/N_s$, with θ_{sk} given above. With PM modules equally skewed, all the T_{cog} harmonics are eliminated except those multiples of N_s.

Adopting a stepped skewing, there is a reduction of the harmonics of back EMF. The reduction for the kth harmonic of back EMF is estimated by means of the corrective factor given by $k_{sk} = \sin(kp\theta_{sk})/(kp\theta_{sk})$.

6.5.1.3 PM Pole Arc Width

The PM pole arc width can be arranged in order to reduce or eliminate some T_{cog} harmonics [18]. According to the simple model of Figure 6.21a, the PM should span almost an integral number of slot pitches. In this way, the positive torque of each PM edge (e.g., the right-hand one) is compensated by the negative torque of the other edge (e.g., the left-hand one). As a confirmation, in [18,19] the optimal PM extension has been computed slightly greater of n times the slot pitch, that is, $(n + 0.14)$ or $(n + 0.17)$, where n is an integer.

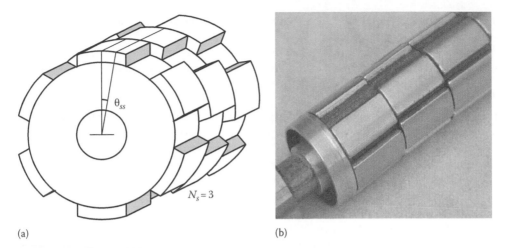

(a) (b)

FIGURE 6.22 Stepped rotor skewing with three modules. (Adapted from Bianchi, N. and Bolognani, S., *IEEE Trans. Ind. Appl.*, 38(5), 1259, 2002.)

As far as the back EMF is concerned, each kth harmonic of the back EMF is reduced by the factor $k_{pm} = \sin(kp\alpha_m)$, where $2\alpha_m$ is the PM pole arc angle.

6.5.1.4 PM Pole Arc with Different Width

A multipole machine can be designed with PMs of different arc width, as shown in Figure 6.23a. In this way, the elementary T_{edge}, shown in Figure 6.21, are distributed along the slot pitch, obtaining a reduction of the total T_{cog}.

6.5.1.5 Notches in the Stator Teeth

A further technique to reduce the T_{cog} consists in introducing in the stator teeth a number N_n of notches in order to obtain dummy slots [20]. They are equally spaced and as wide as the opening of the actual slots. Solutions with $N_n = 1$ and $N_n = 2$, as shown in Figure 6.23b, are mainly adopted. The result is an increased number of interactions between rotor PMs and stator slots and a reduced peak value of the cogging torque. In fact, the N_n equally spaced notches produce additional cogging torque curves, with the same behavior of the original one, but with a displacement $\phi_n = 2\pi/Q(N_n + 1)$ mech. degrees. By adding the original cogging torque with the additional torques, the harmonics multiple of $(N_n + 1)$ are in phase, thus their sum becomes $(N_n + 1)$ times higher. Conversely, the other harmonics are cancelled. The resulting T_{cog} is characterized by a higher frequency and an attenuated peak value. An equivalent

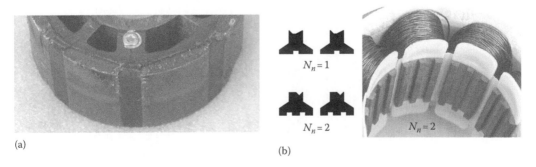

(a) (b)

FIGURE 6.23 Motor design strategies to reduce cogging torque: (a) rotor with PMs of different arc width and (b) notches in the tooth (dummy slots). (Modified from Bianchi, N. and Bolognani, S., *IEEE Trans. Ind. Appl.*, 38(5), 1259, 2002.)

strategy is to introduce dummy teeth in each slot opening [21]. The reason of their effect is similar to that presented above for the dummy slots.

A proper number of notches N_n is obtained when GCD{(N_n + 1), N_p} = 1 as shown in [16]. Conversely, the equality (N_n + 1) = N_p has to be avoided, since it produces an increase of all the T_{cog} harmonics.

6.5.1.6 Shifting of the PMs

For reducing the T_{cog}, it is also possible to modify the position of the PMs on the rotor surface: a sort of "circumferential" skewing with an effect similar to the stepped axial skewing. The technique of shifting the PMs of two adjacent poles to eliminate the T_{cog} harmonic of second order has been firstly presented in [18], then the technique has been refined and generalized in [22]. The general rule states that in a motor with $2p$ poles, the jth PM pole has to be shifted of an angle $\phi_{sh,j} = 2\pi(j - 1)/(2pN_pQ)$ with $j = 1,\ldots, 2p$. The effect of PM shifting on the back EMF harmonics can be estimated by means of the shifting factor, computed in [23].

A sketch of shifted PMs in 4-, 6-, and 8-pole motor is reported in Figure 6.24a, and two photos of a 6-pole rotor with shifted PMs are shown in Figure 6.24b.

6.5.2 Reduction of the Torque Ripple in IPM Motors

Synchronous motors with anisotropic rotor (not only IPM motors but also synchronous reluctance motors) exhibit often a high torque ripple [24]. An example is shown in Figure 6.20a. This ripple is caused by the interaction between the spatial harmonics of electrical loading and the rotor anisotropy. Some techniques presented to reduce the torque ripple of SPM motors can be used; however, some of them are not enough to achieve a smooth torque. Rotor skewing reduces the torque ripple only in part [25], while a slight compensation of the torque harmonics is achieved by shifting the flux barriers from their symmetrical position [22,26].

A reduction of the torque ripple can be achieved by means of a suitable choice of the number of flux barriers with respect to the number of stator slots [25]. The suggested number N_{rfb} of rotor flux barriers

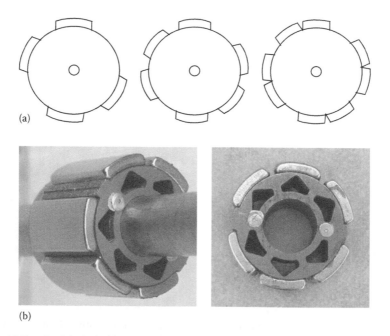

FIGURE 6.24 (a) Sketch of shifted PMs in rotor with 4, 6, and 8 poles and (b) photo of a 6-pole rotor with shifted PMs. (Adapted from Bianchi, N. and Bolognani, S., *IEEE Trans. Ind. Appl.*, 38(5), 1259, 2002.)

per pole pair (whose ends are uniformly distributed along the rotor circumference) is related to the number of stator slots Q so that

$$N_{rfb} = \frac{Q}{2p} \pm 1 \tag{6.24}$$

A different strategy to compensate the torque harmonics of anisotropic motors is based on a two-step design procedure [27,28]:

1. A set of flux barrier geometries is identified so as to cancel a torque harmonic of given order. This means that a harmonic of given order is zero according to the geometry of the rotor flux barriers.
2. Couples of flux barriers belonging to this set are combined together so as the remaining torque harmonics of one flux barrier geometry compensate those of the other geometry. This second step can be achieved in two ways: (**1**) either by forming the rotor with laminations of two different kinds, (**2**) or by adopting two different flux barrier geometries in the same lamination.

Figure 6.25 shows the "Romeo and Juliet" rotor, formed by two different and inseparable kinds of lamination (the first labeled **R** as Romeo, and the second labeled **J** as Juliet). Each lamination has a hole to hold the PM in the same position and with the same size.

Figure 6.26 shows the "Machaon" rotor, formed by laminations with flux barriers of different geometry, large and small, alternatively under the adjacent poles. The name comes from a butterfly with two large and two small wings.

Such solutions yield an appreciable reduction of the torque ripple [76]. Table 6.3 reports a comparison between the average torque (T_{avg}) and torque ripple (ΔT), measured on three IPM motor prototypes with two flux barriers per pole.

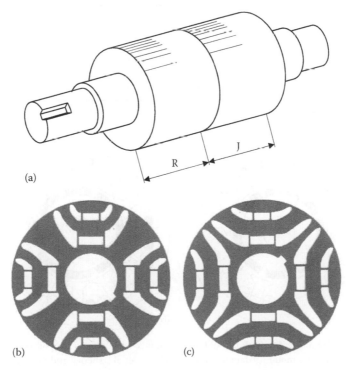

(a)

(b)   (c)

FIGURE 6.25 Photos of the "Romeo and Juliet" laminations. (a) The two-part rotor, (b) R-lamination, and (c) J-lamination. (Modified from Bianchi, N. et al., *IEEE Trans. Ind. Appl.*, 45(3), 921, 2009.)

6.6 Fractional-Slot PM Synchronous Motors

In PM synchronous motors, an alternative to the integral-slot winding is represented by the fractional-slot winding. Among the others, the winding with nonoverlapped coils, that is, with coils wound around a single tooth (coil throw $y_q = 1$), is of particular interest. Two examples of fractional-slot windings are reported in Figure 6.27.

FIGURE 6.26 Photo of the "Machaon" lamination. (Adapted from Bianchi, N. et al., *IEEE Trans. Ind. Appl.*, 45(3), 921, 2009.)

There are several reasons for choosing such a fractional-slot PM motors:

- The length of the end winding is reduced, hence copper weight and Joule losses for given torque are reduced as well. This assumes an important role in applications requiring high efficiency [29].
- The periodicity between stator slots and rotor poles is reduced, so that both cogging torque and torque ripple under load are low [20,30], yielding a smooth torque behavior under various operating conditions [31].
- The synchronous inductance is higher than the corresponding integral-slot winding motor. Thus, in the event of fault, the short-circuit current is limited. This is important in applications requiring high fault tolerance [32,33]. A very high inductance is achieved adopting single-layer windings (see later) and a number of slots lower than the number of poles [34].
- Fractional-slot winding motors are well-suited to be adopted in fault-tolerant motor drives. In the single-layer winding, each slot contains only coil sides of the same phase [35]. Thus,

TABLE 6.3 Torque Comparison at Different Currents among IPM Motor with Classic Geometry, "Romeo and Juliet" (R&J) Motor and "Machaon" Motor (Experimental Results)

	Classic IPM		R&J		Machaon	
I (A)	T_{avg} (Nm)	$\Delta T/T_{avg}$ (%)	T_{avg} (Nm)	$\Delta T/T_{avg}$ (%)	T_{avg} (Nm)	$\Delta T/T_{avg}$ (%)
2.64	2.14	13.1	1.96	4.84	2.182	4.757
2.84	2.39	12.2	2.18	4.91	2.430	4.720
5.30	5.02	11.6	4.63	4.92	5.240	5.784

Source: Bianchi, N. et al., *IEEE Trans. Ind. Appl.*, 45(3), 921, May/June 2009.

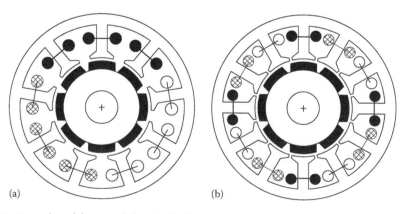

(a)   (b)

FIGURE 6.27 Examples of fractional-slot double-layer motors with nonoverlapped coils. (a) A 9-slot 8-pole motor and (b) a 12-slot 8-pole motor. (Adapted from Bianchi, N. et al., Theory and design of fractional-slot PM machines, Tutorial Course Notes, sponsored by the IEEE-IAS Electrical Machines Committee, *Presented at the IEEE IAS Annual Meeting*, New Orleans, LA, September 23, 2007.)

single-layer windings yield a physical separation between the phases. Furthermore, some configurations exhibit no magnetic coupling between phases [36].

However, the fractional-slot PM motors have not only advantages, in fact

- Some solutions exhibit a low winding factor, that means a low torque density [29,37].
- The MMF space harmonic contents increase heavily, which cause high PM stress, iron saturation, and unbalanced torque [33,38].
- The rotor losses can drastically increase [39]. Then, the solutions with higher armature MMF harmonic contents have to be avoided.

As a consequence, a correct choice of the number of slots and poles, together with the winding arrangement is really important.

6.6.1 Winding Design by Means of the Star of Slots

The fractional-slot windings are designed by means of the star of slots. This is the phasor representation of the main EMF harmonic induced in the coil side of each slot, where "main" is the harmonic of order equal to the number of pole pairs, $\nu = p$.

Let t be the machine periodicity, defined as the greatest common divisor (GCD) between the number of stator slots Q and the number of pole pairs p, which is

$$t = \text{GCD}\{Q, p\} \qquad (6.25)$$

Then, the star of slots is characterized by

1. Q/t spokes
2. Each spoke containing t phasors

The angle between the phasors of two adjacent slots is the electrical angle $\alpha_s^e = p\alpha_s$, where α_s is the slot angle in mechanical radians, that is, $\alpha_s = 2\pi/Q$. The angle between two spokes results in

$$\alpha_{ph} = \frac{2\pi}{(Q/t)} = \frac{\alpha_s^e}{p} t \qquad (6.26)$$

Since electrical angles are considered, the star of slots refers to the equivalent *2-pole machine*. The number given to each phasor corresponds to the number given consecutively to each stator slot. In order to individuate which phasor is to be assigned to each phase, the star of slots is divided into $2m$ equal sectors (where m corresponds to the number of phases). Then, according to which sector they occupy, the phasors and the corresponding coil sides are assigned to the various phases [40].

An example of the star of slots of a 12-slot 10-pole motor is shown in Figure 6.28a. The corresponding phase coils are sketched in Figure 6.28b.

6.6.2 Computation of the Winding Factor

The winding factor (of the main harmonic, i.e., of order $\nu = p$) is an indirect index of the goodness of the winding, since it results to be proportional to the torque density. It is obtained as the product of the distribution factor k_d times the pitch factor k_p, that is, $k_w = k_d k_p$.

The distribution factor k_d is the ratio between the geometrical and the arithmetic sum of the phasors of the same phase. The distribution factor depends only on the number of spokes per phase q_{ph} of the star of slots, given by $q_{ph} = Q/(mt)$.

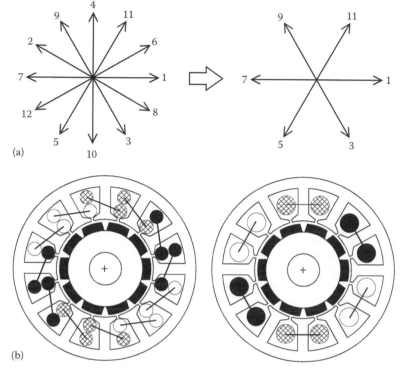

(a)

(b)

FIGURE 6.28 Motor with $Q = 12$, $2p = 10$: thus $t = 1$ odd, $Q/t = 12$ even, and $Q/(2t) = 6$ even. (a) Star of slots and (b) coil distributions. (Adapted from Bianchi, N. et al., Theory and design of fractional-slot PM machines, Tutorial Course Notes, sponsored by the IEEE-IAS Electrical Machines Committee, *Presented at the IEEE IAS Annual Meeting*, New Orleans, LA, September 23, 2007.)

The distribution factor for the main harmonic (i.e., of order $\nu = p$) can be expressed as

$$k_d = \frac{\sin\big((q_{ph}/2)(\alpha_{ph}/2)\big)}{(q_{ph}/2)\sin(\alpha_{ph}/2)} \quad \text{if } q_{ph} \text{ is even}$$

$$k_d = \frac{\sin\big(q_{ph}(\alpha_{ph}/4)\big)}{q_{ph}\sin(\alpha_{ph}/4)} \quad \text{if } q_{ph} \text{ is odd}$$

(6.27)

As an example, referring to the star of slots of the winding shown in Figure 6.27a, it is $q_{ph} = 3$ and $\alpha_{ph} = 2\pi/9$ so that $k_d = 0.959$. For the star of slots of the winding shown in Figure 6.27b, it is $q_{ph} = 1$ and $\alpha_{ph} = 2\pi/3$ so that $k_d = 1$.

The pitch factor is independent of the star of slots and is computed from the coil throw. The coil throw y_q, measured in number of slots, is approximated by $y_q = \text{round}\{Q/(2p)\}$, with the lowest value equal to unity. The pitch factor of the main harmonic is given by

$$k_p = \sin\frac{\sigma_w}{2}$$

(6.28)

where the coil span angle is given by $\sigma_w = (2\pi p y_q)/Q$.

6.6.3 Transformation from Double- to Single-Layer Winding

The single-layer winding with nonoverlapped coil was first proposed for fault-tolerant applications, since it allows a physical separation between the coils. Each coil is wound around a single tooth and separated from the others by a stator tooth. An example is shown in Figure 6.29a, corresponding to the SPM motor shown in Figure 6.28b. Further examples of single-layer fractional-slot windings are described in [41,42].

The single-layer winding can be realized by a transformation from a double-layer winding, as sketched in Figure 6.29b, referring to a 12-slot 10-pole SPM motor. Every other coil of the double-layer winding is removed and reinserted into the stator according to the position of the coils of the same phase. The transformation affects the star of slots of the winding, as illustrated in Figure 6.28a.

There are some geometrical and electrical constraints to the transformation [40]. As regards the geometrical constraints,

1. The number of slots Q must be even.
2. The slot throw y_q must be odd (of course, this constraint is inherently satisfied with the nonoverlapped coil winding, being $y_q = 1$).

As regards the electrical constraints,

1. If Q/t is even, the transformation is always possible. The machine exhibits different performance depending on whether the periodicity t is even or odd.
2. If Q/t is odd, the transformation is possible only if the periodicity t is even.

Table 6.4 summarizes the winding features according to Q and p. The upper part refers to double-layer windings, while the lower part refers to single-layer windings. Table 6.4 highlights the harmonic order (HO) of the armature MMF distribution. For any configuration, the lowest order of the MMF harmonic is the machine periodicity t. In particular, there are no MMF subharmonics when the periodicity t is equal to the number of pole pairs p (i.e., when $Q/p = m$).

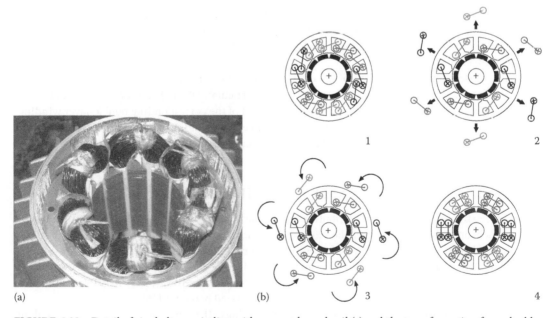

(a)　　　　　(b)

FIGURE 6.29 Detail of single-layer winding with nonoverlapped coil (a) and the transformation from double- to single-layer winding (b). (Modified from Bianchi, N. et al., Theory and design of fractional-slot PM machines, Tutorial Course Notes, sponsored by the IEEE-IAS Electrical Machines Committee, *Presented at the IEEE IAS Annual Meeting*, New Orleans, LA, September 23, 2007.)

TABLE 6.4   Harmonic Order (HO), Distribution Factor (k_d), and Mutual Inductance (M) for Different Combinations of Slots (Q) and Pole Pairs (p) in an m-Phase Fractional-Slot PM Machine

Machine Periodicity $t = \mathrm{GCD}\{Q, p\}$		
Feasibility: Number of Spokes per Phase $Q/(mt)$ Integer		
Q/t even		Q/t odd
Double layer	Adjacent phasors are odd and even alternatively	Superimposed phasors odd or even alternatively
	Superimposed phasors are all odd or all even	
	HO: $(2n-1)t$	HO: nt
	Mutual inductance $M = 0$ when $y_q = 1$	$M \neq 0$
	$Q/(2t)$ even	$Q/(2t)$ odd
	Opposite phasors are both even or both odd	Opposite phasors are one even and the other odd
Transformation from double- to single-layer winding (geometrical constraints: Q even and y_q odd)		
Single layer		(only if t is even)
	k_d increases	k_d unchanged	k_d unchanged
	HO: $(2n-1)t$	HO: nt	HO: $nt/2$
	$M = 0$ remains when $y_q = 1$	$M \neq 0$	$M \neq 0$

Source: Bianchi, N. et al., Theory and design of fractional-slot PM machines, Tutorial Course Notes, sponsored by the IEEE-IAS Electrical Machines Committee, *Presented at the IEEE IAS Annual Meeting*, New Orleans, LA, September 23, 2007.

Note: n is an integer.

With double-layer winding and Q/t even, or with single-layer windings and $Q/(2t)$, the MMF space harmonics are only of odd order, so that the harmonic order can be expressed as $(2n - 1)$ times t [40,43]. Conversely, with the other winding combinations, the order of the MMF space harmonics is both odd and even, so that the harmonic orders are expressed as n times t. Finally, when the double-layer winding is transformed into a single-layer winding and Q/t is odd (the transformation is possible only if t is even), the machine periodicity decreases to $t/2$. Thus, harmonics of lower order (i.e., subharmonics) appear.

Table 6.4 also shows the variation of the distribution factor of the main harmonic, after the transformation from a double- to a single-layer winding. In particular, when $Q/(2t)$ is even, the single-layer winding can exhibit a winding factor even higher than that of the corresponding double-layer winding.

Table 6.4 also shows that, when Q/t is even with a double-layer winding, and when $Q/(2t)$ is even with a single-layer winding, there is no coupling among the phases (i.e., $M = 0$).

6.6.4 Rotor Losses Caused by MMF Space Harmonics

The space harmonics in the MMF distribution, particularly high in fractional-slot PM motors [31,35], move asynchronously with the rotor. Therefore, they induce currents in all rotor conductive parts, producing rotor losses. The amount of rotor losses assumes a peculiar importance in large machines where the number of slots and poles is high, such as in wind turbine PM generators, PM motors in direct drive lift applications, and so on. Some analytical models to compute the rotor losses in SPM motors have been proposed recently [39,44,45]. Some results are summarized hereafter [46].

Figure 6.30a refers to machines with double-layer windings. The lower rotor losses are found along the line of the integral-slot configurations, the bold line of Figure 6.30a, where the number of slots per pole $Q/2p$ is equal to the number of phases $m = 3$. In these configurations, there are only harmonics of odd order multiple of p, with decreasing amplitude, and no third harmonics and subharmonics. Moving

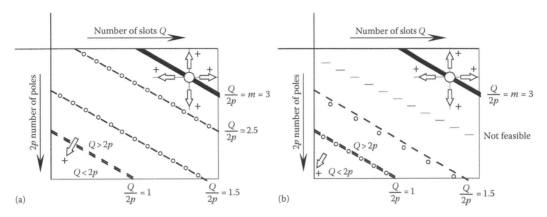

FIGURE 6.30 Map of the rotor losses with double-layer (a) and single-layer (b) windings. (Modified from Bianchi, N. et al., Theory and design of fractional-slot PM machines, Tutorial Course Notes, sponsored by the IEEE-IAS Electrical Machines Committee, *Presented at the IEEE IAS Annual Meeting*, New Orleans, LA, September 23, 2007.)

along all directions from the line $Q/2p = 3$, the rotor losses increase. This is highlighted by the white arrows, departing from the white point drawn on the line (the white points indicate minimum values). Although the rotor losses increase when moving away from the line, the increase is not monotonic: there are some local minima. They are found along the lines characterized by $Q/2p \approx 2.5$ and $Q/2 = 1.5$, that is, the two dashed lines in Figure 6.30a. Along these lines, white points are used to highlight the local minima. Let us note that in the machines with $Q/2p = 1.5$, the machine periodicity returns to be $t = p$, so that there are no MMF subharmonics. Figure 6.30a also shows the border line $Q = 2p$. There are no local minima along this line. On the contrary, the rotor losses continue to increase as the number of slots decreases with respect to the number of poles, as indicated by the white arrow crossing the line $Q/2p = 1$.

Figure 6.30b refers to machines with single-layer winding. The lower rotor losses are found again along the line of the integral-slot configurations, the bold line where $Q/2p = 3$. Of course, the computed rotor losses are the same as those computed with the double-layer winding. As above, along all directions from the line $Q/2p = 3$, the rotor losses increase (white arrows are used again). These losses are generally higher than those computed with the same slot and pole combinations using a double-layer winding. Figure 6.30b also highlights that there are many combinations of slots and poles that do not yield a feasible three-phase winding. In other words, the transformation from double- to single-layer winding is not feasible [40]. Some local minima are found along the line $Q/2p = 1.5$, that is, the dashed line in Figure 6.30b (small circles are drawn in the map). Finally, only with single-layer windings, other rotor losses minima are around the line $Q = 2p$ (highlighted using small circles again). Then, the rotor losses continue to increase as the number of slots decreases with respect to the number of poles, as pointed out by the white arrow.

6.7 Vector Control of PM Motors

For achieving high performance of the PM synchronous motors, it is extremely important to apply appropriate control strategies. Thus, a current-regulated PWM inverter is commonly used to control the current vector of the PM synchronous motor. The main vector control strategies are summarized hereafter.

The voltage equations are given by (6.16). Neglecting the torque ripple, the torque is approximated by (6.19), where flux linkages are expressed as in (6.14). It is

$$\tau = \frac{3}{2} p[\Lambda_m i_q + (L_d - L_q)i_d i_q] \tag{6.29}$$

The first term represents the PM torque and the second term represents the reluctance torque.

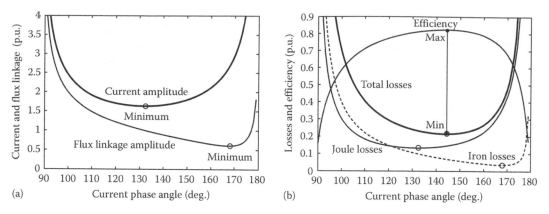

FIGURE 6.31 Stator current, flux linkage, losses, and efficiency as a function of current vector angle α_i^e under constant torque ($\tau_{pu} = 1$) and constant speed ($\omega_{pu} = 1$) condition. (a) Current and flux linkage and (b) losses and efficiency.

Figure 6.31 shows the key characteristic of the motor as a function of the current vector angle α_i^e, defined in Figure 6.17, for given torque and speed. Normalized parameters are used, that is, unity torque $\tau_{pu} = 1$ and unity speed $\omega_{pu} = 1$ are fixed.

Since the parameter variation affects the control performances, the d- and q-axis inductances have to be modeled in the current vector control algorithm as a function of the d- and q-axis current i_d and i_q [47], as indicated by (6.17). However, in the following, constant parameters are considered.

6.7.1 Maximum Torque-per-Ampere Control

For a given torque, there is an optimal operating point in which the current is minimum, as shown in Figure 6.31a. Therefore, a maximum torque-to-current ratio exists. When such a ratio is maximized for any operating condition, the maximum torque-per-Ampere (MTPA) control is achieved [48].

The torque-to-current ratio τ/i is maximized with respect to the current vector angle α_i^e, yielding

$$\cos\alpha_i^e = \frac{-\Lambda_m + \sqrt{\Lambda_m^2 - 8(L_d - L_q)^2 i^2}}{4(L_d - L_q)i} \tag{6.30}$$

Therefore, the relation between d- and q-axis currents for MTPA condition is given by

$$i_d = \frac{\Lambda_m}{2(L_d - L_q)} - \sqrt{\frac{\Lambda_m^2}{4(L_d - L_q)^2} + i_q^2} \tag{6.31}$$

Figure 6.32 shows the MTPA trajectory in the (i_d, i_q) plane. The MTPA trajectory corresponds to the tangent points of the constant torque loci and the constant current circles (e.g., points B_1, B_2, B_3 in Figure 6.32). When the current limit is considered, the maximum available torque is obtained by the MTPA control. The characteristic curves are shown only in the region of $i_d < 0$ and $i_q > 0$, however each characteristic curves are symmetric against the d-axis, thus the current vector in the region of $i_d < 0$ and $i_q < 0$ is used when a negative torque is required.

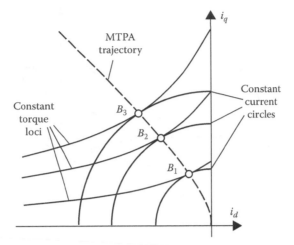

FIGURE 6.32 Current vector trajectory for MTPA control.

6.7.2 Flux-Weakening Control

By increasing the current vector angle α_i^e, the flux linkage λ decreases, yielding the flux-weakening (FW) control [49–51]. The total flux linkage λ is given by

$$\lambda = \sqrt{(\Lambda_m + L_d i_d)^2 + (L_q i_q)^2} \tag{6.32}$$

The d-axis flux linkage λ_d can be adjusted by utilizing the field due to negative d-axis current. This technique allows high speed to be reached. From the condition by which the inner voltage becomes equal to its limit value V_N, the following equation is obtained:

$$\left(\frac{V_N}{\omega}\right)^2 = (\Lambda_m + L_d i_d)^2 + (L_q i_q)^2 \tag{6.33}$$

In the (i_d, i_q) plane, such a relationship defines a family of voltage-limit ellipses. Their size is a function of operating electrical speed ω and their center is located at the point F $(-\Lambda_m/L_d, 0)$ shown in Figure 6.33.

6.7.3 Maximum Torque-per-Voltage Control

For a given torque, there is an optimal operating point minimizing the total flux linkage λ, as shown in Figure 6.31a. This leads to the maximum torque-per-flux (MTPF) control, or, in other words, the maximum torque-per-voltage (MTPV) control [52]. For a given flux linkage λ, the MTPV control is achieved by means of a current vector given as

$$i_d = -\frac{\Lambda_m + \Delta\Lambda_d}{L_d} \tag{6.34}$$

$$i_q = \frac{\sqrt{\lambda^2 - \Delta\Lambda_d^2}}{L_q} \tag{6.35}$$

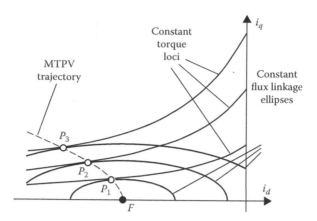

FIGURE 6.33   Current vector trajectory for MTPV control.

where

$$\Delta \Lambda_d = \frac{-L_q \Lambda_m + \sqrt{(L_q \Lambda_m)^2 + 8(L_q - L_d)^2 \lambda^2}}{4(L_q - L_d)^2} \qquad (6.36)$$

The relation between d- and q-axis currents is shown as the MTPV trajectory in Figure 6.33. The MTPV trajectory represents the tangent points of constant torque loci and constant flux linkage ellipses (points P_1, P_2, P_3 in Figure 6.33). When the maximum voltage V_N is reached, the flux linkage λ is decreased along the MTPV trajectory with the increase of speed because $\lambda \approx V_N/\omega$. When the speed tends to infinity, the current vector tends to the center of the ellipses, defined by $i_d = -\Lambda_m/L_d$ and $i_q = 0$.

6.7.4 Maximum Efficiency Control

For a given torque and speed, the minimum copper loss P_J corresponds to the condition of MTPA, while the minimum iron loss P_{Fe} corresponds to the condition of MTPV. Therefore, the total loss $P_{loss} = P_J + P_{Fe}$ are minimized at an optimal current vector angle between the MTPA condition and the MTPF condition, as also shown in Figure 6.31b, leading to the minimum loss (or maximum efficiency) control [10].

6.7.5 Limit Operating Regions

The optimal steady-state current vector is determined under both voltage and current constraints. Neglecting stator resistance, they are given by

$$i = \sqrt{i_d^2 + i_q^2} \le I_N \qquad (6.37)$$

and

$$\lambda = \sqrt{\lambda_d^2 + \lambda_q^2} \le \frac{V_N}{\omega} \qquad (6.38)$$

where the current I_N is the maximum available current of the inverter. The voltage limit V_N is a maximum available output voltage of the inverter, according to the given dc voltage V_{dc}.

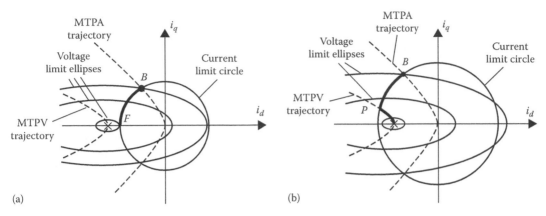

FIGURE 6.34 Selection of optimum current vector for producing maximum torque in consideration of voltage and current constraints. (a) Two operating regions and (b) three operating regions.

Figure 6.34 shows a graphical representation of the control strategies presented above, in the (i_d, i_q) plane. In such a plane, the constant current loci are circles, the constant flux linkage loci are ellipses, and the constant torque loci are hyperbolae. The critical conditions are respectively shown by the current-limit circle and the voltage-limit ellipse in the (i_d, i_q) plane. The current vector satisfying both constraints of voltage and current must be inside of both the current-limit circle and the voltage-limit ellipse.

According to both voltage and current constraints, the optimum current vector producing the maximum torque at any speed is given as follows.

Region I (Constant torque region): Below the base speed ω_B, the maximum torque is produced by the MTPA control. The current vector producing maximum torque is derived from (6.31) and $i = I_N$. This current vector corresponds to the point **B** in Figure 6.34. In this region, $i = I_N$, $\omega\lambda < V_N$, and V reaches its limited value at the base speed ω_B.

Region II (FW, constant volt–ampere region): Above the base speed, the current vector is controlled by the FW control, in which the voltage is kept fixed to $\omega\lambda = V_N$ by utilizing the demagnetizing d-axis armature reaction. The optimum current vector producing the maximum torque at speed $\omega > \omega_B$ is derived from (6.33) and amplitude $i = I_N$.

This current vector corresponds to the intersecting point of the current-limit circle and the voltage-limit ellipse. The current vector angle α_i^e increases as the speed increases. The d-axis current increases toward negative direction and the q-axis current decreases. The current vector trajectory moves along the current-limit circle (bold line in Figure 6.34a). Assuming that $\Lambda_m > L_d I_N$, the FW operation continues up to a maximum speed ω_{max}. The minimum d-axis flux linkage is achieved when i_d reaches $-I_N$ and i_q becomes zero (point **F** in Figure 6.34a), so that $\Lambda_{dmin} = \Lambda_m - L_d I_N$. The torque and power become zero, and the maximum speed results in $\omega_{max} = V_N/\Lambda_{dmin}$.

Region III (FW, decreasing volt–ampere region): If $\Lambda_m < L_d I_N$, the center point of the voltage-limit ellipse is located inside the current-limit circle, as shown in Figure 6.34b. The current vector trajectory moves along the current-limit circle up to the speed ω_p, corresponding to the intersection point **P** between the current-limit circle and the MTPV trajectory. Above ω_p, the optimum current vector is achieved applying the MTPV control.

Figures 6.35 and 6.36 show the curves of torque, power, and d- and q-axis current components versus speed, when the maximum torque control is applied, constrained by voltage and current limit, that is, $v \leq V_N$ and $i \leq I_N$. Normalized parameters are used. The parameters of the two PM synchronous motors are reported in Table 6.5.

Figure 6.35 refers to the case $\Lambda_m - L_d I_N > 0$, therefore only two operating regions exist. Figure 6.36 refers to the case $\Lambda_m - L_d I_N < 0$. In this case, three operating regions exist.

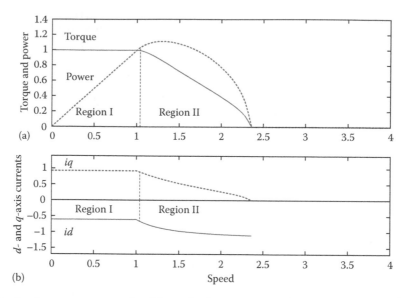

FIGURE 6.35 Maximum torque control with limited voltage and current limit, in case of $\Lambda_m - L_d I_N > 0$ (IPM#1): (a) torque and power and (b) d- and q-axis currents versus speed.

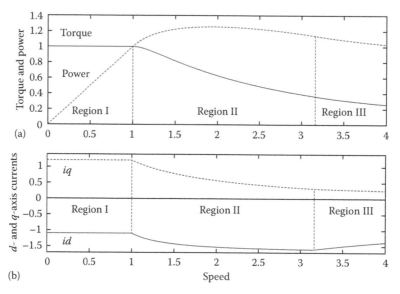

FIGURE 6.36 Maximum torque control with limited voltage and current limit, in case of $\Lambda_m - L_d I_N < 0$ (IPM#2): (a) torque and power and (b) d- and q-axis currents versus speed.

6.7.6 Loss Minimization Control

The optimal current vector for minimizing the total loss at any operating condition can be derived considering both copper and iron losses. The optimum current vector is a function of both torque and speed. Since the iron loss is zero at standstill, where the harmonic losses due to PWM inverter are neglected, the LM trajectory at $\omega = 0$ corresponds to the MTPA trajectory on which the copper loss is minimized. The LM trajectory moves toward negative d-axis current as the speed increases, and it reaches the MTPV trajectory at infinity speed.

TABLE 6.5   IPM Motor Parameters (p.u. Value)

	IPM#1	IPM#2
PM flux linkage	$\Lambda_m = 0.6$	$\Lambda_m = 0.15$
q-Axis inductance	$L_q = 0.933$	$L_q = 0.824$
d-Axis inductance	$L_d = 0.155$	$L_d = 0.206$
Saliency ratio	$\xi = 6$	$\xi = 4$
Resistance	$R = 0.02$	$R = 0.05$
Nominal current	$I_N = 1.11$	$I_N = 1.632$

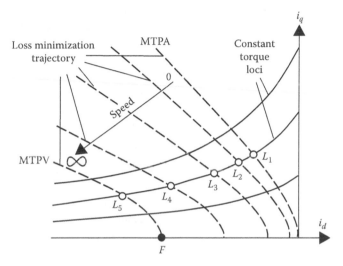

FIGURE 6.37   Current vector trajectory for loss minimization control.

The optimal current vector trajectories at constant speeds are shown in Figure 6.37. When the torque is fixed, the optimal operating point moves along the constant torque locus as the speed increases (e.g., points L_1, L_2, L_3, L_4, L_5 in Figure 6.37). In practice, the optimal d-axis and q-axis currents are numerically calculated or searched experimentally, and then they are stored in a look-up table or modeled by proper functions.

6.8 Fault-Tolerant PM Motors

The fault-tolerant capability of electrical motor drives is an essential feature in applications such as automotive, aeronautic [53], and many others. Even though less stringent, fault tolerance is a positively acknowledged feature also in the industrial environment, due to the related productivity enhancement. A fault-tolerant motor is a motor able to sustain a fault without destroying itself and without propagating the fault.

Some examples are given in the following.

6.8.1 Short-Circuit Fault

In the event of a three-phase short-circuit, $v_d = v_q = 0$. The i_d and i_q currents are computed from (6.16). In the following example, the parameters of IPM motor labeled IPM#1 reported in Table 6.5 are considered.

Figure 6.38 shows the i_d and i_q currents in the (i_d, i_q) plane [77]. The dashed line represents the ellipse trajectory described by the currents when the stator resistance is zero. The solid line refers to the case with a

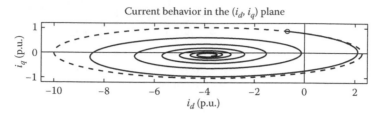

FIGURE 6.38 Short-circuit current trajectory in the (i_d, i_q) plane, at $\omega = 1$. (Adapted from Bianchi, N. et al., *IEEE Trans. Veh. Technol.*, 55(4), 1102, 2006.)

resistance different from zero. The currents move from their initial value I_{d0} and I_{q0} given by the operating point before the fault (and highlighted by the circle in Figure 6.38) toward the steady-state short-circuit value, defined by

$$I_{d,shc} = -\frac{\omega^2 L_q \Lambda_m}{R^2 + \omega^2 L_d L_q} \tag{6.39}$$

and

$$I_{q,shc} = -\frac{\omega \Lambda_m R}{R^2 + \omega^2 L_d L_q} \tag{6.40}$$

It corresponds to $I_{d,shc} = -3.85$ p.u. and $I_{q,shc} = -0.083$ p.u. in Figure 6.38, resulting about 3.5 times the nominal current. It reduces to $I_{d,shc} = -\Lambda_m/L_d$ and $I_{q,shc} = 0$ neglecting the resistance R.

The minimum d-axis current represents the negative current peak that may demagnetize the PM. Neglecting the stator resistance, it is computed as

$$I_{d,min} = -\frac{\Lambda_m}{L_d} - \sqrt{\left(I_{d0} + \frac{\Lambda_m}{L_d}\right)^2 + \left(\frac{L_q}{L_d} I_{q0}\right)^2} \tag{6.41}$$

In the reported example, it is $I_{d,min} = -10$ p.u. Its amplitude is more than nine times the nominal current, that is, higher than 2.5 times $I_{d,shc} = -\Lambda_m/L_d$. Assuming the nominal amplitude of the initial current, the worst case is with q-axis current only, that is, $I_q0 = I_N$ and then $I_{d0} = 0$. In this case, the ideal ellipse exhibits the largest area, and the minimum d-axis current becomes

$$I_{d,min} = \frac{-\Lambda_m - \sqrt{\Lambda_m^2 + (L_q I_N)^2}}{L_d} \tag{6.42}$$

reaching $I_{d,min} = 11.58$ p.u. in the considered example. If the PM flux linkage is higher than the initial q-axis flux linkage, that is, $\Lambda_m \gg L_q I_N$, then the minimum d-axis current can be approximated as $I_{d,min} \approx -2\Lambda_m/L_d$. Conversely, in the case of a reluctance motor, where $\Lambda_m = 0$, the minimum d-axis current becomes $I_{d,min} = -\xi I_N$.

The analysis of the steady-state braking torque is carried out in the synchronous d–q reference frame. The voltage equations are expressed by (6.16) without derivatives and with $v_d = v_q = 0$. The steady-state braking torque [54] results in

$$T_{brk} = -\frac{3}{2} pR\Lambda_m^2 \omega \frac{R^2 + \omega^2 L_q^2}{(R^2 + \omega^2 L_d L_q)^2} \tag{6.43}$$

and the short-circuit current amplitude is obtained by (6.39) and (6.40), resulting in

$$I_{shc} = \frac{\sqrt{(\omega^2 L_q \Lambda_m)^2 + (\omega R \Lambda_m)^2}}{R^2 + \omega^2 L_d L_q}$$ (6.44)

The short-circuit current always increases with the speed ω, approaching Λ_m/L_d. A typical behavior of the braking torque (negative with the motoring convention) and the short-circuit current are shown in Figure 6.39 as a function of the motor speed. The maximum amplitude of T_{brk} is computed by equating the derivative of (6.43), with respect to the speed ω, to zero. The maximum braking torque results in

$$T_{brk}^* = \frac{3}{2} p \frac{\Lambda_m^2}{L_q} f(\xi)$$ (6.45)

and it occurs at the speed

$$\omega^* = \frac{R}{L_q} \sqrt{\chi}$$ (6.46)

where the function $f(\xi)$ is

$$f(\xi) = \sqrt{\chi} \frac{1 + \chi}{\left(1 + (\chi/\xi)\right)^2}$$ (6.47)

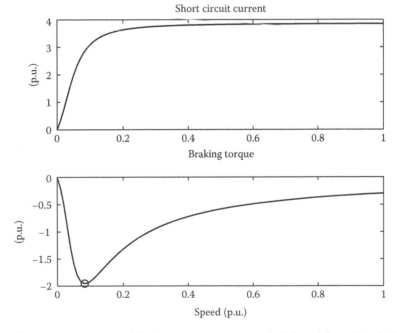

FIGURE 6.39 Short-circuit current and braking torque versus speed. (Adapted from Bianchi, N. et al., *IEEE Trans. Veh. Technol.*, 45(4), 1102, 2006.)

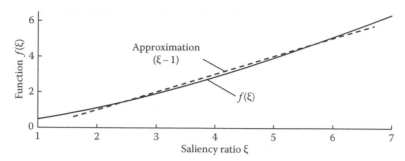

FIGURE 6.40 The function $f(\xi)$, given in (6.47), and its approximating straight line. (Adapted from Bianchi, N. et al., *IEEE Trans. Veh. Technol.*, 55(4), 1102, 2006.)

with

$$\chi = \frac{1}{2}\left[3(\xi-1) + \sqrt{9(\xi-1)^2 + 4\xi}\right] \tag{6.48}$$

It is worth noticing that $f(\xi)$ in (6.47) is a function of the saliency ratio $\xi = L_q/L_d$ only. Its behavior is reported in Figure 6.40. In the range between $\xi = 2$ and $\xi = 6$, such a function can be approximated by the straight line $f(\xi) \approx \xi - 1$, that is, shown in the same Figure 6.40 by a dashed line.

6.8.2 Decoupling between the Phases

In designing fault-tolerant PM motors, it is imperative to consider that any fault does not propagate from a faulty phase to the other healthy phases. Then, a complete decoupling among the phases is proposed [34], including

- An electrical isolation between phases, for example, adopting a full-bridge converter
- A physical separation between phases, for example, adopting motors with fractional-slot winding and nonoverlapped coils
- A magnetic decoupling, for example, adopting fractional-slot motors with a proper combination of slots Q and pole pairs p
- A thermal decoupling, by means of modular solutions, for example, adopting single-layer windings

6.8.3 Multiphase Motor Drives

In multiphase motor drives, the electric power is divided into more inverter legs, reducing the current of each switch [55,56]. In the event of failure of one phase, the remaining healthy phases let the motor to operate properly. Current control strategies are proposed so as to achieve high and smooth torque even without one or more phases [57,58].

Among the others, the two more attractive solutions seem to be

- Five-phase PM motor drive
- Double three-phase PM motor drive

A five-phase PM motor can be designed to reduce the fault occurrence, as well as to operate indefinitely in the presence of fault [59]. Proper current control strategies have been proposed to face the post-fault situation, with a minimum impact of the fault on torque ripple, noise [60–62], and losses [63]. Two five-phase stators with double-layer and single-layer winding, respectively, and unity coil throw are shown in Figure 6.41 [78].

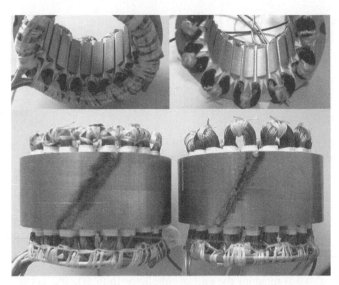

FIGURE 6.41 A 5-phase 20-slot stator with double-layer winding (on left-hand side) and single-layer winding (on right-hand side). (Adapted from Bianchi, N. et al., *IEEE Trans. Ind. Appl.*, 43(4), 960, 2007.)

The alternative to the five-phase motor is represented by the dual three-phase motor. Such a motor includes two identical three-phase windings that are supplied in parallel by two inverters. In the event of a fault of one phase, one inverter is switched off and the healthy three-phase winding is supplied by means of the other inverter, so that the motor continues to operate, even though with reduced power. The advantage of such a solution is that only standard components are used, so that the resulting fault-tolerant motor drive results to be cheaper.

6.9 Sensorless Rotor Position Detection

Among the different techniques for sensorless rotor position detection of PM synchronous motors, the technique described hereafter is based on the high-frequency voltage signal injection. Such a technique is strictly bound to the rotor geometry, requiring a synchronous PM motor with anisotropic rotor, for example, an IPM motor as in Figure 6.1c or an inset motor as in Figure 6.1b The rotor position is detected even at low and zero speed, by elaborating the response of the synchronous PM machine to the high-frequency signal [64,65].

When the high-frequency stator voltage is added to the fundamental voltage, the corresponding high-frequency stator current is affected by the rotor saliency [66,67] and information of the rotor position is extracted from current measurement [68,69]. The two main techniques used to detect the PM rotor position by means of high-frequency signal injection, are briefly summarized hereafter.

Let L_{qh} and L_{dh} be the incremental inductances (also called dynamic or differential inductances), corresponding to the actual operating point.
Then

$$L_{avg} = \frac{L_{qh} + L_{dh}}{2} \quad \text{and} \quad L_{dif} = \frac{L_{qh} - L_{dh}}{2} \tag{6.49}$$

are the average and difference inductances of the high-frequency motor model. The accuracy of the rotor position detection depends on the rotor position and it is strongly affected by saturation and magnetic cross-coupling between d- and q-axis [25,70,71]. The precise magnetic model, described by (6.17), must be used.

6.9.1 Pulsating Voltage Vector Technique

A pulsating voltage vector is superimposed along the estimated d-axis at a constant carrier frequency ω_h. In the estimated synchronous reference frame $\tilde{d} - \tilde{q}$, such a voltage vector is given by

$$\tilde{v}_{dh} = V_h \cos(\omega_h t)$$

$$\tilde{v}_{qh} = 0 \tag{6.50}$$

where the superscript ~ means that the vector is in the estimated reference frame. The corresponding high-frequency current components can be expressed as

$$\tilde{i}_{dh} = \frac{V_h}{\omega_h L_{dh} L_{qh}} \Big[L_{avg} + L_{dif} \cos(2\vartheta_{err}^e) \Big] \sin(\omega_h t)$$

$$\tilde{i}_{qh} = \frac{V_h}{\omega_h L_{dh} L_{qh}} \Big[L_{dif} \sin(2\vartheta_{err}^e) \Big] \sin(\omega_h t) \tag{6.51}$$

where a rotor speed $\omega_m^e = 0$ is fixed for the sake of simplicity. In (6.51), ϑ_{err}^e is the electrical angle error between the estimated $\tilde{d} - \tilde{q}$ and the actual $d-q$ synchronous reference frame.

Equation 6.51 shows that high-frequency component of q-axis current in the estimated rotor reference frame becomes zero when the rotor position angle error is zero. Thus, only q-axis component could be processed using a low-pass filter (LPF), obtaining the rotor position estimation error signal $\varepsilon(\vartheta_{err}^e)$ as

$$\varepsilon(\vartheta_{err}^e) = \text{LPF}\Big[\tilde{i}_{qh} \sin(\omega_h t) \Big]$$

$$= \frac{V_h}{\omega_h} \frac{L_{dif}}{2 L_{dh} L_{qh}} \sin(2\vartheta_{err}^e) \tag{6.52}$$

It can be noted that the error signal is proportional to the sine function of twice the rotor position estimation error. In addition, the signal is proportional to the difference inductance L_{dif}.

6.9.2 Rotating Voltage Vector Technique

Alternatively, a voltage vector rotating at a constant carrier frequency ω_h is superimposed to the fundamental voltage. In the stationary reference frame $\alpha-\beta$, such a voltage vector is given by

$$v_{\alpha\beta h} = V_h e^{j\omega_h t}. \tag{6.53}$$

Neglecting the stator resistance, the corresponding high-frequency current vector is given by

$$\mathbf{i}_{\alpha\beta h} = \frac{L_{avg} v_{\alpha\beta h} - L_{dif} e^{j 2\vartheta_m^e} v_{\alpha\beta h}^*}{j \omega_h L_{dh} L_{qh}}, \tag{6.54}$$

where

   superscript * means the complex conjugate

   ϑ_m^e is the rotor position angle in electrical radians

In order to achieve a signal related to the rotor position angle, the current vector (6.54) is first multiplied by $e^{-j\omega_h t}$ and the result is processed by means of a high-pass filter (HPF), as in a heterodyning scheme [72], yielding

$$\mathrm{HPF}\left[\mathbf{i}_{\alpha\beta h}e^{-j\omega_h t}\right] = j\frac{V_h}{\omega_h}\frac{L_{dif}}{L_{dh}L_{qh}}e^{j2(\vartheta_m^e-\omega_h t)} \tag{6.55}$$

Let $\tilde{\vartheta}_m^e$ be the estimated rotor position angle, the signal (6.55) is multiplied by $e^{j2(\omega_h t-\tilde{\vartheta}_m^e)}$. Then, the rotor position estimation error signal ε corresponds to the real part of such a product, which is

$$\varepsilon = -\frac{V_h}{\omega_h}\frac{L_{dif}}{L_{dh}L_{qh}}\sin[2(\vartheta_m^e - \tilde{\vartheta}_m^e)]. \tag{6.56}$$

Also in this case, the information of the rotor position strongly depends on the difference inductance L_{dif}, so that the error signal disappears when $L_{dh} = L_{qh}$.

6.9.3 Prediction of Sensorless Capability of PM Motors

An accurate magnetic model of the motor is mandatory to predict the capability of the motor for the sensorless rotor position detection. The magnetic model to predict the error signal $\varepsilon(\vartheta_{err}^e)$ is achieved by a set of finite element simulations carried out so as to compute the d- and q-axis flux linkages as functions of the d- and q-axis currents [73]. Then, for a given operating point (defined by the fundamental d- and q-axis currents), a small-signal model is built, defined by the incremental inductances:

$$L_{dd} = \frac{\partial\lambda_d}{\partial i_d} \qquad L_{dq} = \frac{\partial\lambda_d}{\partial i_q}$$
$$L_{qd} = \frac{\partial\lambda_q}{\partial i_d} \qquad L_{qq} = \frac{\partial\lambda_q}{\partial i_q} \tag{6.57}$$

considering both saturation and cross-coupling effect.

When a high-frequency voltage vector is injected along the direction α_v^e (i.e., the d- and q-axis voltage components are $V_h \cos \alpha_v^e$ and $V_h \sin \alpha_v^e$, respectively), the small-signal model described by (6.57) allows to compute the amplitude and the angle of current vector [79]. The phasor diagram is shown in Figure 6.42, including both fundamental components ($\bar{V}_0$ and $\bar{I}_0$) and high-frequency components ($\bar{V}_h$ and $\bar{I}_h$). Such a study is repeated, varying the voltage vector angle α_v^e, so as to estimate the rotor position error signal ε. Let I_{max} and I_{min} be the maximum and minimum of the high-frequency current

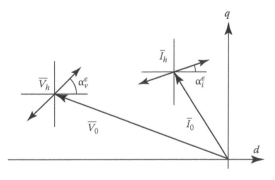

FIGURE 6.42 Phasor diagram with steady-state and high-frequency components. (Modified from Bianchi, N. and Bolognani, S., *IEEE Trans. Ind. Appl.*, 45(4), 1249, 2009.)

(computed with the various voltage vector angles α_v^e), respectively, and α_{Imax}^e the angle where I_{max} is found (defined with respect to the d-axis). Then, the rotor position estimation error signal is computed as

$$\varepsilon(\alpha_v^e) = k_{st} \frac{I_{max} - I_{min}}{2} \sin 2(\alpha_v^e - \alpha_{Imax}^e). \tag{6.58}$$

where

　　α_v^e can be considered as the injection angle

　　$(\alpha_v^e - \alpha_{Imax}^e)$ can be considered as the error signal angle ϑ_{err}^e

Then, α_{Imax}^e corresponds to the angular displacement due to the d–q axis cross-coupling.

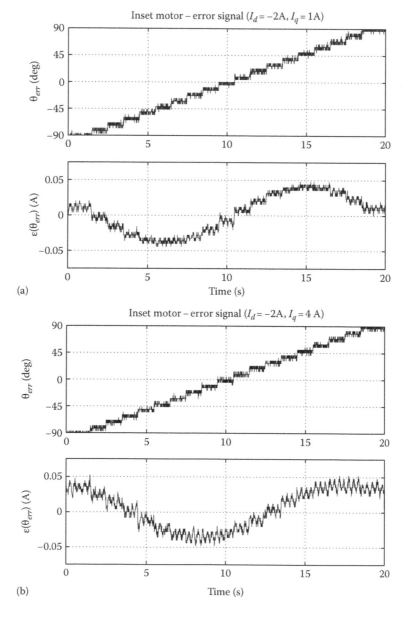

FIGURE 6.43　Rotor position error ϑ_{err}^e and estimation error signal ε with pulsating voltage injection and inset PM motor (experimental test and prediction). (a) Test at low q-axis current, (b) test at high q-axis current,

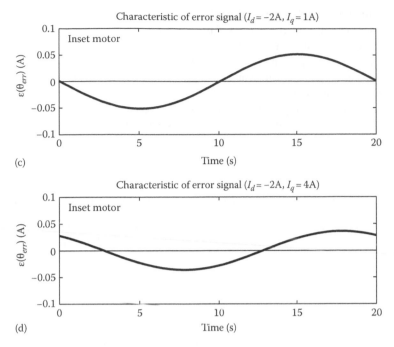

FIGURE 6.43 (continued) (c) prediction at low *q*-axis current, and (d) prediction at high *q*-axis current.

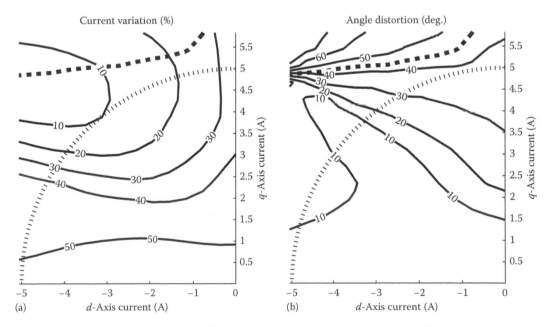

FIGURE 6.44 Contour map of the signal Δ*I*% and of the angle error in the current signal (inset motor). (a) Current variation and (b) angle distortion. (Modified from Bianchi, N. and Bolognani, S., *IEEE Trans. Ind. Appl.*, 45(4), 1249, 2009.)

Figure 6.43 compares experimental and predicted results, referring to the inset motor shown in Figure 6.1b, whose rated current is $\hat{I} = 2.5$ A. The pulsating voltage vector technique has been used, adopting a high-frequency voltage with amplitude $V_h = 50$ V, frequency $f_c = 500$ Hz, and a motor speed $n = 0$ rpm [74]. There is an appreciable agreement in both part and full load. The satisfactory match between predictions and measurements confirms that a PM machine model can be profitably used to predict the sensorless capability of the motor.

Similar results are found using an IPM motor. However, although the waveform of the estimation error signal $\varepsilon(\vartheta_{err}^e)$ is correctly predicted, its amplitude is generally lower than the measured one. This is mainly due to the saturation of the rotor bridges, which has a strong impact on the d-axis inductance, especially with low PM volume.

6.9.4 Contour Map of Rotor Position Error Angle Signal

From the computation presented above, it is possible to draw the contour map of rotor position error signal. Referring to the inset motor tested in Figure 6.43, Figure 6.44a shows the map of the signal $\Delta I\%$, defined as

$$\Delta I\% = 100 \frac{I_{max} - I_{min}}{I_{max} + I_{min}} \tag{6.59}$$

in the (i_d, i_q) plane. The signal $\Delta I\%$ remains in the whole current-limit region (about two times the nominal current), bordered by the dotted circle. The minimum current variation remains higher than 10%. Dashed line highlights the locus where $L_{dh} = L_{qh}$, that is, $L_{dif} = 0$. Figure 6.44b shows the map of the angle of the maximum current with respect to the d-axis. Such an angle corresponds to the angular distortion caused by the cross-coupling effect [75]. The dashed line refers to an error in the rotor position estimation angle of 45 electrical degrees.

References

1. E. Levi, *Polyphase Motors—A Direct Approach to Their Design*. New York: John Wiley & Sons, 1984.
2. V. Honsinger, The fields and parameters of interior type AC permanent magnet machines, *IEEE Transactions on PAS*, 101, 867–876, 1982.
3. J. F. Gieras and M. Wing, *Permanent Magnet Motors Technology: Design and Application*, 2nd edn. New York: Marcel Dekker, 2002.
4. R. M. Bozorth, *Ferromagnetism*, ser. IEEE Magnetics Society. New York: IEEE Press, 1993.
5. J. M. D. Coey, *Rare Heart Iron Permanent Magnet—Monographs on the Physics and Chemistry of Materials*, ser. Oxford Science Publications. Oxford, U.K.: Claredon Press, 1996.
6. G. R. Slemon and A. Straughen, *Electric Machines*. New York: Addison-Wesley, 1980.
7. N. Bianchi and T. Jahns (eds.), Design, analysis, and control of interior PM synchronous machines, ser. IEEE IAS Tutorial Course Notes, *IAS'04 Annual Meeting*, CLEUP, Padova, Italy/Seattle, WA, October 3, 2004 (info@cleup.it).
8. T. Miller, *Brushless Permanent-Magnet and Reluctance Motor Drives*, ser. *Monographs in Electrical and Electronic Engineering*. Oxford, U.K.: Claredon Press/Oxford University Press, 1989.
9. I. Boldea and S. A. Nasar, *Electric Drives*, ser. Power Electronics and Applications *Series*. Boca Raton, FL: CRC Press/Taylor & Francis Group, 1999.
10. S. Morimoto, Y. Tong, Y. Takeda, and T. Hirasa, Loss minimization control of permanent magnet synchronous motor drives, *IEEE Transactions on Industry Electronics*, 41(5), 511–517, September–October 1994.
11. D. White and H. Woodson, *Electromechanical Energy Conversion*. New York: John Wiley & sons, 1959.

12. P. Vas, *Vector Control of AC Machines*, ser. Oxford Science Publications. Oxford, U.K.: Claredon Press, 1990.
13. N. Ida and J. Bastos, *Electromagnetics and Calculation of Fields*. New York: Springer-Verlag, 1992.
14. J. Jin, *The Finite Element Method in Electromagnetics*. New York: John Wiley & Sons, 1992.
15. S. Salon, *Finite Element Analysis of Electrical Machine*. Boston, MA: Kluwer Academic Publishers, 1995.
16. N. Bianchi and S. Bolognani, Design techniques for reducing the cogging torque in surface-mounted PM motors, *IEEE Transactions on Industry Applications*, 38(5), 1259–1265, 2002.
17. T. Jahns and W. L. Soong, Pulsating torque minimization techniques for permanent magnet ac motor drives—A review, *IEEE Transactions on Industrial Electronics*, 43(2), 321–330, April 1996.
18. T. Li and G. Slemon, Reduction of cogging torque in permanent magnet motors, *IEEE Transactions on Magnetics*, 24(6), 2901–2903, 1988.
19. T. Ishikawa and G. Slemon, A method to reduce ripple torque in permanent magnet motors without skewing, *IEEE Transactions on Magnetics*, 29(2), 2028–2031, March 1993.
20. M. Goto and K. Kobayashi, An analysis of the cogging torque of a dc motor and a new technique of reducing the cogging torque, *Electrical Engineering in Japan*, 103(5), 113–120, 1983.
21. K. Kobayashi and M. Goto, A brushless DC motor of a new structure with reduced torque fluctuations, *Electrical Engineering in Japan*, 105(3), 104–112, 1985.
22. N. Bianchi and S. Bolognani, Reducing torque ripple in PM synchronous motors by pole shifting, in *Proceedings of International Conference on Electrical Machines (ICEM)*, Helsinki, Finland, August 2000, pp. 1222–1226.
23. N. Bianchi, S. Bolognani, and A. D. F. Cappello, Back EMF improvement and force ripple reduction in PM linear motor drives, in *Proceedings of the 35th IEEE Power Electronics Specialist Conference (PESC'04)*, Aachen, Germany, June 20–25, 2004, pp. 3372–3377.
24. A. Fratta, G. Troglia, A. Vagati, and F. Villata, Evaluation of torque ripple in high performance synchronous reluctance machines, in *Records of IEEE Industry Application Society Annual Meeting*, Vol. 1, October 1993, Toronto, Canada, 1993, pp. 163–170.
25. A. Vagati, M. Pastorelli, G. Franceschini, and S. Petrache, Design of low-torque-ripple synchronous reluctance motors, *IEEE Transactions on Industry Application*, 34(4), 758–765, July–August 1998.
26. M. Sanada, K. Hiramoto, S. Morimoto, and Y. Takeda, Torque ripple improvement for synchronous reluctance motor using asymmetric flux barrier arrangement, in *Proceedings of the IEEE Industrial Application Society Annual Meeting*, Salt Lake City, UT, October 12–16, 2003.
27. N. Bianchi, S. Bolognani, D. Bon, and M. D. Pré. Torque harmonic compensation in a synchronous reluctance motor, *IEEE Transactions on Energy Conversion*, 23(2), 466–473, June 2008.
28. N. Bianchi, S. Bolognani, D. Bon, and M. D. Pré. Rotor flux-barrier design for torque ripple reduction in synchronous reluctance and PM assisted synchronous reluctance motors, *IEEE Transactions on Industry Applications*, 45(3), 921–928, 2009.
29. N. Bianchi, S. Bolognani, and P. Frare, Design criteria of high efficiency SPM synchronous motors, *IEEE Transactions on Energy Conversion*, 21(2), 396–404, 2006.
30. Z. Q. Zhu and D. Howe, Influence of design parameters on cogging torque in permanent magnet machines, *IEEE Transactions on Energy Conversion*, 15(4), 407–412, December 2000.
31. P. Salminen, Fractional slot permanent magnet synchronous motor for low speed applications, Dissertation, 198, Lappeenranta University of Technology, Lappeenranta, Finland, 2004, ISBN 951-764-982-5 (pdf).
32. B. Mecrow, A. Jack, D. Atkinson, G. Atkinson, A. King, and B. Green, Design and testing of a four-phase fault-tolerant permanent-magnet machine for an engine fuel pump, *IEEE Transactions on Energy Conversion*, 19(4), 671–678, December 2004.
33. F. Magnussen, P. Thelin, and C. Sadarangani, Performance evaluation of permanent magnet synchronous machines with concentrated and distributed winding including the effect of field weakening, in *Proceedings of the Second IEE International Conference on Power Electronics, Machines and Drives (PEMD 2004)*, Vol. 2, Edinburgh, U.K., March 31–April 2, 2004, pp. 679–685.

34. B. Mecrow, A. Jack, and J. Haylock, Fault-tolerant permanent-magnet machine drives, *IEE Proceedings—Electrical Power Applications*, 143(6), 437–442, December 1996.

35. N. Bianchi, M. D. Pré, G. Grezzani, and S. Bolognani, Design considerations on fractional-slot fault-tolerant synchronous motors, *IEEE Transactions on Industry Applications*, 42(4), 997–1006, 2006.

36. N. Bianchi, S. Bolognani, and G. Grezzani, Fractional-slot IPM servomotors: Analysis and performance comparisons, in *Proceedings of the International Conference on Electrical Machines (ICEM'04)*, Vol. CD Rom, paper no. 507, Cracow, Poland, September 5–8, 2004, pp. 1–6.

37. F. Magnussen and C. Sadarangani, Winding factors and joule losses of permanent magnet machines with concentrated windings, in *Proceedings of the IEEE International Electric Machines and Drives Conference (IEMDC'03)*, Vol. 1, Madison, WI, June 2–4, 2003, pp. 333–339.

38. A. D. Gerlando, R. Perini, and M. Ubaldini, High pole number, PM synchronous motor with concentrated coil armature windings, in *Proceedings of International Conference on Electrical Machines (ICEM'04)*, CD-Rom, paper no. 58, Cracow, Poland, September 5–8, 2004, pp. 1–6.

39. N. Schofield, K. Ng, Z. Zhu, and D. Howe, Parasitic rotor losses in a brushless permanent magnet traction machine, in *Proceedings of the Electric Machine and Drives Conference (EMD'97)*, I. C. No. 444, Cambridge, U.K., 1997, September 1–3, 1997, pp. 200–204.

40. N. Bianchi and M. D. Pré, Use of the star of slots in designing fractional-slot single-layer synchronous motors, *IEE Proceedings—Electrical Power Applications*, 153(3), 459–466, May 2006 (Online no. 20050284).

41. J. Cros, P. Viarouge, and A. Halila, Brush dc motors with concentrated windings and soft magnetic composites armatures, in *Conference Record of IEEE Industry Applications Annual Meeting (IAS'01)*, Vol. 4, Chicago, IL, September 30–October 4, 2001, pp. 2549–2556.

42. F. Magnussen and H. Lendenmann, Parasitic effects in PM machines with concentrated windings, in *Conference Record of 40th IEEE Industry Applications Annual Meeting (IAS'05)*, Vol. 2, Kowloon, Hong-Kong, October 2–6, 2005, pp. 1044–1049.

43. N. Bianchi, S. Bolognani, and M. D. Pré, Magnetic loading of fractional-slot three-phase PM motors with non-overlapped coils, in *Conference Record of the IEEE 41st Industry Applications Society Annual Meeting (IAS'05)*, CD-ROM, Tampa, FL, October 8–12, 2006.

44. H. Polinder and M. J. Hoeijmaker, Eddy current losses in segmented surface-mounted magnets of a PM machine, *IEE Proceedings—Electrical Power Applications*, 146(3), 261–266, May 1999.

45. K. Atallah, D. Howe, P. Mellor, and D. Stone, Rotor loss in permanent-magnet brushless AC machines, *IEEE Transactions on Industry Applications*, 36(6), 1612–1617, November/December 2000.

46. N. Bianchi, S. Bolognani, and E. Fornasiero, A general approach to determine the rotor losses in three-phase fractional-slot PM machines, in *Proceedings of the IEEE International Electric Machines and Drives Conference (IEMDC'07)*, Antalya, Turkey, May 2–5, 2007, pp. 634–641.

47. S. Morimoto, M. Sanada, and Y. Takeda, Effects and compensation of magnetic saturation in flux-weakening controlled permanent magnet synchronous motor drives, *IEEE Transactions on Industry Applications*, 30(6), 1632–1637, November–December 1994.

48. T. Jahns, G. Kliman, and T. Neumann, Interior PM synchronous motors for adjustable speed drives, *IEEE Transactions on Industry Applications*, 22(4), 738–747, July/Aug 1986.

49. T. Jahns, Flux-weakening regime operation of an interior permanent magnet synchronous motor drive, *IEEE Transactions on Industry Applications*, 23(3), 681–689, May 1987.

50. B. E. Donald, D. W. Novotny, and T. A. Lipo, Field weakening in buried permanent magnet ac motor drives, *IEEE Transactions on Industry Applications*, 21(2), 398–407, March–April 1987.

51. B. K. Bose, A high-performance inverter-fed drive system of an interior permanent magnet synchronous machine, *IEEE Transactions on Industry Applications*, 24(5), 987–997, November–December 1988.

52. S. Morimoto, Y. Takeda, T. Hirasa, and K. Taniguchi, Expansion of operating limits for permanent magnet motor by current vector control considering inverter capacity, *IEEE Transactions on Industry Applications*, 26(5), 866–871, September–October 1990.

53. J. Haylock, B. Mecrow, A. Jack, and D. Atkinson, Operation of fault tolerant PM drive for an aerospace fuel pump application, *IEE Proceedings—Electrical Power Applications*, 145(5), 441–448, September 1998.

54. T. Jahns, Design, analysis, and control of interior PM synchronous machines, in N. Bianchi, T.M. Jahns (eds.), IEEE IAS Tutorial Course Notes, *IAS Annual Meeting*, CLEUP, Seattle, WA, October 3, 2005, ch. Fault-mode operation, pp. 10.1–10.21 (info@cleup.it).

55. M. Lazzari and P. Ferrari, Phase number and their related effects on the characteristics of inverter-fed induction motor drives, in *Conference Record of IEEE Industry Applications Annual Meeting (IAS'83)*, Vol. 1, Mexico, October 1983, pp. 494–502.

56. T. M. Jahns, Improved reliability in solid state ac drives by means of multiple independent phase-drive units, *IEEE Transactions on Industry Applications*, 16(3), 321–331, May 1980.

57. G. Singh and V. Pant, Analysis of a multiphase induction machine under fault condition in a phase-redundant ac drive system, *Electric Machines and Power Systems*, 28(6), 577–590, December 2000.

58. N. Bianchi, S. Bolognani, and M. D. Pré, Design and tests of a fault-tolerant five-phase permanent magnet motor, in *Proceedings of the IEEE Power Electronics Specialist Conference (PESC'06)*, Jeju, Korea, June 18–22, 2006, pp. 2540–2547.

59. L. Parsa and H. Toliyat, Five-phase permanent-magnet motor drives, *IEEE Transactions on Industry Applications*, 41(1), 30–37, January/February 2005.

60. C. French, P. Acarnley, and A. Jack, Optimal torque control of permanent magnet motors, in *Proceedings of the International Conference on Electrical Machines, ICEM'94*, Vol. 1, Paris, France, September 5–8, 1994, pp. 720–725.

61. T. Gobalarathnam, H. Toliyat, and J. Moreira, Multi-phase fault-tolerant brushless dc motor drives, in *Conference Record of IEEE Industry Applications Annual Meeting (IAS'00)*, Vol. 2, Rome, Italy, October 8–12, 2000, pp. 1683–1688.

62. J. Wang, K. Atallah, and D. Howe, Optimal torque control of fault-tolerant permanent magnet brushless machines, *IEEE Transactions on Magnetics*, 39(5), 2962–2964, September 2003.

63. J. Ede, K. Atallah, J. Wang, and D. Howe, Effect of optimal torque control on rotor loss of fault-tolerant permanent magnet brushless machines, *IEEE Transactions on Magnetics*, 38(5), 3291–3293, September 2002.

64. S. Ogasawara and H. Akagi, An approach to real-time position estimation at zero and low speed for a PM motor based on saliency, *IEEE Transactions on Industry Applications*, 34(1), 163–168, January–February 1998.

65. N. Bianchi, S. Bolognani, and M. Zigliotto, Design hints of an IPM synchronous motor for an effective position sensorless control, in *Proceedings of the IEEE Power Electronics Specialist Conference (PESC'05)*, Recife, Brazil, June 12–16, 2005, pp. 1560–1566.

66. M. Harke, H. Kim, and R. Lorenz, Sensorless control of interior permanent magnet machine drives for zero-phase-lag position estimation, *IEEE Transactions on Industry Applications*, 39(12), 1661–1667, November/December 2003.

67. M. Linke, R. Kennel, and J. Holtz, Sensorless speed and position control of synchronous machines using alternating carrier injection, in *Proceedings of International Electric Machines and Drives Conference (IEMDC'03)*, Madison, WI, June 2–4, 2003, pp. 1211–1217.

68. A. Consoli, G. Scarcella, G. Tutino, and A. Testa, Sensorless field oriented control using common mode currents, in *Proceedings of the IEEE Industrial Applications Society Annual Meeting*, Vol. 3, Rome, Italy, October 8–12, 2000, pp. 1866–1873.

69. J. Jang, S. Sul, and Y. Son, Current measurement issues in sensorless control algorithm using high frequency signal injection method, in *Conference Records of the 38th IEEE Industrial Applications Society Annual Meeting (IAS'03)*, Salt Lake City, UT, October 12–16, 2003.

70. A. Vagati, M. Pastorelli, G. Franceschini, and F. Scapino, Impact of cross saturation in synchronous reluctance motors of transverse-laminated type, *IEEE Transactions on Industry Application*, 36(4), 1039–1046, July–August 2000.

71. P. Guglielmi, M. Pastorelli, and A. Vagati, Impact of cross-saturation in sensorless control of transverse-laminated synchronous reluctance motors, *IEEE Transactions on Industrial Electronics*, 53(2), 429–439, April 2006.

72. Y. Jeong, R. Lorenz, T. Jahns, and S. Sul, Initial rotor position estimation of an IPM synchronous machine using carrier-frequency injection methods, *IEEE Transactions on Industry Applications*, 40(1), 38–45, January/February 2005.

73. N. Bianchi, *Electrical Machine Analysis using Finite Elements*, ser. Power Electronics and Applications Series. Boca Raton, FL: CRC Press/Taylor & Francis Group, 2005.

74. N. Bianchi, S. Bolognani, J.-H. Jang, and S.-K. Sul, Comparison of PM motor structures and sensorless control techniques for zero-speed rotor position detection, *IEEE Transactions on Power Electronics*, 22(6), 2466–2475, November 2007.

75. F. Briz, M. Degner, A. Diez, and R. Lorenz, Measuring, modeling, and decoupling of saturation-induced saliencies in carrier signal injection-based sensorless ac drives, *IEEE Transactions on Industry Applications*, 37(5), 1356–1364, September–October 2001.

76. N. Bianchi, Dai Pré, M., Alberti, L., and Fornasiero, E., Theory and design of fractional-slot PM machines, Tutorial Course Notes, sponsored by the IEEE-IAS Electrical Machines Committee, Presented at the IEEE IAS Annual Meeting, New Orleans, LA, September 23, 2007.

77. N. Bianchi, S. Bolognani, and M. Dai Pré, Design of a fault-tolerant IPM motor for electric power steering, *IEEE Transactions on VT*, 55(4), 1102–1111, 2006.

78. N. Bianchi, S. Bolognani, and M. Dai Pré, Strategies for the fault-tolerant current control of a five-phase permanent-magnet motor", *IEEE Transactions on Industry Applications*, 43(4), 960–970, 2007.

79. N. Bianchi and S. Bolognani, Sensorless-oriented-design of PM Motors, *IEEE Transactions on Industry Applications*, 45(4), 1249–1257, 2009.

7

Switched-Reluctance Machines

Babak Fahimi
*University of Texas
at Arlington*

7.1 Introduction

Switched-reluctance machines (SRM) have resurfaced in the field of adjustable speed motor drives over the past three decades. This has been mainly contributed to enabling technologies and devices offered by power electronics, the semiconductor industry, and cost-effective microprocessors. Being complemented by a rugged structure (see Figure 7.1) suitable for harsh environmental conditions and high-speed applications, a brushless structure that requires minimum maintenance, and a wide constant power region at an affordable cost has turned SRM into a prime candidate for niche industrial and domestic applications. Considering the renewed attention to renewable energy harvesting and advanced electric propulsion of automobiles, SRM drives are expected to attract more attention in years to come [1]. This chapter will provide the reader with a brief historical background, fundamental elements of operation, and state-of-the art in control.

7.2 Historical Background

SRM enjoys a very long history. Elementary versions of the SRM have existed since the early 1800s. The origins of the reluctance machine can be found in the horseshoe electromagnet developed by William Sturgeon in 1824 seen in Figure 7.2, as well as in an improved version of the horseshoe electromagnet developed by Joseph Henry. The improved version of the horseshoe electromagnet was an attempt to convert the single movement device into an actuator with a continuous oscillating motion device.

Many of the early motor designs were of the reluctance motor type and were strongly influenced by the steam engines developed during the same time period. Some of the more interesting machines of that period, which closely relate to the modern SRM, are those invented by Taylor and Davidson in 1935 (see Figure 7.3), and those for Charles Wheatstone developed by William Henely around 1842 (see Figure 7.4). These electromechanical converters operated based on the sequential energizing of the

FIGURE 7.1   Rotor and stator stack of an 8/6 SRM.

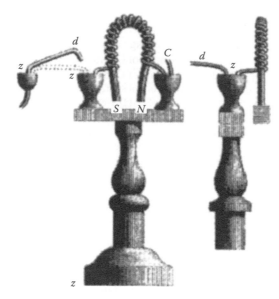

FIGURE 7.2   Horseshoe electromagnet proposed by William Sturgeon. (Courtesy of T. J. E. Miller.)

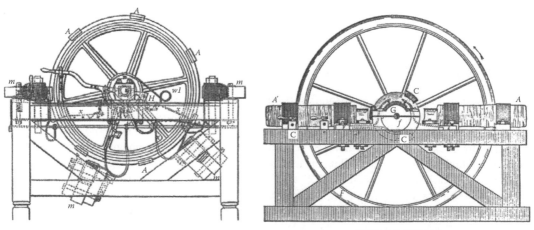

-Davidson's electric motor

FIGURE 7.3   Machines developed by Taylor and Davidson. (Courtesy T. J. E. Miller.)

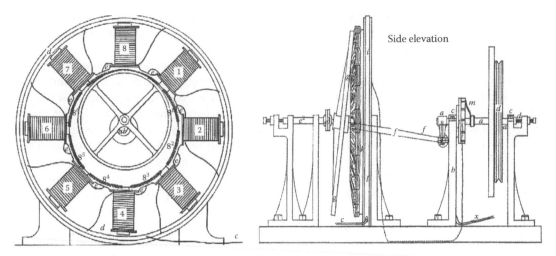

FIGURE 7.4 Machines developed for Charles Wheatstone by William Henely. (Courtesy T. J. E. Miller.)

spatially separated coils. However, the problem of de-energizing the inductors remained an unsolved challenge. Early attempts at using mechanical switches resulted in arcing and sparking, which turned reluctance machines into an entertainment attraction.

With the advent of ac and dc machinery during the later 1800s, the SRM took a backseat, living on in only very specialized applications such as vibration devices like electric bells and some instrumentation mechanisms.

Decades later, with the success of the power electronics and microprocessor-based control, efficient solutions for the early challenge of de-magnetization of the excited stator poles were invented. This gave rise to a new generation of the research, development, and commercialization of the SRM technology into industrial and domestic drives and actuators. The simplicity of the control in SRM drives, as compared to the ac counterparts such as induction and permanent magnet synchronous machines, was an added benefit in the new age of electronically controlled adjustable speed motor drives.

In recent years, the development of cost-effective digital signal controllers with comprehensive set of peripheral devices such as high resolution and ultrafast analog-to-digital converters, pulse width modulation (PWM) hardware on-the-chip and multiple timer configuration along with embedded current and voltage sensors, gate driver, and coupling circuits in semiconductor devices at an affordable cost has opened new opportunities for employment of SRM drives in high impact applications. These enabling technologies have changed the paradigm of the conventional design, where existence of a sinusoidal rotating field would have been considered as a necessity. To the contrary, non-sinusoidal, stepwise magneto-motive fields of SRM are no longer viewed as a handicap, and as a result SRM can compete with induction and permanent magnet synchronous machines on an equal footing, where the focus is on the performance and not conventional design practices.

7.3 Fundamentals of Operation

As a singly excited synchronous machine, SRM generates its electromagnetic torque solely on the principle of reluctance. In most electric machines, an attraction and repletion force between the magnetic fields caused by the armature and field windings forms the dominant part of the torque. In a SRM, the tendency of a polarized rotor pole to align with an excited stator pole is the only source of torque. It must be noted that optimal performance is achieved by proper positioning of the stator winding excitation with respect to the magnetic status of the machine. The magnetic status of the machine can be fully defined by the flux linkages in each phase, the slope of the flux linkages with respect to time, and their respective currents. Therefore, the sensing of the magnetic status of the machine becomes an integral part of the control in a SRM drive. Due to the existence of a one-to-one correspondence between magnetic flux in stator poles and the position of the rotor for any given current, sensing of the position using external sensors has been the practiced method in most developments.

Unlike most ac electric machines, in which the rotating magneto-motive force of the stator portrays a constant magnitude field with a constant angular velocity, the magnetic field of the stator windings in SRM exhibits an impulsive behavior, similar to that of a pair of electrodes forming a capacitor that are periodically charged to a maximum level and then suddenly discharged through an arc. In fact, once a stator phase is charged with a pulse of current, its stored magnetic energy will rise, and eventually at the instant of commutation, all the stored magnetic energy is quickly removed and fed back to the source. At this very moment, the next stator phase will be excited resulting in a frog-leap of the magneto-motive force from one location in the air gap to another. While the resultant rotating field will rotate at the synchronous frequency, the transition from one phase to another phase will occur within a few microseconds. Therefore, one can imagine that the rotating field in a SRM is not sinusoidal (neither with respect to time nor with respect to displacement). In fact, due to quick transitions of the magnetic field, attempts to use the fundamental components of the magnetic field are often inaccurate and inefficient.

A unipolar power inverter is usually used to supply the SRM. The generation of the targeted current profile is performed using a hysteresis or PWM type current controller. Although a square-shaped current pulse is commonly used for excitation in a SRM, different optimal current profiles are sometimes used to mitigate the undesirable effects of excessive torque undulation and audible noise [2]. In fact, SRM drives serve as an outstanding example of advanced motor drive systems where the focus is not on the complicated geometries of the motor. Rather, development of a sophisticated control algorithm and its implementation using a power electronic converter is the focus. The development of control algorithms is facilitated by the recent development of high-performance, cost-effective DSP-based controllers.

SRMs operate on the principle of reluctance torque. To best facilitate the explanation of how the SRM works, it is necessary to look at a more basic structure, namely a simple variable reluctance machine. Figure 7.5 shows a simple C-core with a single winding and a two-pole salient rotor.

If the coil is excited, then according to the Ampere's law (i.e., right hand rule), flux flows in the direction shown by the arrows. Subsequently, the stator and rotor pole faces will become magnetically polarized. Opposite magnetic poles will attract each other, with the tendency to bring them into alignment as much as is possible, in an effort to minimize the reluctance of the system. Thus, in order to achieve this alignment, tangential/normal forces are created on the faces/corners of the stator and rotor poles. In fact it is the normal forces acting on the side of the rotor poles that create the majority of the motional forces.

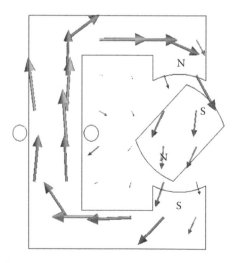

FIGURE 7.5   Simple variable reluctance machine with rotor at 45°.

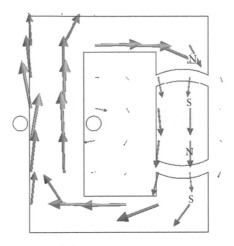

FIGURE 7.6 Simple variable reluctance machine with rotor at 90°.

Electromagnetic torque will be produced due to these motional forces. Since the torque is a product of the system trying to reach a state of minimum reluctance, as well as the fact that the torque only arises due to the excitation source and not from coils or magnets on the rotor, it is termed reluctance torque.

Figure 7.6 shows the system after it has reached the aligned position. This is one of two equilibrium points. The first equilibrium point is obtained at the unaligned position, and is unstable. Without any disturbance, the rotor will remain at this point; however disturbance of this point away from the unaligned position will result in the rotor moving to the aligned position. The aligned position is a stable equilibrium since any disturbance of the rotor from this position will result in the rotor eventually returning to the aligned position.

A SRM, ideally, is a modular combination of several variable reluctance machines that are magnetically linked and are highly dependent on complex switching algorithms and knowledge of the magnetic status in SRM. SRMs can operate as either a motor or a generator [6–9]. In order to obtain motoring action, a stator phase is excited when the rotor is moving from the unaligned position toward the aligned position. Similarly, by exciting a stator phase when the rotor is moving away from aligned toward unaligned position, a generating action will be achieved. By the sequential excitation of the stator phases, a continuous rotation can be achieved. Figure 7.7 illustrates the distribution of the magnetic field during commutations in an 8/6 SRM drive. Notably, the direction of the rotation is opposite to that of the stator excitation. A short flux path in the back-iron of the motor occurs in each electrical cycle. This, in turn, may cause asymmetry in the torque production process.

The proper synchronization of the stator excitation with the rotor position is a key step in the development of an optimal control strategy in SRM drives. Because the magnetic characteristics of the SRM, such as phase inductance or phase flux linkage, portray a one-to-one correspondence with the rotor position, they may be directly used for control purposes. In either case, direct or indirect detection of the rotor position forms an integral part of the control in the SRM drives.

The asymmetric bridge shown in Figure 7.8 is the most commonly used power electronics inverter for a SRM drive. This topology features a unipolar architecture that allows for satisfactory operation in SRM drives. If both switches are closed, the available dc link voltage is applied to the winding. By opening the switches, the negative dc link voltage will be applied to the winding, and freewheeling diodes guarantee a continuous current in the windings. Obviously, by keeping one of the switches closed while the other one is open, the respective freewheeling diode will provide a short-circuited path for the current. This topology can be used effectively to implement PWM-based or hysteresis-based current regulation as demanded by the control system. However, one should notice that at high speeds, the induced EMF in the winding is dominant and does not allow effective control of

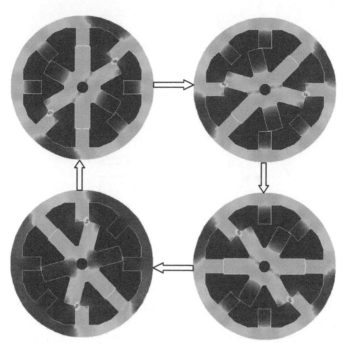

FIGURE 7.7 Illustration of short versus long flux paths for a 8/6 SRM.

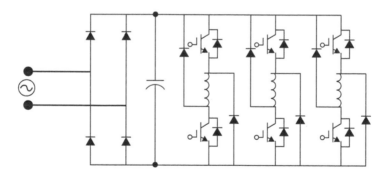

FIGURE 7.8 An asymmetric bridge with the front end rectifier for a 3ϕ SRM drive.

the current waveform. Therefore, current regulation is an issue related only to the low speed mode of operation. During generation, the mechanical energy supplied by the prime mover will be converted into an electrical form manifested by the induced EMF. Unlike the motoring mode of operation, this voltage acts as a voltage source that increases the current in the stator phase, thereby resulting in the generation of electricity.

7.4 Fundamentals of Control in SRM Drives

The control of electromagnetic torque is the main differentiating factor between various types of adjustable speed motor drives. In switched-reluctance motor drives, tuning the commutation instant and profile of the phase current tailors electromagnetic torque. Figure 7.9 depicts the basics of commutation in SRM drives. It can be seen that by properly positioning the current pulse, one can obtain positive (motoring) or negative (generating) modes of operation.

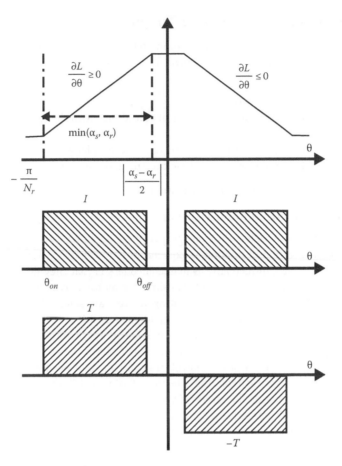

FIGURE 7.9 Commutation in SRM drives.

The induced EMF and electromagnetic torque generated by the SRM drive can be expressed in terms of co-energy, under unsaturated conditions, as follows:

$$E = \frac{\partial^2 W_c}{\partial \theta \partial i} \omega \approx \frac{dL(\theta)}{d\theta} i\omega$$

$$T = \frac{\partial W_c}{\partial \theta} \approx \frac{1}{2} \frac{dL(\theta)}{d\theta} i^2$$

(7.1)

where W_c, L, θ, i, and ω stand for co-energy, phase inductance, rotor position, phase current, and angular speed, respectively.

It must be noted that the nonlinear effects of magnetic saturation are neglected here. It is evident that a positive torque is achieved only if the current pulse is positioned in a region with an increasing inductance profile. Similarly, a generating mode of operation is achieved when the excitation is positioned in a region with a decreasing inductance profile. In order to enhance the productivity of the SRM drive, the commutation instants, (i.e., θ_{on}, θ_{off}), need to be tuned as a function of the angular speed and phase current. To fulfill this goal, the optimization of torque per Ampere is a meaningful objective. Therefore, exciting the motor phase when the inductance has a flat shape should be avoided. At the same time, the phase current needs to be removed well before the aligned position to avoid the generation of negative torque.

7.4.1 Open Loop Control Strategy for Torque

By the proper selection of the control variables, commutation instants, and reference current, an open loop control strategy for SRM drive can be designed. The open loop control strategy is comprised of the following steps:

- Detection of the initial rotor position
- Computation of the commutation thresholds in accordance with the sign of torque, current level, and speed
- Monitoring of the rotor position and selection of the active phases
- A control strategy for regulation of the phase current at low speeds

Each step is explained in detail in Sections 7.4.1.1 through 7.4.1.4.

7.4.1.1 Detection of the Initial Rotor Position

The main task at standstill is to detect the most proper phase for initial excitation. Once this is established, according to the direction of rotation, a sequence of stator phase excitation will be put in place. The major difficulty in using commercially available encoders is that they do not provide a position reference. Therefore, the easiest way to find rotor position for motor startup is to align one of the stator phases with the rotor. This can be achieved by exciting an arbitrary stator phase with an adequate current for a short period of time. Once the rotor is in an aligned position, a reference initial position can then be established. This method requires an initial movement by the rotor, which may not be acceptable in some applications. In these cases, the incorporation of a sensorless scheme at standstill is sought out. Although the explanation of sensorless control strategies for rotor position detection is beyond the extent of this chapter, due to its critical role, the detection of rotor position at standstill is explained here.

To detect rotor position at standstill, a series of voltage pulses with fixed and sufficiently short duration is applied to all phases. By consequent comparison between the magnitudes of the resulting peak currents, the most appropriate phase for conduction is selected. Figure 7.10 shows a set of normalized inductance profiles for a 12/8 SRM drive.

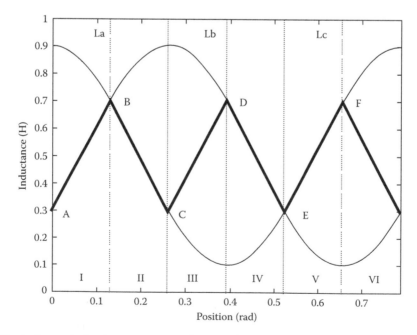

FIGURE 7.10 Assignment of various regions according to inductances in a 12/8 SRM drive.

TABLE 7.1   Detection of Best Phase to Excite at Standstill

Region	Condition	Rotor Angle (Mech.)
I	$I_A < I_B < I_C$	$0 < \theta^* < 7.5°$
II	$I_B < I_A < I_C$	$7.5° < \theta^* < 15°$
III	$I_B < I_C < I_A$	$15° < \theta^* < 22.5°$
IV	$I_C < I_B < I_A$	$22.5° < \theta^* < 30°$
V	$I_C < I_A < I_B$	$30° < \theta^* < 37.5°$
VI	$I_A < I_C < I_B$	$37.5° < \theta^* < 45°$

A full electrical period is divided into six separate regions according to the magnitudes of the inductances. Due to the absence of the induced voltage and small amplitude of currents, one can prove that the following relationship will hold for the magnitudes of measured currents:

$$I_{ABC} = \frac{V_{Bus}\,\Delta T}{L_{ABC}} \tag{7.2}$$

where ΔT, V_{Bus}, and L_{ABC} stand for duration of pulses, dc link voltage, and phase inductances, respectively. Table 7.1 summarizes the detection process for a 12/8 SRM drive. Once the range of position is detected, the proper phase for starting can be easily determined. Furthermore, in each region there exists a phase that offers a linear inductance characteristic. This phase can be used for the computation of rotor position using (7.2).

The flowchart shown in Figure 7.11 summarizes the detection process at standstill.

7.4.1.2 Computation of the Commutation Thresholds

In the next step, the commutation angles for each phase should be computed and stored in memory. If the commutation angles are fixed, computing the thresholds is relatively straightforward. It must be noted that within each electrical cycle, every phase should be excited only once. In addition, a symmetric SRM phase is shifted by

$$\Delta\theta = \frac{(N_s - N_r)360°}{N_s} \tag{7.3}$$

where N_s and N_r stand for the number of stator and rotor poles, respectively. Given a reference for rotor position such as the aligned rotor position with phase-A, one can compute and store the commutation

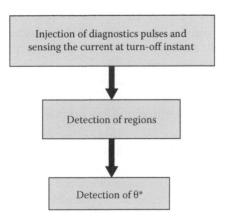

FIGURE 7.11   Detection of rotor position at standstill.

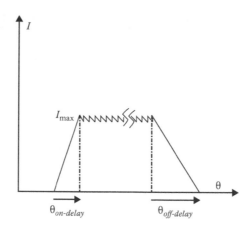

FIGURE 7.12   A typical current pulse at low speeds.

instants for each phase. The commutation thresholds are usually converted into a proper scale, so they can be compared with the value of a counter that tracks the number of incoming pulses from the position sensor. If a particular encoder can generate N pulses per mechanical revolution, then every mechanical degree corresponds to $4N/360$ pulses received by the processor (viz. quadrature pulses provide four rising and falling edges per pulse).

   If optimal performance of the machine is targeted, the effects of rotational speed and current must be taken into account. Figure 7.12 shows a typical current pulse for SRM drive. To achieve optimal control, the delay angles during the turn-on and turn-off process need to be taken into account. By neglecting the effects of motional back-EMF in the neighborhood of commutation, which is a valid assumption as turn-on and turn-off instants occur close to unaligned and aligned position, respectively, one can calculate the delay angles as

$$\theta_{on\text{-}delay} = \frac{\omega L_u}{r} \ln\left(\frac{V}{V - rI_{\max}} \right)$$

$$\theta_{off\text{-}delay} \approx \theta_{on\text{-}delay}\left(\frac{L_a I_{\max}}{L_u} \right)$$

(7.4)

where L_u, L_a, ω, V, and r denote unaligned inductance, aligned inductance, angular speed, bus voltage, and stator phase resistance, respectively. The dependency of the aligned position inductance upon maximum phase current is an indication of the nonlinear effects of saturation that need to be taken into account. As the speed and level of current increases, one needs to adopt the commutation angles using (7.4). As can be seen, the dependency of commutation angle upon the angular speed is linear, while its dependency upon the maximum phase current has a very nonlinear relationship.

7.4.1.3 Monitoring of the Rotor Position and Selection of the Active Phases

Once the previous steps are done, one can start with the main control tasks, namely, enforcing the conduction band and regulating the current. The block diagram depicted in Figure 7.13 shows the structure used in a typical algorithm, which forms the basic control strategy of the SRM drive. Monitoring the rotor position is a relatively easy task with a microcontroller. For the first task in the interrupt service routine, the current value of the rotor position will be compared against the commutation thresholds, and phases that should be on will be identified. In the next step, the current in active phases where an active phase is referred to as a phase that is turned on will be regulated.

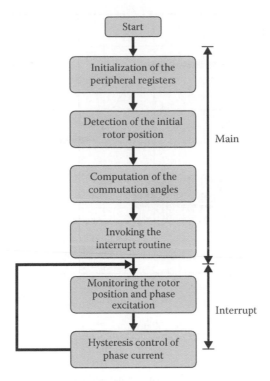

FIGURE 7.13 Block diagram of the basic control in SRM drives.

7.4.1.4 A Control Strategy for Regulation of the Phase Current at Low Speeds

At low speeds where the induced EMF is small, a method for control of the phase current is necessary. In the absence of such routines, the phase current will increase exponentially, possibly damaging the semiconductor devices or motor windings. Hysteresis and PWM control strategies are commonly used for regulating the phase current at low speeds. At higher speeds, the presence of a significantly larger back-EMF limits the growth of the phase current and there is no need for such regulation schemes. The profile of the regulated current depends on the control objective. In most applications, a flat-topped or square-shaped current pulse will be used. Figure 7.14 shows a regulated current waveform along with the gate pulse that is recorded at low speed region.

In order to conduct hysteresis control, the currents in active phases need to be sensed. Once the phase current is sampled, it needs to be converted into digital form. This can be done using the on-chip analog-to-digital converters or external A–D converters. The control rules for a classic two switch per phase inverter shown in Figure 7.8 are given by the following:

- If $I_{\min} \geq I$, then both switches are on. This results in applying the bus voltage across the coil terminals.
- If $I_{\max} \leq I$, then both switches are turned off. This results in applying the negative bus voltage across the coil terminals.
- If $I_{\min} \leq I \leq I_{\max}$, there is no need to make any changes in the status of the switches (i.e., if the switches are on, they remain on and if they are off, they remain off).

By simple comparison between the sampled current and current limits, one can develop a hysteresis control strategy. Since the current is sampled during each interrupt service routine, the time period of the interrupt should be sufficiently small to allow for a tight regulation. Because in most practical cases, only two phases conduct simultaneously, and given the speed of computation in the state-of-the-art microcontrollers, the interrupt service time should be very small.

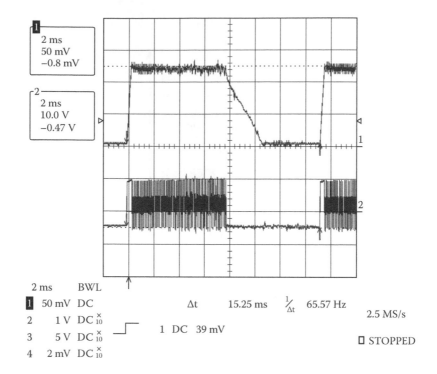

FIGURE 7.14 Phase current waveform and the gating signal without optimization. Reference current = 5.5 A; conduction angle = 180 (electrical) (operating speed = 980 rpm; output power = 120 W).

PWM technique can also be used in control of the phase current in SRM drive. Most application-specific digital signal controllers (i.e., TMS320F2812 from Texas Instruments) offer a fully controllable set of PWM signals via the compare units. The frequency and duty cycle of these PWM pulses can be adjusted at any stage of the program by the setting of two peripheral registers. This valuable feature can easily accommodate the control needs of a three-phase SRM drive system. In the case of a four-phase SRM drive, the fourth PWM signal can be generated by using one of the timer compare outputs.

The block diagram in Figure 7.15 summarizes the various steps along with the peripherals used in an application-specific digital signal controller such as TMS320LF2407. The main inputs to the program consist of commutation angles and the current profile. The quadrature outputs of an encoder have been used to determine the rotor position and angular speed of the drive. The phase currents have been sampled and converted into a digital form, to be used in current control. The output gates have been chosen from the general purpose input/output (GPIO) pins. The interface, conditioning circuit, and buffers are not shown in this picture. The control routine, used for the detection of active phases and hysteresis/PWM control of the phase currents, is combined in the software to form the final gating signal.

Once the basic operation of the SRM drive is established, one can design and develop closed-loop forms of the control. In the following sections, closed-loop torque and speed control routines in the SRM drive are discussed, including four-quadrant operation of the drive.

7.4.2 Closed-Loop Torque Control of the SRM Drive

As SRM technology begins to emerge in the form of a viable candidate for industrial applications, the significance of reliable operation under closed-loop torque, speed, and position control increases. Figure 7.16 depicts a typical cascaded control configuration for SRM drives. The main control block is responsible for

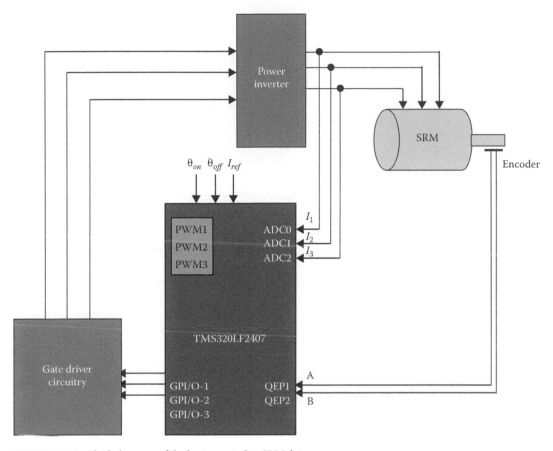

FIGURE 7.15 Block diagram of the basic control in SRM drive system.

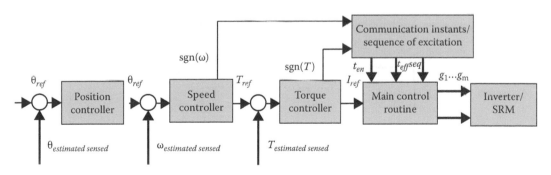

FIGURE 7.16 Cascaded control configuration for a SRM drive system.

generating the gate signals for the power switches. It also performs current regulation and phase commutation functions. In order to perform these tasks, it requires reference current, commutation instants, and a sequence of excitation. The torque controller provides the reference current, while the information regarding the commutation is obtained from a separate block that coordinates motoring, generating, and direction of rotation, as demanded by the various types of control. The various feedback informations are generated using either estimators or transducers.

Depending on the application, an adjustable speed motor drive may operate in various quadrants of the torque/speed plane. For instance, in a water pump application, where control of the output pressure

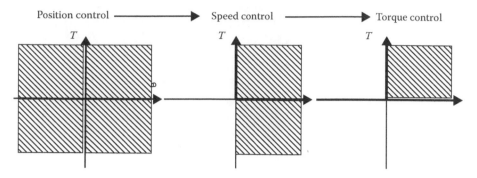

FIGURE 7.17 Minimum requirement of an adjustable speed drive for performing torque, speed, and position control.

is targeted, torque control in one quadrant is sufficient, whereas in an integrated starter/alternator, four-quadrant operation is necessary. Figure 7.17 shows the minimum requirement of an adjustable speed motor drive for performing torque, speed, and position control tasks. A speed controller may issue positive (motoring) or negative (generating) torque commands to regulate the speed. In a similar way, a position controller will ask for positive (clockwise) and negative (counter clockwise) speed commands. The accommodation of such commands will span all four quadrants of operation in the torque/speed plane. As a result, four-quadrant operation is a necessity for many applications in which positioning the rotor is an objective. In order to achieve four-quadrant operation in SRM drives, the direction of rotation in the air gap field needs to be altered. In addition, to generate negative torque during generation mode, the conduction band of the phase should be located in a region with negative inductance slope.

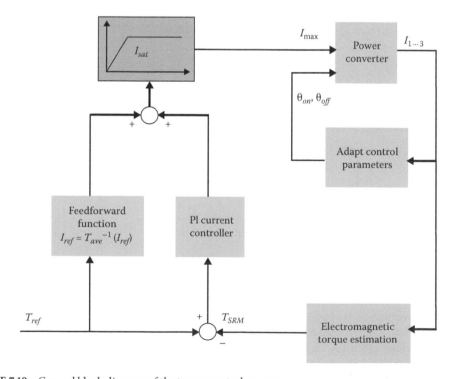

FIGURE 7.18 General block diagram of the torque control system.

Figure 7.18 depicts a general block diagram of the closed-loop torque control system. The main modules in this figure are as follows:

- An estimator for the average/instantaneous electromagnetic torque
- A feed-forward function for fast and convergent tracking of the commanded torque
- A computational block to determine commutation instants according to the sign of demanded torque and magnitude of the phase current

The estimator for average/instantaneous electromagnetic torque is designed based on (7.1). The design also incorporates an analytical model of the phase inductance/flux linkage as shown in the following:

$$L(i,\theta) = L_0(i) + L_1(i)\cos(N_r\theta) + L_2(i)\cos(2N_r\theta) \tag{7.5}$$

where L_0, L_1, and L_2 represent polynomials that reflect the nonlinear effects of saturation. (Derivation of the above formula is explained in Appendix 7.A.) Moreover, the inverse mapping of the torque estimator is used to form a feed-forward function. In the absence of the torque sensor/estimator, this feed-forward function can be used effectively to perform open loop control of the torque. The use of a feed-forward controller accelerates the convergence of the overall torque tracking. The partial mismatch between reference and estimated torque is then compensated via a *PI* controller. It must be noted that the introduction of the measured torque into the control system requires an additional analog-to-digital conversion. Figure 7.19 shows a comparison between the estimated and measured torque in a 12/8 SRM drive at steady state, when responding to a periodic ramp function in closed-loop control. The average torque estimator shows good accuracy. The existence of a 0.4 Nm averaging error is due to the fact that iron and stray losses are not included in the torque estimator. In order to perform this test, a permanent magnet drive acting as an active load was set in a speed control loop running in the same direction at 800 rpm.

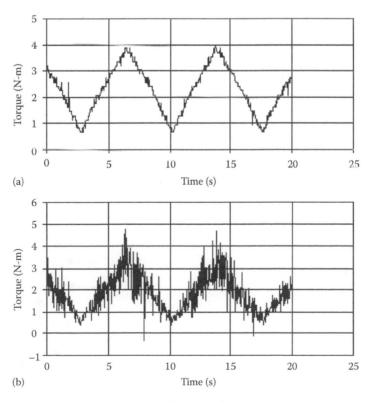

FIGURE 7.19 Comparison between (a) measured and (b) estimated average torque.

As mentioned earlier, operation in all four quadrants of the torque versus speed plane is a requirement for many applications. Given the symmetric shape of the inductance profile with respect to the aligned rotor position, one can expect that for a given conduction band at a constant speed, current waveforms during motoring and generating should be a mirror image of each other. However, one should note that the back-EMF during generation acts as a voltage source resulting in an increase of phase current even after a phase is shut down. This may cause some complications in terms of stability at high speeds. In order to alter the direction of rotation, the only necessary step is to change the sequence of excitation. Notably, the sequence of excitation among stator phases is opposite of the direction of rotation. The transition between two modes needs to be quick and smooth. Upon the receipt of a command requesting a change in direction, the excited phase needs to be turned off to avoid generating additional torque. Regenerative braking should be performed simultaneously. This requires the detection of a phase in which the inductance profile has a negative slope. The operation in generation mode continues until the speed decays to zero or a tolerable near-zero speed. At this time, all the phases will be cleared and a new sequence of excitation can be implemented. Speed reversal during generating is not a usual case because the direction of rotation is dictated by the prime mover. In the case of the speed reversal being initiated by the prime mover, the SRM controller needs to be notified. Otherwise, a mechanism for the detection of rotation direction should be in place. Such a mechanism would detect any unexpected change of mode, i.e., motoring to generating.

7.4.2.1 Closed-Loop Speed Control of the SRM Drive

As it is the next step in developing a high-performance SRM drive, speed control is explained. As shown in Figure 7.20, a cascaded type of control can be used to perform closed-loop speed control. The speed can be sensed using the position information that is already provided by the encoder. Because the SRM is a synchronous machine, one may choose the electrical frequency of excitation for control purposes. The relationship between mechanical and electrical speeds is given by

$$\omega_e = N_r \omega_m \tag{7.6}$$

where N_r is the number of rotor poles. Ultimately, success in performing tightly regulated speed control depends upon the performance of the inner torque control system as depicted in Figure 7.20. It is recommended that a feed-forward function be used to mitigate the initial transients in issuing commands to the torque control system.

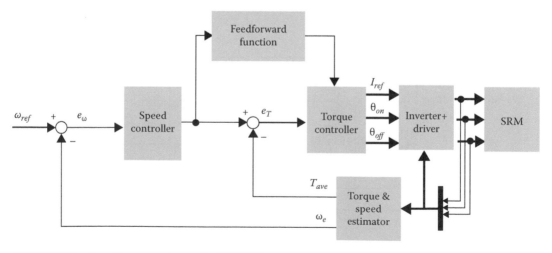

FIGURE 7.20   Closed loop speed control of SRM drive.

7.5 Summary

SRM drives are making their entry into the adjustable speed motor drive market. To take full advantage of their capacities, the development of high-performance control strategies has turned into a necessity. The advent of cost-effective DSP-based controllers provides an opportunity to engineer for this need in an effective way. A successful implementation of these methodologies demands a good understanding of the torque generation process. Basic control methods for the SRM drive have been discussed. These include the principles of design for closed-loop control strategies. More advanced technologies, such as position sensorless and adaptive control [9–12], are also being investigated by many researchers across the globe, and there have been great advances in these areas as well [3–5]. It is expected that developments in better efficiency, fault tolerance, and compactness will come about as a result of these efforts in years to come.

Appendix 7.A

7.A.1 Modeling of Inductance Profile in an 8/6 SRM

The following dynamic equation relates phase voltage, current, and flux linkage as shown in

$$v = ri + \frac{d\lambda}{dt} \tag{7.A.1}$$

where
  v represents the phase voltage
  r is the winding resistance
  i is the winding current
  λ is the flux linkage

Expansion of the equation, neglecting mutual inductance terms yields the following:

$$v = ri + \frac{dL}{d\theta}\frac{d\theta}{dt}i + \frac{di}{dt}L + \frac{dL}{di}\frac{di}{dt} \tag{7.A.2}$$

where
  L is the bulk inductance of the phase, also termed the self-inductance
  θ is the rotor position

The phase voltage equation can be simplified by making the following substitution as shown in (7.A.3), as well as by neglecting saturation, which eliminates the rate of change of the self-inductance L with respect to the phase current.

$$e = \frac{dL}{d\theta}\frac{d\theta}{dt}i \tag{7.A.3}$$

The motional back-EMF is now represented by e, which is a function of the phase current, the rotor speed, and the rate of change of the self-inductance with respect to the rotor position. By substituting (7.A.3) into (7.A.2), the phase voltage equation can be rewritten as

$$v = ri + e + \frac{di}{dt}L \tag{7.A.4}$$

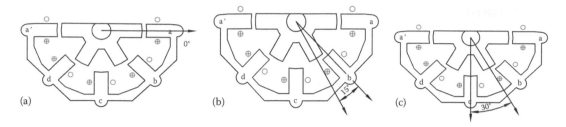

(a)   (b)   (c)

FIGURE 7.A.1 (a) Phase a at aligned position, (b) phases b and d at midway to aligned position, and (c) phase c at unaligned position.

Equation 7.A.4 provides a simpler equation and more clearly shows the role of the self-inductance. The profile of the self-inductance is an important quantity, in that it is very useful for graphically describing the operation of the machine and the placement of current pulses with respect to the position. In the following, the self-inductance profile for an 8/6 SRM is derived, and its relationship to the geometrical uniqueness of the machine is discussed in the context of auto-calibration. Notably, the same methodology can be applied to other machine configurations without the loss of generality.

Figure 7.A.1, shows the basic structure of the 8/6 SRM at three different rotor positions: aligned, midway, and unaligned.

In Figure 7.A.1, the machine is illustrated as being sliced axially, which reflects the symmetry of the stator and rotor structure. As can be seen from Figure 7.A.1, the center of the rotor pole, with respect to a fixed reference position centered on the a-phase stator pole, is at 0°. At the same time that a-phase is aligned, the rotor pole position with respect to the b-phase and d-phase of the machine are at 15°, and are understood to be at the midway position. Finally, analysis of the c-phase stator pole position with respect to the nearest rotor pole shows that the c-phase is at 30°, and is subsequently understood to be unaligned. The 8/6 SRM possesses a unique geometry that allows the self-inductance to be easily modeled using a Fourier series.

Due to the geometric nature of the 8/6 SRM, it is useful to describe the inductance in terms of the aligned, midway, and unaligned positions. These inductances are denoted as L_a, L_m, and L_u respectively. Figure 7.A.2 shows a plot of the self-inductance versus rotor position over one period.

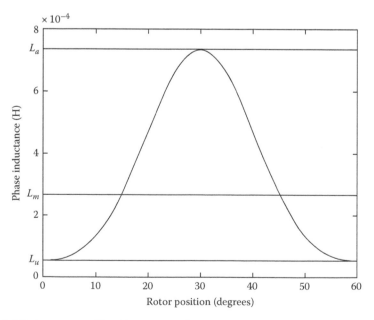

FIGURE 7.A.2 Self-inductance profile with respect to the rotor position.

Electromagnetic torque for a single phase of the machine is proportional to the derivative of the inductance with respect to the rotor angle. Thus from graphical analysis of Figure 7.A.2, it is seen that there are 2 zero torque zones: namely at the aligned and unaligned positions. As stated previously, the inductance of the single phase may be represented by a Fourier series, as follows [13,14]:

$$L(\theta,i) = L_0(i) + L_1(i)f(\theta) + L_2(i)f(2\theta) + \cdots \qquad (7.A.5)$$

In this formulation for inductance, $f(\theta)$ is represented by a smooth basis function that is chosen to be a cosine as is typical for Fourier series formulations. Furthermore, the coefficients of the series expansion are dependent upon the current, thereby allowing the inclusion of the saturation effect. Replacing $f(\theta)$ with $\cos(\theta)$ yields

$$L(\theta,i) = L_0(i) + L_1(i)\cos(\alpha + \phi) + L_2(i)\cos(2(\alpha + \phi)) + \cdots \qquad (7.A.6)$$

where
  $\alpha = N_r\theta$, N_r is the number of rotor poles
  θ is the rotor angle

Although (7.A.6) can be extended to an infinite number of terms, it is beneficial based on judicious examination of the machine's geometric aspects as shown in Figure 7.A.1, to choose only three terms. The angles of interest are listed in Table 7.A.1.

The number of rotor poles for the 8/6 SRM is $N_r = 6$. Replacement of the angle α by $N_r\theta$ at the specified positions listed in Table 7.A.1 yields the following three equations describing the aligned, midway, and unaligned inductances:

$$L_a = L_0(i) + L_1(i)\cos(\phi) + L_2(i)\cos(2\phi)$$

$$L_u = L_0(i) - L_1(i)\cos(\phi) + L_2(i)\cos(2\phi) \qquad (7.A.7)$$

$$L_m = L_0(i) - L_1(i)\sin(\phi) - L_2(i)\cos(2\phi)$$

Since, as stated earlier, the aligned and unaligned positions represent zero torque zones, and since the torque is proportional to the derivative of the self-inductance with respect to the rotor angle under single phase excitation, then the relationship for the derivatives of the aligned and unaligned inductances are as follows:

$$0 = L_1(i)\sin(\phi) + 2L_2(i)\sin(2\phi)$$

$$0 = L_1(i)\sin(\phi) - 2L_2(i)\sin(2\phi) \qquad (7.A.8)$$

By simplifying (7.A.8), it can be shown that $\sin(\phi) = 0$. The value for ϕ may be obtained by noting that it can either be zero or $k\pi$, where k is an integer. Thus by choosing $\phi = 0$, Equation 7.A.7 can be simplified into the following form:

TABLE 7.A.1 Description of Self-Inductance at Specified Positions

Inductance	Angle
L_a = aligned inductance	$\theta = 0°$
L_m = midway inductance	$\theta = 15°$
L_u = unaligned inductance	$\theta = 30°$

$$L_a = L_0(i) + L_1(i) + L_2(i)$$

$$L_u = L_0(i) - L_1(i) + L_2(i) \qquad (7.A.9)$$

$$L_m = L_0(i) - L_1(i) - L_2(i)$$

The ultimate goal of this derivation is to determine the Fourier coefficients L_0, L_1, and L_2, in terms of the quantities listed in Table 7.A.1. With this in mind, (7.A.9) is reformulated into matrix form and subsequently matrix inversion is applied to yield the following relationship:

$$\begin{bmatrix} L_0(i) \\ L_1(i) \\ L_2(i) \end{bmatrix} = \begin{bmatrix} \frac{1}{4} & \frac{1}{2} & \frac{1}{4} \\ \frac{1}{2} & 0 & \frac{-1}{2} \\ \frac{1}{4} & -\frac{1}{2} & \frac{1}{4} \end{bmatrix} \begin{bmatrix} L_a(i) \\ L_m(i) \\ L_u(i) \end{bmatrix}$$

(7.A.10)

It should be noted from Equation 7.A.10, as well as from the physics of the machine, that the unaligned inductance L_u will have no dependency upon the current, thus it can be measured from the machine at a single current level and/or calculated by general formulae for inductance. However, aligned and midway inductances will have current dependency. This dependency upon current is represented by Figures 7.A.3 and 7.A.4, for the aligned and midway inductances. Figure 7.A.5 illustrates the current independency of the unaligned inductance.

There are several choices in representing the inductances L_a and L_m in Equation 7.A.10. One choice that results in a computationally efficient representation is to use polynomials. Equation 7.A.11 is the polynomial representation for the aligned and midway inductances.

$$L_a(i) = \sum_{k=0}^{n} a_k i^k$$

$$L_m(i) = \sum_{k=0}^{n} b_k i^k$$

(7.A.11)

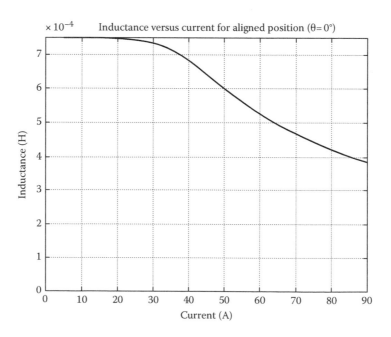

FIGURE 7.A.3 Aligned inductance versus phase current.

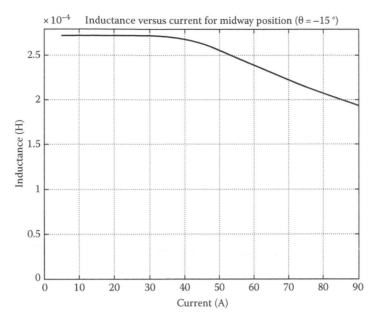

FIGURE 7.A.4 Midway inductance versus phase current.

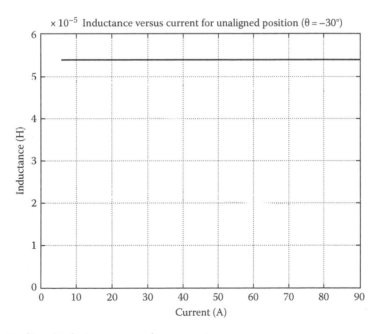

FIGURE 7.A.5 Unaligned inductance versus phase current.

With only five terms in the polynomial representation, a very good fit can be obtained. This is shown in Figure 7.A.6 for the aligned inductance case, and in Figure 7.A.7 for the midway inductance case.

The analysis shows that using the polynomial approximation provides a very good approximation for the Fourier series coefficients in representing inductance and its current dependency [15,16]. Now it is possible to write the Fourier expansion coefficients in terms of the aligned, midway, and unaligned

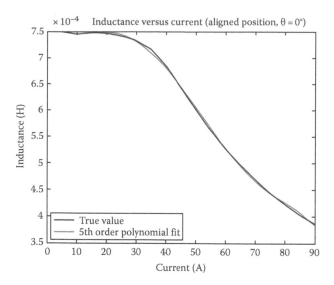

FIGURE 7.A.6   Measured and modeled inductance at aligned position versus current.

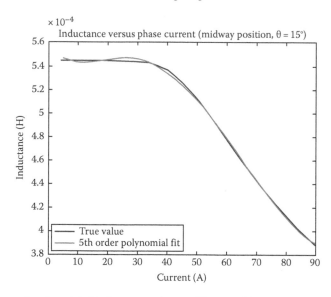

FIGURE 7.A.7   Measured and modeled inductance value at midway position versus current.

inductances using a polynomial fit. Combining Equation 7.A.11 with (7.A.10), the following form for L_0, L_1, and L_2 is obtained:

$$L_0(i) = \left\{ \sum_{k=0}^{n} \left\{ \frac{1}{4} a_k + \frac{1}{2} b_k \right\} i^k \right\} + \frac{1}{4} L_u = \sum_{k=0}^{n} A_k i^k$$

$$L_1(i) = \left\{ \sum_{k=0}^{n} \frac{1}{2} a_k i^k \right\} - \frac{1}{2} L_u = \sum_{k=0}^{n} B_k i^k \qquad (7.A.12)$$

$$L_2(i) = \left\{ \sum_{k=0}^{n} \left\{ \frac{1}{4} a_k - \frac{1}{2} b_k \right\} i^k \right\} + \frac{1}{4} L_u = \sum_{k=0}^{n} C_k i^k$$

Replacement of the Fourier coefficients in Equation 7.A.6 with the result obtained in (7.A.12) yields the following compact formula for the self-inductance of a single phase of the SRM:

$$L(\theta, i) = \sum_{k=0}^{n} \left\{ A_k i^k + B_k i^k \cos(N_r \theta) + C_k i^k \cos(2N_r \theta) \right\} \tag{7.A.13}$$

References

1. K. M. Rahman, B. Fahimi, G. Suresh, A. V. Rajarathnam, and M. Ehsani, Advantages of switched reluctance motor applications to EV and HEV: Design and control issues, *IEEE Transactions on Industry Applications*, 36(1), 119–121, Jan./Feb. 2000.
2. B. Fahimi, G. Suresh, K. M. Rahman, and M. Ehsani, Mitigation of acoustic noise and vibration in switched reluctance motor drive using neural network based current profiling, in *Proceedings of the IEEE 1998 Industry Applications Society Annual Meeting*, St. Louis, MO, Oct. 1998, pp. 715–722.
3. P. P. Acarnley, R. J. Hill, and C. W. Hooper, Detection of rotor position in stepping and switched reluctance motors by monitoring of current waveforms, *IEEE Transactions on Industrial Electronics*, 32(3), 215–222, Aug. 1985.
4. G. Suresh, B. Fahimi, K. M. Rahman, and M. Ehsani, Inductance based position encoding for sensorless SRM drives, in *Proceedings of the 30th IEEE Power Electronics Specialist Conference*, Charleston, SC, July 1999, pp. 832–837.
5. C. C. Chan and Q. Jiang, Study of starting performances of switched reluctance motors, in *Proceedings of the 1995 International Conference on Power Electronics and Motor Drive Systems*, Vol. 1, Singapore, Feb. 1995, pp. 174–179.
6. J. M. Miller, P. J. McClear, and J. H. Lang, Starter-alternator for hybrid electric vehicle: Comparison of induction and variable reluctance machines and drives, in *Proceedings of the 33rd IEEE Industry Application Society Annual Meeting*, Oct. 1998, St. Louis, MO, pp. 513–523.
7. D. A. Torrey, Switched reluctance generators and their control, *IEEE Transactions on Industrial Electronics*, 49(1), 3–14, Feb. 2002.
8. E. Mese, Y. Sozer, J. M. Kokernak, and D. A. Torrey, Optimal excitation of a high speed switched reluctance generator, in *Proceedings of the IEEE 2000 Applied Power Electronics Conference*, New Orleans, LA, 2000, pp. 362–368.
9. B. Fahimi, A. Emadi, and R. B. Sepe, A switched reluctance machine based starter/alternator for more electric cars, *IEEE Transactions on Energy Conversion*, 19(1), 116–124, March 2004.
10. P. Tandon, A. V. Rajarathnam, and M. Ehsani, Self-tuning control of a switched-reluctance motor drive with shaft position sensor, *IEEE Transactions on Industry Applications*, 33(4), 1002–1010, July/Aug. 1997.
11. B. Fahimi, A. Emadi, and R. B. Sepe, Four-quadrant position sensorless control in SRM drives over the entire speed range, *IEEE Transactions on Power Electronics*, 20(1), 154–163, Jan. 2005.
12. M. Ehsani and B. Fahimi, Elimination of position sensors in switched reluctance motor drives: State of the art and future trends, *IEEE Transactions on Industrial Electronics*, 49(1), 40–48, Feb. 2002.
13. B. Fahimi, G. Suresh, J. Mahdavi, and M. Ehsani, A new approach to model switched reluctance motor drive application to dynamic performance prediction, design and control, in *Proceedings of the IEEE Power Electronics Specialists Conference*, Fukuoka, Japan, May 1998, pp. 2097–2102.
14. C. S. Edrington and B. Fahimi, An auto-calibrating model for switched reluctance motor drives: Application to design and control, in *Proceedings of the IEEE 2003 Power Electronics Specialists Conference*, Acapulco, Mexico, June 2003, pp. 409–415.
15. S. Dixon and B. Fahimi, Enhancement of output electric power in switched reluctance generators, in *IEEE International Electric Machines and Drives Conference*, Vol. 2, Madison, WI, June 2003, pp. 849–856.
16. C. S. Edrington, Bipolar excitation of switched reluctance machines, Dissertation at University of Missouri-Rolla, Rolla, MO, 2004.

8

Thermal Effects

Aldo Boglietti
Politecnico di Torino

8.1 Introduction

In the electromagnetic devices, the losses produce an increase of the temperatures. As a consequence, in addition to the electromagnetic design, it is very important a device thermal analysis for well understanding its thermal limit, taking into account the thermal constrains imposed by the insulating material classes. The winding maximum temperatures with respect to the insulation classes are reported in Table 8.1.

It is important to underline that the increase of the winding temperature over the insulation class reduces heavily the insulation life as shown in Figure 8.1.

It is well evident that a correct thermal analysis is essential for a correct electromagnetic device design. In particular, the thermal and electromagnetic designs should have to be developed in parallel, because electromagnetic performances are directly correlated to the thermal conditions. As an example, a 50°C temperature rise in a winding leads to a 20% resistance increase, while a 135°C rise gives 53% resistance increase, with an increase of the copper losses of same amount, when the winding current is constant. With reference to the permanent magnet (PM) motor, in rare earth magnets, any increase in temperature leads to a flux density reduction with a consequent current increase in order to maintain the same output torque. Consequently, the related winding losses will increase with the square of the winding current. The thermal behavior of an electromagnetic device is depending on the adopted cooling system. With reference to rotating electrical machines, the following cooling systems can be found:

- Totally enclosed with natural ventilation "TENV" (typically adopted in servo motors)
- Totally enclosed fan cooled "TEFC" (typically used in induction motors for industrial applications)
- Drip proof radial or axial cooling
- Water jacket cooling system

TABLE 8.1 Winding Maximum Temperatures with Respect to the Insulation Classes

Insulation Class	Winding Temperature Limit [°C]
Class A	105
Class B	130
Class F	155
Class H	180

8.2 Basic Heat Transfer and Flow Analysis

The thermal design requires a good knowledge of the thermal transfer phenomena involved in the electromagnetic device. A short summary of the heat transfers and flow phenomena is reported. The heat transfer exchange is due to conduction, radiation, natural and forced convection. In the following, a short outline of the heat transfer general aspects is reported hereafter, while specific evaluations of rotating electrical machine are reported in Sections 8.5 and 8.6.

8.2.1 Conduction

Conduction is the heat transfer mode in a solid due to a temperature difference between different parts of the material, as shown in Figure 8.2. Conduction is typically present in solid material but it also occurs in liquids and gases even if convection is usually the dominant phenomena in these materials. In the conduction heat transfer, the heat flows from the higher to the lower temperature point due to the vibration of the molecules within the material. As well known, the good electrical conductors are good thermal conductors too.

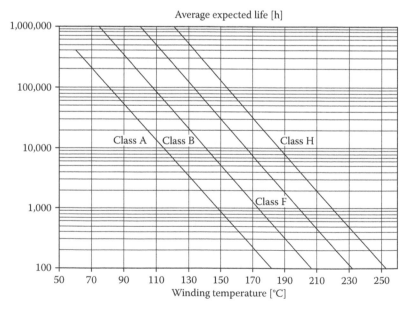

FIGURE 8.1 Expected insulation life as a function of the temperature increase.

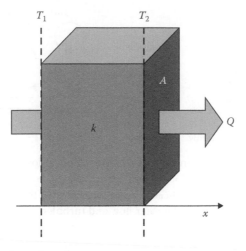

FIGURE 8.2 Heat transfer by thermal conduction.

Fourier's law defines the conduction heat transfer phenomenon:

$$Q = kA \frac{dT}{dx} \tag{8.1}$$

where
  Q [W] is the rate of heat transfer
  A [m²] is the cross-sectional area
  k [W/(m °C)] is material thermal conductivity
  dT/dx [°C/m] is the temperature gradient

For the metals materials k is in the range 10–400 W/(m °C), while solid insulating materials have k values in range 0.1–1 W/(m °C). The air thermal conductivity k_{iar} is equal to 0.026 [W/(m °C)]. When the geometrical and physics characteristics of a homogeneous solid are known, it is possible to compute its thermal resistance using

$$R_{cond} = \frac{L}{kA} \, [°C/W] \tag{8.2}$$

where
  L is the length
  A is the area
  k is the thermal conductivity

The L and A values can obtained from the component geometry.

8.2.2 Convection

Convection is the heat transfer mode between a surface and a fluid. The convection heat transfer can be divided in two main phenomena:

1. Natural convection, where the fluid motion is due to buoyancy forces, because of the density modification of the fluid close to the surface
2. Forced convection, where the fluid motion is due to external forces (imposed for example by a fan)

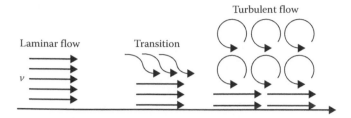

FIGURE 8.3 Laminar and turbulent flow.

Both natural and forced convection can present a laminar flow at lower velocities and turbulent flow when the streamline flow is at higher velocities (see Figure 8.3).

Turbulent flow increases not only the heat transfer rate but also the friction between the fluid and contact surfaces. The transition between laminar flow and turbulent flow is defined on the basis of the Reynolds number:

$$Re = \frac{\rho v L}{\mu} \qquad (8.3)$$

where

ρ [kg/m³] is the fluid density
μ [kg/s m] is the fluid dynamic viscosity
L [m] is the characteristic length of the surface
v [m/s] is the fluid velocity

Newton's law defines the convection heat transfer phenomenon:

$$Q = h_C A (T_1 - T_2) \qquad (8.4)$$

where

Q [W] is the rate of heat transfer
A [m²] is the surface area
h_C [W/(m² °C)] is the convection heat transfer coefficient
$(T_1 - T_2)$ [°C] is the temperature difference between the surface and the fluid

In convection thermal analysis, the correct definition of the heat transfer coefficient h_C is the most difficult problem to be solved. The typical range of the heat transfer coefficient values for the several convection phenomena are reported hereafter:

Air natural convection $h_C = 5$–25 [W/(m² °C)]
Air forced convection $h_C = 10$–300 [W/(m² °C)]
Liquid forced convection $h_C = 50$–5000 [W/(m² °C)]

When the geometrical and physics characteristics of a system are known, it is possible to compute the related thermal resistance using

$$R_{conv} = \frac{1}{A h_C} \, [°C/W] \qquad (8.5)$$

where
A [m²] is the area
h_C is the heat transfer coefficient for convection

Due to the complex nature of the convection, it is often not possible to find directly exact mathematical solutions to the problems. As a consequence, an empirical technique of dimensional analysis, based on experiments and tests, is used in alternative to determine the heat transfer coefficient. In fact, many factors determine the phenomena involved in the convection process between a surface and a fluid, such as shape and size of the solid–fluid boundary, fluid flow characteristics (i.e., turbulent flow), characteristics of the fluid material, etc. This approach uses a set of dimensionless numbers to obtain a functional relationship for h_C and the main fluid physical properties in the flow working conditions. It is important to underline that these dimensionless numbers allow the use of the same formulations with different fluid materials and dimensions not included in the original experiments. The most used dimensionless numbers are the following ones:

Reynolds number: $Re = \rho v L/\mu$, i.e., inertia force/viscous force
Grashof number: $Gr = \beta g \theta \rho^2 L^3/\mu^2$, i.e., buoyancy force/viscous force
Prandtl number: $Pr = c_p \mu/k$, i.e., momentum/thermal diffusivity for a fluid
Nusselt number: $Nu = hL/k$, i.e., convection heat transfer/conduction heat transfer in a fluid

In the previous equation set, the meaning of the used symbol is listed hereafter.

h heat transfer coefficient [W/(m²/°C)]
μ fluid dynamic viscosity [kg/(s.m)]
k thermal conductivity of the fluid [W/(m °C)]
c_p specific heat capacity of the fluid [kJ/(kg °C)]
θ temperature difference between the surface and fluid [°C]
L characteristic length of the surface [m]
β coefficient of cubical expansion of fluid [1/°C]
g gravitational force [m/s²]
v fluid velocity [m/s]
ρ fluid density [kg/m³]

The Reynolds number is used to predict the transition from laminar to turbulent flow in forced convection systems, while the product of the Grashof and Prandtl numbers is used to predict the transition to turbulent flow in systems dominated by natural convection. The natural and forced convection heat transfer coefficients for the geometries involved in electromechanical devices and, in particular, in electrical machine can be found in the technical literature and in books on heat transfer. In presence of a mix between natural and forced convection, the relation $h_{mix}^3 = h_{forced}^3 \pm h_{natural}^3$ is used, where the sign ± takes into account if the two phenomena are in opposing or not.

8.2.3 Radiation

Radiation is the heat transfer mode between two surfaces due to the energy transfer by electromagnetic waves; as a consequence, thank the radiation phenomenon, the heat can be transferred in the vacuum (Figure 8.4).

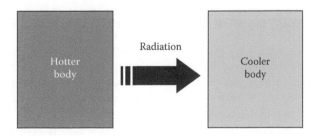

FIGURE 8.4 Heat transfer by radiation.

The amount of the emitted heat depends on the absolute temperature of the body. The Stefan-Boltzmann's law defines the convection heat transfer phenomenon:

$$Q = \sigma A T^4 \tag{8.6}$$

where

Q [W] is the rate of heat transfer
A [m^2] is the surface area of perfectly radiating body
σ [W/(m^2 K^4)] is the Stefan-Boltzmann constant equal to 5.669×10^{-8}
T [K] is the absolute surface temperature

An ideal radiating body (technically defined as "black body") emits at a given temperature the maximum possible energy at all wavelengths. The Stefan-Boltzmann equation defines the energy emission rather than the energy exchange. Since the area A may also absorb radiation from elsewhere, the emitting and absorbing characteristics (called emissivity) and the view that the surfaces have of the other ones (called view factor, taking into account how well one surface is viewed by another one) have to be considered for a correct computation of the heat transfer. The radiation exchange between two surfaces can be computed by

$$Q = \sigma \varepsilon_1 F_{1-2}(T_1^4 - T_2^4), \tag{8.7}$$

where

Q [W] is the rate of heat transfer
A_1 [m^2] is the area of radiating surface 1
T_1 [K] is the absolute temperature of surface 1
T_2 [K] is the absolute temperature of surface 2
ε_1 is the emissivity of surface 1 ($\varepsilon \leq 1$)
F_{1-2} is the view factor (it takes into account how well the surface 2 is viewed by surface 1 ($F_{1-2} \leq 1$))

The heat transfer coefficient of radiation, h_R can be computed by

$$h_R = \frac{\sigma \varepsilon_1 F_{1-2}(T_1^4 - T_2^4)}{T_1 - T_2} \tag{8.8}$$

Consequently, the thermal resistance due to radiation phenomenon is equal to $R_{Rad} = \left(1/A h_R\right)$

8.3 Thermal Analysis and Related Thermal Models

Nowadays, the thermal analysis is typically based on analytical lumped circuits or numerical models. Analytical lumped circuit models have excellent calculation speed, but a correct determination of the thermal resistances is not a simple task. Numerical Computational Fluid Dynamics (CFD) or numerical Finite Element Analysis (FEA) software can be used to accurately predict flow in complex regions and temperature distribution in solid components, respectively. Both the methods suffer of long model setup and computation times, especially when it is virtually impossible to reduce the problem to a two dimensions (2D) problem.

8.4 Numerical Models

8.4.1 Numerical Computational Fluid Dynamics

CFD is used for the determination of the coolant flow rate, speed, and pressure distribution of the cooling fluid inside and outside the device and in cooling passages. In addition, the CFD analysis is very useful to compute the surface heat transfers. These values can be used as starting conditions for subsequent temperature analysis in the active material and in solid structures. The approach by CFD requires modern CFD codes and specialized software available in the market. These softwares are mostly based on the finite volume technique solving Navier-Stokes equations complimented by a selection of validated and proven physical models to solve three-dimensional (3D) laminar or turbulent flow and to obtain heat transfer coefficients with a high degree of accuracy. Both 2D and 3D packages can be found and the choice is depending on the geometry under analysis. CFD analysis using 3D model suffers of a very long model setup and computation times. The typical use of CFD analysis for electrical machines is the following:

- Internal flow either in a through-ventilated machine, where ventilation is driven by a fan or by self-pumping effect of rotor, or in a TEFC motors and generators to assess the air movements that exchange heat from winding endwinding to external endcaps.
- External flow and flow around the enclosure of a TEFC motors and generators.
- Fan design and related performance analysis in order to optimize material, cost, manufacturing processes, space, or access constraints. In fact, fans used in electrical machines often have very poor aerodynamic efficiency and require a low-cost production. CFD offers a great advantage in improving fan design, taking into account its interaction with the cooling circuit.
- Supporting analysis for water flow cooling system both, in electrical machines and power converters.

The use of CFD is devoted and recommended when sophisticated simulations are imposed by the high costs of the prototypes, typically for big motors or generators. It is important to underline that the data obtained using CFD can be usefully adopted to improve the analytical algorithms used in the FEM model or in the thermal resistance evaluation.

8.4.2 Finite Element Analysis

FEA is now a standard tool for electromagnetic analysis and it is more and more used in electromagnetic devices design with both 2D and 3D approaches. Often software packages for electromagnetic analysis include a module for thermal analysis too. At first quick look FEA could seem more accurate than thermal network analysis; however, it has the same problems in the definition of the thermal quantities such as convection heat transfer coefficients and interface gaps. The FEA main advantage is the accurate calculation of conduction heat transfer in complex geometric shapes not approachable with lumped parameters.

8.5 Thermal Analysis Using Thermal Network

The thermal network using lumped thermal parameters is the most used approach for the thermal analysis of electromagnetic devices. This method is based on the following electrothermal equivalence: temperature to voltage, thermal power to current, and thermal resistance to electrical resistance. In thermal network, it is possible to lump together components that have the same temperatures and to connect these components in a single isothermal node in the network. These nodes are separated by thermal resistances, which represent the heat transfer between components. In Figure 8.5, an example of a simplified thermal network for a TEFC induction motor is reported. The method is accurate as the thermal

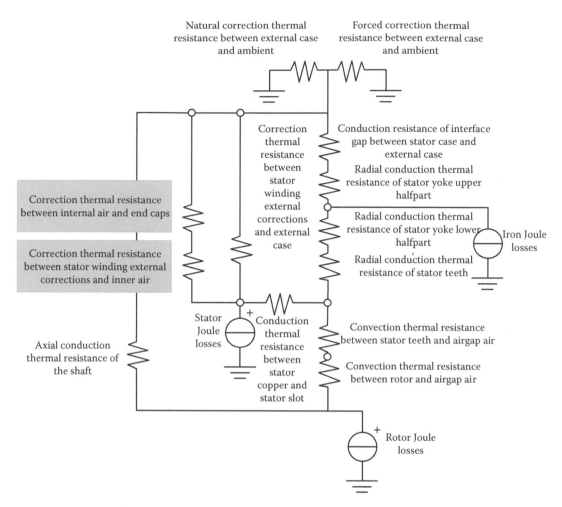

FIGURE 8.5 Simplified thermal network for a TEFC induction motor.

resistances are well determined. The thermal resistances have to represent all the thermal heat transfer phenomena inside and outside the system under analysis. As a consequence, conduction, natural and forced convection, and radiation thermal resistances have to be taken into account. Unfortunately, these thermal resistances are often very complex to be determined, and this is due to the involved geometrical shapes and the physic phenomena. For these reasons, some thermal resistance determinations have to be done using analytical equations often based on the designer experience. The equations used to compute these thermal resistances are summarized in the following. It is important to remark that, from the thermal resistance determination point of view, these methods have general validity and they are not linked to the thermal network complexity.

8.5.1 Conduction Heat Transfer Resistance

Conduction thermal resistances can be simply calculated using the following formulation:

$$R = \frac{L}{kA} \tag{8.9}$$

where
　　L [m] is the path length
　　A [m²] is the path area
　　k [W/(m °C)] is the thermal conductivity of the material

In most cases, L and A can be simply obtained from component geometry, but particular care has to be taken for a correct value of L in the presence of thermal resistances due to interface gap between components. The conductor materials have a thermal conductivity in the range 10–400 W/(m °C), while the insulation material has a thermal conductivity in the range 0.1–1.0 W/(m °C). In Table 8.2, the thermal conductivity of some materials is reported.

8.5.2 Radiation Heat Transfer Resistance

Radiation thermal resistances for a given surface can be simply calculated using

$$R = \frac{1}{h_R A} \tag{8.10}$$

where
　　A [m²] is the surface area
　　h_R [W/(m² °C)] is the heat transfer coefficient

The surface area can be calculated from the surface geometry. The radiation heat transfer coefficient can be calculated using

$$h_R = \sigma \varepsilon F_{1-2} \frac{T_1^4 - T_2^4}{T_1 - T_2} \tag{8.11}$$

The emissivity ε is a function of the material and finish surface for which data are available in most engineering textbooks. The view factor can easily be calculated for simple geometric surfaces such as cylinders and flat plates, but it is more difficult for complex geometries, and the help of specialist books on heat transfer is mandatory.

8.5.3 Convection Heat Transfer Resistance

Convection thermal resistances for a given surface can be calculated using

$$R = \frac{1}{h_C A} \tag{8.12}$$

TABLE 8.2　Thermal Conductivity k in W/(m °C) for Some Materials

Aluminum	237	Zinc	116	Rubber	0.15
Brass	111	Epoxy	0.207	Plastic	0.25
Copper	401	Mica	0.71	Teflon	0.22
Iron	80	Mylar	0.19	Paper	0.15
Iron 1% silicon	42	Nylon	0.242	Air	0.0262
Iron 5% silicon	19	Bakelite	0.19	Water	0.597

The previous equation is basically the same equation as for radiation but with the radiation heat transfer coefficient replaced by the convection heat transfer coefficient, h_C [W/(m² °C)].

The determination of h_C is not easy because convection heat transfer process is due to fluid motion. In natural convection, the fluid motion is due entirely to buoyancy forces linked to the fluid density variations. In a forced convection system, fluid movement is by an external force (e.g., fan, blower, pump). If the fluid velocity is high, then turbulence can be present and in such cases, the mixing of hot and cold air is more efficient with an increase of the heat transfer. However, the turbulent flow produces a larger pressure drop, with a reduction of the fluid volume flow rate. Tested empirical heat transfer correlations, based on dimensionless analysis, are used to predict h_C. The equations for h_C determination can be found in the technical literature for the convection surfaces typically involved in electrical machines and heat sink. A first approximation, for simple geometric shapes, the combined natural convection, and radiation heat transfer coefficient values are in the range 12–14 W/(m² °C).

8.6 Thermal Resistance in Electrical Machines

In the following, some information on appropriated heat transfer coefficient or thermal resistance to be used in electrical machine thermal models is provided. These values take into account complex parts, which cannot be easily determined by the classical relations.

8.6.1 Convection Heat Transfer Resistance

With reference to TENV machines, the equivalent thermal resistance (natural convection and radiation) R_0 [°C/W] between the housing and ambient can be computed in first approximation using

$$R_0 = 0.167 A^{1.039} \tag{8.13}$$

where A [m²] is the total area of the external frame including the fins. This relation can be used in TEFC motor operating in variable speed drive at low speed, where the convection heat transfer overcomes the heat transfer for forced convection.

8.6.2 Radiation Heat Transfer

For inside and outside electrical machine parts, the following average values of the radiation heat transfer coefficients can be used:

8.5 W/(m² °C)	Between copper–iron lamination
6.9 W/(m² °C)	Between endwinding–external cage
5.7 W/(m² °C)	Between external cage–ambient

8.6.3 Equivalent Thermal Conductivity between Winding and Lamination

The thermal behavior of wires positioned inside the slot is a very complex problem because the value of the thermal conductivity k is not simple to be defined. A possible approach to simplify the thermal resistance computation is to use an equivalent thermal conductivity "$k_{cu,ir}$" taking into account, at the same time, the impregnation and insulation system present in the slot. This equivalent thermal conductivity depends on several factors such as material and quality of the impregnation, residual air quantity after

TABLE 8.3 Coefficient
for the Computation of the
Forced Convection Heat
Transfer Coefficient between
End Winding and Endcaps

K_1	K_2	K_3
15.5	0.39	1
15	0.4	0.9
20	0.425	0.7
33.2	0.0445	1
40	0.1	1
10	0.3	1
41.2	0.151	1

the impregnation process. If the equivalent thermal conductivity $k_{cu,ir}$ is known, the thermal resistance between the winding and the stator laminations can be computed using the equation reported in previous section. When the slot fill factor k_f, the slot area A_{slot} [cm²], and axial core length L_{core} [cm] are known, the following relation can be used as a reasonable starting value:

$$k_{cu,ir} = 0.2749 \left[(1 - k_f) A_{slot} L_{core} \right]^{-0.4471} \tag{8.14}$$

The quantity inside the square bracket represents the available net volume inside the slot for the wire/slot insulation and the impregnation. Reasonable values to be used are in the range 0.08–0.04 W/(m² °C)

8.6.4 Forced Convection Heat Transfer Coefficient between End Winding and Endcaps

The thermal resistance between electrical machine endwindings and endcaps due to forced convection can be evaluated by the equation previously reported, where the value of h_C is not so easy to be defined. For totally enclosed machines, the value of h_C can be evaluated using the following formulation:

$$h = k_1 \left[1 + k_2 v^{k3} \right] \tag{8.15}$$

where v [m/s] is the speed air inside the motor endcaps. The three coefficients K_1, K_2, and K_3 are provided by several authors and they are reported in Table 8.3.

8.7 Transient Thermal Analysis Using Thermal Network

In presence of time variations of the heat transfer, heat capacitances have to be added to the previously discussed thermal resistance. Using the same electrothermal equivalence, the heat capacity is equivalent to the electrical capacity and the following thermal equation can be written:

$$P_{th} = C_{th} \frac{dT}{dt} \tag{8.16}$$

where
   P_{th} [W] is the thermal power
   C_{th} [J/°C] is the heat capacity
   T [°C] is the temperature
   t [s] is the time

The heat capacity can be computed by

$$C = \rho V C_{sp} \tag{8.17}$$

where
   ρ [kg/m³] is the density
   V [m³] the considered homogeneous volume
   C_{sp} [J/(kg C)] the specific heat capacity

During the thermal transient conditions, the thermal network is represented by differential equations to be solved using appropriated mathematical methods. Transient thermal analysis are fundamental for predicting the thermal behaviors, such as the instantaneous overheating in electrical machines, electromagnetic devices, and power converter structures when the power losses inside the device are varying in the time due to load with duty cycle.

8.8 Final Considerations

As well evident in this chapter, the thermal analysis in electrical devices is not a simple task, and it requires experience and skill for a correct management and use of the thermal relations and related thermal quantities and coefficients. The support of specialized heat transfer books is quite mandatory in order to avoid wrong approaches and unacceptable results.

Bibliography

1. W.S. Janna, *Engineering Heat Transfer*, Van Nostrand Reinhold (International), London, U.K., 1988.
2. A. Boglietti, A. Cavagnino, D. Staton, M. Shanel, M. Mueller, and C. Mejuto, Evolution and modern approaches for thermal analysis of electrical machines, *IEEE Transactions on Industrial Electronics*, 56(3), 871–882, March 2009.
3. A. Boglietti, A. Cavagnino, M. Lazzari, and M. Pastorelli, A simplified thermal model for variable-speed self-cooled industrial induction motor, *IEEE Transactions on Industry Applications*, 39(4), 945–952, July/August 2003.
4. P. Mellor, D. Roberts, and D. Turner, Lumped parameter thermal model for electrical machines of TEFC design, *IEE Proceedings—B*, 138(5), 205–218, September 1991.
5. N. Jaljal, J.-F. Trigeol, and P. Lagonotte, Reduced thermal model of an induction machine for real-time thermal monitoring, *IEEE Transactions on Industrial Electronics*, 55(10), 3535–3542, October 2008.
6. C. Kral, A. Haumer, and T. Bauml, Thermal model and behavior of a totally-enclosed-water-cooled squirrel-cage induction machine for traction applications, *IEEE Transactions on Industrial Electronics*, 55(10), 3555–3564, October 2008.
7. D. Staton, A. Boglietti, and A. Cavagnino, Solving the more difficult aspects of electric motor thermal analysis, *IEEE Transactions on Energy Conversion*, 20(3), 620–628, September 2005.
8. A. Boglietti and A. Cavagnino, Analysis of the endwinding cooling effects in TEFC induction motors, *IEEE Transactions on Industry Applications*, 43(5), 1214–1222, September–October 2007.
9. A. Boglietti, A. Cavagnino, M. Parvis, and A. Vallan, Evaluation of radiation thermal resistances in industrial motors, *IEEE Transactions on Industry Applications*, 42(3), 688–693, May/June 2006.
10. J. Mugglestone, S.J. Pickering, and D. Lampard, Effect of geometry changes on the flow and heat transfer in the end region of a TEFC induction motor, *Ninth IEE International Conference Electrical Machines & Drives*, Canterbury, U.K., September 1999.
11. C. Micallef, S.J. Pickering, K.A. Simmons, and K.J. Bradley, An alternative cooling arrangement for the end region of a totally enclosed fan cooled (TEFC) induction motor, *IEE Conference Record PEMD 08*, April 3–5, 2008, York, U.K.
12. A. Boglietti, A. Cavagnino, D. Staton, M. Popescu, C. Cossar, and M.I. McGilp, End space heat transfer coefficient determination for different induction motor enclosure types, *IEEE Transactions on Industry Applications*, 45(3), 929–937, May/June 2009.
13. C. Micallef, S.J. Pickering, K.A. Simmons, and K.J. Bradley, Improved cooling in the end region of a strip-wound totally enclosed fan-cooled induction electric machine, *IEEE Transactions on Industrial Electronics*, 55(10), 3517–3524, October 2008.
14. M.A. Valenzuela and J.A. Tapia, Heat transfer and thermal design of finned frames for TEFC variable-speed motors, *IEEE Transactions on Industrial Electronics*, 55(10), 3500–3508, October 2008.

15. D.A. Staton and A. Cavagnino, Convection heat transfer and flow calculations suitable for electric machines thermal models, *IEEE Transactions on Industrial Electronics*, 55(10), 3509–3516, October 2008.
16. A. DiGerlando and I. Vistoli, Thermal networks of induction motors for steady state and transient operation analysis, *Conference Record ICEM 1994*, Paris, France, 1994.
17. E. Schubert, Heat transfer coefficients at end winding and bearing covers of enclosed asynchronous machines, *Elektrie*, 22, 160–162, April 1968.

9

Noise and Vibrations of Electrical Rotating Machines

Bertrand Cassoret
Université d'Artois

Jean-Philippe
Lecointe
Université d'Artois

Jean-François
Brudny
Université d'Artois

9.1 Introduction

The problem of noise and vibrations is important for electrical machines. Indeed, standards in terms of noise are more and more restrictive. The tendency to increase the power of machines for a given size leads to increase in noise and vibrations. That is why the knowledge of phenomena generating acoustic noise is necessary not only to design modern electrical machines, but also to analyze systems with acoustic and vibration problems.

This chapter focuses on noise of AC electrical rotating machines connected to the grid or operating with adjustable-speed drives, with special emphasis on noise of magnetic origin. The objectives are multiple; they aim at answering the following questions: What is the link between the vibrations and the acoustic noise? How can the acoustic spectra of an AC machine be exploited? How can the noise and vibrations be predetermined and their occurrence be avoided?

Section 9.2 presents the various origins of noise of electrical rotating machines. Typical spectra of noisy machines are depicted. Then, in Section 9.3, the phenomenon of noise of magnetic origin is described. The suggested analytical method makes it possible to understand what the contribution of the flux density harmonics on the noisy forces is. In Section 9.4, an analytical mechanical and acoustic modeling for rotating machines is proposed. These developments are important because the forces that produce the noise have various consequences, not only in function of their amplitude and frequency, but also in function of the mechanical structure response. The method allows laying emphasis on the important parameters to supervise for an efficient acoustic design. Finally, Section 9.5 presents an analytical method to determine the flux density harmonics of AC machines.

9.2 Origins of Noise and Vibrations of Electrical Rotating Machines

9.2.1 Mechanical, Aerodynamical, and Magnetic Noises

Noise of electrical rotating machines has essentially three origins: electromagnetic, aerodynamic, and mechanical [1–3].

9.2.1.1 Noise of Mechanical Origin

The noise of mechanical origin comes mainly from the bearings. So, it exists for most of the rotating electrical machines, except in the case of magnetic bearings. The level of this noise, which is tied to frictions, depends on the bearing type and quality, the oiling, and the rotor speed. It is admitted that the noise generated by the plain bearings is widely lower than the other noises. For the roller bearings, the noise depends mainly on the external resonance frequency; those bearings sometimes produce treble sounds, which disappear temporally when a small quantity of grease is injected. Roller bearings lubricated with oil are less noisy [4].

It is also necessary to take into account frictions of brushes, particularly with DC machines because of the non-smooth collector. The sound level due to mechanical frictions increases generally with the square of the speed. Mechanical noises are important only for machines with high rotation speed.

9.2.1.2 Noise of Aerodynamic Origin

Aerodynamic noises are often more higher than mechanical noises. The noise results of air vibrations, the rotating parts create air turbulences and noise. They come from the fan or from active parts of the rotor, which act as a fan (e.g., the ends of the induction machine rotor bars). Obstructions in airflows are a supplementary fact of noise. Ventilation allows convection for cooling; it reduces notably the size of the machines but it creates noise. Thus, there is a compromise to do for the designers between drawing a small-sized machine or a noisy one. Aerodynamic noise increases with the fifth power of the speed. An 80 dB ventilation noise at 1000 revolutions per minute (rpm) reaches 104 dB at 3000 rpm.

9.2.1.3 Noise of Electromagnetic Origin

The level of noise of magnetic origin is variable because it depends on the design, the load, the speed, and the power supply. For low-speed machines, magnetic noise is almost always prevailing. It is generated by electromagnetic forces, which occur between the stator and the rotor. They produce vibrations of the machine, mainly the stator. When the frequencies of the electromagnetic forces are close to the resonance frequencies of the stator, the vibrations and the noise are amplified. The magnetic noise of rotating machines can easily be distinguished from other noises by cutting off the electric supply: the magnetic noise is immediately stopped while aerodynamic and mechanical noises decrease slowly with the speed. Sound spectra show few fine lines typical of magnetic noise.

9.2.2 Examples of Rotating Machine Spectra

The following spectra have been recorded for several types of electrical rotating machines operating at no load. They are obtained with a spectrum analyzer; it displays the FFT of the measured signals which come from

- A microphone, located at 1 m from the surface of the tested machine placed in a semi-anechoic room.
- An accelerometer that measures the stator vibrations amplitudes. Let us point out that the acceleration amplitude results from the product of the vibration amplitude by the square of its angular frequency. That justifies that sometimes, on the acceleration spectra, the lines of high frequency have the highest magnitudes.

9.2.2.1 Example of a 650 W Single-Phase Induction Machine

This cage rotor machine usually works in washing machines. Two kinds of spectra are presented, obtained with the machine normally supplied by the grid or immediately after having cut off the power supply.

9.2.2.1.1 Spectra with the Supplied Machine

Figures 9.1 and 9.2 show respectively the acoustic and vibration spectra for a 50 Hz single-phase supply. On the acoustic spectrum, noise components can be seen at 336, 2,016, 2,160, 3,264, 3,920, 7,550, 7,650, 7,750, and 11,490 Hz. The total noise is 57 dB. The three lines at 7550, 7650, 7750 Hz, spaced at 100 Hz, are typical of magnetic noise.

On the vibration spectrum, lines at 2,016, 3,264, 3,920, 7,550, 7,650, 7,750, and 11,490 Hz can still be observed. As this spectrum includes only noises of magnetic and mechanical origin, it can be concluded that the 2160 Hz component is due to ventilation.

9.2.2.1.2 Spectra after Having Cut Off the Power Supply

Figure 9.3 shows the acoustic spectrum when the rotor is still rotating. It includes only mechanical and aerodynamic noises. The lines at 2,016, 3,920, 7,550, 7,650, 7,750, and 11,490 Hz disappeared. It can be concluded that they have magnetic origin while components at 336, 2160, and 3264 Hz have mechanical or aerodynamic origins. Figure 9.4 presents the corresponding vibration spectrum, which is only

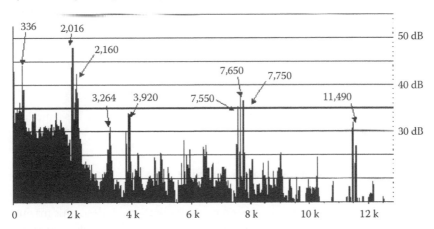

FIGURE 9.1 Spectrum of acoustic pressure level (dBA) of a 650 W single-phase machine.

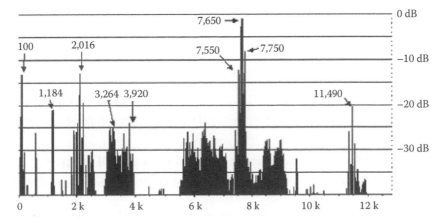

FIGURE 9.2 Vibration spectrum (accelerations) of a 650 W single-phase machine.

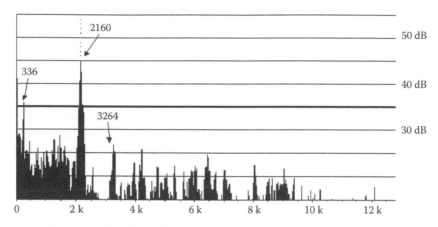

FIGURE 9.3   Aerodynamic and mechanical noises (dBA) of a 650 W single-phase machine.

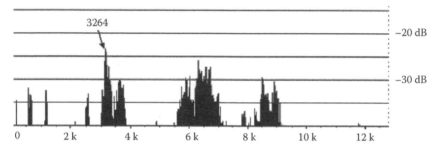

FIGURE 9.4   Mechanical vibrations (accelerations) of a 650 W single-phase machine.

concerned by the mechanical effects. It can be deduced that the 3264 Hz line has a mechanical origin; and the 336 and 2160 Hz probably an aerodynamic origin. Thus, this machine has a low mechanical noise.

9.2.2.2 Example of a Switched Reluctance Machine

The switched reluctance machine (SRM) and, especially, the doubly salient SRM (BDSRM) can be used in many industrial, aerospace, automotive, and domestic applications. Indeed, this machine is easy to manufacture and has low cost because the rotor has no winding and the power electronic controller has few components. In addition to its simple and rugged construction, it has a high efficiency [5]. But vibrations and acoustic noise can be particularly problematic with the SRM because of the stator back-iron deformation induced by radial magnetic forces [6,7]. Figure 9.5 presents the acoustic spectrum of a BDSRM equipped with eight teeth on the stator and six teeth on the rotor; it is supplied with voltage rectangular waveforms and the rotor rotates at 1466.6 rpm. Figure 9.6 shows the radial stator frame vibrations. Both spectra are composed of many thin lines, regularly spaced. The high number of components is explained by the phase currents and voltages, which are not sinusoidal waveforms. They are responsible for harmonics generating noise and vibrations. A major line at 2200 Hz appears clearly on the vibration spectrum; a modal analysis explains it by the presence of a natural resonance of the stator frame around 2100 Hz.

9.2.2.3 Example of a Synchronous Machine Supplied with a PWM Inverter

Let us consider a synchronous AC machine supplied with a Pulse Width Modulation (PWM) inverter, operating at 100 Hz fundamental frequency.

The vibration spectrum measured with an accelerometer (Figure 9.7) shows lines around frequencies that are multiple of the switching frequency f_w (3 kHz). Thus, the PWM switching frequency has an

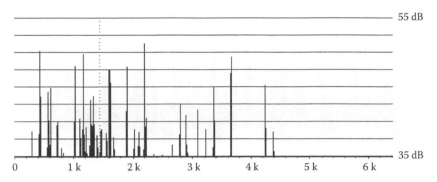

FIGURE 9.5 Spectrum of acoustic pressure level (dBA) of a 8/6 BDSRM machine.

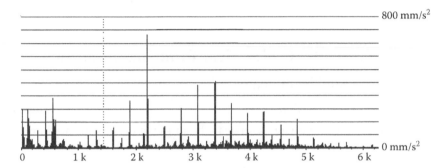

FIGURE 9.6 Vibration spectrum (accelerations, in m/s^2) of a 8/6 BDSRM machine.

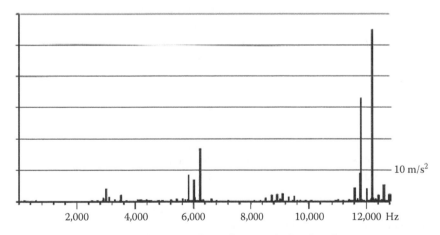

FIGURE 9.7 Vibration spectrum (accelerations) of a machine supplied with a 3 kHz PWM inverter.

obvious influence on noise and vibrations. The best way to obtain a silent machine is to choose a high PWM frequency, so that human ear cannot hear the noise (over 15 kHz); the problem is that the losses of the inverter increase with the frequency.

9.2.2.4 Example of a Saturated Machine

Magnetic saturation can produce noise [8]. The acoustic spectrum shown in Figure 9.8 concerns an industrial three-speed three-phase induction motor with a squirrel cage rotor. Low speed (the synchronous speed is 375 rpm with the 50 Hz grid) is used sporadically, for hoisting, for instance, but a high acoustic level is generated. As this machine has been designed in priority for working at the two

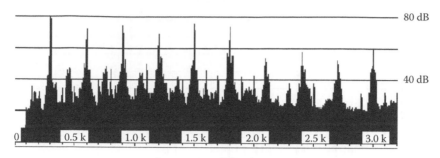

FIGURE 9.8 Noise spectrum (dBA) of a saturated AC machine.

high speeds (two and four poles); the stator magnetic circuit is saturated and consequences on noise are shown on the acoustic spectrum, which is composed of lines of frequency multiple of 300 Hz with particularly high levels.

9.3 Magnetic Noise of AC Electrical Rotating Machines

9.3.1 Description of the Phenomenon

9.3.1.1 Flux Density in the Air Gap

The electrical currents flowing in the wires generate in the air gap of the rotating machines a magnetic field, which acts on the stator and rotor iron. Three kinds of forces appear:

1. Tangential forces, which create torque and rotor rotation.
2. Magnetostrictive forces, negligible for rotating machines (magnetostriction is a property of ferromagnetic materials, which are deformed when they are submitted to a magnetic field; this phenomenon can be important with transformers).
3. Radial Maxwell forces. The radial component of the magnetic flux in the air gap of a magnetic circuit creates a force F_M that tends to attract stator and rotor. Its amplitude per area unit is given by

$$F_M = \frac{b^2}{2\mu_0} \qquad (9.1)$$

where
    b is the flux density at a given point of the stator internal surface
    μ_0 is the vacuum permeability ($4 \times \pi \times 10^{-7}$ H/m)

Those magnetic forces act essentially on the stator by deforming it and by creating vibrations. The rotor is less deformed because of its important rigidity; moreover, its surface is smallest. Then, the rotor vibrations are not taken into account to estimate magnetic noise.

The flux density in the air gap contains a fundamental component and a lot of harmonics generated by

- The spatial distribution of the coils in a finite number of slots, which affects the magnetomotive force (m.m.f.) waveform; those harmonics are called space harmonics
- The variable thickness of the air gap due to the slots, which leads to variable reluctance [9,10]
- The eventual eccentricity of the rotor, creating variable minimal value of the air gap thickness; due to radial forces, manufacturing, or aging of the bearings [11]
- The magnetic saturation of steel sheets, namely at the level of the teeth [8]
- The current harmonics due to the power supply (variable speed drive) [12]

Thus, to design a silent machine, flux density harmonics have to be minimized.

9.3.1.2 Force Waves

Let us consider a p pole pair machine. The flux density b can be expressed as

$$b = \sum_h b_h \tag{9.2}$$

where the harmonic b_h, whose pole pair number is hp, can be written as

$$b_h = \hat{b}_h \cos(\omega_h t - hp\alpha - \psi_h) \tag{9.3}$$

The amplitude $\hat{b}_h$, the angular frequency ω_h (frequency f_h), and the phase angle ψ_h are complex functions of h (e.g., for given h, ω_h can take several values). The angular position of any point in the air gap relatively to a fixed stator reference arbitrary chosen is denoted α. F_M given by (9.1), results from the relationship

$$F_M = \sum_m f_{mM} = \frac{\left(\sum_h b_h\right)^2}{2\mu_0} \tag{9.4}$$

In order to express f_{mM} let us introduce a flux density component with $h'p$ pole pair, which makes it possible to distinguish the different terms. It comes

$$F_M = \frac{1}{2\mu_0}\left[\sum_h \hat{b}_h^2 \cos^2(\omega_h t - hp\alpha - \psi_h) + \sum_h \sum_{h'} \hat{b}_h \hat{b}_{h'} \cos(\omega_h t - hp\alpha - \psi_h)\cos(\omega_{h'} t - h'p\alpha - \psi_{h'})\right] \tag{9.5}$$

Considering the second term (double product), h' and h have to take all the values but h must be different from h'. It can be deduced

$$F_M = \frac{1}{4\mu_0}\left[\sum_h \hat{b}_h^2 \left[1 + \cos(2\omega_h t - 2hp\alpha - 2\psi_h)\right]\right.$$

$$+ \sum_h \sum_{h'} \hat{b}_h \hat{b}_{h'} \left[\cos((\omega_h + \omega_{h'})t - (h+h')p\alpha - (\psi_h + \psi_{h'}))\right.$$

$$\left.\left. + \cos((\omega_h - \omega_{h'})t - (h-h')p\alpha - (\psi_h - \psi_{h'}))\right]\right] \tag{9.6}$$

It appears first that a squared term generates a constant pressure $\hat{f}_{hM} : \hat{f}_{hM} = \hat{b}_h^2/4\mu_0$. This quantity doesn't intervene in the noise definition because only the nonstationary pressure components generate magnetic noise. Let us note f_{mM} such a component, which presents the following general form:

$$f_{mM} = \hat{f}_{mM} \cos(\omega_m t - m\alpha - \psi_m) \tag{9.7}$$

where
    m is the pole pair force number, called mode number
    f_m is the force frequency
    $\omega_m = 2\pi f_m$ is the corresponding angular frequency
    $\hat{f}_{mM}$ is the force component amplitude (N/m²)
    ψ_m is a spatial angle

The forces waves (exactly pressure waves) rotate at ω_m/m angular speed. They generate, at a given point located at the external stator area, vibrations and then, variable air pressure responsible for noise.

Equation 9.6 shows two kinds of forces components f_{mM}: those due to $\hat{b}_h^2$ and those due to the double products $\hat{b}_h\hat{b}_{h'}$. The angular frequencies of the first ones are two times higher than the corresponding magnetic field angular frequencies. The angular frequencies of the second term are the result of the sum and the difference of the angular frequencies of each component. The magnetic noise is generally mainly caused by the second ones [3].

9.3.2 Deformation Modes

Parameter m needs to be considered seriously because it affects the mechanical response of the stator.

- For $m = 0$, the attraction between stator and rotor is uniform along the air gap. Stator vibration is uniform along its circumference at frequency f_m as shown in Figure 9.9: the stator at rest is drawn with full line and with a dotted line when the attraction is maximal.
- $m = 1$ is particular because the attraction between stator and rotor is maximal at one point and minimal at the opposite point. The rotor is off center as shown in Figure 9.10. The maximal attraction point rotates at the angular speed ω_m, creating an unbalanced mass very dangerous for noise and vibrations. This eccentricity leads to air gap thickness and flux density variations. This case is rare.
- For $m \geq 2$, the m points of maximal attraction between stator and rotor cause a deformation of the stator with $2m$ poles, which rotates at angular speed ω_m/m. Figure 9.11 shows deformations for $m = 2$ and 3. As it will be explained later, the deformation amplitude is inversely proportional to m^4.

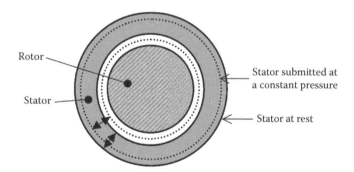

FIGURE 9.9   Stator deformations for $m = 0$.

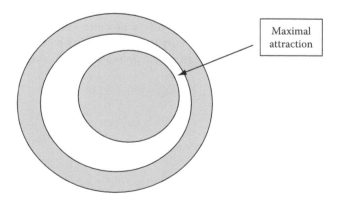

FIGURE 9.10   Rotor displacement for $m = 1$.

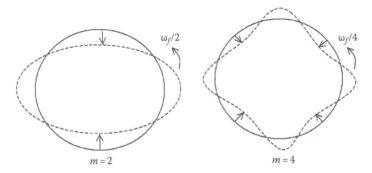

FIGURE 9.11 Stator deformation for $m = 2$ and 4.

9.3.3 Examples

9.3.3.1 15 kW Induction Machine

Let us consider a 15 kW, $p = 3$ induction machine supplied by the 50 Hz grid. For this example, only two components to define b (9.2) will be considered:

- The first one corresponds to the fundamental wave defined for $h = 1$: $\hat{b}_h = 0.7$T, $f_h = 50$ Hz.
- The second one describes a flux density harmonic such as $h' = -1$: $\hat{b}_{h'} = 0.005$T (0.71% of $\hat{b}_h$), $f_{h'} = 3370$ Hz.

The constant pressures take the numerical values: $\hat{f}_{hM} = 97{,}500$ N/m^2, $\hat{f}_{h'M} = 5$ N/m^2.

Equations 9.5 and 9.6 lead to the numerical values given in Table 9.1, which characterize the nonstationary force components (the phase angles are not considered).

The 5 N/m^2 $\hat{f}_{mM}$ component can be neglected. The 100 Hz frequency force has high amplitude and can produce vibrations, but not much noise because its frequency is low for human ear. The two last terms of 1400 N/m^2 amplitudes can generate noise because their amplitudes are sufficiently high, their frequencies are audible, and their mode numbers (0 and 6) are low.

Let us consider the constant pressure f_{hM} that results from $\hat{b}_h$ (97,500 N/m^2). As the considered machine presents a 0.118 m internal radius with a 0.16 m iron length, the internal stator surface area is 0.1186 m^2; it results that a radial force of 11,560 N acts on the stator. As the rated speed is 950 rpm, the rated torque is about 150 N m that leads to a tangential force close to 1270 N. So, the radial force is largely higher than these, allowing the rotation of the rotor.

9.3.3.2 Synchronous Machine Supplied with a PWM Inverter

Let us consider the case of a three-phase, $p = 4$ synchronous machine operating at 50 Hz frequency with the PWM frequency $f_w = 3$ kHz. The aim is to define the m values that concern the 5900, 6000, and 6100 noise lines (see Figure 9.12). Stator current analysis shows preponderant three-phase harmonics currents at 5950 and 6050 Hz, respectively as clockwise and anticlockwise systems (classical result for such an inverter). Each of them generates four density waves, which the most important component correspond to the fundamental term, so a four-pole pair wave.

TABLE 9.1 Example of Pressure Waves

Magnitude, $\hat{f}_{mM}$	$\dfrac{\hat{b}_h^2}{4\mu_0} = 97{,}500$ N/m^2	$\dfrac{\hat{b}_{h'}^2}{4\mu_0} = 5$ N/m^2	$\dfrac{\hat{b}_h^2 \times \hat{b}_{h'}^2}{2\mu_0} = 1{,}400$ N/m^2	
Frequency, f_m	$2f_h = 100$ Hz	$2f_{h'} = 6{,}740$ Hz	$f_h + f_{h'} = 3{,}420$ Hz	$f_h - f_{h'} = 3{,}370$ Hz
Mode, m	$2h \times p = 6$	$2h' \times p = 6$	$P(h + h') = 0$	$p(h - h') = 6$

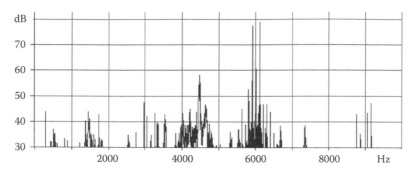

FIGURE 9.12 Acoustic pressure level (dBA) at 1 m of a machine fed by a 3 kHz PWM inverter.

TABLE 9.2 Pressure Waves of a Machine Supplied by a PWM Inverter

	h, h'		h, h''		h', h''	
Magnitude, $\hat{f}_{mM}$	1,950	1,950	1,950	1,950	19.5	19.5
Frequency, f_m	6,000	5,900	6,000	6,100	12,000	100
Mode, m	8	0	8	0	0	8

Let us introduce, as it was done for h', a similar quantity denoted h''. Let us define b using only three components defined as follows:

$$\left. \begin{array}{l} h = 1, f_h = 50 \text{ Hz}, \hat{b}_h = 0.7\text{T} \\ h' = 1, f_{h'} = 5{,}950 \text{ Hz}, \hat{b}_{h'} \approx 0.01 \times \hat{b}_h = 0.007\text{T} \\ h'' = -1, f_{h''} = 6{,}050 \text{ Hz}, \hat{b}_{h''} \approx 0.01 \times \hat{b}_h = 0.007\text{T} \end{array} \right\} \quad (9.8)$$

It can be deduced the constant pressures: $\hat{f}_{hM} = 97{,}500 \text{ N/m}^2$, $\hat{f}_{h'M} = \hat{f}_{h''M} = 9.75 \text{ N/m}^2$.

The f_{mM} quantities resulting from the squared terms present the following characteristics:

$$\left. \begin{array}{l} h = 1, f_m = 100 \text{ Hz}, \hat{f}_{mM} = 97{,}500 \text{ N/m}^2, m = 8 \\ h' = 1, f_m = 11{,}900 \text{ Hz}, \hat{f}_{mM} = 9.75 \text{ N/m}^2, m = 8 \\ h'' = -1, f_m = 12{,}100 \text{ Hz}, \hat{f}_{mM} = 9.75 \text{ N/m}^2, m = -8 \end{array} \right\} \quad (9.9)$$

The f_{mM} components deduced from the double products are presented together in the Table 9.2. It appears that the 5900 and 6100 Hz pressure waves have a 0 mode. The 6000 Hz component is an $m = 8$ mode; it is obtained by adding two pressure waves. The observed noise lines are probably generated by the pressure waves of the Table 9.2.

9.4 Mechanical and Acoustic Modeling

The characteristics of an f_{mM} force component and the stator design make it possible to estimate the vibration amplitude and the corresponding noise. First, it is calculated the amplitude Y_{ms} of the static distortion. Second, the vibration amplitude Y_{md} is determined taking the mechanical resonance frequencies into account. At last, the acoustic noise is estimated. Most of the given mechanical expressions come from the beam theory [2,3].

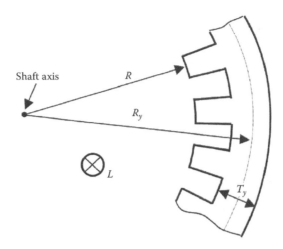

FIGURE 9.13 Notations for the stator frame.

9.4.1 Amplitudes of Static Distortions

9.4.1.1 Static Distortions

For given $\hat{f}_{mM}$, the relationships that define Y_{ms} depend on the m values. They are given by (9.10), (9.11), and (9.12), respectively for $m = 0$, 1, and $m \geq 2$. The following notations are used (Figure 9.13):

- R, internal radius of the stator
- R_y, yoke average radius
- T_y, yoke radial thickness
- L, iron length
- L_s, distance between rotor shaft supports
- d, shaft diameter
- E, elasticity coefficient or Young's modulus: $E = 2.1 \times 10^{11}$ N/m^2 for iron

$$Y_{0s} = \frac{RR_y \hat{f}_{mM}}{ET_y} \tag{9.10}$$

$$Y_{1s} = \frac{4RL_s^3 L \hat{f}_{mM}}{3Ed^4} \tag{9.11}$$

$$Y_{ms} = \frac{12RR_y^3 \hat{f}_{mM}}{ET_y^3 (m^2 - 1)^2} \tag{9.12}$$

For $m \geq 2$, Y_{ms} decreases with m^4. Forces of high modes can have difficulty generating vibrations and noise. Practically, it is not useful to consider forces having mode number higher than 8.

9.4.1.2 Considerations about the Pole Pair Number

In a general way, the yoke width of a AC machine is inversely proportional to p. Ph.L. Alger gives, very roughly, (9.13) and (9.14) [13]:

$$T_y \approx \frac{2R}{5p} \tag{9.13}$$

$$R_y \approx 1.4R \tag{9.14}$$

Replacing T_y and R_y in (9.10) and (9.11), leads to the following quantities:

$$Y_{0s} = \frac{3.5Rp\hat{f}_{mM}}{E} \tag{9.15}$$

$$Y_{ms} = \frac{514.5Rp^3\hat{f}_{mM}}{E(m^2-1)^2} \tag{9.16}$$

It can be observed that the static distortion amplitudes are proportional to R and p or p^3. It is well known that a machine with a high pole number has a large R. So, in the best case, supposing R is constant with p variations, the distortion amplitudes are directly linked to p for $m = 0$ and p^3 for $m \geq 2$. It means that in a $p = 2$ machine, compared to $p = 1$, a force wave with the same characteristics generates deformations at least eight times more important. For $p = 3, 4$, and 5, the deformations are respectively at least 27, 64, and 125 times higher.

Machines with high pole number have a large diameter and a small yoke width. As a consequence, they are less rigid and they can vibrate and create noise more easily.

9.4.2 Resonance Frequencies and Vibration Amplitudes

Every machine has a lot of own frequencies; each of which is associated to a vibration mode. Only one hammer impact can excite these modes. The stroke leads to a noise made of many distinct frequencies corresponding to natural resonance frequencies. Consequently, if a force frequency is close to a resonance frequency, the vibration amplitude increases. Mechanical phenomena are particularly complex and it is difficult to find simple and accurate analytical equations. Following relations of resonance frequencies have been given by Jordan and Timar [2,3]. These general laws are not very accurate but they give an easy location of dangerous zones on the spectrum. Their determination is based on the beam theory [14]. Proposed equations consider a machine as a perfect cylinder without taking into account, for instance, elements like feet that change the natural frequencies [15–17].

9.4.2.1 Resonance Frequencies

Two kinds of resonance frequencies, noted f_m^{s*}, which concerns generally radial vibrations, can be distinguished according to $m = 0$ (9.17), or $m \geq 2$ (9.18):

$$f_0^{s*} = \frac{837.5}{R_y\sqrt{\Delta}} \tag{9.17}$$

$$f_m^{s*} = \frac{f_0^{s*}T_y m(m^2-1)}{2\sqrt{3}R_y\sqrt{m^2+1}} \tag{9.18}$$

where Δ = (weight of yoke + weight of teeth)/weight of yoke. It is difficult to estimate the weight of teeth and yoke, but it is easy to calculate Δ by estimating the surface of elements on a horizontal frame section.

A machine with a high p has a large radius and so a low resonance frequency.

9.4.2.2 Vibration Amplitudes

Amplitude of dynamic vibrations Y_{md} is obtained by multiplying Y_{ms} by a magnification factor η_m depending on frequencies:

$$Y_{md} = \eta_m Y_{ms} \tag{9.19}$$

Introducing $\Delta_f = f_m/f_m^{s*}$, leads to define η_m as following:

$$\eta_m = \left[\left(1 - \Delta_f^2\right)^2 + (2\xi_a\Delta_f) \right]^{-0.5} \tag{9.20}$$

ξ_a is an absorption coefficient difficult to estimate. Generally, for an induction motor, $0.01 < \xi_a < 0.04$. Its value is low and can often be neglected because it interferes only when the force frequency is close to a resonance. In fact, ξ_a avoids that η_m, and so the vibration amplitude, tends to an infinite value, which is physically impossible. The coefficient ξ_a is small if a structure continues to vibrate a long time after a hammer impact (e.g., a bell has a low absorption coefficient).

For low f_m values ($f_m \ll f_m^{s*}$), $\eta_m \rightarrow 1$. For high f_m ($f_m \gg f_m^{s*}$), $\eta_m \rightarrow 0$. So low f_m^{s*} values seem to be better, but they occur with large machines and they are in audible frequencies. The external frame around the magnetic sheets can modify slightly all the equations but it can be neglected in a first approach.

9.4.3 Acoustic Radiations of Electrical Machines

9.4.3.1 Acoustic Notions

9.4.3.1.1 Acoustic Pressure

Vibrations of a material generate vibrations of air particles, so variations of air pressure. If air particle oscillations are time sinusoidal waves y_a of amplitude $\hat{y}_a$ and frequency f_a (angular frequency ω_a), it comes: $y_a = \hat{y}_a \sin(\omega_a t)$. The instantaneous speed v_a is given by $v_a = \omega_a \hat{y}_a \cos(\omega_a t)$. The rms speed is $v_a = \omega_a \hat{y}_a/\sqrt{2}$. Air pressure variations of instantaneous value p_a and rms value P_a (called sound pressure or acoustic pressure), are tied to v_a, expressed in m/s, according to (9.21) where $\underline{Z}$ is the complex acoustic impedance. In the air, in free field (a space without sound reflections), $\underline{Z}$ is a real term such as $Z \approx 415\,\text{kg/m}^2\text{/s}$ (at 20°C and for an atmospheric pressure of 1013 hPa):

$$p_a = \underline{Z} v_a \tag{9.21}$$

Vibration of a particle is transmitted to next particles with the sound speed c (sound wave propagation). As $c = 344\,\text{m/s}$ at 20°C in the air, the wave length λ_a is defined by $\lambda_a = c/f_a$.

9.4.3.1.2 Acoustic Intensity

Pressure and speed of air particles can have different directions, which changes the propagation of the sound wave. The acoustic intensity is a vector; it allows defining the amplitude and also the direction of the sound. Acoustic intensity is the flux of sound energy per area unit. It corresponds to the average rate of sound energy transmitted through a unit area, perpendicular to the direction of travel of the sound. The modulus of this vector, denoted I_a, and measured in W/m², is expressed in

$$I_a = \frac{1}{T_a} \int_0^{T_a} p_a \times v_a \, dt \tag{9.22}$$

In the air, by replacing Z, considering $T_a = 1/f_a$, I_a becomes

$$I_a = 2\pi^2 Z f_a^2 \hat{y}_a^2 \approx 8200 f_a^2 \hat{y}_a^2 \tag{9.23}$$

I_a can also be expressed in function of V_a: $I_a = ZV_a^2$. So, the acoustic intensity is proportional to the square of the rms vibration speed.

9.4.3.1.3 Acoustic Power

Acoustic power, measured in Watt, defines a sound source and doesn't depend on the environment. Sound pressure or acoustic intensity, which is measured at distance of a noisy machine, are different if the machine is, for instance, in a reflective room or outdoors. Standards define the maximal acoustic power of electrical machines. The acoustic power W_a is obtained by integrating I_a through a surface S around the sound source:

$$W_a = \int_S \vec{I_a}\ d\vec{S} \tag{9.24}$$

9.4.3.1.4 Use of Decibels

There is a notable difference between the lowest pressure variation P_{a0} (rms value) that can be heard by a human ear (about $20\,\mu\text{Pa}$), and the ache point (about $100\,\text{Pa}$). Sound is measured in decibels taking P_{a0} as for reference. The level of acoustic pressure $L(P_a)$ is defined as

$$L(P_a) = 10\log\left(\frac{P_a}{P_{a0}}\right)^2 = 20\log\left(\frac{P_a}{P_{a0}}\right) \tag{9.25}$$

The level of acoustic intensity $L(I_a)$ is defined by

$$L(I_a) = 10\log\left(\frac{I_a}{I_{a0}}\right) \tag{9.26}$$

$I_{a0} = 10^{-12}\ \text{W/m}^2$ is the perception threshold of the human ear. In free field, pressure levels and intensity levels are the same. The level of acoustic power $L(W_a)$ of a sound source is given by

$$L(W_a) = 10\log\left(\frac{W_a}{W_{a0}}\right) \tag{9.27}$$

with $W_{a0} = 10^{-12}\ \text{W}$, which is the source power with a uniform acoustic intensity $I_0 = 10^{-12}\ \text{W/m}^2$ through a surface of $1\,\text{m}^2$.

9.4.3.1.5 Human Ear

The constant 10 in the previous equations has been chosen so that a pressure variation of 25%, which is the smallest audible variation for the human ear, leads to a variation of $1\,\text{dB}$.

When many sounds with different frequencies occur together, the resultant acoustic pressure is the square root of the sum of the square for each pressure. Then, if two sounds with different frequencies at the same levels are present, the resulting pressure level is $3\,\text{dB}$ higher than for each sound. But the human ear does not hear so well every frequency. The bandwidth of the human ear goes approximately from 20 to 16,000 Hz. Frequencies that are the best heard are included between 1,000 and 5,000 Hz. Young people can hear higher frequencies than old people. The systems of acoustic measurement can take those phenomena into account with curves noted A, B, C, or D and the unities are dBA, dBB, dBC, or dBD. The most common quantity corresponds to dBA.

9.4.3.2 Acoustic Radiations of Electrical Machines

The f_m and Y_{md} determinations allow the acoustic power and intensity estimations. The acoustic intensity $I_{a(S)}$ at the surface S_e of the machine results from (9.23): considering one force component with a mode number m, it comes

$$I_{a(S)m} = 8200\sigma_m f_m^2 Y_{md}^2 \tag{9.28}$$

σ_m indicates the capacity of the machine, relating to its size, to be a good loudspeaker to emit the sound of λ_{am} wavelength. A large loudspeaker is better to radiate low frequencies. σ_m is difficult to estimate. Some authors consider that the machine is similar to a sphere [3] or a cylinder [13]. In a simplified way, σ_m can be expressed as

$$\sigma_m = 1 - \exp\left(\frac{-\pi D_e}{\lambda_{am}}\right) \tag{9.29}$$

where D_e is the machine external diameter. For D_e large referred to as λ_{am}, σ_m tends to 1.

The $W_{a(S)m}$ acoustic power is the result of the product of $I_{a(S)m}$ by S_e:

$$W_{a(S)m} = I_{a(S)m} S_e \tag{9.30}$$

In decibels, the acoustic power level is

$$LW_{a(S)m} = 10\log\left(\frac{8200\sigma_m f_m^2 Y_{md}^2 S_e}{10^{-12}}\right) \tag{9.31}$$

To calculate the acoustic intensity $I_{a(x)m}$ at a x distance from the sound source, a vibrating sphere in a free field can be considered. As the x radius sphere surface is $4\pi x^2$, it comes

$$I_{a(x)m} = \frac{W_{a(S)m}}{4\pi x^2} = \frac{8200\sigma_m f_m^2 Y_{md}^2 S_e}{4\pi x^2} \tag{9.32}$$

The corresponding acoustic intensity level in dB can be deduced:

$$LI_{a(x)m} = 10\log\left(\frac{I_{a(x)m}}{10^{-12}}\right) = 159.14 + 20\log(f_m Y_{md}) + 10\log(\sigma_m) + 10\log\left(\frac{S_e}{4\pi x^2}\right) \tag{9.33}$$

9.5 Flux Density Harmonics of AC Machines

As explained previously, the magnetic noise is generated by the combination of flux density harmonics. Different ways can be used to determine those harmonics; they are currently estimated by finite element software or by analytical methods. Next paragraph presents such an analytical method for AC machines.

The radial component b of the air gap flux density is obtained by multiplying the ε magnetomotive force applied to the air gap by the Λ per unit area air gap permeance

$$b = \Lambda\varepsilon \tag{9.34}$$

The determination of ε is not a problem when the iron permeability is supposed to be infinite and the eccentricity is neglected. The difficulty consists in the determination of Λ. Approximate expressions exist for a long time in the literature. Timar [2] gives an expression, which neglects the interactions between stator and rotor slots. Alger [13] takes them into account by the mean of only one single term, which corresponds to the fundamental component. It has been shown on several occasions that certain effects, like the magnetic noise [18], are generally mainly tributary of higher-rank components that convey the interactions between stator and rotor slots. This aspect requires presenting an expression of Λ established during the years 1980 [19]. The complete theoretical approach is given in [10]. Let us specify that a similar expression of Λ was presented in 1992 [9] by considering unit slot depths.

9.5.1 Magnetomotive Force Harmonics

Let us consider a one-pole pair three-phase stator with one coil per phase as shown in Figure 9.14.

The windings are connected to a three-phase grid. Every individual coil produces a magnetic flux through the air gap that creates a resultant rotating field. The m.m.f. is the difference of the magnetic potential in the air gap where almost the ampere-turns are consumed.

- For a current i^s, each coil with z^s turns produces, along the air gap, an m.m.f. ε^s, whose amplitude is $\pm z^s i^s/2$, as shown in Figure 9.15 where α is the angular position along the air gap. Denoting h^s the harmonic rank (h^s takes only odd values), the corresponding amplitude is $4z^s i^s/2h^s\pi$.

Aiming to limit harmonic amplitudes, machine designers distribute z^s turns in m^s coils with z^s/m^s turns (m^s is the number of slots per pole and per phase). Then, the m.m.f. results in a sum of rectangular waves as shown in Figure 9.16 for $m^s = 2$. $K_{h^s}^s$, which is the winding distribution factor, defines the decrease of each harmonic. For a three-phase machine, $K_{h^s}^s$ is given by (9.35):

$$K_{h^s}^s = \frac{\sin(h^s\pi/6)}{m^s \sin(h^s\pi/6m^s)} \tag{9.35}$$

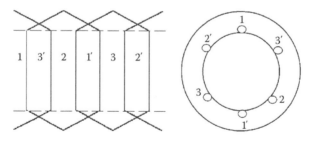

FIGURE 9.14 Windings of a three-phase machine stator.

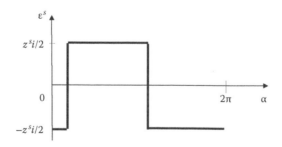

FIGURE 9.15 Magnetomotive force of one coil.

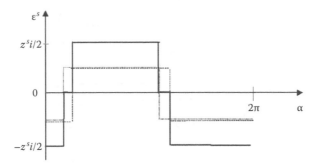

FIGURE 9.16 Magnetomotive force with two slots per pole and per phase.

As the number of slots per pole and per phase can't be infinite (m^s is generally included between two and four), the m.m.f. harmonics exist and they are called space harmonics. If there are p pole pairs, replacing i^s by $I^s\sqrt{2}\cos(\omega t)$, the m.m.f. created by a phase is expressed as

$$\sum_{h^s}\frac{4}{h^s\pi}\frac{z^s}{2}K^s_{h^s}I^s\sqrt{2}\cos(\omega t)\cos(h^s p\alpha)$$

Adding the m.m.f. created by each winding and taking care of their spatial distribution ($2\pi/3$), the air gap m.m.f. created by the single-layer stator windings is

$$\varepsilon^s(\alpha)=H^sI^s\sum_{h^s}G^s_{h^s}\cos(\omega t-h^s p\alpha)\tag{9.36}$$

with

$$G^s_{h^s}=(-1)^{(h^s-1)/2}\frac{K^s_{h^s}}{h^s}$$

and

$$H^s=\frac{3\sqrt{2}z^s}{\pi}.$$

Calculations show that the terms relative to h^s multiple of 3 are null so that $h^s \in [1,-5,7,-11,13,-17,19,-23,\ldots]$. Let us point out that $G^s_1\cong 1$; while $\left|G^s_{h^s}\right|\ll 1$ for $h^s\neq 1$.

9.5.2 Air Gap Permeance Harmonics

The stator wires, whose magnetic permeability is equivalent to the air permeability, are located in slots. An induction machine with a wounded rotor has slots at the rotor (Figure 9.17). For machines with squirrel cages, holes on the rotor can be considered as slots with a low magnetic permeability. Equivalent slots can also be defined for synchronous machines or switched reluctance machines [20].

Therefore, the thickness of air gap is not constant and equal to the minimal air gap width called g. The shape of real slots is rather complex and a simplified developed model is shown in Figure 9.18 [10,21].

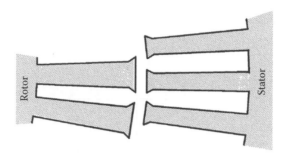

FIGURE 9.17 Stator and rotor slots of an induction machine with a wounded rotor.

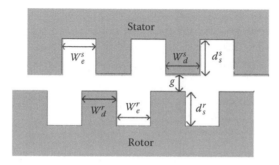

FIGURE 9.18 Simplified model of slots.

The model supposes that magnetic flux lines have a radial direction. The permeance is inversely proportional to the air gap thickness. By developing it in Fourier series, the permeance per area unit $\Lambda(\alpha, \theta)$ (9.37) is obtained [10]:

$$\Lambda(\alpha,\theta) = \mu_0 A_{00} + 2\mu_0 A_{s0} \sum_{k_s=1}^{\infty} f(k_s)\cos(k_s N_t^s \alpha) - 2\mu_0 A_{0r} \sum_{k_r=1}^{\infty} f(k_r)\cos(k_r N_t^r \alpha - k_r N_t^r \theta)$$

$$+ 2\mu_0 A_{sr} \sum_{k_s=1}^{\infty} \sum_{k_r=1}^{\infty} f(k_s)f(k_r)\left\{\cos\left[(k_s N_t^s - k_r N_t^r)\alpha + k_r N_t^r \theta\right]\right.$$

$$\left. + \cos\left[(k_s N_t^s + k_r N_t^r)\alpha - k_r N_t^r \theta\right]\right\} \tag{9.37}$$

where
w_e^s, w_d^s are respectively the width of one stator slot and one stator tooth
d_s^s is the fictitious depth of one stator slot, defined [13,22] as: $d_s^s = w_e^s/5$
r_t^s is the stator slotting ratio: $r_t^s = w_d^s/(w_d^s + w_e^s)$
$f(k_s)$ is the stator slotting function: $f(k_s) = (\sin k_s r_t^s \pi)/2k_s$
$w_e^r, w_d^r, d_s^r, r_t^r, f(k_r)$, are the analogous quantities relative to the rotor
k_s and k_r are integers that take all the values between $-\infty$ and $+\infty$
g is the minimal air gap thickness
g_M is the maximal air gap fictitious thickness: $g_M = g + d_s^s + d_s^r$
g^s and g^r are intermediate fictitious air gap thicknesses expressed by $g + d_s^s$ and $g + d_s^r$, respectively
N_t^s, N_t^r are the total number of stator and rotor slots (or bars)
θ characterizes the angle between a stator and a rotor reference, it depends on time t. For example, for an induction machine: $\theta = (1 - s)\omega t/p + \theta_0$ (s is the slip and θ_0 depends on the loading state of the machine).

The geometric parameters, which characterize the slotting effect, are given by

$$
\left.
\begin{aligned}
A_{00} &= \frac{\left[1 + d_s^s r_t^s / g^r + d_s^r r_t^r / g^s + d_s^s d_s^r (g + g_M) r_t^s r_t^r / g g^s g^r \right]}{g_M} \\[2mm]
A_{s0} &= \frac{2 d_s^s \left[1 + d_s^r (g + g_M) r_t^r / g g^s \right]}{\pi g_M g^r} \\[2mm]
A_{0r} &= \frac{2 d_s^r \left[1 + d_s^s (g + g_M) r_t^s / g d^r \right]}{\pi g_M g^s} \\[2mm]
A_{sr} &= \frac{4 d_s^s d_s^r \left[g_M + g \right]}{\pi^2 g g^s g^r g_M}
\end{aligned}
\right\}
\tag{9.38}
$$

The permeance expression (9.37) contains four groups of terms:

1. A constant term depending on A_{00} (it would be equal to $1/g$ with a constant air gap)
2. Terms depending on A_{s0}, linked to the stator slots
3. Terms depending on A_{0r}, linked to the rotor slots
4. Terms depending on A_{sr}, linked to the interaction between stator and rotor slots

In order to estimate qualitatively, the relative importance of the different terms, the following inequalities can be pointed out: $A_{s0} \cong A_{0r}$; $A_{00} > A_{s0}$ or A_{0r}; A_{s0} or $A_{0r} > A_{sr}$.

The permeance expression (9.37) can be used with all AC machines (induction, synchronous, or switched reluctance machines). From a qualitative point of view, knowing the number of slots is enough to obtain all the harmonics. From a quantitative point of view, the slot dimensions have to be adapted.

9.5.3 Flux Density Harmonics

The stator radial flux density waves in the air gap result from the product $\Lambda(\alpha, \theta) \varepsilon^s(\alpha)$. The same method gives the rotor radial flux density waves. Let us introduce the number of slots per pole pair: $N^r = N_t^r / p$, $N^s = N_t^s / p$.

9.5.3.1 Stator Flux Density Harmonics

Four types of stator flux density harmonics are obtained: stator flux density harmonics independent of the rotor (linked to A_{00} and A_{s0}) and stator flux density dependant on the rotor (linked to A_{0r} and A_{sr}).

9.5.3.1.1 Stator Harmonics Independent of the Rotor

They come from the m.m.f. harmonics (called space harmonics) and from the stator slots:

$$
b_{h^s 0}^s(\alpha, t) = \hat{b}_{h^s 0}^s \cos(\omega t - h^s p \alpha)
\tag{9.39}
$$

$$
\hat{b}_{h^s 0}^s = H^s I^s \mu_0 \left\{ A_{00} G_{h^s}^s + A_{s0} \sum_{\substack{k_s = -\infty \\ k_s \neq 0}}^{+\infty} G_{h_*^s}^s f(k_s), \quad h_*^s = h^s + k_s N^s \right\}
\tag{9.40}
$$

Those harmonics, which rank the same as the m.m.f. ones (1, −5, 7, −11, 13, −17, 19, …) can have an important amplitude depending on the stator slots number. For example, a two-pole pair machine with

$N_t^s = 36$ stator slots, the harmonics, which rank h^s are $-17, 19, -35, 37, \ldots$, have particularly important amplitudes.

9.5.3.1.2 Stator Harmonics Dependent on the Rotor

They come from rotor slots and interaction between both teeth:

$$b_{h^s k_r}^s(\alpha, t) = \hat{b}_{h^s k_r}^s \cos((1 - k_r N^r(1 - s))\omega t - (h^s - k_r N^r)p\alpha - pk_r N^r \theta_0) \tag{9.41}$$

$$\hat{b}_{h^s k_r}^s = H^s I^s \mu_0 f(k_r) \left[A_{0r} G_{h^s}^s + A_{sr} \sum_{\substack{k_s = -\infty \\ k_s \neq 0}}^{+\infty} G_{h_*^s}^s f(k_s), \quad h_*^s = h^s + k_s N^s \right] \tag{9.42}$$

The angular frequency $(1 - k_r N^r(1 - s))\omega$ of those harmonics is not of the grid: it is linked to the rotor slot number and, with an induction machine, to the slip (then to the rotor speed; s is equal to 0 for a synchronous machine). In a general way, these amplitudes are lower than those of the stator flux density harmonics independent of the rotor.

9.5.3.2 Rotor Flux Density Harmonics

With a synchronous or a switched reluctance machine, a good representation of the harmonics in the air gap is obtained by only taking into account the flux density components created by the stator.

In an induction machine with a wounded rotor, the harmonics created by the rotor have the same rank and frequencies as those from the stator. So, a good qualitative representation of the flux density in the air gap is obtained by only taking into account the flux density components created by the stator.

For an induction machine with a cage rotor, the rotor harmonic ranks can be different from those of the stator, new flux density harmonics are created by the rotor and must be taken into account [23,24]. Then two types of rotor flux density harmonics exist, those independent of the stator (linked to A_{00} and A_{0r}) and those dependant on the stator (linked to A_{s0} and A_{sr}). For a real good accuracy, the flux density created by the fundamental current and also the harmonic rotor currents should be considered. Nevertheless, this paragraph only considers the most important flux density harmonics, which are generated by the rotor fundamental current (induced by the fundamental of the stator flux density, corresponding to $h^s = 1$).

9.5.3.2.1 Rotor Harmonics Independent of the Stator (Cage Rotor Induction Machine)

They come from the m.m.f. harmonics (called space harmonics) and from the rotor slots:

$$b_{h^r 0}^r(\alpha, t) = \hat{b}_{h^r 0}^r \cos\left\{ \left[1 + iN^r(1 - s) \right] \omega t - h^r p\alpha + iN^r p\theta_0 - \frac{\pi}{2} - \text{Arg}(\bar{Z}_1) \right\} \tag{9.43}$$

$$\begin{aligned}
\hat{b}_{h^r 0}^r &= H^r I_1^r \mu_0 \left[A_{00} G_{h^r}^r + A_{0r} \sum_{\substack{k_r = -\infty \\ k_r \neq 0}}^{+\infty} G_{h_*^r}^r f(k_r), \quad h_*^r = h^r + k_r N^r \right] \\
H^r &= \frac{N_t^r}{\pi\sqrt{2}} \\
G_{h^r}^r &= \frac{(-1)^{(h^r - 1)/N^r}}{h^r p}
\end{aligned} \tag{9.44}$$

$$h^r = iN^r + 1 \quad (i = 0, \pm 1, \pm 2, \pm 3, \pm 4\ldots), \tag{9.45}$$

I_1^r is the rms value of the rotor fundamental current, varying with the load and $\mathrm{Arg}(\overline{Z}_1)$ its phase angle.

Equation 9.45 shows that the number of rotor bars has an influence on the cage rotor harmonic existence, so a good choice of the slot number is very important to avoid magnetic noise [25,26]. For example, if the number of rotor bars is 34 with $p = 2$, the ranks h^r of the first rotor harmonic are −16 and 18. They will interfere with the stator harmonics −17 and 19 and create force waves with a mode number 2, as shown in (9.6).

9.5.3.2.2 Rotor Harmonics Dependent on the Stator (Cage Rotor Induction Machine)

Those harmonics are tied to the stator slots and to the interaction between stator and rotor slots:

$$b^r_{h^r k_s}(\alpha, t) = \hat{b}^r_{h^r k_s} \cos\left\{ \left[1 + iN^r(1-s) \right]\omega t - (h^r - k_s N^s)p\alpha + iN^r p\theta_0 - \frac{\pi}{2} - \mathrm{Arg}(\overline{Z}_1) \right\} \tag{9.46}$$

$$\hat{b}^r_{h^r k_s} = H^r I_1^r \mu_0 f(k_s) \left\{ A_{s0} G^r_{h^r} + A_{sr} \sum_{\substack{k_r = -\infty \\ k_r \neq 0}}^{+\infty} G^r_{h^r_\star} f(k_r), \quad h^r_\star = h^r + k_r N^r \right\} \tag{9.47}$$

h^r is always given by (9.45). More details can be found in reference [18]. In a general way, these amplitudes are smaller than those of the rotor flux density harmonics independent of the stator.

9.6 Conclusion

The acoustic noise of electrical machines is generally due to aerodynamic phenomena for high-speed machines. In such a case, it is difficult to avoid it. The noise of electromagnetic origin appears often with high pole pair number or when the machine is fed by a PWM inverter, but it can also appear with a bad choice of slot numbers. Designers have to take into account the phenomenon explained in this chapter. The given equations permit to theoretically estimate the vibrations and noise and so to avoid it. For a given machine, all the space and slots harmonics can be estimated. It is necessary to search which combinations can create pressure waves of audible frequency, important amplitude, and low mode number.

Then, it is possible to avoid important noise and vibrations of magnetic origin [27]. When a noisy machine has been made, active noise reduction methods can permit to decrease noise [20,28].

References

1. W.R. Finley. Noise in induction motors—Causes and treatments. *IEEE Transactions on Industry Applications*, 27(6), 1204–1213, November/December.
2. P.L. Timar, A. Fazekas, J. Kiss, A. Miklos, and S.J. Yang. *Noise and Vibration of Electrical Machines*. Elsevier, Amsterdam, the Netherlands, 1989.
3. H. Jordan. *Geräuscharme elektromotoren*. W. Girardet, Essen, Germany, 1950.
4. J. Bonal. *Utilisation industrielle des moteurs à courant alternatif*. Technique & Documentation, Paris, France, 2001.
5. C.-Y. Wu and C. Pollock. Acoustic noise cancellation techniques for switched reluctance drives. *IEEE Transactions on Industry Applications*, 33(2), 477–484, March/April 1997.

6. D.E. Cameron, J.H. Lang, and S.D. Umans. The origin and reduction of acoustic noise in doubly salient variable-reluctance motors. *IEEE Transactions on Industry Applications*, 28(6), 1250–1255, November/December 1992.

7. R.S. Colby, F.M. Mottier, and T.J.E. Miller. Vibration modes and acoustic noise in a four-phase switched reluctance motor. *IEEE Transactions on Industry Applications*, 32(6), 1357–1364, November/December 1996.

8. J.C. Moreira and T.A. Lipo. Modeling of saturated AC machines including airgap flux harmonic components. *IEEE Transactions on Industry Applications*, 28(2), 343–349, March-April 1992.

9. H. Hesse. Air gap permeance in doubly slotted asynchronous machines. *IEEE Transactions on Energy Conversion*, 7(3), 491–499, September 1992.

10. J.F. Brudny. Modelling of induction machine slotting: Resonance phenomenon. *Journal de Physique III*, JP, III, 1009–1023, Mai 1997.

11. S. Ayari, M. Besbes, M. Lecrivain, and M. Gabsi. Effects of the airgap eccentricity on the SRM vibrations. *International Conference on Electric Machines and Drives 1999 (IEMD'99)*, Seattle, WA, May 1999, pp. 138–140.

12. R.J.M. Belmans, D. Verdyck, W. Geysen, and R.D. Findlay. Electro-mechanical analysis of the audible noise of an inverter-fed squirrel cage induction motor. *IEEE Transactions on Industry Applications*, 27(3), 539–544, May/June 1991.

13. Ph.L. Alger. *The Nature of Induction Machines*, 2nd edn. Gordon & Breach Publishers, New York, 1970.

14. S.P. Timoshenko and J.N. Goodier. *Theory of Elasticity*, 3rd edn., International Student Edition, McGraw Hill, New York, 1970.

15. J.Ph. Lecointe, R. Romary, and J.F. Brudny. A contribution to determine natural frequencies of electrical machines. Influence of stator foot fixation, in S. Wiak, M. Dems, and K. Komęza (eds.) *Recent Developments of Electrical Drives*, Springer, Dordrecht, the Netherlands, 2006, pp. 225–236.

16. S. Wanatabe, S. Kenjo, K. Ide, F. Sato, and M. Yamamoto. Natural frequencies and vibration behaviour of motor stators. *IEEE Transactions on Power Apparatus and Systems*, 102(4), 949–956, April 1983.

17. S.P. Verma and A. Balan. Measurements techniques for vibrations and acoustic noise of electrical machines. *Sixth International Conference on Electrical Machines and Drives*, IEE, London, U.K., 1993, pp. 546–551.

18. B. Cassoret, R. Corton, D. Roger, and J.F. Brudny. Magnetic noise reduction of induction machines. *IEEE Transactions on Power Electronics*, 18(2), 570–579, March 2003.

19. J.F. Brudny. Etude quantitative des harmoniques de couple du moteur asynchrone triphasé d'induction. Habilitation thesis, Lille, France, 1991, No. H29.

20. J.Ph. Lecointe, R. Romary, J.F. Brudny, and M. McClelland. Analysis and active reduction of vibration and acoustic noise in the switched reluctance motor. *IEEE Proceedings on Electric Power Applications*, 151(6), 725–733, November 2004.

21. D. Belkhayat, J.F. Brudny, and Ph. Delarue. Fictitous slot model for more precise determination of asynchronous machine torque harmonics. *Proceedings of the IMACS MCTS*, Lille, France, May 1991, pp. 230–235.

22. F.W. Carter. Air-gap induction. *Electrical World and Engineer*, 38(22), 884–888, November 1901.

23. M. Poloujadoff. General rotating m.m.f. theory of the squirrel-cage induction machines with non uniform air-gap and several non sinusoidally distributed windings. *IEEE Transactions on Power Apparatus and Systems*, 95, 583–591, 1982.

24. S. Nandi. Modeling of induction machines including stator and rotor slot effects. *IEEE Transactions on Industry Applications*, 40(4), 1058–1065, July–August 2004.

25. G. Kron. Induction motor slot combinations: Rules to predetermine crawlin, vibration, noise and hooks in the speed torque curves. *AIEE Transactions*, 50, 757–768, 1931.

26. R.P. Bouchard and G. Olivier. *Conception de moteurs asynchrones triphasés*. Presses Internationale Polytechnique, Montreal, Canada, 1997.

27. R. Corton, B. Cassoret, and J.F. Brudny. Prediction and reduction of magnetic noise in induction electrical motors. *Euro-Noise 98*, Vol. 2, Munchen, Germany, October 1998, pp. 1065–1069.

28. B. Cassoret. Active reduction of magnetic noise from induction machines directly connected to the network. PhD thesis, Artois University, Arras, France, 1996.

10

AC Electrical Machine Torque Harmonics

Raphael Romary
Univ Lille Nord de France

Jean-François Brudny
Univ Lille Nord de France

10.1 Introduction

The magnetic noise emitted by electrical machines is a phenomenon generated by the machine itself, because it directly originates from the radial vibrations of the stator frame produced by forces that act on the stator iron in the airgap [1]. The torque harmonics due to the electromagnetic torque time variations can be more disturbing because tangential vibrations can be transmitted to the mechanical load. Thus, the analysis of the effects of torque harmonics is all the more complex as it requires to take into account the characteristics of the mechanical load associated with the considered electrical machine, in particular, the resonance frequencies of the whole structure [2]. In the study of the radial vibrations, the mechanical modeling is easier because it concerns only the stator structure. This is the reason why the studies on torque harmonics generally deal with the mechanical excitation but not with the vibratory analysis.

    The computation of electrical machine magnetic noise requires the knowledge of the radial force repartition at the inner surface of the stator [3]. Thus, a local computation of these magnetic forces allows one to determine the mode of each force component. Concerning the electromagnetic torque, this one results in the integration of the tangential forces on the rotor periphery. These forces are tied to the tangential component of the airgap flux density, which is not given by analytical models. Different ways can be used to determine these forces and consequently the electromagnetic torque [4–7]. In this paper, a simpler global approach will be used, such as the method of magnetic energy derivation [8–9]. Moreover, this method can be simplified if it is associated with the space phasor transformation of the electrical and magnetic variables [10–11].

The first part of this chapter deals with the presentation of the space phasor transformation. This concept is then applied to characterize various three-phase systems: balanced, unbalanced, or non-sine systems. The third part gives the modeling of an AC rotating electrical machine using space phasor variables assuming infinite iron permeability. The fourth part concerns the modeling of an induction machine. The torque harmonic determination in case of a non-sine supply considering a smooth airgap machine is presented in the fifth part. The last part concerns the torque harmonics generated by the variable reluctance effects. In the two last parts, results of numerical applications are compared to those that result in a global modeling using the space phasor to characterize in steady state the electrical machine operating.

10.2 Space Phasor Definition

Let us consider a smooth airgap, two-pole electrical rotating machine. The fixed external part and the interior part in rotation are respectively qualified as stator and as rotor. In order to distinguish the variables according to whether they are relative to the stator or the rotor, they will be labeled with an upper index "s" or "r." For the space phasor definition, only the stator is assumed to be energized by the mean of a three-phase stator symmetrical winding. Each phase q (q = 1, 2, or 3) is constituted of n^s diagonal turn coil. The stator spatial reference, denoted d^s, is assumed to be confounded with the phase 1 axis.

10.2.1 Case of Only One Stator Phase Energized

Let us consider only phase q to be energized with a current i_q^s. This phase, spatially shifted of $\Delta_q = (q-1)2\pi/3$ from d^s, creates a magnetomotive force (mmf) that is responsible for the magnetic flux in the airgap. The f_q^s fundamental airgap mmf, defined in the referential tied to d^s, is given by

$$f_q^s = K^s i_q^s \cos(\alpha^s - \Delta_q) \tag{10.1}$$

where

α^s represents the angular position of an arbitrary point M in the airgap relative to d^s

K^s is a coefficient equal to $K^s = 2n_e^s/\pi$, where n_e^s is the effective coil number obtained by multiplying n^s by the stator fundamental distribution factor

The b_q^s corresponding airgap flux density wave results from $b_q^s = \lambda_{00} f_q^s$. λ_{00} corresponds to the airgap permeance by area unit $\lambda_{00} = \mu_0/g$, where μ_0 is the vacuum permeability ($4\pi 10^{-7}$ H/m) and g the airgap thickness. So, for a smooth airgap machine, b_q^s and f_q^s are tied within a constant.

Let us introduce the $\vec{f_q^s}$ vector to represent the f_q^s sine function. This vector, which presents a $K^s |i_q^s|$ modulus, is oriented along the phase q axis as shown in Figure 10.1, where the phase 2 is concerned. This vector shows the north location of the generated airgap magnetic field. The f_q^s value at M point is obtained projecting $\vec{f_q^s}$ on Ox axis passing by M. In Figure 10.1, this mmf is given by OB. The corresponding flux density is equal to λ_{00} OB.

$\vec{f_q^s}$ can also be expressed as $\vec{f_q^s} = K^s \vec{i_q^s}$. The $\vec{i_q^s}$ vector presents a $|i_q^s|$ modulus and it is also oriented along the phase q axis as shown in Figure 10.1 where OB' is the $\vec{i_q^s}$ projection on Ox. The flux density at M point is given by $\lambda_{00} K^s OB'$. As $\vec{i_q^s}$ makes it possible to characterize the airgap magnetic field spatial repartition, $\vec{i_q^s}$ is defined as a current space phasor.

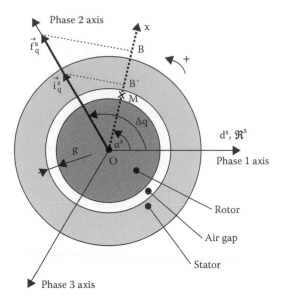

FIGURE 10.1 Current space phasor definition.

As for the time phasor, it is possible to associate a complex quantity $\underline{i}_q^s$ to $\vec{i}_q^s$, that needs to introduce a complex referential $(\mathfrak{R}^s, \mathfrak{I}^s)$ such as the $\mathfrak{R}^s$ real axis is confounded with d^s. In order to distinguish $\underline{i}_q^s$, from the complex quantities tied to the time phasors, $\underline{i}_q^s$ is denoted as a complex vector.

Introducing the complex formulation of the cosinus function, $\underline{i}_q^s$ can be expressed as

$$\underline{i}_q^s = i_q^s e^{j\Delta q} \tag{10.2}$$

It results that

$$f_q^s = K^s \mathfrak{R}^s \left[\underline{i}_q^s e^{-j\alpha^s} \right] \tag{10.3}$$

where $\mathfrak{R}^s \begin{bmatrix} \ \end{bmatrix}$ means that the real part of the complex quantity has to be considered.

10.2.2 Case of a Three-Phase Supply

Let us consider the three-phase stator winging flowed through by a i_q^s three-phase current system. Introducing the complex term "a" such as $a = e^{j\frac{2\pi}{3}}$, Equation 10.2 makes it possible to define the three elementary current space phasors. Nevertheless, from practical reasons, the relationships that define these quantities are adapted according to the considered system phase number. For a three-phase system, a coefficient equal to 2/3 is introduced, that leads to

$$\left.\begin{aligned} \underline{i}_1^s &= \frac{2i_1^s}{3} \\ \underline{i}_2^s &= \frac{2ai_2^s}{3} \\ \underline{i}_3^s &= \frac{2a^2 i_3^s}{3} \end{aligned}\right\} \tag{10.4}$$

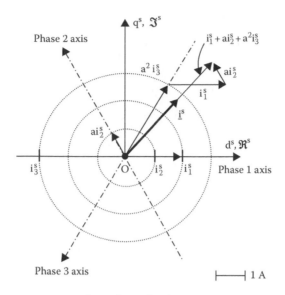

FIGURE 10.2 Three-phase current space phasor determination.

As the f^s resulting airgap fundamental mmf is obtained adding the effects generated by each phase, the quantity $\underline{i}^s$ defined as

$$\underline{i}^s = \sum_q \underline{i}_q^s = \frac{2}{3}\left(i_1^s + ai_2^s + a^2 i_3^s\right) \tag{10.5}$$

leads, according to Equation 10.3, to define f^s as follows:

$$f^s = \frac{3}{2}K^s \Re^s\left[\underline{i}^s e^{-j\alpha^s}\right] \tag{10.6}$$

The Figure 10.2 presents the principle of $\underline{i}^s$ determination taking into account that the i_q^s currents are equal, for this example, to $i_1^s = 2\,\text{A}$, $i_2^s = 1\,\text{A}$, $i_3^s = -3\,\text{A}$.

Let us specify that, thereafter, the trigonometrical direction will be regarded as the positive direction of displacement. The corresponding angular speed will be also counted positively; it will be counted negatively in the opposite direction

10.2.3 Remarks

- Let us point out that the "a" operator is tied to a spatial shift angle but not a time phase one.
- If a p pole pair machine is considered, the previous approach leads to define p complex vectors $\underline{i}^s$ indicating the p airgap north directions. These complex vectors present the same modulus and are spatially shifted of $2\pi/p$. Practically, as the phenomena are the same under each pole pair of the machine, one will represent only one vector under one pole pair as it is done in Figures 10.1 and 10.2. Considering a p pole pair machine does not change the definition of the current space phasor given by (10.5). Only changes the f^s expression that becomes

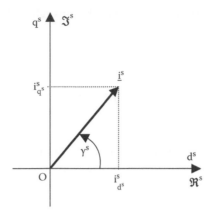

FIGURE 10.3 Current space phasor characterization.

$$f^s = \frac{3}{2} K^s \Re^s \left[\underline{i}^s e^{-jp\alpha^s} \right]$$

(10.7)

The constant K^s is also unchanged on condition that n^s represents the per phase per pole pair coil turn number.

- The definition of the current space phasor has been performed without any assumptions concerning the current waveforms. Consequently, this space phasor can be defined even if the considered variables are non-sine.
- As for all complex quantities, $\underline{i}^s$ can be defined by its polar coordinates or its real and imaginary components as pointed out in Figure 10.3:

$$\left. \begin{aligned} \underline{i}^s &= \left| \underline{i}^s \right| e^{j\gamma^s} \\ \underline{i}^s &= i_{d^s}^s + j i_{q^s}^s \end{aligned} \right\}$$

(10.8)

- If $\underline{i}^s$ is known, it is possible to determine the real variables i_1^s, i_2^s and i_3^s through $\underline{i}^s$ and $\underline{i}^{s*}$, its conjugate quantity. As $a^* = a^2$ and $a^{2*} = a$, it comes $\underline{i}^{s*} = (2/3)(i_1^s + a^2 i_2^s + a i_3^s)$. It results that one can obtain $i_1^s = (\underline{i}^s + \underline{i}^{s*})/2$, $i_2^s = (a^2 \underline{i}^s + a \underline{i}^{s*})$, $i_3^s = (a \underline{i}^s + a^2 \underline{i}^{s*})/2$. These quantities correspond to the $\underline{i}^s$ projections respectively on the phase 1, phase 2, and phase 3 axis.
- $\underline{i}^s$ has been defined from physical considerations. However, equations similar to Equation 10.5 can be used to characterize other space phasors, like voltage or linkage flux ones, although these quantities do not present any particular physical properties.

10.3 Using the Space Phasor for a Three-Phase System Characterization

10.3.1 Three-Phase Sinusoidal Balanced System

Let us consider an ω angular frequency sine, balanced, clockwise, three-phase voltage system: $v_q^s = V^s \sqrt{2} \cos(\omega t - \Delta_q - \varphi_v)$, applied to a balanced load. The line currents can be expressed as $i_q^s = I^s \sqrt{2} \cos(\omega t - \Delta_q - \varphi_v - \varphi_i)$. Let us note that these quantities can be characterized by the time phasors $\bar{v}^s$ and $\bar{I}^s$ whose moduli correspond to the variable rms values. The use of Equation 10.5 leads to express the voltage and current space phasors as follows:

$$
\left.\begin{aligned}
\underline{v}^s &= V^s\sqrt{2}e^{\,j(\omega t-\varphi_v)} \\
\underline{i}^s &= I^s\sqrt{2}e^{\,j(\omega t-\varphi_v-\varphi_i)}
\end{aligned}\right\}
\tag{10.9}
$$

These vectors present constant moduli and rotate at positive ω angular speed.

In case of an anticlockwise voltage system $v_q^s = V^s\sqrt{2}\cos(\omega t + \Delta_q - \varphi_v)$, the i_q^s expression becomes $i_q^s = I^s\sqrt{2}\cos(\omega t + \Delta_q - \varphi_v - \varphi_i)$. The corresponding space phasors that result from the use of Equation 10.5 can be written as

$$
\left.\begin{aligned}
\underline{v}^s &= V^s\sqrt{2}e^{\,-j(\omega t-\varphi_v)} \\
\underline{i}^s &= I^s\sqrt{2}e^{\,-j(\omega t-\varphi_v-\varphi_i)}
\end{aligned}\right\}
\tag{10.10}
$$

These vectors rotate at negative ω angular speed.

Figure 10.4 presents these space phasors for clockwise and anticlockwise systems for $t = 0$, $\varphi_v = \pi/4$, and $\varphi_i = \pi/2$.

The space phasors given in Equations 10.9 and 10.10 are similar to time phasors commonly used to represent sine variables considering the peak values instead of the rms ones. The main difference concerns the rotation that must be considered with the space phasors according to $\underline{i}^s$ has to represent the airgap north magnetic field axis, which rotates at constant speed ω in a direction or the other, following the nature of the three-phase system. For time phasors, this nature does not intervene in their definitions.

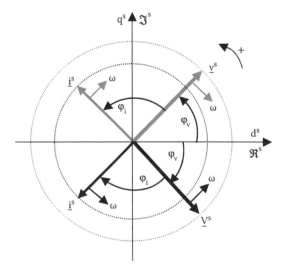

FIGURE 10.4   Locations of the $\underline{i}^s$ and $\underline{V}^s$ space phasors for $t = 0$, $\varphi_v = \pi/4$, and $\varphi_i = \pi/2$ considering a three-phase clockwise system, a three-phase anticlockwise system.

10.3.2 Three-Phase Sinusoidal Unbalanced System

A three-phase unbalanced system is the sum of clockwise, anticlockwise, and homopolar systems. As $1 + a + a^2 = 0$, the homopolar system disappears in the corresponding space phasor that is composed of only two components. They rotate at the same angular frequency in opposite directions and present, in a general way, different moduli.

10.3.3 Case of a Non-Sine System

For sine variables, space phasors are similar to time phasors. The interest of space phasor transformation is that it can be also applied to non-sine variables. The Fourier series decomposition is considered assuming that only odd rank harmonics exist, because of the symmetry that usually appears in the variable time variations. Let us consider again a three-phase stator current system with clockwise fundamental terms. Characterizing with $2k + 1$ the harmonic ranks (k varying from 0 to $+\infty$), the i_q^s current can be expressed as

$$i_q^s = \sum_{k=0}^{+\infty} I_{2k+1}^s \cos\left\{(2k+1)\left[\omega t - (q-1)\frac{2\pi}{3}\right]\right\} \tag{10.11}$$

According to the remark formulated in the previous paragraph concerning the homopolar systems, Equation 10.5 leads to define $\underline{i}^s$ as follows:

$$\underline{i}^s = \sum_{k=-\infty}^{+\infty} \underline{i}_{(6k+1)}^s = \sum_{k=-\infty}^{+\infty} I_{(6k+1)}^s \sqrt{2} e^{j(6k+1)\omega t} \tag{10.12}$$

It can be noticed that this formulation needs that k has to vary from $-\infty$ to $+\infty$. It appears that $\underline{i}^s$ results from the sum of two groups of terms. The first one concerns all the clockwise harmonic components defined for $k \geq 0$ ($6k + 1 = 1, 7, 13, \ldots$), which rotate at $(6k + 1)\omega$ angular frequency in positive direction. The second one is relative to the anticlockwise harmonic components that result from $k < 0$ ($6k + 1 = -5, -11, -17, \ldots$) and that rotate at $(6k + 1)\omega$ angular frequency in negative direction. Figure 10.5 gives an illustration of the stator current harmonic space phasor components. In order to distinguish the terms, the ranks associated to $k \geq 0$, will be noted k_d, those relative to $k < 0$ are denoted k_i.

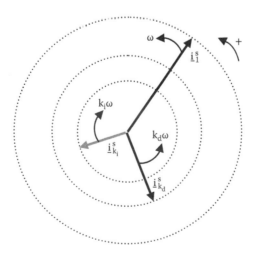

FIGURE 10.5 Case of a non-sine system.

10.4 Preliminary Considerations on the Electrical Rotating Machines

10.4.1 Introduction of a Spatial Referential Tied to the Rotor

As for the stator, a spatial reference d^r, tied to the rotor, can be introduced. One can associate to d^r a complex referential $(\Re^r, \Im^r)$. d^r is spatially shifted from d^s of $\theta = \theta_0 + \Omega t$, where Ω is the angular speed of the rotor. Considering a wound rotor whose windings are constituted of n^r diagonal turn coils crossed by rotor currents, makes it possible, as it was done for the stator, to introduce the current $\underline{i}^r$ space phasor as presented in Figure 10.6a.

$\underline{i}^r$ can be defined by its polar coordinates or its real and imaginary components defined in the referential tied to the rotor:

$$\left. \begin{aligned} \underline{i}^r &= \left| \underline{i}^r \right| e^{j\gamma^r} \\ \underline{i}^r &= i^r_{d^r} + j i^r_{q^r} \end{aligned} \right\} \tag{10.13}$$

The f^r resulting airgap fundamental mmf generated by the rotor can be deduced from (10.6):

$$f^r = A^r K^r \Re^r \left[\underline{i}^r e^{-j\alpha^r} \right] \tag{10.14}$$

where

$K^r = 2n^r_e/\pi$, n^r_e results from n^r by multiplying this quantity by the rotor fundamental distribution factor

A^r is a coefficient that depends on the rotor phase number (2/3 for a three-phase system)

α^r is the angular position of any point M in the airgap relatively to d^r

To obtain the fundamental airgap mmf due to the effects of the stator and of the rotor, it is advisable to sum the effects generated by each armature. Nevertheless, f^s and f^r are not expressed in the same referential. So, one has to define, for example, the rotor mmf in the referential tied to d^s. Let f'^r denote this quantity expressed using the variable α^s. As $\alpha^s = \alpha^r + \theta$, it comes

$$f'^r = A^r K^r \Re^s \left[\underline{i}^r e^{j\theta} e^{-j\alpha^s} \right] \tag{10.15}$$

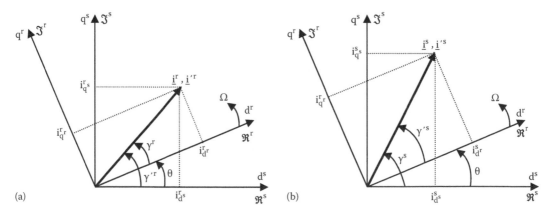

(a)   (b)

FIGURE 10.6 Introduction of spatial rotor referential. (a) For rotor variables and (b) for stator variables.

Let us introduce the space phasor $\underline{i}'^r$, defined as

$$\underline{i}'^r = \underline{i}^r e^{j\theta} \tag{10.16}$$

Then, f'^r can be expressed as

$$f'^r = A^r K^r \mathfrak{R}^s \left[\underline{i}'^r e^{-j\alpha^s} \right] \tag{10.17}$$

The $\underline{i}'^r$ expression given by Equation 10.16 could also be deduced from Figure 10.6a. Its appears that $\underline{i}'^r = |\underline{i}^r| e^{j\gamma'^r}$. As $\gamma'^r = \gamma^r + \theta$, according to the $\underline{i}^r$ definition given by Equation 10.13, one can find again Equation 10.15. Let us point out that $\underline{i}'^r$ can also be expressed using the real and imaginary components as shown in Figure 10.6a:

$$\underline{i}'^r = i^r_{d^s} + j i^r_{q^s} \tag{10.18}$$

In the same way, $\underline{i}^s$ can be expressed in the referential tied to d^r. Let $\underline{i}'^s$ denotes the quantity thus defined. Considering the Figure 10.6b, one can obtain

$$\left. \begin{aligned} \underline{i}'^s &= \underline{i}^s e^{-j\theta} \\ \underline{i}'^s &= |\underline{i}^s| e^{j\gamma'^s} \\ \underline{i}'^s &= i^s_{d^r} + j i^s_{q^r} \end{aligned} \right\} \tag{10.19}$$

10.4.2 Voltage Equations: Instantaneous Power

Let us consider the three-phase stator supplied by a three-phase voltage system. The phase q operating results from the following relationship:

$$v^s_q = r^s i^s_q + \frac{d\psi^s_q}{dt} \tag{10.20}$$

$r^s i^s_q$ is the resistive voltage drop in the phase q, ψ^s_q corresponds to the flux linked by this phase. Multiplying v^s_1 by "1," v^s_2 by "a," v^s_3 by "a²" and adding the three equations thus obtained allows one to express the stator space phasor voltage equation:

$$\underline{v}^s = r^s \underline{i}^s + \frac{d\underline{\psi}^s}{dt} \tag{10.21}$$

Considering the stator instantaneous power, it comes $p^s = \sum_q v^s_q i^s_q$. Developing this expression, taking into account the space phasor definition, leads to

$$p^s = \frac{3}{2} \mathfrak{R}^s \left[\underline{i}^s \underline{v}^{s*} \right] = \frac{3}{2} \mathfrak{R}^s \left[\underline{v}^s \underline{i}^{s*} \right] \tag{10.22}$$

In the same way, one can obtain the rotor space phasor voltage equation

$$\underline{v}^r = r^r \underline{i}^r + \frac{d\underline{\psi}^r}{dt} \tag{10.23}$$

and the rotor instantaneous power

$$p^r = \frac{3}{2}\Re^r\left[\ \underline{i}^r \underline{v}^{r*}\ \right] = \frac{3}{2}\Re^r\left[\ \underline{v}^r \underline{i}^{r*}\ \right] \tag{10.24}$$

Equations 10.21 and 10.23 are expressed in their own referential. In order to exploit them, one has to define these voltage space phasors in the same referential. Let us consider the referential tied to the stator. According the variable change given by (10.16), the following system is obtained:

$$\left.\begin{aligned} \underline{v}^s &= r^s \underline{i}^s + \frac{d\underline{\psi}^s}{dt} \\[2mm] \underline{v}'^r &= r^r \underline{i}'^r + \frac{d\underline{\psi}'^r}{dt} - j\frac{d\theta}{dt}\underline{\psi}'^r \end{aligned}\right\} \tag{10.25}$$

The whole electromechanical system instantaneous power results from

$$p = p^s + p^r = \frac{3}{2}\Re^s\left[\ \underline{v}^s \underline{i}^{s*} + \underline{v}'^r \underline{i}'^{r*}\ \right] \tag{10.26}$$

10.4.3 Electromagnetic Torque Definition

Let us consider an n windings energized electrical rotating machine. Each k winding is supplied by an v_k voltage. Let i_k and ψ_k denote the corresponding winding current and linked flux. The phase k operating results from

$$v_k = r_k i_k + \frac{d\psi_k}{dt} \tag{10.27}$$

As previously mentioned, θ makes it possible to locate the moving part relatively to the stationary one and, consequently, to characterize the variable airgap permeance. So, one can write

$$\psi_k = \psi_k(i_1, i_2, \ldots, i_h, \ldots, i_n, \theta) \tag{10.28}$$

that implies that i_k and θ variables are independent.

Conversely, it becomes

$$i_k = i_k(\psi_1, \psi_2, \ldots, \psi_h, \ldots, \psi_n, \theta) \tag{10.29}$$

So, with this formulation, the ψ_k and θ variables are assumed to be independent.

Considering Equation 10.28, ψ_k can be expressed as

$$\psi_k = \sum_{h=1}^{n} L_{kh}(i_h, \theta)i_h \tag{10.30}$$

It results that Equation 10.27 can be written as follows:

$$v_k = r_k i_k + \frac{d\theta}{dt}\sum_{h=1}^{n} i_h \frac{\partial L_{kh}}{\partial \theta} + \sum_{h=1}^{n} i_h \frac{di_h}{dt}\frac{\partial L_{kh}}{\partial i_h} + \sum_{h=1}^{n} L_{kh}\frac{di_h}{dt} \tag{10.31}$$

For h = k, L_{kk} is denoted as self inductance coefficient including the leakage inductance, for h ≠ k, L_{kh} corresponds to a mutual inductance coefficient between windings k and h.

In the following, a linear magnetic circuit is considered (L_{kh} does not depend on i_h but only on θ). So $\dfrac{\partial L_{kh}}{\partial i_h} = 0$ and Equation 10.31 becomes

$$v_k = r_k i_k + \frac{d\theta}{dt} \sum_{h=1}^{n} i_h \frac{\partial L_{kh}}{\partial \theta} + \sum_{h=1}^{n} L_{kh} \frac{di_h}{dt} \qquad (10.32)$$

Let us consider the total energy dW_t = pdt applied to the system during the time interval dt:

$$\sum_{k=1}^{n} v_k i_k \, dt = \sum_{k=1}^{n} r_k i_k^2 \, dt + \sum_{k=1}^{n} i_k \, d\psi_k \qquad (10.33)$$

Let dW denote the energy applied to the idealized system (initial system without copper losses). dW is expressed as

$$dW = \sum_{k=1}^{n} i_k \, d\psi_k \qquad (10.34)$$

dW is transformed into

- One part, $dW_m = \Gamma_e \, d\theta$, that corresponds to mechanical energy
- Another part, dW_{mag}, that produces a charge of stored magnetic energy:

$$dW = dW_{mag} + dW_m = dW_{mag} + \Gamma_e \, d\theta \qquad (10.35)$$

In order to define the dW_{mag} mathematical formulation, one can consider that the moving part is blocked, so that dθ = 0 and $dW_m = 0$. It results that

$$dW_{mag} = \sum_{k=1}^{n} i_k \, d\psi_k \qquad (10.36)$$

Considering (10.29) makes it possible to define the variables that concern W_{mag}: $W_{mag}(\psi_1, \psi_2, \ldots, \psi_h, \ldots, \psi_n, \theta)$. Therefore, with regards to partial derivatives, dW_{mag} can also be expressed as

$$dW_{mag} = \sum_{k=1}^{n} \frac{\partial W_{mag}}{\partial \psi_k} \, d\psi_k + \frac{\partial W_{mag}}{\partial \theta} \, d\theta \qquad (10.37)$$

Considering (10.34) and (10.35), leads to

$$dW_{mag} = \sum_{k=1}^{n} i_k d\psi_k - \Gamma_e \, d\theta \qquad (10.38)$$

So, by equating coefficients, the second terms of (10.37) and (10.38) lead to

$$\Gamma_e = -\frac{\partial W_{mag}}{\partial \theta} \qquad (10.39)$$

One can introduce the coenergy W'_{mag} defined as

$$W_{mag} + W'_{mag} = \sum_{k=1}^{n} i_k \psi_k \tag{10.40}$$

Using Equation 10.28 makes it possible to characterize W'_{mag}: $W'_{mag}(i_1, i_2, \ldots, i_h, \ldots, i_n, \theta)$. That leads to define dW'_{mag} as

$$dW'_{mag} = \sum_{k=1}^{n} \psi_k di_k \tag{10.41}$$

Similar developments as those realized on W_{mag} allows one to define Γ_e as follows:

$$\Gamma_e = \frac{\partial W'_{mag}}{\partial \theta} \tag{10.42}$$

10.5 Induction Machine Modeling

Let us consider a three-phase, 2-pole pair induction machine. Whatever the rotor (wounded or squirrel cage type), it is supposed to be constituted of a three-phased winding with 180° open coils of n_e^r effective turns for each one. The d^r rotor spatial reference is assumed to be confounded with the rotor phase 1 axis.

So f^s and f'^r are given by (10.6) and (10.17) substituting in this last one 3/2 to A^r. As $K^r/K^s = n_e^r/n_e^s$ the resulting f airgap mmf, obtained adding the effects, can be expressed as

$$f = \frac{3}{2} K^s \Re^s \left[\underline{i}_m e^{-j\alpha^s} \right] \tag{10.43}$$

where the $\underline{i}_m$ magnetizing current is defined as

$$\underline{i}_m = \underline{i}^s + \frac{n_e^r}{n_e^s} \underline{i}'^r \tag{10.44}$$

So, the radial airgap flux density b can be expressed as

$$b = f\Lambda \tag{10.45}$$

where Λ is the permeance per area unit.

In order to express the linked flux space phasors, the stator and rotor phase q linked flux have to be determined. ψ_q^s is given by $\psi_q^s = \psi_{qm}^s + \psi_{ql}^s$, where ψ_{qm}^s is tied to the main effects (flux density wave that crosses the airgap and that results both from stator and rotor effects), and ψ_{ql}^s that comes from the leakage fluxes due to the stator currents. ψ_{qm}^s is given by $\psi_{qm}^s = n_e^s \int_S b \, dS$. S represents the area relating to the coil opening. Considering an area element dS located at angular abscissa α^s and included in a $d\alpha^s$ angle, it comes $dS = RL_a d\alpha^s$. R is the average airgap radius (few different from the internal stator radius or from external rotor one) and L_a is the length of stator and rotor armatures. In these conditions, it can be written as

$$\psi^s_{qm} = n^s_e R L_a \int_{-\frac{\pi}{2}+\Delta_q}^{\frac{\pi}{2}+\Delta_q} b \, d\alpha^s \tag{10.46}$$

Concerning ψ^r_{qm}, it comes

$$\psi^r_{qm} = n^r_e R L_a \int_{-\frac{\pi}{2}+\Delta_q+\theta}^{\frac{\pi}{2}+\Delta_q+\theta} b \, d\alpha^s \tag{10.47}$$

It results that the main $\underline{\psi}^s_m$ and $\underline{\psi}^r_m$ linked flux space phasors, can be expressed as

$$\left.\begin{array}{l} \underline{\psi}^s_m = L^s \underline{i}^s + M \underline{i}'^r \\[2mm] \underline{\psi}^r_m = L^r \underline{i}^r + M \underline{i}'^s \end{array}\right\} \tag{10.48}$$

L^s, L^r, and M take into account the airgap variable permeance. Concerning the leakage fluxes, it is admitted that they are independent of the variable reluctance effects. So, the corresponding leakage flux space phasors are defined as $\underline{\psi}^s_l = l^s \underline{i}^s$, $\underline{\psi}^r_l = l^r \underline{i}^r$, where l^s and l^r are constants. This remark leads to

$$\left.\begin{array}{l} \underline{\psi}^s = (L^s + l^s)\underline{i}^s + M \underline{i}'^r \\[2mm] \underline{\psi}^r = (L^r + l^r)\underline{i}^r + M \underline{i}'^s \end{array}\right\} \tag{10.49}$$

In order to express the electromagnetic torque, let us consider Equations 10.25 and 10.26. It comes

$$dW = \frac{3}{2}\Re^s\left[\underline{i}^{s*} d\underline{\psi}^s + \underline{i}'^{r*} d\underline{\psi}'^r - j\underline{i}'^{r*}\underline{\psi}'^r d\theta\right] \tag{10.50}$$

According to Equation 10.36, the variation of the magnetic energy dW_{mag} can be written as

$$dW_{mag} = \frac{3}{2}\Re^s\left[\underline{i}^{s*} d\underline{\psi}^s + \underline{i}'^{r*} d\underline{\psi}'^r\right]$$

As the system is linear, the integration of dW_{mag} at given θ leads to

$$W_{mag} = \frac{3}{4}\Re^s\left[\underline{i}^{s*}\underline{\psi}^s + \underline{i}'^{r*}\underline{\psi}'^r\right]$$

Then, as $\dfrac{\partial \underline{i}'^{r*}}{\partial\theta} = -j\underline{i}'^{r*}$ and according to Equation 10.39, Γ_e can be expressed as

$$\Gamma_e = -\frac{3}{2}\Re^s\left[\underline{i}^{s*}\frac{\partial\underline{\psi}^s}{\partial\theta} + \underline{i}'^{s*}\frac{\partial\underline{\psi}'^r}{\partial\theta} - j\underline{i}'^{s*}\underline{\psi}'^r\right] \tag{10.51}$$

The use of the variable change previously define, allows the $\underline{\psi}'^r$ characterization $\underline{\psi}'^r = L^r\underline{i}'^r + M\underline{i}^s$. The development of (10.51) leads to

$$\Gamma_e = -\frac{3}{2}\Re^s\left[\underline{i}^{s^*}\underline{i}^s\frac{\partial L^s}{\partial\theta} + \underline{i}'^{r^*}\underline{i}'^r\frac{\partial L^r}{\partial\theta} + \frac{\partial M}{\partial\theta}(\underline{i}'^{r^*}\underline{i}^s + \underline{i}^{s^*}\underline{i}'^r) + j\underline{i}'^r(M\underline{i}^{s^*} + L^r\underline{i}'^{r^*}) - j\underline{i}'^{s^*}\underline{\psi}'^r\right] \tag{10.52}$$

The calculus define the following equality: $(\underline{i}'^{r^*}\underline{i}^s + \underline{i}^{s^*}\underline{i}'^r) = 2\left|\underline{i}'^r\right|\left|\underline{i}^s\right|\cos(\gamma^s - \gamma'^r)$. On the other hand, it appears that $j\underline{i}'^r(M\underline{i}^{s^*} + L^r\underline{i}'^{r^*}) = jM\underline{i}'^r\underline{\psi}'^{r^*} = j\underline{i}'^r\underline{\psi}'^{r^*}$. Introducing the cross product "x," the development of $\Re^s(j\underline{i}'^r\underline{\psi}'^{r^*} - j\underline{i}'^{r^*}\underline{\psi}'^r)$ gives $-2\underline{\psi}'^r$ xi'^{r}. So, the electromagnetic torque can be expressed as

$$\Gamma_e = -\frac{3}{4}\left\{\left|\underline{i}^s\right|^2\frac{\partial L^s}{\partial\theta} + \left|\underline{i}^r\right|^2\frac{\partial L^r}{\partial\theta} + 2\frac{\partial M}{\partial\theta}\left|\underline{i}'^r\right|\left|\underline{i}^s\right|\cos(\gamma'^s - \gamma'^r)\right\} + \frac{3}{2}\underline{\psi}'^r x\underline{i}'^r \tag{10.53}$$

10.6 Case of a Smooth Airgap Induction Machine Modeling

Let us consider a smooth airgap of constant thickness g. Λ is defined as $\Lambda = \Lambda_{00} = \mu_0/g$ ($\mu_0 = 4\pi10^{-7}$ H/m).

10.6.1 Linked Flux Space Phasors

ψ_{qm}^s and ψ_{qm}^r defined by (10.46) and (10.47), become

$$\psi_{qm}^s = n_e^s RL_a\lambda_{00}\int_{-\frac{\pi}{2}+\Delta_q}^{\frac{\pi}{2}+\Delta_q} fd\alpha^s, \quad \psi_{qm}^r = n_e^r RL_a\lambda_{00}\int_{-\frac{\pi}{2}+\Delta_q+\theta}^{\frac{\pi}{2}+\Delta_q+\theta} fd\alpha^s.$$

Developing these terms and using the space phasor definition (see Equation 10.5) lead to

$$\left.\begin{array}{l}\underline{\psi}^s = (L_{00}^s + l^s)\underline{i}^s + M_{00}\underline{i}'^r \\ \underline{\psi}^r = (L_{00}^r + l^r)\underline{i}^r + M_{00}\underline{i}'^s\end{array}\right\} \tag{10.54}$$

where the main cyclic self and mutual inductance coefficients are constants defined as

$$\left.\begin{array}{l}L_{00}^s = 6n_e^{s2}RL_a\Lambda_{00}/\pi \\ M_{00} = 6n_e^s n_e^r RL_a\Lambda_{00}/\pi \\ L_{00}^r = 6n_e^{r2}RL_a\Lambda_{00}/\pi\end{array}\right\} \tag{10.55}$$

Let us introduce the turn ratio $m = n_e^s/n_e^r$, the following correspondences exist between these coefficients: $L_{00}^s = M_{00}m, L_{00}^r = M_{00}/m$, so that $M_{00} = \sqrt{L_{00}^s L_{00}^r}$.

Considering (10.53), the electromagnetic torque is reduced in this case to

$$\Gamma_e = \frac{3}{2}\underline{\psi}'^r x\underline{i}'^r \tag{10.56}$$

10.6.2 Other Formulations for Electromagnetic Torque

Equation 10.56 can also be written as

$$\Gamma_e = \frac{3}{2}\underline{\psi}^r x \underline{i}^r = \frac{3}{2}\left[(L_{00}^r + l^r)\underline{i}^r + M_{00}\underline{i}'^s x \underline{i}^r\right] = -\frac{3}{2}M_{00}\underline{i}'^r x \underline{i}^s = -\frac{3}{2}\left(L_{00}^s \underline{i}^s + M_{00}\underline{i}'^r\right)x\underline{i}^s$$

Introducing the $\underline{\psi}_m$ magnetizing flux space phasor defined as

$$\underline{\psi}_m = L_{00}^s \underline{i}_m \tag{10.57}$$

where the $\underline{i}_m$ magnetizing current space phasor is given by

$$\underline{i}_m = \underline{i}^s + \frac{\underline{i}'^r}{m} \tag{10.58}$$

Γ_e can be expressed as

$$\Gamma_e = -\frac{3}{2}\underline{\psi}_m x \underline{i}^s \tag{10.59}$$

This development shows that the leakage fluxes do not produce torque as it is generally admitted.

10.6.3 Balanced Sinusoidal Three-Phase Supply: Steady-State Operating Mode

For an ω angular pulsation supply, the rotor is short circuited ($\underline{v}^r = 0$) and it rotates at $(1 - s)\omega$ angular speed, where s is the slip. The various space phasors can be expressed as

$$\underline{i}^s = I^s\sqrt{2}e^{j\omega t}, \quad \underline{v}^s = V^s\sqrt{2}e^{j(\omega t + \varphi^s)}, \quad \underline{i}'^r = I^r\sqrt{2}e^{j(\omega t + \vartheta^r)}$$

So, the time phasors can be used to express the voltage system given by (10.25). It comes

$$\left.\begin{array}{l} \overline{V}^s = r^s\overline{I}^s + jl^s\omega\overline{I}^s + jL_{00}^s\omega\overline{I}^s + jM_{00}\omega\overline{I}^r \\[2mm] 0 = r^r\overline{I}^r + jl^r s\omega\overline{I}^r + jL_{00}^r s\omega\overline{I}^r + jM_{00}s\omega\overline{I}^s \end{array}\right\} \tag{10.60}$$

Dividing by s all the terms of the second equation of system (10.60) leads to

$$\left.\begin{array}{l} \overline{V}^s = r^s\overline{I}^s + jl^s\omega\overline{I}^s + jL_{00}^s\omega\overline{I}^s + jM_{00}\omega\overline{I}^r \\[2mm] 0 = \dfrac{r^r}{s}\overline{I}^r + jl^r\omega\overline{I}^r + jL_{00}^r\omega\overline{I}^r + jM_{00}\omega\overline{I}^s \end{array}\right\} \tag{10.61}$$

This procedure makes it possible to assume that all the variables of system (10.61) are of ω angular frequency. Introducing $\overline{I}^{\circ r} = \overline{I}^r/m$, system (10.61) can be rewritten as

$$\left.\begin{array}{l} \overline{V}^s = r^s\overline{I}^s + jl^s\omega\overline{I}^s + jM_{00}\omega m\left(\overline{I}^s - \overline{I}^{\circ r}\right) \\[2mm] 0 = -\dfrac{r^r}{s}m\overline{I}^{\circ r} - jl^r\omega m\overline{I}^{\circ r} - jM_{00}\omega\left(\overline{I}^{\circ r} - \overline{I}^s\right) \end{array}\right\} \tag{10.62}$$

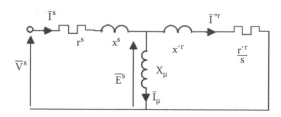

FIGURE 10.7   Single-phase equivalent circuit.

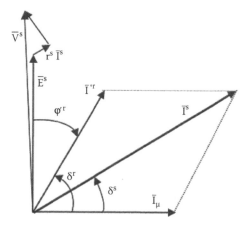

FIGURE 10.8   Corresponding time diagram.

Multiplying the two members of the second equation of system (10.62) by m and noting $r'^r = m^2 r^r$, $l'^r = m^2 l^r$, $L_\mu = mM_{00} = L_{00}^s$, $x^s = l^s \omega$, $x'^r = l'^r \omega$, $X\mu = L\mu\omega$, $\overline{I}\mu = \overline{I}^s - \overline{I}^{or}$, system (10.62) becomes

$$\left. \begin{aligned} \overline{V}^s &= r^s \overline{I}^s + jx^s \overline{I}^s + jX_\mu \overline{I}_\mu \\ \frac{r'^r}{s} \overline{I}^{or} + jx' \overline{I}^{or} &= jX_\mu \overline{I}_\mu \end{aligned} \right\} \tag{10.63}$$

It leads to the classical equivalent single-phase circuit of the Figure 10.7 as to the corresponding time diagram of the Figure 10.8. One can notice that $\overline{I}_N$ current, according to the rotor current change in the way of displacement, as a similar form that $\underline{i}_m$ given by (10.58). On other hand, one can remark that L_μ is defined by L_{00}^s, which corresponds also to the quantity used to characterize the magnetizing flux phasor given by (10.57).

The electromechanical torque deduced from the single-phase equivalent circuit results from the P^r active power transferred to the rotor according to the relationship: $\Gamma_e = P^r/\Omega_s$. Ω_s corresponds to the synchronous speed equal in this case to ω. P^r is given by $P^r = 3E^s I^{or} \cos \varphi'^r$. As $I^{or} \cos \varphi'^r = \overline{I}^s \sin \delta^s$, P^r can be expressed as $P^r = 3E^s I^s \sin \delta^s$. In so far as δ^s is the angle between $-\overline{I}^s$ and $-\overline{I}_\mu$ (and consequently $\overline{\psi}_\mu$), it appears that this Γ_e expression is the same as the one given by (10.57), according to that the space phasor moduli are defined by their amplitude (see Section 10.5.1) and, on the other hand, that the change in the way of displacement of the rotor currents needs to define Γ_e the use of the following relationship:

$$\Gamma_e = \frac{3}{2} \underline{\psi}_m x \underline{i}^s \tag{10.64}$$

10.6.4 Non-Sine Supply: Torque Harmonics

These torque harmonics generated by the supply are of importance especially when their frequencies are close to the mechanical system natural frequencies. That justifies that numerous papers deal with this subject in the literature [12–14], namely concerning their minimization [15–17].

Let us consider a non-sine, three-phase, f frequency, balanced, clockwise, voltage system. The stator voltage space phasor can be expressed as follows:

$$\underline{v}^s = \sum_{k=-\infty}^{+\infty} \underline{v}^s_{(6k+1)} = \sum_{k=-\infty}^{+\infty} V^s_{(6k+1)} \sqrt{2} e^{j(6k+1)\omega t} \tag{10.65}$$

One can deduced from (10.64) that

$$\Gamma_e = \frac{3}{2} \left(\sum_{k=-\infty}^{+\infty} \underline{\psi}_{m(6k+1)} \right) \times \left(\sum_{k'=-\infty}^{+\infty} \underline{i}^s_{(6k'+1)} \right) \tag{10.66}$$

The products of space phasors that present the same angular speed (k = k′) define the Γ_{e0k} mean torques. For k = 0, the operation is similar to this one described in the previous paragraph and the corresponding electromagnetic torque is denoted Γ_{e01}. For k ≠ 0, the corresponding mean torques are qualified as parasitical mean torques.

The products of space phasors such as k ≠ k′ lead to $\Gamma_{ek'k}$ harmonic torques.

In order to express the harmonic and parasitical mean torques, it is advisable to take into account the single-phase induction machine equivalent circuit relative to the harmonics given in Figure 10.9.

For the (6k + 1) voltage harmonic, the synchronous speed is equal to (6k + 1)ω. The corresponding $s_{(6k+1)}$ slip results from

$$s_{(6k+1)} = \frac{(6k+1)\omega - (1-s_1)\omega}{(6k+1)\omega} = \frac{6k + s_1}{6k+1} \tag{10.67}$$

where s_1 represents the slip relative to the fundamental terms. For a normal running, s_1 is of few percent. So, $s_{(6k+1)}$ is close to unity. As $|r'^r + j(6k+1)x'^r| \ll 6(k+1)X_\mu$, $I_{\mu(6k+1)}$ is negligible with regard to $I^{\circ r}_{(6k+1)}$. This particularly has two consequences:

- Γ_{e0k} torques can be neglected comparing to the corresponding Γ_{e01} fundamental quantity.
- The considered harmonic single-phase equivalent circuit becomes the one given in Figure 10.10, it results that $\Gamma_{ek'k}$ can be reduced to Γ_{e1k}.

Let us consider two voltage harmonics defined for $k_d = 6h + 1$ and $k_i = -6h + 1$, where h is an integer that takes only positive values (see Section 10.3.3). Figure 10.11 presents, for given h, the space phasor diagram for t = 0 and t ≠ 0 considering $\underline{\psi}_{m1}$, $\underline{i}^s_{k_d}$, and $\underline{i}^s_{k_i}$. For t = 0, the $\Gamma_{e1(6h)}$ resulting torque harmonic can be expressed as

$$\Gamma_{e1(6h)} = \Gamma_{e1k_d} + \Gamma_{e1k_i} = \frac{3}{2} \left| \underline{\psi}_{m1} \right| \left\{ \left| \underline{i}^s_{k_d} \right| \sin \delta_{k_d} + \left| \underline{i}^s_{k_i} \right| \sin \delta_{k_i} \right\} \tag{10.68}$$

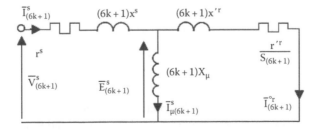

FIGURE 10.9 Single-phase equivalent circuit for the (6k + 1) rank harmonic.

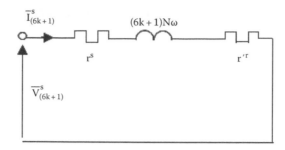

FIGURE 10.10   Simplified single-phase equivalent circuit for the (6k + 1) rank harmonic.

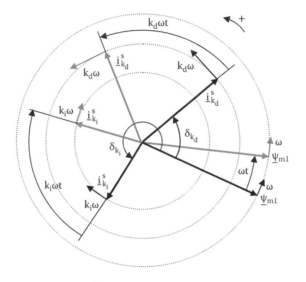

FIGURE 10.11   Harmonic torques: space phasor location. (—) t = 0, (—) t ≠ 0.

At $t \neq 0$, the space phasor moduli do not change, only δ_{k_d} and δ_{k_i} vary. According to space phasor ways of displacement, δ_{k_d} increases by $6h\omega t$ although δ_{k_i} decreases by $6h\omega t$. So, for any time instant t, Equation 10.68 can be written as

$$\Gamma_{el(6h)} = \frac{3}{2} \left| \underline{\psi}_{m1} \right| \left\{ \left| I^s_{(6h+1)} \right| \sqrt{2} \sin(\delta_{(6h+1)} + 6h\omega t) + \left| I^s_{(-6h+1)} \right| \sqrt{2} \sin(\delta_{(-6h+1)} - 6h\omega t) \right\} \tag{10.69}$$

Developing this expression, one can obtain

$$\Gamma_{el(6h)} = \hat{\Gamma}_{el(6h)} \cos(6h\omega t + \beta_{(6h)}) \tag{10.70}$$

where

$$\left. \begin{aligned} \hat{\Gamma}_{el(6h)} &= 3L_\mu I_{m1} \sqrt{I^{s2}_{(-6h+1)} + I^{s2}_{(6h+1)} - 2I^s_{(-6h+1)} I^s_{(6h+1)} \cos(\delta_{(-6h+1)} + \delta_{(6h+1)})} \\ \mathrm{tg}\beta_{(6h)} &= \frac{I^s_{(-6h+1)} \cos\delta_{(-6h+1)} - I^s_{(6h+1)} \cos\delta_{(6h+1)}}{I^s_{(-6h+1)} \sin\delta_{(-6h+1)} + I^s_{(6h+1)} \sin\delta_{(6h+1)}} \end{aligned} \right\} \tag{10.71}$$

These developments point out that the torque harmonics are independent of induction machine load, which means that from an experimental point of view, their determination will be more efficiency at no load [18].

10.6.5 Numerical Application

Let us consider a 2-pole pair induction machine defined through the following parameters: $r^s = 3.6\,\Omega$, $r^r = 1.8\,\Omega$, $l^s = 18\,\text{mH}$, $l^r = 9\,\text{mH}$, $L^s_{00} = 0.4\,\text{H}$, $L^r_{00} = 0.2\,\text{H}$.

The stator is supplied by a three-phase voltage source inverter of input DC voltage $E = 300\,\text{V}$. The phase 1 v^s_1 is represented in Figure 10.12, the supply frequency is 50 Hz. The aim is the $\hat{\Gamma}_{e1(6h)}$ determination for $h = 1$.

The voltage Fourier series decomposition leads to

$$V^s_{2k+1} = \frac{\sqrt{2}E}{\pi(2k+1)}\left[(-1)^k + \sin(2k+1)\frac{\pi}{6}\right]$$

Then it comes $V^s_1 = 202.56\,\text{V}$, $V^s_5 = 40.51\,\text{V}$, $V^s_7 = -28.93\,\text{V}$.

The magnetizing current I_{m1} is deduced from Figure 10.7. Actually, I_{m1} slightly depends on the load because of the voltage drop due to r^s and x^s. Here, the magnetizing current will be calculated for $s = 0$, what leads to $\underline{i}_{m1} = \sqrt{2}I_{m1}e^{j\omega t + \varphi^s_{m1}}$ with $I_{m1} = 1.54\,\text{A}$, and $\varphi^s_{m1} = -88.43°$.

The space phasor currents corresponding to the fifth and seventh current harmonics are deduced from Figure 10.10. They are defined as follows:

$$\underline{i}^s_5 = \sqrt{2}I^s_5 e^{-j[5\omega t + \varphi^s_5]}, \quad \text{with } I^s_5 = 0.707\,\text{A}, \quad \varphi^s_5 = 82.72°$$

$$\underline{i}^s_7 = \sqrt{2}I^s_7 e^{j[7\omega t + \varphi^s_7]}, \quad \text{with } I^s_7 = 0.364\,\text{A}, \quad \varphi^s_7 = 95.2°$$

φ^s_5 and φ^s_7 are the phase angle of $\underline{i}^s_5$ and $\underline{i}^s_7$ relatively to $\underline{v}^s_5$ and $\underline{v}^s_7$, they are tied to δ_5 and δ_7 through

$$\delta_5 = -\varphi^s_5 - \varphi^s_{m1} = 171.15°$$

$$\delta_7 = -\varphi^s_7 - \varphi^s_{m1} = 183.15°$$

Finally, the first equation of (10.71), which has to be multiplied by $p = 2$, leads to $\hat{\Gamma}_{e1(6)} = 1.28\,\text{N·m}$.

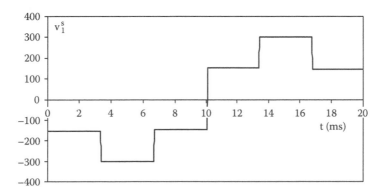

FIGURE 10.12 Single-phase voltage wave form.

10.6.5.1 Simulation Results

Now, the global simulation of the whole system will be performed by solving the voltage equations (10.25) and considering the mechanical equation, $\Gamma_e - \Gamma_r = J\left(d^2\theta/dt^2\right)$. Γ_r corresponds to the torque imposed by the load and J is the polar moment of inertia of the mechanical system. Figures 10.13 and 10.14 give electromagnetic torque and phase 1 current waveforms at no load. An FFT applied to these variables gives $I_1^s = 1.52\,\text{A}$, $I_5^s = 0.717\,\text{A}$, $I_7^s = 0.374\,\text{A}$, $\hat{\Gamma}_{e1(6)} = 1.29\,\text{N·m}$.

The magnitude of the sixth rank harmonic torque obtained by simulation is the same as the value deduced from equivalent circuit taking into account only the fifth and the seventh rank harmonic current.

For a machine running such as $\Gamma_r = 7\,\text{N·m}$, what corresponds to the rated torque, the simulation leads to $I_1^s = 2.51\,\text{A}$, $I_5^s = 0.707\,\text{A}$, $I_7^s = = 0.368\,\text{A}$, $\hat{\Gamma}_{e1(6)} = = 1.24\,\text{N·m}$.

The torque and current waveforms are given in Figures 10.15 and 10.16. It can be noticed that the considered harmonic torque slightly decreases with the load. This is due to the decrease of the magnetizing current in load.

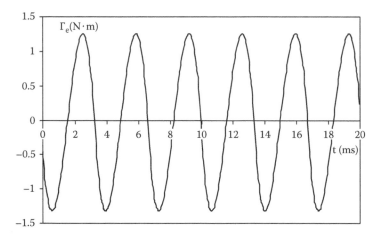

FIGURE 10.13 Torque waveform—no load.

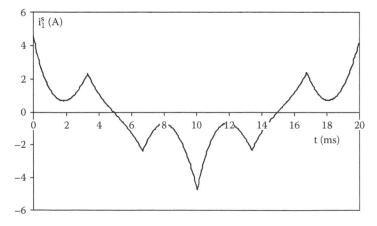

FIGURE 10.14 Current waveform—no load.

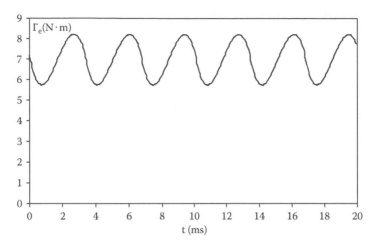

FIGURE 10.15 Torque waveform—rated load.

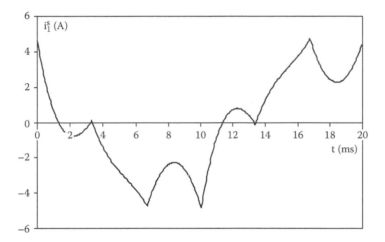

FIGURE 10.16 Current waveform—rated load.

10.7 Reluctant Torques

These reluctant torques are often considered concerning salient synchronous machines or reluctance ones [19–21]. In this chapter, these reluctant torques will be expressed considering an induction machine, taking the slotting effect into account and assuming sinusoidal the stator and rotor currents.

10.7.1 Machine Modeling

The Equations 10.43 through 10.47 are always valid, only Λ, the permeance per area unit, changes. Introducing the integers k_s and k_r, which take all the values between $-\infty$ and $+\infty$, Λ can be written, for a 2-pole machine, as follows [2,22]:

$$\Lambda = \sum_{k_s=-\infty}^{+\infty} \sum_{k_r=-\infty}^{+\infty} \Lambda_{k_s k_r} \cos\left[(k_s N^s + k_r N^s)\alpha^s - k_r N^s \theta\right] \tag{10.72}$$

N^s and N^r denote, respectively, the numbers of stator and rotor slots (or bars). $\Lambda_{k_s k_r}$ is defined as follows:

$$
\left.
\begin{aligned}
\Lambda_{00} &= \mu_0 A_{00} \\
\Lambda_{k_s 0} &= \mu_0 A_{s0} f(k_s) \\
\Lambda_{0k_r} &= \mu_0 A_{0r} f(k_r) \\
\Lambda_{k_s k_r} &= \mu_0 A_{sr} f(k_s) f(k_r)
\end{aligned}
\right\}
\tag{10.73}
$$

where

$$
\left.
\begin{aligned}
A_{00} &= \frac{\left[1 + d_s^s r_t^s/g^r + d_s^r r_t^r/g^s + d_s^s d_s^r (g + g_M) r_t^s r_t^r/gg^s g^r\right]}{g_M} \\
A_{s0} &= 2d_s^s \frac{\left[1 + d_s^r(g + g_M) r_t^r/gg^s\right]}{\pi g_M g^r} \\
A_{0r} &= 2d_s^r \frac{\left[1 + d_s^s(g + g_M) r_t^s/gd^r\right]}{\pi g_M g^s} \\
A_{sr} &= \frac{4d_s^s d_s^r \left[g_M + g\right]}{\pi^2 gg^s g^r g_M}
\end{aligned}
\right\}
\tag{10.74}
$$

The different parameters have the following signification:

- w_s^s, w_t^s are the widths, respectively, of one stator slot and one stator tooth
- d_s^s is the fictitious depth of one stator slot, defined [22] as $d_s^s = w_s^s/5$
- r_t^s is the stator slotting ratio: $r_t^s = w_t^s/(w_t^s + w_s^s)$
- $f(k_s)$ is the stator slotting function: $f(k_s) = (\sin k_s r_t^s \pi)/2k_s$
- w_s^r, w_t^r, d_s^r, r_t^r, $f(k_r)$ are analogous quantities relative to the rotor
- k_s and k_r are the integers that take all the values between $-\infty$ and $+\infty$
- g is the minimal airgap thickness
- g_M is the maximal airgap fictitious thickness: $g_M = g + d_s^s + d_s^r$
- g^s and g^r are the intermediate fictitious airgap thicknesses expressed by $g + d_s^s$ and $g + d_s^r$, respectively

In order to estimate qualitatively the relative importance of the different terms, the following inequalities can be pointed out: $A_{s0} \cong A_{0r}$; $A_{00} > A_{s0}$ or A_{0r}; A_{s0} or $A_{0r} > A_{sr}$.

b, which results from Equation 10.45, is given by

$$
b = \sum_{k_s} \sum_{k_r} b_{k_s k_r}
$$

Using the exponential form and noting $S_{k_s k_r} = k_s N^s + k_r N^r$, Λ can be rewritten as follows:

$$
\Lambda = \frac{1}{2} \sum_{k_s} \sum_{k_r} \Lambda_{k_s k_r} \left(e^{jS_{k_s k_r}\alpha^s} e^{-jk_r N^r \theta} + e^{-jS_{k_s k_r}\alpha^s} e^{jk_r N^r \theta} \right)
\tag{10.75}
$$

So, $b_{k_s k_r}$ can be expressed as

$$
b_{k_s k_r} = \frac{3}{8} K^s \Lambda_{k_s k_r}
\begin{bmatrix}
\left[\underline{i}_m e^{j(S_{k_s k_r} -1)\alpha_s} + \underline{i}_m^* e^{j(S_{k_s k_r} +1)\alpha^s} \right] e^{jk_r N^r \theta} \\[2ex]
+ \left[\underline{i}_m e^{-j(S_{k_s k_r} +1))\alpha_s} + \underline{i}_m^* e^{-j(S_{k_s k_r} -1)\alpha^s} \right] e^{jk_r N^r \theta}
\end{bmatrix}
$$

The use of Equations 10.46 and 10.47 leads to

$$
\psi_{q k_s k_r}^s = \frac{3}{4} K^s \Lambda_{k_s k_r} n_e^s RL
\left\{
\begin{bmatrix}
A_{k_s k_r} \left[\underline{i}_m e^{j(S_{k_s k_r}-1)^{\Delta q}} e^{-jk_r N^r \theta} + \underline{i}_m^* + \underline{i}_m^* e^{-j(S_{k_s k_r}-1)^{\Delta q}} e^{jk_r N^r \theta} \right] \\[2ex]
+ B_{k_s k_r} \left[\underline{i}_m^* e^{j(S_{k_s k_r}+1)^{\Delta q}} e^{-jk_r N^r \theta} + \underline{i}_m e^{-j(S_{k_s k_r}+1)^{\Delta q}} e^{jk_r N^r \theta} \right]
\end{bmatrix}
\right\}
$$

$$
\psi_{q k_s k_r}^r = \frac{3}{4} K^s \Lambda_{k_s k_r} n_e^r RL
\left\{
\begin{bmatrix}
A_{k_s k_r} \left[\underline{i}_m e^{j(S_{k_s k_r}-1)\Delta q} e^{j(k_s N^s-1)\theta} + \underline{i}_m^* e^{-j(S_{k_s k_r}-1))\Delta q} e^{-j(k_s N^s-1)\theta} \right] \\[2ex]
+ B_{k_s k_r} \left[\underline{i}_m^* e^{j(S_{k_s k_r}+1)\Delta q} e^{j(k_s N^s+1)\theta} + \underline{i}_m e^{-j(S_{k_s k_r}+1))\Delta q} e^{-j(k_s N^s+1)\theta} \right]
\end{bmatrix}
\right\}
$$

where

$$
A_{k_s k_r} = \frac{\sin\left[\left(S_{k_s k_r} -1 \right) \pi/2 \right]}{\left(S_{k_s k_r} -1 \right)}
$$

$$
B_{k_s k_r} = \frac{\sin\left[\left(S_{k_s k_r} +1 \right) \pi/2 \right]}{\left(S_{k_s k_r} +1 \right)}
$$

$A_{k_s k_r}$ and $B_{k_s k_r}$ only exist for $S_{k_s k_r} = 2n$, where n is negative, positive, or null integer.
Considering the space phasor definition, one can obtain

$$
\underline{\psi}_{k_s k_r}^s = \frac{1}{2} K^s \Lambda_{k_s k_r} n_e^s RL
\left\{
\begin{bmatrix}
A_{k_s k_r} \left[\underline{i}_m e^{-jk_r N^r \theta} \left(1+a^{2n}+a^{4n} \right) + \underline{i}_m^* e^{jk_r N^r \theta} \left(1+a^{2(1-n)}+a^{4(1-n)} \right) \right] \\[2ex]
+ B_{k_s k_r} \left[\underline{i}_m^* e^{-jk_r N^r \theta} \left(1+a^{2(1+n)}+a^{4(1+n)} \right) + \underline{i}_m e^{jk_r N^r \theta} \left(1+a^{-2n}+a^{-4n} \right) \right]
\end{bmatrix}
\right\}
$$

$$
\underline{\psi}_{k_s k_r}^r = \frac{1}{2} K^s \Lambda_{k_s k_r} n_e^r RL
\left\{
\begin{bmatrix}
A_{k_s k_r} \left[\underline{i}_m e^{j(k_s N^s-1)\theta} \left(1+a^{2n}+a^{4n} \right) + \underline{i}_m^* e^{-j(k_s N^s-1)\theta} \left(1+a^{2(1-n)}+a^{4(1-n)} \right) \right] \\[2ex]
+ B_{k_s k_r} \left[\underline{i}_m^* e^{j(k_s N^s+1)\theta} \left(1+a^{2(1+n)}+a^{4(1+n)} \right) + \underline{i}_m e^{-j(k_s N^s+1)\theta} \left(1+a^{-2n}+a^{-4n} \right) \right]
\end{bmatrix}
\right\}
$$

Different cases have to be considered according to the n values.

Case 1: $(1 + a^{2n} + a^{4n})$ and $(1 + a^{-2n} + a^{-4n})$ are not nil only for $n = 3n'$, n' being an integer that takes all the values between $-\infty$ and $+\infty$. According to the $S_{k_s k_r}$ definition, it comes $S_{k_s k_r} = 6n'$

Case 2: $(1 + a^{2(1-n)} + a^{4(1-n)}) \neq 0$ for $n = 1 - 3n'$, so for $S_{k_s k_r} = 2 - 6n'$

Case 3: $(1 + a^{2(1+n)} + a^{4(1+n)}) \neq 0$ for $n = -1 + 3n'$, so for $S_{k_s k_r} = 2 - 6n'$

When these quantities are not nil, they take the value 3.

    The numerical applications will bring on a three-phase, wound rotor induction machine such as $N^s = 6m^s$ and $N^r = 6m^r$, where m^s and m^r are the stator and rotor per pole per phase slot numbers. So $S_{k_s k_r} = 6(k_s m^s + k_r m^r)$. It results that only the case 1 has to be considered where $n' = (k^s m^s + k^r m^r)$.

- Case of a smooth airgap machine
  In this case, k_s and k_r take the value 0. It results that $S_{00} = 0$ and $A_{00} = B_{00} = 1$. According to the K^s expression it comes

$$\left.\begin{array}{l} \underline{\psi}^s_{00} = 6\dfrac{n^s_e}{\pi} \Lambda_{00} n^s_e RL\underline{i}_m \\[4mm] \underline{\psi}^r_{00} = 6\dfrac{n^s_e}{\pi} \Lambda_{00} n^r_e RL\underline{i}_m \end{array}\right\} \tag{10.76}$$

- Inductance harmonics
  The flux linkage space phasors are reduced to

$$\underline{\psi}^s_{k_s k_r} = \frac{L^s_{00}\underline{i}^s + M_{00}\underline{i}'^r}{2\Lambda_{00}}(-1)^{3(k_s m^s + k_r m^r)}\lambda_{k_s k_r}\left\{\frac{e^{-jk_r N^r \theta}}{1 - 6(k_s m^s + k_r m^r)} + \frac{e^{jk_r N^r \theta}}{1 + 6(k_s m^s + k_r m^r)}\right\} \tag{10.77}$$

$$\underline{\psi}^r_{k_s k_r} = \frac{L^r_{00}\underline{i}^r + M_{00}\underline{i}'^s}{2\Lambda_{00}}(-1)^{3(k_s m^s + k_r m^r)}\lambda_{k_s k_r}\left\{\frac{e^{jk_s N^s \theta}}{1 - 6(k_s m^s + k_r m^r)} + \frac{e^{-jk_s N^s \theta}}{1 + 6(k_s m^s + k_r m^r)}\right\} \tag{10.78}$$

The $\underline{\psi}^s_{k_r}$ quantities, which depend on $e^{jk_r N^r \theta}$ or $e^{-jk_r N^r \theta}$, result on the sum of $\underline{\psi}^s_{k_s k_r}$ on k_s that varies from $-\infty$ to $+\infty$. Concerning the $\underline{\psi}^s_{k_s}$ quantities, which depend on $e^{jk_s N^s \theta}$ or $e^{-jk_s N^s \theta}$, they result on the sum of $\underline{\psi}^s_{k_s k_r}$ on k_r that also varies from $-\infty$ to $+\infty$. So, according to some considerations that concern these sums and that enable to consider only the positive values of k_s and k_r, the following equations can be established:

$$\left.\begin{array}{l} \underline{\psi}^s_{k_r} = (L^s_{k_{r+}} + L^s_{k_{r-}})\underline{i}^s + \left(M_{k_{r+}} + M_{k_{r-}}\right)\underline{i}'^r \\[3mm] \underline{\psi}^r_{k_s} = (L^r_{k_{s+}} + L^r_{k_{s-}})\underline{i}^r + \left(M_{k_{s+}} + M_{k_{s-}}\right)\underline{i}'^s \end{array}\right\} \tag{10.79}$$

where

$$
L_{k_{r+}}^s = \frac{L_{00}^s}{\Lambda_{00}} e^{jk_r N^r \theta} \sum_{k_s = -\infty}^{+\infty} \lambda_{k_s k_r} \left\{ \frac{(-1)^{3(k_s m^s + k_r m^r)}}{1 + 6(k_s m^s + k_r m^r)} \right\}
$$

$$
L_{k_{r-}}^s = \frac{L_{00}^s}{\Lambda_{00}} e^{-jk_r N^r \theta} \sum_{k_s = -\infty}^{+\infty} \lambda_{k_s k_r} \left\{ \frac{(-1)^{3(k_s m^s + k_r m^r)}}{1 - 6(k_s m^s + k_r m^r)} \right\}
$$

$$
L_{k_{s+}}^r = \frac{L_{00}^r}{\Lambda_{00}} e^{jk_s N^s \theta} \sum_{k_r = -\infty}^{+\infty} \lambda_{k_s k_r} \left\{ \frac{(-1)^{3(k_s m^s + k_r m^r)}}{1 - 6(k_s m^s + k_r m^r)} \right\}
$$

$$
L_{k_{s-}}^r = \frac{L_{00}^r}{\Lambda_{00}} e^{-jk_s N^s \theta} \sum_{k_r = -\infty}^{+\infty} \lambda_{bk_r} \left\{ \frac{(-1)^{3(bm^s + k_r m^r)}}{1 + 6(bm^s + k_r m^r)} \right\}
$$

(10.80)

and

$$
M_{k_{r+}} = \frac{n_e^s}{n_e^r} L_{k_{r+}}^s, \quad M_{k_{r-}} = \frac{n_e^s}{n_e^r} L_{k_{r-}}^s, \quad M_{k_{s+}} = \frac{n_e^r}{n_e^s} L_{k_{s+}}^r, \quad M_{k_{s-}} = \frac{n_e^r}{n_e^s} L_{k_{s-}}^r
$$

(10.81)

Considering the linked flux space phasors definition, it comes

$$
\underline{\psi}^s = \left[L_{00}^s + \sum_{k_r=1}^{+\infty} (L_{k_{r+}}^s + L_{k_{r-}}^s) + l^s \right] \underline{i}^s + \left[M_{00} + \sum_{k_r=1}^{+\infty} \left(M_{k_{r+}} + M_{k_{r-}} \right) \right] \underline{i'}^r
$$

$$
\underline{\psi}^r = \left[L_{00}^r + \sum_{k_s=1}^{+\infty} \left(L_{k_{s+}}^r + L_{k_{s-}}^r \right) + l^r \right] \underline{i}^r + \left[M_{00} + \sum_{k_s=1}^{+\infty} \left(M_{k_{s+}} + M_{k_{s-}} \right) \right] \underline{i'}^s
$$

(10.82)

Considering the magnetizing current given by (10.58) makes it possible to define these quantities in the referential tied to d^s:

$$
\underline{\psi}^s = l^s \underline{i}^s + \left[L_{00}^s + \sum_{k_r=1}^{+\infty} \left(L_{k_{r+}}^s + L_{k_{r-}}^s \right) + l^s \right] \underline{i}_m
$$

$$
\underline{\psi'}^r = l^r \underline{i'}^r + \left[L_{00}^r + \sum_{k_s=1}^{+\infty} \left(L_{k_{s+}}^r + L_{k_{s-}}^r \right) + l^r \right] \underline{i}_m
$$

(10.83)

10.7.2 Reluctant Torque Calculation

The torque delivered by the machine is deduced from Equations 10.51 and 10.82.

A numerical application can be performed considering the machine defined in Section 10.6.5. It is a 2-pole pair machine with 18 stator and 24 rotor slots per pole pair ($N^s = 18$, $N^r = 12$, $m^s = 3$, $m^r = 2$). For the reluctant torque computation, other geometrical parameters are necessary to define the quantities $\Lambda_{k_s k_r}$. One gives R = 0.4 m, L_a = 0.15 m, $n_e^s = 123$, $n_e^r = 87$. Other parameters that concern the stator and rotor slots dimensions have to be introduced. They lead to define the following quantities: $r_t^s = 0.4$, $r_t^r = 0.6$, A_{00} = 1829 m^{-1}, A_{s0} = 1651 m^{-1}, A_{0r} = 1192 m^{-1}, A_{sr} = 1461 m^{-1}.

TABLE 10.1   Reluctant Harmonic Torque Magnitude

kr	1	2	3	4
$\hat{\Gamma}_{e(k_r N^r)}$(N·m)	3.42	1.12	0.22	0.11

In order to be in the same conditions than the simulation performed in Section 10.6.5, the machine will be supposed to be fed by a sine three-phase voltage system of 50 Hz, 203.56 V voltage rms. value. The calculation will be performed at no-load condition so that the current $\underline{i}^s$ is equal to the magnetizing current $\underline{i}_{m1}$: $\underline{i}_{m1} = \sqrt{2}I_{m1}e^{j(\omega t + \varphi_{m1}^s)}$ with $I_{m1} = 1.54$ A, and $\varphi_{m1}^s = -88.43°$.

At no load, only the torque components due to the interaction between ψ^s and $\underline{i}^{s^*}$ exist, generating harmonics of $k_r N^r fr$ frequency (fr = (1 − s)f). The obtained magnitudes $\hat{\Gamma}_{e(k_r N^r)}$ are given in Table 10.1.

One can notice that the magnitude for $k_r = 1$ is higher than this obtained for the sixth harmonic torque determined in Section 10.6.5 in case of non-sine supply ($\hat{\Gamma}_{e1(6)} = 1.24$ N · m)

10.8 Conclusion

The tangential vibrations of electrical rotating machines originate from the variations of its electromagnetic torque. These variations can be associated to torque harmonic components. Two kinds of torque harmonics can be defined: those due to the supply and those due to the reluctant effects. A method to calculate these torque harmonics has been presented. It is based on the space phasor definition that enables to present very synthetic expression of the electromagnetic torque. The space phasor transformation also provides simple voltage equations. Application to induction machines has been performed and the numerical applications show that the torque harmonics have significant magnitudes compared to the rated torque. Moreover, the reluctant torque harmonics are higher than these due to the supply (in case of voltage source inverter supply). A technique to reduce the reluctant torques consists in skewing the rotor [23].

References

1. Ph.L. Alger. *The Nature of Induction Machines*, 2nd edn. Gordon & Breach Publishers, New York/London, U.K./Paris, France, 1970.
2. J.F. Brudny. Etude quantitative des harmoniques de couple du moteur asynchrone triphasé d'induction. Habilitation thesis, Lille, France, 1991, NH29.
3. P.L. Timar, A. Fazekas, J. Kiss, A. Miklos, and S.J. Yang. *Noise and Vibration of Electrical Machines*. Elsevier, Amsterdam, the Netherlands, 1989.
4. T. Tarhuvud and K. Riechert. Accuracy problems of force and torque calculation in FE-systems. *IEEE Transactions on Magnetics*, 24, 443–446, 1988.
5. W. Muller. Comparison of different methods of force calculation. *IEEE Transactions on Magnetics*, 26, 1058–1061, 1990.
6. N. Sadowski, Y. Lefevre, M. Lajoie-Mazenc, and J. Cros. Finite element torque calculation in electrical machine while considering the movement. *IEEE Transactions on Magnetics*, 38, 1410–1413, 1992.
7. M. Marinescu and N. Marinescu. Numerical computation of torques in permanent magnet motors by Maxwell stresses and energy method. *IEEE Transactions on Magnetics*, 24, 463–466, 1988.
8. D.O'Kelly. *Performance and Control of Electrical Machines*. McGraw-Hill Book Company, Maidenhead, U.K., 1991.
9. M. Jufer. *Electromecanique*, Vol. 9. Presses Polytechniques et Universitaires Romandes, Lausanne, Switzerland, 1995.

10. W. Leonhard. 30 tears space vectors, 20 years field orientation, 10 years digital signal processing with controlled AC-drives, a review (part 1). *European Power Electronics and Drives Journal*, 1, 13–31, July 1991.

11. P. Vas. *Vector Control of AC Machine*. Oxford Science Publication, Oxford, U.K., 1990.

12. T.M. Jahns. Torque production in permanent magnet motors drives with rectangular current excitation. *IEEE Transactions on Industry Applications*, 20, 803–813, 1984.

13. S.M. Abdulrahman, J.G. Kettleborough, and I.R. Smith. Fast calculation of harmonic torque pulsations in a VSI/induction motor drive. *IEEE Transactions on Industrial Electronics*, 40(6), 561–569, 1993.

14. J.P.G. De Abreu, J.S. De Sa, and C.C. Prado. Harmonic torques in three-phase induction motors supplied by nonsinusoidal voltages. *Eleventh International Conference on Harmonics and Quality of Power*, Sept. 2004, pp. 652–657.

15. H. Le Huy, R. Perret, and R. Feuillet. Minimization of torque ripple in brushless DC motor drives. *IEEE Transactions on Industry Applications*, 22, 748–755, 1986.

16. J.J. Spangler. A low cost, simple torque ripple reduction technique for three phase inductor motors. *Proceedings of the Applied Power Electronics Conference and Exposition (APEC 2002)*, Vol. 2, Dallas, TX, March 2002, pp. 759–763.

17. M. Elbuluk. Torque ripple minimization in direct torque control of induction machines. *Proceedings of the 38th IAS Annual Meeting*, Vol. 1, Salt Lake City, UT, Oct. 2003, pp. 11–16.

18. J.M.D. Murphy and F.G. Turnbull. *Power Electronic Control of AC Motors*. Pergamon Press, Oxford, U.K., 1988.

19. R. Romary, D. Roger, and J.F. Brudny. A current source PWM inverter used to reject harmonic torques of the permanent magnet synchronous machine. *SPEEDAM 1994*, Taormina, Italie, Juin 1994, pp. 115–120.

20. J. Zhao, M.J. Kamper, and F.S. Van der Merwe. On-line control method to reduce mechanical vibration and torque ripple in reluctance synchronous machine drives. *Proceedings of the IECON 97*, Vol. 1, New Orleans, LA, Nov. 1997, pp. 126–131.

21. T.J.E. Miller. *Brushless, Permanent-Magnet and Reluctance Motor Drives*. Clarendon Press, Oxford, U.K., 1989.

22. F.W. Carter. Air-gap induction. *Electrical World and Engineer*, 38(22), 884–888, November 1901.

23. R. Romary and J.F. Brudny, A skew shape rotor to optimize magnetic noise reduction of induction machine. *ICEM 98*, Istanbul, Turkey, 1998, pp. 1756–1760.

III

Conversion

11
Three-Phase AC–DC Converters

Mariusz
Malinowski
*Warsaw University
of Technology*

Marian P.
Kazmierkowski
*Warsaw University
of Technology*

11.1 Overview

11.1.1 Introduction

Currently, an increasing part of generated electric energy is converted through AC–DC converters, before it is consumed in the final load. The majority of systems apply a diode rectifier (Figure 11.1). Diode rectifiers are simple, reliable, robust, and cheap. However, diodes rectifiers enable only unidirectional power flow and cause a high level of harmonic input currents; moreover the performance varies significantly with load [1]. Performance can be improved by the application of input inductance but it demands installation of a bulky three-phase choke (Figure 11.2).

During the last years, PWM converters have drastically increased their importance on the market of AC–DC conversion [2]. Two technology breakthroughs of the electronic industry have enabled this remarkable development:

- The introduction of IGBTs on the market enabled the manufacturing of reliable, robust, and low-cost PWM converter modules.
- The introduction of low-cost microprocessors (e.g., digital signal processors—DSPs) and FPGA for real-time applications allowed the successful implementation of complex vector control schemes for PWM converters.

AC–DC voltage source converters (VSC) are widely used in industrial AC drives as active front end (AFE), DC-power supply, power quality improvement, and harmonic compensation (active filter) equipments. Lately, AC–DC converters have become a very important part of an AC–DC–AC line interfacing converters in renewable and distributed energy systems. Three-phase PWM converters are increasingly

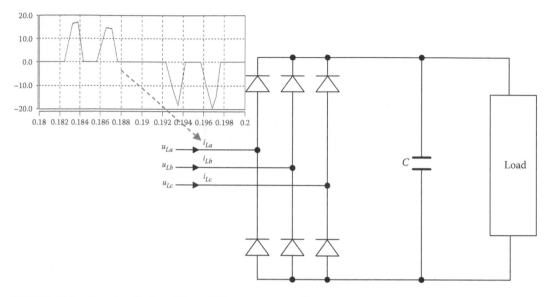

FIGURE 11.1   Six-pulse diode rectifier with line current waveform.

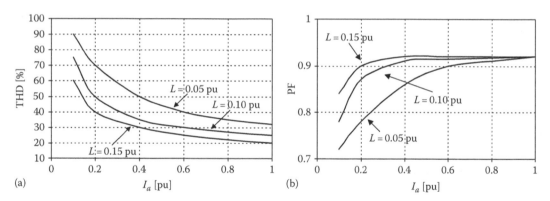

FIGURE 11.2   (a) THD [%] of six-pulse diode rectifier phase current versus variation of input inductance and load and (b) power factor (PF) of six-pulse diode rectifier versus variation of input inductance and load.

applied if a substantial amount of energy should be fed into the grid. Furthermore, it offers characteristics like a low harmonic distortion of the line currents (compliance to IEEE 519), a regulation of the input power factor, an adjustment and stabilization of the DC-link voltage, and, depending on the requirements, possibly a reduction of the DC filter capacitor size for certain applications.

Three-phase PWM AC–AD converter is connected to the grid through inductor L or LCL filter (high-performance application), which are an integral part of the circuit (Figure 11.3).

Two-level converters for small power (Figure 11.3a) and three-level converters for high power (Figure 11.3b) are typical PWM topologies used by industry. Three-level converters require more complex modulation but they has, in comparison to two-level converters, reduced voltage stress on every switch, better power quality (lower current and voltage THD), and about 30% smaller LCL filter [3].

11.1.2 Control Strategies

Another important technology is innovations in the field of converter control principles. Various control strategies have been proposed in recent works on this type of PWM converter [4–9]. A well-known method of indirect active and reactive power control is based on current vector orientation with respect

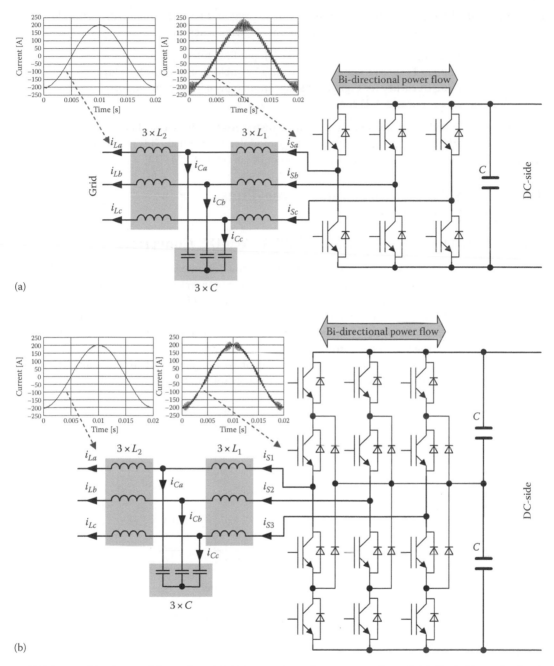

FIGURE 11.3 Topologies of three-phase AC–DC converter with current waveforms (LCL filter): (a) two-level and (b) three-level.

to the line voltage vector (voltage-oriented control—VOC) [4–6,9]. VOC guarantees high dynamics and static performance via internal current control loops. However, the final configuration and performance of the VOC system largely depends on the quality of the applied current control strategy [10]. Another less-known method based on instantaneous direct active and reactive power control is called direct power control (DPC) [7,8]. Both mentioned strategies do not perform sinusoidal current when the line voltage is distorted. Only a DPC strategy based on virtual flux instead of the line voltage vector orientation called VF-DPC, provides sinusoidal line current and lower harmonic distortion [4,11,12]. However, among the well-known disadvantages of the VF-DPC scheme are

- Variable switching frequency (difficulties of LC input EMI filter design)
- Violation of polarity consistency rules (to avoid ±1 switching over dc link voltage)
- High sampling frequency for digital implementation of hysteresis comparators
- Fast microprocessor and A/D converters

Therefore, it is difficult to implement VF-DPC in industry. All the above drawbacks can be eliminated when, instead of the switching table, a PWM voltage modulator is applied. This is realized in DPC with constant switching frequency using space vector modulation (DPC-SVM) [13].

11.2 Control Techniques for Three-Phase PWM AC–DC Converters

11.2.1 Basic Operation Principles of PWM AC–DC Converters

Figure 11.4a shows a single-phase representation of the circuit presented in Figure 11.4b. L and R represent the line inductor, $\underline{u}_L$ is the line voltage, and $\underline{u}_S$ is the bridge converter voltage, controllable from the DC-side. The magnitude of $\underline{u}_S$ depends on the modulation index and DC voltage level [4].

The line current i_L is controlled by the voltage drop across the inductance L interconnecting two voltage sources (line and converter). It means that the inductance voltage u_I equals the difference between the line voltage u_L and the converter voltage u_S. When we control phase angle ε and amplitude of converter voltage u_S, we control indirectly the phase and amplitude of line current. In this way, average value and sign of DC current is subject to control what is proportional to active power

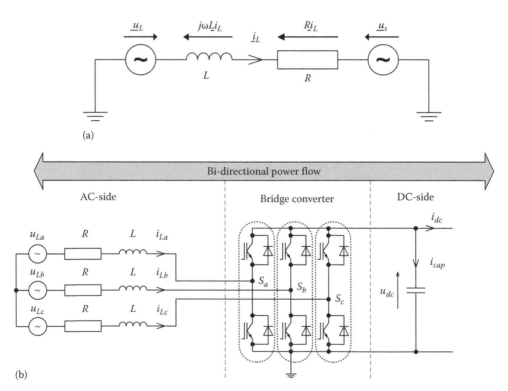

FIGURE 11.4 Simplified representation of three-phase PWM AC–DC converter: (a) single-phase representation of the circuit and (b) main circuit.

conducted through converter. The reactive power can be controlled independently with shift of fundamental harmonic current I_L in respect to voltage U_L.

11.2.2 Mathematical Description of the PWM AC–DC Converters

The basic relationship between vectors of the PWM rectifier is presented in Figure 11.5. For three-phase, line voltage and the fundamental line current is

$$u_{La} = E_m \cos \omega t \tag{11.1a}$$

$$u_{Lb} = E_m \cos\left(\omega t + \frac{2\pi}{3}\right) \tag{11.1b}$$

$$u_{Lc} = E_m \cos\left(\omega t - \frac{2\pi}{3}\right) \tag{11.1c}$$

$$i_{La} = I_m \cos(\omega t + \phi) \tag{11.2a}$$

$$i_{Lb} = I_m \cos\left(\omega t + \frac{2\pi}{3} + \phi\right) \tag{11.2b}$$

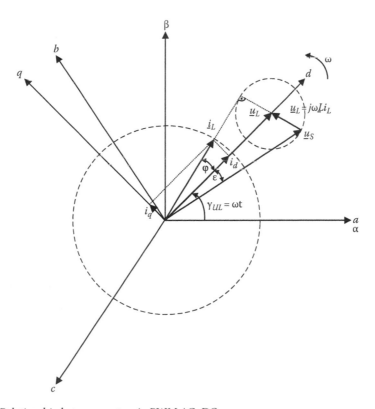

FIGURE 11.5 Relationship between vectors in PWM AC–DC.

$$i_{Lc} = I_m \cos\left(\omega t - \frac{2\pi}{3} + \phi\right) \tag{11.2c}$$

where E_m (I_m) and ω are amplitude of the phase voltage (current) and angular frequency, respectively.

Line-to-line input voltages of PWM converter can be described as

$$u_{Sab} = (S_a - S_b) \cdot u_{dc} \tag{11.3a}$$

$$u_{Sbc} = (S_b - S_c) \cdot u_{dc} \tag{11.3b}$$

$$u_{Sca} = (S_c - S_a) \cdot u_{dc} \tag{11.3c}$$

where

$S_x = 1$ when the upper transistor of one leg in converter is switched ON ($x = a, b, c$)
$S_x = 0$ when the lower transistor of one leg in converter is switched ON ($x = a, b, c$)

The phase voltages are equal:

$$u_{Sa} = f_a \cdot u_{dc} \tag{11.4a}$$

$$u_{Sb} = f_b \cdot u_{dc} \tag{11.4b}$$

$$u_{Sc} = f_c \cdot u_{dc} \tag{11.4c}$$

where

$$f_a = \frac{2S_a - (S_b + S_c)}{3} \tag{11.5a}$$

$$f_b = \frac{2S_b - (S_a + S_c)}{3} \tag{11.5b}$$

$$f_c = \frac{2S_c - (S_a + S_b)}{3} \tag{11.5c}$$

The f_a, f_b, f_c are assumed 0, ±1/3, and ±2/3, respectively.

11.2.2.1 Model of Three-Phase PWM AC–DC Converters in Natural Coordinates (*abc*)

The voltage equations for balanced three-phase system without the neutral connection can be written as (Figure 11.5)

$$\underline{u}_L = \underline{u}_I + \underline{u}_S \tag{11.6}$$

$$\underline{u}_L = R\underline{i}_L + \frac{di_L}{dt}L + \underline{u}_S \tag{11.7}$$

$$\begin{bmatrix} u_{La} \\ u_{Lb} \\ u_{Lc} \end{bmatrix} = R \begin{bmatrix} i_{La} \\ i_{Lb} \\ i_{Lc} \end{bmatrix} + L \frac{d}{dt} \begin{bmatrix} i_{La} \\ i_{Lb} \\ i_{Lc} \end{bmatrix} + \begin{bmatrix} u_{Sa} \\ u_{Sb} \\ u_{Sc} \end{bmatrix} \tag{11.8}$$

Other current equations determine relations among phase currents (i_{La}, i_{Lb}, i_{Lc}), load current (i_{dc}), and dc-link capacitor current (i_{cap})

$$C \frac{du_{dc}}{dt} = i_{cap} = S_a i_{La} + S_b i_{Lb} + S_c i_{Lc} - i_{dc} \tag{11.9}$$

For example, one of eight possible converter switching states ($S_a = 1$, $S_b = 0$, $S_c = 0$) gives the following current equation $i_{cap} = i_{La} - i_{dc}$.

The combination of Equations 11.4 through 11.9 can be represented as three-phase block diagram [14].

11.2.2.2 Model of PWM AC–DC Converters in Stationary Coordinates (α-β)

The voltage equations for balanced three-phase system without the neutral connection can be simplified by representation in fixed rectangular α-β coordinate system, where each vector is described by only two variables. The α-axis (real) and the *a*-axis have the same orientation but the β-axis (imaginary) leads the *a*-axis with 90° (Figure 11.5). It resolves each vector into real and imaginary parts $x_{\alpha\beta} = x_\alpha + jx_\beta$, where x_α and x_β are obtained by applying Clarke transformation defining relationship between the three-phase system *a-b-c* and the stationary reference frame α-β:

$$\begin{bmatrix} x_\alpha \\ x_\beta \end{bmatrix} = \sqrt{\frac{2}{3}} \begin{bmatrix} 1 & -1/2 & -1/2 \\ 0 & \sqrt{3}/2 & -\sqrt{3}/2 \end{bmatrix} \begin{bmatrix} x_a \\ x_b \\ x_c \end{bmatrix}. \tag{11.10}$$

Then, Equations 11.8 and 11.9 can be written as

$$\begin{bmatrix} u_{L\alpha} \\ u_{L\beta} \end{bmatrix} = R \begin{bmatrix} i_{L\alpha} \\ i_{L\beta} \end{bmatrix} + L \frac{d}{dt} \begin{bmatrix} i_{L\alpha} \\ i_{L\beta} \end{bmatrix} + \begin{bmatrix} u_{S\alpha} \\ u_{S\beta} \end{bmatrix} \tag{11.11}$$

and

$$C \frac{du_{dc}}{dt} = (i_{L\alpha} S_\alpha + i_{L\beta} S_\beta) - i_{dc} \tag{11.12}$$

where

$$S_\alpha = \frac{1}{\sqrt{6}} (2S_a - S_b - S_c)$$

$$S_\beta = \frac{1}{\sqrt{2}} (S_b - S_c)$$

11.2.2.3 Model of PWM AC–DC Converters in Synchronous Rotating Coordinates (*d-q*)

The voltage equations can be even represented in coordinate system, where vectors are transformed from stationary α-β to synchronously rotating, with line voltage vector, *d-q* coordinate system. The angle between the real axis α of the stationary system and real axis *d* of the new rotating system can be described as $\gamma_{UL} = \omega t$. Then, vectors can be transformed from α-β in the synchronous *d-q* coordinates with the help of simple trigonometrical relationship:

$$\begin{bmatrix} k_d \\ k_q \end{bmatrix} = \begin{bmatrix} \cos\gamma_{UL} & \sin\gamma_{UL} \\ -\sin\gamma_{UL} & \cos\gamma_{UL} \end{bmatrix} \begin{bmatrix} k_\alpha \\ k_\beta \end{bmatrix}. \tag{11.13}$$

It gives

$$u_{Ld} = Ri_{Ld} + L\frac{di_{Ld}}{dt} - \omega Li_{Lq} + u_{Sd} \tag{11.14a}$$

$$u_{Lq} = Ri_{Lq} + L\frac{di_{Lq}}{dt} + \omega Li_{Ld} + u_{Sq} \tag{11.14b}$$

$$C\frac{du_{dc}}{dt} = (i_{Ld}S_d + i_{Lq}S_q) - i_{dc} \tag{11.15}$$

where

$$S_d = S_\alpha \cos\omega t + S_\beta \sin\omega t$$

$$S_q = S_\beta \cos\omega t - S_\alpha \sin\omega t$$

A block diagram of *d-q* model is presented in Figure 11.6.

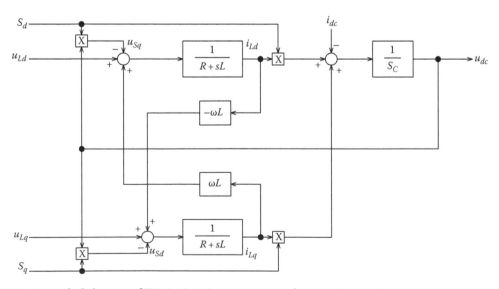

FIGURE 11.6 Block diagram of PWM AC–DC converter in synchronous *d-q* coordinates.

11.2.3 Line Voltage, Virtual Flux, and Instantaneous Power Estimation

11.2.3.1 Line Voltage Estimation

An important requirement for a voltage estimator is to estimate the voltage correct, also under unbalanced conditions and pre-existing harmonic voltage distortion. Not only the fundamental component should be estimated correct, but also the harmonic components and the voltage unbalance. It gives a higher total power factor [7]. It is possible to calculate the voltage across the inductance by the current differentiating. The line voltage can then be estimated by adding reference of the rectifier input voltage to the calculated voltage drop across the inductor [15]. However, this approach has the disadvantage that the current is differentiated and noise in the current signal is gained through the differentiation. To prevent this, a voltage estimator based on the power estimator of [7] can be applied, where the active power p is described as *scalar product* between the three-phase voltages drop across the inductor and line currents and reactive power q as *vector product* between them:

$$p_I = \underline{u}_{I(abc)} \times \underline{i}_{L(abc)} = u_{Ia}i_{La} + u_{Ib}i_{Lb} + u_{Ic}i_{Lc} \tag{11.16a}$$

$$q_I = \underline{u}_{I(abc)} \times \underline{i}_{L(abc)} = u'_{Ia}i_{La} + u'_{Lb}i_{Lb} + u'_{Lb}i_{Lc} \tag{11.16b}$$

$u'_{Ia}, u'_{Ib}, u'_{Ic}$ is 90° lag of u_{Ia}, u_{Ib}, u_{Ic}, respectively. The same equations can be described in matrix form as

$$\begin{bmatrix} p_I \\ q_I \end{bmatrix} = \begin{bmatrix} u_{Ia} & u_{Ib} & u_{Ic} \\ u'_{Ia} & u'_{Ib} & u'_{Ic} \end{bmatrix} \begin{bmatrix} i_{La} \\ i_{Lb} \\ i_{Lc} \end{bmatrix} \tag{11.16c}$$

where

$$\begin{bmatrix} u'_{Ia} \\ u'_{Ib} \\ u'_{Ic} \end{bmatrix} = \frac{1}{\sqrt{3}} \begin{bmatrix} u_{Ic} - u_{Ib} \\ u_{Ia} - u_{Ic} \\ u_{Ib} - u_{Ia} \end{bmatrix} = \frac{1}{\sqrt{3}} \begin{bmatrix} u_{Icb} \\ u_{Iac} \\ u_{Iba} \end{bmatrix}. \tag{11.16d}$$

Using Equations 11.16a through d, the estimated active and reactive power across the inductance can be expressed as

$$p_I = L\left(\frac{di_{La}}{dt} i_{La} + \frac{di_{Lb}}{dt} i_{Lb} + \frac{di_{Lc}}{dt} i_{Lc} \right) = 0 \tag{11.17a}$$

$$q_I = \frac{3L}{\sqrt{3}} \left(\frac{di_{La}}{dt} i_{Lc} - \frac{di_{Lc}}{dt} i_{La} \right). \tag{11.17b}$$

where

$$L \frac{di_{La}}{dt} = u_{Ia}$$

$$L \frac{di_{Lb}}{dt} = u_{Ib}$$

$$L \frac{di_{Lc}}{dt} = u_{Ic}$$

Since powers are DC-values, it is possible to prevent the noise of the differentiated current by use of a simple low-pass filter. This ensures a robust and noise-insensitive performance of the voltage estimator.

Then, based on instantaneous power theory proposed by Akagi et al. [16], when the three-phase voltages and currents are transformed into α-β coordinates,

$$\begin{bmatrix} p_I \\ q_I \end{bmatrix} = \begin{bmatrix} i_{L\alpha} & i_{L\beta} \\ -i_{L\beta} & i_{L\alpha} \end{bmatrix} \begin{bmatrix} u_{I\alpha} \\ u_{I\beta} \end{bmatrix} \tag{11.18}$$

the estimated voltages across the inductance after transformation of (11.18) are

$$\begin{bmatrix} u_{I\alpha} \\ u_{I\beta} \end{bmatrix} = \frac{1}{i_{L\alpha}^2 + i_{L\beta}^2} \begin{bmatrix} i_{L\alpha} & -i_{L\beta} \\ i_{L\beta} & i_{L\alpha} \end{bmatrix} \begin{bmatrix} 0 \\ q_I \end{bmatrix} \tag{11.19}$$

It should be noted that in this special case, it is only possible to estimate the reactive power in the inductor.

The estimated line voltage $u_{L(est)}$ can now be found by adding the voltage reference of the PWM rectifier to the estimated inductor voltage [6].

$$\underline{u}_{L(est)} = \underline{u}_S + \underline{u}_I \tag{11.20}$$

11.2.3.2 Virtual Flux Estimation

It is possible to replace the AC-line voltage sensors with a virtual flux estimator, what gives technical and economical advantages to the system as: simplification, isolation between the power circuit and control system, reliability, and cost effectiveness.

The integration of the voltages leads to a virtual flux (VF) vector $\underline{\Psi}_L$, in stationary α-β coordinates [4,12]:

$$\underline{\Psi}_L = \begin{bmatrix} \Psi_{L\alpha} \\ \Psi_{L\beta} \end{bmatrix} = \begin{bmatrix} \int u_{L\alpha} \, dt \\ \int u_{L\beta} \, dt \end{bmatrix} \tag{11.21}$$

where

$$\underline{u}_L = \begin{bmatrix} u_{L\alpha} \\ u_{L\beta} \end{bmatrix} = \sqrt{\frac{2}{3}} \begin{bmatrix} 1 & 1/2 \\ 0 & \sqrt{3}/2 \end{bmatrix} \begin{bmatrix} u_{Lab} \\ u_{Lbc} \end{bmatrix} \tag{11.22}$$

When we establish, that [6]

$$\underline{u}_L = \underline{u}_S + \underline{u}_I \tag{11.23}$$

then, similarly to (11.23), a virtual flux equation can be presented as [12]

$$\underline{\Psi}_L = \underline{\Psi}_S + \underline{\Psi}_I \tag{11.24}$$

Based on the measured DC link voltage u_{dc} and the duty cycles of modulator D_a, D_b, D_c, the virtual flux $\underline{\Psi}_L$ components are calculated in stationary (α-β) coordinates system as follows:

$$\Psi_{L\alpha} = \int \left(\sqrt{\frac{2}{3}} u_{dc} \left(D_a - \frac{1}{2} (D_b + D_c) \right) \right) dt + Li_{L\alpha} \tag{11.25a}$$

$$\Psi_{L\beta} = \int \left(\frac{1}{\sqrt{2}} u_{dc}(D_b - D_c) \right) dt + Li_{L\beta} \tag{11.25b}$$

11.2.3.3 Instantaneous Power Calculation Based on Voltage Estimation

The instantaneous values of active (p) and reactive power (q) in AC voltage sensorless system are estimated by Equations 11.26. First part of both equations represents power in the inductance and the second part is the power of the PWM AC–DC converter [7]:

$$p = L \left(\frac{di_{La}}{dt} i_{La} + \frac{di_{Lb}}{dt} i_{Lb} + \frac{di_{Lc}}{dt} i_{Lc} \right) + u_{dc}(S_a i_{La} + S_b i_{Lb} + S_c i_{Lc}) \tag{11.26a}$$

$$q = \frac{1}{\sqrt{3}} \left\{ 3L \left(\frac{di_{La}}{dt} i_{Lc} - \frac{dt_{Lc}}{dt} i_{La} \right) - u_{dc} \left[S_a(i_{Lb} - i_{Lc}) + S_b(i_{Lc} - i_{La}) + S_c(i_{La} - i_{Lb}) \right] \right\} \tag{11.26b}$$

As can be seen in (11.26), the forms of equations have to be changed according to the switching state of the converter, and both equations require the knowledge of the line inductance L.

11.2.3.4 Instantaneous Power Calculation Based on Virtual Flux Estimation

The measured line currents i_{La}, i_{Lb} and the estimated virtual flux components $\Psi_{L\alpha}$, $\Psi_{L\beta}$ are used to the power estimation [4,11]. Using (11.23), the voltage equation can be written as (in practice, R can be neglected)

$$\underline{u}_L = L \frac{di_L}{dt} + \frac{d}{dt} \underline{\Psi}_S = L \frac{di_L}{dt} + \underline{u}_S \tag{11.27}$$

Using complex notation, the instantaneous power can be calculated as follows:

$$p = Re(\underline{u}_L \cdot \underline{i}_L^*) \tag{11.28a}$$

$$q = Im(\underline{u}_L \cdot \underline{i}_L^*) \tag{11.28b}$$

where * denotes conjugate of the line current vector. The line voltage can be expressed by the virtual flux as

$$\underline{u}_L = \frac{d}{dt} \underline{\Psi}_L = \frac{d}{dt} (\Psi_L e^{j\omega t}) = \frac{d\Psi_L}{dt} e^{j\omega t} + j\omega \Psi_L e^{j\omega t} = \frac{d\Psi_L}{dt} e^{j\omega t} + j\omega \underline{\Psi}_L \tag{11.29}$$

where $\underline{\Psi}_L$ denotes the space vector and Ψ_L its amplitude. For virtual flux-oriented quantities, in α-β coordinates and using (11.28) and (11.29):

$$\underline{u}_L = \frac{d\Psi_L}{dt} \bigg|_{\alpha} + j \frac{d\Psi_L}{dt} \bigg|_{\beta} + j\omega (\Psi_{L\alpha} + j\Psi_{L\beta}) \tag{11.30}$$

$$\underline{u}_L \underline{i}_L^* = \left\{ \left. \frac{d\Psi_L}{dt} \right|_\alpha + j \left. \frac{d\Psi_L}{dt} \right|_\beta + j\omega(\Psi_{L\alpha} + j\Psi_{L\beta}) \right\}(i_{L\alpha} - ji_{L\beta}) \tag{11.31}$$

That gives

$$p = \left\{ \left. \frac{d\Psi_L}{dt} \right|_\alpha i_{L\alpha} + \left. \frac{d\Psi_L}{dt} \right|_\beta i_{L\beta} + \omega(\Psi_{L\alpha}i_{L\beta} - \Psi_{L\beta}i_{L\alpha}) \right\} \tag{11.32a}$$

and

$$q = \left\{ - \left. \frac{d\Psi_L}{dt} \right|_\alpha i_{L\beta} + \left. \frac{d\Psi_L}{dt} \right|_\beta i_{L\alpha} + \omega(\Psi_{L\alpha}i_{L\alpha} + \Psi_{L\beta}i_{L\beta}) \right\} \tag{11.32b}$$

For sinusoidal and balanced line voltage, the derivatives of the flux amplitudes are zero. The instantaneous active and reactive powers can be computed as [11]

$$p = \omega \cdot (\Psi_{L\alpha}i_{L\beta} - \Psi_{L\beta}i_{L\alpha}) \tag{11.33a}$$

$$q = \omega \cdot (\Psi_{L\alpha}i_{L\alpha} + \Psi_{L\beta}i_{L\beta}). \tag{11.33b}$$

11.2.4 Voltage-Oriented Control

The conventional control system uses closed-loop current control in rotating reference frame, the VOC scheme is shown in Figure 11.7. A characteristic feature for this current controller is processing of signals in two coordinate systems. The first is stationary α-β and the second is synchronously rotating d-q coordinate system. Three-phase measured values are converted to equivalent two-phase system α-β and then are transformed to rotating coordinate system in a block α-β/d-q (11.13).

Thanks to this type of transformation, the control values are DC signals. An inverse transformation d-q/α-β is achieved on the output of control system and it gives a result, the AC–DC converter reference signals in stationary coordinate:

$$\begin{bmatrix} u_{S\alpha} \\ u_{S\beta} \end{bmatrix} = \begin{bmatrix} \cos\gamma_{UL} & -\sin\gamma_{UL} \\ \sin\gamma_{UL} & \cos\gamma_{UL} \end{bmatrix} \begin{bmatrix} u_{Sd} \\ u_{Sq} \end{bmatrix} \tag{11.34}$$

For both coordinate transformation, the angle of the voltage vector γ_{UL} is defined as

$$\sin\gamma_{UL} = \frac{u_{L\beta}}{\sqrt{(u_{L\alpha})^2 + (u_{L\beta})^2}} \tag{11.35a}$$

$$\cos\gamma_{UL} = \frac{u_{L\alpha}}{\sqrt{(u_{L\alpha})^2 + (u_{L\beta})^2}}. \tag{11.35b}$$

In voltage oriented d-q coordinates, the AC line current vector $\underline{i}_L$ is split into two rectangular components $\underline{i}_L = [i_{Ld}, i_{Lq}]$ (Figure 11.8). The component i_{Lq} determinates reactive power, whereas i_{Ld} decides

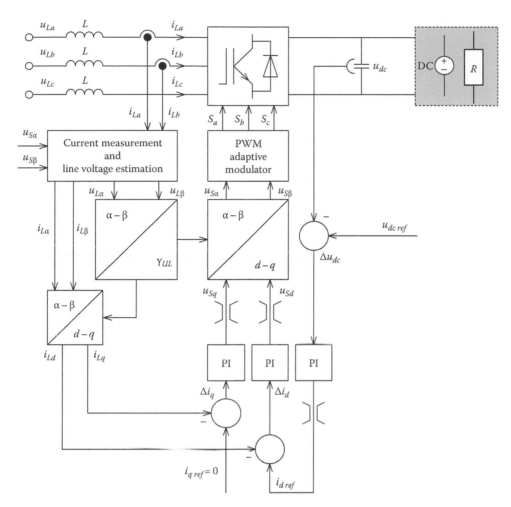

FIGURE 11.7 Block scheme of AC voltage sensorless VOC.

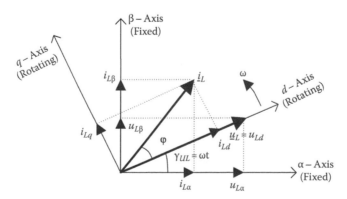

FIGURE 11.8 Coordinate transformation of line current and line voltage from stationary α–β coordinates to rotating *d-q* coordinates.

about active power flow. Thus, the reactive and the active power can be controlled independently. The UPF condition is met when the line current vector, $\underline{i}_L$, is aligned with the line voltage vector, $\underline{u}_L$.

The voltage equations in the *d-q* synchronous reference frame in accordance with Equations 11.14 and with assumption that the *q*-axis current is set to zero in all condition for unity power factor control while the reference current i_{Ld} is set by the DC-link voltage controller and controls the active power flow between the grid and the DC-link can be reduced to ($R \approx 0$) [4]

$$u_{Ld} = L\frac{di_{Ld}}{dt} + u_{Sd} - \omega \cdot L \cdot i_{Lq} \tag{11.36a}$$

$$0 = L\frac{di_{Lq}}{dt} + u_{Sq} + \omega \cdot L \cdot i_{Ld} \tag{11.36b}$$

Assuming that the *q*-axis current is well regulated to zero, the following equations hold true:

$$u_{Ld} = L\frac{di_{Ld}}{dt} + u_{Sd} \tag{11.37a}$$

$$0 = u_{Sq} + \omega \cdot L \cdot i_{Ld} \tag{11.37b}$$

As current controller, the PI-type can be used. However, the PI current controller has no satisfactory tracing performance, especially, for the coupled system described by Equations 11.36. Therefore, for high-performance application with accuracy current tracking at dynamic state, the decoupled controller diagram for the PWM AC–DC converter should be applied to what is shown in Figure 11.9 [4]:

$$u_{Sd} = \omega L i_{Lq} + u_{Ld} + \Delta u_d \tag{11.38a}$$

$$u_{Sq} = -\omega L i_{Ld} + \Delta u_q \tag{11.38b}$$

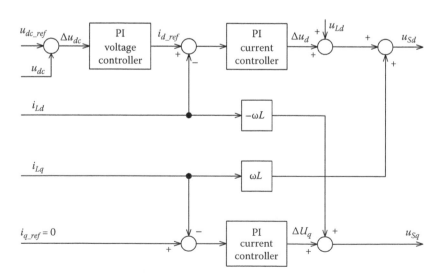

FIGURE 11.9 Decoupled current control of PWM AC–DC converter.

where Δ is the output signals of the current controllers:

$$\Delta u_d = k_p \left(i_{d\_ref} - i_{Ld} \right) + k_i \int \left(i_{d\_ref} - i_{Ld} \right) dt \qquad (11.39a)$$

$$\Delta u_q = k_p = \left(i_{q\_ref} - i_{Lq} \right) + k_i \int \left(i_{q\_ref} - i_{Lq} \right) dt \qquad (11.39b)$$

The output signals from PI controllers after $dq/\alpha\beta$ transformation (Equation 11.24) are used for switching signals generation by a space vector modulator (SVM).

11.2.5 Virtual Flux-Based Direct Power Control

Figure 11.10 shows configuration of virtual flux-based direct power control (VF-DPC), where the commands of reactive power q_{ref} (set to zero for unity power factor) and active power p_{ref} (delivered from the outer PI-DC voltage controller) are compared with the estimated q and p values (Equations 11.33a and b), in reactive and active power hysteresis controllers, respectively. The digitized variables d_p, d_q and the line voltage vector position $\gamma_{UL} = arc\ tg\ (u_{L\alpha}/u_{L\beta})$ form a digital word, which by accessing the address of the look-up table, selects the appropriate voltage vector according to the switching table [4].

However, disturbances supcrimposed onto the line voltage influence directly the line voltage vector position in control system. Sometimes, this problem is overcome by phase-locked loops (PLLs) only, but the quality of the controlled system depends on how effectively the PLLs have been designed. Therefore, it is easier to replace angle of the line voltage vector γ_{UL} by angle of VF vector $\gamma_{\psi L} = arc\ tg\ (\psi_{L\alpha}/\psi_{L\beta})$, because $\gamma_{\psi L}$ is less sensitive than γ_{UL} to disturbances in the line voltage, thanks to the natural low-pass behavior of the integrators in estimator (Equations 11.25a and b). For this reason, it is not necessary to implement PLLs to achieve robustness in the flux-oriented scheme.

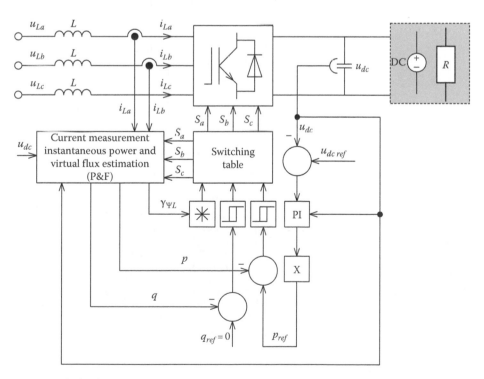

FIGURE 11.10 Block scheme of VF-DPC.

11.2.6 Direct Power Control–Space Vector Modulated

The DPC-SVM with constant switching frequency uses closed-loop power control, what is shown in Figure 11.11a [13,17]. The commanded reactive power q_{ref} (set to zero for unity power factor operation) and (delivered from the outer PI-DC voltage controller) active power p_{ref} (power flow between the supply and the DC-link) values are compared with the estimated q and p values (Equations 11.33a and b),

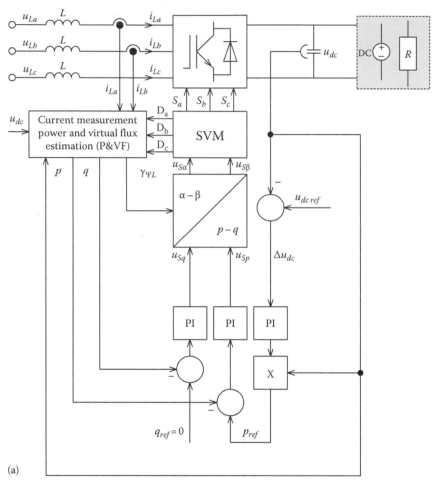

(a)

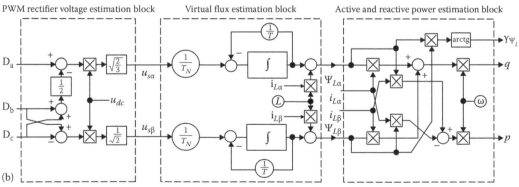

(b)

FIGURE 11.11 (a) Block scheme of DPC-SVM and (b) block scheme of DPC-SVM estimators (P&VF). (From Malinowski, M. et al., *IEEE Trans. Ind. Elect.*, 51(2), 447, 2004. With permission.)

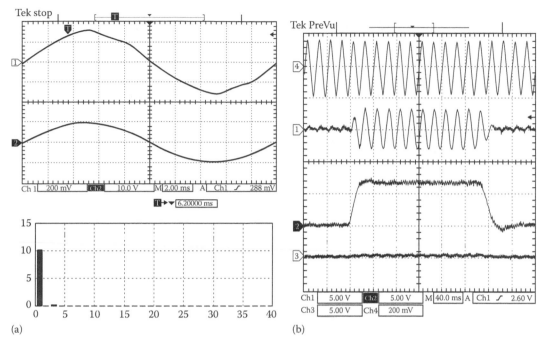

(a) (b)

FIGURE 11.12 Basic signal waveforms for DPC-SVM: (a) steady state. From the top: distorted line voltage, line currents (10 A/div), and harmonic spectrum of line current (THD = 2.6%). (b) Transient of the step change of the load. From the top: line voltages, line currents (10 A/div), active and reactive power. (From Malinowski, M. et al., *IEEE Trans. Ind. Elect.*, 51(2), 447, 2004. With permission.)

respectively. The errors are delivered to PI controllers, where the variables are DC quantities, what eliminates steady state error. The output signals from PI controllers after transformation are described as

$$
\begin{bmatrix} u_{S\alpha} \\ u_{S\beta} \end{bmatrix} = \begin{bmatrix} -\sin\gamma_{\Psi L} & -\cos\gamma_{\Psi L} \\ \cos\gamma_{\Psi L} & -\sin\gamma_{\Psi L} \end{bmatrix} \begin{bmatrix} u_{Sp} \\ u_{Sq} \end{bmatrix}
\tag{11.40}
$$

where

$$
\sin\gamma_{\Psi L} = \frac{\Psi_{L\beta}}{\sqrt{(\Psi_{L\alpha})^2 + (\Psi_{L\beta})^2}}
\tag{11.41a}
$$

$$
\cos\gamma_{\Psi L} = \frac{\Psi_{L\alpha}}{\sqrt{(\Psi_{L\alpha})^2 + (\Psi_{L\beta})^2}}.
\tag{11.41b}
$$

are used for switching signals generation by SVM. Block scheme of all DPC-SVM estimators and basic signals waveforms are shown in Figures 11.11b and 11.12.

11.2.7 Active Damping

The reduction of the current harmonics around switching frequency and the multiple of switching frequency is an important point to get high-performance PWM AC–DC converter, which fulfills standards (IEEE 519-1992, IEC 61000-3-2/IEC 61000-3-4). Large value of input inductance allows achieving this goal; however, it reduces dynamics and operation range of AC–DC converter [4]. Therefore, simple inductance is replaced by, third-order low-pass LCL filter [3,18] (Figure 11.13). In this solution, the

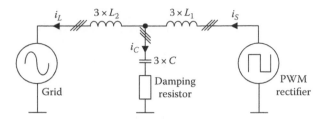

FIGURE 11.13 Equivalent circuit of three-phase PWM AC–DC converter with LCL filter. (From Malinowski, M. and Bernet, S., *IEEE Trans. Ind. Elect.*, 55(4), 1876, 2008. With permission.)

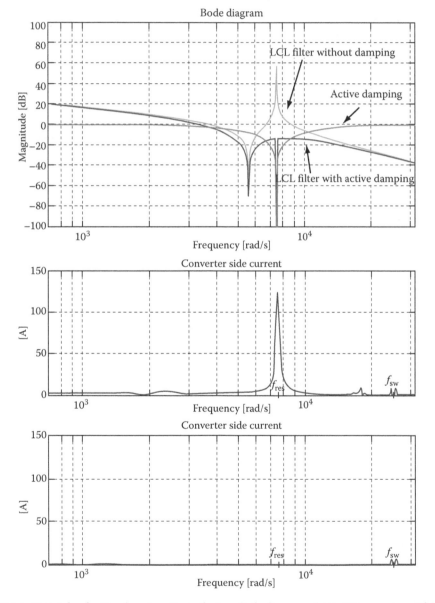

FIGURE 11.14 Principle of active damping. From the top: Bode diagram of AC–DC converter with LCL filter, harmonic spectrum of current with resonance effect, harmonic spectrum of current with active damping. (From Malinowski, M. and Bernet, S., *IEEE Trans. Ind. Elect.*, 55(4), 1876, 2008. With permission.)

current ripple attenuation is very effective even for small inductance size, because capacitor impedance is inversely proportional to the frequency of current and creates a low impedance path for higher harmonics. However, LCL can bring even undesired resonance effect (stability problems), caused by zero impedance for some higher-order harmonics of current. Unstable system can be stabilized using a damping resistor, so called *passive damping*. This solution despite advantages such as simplicity and reliability, due to which it is widely used in industry, has a main drawback: increase of losses and hence reduction of efficiency. Therefore, nowadays, a tendency to replace passive with *active damping* (AD) may be observed. AD is implemented by the modification of control algorithm, which stabilizes the system without increasing losses. Basic idea may be explained easily in frequency domain (Figure 11.14). Addition of AD algorithm introduces a negative peak that compensates for the positive one caused by presence of LCL filter [18]. In this section, a few different AD methods applied for VOC are shortly presented.

11.2.7.1 AD Based on Lead-Lag Compensator

General block scheme of the VOC with additional lead-lag element $L(s) = k_d(T_d s + 1)/(\alpha T_d s + 1)$ is presented in Figure 11.15. Measured capacitor voltages after transformation from stationary $\alpha\beta$ to synchronous dq rotating system (V_{C_d}, V_{C_q}) are delivered to lead-lag compensator. Then, output signals (V_{CR_d}, V_{CR_q}) are subtracted from modulator input signals (u_{Sd}, u_{Sq}). Proper effects of AD may be achieved only when correct system tuning has been carried out [19].

11.2.7.2 AD Based on Virtual Resistor

This method based on idea that real damping resistor, which is in series with capacitor, through simple block transformation, can be replaced by additional differential control block realizing function of "Virtual Resistor" [20]. Figure 11.16 presents simple block scheme of the VOC with "Virtual Resistor," where measured capacitor currents after transformation from stationary $\alpha\beta$ to synchronous dq rotating system are

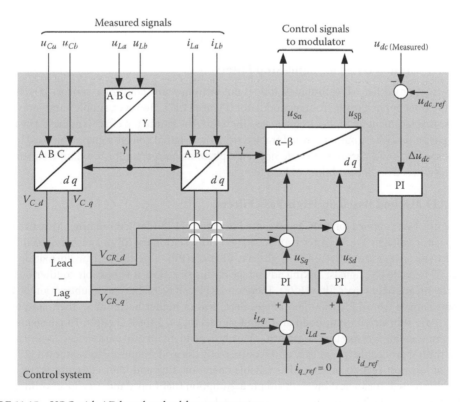

FIGURE 11.15 VOC with AD based on lead-lag compensator.

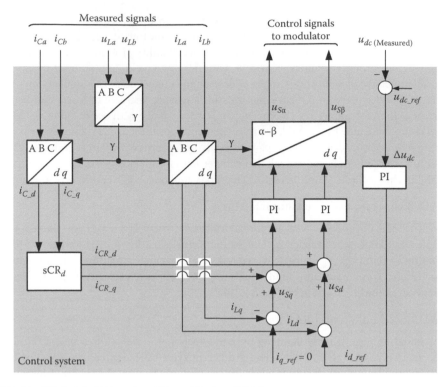

FIGURE 11.16   VOC with AD based on "Virtual Resistor" block.

differentiated and the output (i_{CR_d}, i_{CR_q}) is delivered to reference current signals (i_{d_ref}, i_{q_ref}). The main drawback of this method is the necessity of additional current sensors, which is difficult to replace by estimator.

11.2.7.3 AD Method Based on Band-Stop Filters

In contrary to many other AD techniques, it does not introduce any additional sensors [21]. It is based on band-stop filter applied in the front of modulator (Figure 11.17). However, simple band-stop filter can cause phase displacement for higher frequencies. Therefore, the band-stop effect is achieved by means of two band-pass filters whose outputs are subtracted from the original voltage signals. Imposed band-stop filters are based on band-pass ones that do not cause phase displacement for resonant frequency. Filter tuning is easy, based on known values of LCL filter.

11.2.7.4 AD Method Based on High-Pass Filters

Block scheme of AD applied for VOC is presented in Figure 11.18 (dashed red line). Measured or estimated capacitor voltages $V_{C\alpha}$, $V_{C\beta}$ after transformation from stationary $\alpha\beta$ to synchronous dq rotating coordinates (V_{C_d}, V_{C_q}) are delivered to high-pass filters (HPF). Due to the transformation ($\alpha\beta/dq$), 50 Hz signals become DC signals. Taking advantage of this situation, it is possible to filter out the first harmonic by means of low-pass filter (LPF). Therefore, HPF can be realized as a subtraction of V_{C_d}, V_{C_q} and outputs signals from LPF, what guarantees delay less in higher harmonics. Then, output signals (V_{CR_d}, V_{CR_q}) are subtracted from modulator input signals (u_{Sd}, u_{Sq}) for AD effect. In consequence, this method becomes really perspective, because it is hardly dependant on grid parameters, as well as the AD algorithm tuning procedure is very easy and requires only the grid frequency to be known [22].

In view of low-cost realization as well as reliable operation, line and filter-capacitor voltage sensors should be eliminated. It can be done with simple assumption that impedance of capacitor in LCL filter for low frequency is very high: ($i_C \approx 0$, $i_L \approx i_S$, $L_{12} = L_1 + L_2$).

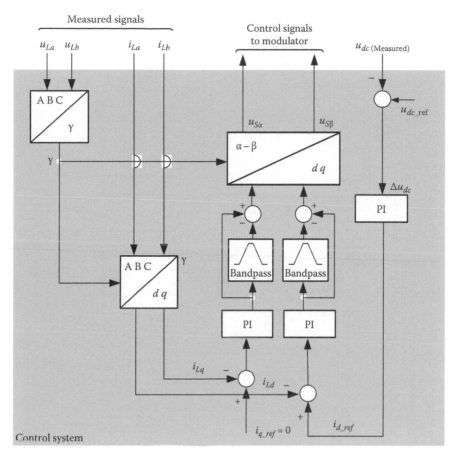

FIGURE 11.17 Control structure of VOC with active damping algorithm based on two band-pass filters.

With these assumptions, the line voltage can be estimated by summing rectifier's voltage with voltage drop across inductors. The calculation of voltage drop across the inductance can be done by, as described in previous section, differentiation of current or estimation based on the power theory [6]:

$$\underline{u}^{*}_{Line} = \underline{u}_{S} + \underline{u}_{L12}. \tag{11.42}$$

In the case of filter-capacitor voltage estimation, calculations are realized in a similar way, but the rectifier's voltage is summed with only voltage drop across inductance L_1 [22]:

$$\underline{u}_{C} = \underline{u}_{S} + \underline{u}_{L1}. \tag{11.43}$$

Taking in account (11.43) through (11.45):

$$p_{L1} = L_1 \left(\frac{di_{La}}{dt} i_{La} + \frac{di_{Lb}}{dt} i_{Lb} + \frac{di_{Lc}}{dt} i_{Lc} \right) = 0 \tag{11.44a}$$

$$q_{L1} = \frac{3L_1}{\sqrt{3}} \left(\frac{di_{La}}{dt} i_{Lc} - \frac{di_{Lc}}{dt} i_{La} \right) \tag{11.44b}$$

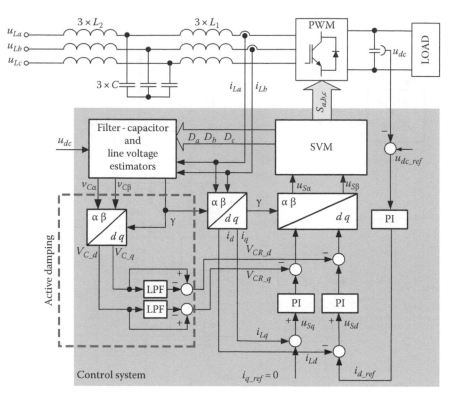

FIGURE 11.18   Basic scheme of VOC with active damping and voltage estimators.

$$
\begin{bmatrix} u_{L1\alpha} \\ u_{L1\beta} \end{bmatrix} = \frac{1}{i_{L\alpha}^2 + i_{L\beta}^2} \begin{bmatrix} i_{L\alpha} & -i_{L\beta} \\ i_{L\beta} & i_{L\alpha} \end{bmatrix} \begin{bmatrix} 0 \\ q_{L1} \end{bmatrix}
\tag{11.45}
$$

we get equations describing estimated filter-capacitor voltage:

$$
u_{C\alpha}^* = \left(\sqrt{\frac{2}{3}}\, u_{dc}\left(D_a - \frac{1}{2}(D_b + D_c) \right) \right) + u_{L1\alpha}
\tag{11.46a}
$$

$$
u_{C\beta}^* = \left(\frac{1}{\sqrt{2}}\, u_{dc}(D_b - D_c) \right) + u_{L1\beta}
\tag{11.46b}
$$

Basic waveforms are shown in Figure 11.19, what proves that estimators correctly estimate line voltage and capacitor voltage (AD activated at 0.08 s). It shows that resonance is visible in estimated capacitor voltage. Therefore, signals delivered from filter-capacitor estimators to AD block can correctly attenuate existing oscillations.

11.2.8 Summary of Control Schemes for PWM AC–DC Converters

The advantages and features of control schemes for PWM AC–DC converter are summarized in Table 11.1.

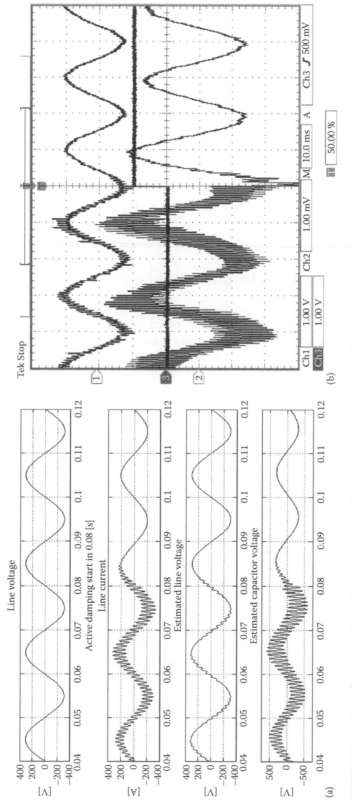

FIGURE 11.19 Sensorless *VOC* with AD based on high-pass filters: (a) simulated waveforms of line voltage u_{Line}, line current i_{Line}, estimated line voltage u'_{Line}, and estimated capacitor voltage $u_C^\star$ (switch-on of active damping function in 0.08 [s]). (b) Experimental waveforms of line voltage u_{Line}, signal showing when AD is activated, line current i_{L2} (From Malinowski, M. and Bernet, S., *IEEE Trans. Ind. Elect.*, 55(4), 1876, 2008. With permission.).

TABLE 11.1 Advantages and Features of Control Schemes for PWM AC–DC Converter

		VOC	VF-DPC	DPC-SVM
Power control—indirect		Yes	No	No
Power control—direct		No	Yes	Yes
Modulation techniques	SVM	Yes	No	Yes
	Switching table	No	Yes	No
Line voltage orientation		Yes	No	No
Virtual flux orientation		No	Yes	Yes
Decoupling block		Yes	No	No
Low algorithm complexity		No	Yes	No
Low computation intensity		Yes	No	Yes
Constant switching frequency		Yes	No	Yes
Low sensitivity to line inductance variation		Yes	No	Yes
Low sensitivity to line voltage distortion	THD of line current	No	Yes	Yes
	Power factor	Yes	No	No

11.3 Summary and Conclusion

This chapter has reviewed most popular three-phase voltage source AC–DC bridge converters. Uncontrolled diode rectifiers should be used in simple and low-cost application, where the regeneration of energy from DC-side is not required and power quality is not a crucial issue. More expensive PWM-controlled AC–DC converters provide very high control performance, operation at unity power factor (UPF) condition, low harmonic distortion of line currents, possibility fed energy into the grid, and reduction of passive components.

Various control techniques for control of PWM AC–DC converters have been discussed. In the VOC scheme, the UPF condition is enforced by aligning the direct component of the reference voltage vector with the line current vector. The VOC scheme is simple to implement in cheap microcontrollers but does not give good results at significantly distorted line voltage. In the direct power control (DPC) schemes associated with virtual flux (VF) oriented control techniques, bang–bang controllers in the active and reactive power loops are employed for the selection of the next state of the rectifier. The VF-DPC scheme is very good but it demands very sophisticated control platform and—because of variable switching frequency—results in difficulties of LCL input filter design. The DPC-SVM system constitutes a viable alternative to the other control strategies thanks to advantages like: simple control algorithm (inexpensive microcontroller), constant switching frequency (easy LCL input filter design and active damping), and sinusoidal line currents (low THD) for slightly distorted line voltage.

The common tendency in PWM AC–DC control systems is the elimination of line side voltage sensors and replacing them by appropriate estimators.

Also, the stability problem of converter with LCL line side filter can be effectively solved using AD algorithm (see Section 11.2.7).

It is believed that thanks to continuous developments in power semiconductor components and digital signal processing, the voltage source PWM AC–DC converters will have a strong impact on power conversion, especially in renewable and distributed energy systems.

List of Symbols

Abbreviations

AD	Active damping
ASD	Adjustable speed drives
DPC	Direct power control

DPC-SVM	Direct power control-space vector modulated
DSP	Digital signal processor
IGBT	Insulated gate bipolar transistor
PFC	Power factor correction
PI	Proportional integral (controller)
PLL	Phase locked loop
PWM	Pulse-width modulation
SVM	Space vector modulation
THD	Total harmonic distortion
UPF	Unity power factor
VF-DPC	Virtual flux-based direct power control
VOC	Voltage-oriented control
VSI	Voltage source inverter

General Symbols

f	Frequency
I	Current
J	Imaginary unit
T	Instantaneous time
u, v	Voltage
u_{dc}	DC link voltage
i_{dc}	DC link current
S_a, S_b, S_c	Switching state of the converter
D_a, D_b, D_c	Duty cycles of modulator
C	Capacitance
L	Inductance
R	Resistance
ω	Angular frequency
$\cos \varphi$	Fundamental power factor
P	Instantaneous active power
Q	Instantaneous reactive power
$\underline{u}_L$	Line voltage vector
$u_{L\alpha}$	Line voltage vector components in the Stationary α, β coordinates
$u_{L\beta}$	Line voltage vector components in the Stationary α, β coordinates
u_{Ld}	Line voltage vector components in the synchronous d, q coordinates
u_{Lq}	Line voltage vector components in the synchronous d, q coordinates
$\underline{i}_L$	Line current vector
$i_{L\alpha}$	Line current vector components in the stationary α, β coordinates
$i_{L\beta}$	Line current vector components in the stationary α, β coordinates
i_{Ld}	Line current vector components in the synchronous d, q coordinates
i_{Lq}	Line current vector components in the synchronous d, q coordinates
$\underline{u}_S$	Converter voltage vector
$u_{S\alpha}$	Converter voltage vector components in the stationary α, β coordinates
$u_{S\beta}$	Converter voltage vector components in the stationary α, β coordinates
u_{Sd}	Converter voltage vector components in the synchronous d, q coordinates
u_{Sq}	Converter voltage vector components in the synchronous d, q coordinates
$\underline{\psi}_L$	Virtual line flux vector
$\psi_{L\alpha}$	Virtual line flux vector components in the stationary α, β coordinates
$\psi_{L\beta}$	Virtual line flux vector components in the stationary α, β coordinates

ψ_{Ld}	Virtual line flux vector components in the synchronous d, q coordinates
ψ_{Lq}	Virtual line flux vector components in the synchronous d, q coordinates

Indices

a, b, c	Phases of three-phase system
d, q	Direct and quadrature component
α, β	Alpha, beta components
ref, c	Reference
rms	Root mean square value
m	Amplitude
est	Estimated
L	Grid
C, cap	Capacitor
S	Converter
I	Inductance

References

1. B. Wu, L. Li, and S. Wei, Multipulse diode rectifiers for high power multilevel inverter fed drives, in *Proceedings of the Power Electronics Congress,* 2004, CIEP 2004, pp. 9–14.
2. H. Kohlmeier, O. Niermeyer, and D. Schroder, High dynamic four quadrant AC-motor drive with improved power-factor and on-line optimized pulse pattern with PROMC, in *Proceedings of the EPE Conference*, Brussels, Belgium, 1985, pp. 3.173–3.178.
3. R. Teichmann, M. Malinowski, and S. Bernet, Evaluation of three-level rectifiers for low voltage utility applications, *IEEE Transactions on Industrial Electronics*, 52(2), 471–482, April 2005.
4. M. Malinowski, Sensorless control strategies for three-phase PWM rectifiers, PhD thesis, Warsaw University of Technology, Warsaw, Poland, 2001.
5. M. P. Kazmierkowski, R. Krishnan, and F. Blaabjerg, *Control in Power Electronics*, Academic Press, San Diego, CA/London, U.K., 2002.
6. S. Hansen, M. Malinowski, F. Blaabjerg, and M. P. Kazmierkowski, Control strategies for PWM rectifiers without line voltage sensors, in *Proceedings of the IEEE-APEC Conference*, Vol. 2, New Orleans, LA, 2000, pp. 832–839.
7. T. Noguchi, H. Tomiki, S. Kondo, and I. Takahashi, Direct Power Control of PWM converter without power-source voltage sensors, *IEEE Transactions on Industry Application*, 34(3), 1998, pp. 473–479.
8. T. Ohnishi, Three-phase PWM converter/inverter by means of instantaneous active and reactive power control, in *Proceedings of the IEEE-IECON Conference*, Krobe, Japan, 1991, pp. 819–824.
9. B. T. Ooi, J. W. Dixon, A. B. Kulkarni, and M. Nishimoto, An integrated AC drive system using a controlled current PWM rectifier/inverter link, in *Proceedings of the IEEE-PESC Conference*, Vancouver, Canada, 1986, pp. 494–501.
10. M. P. Kazmierkowski and L. Malesani, Current control techniques for three-phase voltage-source PWM converters: A survey, *IEEE Transactions on Industrial Electronics*, 45(5), 691–703, 1998.
11. M. Malinowski, M. P. Kaźmierkowski, S. Hansen, F. Blaabjerg, and G. D. Marques, Virtual flux based direct power control of three-phase PWM rectifiers, *IEEE Transactions on Industry Applications*, 37(4), 1019–1027, 2001.
12. M. Weinhold, A new control scheme for optimal operation of a three-phase voltage dc link PWM converter, in *Proceedings of the PCIM Conference*, Nurberg, Germany, 1991, pp. 371–383.
13. M. Malinowski, M. Jasinski, and M. P. Kaźmierkowski, Simple direct power control of three-phase PWM rectifier using space-vector modulation (DPC-SVM), *IEEE Transactions on Industrial Electronics*, 51(2), 447–454, April 2004.

14. V. Blasko and V. Kaura, A new mathematical model and control of a three-phase AC-DC voltage source converter, *IEEE Transactions on Power Electronics*, 12(1), 116–122, January 1997.

15. T. Ohnuki, O. Miyashida, P. Lataire, and G. Maggetto, A three-phase PWM rectifier without voltage sensors, in *Proceedings of the EPE Conference*, Trondheim, Norway, 1997, pp. 2.881–2.886.

16. H. Akagi, Y. Kanazawa, and A. Nabae, Instantaneous reactive power compensators comprising switching devices without energy storage components, *IEEE Transactions on Industry Applications*, 20(3), 625–630, May/June 1984.

17. M. Malinowski and M. P. Kaźmierkowski, Simple direct power control of three-phase PWM rectifier using space vector modulation—A comparative study, *EPE Journal*, 13(2), 28–34, 2003.

18. M. Liserre, F. Blaabjerg, and S. Hansen, Design and control of an LCL-filter based three-phase active rectifier, in *Industry Applications Conference (IAS'01)*, Vol. 1, Chicago, IL, 2001, pp. 299–307.

19. V. Blasko and V. Kaura, A novel control to actively damp resonance in input LC filter of a three phase voltage source converter, in *Eleventh Annual Applied Power Electronics Conference and Exposition (APEC'96), Conference Proceedings*, Vol. 2, San Jose, CA, March 3–7, 1996, pp. 545–551.

20. P. K. Dahono, A control method for DC-DC converter that has an LCL output filter based on new virtual capacitor and resistor concepts, in *IEEE 35th Annual Power Electronics Specialists Conference (PESC '04)*, Vol. 1, Aachen, Germany, June 20–25, 2004, pp. 36–42.

21. M. Liserre, A. Dell'Aquila, and F. Blaabjerg, Genetic algorithm-based design of the active damping for an LCL-filter three-phase active rectifier, *IEEE Transactions on Power Electronics*, 19(1), 76–86, January 2004.

22. M. Malinowski and S. Bernet, A simple voltage sensorless active damping scheme for three-phase PWM converters with an *LCL* filter, *IEEE Transactions on Industrial Electronics*, 55(4), 1876–1880, April 2008.

12

AC-to-DC Three-Phase/ Switch/Level PWM Boost Converter: Design, Modeling, and Control

Hadi Y. Kanaan
St. Joseph University

Kamal Al-Haddad
*École de Technologie
Supérieure*

12.1 Introduction

AC-to-DC family of converters constitutes the interface circuit between the network and the loads. These converters are known by single-phase or three-phase active rectifiers. They do play an important role in controlling the energy transfer from the utility to the load and vice versa. With the ever increase of power quality requirement at the point of common coupling with the network, AC-to-DC converters are nowadays required to achieve different tasks such as: provide high input power factor, low line current distortion [1–3], fixed output voltage and robustness to load, and utility voltage unbalances. Several topologies that satisfy these requirements have been studied [4–6]. Among these structures, one can recall the six-switch rectifier shown in Figure 12.1, which is the most conventionally used topology for bidirectional power flow applications [7,8]. It is characterized by high performance in terms of input power factor and DC voltage regulation, but at the cost of a high number of hard-switching devices,

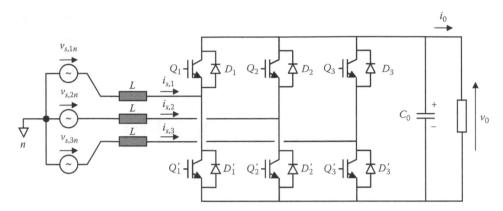

FIGURE 12.1   Six-switch rectifier.

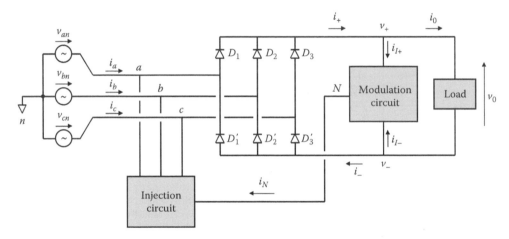

FIGURE 12.2   Basic structure of a three-phase current-injection rectifier.

yielding relatively high-power losses and consequently a relatively low efficiency. To a certain extent, this topology has no competitor in the applications where a bidirectionality of the power flow is required.

For unidirectional power flow application, another approach for designing a high-power-factor three-phase rectifiers consists of using the current-injection principle (Figure 12.2). In such a case, the rectifier would be a combination of three blocks: the conventional diode bridge that embeds the rectifying process, a modulation circuit that is aimed for current wave-shaping and DC voltage regulation, and, finally, an injection circuit where the major role is to compensate the intermittencies and, thus, avoid the irregularities in the line currents waveforms [9–24]. Most of the rectifiers found are used as a combination of an active modulation circuit consisting of a dual boost and a passive injection circuit, such as coupled inductors, transformers, and series inductor-capacitor connections tuned at the third harmonic, all suffering from bulkiness, additional costs, and power losses. In an attempt to increase the efficiency and power density of current-injection rectifiers, a topology, based on a totally active injection circuit consisting of three star connected four-quadrant switches, has been proposed [19]. However, the reliability of this structure is dramatically affected by a slight unbalance of the three-phase voltage source.

This chapter is dedicated to the study of a powerful, simple, and promising topology of a high power factor three-phase, three-switch, three-level pulse-width modulated (PWM) boost rectifier [25,26]. This rectifier exhibits high interests among power electronics researchers and engineers, and is increasingly used in medium unidirectional power applications that require rectification with low harmonic distortion. Besides its topological advantages, namely low number of high-frequency active switches, high efficiency, low design costs, and low voltage stresses, this converter is also known for its low control complexity and low sensing efforts regarding the control system design and implementation [27,28]. Tremendous work has been carried out on the design, modeling, and control of such converter by means of conventional or modern control methods to enhance its performance in terms of current distortion, DC voltage regulation, transients, power density, efficiency, cost, and reliability and robustness toward external disturbances [29–37].

The chapter is divided into four sections. First, some basic issues concerning the modeling of switch-mode converters with fixed-switching-frequency are addressed in Section 12.2. Then, in order to understand in a simple manner the operation of the three-phase Vienna rectifier, the single-phase version of the converter is considered and studied in detail in Section 12.3. Some design criteria or constraints for ensuring the current modulation ability of this topology are deduced, which constitute the basic frame on which the study of the three-phase structure is established. In Section 12.4, the rectifier sequences of operation are presented and a corresponding state model is derived for a continuous current mode (CCM) and fixed-switching-frequency operation. The modeling approach uses the state-space averaging technique, and the averaging process is applied on two time intervals—the switching period for average current evaluation and the mains period for average voltage computation. A basic mathematical model of the converter is first established. A simplified time-invariant model is then deduced using rotating Park transformation, and corresponding transfer functions are calculated through a small-signal linearization process. The steady-state regime is analyzed on the basis of the obtained model, and converter design criteria is consequently discussed.

Finally, in Section 12.5, a comparative evaluation of two multiple-loops duty-cycle-based control schemes is presented. The control laws are both elaborated on the basis of a state-space averaged model of the converter, expressed in the synchronous rotating frame. On the one hand, a control scheme that uses linear regulators is designed by using the small-signal transfer functions of the converter's model, which was linearized on the neighborhood of a suitable steady-state operating point. On the other hand, a nonlinear control scheme that uses the input–output feedback linearization approach is also designed in order to satisfy the same requirements as for the linear control system. For comparison purpose, both control schemes are implemented numerically using the Simulink® tool of MATLAB®, and simulation experiments are carried out in order to test and verify the tracking and regulation performance of each control law as well as their robustness toward load or mains source disturbance. For the same operating conditions, the performance of the two control laws are analyzed and compared in terms of line currents total harmonic distortion (THD) and DC voltage regulation.

12.2 Overview on Modeling Techniques Applied to Switch-Mode Converters

Reliable numerical models for power converters are increasingly required for both academic and industrial purposes. The aim is to elaborate highly precise mathematical representations of widely used converters. The usefulness of such virtual models could be emphasized in several aspects. More specifically, they allow

1. A systematic design of well-tuned control systems that improve the time response of the converter
2. A preevaluation of the operating regime as well as the analysis of the static and dynamic performance of the converter
3. A better selection of the system parameters and components
4. Fast simulations, which make these models suitable for real-time applications such that the hardware in the loop (HIL) and the power hardware in the loop (PHIL) techniques, widely used in the industry to test hardware controllers before being integrated into the real plants
5. In avoiding the elaboration of a real laboratory prototype, which can be costly and both time- and effort-consuming

Three modeling techniques for representing switch-mode power converters already exit. The first approach is the circuit averaging method [38], which is based on topological manipulations applied to a *N*-states converter. It consists, more particularly, of replacing each semiconductor by either a controlled source voltage or a controlled source current, depending on its topological position in the converter circuit.

The second approach concerns the state-space averaging method [39,40], which is based on analytical manipulations using the different state representations of a converter. It consists to determine, first, the linear state model for each possible configuration of the circuit and, then, to combine all these elementary models into a single and unified one, through a weighted sum. The weights are the occurrence degree of all the possible configurations.

As far as low-frequency modeling is concerned, both techniques described above are quite similar and give identical results. They are quite simple and may be useful as far as the required behavior of the converter is limited to the low-frequency region. For instance, they could well be used in the design of the controllers [41–43]. On the other hand, they give no information concerning the induced high-frequency phenomena and, therefore, they do not present any credibility for electromagnetic compatibility analysis. Despite the hard limitations concerning their applicability, these methods provide simple and time saving tools for simulating power converters. They also allow a real-time analysis of these converters, widely recommended in academic and research fields.

The third approach is based on the switching-function concept [44,45]. Although it is more complex and time-consuming than the averaging technique, this method does not neglect the high-frequency operating regime and allows, consequently, to study, on the one hand, the effects of the switching phenomenon on the waveforms of the converter's variables and, on the other hand, to analyze the electromagnetic compatibility of the converter.

In this section, all three modeling approaches are described and applied to a conventional single-phase boost-type full-bridge AC–DC power converter that operates with a fixed-switching-frequency. The converter's topology is depicted in Figure 12.3. It consists of two inverter legs with a current smoothing inductor at the AC side and a filtering capacitor at the DC side. The voltage source is assumed to be an ideal sine wave. The DC load is a pure resistor. The couple of switches in the upper or lower level, and those belonging to a same leg, are controlled complementarily. The gate signals are delivered by a PWM carrier-based circuit that operates with a fixed–switching-frequency. The DC voltage is controllable, and is assumed to be higher than the peak value of the AC source voltage. In this case, the ability of source current wave-shaping is maintained over the entire mains cycle.

12.2.1 Average Modeling Techniques

Average modeling techniques are based principally on replacing all the system variables by their mean value over a switching period and ignoring, thus, their high-frequency components. Their application is particularly suitable for high-switching frequency converters operating in a continuous mode, where it can be assumed that the time variations of all the capacitors voltages and inductors currents are of constant slope, or even negligible, on a switching period. In this case, the elaboration of the mathematical

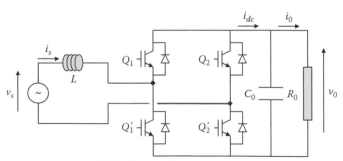

FIGURE 12.3   Single-phase boost-type full-bridge rectifier.

model of the converter, known as the averaged or low-frequency model, seems quite simple and straight-forward. It is worthy to note that these techniques can be applied to the discontinuous mode case [46], but the obtained models are generally highly nonlinear and not suited for control design.

There are two strategies used for the derivation of the averaged model of a fixed-frequency converter. They are presented in the following.

12.2.1.1 Circuit Averaging Technique

The circuit averaging approach [38] is based on topological manipulations applied to an N-configurations converter. It consists, more particularly, of replacing each semiconductor or group of semiconductors by either a controlled source voltage or a controlled source current, depending on its topological position in the converter circuit. Denoting by x_i, $i = 1, 2, ..., N$, the value taken by a variable x (that might be the voltage or current of any switch or group of switches in the converter) at a configuration i, and by d_i the occurrence degree of this configuration in a switching period T_S, the circuit averaging method consists of replacing, therefore, the instantaneous variable x by its averaged value over T_S:

$$x \rightarrow \sum_{i=1}^{N} d_i x_i \tag{12.1}$$

In order to illustrate this technique, let us apply it to the boost rectifier of Figure 12.3. In most applications where power factor improvement is required, this topology operates in the continuous mode and, therefore, presents only two configurations—the first one corresponds to the switching-on-state of switches Q_1 and Q_2' (Figure 12.4a), whereas the second configuration corresponds to the switching-on of Q_1' and Q_2 (Figure 12.4b). Whatever the state of the switches might be, the time variation laws of the source current i_s and the DC voltage v_0 are always given by the following equations:

$$L\frac{di_s}{dt} = v_s - v_{AB}$$

$$C_0\frac{dv_0}{dt} + \frac{v_0}{R_0} = i_{dc}, \tag{12.2}$$

where
  v_s denotes the source voltage
  v_{AB} the voltage at the AC side of the rectifier
  i_{dc} the current delivered at the DC side of the rectifier
  R_0 the load resistor

In the following, the voltage ripple at the load side (on the output capacitor) and the current ripple at the input side (in the inductor) at the switching frequency are neglected. This assumption is justified by a suitable choice of the reactive elements. In such a case, the values of the rectifier input voltage v_{AB} and output

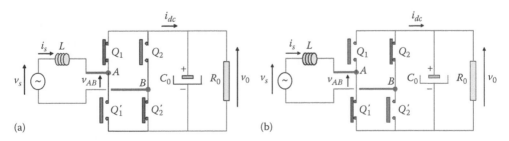

FIGURE 12.4 Circuit configurations: (a) Q_1 and Q_2' are switched-on, (b) Q_1' and Q_2 are switched-on.

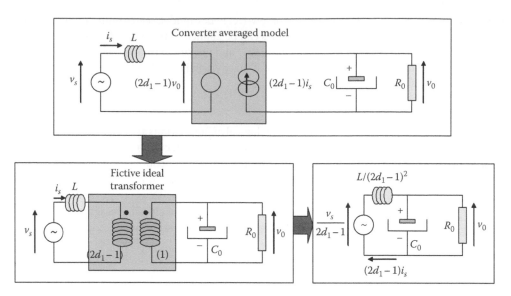

FIGURE 12.5 Circuit averaging applied to the boost rectifier.

current i_{DC} are practically constant in each configuration; they are equal, respectively, to v_0 and i_s during d_1T_S, where d_1 denotes the duty cycle of switch Q_1, and to $(-v_0)$ and $(-i_s)$ during the rest of the switching period. Applying the circuit averaging technique to the converter signifies replacing the instantaneous values of v_{AB} and i_{DC} by their average value calculated over the switching period T_S (Figure 12.5). It yields

$$v_{AB} \# (2d_1 - 1)v_0$$
$$i_{dc} \# (2d_1 - 1)i_s$$

(12.3)

Replacing the averaged expressions (12.3) of v_{AB} and i_{DC} into Equation 12.2 yields a one-input-two-outputs bilinear system represented by

$$L\frac{di_s}{dt} \# v_s - (2d_1 - 1)v_0$$
$$C_0\frac{dv_0}{dt} + \frac{v_0}{R_0} \# (2d_1 - 1)i_s$$

(12.4)

In a control system view, the duty cycle d_1 is considered as the control input, v_0 and i_s the state variables and v_s the disturbance signal. Note that, in system (12.4), only the low-frequency components (at the left-side of the switching frequency f_S) of the variables are taken into account. The harmonics at frequencies higher than f_S are neglected.

The circuit averaging technique applied to a boost rectifier is illustrated in Figure 12.5. An ideal transformer could be used to represent, in a more convenient manner, the coupling between the source and the load. The transformer can be omitted by having all the circuit elements put on the primary or the secondary of the transformer. A final equivalent circuit with a minimized number of components is thus obtained as shown in Figure 12.5.

The modeling approach described above is quite simple and may be useful as far as the required behavior of the converter is limited to the low-frequency region. For instance, it could well be used in the design of the controllers [41–43]. On the other hand, it gives no information concerning the induced high-frequency phenomena and, therefore, it does not present any credibility for electromagnetic compatibility analysis. Furthermore, as it can be noticed from the example, the complexity of this method to a given topology increases considerably with the number of switches or group of switches. Therefore, its application is generally limited to simple topologies with reduced number switches.

12.2.1.2 State-Space Averaging Technique

The state-space averaging method [39,40] is based on analytical manipulations using the different state representations of a converter. It is summarized by the diagram in Figure 12.6 for an N-states converter, where $\mathbf{x}$ denotes the state vector, $\mathbf{y}$ the output vector, and $\mathbf{v}$ the disturbance vector. This modeling technique consists to determine, first, the linear state model for each possible configuration of the circuit and, then, to combine all these elementary models into a single and unified one, through a weighted sum. The weights are the occurrence degree of all the possible configurations.

For instance, let us consider again the continuous mode operation of the boost rectifier depicted in Figure 12.3. Recall that this topology has two stable configurations presented in Figure 12.4a and b. The first one (Figure 12.4a) can be represented by the following state-space model:

$$\dot{\mathbf{x}} = \mathbf{A}_1 \mathbf{x} + \mathbf{E}_1 v_s, \tag{12.5}$$

where

$\mathbf{x} = [i_s, v_0]^T$ is the state vector
$\mathbf{A}_1$ the state matrix
$\mathbf{E}_1$ the disturbance matrix given as

$$\mathbf{A}_1 = \begin{bmatrix} 0 & -\dfrac{1}{L} \\ \dfrac{1}{C_0} & -\dfrac{1}{R_0 C_0} \end{bmatrix} \quad \text{and} \quad \mathbf{E}_1 = \begin{bmatrix} \dfrac{1}{L} \\ 0 \end{bmatrix}$$

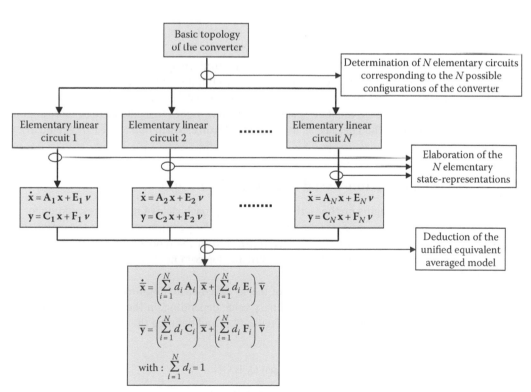

FIGURE 12.6   Diagram representation of the state-space averaging technique.

The second configuration is represented in the state-space as

$$\dot{\mathbf{x}} = \mathbf{A}_2\mathbf{x} + \mathbf{E}_2 v_s \tag{12.6}$$

with

$$\mathbf{A}_2 = \begin{bmatrix} 0 & \dfrac{1}{L} \\ -\dfrac{1}{C_0} & -\dfrac{1}{R_0 C_0} \end{bmatrix} \quad \text{and} \quad \mathbf{E}_2 = \begin{bmatrix} \dfrac{1}{L} \\ 0 \end{bmatrix} = \mathbf{E}_1$$

Denoting by d_1 the duty cycle of switch Q_1, the state-space averaged model of the boost converter is obtained as follows:

$$\dot{\mathbf{x}} \# \mathbf{A}\mathbf{x} + \mathbf{E}v_s \tag{12.7}$$

with

$$\mathbf{A} = d_1\mathbf{A}_1 + (1-d_1)\mathbf{A}_2 = \begin{bmatrix} 0 & -\dfrac{2d_1-1}{L} \\ \dfrac{2d_1-1}{C_0} & -\dfrac{1}{R_0 C_0} \end{bmatrix} \tag{12.8}$$

and

$$\mathbf{E} = d_1\mathbf{E}_1 + (1-d_1)\mathbf{E}_2 = \mathbf{E}_1 = \mathbf{E}_2 = \begin{bmatrix} \dfrac{1}{L} \\ 0 \end{bmatrix} \tag{12.9}$$

It can be noticed that model (12.7) is equivalent to model (12.4), given by the circuit averaging approach. As far as the low-frequency modeling is concerned, the two techniques are quite similar and give identical results. Here again, the study of the operation regime at the switching frequency is not possible due to the averaging process. However, the state-space averaging technique is considered to be more popular than the circuit averaging method because of its systematic feature, which makes it easily extendable to more complex topologies.

12.2.2 Switching-Function-Based Modeling Technique

Contrarily to the former averaging methods, this modeling approach yields a state representation for the converter, which is valid in the entire frequency range. So, both operation regimes at the mains (low) frequency and the switching (high) frequency are considered in the modeling process, which makes this approach more accurate and, therefore, more suitable for computer simulations and, especially, real-time applications.

This modeling technique is based on the use of the so-called switching-function that is associated to a switch or group of switches, which gives a binary value depending on the state of these switches. By recalling the example of the boost rectifier of Figure 12.3, and defining the switching-function s_1 of switch Q_1 as:

$$s_1 = \begin{cases} 1 & \text{when } Q_1 \text{ is ON} \\ 0 & \text{when } Q_1 \text{ is OFF} \end{cases} \tag{12.10}$$

we may set:

$$v_{AB} = (s_1 - \bar{s}_1)v_0$$

$$i_{dc} = (s_1 - \bar{s}_1)i_s,$$

(12.11)

where $\bar{s}_1$ denotes the logical complement of s_1. Replacing expressions (12.11) into (12.2) yields:

$$L\frac{di_s}{dt} = v_s - (s_1 - \bar{s}_1)v_0$$

$$C_0\frac{dv_0}{dt} + \frac{v_0}{R_0} = (s_1 - \bar{s}_1)i_s$$

(12.12)

Note that model (12.12) is more general than the one given by (12.4) or (12.7). In fact, models (12.4) and (12.7) are straightforwardly obtained from (12.12) by replacing the switching-function s_1 and its complement $\bar{s}_1$ by their, respective, average values d_1 and $(1 - d_1)$ over the switching period T_s. Hence, and contrarily to the former ones, model (12.12) allows to consider, in addition to the averaged or low-frequency behavior of the converter, the effects of the switching process on the system.

12.3 Study of a Basic Topology: The Single-Phase, Single-Switch, Three-Level Rectifier

For a better understanding of the principles and law that govern the operation of the three-phase/switch/level (or Vienna) rectifier, it is convenient at this beginning stage to consider only the simple single-phase version of this topology and to develop its equations in order to deduce, first, some design features and constraints related to its variables structural devices and to evaluate, secondly, its performance and limitations.

The single-phase, single-switch, three-level boost rectifier is presented in Figure 12.7. The Q is a four-quadrant switch, required to allow the reversibility of both current and voltage due to the bipolarity of the source. In fact, as illustrated in Figure 12.8, this converter can be viewed as no other than the association of two DC–DC boost converters that operate in a complementary manner, i.e., during either the positive or negative half-wave of the source voltage or current. The two transistors are combined together to form the four-quadrant switch. In a normal operation of the converter, the total DC output voltage v_0 is equally divided across the split capacitors (i.e., $v_{0,h} \approx v_{0,l} \approx v_0/2$). This voltage balancing, as well as the total voltage level, can be both adjusted through a properly designed feedback control scheme. The resistors $R_{0,h}$ and $R_{0,l}$ represent, respectively, the upper and lower DC loads connected to each output capacitor. The load is said to be balanced if $R_{0,h} = R_{0,l}$.

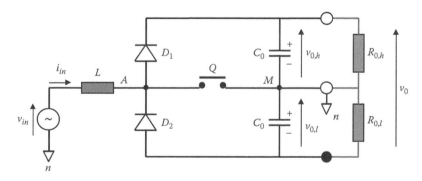

FIGURE 12.7   Single-phase, single-switch, three-level boost rectifier.

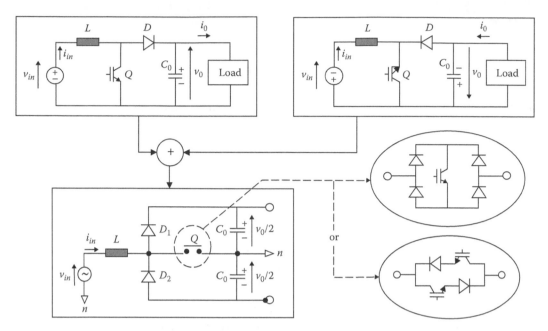

FIGURE 12.8 Design of the rectifier by the association of two complementary DC–DC boost converters.

The converter in Figure 12.7 is a two-quadrant converter that operates normally only when the source voltage and current have the same polarity (both should be either positive or negative), as is be demonstrated next. Consequently, the power flow is always unidirectional, transmitted from the AC source to the DC load. Therefore, it appears that this topology could be suitable for power factor improvement in single-phase applications since, there, the AC-side voltage and current are required to be proportional (both sine-waves with zero phase margin), but not applicable for power conditioning or compensation since, in these specific applications, the bidirectionality of the power flow in the converter is mandatory.

In power factor correction applications, the source current i_{in} should track in an average a sine-wave shape that is proportional to the source voltage v_{in}. It could be assumed, therefore, if the control algorithm is properly selected and if the high-frequency ripple in the source current is significantly reduced through an adequate choice of the inductor, the converter will operate in a CCM in the inductor except in local regions around the zero-crossings of the source voltage or current, where the discontinuous current mode (DCM) could take place. Since these time intervals are in practice negligible compared to the mains period, especially at medium and high loads, only the CCM operation could be considered in the study, and the converter would have only three possible configurations, depending on the state of the main switch Q and the sign of the source current i_{in}, as illustrated in Figure 12.9.

Following these considerations, the input current variations are governed by the following state equation:

$$L\frac{di_{in}}{dt} = \begin{cases} v_{in} & \text{if } Q \text{ is ON} \\ v_{in} - \dfrac{v_0}{2}\text{sgn}(i_{in}) & \text{if } Q \text{ is OFF,} \end{cases} \tag{12.13}$$

where sgn denotes the sign function defined as

$$\forall x, \text{sgn}(x) = \begin{cases} -1 & \text{if } x < 0 \\ +1 & \text{if } x \geq 0 \end{cases} \tag{12.14}$$

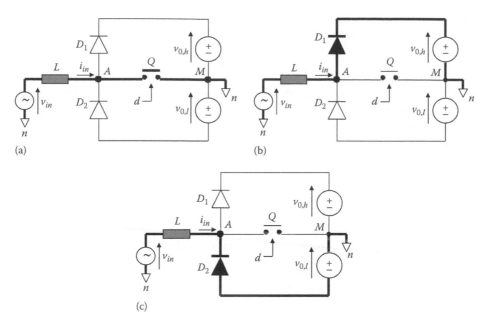

FIGURE 12.9 Possible configurations of the rectifier in a CCM operation: (a) Q is ON, (b) Q is OFF and $i_{in} > 0$, and (c) Q is OFF and $i_{in} < 0$.

It can be noticed, from Equation 12.13, that for $v_{in} > 0$ and $i_{in} < 0$, the source current i_{in} always increases whatever the state of Q is (note that the output voltage v_0 is always positive). Similarly, for $v_{in} < 0$ and $i_{in} > 0$, it decreases continually. Hence, in both cases, the source current tends to have the same polarity as the source voltage. Furthermore, the source current could not be modulated and, thus, has no tracking ability if it does not have the same polarity as the source voltage. In other words, in order to ensure the current modulation ability, the converter should operate only in the two quadrants, where v_{in} and i_{in} are both positive or both negative. The instantaneous input power $p_{in} = v_{in} \cdot i_{in}$ should be, thus, always positive, which limits the use of this topology to the applications where the bidirectionality of the active power flow is not necessary.

Another condition for preserving the current modulation ability is to choose a suitable value of v_0 that would allow a sign permutation of the current slope (di_{in}/dt) whenever the main switch Q changes its state. It yields that:

$$v_0 > 2 \cdot v_{in}(t), \quad \forall t \tag{12.15}$$

Any value of the DC output voltage greater than twice the input voltage peak value is theoretically convenient.

On the other hand, depending on the state of Q and the sign of i_{in}, the anode potential of diode D_1 (at point A in Figure 12.7) has three possible values: 0, $v_0/2$, and $-v_0/2$. For this reason, this topology is commonly said to be a three-level device. This same concept has been used in the elaboration of the three-phase/switch/level (or Vienna) rectifier, which will be studied later.

For a more detailed analysis on how the current tracking is performed, how a unity power factor could be obtained, and how the corresponding limitations and design criteria are considered, let us first consider an ideal source voltage expressed as

$$v_{in}(t) = \hat{v}_{in} \cdot \sin(\omega t), \quad \forall t, \tag{12.16}$$

where

$\hat{v}_{in}$ is the peak value of the source voltage
ω its angular frequency
t the time variable

In order to get a unity power factor operation, the input current i_{in} should track on an average a reference i_{in}^* that should be proportional to v_{in}, i.e.,

$$i_{in}^*(t) = \hat{i}_{in}^* \cdot \sin(\omega t), \quad \forall t \tag{12.17}$$

For a proper design of the control system that is aimed to ensure current tracking, the real input current i_{in} converges after a limited transient regime toward its reference i_{in}^*. Under these circumstances and during the first positive half-wave located between 0 and π/ω, where the input current i_{in} could be assumed always positive, Equation 12.13 becomes

$$L\frac{di_{in}}{dt} = \begin{cases} v_{in} & \text{if } Q \text{ is ON} \\ v_{in} - \dfrac{v_0}{2} & \text{if } Q \text{ is OFF} \end{cases} \tag{12.18}$$

In order to preserve the current modulation ability and to allow the input current to track on an average its reference, the following two conditions must be satisfied simultaneously:

1. The sign of the input current slope must change at each commutation of the main switch Q; it yields condition (12.15) by noticing from (12.16) the positive value of v_{in} between 0 and π/ω.
2. The input current must vary faster than its reference, i.e.,

$$\left|\frac{di_{in}(t)}{dt}\right| > \left|\frac{di_{in}^*(t)}{dt}\right|, \quad \forall t \tag{12.19}$$

or, using expressions (12.17) and (12.18) into (12.19) and taking account of (12.15)

$$\min\left(\frac{v_{in}(t)}{L}, \frac{(v_0/2) - v_{in}(t)}{L}\right) > \hat{i}_{in}^* \, \omega \cdot \left|\cos(\omega t)\right|, \quad \forall t \tag{12.20}$$

Condition (12.15) is satisfied if the DC output voltage is chosen to be greater than twice the peak value of the source voltage, i.e.,

$$v_0 > 2 \cdot \hat{v}_{in} \tag{12.21}$$

As for condition (12.20), two remarks deserve to be pointed out:

1. At the zero-crossings of the source voltage v_{in}, which take place at $\omega t = k\pi$ for any integer k, condition (12.20) cannot be satisfied for any choice of L, since the left-hand term would be zero and the right-hand one is always positive. At these instants, the modulation ability is lost and the input current cannot reach its reference. This temporary loss of current tracking ability, which is inevitable for this particular converter, is called the *detuning phenomenon* (see Figure 12.10). The detuning angle γ, which represents the duration of that phenomenon after each zero-crossing of the source voltage, can be easily calculated. It is expressed as

$$\gamma = 2\tan^{-1}\left(\frac{L\omega \hat{i}_{in}^*}{\hat{v}_{in}}\right) \tag{12.22}$$

The limitation of angle γ sets a criterion for the design of the AC inductor L, i.e.,

$$L \leq \frac{\hat{v}_{in}}{\hat{i}_{in}^*\omega}\tan\left(\frac{\gamma_M}{2}\right), \tag{12.23}$$

where γ_M denotes the maximum admissible value for γ.

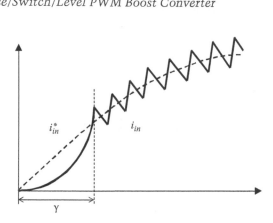

FIGURE 12.10   The detuning phenomenon of the converter.

2. Outside the detuning regions, condition (12.20) can be reduced to

$$L < \frac{\hat{v}_{in}}{\hat{i}_{in}^{*} \omega} \tag{12.24}$$

assuming

$$v_0 > 2\, \hat{v}_{in}(1 + \sin\gamma) \tag{12.25}$$

or

$$v_0 > 2\, \hat{v}_{in}\left(1 + \frac{2L\omega\, \hat{i}_{in}^{*}\, \hat{v}_{in}}{\hat{v}_{in}^2 + \left(L\omega\, \hat{i}_{in}^{*}\right)^2}\right), \tag{12.26}$$

which is easily satisfied by a proper adjustment of v_0 through a control loop. Note that, in practice, the detuning angle γ is relatively small. For instance, if the source voltage RMS-value is 120 V, the load power is 1 kW, the mains frequency is 60 Hz and the inductor's value is 4 mH, the detuning angle would be around 12° or $\pi/15$ radians, which corresponds to only 6.7% of the mains half-period. The minimum value permissible for the DC output voltage, set by condition (12.25) or (12.26), is increased by 21% in such a case, compared to the limit given by (12.21).

Another constraint considered in the design of the inductor is the limitation of the input current ripple at the switching frequency. Figure 12.11 shows the input current variation during a switching period T_S inside the positive half-cycle of the source voltage, where it is assumed that condition (12.25) always stands, the converter operates with a fixed-switching-frequency (through a carrier-based PWM), and the switching period is too small with respect to the mains period such that the slope of the current reference i_{in}^{*} would be considered as constant. The high-frequency ripple of the input current can be expressed as

$$\Delta i_{in} = dT_S(\tan\beta - \tan\alpha) \tag{12.27}$$

or

$$\Delta i_{in} = (1 - d)T_S(\tan\delta + \tan\alpha), \tag{12.28}$$

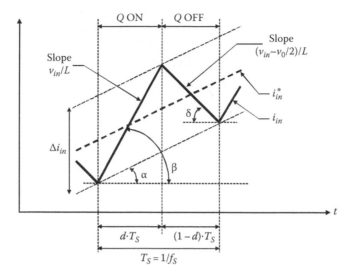

FIGURE 12.11 Input current in a switching period during the positive half-cycle of the source voltage.

where d represents the duty cycle, and

$$\tan\alpha = \frac{di^{*}_{in}}{dt} \quad \tan\beta = \frac{v_{in}}{L} \quad \tan\delta = \frac{(v_0/2) - v_{in}}{L}$$

Combining (12.27) with (12.28) yields

$$d(t) = \frac{\tan\alpha + \tan\delta}{\tan\beta + \tan\delta} = 1 - 2\frac{v_{in} - L\,di^{*}_{in}/dt}{v_0} \tag{12.29}$$

and

$$\Delta i_{in}(t) = \frac{2}{Lf_S} \cdot \frac{(v_0/2 - v_{in} + L\,di^{*}_{in}/dt)\cdot(v_{in} - L\,di^{*}_{in}/dt)}{v_0} \tag{12.30}$$

The current ripple has a maximum value when

$$v_{in} - L\frac{di^{*}_{in}}{dt} = \frac{v_0}{4}, \tag{12.31}$$

i.e.,

$$\omega t = \sin^{-1}\left(\frac{v_0}{4\sqrt{\hat{v}^2_{in} + \left(L\omega\,\hat{i}^{*}_{in}\right)^2}}\right) + \frac{\gamma}{2} \tag{12.32}$$

Replacing (12.31) or (12.32) into (12.30), we obtain

$$(\Delta i_{in})_{max} = \frac{v_0}{8Lf_S}, \tag{12.33}$$

which should be less than $(\Delta i_{in})_{\text{admissible}}$. It yields

$$L > \frac{v_0}{8 f_S (\Delta i_{in})_{\text{admissible}}} \tag{12.34}$$

Finally, in order to ensure high quality current tracking, as well as convenient reduction of the high-frequency ripple in the source current, the value to be chosen for L should satisfy (12.23), (12.24), and (12.34), i.e.,

$$\frac{v_0}{8 f_S (\Delta i_{in})_{\text{admissible}}} < L \le \frac{\hat{v}_{in}}{\hat{i}^*_{in}\omega} \tan\left(\frac{\gamma_M}{2}\right) \tag{12.35}$$

considering, in practice, that γ_M is chosen to be much less than $\pi/2$.

The criterion for choosing the split DC capacitors, both having the common value C_0, is derived from the averaged model of the converter. Assuming a fixed-switching-frequency operation, where the switch gate signal is generated by a carrier-based PWM controller, the averaged model of the converter is presented as follows in the case of two resistive loads applied to both output capacitors (as in Figure 12.7):

$$L\frac{d\bar{i}_{in}}{dt} = \begin{cases} \bar{v}_{in} - (1-d)\,\bar{v}_{0,h} & \text{if } \bar{i}_{in} > 0 \\ \bar{v}_{in} + (1-d)\,\bar{v}_{0,l} & \text{if } \bar{i}_{in} < 0 \end{cases} \tag{12.36a}$$

$$C_0\frac{d\bar{v}_{0,h}}{dt} = \begin{cases} (1-d)\,\bar{i}_{in} - \dfrac{\bar{v}_{0,h}}{R_{0,h}} & \text{if } \bar{i}_{in} > 0 \\ -\dfrac{\bar{v}_{0,h}}{R_{0,h}} & \text{if } \bar{i}_{in} < 0 \end{cases} \tag{12.36b}$$

$$C_0\frac{d\bar{v}_{0,l}}{dt} = \begin{cases} -\dfrac{\bar{v}_{0,l}}{R_{0,l}} & \text{if } \bar{i}_{in} > 0 \\ -(1-d)\,\bar{i}_{in} - \dfrac{\bar{v}_{0,l}}{R_{0,l}} & \text{if } \bar{i}_{in} < 0 \end{cases} \tag{12.36c}$$

$\bar{v}_{in}, \bar{i}_{in}, \bar{v}_{0,h}, \bar{v}_{0,l}$ denote, respectively, the averaged values, evaluated on a switching period, of the source voltage, the source current, the upper DC voltage, and the lower DC voltage. For a balanced load (i.e., $R_{0,h} = R_{0,l} = R_0$), and knowing that $\bar{v}_0 = \bar{v}_{0,h} + \bar{v}_{0,l}$, $\bar{v}_0$ being the averaged total DC voltage, Equations 12.36b and 12.36c can be combined into

$$C_0\frac{d\bar{v}_0}{dt} = \begin{cases} (1-d)\bar{i}_{in} - \dfrac{\bar{v}_0}{R_0} & \text{if } \bar{i}_{in} > 0 \\ -(1-d)\bar{i}_{in} - \dfrac{\bar{v}_0}{R_0} & \text{if } \bar{i}_{in} < 0 \end{cases} \tag{12.36d}$$

In the steady regime, $\bar{i}_{in}$ follows its reference i^*_{in}, and the averaged total output voltage $\bar{v}_0$ is equal to a desired constant value V_0. Therefore, by using expressions (12.16) and (12.17) into (12.36a), the following expression of the duty cycle is obtained:

$$d = \begin{cases} 1 - 2\dfrac{\hat{v}_{in}\sin(\omega t) - L\omega\,\hat{i}^*_{in}\cos(\omega t)}{V_0} & \text{for } 0 < t < \dfrac{\pi}{\omega} \\[4mm] 1 + 2\dfrac{\hat{v}_{in}\sin(\omega t) - L\omega\,\hat{i}^*_{in}\cos(\omega t)}{V_0} & \text{for } \dfrac{\pi}{\omega} < t < \dfrac{2\pi}{\omega}, \end{cases} \tag{12.37}$$

which, for $\bar{i}_{in} > 0$, is similar to Equation 12.29. Replacing (12.37) into (12.36d) yields

$$\bar{v}_0 = V_0 - \frac{V_0}{2 R_0 C_0 \cos(\gamma/2)} \sin\left(2\omega t - \frac{\gamma}{2} \right) \qquad (12.38)$$

The voltage ripple is then deduced as

$$\rho_0 = \frac{\Delta \bar{v}_0}{V_0} = \frac{1}{R_0 C_0 \cos(\gamma/2)} \qquad (12.39)$$

and, if $\rho_{0,max}$ is the maximum admissible value for the voltage ripple, the value of the DC capacitors should satisfy

$$C_0 > \frac{1}{R_0 \rho_{0,max} \cos(\gamma/2)} \qquad (12.40)$$

12.4 Design and Average Modeling of the Three-Phase/Switch/Level Rectifier

In this section, a simple mathematical model of the three-phase, three-switch, three-level, fixed-frequency PWM rectifier (or more simply the Vienna rectifier) operating in continuous current mode is developed from a control design perspective. The model is elaborated using the state-space averaging technique, commonly used in PWM DC–DC converters modeling problems [39,40], and presented in Section 12.2. Recall that this modeling approach is so far valid as long as the input and state variables of the converter vary slowly in time. Furthermore, other modeling techniques, such that the averaging technique that is based on equivalent circuit manipulations [38] and the Fourier analysis–based modeling approach [44], already exist in the literature and could be used for the same purpose. Albeit their differences, they all yield at the same low-frequency representation of the converter.

The basic model first obtained for the converter is a nonlinear fifth-order time-varying system, and the elaboration and implementation of a corresponding suitable control law seem highly difficult. Thus, in order to simplify the eventual control design procedure, a fourth-order time-invariant model is elaborated by applying to the former two transformations—a three-axis/two-axis frame transformation [44], known as Park transformation, and an input vector nonlinear transformation [32]. Finally, a small-signal linearization of the model around its static point is elaborated in order to deduce the corresponding transfer functions, on the basis of which frequency-domain linear control design could be carried out.

The reliability of the proposed model is investigated through numerical results using the MATLAB and Simulink simulation tool. A digital version of the converter has been integrated using the switching-function approach. The model parameters are shown to track their theoretically estimated values.

12.4.1 Topology and Operation of the Three-Phase/Switch/Level Rectifier

The scheme of the Vienna rectifier is illustrated in Figure 12.12a. It consists of three identical legs, each one having one high-frequency controlled switch and six diodes. The three legs operate in the same manner but are shifted by $2\pi/3$ and $4\pi/3$ in time. From an operational view, this structure is equivalent to the simplified circuit representation given in Figure 12.12b, which will be considered later for developing the converter model and control schemes. This equivalent topology consists of three single-switch legs associated to each phase. Q_1, Q_2, and Q_3 are four-quadrants switches; they are controlled in order

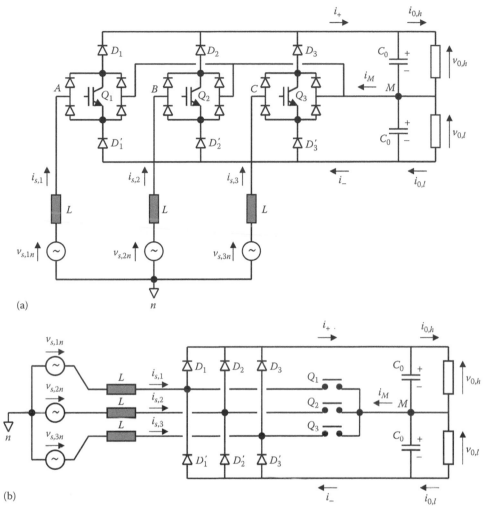

FIGURE 12.12 Three-phase/switch/level rectifier: (a) Vienna rectifier and (b) equivalent topology.

to ensure line current shaping at the input, DC voltage regulation and middle point stabilization at the output. In order to simplify the analysis, the converter of Figure 12.12b can be seen as an association of three identical bidirectional boost converters, as the one presented in Figure 12.13 for phase 1. Note that the single-phase equivalent topology presented in Figure 12.13 is similar to the one already given in Figure 12.7.

Referring to Figure 12.13, we may write the following equation for phase 1

$$v_{s,1n} = L\frac{di_{s,1}}{dt} + v_{M,n} + v_{AM}, \qquad (12.41)$$

where

$v_{s,1n}$ is the phase-to-neutral voltage

$i_{s,1}$ is the phase current

$v_{M,n}$ is the middle point voltage with respect to the mains neutral

v_{AM} is the switch voltage defined as $v_{0,h}$ and $v_{0,l}$ being the upper and lower output voltages, respectively

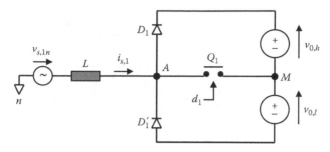

FIGURE 12.13   Single-phase equivalent circuit.

$$v_{AM} = \begin{cases} 0 & \text{if } Q_1 \text{ is turned-on} \\ v_{0,h} & \text{if } Q_1 \text{ is turned-off and } i_{s,1} > 0 \\ -v_{0,l} & \text{if } Q_1 \text{ is turned-off and } i_{s,1} < 0 \end{cases} \qquad (12.42)$$

Hence, we may express v_{AM} as follows:

$$v_{AM} = (1 - s_1) \cdot \left[v_{0,h} \cdot \theta(i_{s,1}) - v_{0,l} \cdot \overline{\theta(i_{s,1})} \right], \qquad (12.43)$$

where
   θ is the threshold function
   $\bar{\theta}$ its logic complement
   s_1 is the switching-function defined as

$$s_1 = \begin{cases} 0 & \text{if } Q_1 \text{ is turned-off} \\ 1 & \text{if } Q_1 \text{ is turned-on} \end{cases} \qquad (12.44)$$

In the same way, we can write for the other two phases the following equations:

$$v_{s,2n} = L \frac{di_{s,2}}{dt} + v_{M,n} + v_{BM} \qquad (12.45)$$

and

$$v_{s,3n} = L \frac{di_{s,3}}{dt} + v_{M,n} + v_{CM}, \qquad (12.46)$$

where

$$v_{BM} = (1 - s_2) \cdot \left[v_{0,h} \cdot \theta(i_{s,2}) - v_{0,l} \cdot \overline{\theta(i_{s,2})} \right] \qquad (12.47)$$

and

$$v_{CM} = (1 - s_3) \cdot \left[v_{0,h} \cdot \theta(i_{s,3}) - v_{0,l} \cdot \overline{\theta(i_{s,3})} \right] \qquad (12.48)$$

s_2 and s_3 being the switching-functions corresponding to Q_2 and Q_3, respectively.

In the nominal steady-state regime with a balanced load, $v_{0,h}$ and $v_{0,l}$ are equal to $v_0/2$, where $v_0 = v_{0,h} + v_{0,l}$ is the overall output voltage. We may, thus, rewrite Equations 12.43, 12.47, and 12.48 as follows:

$$v_{AM} \cong \frac{v_0}{2} \text{sgn}(i_{s,1})(1 - s_1) \tag{12.49a}$$

$$v_{BM} \cong \frac{v_0}{2} \text{sgn}(i_{s,2})(1 - s_2) \tag{12.49b}$$

$$v_{CM} \cong \frac{v_0}{2} \text{sgn}(i_{s,3})(1 - s_3), \tag{12.49c}$$

where sgn denotes the signum function.

Furthermore, assuming that the utility voltages are balanced sine waves, and that the neutral is disconnected, it follows

$$v_{s,1n}(t) + v_{s,2n}(t) + v_{s,3n}(t) = 0, \quad \forall t \tag{12.50}$$

and

$$i_{s,1}(t) + i_{s,2}(t) + i_{s,3}(t) = 0, \quad \forall t \tag{12.51}$$

Using identities (12.50) and (12.51) in Equations 12.41, 12.45, and 12.46 yields

$$v_{M,n} = -\frac{1}{3}(v_{AM} + v_{BM} + v_{CM}) \tag{12.52}$$

which can be rewritten using expressions (12.49)

$$v_{M,n} = -\frac{v_0}{6} \sum_{k=1}^{3} \text{sgn}(i_{s,k})(1 - s_k) \tag{12.53}$$

The values of $v_{M,n}$ are given in Table 12.1 with respect to the switching states s_k and the sign of line currents $i_{s,k}$, $k \in \{1,2,3\}$. Thus, the value of $v_{M,n}$ depends only on the output voltage v_0. Referring to Equations 12.41, 12.45, and 12.46, it is noticed that, in order to ensure line current wave shaping, the following two conditions must always be respected:

$$|v_{s,kn}(t)| > v_{M,n}(t) \cdot \text{sgn}[i_{s,k}(t)], \quad \forall t \quad \text{and} \quad \forall k \in \{1, 2, 3\} \tag{12.54}$$

$$|v_{s,kn}(t)| < v_{M,n}(t) \cdot \text{sgn}[i_{s,k}(t)] + \frac{v_0}{2}, \quad \forall t \quad \text{and} \quad \forall k \in \{1, 2, 3\} \tag{12.55}$$

TABLE 12.1   Values of $v_{M,n}$ with Respect to the Switching States and the Sign of Line Currents

Conditions	Switching Functions s_1, s_2 and s_3							
	111	110	101	011	100	001	010	000
$i_{s,1} > 0, i_{s,2} < 0, i_{s,3} > 0$	0	$-v_0/6$	$v_0/6$	$-v_0/6$	0	0	$-v_0/3$	$-v_0/6$
$i_{s,1} > 0, i_{s,2} < 0, i_{s,3} < 0$	0	$v_0/6$	$v_0/6$	$-v_0/6$	$v_0/3$	0	0	$v_0/6$
$i_{s,1} > 0, i_{s,2} > 0, i_{s,3} < 0$	0	$v_0/6$	$-v_0/6$	$-v_0/6$	0	$-v_0/3$	0	$-v_0/6$
$i_{s,1} < 0, i_{s,2} > 0, i_{s,3} < 0$	0	$v_0/6$	$-v_0/6$	$v_0/6$	0	0	$v_0/3$	$v_0/6$
$i_{s,1} < 0, i_{s,2} > 0, i_{s,3} > 0$	0	$-v_0/6$	$-v_0/6$	$v_0/6$	$-v_0/3$	0	0	$-v_0/6$
$i_{s,1} < 0, i_{s,2} < 0, i_{s,3} > 0$	0	$-v_0/6$	$v_0/6$	$v_0/6$	0	$v_0/3$	0	$v_0/6$

Conditions (12.54) and (12.55) limit the choice of the output voltage value in the range

$$\frac{3}{2}V_S\sqrt{6} < v_0 < 3V_S\sqrt{6} \tag{12.56}$$

i.e., between $3.68V_S$ and $7.34V_S$, where V_S is the RMS-value of the phase-to-neutral mains voltage.

At the output side, the converter is represented by the following state equations:

$$C_0\frac{dv_{0,h}}{dt} = \sum_{k=1}^{3}(1-s_k)\cdot i_{s,k}\cdot\theta(i_{s,k}) - i_{0,h} \tag{12.57}$$

$$C_0\frac{dv_{0,l}}{dt} = -\sum_{k=1}^{3}(1-s_k)\cdot i_{s,k}\cdot\overline{\theta(i_{s,k})} - i_{0,l} \tag{12.58}$$

12.4.2 State-Space Average Modeling of the Three-Phase/Switch/Level Rectifier

12.4.2.1 Basic Model

The modeling approach applied to the converter in Figure 12.12 is based on the state-space averaging technique [39,40]. In this method, all variables are averaged on a sampling period T_S. Including Equations 12.43, 12.47, 12.48, and 12.52 in the system Equations 12.41, 12.45, and 12.46, the equivalent average model of the converter, viewed on the AC side, is as follows:

$$\begin{bmatrix} v_{s,1n} \\ v_{s,2n} \\ v_{s,3n} \end{bmatrix} = L\frac{d}{dt}\begin{bmatrix} i_{s,1} \\ i_{s,2} \\ i_{s,3} \end{bmatrix} + \Gamma\left(\frac{v_0}{2}\mathbf{SGN} + \frac{\Delta v_0}{2}\mathbf{I}_3\right)\begin{bmatrix} 1-d_1 \\ 1-d_2 \\ 1-d_3 \end{bmatrix} \tag{12.59}$$

where

$$\Gamma = \begin{bmatrix} \dfrac{2}{3} & \dfrac{-1}{3} & \dfrac{-1}{3} \\ \dfrac{-1}{3} & \dfrac{2}{3} & \dfrac{-1}{3} \\ \dfrac{-1}{3} & \dfrac{-1}{3} & \dfrac{2}{3} \end{bmatrix} \quad \text{and} \quad \mathbf{SGN} = \begin{bmatrix} \operatorname{sgn}(i_{s,1}) & 0 & 0 \\ 0 & \operatorname{sgn}(i_{s,2}) & 0 \\ 0 & 0 & \operatorname{sgn}(i_{s,3}) \end{bmatrix}$$

$\Delta v_0 = v_{0,h} - v_{0,l}$, and $\mathbf{I_3}$ is the third-order identity matrix. d_1, d_2, and d_3 are the duty cycles of switches Q_1, Q_2, and Q_3 respectively. Note that system (12.59) is a time-varying model that depends on the sign of the line currents $i_{s,1}$, $i_{s,2}$, and $i_{s,3}$. Therefore, it is not suitable for a stationary control design process. In order to overcome this drawback, the following input transformation is proposed:

$$d'_k = (1 - d_k)\left[\mathrm{sgn}(i_{s,k}) + \frac{\Delta v_0}{v_0} \right], \quad \forall k \in \{1,2,3\} \tag{12.60}$$

Adding Equation 12.60 to system (12.59) yields

$$\mathbf{v_s} = L\frac{d\mathbf{i_s}}{dt} + \frac{v_0}{2}\mathbf{\Gamma d'}, \tag{12.61}$$

where
$\mathbf{v_s} = [v_{s,1n}, v_{s,2n}, v_{s,3n}]^T$ is the input voltage vector
$\mathbf{i_s} = [i_{s,1}, i_{s,2}, i_{s,3}]^T$ the input current vector
$\mathbf{d'} = [d'_1, d'_2, d'_3]^T$ the new control vector

Furthermore, at the load level, the average model of the converter is viewed as

$$C_0\frac{dv_{0,h}}{dt} + i_{0,h} = i_+ = \frac{1}{2}\sum_{k=1}^{3}(1 - d_k)i_{s,k}\left[1 + \mathrm{sgn}(i_{s,k})\right] \tag{12.62a}$$

$$C_0\frac{dv_{0,l}}{dt} + i_{0,l} = i_- = -\frac{1}{2}\sum_{k=1}^{3}(1 - d_k)i_{s,k}\left[1 - \mathrm{sgn}(i_{s,k})\right], \tag{12.62b}$$

where
$i_{0,h}$ and $i_{0,l}$ are the upper and lower output currents
i_+ and i_- the DC-side currents of the diode bridge

Introducing the overall output voltage v_0, the output voltage unbalance Δv_0 and transformation of (12.60) into Equations 12.62a and 12.62b yields

$$C_0\frac{dv_0}{dt} + i_{0,h} + i_{0,l} \cong \sum_{k=1}^{3}d'_k i_{s,k}\left[1 - \frac{\Delta v_0}{v_0}\mathrm{sgn}(i_{s,k})\right] \tag{12.63a}$$

$$C_0\frac{d(\Delta v_0)}{dt} + i_{0,h} - i_{0,l} \cong \sum_{k=1}^{3}d'_k i_{s,k}\left[\mathrm{sgn}(i_{s,k}) - \frac{\Delta v_0}{v_0}\right] \tag{12.63b}$$

In the derivation of Equations 12.63a and 12.63b, it was assumed that $\Delta v_0/v_0 \ll 1$. Equations 12.61 and 12.63 represent the basic low-frequency model of the converter in the stationary-frame. Although Equation 12.61 is time-invariant, the subsystem (12.63) is not. Nevertheless, knowing that the output voltages variations are relatively slow with respect to the mains frequency, it seems more convenient, from a control design perspective, to consider their average on a mains period T_0 instead of the one computed on a sampling period T_S. Furthermore, the basic model can be significantly reduced by applying Park's transformation, as discussed in the next subsection.

12.4.2.2 Frame Transformation

The model defined by Equations 12.61 and 12.63 can be expressed in a new rotating frame using Park's transformation. The Park's matrix is defined as [44]

$$
\mathbf{K} = \frac{2}{3}
\begin{bmatrix}
\sin(\omega_0 t) & \sin\left(\omega_0 t - \dfrac{2\pi}{3}\right) & \sin\left(\omega_0 t - \dfrac{4\pi}{3}\right) \\[2mm]
\cos(\omega_0 t) & \cos\left(\omega_0 t - \dfrac{2\pi}{3}\right) & \cos\left(\omega_0 t - \dfrac{4\pi}{3}\right) \\[2mm]
\dfrac{3}{2} & \dfrac{3}{2} & \dfrac{3}{2}
\end{bmatrix},
\tag{12.64}
$$

where ω_0 is the mains angular frequency. Defining the new vectors $\mathbf{v}_s^r$, $\mathbf{i}_s^r$, and $\mathbf{d}''$ as follows

$$
\mathbf{v}_s^r \overset{\Delta}{=} \begin{bmatrix} v_{s,d} & v_{s,q} & v_{s,0} \end{bmatrix}^T = \mathbf{K}\mathbf{v}_s
\tag{12.65a}
$$

$$
\mathbf{i}_s^r \overset{\Delta}{=} \begin{bmatrix} i_{s,d} & i_{s,q} & i_{s,0} \end{bmatrix}^T = \mathbf{K}\mathbf{i}_s
\tag{12.65b}
$$

$$
\mathbf{d}'' \overset{\Delta}{=} \begin{bmatrix} d_d' & d_q' & d_0' \end{bmatrix}^T = \mathbf{K}\mathbf{d}'
\tag{12.65c}
$$

Equations 12.61 and 12.63 can be arranged as

$$
v_{s,d} = L\frac{di_{s,d}}{dt} - L\omega_0 i_{s,q} + \frac{v_0}{2} d_d'
\tag{12.66a}
$$

$$
v_{s,q} = L\frac{di_{s,q}}{dt} + L\omega_0 i_{s,d} + \frac{v_0}{2} d_q'
\tag{12.66b}
$$

$$
C_0\frac{dv_0}{dt} + i_{0,h} + i_{0,l} = \frac{3}{2}\left(d_d' i_{s,d} + d_q' i_{s,q}\right) - \frac{\Delta v_0}{v_0}(\mathbf{d}'')^T\left[\mathbf{K}\,\mathbf{SGN}^{-1}\,\mathbf{K}^T\right]^{-1}\mathbf{i}_s^r
\tag{12.66c}
$$

$$
C_0\frac{d(\Delta v_0)}{dt} + i_{0,h} - i_{0,l} = -\frac{3}{2}\frac{\Delta v_0}{v_0}\left(d_d' i_{s,d} + d_q' i_{s,q}\right) + (\mathbf{d}'')^T\left[\mathbf{K}\,\mathbf{SGN}^{-1}\,\mathbf{K}^T\right]^{-1}\mathbf{i}_s^r
\tag{12.66d}
$$

The voltage and current zero sequence components $v_{s,0}$ and $i_{s,0}$ are eliminated, as shown by (12.50) and (12.51). A time-invariant equivalent model of the converter can be elaborated by averaging the term $[\mathbf{K}\,\mathbf{SGN}^{-1}\,\mathbf{K}^T]^{-1}$ over the mains period T_0, as mentioned above. Furthermore, the generality of the modeling approach would not be lost if we assume balanced sine-wave line currents. It follows

$$
\left\langle \left[\mathbf{K}\cdot\mathbf{SGN}^{-1}\cdot\mathbf{K}^T\right]^{-1}\right\rangle_{T_0} \cong \alpha\cdot
\begin{bmatrix}
0 & 0 & \cos\phi \\
0 & 0 & \sin\phi \\
\cos\phi & \sin\phi & 0
\end{bmatrix},
\tag{12.67}
$$

where φ denotes the phase shift between the phase voltage and the corresponding line current. Here, α is a parameter estimated at

$$\alpha \cong \frac{2}{\pi} \tag{12.68}$$

In a unity-power-factor operating mode, φ equals zero and, hence, we may rewrite system (12.66) as follows:

$$v_{s,d} = L\frac{di_{s,d}}{dt} - L\omega_0 i_{s,q} + \frac{v_0}{2}d'_d \tag{12.69a}$$

$$v_{s,q} = L\frac{di_{s,q}}{dt} + L\omega_0 i_{s,d} + \frac{v_0}{2}d'_q \tag{12.69b}$$

$$C_0\frac{d(\Delta v_0)}{dt} = \alpha d'_0 i_{s,d} - \frac{3}{2}\cdot\frac{\Delta v_0}{v_0}\cdot\left(d'_d i_{s,d} + d'_q i_{s,q}\right) - i_{0,h} + i_{0,l} \tag{12.69c}$$

$$C_0\frac{dv_0}{dt} = \frac{3}{2}\cdot\left(d'_d i_{s,d} + d'_q i_{s,q}\right) - \alpha\frac{\Delta v_0}{v_0}d'_0 i_{s,d} - i_{0,h} - i_{0,l} \tag{12.69d}$$

The system (12.69) represents a low-frequency fourth-order time-invariant continuous nonlinear state model of the converter, that has $i_{s,d}$, $i_{s,q}$, v_0, and Δv_0 as state variables, d'_d, d'_q, and d'_0 as control inputs, $v_{s,d}$ and $v_{s,q}$ as disturbance inputs. Note that the zero-sequence components of the mains voltage and current, respectively, $v_{s,0}$ and $i_{s,0}$, are identically zero (assuming a balanced three-phase source voltage and a nonconnected neutral point) and are, therefore, neglected.

12.4.3 Desired Steady-State Operating Regime

In the following, the theoretical expressions and waveforms of all system variables are established in the desired steady-state regime assuming:

1. A balanced three-phase voltage source, i.e.,

$$v^*_{s,1n}(t) = V^*_S\sqrt{2}\sin(\omega_0 t)$$

$$v^*_{s,2n}(t) = V^*_S\sqrt{2}\sin\left(\omega_0 t - \frac{2\pi}{3}\right) \tag{12.70}$$

$$v^*_{s,3n}(t) = V^*_S\sqrt{2}\sin\left(\omega_0 t - \frac{4\pi}{3}\right),$$

where V^*_S is the desired steady-state RMS-value of the mains phase-to-neutral voltage.

2. A unity power factor operating condition, i.e.,

$$i^*_{s,1}(t) = I^*_S\sqrt{2}\sin(\omega_0 t)$$

$$i^*_{s,2}(t) = I^*_S\sqrt{2}\sin\left(\omega_0 t - \frac{2\pi}{3}\right) \tag{12.71}$$

$$i^*_{s,3}(t) = I^*_S\sqrt{2}\sin\left(\omega_0 t - \frac{4\pi}{3}\right),$$

where I^*_S is the desired steady-state RMS-value of the line currents.

3. A load balance, i.e.,

$$v_{0,h}^* = v_{0,l}^* = \frac{V_0^*}{2} \quad \text{and} \quad i_{0,h}^* = i_{0,l}^* = I_0^* \tag{12.72}$$

where, V_0^* and I_0^* are, respectively, the desired constant values for the output voltage and current. The asterisks in expressions (12.70) through (12.72) characterize the desired regime. Applying Park's transformation to the voltage and current expressions (12.70) and (12.71) yield time-invariant vectors expressed in the rotating frame as

$$\mathbf{v_s^r}^* \overset{\Delta}{=} \begin{bmatrix} v_{s,d}^* \\ v_{s,q}^* \\ v_{s,0}^* \end{bmatrix} = \begin{bmatrix} V_S^* \sqrt{2} \\ 0 \\ 0 \end{bmatrix} \quad \text{and} \quad \mathbf{i_s^r}^* \overset{\Delta}{=} \begin{bmatrix} i_{s,d}^* \\ i_{s,q}^* \\ i_{s,0}^* \end{bmatrix} = \begin{bmatrix} I_S^* \sqrt{2} \\ 0 \\ 0 \end{bmatrix} \tag{12.73}$$

Integrating expressions (12.73) into system (12.69), the steady-state values of the control inputs are obtained as follows:

$$d_d'^* = \frac{2V_S^* \sqrt{2}}{V_0^*}$$

$$d_q'^* = -\frac{2L\omega_0 I_S^* \sqrt{2}}{V_0^*} \tag{12.74}$$

$$d_0'^* = 0$$

In addition, the power conservation law is verified, i.e.,

$$3V_S^* I_S^* = V_0^* I_0^* \tag{12.75}$$

Referring to the stationary-frame, the steady-state control inputs are expressed as follows:

$$\begin{bmatrix} d_1'^* \\ d_2'^* \\ d_3'^* \end{bmatrix} \overset{\Delta}{=} \mathbf{K}^{-1} \begin{bmatrix} d_d'^* \\ d_q'^* \\ d_0'^* \end{bmatrix} = \begin{bmatrix} \hat{d}' \sin(\omega_0 t - \phi) \\ \hat{d}' \sin\left(\omega_0 t - \phi - \frac{2\pi}{3}\right) \\ \hat{d}' \sin\left(\omega_0 t - \phi - \frac{4\pi}{3}\right) \end{bmatrix}, \tag{12.76}$$

where

$$\hat{d}' = \frac{2V_S^* \sqrt{2}}{V_0^* \cos\phi} \tag{12.77}$$

and

$$tg\,\phi = \frac{L\omega_0 I_S^*}{V_S^*} \tag{12.78}$$

Using Equation 12.60, we may set therefore

$$d_k^* = 1 - d_k'^* \, \text{sgn}(i_{s,k}^*) = \begin{cases} 1 - \hat{d}' \sin\left[\omega_0 t - \phi - 2(k-1)\dfrac{\pi}{3}\right], & 2(k-1)\dfrac{\pi}{3} < \omega_0 t < \pi + 2(k-1)\dfrac{\pi}{3} \\[2mm] 1 + \hat{d}' \sin\left[\omega_0 t - \phi - 2(k-1)\dfrac{\pi}{3}\right], & \pi + 2(k-1)\dfrac{\pi}{3} < \omega_0 t < 2\pi + 2(k-1)\dfrac{\pi}{3} \end{cases} \tag{12.79}$$

for $k \in \{1,2,3\}$. Expressions (12.79) show that the duty cycles d_1^*, d_2^*, and d_3^* vary periodically, with a $T_0/2$ period. It also emphasizes the control saturation phenomenon, which takes place periodically and has a duration angle ϕ. It is noticed from expression (12.78) that, in order to reduce the undesirable effects of the control saturation, the inductor value L has to be minimized.

Furthermore, following Equations 12.53 and 12.60, the averaged middle point voltage may be expressed as

$$v_{M,n} \cong -\frac{v_0}{6} \sum_{k=1}^{3} \left[1 - \frac{\Delta v_0}{v_0} \text{sgn}(i_{s,k})\right] d_k' \tag{12.80}$$

Using Equation 12.76, it follows in steady-state regime

$$v_{M,n}^* = -\frac{V_0^*}{6} \sum_{k=1}^{3} d_k'^* \equiv 0 \tag{12.81}$$

Concerning the steady-state expressions of DC-side currents i_+ and i_-, they could be easily established as

$$i_+^* = \frac{1}{2} \sum_{k=1}^{3} d_k'^* i_{s,k}^* \left[1 + \text{sgn}(i_{s,k}^*)\right] \tag{12.82a}$$

and

$$i_-^* = \frac{1}{2} \sum_{k=1}^{3} d_k'^* i_{s,k}^* \left[1 - \text{sgn}(i_{s,k}^*)\right] \tag{12.82b}$$

Integrating Equations 12.71 and 12.79 into (12.82), it yields after some manipulation to the expressions given in (12.83) for current i_+^*. The expressions corresponding to current i_-^* are obtained by applying a phase shifting of π to the former ones. Note that currents i_+^* and i_-^* have practically a third harmonic sine wave shape, as illustrated in Figure 12.14.

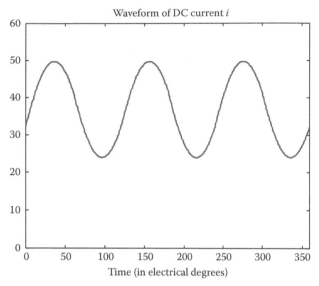

FIGURE 12.14 Steady-state waveform of DC current i_-^*.

$$
i_+^* = \begin{cases}
\dfrac{\sqrt{2}}{2}\,\hat{d}'I_S^*\left[2\cos\phi + \cos\left(2\omega_0 t - \phi - \dfrac{4\pi}{3}\right)\right], & 0 < \omega_0 t < \dfrac{\pi}{3} \\[3mm]
\dfrac{\sqrt{2}}{2}\,\hat{d}'I_S^*\left[\cos\phi - \cos(2\omega_0 t - \phi)\right] & \dfrac{\pi}{3} < \omega_0 t < \dfrac{2\pi}{3} \\[3mm]
\dfrac{\sqrt{2}}{2}\,\hat{d}'I_S^*\left[2\cos\phi + \cos\left(2\omega_0 t - \phi - \dfrac{2\pi}{3}\right)\right] & \dfrac{2\pi}{3} < \omega_0 t < \pi \\[3mm]
\dfrac{\sqrt{2}}{2}\,\hat{d}'I_S^*\left[\cos\phi - \cos\left(2\omega_0 t - \phi - \dfrac{4\pi}{3}\right)\right] & \pi < \omega_0 t < \dfrac{4\pi}{3} \\[3mm]
\dfrac{\sqrt{2}}{2}\,\hat{d}'I_S^*\left[2\cos\phi + \cos(2\omega_0 t - \phi)\right] & \dfrac{4\pi}{3} < \omega_0 t < \dfrac{5\pi}{3} \\[3mm]
\dfrac{\sqrt{2}}{2}\,\hat{d}'I_S^*\left[\cos\phi - \cos\left(2\omega_0 t - \phi - \dfrac{2\pi}{3}\right)\right] & \dfrac{5\pi}{3} < \omega_0 t < 2\pi
\end{cases}
\tag{12.83}
$$

12.4.4 Design Criteria

12.4.4.1 Design of the Inductors

In order to ensure current wave-shaping in steady-state regime, the common value of mains series inductors have to satisfy the following conditions, as described in Figure 12.15:

$$
\begin{aligned}
v_{s,kn}^* - v_{M,n} > L\frac{di_{s,k}^*}{dt}, \quad \text{when } i_{s,k}^* > 0 \\[3mm]
v_{s,kn}^* - v_{M,n} - \frac{V_0^*}{2} < L\frac{di_{s,k}^*}{dt}
\end{aligned}
\tag{12.84}
$$

$$
\begin{aligned}
v_{s,kn}^* - v_{M,n} < L\frac{di_{s,k}^*}{dt}, \quad \text{when } i_{sk}^* > 0 \\[3mm]
v_{s,kn}^* - v_{M,n} + \frac{V_0^*}{2} > L\frac{di_{s,k}^*}{dt}
\end{aligned}
\tag{12.85}
$$

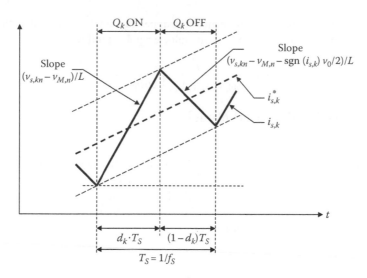

FIGURE 12.15 Line current wave-shaping.

for each $k \in \{1,2,3\}$. The value of $v_{M,n}$, corresponding to each case, is given in Table 12.1. After some mathematical developments, we obtain the following condition:

$$L < \min\left(\frac{V_0^* \sqrt{2} - 3V_S^* \sqrt{3}}{6\omega_0 I_S^*}, \frac{6V_S^* \sqrt{3} - V_0^* \sqrt{2}}{6\omega_0 I_S^*} \right) \tag{12.86}$$

The range of the inductor value L is thus maximized if

$$V_0^* = \frac{9}{4} V_S^* \sqrt{6} \cong 5.51 V_S^* \tag{12.87}$$

Furthermore, the inductors are also designed for current ripple limitation. In this perspective, reasoning around the peak value of the line currents yields

$$L > \frac{1}{f_S (\Delta i_s)_{max}} \left(2V_S^* \sqrt{2} - \frac{V_0^*}{4} - \frac{6V_S^{*2}}{V_0^*} \right), \tag{12.88}$$

where
f_S is the switching frequency
$(\Delta i_s)_{max}$ the acceptable current ripple

Finally, the inductor's value is chosen accordingly to conditions (12.86) and (12.88).

12.4.4.2 Design of the Capacitors

The two DC-side capacitors of the converter are designed in the low-frequency domain. Referring to the expression (12.83), the magnitude of the DC-side upper current ripple can be obtained as

$$\left(\Delta \hat{i}_+^* \right) = \frac{V_S^* I_S^*}{V_0^*} \left(\frac{2}{\cos \phi} - 1 \right) \tag{12.89}$$

Assuming that the totality of the AC-component of the upper current i_+^* is derived by the upper capacitor, and denoting by $(\Delta v_0)_{max}$ the admissible output voltage ripple, it follows

$$C_0 > \frac{2V_s^* I_s^*}{3\omega_0 V_0^* (\Delta v_0)_{max}} \left(\frac{2}{\cos \phi} - 1 \right)$$

(12.90)

12.4.5 State-Space Small-Signal Model

12.4.5.1 Static Point

In the (d, q) frame, the calculation of the static point is carried out by setting all the time-derivatives in Equations 12.69a through 12.69d to zero. Assuming that the converter operates near a unity power factor condition, the steady-state space-vector of the line currents is considered proportional to the space-vector of the mains line-to-neutral voltage, both oriented with respect to the d-axis. These considerations yield the following nominal static point:

$$V_{s,d} = V_s \sqrt{2}$$

$$I_{s,d} = I_s \sqrt{2}$$

$$V_{s,q} = I_{s,q} = \Delta V_0 = 0$$

$$D_d' = \frac{2V_s \sqrt{2}}{V_0}$$

(12.91)

$$D_q' = -\frac{2L\omega_0 I_s \sqrt{2}}{V_0}$$

$$D_0' = \frac{I_{0,h} - I_{0,l}}{\alpha I_s \sqrt{2}}$$

where
  V_s denotes the RMS-value of the source line-to-neutral voltage
  I_s the RMS-value of the line current
  V_0 the steady-state value of the total DC output voltage

In addition, by assuming a balanced purely resistive DC load ($R_{0,h} = R_{0,l} = R_0$), we get

$$i_{0,h} = \frac{v_0 + \Delta v_0}{2R_0}$$

(12.92)

$$i_{0,l} = \frac{v_0 - \Delta v_0}{2R_0}$$

Thus,

$$I_{0,h} = I_{0,l} = \frac{V_0}{2R_0}$$

(12.93)

and

$$D_0' = 0$$

(12.94)

It is obvious that the static values of the converter's input and state variables given in this subsection are similar to their steady-state values obtained in the desired regime (refer to Section 12.4.3). This result can be easily predicted by the fact that, in the (d, q) frame, all the state variables tend in steady-state toward constant values.

12.4.5.2 Time-Domain Small-Signal Model

The small-signal linearization of a system consists of representing each time-variable $z(t)$ by a superposition of two terms: (1) its desired steady-state value z^* and (2) a time-varying signal $z_{\sim}(t)$ representing the assumed small variation of the variable in the neighborhood of its steady-state value.

Applying this first-order linearization process to the converter model represented by Equations 12.69a through 12.69d and 12.92, around the desired steady-state point represented by (12.91), (12.93), and (12.94), yields the following linear state model:

$$\dot{\mathbf{x}}_{\sim} = \mathbf{A} \cdot \mathbf{x}_{\sim} + \mathbf{B} \cdot \mathbf{d}_{\sim} + \mathbf{E} \cdot \mathbf{v}_{\sim}, \tag{12.95}$$

where

$\mathbf{x}_{\sim} = [i_{s,d\rightarrow}, i_{s,q\rightarrow}, (\Delta v_0)_{\sim}, v_{0\sim}]^T$ is the state vector

$\mathbf{d}_{\sim} = [d'_{d\sim}, d'_{q\sim}, d'_{0\sim}]^T$ is the control or input vector

$\mathbf{v}_{\sim} = [v_{s,d\rightarrow}, v_{s,q\rightarrow}]^T$ is the disturbance vector

$\mathbf{A}, \mathbf{B},$ and $\mathbf{E}$ respectively are the state matrix, the control matrix, and the disturbance matrix, defined as

$$\mathbf{A} = \begin{bmatrix} 0 & \omega_0 & 0 & -\dfrac{V_S\sqrt{2}}{LV_0} \\[2ex] -\omega_0 & 0 & 0 & \dfrac{\omega_0 I_S\sqrt{2}}{V_0} \\[2ex] 0 & 0 & -\dfrac{2}{R_0 C_0} & 0 \\[2ex] \dfrac{3V_S\sqrt{2}}{C_0 V_0} & -\dfrac{3L\omega_0 I_S\sqrt{2}}{C_0 V_0} & 0 & -\dfrac{1}{R_0 C_0} \end{bmatrix}$$

$$\mathbf{B} = \begin{bmatrix} -\dfrac{V_0}{2L} & 0 & 0 \\[2ex] 0 & -\dfrac{V_0}{2L} & 0 \\[2ex] 0 & 0 & \dfrac{\alpha I_S\sqrt{2}}{C_0} \\[2ex] \dfrac{3I_S\sqrt{2}}{2C_0} & 0 & 0 \end{bmatrix} \quad \text{and} \quad \mathbf{E} = \begin{bmatrix} \dfrac{1}{L} & 0 \\[2ex] 0 & \dfrac{1}{L} \\[2ex] 0 & 0 \\[2ex] 0 & 0 \end{bmatrix}$$

12.4.5.3 Transfer Functions

The frequency-domain representation of the converter is obtained by applying the Laplace transform to the state Equations 12.95. It yields

$$\mathbf{X}(s) = (s\mathbf{I}_4 - \mathbf{A})^{-1}\mathbf{B} \cdot \mathbf{D}(s) + (s\mathbf{I}_4 - \mathbf{A})^{-1}\mathbf{E} \cdot \mathbf{V}(s), \tag{12.96}$$

where

$\mathbf{I}_4$ denotes the 4-by-4 identity matrix

s is the Laplace operator

$\mathbf{X}(s) = [I_{s,d}(s), I_{s,q}(s), \Delta V_0(s), V_0(s)]^T$, $\mathbf{D}(s) = [D'_d(s), D'_q(s), D'_0(s)]^T$, and $\mathbf{V}(s) = [V_{s,d}(s), V_{s,q}(s)]^T$ are the Laplace transforms of vectors $\mathbf{x}_{\sim}, \mathbf{d}_{\sim},$ and $\mathbf{v}_{\sim}$, respectively

The development of expression (12.96) leads to the following input–output transfer functions:

$$G_{dd}(s) \equiv \left. \frac{I_{s,d}(s)}{D'_d(s)} \right|_{\substack{D'_q=0 \\ D'_0=0 \\ V_{s,d}=0 \\ V_{s,q}=0}} = -\frac{V_0}{2L} \cdot \frac{s(s+\omega_{z1})}{s^3 + \omega_{p1}s^2 + \omega_{p2}^2 s + \omega_{p3}^3} \tag{12.97a}$$

$$G_{dq}(s) \equiv \left. \frac{I_{s,d}(s)}{D'_q(s)} \right|_{\substack{D'_d=0 \\ D'_0=0 \\ V_{s,d}=0 \\ V_{s,q}=0}} = -\frac{V_0}{2L} \cdot \frac{\omega_0(s+\omega_{z1})}{s^3 + \omega_{p1}s^2 + \omega_{p2}^2 s + \omega_{p3}^3} \tag{12.97b}$$

$$G_{d0}(s) \equiv \left. \frac{I_{s,d}(s)}{D'_0(s)} \right|_{\substack{D'_d=0 \\ D'_q=0 \\ V_{s,d}=0 \\ V_{s,q}=0}} = 0 \tag{12.97c}$$

$$G_{qd}(s) \equiv \left. \frac{I_{s,q}(s)}{D'_d(s)} \right|_{\substack{D'_q=0 \\ D'_0=0 \\ V_{s,d}=0 \\ V_{s,q}=0}} = \left(\frac{V_0}{2L} + \frac{3I_S^2}{C_0 V_0} \right) \cdot \frac{\omega_0(s+\omega_{z2})}{s^3 + \omega_{p1}s^2 + \omega_{p2}^2 s + \omega_{p3}^3} \tag{12.97d}$$

$$G_{qq}(s) \equiv \left. \frac{I_{s,q}(s)}{D'_q(s)} \right|_{\substack{D'_d=0 \\ D'_0=0 \\ V_{s,d}=0 \\ V_{s,q}=0}} = -\frac{V_0}{2L} \cdot \frac{s^2 + \omega_{z3}s + \omega_{z4}^2}{s^3 + \omega_{p1}s^2 + \omega_{p2}^2 s + \omega_{p3}^3} \tag{12.97e}$$

$$G_{q0}(s) \equiv \left. \frac{I_{s,q}(s)}{D'_0(s)} \right|_{\substack{D'_d=0 \\ D'_q=0 \\ V_{s,d}=0 \\ V_{s,q}=0}} = 0 \tag{12.97f}$$

$$G_{\Delta 0d}(s) \equiv \left. \frac{\Delta V_0(s)}{D'_d(s)} \right|_{\substack{D'_q=0 \\ D'_0=0 \\ V_{s,d}=0 \\ V_{s,q}=0}} = 0 \tag{12.97g}$$

$$G_{\Delta 0q}(s) \equiv \left. \frac{\Delta V_0(s)}{D'_q(s)} \right|_{\substack{D'_d=0 \\ D'_0=0 \\ V_{s,d}=0 \\ V_{s,q}=0}} = 0 \tag{12.97h}$$

$$G_{\Delta 00}(s) \equiv \left. \frac{\Delta V_0(s)}{D'_0(s)} \right|_{\substack{D'_d=0 \\ D'_q=0 \\ V_{s,d}=0 \\ V_{s,q}=0}} = \frac{\alpha I_S \sqrt{2}}{C_0} \cdot \frac{1}{s+\omega_{p4}} \tag{12.97i}$$

$$G_{0d}(s) \equiv \left. \frac{V_0(s)}{D'_d(s)} \right|_{\substack{D'_q=0 \\ D'_0=0 \\ V_{s,d}=0 \\ V_{s,q}=0}} = \frac{3I_S \sqrt{2}}{2C_0} \cdot \frac{s(s-\omega_{z5})}{s^3 + \omega_{p1}s^2 + \omega_{p2}^2 s + \omega_{p3}^3} \tag{12.97j}$$

$$G_{0q}(s) \equiv \left. \frac{V_0(s)}{D_q'(s)} \right|_{\substack{D_d'=0 \\ D_0'=0 \\ V_{s,d}=0 \\ V_{s,q}=0}} = \frac{3I_S\sqrt{2}}{2C_0} \cdot \frac{\omega_0(s-\omega_{z5})}{s^3 + \omega_{p1}s^2 + \omega_{p2}^2 s + \omega_{p3}^3} \qquad (12.97k)$$

$$G_{00}(s) \equiv \left. \frac{V_0(s)}{D_0'(s)} \right|_{\substack{D_d'=0 \\ D_q'=0 \\ V_{s,d}=0 \\ V_{s,q}=0}} = 0 \qquad (12.97l)$$

with

$$\omega_{z1} = 2\omega_{z3} = 2\omega_{p1} = \omega_{p4} = \frac{2}{R_0 C_0} \qquad \omega_{z2} = \frac{V_0^2}{R_0 C_0 V_0^2 + 6R_0 L I_S^2}$$

$$\omega_{z4} = \frac{V_S}{V_0}\sqrt{\frac{6}{LC_0}} \qquad \omega_{z5} = \frac{V_S}{LI_S}$$

$$\omega_{p2} = \sqrt{\omega_0^2 + \frac{6V_S^2}{LC_0 V_0^2} + \frac{6L\omega_0^2 I_S^2}{C_0 V_0^2}} \qquad \omega_{p3} = \sqrt[3]{\frac{\omega_0^2}{R_0 C_0}}$$

It is on the basis of these transfer functions that the proposed control scheme, which ensures unity power factor as well as DC voltage stabilization, will be designed.

Furthermore, the poles of the converter (which are the roots of the fourth-degree polynomial characteristic) are given as follows:

$$p_1 = -\omega_{p4} = -\frac{2}{R_0 C_0}$$

$$p_2 = \sigma + \tau - \frac{\omega_{p1}}{3}$$

$$p_3 = -\frac{1}{2}(\sigma + \tau) - \frac{\omega_{p1}}{3} + j\frac{\sqrt{3}}{2}(\sigma - \tau)$$

$$p_4 = -\frac{1}{2}(\sigma + \tau) - \frac{\omega_{p1}}{3} - j\frac{\sqrt{3}}{2}(\sigma - \tau)$$

where

$$\sigma = \sqrt[3]{\rho + \sqrt{\mu^3 + \rho^2}}$$

$$\tau = \sqrt[3]{\rho - \sqrt{\mu^3 + \rho^2}}$$

$$\rho = \frac{1}{54}\left(9\omega_{p1}\omega_{p2}^2 - 27\omega_{p3}^3 - 2\omega_{p1}^3\right)$$

$$\mu = \frac{1}{9}\left(3\omega_{p2}^2 - \omega_{p1}^2\right)$$

Similarly, the disturbance transfer functions can also be derived from (12.96), and are obtained as follows:

$$F_{dd}(s) \equiv \left.\frac{I_{s,d}(s)}{V_{s,d}(s)}\right|_{\substack{D_d'=0 \\ D_q'=0 \\ D_0'=0 \\ V_{s,q}=0}} = \frac{1}{L}\cdot\frac{s^2 + \omega_{z3}s + \omega_{z6}^2}{s^3 + \omega_{p1}s^2 + \omega_{p2}^2 s + \omega_{p3}^3} \tag{12.98a}$$

$$F_{dq}(s) \equiv \left.\frac{I_{s,d}(s)}{V_{s,q}(s)}\right|_{\substack{D_d'=0 \\ D_q'=0 \\ D_0'=0 \\ V_{s,d}=0}} = \frac{1}{L}\cdot\frac{\omega_0(s + \omega_{z1})}{s^3 + \omega_{p1}s^2 + \omega_{p2}^2 s + \omega_{p3}^3} \tag{12.98b}$$

$$F_{qd}(s) \equiv \left.\frac{I_{s,q}(s)}{V_{s,d}(s)}\right|_{\substack{D_d'=0 \\ D_q'=0 \\ D_0'=0 \\ V_{s,q}=0}} = -\frac{1}{L}\cdot\frac{\omega_0 s}{s^3 + \omega_{p1}s^2 + \omega_{p2}^2 s + \omega_{p3}^3} \tag{12.98c}$$

$$F_{qq}(s) \equiv \left.\frac{I_{s,q}(s)}{V_{s,q}(s)}\right|_{\substack{D_d'=0 \\ D_q'=0 \\ D_0'=0 \\ V_{s,d}=0}} = \frac{1}{L}\cdot\frac{s^2 + \omega_{z3}s + \omega_{z4}^2}{s^3 + \omega_{p1}s^2 + \omega_{p2}^2 s + \omega_{p3}^3} \tag{12.98d}$$

$$F_{\Delta 0d}(s) \equiv \left.\frac{\Delta V_0(s)}{V_{s,d}(s)}\right|_{\substack{D_d'=0 \\ D_q'=0 \\ D_0'=0 \\ V_{s,q}=0}} = 0 \tag{12.98e}$$

$$F_{\Delta 0q}(s) \equiv \left.\frac{\Delta V_0(s)}{V_{s,q}(s)}\right|_{\substack{D_d'=0 \\ D_q'=0 \\ D_0'=0 \\ V_{s,d}=0}} = 0 \tag{12.98f}$$

$$F_{0d}(s) \equiv \left.\frac{V_0(s)}{V_{s,d}(s)}\right|_{\substack{D_d'=0 \\ D_q'=0 \\ D_0'=0 \\ V_{s,q}=0}} = \frac{3V_S\sqrt{2}}{LC_0 V_0}\cdot\frac{s + \omega_{z7}}{s^3 + \omega_{p1}s^2 + \omega_{p2}^2 s + \omega_{p3}^3} \tag{12.98g}$$

$$F_{0q}(s) \equiv \left. \frac{V_0(s)}{V_{s,q}(s)} \right|_{\substack{D'_d=0 \\ D'_q=0 \\ D'_0=0 \\ V_{s,d}=0}} = -\frac{3\omega_0 I_S \sqrt{2}}{C_0 V_0} \cdot \frac{s - \omega_{z5}}{s^3 + \omega_{p1}s^2 + \omega_{p2}^2 s + \omega_{p3}^3} \tag{12.98h}$$

where

$$\omega_{z6} = \frac{I_S \omega_0}{V_0} \sqrt{\frac{6L}{C_0}} \quad \omega_{z7} = \frac{L\omega_0^2 I_S}{V_S}$$

12.4.6 Simulation Results

In order to highlight the steady-state and transient performances of the proposed control scheme, a simulation work is carried using the MATLAB and Simulink tool. A numerical version of the converter is, hence, implemented and two resistors, denoted, respectively, by $R_{0,h}$ and $R_{0,l}$, are used to represent the upper and lower DC loads. The numerical values of all parameters and operating conditions of the converter are presented in Table 12.2.

Figures 12.16 and 12.17 show the dependency of the model parameter α on the DC load power and the unbalance factor σ defined as

$$\sigma = \left| \frac{R_{0,h} - R_{0,l}}{R_{0,h} + R_{0,l}} \right| \tag{12.99}$$

Note that, for a given σ, the parameter α does not vary significantly with the load power, even for heavy unbalance conditions. Contrarily, the dependency of α on σ for a given load power is quite noticeable and has to be considered in the design of robust or adaptive control circuit with high dynamic performance. This dependency was not taken into account in the development of the theoretical expression (12.68), where a balanced DC load has been considered. However, it is clear that, for small values of the unbalance factor σ, the theoretical and numerical values of the parameter α are quite similar.

TABLE 12.2 System Parameters and Operating Conditions

Phase-to-neutral voltage RMS-value	$V_S = 120\,\text{V}$
Desired total output voltage	$V_0 = 700\,\text{V}$
Nominal load power	$P_0 = 25\,\text{kW}$
Mains frequency	$f_0 = 60\,\text{Hz}$
Switching frequency	$f_S = 50\,\text{kHz}$
AC-side inductors	$L = 1\,\text{mH, each}$
DC-side capacitors	$C_0 = 1\,\text{mF, each}$
Current feedback gains	$K_i = 0.05\,\Omega$
Voltage feedback gains	$K_v = 5/700$
PWM dynamic gain	$K_{PWM} = 1$

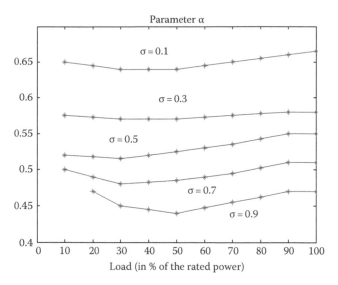

FIGURE 12.16 Parameter α with respect to the load and unbalance factor.

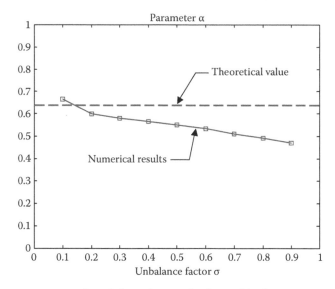

FIGURE 12.17 Parameter α versus the unbalance factor σ for the rated load.

12.5 Averaged Model-Based Multi-Loop Control Techniques Applied to the Three-Phase/ Switch/Level PWM Boost Rectifier

In most applications that use the Vienna rectifier, hysteretic-based control was implemented in order to ensure line current wave shaping [25]. However, this control technique suffers from a major drawback, namely a time-varying switching frequency, which may reduce, on the one hand, the reliability of the converter due to an eventual excess of power losses in the switching devices, and on the other hand, cause filtering difficulties due to a wide spread harmonic spectrum for the line currents. In order to avoid these inconveniences, fixed-frequency PWM control is considered. But the application of such control technique requires the knowledge of the dynamics of the converter. A low-frequency state-space model of the converter can be derived by applying the well-known state-space averaging technique [39]. This approach is widely used in the modeling process of switch-mode converters. The existence of such model allows the systematic development of control laws offered by the classic or modern control theory.

Applying forward the state-space-averaging approach to the converter yields a nonlinear fifth-order time-varying model. Therefore, the elaboration and implementation of a corresponding suitable control law become very difficult tasks. Thus, in order to simplify the control design procedure, a fourth-order time-invariant model is elaborated in [32] by applying to the former two transformations—a three-to-two-axis transformation using the synchronous rotating frame and an input vector nonlinear transformation. This state-space representation is used for the design of a multiple-loops nonlinear controller that uses the input/output feedback linearization approach. The implemented control scheme offered high steady-state and dynamic performance, especially in terms of line current THD, DC voltage regulation, and robustness toward load or mains voltage disturbances, but at the expense of a high control and sensing effort, and a low robustness toward structural parameters variations.

The control scheme can be considerably simplified if single-input-single-output (SISO) linear regulators are used. For this purpose, a small-signal representation of the converter must be derived, and the corresponding transfer functions must be computed. A linear multiple-loop control system is then designed by neglecting the cross-coupling between the input and output variables of the converter. This assumption allows in designing each control loop independently from the others. Although its simplicity, it will be shown that the linear control scheme thus obtained suffers from an instability that occurs inevitably at low power.

In this section, a comparative evaluation of the two control approaches described above is established. The comparison is based on simulation experiments carried out on a numerical versions of the converter associated to each control algorithm. The line current shaping and output DC voltage regulation are evaluated and analyzed in both balanced and unbalanced operating conditions, at full or partial load.

12.5.1 Linear Control Design

The proposed multiple-loops linear control system is presented in Figure 12.18. The letter **K** stands for the stationary-frame/synchronous-frame transformation. The current references $i_{s,d,ref}$ and $i_{s,q,ref}$ are generated as follows

$$i_{s,d,ref} = \frac{I_S\sqrt{2}}{\sqrt{v_{s,d}^2 + v_{s,q}^2}} v_{s,d}$$

$$i_{s,q,ref} = \frac{I_S\sqrt{2}}{\sqrt{v_{s,d}^2 + v_{s,q}^2}} v_{s,q}$$

$$(12.100)$$

K_i and K_v are, respectively, the current and voltage loops feedback-scaling gains. The linear inner regulators $H_{i,d}(s)$, $H_{i,q}(s)$, $H_{\Delta v}(s)$, and outer regulator $H_v(s)$ are designed, as indicated in Figure 12.19, by using the independent multiple feedback looping approach. In other words, all the cross-coupling transfer functions between the control inputs and the system outputs are neglected. In addition, for the sake of simplicity, the effects of the disturbance source voltages $v_{s,d}$ and $v_{s,q}$ are neglected and, consequently, their corresponding transfer functions are not considered.

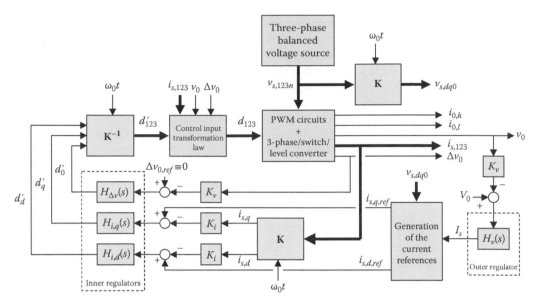

FIGURE 12.18   Block diagram of the linear control scheme.

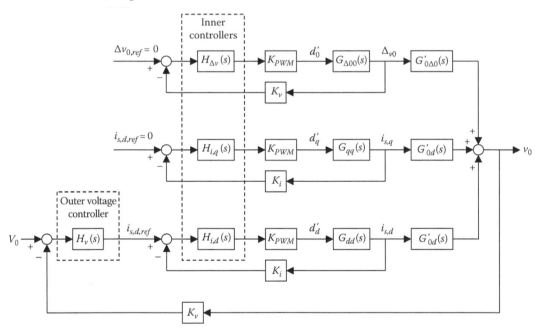

FIGURE 12.19   Design of the multi-loops linear regulator.

The factor K_{PWM} that appears in the inner loops represents the dynamic gain introduced by the pulse-width-modulators. The structure and parameters of the regulators are chosen on the basis of the pole-zero compensation method, in order to ensure an optimal second-order behavior to the corresponding closed-loop (i.e., a damping factor equal to 0.707). Considering the numerical values given in Table 12.2, it yields for the inner loops

$$H_{i,d}(s) = -\frac{1{,}128(s+44.72)^2}{s^2(s+6{,}283)} \tag{12.101}$$

$$H_{i,q}(s) = -\frac{1,128(s+44.72)}{s(s+6,283)} \tag{12.102}$$

$$H_{\Delta v}(s) = \frac{44,200(s+204)}{s(s+6,283)} \tag{12.103}$$

Furthermore, the outer loop is designed to be slower enough than the inner ones in order to ensure high stability to the control system. In this case, only $G'_{0d}(s)$ is considered in the calculation of $H_v(s)$. It is given as

$$G'_{0d}(s) \equiv \left.\frac{V_0(s)}{I_{s,d}(s)}\right|_{\substack{I_{s,q}=0 \\ \Delta V_0=0 \\ V_{s,d}=0 \\ V_{s,q}=0}} = \frac{G_{0d}(s) \cdot G_{qq}(s) - G_{0q}(s) \cdot G_{qd}(s)}{G_{dd}(s) \cdot G_{qq}(s) - G_{dq}(s) \cdot G_{qd}(s)} \tag{12.104}$$

or, employing (12.97)

$$G'_{0d}(s) = -\frac{3LI_S\sqrt{2}}{C_0V_0} \cdot \frac{s-\omega_{z5}}{s+\omega_{p4}} \tag{12.105}$$

It yields, after calculations,

$$H_v(s) = \frac{19(s+204)}{s(s+62.83)} \tag{12.106}$$

Note that $G'_{0q}(s)$ and $G'_{0\Delta 0}(s)$ do not interfere in the calculation of $H_v(s)$ and, therefore, are not considered.

12.5.2 Nonlinear Control Design

The proposed nonlinear control scheme is presented in Figure 12.20. It consists, first, of an inner multiple-input-multiple-output (MIMO) feedback loop for current wave-shaping and DC voltage balance adjustment and, secondly, of a SISO outer feedback loop for DC voltage regulation. Both inner and outer

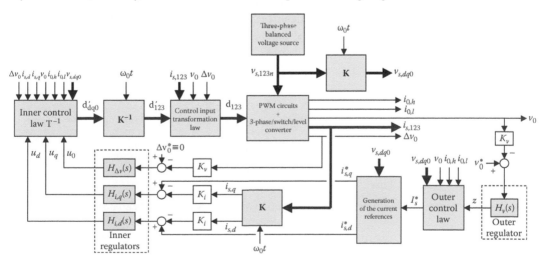

FIGURE 12.20 Nonlinear control scheme.

control laws are elaborated on the basis of the nonlinearity compensation technique [47] applied to the three-input-four-output system given by (12.69). The design procedure of the control system is described in the following subsections.

12.5.2.1 Inner Control Law

The design of the inner control law is based on applying the nonlinearity compensation technique to the subsystem described by the Equations 12.69a through 12.69c. The major purpose of this strategy is to find a multivariable nonlinear function T that transforms the original subsystem into a linear decoupled one. Such function is given by

$$\mathbf{u}_{dq0} = \mathbf{T}\left(\mathbf{v}_{s,dq0}, \mathbf{i}_{s,dq0}, \mathbf{d}'_{dq0}, v_0, \Delta v_0, i_{0,h}, i_{0,l}\right) = \begin{bmatrix} \dfrac{2v_{s,d} + 2L\omega_0 i_{s,q} - v_0 d'_d}{2L} \\[2ex] \dfrac{2v_{s,q} - 2L\omega_0 i_{s,d} - v_0 d'_q}{2L} \\[2ex] \dfrac{2\alpha v_0 i_{s,d} d'_0 - 3\Delta v_0\left(i_{s,d} d'_d + i_{s,q} d'_q\right) - 2v_0(i_{0,h} - i_{0,l})}{2C_0 v_0} \end{bmatrix}, \quad (12.107)$$

where $\mathbf{u}_{dq0} = [u_d, u_q, u_0]^T$ denotes the new inner input vector of the linearized system represented by the following canonical form

$$\frac{di_{s,d}}{dt} = u_d$$

$$\frac{di_{s,q}}{dt} = u_q \qquad\qquad (12.108)$$

$$\frac{d(\Delta v_0)}{dt} = u_0$$

Here again, the inner regulators $H_{i,d}(s)$, $H_{i,q}(s)$, and $H_{\Delta v}(s)$ of each single loop are designed so that the corresponding closed-loop transfer function is equal to a low-pass optimal transfer function filter. The calculation of the regulators parameters should also take account of two additional design criteria. The first one considers that the inner current loops are relatively much faster than the outer one. The second criterion is that the regulators should attenuate the high-frequency components of the controlled variables that are inherently generated by the switching process. Considering again the numerical values given in Section 12.5.1, it yields

$$H_{i,d}(s) = H_{i,q}(s) = \frac{2 \times 10^5 \pi \sqrt{2}}{1 + s/2 \times 10^4 \pi \sqrt{2}} \qquad\qquad (12.109)$$

$$H_{\Delta v}(s) = \frac{7000 \pi \sqrt{2}}{1 + s/100\pi\sqrt{2}} \qquad\qquad (12.110)$$

12.5.2.2 Outer Control Law

Once the inner control law is implemented, the original MIMO system (12.69) can be reduced to a SISO model given by

$$C_0 \frac{dv_0}{dt} = \frac{3}{2}\left(d'^*_d i^*_{s,d} + d'^*_q i^*_{s,q}\right) - i_{0,h} - i_{0,l} \qquad\qquad (12.111)$$

$d_d'^*$, $d_q'^*$, $i_{s,d}^*$, and $i_{s,q}^*$ are, respectively, the desired values of d_d', d_q', $i_{s,d}$, and $i_{s,q}$. They are given by

$$d_d'^* = \frac{2}{v_0}\left(v_{s,d} + L\omega_0 i_{s,q}^*\right)$$

$$d_q'^* = \frac{2}{v_0}\left(v_{s,q} - L\omega_0 i_{s,d}^*\right)$$

$$i_{s,d}^* = \frac{I_S^*}{V_S} v_{s,d}$$

(12.112)

$$i_{s,q}^* = \frac{I_S^*}{V_S} v_{s,q},$$

where
   V_S denotes the RMS-value of the mains voltage
   I_S^* is the desired RMS-value of the line current

Using expressions (12.112) into Equation 12.111, and knowing that

$$v_{s,d}^2 + v_{s,q}^2 = 2V_S^2, \quad \forall t$$

(12.113)

we get

$$C_0 \frac{dv_0}{dt} = \frac{6}{v_0} V_S I_S^* - i_{0,h} - i_{0,l}$$

(12.114)

Equation 12.114 represents a nonlinear SISO system, having I_S^* as an input and v_0 as an output. By introducing a new input variable z given by

$$z = \frac{3\sqrt{2}\sqrt{v_{s,d}^2 + v_{s,q}^2}\, I_S^* - v_0(i_{0,h} + i_{0,l})}{C_0 v_0}$$

(12.115)

the system becomes equivalent to the following minimized canonical form

$$\frac{dv_0}{dt} = z$$

(12.116)

The design of the linear regulator $H_v(s)$ follows the same considerations indicated previously. It yields

$$H_v(s) = \frac{2800\pi\sqrt{2}}{1 + s/40\pi\sqrt{2}}$$

(12.117)

12.5.3 Simulation Results

In order to highlight the steady-state and transient performances of the proposed control schemes, a simulation work is carried using the MATLAB and Simulink tool. Numerical versions of the systems depicted in Figures 12.18 and 12.20 are, hence, implemented and two resistors, denoted, respectively, by $R_{0,h}$ and $R_{0,l}$, are used to represent the upper and lower DC loads. The numerical values of all parameters and operating conditions of the converter are as selected in Section 12.5.1.

Figure 12.21 shows the source currents and DC output voltages in the steady-state regime, where a balanced nominal load is considered. A practically unity power factor operation is obtained for both control schemes as noticed. Figures 12.22 and 12.23 illustrate, respectively, the response of the linear and nonlinear control systems to a sudden variation of the DC load ($R_{0,l}$ is increased by 100% at $t = 0.05\,\mathrm{s}$). The given time-response of the duty cycle d_1 of switch Q_1 lets to appear a control saturation phenomenon for both control schemes. Finally, Figure 12.24 presents the variation of the source current THD for both control schemes, with respect to the load power and the load unbalance factor defined in (12.99).

The impacts of input voltage disturbances on the system performance are only analyzed in the case of the nonlinear control, since it generally gives better results than the linear one. Figure 12.25a and b shows the response of the controlled converter to a sudden shortening then reestablishment of the source voltage in terms of line currents $i_{s,1}$, $i_{s,2}$, $i_{s,3}$, and DC output voltages $v_{0,h}$ and $v_{0,l}$. Figure 12.25c

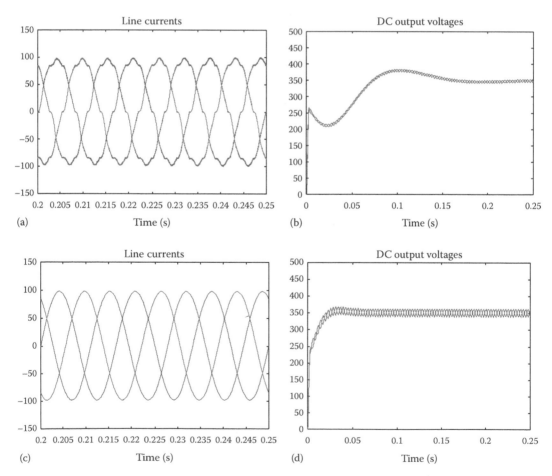

FIGURE 12.21 Line currents and DC voltages for the linear (a and b) and nonlinear (c and d) control schemes under normal operating conditions.

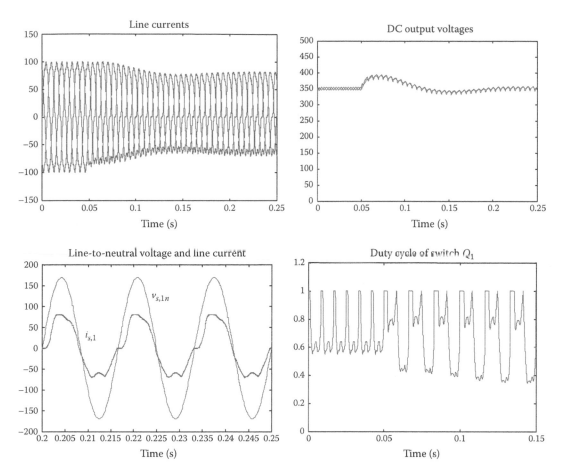

FIGURE 12.22 Effects of a load variation on the line currents and DC voltages for the linear control scheme.

and d represents the system response to a sudden increase then decrease of the mains voltage. The same mains voltage disturbances are applied to the converter under unbalanced DC load operating conditions. The corresponding results are shown in Figures 12.26 and 12.27 for a small ($\sigma = 0.33$) and severe ($\sigma = 0.82$) load unbalance, respectively.

12.5.4 Comparative Evaluation

By inspecting Figures 12.21 through 12.24, we can deduce the following conclusions:

1. During the rated and balanced operating conditions (Figure 12.21), the nonlinear control offers a better current shaping than the linear one; however, the low-frequency ripple in the DC output voltages is more noticeable.
2. At low power (lower than 30% of the rated power), the linear control system becomes unstable, and a low-frequency component (around 8 Hz) appears in the waveforms of the DC voltages and line currents (see Figure 12.23a). Furthermore, the current THD offered by the linear control system is severely deteriorated when the load unbalance increases significantly. This is due to the fact that the set point chosen in the linearization process is varying, which has a major impact on the placement of the poles and zeros of the transfer functions and, consequently, affects considerably the credibility of the small-signal model of the converter. The regulators calculated in Section 12.5.1 are,

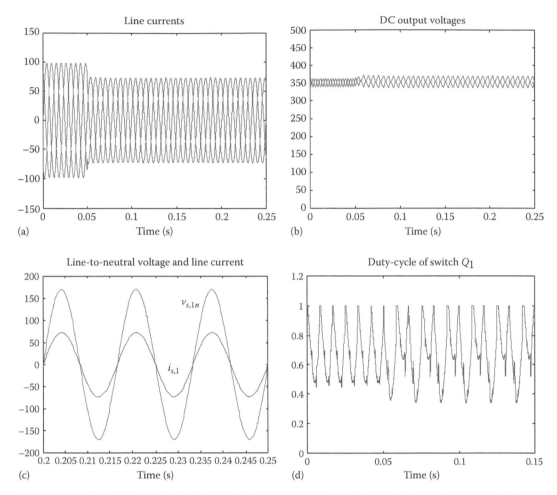

FIGURE 12.23 Effects of a load variation on the line currents and DC voltages for the nonlinear control scheme.

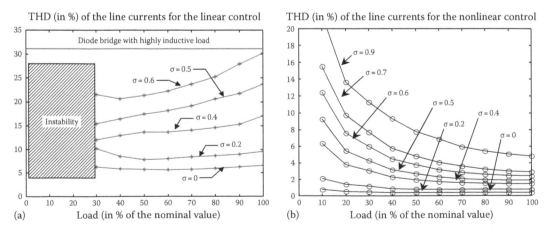

FIGURE 12.24 Current THD with respect to the load characteristics for the linear (a) and nonlinear (b) control systems.

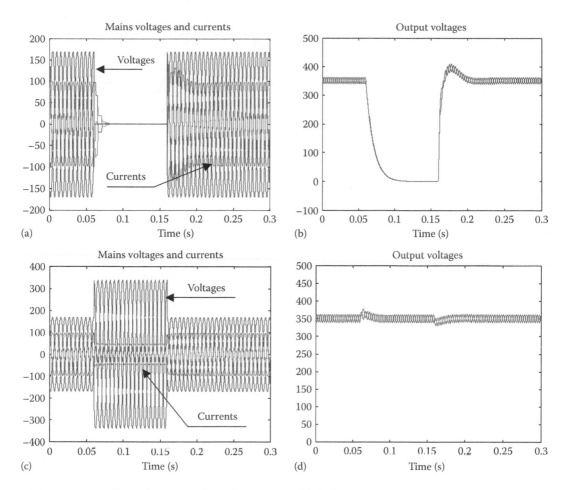

FIGURE 12.25 Effects of a mains voltage shortening on (a) the line currents and (b) the DC output voltages. Effects of a sudden mains overvoltage on (c) the line currents and (d) the DC output voltages. Case of the nonlinear control strategy with a balanced DC nominal load.

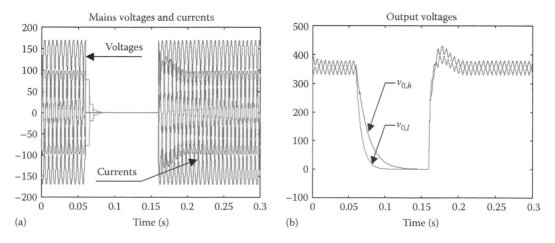

FIGURE 12.26 Effects of a mains voltage shortening on (a) the line currents and (b) the DC output voltages.

(*continued*)

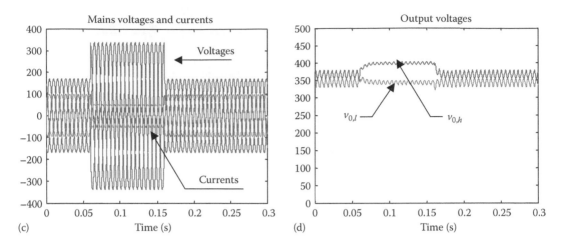

FIGURE 12.26 (continued) Effects of a sudden mains overvoltage on (c) the line currents and (d) the DC output voltages. Case of the nonlinear control strategy with an unbalanced DC nominal load ($\sigma = 0.33$).

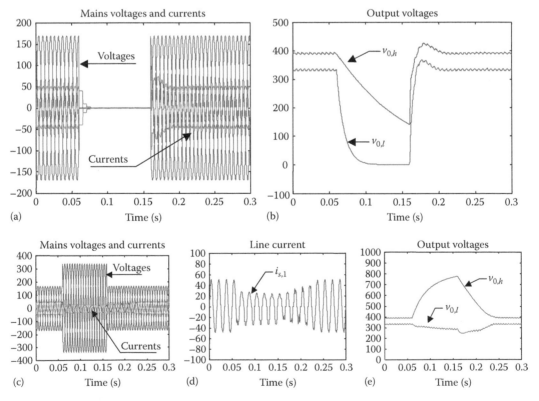

FIGURE 12.27 Effects of a mains voltage shortening on (a) the line currents and (b) the DC output voltages. Effects of a sudden mains overvoltage on (c) the line currents, (d) the current in phase 1 and (e) the DC output voltages. Case of the nonlinear control strategy with a highly unbalanced DC load ($\sigma = 0.82$).

thus, no longer tuned to the converter, and the stability is severely affected. Note, in addition, that the structural parameter α that appears in the converter model given by (12.69) is not quite independent from the operating conditions and does also vary. For high values of σ (>0.6), the current THD exceeds the critical value of 31% given by a classical three-phase diode bridge connected to a highly inductive DC load.

3. In response to a sudden load variation, the nonlinear control scheme offers better dynamics than the linear one. As seen in Figures 12.22 and 12.23, the nonlinear control system responds in less than 10 ms without voltage overstepping, whereas, for the linear controller, the time response exceeds 10 ms and the DC voltages attain temporarily 110% of their steady-state values. Furthermore, the control saturation problem (detected in the waveforms of the duty cycle d_1) is more noticeable in the case of the linear control system. This phenomenon, in addition to the discontinuities that the duty cycle presents, affects considerably the validity of the mathematical model (12.69) on which the design of the regulators was based.

12.6 Conclusion

In this chapter, after addressing some basic issues concerning the modeling of switch-mode converters with fixed-switching-frequency, the authors presented a comprehensive method to facilitate the understanding of the Vienna rectifier operation. At first the single-phase topology was considered and studied. Some design criteria or constraints are deduced, which constitutes the basic frame on which the study of the three-phase structure is established. The converter sequences of operation is presented and a corresponding state model is derived for a CCM and fixed-switching-frequency operation. A basic mathematical model of the converter is established, and a simplified time-invariant model is, thereafter, deduced using rotating Park transformation, that led to transfer functions calculation through a small-signal linearization process. Finally, the results of two control approaches, namely multiple-loops linear control design theory (applied on the small-signal model) and the input–output feedback linearization (to compensate the nonlinearity and cross-coupling), are presented. The obtained results emphasize a high performance of both control laws in terms of line current THD, DC voltage regulation, and stability near full and balanced load operating conditions.

References

1. *IEEE Recommended Practices and Requirements for Harmonic Control in Electric Power Systems*, IEEE Std. 519, Institute of Electrical and Electronics Engineers, June 1992.
2. IEC Subcommittee 77A, Disturbance in supply systems caused by household appliance and similar electrical equipment, Part 2: Harmonics, IEC 555-2 (EN 60555-2), September 1992.
3. T. S. Key and J.-S. Lai, Comparison of standards and power supply design options for limiting harmonic distortion in power systems, *IEEE Trans. Ind. Appl.*, 29(4), 688–695, July/August 1993.
4. M. Rastogi, R. Naik, and N. Mohan, A comparative evaluation of harmonic reduction techniques in three-phase utility interface of power electronic loads, *IEEE Trans. Ind. Appl.*, 30(5), 1149–1155, September/October 1994.
5. H. Mao, F. C. Y. Lee, D. Boroyevich, and S. Hiti, Review of high-performance three-phase power-factor correction circuits, *IEEE Trans. Ind. Electron.*, 44(4), 437–446, August 1997.
6. J. W. Kolar and H. Ertl, Status of the techniques of three-phase rectifier systems with low effects on the mains, in *Proceedings of 21st INTELEC*, Copenhagen, Denmark, June 6–9, 1999.
7. R. Wu, S. B. Dewan, and G. R. Slemon, A PWM AC to DC converter with fixed switching frequency, *IEEE Trans. Ind. Appl.*, 26(5), 880–885, 1990.
8. W.-C. Lee, D.-S. Hyun, and T.-K. Lee, A novel control method for three-phase PWM rectifiers using a single current sensor, *IEEE Trans. Power Electron.*, 15(5), 861–870, September 2000.

9. S. Kim, P. N. Enjeti, P. Packebush, and I. J. Pitel, A new approach to improve power factor and reduce harmonics in a three-phase diode rectifier type utility interface, *IEEE Trans. Ind. Appl.*, 30(6), 1557–1564, November/December 1994.

10. W. B. Lawrance and W. Mielczarski, Harmonic current rejection in a three-phase diode bridge rectifier, *IEEE Trans. Ind. Electron.*, 39, 571–576, December 1992.

11. P. Pejovic and Z. Janda, An analysis of three-phase low harmonic rectifiers applying the third-harmonic current injection, *IEEE Trans. Power Electron.*, 14(3), 397–407, May 1999.

12. N. Mohan, M. Rastogi, and R. Naik, Analysis of a new power electronics interface with approximately sinusoidal 3-phase utility currents and a regulated DC output, *IEEE Trans. Power Deliv.*, 8(2), 540–546, April 1993.

13. R. Naik, M. Rastogi, and N. Mohan, Third-harmonic modulated power electronics interface with 3-phase utility to provide a regulated DC output and to minimize line current harmonics, *IEEE/IAS Annual Meeting Conference*, Houston, TX, 689–694, October 4–9, 1992.

14. R. Naik, M. Rastogi, and N. Mohan, Third-harmonic modulated power electronics interface with three-phase utility to provide a regulated DC output and to minimize line-current harmonics, *IEEE Trans. Ind. Appl.*, 31(3), 598–602, May/June 1995.

15. M. Rastogi, N. Mohan, and C. P. Henze, Three-phase sinusoidal current rectifier with zero-current switching, *IEEE Trans. Power Electron.*, 10(6), 753–759, November 1995.

16. R. Naik, M. Rastogi, N. Mohan, R. Nilssen, and C. P. Henze, A magnetic device for current injection in a three-phase, sinusoidal-current utility interface, *IEEE/IAS Annual Meeting*, Toronto, Ontario Canada, October 1993, vol. 2, pp. 926–930, 1993.

17. R. Naik, N. Mohan, M. Rogers, and A. Bulawka, A novel grid interface, optimized for utility-scale applications of photovoltaic, wind-electric, and fuel-cell systems, *IEEE Trans. Power Deliv.*, 10(4), 1920–1926, October 1995.

18. P. Pejovic and Z. Janda, Optimal current programming in three-phase high-power-factor rectifier based on two boost converters, *IEEE Trans. Power Electron.*, 13(6), 1152–1163, November 1998.

19. J. C. Salmon, Operating a three-phase diode rectifier with a low-input current distortion using a series-connected dual boost converter, *IEEE Trans. Power Electron.*, 11(4), 592–603, July 1996.

20. A. M. Cross and A. J. Forsyth, A high-power-factor, three-phase isolated AC-DC converter using high-frequency current injection, *IEEE Trans. Power Electron.*, 18(4), 1012–1019, July 2003.

21. N. Vazquez, H. Rodriguez, C. Hernandez, E. Rodriguez, and J. Arau, Three-phase rectifier with active current injection and high efficiency, *IEEE Trans. Ind. Electron.*, 56(1), 110–119, January 2009.

22. C. Qiao and K. M. Smedley, A general three-phase PFC controller for rectifiers with a series-connected dual-boost topology, *IEEE Trans. Ind. Appl.*, 38(1), 137–148, January/February 2002.

23. B. M. Saied and H. I. Zynal, Minimizing current distortion of a three-phase bridge rectifier based on line injection technique, *IEEE Trans. Power Electron.*, 21(6), 1754–1761, November 2006.

24. J.-I. Itoh and I. Ashida, A novel three-phase PFC rectifier using a harmonic current injection method, *IEEE Trans. Power Electron.*, 23(2), 715–722, March 2008.

25. J. W. Kolar and F. C. Zach, A novel three-phase utility interface minimizing line current harmonics of high-power telecommunications rectifier modules, *IEEE Trans. Ind. Electron.*, 44(4), 456–467, August 1997.

26. J. W. Kolar and F. C. Zach, A novel three-phase three-switch three-level unity power factor PWM rectifier, in *Proceedings of the 28th Power Conversion Conference*, pp. 125–138, Nuremberg, Germany, June 28–30, 1994.

27. J. W. Kolar, F. Stögerer, J. Miniböck, and H. Ertl, A new concept for reconstruction of the input phase currents of a three-phase/switch/level PWM (Vienna) rectifier based on neutral point current measurement, *IEEE 31st Annual Power Electronics Specialists Conference (PESC'00)*, Galway, Ireland, June 2000, vol. 1, pp. 139–146, 2000.

28. C. Qiao and K. M. Smedley, Three-phase unity-power-factor star-connected switch (VIENNA) rectifier with unified constant-frequency integration control, *IEEE Trans. Power Electron.*, 18(4), 952–957, July 2003.

29. T. Nussbaumer and J. W. Kolar, Comparison of 3-phase wide output voltage range PWM rectifiers, *IEEE Trans. Ind. Electron.*, 54(6), 3422–3425, December 2007.

30. C.-M. Young, C.-C. Wu, and C.-H. Lu, Constant-switching-frequency control of three-phase/ switch/level boost-type rectifiers without current sensors, *IEEE Trans. Ind. Electron.*, 50(1), 246–248, February 2003.

31. J. Minibock and J. W. Kolar, Novel concept for mains voltage proportional input current shaping of a VIENNA rectifier eliminating controller multipliers, *IEEE Trans. Ind. Electron.*, 52(1), 162–170, February 2005.

32. H. Y. Kanaan, K. Al-Haddad, and F. Fnaiech, Modelling and control of a three-phase/switch/level fixed-frequency PWM rectifier: State-space averaged model, *IEE Proc. Electr. Power Appl.*, 152(03), 551–557, May 2005.

33. H. Y. Kanaan, K. Al-Haddad, and F. Fnaiech, A study on the effects of the neutral inductor on the modeling and performance of a four-wire three-phase/switch/level fixed-frequency rectifier, *J. Math. Comput. Simul. (IMACS)*, 71(4–6), 487–498, June 2006 (Special issue on Modeling and Simulation of Electric Machines, Converters and Systems).

34. N. Bel Hadj-Youssef, K. Al-Haddad, H. Y. Kanaan, and F. Fnaiech, Small-signal perturbation technique used for DSP-based identification of a three-phase three-level boost-type Vienna rectifier, *IET Proc. Electr. Power Appl.*, 1(2), 199–208, March 2007.

35. N. Bel Haj Youssef, K. Al-Haddad, and H. Y. Kanaan, Real-time implementation of a discrete nonlinearity compensating multiloops control technique for a 1.5 kW three-phase/switch/level Vienna converter, *IEEE Trans. Ind. Electron.*, 55(3), 1225–1234, March 2008.

36. N. Bel Haj Youssef, K. Al-Haddad, and H. Y. Kanaan, Large signal modeling and steady-state analysis of a 1.5 kW three phase/switch/level (Vienna) rectifier with experimental validation, *IEEE Trans. Ind. Electron.*, 55(3), 1213–1224, March 2008.

37. N. Bel Haj Youssef, K. Al-Haddad, and H. Y. Kanaan, Implementation of a new linear control technique based on experimentally validated small-signal model of three-phase three-level boost-type Vienna rectifier, *IEEE Trans. Ind. Electron.*, 55(4), 1666–1676, April 2008.

38. G. W. Wester and R. D. Middlebrook, Low-frequency characterization of switched DC-to-DC converters, in *Proceedings of the IEEE Power Processing and Electronics Specialists Conference*, Atlantic City, NJ, May 22–23, 1972.

39. R. D. Middlebrook and S. Cuk, A general unified approach to modeling switching-converter power stages, in *Proceedings of the IEEE Power Electronics Specialists Conference*, Cleveland, OH, June 8–10, 1976.

40. S. R. Sanders, J. M. Noworolski, X. Z. Liu, and G. C. Verghese, Generalized averaging method for power conversion circuits, *IEEE Trans. Power Electron.*, 6(2), 251–259, April 1991.

41. J. P. Noon, UC3855A/B high performance power factor preregulator, Unitrode Corporation, Merrimack, NH, Unitrode Application Notes, Section U-153, pp. 3.460–3.479, 1998.

42. H. Kanaan, K. Al-Haddad, R. Chaffaï, and L. Duguay, Susceptibility and input impedance evaluation of a single phase unity power factor rectifier, in *Proceedings of Seventh IEEE ICECS'2K Conference*, Beirut, Lebanon, December 17–20, 2000.

43. H. Kanaan and K. Al-Haddad, A comparative evaluation of averaged model based linear and nonlinear control laws applied to a single-phase two-stage boost rectifier, in *Proceedings of RTST 2002 Conference*, Beirut & Byblos, Lebanon, March 4–6, 2002.

44. R. Wu, S. B. Dewan, and G. R. Slemon, Analysis of an AC-to-DC voltage source converter using PWM with phase and amplitude control, *IEEE Trans. Ind. Appl.*, 27(2), 355–364, March/April 1991.

45. H. Y. Kanaan and K. Al-Haddad, A comparison between three modeling approaches for computer implementation of high-fixed-switching-frequency power converters operating in a continuous mode, in *Proceedings of CCECE'02*, vol. 1, pp. 274–279, Winnipeg, Canada, May 12–15, 2002.

46. J. Sun, D. M. Mitchell, M. F. Greuel, P. T. Krein, and R. M. Bass, Averaged modeling of PWM converters operating in discontinuous conduction mode, *IEEE Trans. Power Electron.*, 16(4), 482–492, July 2001.

47. J.-J. E. Slotine and W. Li, *Applied Nonlinear Control*, Prentice-Hall, Englewood Cliffs, NJ, 1991.

13

DC–DC Converters

István Nagy
*Budapest University of
Technology and Economics*

Pavol Bauer
*Delft University
of Technology*

13.1 Introduction

The DC–DC converters illustrated in Figure 13.1 are used to interface two DC systems and control the flow of power between them. Their basic function in a DC environment is similar to that of transformers in AC systems. Unlike in transformers, the ratio of the input to the output, either voltage or current, can continuously be varied by the control signal and this ratio can be higher or lower than unity.

DC–DC converters are called choppers in high-power applications. They are used for DC motor control, for example, in battery-supplied vehicles and in different applications such as in electric cars, airplanes, and spaceships, where onboard-regulated DC power supplies are required. In general, DC–DC converters are employed as power supplies in sensors, controllers, transducers, computers, commercial electronics, electronic instruments, as well as a variety of technologies that include plasma, arc, electron beam, electrolytic, nuclear physics, solar energy conversion, wind energy conversion, and the like. The power levels encountered in DC–DC converters range from (1) less than one watt, in DC–DC converters within battery-operated portable equipment; (2) tens, hundreds, or thousands of watts in power supplies for computers and office equipment; (3) kilowatts to megawatts in variable speed motor drives; and (4) roughly 100 MW in the DC transmission lines, for example, offshore wind farms.

The DC–DC converters are constructed of electronic switches and sometimes include inductive and capacitive components, all of which are normally followed by a low-pass filter. If the filter corner frequency is sufficiently lower than the switching frequency, then the filter essentially passes only the DC component. There are a number of classifications for these converters that are dependent upon the input impedance, $\bar{Z}_i$, of the low-pass filter, as shown in Figure 13.2 (Rashid, 1993). The converter is either output current sourced, in which case $\bar{Z}_i \cong j\omega L$, or output voltage sourced such that $\bar{Z}_i \cong -(j/\omega C)$, in which case either the output current or voltage is designed to be ripple free, i.e., constant in one switching cycle. Some DC–DC converters permit power flow in only one direction, others implement bidirectional power flow. Depending upon the direction of the output current and voltage, the converters can

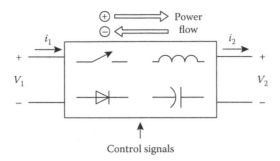

FIGURE 13.1   DC–DC converters.

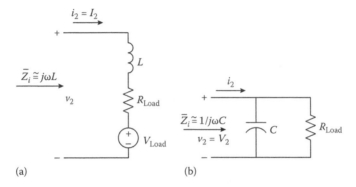

FIGURE 13.2   Basic low-pass filters for output current sourced (a) and voltage sourced (b) converter.

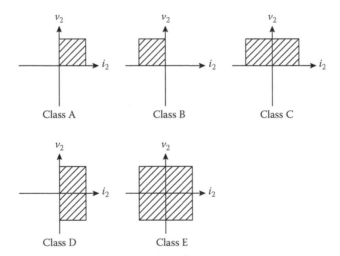

FIGURE 13.3   Unidirectional (class A and class B) and bidirectional (class C, class D, and class E) power flow.

be classified into five classes as shown in Figure 13.3. One-quadrant (classes A and B), two-quadrant (classes C and D) and four-quadrant operation can be realized.

Hard switched and soft switched or resonant converters exhibit another classification. In the first (second) group the power loss is high (low) in switching as a result of the nonzero voltage and current (zero voltage and/or current) on the switches at the initialization of the switching action.

The step-down or buck converter can only reduce, while the step-up or boost converter can only increase the average output voltage in comparison with the input voltage. The step-up/down or buck-and-boost converter produces an output voltage that is either lower or higher than the input voltage. DC–DC converters are built with and without electrical isolation. The former usually incorporate both a DC–AC and an AC–DC converter in cascade as well as a transformer at the terminals of the AC signals for electrical isolation. The transformer turns ratio is also utilized for bridging a larger gap between the input and output voltage.

There is (not) a direct path between the input and output terminals in the direct (indirect) converter. Although these converters may operate in either a continuous or discontinuous current conduction mode, only the continuous current conduction mode will be discussed in this chapter.

13.2 Switch Mode Conversion Concept

The ripple-free DC voltage shown in Figure 13.4a or the ripple-free current shown in Figure 13.4b is periodically chopped by the switch S. By changing the duty ratio $D = T_{ON}/T$, the average value of either waveform can be varied continuously. The ratio of the switching frequency $f_s = 1/T$ to the frequency of the external signals is large enough to remove the switching frequency component from the signals.

13.3 Output Current Sourced Converters

A typical load circuit is given in Figure 13.2a. The input voltage v_1 and the load current i_2 are assumed to be ripple free in all cases. The circuit configurations and the time functions for the output voltage v_2 and the input current i_1 are illustrated in Figures 13.5 and 13.6. The voltage ratio in class A (class B) is $V_2/V_1 = D(V_2/V_1 = 1 - D)$. If switch S_p (S_n) is turned on and off while the other switch remains off, the circuit configuration for class C operates like class A (B) in the first (second) quadrant. If the load is connected across the terminals of the positive switch S_p–D_p as shown in Figure 13.5c by the dotted line, the converter operates either in the first or third quadrants. Classes D and E can be operated with either bipolar or unipolar voltage switching. In the first case, two switches located diagonally in the circuit diagram are simultaneously turned on and off as a pair (see Figure 13.5e and Figure 13.6b). Operation with unipolar voltage switching is achieved by shifting the turn-on-off process in these switches by half a cycle (see Figure 13.5f and Figure 13.6c). Figure 13.6 shows the time functions for both bipolar (Figure 13.6b) and unipolar (Figure 13.6c) voltage switching in all four quadrants. The conducting device is either the switch turned on or its antiparallel diode. The turned-on switches and the conducting diodes,

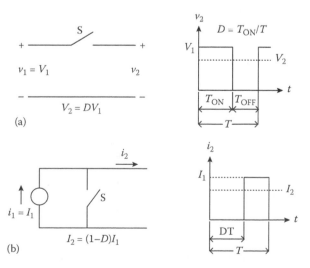

FIGURE 13.4 Switch mode conversion concept.

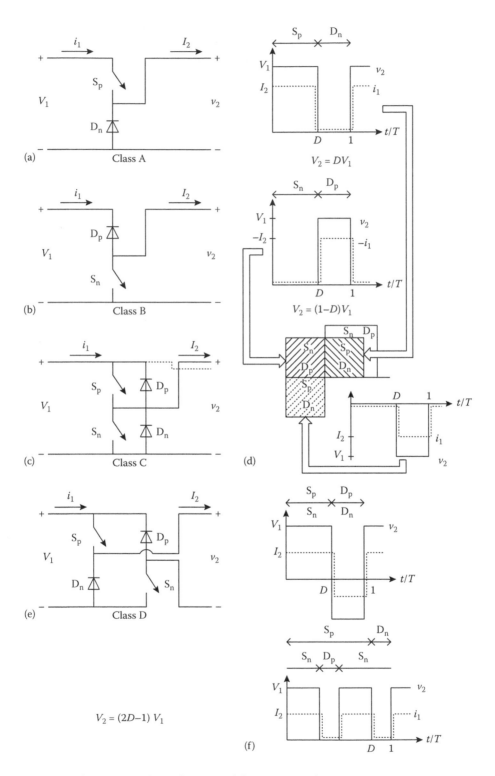

FIGURE 13.5 Configurations and time functions of class A, B, C, and D converters.

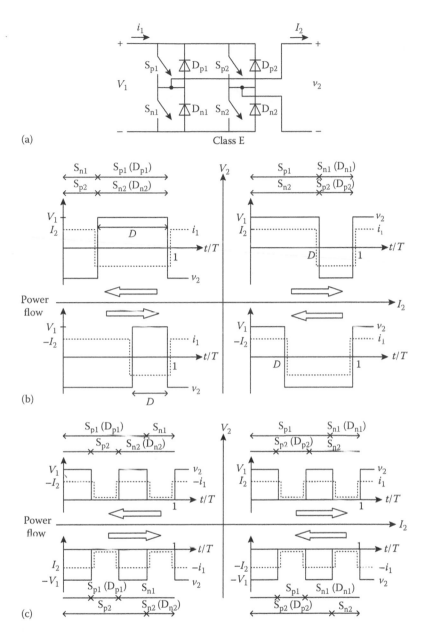

FIGURE 13.6 Configuration of class E converter (a), time function of bipolar (b), and voltage switching (c).

in bracket, are shown in Figure 13.6b and c. The conducting diode is always indicated along with the switch that is turned on. Both for bipolar and unipolar voltage switching, the average output voltage is $V_2 = (2D - 1)V_1$, where D is the duty ratio of switch S_{p1}, S_{n2}.

Assuming the switches have an identical switching frequency, the unipolar voltage switching produces better output voltage input current waveforms as well as a better frequency response, since the "effective" switching frequency of the two waveforms is doubled and the ripple amplitude is halved.

Class E can be converted to a class C or class D configuration by appropriate control, for example, continuously turning on S_{n2}. The antiparallel-connected S_{n2} and D_{n2} constitute a short circuit, and S_{p2} and D_{p2} are equivalent to an open circuit. First- and second-quadrant operations are achieved with the waveforms shown in Figure 13.6a and b. On the other hand, by continuously turning on S_{n1}, in which case switch S_{n1} and D_{n1} constitute a short circuit and S_{p1} and D_{p1} are equivalent to an open circuit, third-, and fourth-quadrant operations are accomplished.

Since the converters are ideally lossless, the current ratio for all configurations is $I_2/I_1 = V_2/V_1$.

13.4 Output Voltage Sourced Converters

In what follows, it will be assumed that L and C are large enough to eliminate switching frequency components from the terminal variables v_1, i_1 and v_2, i_2. Furthermore, the relation between the average input and output voltage can be derived by using the simple fact that the time integral of the inductor voltage v_2 over one period must be zero.

13.4.1 Direct Converters

The circuit configurations and the time functions for the buck (step-down) and boost (step-up) converters are shown in Figure 13.7 (Mohan et al., 2002). By turning on the switch S in interval DT in the buck converter, the diode becomes reverse biased and the input supplies energy to both the load and the inductor L. If the same action is repeated in the boost converter, energy is supplied only to the inductor L. If switch S is turned off in interval $(1 - D)$, the inductor current flows through the diode in the buck converter transferring some of its stored energy to the load, while in the boost converter the energy is forced toward the output both from the inductor and the input through the diode as a result of the inductor current even though $V_2 > V_1$.

13.4.2 Indirect Converters

The circuits and time functions for the buck-and-boost (step-up/down) and Ćuk converters are shown in Figure 13.8. Note that the polarity of the output voltage is negative. Turning on the switch S in interval D reverse biases the diode. In Figure 13.8a, energy is supplied from the input to the inductor L and from the capacitor C to the load. In Figure 13.8b, energy is supplied from the input to inductor L_1, and from capacitor C to the load as well as inductor L_2. This converter operates via capacitive energy transfer. As illustrated in Figure 13.8b, capacitor C is connected through L_1 to the input source while the diode is conducting, and source energy is stored in C. The diode conducts current, if the switch S is turned off in interval $(1 - D)$. When the switch S is conducting, this energy is released through L_2 to the load.

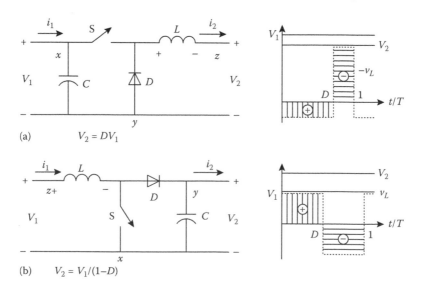

FIGURE 13.7 Configuration and time function of buck (step-down) (a) and boost (step-up) (b) converters.

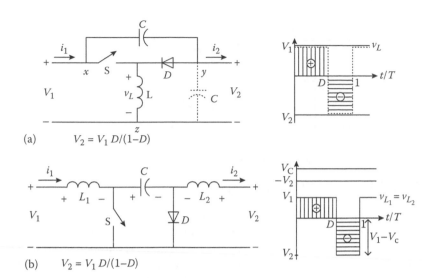

FIGURE 13.8 Configuration and time function of buck-and-boost (a) and Čuk (b) converters for $D = 0.5$.

Also, in Figure 12.21a, energy is supplied from the input to the capacitor C. In Figure 13.8b, energy is supplied to the capacitor C from the input and inductor L_1. The relation V_2/V_1 is the same for the buck-and-boost and Čuk converters. The output voltage V_2 can be either smaller or larger than V_1.

The capacitor C can be placed either between terminals x and y, or y and z without changing the operation of the buck-boost converter. In both cases, the voltages v_1, v_2, and v_{xy} are ripple free.

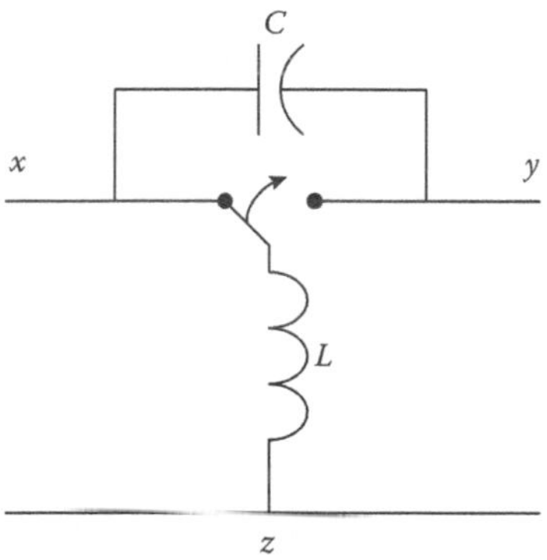

13.5 Fundamental Topological Relationships

The basic circuit, the so-called canonical switching cell (CSC), that is common to buck, boost, and buck-and-boost converters is shown in Figure 13.9 (Kassakian et al., 1992). It uses a double-throw switch that satisfies the condition that the two switches—transistor and diode in the four converters—be neither on nor off simultaneously. The CSC is the basic building block for a large number of DC–DC converters in addition to those discussed in the previous section. The different converter configurations, i.e., buck, boost, and buck-and-boost, are dependent upon both the way in which the CSC is connected to the external system and the implementation of switches. It can be shown that the Čuk converter can easily be derived from CSC as well (Kassakian et al., 1992).

FIGURE 13.9 Canonical switching cell (CSC).

13.6 Bidirectional Power Flow

Power can flow only from left to right in the configurations discussed in Section 13.2.2. However, bidirectional power flow is required in some applications. Figure 13.10 shows the implementation of the switches within the CSC for bidirectional power flow under the conditions that the polarity of the two external voltages (currents) can (cannot) change. Assuming $V_1 > 0$ and $V_2 > 0$, ($V_1 < 0$ and $V_2 < 0$), transistor S_2 (S_1) can be kept continuously on. Control is achieved by switching the other transistor. When S_2 (S_1) is continuously on, the configuration works as a buck (boost) converter and the power flows from left to right (right to left). The converter can operate in quadrant I and IV like class D converters.

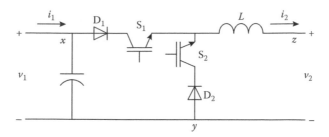

FIGURE 13.10   Configuration providing bidirectional power flow.

13.7 Isolated DC–DC Converters

Each of the basic converters can only accommodate one input and one output with input and output sharing a common reference line. To overcome these limitations, an isolation transformer is added to the DC–DC converters. An additional benefit achieved through the application of a transformer is the reduction of component stresses when the conversion ratio V_2/V_1 is far from unity. Isolated DC–DC converters can be classified according to the core excitation of their transformer:

- In unidirectional core excitation, the flux density B and the magnetic field strength H can be of only one polarity. For example, the forward converter that is derived from the buck converter and the flyback converter derived from the buck-and-boost converter belong to this group. They are called "single-ended" converters, also, because power is forwarded through the transformer in only one polarity of the primary voltage.
- In the bidirectional core excitation, B and H can have both positive and negative polarity. Push-pull, half-bridge, and full-bridge inverter topologies belong to this group. They are called "double-ended" converters, as well, because power is forwarded through the transformer in both polarities of the primary voltage.

13.7.1 Single-Ended Forward Converter

The basic configuration for this converter and the associated time functions are shown in Figure 13.11. The losses and the leakage inductance of the transformer are neglected and it is modeled by an ideal transformer with the turns ratio $N{:}1$ and magnetizing inductance L_m.

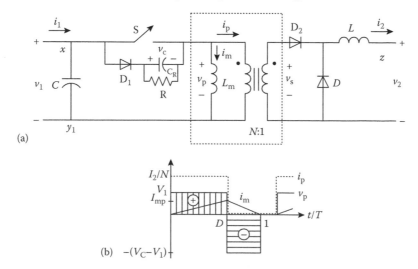

FIGURE 13.11   Configuration (a) and time functions (b) of forward converter.

By ignoring the magnetizing current I_m, i.e., $L_m = \infty$ and assuming $N = 1$, the operation of the configuration in Figure 13.1 is the same as that of the buck converter (Figure 13.7a). By changing N, the voltage ratio is simply altered.

The magnetizing current cannot be ignored in a practical forward converter. Assuming the magnetizing current, $i_m(0) = 0$, at the beginning of the period (Figure 13.11b), the DC voltage $v_p = V_1$ in interval DT during the on time of switch S, causes magnetizing current i_m and the flux density to increase in a linear fashion, reaching their peaks at $t = DT$. Power is delivered through the transformer and diode D_2 to the load and inductor L. By turning off switch S, current i_m is diverted from S to clamping circuit consisting of D_1, R, and C_R. Assuming an approximately constant clamping voltage $v_c = V_c > V_1$, the primary voltage of the transformer is $-(V_c - V_1) < 0$ for time $t \geq DT$ and the current begins decreasing linearly. Diodes D_2 and D become reverse and forward biased, respectively. In steady state, the magnetizing current, i_m, must reach zero prior to, or at time $t = T$, and the core is reset. The required maximum value, the duty ratio D_{max}, determines the minimum value of the clamping voltage $V_{c,min}$, since the relationship for the voltage–time area, i.e., $(V_c - V_1)(1 - D)T \geq V_1 DT$ must be satisfied. The higher the value of D_{max}, the bigger $V_{c,min}$ must be.

The energy stored in the magnetizing inductance by $i_m = I_{mp}$ is partially dissipated in the resistance R. At high power, the resistance R can be replaced by a DC–DC converter to recover the magnetizing energy. The clamping function can be implemented with a Zener diode or the addition of a tertiary winding on the transformer. In the latter case, the winding has to be connected in series with a diode, either across the input or output terminals of the converter in such a way that the magnetizing energy is supplied back to the input or output circuit during the off interval of switch S.

13.7.2 Single-Ended Hybrid-Bridge Converter

In contrast to the single switch and the clamping circuit of the forward converter shown in Figure 13.11, the single-ended hybrid-bridge converter has two switches turned on and off simultaneously and two diodes performing the clamping function on the primary side of the transformer as shown in Figure 13.12. Otherwise this circuit is the same as that of the forward converter. The two converters operate in a similar manner as shown in Figures 13.11b and 13.12b, and the transformer core is excited unidirectionally. However, the magnetizing current i_m is flowing through diode D_1 and D_2 in the off interval and the primary voltage is clamped at $v_p = V_1$. i_m decays to zero at $t = 2DT$ and the maximum value of the duty ratio is $D_{max} = 0.5$.

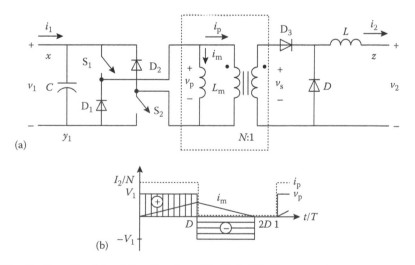

FIGURE 13.12 Configuration (a) and time functions (b) of hybrid-bridge converter.

13.7.3 Flyback Converter

The circuit, which employs the same transformer as that used in the forward converter, and the time functions shown in Figure 13.13 reveal the basic similarity between the flyback converter and the buck-and-boost converter shown in Figure 13.8a. Ignoring the leakage inductances of the transformer, its operation is identical to that of the nonisolated buck-and-boost converter except for the transformer effect. Unlike the forward converter, the transformer magnetizing inductance stores energy during the on interval of switch S. This energy is transferred during the off interval through the transformer and the diode D to the load.

Flyback converters are applied in television receivers with a very high turns ratio to produce a high voltage "to flyback" the horizontal beam on the screen in order to start the next line, The flyback converter is single-ended and the transformer core is excited unidirectionally.

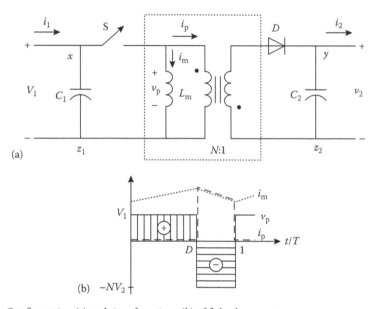

FIGURE 13.13   Configuration (a) and time functions (b) of flyback converter.

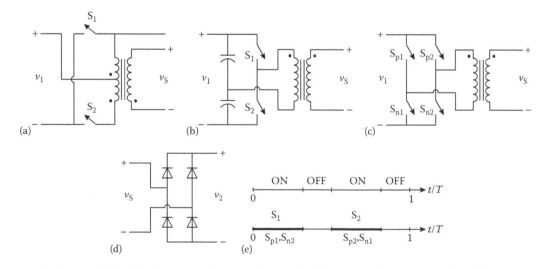

FIGURE 13.14   Push-pull (a), half-bridge (b), full-bridge (c), double-ended (d) converters and their control (e).

13.7.4 Double-Ended Isolated Converters

The transformer core for these types of converters is excited bidirectionally. This group of converters includes the push-pull shown in Figure 13.14a, the half-bridge shown in Figure 13.14b, and the full-bridge shown in Figure 13.14c. All three converters generate a high-frequency AC voltage without any DC component across the primary of the transformer by turning the switches on-and-off periodically according to the pattern shown in Figure 13.14d. The AC voltage, v_s, is rectified by a diode bridge in all three converters as shown in Figure 13.14e.

13.8 Control

There are basically three control methods as illustrated in Table 13.1 (Severns and Blomm, 1985).

TABLE 13.1 Control Methods of Converters

Constant	Controlled
Period T	T_{ON}, T_{OFF} or duty ratio T_{ON}/T
Pulse width T_{ON}	T, T_{OFF} or frequency $1/T$
Pulse pause T_{OFF}	T, T_{ON} or frequency $1/T$

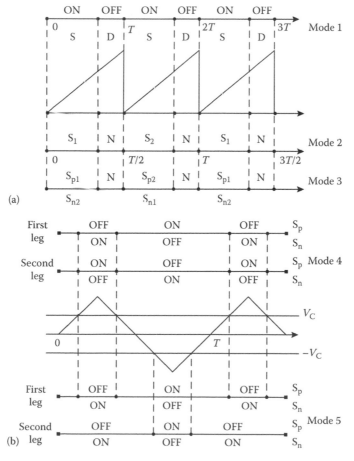

FIGURE 13.15 Control modes of converters. Pulse-width modulation Mode 1, 2, 3 (a) Mode 4 and 5 (b).

The first control method is referred to as pulse-width modulation (PWM). Five PWM control modes of DC–DC converters are shown in Figure 13.15. There is only one controlled switch in mode 1, two switches in mode 2 and four switches in modes 3, 4, and 5. Control mode 1 is applied in class A, B, and C as well as in the buck, boost, buck-and-boost, Čuk, single-ended forward, and flyback converters. Control mode 2 is applied in isolated converters, single-ended hybrid-bridge, push-pull, and half-bridge converters. Mode 3 is used in isolated double-ended full-bridge converters. Modes 4 and 5 are applied in class D and E converters as well as the nonisolated full-bridge converter for bipolar and unipolar voltage switching, respectively.

Note the basic difference between control modes 2 and 3, and modes 4 and 5. In modes 2 and 3 there are intervals when none of the controlled switches are turned on. In modes 4 and 5, one controlled switch is always on in each leg. In other words, two switches are never off nor on simultaneously in one leg. Switch $S_p(S_n)$ in the first leg is controlled together with $S_n(S_p)$ in the second leg in mode 4. On the other hand, the control of switch $S_p(S_n)$ in the first leg is shifted by half a cycle to the control of switch $S_n(S_p)$ in the second leg in mode 5.

References

Kassakian, J. G., Schlecht, M. E., and Verghese, G. C. 1992. *Principles of Power Electronics*, Addison-Wesley, Reading, MA.

Mohan, N., Undeland, T. M., and Robinsons, W. E. 2002. *Power Electronics*, John Wiley & Sons, New York.

Rashid, M. H. 1993. *Power Electronics*, Prentice-Hall International, London, U.K.

Severns, R. R. and Blomm, G. E. 1985. *Modern DC-to-DC Switchmode Power Converter Circuits*, Van Nostrand Reinhold Electrical/Computer Science and Engineering Series, Van Nostrand Reinhold, New York.

14

DC–AC Converters

Samir Kouro
Ryerson University

José I. León
University of Sevilla

Leopoldo Garcia
Franquelo
University of Sevilla

José Rodríguez
*Universidad Tecnica
Federico Santa Maria*

Bin Wu
Ryerson University

14.1 Introduction

Static power converters that adapt DC voltages and currents to AC waveforms are usually known as inverters. Their main function is to generate from one or multiple DC sources an AC switched pattern output waveform, with a fundamental component with adjustable phase, frequency, and amplitude to meet the needs of a particular application. A generic block diagram describing this function of the inverter is shown in Figure 14.1 for a generic DC-variable x_{dc}, usually voltage or current. Note that A_{dc} is the fixed amplitude of x_{dc}, while $A_{ac}, f,$ and θ represent the adjustable amplitude, frequency, and phase of the fundamental component of the switched AC-variable $(x_{ac}\{f_1\})$, respectively. This conversion is achieved by the proper control, better known as modulation, of the static power switches that interconnect the DC source to the AC load using the different configurations or conduction states provided by the switches arrangement or topology.

The DC sources can be either current or voltage sources, dividing the inverter family into two main groups: current source inverters (CSI) and voltage source inverters (VSI), as shown in Figure 14.2. The DC source is usually composed of a rectifier followed by an energy storage or filter stage known as DC link (this conversion concept is known as indirect conversion, AC–DC/DC–AC). Typical DC links are inductors and capacitors used for CSI and VSI, respectively. Less common are direct conversion applications where other DC sources are used like batteries, photovoltaic modules, and fuel cells. Figure 14.2 further classifies the different type of CSI and VSI topologies depending on their typical power range of application. While CSI have been dominating in the medium-voltage high-power range with the pulse-width modulated CSI (PWM-CSI) and the load-commutated inverter (LCI) [1], voltage source are widely found in low- and medium-power applications with single-phase and three-phase two-level VSI. Recently, VSI have also become attractive in the medium-voltage high-power market with multilevel converter topologies [2].

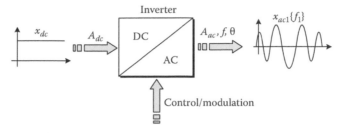

FIGURE 14.1 Inverter operating principle.

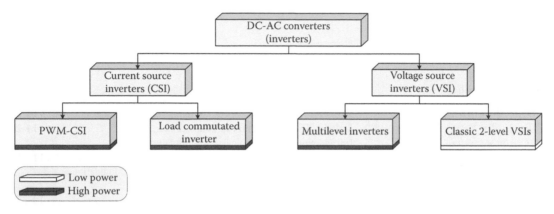

FIGURE 14.2 Inverter topology classification according to type of source (voltage or current) and power range.

This chapter describes the most common inverter topologies and modulation schemes found in industry. Special attention is given to basic concepts, operating principles, and figures of merits like efficiency, power quality, power range, and implementation complexity. The chapter is organized as follows: Section 14.2 is focused on VSI introducing the most common topologies and modulation methods. Section 14.3 addresses the multilevel converters specially designed for medium-voltage high-power applications. Finally, Section 14.4 is focused on CSI and their modulation methods.

14.2 Voltage Source Inverters

14.2.1 Introduction

VSI use a constant voltage source usually provided by a voltage source rectifier and a capacitive DC link, to generate a switched voltage waveform at the output with a fundamental voltage component with adjustable frequency, phase, and amplitude that matches a desired reference voltage. On the other hand, the inverter output current is defined by the load, which is usually very sinusoidal for inductive loads such as motor drives; otherwise output filters are used.

VSI are the most common power conversion systems in DC–AC powered applications, particularly in low and medium power, either in single- or three-phase systems with the classic two-level topologies. Currently, they have also an important presence in the high-power-medium-voltage market (which have been dominated by CSI topologies), with the development of multilevel converters. VSI are widely used in single-phase ac power applications like uninterruptable power supplies (UPS), class-D audio power amplifiers, domestic appliances (washing machines, air conditioning, etc.), photovoltaic power conversion; and in three-phase systems such as adjustable speed drives, pumps, compressors, fans, conveyors, industrial robots, active filters, elevators, mills, mixers, crushers, paper machines, cranes, flexible ac transmission systems (FACTS), train traction, shovels, electric vehicles, wind power conversion and mining haul trucks, to name a few. They cover such a wide power range; they can be found in sizes from cubic millimeters as signal amplifiers in cell phones to cubic meters to drive fans in the cement industry.

The following sections present the operating principles and concepts related to the most common VSI topologies and their corresponding modulation schemes found in industry.

14.2.2 VSI Topologies

14.2.2.1 Half-Bridge VSI (Single-Phase)

The half bridge is a two-level single-phase inverter whose power circuit is illustrated in Figure 14.3a. It is composed of one inverter leg containing two semiconductor switches (T_1 and T_2) with antiparallel connected freewheeling diodes (D_1 and D_2) used to provide a negative current path through the switch when required. The inverter also features two capacitors in the DC link that split the total DC link voltage to provide a 0 V midpoint connection for the load, also known as neutral point (denoted as node O in Figure 14.3). The load is connected between this node and the inverter leg output phase node a. Note that isolated gate bipolar transistors (IGBT) are used as power switches in Figure 14.3b for illustrative purposes, but could be any other power semiconductor (metal oxide semiconductor field effect transistor [MOSFET], gate turn-off thyristor [GTO], integrated gate-commutated thyristor [IGCT], etc.), although MOSFETs and IGBTs are the most used in this topology due to power range and application field [3]. The positive and negative bus bars of the inverter are denoted by P and N, respectively. It is worth mentioning that the DC-side capacitors do not correspond to the DC voltage source, since they cannot provide active power. Instead, the DC-side source is represented by the constant voltage V_{dc} at the open-end input nodes, and could be provided by any DC source (rectifier, batteries, fuel cells, etc.). This way of illustrating the DC source is kept throughout the chapter for generality.

The inverter is controlled by a binary gate signal $S_a \in \{1,0\}$, where 1 represents the "on" state of the switch (switch is conducting) and 0 the "off" state (switch is open). As can be seen from Figure 14.3, the upper switch T_1 is controlled by S_a, while the lower switch T_2 is controlled using its logic complement $\overline{S}_a$. This alternate control is necessary to avoid simultaneous conduction of T_1 and T_2, since it would short-circuit the DC link, or to avoid both switches to be open generating undefined output voltages [3]. Hence the gate signal S_a defines two switching states: when $S_a = 1$ the inverter output node a is connected to the positive busbar P, resulting in a positive output voltage $v_{ao} = V_{dc}/2$; and when $S_a = 0$ the inverter output node a is connected to the negative busbar N, resulting in a negative output voltage $v_{ao} = -V_{dc}/2$. The fact that there are only two possible output voltages is why this VSI is classified as a two-level inverter. The alternation between these two switching states over different periods of time, process called modulation, is how the DC voltage V_{dc} is converted into an AC switched waveform,

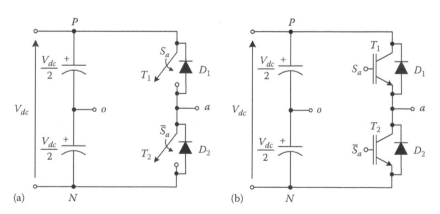

FIGURE 14.3 Half-bridge inverter power circuit: (a) with generic semiconductor switches and (b) featuring IGBTs.

achieving the desired operation of the converter. It is worth to mention that in practice, the commutation of a power device is not instantaneous; therefore a dead time has to be added before a turn-on (change from 0 to 1), to avoid two switches to be conducting simultaneously which would short-circuit the DC-link capacitor. The dead time usually is just a little bit larger than the turn-off commutation time of the switch, and therefore depends on the semiconductor type and power rating. In case of the IGBT, the dead time is usually a couple of micro seconds.

Although S_a is a binary signal leading to two different switching states, there will be four different conduction states depending on the load current polarity, which determines which semiconductor device is conducting the current (the power transistor or the freewheeling diode). These four conduction states together with the switching states are illustrated in Figure 14.4 and listed in Table 14.1. The qualitative example shown in Figure 14.4 presents a hypothetical AC square-wave operation of the inverter feeding a highly inductive load as an example to illustrate the different conduction states, and it does not illustrate a real current waveform for the given voltage. For example, the equivalent circuit for the two conduction states obtained for a negative and positive load current i_a when generating the switching state $S_a = 1$, are illustrated in Figure 14.4a and b respectively. Note the active part of the circuit is highlighted to show the current path. In the first, the negative current is conducted from the load to the upper DC-link capacitor through the freewheeling diode, while in the second case, the positive current flows from the capacitor to the load through the power transistor. Both conduction states correspond to the same switching state, with output voltage $v_{ao} = V_{dc}/2$. For the next topologies analyzed in this chapter, only the switching states

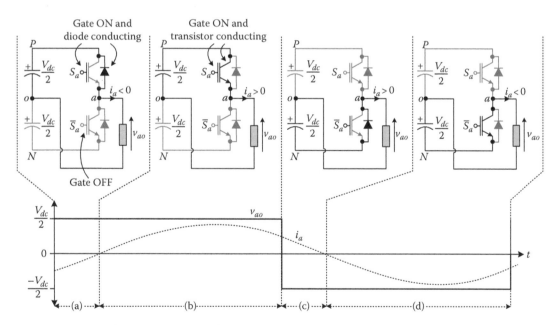

FIGURE 14.4 Half-bridge inverter conduction states when (a) $v_{ao} = V_{dc}/2$ and $i_a < 0$, (b) $v_{ao} = V_{dc}/2$ and $i_a > 0$, (c) $v_{ao} = -V_{dc}/2$ and $i_a < 0$, and (d) $v_{ao} = -V_{dc}/2$ and $i_a < 0$.

TABLE 14.1 Half-Bridge Switching and Conduction States

Switching State	Gate Signal, S_a	Output Voltage, v_{ao}	Conduction State	Output Current, i_a	Semiconductor Conducting
1	1	$V_{dc}/2$	(a)	<0	D_1
			(b)	>0	T_1
2	0	$-V_{dc}/2$	(c)	>0	D_2
			(d)	<0	T_2

will be considered, since it has direct relation to the generated output voltage. The conduction states will be neglected due to less relevance in the analysis of the modulation principles.

It is important for designing considerations to notice that, when not conducting, the semiconductor switches in this topology are blocking the complete DC-link voltage V_{dc}. Therefore, this topology is more common in applications in the low voltage range with a maximum of operation defined by the semiconductor technology being used.

14.2.2.2 H-Bridge VSI (Single-Phase)

Another popular single-phase DC–AC power converter, is the H-bridge VSI. Basically, the H-bridge is composed of two half-bridge inverter legs connected in parallel to provide two output nodes a and b to connect the load between them, as shown in Figure 14.5. Since the load is connected between the inverter legs (giving this converter its name), the midpoint in the DC link is no longer necessary; hence only one capacitor is required. Each leg has its own binary control signal $S_{a,b} \in \{1,0\}$, where 1 represents the "on" state of the switch (switch is conducting) and 0 the "off" state (switch is open). As it happens with the half-bridge, the semiconductor switches in one inverter leg are controlled with complementary signals to avoid both conducting at the same time and short-circuit the DC link, and to avoid both switches to be simultaneously open leading to undefined output voltages [3]. Since the inverter is controlled with two binary signals, it features $2^2 = 4$ different switching states defined by (S_a, S_b), which are illustrated in Figure 14.6 highlighting the corresponding active parts of the circuit. For example, consider the switching state (1,0) shown in Figure 14.6a. The output of leg a is connected to the positive

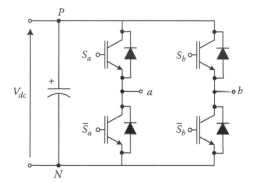

FIGURE 14.5 H-bridge inverter power circuit (IGBT based).

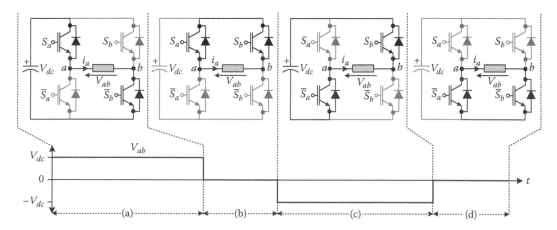

FIGURE 14.6 H-bridge inverter switching states: (a) $v_{ab} = V_{dc}$, (b) $v_{ab} = 0$, (c) $v_{ab} = -V_{dc}$, and (d) $v_{ab} = 0$.

TABLE 14.2   H-Bridge Switching States

Switching State	Gate Signal, S_a	Gate Signal, S_b	Output Voltage, v_{ab}
(a)	1	0	V_{dc}
(b)	1	1	0
(c)	0	1	$-V_{dc}$
(d)	0	0	0

bar P, while the output of leg b is connected to the negative bar N, generating an output voltage $v_{ab} = V_{dc}$. A general expression for the output voltage is

$$v_{ab} = (S_a - S_b)V_{dc}, \quad S_{a,b} \in \{0,1\}. \tag{14.1}$$

By replacing in (14.1) the different binary combinations of the gate signals, it is easy to obtain the different output voltage levels listed in Table 14.2 and illustrated in Figure 14.6.

Note that two of them, (1,1) and (0,0), generate both a zero voltage level. This feature is called voltage-level redundancy and can be used for other control purposes since it does not affect the voltage level generated at the load side. Hence, there are three different output voltage levels $\{V_{dc}, 0, -V_{dc}\}$ compared to the two-level half bridge. This is why the H-bridge is classified as a three-level topology and can be considered as a multilevel inverter [4].

When not conducting, each power semiconductor blocks the total DC-link voltage V_{dc}. Therefore, like the half bridge, this topology is also restricted to low-voltage applications, limited by the semiconductor technology being used. Nevertheless, as will be discussed later, the H-bridge can be used as a basic module for larger multilevel converters, with more levels and higher voltage operation, suitable for medium-voltage applications [4].

14.2.2.3 Full-Bridge VSI (Three-Phase VSI)

The three-phase VSI is composed by the parallel connection of three inverter legs like the one used in the half- and H-bridge, as is shown in Figure 14.7. Therefore, its operation is very similar. Each leg has its own binary control signal $S_{a,b,c} \in \{1,0\}$, where 1 represents the "on" state of the switch (switch is conducting) and 0 the "off" state (switch is open). As with the half bridge, the semiconductor switches in one inverter leg are controlled with complementary signals to avoid both conducting at the same time and

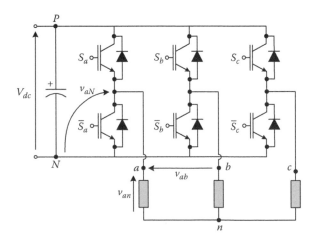

FIGURE 14.7   Full-bridge three-phase VSI power circuit (shown with Y-connected load, and IGBTs).

short-circuit the DC link, and to avoid both switches to be simultaneously open and produce undefined output voltages. Hence, when $S_x = 1$, phase x output node is connected to the positive bar, generating a phase output voltage $v_{xN} = V_{dc}$, while with $S_x = 0$, phase x output node is connected to the negative bar, generating a phase output voltage $v_{xN} = 0$. Therefore, this inverter is classified as a two-level inverter. All the inverter phase output voltages can be obtained by

$$v_{xN} = S_x V_{dc}, \quad S_x \in \{0,1\}, \quad x = a,b,c. \tag{14.2}$$

Since the inverter is controlled with three binary signals, it features $2^3 = 8$ different switching states (S_a, S_b, S_c), which are listed in Table 14.3 with the corresponding phase output voltages. The space vectors for each switching state listed in Table 14.3 will be introduced later in this chapter.

Note that a difference compared to the half- and H-bridge inverter is the fact that the phase output voltage is a switched AC waveform between 0 and V_{dc}, i.e, unlike the previous single-phase inverters that connect the load to a midpoint, the three-phase VSI output voltage has a DC component equal to $V_{dc}/2$. However, this DC offset is common to the three phases and it is eliminated with the three-phase connection and does not appear in the line–line and load voltages [5].

When not conducting, each power semiconductor blocks the total DC-link voltage V_{dc}. Therefore, like the half- and H-bridge, this topology is also restricted to low-voltage applications, where it is currently the dominating topology in industry. Nevertheless, HV-IGBT-, GTO-, and IGCT-based inverters, or inverters with several IGBTs in series (for higher voltage) or in parallel (for larger currents), have made this topology available even in medium-voltage and high-power applications. In this power range, this topology has the main drawback of very high voltage derivatives (dv/dt) that force the need of filters to produce motor friendly waveforms [6]. Also, the high switching frequencies needed to avoid low-order harmonics in the load current are not suitable in the high-power range due to switching losses, unless special modulation techniques are used as will be discussed later in this chapter.

Finally, it should be noticed that multiphase converters can be built adding new legs to the topology shown in Figure 14.7. Multiphase variable speed drives have become very attractive in some specific

TABLE 14.3 Two-Level Three-Phase VSI Switching States

Switching State	Gating Signals			Output Voltage			Space Vector
	S_a	S_b	S_c	v_{aN}	v_{bN}	v_{cN}	v_s
1	0	0	0	0	0	0	$V_0 = 0$
2	1	0	0	V_{dc}	0	0	$V_1 = \dfrac{2}{3} V_{dc}$
3	1	1	0	V_{dc}	V_{dc}	0	$V_2 = \dfrac{2}{3} V_{dc} e^{j(\pi/3)}$
4	0	1	0	0	V_{dc}	0	$V_3 = \dfrac{2}{3} V_{dc} e^{j(2\pi/3)}$
5	0	1	1	0	V_{dc}	V_{dc}	$V_4 = \dfrac{-2}{3} V_{dc}$
6	0	0	1	0	0	V_{dc}	$V_5 = \dfrac{2}{3} V_{dc} e^{j(4\pi/3)}$
7	1	0	1	V_{dc}	0	V_{dc}	$V_6 = \dfrac{2}{3} V_{dc} e^{j(5\pi/3)}$
8	1	1	1	V_{dc}	V_{dc}	V_{dc}	$V_7 = 0$

application areas such as electric ship propulsion, locomotive traction, and military applications due to higher power capability and improved reliability (fault-tolerant application) [39].

14.2.3 Modulation Methods

As discussed in the previous section, static power converters like the VSI generate constant output voltage levels. Hence, in order to generate an arbitrary voltage waveform, the inverter has to be controlled alternating the available voltage levels or vectors in such a way that the time average of the switched voltage waveform, or its fundamental component, approximates the desired voltage reference. This process is called modulation, and over the years several different methods have been proposed and applied in industry [5]. They have different operating principles, implementation, and performance, and the selection of a particular one is in direct relation to the type of application, its power range, and its dynamic requirements. This section describes the most common modulation schemes found for classic VSI.

14.2.3.1 Square-Wave Operation

The square-wave operation is the most basic and easy to implement modulation scheme for the VSI [7]. Like its name suggests, the main idea is to generate an AC square output waveform with the desired frequency. Figure 14.8a shows the voltage waveforms produced by this modulation scheme in a three-phase two-level VSI. The inverter's phase *a* output voltage v_{aN}, alternates between the V_{dc} and 0 voltage levels every half fundamental cycle. This is achieved by a very simple control strategy based on the comparison between the reference voltage v^* and zero, as shown in the block diagram illustrated in Figure 14.8b. Hence, when the voltage reference is positive V_{dc} is generated, and when negative, zero is generated. The

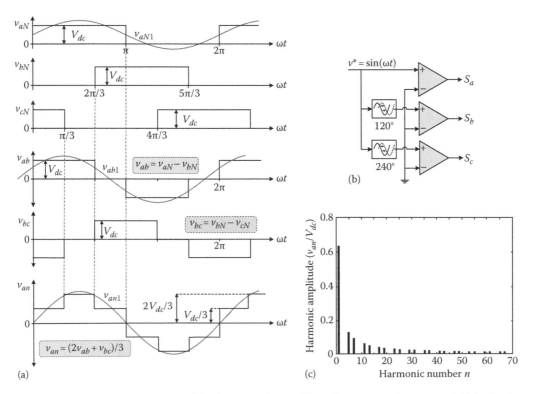

FIGURE 14.8 Square-wave operation: (a) voltage waveforms, (b) implementation diagram, and (c) load voltage harmonics.

other phases are controlled similarly with $2\pi/3$ phase shifts between the references of each phase. Note that the line–line voltages, for example, v_{ab} is equal to the difference of both phase voltages $v_{ab} = v_{aN} - v_{bN}$, eliminating the phase-voltage DC offset, creating the ac waveform. When considering a star-connected load, the load voltages can also be obtained through Kirchoff voltage law (KVL) of the line–line voltages, which for phase a is $v_{an} = (2v_{ab} + v_{bc})/3$, which is also illustrated in Figure 14.8a. Note that in this case the load voltage has a four-level stepped waveform that approximates better a sinusoidal waveform, improving the harmonic distortion compared with the inverter phase output voltage. The harmonic content of this voltage waveform is given by its Fourier series representation

$$v_{an} = \frac{2V_{dc}}{\pi}\left[\sin(\omega t) + \frac{1}{5}\sin(5\omega t) + \frac{1}{7}\sin(7\omega t) + \frac{1}{11}\sin(11\omega t) + \frac{1}{13}\sin(13\omega t) + \cdots\right], \qquad (14.3)$$

and is plotted in the spectrum shown in Figure 14.8c. The total harmonic distortion of the line-to-line voltage is 31%.

Note that the square-waveform operation for the single-phase half-bridge has the same control diagram of one phase only of those shown in Figure 14.8b, and the inverter output voltage is equal to the load voltage and corresponds to the first waveform in Figure 14.8a but without the DC offset, i.e., commutating between $\pm V_{dc}/2$. For the single-phase H-bridge, two phases of the control diagram of Figure 14.8b are used. In this case, the inverter line–line voltage is equal to the load voltage and corresponds to the v_{ab} waveform shown in Figure 14.8a. Both single-phase solutions have even worse THD than the three-phase case and have little practical use. Even the three-phase square-wave operation is nowadays considered obsolete and is only used in reduced dynamic performance systems. The low power quality is the price to pay for implementation simplicity and efficiency, since devices switch at fundamental switching frequency. This method is also used for low-cost systems.

14.2.3.2 Sinusoidal PWM: Bipolar PWM and Unipolar PWM

Sinusoidal pulse-width modulation (PWM), also known as carrier-based modulation methods, are perhaps the most widely developed and applied modulation schemes for power converters in industry [5]. As will be discussed in this section, the main reasons are its simple implementation, online operation, and good power quality. On the weak side is the need of higher switching frequencies that affect the system efficiency by introducing more switching losses, which is not always suitable for high-power applications. On the other hand, if the switching frequency is low, the size, volume, and economical cost of the necessary filters (mainly inductances) increase. Therefore, a trade-off between the power losses and the filter design cost has to be done. Nevertheless, it has a wide range of application field, from switching power supplies, digital audio amplifiers to high-performance variable speed drives.

The basic idea behind PWM, is to alternate between the different switching states of the inverter in such a way that the time average of the switched voltage waveform equals the desired reference. Since the output voltage levels of the inverter are fixed, the modulation is performed by changing the width of the pulses, also known as duty cycle, or duty ratio, or dwell times. This was originally achieved with analog circuitry by comparing the reference with triangle carrier signals that span in amplitude over the whole modulation range (also known as naturally sampled PWM). Today digital implementations sample and hold the reference value over the modulation period (also known as regularly sampled PWM), which is then used to compute the dwell times by comparing to the carrier waveform or via a simple average value algorithm.

Sinusoidal PWM can be classified into three different categories: bipolar, unipolar, and multicarrier PWM. For the first one, the output voltage switches between the negative and positive output voltages, while in unipolar the output voltage switches between zero and the positive output voltage or between zero and the negative output voltage of the inverter. Multicarrier PWM strategies are used for multilevel converters, and they are discussed in Sections 14.3.3.2.1 and 14.3.3.2.2.

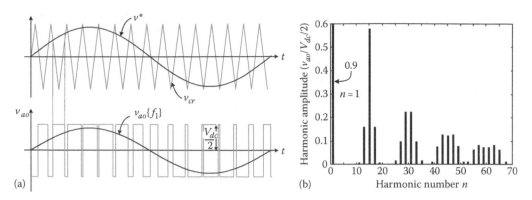

FIGURE 14.9   Sinusoidal bipolar PWM for single-phase half-bridge inverters. (a) Switching waveform generation and (b) harmonic spectrum of the phase voltage (in pu).

14.2.3.2.1 Bipolar PWM for Single-Phase Half-Bridge Inverters

The triangular carrier signal v_{cr} along the reference voltage or modulating signal v^* are illustrated in Figure 14.9. By simple comparison when the reference signal is over the carrier signal ($v^* \geq v_{cr}$) the inverter gate signal S_a defined in Figure 14.3 is set to logic "1" turning on the upper power switch T_1 which connects the output node to the positive bar, generating $v_{ao} = V_{dc}/2$. On the contrary, when $v^* < v_{cr}$, the gate signal $S_a = 0$ and the output node is connected to the negative bar generating $v_{ao} = -V_{dc}/2$. Since the triangle waveform is linear with respect to time, the instants in which those changes occur and consequently the width of the pulses will be proportional to the instantaneous reference signal amplitude, achieving the desired time average. The faster the carrier frequency with respect to the modulating signal, the better the pulses will approximate the time average, hence better reference tracking. However, this comes at expense of higher switching frequency, affecting efficiency, which constitutes a design constraint.

There are two useful concepts to better understand and analyze PWM methods: the amplitude modulation index m_a and the frequency modulation index m_f [1]. The amplitude modulation index defines the relation between the fundamental component amplitude and the switched AC waveform amplitude (usually the DC-link voltage); hence for a single-phase half-bridge it is defined as

$$m_a = \frac{\hat{v}_{ao}\{f_1\}}{V_{dc}/2}, \tag{14.4}$$

where f_1 is the fundamental frequency. Since the carrier signal is defined in such a way it covers the whole modulation range, it is proportional to the inverter DC-link voltage. In the same way the output fundamental component is the time average of the reference voltage. Hence, the amplitude modulation index can also be defined as

$$m_a = \frac{\hat{v}^*}{\hat{v}_{cr}}. \tag{14.5}$$

On the other hand, the frequency modulation index is the relation between the reference voltage frequency and the carrier frequency

$$m_f = \frac{f_{cr}}{f_1}. \tag{14.6}$$

The frequency modulation index is useful to know where the harmonic content of the switched or PWM waveform are located in the corresponding spectrum, as shown in Figure 14.9. Note that for the qualitative case illustrated in Figure 14.9, the carrier signal has 15 cycles compared to the reference, hence $m_f = 15$. Therefore, the main or strongest harmonic component is located at $n = 15$. Each main harmonic is accompanied by a group of sideband harmonics located at $m_f \pm 2$ and $m_f \pm 4$, therefore in the spectrum of Figure 14.9b, some lower order harmonics located at $n = 11$ and $n = 13$ are visible. These harmonics appear due to a convolution effect between the fundamental and carrier frequency during the comparison performed by PWM. In addition, because of the same effect sidebands appear around multiples of the central frequency located at $2m_f$, $3m_f$, etc. Note that the fundamental component has been truncated in the spectrum shown in Figure 14.9 to highlight the harmonics present in the output voltage. It is worth mentioning that the carrier frequency f_{cr} is in this case equal to the device switching frequency f_{sw} and the output voltage switching pattern frequency f_{ao}.

14.2.3.2.2 Unipolar PWM for Single-Phase H-Bridge Inverters

Unipolar PWM is specially used for the single-phase H-bridge. In this case, the comparison between one reference and one carrier as done in bipolar PWM is not enough, since this binary output would leave one of the levels undefined. Basically, unipolar PWM is a combination of two bipolar PWM modulations, one for each leg of the H-bridge shown in Figure 14.5. The main difference is that the carrier for the second leg is shifted in 180° or is in opposite phase, and the comparison logic is inverse, i.e., when the reference is above the carrier, the gating signal is zero instead of one. This operating principle is shown in Figure 14.10, where both carriers are shown together with the reference. Each leg output voltage is also shown and consists of a bipolar switched waveform. Since the load is connected between the legs or line–line, the output voltage is $v_{ab} = v_{aN} - v_{bN}$, which eliminates the DC offset and produces the negative voltage levels. Note that because each leg is modulated using the carrier frequency f_{cr} and they are phase shifted in 180°, the resulting waveform has the double of commutations per cycle. This is also why in the spectrum no dominant harmonics appear at m_f but at $2m_f$ and integer multiples of it ($4m_f$, $6m_f$, etc.). It is worth mentioning that the carrier frequency f_{cr} is in this case also equal to the device switching frequency f_{sw}, but due to the H-bridge connection of the load, a multiplicative effect is obtained in the output voltage switching pattern frequency $f_{ao} = 2f_{cr}$. This is a positive from power quality point of view, since harmonics are shifted to $2m_f$, without increasing the device switching frequency [8]. In fact, the device switching frequency could be reduced to the half, and still achieve same power quality as with bipolar PWM. Nevertheless, this does not mean an improvement in efficiency with respect to the half-bridge case, since now two legs are switching instead of one.

14.2.3.2.3 Bipolar PWM for Three-Phase VSIs

This is the three-phase extension of the same bipolar PWM used for the half-bridge. The only difference is that the reference signals are phase shifted in 120° among each other to obtain a balanced three-phase

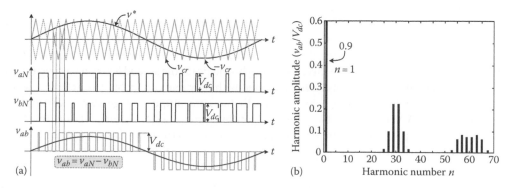

FIGURE 14.10 Sinusoidal unipolar PWM for single-phase H-bridge inverters. (a) Switching waveform generation and (b) harmonic spectrum of the line-to-line voltage (in pu).

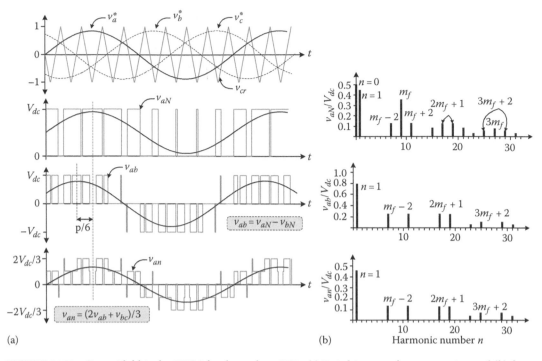

FIGURE 14.11 Sinusoidal bipolar PWM for three-phase VSIs. (a) Switching waveform generation and (b) from top to bottom, harmonic spectrum of the converter phase output voltage, the line-to-line voltage and the load phase voltage (in pu).

voltage at the output. Figure 14.11 shows the carrier and reference signals with the resulting inverter phase a output voltage, the line–line voltage, and the load voltage obtained with this modulation. Note that the line–line voltage has three levels and no DC offset because of the same reason than in unipolar PWM. The only difference is that because there is only one carrier, the harmonics are not shifted to $2m_f$ as with unipolar PWM. Note that because the carriers are equal for all phases (have no phase shifts), the dominant harmonic at m_f is eliminated in three-phase connection, since it is common to all the phases, therefore it does not appear in the line–line voltage, reducing the load voltage THD. In addition, the three-level line–line voltages when connected to a Y-load combine together to form a five-level voltage waveform, which also reflects a better THD and reduces dv/dt. Note that for the qualitative example of Figure 14.11, a carrier frequency of nine times the fundamental is considered ($m_f = 9$) for illustrative purposes to be able to appreciate the commutations. However, in practice, higher carrier frequencies are used ($m_f > 20$), especially in low-power applications.

The implementation block diagrams for the three PWM methods exposed in this section are illustrated in Figure 14.12. Note that the three-phase VSI has exactly the same control scheme as the half bridge but repeated three times for the different reference signals, while the unipolar implementation features the additional carrier and the inverter comparison logic for the second leg, as mentioned earlier. Unipolar PWM can also alternatively be implemented with only one carrier but two reference signals (one in opposite phase to the other), achieving exactly the same output voltage. A non-carrier-based implementation of PWM is also possible; a simple algorithm can be used to calculate the on time (t_{on}) of each leg, i.e., the time portion of the modulation period T_m in which the phase output node is connected to the positive bar of the inverter, given by

$$t_{on} = \frac{v^*}{V_{dc}/2} T_m = m_a T_m. \tag{14.7}$$

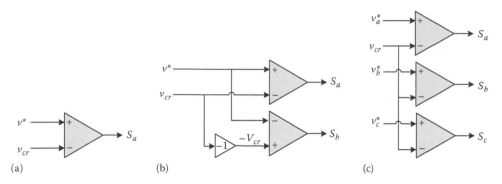

FIGURE 14.12 Implementation block diagram for PWM: (a) half-bridge bipolar PWM, (b) H-bridge unipolar PWM, and (c) three-phase VSI bipolar PWM.

Then, the 0 gating signal is generated during $(T_m - t_{on})/2$, followed by 1 during t_{on} and finally completed with 0 for the other half of $(T_m - t_{on})/2$. In this way, if the modulation period T_m is considered equal to the carrier signal period $T_{cr} = 1/f_{cr}$ the exact same results are obtained.

Note that it is important to divide the 0 state in two, and apply one before and after the 1 state to achieve a symmetrical or center-weighted PWM pulse pattern as the one achieved with a triangular carrier. Although the order in which the states are generated has no impact on the average value generated over T_m, it does have importance for practical implication in digital platforms and feedback purpose. For example, consider the current control of an inverter feeding an RL load: if the 0 state would not be divided in two, and is completely generated and then followed by the 1 state, it would correspond to a sawtooth carrier PWM. In this case, no synchronous sampling of the current would be possible, and the current value fed back into the control loop would have a time average error that can affect the overall system [5]. This can be observed in Figure 14.13 where the sawtooth and the triangular carrier implementation are compared. The triangular carrier produces a centered pulse in T_m, which allows the real current $i_a(t)$ to be crossing its average value during the sample time. In this way, the sampled current $i_a(k)$ used for measurement and feedback better approximates the real current $i_a(t)$ compared to the sawtooth case shown in Figure 14.13b.

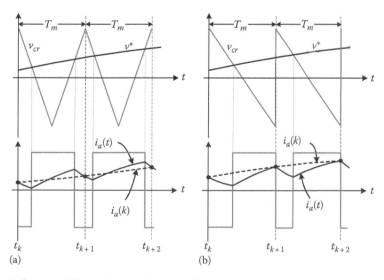

FIGURE 14.13 Influence of the carrier signal over synchronous current sampling: (a) triangular carrier PWM and (b) sawtooth carrier PWM (right weighted).

The qualitative examples shown in Figures 14.9 through 14.11 have carrier frequencies of integer multiples of the fundamental frequency, and are in phase with the sinusoidal references; this is also known as synchronous PWM, which generates the characteristic harmonics shown in the respective spectra with symmetrical sidebands [8]. In practice, for variable frequency applications like adjustable speed drives, the carrier signal is fixed and therefore is not necessarily in phase and has not necessarily an integer multiple of the fundamental frequency, producing slight variations in the characteristic harmonics. This is known as asynchronous PWM. Special care must be taken when the frequency index m_f is low ($m_f < 20$) and non-integer, as low-order harmonics can appear in the converter output voltage spectrum. In practice, asynchronous PWM can be used only when large m_f can be applied, because of the low amplitude of the low-order harmonics.

14.2.3.3 Space Vector Modulation

The space vector modulation (SVM) algorithm is basically also a PWM strategy with the difference that the switching times are computed based on the three-phase space vector representation of the reference and the VSI switching states [5], rather than the per-phase amplitude in time representation of previous analyzed methods. Therefore space vector–based modulation methods have only real purpose for three-phase inverters.

The voltage space vector of a VSI can be defined in the α–β complex plane by

$$v_s = \frac{2}{3}[v_{aN} + a v_{bN} + a^2 v_{cN}], \tag{14.8}$$

where

$$a = -\frac{1}{2} + j\frac{\sqrt{3}}{2}$$

v_{aN}, v_{bN}, and v_{cN} are the inverter phase output voltages.

It can be demonstrated that the space vector can be computed also using the load voltages v_{an}, v_{bn}, and v_{cn} without any difference, since the common mode voltage v_{nN} that relates both voltages ($v_{aN} = v_{an} + v_{nN}$) is common to the three phases and when multiplied by $(1 + a + a^2)$ is eliminated in the space vector transformation given in (14.8).

As seen previously, the inverter phase output voltages are defined by the gating signals according to (14.2). By replacing (14.2) in (14.8), the voltage space vector can then be defined using the gating signals S_a, S_b, and S_c, which leads to

$$v_s = \frac{2}{3} V_{dc} \left[S_a + a S_b + a^2 S_c \right] \tag{14.9}$$

Replacing in (14.9) all the binary combinations of the gating will lead to $2^3 = 8$ space vectors, which are listed in Table 14.3. Note that there are only seven different vectors, as vectors V_0 and V_7 result both in zero. These are also called non-active vectors since they produce zero voltage level at the load while the current freewheels via the active switches or the antiparallel diodes without interacting with the DC link. These vectors can be plotted in the α–β complex plane, resulting in the VSI voltage space vector states representation illustrated in Figure 14.14a. From Table 14.3 and Figure 14.14a, it is clear that all the active space vectors (i.e., excluding the zero vectors V_0 and V_7) have the same magnitude

$$|V_k| = \frac{2}{3} V_{dc}, \quad \text{with } k = 1, \ldots, 6 \tag{14.10}$$

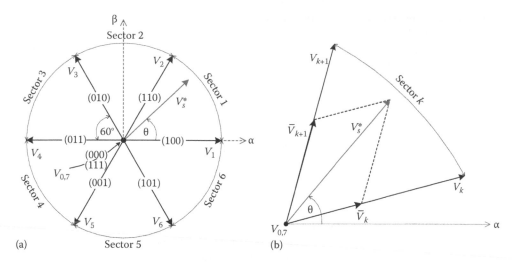

FIGURE 14.14 (a) Space vectors generated by a three-phase VSI and (b) SVM operating principle for a generic sector k.

and different angles, which are rotated in $\pi/3$ with respect to each other

$$\angle\{V_k\} = (k-1)\frac{\pi}{3}, \quad \text{with } k = 1,\dots,6 \tag{14.11}$$

Each adjacent pair of active vectors define an area in the α–β plane, dividing it in six sectors. The voltage reference space vector V_s^* can be also computed by (14.8), and the resulting vector can be mapped in the α–β plane, falling in one of the sectors. For balanced three-phase sinusoidal references, as is usual in power converter systems in steady state, the resulting reference vector is a fixed amplitude rotating space vector with the same amplitude and angular speed (ω) of the sinusoidal references, with an instantaneous position with respect to the real axis α given by $\theta = \omega t$.

The main idea behind the working principle is to generate over a modulation period T_m, a time average equal to the regularly sampled reference vector (amplitude and angular position) [5]. Hence, the problem is reduced to finding the duty cycles (on and off times) of the zero vector and the two active vectors that define the sector in which the reference is located. Consider the generic case of sector k in Figure 14.14b, then the time average over a modulation period can be defined by

$$V_s^* = \frac{1}{T_m}(t_k V_k + t_{k+1} V_{k+1} + t_0 V_0) \tag{14.12}$$

$$T_m = t_k + t_{k+1} + t_0, \tag{14.13}$$

where t_k/T_s, t_{k+1}/T_s, and t_0/T_s are the duty cycles of the respective vectors. Using trigonometric relations it can be easily found that

$$|\bar{V}_k| = \frac{t_k}{T_m}|V_k| = |V_s^*|\left\{\cos(\theta - \theta_k) - \frac{\sin(\theta - \theta_k)}{\sqrt{3}}\right\}, \tag{14.14}$$

$$|\bar{V}_{k+1}| = \frac{t_{k+1}}{T_m}|V_{k+1}| = 2|V_s^*|\frac{\sin(\theta - \theta_k)}{\sqrt{3}}, \tag{14.15}$$

where θ_k is the angle between the α axis and the current space vector k. Since all the space vectors have the same amplitude $|V_k| = |V_{k+1}| = 2V_{dc}/3$, they can be replaced in (14.14) and (14.15). Then the only unknown variables left in (14.14) and (14.15) are t_k and t_{k+1}. Thus, the following set of equations to solve the duty cycles can be obtained:

$$t_k = \frac{3T_m\,|V_s^*|}{2V_{dc}}\left\{\cos(\theta - \theta_k) - \sin\frac{\theta - \theta_k}{\sqrt{3}}\right\} \tag{14.16}$$

$$t_{k+1} = \frac{3T_m\,|V_s^*|}{V_{dc}}\frac{\sin(\theta - \theta_k)}{\sqrt{3}} \tag{14.17}$$

$$t_0 = T_m - t_k - t_{k+1} \tag{14.18}$$

Note that (14.18) is simply obtained from (14.13) once the two nonzero vector duty cycle times have been computed, in order to complete the modulation period T_m. This generic sector solution described earlier can be easily applied to any sector replacing the numeric k index ($k = 1, \ldots, 6$).

The final stage in the SVM algorithm is to generate an appropriate switching sequence of the modulating vectors and their duty cycles. As explained in carrier-based PWM, it is desirable to have a center-weighted PWM sequence, i.e., centered switching pulses over T_m to achieve a synchronous operation of the inverter. Since in terms of average value there is no difference in relation which vector is generated first or last, other issues can be addressed in the definition of the switching sequence. Particularly, efficiency can be taken into account trying to decrease the number of commutations, thus reducing switching losses [1,5].

A popular vector generation sequence with a center-weighted pulse pattern is illustrated in Figure 14.15a and b depending on if the reference vector is located in an even or odd sector. The zero vector is divided into four segments and generated using both zero vector possibilities V_0 and V_7. In the particular case shown in Figure 14.15, V_7 has been selected to start and finish the sequence, while V_0 is used for the middle pulse. This sequence can be reversed from the center to the sides (V_7 in the middle and V_0 at

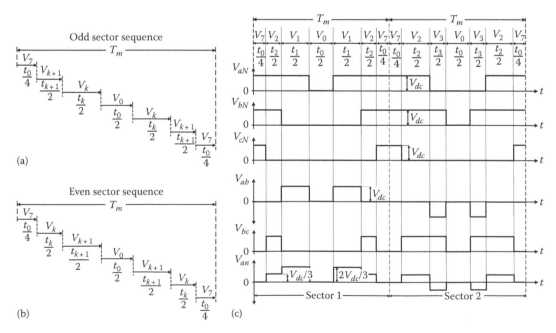

FIGURE 14.15 Center weighted pulse pattern space vector generation sequence: (a) for odd sector, (b) for even sector, and (c) example for sector 1 and 2 transition.

both ends); this is equivalent to changing the polarity of the carrier signal in PWM, and does not affect the output voltage THD. Note that the difference between the odd and even sector is a swap in which active vector is generated first, which is necessary in order to keep a center pulse pattern. This is made more clear in a qualitative example for sector 1 and 2 transition illustrated in Figure 14.15c, where the vector sequence can be tracked down to all inverter phase output voltages (v_{aN}, v_{bN}, and v_{cN}), the line–line voltages ($v_{ab} = v_{aN} - v_{bN}$ and $v_{bc} = v_{bN} - v_{cN}$) and the phase load voltage ($v_{an} = [2v_{ab} - v_{bc}]/3$). Note how each voltage has a symmetrical waveform within each modulation period T_m.

Another state-of-the-art sequence known as discontinuous SVM [9] has attractive features in terms of reduction of the switching frequency. This sequence takes in advantage the fact that a phase of the inverter can be kept on a fixed switching state for two sectors, or equivalently for $2\pi/3$, i.e., without switching during a third of the fundamental cycle. From Figure 14.15a, it is clear that if you consider only $V_0 = (0,0,0)$ as zero vector, the following relations hold:

- The phase c component of all vectors generated in sector 1 and 2 is always 0.
- The phase a component of all vectors generated in sector 3 and 4 is always 0.
- The phase b component of all vectors generated in sector 5 and 6 is always 0.

In the same way, considering only $V_1 = (1,1,1)$ as zero vector, the following relations hold:

- The phase a component of all vectors generated in sector 6 and 1 is always 1.
- The phase b component of all vectors generated in sector 2 and 3 is always 1.
- The phase c component of all vectors generated in sector 4 and 5 is always 1.

By considering one of both cases, it is possible to define a sequence in which one phase can be kept fixed during both corresponding sectors. Figure 14.16 and 14.17 show the vector sequences for odd and even sectors to be considered when using (0,0,0) and (1,1,1) as boundary vectors, respectively. Note that choosing between one or another is equivalent to changing the polarity of the carrier in traditional PWM, and therefore does not affect the output voltage. From Figures 2.14 and 1.15c, it is clear how a phase of the inverter is kept fixed on 0 and 1, respectively, without switching. This strongly reduces the number of commutations and improves efficiency, compared to the 7-segment sequence shown in Figure 14.15.

The inverter output phase voltages, the line–line voltage, load voltage, and load current for discontinuous SVM using V_0, are shown in Figure 14.18. Note how each phase is kept at zero voltage level during $2\pi/3$ of the whole fundamental cycle.

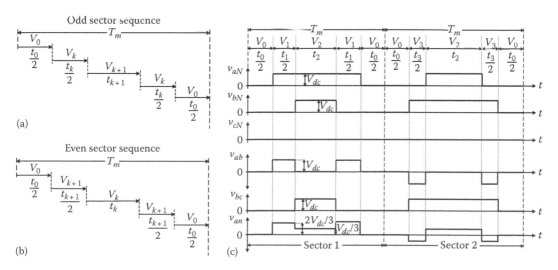

FIGURE 14.16 Discontinuous SVM sequence: (a) for odd sector, (b) for even sector, and (c) example for sector 1 and 2 transition.

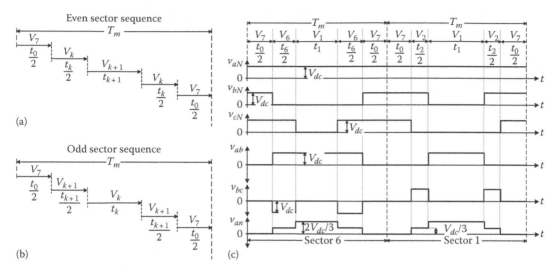

FIGURE 14.17 Discontinuous SVM sequence: (a) for even sector, (b) for odd sector, and (c) example for sector 6 and 1 transition.

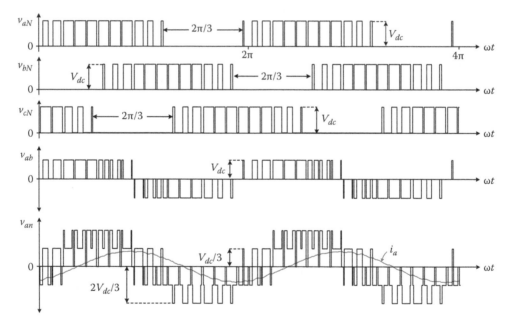

FIGURE 14.18 SVM voltage and current waveforms (inverter phase voltages, line–line voltage, and load voltage and current).

14.2.3.4 Over-Modulation and Zero-Sequence Injection

The voltage reference used for modulation using carrier-based PWM, in order to be properly modulated, needs to be always within the modulation range of the carrier signals. In practice, this means

$$m_a = \frac{\hat{v}^*}{\hat{v}_{cr}} \leq 1. \tag{14.19}$$

If the amplitude of the reference is higher than the amplitude of the carrier signal, the generated pulses can no longer warranty the time-average equivalence, and the linearity of the modulation is

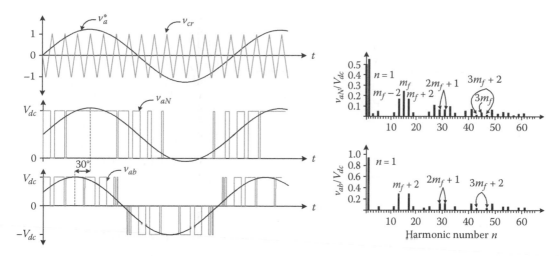

FIGURE 14.19 Over-modulation concept in a VSI.

lost, producing saturation. This concept, called over-modulation is illustrated in Figure 14.19. Because the fundamental component is not properly modulated, the control loop providing the voltage reference will be affected. Moreover, from the voltage spectra shown in Figure 14.19, it is clear that over-modulation introduces undesirable low-order harmonics in the output voltage that will not be filtered by the load, and will appear in the load current. These harmonics will be fed back into the control loop, affecting overall performance [1,5].

On the other hand over-modulation has as positive counterpart the fact that higher amplitude fundamental components can be generated by the inverter with $m_a > 1$, utilizing the same DC-link voltage to obtain higher load voltages, hence higher power, for the same rated converter. To overcome the loss of linearity, zero sequence signals can be injected to over-modulating reference voltages in such a way that the modified reference is kept within the modulating range of the carriers. Since zero sequence signals are canceled in three-phase connection, they will not appear in the line–line and load voltages, delivering the over-modulated reference. Therefore, this principle can only be applied for three-phase VSIs. The two most popular zero sequence signals are the third harmonic and the min–max sequence [8].

14.2.3.4.1 *Third Harmonic Injection*

Figure 14.20 shows a traditional bipolar PWM for phase a of the inverter, in which the reference voltage v_a^* is in over-modulation. A third harmonic signal v_{a3}^*, which is also in phase with the reference voltage is added to form a new reference voltage $\tilde{v}_a^* = v_a^* + v_{a3}^*$, that is completely included in the carrier range, and therefore is not over-modulated. As expected, the inverter phase output voltage v_{aN} shown in Figure 14.20 and in the corresponding spectrum, contains the $V_{dc}/2$ DC offset, the characteristic carrier harmonics and their sidebands, the desired fundamental component and the third harmonic, which has also been modulated by the carrier. However, in the line–line voltage spectrum, the third harmonics have disappeared, leaving only the desired fundamental component. This can be demonstrated by deriving the line–line voltages analytically. Consider the modified references for phases a and b, defined by

$$\tilde{v}_a^* = v_a^* \sin(\omega t) + v_{a3}^* \sin(3\omega t) \tag{14.20}$$

$$\tilde{v}_b^* = v_a^* \sin\left(\omega t - \frac{2\pi}{3}\right) + v_{a3}^* \sin\left(3\left(\omega t - \frac{2\pi}{3}\right)\right) \tag{14.21}$$

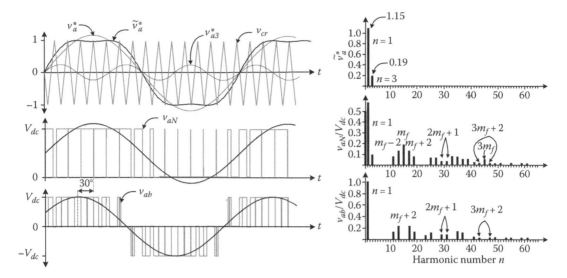

FIGURE 14.20 Third harmonic injection operating principle, waveforms, and spectra.

Then after modulation, the corresponding switched inverter phase output voltages can be expressed by

$$v_{aN} = v_a^* \sin(\omega t) + v_{a3}^* \sin(3\omega t) + v_{hf} \tag{14.22}$$

$$v_{bN} = v_a^* \sin\left(\omega t - \frac{2\pi}{3}\right) + v_{a3}^* \sin\left(3\left(\omega t - \frac{2\pi}{3}\right)\right) + v_{hf}, \tag{14.23}$$

where v_{hf} are the high-frequency components grouping all characteristic harmonics. Then the line–line voltage $v_{ab} = v_{aN} - v_{aN}$ can be computed from (14.22) and (14.23), resulting in

$$v_{ab} = v_a^*\left[\sin(\omega t) - \sin\left(\omega t - \frac{2\pi}{3}\right)\right] + v_{a3}^* \underbrace{\frac{[\sin(3\omega t) - \sin(3\omega t - 2\pi)]}{0}} + v_{hf} \tag{14.24}$$

$$v_{ab} = \sqrt{3}v_a^* \sin(\omega t) + v_{hf} \tag{14.25}$$

From (14.25), it is clear that no triple harmonic appear in the line–line voltage, and consequently in the load voltage and current.

An important aspect to consider is the limit of over-modulation the fundamental component can incur, and the corresponding amplitude the third harmonic needs to have to achieve the necessary compensation. For this consider the modified reference of (14.20). To analyze the maximal value, (14.20) can be derived with respect to ωt and equaled to zero:

$$\frac{d\tilde{v}_a^*}{d\omega t} = v_a^* \cos(\omega t) + 3v_{a3}^* \cos(3\omega t) = 0. \tag{14.26}$$

Since the maximal value can only occur at $\omega t = \pi/3$ when the third harmonic is crossing zero, this value can be considered into (14.26), which yields

$$v_{a3}^* = \frac{1}{6}v_a^*. \tag{14.27}$$

Replacing (14.27) in (14.20) and by considering that at $\omega t = \pi/3$ the value of $\tilde{v}_a^*$ has to be equal to the carrier maximum value, i.e., $\tilde{v}_a^*(\pi/3) = \hat{v}_{cr} = 1$, the following solution is obtained:

$$\hat{v}_a^*\left(\frac{\pi}{3}\right) = v_a^* \sin\left(\frac{\pi}{3}\right) + \frac{1}{6}v_a^* \underbrace{\sin\left(\frac{3\pi}{3}\right)}_{0} = 1 \tag{14.28}$$

$$\Rightarrow v_a^* = \frac{2}{\sqrt{3}} = 1.1547. \tag{14.29}$$

Replacing (14.29) in (14.27) yields to

$$v_{a3}^* = 0.19245. \tag{14.30}$$

Summarizing, the maximum value the fundamental component of the reference can have is an additional 15.47%, and the necessary third harmonic component will be 1/6 part of that. This case (the maximum permitted over-modulation) is the one illustrated in Figure 14.20.

One of the disadvantages of the third harmonic injection method is that the injection has to be synchronized and the reference voltage amplitude must be known in order to compute the amplitude of the third harmonic to be injected. This makes this method unfeasible for variable speed and closed-loop operation. In those cases, min–max zero sequence injection is a better choice.

14.2.3.4.2 Min–Max Injection

The min–max signal is a zero sequence signal composed only by odd triple harmonics (mainly third and ninth) [5]. Therefore, this method only can be used for three-phase inverters where this additional zero sequence signal will be canceled in the line–line voltages. The purpose of the signal is to lower the amplitude of the reference such that it can be completely included in the carrier signal modulating range. The min–max signal is defined by

$$v_{mm}(t) = \frac{\min\{v_a^*(t), v_b^*(t), v_c^*(t)\} + \max\{v_a^*(t), v_b^*(t), v_c^*(t)\}}{2} \tag{14.31}$$

The modified reference signals are $\tilde{v}_x^*(t) = v_x^*(t) - v_{mm}(t)$, where x stands for the three phases (a,b,c). Note that an important difference with third-harmonic injection is the fact that $v_{mm}(t)$ is time-dependent, and therefore can be computed online, regardless of the phase of the references; thus it can be used for variable speed and closed-loop operation. The three over-modulating reference signals, min and max components and the min–max sequence are illustrated in Figure 14.21a. The modified reference signal $\tilde{v}_a^*$ for phase a is shown in Figure 14.21b.

As happens with the third harmonic injection, it can be demonstrated that the maximum amount of over-modulation permitted for the references is also a 15.47%. Figure 14.22 shows a bipolar PWM

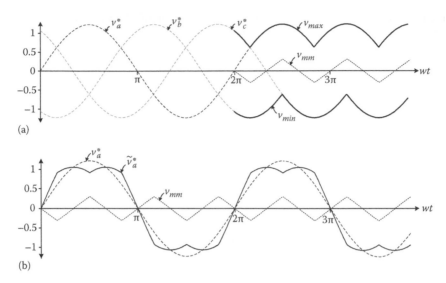

FIGURE 14.21   (a) Generation of the min–max sequence (v_{mm}) for three-phase reference voltages and (b) modified or injected reference waveform ($\tilde{v}_a^*$).

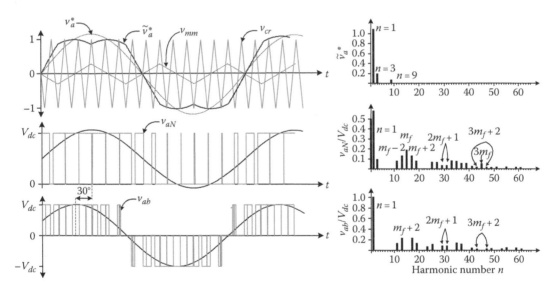

FIGURE 14.22   Min–max zero sequence injection operating principle, waveforms, and spectra.

implementation including min–max injection, considering the maximum admissible amplitude for the reference. Note that the modified reference $\tilde{v}_a^*$ is completely included in the carrier range. From its spectrum, it is clear that it includes the over-modulating fundamental component along with third and ninth harmonics given by v_{mm}. The resulting inverter phase output voltage does present the low-order injected harmonics, but it also includes the fully modulated fundamental component. In the line–line voltage, the min–max harmonics are eliminated as expected. The demonstration of this cancelation is similar to the one performed for third harmonic injection and will not be included here.

   It is worth mentioning that bipolar PWM with min–max sequence injection produces exactly the same pulse pattern than the one achieved with SVM considering the center-weighted seven-segment vector sequence of Figure 14.15 [8]. This means that SVM achieves a 15.47% over-modulation capability and better use of the inverter rating without the need of zero sequence injections vs. carrier-based PWM methods. Nevertheless, considering the easy implementation of carrier-based PWM, and that PWM signals are available in most digital platforms, the slight modification of the reference signal using min–max is a very simple way to implement something equivalent to center-weighted SVM, without the complex algorithm, calculations, and vector generation sequence. This is why carrier-based PWM with min–max is considered a standard nowadays.

14.2.3.5 Selective Harmonic Elimination

In the megawatt range, switching losses caused during the commutation of a power device, especially when the inverter contains freewheeling diodes responsible for large reverse recovery currents during the commutation, can lead to high-energy losses in long-term operation [10]. In addition, it requires larger and more sophisticated heat dissipation systems, usually air and water cooled. Therefore, high-switching frequency modulation methods, like those based on PWM or SVM are not suitable. Unfortunately, lowering the carrier frequency in PWM (or the modulation period in SVM) imposes a trade-off between efficiency improvement and power quality reduction, since linearity is lost in the modulation (carrier gets slow compared to the reference), and low-order sideband harmonics appear, which cannot be filtered by the load. This results in higher load current THD.

As a solution to the aforementioned problem, selective harmonic elimination (SHE) has been developed mainly targeted for high-power applications [1,5,11]. Basically, SHE is a PWM strategy where the commutation angles are predefined and precalculated in order to eliminate low-order harmonics and keep fundamental component tracking. To achieve this, the Fourier series of the predefined waveform is used to equal each non-desired low-order harmonic to zero, hence the name, and in additionally match the fundamental component with the desired modulation index given by the reference. Figure 14.23 shows a predefined SHE voltage waveform for a half- and H-bridge inverter, known as bipolar SHE and unipolar SHE, respectively. Both waveforms are depicted with five switching angles per quarter fundamental cycle ($\theta_{1,\dots,5}$). The three-phase full bridge VSI case is shown with three angles per quarter cycle in Figure 14.24 together with the line–line and load voltage, and their respective spectra.

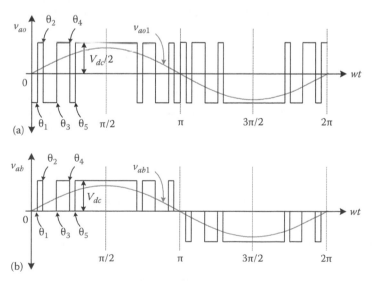

FIGURE 14.23 Five-angle SHE waveform: (a) half-bridge inverter and (b) H-bridge inverter.

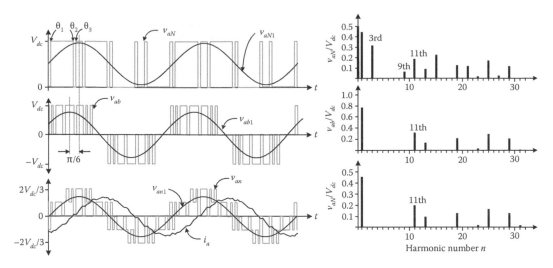

FIGURE 14.24 Full-bridge VSI three-angle SHE voltage waveforms and spectra (inverter phase voltage, line–line voltage, and load phase voltage).

To explain the operating principle, consider the three-angle case of Figure 14.24. The Fourier series for the switched voltage waveform is given by

$$v_{aN}(t) = \frac{V_{dc}}{2} + \sum_{n=1}^{\infty} b_n \sin(n\omega t), \tag{14.32}$$

$$b_n = \frac{4}{\pi} \int_0^{\pi/2} v_{aN}(\omega t)\sin(n\omega t)\,d\omega t, \tag{14.33}$$

where n is the harmonic number ($n = 1, 3, 5, \ldots$). There are no even order harmonics due to half-wave symmetry. By replacing the angles in (14.33), the following coefficients are obtained

$$b_n = \frac{4V_{dc}}{\pi n}\Big[\cos(n\theta_1) - \cos(n\theta_2) + \cos(n\theta_3)\Big] \tag{14.34}$$

Since there are no even order harmonics, and third-order harmonics and its multiples (called zero sequence signals) are cancelled by the three-phase connection of a balanced load, it is usual to eliminate the 5th and 7th harmonic. This is achieved by replacing $n = 5$ and $n = 7$ in (14.34), and forcing the coefficients to zero:

$$b_5 = 0 = \Big[\cos(5\theta_1) - \cos(5\theta_2) + \cos(5\theta_3)\Big] \tag{14.35}$$

$$b_7 = 0 = \Big[\cos(7\theta_1) - \cos(7\theta_2) + \cos(7\theta_3)\Big] \tag{14.36}$$

To complete the set of equations, the fundamental is forced to obtain the desired modulation index

$$b_1 = M\frac{V_{dc}}{2} = \frac{4V_{dc}}{\pi}\Big[\cos(\theta_1) - \cos(\theta_2) + \cos(\theta_3)\Big]. \tag{14.37}$$

The three angles and three equations form the nonlinear system to be solved. Note that the addition of an additional coefficient to eliminate another harmonic is not possible since two equations would be linear dependent. The only way to eliminate more harmonics is to add more angles, increasing the complexity of the system. The general rule is that with k angles, $k - 1$ harmonics can be eliminated while keeping control of the fundamental component. Figure 14.24 shows a particular solution for a modulation index close to one. Note how the 5th and 7th harmonics are eliminated in the inverter phase voltage while the 3rd and 9th still appear. Nevertheless, as mentioned before they are eliminated through the three-phase connection of the load and do not appear in the line–line and load voltages, as can be corroborated by their spectra. In addition the load current is shown, and appears highly sinusoidal despite the low switching frequency of the inverter. If more angles are considered, a natural choice for elimination would be the 11th, 13th, 17th, and so on, since no even harmonics or multiples of three need to be eliminated. For the single-phase case, this does not hold and triple harmonics also need to be eliminated.

It is worth mentioning that the set of equations cannot be solved online or analytically, being this the main disadvantage of SHE. Hence, all the switching patterns have to be pre-calculated off-line and stored in lookup tables. Many types of algorithms are used to solve these equations, mainly based on iterative numerical techniques, such as genetic algorithms [12]. A typical five-angle solution for the whole modulation index range is shown in Figure 14.25a. Usually this solution is stored in lookup tables, which are accessed using the modulation index given by the voltage reference. Then, the angles are converted into time by using a triangle waveform with amplitude $\pi/2$ and with the double of the desired fundamental frequency ω in [rad/s]. This implementation strategy is illustrated in Figure 14.25b.

To illustrate the effectiveness of SHE, consider the following example: A five-angle SHE waveform produces an average device switching frequency f_{sw} of 11 times the fundamental (number of pulses per cycle). Thus, for a 50 Hz fundamental frequency $f_{sw} = 550\,\text{Hz}$. This waveform in a three-phase connection

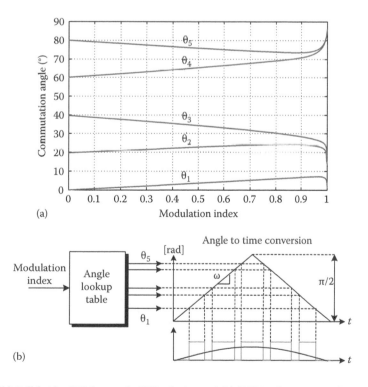

FIGURE 14.25 (a) Full-bridge VSI five-angle SHE solution and (b) SHE implementation diagram.

would generate the 17th as the first harmonic equivalent to 850 Hz. On the contrary, a 550 Hz carrier PWM would have central harmonics at 550 Hz, and significant low-order sidebands at 450 Hz, almost two times lower than SHE. Nevertheless, the main drawback of SHE is the fact it is computed off-line and stored in lookup tables, which are inherently discontinuous in nature. In addition, the angles are computed assuming pure sinusoidal waveform operation in steady state, hence in dynamic operation for variable frequency and amplitude the angles are no longer optimal and low-order harmonics do appear. These harmonics are fed back in closed-loop operation, affecting the system performance [13]. Hence, SHE is not recommended for high-performance variable speed motor drives.

A modulation method that can combine low switching frequency and high bandwidth closed-loop operation still remains an important subject of development in power electronics.

14.3 Multilevel Voltage Source Converters

14.3.1 Introduction

Several applications demand higher power due to economy of scale and efficiency reasons. To reach high power levels, VSI are needed to increase their voltage operation above the limits imposed by the semiconductor technology. The series connection of devices in the two-level topologies analyzed in previous section can increase the voltage rating of the inverter, but high dv/dt's are obtained. Moreover, the voltage distribution among the series-connected devices is not even due to mismatch between devices, which leads to derating of the devices and less reliability due to possible overvoltages across one device. This is why current source topologies were the only alternative for high-power applications during several decades. Multilevel inverters were specially developed to enable voltage source topologies reach higher voltages [28]. Instead of connecting several power switches in series, multilevel inverter structures arrange the semiconductors with additional DC-link capacitors that subdivide the total converter voltage rating to the blocking limit of the semiconductor. The additional dc-link capacitors, hence dc-sources, are connected to the load in sequences through the different switching states of the semiconductor arrangement, enabling the possibility not only to increase the voltage, but also to generate more voltage levels at the output, improving the quality of the generated voltage waveform. Figure 14.26 shows the difference between the concept of the two-level and the multilevel voltage waveform (a nine-level example is shown).

From Figure 14.26, a clear improvement in the voltage waveform can be noticed in terms of THD reduction, and the dv/dt for the same voltage rating is a $1/(k-1)$ fraction of the two-level waveform, for a k-level inverter (1/8 in the case of the nine-level inverter shown in Figure 14.26b). This also means that

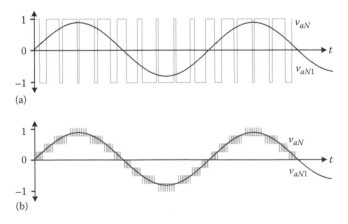

FIGURE 14.26 Multilevel inverter output voltage: (a) two-level and (b) nine-level.

the voltage rating can be increased $k - 1$ times given a specific semiconductor blocking voltage limit, which effectively increases the nominal power of the converter.

These properties make multilevel converters very attractive for high-power applications (1–50 MW) that reach the medium-voltage level (2.3–10 kV) like pumps, fans, conveyors, high speed traction, and ship propulsion to name a few. Currently several topologies have found industrial acceptance and are commercialized by several medium-voltage converter manufacturers [38]. Despite the level of maturity these inverters have reached, multilevel converters have a more complex circuit structure, with more semiconductors, hence more switching states or control options, which all together introduce several technical challenges. Nevertheless, these extra switching states also enable a great number of possibilities and additional degrees of freedom. This is why new topologies and modulation methods are still very actively under research and development.

This section presents a brief overview of the most common multilevel converter topologies and multilevel modulation methods used in industry.

14.3.2 Multilevel Converter Topologies

There are a large number of multilevel converter topologies reported in the literature [14,38], but the most common multilevel converter topologies in industry are diode-clamped converters also called neutral-point-clamped (NPC) converters, cascaded H-bridge (CHB) converters, and flying capacitor (FC) converters. A classification of multilevel converter topologies is represented in Figure 14.27, which includes these three converters along with more recently introduced topologies, some of them derived from the conventional multilevel inverters. Since the NPC, CHB, and FC have been successfully introduced as commercial products for more than a decade they will be covered in further detail in this section, while the other topologies (some of them not available in industry) will be briefly addressed and referenced for further reading.

14.3.2.1 Neutral-Point-Clamped Inverters

The NPC multilevel inverter was introduced in the early 1980s [15]. This converter was based on a modification of the classical three-phase two-level converter topology. In the conventional two-level converter (see Figure 14.7), each power semiconductor must withstand a voltage stress equal to V_{dc}. The

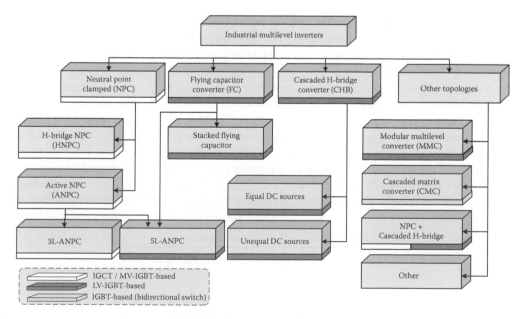

FIGURE 14.27 Industrial multilevel converter classification.

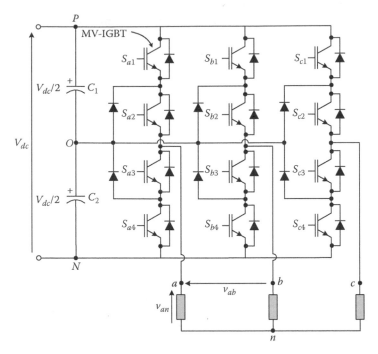

FIGURE 14.28 Three-phase three-level diode-clamped converter also called NPC converter.

modification to get the three-level NPC converter adds two extra semiconductors per phase and two clamping diodes that divide the DC link in half. Using this new topology, each switching device blocks at the most a voltage equal to $V_{dc}/2$. So, if these semiconductors have the same characteristics than those used in the two-level converter, the DC-link voltage can be theoretically doubled leading to double the nominal power of the converter. In Figure 14.28, the three-phase three-level NPC, also called diode-clamped converter is represented. In this topology, the total DC-link voltage (V_{dc}) has to be equally shared between capacitors C_1 and C_2.

In order to avoid short-circuit of the DC-link capacitors, there are only three possible switching states for the NPC, which are summarized in Table 14.4. These three switching states produce three output phase voltages with respect to neutral point O (the middle point of the dc-link), which is why the NPC is referred as a three-level inverter.

The NPC topology can be extended to achieve more levels in the output phase voltages, for which more capacitors are connected in series in the dc-link, and additional switches and clamping diodes are used to clamp the switches to each capacitor [38]. Therefore, in this case it is more correctly referred as a diode-clamped converter, since there is not only one clamping node at the dc-link side, and even not necessarily a neutral point with zero volt potential (this is the case for even number levels). However, the NPC topology with a high number of levels has a high unequal distribution of the losses among semiconductors which forces a derating of the power devices, and also a reduction of the lifetime of the power semiconductors. On the other hand, although all the power switches share

TABLE 14.4 Three-Level NPC Switching States

S_{a1}	S_{a2}	S_{a3}	S_{a4}	Phase Voltage, v_{ao}
1	1	0	0	$V_{dc}/2$
0	1	1	0	0
0	0	1	1	$-V_{dc}/2$

Note: Only phase *a* given.

the same blocking voltages, the clamping diodes do not share it equally. Note that frequently these power converters use the top-rated devices in the market, so clamping diodes for higher number of levels will require the connection of several diodes in series. These problems and other such as the dc voltage balance of the capacitors have prevented the industrial implementation of the NPC topology with more than three levels.

Three-level NPC topology has become very popular in industry and academic research all over the world. As some commercial examples, converters such as the ACS1000 (ABB), MV Simovert (Siemens), TMdrive-70 (TMEIC-GE), Silcovert-TN (Ansaldo), MV7000 (Converteam) and IngeDrive MV500 (IngeTeam) to name a few are current commercial three-level NPC solutions. The NPC can be found in industry featuring IGCT, IEGT, and medium-voltage IGBT (MV-IGBT), like the one shown in Figure 14.28.

14.3.2.2 Flying Capacitor Inverters

Multilevel FC converter topology was developed in the 1990s and it uses several floating capacitors instead of clamping diodes, to share the voltage stress among devices, and to achieve different voltage levels in the output voltage [16]. In Figure 14.29, a three-phase FC is shown. Depending on the voltage of the floating capacitors, the number of voltage levels change. In the converter represented in Figure 14.29, if the floating capacitor voltages are equal to $v_{a1} = v_{b1} = v_{c1} = V_{dc}/2$, the number of output voltage levels is three, as is summarized in Table 14.5. Other voltage ratios can be used for the floating capacitors, increasing the number of levels. However, this makes the voltage balancing of the capacitors more difficult and imposes different blocking voltage among the devices, and therefore does not find industrial acceptance.

The FC converter shown in Figure 14.29 can be represented in a different way in order to show its high modularity. In fact, each phase of the FC topology is formed by several basic power cells connected in cascade, as is shown in Figure 14.30. Each cell is composed of a pair of switches and one capacitor. It is clear that both power semiconductors of each power cell are controlled with opposite signals to avoid a short-circuit of the capacitors. Therefore, only one control signal has to be used in order to trigger the power semiconductors of each cell of the FC.

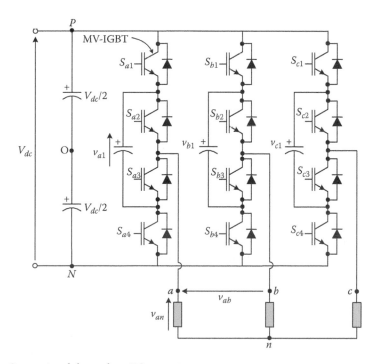

FIGURE 14.29 Conventional three-phase FC converter.

TABLE 14.5   Three-Level FC Switching States

S_{a1}	S_{a2}	S_{a3}	S_{a4}	Phase Voltage, v_{ao}
1	1	0	0	$V_{dc}/2$
1	0	1	0	0
0	1	0	1	0
0	0	1	1	$-V_{dc}/2$

Note: Only phase *a* given.

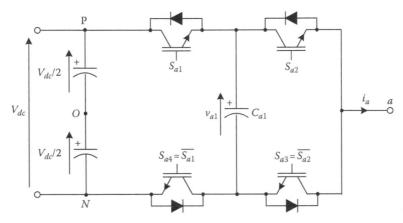

FIGURE 14.30   Two-cell FC converter topology. If v_{a1} is equal to $V_{dc}/2$, this is the three-level FC topology.

The FC topology can be extended achieving more levels in the output phase voltages only connecting more power cells in series. In general, for a multilevel FC with *m* cells, several FC voltages can be considered $(v_{a1}:v_{a2}:v_{a3}:\ldots:v_{a(m-1)})$ However, the conventional FC topology has floating capacitors with voltage ratios equal to $m - 1:\ldots:2:1$ what means that the voltage of the floating capacitor *j* is equal to $v_{aj} = jV_{dc}/m$. Using this conventional dc voltage ratio, the number of levels of a *m*-cell FC is equal to $m + 1$. Nowadays, the FC topology has a reduced industrial presence but some commercial products can be found (converter ALSPA VDM6000 by Alstom).

14.3.2.3 Multilevel Cascaded H-Bridge Inverters

The multilevel cascaded H-bridge converter (usually called CHB converter) is formed by the series connection of several H-bridges with their corresponding independent voltage sources [17]. In Figure 14.5, a conventional H-bridge VSI was shown. This circuit can be considered as the basic cell to develop multilevel CHB converters and its operating principles were introduced in Section 14.2.2.2. A CHB is easily built connecting several H-bridge cells in series, like the two-cell CHB shown in Figure 14.31. In this way, the CHB topology is able to reach higher voltage levels by just adding H-bridge cells in series. This high modularity feature is very attractive to reach medium voltages up to 10 or even 13 kV in some industrial applications. This is why the CHB is found in practical applications up to nine cells in series [43]. Because the voltage is shared among so many cells and devices, low-voltage IGBT (LV-IGBT) are used. However, the main drawback of this topology is that each basic cell needs an independent voltage source, V_{dc1} and V_{dc2} in Figure 14.31, which are usually equal (V_{dc}). These isolated dc-sources are commonly provided by a multi-secondary transformer with diode rectifiers. The secondaries of the transformer are shifted in phase (zigzag transformers) so that together with the diode rectifiers a multipulse rectifier configuration is achieved that enables input current harmonics mitigation. Hence, the transformer can be seen as a drawback in terms of design and implementation complexity and additional cost, but on the other hand introduces improved power quality.

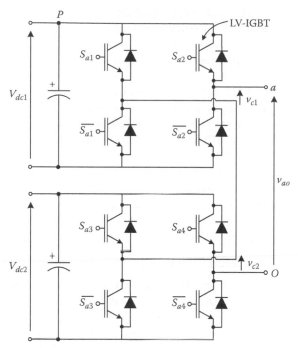

FIGURE 14.31 Two-cell CHB converter.

The conventional CHB assumes that all the dc voltage sources V_{dci} have exactly the same values, this corresponds to the CHB with equal dc sources of the classification in Figure 14.27. Assuming this conventional dc voltage ratio and considering a two-cell cascaded converter, like the one shown in Figure 14.31, the possible switching states are listed in Table 14.6. The two-cell achieves five possible output voltages and, therefore it is a five-level converter. Many of the switching states generate the same

TABLE 14.6 Five-Level CHB Switching States with Equal dc-Sources ($V_{dc1} = V_{dc2}$)

Cell 1		Cell 2		Cell 1 Voltage	Cell 2 Voltage	Phase Voltage
S_{a1}	S_{a2}	S_{a3}	S_{a4}	v_{c1}	v_{c2}	$v_{ao} = v_{c1} + v_{c2}$
1	0	1	0	V_{dc}	V_{dc}	$2V_{dc}$
1	0	0	0	V_{dc}	0	V_{dc}
1	0	1	1	V_{dc}	0	V_{dc}
0	0	1	0	0	V_{dc}	V_{dc}
1	1	1	0	0	V_{dc}	V_{dc}
0	0	0	0	0	0	0
1	1	0	0	0	0	0
0	0	1	1	0	0	0
1	1	1	1	0	0	0
1	0	0	1	V_{dc}	$-V_{dc}$	$V_{dc}, -V_{dc}$
0	1	1	0	$-V_{dc}$	V_{dc}	$-V_{dc}, V_{dc}$
0	1	0	0	$-V_{dc}$	0	$-V_{dc}$
0	1	1	1	$-V_{dc}$	0	$-V_{dc}$
0	0	0	1	0	$-V_{dc}$	$-V_{dc}$
1	1	0	1	0	$-V_{dc}$	$-V_{dc}$
0	1	0	1	$-V_{dc}$	$-V_{dc}$	$-2V_{dc}$

output voltage level (voltage level redundancy), which increases overproportionally to the amount of cells. In general terms, the number of different voltage levels generated by a CHB with k cells is $2k + 1$.

Different dc voltage source ratios can be applied in order to achieve more voltage levels in the output voltage [18]. These converters are known as CHB with unequal dc sources or asymmetric CHB, as shown in classification of Figure 14.27.

Depending on the dc voltage ratio, up to nine levels can be obtained using a two-cell CHB topology shown in Figure 14.31. In general terms, a voltage ratio in multiples of three between each cell of the CHB ($V_{dc(I+1)} = 3V_{dci}$) eliminates all the voltage-level redundancies, maximizing the number of generated voltage levels. In this case, a k-cell CHB will generate 3^k levels in the output voltage. Compared to a CHB with equal dc sources, a 4-cell asymmetric converter will generate $3^4 = 81$ levels compared to $2 \cdot 4 + 1 = 9$ levels of the symmetric CHB. However, like with the FC, the modularity is lost since different blocking voltages appear among the semiconductors of the different cells.

The CHB with equal dc sources is recently achieving a high industrial impact and commercial products such as MVD Perfect Harmony (Siemens), Tmdrive-MV (TMEIC-GE), LSMV VFD (LS Industrial Systems), AS7000 (ArrowSpeed), and FSDrive-MV1S (Yaskawa) to name a few can be currently found.

14.3.2.4 Other Multilevel Inverter Topologies

Other topologies, usually derived from the classic multilevel converter topologies (NPC, CHB, and FC), have been introduced in the literature. Among them, one derived from the three-level NPC and called three-level active NPC (ANPC) topology has been developed trying to improve the NPC features. As can be observed in Figure 14.32, this topology replaces the clamping diodes by clamping switches offering the possibility to equalize the losses in the overall converter (which is a drawback of the conventional three-level NPC) enabling a substantial increase in the power rating of the converter. A detailed analysis on the loss distribution and how to control it through the new switching states provided by the additional clamping switches is performed in [40].

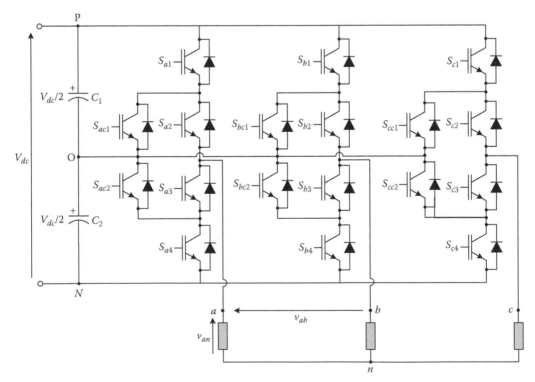

FIGURE 14.32 Three-level ANPC converter.

Another interesting hybrid topology called NPC-CHB has been recently introduced [41] and it is formed by a three-level NPC connected in series with single-phase H-bridge cells (usually one or two cells). In this topology, normally the H-bridges dc sides are floating capacitors without any dc voltage supply. Therefore, the addition of the H-bridge cells increases the number of voltage levels, but does not increase the active power rating of the overall converter. The functionality of the H-bridge cells is the active filtering and the enhancement of the output voltage harmonic distortion.

Finally, another hybrid multilevel converter, which is the current focus of industrial interest is the modular multilevel converter (M2C or MMC), particularly for HVDC systems [42]. The MMC is derived from the CHB topology and is usually formed by single-phase half-bridges with floating dc sides connected in series. However, in the MMC case, the phase leg is divided in two equal parts to be able to generate equal number of positive and negative levels at the ac side. Normally, some inductor is connected at the output of each leg to protect during transitory short circuits. Also, floating H-bridges have been used as power cells to implement the MMC topology. As the number of cells of the MMC used to be high, the output voltage present a high number of levels improving its harmonic content.

Another multilevel topology that has found industrial application is an hybrid between the NPC and an H-bridge. Basically the topology connects in parallel two three-level NPC legs to form a five-level H-bridge, called H-NPC. This topology also needs isolated dc-sources for the H-bridge of each phase. This topology is commercialized by two major manufacturers (ACS5000 by ABB and Dura-Bilt5i MV by TMEIC-GE).

14.3.3 Modulation Techniques for Multilevel Inverters

The most common modulation techniques for multilevel converters have been extended from modulation methods used in conventional two-level VSIs. In general, as the multilevel converters are specially well suited for high-power applications, usually the modulation technique is focused on the minimization of the switching losses mainly by reducing the switching frequency. The most common modulation techniques for multilevel converters are divided in two main groups: techniques based on the space vector concept and techniques based on the voltage levels. A classification of the most common modulation techniques for multilevel converters is introduced in Figure 14.33. Furthermore, as is also shown in

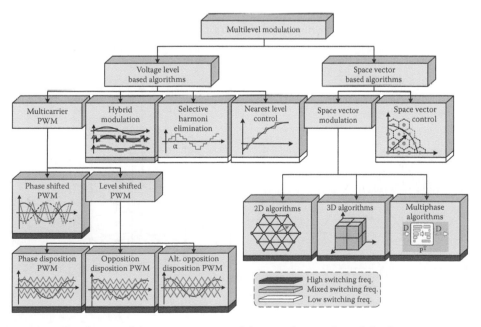

FIGURE 14.33   Classification of the most common modulation techniques for multilevel converters.

Figure 14.33, the modulation techniques can be classified depending on the switching frequency of the method (high, mixed, or low switching frequency).

14.3.3.1 Space Vector–Based Modulation Techniques for Multilevel Inverters

In general, the space-vector modulation (SVM) techniques can be classified into three main groups depending on the number of phases of the converter and depending on the coordinates used to plot the space vectors generated by the converter, also called control region of the converter.

14.3.3.1.1 SVM Techniques Based on Two-Dimensional Control Regions

The conventional way to introduce a SVM technique for a three-phase converter is based on a two-dimensional (2D) representation of the control region of the converter using the α–β plane. As an example, the 2D control region of a three-phase three-level converter is represented in Figure 14.34a. In these SVM techniques, the main purpose of the algorithm is to determine the three space vectors, their duty cycle or switching times and the switching sequence in which they will be generated to approximate in average the reference voltage vector over the modulation period. In [19], a summary of the most used 2D SVM techniques is presented. Usually, the space vectors and switching times are determined by geometrical-based calculations, like the ones analyzed for the two-level SVM in previous section. This is usually done between the projections of the reference space vector over the triangle formed by the closest voltage space vectors generated by the converter to the reference vector in a given moment. The final switching sequence and the switching times are easily calculated using simple mathematical equations [20].

14.3.3.1.2 SVM Techniques Based on Three-Dimensional Control Regions

SVM techniques designed taking into account the control region represented in the α–β plane are well suited for converter topologies where the zero sequence voltage and the zero sequence current are zero and therefore their γ components are zero. However, some converter topologies such as the three-phase four-wire and four-phase four-wire topologies present zero sequence voltages and zero sequence currents in general different to zero. In these cases, SVM techniques have to consider the three components α, β, and γ to carry out the modulation without errors. The easiest way to design a SVM technique using three components is to use the natural coordinates *abc* because, in this case, the control region

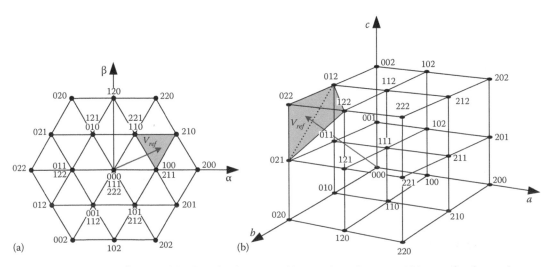

FIGURE 14.34   Control region of the three-level converter (a) using the α–β plane and (b) using the *abc* coordinates.

is formed by regular volumes simplifying the necessary calculations. This concept is shown in Figure 14.34b, where the control region of a three-phase three-level converter is represented using the *abc* frame. It can be noticed that the 3D-based SVM techniques are an extension of the 2D-based SVM techniques, and they can be applied without restriction to any power converter topology with or without zero sequence components.

For instance, a very simple 3D SVM using *abc* coordinates is introduced in [21] and is based on a normalization of the reference voltage and a simple geometrical search of the tetrahedron where the normalized reference vector is located. In this case, the final switching sequence formed by the four nearest state vectors and their corresponding switching times are also calculated with simple mathematical expressions.

14.3.3.1.3 SVM Techniques for Multilevel Multiphase Converters

The previous SVM techniques including the 2D-SVM and the 3D-SVM can be only applied to three-phase multilevel converters. The graphical representation of the control region, present on three phase converters, is lost when the number of phases increases. For a higher number of phases, new modulation techniques have been introduced in [22]. This method solves the modulation problem of multiphase multilevel converters by using matrix calculations in order to determine the switching sequence and the switching times to generate the reference voltage for each phase. Using this multiphase SVM method, in the first step, a normalization of the reference voltage is done and all the calculations are written in matrix format. Using this modulation technique, the multiphase multilevel modulation problem is reduced to a multiphase two-level problem using simple calculations determining a normalized two-level reference vector. This technique is simple but it should be noticed that the number of cases highly increases when the number of phases is increased.

14.3.3.1.4 Space Vector Control

In the space vector control (SVC), the basic idea is to take advantage of converters with high number of voltage vectors (for inverters of at least seven levels), by simply approximating the reference to the closest voltage vector that can be generated. SVC was in [23], as an alternative to the SVM techniques to provide a high performance using a low switching frequency. SVC is not actually a modulation method because the reference vector is not achieved by the switching of the converter averaged over a switching period. Using SVC, the reference vector is only approximated, generating an error which is small in the case of converters with high amounts of levels due to the dense SVC region. Because of the approximation error, some low-order harmonic distortion does appear in the output voltage; however, this drawback comes with the great benefit of very low switching frequency, which improves the efficiency of the converter.

14.3.3.2 Voltage Level–Based Modulation Techniques for Multilevel Inverters

14.3.3.2.1 Multicarrier Level-Shifted PWM

Level-shifted PWM (LS-PWM) is the natural extension of bipolar PWM for multilevel inverters [8]. Bipolar PWM uses one carrier signal, which is compared to the reference to decide between two different voltage levels, typically the positive and negative busbars of a VSI. By generalizing this idea, for an *k*-level inverter, *k* − 1 carriers are needed arranging them in vertical shifts. Since each carrier is associated to two levels, the same principle of bipolar PWM can be applied. The *k* − 1 carriers span the whole amplitude range that can be generated by the converter.

The carriers can be arranged in vertical shifts, with all the signals in phase with each other, called phase disposition (PD-LS-PWM), with all the positive carriers in phase with each other and in opposite phase of the negative carriers, known as phase opposition disposition (POD-LS-PWM), and finally alternate phase opposition disposition (APOD-LS-PWM) which is obtained by alternating the phase between adjacent carriers. The LS-PWM modulation is especially useful for NPC converters, since each carrier can be easily associated to two power switches of the converter.

The LS-PWM technique leads to high-quality line voltages since all the carriers are in phase when compared to other multicarrier PWM methods. In addition since it is based on the output voltage levels of an inverter, this principle can be easily adapted to any multilevel converter topology. However, this method is not preferred for CHB and FC, since it causes an uneven power distribution among the different cells. This generates input current distortion in the CHB and a capacitor unbalance in the FC.

14.3.3.2.2 Multicarrier Phase-Shifted PWM

Phase-shifted PWM (PS-PWM) is a multicarrier PWM method, and is a natural extension of PWM especially suited to be applied to the FC and CHB [8] converters, mainly due to the modularity of these topologies. Each cell is modulated independently using unipolar or bipolar PWM, for the CHB and FC, respectively, with the same reference signal. A phase shift is introduced across the carrier signals of each cell in order to produce the stepped multilevel waveform. The lowest output voltage distortion is achieved with a $180°/k$ or $360°/k$ phase shifts between the carriers, for a k-cell CHB or k-cell FC converters, respectively. This is because the FC power cells have two-level output voltages, compared to the three output levels of the H-bridges of the CHB.

Using the PS-PWM technique for a conventional symmetrical k-cell multilevel converter, the inverter has k times the nominal power of each cell. In addition, the frequency of the inverter output voltage switching pattern is also k times higher, without increasing the average switching frequency. This is produced by a multiplicative effect in the output switching pattern, due to the series connection of the cells and the phase shifts introduced in the carriers. The PS-PWM technique produces an even usage of the different power cells of the CHB leading to an equal power distribution and power losses between the H-bridges. The same happens with the FC, but in relation to the natural balancing of the capacitor voltages what is naturally achieved by using the PS-PWM. However, NPC inverters cannot operate with PS-PWM, since it has no modular structure, and therefore the carriers cannot be associated to a particular cell or treated independently.

14.3.3.2.3 Hybrid Modulation

The hybrid modulation [24] is specially conceived for the asymmetrical CHB converters (with unequal dc sources). The basic idea is to take advantage of the different power rates among the cells to reduce switching losses and improve the converter efficiency. In the high-power cells, a square-wave modulation is applied, while the lowest voltage power cell uses a conventional PWM technique. In this way, the high-power cells switch at fundamental frequency, generating very low switching losses, while the quality of the output voltage is achieved by using traditional PWM in the lowest voltage cell.

14.3.3.2.4 Selective Harmonic Elimination and Selective Harmonic Mitigation

The switching losses limit the maximum switching frequency that can be used in a multilevel converter. In order to reduce the harmonic distortion of the output waveforms, modulation strategies such as the well-known SHE technique has been extended to be applied to multilevel converters. As in the conventional two-level case, the SHE technique applied to multilevel converters can set the amplitude of the fundamental harmonic and make zero the amplitude of $k - 1$ desired harmonics if k switching angles are used per quarter of period [25].

It is important to notice that other harmonic-based modulation techniques have been introduced recently in order to take into account actual grid codes imposed by the electric providers all over the world. For example, the selective harmonic mitigation (SHM) modulation is based on the idea that it is not necessary to completely eliminate the amplitude of the harmonics. They just have to be reduced below the levels imposed by grid codes [26]. The SHM technique generates output waveforms that can completely fulfill specific grid codes with a lower switching frequency than the conventional SHE. The SHM technique has been applied to a three-level converter but, as happens with the SHE technique, it can also be applied to converters with any number of levels independent of the specific converter topology.

14.3.3.2.5 Nearest Level Control

The nearest level control (NLC) [27] also known as the round method, is somehow, the time domain counterpart of SVC for single-phase systems. Basically, the same principle is applied, by selecting the nearest voltage level that can be generated by the inverter to the desired output voltage reference. The main advantage is that the algorithm is extremely simple and the output voltage level selection is reduced to a simple expression. In case of a multilevel converters with output voltage steps equal to V_{dc} the output voltage level v_o for a desired reference voltage v_o^* is determined by

$$v_o = V_{dc} \cdot \text{Round} \left\{ \frac{v_o^*}{V_{dc}} \right\}. \tag{14.38}$$

The nearest integer function or round function, is defined such that round $\{x\}$ is the integer closest to x. Since this definition is ambiguous for half-integers, the additional convention is that half-integers are always rounded to even numbers.

Note that, as in the SVC case, the NLC is not strictly a modulation technique because the reference voltage is only approximated by the closest voltage level. In the NLC case, the maximum approximation error is $V_{dc}/2$. The voltage waveform is very similar to the one obtained with the SHE methods. However, this method does not eliminate specific harmonics, and therefore has to be used in inverters with a high number of levels to avoid important values of low-order harmonics at the output as happens with SVC.

14.4 Current Source Inverters

14.4.1 Introduction

CSI, unlike VSIs, use a constant current source usually provided by a controlled current source rectifier (CSR) and an inductive dc-link, to generate a switched current waveform at the output with adjustable frequency, phase, and amplitude. Hence the output voltage is defined by the current load, which is usually very sinusoidal for resistive-inductive loads such as motor drives, and is not switched as in VSIs; thus dv/dts are very motor friendly. Furthermore, the current is not defined by the load and is always controlled; hence, this topology offers inherent over-current and short-circuit control. The main drawbacks of CSIs are the current harmonics, which need to be mitigated with capacitive output filters (although they together with the inductive load also contribute to the highly sinusoidal voltages), and the reduced dynamic performance since the current amplitude is controlled by the input rectifier with large dc chokes [1].

From the DC/AC classification in Figure 14.2, it can be seen there are two basic CSI topologies: the PWM-CSI and the load-commutated CSI (LCI-CSI). The first one features IGCTs (in replacement of GTOs) as power semiconductors for hard-switched modulation methods, while the latter uses SCR devices with load-dependent commutation, which is more efficient and reliable although slower and with reduced dynamic performance. Therefore the LCI has been mainly used in very high-power synchronous motor drives operated at a leading power factor, in applications up to several tens of megawatt. The following sections describe the power circuit structures, operating principles, modulation methods, and applications of CSIs.

14.4.2 PWM-CSI

14.4.2.1 Operating Principle

Although single-phase CSI are conceptually possible, in this chapter, only three-phase CSI topologies and control methods will be analyzed, due to their application field in industry, which are mainly three-phase systems [29–32]. The CSI power circuit is illustrated in Figure 14.35, featuring symmetric gate-commutated thyristor (SGCT) power semiconductors, which is currently the industry standard [33],

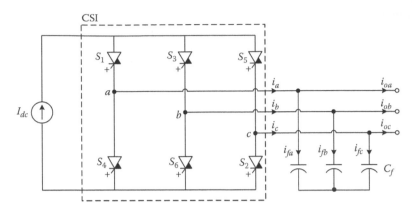

FIGURE 14.35   Basic PWM-CSI power circuit featuring IGCT semiconductors.

although many of the operating CSIs commissioned in past decades are working with GTOs [34]. As can be seen, the CSI has mainly three parts, the controlled current source, the converter full bridge, and the output capacitive filter.

As mentioned in the introduction, the output filter is in part responsible for the output power quality (more sinusoidal voltage and current waveforms), but more important it provides a current path for the hard-switched inductive load current. In this way, device damaging over voltages at the inverter output, due to the high *di/dt*, are overcome while improving power quality.

The controlled current source is usually provided by a controlled CSR and a pair of large inductive reactors or dc chokes, as can be seen in Figure 14.36. Typical rectifiers for CSIs are full-bridge SRC rectifiers, which can also be connected in series with multipulse transformer configurations, depending on the power level needs and input power quality requirements (grid codes compliance) [1]. Another common rectifier is the PWM-CSR with the same structure as the PWM-CSI of Figure 14.35 connected together in a back-to-back configuration [35]. In this case, series connection of IGCT devices can be used to reach higher blocking voltages in medium-voltage applications (at rectifier and inverter side), while the capacitive input filter is used to meet grid codes. To overcome possible resonances between the rectifier line current harmonics and the filter plus grid, active damping methods can be applied [36]. The purpose of the CSR is to keep the current controlled and constant at a desired reference I_{dc}^* (therefore the large dc chokes), although this reference can change depending on the required current amplitude of the inverter output. Hence the inverter is only in charge of controlling the phase and frequency of the ac current. This differs from VSIs, which have a fixed DC-link voltage, and are responsible of controlling phase, frequency, and amplitude of the output voltage.

In VSIs, the two semiconductors in one inverter leg operate alternately to avoid capacitive DC-link short circuit. In the same sense, CSIs also require switching restrictions to operate appropriately. Since the inverter bridge is connected to a constant current source (inductive DC link), there must be always a path for I_{dc}, hence at least one of the upper and one of the lower IGCTs must be turned on at any time

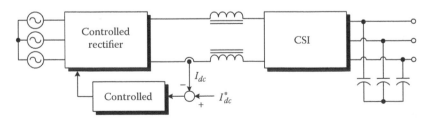

FIGURE 14.36   Controlled dc-link current source (generic circuit), used to feed the CSI.

TABLE 14.7 PWM-CSI Switching States

Switching State	Upper Gating Signals			Lower Gating Signals			Output Currents			Space Vectors
	S_1	S_3	S_5	S_4	S_6	S_2	i_a	i_b	i_c	
1	1	0	0	1	0	0	0	0	0	$I_{0a}=I_0=0$
2	0	1	0	0	1	0	0	0	0	$I_{0b}=I_0=0$
3	0	0	1	0	0	1	0	0	0	$I_{0c}=I_0=0$
4	1	0	0	0	1	0	I_{dc}	$-I_{dc}$	0	$I_1=\dfrac{2}{\sqrt{3}}I_{dc}e^{j(-\pi/6)}$
5	1	0	0	0	0	1	I_{dc}	0	$-I_{dc}$	$I_2=\dfrac{2}{\sqrt{3}}I_{dc}e^{j(\pi/6)}$
6	0	1	0	0	0	1	0	I_{dc}	$-I_{dc}$	$I_3=\dfrac{2}{\sqrt{3}}I_{dc}e^{j(\pi/2)}$
7	0	1	0	1	0	0	$-I_{dc}$	I_{dc}	0	$I_4=\dfrac{-2}{\sqrt{3}}I_{dc}e^{j(5\pi/6)}$
8	0	0	1	1	0	0	$-I_{dc}$	0	I_{dc}	$I_5=\dfrac{-2}{\sqrt{3}}I_{dc}e^{j(7\pi/6)}$
9	0	0	1	0	1	0	0	$-I_{dc}$	I_{dc}	$I_6=\dfrac{2}{\sqrt{3}}I_{dc}e^{j(-\pi/2)}$

($S_1 + S_3 + S_5 \geq 1$ and $S_2 + S_4 + S_6 \geq 1$). In addition, the output terminals of the inverter bridge are connected to a capacitive filter, thus to avoid the capacitors short circuit, at the most one upper and one lower IGCT are allowed to be on at any time ($S_1 + S_3 + S_5 \leq 1$ and $S_2 + S_4 + S_6 \leq 1$). This restriction also prevents I_{dc} from splitting into two undefined line currents (determined by load conditions), since this would affect the proper operation of the modulation stage. By intersecting both restrictions it is clear that one upper and one lower power switch must be always conducting ($S_1 + S_3 + S_5 = 1$ and $S_2 + S_4 + S_6 = 1$). This restriction provides a positive and negative current path in the inverter, hence there is no need for antiparallel diodes as it is in VSIs, resulting in less semiconductors and simpler topology structure. Considering all the admissible switching states of the inverter, there are nine possible switching combinations, which generate seven different three-phase line currents, which are listed in Table 14.7. The three-phase output currents can be represented with one space vector, consequently resulting in seven different space vectors ($I_0, I_1,\ldots, I_6$), as will be discussed later.

From the different switching states shown in Table 14.7, it can be appreciated that the CSI can generate three different output current levels in each phase: $-I_{dc}$, 0, and I_{dc}. It is then the job of the modulation stage to make the inverter alternate appropriately between these constant current levels, to deliver an AC switched current waveform with the desired phase and frequency fundamental component to the load. There are four well-established modulation methods for CSIs: square-wave modulation, trapezoidal PWM (T-PWM), SHE [11], and SVM [37], which are analyzed in the next section.

14.4.3 PWM-CSI Modulation Methods

14.4.3.1 Square-Wave Modulation

In the square-wave operation analyzed previously for VSI, the voltage waveform featured only two levels, since the waveform was defined from the inverter output to the negative bar. Nevertheless, the output line voltages do have three levels like the output line currents in a CSI. Moreover the line voltage waveform obtained with square-wave operation in VSIs, if seen as a line current, fulfills the switching restrictions of a CSI, and therefore the same pattern can be defined, as illustrated in Figure 14.37a. Note that at any time only one phase is conducting $-I_{dc}$ and one I_{dc} as supposed to be.

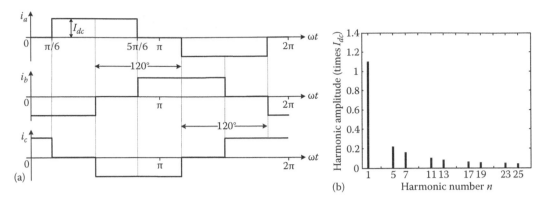

FIGURE 14.37 Square-wave modulation for CSIs: (a) current waveforms and (b) line current spectrum.

The main advantage of this method is its extreme simplicity, since phase and frequency can be easily controlled by converting the fixed switching angles ($\pi/6$ and $5\pi/6$ for phase a) to time, while the amplitude of the ac current is controlled by the SCR. In fact, this is an advantage over square-wave operation in VSIs, where no control of the fundamental is possible, hence here called "modulation" instead of "operation." Another important advantage is that each power semiconductor switches at fundamental switching frequency, i.e., the device switching frequency $f_{sw} = f_1$, resulting in a very efficient modulation scheme due to lower switching losses. These advantages come at expense of poor power quality as the defined current waveform has low-order harmonics, which cannot be fully mitigated by the output filter. From the Fourier series of the line currents shown in Figure 14.37a, each harmonic component h_n can be computed as $2\sqrt{3}\, I_{dc}/n\pi$, where $n = 6k \pm 1$, $k \in N$. Considering these harmonics, the line current spectrum is given in Figure 14.37b, normalized by the dc current amplitude I_{dc}. Note that the fundamental component is always 1.1 I_{dc} since there are no additional or variable switching angles. The low-order harmonics introduced by this modulation method can excite the capacitive filter and inductive motor resonances and therefore has limited application in adjustable speed drives.

14.4.3.2 Trapezoidal PWM

Traditional carrier-based PWM with sinusoidal references cannot be directly extended to CSIs, due to the switching constraints defined earlier. Therefore, a trapezoidal reference waveform is used instead and compared to a modified carrier waveform, which can be appreciated for phase a in Figure 14.38a. Note that the reference signal for each semiconductor v_{mi} ($i = 1, \dots, 6$) are active in modulation only during one-half of the fundamental cycle, which can be the positive or negative semi-cycles, depending if they conduct I_{dc} or $-I_{dc}$. In addition, the active semi-cycle is divided into three segments of $\pi/3$. To meet the switching constraints, the central $\pi/3$ segment is always "on" for the active phase. During this central segment, the other two phases are in their first and last segment, respectively, featuring complementary linear carriers PWM, to alternate between each other the necessary current return path for the active phase. This also introduces the commutations to produce the switched current waveform and reduce harmonics compared to square-wave modulation. The three-phase output line currents obtained with T-PWM are shown in Figure 14.38a.

The fact that a device only commutates during 1/3 of the whole fundamental cycle strongly reduces switching losses. Generally the carrier frequency f_{cr} is an even multiple of the fundamental frequency ($f_{cr} = kf_1$, with k even), then the device switching frequency can be computed as $f_{sw} = (1 + k/3)\, f_1$. For the example illustrated in Figure 14.38a, $k = 18$, then $f_{sw} = 7f_1$. Considering a 50 Hz fundamental frequency in this example, a device switching frequency of 350 Hz would be obtained, which is very low and useful for high-power applications.

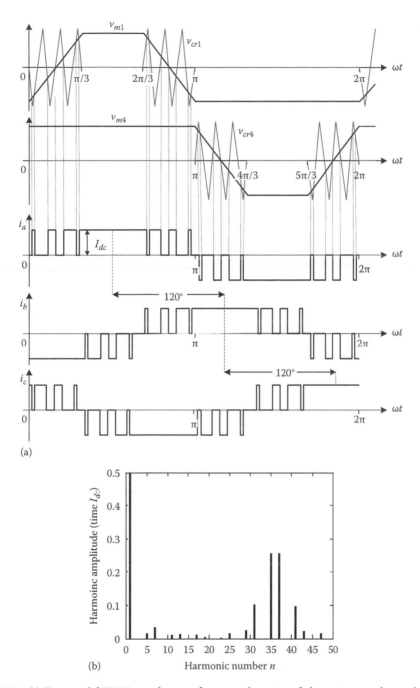

(a)

(b)

FIGURE 14.38 (a) Trapezoidal PWM waveforms: reference and carriers of phase a inverter leg, and three-phase output currents. (b) Output current spectrum for modulation index 0.85 and carrier $f_{cr} = 36f_1$.

Another difference with traditional carrier-based PWM is that there is no linearity between the modulation index and the amplitude of the fundamental component i_1. This is due the fact that the central segment of $\pi/3$ is not modulated, being always in "on" state, thereby forcing a high amplitude for the fundamental frequency component. In fact, when changing the modulation index through all its range (from 0 to 1), the peak amplitude for the current fundamental component varies only within $0.9314I_{dc} \leq \hat{i}_1 \leq 1.0465I_{dc}$. This is no problem, since the current amplitude is controlled by the CSR.

The advantage of T-PWM compared to square-wave modulation is that the additional commutations reduce significantly the output current low-order harmonics. Figure 14.38b shows the current spectrum for a T-PWM waveform obtained for a 0.85 modulation index and a carrier frequency 36 times the fundamental frequency. Note that most of the harmonic energy content appears as sidebands around the carrier frequency in $n = (f_{cr}/f_1) \pm 1$ and $n = (f_{cr}/f_1) \pm 5$. Compared to the square-wave modulation, 5th and 7th harmonics have been reduced around 90% and 70%, respectively. They do not disappear completely compared to traditional carrier-based PWM used in VSIs, since the trapezoidal output current low-order harmonics are not completely filtered. In practice, carrier frequencies should be $f_{cr} \geq 18f_1$ in order to have admissible 5th and 7th harmonics. A more detailed harmonic analysis of this method is presented in [1].

14.4.3.3 Selective Harmonic Elimination for CSI

SHE seen in the previous section for VSIs can also be applied to CSIs, and has no conceptual differences apart from incorporating the switching constraints into the SHE waveform. These can be easily achieved by fixing and make dependent some angles to provide certain symmetry in the switching pattern and in this way avoid overcrossing of "on" or "off" states of two phases, or the overcrossing from a zero current level in the three phases. Figure 14.39a shows the three-phase line currents generated by SHE considering five switching angles, from which one is fixed in $\pi/6$ and two are dependent on the only two variable angles θ_1 and θ_2. The dependent angles are set to $\pi/3 - \theta_1$ and $\pi/3 - \theta_2$ to introduce the symmetry in the waveform. Any additional angle θ_k, has to be placed between 0 and $\pi/6$, and directly defines a dependent angle in $\pi/3 - \theta_k$.

Since not all the angles can be defined at will, a reduction of the degrees of freedom of the SHE pattern is introduced compared to the VSI case, hence less harmonics can be eliminated. For example, the waveform illustrated in Figure 14.39a has five switching angles in the first quarter cycle, but only two of them are controllable, compared to the five independent angles in SHE for VSIs. In this case, only one harmonic can be eliminated while controlling the fundamental component, or two harmonics can be eliminated without fundamental component control. Unlike with SHE in VSIs, where one angle is reserved for the fundamental component control, in CSI the angles are fully devoted to harmonic elimination, since the CSR can be used externally to control the current amplitude.

The Fourier series for the switched current waveform illustrated in Figure 14.39a is given by

$$i_a(t) = \sum_{n=1}^{\infty} b_n \sin(n\omega t) \tag{14.39}$$

$$b_n = \frac{4}{\pi} \int_0^{\pi/2} i_a(\omega t) \sin(n\omega t) d\omega t, \tag{14.40}$$

where n is the harmonic number ($n = 1, 3, 5, \ldots$). For the two-angle example given in Figure 14.39a, solving (14.40) considering I_{dc} constant leads to

$$b_n = \frac{4I_{dc}}{\pi n} \left\{ \cos(n\theta_1) + \cos\left[n\left(\frac{\pi}{3} - \theta_1 \right) \right] - \cos(n\theta_2) - \cos\left[n\left(\frac{\pi}{3} - \theta_2 \right) \right] + \cos\left(n\frac{\pi}{6} \right) \right\} \tag{14.41}$$

If the 5th and 7th harmonic needs to be eliminated, (14.41) turns into the following set of equations:

$$b_5 = 0 = \left\{ \cos(5\theta_1) + \cos\left[5\left(\frac{\pi}{3} \right) - \theta_1 \right] - \cos(5\theta_1) - \cos\left[5\left(\frac{\pi}{3} - \theta_2 \right) \right] + \cos\left(5\frac{\pi}{6} \right) \right\}$$

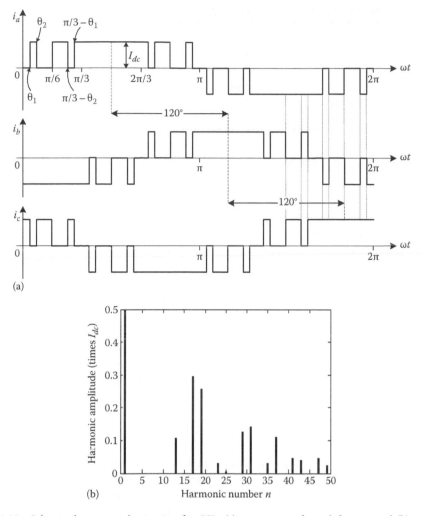

FIGURE 14.39 Selective harmonic elimination for CSIs: (a) current waveform definition and (b) current spectrum for a three independent angles SHE, harmonics eliminated up to the 11th.

$$b_7 = 0 = \left\{ \cos(7\theta_1) + \cos\left[7\left(\frac{\pi}{3} - \theta_1\right)\right] - \cos(7\theta_2) - \cos\left[7\frac{\pi}{3} - \theta_2\right] + \cos\left(7\frac{\pi}{6}\right) \right\}$$

This set of equations cannot be solved online, being this the main disadvantage of SHE. Hence, all the switching patterns have to be pre-calculated off-line and stored in lookup tables. Many types of algorithms are used to solve these equations, mainly based on iterative numerical techniques, such as genetic algorithms. The advantage for CSIs is that this algorithm does not have to be performed several times for different modulation indexes, since the amplitude is controlled by the CSR, reducing considerably design time. Also for this reason, no lookup table interpolations need to be performed for those current amplitude values which are not pre-calculated, as it is necessary for SHE in VSIs. Solutions for the SHE problem from one up to four angles are given in Table 14.8, considering the elimination from the 5th up to the 13th harmonics.

A detailed table, with more solutions of the SHE problem and different combinations of eliminated harmonics can be found in [1]. The output current spectrum for a SHE waveform with three independent switching angles is shown in Figure 14.39b, which corresponds to the third solution given in Table 14.8.

TABLE 14.8   Selective Harmonic Elimination Switching Angles for CSI

Eliminated Harmonics	Switching Angles			
	θ_1	θ_2	θ_3	θ_4
5th	18.0°	—	—	—
5th, 7th	7.93°	13.75°	—	—
5th, 7th, 11th	2.24°	5.600°	21.26°	—
5th, 7th, 11th, 13th	0.00°	1.600°	15.14°	20.26°

14.4.3.4 Space Vector Modulation for CSI

As with previous modulations schemes, SVM used in VSIs can also be extended for CSIs, providing that the switching constraints mentioned earlier are satisfied. The SVM algorithm is basically also a PWM strategy with the difference that the switching times are computed based on the three-phase space vector in time representation of the reference and the CSI switching states, rather than the per-phase amplitude in time representation of previous analyzed methods.

The current space vector can be defined in the α–β complex plane by

$$I_s = \frac{2}{3}\left[i_a + ai_b + a^2 i_c\right], \tag{14.42}$$

where $a = -1/2 + j\sqrt{3}/2$. As seen previously, the line of a CSI can be currents that have three different values $-I_{dc}$, 0, and I_{dc}, depending on the switching states. According to the definition of the gating signals, $S_1, S_2, \ldots, S_6$ in Figure 14.35 and the switching states in Table 14.7, the following relations can be obtained for each phase current:

$$i_a = (S_1 - S_4)I_{dc}; \quad i_b = (S_3 - S_6)I_{dc}; \quad i_c = (S_5 - S_2)I_{dc} \tag{14.43}$$

For example, when S_1 is 1 and S_4 is 0, the dc current is passed through the upper switch to the load, generating a positive phase current $i_a = I_{dc}$. On the contrary, if S_1 is 0 and S_4 is 1, the lower switch is conducting the negative current from the load back to the dc source, and the phase current is $i_a = -I_{dc}$. Finally, if S_1 and S_4 are both 1(0), phase a is bypassed (opened), resulting in $i_a = 0$. Note that the four binary combinations in the previous example for phase a are consistent with the first term in (14.43). By replacing (14.43) into (14.42), the current space vector is then defined by the gating signals $S_1, S_2, \ldots, S_6$, with the expression

$$I_s = \frac{2}{3}I_{dc}\left[(S_1 - S_4) + a(S_3 - S_6) + a^2(S_5 - S_2)\right] \tag{14.44}$$

Replacing all the admissible switching states in (14.44), leads to the nine space vectors listed in Table 14.7. Note that seven are different, as vectors I_{0a}, I_{0b}, and I_{0c} are the same zero vector I_0, also called bypass vectors, since the DC-link current freewheels via one leg of the inverter bridge without interacting with the load (the lowercase letter indicates which phase leg of the inverter is being bypassed). These vectors can be plotted in the α–β complex plane, resulting in the CSI current space vector representation illustrated in Figure 14.40a. Note that all the vectors (excluding the zero vectors) have the same amplitude $|I_k| = I_{dc}2/\sqrt{3}$ (with $k = 1, \ldots, 6$) and are rotated in $\pi/3$ respect each other, and are usually called active vectors. Each adjacent pair of active vectors define an area in the α–β plane, dividing it in six sectors. The current reference space vector I_s^* can be also computed by (14.42), and the resulting vector can be mapped in the α–β plane, falling in one of the sectors. For balanced three-phase sinusoidal references, as is usual in power converter systems, the resulting reference vector is a fixed amplitude rotating space

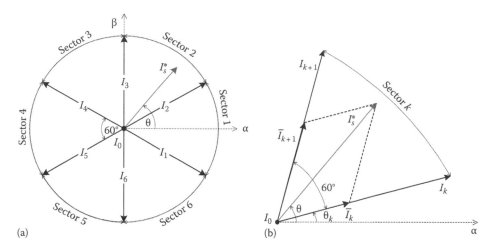

FIGURE 14.40 (a) Current space vectors generated by a CSI and (b) SVM operating principle for generic sector k.

vector with the same amplitude and angular speed (ω) of the sinusoidal references, with an instantaneous position with respect to the real axis α given by $\theta = \omega t$.

As with SVM used in VSIs, the main idea behind the working principle is to generate over a modulation period T_s, a time average equal to the regularly sampled reference vector (amplitude and angular position). Hence, the problem is reduced to finding the duty cycles (on and off times) of the zero vector and the two vectors that define the sector in which the reference is crossing. Consider the generic case of sector k in Figure 14.40b; then the time average over a modulation period can be defined by

$$I_s^* = \frac{1}{T_s}(t_k I_k + t_{k+1} I_{k+1} + t_0 I_0) \tag{14.45}$$

$$T_s = t_k + t_{k+1} + t_0, \tag{14.46}$$

where t_k/T_s, t_{k+1}/T_s, and t_0/T_s are the duty cycles of the respective vectors. Using trigonometric relations, it can be easily found that

$$|\overline{I}_k| = \frac{t_k}{T_s}|I_k| = |I_s^*| \left\{ \cos(\theta - \theta_k) - \frac{\sin(\theta - \theta_k)}{\sqrt{3}} \right\}, \tag{14.47}$$

$$|\overline{I}_{k+1}| = \frac{t_{k+1}}{T_s}|I_{k+1}| = 2|I_s^*| \frac{\sin(\theta - \theta_k)}{\sqrt{3}}, \tag{14.48}$$

where θ_k is the angle between the α axis and the current space vector k. Since all the space vectors have the same amplitude $|I_k| = |I_{k+1}| = I_{dc} 2/\sqrt{3}$, they can be replaced in (14.47) and (14.48). Then the only unknown variables left in (14.47) and (14.48) are t_k and t_{k+1}. Thus, the following set of equations to solve the duty cycles can be obtained

$$t_k = \frac{T_s|I_s^*|}{2I_{dc}} \left\{ \sqrt{3}\cos(\theta - \theta_k) - \sin(\theta - \theta_k) \right\} \tag{14.49}$$

$$t_{k+1} = \frac{T_s \, | I_s^* |}{I_{dc}} \sin(\theta - \theta_k) \tag{14.50}$$

$$t_0 = T_s - t_k - t_{k+1} \tag{14.51}$$

Note that (14.51) is simply obtained from (14.46) once the two nonzero vector duty cycle times have been computed, in order to complete the modulation period T_s.

This generic sector solution described earlier can be easily applied to any sector replacing the numeric k index ($k = 1,...,6$). The final stage in the SVM algorithm is to generate an appropriate switching sequence between the modulating vectors and their duty cycles. Since the switching constraints are inherently satisfied because only admissible switching states are considered, the switching sequence can be used to improve other aspects of the modulation. Particularly, efficiency can be taken into account trying to decrease the number of commutations, thus reducing switching losses (useful in CSIs due to their application field in high-power drives). A popular switching sequence that only requires three device turn "on" and turn "off" per modulation period is first I_k, then I_{k+1}, and finally I_0. The reduction of the commutations is achieved by the proper use of the different bypass vectors available as redundancies of I_0 (I_{0a}, I_{0b}, and I_{0c}), depending on the sector in which the VSI is operating, as given in Table 14.9. Considering this switching sequence, the device average switching frequency is given by $f_{sw} = 1/2T_s$. Reducing the modulation period, T_s is analogous to using higher carrier frequencies in traditional carrier-based PWM, hence it will produce higher switching losses. Furthermore, it will depend on the computational power of the digital platform used to implement the algorithm.

One of the advantages of SVM for CSIs is that the fundamental component of the switched output current is controlled directly by the inverter instead by the CSR. This is due to the fact that the current waveform is always defined, since only admissible switching states are considered for the space vector representation, hence for modulation. In addition, the amplitude of the fundamental component is included in the duty cycle calculation, and therefore is directly controlled. This results in superior dynamic performance in comparison to square-wave modulation, SHE and T-PWM, where the fundamental component is tracked slowly due to the large dc chokes used as dc link, resulting in a large time constant for the control of I_{dc} by the CSR. Instead, I_{dc} is kept fixed for SVM.

Figure 14.41a shows a typical output current waveform obtained with SVM for a sinusoidal reference with amplitude I_{dc} and frequency f_1, using a sample period $T_s = 1/18f_1$, resulting in an average switching frequency of $f_{sw} = 9f_1$. The corresponding spectrum is illustrated in Figure 14.41b. A more detailed harmonic analysis of this method is presented in [1].

Note that SVM in CSIs does not eliminate completely the low-order harmonics, compared to the SVM applied in VSIs. In fact SHE and T-PWM provide better power quality, being SHE the best of all in this aspect. On the other hand, SVM has the best dynamic performance of all the methods discussed here for CSIs. T-PWM lies somewhere in between SHE and SVM, combining partially the favorable characteristics of these methods. Square-wave modulation has the worst power quality and poor dynamic behavior among the modulation methods, nevertheless, it is the most efficient and easiest for

TABLE 14.9 SVM Vector Sequence for PWM-CSI Depending on the Sector

Vector Sequence	Sector					
	1	2	3	4	5	6
I_k	I_1	I_2	I_3	I_4	I_5	I_6
I_{k+1}	I_2	I_3	I_4	I_5	I_6	I_1
0	I_{0a}	I_{0c}	I_{0b}	I_{0a}	I_{0c}	I_{0b}

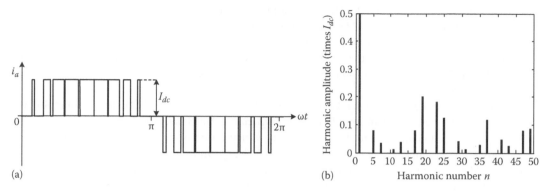

FIGURE 14.41 SVM for CSIs: (a) line output current waveform and (b) line output current spectrum.

implementation. The choice between one particular modulation method over the other will depend on the application and its specific requirements.

Currently, most CSI-driven applications are aimed at high-power motor drives, where efficiency is an important issue. Therefore, a combination of T-PWM and SHE is preferred. The first is used for fundamental frequencies up to 30 Hz, and the later is used from that frequency and above. In addition, the carrier frequencies are also varied in T-PWM proportionally to the increase or decrease of the fundamental frequency. In the same way, the number of angles used in SHE is modified proportionally to the variations in f_1 to keep the switching frequencies low. In this way, the device average switching frequency is kept under 500 Hz in typical CSI high-power applications.

References

1. B. Wu, *High-Power Converters and AC Drives*, Wiley-IEEE Press, Piscataway, NJ, 2006.
2. J. Rodriguez, J. S. Lai, and F. Z. Peng, Multilevel inverters: A survey of topologies, controls, and applications, *IEEE Transactions on Industrial Electronics*, 49(4), 724–738, August 2002.
3. N. Mohan, T. M. Undeland, and W. P. Robbins, *Power Electronics: Converters, Applications, and Design*, 3 edn, Wiley, Hoboken, NJ, October 10, 2002.
4. J. Rodriguez, S. Bernet, B. Wu, J. O. Pontt, and S. Kouro, Multilevel voltage-source-converter topologies for industrial medium-voltage drives, *IEEE Transactions on Industrial Electronics*, 54(6), 2930–2945, December 2007.
5. J. Holtz, Pulsewidth modulation for electronic power conversion, *Proc. IEEE*, 82(8), 1194–1214, Aug. 1994.
6. D. Busse, J. Erdman, R. Kerkman, D. Schlegel, and G. Skibinski, Bearing currents and their relationship to PWM drive, *IEEE Transactions on Power Electronics*, 12, 243–252, March 1997.
7. B. K. Bose, *Modern Power Electronics and AC Drives*, Prentice Hall PTR, Upper Saddle River, NJ, October 22, 2001.
8. D. G. Holmes and T. A. Lipo, *Pulse Width Modulation for Power Converters: Principles and Practice*, 1st edn., Wiley-IEEE Press, Piscataway, NJ, October 3, 2003.
9. L. Asiminoaei, P. Rodríguez, and F. Blaabjerg, Application of discontinuous PWM modulation in active power filters, *IEEE Transactions on Power Electronics*, 23(4), 1692–1706, July 2008.
10. S. Kouro, M. A. Perez, H. Robles, and J. Rodríguez, Switching loss analysis of modulation methods used in cascaded H-bridge multilevel converters, in *39th IEEE Power Electronics Specialists Conference (PESC08)*, Rhodes, Greece, June 15–19, 2008, pp. 4662–4668.
11. J. R. Espinoza, G. Joós, J. I. Guzmán, L. A. Morán, and R. P. Burgos, Selective harmonic elimination and current/voltage control in current/voltage-source topologies: A unified approach, *IEEE Transactions on Industrial Electronics*, 48(1), 71–81, February 2001.

12. B. Ozpineci, L. Tolbert, and J. Chiasson, Harmonic optimization of multilevel converters using genetic algorithms, *IEEE Power Electronics Letters*, 3(3), 92–95, September 2005.

13. S. Kouro, B. La Rocca, P. Cortes, S. Alepuz, B. Wu, and J. Rodriguez, Predictive control based selective harmonic elimination with low switching frequency for multilevel converters, in *Proceedings of the 2009 IEEE Energy Conversion Congress and Exposition (ECCE 2009)*, San Jose, CA, September 20–24, 2009.

14. L. G. Franquelo, J. Rodriguez, J. I. Leon, S. Kouro, R. Portillo, and M. M. Prats, The age of multilevel converters arrives, *IEEE Industrial Electronics Magazine*, 2(2), 28–39, June 2008.

15. A. Nabae, I. Takahashi, and H. Akagi, A new neutral-point-clamped PWM inverter, *IEEE Transactions on Industry Applications*, 17(5), 518–523, September 1981.

16. T. A. Meynard and H. Foch, Multi-level conversion: High voltage choppers and voltage-source inverters, in *23rd Annual IEEE Power Electronics Specialists Conference, 1992 (PESC '92)*, Vol. 1, Toledo, Spain, June-29–July 3, 1992, pp. 397–403.

17. M. Marchesoni, M. Mazzucchelli, and S. Tenconi, A non conventional power converter for plasma stabilization, in *19th Annual IEEE Power Electronics Specialists Conference, 1988 (PESC '88)*, Vol. 1, Kyoto, Japan, April 11–14, 1988, pp. 122–129.

18. C. Rech and J. R. Pinheiro, Hybrid multilevel converters: Unified analysis and design considerations, *IEEE Transactions on Industrial Electronics*, 54(2), 1092–1104, April 2007.

19. A. M. Massoud, S. J. Finney, and B. W. Williams, Systematic analytical based generalised algorithm for multilevel space vector modulation with a fixed execution time, *IET Power Electronics*, 1(2), 175–193, June 2008.

20. N. Celanovic and D. Boroyevich, A fast space-vector modulation algorithm for multilevel three-phase converters, *IEEE Transactions on Industrial Applications*, 37(2), 637–641, March/April 2001.

21. M. M. Prats, L. G. Franquelo, R. Portillo, J. I. Leon, E. Galvan, and J. M. Carrasco, A 3-D space vector modulation generalized algorithm for multilevel converters, *IEEE Power Electronics Letters*, 1(4), 110–114, December 2003.

22. O. Lopez, J. Alvarez, J. Doval-Gandoy, and F. D. Freijedo, Multilevel multiphase space vector PWM algorithm, *IEEE Transactions Industrial Electronics*, 55(5), 1933–1942, May 2008.

23. J. Rodriguez, L. Moran, P. Correa, and C. Silva, A vector control technique for medium-voltage multilevel inverters, *IEEE Transactions on Industrial Electronics*, 49(4), 882–888, August 2002.

24. M. D. Manjrekar, P. K. Steimer, and T. A. Lipo, Hybrid multilevel power conversion system: A competitive solution for high-power applications, *IEEE Transactions on Industry Applications*, 36(3), 834–841, May 2000.

25. Z. Du, L. M. Tolbert, and J. N. Chiasson, Active harmonic elimination for multilevel converters, *IEEE Transactions Power Electronics*, 21(2), 459–469, March 2006.

26. L. G. Franquelo, J. Napoles, R. Portillo, J. I. Leon, and M. A. Aguirre, A flexible selective harmonic mitigation technique to meet grid codes in three-level PWM converters, *IEEE Transactions on Industrial Electronics*, 54(6), 3022–3029, December 2007.

27. M. Perez, J. Rodriguez, J. Pontt, and S. Kouro, Power distribution in hybrid multi cell converter with nearest level modulation, in *IEEE International Symposium on Industrial Electronics (ISIE 2007)*, Vigo, Spain, June 4–7, 2007, pp. 736–741.

28. L. G. Franquelo, J. I. Leon, and E. Dominguez, New trends and topologies for high power industrial applications: The multilevel converters solution, *IEEE International Conference on Power Engineering, Energy and Electrical Drives, 2009 (POWERENG '09)*, Lisbon, Portugal, March 18–20, 2009.

29. B. Wu, J. Pontt, J. Rodríguez, S. Bernet, and S. Kouro. Current-source converter and cycloconverter topologies for industrial medium-voltage drives, *IEEE Transactions on Industrial Electronics*, 55(7), 2786–2797, July 2008.

30. M. Salo and H. Tuusa, A vector-controlled PWM current-source-inverter fed induction motor drive with a new stator current control method, *IEEE Transactions on Industrial Electronics*, 52(2), 523–531, 2005.

31. P. Cancelliere, V. D. Colli, R. Di Stefano, and F. Marignetti, Modeling and control of a zero-current-switching DC/AC current-source inverter, *IEEE Transactions on Industrial Electronics*, 54(4), 2106–2119, August 2007.

32. B. Wu, S. Dewan, and G. Slemon, PWM-CSI inverter induction motor drives, *IEEE Transactions on Industry Applications*, 28(1), 64–71, 1992.

33. N. R. Zargari, S. C. Rizzo, Y. Xiao, H. Iwamoto, K. Satoh, and J. F. Donlon, A new current-source converter using a symmetric gate-commutated thyristor (SGCT), *IEEE Transactions on Industry Applications*, 37(3), 896–903, 2001.

34. P. Espelage, J. M. Nowak, and L. H. Walker, Symmetrical GTO current source inverter for wide speed range control of 2300 to 4160 Volts, 350 to 7000HP induction motors, *IEEE Industry Applications Society Conference (IAS)*, Pittsburgh, PA, 1988, pp. 302–307.

35. S. Rees, New cascaded control system for current-source rectifiers, *IEEE Transactions on Industrial Electronics*, 52(3), 774–784, 2005.

36. J. Wiseman, B. Wu, and G. S. P. Castle, A PWM current source rectifier with active damping for high power medium voltage applications, *IEEE Power Electronics Specialist Conference (PESC)*, Cairns, Australia, 2002, pp. 1930–1934.

37. J. Ma, B. Wu, and S. Rizzo, A space vector modulated CSI-based ac drive for multimotor applications, *IEEE Transactions on Power Electronics*, 16(4), 535–544, 2001.

38. J. Rodriguez, L. G. Franquelo, S. Kouro, J. I. Leon, R. Portillo, and M. M. Prats, Multilevel converters: An enabling technology for high power applications, *Proceedings of the IEEE*, 97(11), 1786–1817, November 2009.

39. E. Levi, Multiphase electric machines for variable-speed applications, *IEEE Transactions on Industrial Electronics*, 55(5), 1893–909, May 2008.

40. T. Bruckner, S. Bernet, and H. Guldner, The active NPC converter and its loss-balancing control, *IEEE Transactions on Industrial Electronics*, 52(3), 855–868, June 2005.

41. P. Steimer and M. Veenstra, Converter with additional voltage addition or substraction at the output, U.S. Patent No. 6,621,719 B2, Filed 17 April 2002 Granted 16 September 2003.

42. R. Marquardt, Stromrichterschaltungen mit verteilten energiespeichern, German Patent No. DE10103031A1, Filed 25 July 2002 Issued 24 January 2001.

43. S. Kouro, M. Malinowski, K. Gopakamar, J. Pou, L. G. Franquelo, B. Wu, J. Rodriguez, M. A. Pérez, and J. I. Leon, Recent advances and industrial applications of multilevel converter, *IEEE Transactions on Industrial Electronics*, 57(8), 2553–2580, 2010.

15

AC/AC Converters

Patrick Wheeler
University of Nottingham

15.1 Matrix Converters

15.1.1 Introduction

The matrix (or direct) converter offers "a more silicon solution" for AC/AC power conversion. The topology consists of an array of bidirectional switches arranged so that any of the output lines of the converter can be connected to any of the input lines. Figure 15.1 shows a typical three-phase to three-phase matrix converter, with nine bidirectional switches. The switches allow any input phase to be connected to any output phase. The output waveform is then created using a suitable pulse width modulation (PWM) pattern similar to a normal inverter, except that the input is a three-phase supply instead of a fixed DC voltage.

When the matrix converter topology was first published in 1976, it was termed a Forced Commutated Cycloconverter [1]. At this time, no fully controllable power semiconductor devices were available, so the early prototype circuits relied on forced-commutated thyristors. When BJTs became available, the matrix converter began to be considered as a viable alternative to a diode bridge/inverter arrangement [2,3]. However, issues regarding device count and current commutation relegated the matrix converter to a decade as an academic curiosity. More recently, the reducing cost of semiconductors and the resolution of the practical problems means that the topology has become a contender in some applications.

The matrix converter has many advantages over traditional topologies. The topology is inherently bidirectional, so can regenerate energy back to the supply. The converter draws sinusoidal input currents and, depending on the modulation technique, it can be arranged that unity displacement factor is seen at the supply side irrespective of the type of load [4]. The size of the power circuit has the potential to be greatly reduced in comparison to conventional technologies since there are no large capacitors or inductors to store energy.

In terms of device count, a comparison can be made between the matrix converter and a back-to-back inverter, which has the same functional characteristics of bidirectional power flow and sinusoidal input currents. It can be shown that the DC link capacitor and input inductors associated with the back-to-back inverter circuit are replaced with the extra six switching devices and a small high-frequency filter in the matrix converter solution. It can also be shown that the device losses from a matrix converter are similar to those in the equivalent diode bridge/back-to-back inverter circuit.

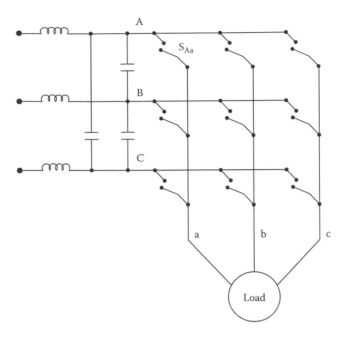

FIGURE 15.1   The matrix converter circuit.

One often-cited problem with the matrix converter is the fundamental maximum voltage transfer ratio of 86%. This limitation results from the fact that there is no energy storage in the circuit and hence the target output voltage waveforms must fall within the envelope of the input voltages. Any attempt to exceed this limit will result in unwanted low-frequency components in the input and output waveforms [5]. This limitation is only a problem if the design engineer does not have control of the design of the load; for example, in an application where the matrix converter is used as a motor drive, a machine can simply be designed to work with the slightly reduced output voltage.

15.2 Matrix Converter Concepts

The output waveforms of the matrix converter are formed by selecting each of the input phases in sequence for defined periods of time. Typical output voltage and input current waveforms for a very low switching frequency are shown in Figure 15.2. The output voltage consists of segments made up from the three input voltages instead of the two fixed DC levels found in an inverter. For this reason, the harmonic spectrum of the output voltage is slightly richer in harmonics around the switching frequency.

The input current consists of segments of the three output currents plus blank periods during which the output current freewheels through the switch matrix. This waveform can then be filtered with a small input filter to provide a good-quality input current. For this input filter to be small, and for the advantages of the matrix converter to be fully realized, it is necessary to have a relatively high switching frequency, usually at least 15 times greater than the highest input or output frequency.

15.2.1 Power Circuit Implementation

In order to construct any form of matrix converter power circuit, a bidirectional switch element is required. The bidirectional switch must be able to conduct current in both directions when turned on and block voltage in both directions when turned off. A suitable bidirectional switch function is not currently available as a single semiconductor device; therefore, bidirectional switches have to be built from discrete devices and packaged in a suitable form for the matrix converter topology. In common

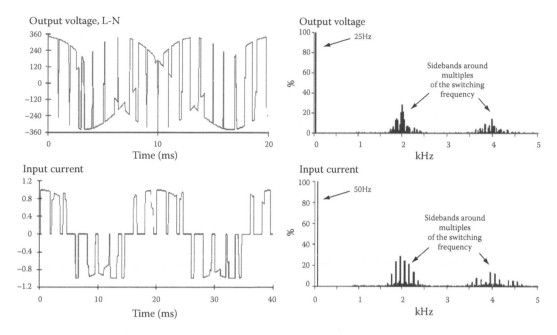

FIGURE 15.2 Examples of typical matrix converter waveforms and spectra.

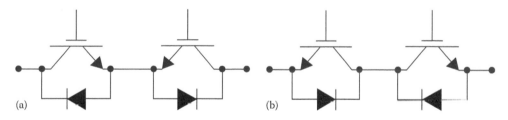

FIGURE 15.3 (a) Common emitter configuration. (b) Common collector configuration. Bidirectional switch cells using anti-parallel IGBTs and diodes.

with other converter topologies, the IGBT is usually the semiconductor device of choice for the matrix converter construction.

There are a number of options for the construction of these bidirectional switch cells. The bidirectional switch can be realized using a pair of anti-parallel IGBTs with series diodes, as shown in Figure 15.3. As shown, the IGBTs can be arranged in either common collector or common emitter configurations.

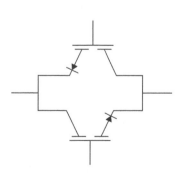

FIGURE 15.4 Bidirectional switch cell using reverse blocking IGBTs.

These configurations can be selected in order to minimize the number of gate drive power supplies required, depending on the choice of packaging option [5]. Both of these options allow the direction of the possible current flow in the switch to be controlled, a useful facility in for solving the current commutation problem.

Bidirectional switch cells can also be formed using reverse blocking IGBTs, as shown in Figure 15.4. This configuration reduces the number of semiconductor devices in the conduction path, but the reduced switching performance caused by the trade-off in the reverse recovery characteristics leads to higher switching devices with currently available devices.

To facilitate the efficient construction of a matrix converter power circuit, it is necessary to package the bidirectional switch cells in a

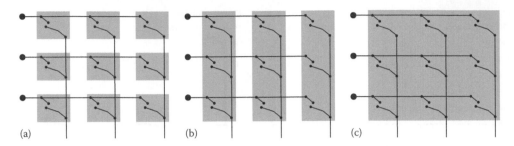

FIGURE 15.5 Module arrangements for matrix converter bidirectional switch cells. (a) One module per switch, (b) one module per output leg, and (c) one module per converter.

suitable form. There are three basic options for this packaging. These options, shown in Figure 15.5, are to arrange the devices so that each module contains

- One bidirectional switch cell (high power, >200 A)
- The switches for one converter output leg (medium power, 50–600 A)
- All the switches required for a complete converter (low power, <100 A)

The choice between these options can be made based on the power level of the matrix converter under consideration and the packaging technologies to be deployed. All these options are already available from device manufacturers either as standard modules or special orders.

To enable the current path to be past between the bidirectional switches, a current commutation strategy must be deployed. This strategy is needed as there are no natural freewheeling paths available in the matrix converter power circuit, unlike in an inverter power circuit. To achieve the safe commutation of current between the switches when a state change is requested by the converter modulation, a safe sequence of switching of the devices can be used. To ensure that the sequence used is safe, most current commutation methods use either the relative magnitude of the input voltages or the direction of the output current to determine the safe device states and to derive the required switching sequence. These safe commutation sequences ensure that there is never a short circuit across the input lines or open circuit on the converter output lines during the commutation, since either of these conditions could lead to the destruction of the converter through over current or voltage, respectively.

The commutation of the current in a matrix converter is considered for each output leg independently as there is no connection between the current commutation paths. For the purpose of modulation, we are interested in passing the current path between two bidirectional switches, as shown in Figure 15.6a. Considering two bidirectional switches, it is possible to find all the possible states of the four devices, given in the format of Equation 15.1, depending on either the output current direction of the relative magnitude or the two input voltages, as shown in Figure 15.6b. It is then possible to draw a plan of all the device states and to link these states where the state of only one device is changed to form a commutation map, as shown in Figure 15.7. This map represents all the possible current commutation strategies that rely on output current direction or relative input voltage magnitude.

Two common current commutation strategies for matrix converters are the output current direction and the relative input voltage-based commutation techniques. The output current direction-based technique, shown in Figure 15.8a, relies on knowledge of the output current direction in order to determine a safe sequence of operation for the devices. When a change in switch state is requested by the converter modulation, the device in the outgoing switch, which is not in the current conduction path is turned off, α_1 in Figure 15.8a. The device in the incoming switch, which will carry the current, can then be turned on, β_2. No short circuit between the input lines has been created because both devices can only conduct in the same direction and a path for the load current is maintained. The second device in the outgoing switch, α_2, can then be turned off followed by the turning on of the second device in the incoming switch, β_1, to complete the sequence.

$$\begin{bmatrix} \alpha_1 & \alpha_2 \\ \beta_1 & \beta_2 \end{bmatrix}$$

where

$\alpha_x, \beta_y = 1$ denotes a device which is on

$\alpha_x, \beta_y = 0$ denotes a device which is off

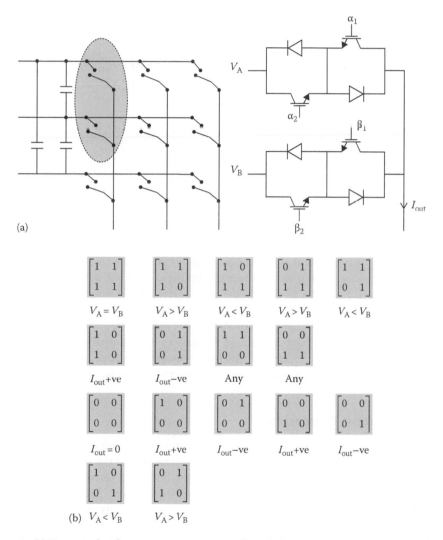

(b) $V_A < V_B$ $V_A > V_B$

FIGURE 15.6 (a) Two switches for current commutation. (b) Safe device states with conditions. Current commutation for two bidirectional switches.

For the relative input voltage-based technique, the principle is that during the commutation, all the devices not required for blocking an input short circuit can be turned on. If it is assumed that $V_A > V_B$ in Figure 15.8b, then the first step in the sequence is to turn on device β_2 in the incoming switch without causing a short circuit of the input lines. It is then possible to turn off device α_2 as there will still be a current path in both directions. Once α_2 has been turned off, device β_1 can be turned on and finally the sequence completed by turning off device α_1. This complete sequence is shown in Figure 15.8b.

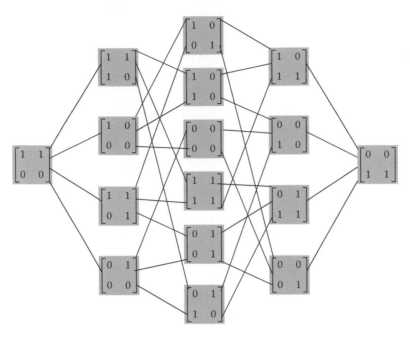

FIGURE 15.7   Current commutation paths for two bidirectional switches.

One disadvantage of this technique is that the commutation sequence takes longer than for the output current-based technique. Also, if an error is made, the input lines will suffer a short circuit rather than the output line open circuit for an error in the input current-based sequence. It is possible to protect against an output line open circuit, but not an input line short circuit. The advantage of the input voltage-based technique is that the input voltage magnitudes must be measured for the modulation of the converter, whereas the output current direction will have to be measured using a dedicated circuit. Both these basic techniques can simply be extended to a converter with three input phases and form the fundamental principles used in the majority of practical matrix converters.

15.2.2 Power Circuit Protection

In order to protect the matrix converter power circuit from overvoltages, a diode clamp circuit is used, as shown in Figure 15.9. These overvoltages can appear when all the switches in the converter are instantaneously turned off and the current in the load is suddenly interrupted, for example, under an output overload condition. The energy stored in the motor inductance has to be safely discharged, and a small clamp circuit can perform this function. The clamp circuit typically uses 12 fast diodes, consisting of two diode bridges connected to the input and output terminals, and a small DC link capacitor. The capacitor is sized to ensure that the maximum energy stored in the inductance of the load does not cause a capacitor voltage above the rating of the semiconductor devices in the matrix converter power circuit.

15.2.3 Modulation Algorithms

There are a number of possible modulation techniques for matrix converters, and this has been a very popular topic for researchers. In this chapter, the two most popular techniques are considered, the Venturini Modulation strategy [4] and Space Vector Modulation [6].

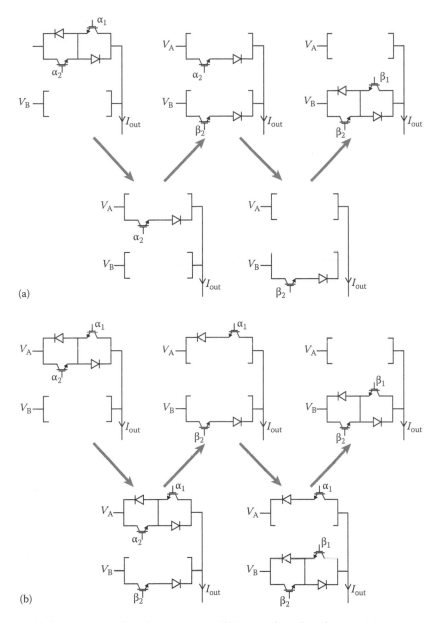

FIGURE 15.8   (a) Output current-based commutation. (b) Input voltage-based commutation.

To examine the basic modulation problem for the matrix converter, a set of input voltages and output currents can be assumed:

$$
\mathbf{v_i} = V_{im}
\begin{bmatrix}
\cos(\omega_i t) \\
\cos\left(\omega_i t + \dfrac{2\pi}{3}\right) \\
\cos\left(\omega_i t + \dfrac{4\pi}{3}\right)
\end{bmatrix},
\quad
\mathbf{i_o} = I_{om}
\begin{bmatrix}
\cos(\omega_o t + \phi_o) \\
\cos\left(\omega_o t + \phi_o + \dfrac{2\pi}{3}\right) \\
\cos\left(\omega_o t + \phi_o + \dfrac{4\pi}{3}\right)
\end{bmatrix}
\tag{15.1}
$$

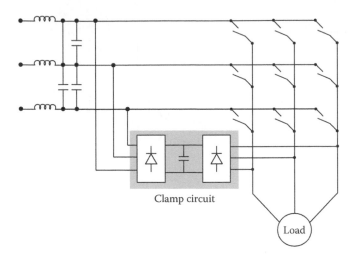

FIGURE 15.9   The matrix converter clamp circuit.

A modulation matrix, $M(t)$, can then be found such that

$$\mathbf{v_o} = qV_{im} \begin{bmatrix} \cos(\omega_o t) \\ \cos\left(\omega_o t + \dfrac{2\pi}{3}\right) \\ \cos\left(\omega_o t + \dfrac{4\pi}{3}\right) \end{bmatrix}, \quad \mathbf{i_i} = q\cos(\phi_o)I_{om} \begin{bmatrix} \cos(\omega_i t + \phi_i) \\ \cos\left(\omega_i t + \phi_i + \dfrac{2\pi}{3}\right) \\ \cos\left(\omega_i t + \phi_i + \dfrac{4\pi}{3}\right) \end{bmatrix}, \quad (15.2)$$

where q is the voltage gain between the output and input voltages.

There are two basic solutions to this problem:

$$\mathbf{M1} = \frac{1}{3} \begin{bmatrix} 1+2q\cos(\omega_m t) & 1+2q\cos\left(\omega_m t - \dfrac{2\pi}{3}\right) & 1+2q\cos\left(\omega_m t - \dfrac{4\pi}{3}\right) \\ 1+2q\cos\left(\omega_m t - \dfrac{4\pi}{3}\right) & 1+2q\cos(\omega_m t) & 1+2q\cos\left(\omega_m t - \dfrac{2\pi}{3}\right) \\ 1+2q\cos\left(\omega_m t - \dfrac{2\pi}{3}\right) & 1+2q\cos\left(\omega_m t - \dfrac{4\pi}{3}\right) & 1+2q\cos(\omega_m t) \end{bmatrix} \quad (15.3)$$

$$\omega_m = (\omega_o - \omega_i)$$

and

$$\mathbf{M2} = \frac{1}{3} \begin{bmatrix} 1+2q\cos(\omega_m t) & 1+2q\cos\left(\omega_m t - \dfrac{2\pi}{3}\right) & 1+2q\cos\left(\omega_m t - \dfrac{4\pi}{3}\right) \\ 1+2q\cos\left(\omega_m t - \dfrac{2\pi}{3}\right) & 1+2q\cos\left(\omega_m t - \dfrac{4\pi}{3}\right) & 1+2q\cos(\omega_m t) \\ 1+2q\cos\left(\omega_m t - \dfrac{4\pi}{3}\right) & 1+2q\cos(\omega_m t) & 1+2q\cos\left(\omega_m t - \dfrac{2\pi}{3}\right) \end{bmatrix} \quad (15.4)$$

$$\omega_m = -(\omega_o + \omega_i)$$

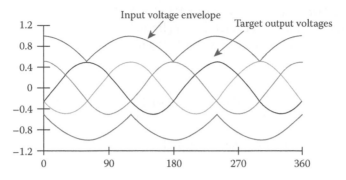

FIGURE 15.10 Waveforms—maximum voltage ratio of 50%.

The first solution gives the same phase displacement at the input and output ports, whereas the second solution gives a reversed phase displacement. Combining the two solutions, therefore, provides a method of controlling the input displacement factor control.

This method is a direct transfer function approach; during each switch sequence time (t_{seq}), the average output voltage is equal to the demanded voltage. For this to be possible, it is clear that the target voltages must fit within the input voltage envelope for any output frequency, limiting the maximum voltage ratio to 50%, as shown in Figure 15.10.

The voltage transfer ratio can be increased to 87% by adding common-mode voltages with frequencies equal to the third harmonics of the input and out to the target outputs. The common-mode voltages have no impact on the line-to-line voltages, but do allow better use to be made of the input voltage envelope as shown in Figure 15.11. It should be noted that a voltage ratio of 87% is the intrinsic maximum for any modulation method where the target output voltage equals the mean output voltage during each switching sequence:

$$\mathbf{v_o} = qV_{im} \begin{bmatrix} \cos(\omega_o t) - \frac{1}{6}\cos(3\omega_o t) + \frac{1}{2\sqrt{3}}\cos(3\omega_i t) \\ \cos\left(\omega_o t + \frac{2\pi}{3}\right) - \frac{1}{6}\cos(3\omega_o t) + \frac{1}{2\sqrt{3}}\cos(3\omega_i t) \\ \cos\left(\omega_o t + \frac{4\pi}{3}\right) - \frac{1}{6}\cos(3\omega_o t) + \frac{1}{2\sqrt{3}}\cos(3\omega_i t) \end{bmatrix} \tag{15.5}$$

Calculating the switch timings directly from the above equations is cumbersome for a practical implementation. Venturini's optimum method employs the common-mode addition technique defined in

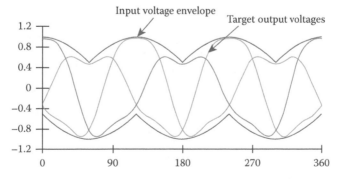

FIGURE 15.11 Illustrating voltage ratio improvement to 87%.

Equation 15.5 to achieve a maximum voltage ratio of 87%. The formal statement of the algorithm, including displacement factor control, is rather complex and appears unsuited for real-time implementation. In fact, if unity input displacement factor is required, then the algorithm can be more simply stated in the form of (15.6):

$$m_{Kj} = \frac{1}{3}\left[1 + \frac{2v_K v_j}{V_{im}^{2}} + \frac{4q}{3\sqrt{3}}\sin(\omega_i t + \beta_K)\sin(3\omega_i t)\right] \quad \text{for } K = A,B,C \quad \text{and} \quad j = a,b,c$$

$$\beta_K = 0, \frac{2\pi}{3}, \frac{4\pi}{3} \quad \text{for } K = A,B,C \text{ respectively} \tag{15.6}$$

Note that the target output voltages, v_j, include the common-mode addition defined in (15.5).

Space vector modulation (SPVM) is well known and established in conventional PWM inverters. Its application to matrix converters is conceptually the same and the concept of space vectors can be applied to output voltage and input current control. The target output voltage space vector of the matrix converter is defined in terms of the line-to-line voltages:

$$\mathbf{V}_o(t) = \frac{2}{3}(v_{ab} + av_{bc} + a^2 v_{ca}) \quad \text{where } a = \exp\left(\frac{j2\pi}{3}\right) \tag{15.7}$$

In the complex plane, $\mathbf{V}_o(t)$ is a vector of constant rotating at angular frequency (Figure 15.12). In the SPVM, $\mathbf{V}_o(t)$ is synthesized by time averaging from a selection of adjacent vectors in the set of converter output vectors in each sampling period. For a matrix converter, the selection of vectors is by no means unique and a number of possibilities exist, leading to many publications.

The 27 possible output vectors, given in Table 15.2, for a three-phase matrix converter can be classified into three groups with the following characteristics:

- Group I: each output line is connected to a different input line. Output space vectors are constant in amplitude, rotate at the supply frequency.

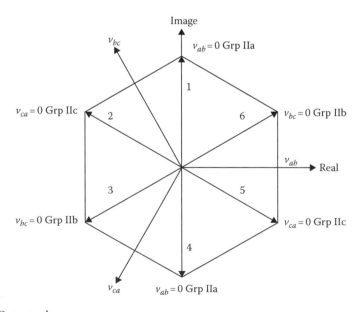

FIGURE 15.12 Output voltage space vectors.

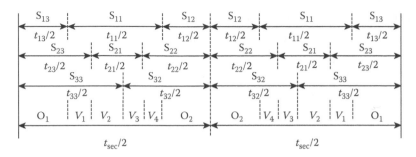

FIGURE 15.13 Possible way of allocating states within switching sequence.

- Group II: two output lines are connected to one input line, the remaining output line is connected to one of the other input lines. Output space vectors have varying amplitude and fixed direction occupying one of six positions regularly spaced 60° apart.
- Group III: all output lines are connected to a common input line. Output space vectors have zero amplitude.

In the SPVM, the group I vectors are not normally used and the desired output is synthesized from the group II active vectors and the group III zero vectors, in a similar way to techniques used for inverters. However, in a matrix converter, the input current must be considered as well as the output voltage vectors. The time weighting for the vectors can then be calculated. There is no unique way for distributing the times within the switching sequence, but one popular method is shown in Figure 15.13.

15.2.4 Two-Stage Matrix Converters (Sparse)

In addition to the standard form of the matrix converter, there has been recent interest in alternative two-stage direct converter topologies. This family of direct converter topologies are usually often referred to as sparse matrix converters, but this term is often improperly used, as discussed below. The basic form of the two-stage direct converter consists of a three-phase to two-phase matrix converter followed by a standard inverter bridge, as shown in Figure 15.14.

The three-phase to two-phase matrix converter is used to create a switched "DC" link voltage. The three-phase to two-phase matrix converter must be modulated in a way that ensures a positive "DC" link voltage in order to avoid a short circuit condition through the diodes of the inverter bridge. This "DC" link voltage is then switched by the inverter bridge to give the desired output waveform.

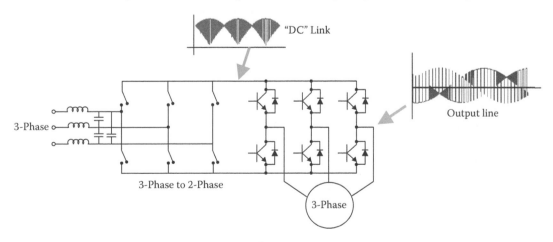

FIGURE 15.14 The two-stage direct converter topology.

TABLE 15.1   Matrix Converter Vectors

Vector Number	Conducting Switches			Output Phase Voltages			Output Line-to-Line Voltages			Input Line Currents		
				v_a	v_b	v_c	v_{ab}	v_{bc}	v_{ca}	I_A	I_B	I_C
+1	S_{Aa}	S_{Bb}	S_{Bc}	v_A	v_B	v_B	v_{AB}	0	$-v_{AB}$	I_a	I_b+I_c	0
−1	S_{Ba}	S_{Ab}	S_{Ac}	v_B	v_A	v_A	$-v_{AB}$	0	v_{AB}	I_b+I_c	I_a	0
+2	S_{Ba}	S_{Cb}	S_{Cc}	v_B	v_C	v_C	v_{BC}	0	$-v_{BC}$	0	I_a	I_b+I_c
−2	S_{Ca}	S_{Bb}	S_{Bc}	v_C	v_B	v_B	$-v_{BC}$	0	v_{BC}	0	I_b+I_c	I_a
+3	S_{Ca}	S_{Ab}	S_{Ac}	v_C	v_A	v_A	v_{CA}	0	$-v_{CA}$	I_b+I_c	0	I_a
−3	S_{Aa}	S_{Cb}	S_{Cc}	v_A	v_C	v_C	$-v_{CA}$	0	v_{CA}	I_a	0	I_b+I_c
+4	S_{Ba}	S_{Ab}	S_{Bc}	v_B	v_A	v_B	$-v_{AB}$	v_{AB}	0	I_b	I_a+I_c	0
−4	S_{Aa}	S_{Bb}	S_{Ac}	v_A	v_B	v_A	v_{AB}	$-v_{AB}$	0	I_a+I_c	I_b	0
+5	S_{Ca}	S_{Bb}	S_{Cc}	v_C	v_B	v_C	$-v_{BC}$	v_{BC}	0	0	I_b	I_a+I_c
−5	S_{Ba}	S_{Cb}	S_{Bc}	v_B	v_C	v_B	v_{BC}	$-v_{BC}$	0	0	I_a+I_c	I_b
+6	S_{Aa}	S_{Cb}	S_{Ac}	v_A	v_C	v_A	$-v_{CA}$	v_{CA}	0	I_a+I_c	0	I_b
−6	S_{Ca}	S_{Ab}	S_{Cc}	v_C	v_A	v_C	v_{CA}	$-v_{CA}$	0	I_b	0	I_a+I_c
+7	S_{Ba}	S_{Bb}	S_{Ac}	v_B	v_B	v_A	0	$-v_{AB}$	v_{AB}	I_c	I_a+I_b	0
−7	S_{Aa}	S_{Ab}	S_{Bc}	v_A	v_A	v_B	0	v_{AB}	$-v_{AB}$	I_a+I_b	I_c	0
+8	S_{Ca}	S_{Cb}	S_{Bc}	v_C	v_C	v_B	0	$-v_{BC}$	v_{BC}	0	I_c	I_a+I_b
−8	S_{Ba}	S_{Bb}	S_{Cc}	v_B	v_B	v_C	0	v_{BC}	$-v_{BC}$	0	I_a+I_b	I_c
+9	S_{Aa}	S_{Ab}	S_{Cc}	v_A	v_A	v_C	0	$-v_{CA}$	v_{CA}	I_a+I_b	0	I_c
−9	S_{Ca}	S_{Cb}	S_{Ac}	v_C	v_C	v_A	0	v_{CA}	$-v_{CA}$	I_c	0	I_a+I_b

With this topology, it is possible to create all the same output vectors as used in a standard matrix converter with the exception of the rotating vectors. The rotating vectors are the vectors where each output of a standard matrix converter is connected to a different input phase, shown in Table 15.1. In most common modulation techniques for the standard matrix converter, the rotating vectors are not used.

If the operation of the three-phase to two-phase matrix converter is analyzed in detail, it can be seen that not all the devices in the matrix converter are actually required. It is possible to remove three IGBTs, as shown in Figure 15.15b, and retain the full functionality of the complete converter. This topology is normally referred to as a sparse matrix converter. A further reduction in device count is possible if only unidirectional power flow is required, as shown in Figure 15.15c. The outer IGBTs in the three-phase to two-phase matrix converter are not required in unidirectional power flow situation, leading to a topology that has been termed the Very Sparse Matrix Converter. The semiconductor device count for these circuits is compared with the standard matrix converter and a back-to-back inverter arrangement in Table 15.2.

To overcome some of the limitations of all the direct converter topologies, it is possible to add some energy storage into the "DC" link using an H-bridge, as shown in Figure 15.15d. With this topology, it is possible to increase the maximum output voltage and compensate for input voltage waveform distortion, but this is usually at the expense of input current waveform quality.

15.2.5 Applications

The range of published practical implementations has demonstrated the technology readiness of matrix converters for motor drive applications. These range from a 2 kW matrix converter using silicon carbide devices and switching at 150 kHz for aerospace applications built at ETH in Zürich, Switzerland to a 150 kVA matrix converter using 600A IGBTs built at the U.S. Army research labs in collaboration with the University of Nottingham, United Kingdom.

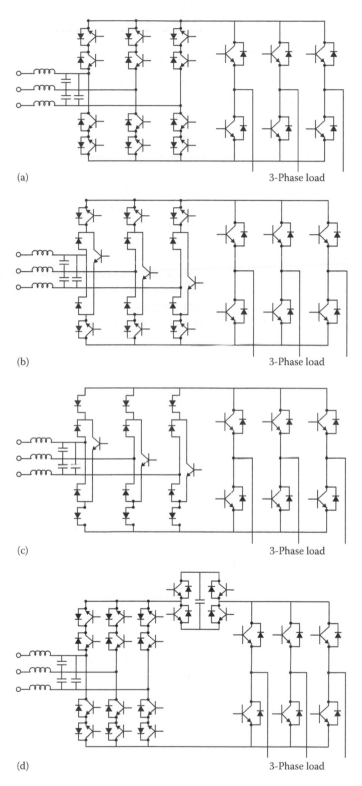

(a)

3-Phase load

(b)

3-Phase load

(c)

3-Phase load

(d)

3-Phase load

FIGURE 15.15 (a) The two-stage direct power converter. (b) The sparse converter. (c) The very sparse matrix converter. (d) Hybrid two-stage direct power converter.

TABLE 15.2 Comparison of Topology Device Count

	Matrix Converter	Two-Stage Direct Converter	Sparse Direct Converter	Very Sparse Direct Converter	Back-to-Back Inverter	Hybrid Two-Stage Direct Converter
IGBTs	18	18	15	9	12	22
Diodes	18	18	18	18	12	22
Electrolytic Capacitors	0	0	0	0	Large	Small

Most of leading-edge research in matrix converters is now focused on potential applications. Many potential applications exist where power density carries a premium, such as integrated motor drives, lifts and hoists, aerospace applications, and marine propulsion. It is in these high-value industries where the advantages become very significant, that the matrix converter will probably first find its first commercial applications. As the price of semiconductors continues to fall, the matrix converter is also becoming a more attractive future alternative to the back-to-back inverter in applications where sinusoidal input currents or true bidirectional power flow are required. The matrix converter could be an ideal converter topology to utilize future technologies such as high-temperature silicon carbide devices. These devices will operate at temperatures up to 300°C, so the lack of large electrolytic capacitors, as normally used in an inverter, would again be a significant advantage.

Acknowledgments

The author would like to thank Prof. Jon Clare and Dr. Lee Empringham for their contribution to the ongoing research effort into matrix converters at the University of Nottingham and for their ideas, input, and contribution to the content of this chapter.

References

1. Gyugi, L. and Pelly, B., *Static Power Frequency Changers: Theory, Performance and Applications*, John Wiley & Sons, New York, 1976.
2. Daniels, A. and Slattery, D., New power converter technique employing power transistors, *IEE Proc.* 25(2), 146–150, February 1978.
3. Venturini, M., A new sine wave in sine wave out, conversion technique which eliminates reactive elements, *Proceedings of the POWERCON 7*, San Diego, CA, 1980, pp. E3/1–E3/15.
4. Alesina, A. and Venturini, M.G.B., Analysis and design of optimum-amplitude nine-switch direct AC-AC converters, *IEEE Trans. Power Electron.* 4(1), 101–112, January 1989.
5. Wheeler, P.W., Rodriguez, J., Clare, J.C., and Empringham L., Matrix converters: A technology review, *IEEE Trans. Ind. Electron.* 49(2), 276–288, 2002.
6. Apap, M., Wheeler, P.W., Clare, J.C., and Bradley, K.J., Analysis and comparison of AC-AC matrix converter control strategies, *IEEE Power Electronics Specialists Conference*, Acapulco, Mexico, June 2003.

16

Fundamentals of AC–DC–AC Converters Control and Applications

Marek Jasiński
Warsaw University
of Technology

Marian P.
Kazmierkowski
Warsaw University
of Technology

16.1 Introduction

AC–DC–AC converters are part of a group of AC/AC converters. Generally, AC/AC converters take power from one AC system and deliver it to another with waveforms of different amplitude, frequency, and phase. Those systems can be single phase or three phase. The major application of voltage source AC/AC converters are adjustable speed drives (ASDs) [5,18,19,41] and adjustable speed generators (ASGs) (variable speed generation systems).

The widely voltage source AC/AC converters utilize a DC-link between the two AC systems, as presented in Figure 16.1a,b, and provide direct power conversion, as in Figure 16.1c.

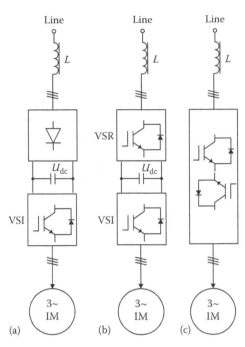

FIGURE 16.1 Chosen AC/AC converters for ASDs: (a) with diode rectifier, (b) with VSR, and (c) direct converter (matrix or cycloconverter). VSI, voltage source inverter; IM, induction machine; PWM, pulse width modulation [Data from 11].

In AC–DC–AC converter, the input AC power is rectified into a DC waveform and then is inverted into the output AC waveform. A capacitor (and/or inductor) in DC-link stores the instantaneous difference between the input and output powers. AC–DC and DC–AC converters can be controlled independently.

The matrix converter (cycloconverter) avoids the intermediate DC-link by converting the input AC waveforms directly into the desired output waveforms (Figure 16.1c) [16].

Although a three-phase induction machine was introduced more than 100 years ago, the research and development in this area is still ongoing. Moreover, there has been a remarkable growth in the development of new power semiconductor devices and power electronics converters in the last 20/30 years, and has not ceased to grow. The introduction of insulated gate bipolar transistors (IGBTs) in the mid-1980s was an important milestone in the history of power semiconductor devices. Similarly, digital signal processors (DSPs) developed in 1990s were a milestone in implementation and applications of advanced control strategies for power converter drives. As a result, ASD systems are widely used in applications such as pumps, fans, paper and textile mills, elevators, electric vehicles, and underground traction, home appliances, wind generation systems (ASG), servo drives and robotics, computer peripherals, steel and cement mills, and ship propulsion [5]. Nowadays, most ASD systems consist of uncontrollable diode rectifier (Figure 16.1a) or a line-commutated phase-controlled thyristor bridge. Although both these converters offer a high reliability and simple structure, they also have serious disadvantages. The DC-link voltage of the diode rectifier is uncontrolled and pulsating; therefore, bulky DC-link capacitor and usually DC-choke are needed. Moreover, the power flow is unidirectional and the input current (line current) is strongly distorted. The last drawback is very important because of standard regulations such as IEEE Std 519-1992 in the United States and IEC 61000-3-2/IEC 61000-3-4 in the European Union. Even small power ASD can cause a total harmonics distortion (THD) problem for a supply line when a large number of nonlinear loads are connected to one point of common coupling (PCC). Table 16.1 lists the harmonic current limits based on the size of a load with respect to the size of line power supply.

TABLE 16.1   Current Distortion Limits for General Distribution Systems (Up to 69 kV)

Maximum Harmonic Current Distortion in Percent of (15 or 30 min Demand) I_{Lm}						
Individual Harmonic Order (Odd Harmonics)						
I_{SC}/I_{Lm}	<11	$11 \leq h < 17$	$17 \leq h < 23$	$23 \leq h < 35$	$35 \leq h$	TDD
$I_{SC}/I_{Lm} < 20$	4.0	2.0	1.5	0.6	0.3	5.0
$20 < I_{SC}/I_{Lm} < 50$	7.0	3.5	2.5	1.0	0.5	8.0
$50 < I_{SC}/I_{Lm} < 100$	10.0	4.5	4.0	1.5	0.7	12.0
$100 < I_{SC}/I_{Lm} < 1000$	12.0	5.5	5.0	2.0	1.0	15.0
$I_{Sn}/I_{Lm} > 1000$	15.0	7.0	6.0	2.5	1.4	20.0

Note: TDD is the total demand distortion (root-sum-square—RSS).
Source: Modified from IEEE Std 519–1992, IEEE Recommended Practices and Requirements for Harmonic Control in Electrical Power Systems, The Institute of Electrical and Electronics Engineers Inc., USA, 1993.

TABLE 16.2   Voltage Distortion Limits

Bus Voltage at PCC	Individual Voltage Distortion [%]	
69 kV and below	3.0	5.0

Source: Modified from IEEE Std 519–1992, IEEE Recommended Practices and Requirements for Harmonic Control in Electrical Power Systems, The Institute of Electrical and Electronics Engineers Inc., USA, 1993.

The recommended voltage distortion limits, usually expressed by THD index, is shown in Table 16.2, where THD is total (root-sum-square (RSS)) harmonic voltage in percent of nominal fundamental frequency voltage. This term has come into common usage to define either voltage or current distortion factor (DF) (Equation 16.1). The DF is the ratio of the RSS of the harmonic content to the root-mean-square (RMS) value of the fundamental quantity, expressed as a percent of the fundamental [13]:

$$\text{THD} = \sqrt{\frac{\sum_{h=2}^{50} U_{L(h)}^2}{U_{L(1)}^2}} 100\% \tag{16.1}$$

Some types of electronic receivers can be affected by transmission of AC supply harmonics through the equipment power supply or by electromagnetic coupling of harmonics into equipment components (electromagnetic interference (EMI) problem). Computers and associated equipment such as programmable controllers frequently require AC sources that have no more distortion than a 5% THD, with the largest single harmonic being no more than 3% of the fundamental. Higher levels of harmonics result in erratic, sometimes subtle malfunctions of the equipment that can, in some cases, have serious consequences. Also, instruments can be affected similarly. Perhaps, the most serious of these are malfunctions in medical instruments. Consequently, many medical instruments are provided with special power electronics devices (line-conditioners). Here is a wide application field, especially, for AC–DC–AC converters, such as uninterruptible power supply (UPS) systems.

Therefore, a lot of methods for elimination of harmonics distortion in the power system are developed and implemented [31]. Moreover, several blackouts in recent years (United States and Canada (New York, Detroit, Toronto) in August 2003, Russia (Moscow) in May 2005, United States (Los Angeles) in September 2005, and high prices of the oil shows that the idea of "clean power" is more and more up to date.

Harmonics reduction methods can be divided into two main groups (Figure 16.2):

1. Passive filters and active filters: harmonics reduction of the already installed nonlinear loads
2. Multi-pulse rectifiers and VSR (active rectifiers): power-grid friendly converters (with limited THD) [22].

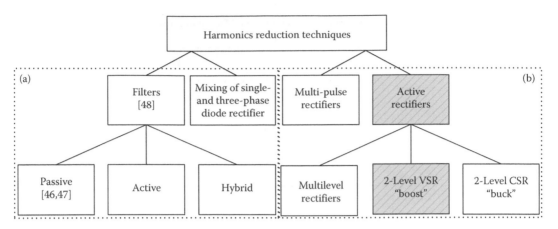

FIGURE 16.2 Harmonics reduction techniques; where CSR is current source rectifier.

Furthermore, the energy saving is important because VSR assures regenerating braking with energy-saving capability as well as after minor modification, active filtering function can be implemented [1].

Typical application of the VSR is shown in Figure 16.1b. Thanks to systematical cost reduction of the IGBTs and DSPs, there have appeared on the market serially produced VSR from few kVA up to MVA. An individual VSR can provide the DC-link voltage to several VSI-fed IM (for cost reduction) [43]. Moreover, VSR can compensate the nonlinear load current connected in parallel with VSR to PCC.

This chapter focuses on three-phase AC–DC–AC converter consisting of two identical voltage source converters (VSCs), IGBT bridges, as in Figure 16.1b. The first of them (at the line side) works as a voltage source rectifier (VSR) feeding the DC-link circuit in motoring mode or feeding the grid in generating mode, whereas the second (at the machine side) operates as a voltage source inverter (VSI) (also called machine side converter (MSC)) feeding induction machine (IM) in motoring mode or taking energy from the IM in generating mode. Sometimes, VSR is called active rectifier, PWM rectifier or active front-end, or grid side converter (GSC).

Generally, the high-performance frequency-controlled AC–DC–AC converter–fed IM drive should offer the following features and abilities:

On the MSC side

- Four-quadrant operation
- Fast flux and torque response
- Maximum output torque available in wide range of speed operation
- Constant switching frequency
- Unipolar voltage PWM, thus lower switching losses
- Low flux and torque ripple
- Wide range of speed control
- Robustness to parameter variations

On the GSC

- Bidirectional power flow
- Nearly sinusoidal input current (low THD typically below 5%)
- Controllable reactive power (up to unity power factor (UPF))
- Controllable DC-link voltage (well stabilized at desired level)
- Reduction of DC-link capacitor and DC voltage fluctuation
- Insensitivity to line voltage variations [26,35,40]
- Reduction of transformer and cable cost due to UPF

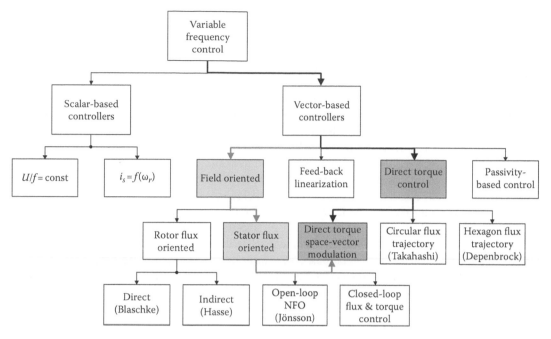

FIGURE 16.3   Classification of IM control methods; where NFO is the natural field orientation [Data from 45].

These features depend mainly on the applied control strategy. The main goal of the chosen control strategy is to provide optimal parameters of ASD concurrently with reduction of the cost and maximal simplification of the whole system. Moreover, robustness of the control system is very important.

IM control methods can be divided into *scalar* and *vector control*. The general classification of the variable frequency methods is presented in Figure 16.3. Following the definition from [19], we may say that "in scalar control, which is based on relationships valid for steady state, only magnitude and frequency (angular speed) of voltage, current and flux linkage space vectors are controlled. As result, the scalar control does not act on space vector position during transients. Contrarily, in vector control, which is based on relations valid for dynamic states, not only magnitude and frequency (angular speed) but also instantaneous positions of voltage, current, and flux space vectors are controlled. Thus, the vector control acts on the positions of the space vectors and provides their correct orientation both in steady state and during transients."

Therefore, vector control is a general control concept that can be implemented in many different ways. The most known method, called *field oriented control* (FOC) [4,7,31] or vector control [5] has been proposed by Hasse (indirect FOC) and Blaschke (direct FOC) [3] (see also [5,18,41]), and gives the induction machine good performance. In the FOC, the IM equations are transformed into rotor flux vector oriented coordinate system. In rotor flux vector oriented coordinates (assumed constant rotor flux amplitude), there is a linear relationship between current vector components and machine torque. Moreover, like in a DC machine, the flux reference amplitude is reduced in the field-weakening range in order to limit the stator voltage typically at higher than nominal speed. IM equations represented in the flux vector oriented coordinates have a good physical basis because they correspond to the decoupled torque generation in separately excited DC machine. Nevertheless, from the theoretical point of view, another type of mathematical transformations can be chosen to achieve decoupling and linearization of IM equations. Those methods are known as *modern nonlinear control* [16]. Marino et al. and Krzeminski (see Kazmierkowski et al. [19]) have proposed a nonlinear transformation of the machine state variables so that, in the new coordinates, the speed and rotor flux amplitude are decoupled by feedback; the method is called *feedback linearization control* (FLC). Also, a method based on the variation theory and energy shaping, called *passivity-based control* (PBC) has been recently investigated [17].

In the mid-1980s, there was a trend toward the standardization of the control systems on the basis of the FOC methodology. However, Depenbrock, Takahashi, and Nogouchi [36] have presented a new strategy, which abandons an idea of mathematical coordinate transformation and the analogy with DC machine control. These authors proposed to replace the averaging based decoupling control with the instantaneous bang-bang control, which very well corresponds to on–off operation of the VSI semiconductor power devices. These strategies are known as *direct torque control* (DTC). Among the main advantages of DTC scheme are simple structure, good dynamic behavior, and that it is inherently a motion-sensorless control method. However, it has very important drawbacks, i.e., variable switching frequency, high torque pulsation, unreliable startup, and low-speed operation performance. Therefore, to overcome these disadvantages, a space vector modulator (SVM) was introduced to DTC structure [9] giving DTC–SVM control scheme. In this method, disadvantages of the classical DTC are eliminated.

However, it should be pointed that no commonly shared terminology exists regarding DTC and DTC–SVM. From the formal considerations, DTC-SVM can also be called as *stator field oriented control* (SFOC). In this chapter, DTC and DTC-SVM scheme refer to control schemes operating with closed torque and flux loops without current controllers.

Control of the VSR can be considered as a dual problem with vector control of an induction machine. The simple scalar control is based on current regulation in three-phase system (AC waveforms) [5] (Figure 16.4).

Like for IM, vector control of VSR is a general control philosophy that can be implemented in many different ways. The most popular method, known as *voltage oriented control* (VOC) [25,28] gives high dynamic and static performances via internal current control loops. In the VOC, the VSR equations are transformed in a line voltage vector oriented coordinate system. In line voltage vector oriented coordinates, there is a linear relationship between current vector control components and power flow. To improve the robustness of VOC scheme, a virtual flux (VF) concept was introduced by Duarte. However, from the theoretical point of view, other types of mathematical coordinate transformations can be defined to achieve decoupling and linearization of the VSR equations. This has originated the methods known as *nonlinear control*. Jung et al. [15] and Lee et al. [23] have proposed a nonlinear

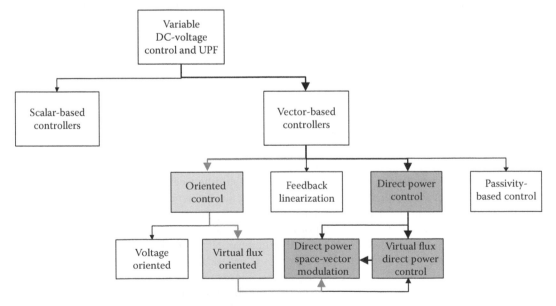

FIGURE 16.4 Classification of VSR control methods.

transformation of VSR state variables so that, in the new coordinates, the DC-link voltage and line current are decoupled by feedback; this method is called also FLC like for induction machine. Moreover, a PBC, as for IM, was also investigated in respect to VSR [19].

In the mid-1990s, Manninen [30] and in the second part of 1990s, Nogouchi at al. [33] have expanded the idea of DTC for VSR called *direct power control* (DPC). From that time, it has been continuously improved. However, these control principles are very similar to DTC schemes for IM and have the same drawbacks. Therefore, to overcome that disadvantages a space vector modulator (SVM) [12] was introduced to DPC structure giving new DPC–SVM control scheme [29]. Hence, presented DPC-SVM and DTC-SVM joins important advantages of SVM (e.g., constant switching frequency, unipolar voltage pulses), with advantages of DPC, and DTC (e.g., simple and robust structure, lack of internal current control loops, good dynamics, etc.). However, when control structure of the VSR operates independently from control of the IM, the DC-link voltage stabilization is not sufficiently fast and, as a consequence a large DC-link capacitor is required for instantaneous power balancing. Therefore, for speed up the DC-link voltage dynamic an additional active power feedforward (PF) loop from the VSI-fed IM side to VSR-fed DC-link control is required. As result a direct power and torque control with space vector modulation (DPTC-SVM) scheme was obtained [14]. This new control scheme with PF loop allows significant reduction of the DC-link capacitor keeping fast instantaneous power balancing. It seems to be very attractive for industrial application. Therefore, this chapter is devoted to analysis and study of DPTC-SVM scheme with PF loop (Figure 16.4).

16.2 Mathematical Model of the VSI-Fed Induction Machine

To present basic control methods of VSI-fed IM (see Figures 16.5 and 16.6), the space vector–based IM mathematical model will be presented and discussed in this section. The fundamental wave IM model is developed under the following idealized assumptions [18]:

- The object is a symmetrical, three-phase machine.
- Only the basic harmonics are considered while the higher harmonics of the spatial field distribution and magnetomotive force (MMF) in the air gap are disregarded.
- The spatially distributed stator and rotor windings are represented by a virtual so-called concentrated coil.

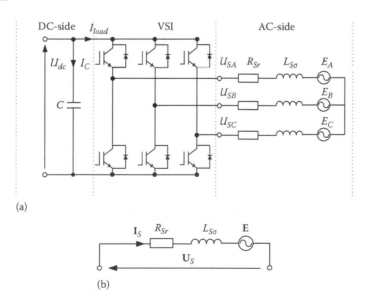

(a)

(b)

FIGURE 16.5 VSI with IM equivalent circuit: (a) three-phase system and (b) single-phase equivalent circuit.

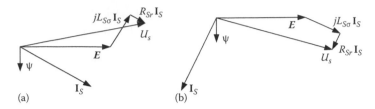

FIGURE 16.6 Pictorial phasor diagrams for VSI-fed IM drive: (a) motoring and (b) regenerating.

- The effects of anisotropy, magnetic saturation, iron losses, and eddy currents are neglected.
- The coil resistances and reactances are assumed to be constant.
- The current and voltage are taken to be sinusoidal (in many cases, especially when considering steady states).

16.2.1 IM Mathematical Model in Rotating Coordinate System with Arbitrary Angular Speed

The model of the IM in natural *ABC* coordinates is very complicated. Therefore, in order to reduce the set of equations from 12 to 4, the complex space vectors are used. Moreover, based on transformation into a common rotating coordinate system with arbitrary angular speed Ω_K and referring rotor quantities to the stator circuit, a following set of equations can be written [18]:

Voltage equations:

$$\mathbf{U}_{SK} = R_S\mathbf{I}_{SK} + \frac{d\boldsymbol{\Psi}_{SK}}{dt} + j\Omega_K\boldsymbol{\Psi}_{SK}, \tag{16.2}$$

$$\mathbf{U}_{rK} = R_r\mathbf{I}_{rK} + \frac{d\boldsymbol{\Psi}_{rK}}{dt} + j\left(\Omega_K - p_b\Omega_m\right)\boldsymbol{\Psi}_{rK}, \tag{16.3}$$

Flux-currents equations:

$$\boldsymbol{\Psi}_{SK} = L_S\mathbf{I}_{SK} + L_M\mathbf{I}_{rK}, \tag{16.4}$$

$$\boldsymbol{\Psi}_{rK} = L_r\mathbf{I}_{rK} + L_M\mathbf{I}_{SK}, \tag{16.5}$$

And motion equation:

$$\frac{d\Omega_m}{dt} = \frac{1}{J}\left[p_b\frac{m_S}{2}\mathrm{Im}\left(\boldsymbol{\Psi}_{SK}^*\mathbf{I}_{SK}\right) - M_L\right] \tag{16.6}$$

16.3 Operation of Voltage Source Rectifier

VSR can be described in different coordinate systems. The basic scheme of the VSR with AC input choke and output DC side capacitor is shown in Figure 16.7a, while Figure 16.7b shows its a single-phase representation, where $\mathbf{U}_L$ is a line voltage space vector, $\mathbf{I}_L$ is a line current space vector, $\mathbf{U}_p$ is the VSR input voltage space vector, and $\mathbf{U}_i$ is a space vector of voltage drop on the input (AC line side) choke L and it resistance R.

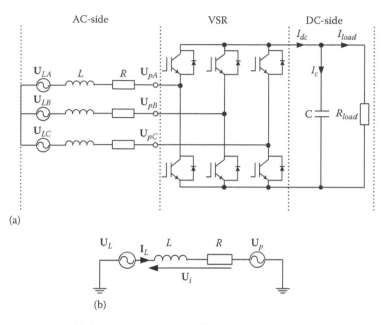

(a)

(b)

FIGURE 16.7 VSR topology: (a) three-phase system and (b) single-phase equivalent circuit.

The $\mathbf{U}_p$ voltage is controllable and depends on switching signals pattern and DC-link voltage level. Thanks to control magnitude and phase of the $\mathbf{U}_p$ voltage, the line current can be controlled by changing the voltage drop on the input choke, $\mathbf{U}_i$. Therefore, inductances between line and AC side of the VSR are indispensable. They create a current source and provide boost feature of the VSR. By controlling the converter AC side voltage in its phase and amplitude $\mathbf{U}_p$, the phase and amplitude of the line current vector $\mathbf{I}_L$ is controlled indirectly.

Further, in Figure 16.8 are shown both motoring and regenerating phasor diagrams of VSR. From this figure, it can be seen that the magnitude of $\mathbf{U}_p$ is higher during regeneration than in rectifying mode. With assumption of a stiff line power (i.e., $\mathbf{U}_L$ is a pure voltage source with zero internal impedance), terminal voltage of VSR $\mathbf{U}_p$ can differ up to about 3% between motoring and regenerating modes.

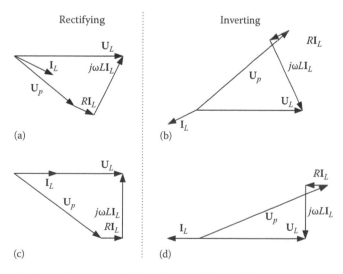

FIGURE 16.8 Pictorial phasor diagrams for VSR: (a, b) non-UPF; (c, d) UPF operation.

16.3.1 Operation Limits of the Voltage Source Rectifier

Figure 16.9 indicates that for VSR is a load current limit for fixed line and DC-link voltage, as well as input choke. Beyond that limit, the VSR is not able to operate and maintain a UPF requirement. Lower line inductance and higher voltage reserve (between the line voltage and the DC side voltage) can increase that limits. However, there is a limitation for the minimum DC-link voltage defined as

$$U_{dc} > \sqrt{2}\sqrt{3}U_{LRMS} \tag{16.7}$$

This limitation is introduced by freewheeling diodes in VSR, which operate as a diode rectifier. However, in the literature exists other limitation [34] which takes into account the input power (value of the current) of the VSR.

Let consider that commanded value of the line current differs from actual current by ΔI_{Lxy}:

$$\Delta I_{Lxy} = I_{Lxyc} - I_{Lxy} \tag{16.8}$$

The direction and velocity of the line current vector changes are described by derivative of that current $L(dI_{Lxy}/dt)$. It can be represented by equations in synchronous rotating xy coordinates:

$$L\frac{dI_{Lxy}}{dt} + R\left(I_{Lxyc} - \Delta I_{Lxy}\right) = U_{Lxy} - U_{dc}S_{1xy} + j\omega_L L\left(I_{Lxyc} - \Delta I_{Lxy}\right) \tag{16.9}$$

With assumptions that resistance of the input chokes $R \cong 0$ and actual current is close to commanded value ($\Delta I_{Lxy} \cong 0$), the above equation can be simplified to

$$L\frac{dI_{Lxy}}{dt} = U_{Lxy} - U_{dc}S_{1xy} + j\omega_L L I_{Lxyc} \tag{16.10}$$

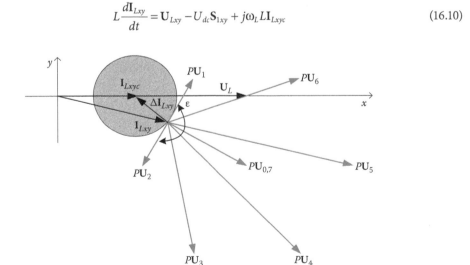

FIGURE 16.9 Error area of the line current vector. (Adapted from Sikorski, A., *Problemy dotyczące minimalizacji strat łączeniowych w przekształtniku AC-DC-AC-PWM zasilającym maszynę indukcyjną*, Politechnika Białostocka, Rozprawy Naukowe nr. 58, Białystok, Poland, pp. 217, 1998 (in Polish).)

Based this equation, the direction and velocity of the line current vector changes depends on

- Values of input chokes L
- Line voltage vector $\mathbf{U}_{Lxy}$
- Line current vector $\mathbf{I}_{Lxy}$
- Value of the DC-link voltage U_{dc}
- Switching states of the VSR $\mathbf{S}_{1xy}$

Command current $\mathbf{I}_{Lxyc}$ is in phase with line voltage vector $\mathbf{U}_{Lxy}$ and it lies on the axis x. The difference between actual current $\mathbf{I}_{Lxy}$ and commanded $\mathbf{I}_{Lxyc}$ is defined by Equation 16.8 and is illustrated in Figure 16.9.

Full current control is possible when the current is kept in desired error area (16.9). Critical operation of the VSR is when the angle achieves $\varepsilon = \pi$. Figure 16.9 shows that for such case ε created by PU_1, PU_2, $\mathbf{U}_{pc1}$, and $\mathbf{U}_{pc2}$ vectors, are the arms of the equilateral triangle. Therefore, based on the equation for its altitude, the boundary condition can be defined as

$$\left|\mathbf{U}_{Lxy} + j\omega_L L \mathbf{I}_{Lxyc}\right| = \frac{\sqrt{3}}{2}\left|\mathbf{U}_{pxy}\right| \tag{16.11}$$

Assuming that $\mathbf{U}_{Lxy} = U_{Lm}$, $\mathbf{I}_{Lxyc} = I_{Lmc}$, and $\mathbf{U}_{pxy} = (2/3)U_{dc}$, the following expression can be derived:

$$\sqrt{U_{Lm}^2 + (\omega_L L I_{Lmc})^2} = \frac{\sqrt{3}}{2}\frac{2}{3}U_{dc} \tag{16.12}$$

After rearranging, one obtains dependence for minimum DC-link voltage:

$$U_{dc\_min} = \sqrt{3\left(U_{Lm}^2 + (\omega_L L I_{Lmc})^2\right)} \tag{16.13}$$

(For example with parameters as $U_{Lm} = 230\sqrt{2}$ V, $\omega_L = 2\pi 50$, $L = 0.01$ H, $I_{Lmc} = 10$ A, then $U_{dc\_min} \geq 566$ V). Based on this relation, the maximum value of the input inductance can be calculated as

$$L_m = \frac{\sqrt{1/3\, U_{dc}^2 - \left(U_{Lm}\right)^2}}{\omega_L I_{Lmc}} \tag{16.14}$$

(For example, with parameters as $U_{Lm} = 230\sqrt{2}$ V, $\omega_L = 2\pi 50$, $I_{Lmc} = 10$ A, and $U_{dc\_min} = 566$ V then maximum input line inductance is $L_m = 0.01$ H.)

16.3.2 VSR Model in Synchronously Rotating xy Coordinates

A two-phase model in synchronously rotating xy coordinates using the complex space vector notation can be expressed as

$$L\frac{d\mathbf{I}_{Lxy}}{dt} + R\mathbf{I}_{Lxy} = \mathbf{U}_{Lxy} - U_{dc}\mathbf{S}_{1xy} + j\omega_L L\mathbf{I}_{Lxy} \tag{16.15}$$

$$C\frac{dU_{dc}}{dt} = \frac{3}{2}\mathrm{Re}\left[\mathbf{I}_{Lxy}\mathbf{S}_{1xy}^*\right] - I_{load} \tag{16.16}$$

16.4 Vector Control Methods of AC–DC–AC Converter–Fed Induction Machine Drives: A Review

VSI: First publications about inverter vector control (FOC) was published 30 years ago [3], and from that time it has been widely used in industry. As mentioned, the FOC can be divided into direct field oriented control (DFOC) and indirect field oriented control (IFOC). The second one seems to be more attractive because of lack of the flux estimator. Thanks to this ability, it is easier in implementation. Therefore, for further consideration, IFOC is chosen.

DTC was proposed by Takahashi [36].

VSR: Control of the VSR can be considered as a dual problem with vector control of an induction machine.

Besides of classification as in Section 16.1, control techniques for VSR can be classified in respect to voltage and VF bases. Overall, four types of techniques can be distinguished:

- Voltage oriented control (VOC)
- Voltage-based direct power control (DPC)
- Virtual flux oriented control (VFOC)
- Virtual flux-based direct power control (VF-DPC)

All these methods are very well described in the literature [19,28], where superiority of VF-based methods is clearly shown. Therefore, only VF-based method will be described.

This chapter has been performed with two main goals: presenting theoretical background of each control technique and brief comparison (Figure 16.10).

16.4.1 Field Oriented Control and Virtual Flux Oriented Control

VSI: The block diagram of the IFOC is presented in Figure 16.11. The commanded electromagnetic torque M_{ec}, is delivered from outer *PI* speed controller, based on mechanical speed error e_{Ω_m}.

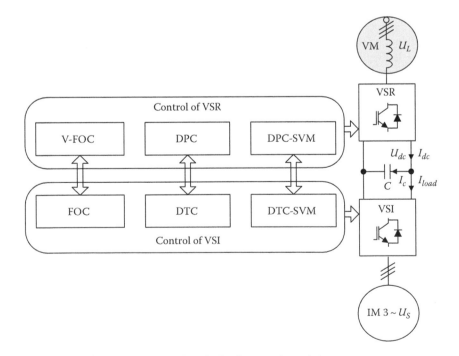

FIGURE 16.10   Relationship between control methods of VSR and VSI-fed IM.

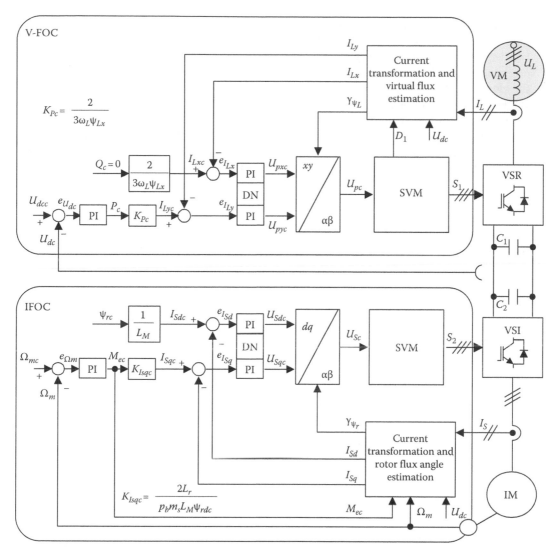

FIGURE 16.11 Virtual flux oriented control (V-FOC) and indirect field oriented control (IFOC); where DN is decoupling network.

Then, command values I_{Sdc} and I_{Sqc} are compared with actual values of current component I_{Sd} and I_{Sq}, respectively. It should be stressed that (for steady state) I_{Sd} is equal to the magnetizing current, while the torque in both dynamic and steady states is proportional to I_{Sq}. The current errors $e_{I_{Sd}}$ and $e_{I_{Sq}}$ are fed to two *PI* controllers, which generate commanded stator voltage components U_{Sqc}, and U_{Sdc}, respectively. Further, commanded voltages are converted from rotating *dq* coordinates into stationary αβ coordinates using rotor flux vector position angle γ_{Ψ_r}. So obtained voltage vector $\mathbf{U}_{Sc}$ is delivered to space vector modulator (SVM), which generates appropriate switching states vector $\mathbf{S}_2(S_{2A}, S_{2B}, S_{2C})$ for control power transistors of the VSI.

VSR: VOC guarantees high dynamics and static performance via an internal current control loop. It has become very popular and has consequently been developed and improved. Therefore, VOC is a basis for V-FOC, which is shown in Figure 16.11.

The goal of the control system is to maintain the DC-link voltage U_{dc}, at the required level, while currents drawn from the power system should be sinusoidal like and in phase with line voltage to satisfy the

UPF condition. The UPF condition is fulfilled when the line current vector $\mathbf{I}_L = I_{Lx} + jI_{Ly}$, is aligned with the phase voltage vector $\mathbf{U}_L = U_{Lx} + jU_{Ly}$, of the line.

The idea of VF has been proposed to improve the VSR control under distorted and/or unbalanced line voltage conditions, taking the advantage of the integrator's low-pass filter behavior [30].

Therefore, a rotating reference frame aligned with Ψ_L is used. The vector of VF lags the voltage vector by 90°. For the UPF condition, the command value of the direct component current vector I_{Lxc} is set to zero. Command value of the I_{Lyc} is an active component of the line current vector. After comparison, commanded currents with actual values, the errors are delivered to PI current controllers. Voltages generated by the controllers are transformed to $\alpha\beta$ coordinates using VF position angle $\gamma_{\Psi L}$. Switching signal vector $\mathbf{S}_1$ for the VSR is generated by a space vector modulator.

16.4.2 Direct Torque Control and VF-Based Direct Power Control

VSI: The block diagram of the method is presented in Figure 16.12. The commanded electromagnetic torque M_{ec} is delivered from outer PI speed controller. Then, M_{ec} and commanded stator flux Ψ_{Sc}

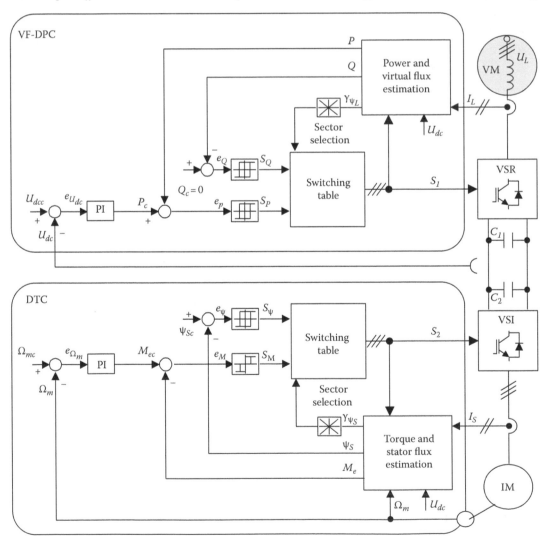

FIGURE 16.12 Conventional switching table-based DPC and DTC.

amplitudes are compared with estimated values of M_e and Ψ_S, respectively. The torque e_M and flux e_ψ errors are fed to two hysteresis comparators.

From predefined switching table, based on digitized error signals S_M and S_ψ, and the stator flux position γ_{Ψ_S} the appropriate voltage vector is selected. The outputs from the predefined switching table are switching states $\mathbf{S}_2$ for the VSI. Then, voltage space vector plane for the DTC needs to be divided into six sectors as in Figure 16.12. The sectors could be defined in different manner [18].

DTC is based on controlling the stator flux vector position in respect to rotor flux vector position based on the expression

$$M_e = p_b \frac{m_S}{2} \frac{L_M}{L_r} \frac{1}{\sigma L_S} \Psi_r \Psi_S \sin \gamma_\Psi, \tag{16.17}$$

where the angle between stator and rotor flux vectors is defined as

$$\gamma_\Psi = \gamma_{\Psi_S} - \gamma_{\Psi_r} \tag{16.18}$$

From Equation 16.17, it can be seen that the electromagnetic torque depends on amplitudes of stator and rotor fluxes and angle between them γ_Ψ. Thanks to long rotor time constant, the angle γ_Ψ can be controlled by fast change of stator flux vector position. Under assumption that the stator resistance R_s is zero, the stator flux can be easily expressed as a function of a stator voltage:

$$\frac{d\Psi_S}{dt} = \mathbf{U}_S. \tag{16.19}$$

Or in the form:

$$\Psi_S = \int \mathbf{U}_S \, dt, \tag{16.20}$$

VSR: From Figure 16.12, it can be seen that there are two power loops: for active P, and reactive Q ones. Command active power P_c is controlled by DC-link voltage loop, while the command reactive power Q_c is given from the outside of the control scheme. Usually reactive power is set to be zero, to obtain a UPF operation. The DC-link voltage is maintaining to be constant by appropriate active power adjustment. Estimated values of the active power P and reactive power Q are compared with commanded values. The power errors e_P and e_Q are input signal to hysteresis comparators. At the output of the comparators are digitized signals S_P and S_Q.

In classical (voltage based) DPC from predefined switching table, based on signals S_P, and S_Q, and position of the line voltage γ_{U_L}, the appropriate voltage vector is selected. In VF-based VF-DPC instead of γ_{U_L}, the position of the VF γ_{Ψ_L} are utilized in control algorithm. The outputs from the predefined switching table are the switching states $\mathbf{S}_1$ for the VSR.

The instantaneous active power P is a scalar product between the line voltages and currents instantaneous space vectors, whereas the instantaneous reactive power Q is a vector product between them, and they can be expressed in complex form as

$$P = \frac{3}{2}\mathrm{Re}\{\mathbf{U}_L\mathbf{I}_L^*\} = \frac{3}{2}\left(U_{L\alpha}I_{L\alpha} + U_{L\beta}I_{L\beta}\right) = \frac{3}{2}\mathbf{U}_L \times \mathbf{I}_L \tag{16.21}$$

$$Q = \frac{3}{2}\mathrm{Im}\{\mathbf{U}_L\mathbf{I}_L^*\} = \frac{3}{2}\left(U_{L\beta}I_{L\alpha} - U_{L\alpha}I_{L\beta}\right) = \frac{3}{2}\mathbf{U}_L \times \mathbf{I}_L \tag{16.22}$$

There is a possibility to estimate the line voltages by adding the input voltage $U_p = U_{dc}\mathbf{S}_1$ of the VSR to the voltages drops on the input choke U_I. Therefore, active and reactive power of the line can be calculated in line voltage sensorless manner as follows:

$$P = \left(U_{dc}S_A + L\frac{dI_{LA}}{dt} \right)I_{LA} + \left(U_{dc}S_B + L\frac{dI_{LB}}{dt} \right)I_{LB} + \left(U_{dc}S_C + L\frac{dI_{LC}}{dt} \right)I_{LC} \tag{16.23}$$

$$Q = \frac{1}{\sqrt{3}}\left\{ 3L\left(\frac{dI_{LA}}{dt}I_{LC} - \frac{dI_{LC}}{dt}I_{LA} \right) + \left(-U_{dc}\left[S_A(I_{LB} - I_{LC}) + S_B(I_{LC} - I_{LA}) + S_C(I_{LA} - I_{LB}) \right] \right) \right\} \tag{16.24}$$

Such calculated power can be used as a feedback signal for DPC scheme. Please consider that power losses on the resistance of the input choke R are neglected because they have low value in comparison to total active power.

Unfortunately, such calculation causes some problems in DSP implementation. The differential operations of the currents are performed on the basis of finite differences and gives very noisy signals. So, to suppress the current ripples, a relatively large inductance is needed. Moreover, calculation of finite differences of the currents should be as accurate as possible (about ten times per a switching period) and should be avoided at the moment of the switching [33].

To avoid this problem, a VF of the line has been introduced in [28,30]. The voltage in the line can be expressed by the formula

$$\frac{d\mathbf{\Psi}_L}{dt} = \mathbf{U}_L \tag{16.25}$$

After integration, the VF can be expressed as

$$\mathbf{\Psi}_L = \int \mathbf{U}_L \, dt + \mathbf{\Psi}_{L0} \tag{16.26}$$

Further, when the frequency of the rotating VF is constant, also the length of VF and voltage are proportional to each other. Moreover, a phase position between VF and voltage is 90° (lagging).

From analogy with IM, the instantaneous active power can be expressed as

$$P = M\omega_L \tag{16.27}$$

where M is an instantaneous virtual torque (VT) and can be expressed as [18]

$$M = \frac{3}{2}\mathrm{Im}\left\{ \mathbf{\Psi}_L^* \mathbf{I}_L \right\} \tag{16.28}$$

Then instantaneous active power is described by

$$P = \frac{3}{2}\mathrm{Im}\left\{ \mathbf{\Psi}_L^* \mathbf{I}_L \right\}\omega_L \tag{16.29}$$

Moreover, instantaneous reactive power can be derived from the following equation:

$$Q = \frac{3}{2}\mathrm{Re}\left\{ \mathbf{\Psi}_L^* \mathbf{I}_L \right\}\omega_L \tag{16.30}$$

After calculation in stationary αβ coordinates, instantaneous active and reactive power can be calculated as

$$P = \frac{3}{2}\omega_L \left(\Psi_{L\alpha} I_{L\beta} - \Psi_{L\beta} I_{L\alpha} \right) \tag{16.31}$$

$$Q = \frac{3}{2}\omega_L \left(\Psi_{L\alpha} I_{L\alpha} + \Psi_{L\beta} I_{L\beta} \right) \tag{16.32}$$

16.4.3 Direct Torque Control with Space Vector Modulation and Direct Power Control with Space Vector Modulator

VSI: To avoid the drawbacks of switching table-based DTC instead of hysteresis controllers and switching table, the *PI* controllers with the SVM block were introduced like in IFOC. Therefore, DTC with SVM (DTC–SVM) joins DTC and IFOC features in one control structure, as in Figure 16.13.

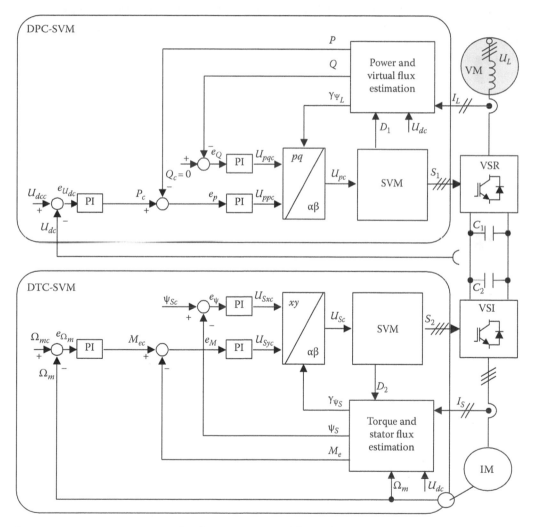

FIGURE 16.13 Direct power control with space vector modulation (DPC-SVM) and direct torque control with space vector modulation (DTC-SVM).

The commanded electromagnetic torque M_{ec} is delivered from outer PI speed controller (Figure 16.13). Then, M_{ec} and commanded stator flux Ψ_{Sc} amplitudes are compared with estimated actual values of M_e and Ψ_S. The torque e_M and flux e_ψ errors are fed to two PI controllers. The output signals are the command stator voltage components U_{Syc}, and U_{Sxc} respectively.

Further, voltage components in rotating xy system of coordinates are transformed into $\alpha\beta$ stationary coordinates using γ_{Ψ_s} flux position angle. Obtained voltage vector $\mathbf{U}_{Sc}$ is delivered to space vector modulator (SVM), which generates appropriate switching states vector $\mathbf{S}_2(S_{2A}, S_{2B}, S_{2C})$ for the VSI.

VSR: Direct power control with space vector modulation (DPC-SVM) [29] guarantees high dynamics and static performance via an internal power control loops. It is not well known in the literature. This method joins the concept of DPC and V-FOC. The active and reactive power is used as control variables instead of the line currents.

The DPC-SVM with constant switching frequency uses closed active and reactive power control loops (Figure 16.13). The command active power P_c are generated by outer DC-link voltage controller, whereas command reactive power Q_c is set to zero for UPF operation. These values are compared with the estimated P and Q values, respectively. Calculated errors e_p and e_Q are delivered to PI power controllers. Voltages generated by power controllers are DC quantities, what eliminates steady-state error (PI controller features), as well as in V-FOC. Then, after transformation to stationary $\alpha\beta$ coordinates, the voltages are used for switching signals generation by SVM block. The proper design of the power controller parameters is very important. Therefore, analysis and synthesis will be described in the followed section.

Based on discussion presented in this chapter, a brief comparison of control techniques for AC–DC–AC converter–fed IM drives is given in Table 16.3.

Among discussed control methods, DTC-SVM and DPC-SVM seem to be most attractive for further consideration. Because these methods connect well-known advantages of FOC and V-FOC with attractiveness of novel strategies such as hysteresis-based DTC and DPC. Therefore, for further considerations, a shorter name will be used for common control method of full-controlled AC–DC–AC converter: DPTC-SVM.

TABLE 16.3   Comparison of Control Techniques for AC–DC–AC Converter–Fed IM Drives

Feature	IFOC/V-FOC	DTC/VF-DPC	DPC-SVM/DTC-SVM
Constant switching frequency	Yes (5 kHz)	No	Yes (5 kHz)
SVM blocks	Yes	No	Yes
Coordinates transformation	Yes	No	Yes (only one)
Direct control of (VSI side)	Stator currents	Torque, stator flux	Torque, stator flux
Estimation of (VSI side)	Rotor flux angle	Torque, stator flux	Torque, stator flux
Coordinates orientation (VSI side)	Rotor flux	Stator flux	Stator flux
Direct control of (VSR side)	Line currents	Line powers	Line powers
Estimation of (VSR side)	Virtual flux	Powers, virtual flux	Powers, virtual flux
Coordinates orientation (VSR side)	Virtual flux	Virtual flux	Virtual flux
Line voltage sensorless	Yes	Yes	Yes
Sampling frequency	5 kHz	50 kHz	5 kHz
Independence from rotor parameters; universal for IM and PMSM	No	Yes	Yes

16.5 Line Side Converter Controllers Design

Because abbreviations VSR and VSI determine a energy-flow direction, it is better to describe these converters as line side converter (LSC) (historically VSR) and MSC (historically VSI).

16.5.1 Line Current and Line Power Controllers

The model presented in Section 16.3 is very convenient to use in synthesis and analysis of the current regulators for VSR. However, presence of coupling requires an application of *decoupling network* (DN), as in Figure 16.14.

Hence, it can be clearly seen that decoupled command rectifier voltage $\mathbf{U}_{pxyc} = U_{dc}\mathbf{S}_{xy}$ would be generated as follows:

$$U_{pxc} = U_{Lx} - L\frac{dI_{Lx}}{dt} - RI_{Lx} + \omega_L L I_{Ly} \tag{16.33}$$

$$U_{pyc} = U_{Ly} - L\frac{dI_{Ly}}{dt} - RI_{Ly} - \omega_L L I_{Lx} \tag{16.34}$$

Decoupling for the x and y axes reduces the synchronous rotating current control plant to a first-order delay.

It simplifies the analysis and enables the derivation of analytical expressions for the parameters of current regulators.

Control structure will operate in discontinuous environment (complete simulation model, and implementation in DSP) therefore, is necessary to take into account the sampling period T_S. It can be done by sample and hold (S&H) block. Moreover, the statistical delay of the PWM generation $T_{PWM} = 0.5T_S$ should be taken into account (block VSC). In the literature [4,25], the delay of the PWM is approximated from zero to two sampling periods T_S. Further, $K_C = 1$ is the VSC gain, τ_0 is a dead time of the VSC ($\tau_0 = 0$ for ideal converter).

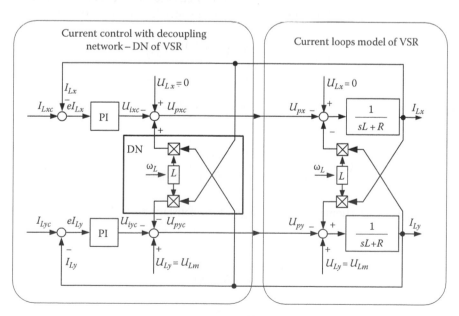

FIGURE 16.14 Current control with DN of VSR controlled by V-FOC.

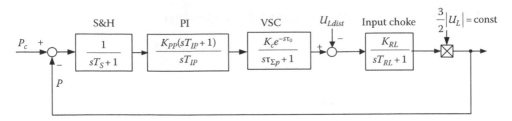

FIGURE 16.15 Block diagram for a simplified active power control loop in the synchronous rotating reference frame.

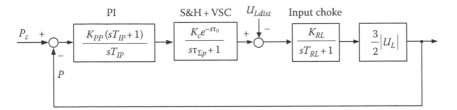

FIGURE 16.16 Modified block diagram of Figure 16.15.

In Figure 16.15, a block diagram for a simplified power control loop in the synchronous *xy* rotating coordinates is presented. Since the same block diagram applies to both *P* and *Q* power controllers, description only for *P* active power control loop will be presented.

The model of Figure 16.15 can be modified, as shown in Figure 16.16, where sum of the small time constants is defined by

$$\tau_{\Sigma p} = T_S + T_{PWM} \tag{16.35}$$

Please note that $\tau_{\Sigma p}$ is a sum of small time constants, T_{RL} is a large time constant of the input choke. From several methods of design, symmetry optimum (SO) is chosen because it has good response to a disturbance U_{Ldist} step. For $U_L = const.$ the following open-loop transfer function can be derived:

$$G_{OP}(s) = \frac{K_{RL}K_{PP}\left(1 + sT_{IP}\right)}{sT_{IP}\left(s\tau_{\Sigma p} + 1\right)\left(sT_{RL} + 1\right)} \frac{3}{2}\left|U_L\right| \tag{16.36}$$

With simplification $(sT_{RL} + 1) \approx sT_{RL}$ [18] gives following closed-loop transfer function for power control loop:

$$G_{ZP}(s) = \frac{K_{RL}K_{PP}\left(1 + sT_{IP}\right)}{K_{RL}K_{PP}\left(1 + sT_{IP}\right) + s^2 T_{IP}T_{RL} + s^3 T_{IP}\tau_{\Sigma p}T_{RL}} \frac{3}{2}\left|U_L\right| \tag{16.37}$$

For this relation, the proportional gain and integral time constant of the PI current controller can be calculated as

$$K_{PP} = \frac{T_{RL}}{2\tau_{\Sigma p}K_{RL}} \frac{2}{3\left|U_L\right|} \tag{16.38}$$

$$T_{IP} = 4\tau_{\Sigma p} \tag{16.39}$$

which, substituted in Equation 16.36, yields open-loop transfer function of the form:

$$G_{OP}(s) = \frac{K_{RL}K_{PP}\left(1+sT_{IP}\right)}{sT_{IP}\left(s\tau_{\Sigma p}+1\right)\left(sT_{RL}+1\right)}\frac{3}{2}\left|U_L\right| \tag{16.40}$$

$$G_{OP}(s) = \frac{2T_{RL}}{2\tau_{\Sigma p}K_{RL}3\left|U_L\right|}\frac{K_{RL}\left(1+s4\tau_p\right)}{s4\tau_{\Sigma p}\left(s\tau_{\Sigma p}+1\right)\left(sT_{RL}+1\right)}\frac{3}{2}\left|U_L\right|$$

$$\approx \frac{T_{RL}}{2\tau_{\Sigma i}}\frac{\left(1+s4\tau_{\Sigma p}\right)}{s4\tau_{\Sigma p}\left(s\tau_{\Sigma p}+1\right)sT_{RL}} = \frac{\left(1+s4\tau_{\Sigma p}\right)}{s^2 8\tau_{\Sigma p}^2 + s^3 8\tau_{\Sigma p}^3}. \tag{16.41}$$

For the closed-loop transfer function:

$$G_{CP}(s) = \frac{1+s4\tau_{\Sigma p}}{1+s4\tau_{\Sigma p}+s^2 8\tau_{\Sigma p}^2 + s^3 8\tau_{\Sigma p}^3}. \tag{16.42}$$

Tuning of the regulators based on Equations 16.38 and 16.39 gives power tracking performance with more then 40% overshoot, as shown in Figure 16.18 caused by the forcing element in the numerator (Equation 16.42). Therefore, for decreasing the overshot (compensate for the forcing element in the numerator), a first order prefilter on the reference signal can be used:

$$G_{pfp}(s) = \frac{1}{1+sT_{pfp}} \tag{16.43}$$

where T_{pfp} usually equals to a few $\tau_{\Sigma p}$. On further investigation, a time delay of the prefilter is set to $4\tau_{\Sigma p}$. Therefore, Equation 16.42 takes a form:

$$G_{CPf}(s) = G_{CP}(s)G_{pfp}(s) = \frac{1}{1+s4\tau_{\Sigma p}+s^2 8\tau_{\Sigma p}^2 + s^3 8\tau_{\Sigma p}^3}. \tag{16.44}$$

Hence, the block diagram of the control loop takes a form, as in Figure 16.17. Relation (16.44) can be approximated by first-order transfer function as

$$G_{Cpf}(s) \cong \frac{1}{1+s4\tau_{\Sigma p}}. \tag{16.45}$$

Comparison of step answer in control loop without (a) and with (b) prefilter is shown in Figures 16.18 and 16.19. The first one shows response (in MATLAB® and Simulink®) to a step change of active power reference step at time $t = 0.1$ [s] whereas in $t = 0.11$ [s] the disturbance step is applied.

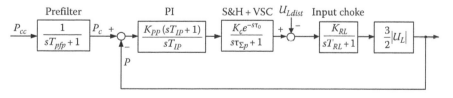

FIGURE 16.17 Power control loop with prefilter.

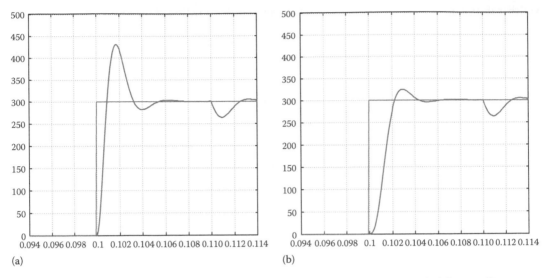

FIGURE 16.18 Active power tracking performance (simulated in MATLAB® and Simulink®) controller parameters designed according to SO: (a) without prefilter and (b) with prefilter.

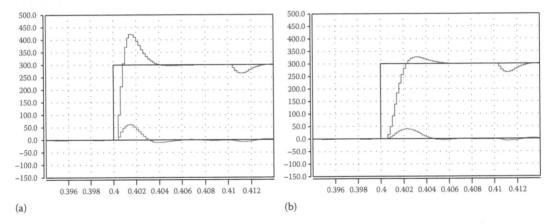

FIGURE 16.19 Active power tracking performance (simulated in Saber) without decoupling: (a) without prefilter and (b) with prefilter. From the top: command and estimated active power, command and estimated reactive power.

Discrete simulations (in Saber) show that the answer is little bit different.

Figure 16.19 presents response to a step change of reference ($t = 0.4$ s) in complete Saber model. At time $t = 0.41$ s, disturbance step is applied. The difference is caused because of presence of the nonlinear coupling.

Therefore, decoupling in power control feedbacks should be introduced. Hence, it could be clearly seen that command line voltages should be generated as follows:

$$L\frac{dI_{Lx}}{dt} + RI_{Lx} + \omega_L L I_{Ly} + U_{dc}S_x = U_{Lxc} \tag{16.46}$$

$$L\frac{dI_{Ly}}{dt} + RI_{Ly} - \omega_L L I_{Lx} + U_{dc}S_y = U_{Lyc} \tag{16.47}$$

The step answer of the system with implemented decoupling in power control loop is presented in Figure 16.20 (reference step at $t = 0.4$ s and disturbance step at $t = 0.41$ s).

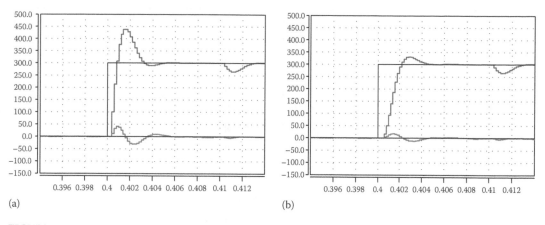

(a) (b)

FIGURE 16.20 Active power tracking performance (simulated in Saber) with decoupling: (a) without prefilter and (b) with prefilter. From the top: command and estimated active power, command and estimated reactive power.

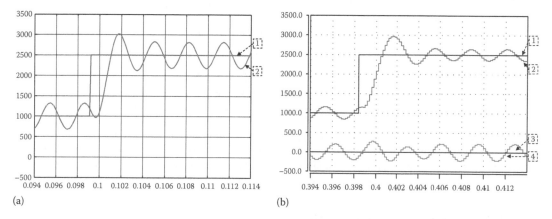

(a) (b)

FIGURE 16.21 Active power tracking performance (simulated) with prefilter: (1) command active power, (2) estimated active power, (3) command reactive power, (4) estimated reactive power. (a) Simulation in MATLAB® and Simulink® and (b) simulation in Saber.

The response is closer to ideal one obtained in MATLAB. However, there is still difference. It is caused by not fully decoupled signals and effect of sampling time T_s of discrete control system.

For better comparison with experimental results, the test under distorted line voltage was performed. Command power has been changed from 1 to 2.5 kW as on the real system. The simulation result for this case is shown in Figure 16.21.

Please take into account the oscillations in Figure 16.21. There are generated by modeled line voltage distortion (THD$_{U_L}$ = 4% of fifth harmonics). This harmonics after coordinate transformation to rotating coordinates gives AC components with frequency six times higher than line voltage frequency (300 Hz) and with amplitude U_{m6} = 6.9 V. Hence, a question arises: how the sampling frequency takes impact on the control parameters and on the design of the power controllers? Therefore, take into consideration the following simulated results presented in Figures 16.22 and 16.23. Figure 16.22 shows active and reactive power tracking performance at different sampling frequency: (a, b) f_s = 2.5 kHz, and (c, d) f_s = 5 kHz. Figure 16.23 presents the same results at different sampling frequency: (a, b) f_s = 10 kHz, and (c, d) f_s = 20 kHz, respectively.

Parameters of the power controllers derived according to SO for different values of sampling frequency are shown in Table 16.4.

Based on this comparison, it can be concluded that for higher sampling frequency distortion with fifth harmonics decay. Therefore, even for distorted line voltage, the line current will be very close to sinusoidal.

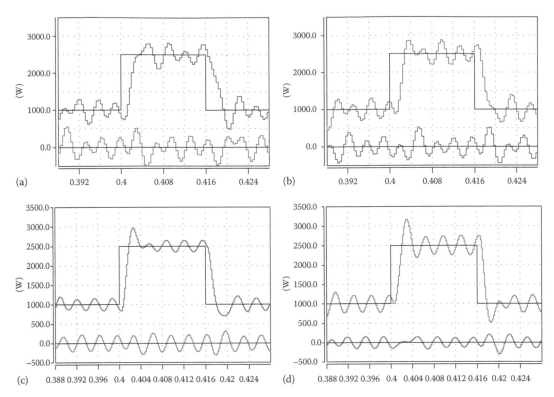

FIGURE 16.22 Active and reactive power tracking performance (simulated in Saber) at different sampling frequencies: f_s = 2.5 kHz (a) active power step and (b) reactive power step; f_s = 5 kHz (c) active power step and (d) reactive power step. From the top: (a,c) command and estimated active power, command and estimated reactive power; (b,d) command and estimated reactive power, command and estimated active power.

16.5.2 DC-Link Voltage Controller

For DC-link voltage controller design, the inner current or power control loop can be modeled with the first-order transfer function (Section 16.5.1).

The power control loop of VSR can be approximated in further consideration by first order block with equivalent time constant T_{IT}.

$$G_{pz}(s) = \frac{1}{1 + sT_{IT}}$$ (16.48)

where $T_{IT} = 2\tau_{\Sigma p}$ for power controllers designed by MO criterion or $T_{IT} = 4\tau_{\Sigma p}$ for power controllers designed by SO criterion. Therefore, the DC-link voltage control loop can be modeled as in Figure 16.24.

The block diagram of Figure 16.24 can be modified as shown in Figure 16.25.

For simplicity, it can be assumed that

$$T_{UT} = T_U + T_{IT}$$ (16.49)

where
  T_U is DC-link voltage filter time constant
  T_{UT} is a sum of small time constants
  CU_{dcc} is an equivalent of integration time constant

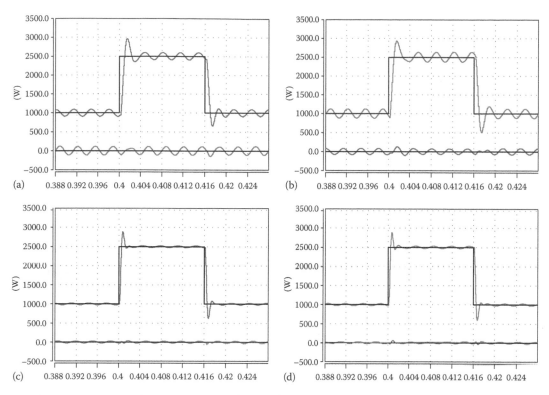

FIGURE 16.23 Active and reactive power tracking performance (simulated in Saber) at different sampling frequency: $f_s = 10\,\text{kHz}$ (a) active power step and (b) reactive power step; $f_s = 20\,\text{kHz}$ (c) active power step and (d) reactive power step. From the top: (a, c) command and estimated active power, command and estimated reactive power; (b, d) command and estimated reactive power, command and estimated active power.

TABLE 16.4 Parameters of Active and Reactive Power Controllers

$U_{LRMS} = 141\,\text{V}$		
f_s [kHz]	K_{PP}	T_{IP} [s]
2.5	0.0279	0.0024
5	0.0557	0.0012
10	0.11	0.0006
20	0.22	0.0003
50	0.44	0.00015

Therefore, the open-loop transfer function can be derived:

$$G_{Uo}(s) = \frac{K_{PU}\left(sT_{IU} + 1\right)}{sT_{IU}\left(sT_{UT} + 1\right)sCU_{dcc}} \qquad (16.50)$$

This gives the following closed-loop transfer function:

$$G_{Uz}(s) = \frac{K_{PU}\left(sT_{IU} + 1\right)}{K_{PU} + sT_{IU}K_{PU} + s^2T_{IU}CU_{dcc} + s^3T_{IU}CT_{UT}U_{dcc}} \qquad (16.51)$$

The method of symmetrical optimum is used to synthesize the DC-link voltage controller. Therefore, square of the module of Equation 16.51 takes a form:

$$G_{Uz}(\omega) = \frac{K_{PU}^2\left(\omega^2T_{IU}^2 + 1\right)}{M_z(\omega)} \qquad (16.52)$$

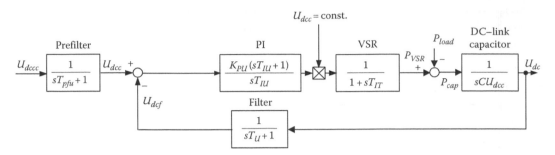

FIGURE 16.24   Block diagram for a simplified DC-link voltage control loop.

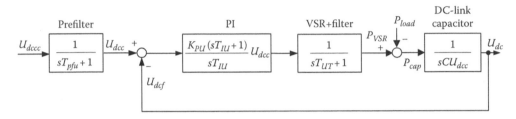

FIGURE 16.25   Modified block diagram of Figure 16.24.

where

$$M_z(\omega) = K_{PU}^2 + \omega^2 T_{IU} K_{PU}\left(T_{IU} K_{PU} - 2CU_{dcc}\right) + \omega^4 T_{IU}^2 CU_{dcc}\left(CU_{dcc} - 2K_{PU}T_{UT}\right)$$

$$+ \omega^6 \left(T_{IU} CU_{dcc} T_{UT}\right)^2$$

ω is the frequency domain

Hence, proportional gain K_{PU} and integral time constant T_{IU} of the DC-link voltage controller can be calculated as follows:

$$K_{PU} = \frac{C}{2T_{UT}} U_{dcc} \tag{16.53}$$

$$T_{IU} = 4T_{UT} \tag{16.54}$$

Please take into consideration that value of the DC-link voltage filter time constant T_U has to be determined. Theoretically, it could be equal to one sampling period T_S. However, in practice the low-pass filter in DC-link voltage loop is needed. Therefore, for further consideration, $T_U = 0.003$ [s] (results in Figure 16.26).

16.6 Direct Power and Torque Control with Space Vector Modulation

DPC-SVM and DTC-SVM seem to be most attractive for control of the AC–DC–AC converter. When both methods are joined for control of the AC–DC–AC converter, DPTC-SVM is obtained.

    In this chapter, DPTC-SVM scheme of AC–DC–AC converter-fed IM drive will be considered and power flow between VSR and VSI side will be also analyzed. Some techniques for reduction of the DC-link capacitor will be described. When the active rectifier DC-link current I_{dc} is equal to the

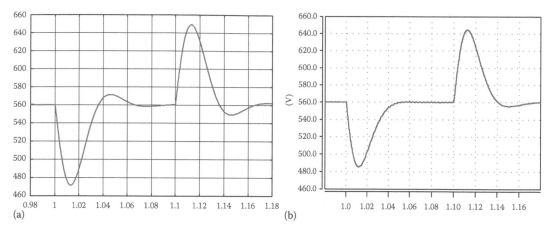

FIGURE 16.26 Voltage disturbance compensation performance (simulated) for controller parameters calculated according SO. Transient in load from zero to nominal (at $t = 1.0$ s) and from nominal to zero (at $t = 1.1$ s). (a) Simulation in MATLAB® and Simulink® and (b) simulation in Saber.

DC-link inverter current I_{load} in the AC–DC–AC converter, no current will flow through the DC-link capacitor. As a result, DC-link voltage will be constant.

However, in spite of very good dynamics behaviors of DPTC-SVM scheme, the control of the DC-link voltage can be improved [8,24]. Therefore, active PF from inverter side to rectifier side is introduced. The PF delivers information about machine states directly to active power control loop of the VSR. Thanks to faster control of power flow between VSR and VSI, the fluctuation of the DC-link voltages will decrease. So, the size of the DC-link capacitor can be significantly reduced (because voltage fluctuation is reduced).

16.6.1 Model of the AC–DC–AC Converter–Fed Induction Machine Drive with Active Power Feedforward

In Figure 16.27 is shown simplified diagram of the AC–DC–AC converter, which consists of VSR-fed DC-link and VSI-fed IM. Both VSR and VSI are IGBT bridge converters.

The mathematical models of the VSR and VSI are given in Sections 16.2 and 16.3, and description of the DPC-SVM and DTC-SVM have been presented in Section 16.4. Here, the whole system with PF will be discussed and studied.

Note again that the coordinates system for control of the VSR is oriented with VF vector. Therefore, I_{Lxc} is set to zero to meet the UPF condition. With this assumption, the VSR input power can be calculated as

$$P_{VSR} = \frac{3}{2}\left(I_{Lx}U_{px} + I_{Ly}U_{py}\right) = \frac{3}{2}I_{Ly}U_{py} \tag{16.55}$$

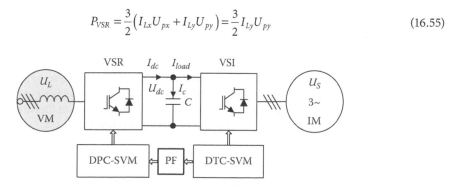

FIGURE 16.27 Configuration of the AC–DC–AC converter–fed induction machine drive with PF loop.

Under steady-state operation, I_{Ly} = const. and, with assumption that resistance of the input choke is $R = 0$, the following equation can be written:

$$P_{VSR} = \frac{3}{2} I_{Ly} U_{Ly}$$ (16.56)

On the other hand, the power consumed/produced by the VSI-fed IM is defined by

$$P_{VSI} = \frac{3}{2} \left(I_{Sx} U_{Sx} + I_{Sy} U_{Sy} \right)$$ (16.57)

Another form of the above equations can be derived based on power (Equation 16.29) where is clearly seen that the active power of the VSR is proportional to the virtual torque (VT). Therefore, Equation 16.55 can be written as

$$P_{VSR} = \frac{3}{2} \omega_L \left(\Psi_{Lx} I_{Ly} - \Psi_{Ly} I_{Lx} \right) = \frac{3}{2} \omega_L \Psi_{Lx} I_{Ly}$$ (16.58)

On the VSI-fed IM side electromagnetic power of the machine is defined by

$$P_e = M_e \Omega_m$$ (16.59)

Taking into account Equation 16.6 yields

$$P_e = p_b \frac{m_S}{2} \Omega_m \Psi_{Sx} I_{Sy}$$ (16.60)

Moreover, it can be assumed (neglecting power losses) that electromagnetic power of the IM is equal to an active power delivered to the machine $P_e = P_{VSI}$, hence,

$$P_{VSI} = p_b \frac{m_S}{2} \Omega_m \Psi_{Sx} I_{Sy}$$ (16.61)

But this is not sufficient assumption because of power losses P_{losses} in the real system, so it should be written as

$$P_{VSI} = p_b \frac{m_S}{2} \Omega_m \Psi_{Sx} I_{Sy} + P_{losses}$$ (16.62)

Further, please consider a situation at standstill ($\Omega_m = 0$) when nominal torque is applied. In such a case, the electromagnetic power will be zero but the IM power P_{VSI} will have a significant value. Estimation of this power is quite difficult, because the parameters of the IM and power switches are needed. Hence, for simplicity of the control structure, a power estimator based on command stator voltage $\mathbf{U}_{Sc}$ and actual current $\mathbf{I}_s$ will be taken into consideration:

$$P_{VSI} = \frac{3}{2} \left(I_{Sx} U_{Sxc} + I_{Sy} U_{Syc} \right)$$ (16.63)

16.6.2 Analysis of the Power Response Time Constant

Based on Equation 16.48 the time constant delay of the VSR response T_{IT} is determined. With assumption that power losses of the converters can be neglected, power tracking performance can be expressed by

$$P_{VSR}(s) = \frac{1}{1 + sT_{IT}} P_{VSRc} \tag{16.64}$$

Similarly, for the VSI, it can be written as

$$P_{VSI}(s) = \frac{1}{1 + sT_{IF}} P_{VSIc} \tag{16.65}$$

where T_{IF} is the equivalent time constant of the VSI step response.

16.6.3 Energy of the DC-Link Capacitor

The DC-link voltage can be described as (for more detail, see Section 16.3):

$$\frac{dU_{dc}}{dt} = \frac{1}{C}\left(I_{dc} - I_{load}\right), \tag{16.66}$$

So

$$U_{dc} = \frac{1}{C}\int \left(I_{dc} - I_{load}\right)dt, \tag{16.67}$$

Assuming the initial condition as in steady state, hence, the actual DC-link voltage U_{dc} is equal to commanded DC-link voltage U_{dcc}. Therefore, Equation 16.67 can be rewritten as

$$U_{dc} = \frac{1}{CU_{dcc}}\int \left(U_{dcc}I_{dc} - U_{dcc}I_{load}\right)dt = \frac{1}{CU_{dcc}}\int \left(P_{dc} - P_{load}\right)dt, \tag{16.68}$$

where $P_{dc} - P_{load} = P_{cap}$; therefore, the above equation can be written as

$$U_{dc} = \frac{1}{CU_{dcc}}\int P_{cap}dt, \tag{16.69}$$

If the power losses of the VSR and VSI are neglected (for simplicity), the energy storage variation of the DC-link capacitor will be the integral of the difference between the input power P_{VSR} and the output power P_{VSI}. Therefore, it can be written as

$$P_{VSR} = P_{cap} + P_{VSI}, \tag{16.70}$$

From this equation, it can be concluded that for proper (accurate) control of the VSR power P_{VSR}, the command power P_{VSRc} should be as follows:

$$P_{VSRc} = P_{capc} + P_{VSIc}, \tag{16.71}$$

where

$P_{capc} = P_c$ denotes power of the DC-link voltage feedback control loop

P_{VSIc} denotes the instantaneous active PF signal

The command output power can be estimated based on different methods that provide additional time constant T_2 [8,10,15,20,21,24,39]. Hence,

$$P_{VSIc2}(s) = \frac{1}{1 + sT_2} P_{VSIc}.$$ (16.72)

Moreover, it should be stressed here that the first-order filter with time constant T_U should be added to DC-link voltage feedback, which strongly delays the signal P_c (see Section 16.5.2):

$$U_{dcf}(s) = \frac{1}{1 + sT_U} U_{dc}$$ (16.73)

This delay is taken into account in DC-link voltage controller design. Hence

$$P_c(s) = \frac{K_{PU}(sT_{IU} + 1)}{sT_{IU}} eU_{dcf}U_{dcc}$$ (16.74)

where

$$eU_{dcf} = U_{dcc} - U_{dcf}$$

Therefore, Equation 16.71 can be rewritten as

$$P_{VSRc}(s) = P_c + P_{VSIc2},$$ (16.75)

Substituting Equations 16.72 and 16.74 into Equation 16.75 yields

$$P_{VSRc}(s) = \frac{K_{PU}(sT_{IU} + 1)}{sT_{IU}} eU_{dcf}U_{dcc} + \frac{1}{1 + sT_2} P_{VSIc},$$ (16.75a)

From Equations 16.64 and 16.65, the open-loop transfer function of the input power (of the VSR) and output power (of the VSI) can be written as

$$G_{VSRo}(s) = \frac{P_{VSR}}{P_{VSRc}} = \frac{1}{1 + sT_{IT}}$$ (16.76)

$$G_{VSIo}(s) = \frac{P_{VSI}}{P_{VSIc}} = \frac{1}{1 + sT_{IF}}$$ (16.77)

Based on these equations, the analytic model of the AC–DC–AC converter–fed IM drive with active PF can be defined as in Figure 16.28. Such a system can be described by open-loop transfer function as

$$G_{Ao}(s) = \frac{U_{dc}}{M_{ec}}$$ (16.78)

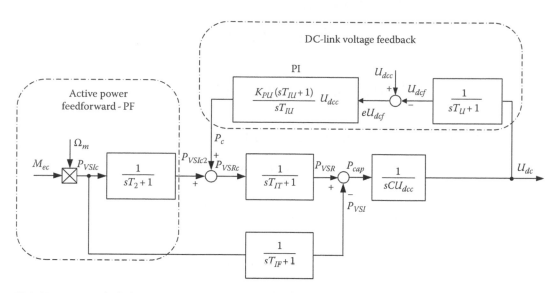

FIGURE 16.28 Block diagram of the AC–DC–AC converter–fed IM with active PF.

Assuming initial steady-state operation, $\Omega_m = \Omega_{mc}$ = const. and $U_{dc} = U_{dcc}$ = const., the transfer function of the AC–DC–AC converter–fed IM drive can be described.

Based on above considerations, it can be concluded that introduced active PF has no negative impact on the line current. This can be derived analytically. Please consider rotating reference frame concurrently with VF. With assumption that the system is decoupled, and meets the UPF condition, i.e., $I_{Lx} = 0$, the following equations can be derived:

$$L\frac{dI_{Ly}}{dt} = U_{Ly} - U_{py} - RI_{Ly} \tag{16.79}$$

$$C\frac{dU_{dc}}{dt} = I_{Ly}\frac{U_{py}}{U_{dc}} - I_{load} \tag{16.80}$$

$$L\frac{dI_{Ly}}{dt} = \frac{K_{Pi1}}{T_{Ii1}}\left(I_{Lyc} - I_{Ly}\right) \tag{16.81}$$

$$L\frac{dI_{Lyc}}{dt} = \frac{K_{PU}}{T_{IU}}\left(U_{dcc} - U_{dc}\right) \tag{16.82}$$

$$I_{Lyc} = K_{PU}\left(U_{dcc} - U_{dc}\right) \tag{16.83}$$

Then, VSR voltage

$$U_{py} = K_{Pi1}\Big[K_{PU}\left(U_{dcc} - U_{dc}\right) + I_{PF} - I_y\Big] \tag{16.84}$$

where I_{PF} is the current proportional to signal of the active PF.
Hence, during steady-state operation, the above equations yield

$$0 = U_{Ly} - K_{Pi1}\Big[K_{PU}\left(U_{dcc} - U_{dc}\right) + I_{PF} - I_y\Big] - RI_{Ly} \tag{16.85}$$

$$0 = I_{Ly} \frac{1}{U_{dc}} \left\{ K_{Pi1} \left[K_{PU} \left(U_{dcc} - U_{dc} \right) + I_{PF} - I_y \right] \right\} - I_{load} \tag{16.86}$$

$$0 = \frac{K_{Pi1}}{T_{Ii1}} \left(K_{PU} \left(U_{dcc} - U_{dc} \right) - I_{Ly} \right) = \left(K_{PU} \left(U_{dcc} - U_{dc} \right) - I_{Ly} \right) \tag{16.87}$$

Substituting Equations 16.87 into Equations 16.85 and 16.86:

$$0 = U_{Ly} - K_{Pi1} I_{PF} - R I_{Ly} \tag{16.88}$$

$$0 = I_{Ly} \frac{1}{U_{dcc}} K_{Pi1} I_{PF} - I_{load} \Rightarrow K_{Pi1} I_{PF} = \frac{I_{load} U_{dcc}}{I_{Ly}} \tag{16.89}$$

Based on Equation 16.89 the current from active PF can be eliminated:

$$0 = U_{Ly} - \frac{I_{load} U_{dcc}}{I_{Ly}} - R I_{Ly} \tag{16.90}$$

From Equation 16.90, it can be seen that steady states do not depends on active PF. Further, based on Equation 16.90 a stationary error elimination condition can be defined as

$$R I_{Ly}^2 - U_{Ly} I_{Ly} - I_{load} U_{dcc} = 0 \tag{16.91}$$

If the solutions are real, the stationary error is eliminated. It takes place for

$$I_{Ly1/2} = \frac{U_{Ly} \pm \sqrt{U_{Ly}^2 - 4R I_{load} U_{dcc}}}{2R} \tag{16.92}$$

when

$$I_{load} U_{dcc} < \frac{U_{Ly}^2}{4R} \tag{16.93}$$

16.7 DC-Link Capacitor Design

In this chapter, methods of passive components for AC–DC–AC converter design are discussed. There are input filter (L or LCL) and the DC-link capacitor, which have strong effect on size, weight, and final price of AC–DC–AC converter.

To minimize the reactive filter requirements of a VSR circuit, certain terminal constraints are needed: input and output voltages, THD of the input and output current (I_L, I_{dc} respectively). The filter requirements are usually defined in terms of filter cost and/or size. Here, the selection of DC-link capacitor is discussed while L and LCL filter is out of the scope of the thesis. However, the design of L as well as LCL filter is important and it has been well discussed in many publications, e.g., [25,37,38].

The advantages of Al electrolytic capacitors that have led to their wide application range, are their high volumetric efficiency (i.e., capacitance per unit volume), which enables the production of capacitors with up to 1 F capacitance, and the fact that an Al electrolytic capacitor provides a high ripple current capability together with a high reliability and an excellent price/performance ratio. However, optimization and minimization of the DC-link capacitor volume is very important for the cost, power density, and reliability of the product. Therefore, advanced control strategies have been investigated for better control of the DC-link voltage.

16.7.1 Ratings of the DC-Link Capacitor

Beside variety design criteria [32,42] of the DC-link capacitor, the minimum capacitance value is designed to limit the DC-link voltage ripple at a specified level, typically ΔU_{dc} is 1% or 2% of U_{dc}. Therefore, peak-to-peak voltage ripple in DC-link is adopted as the design criterion for the DC-link capacitor sizing.

With assumption of a balanced tree-phase line and neglecting the power losses in the power switches, the VSR's DC-link part can be described as

$$C\frac{dU_{dc}}{dt} = I_{dc} - I_{load} = \sum_{k=A}^{C} I_{Lk}S_k - I_{load} \approx I_{LA}S_A + I_{LB}S_B + I_{LC}S_C - \frac{P_{load}}{U_{dc}} \tag{16.94}$$

For a given allowable peak ripple voltage and switching frequency, the minimum capacitor for the converter in Figure 16.27 can be found from [6].

$$C_{min\_VSR1} = P_{load\_max}\frac{\sqrt{2} + \left(\sqrt{3}U_{LL}/U_{dc}\right)}{2\sqrt{3}\Delta U_{dc}f_s U_{LL}}, \tag{16.95}$$

where
  U_{LL} is a line-to-line voltage
  P_{load_max} is the maximal load power
  ΔU_{dc} is the specified peak-to-peak voltage ripple in DC-link during steady states

Another approach of the DC-link capacitors design takes into account the following considerations:

- The voltage ripple, due to the high-frequency components of the modulated DC-link currents of both converters (i.e. VSR and VSI), have to remain within desired limits.
- When all switches of VSR are off, the inductor's energy flows into the capacitor, increasing its voltage.
- The capacitor energy has to sustain the output power demand in a period of the time delay of the DC-link voltage control loop.

The first and the second point of considerations are less important, while the third one practically determines the capacitor value. Assuming time delay of the DC-link voltage control loop T_{UT} and variation of the maximal load power ΔP_{load_max}, the energy exchanged by the DC-link capacitor can be estimated as

$$\Delta W_{dc} = \Delta P_{load\_max}T_{UT} \tag{16.96}$$

where T_{UT} is defined in Section 16.3.

From this equation, the maximal DC-link voltage variation during transient is expressed by

$$\Delta U_{dc\_max} = \frac{\Delta W_{dc}}{C_{min\_VSR2}U_{dc}} \tag{16.97}$$

Considering the maximal voltage variation during transient ΔU_{dc_max} and rearranging an Equation 16.97 the minimal capacitance can be calculated as [27]

$$C_{min\_VSR2} = \frac{T_{UT}\Delta P_{load\_max}}{U_{dc}\Delta U_{dc\_max}} \tag{16.98}$$

Other expression for minimum capacitance can be derived, given the maximum power step modeled by maximal load power in square $P_{load\_max}^2$ and the maximum allowable DC-link voltage variation ΔU_{dc_max} [2].

$$C_{min\_VSR3} = \frac{2LP_{load\_max}^2}{U_{dc}\Delta U_{dc\_max}U_L^2} \tag{16.99}$$

Because of simplification ($P_{load\_max}^2$, T_{UT}, and T_S are not taken into account) used in Equation 16.99, the calculated capacitance is slightly too large.

For a comparison, the DC-link capacitor value of AC–DC–AC converter with diode rectifier is given by [6]:

$$C_{di} = P_{load\_max}\frac{\pi^2}{54\sqrt{2}\Delta U_{dc}f_L U_{LL}} \tag{16.100}$$

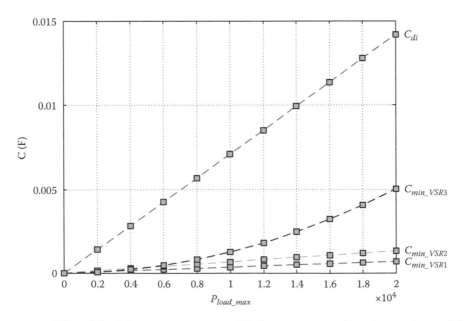

FIGURE 16.29 Value of the DC-capacitor versus rated load power up to 20 kW: C_{di}, capacitance defined by Equation 16.100; C_{min_VSR1}, capacitance defined by Equation 16.95; C_{min_VSR2}, capacitance defined by Equation 1.98; and C_{min_VSR3}, capacitance defined by Equation 1.99.

Figure 16.29 shows values of the DC-link capacitor in AC–DC–AC converter with diode rectifier and with VSR. It can be seen that difference between capacitances raised proportional to power rating. Therefore, only for higher power the installation of the VSR is more cost-effective.

Therefore, it can be concluded that for given nominal power, the DC-link capacitor value depends only on the switching pattern and quality of the applied control methods. In considered case the switching frequency is equal to the sampling frequency. For higher T_s, the DC-link capacitor can be smaller because the DC-link voltage error is significantly reduced, as shown in Figure 16.30. However, the switching frequency is limited by the switching losses of the devices used in the VSC.

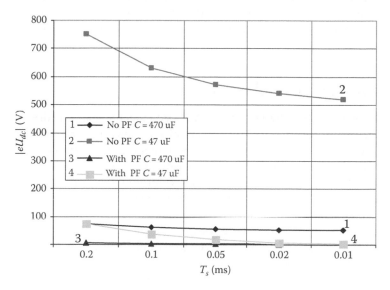

FIGURE 16.30 Module of the DC-link voltage fluctuation $|eU_{dc}|$ as a function of sampling time T_s, during load change from 0 to 3 kW with command DC-link voltage $U_{dcc} = 560$ V. Simulated results in Saber model of DPTC-SVM.

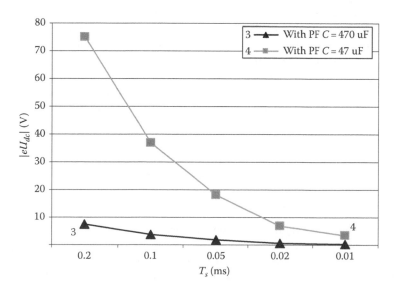

FIGURE 16.31 Module of the DC-link voltage fluctuation $|eU_{dc}|$ as a function of sampling time T_s, during load change from 0 to 3 kW with command DC-link voltage $U_{dcc} = 560$ V for a case when in control structure the active PF is implemented. Simulated results in Saber model of DPTC-SVM.

Hence, to further reduction of the DC-link filter, the device losses should be made independent of switching frequency. It can be realized by the use of soft-switched VSC such as the resonant DC-link converter.

From Figure 16.30, it can be concluded that for ideal control methods, with sufficiently high sampling time, the DC-link voltage fluctuation during transient for reduced capacitor can be even smaller (case with active PF [PF C = 47 μF]) than for higher value of the capacitance (case without active PF [no PF C = 470 μF]). Moreover, taking into consideration only the cases with PF, it can be seen that the fluctuations of DC-link voltage during transient for C = 470 μF and for C = 47 μF are almost equal for higher sampling time (Figure 16.31). Based on this, the postulate can be formed as follows: stabilization of the DC-link voltage in AC–DC–AC converter with transistor bridge rectifier (VSR) mainly depends on the quality of the used DC-link voltage and input/output power flow control methods.

16.8 Summary and Conclusion

In this chapter, a line power friendly AC–DC–AC VSC-fed induction machine drives have been studied and investigated under aspect of DSP-based control and industrial serial manufacturing. Such a control system should meet following requirements: operation with UPF, low THD of the line current, very good stabilization of the DC-link voltage, four-quadrant operation, and be universal for different types of drive application (open-loop control, sensorless control, and control with motion sensor). Also, for serial manufacturing, it is important to have simple reliable and low-cost solution with repeatable parameters of drive.

After literature studies, two control strategies were chosen: DTC with space vector modulation (DTC-SVM) for the machine side DC–AC converter (VSI) and direct power control with space vector modulation DPC-SVM for the line side AC–DC converter (VSR). As results, the novel scheme for control of AC–DC–AC converter–fed induction machine (IM) called (see Figure 16.13) has been proposed.

After simulation study, it has been shown that when control structure of the VSR-fed DC-link operates independently from control of the VSI-fed IM, the DC-link voltage stabilization is not sufficiently fast to balance instantaneous DC-link power flow. Therefore, a high value of DC-link capacitor is required. However, it could be significantly reduced if a new elaborated active PF estimator has been introduced to whole control structure. As shown in Section 1.6 of this chapter, thanks to PF loop, the value of the DC-link capacitor has been reduced 10 times while good stabilization of DC-link voltage in both dynamic and stationary states as well as low line current THD are kept. However, it should be stressed that the system has to be equipped with fast PF loop and/or DC-link chopper.

For proper operation of the closed control loops in DPTC-SVM scheme, the digital PI controllers design is very important. Therefore, in Section 16.5 synthesis of active and reactive power as well as DC-link voltage controller is discussed. This design is based on continuous transfer function approach (symmetric optimum [SO]), because fast sampling used in the system (dominated time constant—line inductor T_{RL}—is much higher as sampling time T_S), and has been confirmed by both simulated and experimental results. The flux and torque control design has been implemented similarly.

The main features and advantages of developed DPTC-SVM scheme (Figure 16.13) for AC–DC–AC converter–fed IM can be summarized as follows:

- Four-quadrant operation (bidirectional power flow)
- Fast flux and torque response
- Available maximum output torque in wide range of speed operation region
- Constant switching frequency and unipolar voltage pulses (thanks to use SVM)
- Low flux and torque ripple
- Robustness to rotor parameter variations
- Motion sensorless operation

- Nearly sinusoidal line current (low THD typically below 5%)
- Direct adjustable reactive power (including UPF)
- Controllable DC-link voltage (well stabilized at desired level)
- AC side voltage sensors are eliminated (only DC-link voltage sensor is required)
- Wide-speed range operation of the IM connected to the output of the VSI
- Reduction of DC-link capacitor, thanks to active PF (gives possibility to use a foil capacitor)
- Easy operation in open- or closed-loop fashion, because (separately SVM blocks are used)

References

1. H. Akagi, New trends in active filters for power conditioning, *IEEE Transactions on Industry Applications*, 32, 1312–1332, 1996.
2. M. Alakula and J. E. Persson, Vector Controlled AC/AC converters with a minimum of energy storage, In *Proceedings of the IEEE Conference*, London, U.K., 1994, pp. 1130–1134.
3. F. Blaschke, A new method for the structural decoupling of A.C. induction machines, in *Proceedings of the Conference Record IFAC*, Duesseldorf, Germany, October 1971, pp. 1–15.
4. V. Blasko and V. Kaura, A new mathematical model and control of a tree-phase AC-DC voltage source converter, *IEEE Transactions on Power Electronics*, 12(1), 116–123, January 1997.
5. B. K. Bose, *Modern Power Electronics and AC Drives*, Prentice-Hall, Upper Saddle River, NJ, 2002.
6. M. Cichowlas, PWM rectifier with active filtering, PhD thesis, Warsaw University of Technology, Warsaw, Poland, 2004.
7. R. W. De Doncker and D. W. Novotny, The universal field oriented controller, *IEEE Transactions on Industry Applications*, 30(1), 92–100, January/February 1994.
8. B.-G. Gu and K. Nam, A DC link capacitor minimization method through direct capacitor current control, in *Conference Record of the Industry Applications Conference, 2002, 37th IAS Annual Meeting*, Pittsburgh, Vol. 2, October 13–18, 2002, pp. 811–817.
9. T. G. Habatler, F. Profumo, M. Pastorelli, and L. M. Tolbert, Direct torque control of induction machines using space vector modulation, *IEEE Transactions on Industrial Applications*, 28(5), 1045–1053, September/October 1992.
10. T. G. Habatler and D. M. Divan, Rectifier/inverter reactive component minimalization, *IEEE Transactions on Industrial Applications*, 25(2), 307–316, March/April 1989.
11. D. G. Holmes and T. A. Lipo, *Pulse Width Modulation for Power Converters, Principles and Practice*, Wiley-Interscience, Hoboken, NJ/IEEE Press, Piscataway, NJ, 2003.
12. J. Holtz, Pulsewidth modulation for electronics power conversion, in *Proceedings of the IEEE*, 82(8), 1194–1214, August 1994.
13. IEEE, IEEE recommended practices and requirements for harmonic control in electrical power systems, IEEE Std 519-1992, The Institute of Electrical and Electronics Engineers, New York, 1993.
14. M. Jasinski, Direct power and torque control of AC/DC/AC converter-fed induction motor drives, PhD-thesis, Warsaw University of Technology, Warsaw, Poland, 2005.
15. J. Jung, S. Lim, and K. Nam, A feedback linearizing control scheme for a PWM converter-inverter having a very small DC-link capacitor, *IEEE Transactions on Industry Applications*, 35(5), 1124–1131, September/October 1999.
16. J. G. Kassakian, M. F. Schlecht, and G. C. Verghese, *Principles of Power Electronics*, Addison-Wesley Publishing Company, Reading, MA, 1991, p. 738.
17. M. P. Kazmierkowski and L. Malesani, Current control techniques for three-phase voltage-source PWM converters: A survey, *IEEE Transactions on Industrial Electronics*, 45(5), 691–703, October 1998.
18. M. P. Kazmierkowski and H. Tunia, Automatic control of converter-fed drives, Elsevier, Amsterdam, the Netherlands/London, U.K., New York/Tokyo, Japan/PWN Warszawa, Poland, 1994, p. 559.

19. M. P. Kazmierkowski, R. Krishnan, and F. Blaabjerg, *Control in Power Electronics*, Academic Press, San Diego, CA/London, U.K., 2002, pp. 579.

20. J. S. Kim and S. K. Sul, New control scheme for ac–dc–ac converter without dc link electrolytic capacitor, in *Proceedings of the IEEE PESC'93*, Seattle, WA, 1993, pp. 300–306.

21. S. Kim, S.-K. Sul, and T. A. Lipo, AC/AC power conversion based on matrix converter topology with unidirectional switches, *IEEE Transactions on Industry Applications*, 36(1), 139–145, January/February 2000.

22. J. W. Kolar, H. Ertl, K. Edelmoser, and F. C. Zach, Analysis of the control behaviour of a bidirectional three-phase PWM rectifier system, in *Proceedings of the EPE 1991 Conference*, Florence, Italy, 1991, pp. (2-095)–(2-100).

23. D.-C. Lee, K.-D. Lee, and G.-M. Lee, Voltage control of PWM converters using feedback linearization, in *Industry Applications Conference, 1998. 33rd IAS Annual Meeting. The 1998 IEEE*, Vol. 2, St Louis, MO, October 12–15, 1998, pp. 1491–1496.

24. J. Ch. Liao and S. N. Yen, A novel instantaneous power control strategy and analytic model for integrated rectifier/inverter systems, *IEEE Transactions on Power Electronics*, 15(6), 996–1006, November 2000.

25. M. Liserre, Innovative control techiques of power converters for industrial automation, Politecnico di Bari, PhD thesis, Politecnico di Bari, Bari, Italy, 2001.

26. K. J. P. Macken, M. H. J. Bollen, and R. J. M. Belmans, Mitigation of voltage dips through distributed generation systems, *IEEE Transactions on Industrial Applications*, 40(6), 1686–1693, November/December 2004.

27. L. Malesani, L. Rossetto, P. Tenti, and P. Tomasin, AC-DC-AC PWM converter with minimum energy storage in the dc link, in *Applied Power Electronics Conference and Exposition, 1993 (APEC '93), Eighth Annual Conference Proceedings 1993*, San Diego, CA, March 7–11, 1993, pp. 306–311.

28. M. Malinowski, Sensorless control strategies for three-phase PWM rectifiers, PhD thesis, Warsaw University of Technology, Warszawa, Poland, 2001.

29. M. Malinowski, M. Jasinski, and M. P. Kazmierkowski, Simple direct power control of three-phase PWM rectifier using space-vector modulation (DPC-SVM), *IEEE Transactions on Industrial Electronics*, 51(2), 447–454, April 2004.

30. V. Manninen, Application of direct torque control modulation technology to a line converter, in *Proceedings of the EPE 1995 Conference*, Seville, Spain, 1995, pp. 1.292–1.296.

31. N. Mohan, T. M. Undeland, and W. P. Robbins, *Power Electronics: Converters, Applications, and Design*, John Wiley & Sons, Singapore, 1989, p. 667.

32. L. Moran, P. D. Ziogas, and G. Joos, Design aspects of synchronous PWM rectifier-inverter systems under unbalanced input voltage conditions, *IEEE Transactions on Industry Applications*, 28(6), 1286–1293, November/December 1992.

33. T. Noguchi, H. Tomiki, S. Kondo, and I. Takahashi, Direct power control of PWM converter without power-source voltage sensors, *IEEE Transactions on Industry Applications*, 34(3), 473–479, 1998.

34. A. Sikorski, Problemy dotyczące minimalizacji strat łączeniowych w przekształtniku AC-DC-AC-PWM zasilającym maszynę indukcyjną, Politechnika Białostocka, Rozprawy Naukowe nr. 58, Białystok, Poland, 1998, pp. 217 (in Polish).

35. K. Stockman, M. Didden, F. D'Hulster, and R. Belmans, Embedded solutions to protect textile processes against voltage sags, *IEEE Industrial Applications Magazine*, 10(5), 59–65, September/October 2004.

36. T. Takahashi and T. Noguchi, A new quick-response and high efficiency control strategy of an induction machine, *IEEE Transactions on Industrial Applications*, 22(5), 820–827, September/October 1986.

37. R. Teodorescu, F. Blaabjerg, M. Liserre, and A. Dell'Aquila, A stable three-phase LCL-filter based active rectifier without damping, *Conference Record of the Industry Applications Conference, 2003, 38th IAS Annual Meeting*, Vol. 3, Salt Lake City, UT, October 12–16, 2003, pp. 1552–1557.

38. E. Twining and D. G. Holmes, Grid current regulation of a three-phase voltage source inverter with LCL input filter, *IEEE Transactions on Power Electronics*, 18(3), 888–895, May 2003.

39. R. Uhrin and F. Profumo, Performance comparison of output power estimators used in AC-DC-AC converters, in *20th International Conference on Industrial Electronics, Control and Instrumentation, 1994 (IECON'94)*, Vol. 1, Bologna, Italy, Sept. 5–9, 1994, pp. 344–348.

40. A. van Zyl, R. Spee, A. Faveluke, and S. Bhowmik, Voltage sag ride-through for adjustable-speed drives with active rectifiers, *IEEE Transactions on Industrial Applications*, 34(6), 1270–1276, November/December 1998.

41. P. Vas, *Sensorless Vector and Direct Torque Control*, Oxford University Press, New York, 1998, p. 729.

42. M. Winkelnkemper and S. Bernet, Design and optimalization of the DC-link capacitor of PWM voltage source inverter with active frontend for low-voltage drives, in *Proceedings of the EPE 2003 Conference*, Toulouse, France, 2003.

43. K. Xing, F. C. Lee, J. S. Lai, Y. Gurjit, and D. Borojevic, Adjustable speed drive neutral voltage shift and grounding issues in a DC distributed system, in *Proceedings of the Annual Meeting of the IEEE-IAS*, New Orleans, LA, 1997, pp. 517–524.

44. IEEE Std 519–1992, IEEE Recommended Practices and Requirements for Harmonic Control in Electrical Power Systems, The Institute of Electrical and Electronics Engineers Inc., USA, 1993.

45. M. P. Kazmierkowski, R. Krishnan, and F. Blaabjerg. *Control in Power Electronics*, Academic Press, San Diego, CA, 2002, pp. 579.

46. A. R. Prasad, P. D. Ziogas, and S. Manias, An Active Power Factor Correction Technique for Three-Phase Diode Rectifiers, *IEEE Transactions on Power Electronics*, 6(1), 83–92, January 1991

47. A. R. Prasad, P. D. Ziogas, and S. Manias, Passive input current waveshaping method for three-phase diode rectifiers, *IEEE Transactions Proceedings-B.*, 139(6), 512–520, November 1992.

48. R. Stzelecki and H. Supronowicz, *Power Factor Correction in AC Current Systems*, OWPW, 2000, p. 451. (in Polish)

17

Power Supplies

Francisco Javier
Azcondo
University of Cantabria

17.1 Introduction

The term "power supply" refers to power conversion systems in the following categories: (1) AC-DC rectifiers, which are usually single- or three-phase active rectifiers; (2) DC-DC converters, which are buck, boost, or buck–boost-derived converters in hard or soft-switched (resonant converters) operation; and (3) DC-AC inverters, e.g., uninterruptible powers supplies (UPS), AC power sources, and motor drives. Recent applications have opened up new categories of power supplies such as electronic ballasts, pulsating power supplies, and high-voltage capacitor charging power supplies.

Commercial power supply catalogs have DC output power ranges from 10 W to 15 kW at output voltages from 3.3 V to 15 kV and AC outputs in UPS reaching 60 kVA. General purpose DC power supplies may include single- or multi-output terminals at fixed or regulated voltage.

This initial classification suggests that the technical solution for the power conversion systems satisfy three groups of specifications. First, the input power source requirements should be fulfilled, which implies complying with the source voltage levels, its current limit, dynamic capabilities, and the applicable standards. Second, the load requirements should be satisfied, i.e., steady-state accuracy and ripple limits and also dynamic behavior; and third, the maximum efficiency should be obtained, reaching the maximum power density, in terms of the ratio power transformed and the equipment weight and volume, while complying with the electromagnetic and radio-frequency interference (EMI and RFI) regulations.

Although linear power supplies still have applications in areas where the load specifications are very restrictive, e.g., laboratories, very high fidelity audio, etc., switched-mode conversion techniques cover most of the practical specifications and are gaining new areas of applications as the performance of power devices and magnetic components increase and the control techniques become more sophisticated and obtain faster responses.

This chapter is focused on modern switched-mode power stages and control techniques oriented to comply with the three groups of specifications. Basic blocks of power supply systems, with the utility line

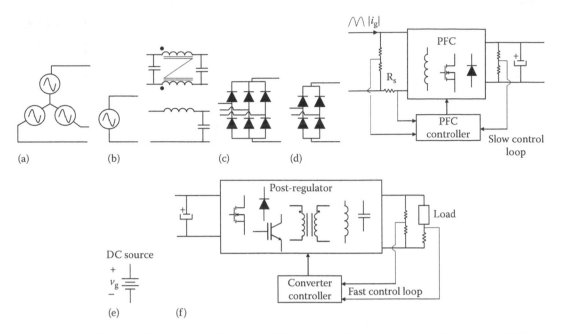

FIGURE 17.1 Elements of the power supply system. (a) Multi- or single-phase source, (b) Common and differential mode filters, (c) multi- or single-phase passive rectifiers, (d) active PFC stage, (e) DC source, and (f) Postregulator DC—load conversion stage.

as the power source or, alternatively, a DC power source such as a battery or a solar panel, are depicted in Figure 17.1. Here the AC-DC power factor correction (PFC) stage makes the system meet the utility standards and minimize the input current i_g. In principle, the PFC may require input current and output voltage control. In a standard two-stages approach, the PFC section is connected to the utility through an input filter to prevent the injection of high frequency (HF) noise, while the subsequent converter, DC-DC or DC-AC, operating in voltage or current mode, is designed to meet the load specifications.

The switched conversion technique is based on the principle that the switch combines regulation capabilities and ideal zero losses. In Figure 17.2, a basic first quadrant v versus i conversion technique is presented, where the average output voltage over a switching period T, $\langle v \rangle_T$, is controlled by the duty cycle, d, and the average input voltage $\langle v_g \rangle_T$,

$$\langle v \rangle_T = d \langle v_g \rangle_T \tag{17.1}$$

The chapter organization is as follows. After the introduction, Section 17.2 is dedicated to single-phase rectifiers. Section 17.3 is focused on DC input power conversion stages, where key components

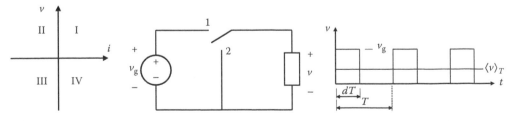

FIGURE 17.2 Left: The four operation quadrants. Middle: first quadrant voltage chopper. Right: controlled average output voltage.

and conversion techniques are detailed according to power levels and load requirements. Section 17.4 introduces the latest advances available for power supply designers, and then finalizing with some concluding remarks.

17.2 Single-Phase Rectifiers

Front-end AC line to DC output power converters are designed when it is necessary to comply with the standard that limits the power factor and current harmonic content. The IEC 1000-3-2 (EN 61000-3-2) is the main reference in technical literature for equipment that require phase current lower than 16 A [LHCE95]. Equipment classification and the corresponding harmonic limits are summarized in Figure 17.3.

A variety of solutions that meet the standard specifications were reported in the 1990s [GCPAU03]. At present, new diodes and transistors are rated to higher voltage and current stress and achieve better switching performance. New magnetic materials are used to build inductors with DC bias current in switch mode power supplies, obtaining higher energy storage capability with smaller size than gapped ferrite cores. Nonlinear and self-tuning control strategies simplify and improve the dynamics of the inner and outer current control loops. All of them are elements that achieve advances in the design of PFC stages.

A few examples of PFC techniques are presented next. They are classified into noncontrolled and controlled input current.

17.2.1 PFC Stages with No Inner Input Current Control Loop

PFC stages extend the time in which the current is absorbed from the mains during each semi-period of the mains' frequency. This can be achieved by using passive solutions based on line inductive filters or valley fill circuits, as is shown in Figure 17.4 [BAB02].

The valley fill circuit is applied in low power conversion stages. Its output voltage has 50% ripple that should be compensated by the subsequent stage.

Boost converter topology is preferable for active PFC, because it offers the highest switch utilization ratio, including an inductor that produces continuous line current, when operating in a continuous

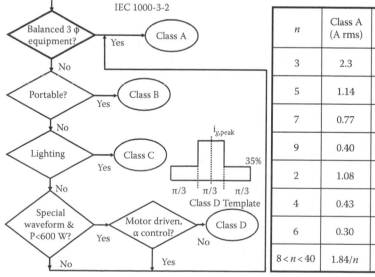

n	Class A (A rms)	Class B (A rms)	Class C (% fun.)	Class A (mA/W)
3	2.3	3.45	30PF	3.4
5	1.14	1.71	10	1.9
7	0.77	1.155	7	1.0
9	0.40	0.60	5	0.5
2	1.08	1.62	2	—
4	0.43	0.645	—	—
6	0.30	0.45	—	—
$8 < n < 40$	$1.84/n$	$2.76/n$	—	—

FIGURE 17.3 Equipment classification and harmonic limits defined by IEC 1000-3-2. (From IEC, Limits for harmonic current emissions (equipment input current <16 A per phase), IEC 1000/3/2 Int. Std., 1995. With permission.)

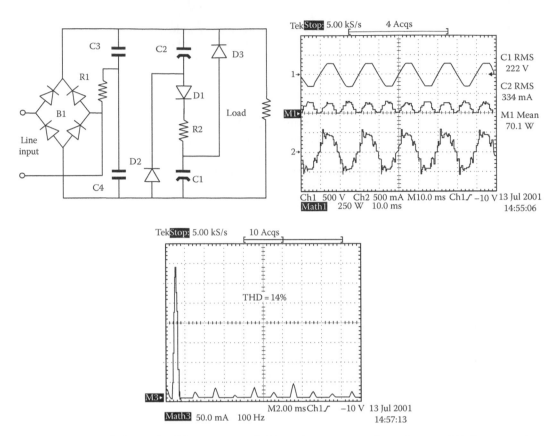

FIGURE 17.4 Top left: A valley fill circuit example. Top right: upper trace line voltage, middle line power, bottom line current. Bottom: line current spectrum. (From Branas, C. et al., Evaluation of an electronic ballast for HID lamps with passive power factor correction, in *Proceedings of the 28th Annual Conference of the Industrial Electronics Society (IECON 2002)*, Sevilla, Spain, November 2002, pp. 371–376. With permission.)

conduction mode. Typical boost converter output voltage is $V_o = 400$ V when connection to the European utility line is included in the specification. The effect of the boost converter in PFC applications is generalized by use of a high-impedance network (HIN) [VSSJ07] placed between the input rectifier and the capacitor that supplies the input DC voltage to the DC–DC converter, as shown in Figure 17.5. This solution leads to size reduction in single-stage PFC, in comparison with the two stage solution (PFC + postregulator), when the target is not unity power factor but to comply with the regulations.

There are applications where buck–boost type topologies are preferred because the control is very simple in discontinuous conduction mode (DCM) or because variable output voltage is required. Buck converters can also be controlled as PFCs when the output voltage is lower than the line voltage during the greater part of each semi-period of the mains' frequency. As an example, Figure 17.6 shows a buck–boost converter operating as a resistor emulator in DCM at constant switching period, T, and duty cycle, D. The input current, i_g, peak input current, i_{gpk}, and average input current, $\langle i_g \rangle_T$, are given in (17.2) and (17.3):

$$i_{gpk} \cong \langle v_g \rangle_T \frac{DT}{L} \tag{17.2}$$

$$\langle i_g \rangle_T = \langle v_g \rangle_T \frac{D^2 T}{2L} \tag{17.3}$$

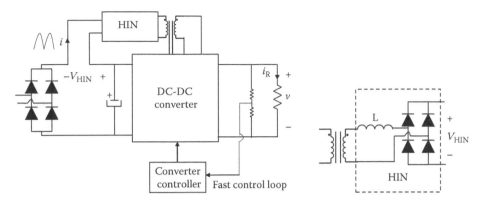

FIGURE 17.5 Left: Single-stage PFC with HIN. Right: example of HIN. (From Villarejo, J.A. et al., *IEEE Trans. Ind. Electron.*, 54(3), 1472, June 2007. With permission.)

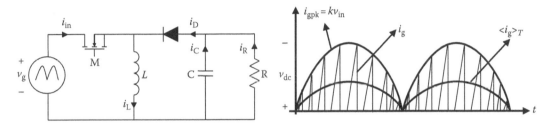

FIGURE 17.6 Left: buck–boost topology. Right: input current, input current envelope, and average input current.

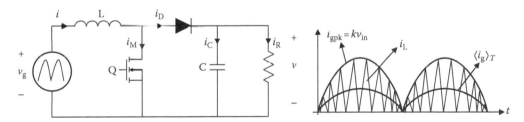

FIGURE 17.7 Left: boost topology. Right: operation at the DCM-CCM boundary condition. Input current, input current envelope, and average input current.

Discontinuous conduction mode produces zero current turn-on in the transistor. Additionally, it requires low inductance (low inductor size), finding the application when the resulting current ripple (peak current) is not excessive. A widely applied solution for a PFC circuit fixes the operation mode at the boundary between CCM and DCM, which is a reasonable solution up to 200 W. Figure 17.7 illustrates this operation mode for a boost converter. In this case, zero current detection is required and the switching period is not constant. The resistor emulator behavior is obtained if constant on-time is imposed as if can be derived from the input current, i_g, peak input current, i_{gpk}, and average input current, $\langle i_g \rangle_T$ in Equations 17.4 and 17.5:

$$i_{gpk} \cong \langle v_g \rangle_T \frac{t_{on}}{L} \tag{17.4}$$

$$\langle i_g \rangle_T = \langle v_g \rangle_T \frac{dT}{2L} \tag{17.5}$$

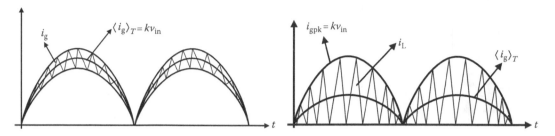

FIGURE 17.8 Current control for PFC applications. Left: average or hysteretic control in CCM. Right: peak current mode control in boundary CCM-DCM operation.

17.2.2 PFC Stages with Inner Control Current Loop

Input current control is intended to guarantee resistor emulator behavior—the inner current loop is combined with an outer loop that consists of either an output voltage control with output power limit or an output power control with output voltage limit. Depending on the power conversion rate, commercial solutions can be found for the boundary between DCM and CCM as well as CCM operations [VSSJ07, PB09]. Current controllers make either the peak or average current over a switching period follow the sinusoidal utility voltage while line period current amplitude is controlled by the outer loop. Figure 17.8 shows the effect of the input current control in CCM either by average current or hysteretic controller.

Average current control can be performed at constant switching frequency operation, which is a desirable characteristic to make the inductor design easier and to predict RFI and EMI. However, inherent dynamics of the average control are limited in a bandwidth well below the switching frequency.

Nonlinear controllers [MJE96] combine constant switching frequency operation with fast response and simplicity. They can be applied to different topologies [LS98, ZM98, SLARF08] and can be implemented in a digital circuit, unifying of the power supply controller, as is depicted in Figure 17.9.

Nonlinear-carrier controllers [SC88] compare a carrier signal with the variable under control, in this case, input current, to determine the switching instant with no switching period delay. Figure 17.10 shows the case for a boost converter. The turn-off instant corresponds to

$$V_{m}\left(1-d\right)=r_{s}i_{\mathrm{Lpk}},\tag{17.6}$$

where

  r_s is the current sensor
  V_m is the maximum carrier signal value controlled by the outer loop

and, therefore, the peak current follows the input voltage in each switching period for the boost converter case.

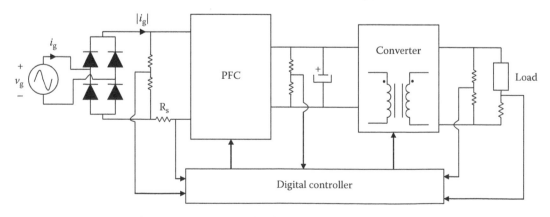

FIGURE 17.9 Power supply architecture with unified digital controller.

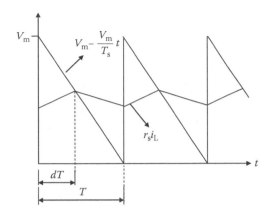

FIGURE 17.10 Carrier signal and sampled current for nonlinear-carrier control PFC boost converter.

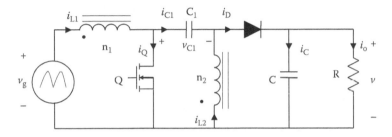

FIGURE 17.11 Single-ended primary inductor converter (SEPIC) with coupled inductor PFC circuit.

Nonlinear-carrier control can be modified to be adapted to other topologies. For example, in a buck–boost type converter, e.g., SEPIC (see Figure 17.11), the turnoff instant is given by

$$V_m - V_m \frac{t}{T} = r_s i_{Lpk} \frac{t}{T} \quad \text{and, therefore,} \quad V_m \frac{1-d}{d} = r_s i_{Lpk}, \tag{17.7}$$

in order to achieve such that the peak current follows the input voltage in each switching period.

17.3 DC-to-Load Power Conversion

Derivation of the different power converter topologies is obtained using methods such as switch matrix circuits [K88], basic three-terminal cells, and topology manipulation starting from the buck converter [EM01].

The derivation of power converters topologies from a switch matrix circuit is a top-down approach. The buck converter is a particular case of the four-quadrant direct switch matrix circuit, shown in Figure 17.12. Switch 11 is transistor M, switch 21 is diode D, switch 12 is permanently off and switch 22 is permanently on. Direct converters include energy storage elements at the input and output only, while in indirect converters the power flow does not go directly from input to output but through an intermediate energy storage element. They can be derived by cascaded connection of direct converters. Switched-mode converters produce a given energy transfer from input (source) to output storage element in a switching period. Therefore, they can be analyzed as discrete energy processors [MZ07, LYR05].

The 3D canonical cell [KSV91], from which the switch inductor or switch capacitor cells are derived, as shown in Figure 17.13, is the common element in the basic conversion topologies shown in Figure 17.14. Connections a-A, b-B, and c-C give the buck converter, connections b-A, a-C, and c-B result in the boost

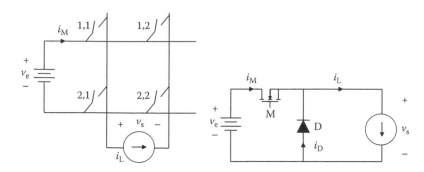

FIGURE 17.12 Left: Two input-two output lines direct matrix converter. Right: derived buck converter. (From Krein, P.T., *Elements of Power Electronics*, Oxford University Press, Oxford, U.K., 1988. With permission.)

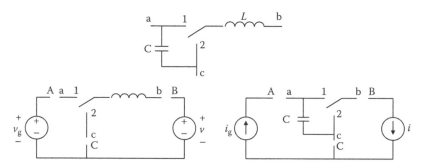

FIGURE 17.13 Three-terminal cells. Above: canonical cell. Below left: switch inductor cell. Below right: switch capacitor cell. (From Erickson, R.W. and Maksimovic, D., *Fundamentals of Power Electronics*, 2nd edn., Kluwer Academic Publishers, Secaucus, NJ, 2001. With permission.)

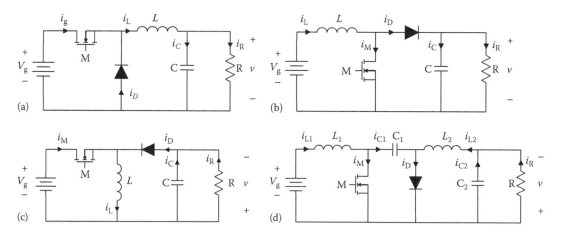

FIGURE 17.14 Basic DC-DC converter topologies: (a) buck, (b) boost, (c) buck–boost, and (d) Ćuk.

converter, while the connection c-A, a-B, and b-C give indirect converters, a buck–boost converter for the switch inductor cell, and a Ćuk converter for the switch capacitor cell.

Power converters can also be derived from the manipulation of the buck converter (bottom-up approach). Inversion of source and load transforms the buck converter into a boost converter, cascaded connection of a buck and a boost converters result in a buck–boost converter, cascaded boost and buck converters give a Ćuk converter, a boost and buck–boost converters form a SEPIC, etc.... Bipolar converters are the result of the differential connection of basic converters to the load. Inverters (DC-AC converters) are bipolar converters, where the controller modifies its reference in each switching period depending on the target output voltage.

Steady-state analysis of the converter topologies under duty cycle control, D, when they operate in CCM at constant switching frequency, gives the well-known input to output voltage ratio $V/V_g = D$, $V/V_g = 1/1 - D$, and $V/V_g = D/1 - D$ for the buck, boost, and indirect converters, respectively.

17.3.1 Isolation

The reasons for introducing galvanic isolation in power converters are (1) safety and grounding requirements, (2) output voltage polarity selection, (3) increase in the resolution of the regulation parameter, (4) low cost implementation of multi-output power supply, and (5) series or parallel association of power outputs to increase output voltage or current.

Output voltage manipulation requires isolation from the utility and terminal grounding. Applications of power supplies in industrial processes, e.g., welding require ground connection of one output terminal, either positive or negative. Positive and negative voltage is often required to supply electronic circuits. Direct regulation of standard low output voltage, e.g., 12 or 5 V, from the rectified utility line would result in a very limited utilization of the duty cycle range. A previous step-down voltage transformation increases the control resolution and improves the switch utilization ratio. An HF switching frequency generates HF voltages that, as a result of the Faraday's law, are transformed using smaller-sized magnetic cores than those required when the low frequency line voltage is transformed.

As an illustrative example, in Figure 17.15, two transformers are compared, the 1 kVA, 50 Hz one is 125 mm high, 150 mm wide, 170 mm deep, and weighs 14 kg!!, while the 1.3 kVA, 125 kHz planar transformer is 22 mm high, 64 mm wide, and 106 mm deep weighing 0.45 kg.

Figure 17.16 shows the simplified schematic of a multiple output DC-DC converter, in this case a forward converter, derived from a nonisolated buck converter. Regulation of only one output voltage is required to fix the voltage in every output as long as CCM operation is guaranteed.

Isolated converters, such as the forward converter, which is derived from the buck converter, or the flyback converter, derived from the buck–boost converter, both shown in Figure 17.17, use only the first quadrant of the B-H magnetic curve.

The transformer of the forward converter requires resetting the magnetic flux in every switching period to prevent magnetic core saturation. This task is performed by an extra winding that, when connected to the supply voltage during part of the off-time, generates a voltage drop across the transformer primary winding, whose polarity is opposite to the primary voltage during the

FIGURE 17.15 1 kVA, 50 Hz (left) vs. 1.3 kVA, 125 kHz planar transformer (right).

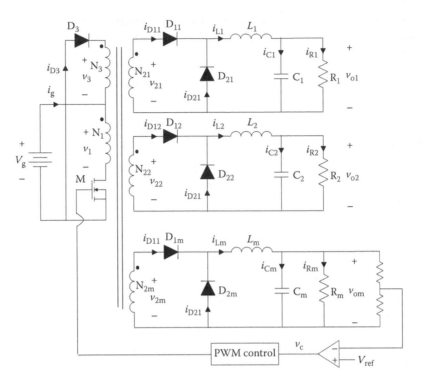

FIGURE 17.16   Multiple outputs forward converter.

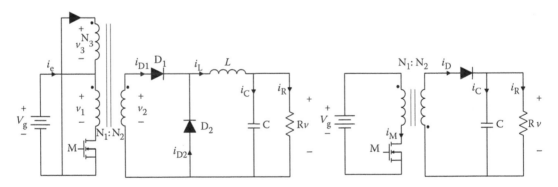

FIGURE 17.17   Left: forward converter. Right: flyback converter.

on-time. The flyback converter is obtained by substituting the inductor of the buck–boost converter by a coupled inductor.

Bidirectional core excitation increases the power conversion density. Examples of isolated converters with the same switch utilization ratio that use two quadrants of the B-H magnetic curve are shown in Figure 17.18. They are derived from the buck or boost topologies. In a first stage, an HF AC voltage or current is generated, which is then transformed and finally rectified for the supply to the DC load.

As a first approach, the magnetic component design sequence tries to minimize the inductor or transformer size, preventing core saturation and excessive losses that are originated both in the magnetic core, i.e., area enclosed by the B-H hysteresis loop and Eddy currents, and in the wire, due to its resistance and skin, proximity, and gap effects [L04, BV05, K09, APCU07].

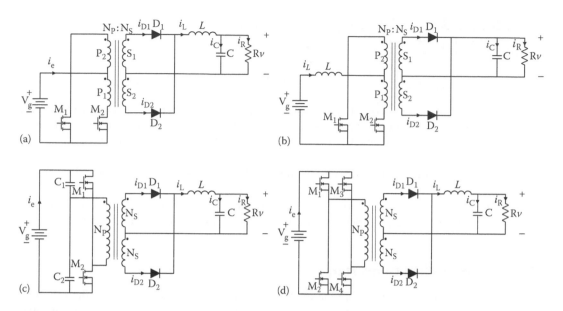

FIGURE 17.18 (a) Buck-derived push–pull, (b) boost-derived push–pull, (c) buck-derived half-bridge, and (d) buck-derived full-bridge.

17.3.2 Switched Capacitor Converters

Magnetic components make the integration of power supplies difficult, this being a limitation when attempting to reduce size and cost. There are commercial monolithic circuits that integrate active components, i.e., switches and controllers, and require the connection of the reactive components to complete the power supply. Fully or partially integrated in a monolithic circuit switched capacitor (SC) power converters can also be found in low power applications [I01].

As is stated in [M97], the conversion ratio of a two-phase SC converter with no load, based on Makowski cells is

$$M[k] = \frac{V}{V_g} = \frac{P}{Q},$$ (17.8)

where

$$\text{Max}[\text{Abs}[P], \text{Abs}[Q]] \le F_k, \quad \text{Min}[\text{Abs}[P], \text{Abs}[Q]] \le 1$$ (17.9)

k is the total number of capacitors, including the one at the output
F_k is the kth Fibonacci number

Therefore, step-up and -down conversion and positive and negative polarity are obtained depending on the states of the different switches in each phase.

Two examples of four-capacitor converters presented in [MM95] are reproduced in Figure 17.19. The circuit shown in Figure 17.19b is obtained by the inversion of source and load of the converter in Figure 17.19a. In phase one, switches 1 are *on* and switches 2 are *off* and in phase two, switches 2 are *on* and switches 1 are *off*.

If, after charged, capacitors are only discharged a little bit, then during phase one the converter in Figure 17.19a charges C1 to V_g, V_g and C1 charges C2 to 2 V_g in phase 2, therefore, C2 and V_g charges C3

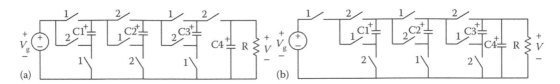

FIGURE 17.19   Two SC converters with different switch control: (a) $V/V_g = 5$ and (b) $V/V_g = 1/5$. (From Makowski, M. S. and Maksimovic, D., Performance limits of switched-capacitors DC-DC converters, *Proceedings of the Power Electronics Specialist Conference*, Atlanta, GA, June 18–22, vol. 2, pp. 1215–1221, 1995. With permission.)

to 3 V_g in phase one, which means C4 is charged up to 5 V_g by C3 and C2 in phase two ($V/V_g = 5$). The opposite sequence for the converter in Figure 17.19b results in $V_g/V = 5$.

Improvements in switching transitions, integrated circuit implementation, and regulation capabilities to reduce the output load dependence are the subject of research efforts and industrial advances in low power supplies based on SC technology.

17.3.3 Soft-Switched Converters

Power losses in switched devices are originated by on-conduction and switching losses. The switching frequency dependence of the switching losses limits the size reduction in inductors, capacitors, and transformers that the increase of the switching frequency could achieve.

Zero voltage switching (ZVS) occurs when, at the beginning of a switch transition, the voltage across the switch is zero, while zero current switching (ZCS) occurs when, at the beginning of a switch transition, the current through the switch is zero. Soft-switched transitions result in ideal zero switching losses.

Resonant, quasi-resonant, and multi-resonant converters are topologies that achieve ZVS, ZCS, or both, allowing the increase of the switching frequency. Therefore, the losses in magnetic elements and switch conduction dictate the limit of the converter size reduction.

Soft-switched transitions are accomplished by modifying the squared profile of voltages and currents into sinusoidal or partially sinusoidal shapes and introducing some delay in the zero crossing time by adding extra inductors, capacitors, or resonant (LC) tanks. For a given constant transient time, the area enclosed by the switched current versus switched voltage ($i_D(v_{DS})$) represents energy losses. Using the function ($i_D(v_{DS})$) hard-switching and soft-switching are compared in Figure 17.20.

Soft turnon or turnoff is obtained by connecting L and C components in series and parallel, respectively, to classic switches, generating resonant switches [L88] (see Figure 17.21) or by designing specific switching arrangements that include additional switches. Examples of new families of converters generated by replacing the traditional switch by a resonant switch arrangement are found in the literature [VCVFF96].

By inserting resonant tanks (LC networks) between the switches and load in classic topologies, square waveforms are essentially transformed into their fundamental modes. If the switching frequency is higher than the resonant frequency ZVS turnon is obtained and if the switching frequency is lower

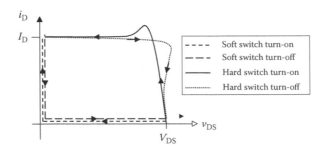

FIGURE 17.20   Current vs. voltage hard- and soft-switched transitions.

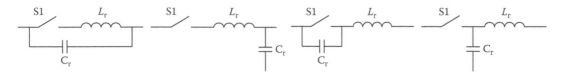

FIGURE 17.21 Left: ZC resonant switches. Right: ZV resonant switches. (From Lee, F.C., *Proc. IEEE*, 76(4), 377, April 1988. With permission.)

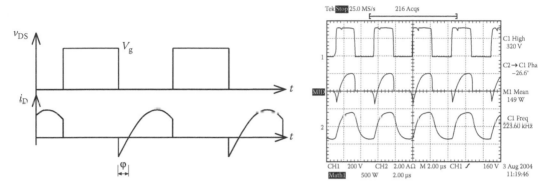

FIGURE 17.22 Operation above resonance. Left: drain-to-source voltage (v_{DS}) and drain current (i_D). Right: Experimental waveforms in a class D LCC parallel series resonant converter. Above: midpoint voltage. Below: inductor current. Middle: product of midpoint voltage by inductor current (midpoint power).

than the resonant frequency ZCS turnoff is obtained. Operation over the resonant frequency, as shown in Figure 17.22, is preferred when power MOSFETs are used as switches, especially in bridge topologies because of the poor switching performance of the diode integrated in the MOSFET structure. The different resonant networks are the origin of the so-called resonant converters (RC) [S88, KC95]. In Figure 17.23, different arrangements of LC networks are shown, resulting in the following resonant converter circuits: LC series RC, LC parallel RC, LCC series–parallel RC, LCC parallel series RC, and LLC RC. The load R_{AC} represents either a possible final load, the converter resulting in a DC to high-frequency AC converter, or the subsequent transformer and rectifier to complete an isolated DC-DC converter.

Designs of RC are oriented not only in guaranteeing soft-switch, but also in maximizing the power factor in the resonant tank, i.e., minimizing the angle φ in Figure 17.22, in order to also minimize the amplitude of the resonant current that causes on-losses, and its value at the turnoff instant that causes switching losses.

Envelope variable models [WHE91,YZGE03] are the resonant converter counterparts of the averaged models for the square wave converters, which are used to determine their dynamics and design the controllers.

Apart from reducing switching losses, the RCs find applications for their electrical properties. Series/parallel resonant converters are particular cases of the LCC and LLC converters. In Figure 17.24, large signal steady-state behavior for different switching frequencies of different variables of the LCC series–parallel RC are shown [CBA10]. The following characteristics can be observed: At the series resonant frequency, ω_s, (Figure 17.24a), the converter is an input voltage controlled voltage source, it being open load protected. At the unloaded resonant frequency, ω_o, (Figure 17.24b) the converter is an input voltage controlled current source, it being short circuit protected. Figure 17.24c shows a frequency, ω_L, where the converter is a resistor emulator, since the input current has no load dependence. Figure 17.24d shows areas of switching frequency between ω_s and ω_o, where the converter behaves close to an input voltage controlled power source [AZB07].

Because of the inherent high output impedance, Z_o, properties, as current or power sources, $Z_o = \infty$ and $Z_o = R_{load}$, respectively, resonant converters find applications as electronic ballast for the control of

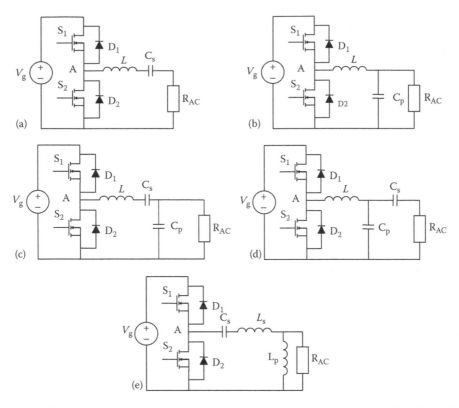

FIGURE 17.23 Different RC converters: (a) series RC, (b) parallel RC, (c) series–parallel RC, (d) parallel series RC, and (e) LLC RC.

discharge lamps [A01], ozone generators [AGCRC05], electrical discharge machining, arc welding [FR98, MMRTMP95], etc. Recent research has focused on the LLC converter, since the resonant tank along with the subsequent transformer can be integrated in a single magnetic device [C07].

17.4 Trends

Challenges for power supply designers are, among others, to achieve higher power density, control speed, and adaptability to load changes.

17.4.1 New Devices and Magnetic Cores

New commercial power devices simplify the power circuit, and increase the efficiency, indicating a step ahead in bringing their performance close to an ideal switch. CoolMOS™ transistors increase their on-resistance linearly with the voltage blocking capabilities, BV_{BR}, while traditional MOSFETs' on-resistance depends on $BV_{BR}{}^{2.5}$. Those devices can block up to 900 V. Trench MOSFETs® offer lower on-resistance and gate charge than previous synchronous MOSFETs, allowing the increase of the switching frequency and reducing on-losses.

SiC power device technology reduces the n^- layer, since the breakdown electric-field, $E_{BR} = 2 \times 10^6$ V/cm, is several times the $E_{BR} = 3 \times 10^5$ V/cm for Si devices. GaN has a breakdown electric field that is about 10 times higher than silicons. Availability of 600 V SiC Schottky diodes, with superior switching performance and higher junction temperature than fast Si diodes, has improved the PFC efficiency, switching frequency, and size [SBCCP03].

High magnetic flux saturation and soft saturation magnetic cores with integrated gaps provide higher energy storage capability than with gapped ferrites, reducing size, and increasing the load range in CCM.

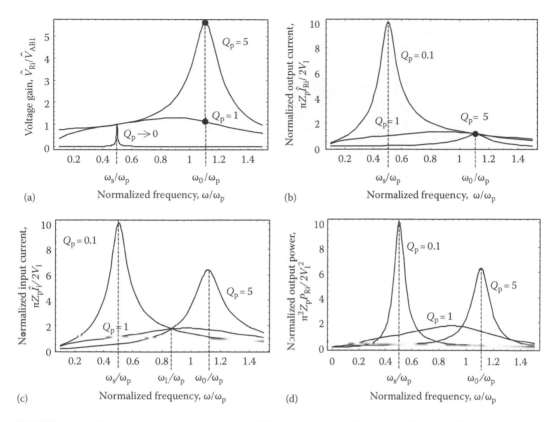

FIGURE 17.24 Large signal steady-state analysis of the LCC series–parallel RC as a function of the load (parallel quality factor, Q_p) and the switching frequency, ω. (From Casanueva, R. et al., *IEEE Trans. Ind. Elec.*, 57, 3355, 2010. With permission.)

17.4.2 Parallel Operation

In most cases, the original power source is a voltage source, e.g., utility voltage and batteries. At the same time, most of the specified power supplies outputs are output voltage. Even in the case of control of gaseous discharges, the arc voltage is constant as a first approach, depending on the electrode separation. Therefore, a power increase usually means higher input current demand and higher output current availability. Paralleling optimized power converter modules leads to several advantages. Besides increasing the power conversion capabilities, power density can be increased, since the interleaving synchronization of different paralleled sections shifts the current harmonics to higher frequencies reducing the filter requirements [GZCC06] and leading to dynamic improvements. In the case of resonant converters, paralleled operation introduces a new control parameter, the overlap among phases [BAC08], resulting in a simple and efficient control technique at constant switching frequency. Thermal distribution is another advantage of the parallel operation of power modules because it reduces the heatsink needs and improves their reliability. Figure 17.25 shows a 1 kW penta-phase class D series–parallel LCC resonant converter laboratory prototype [BACD07], whose operation frequency is 125 kHz, and the midpoint voltages of each phase. In this case, the displacement of one phase is used to regulate the output voltage.

17.4.3 Energy Process Savings

Each energy conversion stage leads to total efficiency reduction but it is accepted as necessary to meet the power supply specifications. Classical and more expensive solutions cascade different passive or active

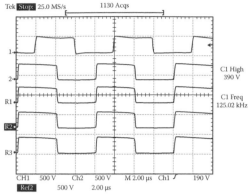

FIGURE 17.25 Left: penta-phase LCsCp resonant inverter laboratory prototype. Right: midpoint voltages with one phase displacement regulation. (From Branas, C. et al., Penta phase series parallel LCsCp resonant inverter to drive 1 kW HPS lamps, in *Conference Record of the 47th Annual Meeting*, New Orleans, September 2007, pp. 839–845. With permission.)

stages in order to shape waveforms, store energy, remove harmonics, modify voltage, and current levels, and achieve the required response under source or load variations. The integration of power stages to complete two or more groups of specifications may improve the power density figures as is the case in indirect converters, but it does not mean a reduction of the power processing stages. The power conversion techniques and power electronics elements presented so far are focused on increasing the efficiency of each energy transformation stage. Overall efficiency also increases when 100% of the energy is not transformed in each stage. For an *n* transformation stages power supply, the total efficiency, η, can be expressed as

$$\eta = \sum_m k_m \left(\prod_{i=j}^{n} \eta_i \right), \tag{17.10}$$

where k_m is the *m*th part of the source energy that is processed through the *j* to *n* stages. The reduction of the amount of energy processed also leads to improvements of the dynamics performance.

An example of partial power processing is the two-input buck converter (TIbuck). Figure 17.26 shows one of the circuits analyzed in [SVHNF99] along with the schematic of the TIbuck differentiating the two structures of the power converter.

Based on the principle of cascaded and parallel energy processing applied to power supplies that integrates the PFC and the postregulator in a single-stage, a review of derived topologies is presented in [QS01]. As a general rule, the energy portion required to achieve fast output response is transformed by cascaded power converter stages, while efficiency improvements are achieved by transferring energy directly to the load (bypassing conversion stages), although parallel energy processing may require more semiconductor switches and a complicated control circuit.

17.4.4 Digital Modeling and Control

Direct digital converter modeling [MZ07] overcome bandwidth limitations of averaged continuous-time modeling and obtains exact small-signal discrete models as it takes into account the sampling, modulator effects, and delays in a digitally controlled converter, facilitating the direct digital compensator ($G_c(z)$) design (see Figure 17.27). Nonlinear controllers [RS08] such as current programmed and voltage regulator, v_2, techniques that find predictive implementation in digital circuits [TMT08] increase the converters' response speed and include additional power management capabilities [WD06]. New self-tuning controllers [PCME03] are capable of adapting their operation to the input voltage conditions and filter components, reducing the design effort.

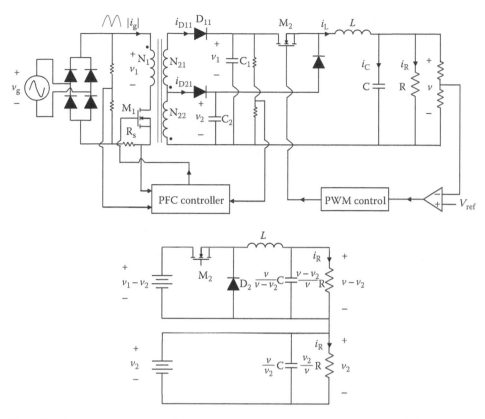

FIGURE 17.26 Top: two-outputs power factor corrector based on a flyback topology with a TIbuck postregulator. Right: detail of the TIbuck structure. (From Sebastian, J. et al., *IEEE Trans. Ind. Electron.*, 46(3), 569, June 1999. With permission.)

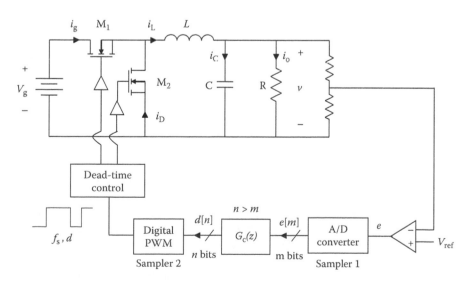

FIGURE 17.27 Block diagram of a digitally controlled buck converter as presented in [LYR05]. (From Prodic, A. et al., *IEEE Trans. Power Electron.*, 18(1), 420, January 2003. With permission.)

17.5 Conclusions

Power supplies are designed to comply with the specifications demanded by the different electrical sources and loads. When the original power source is the utility line, active PFC action is usually required to meet the standards and to provide solutions for universal operation. Switched-mode power converters are gaining areas of application because they fulfill the weight, size and cost reduction, and efficiency increase challenges, even in very low power or high output quality power supplies. Power supply size is mainly dependent on the power dissipation needs and the magnetic component size. Magnetic transformers are a key element in meeting the power supply specifications since they provide electrical isolation to the output, improve the controllability, and simplify the design of multi-output power supplies. In low power nonisolated applications, converters based on SCs are proposed for maximum integration in monolithic circuits. Switching losses that limit the maximum switching frequency can be reduced and even eliminated with soft-switched converters. These converters also present inherent properties at given switching frequencies that makes them suitable for driving special loads such as gaseous discharges.

New technologies cover the main necessities of the power supply design: power devices, magnetic materials, converter architectures, and control strategies, all help in the quest for higher efficiency, faster response, and smarter converters.

References

[A01] J.M. Alonso, Electronic ballasts Chapter 21, *Power Electronics Handbook*, M.H. Rashid (ed.), pp. 507–532, Academic Press, Orlando, FL, 2001.

[AGCRC05] J.M. Alonso, J. Garcia, A.J. Calleja, J. Ribas, J. Cardesin, Analysis, design, and experimentation of a high-voltage power supply for ozone generation based on current-fed parallel-resonant push-pull inverter, *IEEE Transactions on Industry Applications* 41(5), 1364–1372, September–October 2005.

[APCU07] R. Asensi, R. Prieto, J.A. Cobos, J. Uceda, Modeling high-frequency multiwinding magnetic components using finite-element analysis, *IEEE Transactions on Magnetics,* 43(10), 3840–3850, October 2007.

[AZB07] F.J. Azcondo, R. Zane, C. Branas, Design of resonant inverters for optimal efficiency over lamp life in electronic ballast with phase control, *IEEE Transactions Power Electronics* 22 (3, Part Special Section on Lighting Applications), 815–823, May 2007.

[BAB02] C. Brañas, F.J. Azcondo, and S. Bracho, Evaluation of an electronic ballast for HID lamps with passive power factor correction, in *Proceedings of the 28th Annual Conference of the Industrial Electronics Society (IECON 2002)*, Sevilla, Spain, November 2002, pp. 371–376.

[BAC08] C. Branas, F.J. Azcondo, R. Casanueva, A generalized study of multiphase parallel resonant inverters for high-power applications, *IEEE Transactions on Circuits and Systems I: Regular Papers*, 55(7), 2128–2138, August 2008.

[BACD07] Branas, C. et al., Penta phase series parallel LCsCp resonant inverter to drive 1 kW HPS lamps, in *Conference Record of the 47th Annual Meeting*, New Orleans, September 2007, pp. 839–845.

[BV05] A.V. den Bossche, V.C. Valchev, *Inductors and Transformers for Power Electronics*, CRC Press, Boca Raton, FL, 2005.

[C07] H. Choi, Analysis and design of LLC resonant converter with integrated transformer, in *Proceedings of the IEEE Applied Power Electronics Conference (APEC)*, Palm Springs, CA, February 25–March 1, 2007, pp. 1630–1635.

[CBA10] R. Casanueva, C. Brañas, and F.J. Azcondo, Teaching resonant converters: Properties and applications for variable loads, *IEEE Tranctions on Industrial Electronics*, 57, 3355–3363, October 2010.

[EM01] R.W. Erickson, D. Maksimovic, *Fundamentals of Power Electronics*, 2nd edn., Kluwer Academic Publishers, Secaucus, NJ, 2001.

[FR98] J.A. Ferreira, J.A. Roux, A series resonant converter for arc-striking applications, *IEEE Transactions on Industrial Electronics,* 45(4), 585–592, August 1998.

[GCPAU03] O. Garcia, J.A. Cobos, R. Prieto, P. Alou, J. Uceda, Single phase power factor correction: a survey, *IEEE Transactions on Power Electronics,* 18(3), 749–755, May 2003.

[GZCC06] O. García, P. Zumel, A. de Castro, J.A. Cobos, Automotive DC-DC bidirectional converter made with many interleaved stages, *IEEE Transactions on Power Electronics,* 21(3), 578–586, May 2006.

[I01] A. Ioinovici, Switched-capacitor power electronics circuits, *IEEE Circuits and Systems Magazine,* 1(3), 37–420, July 2001.

[K09] M.K. Kazimierczuk, *High-Frequency Magnetic Components,* John Wiley & Sons, Chichester, U.K., 2009.

[K88] P.T. Krein, *Elements of Power Electronics.* Oxford University Press, Oxford, U.K., 1988.

[KC95] M. K. Kazimierzuk, D. Czarkowski, *Resonant Power Converters,* Wiley, New York, 1995.

[KSV91] J.G. Kassakian, M. Schlecht, G. Verghese, *Principles of Power Electronics,* Addison-Wesley Publishing Company, Reading, MA, 1991.

[L04] Colonel Wm. T. McLyman, *Transformer and Inductor Design Handbook,* 3rd edn., Marcel Dekker, Inc., New York, 2004.

[L88] F.C. Lee, High-frequency quasi-resonant converter technologies, *Proceedings of the IEEE,* 76(4), 377–390, April 1988.

[LHCE95] IEC, Limits for harmonic current emissions (equipment input current <16A per phase), IEC 1000/3/2 Int. Std., 1995.

[LS98] Z. Lai, K.M. Smedley, A family of continuous-conduction-mode power-factor-correction controllers based on the general pulse-width modulator, *IEEE Transactions on Power Electronics,* 13(3), 501–510, May 1998.

[LYR05] F.L. Luo, H.Y.M. Rashid, *Digital Power Electronics and Applications,* Elsevier Academic Press, Oxford, U.K., 2005.

[M97] M.S. Makowski, Realizability conditions and bounds on synthesis of switched-capacitor DC-DC voltage multiplier circuits, *IEEE Transactions on Circuits and Systems I: Fundamental Theory and Applications,* 44(8), 684–691, August 1997.

[MJE96] D. Maksimovic, Y. Jang, R.W. Erickson, Nonlinear-carrier control for high-power-factor boost rectifiers, *IEEE Transactions on Power Electronics,* 11(4), 578–584, July 1996.

[MM95] M.S. Makowski, D. Maksimovic, Performance limits of switched-capacitor DC-DC converters, in *Proceedings of the IEEE Power Electronics Specialist Conference,* Atlanta, GA, June 18–22, 1995, vol. 2, pp. 1215–1221.

[MMRTMP95] L. Malesani, P. Mattavelli, L. Rossetto, P. Tenti, W. Marin, A. Pollmann, Electronic welder with high-frequency res sonant inverter, *IEEE Transactions on Industry Applications,* 31(2), 273–279, April 1995.

[MZ07] D. Maksimovic, R. Zane, Small-signal discrete-time modeling of digitally controlled PWM converters, *IEEE Transactions on Power Electronics,* 22(6), 2252–2256, November 2007.

[PB09] A.I. Pressman, K. Billings. *Switching Power Supply Design,* 3rd edn., Mc Graw Hill, New York, 2009.

[PCME03] A. Prodic, J. Chen, D. Maksimovic, R.W. Erickson, Self-tuning digitally controlled low-harmonic rectifier having fast dynamic response, *IEEE Transactions on Power Electronics,* 18(1), 420– 428, January 2003.

[QS01] C. Quiao, K.M. Smedley, A topology survey of single stage power factor corrector with a boost type input-current-sharper, *IEEE Transactions on Power Electronics,* 16(3), 360–368, May 2001.

[RS08] R. Redl, T. Schiff, A new family of enhanced ripple regulators for power-management applications, in *Proceedings of the Power Electronics Control and Intelligent Motion PCIM 2008,* Nürnberg, Germany, May 27–29, 2008.

[S88] R.L. Steigerwald, A comparison of half-bridge resonant converter topologies, *IEEE Transactions on Power Electronics,* 3(2), 174–182, April 1988.

[SBCCP03] G. Spiazzi, S. Buso, M. Citron, M. Corradin, R. Pierobon, Performance evaluation of a Schottky SiC power diode in a boost PFC application, *IEEE Transactions on Power Electronics,* 18(6), 1249–1253, November 2003.

[SC88] K.M. Smedley, S. Cuk, One-cycle control of switching converters, *IEEE Transactions on Power Electronics*, 10(6), 625–633, April 1988.

[SLARF08] J. Sebastian, D.G. Lamar, M. Arias, M. Rodriguez, A. Fernandez, The voltage-controlled compensation ramp: A new waveshaping technique for power factor correctors, in *Proceedings of the IEEE Applied Power Electronics Conference (APEC)*, Austin, TX, pp. 722–728, February 24–28, 2008.

[SVHNF99] J. Sebastian, P.J. Villegas, M. Hernando, F. Nuño, F. Fernandez Linera, Average-current-mode control of two-input buck postregulators used in power-factor correctors, *IEEE Transactions on Industrial Electronics*, 46(3), 569–576, June 1999.

[TMT08] D. Trevisan, P. Mattavelli, P. Tenti, Digital control of single-inductor multiple-output step-down DC–DC converters in CCM, *IEEE Transactions on Industrial Electronics*, 55(9), 3476–3483, September 2008.

[VCVFF96] M.S. Vilela, E.A.A. Coelho, J.B. Vieira Jr., L.C. de Freitas, V.J. Farias, A family of PWM soft-switching converters without switch voltage and current stresses, in *Proceedings of the IEEE International Symposium on Circuits and Systems*, Atlanta, GA, May 12–15, 1996, vol. 1, pp. 533–536.

[VSSJ07] J.A. Villarejo, J. Sebastian, F. Soto, E. de Jodar, Optimizing the design of single-stage power-factor correctors, *IEEE Transactions on Industrial Electronics*, 54(3), 1472–1482, June 2007.

[WD06] R.V. White, D. Durant, Understanding and using PMBus™ data formats, in *Proceedings of the IEEE Applied Power Electronics Conference (APEC)*, Austin, TX, March 2006, pp. 834–840.

[WHE91] A.F. Witulski, A.F. Hernandez, R.W Erickson, Small signal equivalent circuit modeling of resonant converters, *IEEE Transactions on Power Electronics*, 6(1), 11–27, January 1991.

[YZGE03] Y. Yin, R. Zane, J. Glaser, R.W. Erickson, Small-signal analysis of frequency-controlled electronic ballast, *IEEE Transactions on Circuit and Systems-I: Fundamental Theory and Applications*, 50(8), 1103–1110, August 2003.

[ZM98] R. Zane, D. Maksimovic, Nonlinear-carrier control for high-power-factor rectifiers based on up-down switching converters, *IEEE Transactions on Power Electronics*, 13(2), 213–221, March 1998.

18

Uninterruptible Power Supplies

Josep M. Guerrero
Technical University
Catalonia

Juan C. Vasquez
Technical University
Catalonia

18.1 Introduction

Uninterruptible power supply (UPS) systems become more and more important due to the growing critical loads, such as telecommunication systems, computer sets, and hospital equipments. In the last few years, an increasing number of publications about UPS systems research have appeared, and, at the same time, different kinds of industrial UPS units have been introduced in the market. Furthermore, the development of novel energy storage systems, power electronic topologies, fast electrical devices, high-performance digital processors, and other technological advances yield new opportunities for UPS systems [King03, Bekiarov02].

New electrical energy concepts like distributed generation (DG) and microgrids require storage energy systems to be able to manage the energy near the consumption points. In other words, distributed UPS systems are becoming important to integrate variable renewable energy like photovoltaic or wind turbines [Guerrero07]. The use of power electronics helps to control the parts that compound a UPS system, improving power quality and reliability. Nowadays, UPS scenario is very wide not only because there is a broad range of power rating, but also because there exists different kinds of storage energy systems. In addition, the digital signal processors are making real control techniques that could not be implemented in the past, allowing new power system configurations.

The UPS systems field is a multidisciplinary area, which encompasses power stage topologies, control techniques, technological storage solutions, and complex power systems, among others. In the light of worldwide interest among engineers, manufacturers, researchers, and users of UPS systems, this chapter is provided. This chapter is organized according to the following topics: classification of UPS systems, storage energy systems, distributed UPS, and microgrids based on UPS systems.

18.2 Classification of UPS Systems

A UPS is a device that maintains a continuous supply of electric power to the connected equipment by supplying power from a separate source when the utility main is not available. The UPS is normally inserted between the commercial utility mains and the critical loads. When a power failure or abnormality occurs, the UPS will effectively switch from utility power to its own power source almost instantaneously. There are a large variety of power rated UPS units: from units that provide backup to a single computer without a monitor of around 300 W to units that provide power to entire data centers or buildings of several megawatts, which typically work together with generators.

UPS systems are generally classified as static, which use power electronic converters with semiconductor devices, and rotary (or dynamic), which use electromechanical engines such as motors and generators. The combination of both static and rotary UPS systems is often called hybrid UPS systems [Kusko96].

Rotary UPS systems have been around for long time and their power rating reaches several megawatts [Dugan03]. Figure 18.1 shows a configuration of a rotary UPS consisting of a motor-generator set with heavy flywheels and engines. The concept is very simple: a motor powered by the utility drives a generator that powers the critical load. The flywheels located on the shaft provide greater inertia in order to increase the ride-through time. In the case of line disturbances, the inertia of the machines and the flywheels maintains the power supply for several seconds. These systems, due to their high reliability, are still in use and new ones are being installed in industrial settings. Although this kind of UPS is simple in concept, it has some drawbacks such as the losses associated with the motor-generation set, the noise of the overall system, and the need for maintenance. In order to reduce such losses, an off-line configuration is often proposed, as shown in Figure 18.2. Under normal operation, the synchronous machine is used to compensate reactive power. When the utility fails, the static switch opens and the synchronous machine starts to operate as a generator, injecting both active and reactive power. While the flywheel provides the stored energy, the diesel engine has time to start.

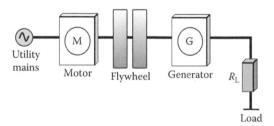

FIGURE 18.1   Block diagram of a rotary UPS consisted of an M-G set with flywheel.

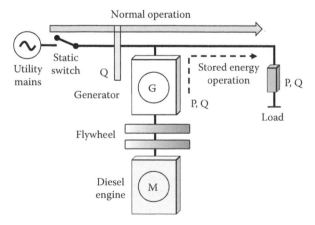

FIGURE 18.2   Off-line UPS with diesel engine backup.

About dynamic or rotary UPS systems, we have to bear in mind that, if a clutch is not installed, the diesel engine has to run in all operation modes, in a stand-by power supply fashion. However, if a clutch is installed, e.g., Piller rotary UPS or Eurodiesel Ltd, the generator can be normally stopped, and when a loss of the mains occurs, the flywheel has enough kinetic energy, in the range of few seconds, to allow the diesel generator to electronically start. At this moment, the electromagnetic clutch is engaged and the flywheel energy is used to raise the generator to the proper speed.

On the other hand, flywheel UPS systems are designed to work in the range of seconds, bearing in mind that if we increase the power drawn from the generator, the time drops down. In that example, the authors show a design of a flywheel UPS to deliver 85 kW for supporting near 40 s of ride-through. In addition, in the industry we can find commercial flywheel UPS systems, e.g., Socomec Sicon. This commercial UPS is able to work in parallel with other modules and with the DC-link of a static UPS. The flywheel is magnetically levitated in a vacuum, spinning at up to 54,000 rpm. The system delivers 190 kW with 13 s of ride-through at the rated power.

Furthermore, the combination of rotary UPS systems with power electronic converters results in hybrid systems, as shown in Figure 18.3. The variable speed drive, consisting of an AC/AC converter, regulates the optimum speed of the flywheel associated with the motor. The written-pole generator produces a constant line frequency as the machine slows down, provided that the rotor is spinning at speeds between 3150 and 3600 rpm. The flywheel inertia allows the generator rotor to keep spinning above 3150 rpm when the utility fails [Dugan03].

Static UPS systems are based on power electronic devices. The continuous development of devices such as insulated gate bipolar transistors (IGBTs) allows high frequency operation, which results in a fast transient response and low total harmonic distortion (THD) in the output voltage. According to the international standards IEC 62040-3 and ENV 500091-3, UPS systems can be classified into three main categories [Karve00, Bekiarov02]:

- *Off-line* (*passive stand-by or line-preferred*). Figure 18.4a shows the configuration of an off-line UPS, also known as line-preferred UPS or passive standby. It consists of a battery set, a charger, and a switch, which normally connects the mains to the load and to the batteries, so that these remain charged (normal operation). However, when the utility power fails or meets abnormal function, the static switch connects the load to the inverter in order to supply the energy from the batteries (stored energy operation). The transfer time from the normal operation to the stored energy operation is generally less than 10 ms, which does not affect typical computer loads. With this configuration, the UPS simply transfers utility power through to the load when either a power failure, sag, or spike occurs; at the same time, the UPS switches the load onto battery power and disconnects the utility power until it returns to an acceptable level. Off-line UPS systems completely solve problems 1–3. However, for the power problems 4–9, they only can be solved by switching to stored energy operation. In this situation, the batteries will be discharged even though the line voltage is presented [Tsai03]. Off-line UPSs are commonly rated at 600 VA for small personal computers and home applications.
- *Online* (*double conversion or inverter-preferred*). Figure 18.4b depicts the configuration of an online UPS, also known as double conversion UPS [Liang04]. During normal or even

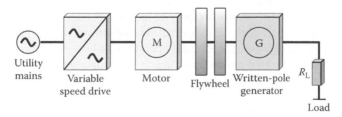

FIGURE 18.3 Hybrid UPS system.

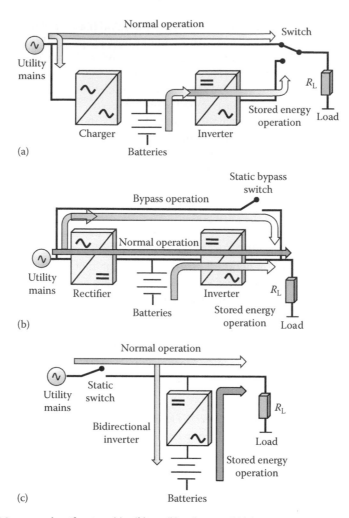

FIGURE 18.4   UPS system classification: (a) off-line, (b) online, and (c) line-interactive.

abnormal line conditions, the inverter supplies energy from the mains through the rectifier, which charges the batteries continuously and can also provide power factor correction. When the line fails, the inverter still supplies energy to the loads but from the batteries. As a consequence, no transfer time exists during the transition from normal to stored energy modes. In general, this is the most reliable UPS configuration due to its simplicity (only three elements), and the continuous charge of the batteries, which means that they are always ready for the next power outage. This kind of UPS provides total independence between input and output voltage amplitude and frequency, and, thus, a high output voltage quality can be obtained. When an overload occurs, the bypass switch connects the load directly to the utility mains, in order to guarantee the continuous supply of the load, thereby avoiding the damage to the UPS module (bypass operation). In this situation, the output voltage must be synchronized with the utility phase, otherwise the bypass operation will not be allowed. Typical efficiency is up to 94%, which is limited due to the double conversion effect. Online UPSs are typically used in environments with sensitive equipment or environments. Almost all commercial UPS units of 5 kVA and above are online.

- *Line-interactive*. Figure 18.4c illustrates the line-interactive UPS configuration, which can be considered as a midway between the online and the off-line configurations [Jou04]. It consists of a single

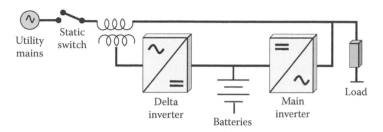

FIGURE 18.5   Series–parallel line-interactive UPS or delta-conversion UPS.

bidirectional converter that connects the batteries to the load. Under normal operation, the mains supplies the load, and the batteries can be charged through the bidirectional inverter, acting as a DC/AC converter. It may also have active power filtering capabilities. When there is a failure in the mains, the static switch disconnects the load from the line and the bidirectional converter acts as an inverter, supplying energy from the batteries. The main advantages of the line-interactive UPS are the simplicity and the lower cost in comparison to the online UPS. Line-interactive units typically incorporate an automatic voltage regulator (AVR), which allows the UPS to effectively step up or step down the incoming line voltage without switching to battery power. Thus, the UPS is able to correct most long-term overvoltages or undervoltages without draining the batteries. Another advantage is that it reduces the number of transfers to battery, which extends the lifetime of the batteries. However, it has the disadvantage that under normal operation it is not possible to regulate output voltage frequency. Line-interactive UPS units typically rate between 0.5 and 5 kVA for small server systems. Typical efficiency is about 97% when there are no problems in the line.

Figure 18.5 shows a special kind of line-interactive UPS, known as series–parallel or delta-conversion UPS [Silva02], which consists of two inverters connected to the batteries: the delta inverter (rated at 20% of the nominal power), connected through a series transformer to the utility, and the main inverter (fully rated at 100% of the nominal power), connected directly to the load. This configuration achieves power factor correction, load harmonic current suppression, and output voltage regulation. The delta inverter works as a sinusoidal current source in phase with the input voltage. The main inverter works as a low-THD sinusoidal voltage source in phase with the input voltage. Usually, only a small portion of the nominal power (up to 15%) flows from the delta to the main inverter, achieving high efficiency. Nevertheless, this configuration needs complex control algorithms. In addition, unlike in online UPSs, there is no continuous separation of load and utility mains. The delta-conversion UPS systems provide protection from all line problems except for frequency variations.

18.3 Storage Energy Systems

One of the problems to be solved by future UPS systems is how to store the energy. This question raises several solutions which can be used alone or combined. Some of the storage energy technologies are summarized as follows [Roberts05]:

- *Battery energy storage system (BESS)*. Typical UPS systems use chemical batteries to store energy. Rechargeable batteries such as valve-regulated lead-acid (VRLA) or nickel-cadmium (Ni-Cd) are the most popular due to their availability and reliability. A lead-acid battery reaction is reversible, allowing the battery to be reused. There are also some advanced sodium/sulfur, zinc/bromine, and lithium/air batteries that are nearing commercial readiness and offer promise for future utility application. On the other hand, flow batteries store and release energy by means of a reversible electrochemical reaction between two electrolyte solutions. There are four main flow battery technologies: polysulfide bromide (PSB), vanadium redox (VRB), zinc bromine (ZnBr), and hydrogen bromine (H-Br) batteries. However, batteries contain heavy metals, such as Cd and Hg, may cause environmental pollution. A large majority of UPS designs use a characteristic constant-voltage charging system with current limit.

- *Flywheels.* This system is essentially a dynamic battery that stores energy mechanically in the form of kinetic energy by spinning a mass around an axis. The electrical input spins the flywheel rotor and keeps it spinning until called upon to release the stored energy through a generator, such as a reluctance motor generator [Lawrence03]. Sometimes the flywheel is enclosed in a vacuum or in gas helium in order to avoid friction losses. The amount of energy available and its duration is governed by the mass and speed of the flywheel. There are two available types of flywheels: low-speed (less than 40,000 rpm) flywheels, which are based on steel rotors, and high-speed (between 40,000 and 60,000 rpm) flywheels, which use carbon fiber rotors and magnetic bearings. Flywheels provide 1–30 s of ride-through time. In addition, the combination of modern power electronics and low-speed flywheels can provide protection against multiple power-line disturbances.

- *Superconducting magnetic energy storage* (*SMES*). This system stores electrical energy in a superconducting coil. The resistance of a superconductor is zero so the current flows without reduction in magnitude. The variable current through the superconducting coil is converted to a constant voltage, which can be connected to an inverter. The superconducting coil as illustrated in Figure 18.6 is made of niobium titanium (NbTi) and it is cooled to 4.2 K by liquid helium [Mito06]. Typical power rates for this application are up to 4 MVA.

- *Fuel cells* (*FC*). These devices convert the chemical energy of the fuel directly into electrical energy. They are good energy sources to provide reliable power at a steady-state. However, due to their slow internal electrochemical and thermodynamic characteristics, they cannot respond to the electrical transients as fast as it is desirable. This problem can be solved by using supercapacitors or BESS in order to improve the dynamic response of the system [Nehrir06]. Fuel cells can be

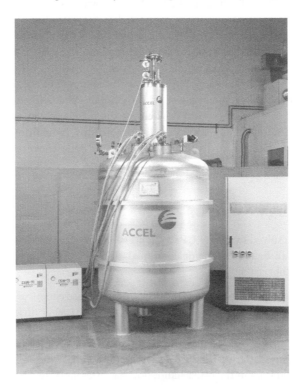

FIGURE 18.6 SMES with 0.6 kWh capacity from Accel. (Courtesy of ACCEL Instruments GmbH.)

classified into proton exchange membrane (PEMFC), solid oxide (SOFC), and molten carbonate (MCFC). PEMFC are more suitable for UPS applications since they are compact, lightweight, and provide high power density at room temperature, while SOFC and MCFC require much higher temperature between 800°C and 1000°C for optimal operation.

- *Compressed air energy storage (CAES)*. This technology uses an intermediary mechanical-hydraulic conversion, also called liquid-piston principle [Lemofouet06]. These devices are raising interest, since they do not generate any waste. They can also be integrated together with a cogeneration system, due to the thermal processes associated with the compression and the expansion of gas. Their efficiency can also be optimized by using power electronics or combining CAES with other storage systems.

18.4 Distributed UPS Systems

With the objective to further increase the reliability of UPS systems, the use of several UPS units connected in parallel is an interesting option. The advantages of a paralleled UPS system over one centralized unit are flexibility to increase the power capability, enhanced availability, fault tolerance with $N + 1$ modules (N modules supporting the load plus one reserve stand-by module), and ease of maintenance due to the redundant configuration [Sears01].

Parallel operation is a special feature of high-performance industrial UPS systems. The parallel connection of UPS inverters is a challenging problem, which is more complex than paralleling DC sources, since every module must share the load properly while staying synchronized. In theory, if the output voltage of every module has the same amplitude, frequency, and phase, the current load could be equally distributed. However, due to the physical differences between the modules and the line-impedance mismatches, the load will not be properly shared. This fact leads to a circulating current among the units, as shown in Figure 18.7. Circulating current is especially dangerous at no-load or light-load conditions, since one or several modules can absorb active power operating in rectifier mode. This increases the DC-link voltage level, which can result in damage to the DC capacitors or a shutdown due to overload. Generally speaking, a paralleled UPS system must achieve the following features:

- The same output voltage amplitude, frequency, and phase
- Equal current sharing between the units
- Flexibility to increase the number of units
- Plug and play operation at any time (hot-swap operation capability)

The fast development of digital signal processors (DSP) has brought about an increase in control techniques for the parallel operation of UPS inverters. These control schemes can be classified into two

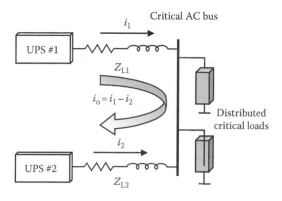

FIGURE 18.7 Circulating current concept.

main groups with regards to the use of control wire interconnections. The first one is based on active load-sharing techniques, which can be classified as follows [Shanxu99, Kawabata88], (see Figure 18.8)

1. *Centralized control*: The total load current is divided by the number of modules N, so that this value becomes the current reference of each module. An outer control loop in the central control adjusts the load voltage. This system is normally used in common UPS equipment with several output inverters connected in parallel [Holtz90].
2. *Master-slave*: The master module regulates the load voltage. Hence, the master current fixes the current references of the rest of the modules (slaves), [Broeck98]. The master can be fixed by the module that brings the maximum rms or crest current, or can be a rotating master. If the master unit fails, another module will take the role of master in order to avoid the overall failure of the system. This system is often adopted when using different UPS units mounted into a rack.
3. *Circular chain control* (3C): The current reference of each module is taken from the above module, forming a control ring [Wu00]. Note that the current reference of the first unit is obtained from that of the last unit. The approach is interesting for distributed power systems based on AC-power rings [Chandorkar00].
4. *Average load sharing*: The current of all modules is averaged by means of a common current bus [Tao03]. The average current of all the modules is the reference for each individual one. This control scheme is highly reliable due to the real democratic conception, in which no master-slave philosophy is present. Also, the approach is highly modular and expandable, making it interesting for industrial UPS systems. In general, this scheme is the most robust and useful of the above controllers.

In general, the last two control schemes require that the modules share two signals—the output voltage reference phase (which can be achieve by a dedicated line or by using a phase-locked loop (PLL) circuit, to synchronize all UPS modules) and the current information (a portion of the load current, master current, or the average current). In a typical UPS application, the reference voltage is synchronized either with the external bypass utility line or to an internal oscillator signal when this is not present. Another possibility is to use active and reactive power information instead of the current. Thus, we use active and reactive power to adjust the phase and the amplitude of each module, but using the same three control schemes [Guerrero04]. Although these controllers achieve both good output voltage regulation and equal current sharing, the need for intercommunication lines among modules reduces the flexibility of the physical location and its reliability, since a fault in one line can result in the shutdown of the system. In order to improve reliability and avoid noise problems in the control lines, digital communications by using a CAN bus or other digital buses are proposed. In this sense, low bandwidth communications can be performed when using active and reactive average power instead of instantaneous output currents.

The second kind of control scheme for the parallel operation of UPSs is mainly based on the droop method (also called independent, autonomous or wireless control). This concept stems from the power system theory, in which a generator connected to the utility line drops its frequency when the power required increases [Tuladhar00]. In order to achieve good power sharing, the control loop makes tight adjustments over the output voltage frequency and the amplitude of the inverter, thus compensating for the active and reactive power unbalances. The droop method achieves higher reliability and flexibility in the physical location of the modules, since it uses only local power measurements. Nevertheless, the conventional droop method shows several drawbacks that limit its application, such as [Guerrero04]: slow transient response, trade-off between the power sharing accuracy and the frequency and voltage deviations, unbalance harmonic current sharing, and high dependency on the inverter output impedance.

Another drawback of the standard droop method is that the power sharing is degraded if the sum of the output impedance and the line impedance is unbalanced. To solve this, interface inductors can be included between the inverter and the load bus, as depicted in Figure 18.9, but they are heavy and bulky. As an alternative, novel control loops that fix the output impedance of the units by emulating lossless resistors or reactors have been proposed [Guerrero05].

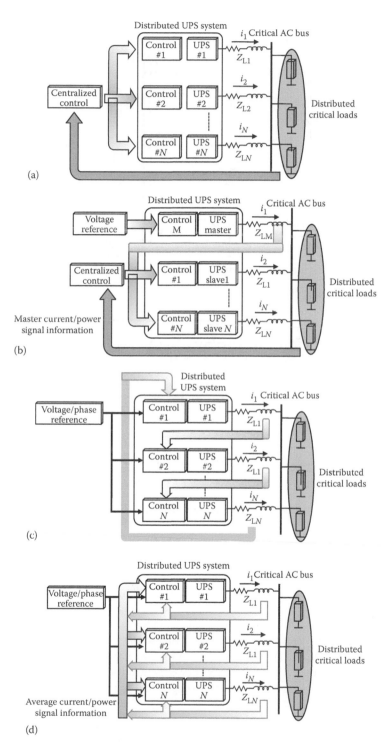

FIGURE 18.8 Active load-sharing control schemes for the parallel operation of distributed UPS systems: (a) centralized control, (b) master–slave control, (c) current chain control, and (d) average load sharing.

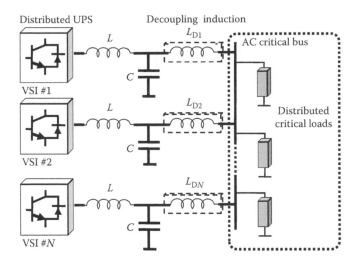

FIGURE 18.9   Equivalent circuit of a distributed UPS system.

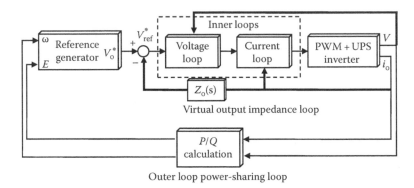

FIGURE 18.10   Block diagram of the closed-loop system with the virtual output impedance path.

Usually, the inverter output impedance is considered to be inductive, which is often justified by the high inductive component of the line impedance and the large inductor of the output filter. However, this is not always true, since the closed-loop output impedance also depends on the control strategy, and the line impedance is predominantly resistive for low voltage cabling. The output impedance of the closed-loop inverter affects the power sharing accuracy and determines the droop control strategy. Furthermore, the proper design of this output impedance can reduce the impact of the line-impedance unbalance. Figure 18.10 illustrates this concept in relation to the rest of the control loops. The output impedance angle determines, to a large extent, the droop control law. Table 18.1 shows the parameters that can be used to control the active and reactive power flow in the functioning of the output impedance. Figure 18.11 shows the droop control functions depending on the output impedance [Guerrero06].

TABLE 18.1   Output Impedance Impact over Power Flow Controllability

Output Impedance	Inductive (90°)	Resistive (0°)
Active power (P)	Frequency (ω)	Amplitude (E)
Reactive power (Q)	Amplitude (E)	Frequency (ω)

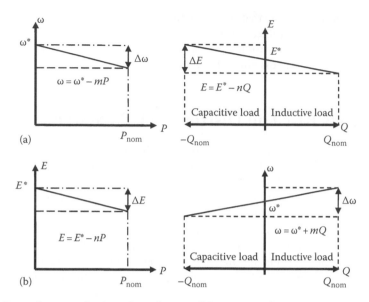

FIGURE 18.11 Droop functions for the independent parallel operation of UPSs.

On the other hand, the droop method has been studied extensively in parallel DC converters. In these cases, resistive output impedance is enforced easily by subtracting a proportional term of the output current from the voltage reference. The resistive droop method can be applied to parallel UPS inverters. The advantages of such an approach are the following: (1) the overall system is more damped; (2) it provides automatic harmonic current sharing; and (3) phase errors barely affect active power sharing.

However, although the output impedance of the inverter can be well established, the line impedance is unknown, which can result in an unbalanced reactive power flow. This problem can be overcome by injecting high frequency signals through power lines [Tuladhar00] or by adding external data communication signals [Marwali04]. Some control solutions are also presented to reduce the harmonic distortion of the output voltage when supplying nonlinear loads by introducing harmonic sharing loops. This solution consists in adding into the virtual impedance loop a bank of band pass filters, which extracts current harmonic components in order to droop the output voltage reference proportionally to these current harmonics [Guerrero07]. Figure 18.12 shows the behavior of a two-parallel-UPS system

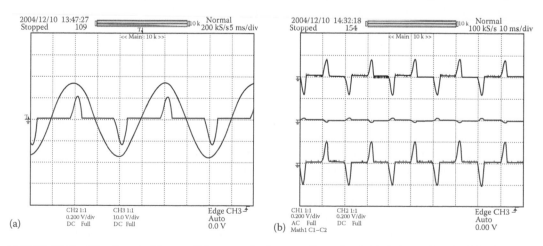

FIGURE 18.12 Waveforms of the parallel system sharing a nonlinear load. (a) Output voltage and load current (*X*-axis: 5 ms/div, *Y*-axis: 40 A/div). (b) Output currents and circulating current (*X*-axis: 10 ms/div, *Y*-axis: 20 A/div).

when sharing a nonlinear load. Note that the circulating current is very low due to the good load-sharing capability when supplying nonlinear loads.

The mentioned autonomous control for parallel UPS systems is expanding in the market, which highlights its applicability in real distributed power systems.

18.5 Microgrids Based on Distributed UPS Systems

In the following years, the electric grid will evolve from the current very centralized model toward a more distributed one. At the present time the generation, consumption, and storage points are very far away from each other. Under these circumstances, relatively frequent failures of the electric supply and important losses take place in the transport and distribution of energy; therefore, it can be stated that the efficiency of the supply system is low.

Both the electric companies and the governments are aiming at an electric grid, formed in a certain proportion by distributed generators, where the consumption points are near the generation points, avoiding high losses in the transmission lines and reducing the rate of shortcomings. Summing up, it is pursued that the generation of small quantities of electric power by the users (this concept is called microgeneration in the origin), considering them not only as electric power consumers but also as responsible for the generation, becomes in this way an integral part of the grid.

A microgrid could be defined as the sum of microgenerators, energy storage systems, and loads operating as a single system. Most of the generators should be controlled by power electronic equipment to provide the system with enough flexibility. Under this perspective, and admitting that the future electric net will be formed by both centralized generation and DG, it should be considered that in the event of failure of the public distribution grid, a microgrid should be disconnected, being able to work autonomously, managing the generation, storage, and consumption of energy.

In this context, it is necessary to develop a new concept of flexible grid, i.e., with a reconfiguration capability for operation with or without connection to the mains. The future microgrids should incorporate supervision and control systems that allow the efficient management of various kinds of energy generators, such as photovoltaic panels and small wind generators, energy storage systems, and local loads. Hence, the management of all the previously described elements interacting with the public electric grid is an important issue. In this case, when the microgrid detects an important failure in the mains, it will be able to disconnect and work in an autonomous way. Similarly, when it works isolated from the grid, it will be monitoring the public mains in order to reconnect when the suitable conditions are met. The intelligent microgrid, through monitoring, management, and control elements, will be capable of reconfiguring its operation modes and making decisions in real time.

In the last years several ideas have been proposed about the control and management of microgrids. One of the problems that are currently under study is the soft transition of the microgrid from isolated mode to grid connection mode and vice versa. Further issues of special relevance in the research about microgrids are the following:

- *New control methods* based on the calculation of the active and reactivate powers [Villeneuve04] with the objective of reducing the circulating currents among the power converters connected to a microgrid. The research in this field has produced new works aiming to improve the dynamic and static benefits of these controllers [Guerrero04, Guerrero05, Guerrero06, Guerrero07].
- *The robustness to voltage sags* while guaranteeing the nominal voltage in the microgrid, and the detection of island operation through estimation of the electric grid impedance [Guerrero09].
- *Microgrid energy management* considering the energy storage systems and the control of the energy flows in both operation modes (with and without connection to the public grid).
- *New functionalities* for the microgrid connected inverters achieving their optimal operation as a function of the surrounding conditions. An example is the operation of the inverters, depending

on the microgrid status, as parallel active filters capable of correcting the current harmonics generated by nonlinear loads connected to the microgrid [Wekesa02, Borup01].

- The start-up process of the power converters in a microgrid (*black-start*), or the loads disconnection in the case of island operation when low energy levels are available from the sources, to give some examples, are the key aspects to consider in the management and control of these systems [Degner04].

- In each power conversion system in the microgrid it can be distinguished between aspects related to the input stage (source of renewable energy) and to the output or microgrid connection stage. In the case of wind energy systems or photovoltaic systems, numerous proposals have been carried out about control algorithms for the maximum power tracking of the energy source. The microgrid connection inverter of the power converters is one of the key points to make the microgrid concept feasible [Tsikalakis08].

- *Grid synchronization techniques* [Blaabjerg06, Svensson01], such as PLL and grid estimators [Karimi04], that allow the calculation of the grid phase and frequency starting from the measurement of the supply voltage(s). Their correct operation under imperfect conditions (unbalances, sags, distortion) is an outstanding problem described in the scientific literature.

Hence, microgrids are becoming a reality in a scenario in which renewable energy, DG, and distributed storage systems can be conjugated and also integrated into the grid. These concepts are growing up due not only to environmental aspects but also to social, economical, and political interests. The variable nature of some renewable energy systems such as photovoltaic or wind energy relies on natural phenomenon like sunshine or wind. Consequently, it is difficult to predict the power that can be obtained through these prime sources, and the peaks of power demand do not coincide necessarily with the generation peaks.

Hence, storage energy systems are required if we want to supply the local loads in an UPS fashion. Some small and distributed energy storage systems, such as flow batteries, fuel cells, flywheels, superconductor inductors, or compressed air devices, can be used for this purpose.

The DG concept is growing in importance, pointing out that the future utility line will be formed by distributed energy resources and small grids (minigrids or microgrids), interconnected between them. In fact, the responsibility of the final user is to produce and store part of the electrical power of the whole system. Hence, microgrid can export or import energy to the utility through the point of common coupling (PCC). And, when there is a utility failure, the microgrid still can work as an autonomous grid. As a consequence, these two classical applications—grid-connected and islanded operations—can be used in the same application. In this sense, the droop control method can be a good solution to connect in parallel several inverters in island mode. However, although it has been investigated and improved, this method by itself is not suitable for the coming flexible microgrids. Further, although there are line-interactive UPS in the market, some UPS systems, able to operate in parallel autonomously forming a microgrid, are under development.

A flexible microgrid has to be able to import/export energy from/to the grid, control the active and reactive power flows, and manage the energy storage. Figure 18.13 shows a microgrid, including small generators, storage devices, and local critical and noncritical loads, which can operate both connected to the grid or autonomously in island mode. In this way, the power sources (photovoltaic arrays, small wind turbines, or fuel cells) or storage devices (flywheels, superconductor inductors, or compressed air systems) use electronic interfaces between them and the microgrid. Usually, these interfaces are AC/AC or DC/AC power electronic converters, also called inverters.

Traditionally, inverters have two separate operation modes, acting as a current source, if they are connected to the grid or as a voltage source, if they work autonomously. In this last case, the inverters must be disconnected from the grid when a grid fault occurs, for security reason and to avoid islanding operation. However, in order to impulse the use of decentralized generation of electrical power, the DG, and the implantation of the microgrids, islanding operation should be accepted if the user is completely disconnected to the grid. In this case, the microgrid could operate as an autonomous grid, using the following three control levels [Guerrero09].

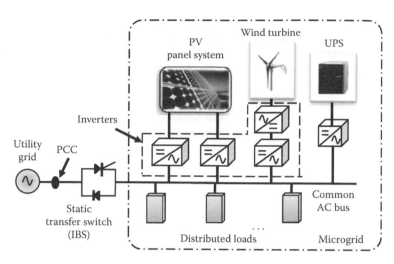

FIGURE 18.13   Diagram of a flexible microgrid.

1. *Primary control*: The inverters are programmed to act as generators by including virtual inertias through the droop method, which ensures that the active and the reactive powers are properly shared between the inverters.

2. *Secondary control*: The primary control achieves power sharing by sacrificing frequency and amplitude regulation. In order to restore the microgrid voltage to nominal values, the supervisor sends proper signals by using low bandwidth communications. This control can also be used to synchronize the microgrid with the main grid before they have to be interconnected, facilitating the transition from islanded to grid-connected mode.

3. *Tertiary control*: The set points of the microgrid inverters can be adjusted, in order to control the power flow, in global (the microgrid imports/exports energy) or local terms (hierarchy of spending energy). Normally, the power flow priority depends on economic issues. Economic data must be processed and used to make decisions in the microgrid.

Figure 18.13 shows the schematic diagram of a microgrid. In this example, it consists of several photovoltaic strings connected to a set of line-interactive UPSs forming a local AC microgrid, which can be connected to the utility mains through an intelligent bypass switch (IBS). The IBS continuously monitors both its sides—the mains and the microgrid. If there is a fault in the mains, the IBS will disconnect the microgrid from the grid, creating an energetic island. When the main is restored, all UPS units are advised by the IBS to synchronize with the mains to properly manage the energy reconnection.

The microgrid has two main possible operation modes: grid-connected and islanded mode. The transitions between both modes and the connection or disconnection of UPS modules should be made seamlessly (hot-swap or plug and play capability). In this sense, the droop control method can be used pretty well for islanding microgrids. Taking into account the features and limitations of the droop method, the control structure of the microgrid should allow operating in both grid-connected and islanded modes. In this case, the operation of the inverters is autonomous as a contrary as other microgrid configurations, which use master-slave principles. Only low bandwidth communications are required in order to control the microgrid power flow and synchronization with the utility grid.

18.6 Droop Method Concept

As explained before, the droop method is often proposed with the objective of connecting several parallel inverters without control intercommunications. The applications of such a kind of control are typically industrial UPS systems or islanding microgrids. The conventional droop method is based on the principle that the phase and the amplitude of the inverter can be used to control active and reactive power flows. Hence, the conventional droop method can be expressed as follows [Chandorkar94]

$$\omega = \omega^* - mP \tag{18.1}$$

$$E = E^* - nQ \tag{18.2}$$

where
    E is the amplitude of the inverter output voltage
    ω is the frequency of the inverter
    ω^* and E^* are the frequency and amplitude at no-load
    m and n are the proportional droop coefficients

The active and reactive power flowing from an inverter to a grid through an inductor can be expressed as follows [Bergen 86]

$$P = \left(\frac{EV}{Z} \cos\phi - \frac{V^2}{Z} \right) \cos\theta + \frac{EV}{Z} \sin\phi \sin\theta \tag{18.3}$$

$$Q = \left(\frac{EV}{Z} \cos\phi - \frac{V^2}{Z} \right) \sin\theta - \frac{EV}{Z} \sin\phi \cos\theta \tag{18.4}$$

where
    Z and θ are the magnitude and the phase of the output impedance
    V is the common bus voltage
    ϕ is the phase angle between the inverter output voltage and the microgrid voltage

Notice that there is no decoupling between $P - \omega$ and $Q - E$. However, it is very important to keep in mind that the droop method is based on two main assumptions:

- *Assumption 1*: The output impedance is pure inductive ($Z = X$) and $0 = 90°$. By using (18.3) and (18.4), it yields,

$$P = \frac{EV}{X} \sin\phi \tag{18.5}$$

$$Q = \frac{EV}{X} \cos\phi - \frac{V^2}{X} \tag{18.6}$$

This is often justified due to the large inductor of the filter inverter and to the impedance of the power lines. However, the inverter output impedance depends on the control loops, and the impedance of the power lines is mainly resistive in low voltage applications. This problem can be overcome by adding an output inductor, resulting in an *LCL* output filter, or by programming a virtual output impedance through a control loop.
- *Assumption 2*: The angle ϕ is small, we can derive that $\sin\phi \approx \phi$ and $\cos\phi \approx 1$, and consequently

$$P \approx \frac{EV}{X} \sin\phi \tag{18.7}$$

$$Q \approx \frac{V}{X}(E - V) \tag{18.8}$$

Note that, taking into account these considerations, P and Q are linearly dependents on ϕ and E. This approximation is true if the output impedance is not too large as in most practical cases.

In droop method, each unit uses frequency instead of phase to control the active power flows, because they do not know the initial phase value of the other units. However, the initial frequency at no-load can be easily fixed as ω^*. As a consequence, the droop method has an inherent trade-off between the active power sharing and the frequency accuracy, thus resulting in frequency deviations. In [Chandorkar94], frequency restoration loops were proposed to eliminate these frequency deviations. However, in general, it is not practical, since the system becomes unstable, due to inaccuracies in inverters output frequency that leads to increasing circulating currents.

18.7 Communications

The droop method does not need any communication link between the UPS inverters. This can be interesting when having islanded inverters that have to share the total load. However, it has several problems when trying to apply to:

- *Online distributed UPS system*: In this case, the UPS inverters must be synchronized in phase with the utility mains when it is present. One additional loop can adjust the frequency and the phase in a PLL fashion. Communications can reduce this problem. Furthermore, little measurement phase errors results in large circulating current between the inverters. In addition, it is necessary to communicate to the UPS units if one of the static bypass switches is turned on, among other emergency settings.
- *Line-interactive distributed UPS system*: When the utility main is disconnected, the UPS units have good balance; however, they must be resynchronized to the utility grid when the fault is cleared. Some authors propose just waiting to match the grid phase or to overload the UPS unit more closed to the utility switch. Both solutions are not reliable and hazards can make the system shutdown.
- *Large area UPS system*: In applications like a microgrid, the units can be located at distant points. Consequently, the power lines can be highly unbalanced, and the measurement errors can contribute to produce high circulating currents.

All these problems can be overcome by using communications. Combining low bandwidth communications with droop method can be a high-performance solution for a true distributed UPS system.

18.8 Virtual Output Impedance

It is known that line impedance has a considerably effect on the power sharing accuracy of the P/Q droop method. Alternatively or complementary to the use of signal communications, it is often used as a fast control loop, called virtual output impedance, which can be used to fix the output impedance of the inverter.

This impedance should be larger than the combined values of the output impedance of the UPS inverter plus the maximum power line impedance. The implementation of the virtual output impedance can by done by using the following expression [Guerrero05]

$$V^* = V_{ref}^* - i_o Z_o(s) \tag{18.9}$$

where
    $Z_o(s)$ is the transfer function of the virtual output impedance
    V_{ref}^* is the voltage reference calculated by the P/Q-sharing loop
    V^* is the output voltage provided to the inner control loops

Figure 18.14 shows the block diagram of a droop controller with the virtual output impedance loop.

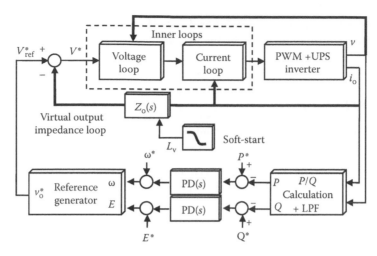

FIGURE 18.14 Block diagram of the inverter control loops.

The output impedance value must be selected following similar way as m and n coefficients, according to the nominal apparent power S_i of each UPS unit i

$$Z_{o1}S_1 = Z_{o2}S_2 = \cdots = Z_{oN}S_N \tag{18.10}$$

At this point, we should bear in mind that the output impedance have become a control variable of our system. Another practical issue is the desirable hot-swap or *plug and play capability*, which consists in a seamless operation of the UPS inverter when it is connected suddenly to the common AC bus. The output current peak in such a situation is expressed in [Guerrero05] as

$$I_{\text{pk}} \approx \frac{E}{X} \cdot \Delta\phi \tag{18.11}$$

where $\Delta\phi$ is the PLL error.

In order to reduce this initial current peak, we can reduce the PLL error to a limited small angle, but this is still not enough because this error is difficult to control, due to the fact that the PLL accuracy depends on the sensor errors and on other nonideal parameters. Bearing in mind that the output impedance is a new adjustable control parameter, by following (18.11), we can deduce that another way to reduce the current peak is to increase the output inductance L_D. Hence, a soft-start operation of the output impedance is proposed to alleviate this initial transient peak, achieving a seamless connection of the inverter to the common bus (hot-swap operation) [Guerrero06]

$$L_D^* = L_{\text{Df}}^* + \left(L_{\text{Do}}^* - L_{\text{Df}}^*\right)e^{-t/T_{\text{ST}}} \tag{18.12}$$

where
L_{Do}^* and L_{Df}^* are the initial and final values of the output impedance
T_{ST} is the time constant of the soft-start operation

The soft-start operation consists in connecting the inverter to the common bus using a high output impedance and reducing it slowly toward the nominal value. This way, the initial current peak can be avoided in spite of the PLL error. A proportional controller detects the error signal of the DC-link voltage. The control algorithm increases the sine reference value of this inverter module to stop this energy feedback.

18.9 Microgrid Control

It is desirable that the control structure of the microgrid allows its operation in grid-connected and islanded modes, and enables the soft transition between both modes [Guerrero09].

- *Grid-connected operation.* The microgrid is connected to the grid through an IBS. In this case, all UPSs have been programmed with the same droop function

$$\omega = \omega^* - m(P - P^*) \tag{18.13}$$

$$E = E^* - n(Q - Q^*) \tag{18.14}$$

  where P^* and Q^* are the desired active and the reactive powers. Normally, P^* should coincide with the nominal active power of each inverter and $Q^* = 0$.

However, we have to distinguish between two possibilities: importing energy from the grid or exporting energy to the grid. The first scenario, in which the total load power is not fully supplied by the inverters, the *IBS* must adjust P^* by using low bandwidth communications to absorb the nominal power from the grid in the *PCC*. This is done with small increments and decrements of P^* as function of the measured grid power, by using a slow *PI* controller, as follows:

$$P^* = k_p\left(P_g^* - P_g\right) + k_i \int \left(P_g^* - P_g\right) dt + P_i^* \tag{18.15}$$

where
  P_g and P_g^* are the measured and the reference active power of the grid
  P_i^* is the nominal power of the inverter i

This way, the UPSs with low battery level can switch to charger mode by using $P^* < 0$. Similarly, we proposed that reactive power control law can be defined as

$$Q^* = k_p'\left(Q_g^* - Q_g\right) + k_i'\int \left(P_g^* - P_g\right) dt + Q_i^* \tag{18.16}$$

where
  Q_g and Q_g^* are the measured and the reference reactive power of the grid
  Q_i^* is the nominal reactive power

The second scenario occurs when the power of the prime movers (e.g., PV panels) is much higher than those required by the loads, and the batteries are fully charged. In this case, the *IBS* may enforce to inject the rest of the power to the grid. Moreover, the *IBS* have to adjust the power references.

- *Islanded operation.* When the grid is not present, the *IBS* disconnects the microgrid from the main grid, starting the autonomous operation. In such a case, the droop method is enough to guarantee proper power sharing between the UPSs. However, the power sharing should take into account the batteries charging level of each module. In this case, the droop coefficient

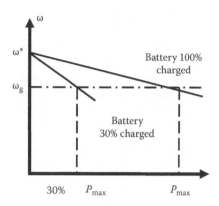

FIGURE 18.15 Droop characteristic as a function of the batteries charge level.

m can be adjusted to be inversely proportional to the charge level of the batteries, as shown in Figure 18.15,

$$m = \frac{m_{min}}{\alpha} \tag{18.17}$$

where
m_{min} is the droop coefficient at full charge
α is the level of charge of the batteries ($\alpha = 1$ when fully charged and $\alpha = 0.001$ when empty)

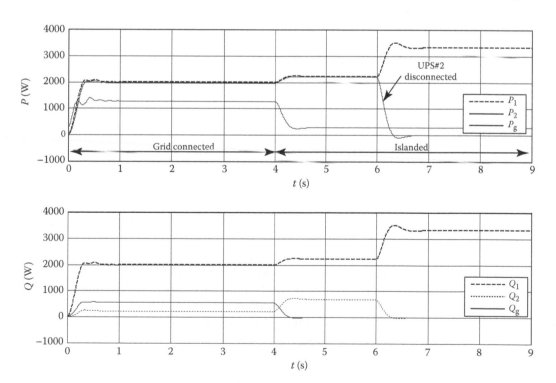

FIGURE 18.16 Active and reactive power transients between grid-connected and islanded modes. (*Y*-axis: $P = 1\,\text{kW/div}$, $Q = 1\,\text{kVAr/div}$).

- *Transitions between grid-connected and islanded operation.* When the *IBS* detects some fault in the grid, it disconnects the microgrid from the grid. In this situation, the *IBS* can readjust the power reference to the nominal values, but this action is not mandatory. Instead, the *IBS* can measure the frequency and the amplitude of the voltage inside the microgrid, and move the set points (P^* and Q^*) in order to avoid the corresponding frequency and amplitude deviations of the droop method. In contrast, when the microgrid is working in islanded mode and the *IBS* detects that the voltage outside of the microgrid is stable and fault free, it can resynchronize the microgrid with the frequency, amplitude, and phase of the grid in order to reconnect the microgrid to the grid seamlessly.

Figure 18.16 shows the active and reactive powers of a two-UPS microgrid sharing power with the grid, the transition to islanded operation, and the disconnection of UPS#2. The system starts being connected to the grid, with $P_g^* = 1000$ W and $Q_g^* = 0$ Var. At $t = 4$ s, the system is disconnected to the grid, and the two-UPS units operate in island mode sharing the overall load. At $t = 6$ s, UPS#2 is disconnected and UPS#1 supplies all the power to the microgrid. Notice the proper transient response, as well as the good power regulation of the system.

18.10 Conclusion

In the coming years, the penetration of DG systems will cause a change of paradigm from centralized electrical generation. It is expected that the utility grid will be formed by a number of interconnected UPSs-based microgrids. However, the onsite generation near the consumption points can be a problem, if we are not able to manage the energy by means of novel kinds of UPSs. One of the problems is that classic renewable energy sources such as photovoltaic and wind energy are variable, since they rely on natural phenomena like sun and wind. In order to accommodate these variable sources to the energy demanded by the loads, it is necessary to regulate the energy flow adequately. On the other hand, the interactivity with the grid and the islanded operation will be the requirements for these new UPSs. In addition, the use of technologies such as compressed air energy devices, regenerative fuel cells and fly-wheel systems will be integrated together with renewable energy sources in order to ensure the continuous and reliable electrical power supply.

References

[Bekiarov02] S. B. Bekiarov and A. Emadi, Uninterruptible power supplies: Classification, operation, dynamics, and control, in *Proceedings of the IEEE APEC'02*, Dallas, TX, 2002, pp. 597–604.

[Bergen86] A. R. Bergen, *Power Systems Analysis*, Prentice-Hall, Englewood Cliffs, NJ, Ed, 1986.

[Blaabjerg06] F. Blaabjerg, R. Teodorescu, M. Liserre, and A. V. Timbus, Overview of control and grid synchronization for distributed power generation systems, *IEEE Trans. Ind. Electron.*, 53, 1398–1409, October 2006.

[Borup01] U. Borup, F. Blaabjerg, and P. N. Enjeti, Sharing of nonlinear load in parallel-connected three-phase converters, *IEEE Trans. Ind. Appl.*, 37(6), 1817–1823, November/December 2001.

[Broeck98] H. van der Broeck and U. Boeke, A simple method for parallel operation of inverters, in *Proceedings of the IEEE INTELEC'98 Conference*, San Francisco, CA, 1998, pp. 143–150.

[Chandorkar94] M. C. Chandorkar, D. M. Divan, Y. Hu, and B. Barnajee, Novel architectures and control for distributed UPS systems, in *Proceedings of the IEEE APEC'94*, Orlando, FL, 1994, pp. 683–689.

[Degner04] T. Degner, P. Taylor, D. Rollinson, A. Neris, and S. Tselepis, Interconnection of solar powered mini-grids—A case study for Kythnos Island, in *Proceedings of the European Photovoltaic Solar Energy Conference and Exhibition*, Bangkok, Thailand, 2004, pp. 1–4.

[Dugan03] R. C. Dugan, M. F. McGranaghan, S. Santoso, and H. W. Beaty, *Electrical Power System Quality*, New York: McGraw-Hill, 2003.

[Guerrero04] J. M. Guerrero, L. García de Vicuña, J. Matas, M. Castilla, and J. Miret, A wireless controller to enhance dynamic performance of parallel inverters in distributed generation systems, *IEEE Trans. Power Electron.*, 19(5), 1205–1213, September 2004.

[Guerrero05] J. M. Guerrero, L. García de Vicuña, J. Matas, M. Castilla, and J. Miret, Output impedance design of parallel-connected UPS inverters with wireless load-sharing control, *IEEE Trans. Ind. Electron.*, 52(4), 1126–1135, August 2005.

[Guerrero06] J. M. Guerrero, J. Matas, L. Garcia de Vicuña, M. Castilla, and J. Miret, Wireless-control strategy for parallel operation of distributed-generation inverters, *IEEE Trans. Ind. Electron.*, 53(5), 1461–1470, October 2006.

[Guerrero07] J. M. Guerrero, J. Matas, L. García de Vicuña, M. Castilla, and J. Miret, Decentralized control for parallel operation of distributed generation inverters using resistive output impedance, *IEEE Trans. Ind. Electron.*, 54(2), 994–1004, April 2007.

[Guerrero09] J. M. Guerrero, J. C. Vasquez, J. Matas, M. Castilla, and L. Garcia de Vicuna, Control strategy for flexible microgrid based on parallel line-interactive UPS systems, *IEEE Trans. Ind. Electron.*, 56(3), 726–736, March 2009.

[Holtz90] J. Holtz and K. H. Werner, Multi-inverter UPS system with redundant load sharing control, *IEEE Trans. Ind. Electron.*, 37(6), 506–513, December 1990.

[Jou04] H.-L. Jou, J.-C. Wu, C. Tsai, K.-D. Wu, and M.-S. Huang, Novel line-interactive uninterruptible power supply, *IEE Proc.-Electron. Power Appl.*, 151(3), 359–364, May 2004.

[Karimi04] M. K. Ghartemani and M. R. Iravani, Method for synchronization of power electronic converters in polluted and variable-frequency environments, *IEEE Trans. Power Syst.*, 19(3), 1263–1270, August 2004.

[Karve00] S. Karve, Three of a kind, *IEE Rev.—Power Syst.*, 46(2), 27–31, 2000.

[Kawabata88] T. Kawabata and S. Higashino, Parallel operation of voltage source inverters, *IEEE Trans. Ind. Appl.*, 24(2), 281–287, March/April 1988.

[King03] A. King and W. Knight, *Uninterruptible Power Supplies and Standby Power Systems*. New York: McGraw-Hill, 2003.

[Kusko96] A. Kusko and S. Fairfax, Survey of rotary uninterruptible power supplies, in *Proceedings of the IEEE Telecommunications and Energy Conference*, Boston, MA, 1996, pp. 416–419.

[Lawrence03] R. G. Lawrence, K. L. Craven, and G. D. Nichols, Flywheel UPS, *IEEE Ind. Appl. Mag.*, 9, 44–50, May/June 2003.

[Lemofouet06] S. Lemofouet and A. Rufer, A hybrid energy storage system based on compressed air and supercapacitors with maximum efficiency point tracking (MEPT), *IEEE Trans. Ind. Electron.*, 53(4), 1105–1115, August 2006.

[Liang04] T.-J. Liang and J.-L. Shyu, Improved DSP-controlled online UPS system with high real output power, *IEE Proc.-Electron. Power Appl.*, 151(1), 121–127, January 2004.

[Marwali04] M. N. Marwali, J.-W. Jung, and A. Keyhani, Control of distributed generation systems—Part II: Load sharing control, *IEEE Trans. Power Electron.*, 19(6), 1551–1561, November 2004.

[Mito06] T. Mito et al., Validation of the high performance conduction-cooled prototype LTS for UPS-SMES, *IEEE Trans. Appl. Supercond.*, 19(2), 608–611, June 2006.

[Nehrir06] M. H. Nehrir, C. Wang, and S. R. Shaw, Fuel cells: promising devices for distributed generation, *IEEE Power Energy Mag.*, 4(1), 47–53, January/February 2006.

[Roberts05] B. Roberts and J. McDowall, Commercial successes in power storage, *IEEE Power Energy Mag.*, 3, 24–30, March/April 2005.

[Sears01] J. Sears, High-availability power systems: Redundancy options, Power Pulse, Darnell.Com Inc., Angel, CA, 2001.

[Shanxu99] D. Shanxu, M. Yu, X. Jian, K.Yong, and C. Jian, Parallel operation control technique of voltage source inverters in UPS, in *Proceedings of the IEEE PEDS'99*, Hong Kong, China, 1999, pp. 883–887.

[Silva02] S. A. O. da Silva, P. F. Donoso-Garcia, P. C. Cortizo, and P. F. Seixas, A three-phase line interactive UPS system implementation with series-parallel active power-line conditioning capabilities, *IEEE Trans. Ind. Appl.*, 38(6), 1581–1590, November/December 2002.

[Svensson01] J. Svensson, Synchronization methods for grid-connected voltage source converters, in *Proceedings of the IEEE Generation, Transmission, Distribution*, vol. 148, May 2001, pp. 229–235.

[Tao03] J. Tao, H. Lin, J. Zhang, and J. Ying, A novel load sharing control technique for paralleled inverters, in *Proceedings of the IEEE PESC'03 Conference*, Acapulco, México, 2003, pp. 1432–1437.

[Tsai03] M. T. Tsai and C. H. Liu, Design and implementation of a cost-effective quasi line-interactive UPS with novel topology, *IEEE Trans. Power Electron.*, 18(4), 1002–1011, July 2003.

[Tsikalakis08] A. G. Tsikalakis and N. D. Hatziargyriou, Centralized control for optimizing microgrids operation, *IEEE Trans. Energy Conversion*, 23(1), 241–248, March 2008.

[Tuladhar00] A. Tuladhar, H. Jin, T. Unger, and K. Mauch, Control of parallel inverters in distributed AC power systems with consideration of line impedance, *IEEE Trans. Ind. Appl.*, 36(1), 131–138, January/February 2000.

[Villeneuve04] P. L. Villeneuve, Concerns generated by islanding, *IEEE Power Energy Mag.*, 2, 49–53, May/June 2004.

[Wekesa02] C. Wekesa and T. Ohnishi, Utility interactive AC module photovoltaic system with frequency tracking and active power filter capabilities, in *Proceedings of the IEEE-PCC'02 Conference*, Osaka, Japan, 2002, pp. 316–321.

[Wu00] T. F. Wu, Y.-K. Chen, and Y.-H. Huang, 3C strategy for inverters in parallel operation achieving an equal current distribution, *IEEE Trans. Ind. Electron.*, 47(2), 273–281, April 2000.

19

Recent Trends in Multilevel Inverter

K. Gopakumar
Indian Institute of Science

19.1 Introduction

Multilevel inverters are finding increased attention in industry as the preferred choice of electronic power conversion for high power applications. It is well suited for applications in a variety of industries involving transportation and energy management. Many research works have been published in recent times stressing the growing importance of multilevel converters [1–4]. This chapter deals with some of the recent trends in this field. It is organized as follows. After a brief introduction of the basics, the most common topologies of multilevel inverter are discussed. The main emphasis of this chapter is to introduce the readers to the many new emerging topologies of multilevel inverters for addressing various operational issues. A simple pulse width modulation (PWM) technique for multilevel inverters is also explained in detail. The future trend of multilevel inverters is discussed in the concluding section.

19.2 Basics of Multilevel Inverter

The basic operation of a multilevel inverter can be understood from Figure 19.1. It consists of a number of voltage sources, in series, and a selector switch. Depending on the position of the switch, a particular magnitude of voltage can be applied to the load. The voltage of point A with respect to the reference point is defined as the *pole voltage*. If there are n dc sources connected in series, then at any instant of time, the pole voltage can assume one of n voltage magnitudes. These voltage magnitudes are defined as *levels* and, thus, n levels of voltage can be impressed at the pole. The term *multilevel* is a term indicating that, at any instant of time, one out of many voltage levels can be realized at the pole. Depending on the number of dc sources, a 2-level, 3-level, or in general an *n*-level inverter can be formed. If the dc sources are equal, then all the levels are equal in magnitude; however, for a general case, the levels may be unequal in magnitude.

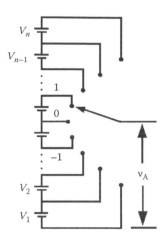

FIGURE 19.1   Basic operation of multilevel inverter.

For most of the multilevel inverter fed applications involving drives or the utility, a sinusoidal fundamental voltage needs to be fabricated by switching fixed dc sources. The typical pole voltages of a 2-level, 3-level, and a 5-level inverter is shown in Figure 19.2, where sinusoidal pulse width modulation (SPWM) is used. As the number of levels increase, the pole voltage approaches closer to the desired sine wave. This is one of the main advantages of the multilevel inverter. If the switches of the 5-level inverter are switched at the same frequency as that of the 2-level inverter, then the output voltage is closer to a sinusoid in case of the 5-level inverter. This means that the harmonic content in the waveform will reduce, or get shifted to higher side of the spectrum, where it can be easily filtered out. The higher the switching frequency, the better will be the quality of the sine wave produced. However, unlike low power applications, switching frequency in multilevel inverters is always restricted. This is because, multilevel converters are mostly used with high voltage dc sources, where high switching frequency of devices cause large d*v*/d*t* stress on the devices and wave reflections in cables, apart from higher switching losses. In general, a design trade-off is made in selecting the switching frequency and the quality of the output voltage produced in case of multilevel inverter driven applications.

Various operational issues of multilevel inverter can be understood from the voltage space vector diagram, which is introduced here. A voltage space vector diagram, or simply a space vector diagram is the 2D representation of voltage phasors in space in the α–β plane. The space vector diagram of a 3-phase, 2-level inverter is shown in Figure 19.3a, where the $2^3 = 8$ switching states are distributed in the space vector plane. In case of 3-level and 5-level inverters (Figure 19.3b and c), the number of locations increases to $3^3 = 27$ and $5^3 = 125$, respectively, leading to an increased density of space vectors in the diagram. For realizing the rotating reference vector, three space vectors enclosing the tip of the reference vector is switched. The instantaneous error between the switching vectors and the tip of the reference vector determines the magnitude of harmonics in the phase voltage. Greater is the error, greater will be the percentage of harmonics in the output voltage. Clearly, the 5-level inverter with higher density of space vectors will produce much less harmonics compared to a 2-level inverter.

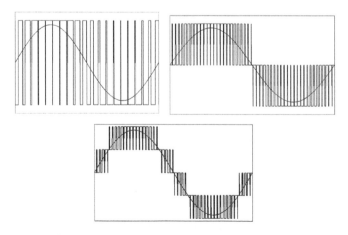

FIGURE 19.2   Pole voltages of 2-level, 3-level, and 5-level inverters.

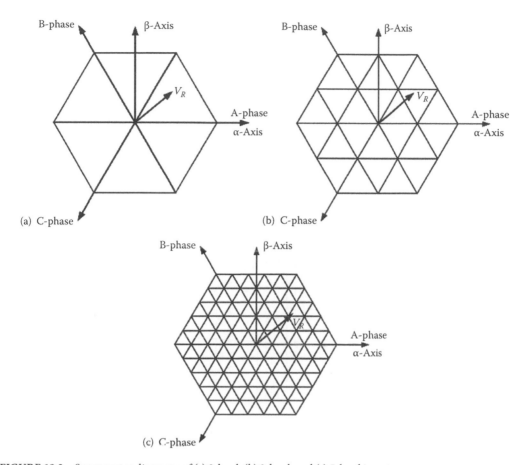

FIGURE 19.3 Space vector diagrams of (a) 2-level, (b) 3-level, and (c) 5-level inverters.

19.3 Topologies for Multilevel Inverter

Presently, a number of circuit topologies exist for multilevel converters. Depending on the application requirements, any one of the topologies may prove to be economical and reliable than others. Among the established ones, three topologies are more popular. They are the neutral point clamped (NPC) topology, cascaded H-bridge (CHB) topology, and the flying capacitor (FC) topology. A comparative performance of these topologies will be presented here. Apart from this, some new topologies like open-end winding structure and cascaded structure for drives application will also be discussed.

19.3.1 Neutral Point Clamped Inverter

The NPC inverter, also known as the diode clamped inverter was first proposed by Nabae et al. [5] and has become the most popular topology for multilevel inverters in recent times. The 3-level and 5-level NPC inverters are shown in Figure 19.4.

For a balanced 3-phase system, the phase voltages and pole voltages are related by the following equations, where v_{xo} and v_{xn} ($x = a, b, c$) are, respectively, the pole and phase voltages of an inverter.

$$\begin{bmatrix} v_{AN} \\ v_{BN} \\ v_{CN} \end{bmatrix} = \begin{bmatrix} 2/3 & -1/3 & -1/3 \\ -1/3 & 2/3 & -1/3 \\ -1/3 & -1/3 & 2/3 \end{bmatrix} \begin{bmatrix} v_{AO} \\ v_{BO} \\ v_{CO} \end{bmatrix} \tag{19.1}$$

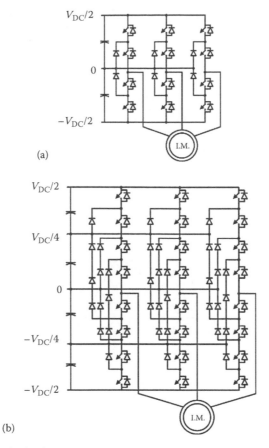

FIGURE 19.4   Three-level and 5-level NPC inverters.

The inverter voltage space vector is represented in terms of the 3-phase voltages as,

$$V_R = v_{AN} + v_{BN}e^{j120} + v_{CN}e^{j240} \tag{19.2}$$

For all the $3^3 = 27$ combinations, the voltage space vectors are plotted in Figure 19.5. Some of the locations, e.g., vectors on the inner hexagon, have more than one unique set of pole voltage combination to achieve the same space vector. If a space vector can be realized by more than one set of pole voltage

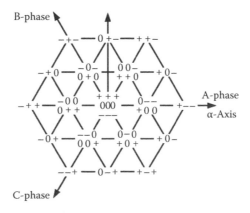

FIGURE 19.5   Space vector diagram of a 3-level NPC inverter showing the switching state combinations.

combinations, then such a space vector is said to have a switching state multiplicity. In this case, the vectors on the inner hexagon have a multiplicity of two. These switching state multiplicities are used to address various operational issues of the NPC inverter, the most common being the control the mid-point voltage of the dc bus so that both the capacitors share half the total dc bus voltage [6].

This topology has become the most popular topology for 3-level inverters, since it offered a simple power circuit to extend the voltage and power ranges of the existing 2-level voltage source inverter (VSI). All semiconductors are operated at a commutation voltage of half the dc-link voltage. The simple transformer rectifier structure at the front end along with a single common dc bus made this converter of particular interest for medium voltage applications.

The main drawback of 3-level NPC inverters is the uneven loss distribution in the devices. In a fundamental cycle, the conduction period of the inner devices is more than the outer devices. This causes unequal losses in devices in a leg. To mitigate this problem, the switching pattern of the devices should rotate. It has been suggested to use active switches in place of the clamping diodes to control the uneven loss distribution of the NPC [7]. The fluctuation of the dc bus midpoint voltage is also another problem, but there are techniques to minimize this problem [1,2].

19.3.2 Cascaded H-Bridge Inverter

This topology can be easily understood from the working knowledge of the basic four quadrant converter, also known as H-bridge converter. Referring to Figure 19.6, each H-bridge cell can produce three levels of voltages at the output. A number of H-bridge cells can be cascaded together to form a CHB structure and a 5-level CHB structure is shown in Figure 19.7. Each H-bridge cell requires four switches and one dc-link formed by a rectifier or a battery. Usually the rectifier is a 3-phase uncontrolled bridge type; however, an active front end can be used for regenerative applications.

Many switching state multiplicities are produced in CHB cell. These multiplicities, together with efficient PWM techniques can be effectively used for different control purposes like even distribution of losses in all the devices and improvement of input current drawn from the converter [1]. The drawback of this inverter structure is the necessity of a number of isolated power supplies, each of which feeds a power cell. Each dc supply requires a transformer, rectifier, and a capacitor adding to the overall size, weight, and cost of the system. However, these dc supplies are normally obtained from multipulse diode rectifiers using star-delta or star-delta-zigzag transformers, thus, improving the input current waveform drawn from the supply.

In order to produce more number of levels in the output voltage with the CHB topology, unequal dc sources can be used. For example, if the two cells of the CHB inverter of Figure 19.7 has dc bus voltages of E and 3E, then nine voltage levels can be produced at the pole, i.e., 4E, 3E, 2E, E, 0, −E, −2E, 3E, and −4E [8,9]. If both the dc sources are equal, then only five levels of voltages are possible. Thus, by using the same topology more number of levels in the phase voltage is possible. The modularity of the inverter is lost in this case, switching redundancies are less, and devices of different ratings have to be

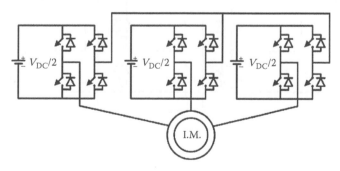

FIGURE 19.6 Three-level H-bridge inverter.

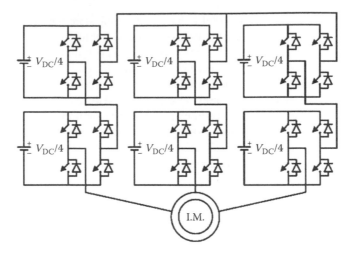

FIGURE 19.7   Cascaded 5-level H-bridge inverter.

used. However, these disadvantages may be acceptable in view of the quality of the output waveform produced, which approaches closer to a sine wave.

19.3.3 Flying Capacitor Inverter

The third topology that has received attention in recent times and has been commercially produced is the FC inverter [1–4]. A 4-level FC topology is shown in Figure 19.8. Similar to the CHB inverter, the FC topology has a number of multiplicities for each space vector.

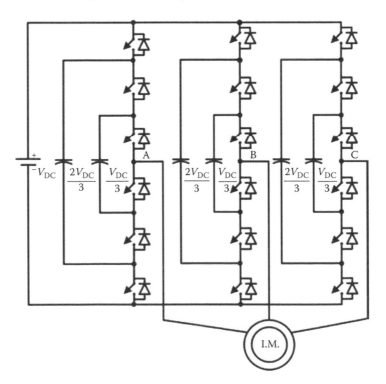

FIGURE 19.8   Four-level FC inverter topology.

It is essential that capacitor voltages are maintained at a desired value before any particular switching state is applied. This also includes the initial precharging of the capacitors [10]. Many of the switching state multiplicities for a space vector are utilized to balance the capacitor voltages of FC. As the load current flows through the capacitor, a quick charge/discharge of the capacitor voltage can happen. A complementary switching state in terms of capacitor voltage restoration is used alternately to maintain the voltage level of capacitors. However, this causes an increase in the switching frequency of the inverter. At the same time, bulky capacitors are placed in the circuit. Typical switching frequencies of about 1 kHz are usually applied in this type of converter [1]. This helps to take corrective action on the capacitor voltages in a shorter duration of time.

19.4 Operational Issues of Multilevel Inverter

One of the most important operational issues of multilevel inverter is the capacitor voltage balancing in the dc bus. In case of NPC converters, capacitor voltage balancing problem has been studied in details and many suggestions have been reported. In case of FC topology, capacitor voltage problem is minimized using the switching state multiplicities. Additionally, extra circuit is needed for precharging the capacitors to the desired level. Capacitor voltage balancing problem and its mitigation are more complex for higher level NPCs and FC topologies, since complementary switching states need to be selected that would balance the capacitor voltages. At the same time, under abnormal or faulty conditions, the capacitor voltages may also deviate from their normal permissible limits. Under such situations, the voltages are restored using a particular set of switching states repeatedly without disturbing the fundamental output voltage.

The common mode voltage elimination for drives application is another important operational issue for multilevel inverters. The common mode voltage is defined as the average of three pole voltages of three phases of an inverter, i.e., $v_{CM} = (v_{AO} + v_{BO} + v_{CO})/3$. With a high switching frequency of the inverter, this voltage gets a low impedance path through the stator winding to the stator and rotor iron via parasitic capacitances (Figure 19.9). The leakage currents flowing to the rotor or through the stator iron builds up motor shaft voltage. When this shaft voltage exceeds the breakdown strength of the lubrication oil, it can cause premature bearing failure due to flashovers. It has been observed that the common mode voltage produced by inverters is one of the primary reasons of motor failures in recent times. Although the common mode voltage is reduced when multilevel inverters are used, many new switching strategy and PWM techniques have been developed to completely eliminate the common mode voltage produced by multilevel inverters. With a total elimination of the common mode voltage in an open-end motor drive, both side inverters can use the same dc-link. Some of these topologies are discussed in detail in the next section.

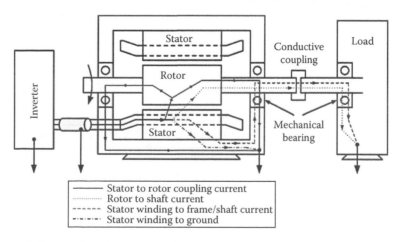

FIGURE 19.9 Path showing flow of common mode current in motor.

19.5 New Trends in Multilevel Topologies for Induction Motor Drives

Although multilevel inverters have many benefits over conventional 2-level inverters, the market penetration of multilevel inverters is still low. As such, many new topologies are still being developed and reported in literature in the field of multilevel inverters. These topologies try to optimize the various operational issues like device count, control, signal improvement, reliability, and fault tolerance capability of an inverter. Some of these new topologies include the cascaded inverter connection, open-end induction motor configuration, and asymmetric dc-link structure. These topologies will be explained in the following section with a particular emphasis on how they address the various operational issues of the inverter.

19.6 Inverters Feeding an Open-End Winding Drive

Strictly speaking, this is not a new inverter topology, but by using conventional inverters and opening the machine neutral, different space vector diagrams can be formed [11]. Nevertheless, this falls under the category of new class of inverter fed drive systems. In a conventional induction motor, one end of the stator windings is connected to the supply while the other end is shorted. In an open-end winding machine, the shorting is removed and the windings are connected to two voltage sources feeding from both the sides of the windings. The total loading on the machine can, therefore, be shared by the two inverters.

For an open-ended induction motor fed from one 2-level inverter on each side, every switching state for a particular space vector is generated by a combination of six voltages (v_{AO}, v_{BO}, v_{CO}, $v_{A'O'}$, $v_{B'O'}$, $v_{C'O'}$), where v_{AO} and $v_{A'O'}$ are, respectively, the "A" phase pole voltages of INV1 and INV2 (Figure 19.10). The phase voltage of the open-end induction motor is given by,

$$v_{AA'} = v_{AO} - v_{A'O'}, \quad v_{BB'} = v_{BO} - v_{B'O'}, \quad v_{CC'} = v_{CO} - v_{C'O'} \tag{19.3}$$

and the resultant space vector is given as,

$$V_R = v_{AA'} + v_{BB'}e^{j120°} + v_{CC'}e^{j240°} \tag{19.4}$$

A 3-level space vector diagram can be thus formed by feeding an open-end winding induction motor by two 2-level inverters [12]. Since, each 2-level inverter produces 8 switching states, a total of (8 × 8 =) 64 switching states are possible here, while a conventional 3-level inverter has 27 switching states. Compared to the 3-level NPC inverter and the FC topology, this topology does not require any capacitor voltage balancing technique to be applied to the inverter. At the same time, unlike the NPC converter, this converter

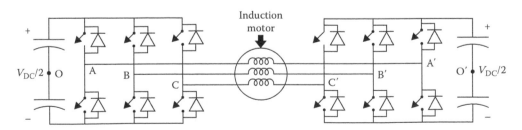

FIGURE 19.10 Open-end induction motor drive.

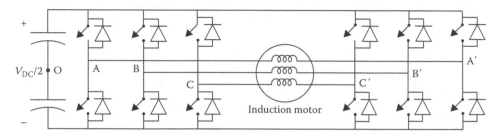

FIGURE 19.11 Open-end induction motor drive with single dc bus.

is devoid of the six clamping diodes. Thus, the number of conducting devices in series is reduced causing an overall reduction of losses. By properly switching the two inverters, the load sharing and the loss distribution in the two inverters can be made equal. Two isolated dc voltage sources are required in this case to restrict the zero sequence currents to circulate, however compared to the CHB inverter, the number of isolated dc sources is less in this topology. In many cases, they are supplied from star-delta transformers causing an improvement in the input current drawn by this converter.

It is even possible to reduce the number of dc sources to one, with a deterioration of the dc bus utilization of the inverters [13]. The circuit configuration is shown in Figure 19.11. This is achieved by making the common mode voltage absent in both the inverters. The common mode voltage is generated by the inverters due to the various switching states. In the 3-level space vector diagram of an open-end induction motor drive (Figure 19.12), six space vectors are shown, which produce zero common mode voltage across the phase winding. This is achieved by observing that for these space vectors, the difference of common mode voltage produced by the two inverters is zero. If the two dc sources are shorted as in Figure 19.11 and any space vector other than these six are switched, then it will result in a huge zero sequence circulating current flow through the winding. However, switching on the space vector diagram with six locations, as shown in Figure 19.12, will result in a decrease in the dc bus utilization and a switching strategy similar to a 2-level inverter.

Full dc bus utilization is proposed in [14], where four bidirectional switches are used along with the two inverters (Figure 19.13). During switching of the space vector locations that produce zero sequence voltage, the bidirectional switches are turned off to deny the zero sequence current path. Otherwise, they are turned on. In this way, by using only one dc source and without any capacitor unbalance problem, a 3-level space vector diagram is achieved with a full dc bus utilization.

The opening of the stator neutral introduces another degree of freedom in controlling the voltage impressed on the machine phase winding. One side of the inverter can be operated at low switching

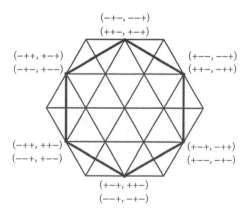

FIGURE 19.12 Three-level space vectors of an open-end induction motor drive that does not produce common mode voltage.

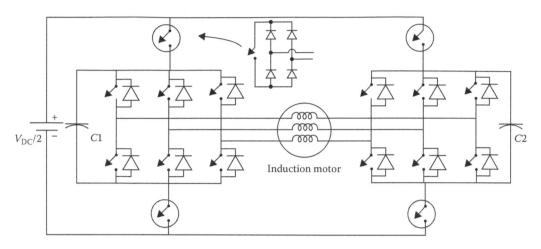

FIGURE 19.13   Open-end winding induction motor drive with full dc bus utilization.

frequency, thereby restricting the switching losses while the other side inverter can be used to cancel the harmonics generated by this inverter. In this way, a highly sinusoidal voltage can be impressed on the machine phase. Note that, the open-end structure may not be limited to drives only. The concept can easily be extended to an open-end transformer also, for front end applications.

19.7 Multilevel Inverter Configurations Cascading Conventional 2-Level Inverters

A 3-level inverter scheme is presented in [15], by cascading two 2-level inverters as shown in Figure 19.14. The pole voltage v_{A2O} can be either $V_{DC}/2$, 0, or $-V_{DC}/2$, by turning on the switches of INV1 and INV2, as shown in Table 19.1. This 3-level inverter can also be used as a 2-level inverter in the lower speed range by clamping one of the 2-level inverters. The advantage here is the absence of the clamping diodes as required in the 3-level NPC inverters and the inverter power-bus structure is very simple as conventional 2-level inverters are used. It can be seen from Figure 19.14 that the voltage rating of the bottom switches in each leg has to withstand a full dc-link voltage and must be rated for a full V_{DC}. Many cascaded inverter structures

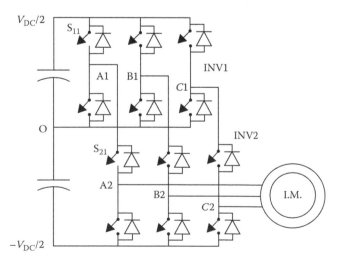

FIGURE 19.14   Three-level inverter scheme cascading conventional 2-level inverters.

TABLE 19.1 Generation of Three Different Voltage Levels Based on State of the Switches for A-Phase (Figure 19.14)

Pole Voltage v_{A2O}	Voltage Level	State of the Switch	
		S_{11}	S_{21}
$V_{DC}/2$	+	1	1
0	0	0	1
$-V_{DC}/2$	–	0	0

have been reported [16–18], and together with open-end winding configuration they can produce different inverter circuits with much functionality, one of which is explained next.

One of the most optimized topology for multilevel inverter is presented in [19] (Figures 19.15 and 19.16). This is a 5-level inverter topology with capacitor voltage balancing and common mode voltage elimination and is a combination of different concepts discussed earlier. There are two 3-level NPC inverters and two 2-level inverters feeding an open-end motor. Two 3-level NPC inverters feeding the open-end motor will produce 5-level space vector diagram, while the presence of two additional 2-level inverters can achieve up to 9-level space vector diagram. However, in order to eliminate the common mode voltage at the machine terminals as well as across the phase, the 9-level structure is reduced to a 5-level structure. The common mode voltage elimination is achieved by selecting inverter switching state combinations, from both ends of the drive, which will not produce the triplen order voltage. This can be easily verified by summing the three pole voltage levels. For example, a switching combination of (10–1, 2′–1′ –1′) will not have any triplen order, since the sum of pole voltages from the left (10–1)

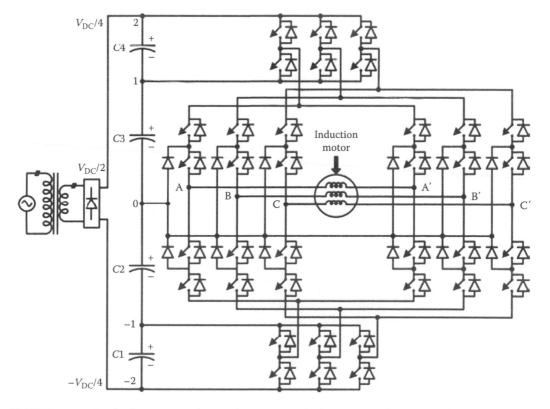

FIGURE 19.15 Five-level inverter topology with common mode elimination and capacitor voltage balancing.

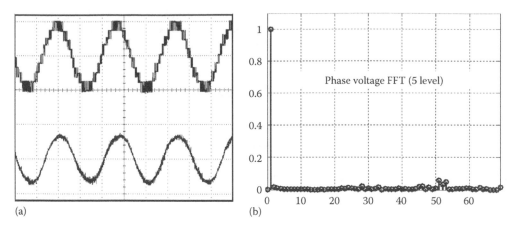

(a)   (b)

FIGURE 19.16 Experimental results of 5-level inverter operation. (a) Phase voltage (top) and phase current (bottom). [Y-axis: 1 div = 200 V and 1 div = 5 A, X-axis: 1 div = 10 ms]. (b) Normalized harmonic spectrum for phase voltage.

and right side (2′ −1′ −1′) will add up to zero. This also permits the use of the same dc-link for both the inverters, and with a proper switching strategy, the load and losses on these inverters can be made equal. Additionally, the cascaded inverter structure helps in sharing the 2-level inverters by both side of the motor, thereby reducing the switch count. In this scheme, capacitor voltage balancing for all the four capacitors of the dc bus is achieved both in steady and transient states. The capacitor voltages can also be restored back to their nominal values after a fault/abnormal condition. All the devices have rating of $V_{DC}/8$. The phase voltage and current waveforms for 5-level operation are shown in Figure 19.16. The normalized harmonic spectrum of phase voltage shows that the motor can be fed from a nearly sinusoidal voltage. The strategy for capacitor voltage balancing in this inverter requires a little explanation. It involves the use of redundant switching states in the space vector diagram that causes opposite effect on the voltage level of capacitors. Figure 19.17 shows the motor winding connection for a particular space vector in the 5-level space vector diagram, which has two multiplicities (01−1, −110) and (1−10, 0−11). These levels are also marked in the dc bus. For any switching state combination, the phase winding is electrically connected between the left and right level points on the dc bus. The assumed positive direction of current is also shown in the figure.

Figure 19.17a shows the switching state (01−1, −110). Here, current $(i_A + i_C)$ will flow through capacitors C4, C3, and C1. The magnitude of current flowing through these three capacitances will be different to that flowing through C2, which will cause unbalance of voltage across C2. Figure 19.17b shows the other switching state for the same space vector. Here, the same current flows through capacitors C1, C2, and C4 causing voltage unbalance across C3. If the magnitude of current remains constant during these two switching states, then the same duration of switching time will cause equal voltage unbalance across C2 and C3. On the other hand, the voltage across C1 and C4 will be equally affected by the two states. So, any unbalance of voltage across C2 and C3 can be restored back to the nominal value, if these two switching states (1−10, 0−11) and (01−1, −110) are used in consecutive sampling periods for the same duration of time, the current remaining constant during this period. The same concept can be extended to maintain the voltage across the other capacitors and a detailed explanation is given in [19]. One experimental result for deliberate capacitor voltage unbalancing and subsequent balancing by corrective switching states is shown in Figure 19.17b.

Many new topologies and inverter configurations have been developed for multilevel inverter fed drive applications. One such application that is of particular interest is feeding the open-end motor with asymmetric dc sources. In fact, by appropriately choosing the dc bus voltage ratios, the total set of 5th and 7th order harmonics can be totally eliminated from the phase voltage leaving the next set of harmonics at 11th and 13th order. This is introduced next.

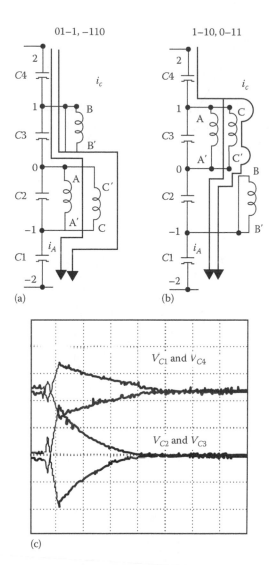

FIGURE 19.17 (a, b) Connection of the motor winding to the capacitors for a space vector having two multiplicities. (c) Experimental results for deliberate capacitor voltage unbalancing and subsequent balancing [*Y*-axis: 1 div – 10 V, *X*-axis: 1 div = 1 s].

19.8 12-Sided Space Vector Structure

As mentioned earlier, the requirement for high power applications is to produce a voltage as close to a sine wave with minimum switching frequency of the inverters. These two are contradictory requirements. With conventional PWM techniques, the switching frequency of the inverters is kept high to shift the harmonics around the switching frequency and its sidebands. But high switching frequency produces higher d*v*/d*t* stress on the devices and wave reflection apart from higher losses. Selection of an optimum switching frequency for high power applications is often a matter of debate. In this respect, a 12-sided polygonal space vector structure is very desirable, that eliminates all the $6n \pm 1$, n = odd harmonics from the phase voltage.

The 12-sided polygonal space vector diagram was first introduced in [20] (Figure 19.18). In this case, two 2-level inverters feed an open-end induction motor, but these two inverters are supplied from two

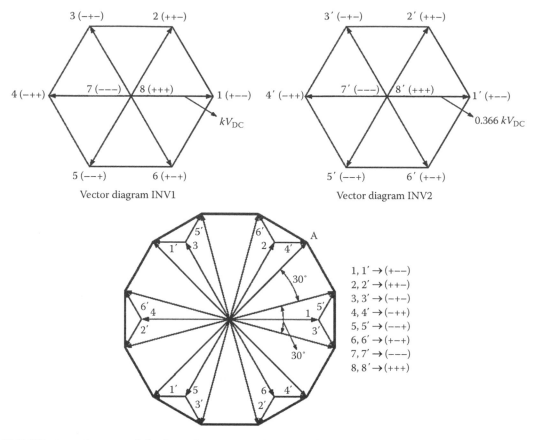

FIGURE 19.18 Basic 12-sided polygonal space vector diagram realization.

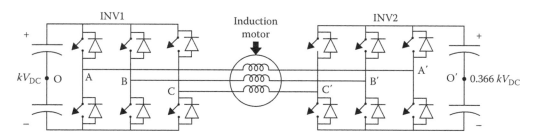

FIGURE 19.19 Power circuit realization.

isolated dc sources of magnitudes in the ratio 1:0.366 (Figure 19.19). Because of the asymmetry in the dc-link, the hexagonal space vector diagram can be modified to form a 12-sided polygonal (*dodecagonal*) space vector diagram. Each of the 2-level inverters produces its own hexagonal space vector diagram with unequal radius. When two such diagrams are combined and the space vectors are switched in a particular sequence, then they form the vertices of a 12-sided polygon. Note that, because of the open-end winding structure, the resultant dodecagonal vectors are obtained by subtracting the space vectors of INV2 from INV1. For realizing one such location (e.g., point A) on the vertex of the polygon, the space vectors that are chosen are 2 from INV1 and 4′ from INV2 (Figure 19.18). The value of "k" decides the radius of the space vector diagram. By choosing an appropriate value of k, the radius of the dodecagonal space vector diagram can be made equal to the hexagonal one. For realizing the volt-second balance in a switching cycle, space vectors from the vertices of the dodecagon and the zero vector at the center are chosen. This method of switching space vectors on a dodecagon will not produce any harmonics in the phase voltage of the order $6n \pm 1$, $n =$ odd. By increasing the number of samples in a sector, it is also possible to suppress the lower order harmonics and a nearly sinusoidal voltage can be obtained. At the same time, a dodecagon is closer to a circle than a hexagon; so the linear modulation range is extended by about 6.6% compared to the hexagonal case. For a 50 Hz rated frequency operation, under constant V/f ratio, the linear modulation can be achieved up to a frequency of 48.3 Hz. Also, the harmonics of the order $6n \pm 1$, $n =$ odd are absent in the overmodulation region, thereby avoiding special current compensated schemes required in the overmodulation region [21]. The maximum voltage is obtained at the end of overmodulation region, where the phase voltage becomes a 12-step waveform. One experimental result of phase voltage, its normalized harmonic spectrum, and phase current is shown in Figure 19.20. Note that in the harmonic spectrum, all the $6n \pm 1$, $n =$ odd harmonics are eliminated.

Instead of an open-end induction motor, the same space vector diagram can be realized by a conventional motor but with a 4-level inverter having asymmetric dc-links (Figure 19.21) [22]. The overall configuration of the inverter consists of cascaded combination of three 2-level inverters, fed from asymmetric isolated dc voltages. Two inverters are supplied with dc voltage of $0.366kV_{DC}$ (INV1 and INV3) while the third one is fed with a dc supply of $0.634kV_{DC}$ (INV2). This voltage ratio can be realized by a combination of star-delta transformers, since $0.634kV_{DC}:0.366kV_{DC} = \sqrt{3}:1$. The value of "$k$" can be chosen to match the radius of the dodecagonal space vector diagram equal to that of the hexagonal one. Based on the bottom rail of INV3, the pole voltage of this inverter structure can have four levels at the output. These are 0, $0.366kV_{DC}$, kV_{DC}, and $1.366kV_{DC}$. They are, respectively, defined as levels 0, 1, 2, and 3 in this discussion.

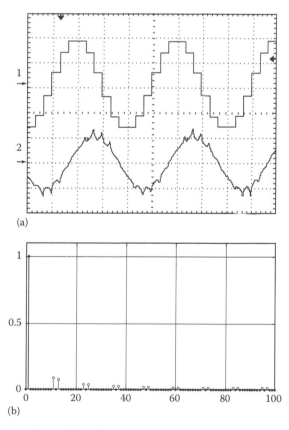

(a)

(b)

FIGURE 19.20 (a) Experimental waveform of 12-step operation (1) phase voltage and (2) phase current; (b) normalized harmonic spectrum of phase voltage.

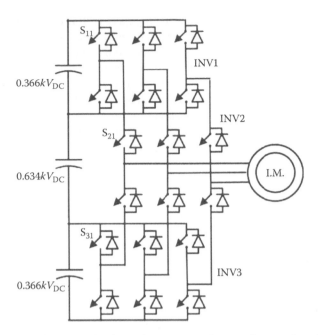

FIGURE 19.21 Power circuit for 12-sided polygonal space vector diagram for a single-end induction motor.

The status of different switches for realizing these voltage levels is tabulated in Table 19.2. Note that, the switches in each leg of the constituent inverters are operated complementary to each other to prevent short circuit of the dc bus.

The same point A on the 12-sided polygon (Figure 19.18) is realized in this topology by the switching combination of (320), where the three numbers represent the level of voltages applied in the three phases. It is also possible to use a common mode eliminated inverter configuration, thereby eliminating the need for isolated sources for open-end structure [23].

TABLE 19.2 Switch Status for Different Levels of Pole Voltage

Pole Voltage	Level	S_{11}	S_{21}	S_{31}
$1.366kV_{DC}$	3	1	1	1
$1kV_{DC}$	2	0	1	1
$0.366kV_{DC}$	1	0	0	1
$0V_{DC}$	0	0	0	0

Although the 12-sided space vector diagram eliminates a set of harmonics from the phase voltage, yet the switching happens similar to 2-level inverter. For realizing the reference, space vectors on the vertices of the 12-sided polygon are chosen. The 12-sided polygonal space vector diagram has been extended to multilevel 12-sided diagram using conventional 3-level NPC inverters [24]. This helps to achieve the space vector diagram using devices of half the rating.

In a recent work, the α–β plane is divided by six concentric dodecagons (Figure 19.22) [25].This space vector diagram is developed by using the switching combinations of two 3-level NPC inverters (with dc-links of magnitude 1:0.366) (Figure 19.23). The inverters can produce three levels of voltages viz. 0, 1, and 2 from INV1 and 0′, 1′, and 2′ from INV2 with respect to their lower dc rail. Out of all the possible switching state combination from both the inverters, there exist space vectors that lie on six concentric dodecagons. These are used for switching in the present work. Because of the presence of six dodecagons, the switching vectors are near to each other, thus the harmonics in the output voltage is minimized. At the same time, all the $6n \pm 1$, n = odd harmonics are absent in the phase voltage throughout the modulation range, including overmodulation region. Thus, very high quality of sine wave can be produced in the phase windings using these space vectors at a reduced switching frequency. The dodecagons have their radius in the ratio 1: $\cos(\pi/12)$: $\cos(2\pi/12)$: $\cos(3\pi/12)$: $\cos(4\pi/12)$: $\cos(5\pi/12)$. Note that, as the radii increases, the dodecagons become closer to each other. This will help to suppress the harmonics more at higher voltage levels. For realizing the reference vector in a switching cycle, three nearest space vectors on adjacent

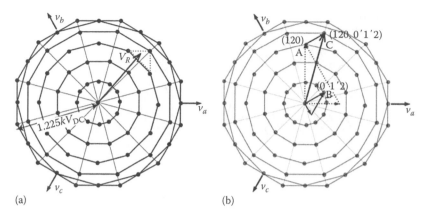

FIGURE 19.22 (a) Multilevel 12-sided space vector diagram and (b) switching state combinations from INV1 and INV2 for realizing space vector.

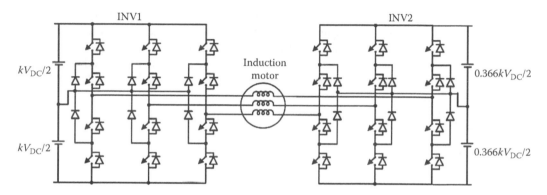

FIGURE 19.23 Power circuit realization of the space vector diagram of Figure 19.22.

dodecagons are used for switching (Figure 19.22a). One switching combination for realizing a space vector is illustrated in Figure 19.22b. Space vector A is generated by INV1 by switching states (120). Space vector B is generated by INV2 by switching state (0'1'2'), measured from INV1 side. When they are combined, space vector C (120, 0'1'2') is formed. Other switching state multiplicities for the entire space vector diagram are presented in [25]. At all modulation indices, harmonics of the order $6n \pm 1$, n = odd are absent from the phase voltage including overmodulation region.

19.9 PWM Strategies for Multilevel Inverter

The PWM strategies for 2-level inverter have been extended for switching of multilevel inverters. In a 2-level inverter, a modulating sine wave is compared with a high frequency triangular wave. Depending on the instantaneous magnitudes of the sine wave and the triangular wave, the upper or lower switch in a single-phase leg is switched ON or OFF. This is called the sine-triangle PWM. Instead of a pure modulating sine wave, different offsets can be added to the sine wave to increase the dc bus utilization of the inverter [26]. The same concept has been extended to the multilevel inverter switching. Here, instead of one triangular carrier, different triangular carriers are used [27]. In level shifted PWM, there are $(n - 1)$ identical carriers for an n-level inverter. The carriers are vertically disposed so that they occupy adjacent vertical bands (Figure 19.24).

One of the most popular PWM strategies for multilevel inverters is the space vector PWM (SVPWM). The SVPWM technique involves calculation of duty cycle of individual switches in an inverter using

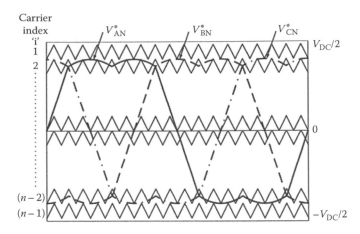

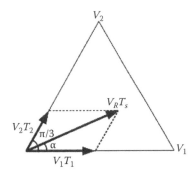

FIGURE 19.24   PWM strategy for multilevel inverters involving multiple carriers.

FIGURE 19.25   Traditional SVPWM based on volt-sec balance.

the concept of volt-second balance in a switching cycle. The hexagonal space vector diagram is divided into six triangular areas (Figure 19.25). For realizing the reference vector, space vectors on the enclosing triangle are chosen to maintain the volt-second balance in a switching cycle. This is mathematically expressed as,

$$V_R^* T_s = V_1 T_1 + V_2 T_2 + V_0 T_0 \tag{19.5}$$

The timing durations for which the active vectors and the zero vectors need to be switched are given by,

$$T_1 = \frac{V_R^*}{V_{DC}} T_s \frac{\sin(60° - \alpha)}{\sin 60°}, \quad T_2 = \frac{V_R^*}{V_{DC}} T_s \frac{\sin \alpha}{\sin 60°}, \quad \text{and} \quad T_0 = T_s - (T_1 + T_2) \tag{19.6}$$

where
  V_R^* = amplitude of reference voltage space phasor ($\mathbf{V_R^*}$)
  V_{DC} = dc-link voltage of inverter
  α = angle of $\mathbf{V_R^*}$ with respect to $\mathbf{V}_1$ in degrees

Historically, SVPWM was developed as a technique different from carrier based PWM; but at a later stage, it was found out to be an extension of the carrier based PWM technique. This is explained next.

For 3-phase balanced sinusoidal waveforms, the instantaneous magnitudes of the phase voltages contain the timing information presented by the equations above. For this, the sampled values are multiplied by the sampling time to transform the voltages into the time domain. Figure 19.26a shows the three sinusoidal modulating waveforms, normalized with respect to the dc bus voltage. The timing information is obtained from these waveforms by multiplying these values by the sampling time T_s, i.e., $T_a = (v_a/V_{DC})^*T_s$, etc. The timing values are plotted in Figure 19.26b, where timing durations are calculated from the zero axis. Timing values may be negative at this point, and at a later stage will be made positive. Here, the sampling instant is chosen when the reference vector lies in sector 1. It is to be noted that the expressions for T_1 and T_2 described by the above equation above

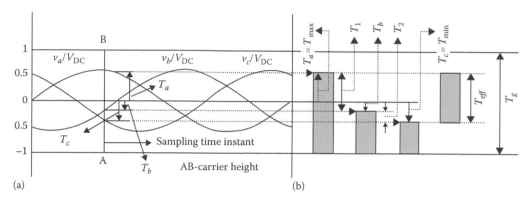

FIGURE 19.26 Sampling of reference phase voltages in sector-1 and the generation of equivalent time signals. (a) Modified reference phase voltages. (b) Equivalent time signal generation for the phases in sector-1.

is similar to the line voltage expressions of the waveforms in sector 1, with a certain gain factor. As such, the difference between the maximum and middle, and middle and minimum sampled waveforms directly gives the timing information of T_1 and T_2. At this sampling instant, $T_{max} = T_a$ and $T_{min} = T_c$, with the instantaneous value of T_c being negative. As such, the effective period defined as $T_{eff} = (T_1 + T_2)$ should be $T_{max} - T_{min}$. The next step is to make the zero periods equal at the start and end of a switching cycle. This is the most important requirement for SVPWM, which ensures 15% more dc utilization compared to sine-PWM [28]. To make the zero periods equal, some offset needs to be added to the timing diagram, which in terms of voltage indicates an additional common mode voltage addition. Because of the isolated neutral of the motor, this additional common mode voltage does not produce any circulating current. To make the zero periods equal, the T_{eff} period shown in Figure 19.26b needs to be placed exactly at the center of the switching period T_s. This implies calculation of the zero period as $T_0 = T_s - T_{eff}$ and adding $T_0/2$ to all the timing waveforms. Additionally, since instantaneous value of T_{min} is negative, a further offset of $(-T_{min})$ is added to all the timing waveforms so that all the three timings now become positive. This addition of $(-T_{min})$ to all the waveforms is particularly suitable for DSP implementation, where all the numbers are necessarily positive.

The space vector PWM switching for multilevel inverters also follows the same principle as above, but there is a slight change in the offset addition. The space vector diagram of a multilevel inverter consists of a number of identical hexagons which are placed in the $\alpha-\beta$ vector plane at different distances from the center. The strategy of switching in a multilevel SVPWM involves mapping of the outer hexagons to the center one, and then using the volt-second balance concept of a 2-level SVPWM. From the modulating waveform point of view, this means adding an offset to the modulating waveform. For example, the original modulating waveform for a 3-level inverter for any phase is shown in Figure 19.27a. After the addition of the offset, which is called the first offset, the modulating waveform becomes as shown in Figure 19.27b. One additional constraint here is the placement of start and end vectors equally at the start and end of the cycle. This is accomplished by giving an additional offset, which is called second offset, to the modulating waveforms to place the zero vector times equally at the start and the end, which is explained in the next paragraph. Note that, after the outer hexagons are mapped into the inner one, the space vectors for the zero periods are no longer the (+++) or (− − −) or (000). In fact, the zero vectors are now at the center of the outer hexagon which is being mapped.

The 3-level modulating waveforms after the addition of $T_{offset1}$ is shown in Figure 19.28a. Note that the upper and lower modulating waveforms are equally placed from the top and bottom boundary of $V_{DC}/2$ and $-V_{DC}/2$. The pole voltages under this condition are also shown. The logic for switching the pole voltages is as follows: If the modulating wave is positive and greater than the triangle, then the pole voltage assumes $V_{DC}/2$; otherwise it is at 0. If the modulating wave is negative, and greater than the triangle,

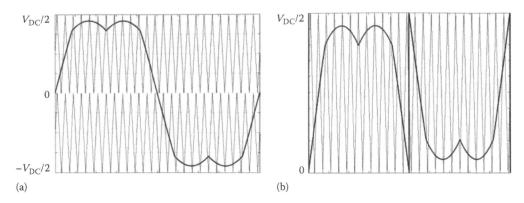

(a)   (b)

FIGURE 19.27   Reference waveform generation for SVPWM of 3-level inverter (a) before addition of first offset and (b) after addition of first offset.

then the pole voltage assumes 0; otherwise it is at $-V_{DC}/2$. With this logic, it is found that the two periods at the start and end of a cycle are not equal. The start and end periods are called the zero periods, but they do not signify the presence of the zero vector, as explained above. So an additional offset called $T_{offset2}$ is needed to make the zero periods equal at the start and end of a cycle, as shown in Figure 19.28b.

The whole algorithm for generation of the active and zero vector timings is given below. The details of this algorithm including the overmodulation region are given in [29].

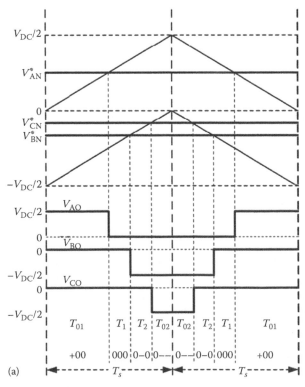

FIGURE 19.28   Placement of zero vectors equally at the start and end of a cycle (a) before equal placement of zero vectors.

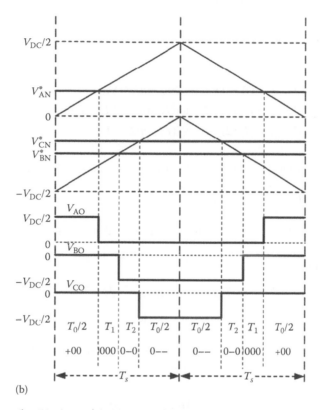

FIGURE 19.28 (continued)   (b) after equal placement of zero vectors.

R ead the sampled amplitudes of v_{AN}, v_{BN}, and v_{CN} for the current sampling interval. Determine the time equivalents of phase voltages, i.e., T_{as}, T_{bs}, and T_{cs}

$$T_{as} = v_{AN} \times \frac{T_s}{(n-1)}, \quad T_{bs} = v_{BN} \times \frac{T_s}{V_{DC}/(n-1)}, \quad T_{cs} = v_{CN} \times \frac{T_s}{V_{DC}/(n-1)},$$

where n is the number of levels.

Find out $T_{offset1}$ as $T_{offset1} = -(T_{max} + T_{min})/2$, where T_{max}, T_{min} are the maximum and minimum of T_{as}, T_{bs}, and T_{cs}.

Determine T_{as}^*, T_{bs}^*, and T_{cs}^* as $T_{as}^* = T_{as} + T_{offset1}$, $T_{bs}^* = T_{bs} + T_{offset1}$, $T_{cs}^* = T_{cs} + T_{offset1}$. Determine the carrier indices I_a, I_b, and I_c for A, B, and C phases, respectively.

Determine T_{a_cross}, T_{b_cross}, and T_{c_cross}

$$T_{a\_cross} = T_{as}^* + ((I_a - (n-1)/2) * T_s)$$
$$T_{b\_cross} = T_{bs}^* + ((I_b - (n-1)/2) * T_s) \quad \text{for } n \text{ odd}$$
$$T_{c\_cross} = T_{cs}^* + ((I_c - (n-1)/2) * T_s)$$

(continued)

(continued)

Or

$$T_{a\_cross} = (T_s/2) + T_{as}^* + ((I_a - (n/2)) * T_s)$$

$$T_{b\_cross} = (T_s/2) + T_{bs}^* + ((I_b - (n/2)) * T_s) \quad \text{for } n \text{ even.}$$

$$T_{c\_cross} = (T_s/2) + T_{cs}^* + ((I_c - (n/2)) * T_s)$$

Sort T_{a_cross}, T_{b_cross}, and T_{c_cross} to determine T_{first_cross}, T_{second_cross}, and T_{third_cross}
The maximum of T_{a_cross}, T_{b_cross}, and T_{c_cross} is T_{third_cross}
The minimum of T_{a_cross}, T_{b_cross}, and T_{c_cross} is T_{first_cross}
And the remaining is T_{second_cross}
Assign first_cross phase, second_cross phase, and third_cross phase according to the phase
that determines T_{first_cross}, T_{second_cross}, and T_{third_cross}
Calculate $T_{offset2}$ as

$$T_0 = T_s - T_{middle} \quad \therefore \quad T_{offset2} = T_0/2 - T_{first\_cross}$$

Determine the timing durations for actual gate signals as (T_{ga}, T_{gb} and T_{gc})

$$T_{ga} = T_{a\_cross} + T_{offset2}, \quad T_{gb} = T_{b\_cross} + T_{offset2}, \quad T_{gc} = T_{c\_cross} + T_{offset2}$$

19.10 Future Trends in Multilevel Inverter

The future of multilevel inverters is promising. The modern trend in multilevel inverters indicates the preference of using multiple numbers of power devices of smaller rating instead of one device of large rating. This factor is particularly important in harsh environment applications, where concerns of reliability and fault tolerance are of utmost importance. Although newer devices of higher voltage rating are continuously being developed, but usage of multiple number of low voltage devices with time tested and proven technology have been found to be more attractive. This has helped in the growth of multilevel inverters as the choice for high power applications. With improvement in device technology, the size and cost of devices is going to come down in near future. This will help to accommodate larger number of cheaper devices in a small package, thereby creating a structure similar to Power VLSI. Many of the multilevel converters of higher levels having a large number of devices will become economically feasible at that time.

Many new applications fields for multilevel inverters are emerging. In particular, multilevel inverters for power transmission and distribution have not been fully explored yet. Similarly, for interfacing renewable energy sources with the utility, multilevel converters can also be effectively used. On the drives side, newer designs of machine and multilevel inverter fed multiphase machines, using multiple numbers of low voltage and low current devices, is also an exciting field of research.

References

1. J. Rodriguez, S. Bernet, B. Wu, J. O. Pontt, and S. Kouro, Multilevel voltage-source-converter topologies for industrial medium-voltage drives, *IEEE Trans. Ind. Electron.*, 54(6), 2930–2945, Dec. 2007.
2. R. Klug and N. Klaassen, High power medium voltage drives: Innovations, portfolio, trends, *Proceedings of the Conference Rec. EPE*, vol. 34, Lausanne, Switzerland, Sept. 2005.

3. J. Rodriguez, J. O. Pontt, P. Lezana, and S. Kouro, Tutorial on multilevel converters, *Proceedings of the Conference Rec. PELINCEC 2005*, Warsaw, Poland, Oct. 2005.

4. L. G. Franquelo, J. Rodriguez, J. I. Leon, S. Kauro, R. Portillo, and M. A. M. Prats, The age of multilevel converters arrives, *IEEE Ind. Electron. Mag.*, 49, 28–39, June 2008.

5. A. Nabae, I. Takahashi, and H. Akagi, A new neutral point clamped PWM inverter, *IEEE Trans. Ind. Appl.*, 17, 518–522, 1981.

6. N. Celanovic and D. Boroyevich, A comprehensive study of neutral point voltage balancing problem in three-level-neutral-point-clamped voltage source PWM inverters, *IEEE Trans. Power Electron.*, 15, 242–249, Mar. 2000.

7. P. Barbosa, P. Steimer, J. Steinke, M. Winkelnkemper, and N. Celanovic, Active neutral point clamped (ANPC) multilevel converter technology, *Proceedings of the Conference Rec. EPE 2005*, Dresden, Germany, 2005.

8. A. Rufer, M. Veenstra, and K. Gopakumar, Asymmetrical multilevel converters for high resolution voltage phasor generation, *Proceedings of the Conference Rec. EPE 1999*, Salt Lake City, UT, pp. 1–10, 1999.

9. M. Veenstra and A. Rufer, Control of a hybrid asymmetric multilevel inverter for competitive medium-voltage industrial drives, *IEEE Trans. Ind. Electron.*, 41(2), 655–664, Mar.–Apr. 2005.

10. T. A. Meynard and H. Foch, Electronic device for electrical energy conversion between a voltage source and a current source by means of controllable switching cells, *IEEE Trans. Ind. Electron.*, 49(5), 955–964, Oct. 2002.

11. H. Stemmler and P. Geggenbach, Configurations of high power voltage source inverter drives, *Proceedings of the Conference Rec. EPE 1993*, vol. 5, Brighton, U.K., pp. 7–14, 1993.

12. E. G. Shivakumar, K. Gopakumar, and V. T. Ranganathan, Space vector PWM control of dual inverter fed open-end winding induction motor drive, *EPE J.*, 12(1), 9–18, Feb. 2002.

13. V. T. Somasekhar, K. Gopakumar, E. G. Shivakumar, and S. K. Sinha, A space vector modulation scheme for a dual two level inverter fed open-end winding induction motor drive for the elimination of zero sequence currents, *EPE J.*, 12(2), 26–36, May 2002.

14. V. T. Somasekhar, K. Gopakumar, and M. R. Baiju, Dual two-level inverter scheme for an open-end winding induction motor drive with a single DC power supply and improved DC bus utilization, *IEE Proc. Electron. Power Appl.*, 151(2), 230–238, Mar. 2004.

15. V. T. Somasekhar and K. Gopakumar, Three-level inverter configuration cascading two two-level inverters, *IEE Proc.-Electron. Power Appl.*, 150(3), 245–254, May 2005.

16. K. A. Corzine, S. D. Sudhoff, and C. A. Whitcomb, Performance characteristics of a cascaded two-level inverter, *IEEE Trans. Energy Conversion*, 14(3), 433–439, Sept. 1999.

17. K. A. Corzine, M. W. Wielebski, F. Z. Peng, and J. Wang, Control of cascaded multilevel inverters, *IEEE Trans. Power Electron.*, 19(3), 732–738, May 2004.

18. P. Xiao, G. K. Venayagamoorthy, and K. A. Corzine, Seven-level shunt active power filter for high-power drive systems, *IEEE Trans. Ind. Electron.*, 24(1), 6–13, Jan. 2009.

19. G. Mondal, F. Sheron, A. Das, K. Sivakumar, and K. Gopakumar, A DC-link capacitor voltage balancing with CMV elimination using only the switching state redundancies for a reduced switch count multi-level inverter fed IM Drive, *EPE J.*, 19(1), 5–15, Mar. 2009.

20. K. K. Mohapatra, K. Gopakumar, V. T. Somasekhar, and L. Umanand, A harmonic elimination and suppression scheme for an open-end winding induction motor drive, *IEEE Trans. Ind. Electron.*, 50(6), 1187–1198, Dec. 2003.

21. M. Khambadkone and J. Holtz, Compensated synchronous PI current-controller in overmodulation range with six-step operation of space vector modulation based vector controlled drives, *IEEE Trans. Ind. Electron.*, 50(6), 1187–1198, 2003.

22. S. Lakshminarayanan, R. S. Kanchan, P. N. Tekwani, and K. Gopakumar, Multilevel inverter with 12-sided polygonal voltage space vector locations for induction motor drive, *IEE Proc. Electr. Power Appl.*, 153(3), 411–419, May 2006.

23. S. Lakshminarayanan, G. Mondal, P. N. Tekwani, K. K. Mohapatra, and K. Gopakumar, Twelve-sided polygonal voltage space vector based multilevel inverter for an induction motor drive with common-mode voltage elimination, *IEEE Trans. Ind. Electron.*, 54(5), 2761–2768, Oct. 2007.

24. A. Das, K. Sivakumar, R. Ramchand, C. Patel, and K. Gopakumar, A pulse width modulated control of induction motor drive using multilevel 12-sided polygonal voltage space vectors, *IEEE Trans. Ind. Electron.*, 56(7), 2441–2449, July 2009.

25. A. Das, K. Sivakumar, R. Ramchand, C. Patel, and K. Gopakumar, A high resolution pulse width modulation technique using concentric multilevel dodecagonal voltage space vector structures, *Proceedings of the Conference Rec. ISIE 2009*, Montreal, Canada, 2009.

26. Dae-Woong Chung, Joohn-Sheok Kim, and Seung-Ki Sul, Unified voltage modulation technique for real-time three phase power conversion, *IEEE Trans. Ind. Appl.*, 34(2), 374–380, 1998.

27. G. Carrara et al., A new multilevel PWM method: A theoretical analysis, *IEEE Trans. Power Electron.*, 7(3), 497–505, July 1992.

28. D. G. Holmes, The significance of zero space vector placement for carrier-based PWM schemes, *IEEE Trans. Ind. Appl.*, 32(5), 1122–1129, Sept. 1996.

29. R. S. Kanchan, M. Baiju, K. K. Mohapatra, P. P. Ouseph, and K. Gopakumar, Space vector PWM signal generation for multi-level inverters using only the sampled amplitudes of reference phase voltages, *IEE Proc. Electr. Power Appl.*, 152(2), 297–309, Apr. 2005.

20

Resonant Converters

István Nagy
*Budapest University
of Technology and
Economics*

Zoltán Sütö
*Budapest University
of Technology and
Economics*

20.1 Introduction

Resonant converters connect a DC system to an AC or another DC system and control both the power transfer between them and the output voltage or current [2,4]. They are used in such applications as induction heating, very-high-frequency DC–DC power supplies, sonar transmitters, ballasts for fluorescent lamps, power supplies for laser cutting machines, and ultrasonic generators [3,5].

There are some common features characterizing the behavior of most, or at least, some of them. DC–DC and DC–AC converters have two basic shortcomings when their switches are operating in switch mode. During the turn-on and turn-off time, high current and voltage appear simultaneously in and across the switches producing high power losses in them, that is, high switching stresses. The power loss increases linearly with the switching frequency. To ensure reasonable efficiency of the power conversion, the switching frequency has to be kept under a certain maximum value. The second shortcoming in a switching mode operation is the electromagnetic interference (EMI) generated by the large dv/dt and di/dt values of the switching variables. The drawbacks have been accentuated by the trend that is pushing the switching frequency to higher and higher range in order to reduce the converter size and weight.

The resonant converters can minimize the shortcomings. The switches in resonant converters create a square-wave-like voltage or current pulse train with or without a DC component. A resonant L–C circuit is always incorporated. Its resonant frequency could be close to the switching frequency or could deviate substantially. If the resonant L–C circuit is tuned to approximately the switching frequency, the unwanted harmonics are removed by the circuit. In both cases, the variation of the switching frequency is one of the means for controlling the output power and voltage.

The advantages of resonant converters are derived from their L–C circuit and they are as follows: sinusoidal-like wave shapes, inherent filter action, reduced dv/dt and di/dt and EMI, facilitation of the turn-off process by providing zero current crossing for the switches, and output power and voltage control by changing the switching frequency. In addition, some resonant converters, e.g., quasi-resonant converters, can accomplish zero current and/or zero voltage across the switches at the switching instant and reduce substantially the switching losses. The literature categorizes these converters as hard-switched and soft-switched converters. Unlike hard-switched converters, the switches in soft-switched converters, quasi-resonant converters, and some resonant converters are subjected to much lower switching stresses. Note that not all resonant converters offer zero current switching (ZCS) and/or zero voltage switching (ZVS), that is, reduced switching power losses. In return for these advantageous features, the switches are subjected to higher forward currents and reverse voltages than they would encounter in a nonresonant configuration of the same power. The variation in the operation frequency can be another drawback.

First, a short review of the two basic resonant circuits, series and parallel, are given. Then the following three types of resonant converters are discussed:

- Load-resonant converters
- Resonant-switch converters
- Resonant DC link converters

The description of the dual-channel resonant DC–DC converter family concludes the chapter [1].

20.2 Survey of the Second-Order Resonant Circuits

The parallel resonant circuit is the dual of the series resonant circuit (Figure 20.1). The series (parallel) circuit is driven by a voltage (current) source. The analog variables for the voltages and currents are the corresponding currents and voltages (Figure 20.1). Kirchhoff's voltage law for the series circuit

$$v_i = v_L + v_R + v_C = i_i \left(sL + R + \frac{1}{sC} \right)$$ (20.1)

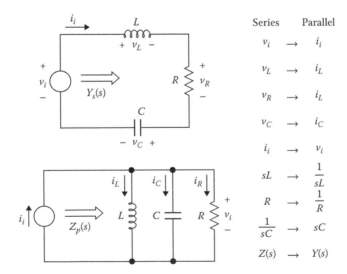

FIGURE 20.1   Dual circuits.

and Kirchhoff's current law for the parallel circuit

$$i_i = i_L + i_R + i_C = v_i \left(\frac{1}{sL} + \frac{1}{R} + sC \right) \tag{20.2}$$

have to be used. The analog parameters for the impedances are the corresponding admittances (Figure 20.1). The input current for the series circuit is

$$i_i = Y_s(s)v_i = \frac{1}{Z_s(s)} v_i \tag{20.3}$$

and the input voltage for the parallel circuit is

$$v_i = Z_p(s)i_i \tag{20.4}$$

where the input admittance

$$Y_s(s) = \frac{1}{R} \frac{2\xi_s Ts}{1 + 2\xi_s Ts + T^2 s^2} \tag{20.5}$$

and the input impedance

$$Z_p(s) = R \frac{2\xi_p Ts}{1 + 2\xi_p Ts + T^2 s^2} \tag{20.6}$$

The time constant T and the damping factor ξ together with some other parameters are given in Table 20.1. ξ must be smaller than unity in (20.5) and (20.6) to have complex roots in the denominators, that is, to obtain oscillatory response.

When v_i is a unit step function, $v_i(s) = 1/s$, the time response of the voltage across R in the series resonance circuit from (20.3) and (20.5) is

$$Ri_i(t) = 2\xi_s T \underbrace{\frac{1}{T\sqrt{1-\xi_s^2}} e^{-\xi_s t/T} \sin\left(\frac{t\sqrt{1-\xi_s^2}}{T} \right)}_{f(t/T)} \tag{20.7}$$

TABLE 20.1   Parameters

	Series	Parallel
Time constant	$T = \sqrt{LC}$	$T = \sqrt{LC}$
Resonant angular frequency	$\omega_0 = 2\pi f_0 = \dfrac{1}{T}$	$\omega_0 = 2\pi f_0 = \dfrac{1}{T}$
Damping factor	$\xi_s = \dfrac{1}{2} \dfrac{R}{\omega_0 L} = \dfrac{1}{2}\omega_0 CR$	$\xi_p = \dfrac{1}{2} \dfrac{\omega_0 L}{R} = \dfrac{1}{2} \dfrac{1}{\omega_0 CR}$
Characteristic impedance	$Z_0 = \sqrt{\dfrac{L}{C}}$	$Z_0 = \sqrt{\dfrac{L}{C}}$
Damped resonant angular frequency	$\omega_d = \omega_0\sqrt{1-\xi_s^2}$	$\omega_d = \omega_0\sqrt{1-\xi_p^2}$
Quality factor	$Q_s = \dfrac{1}{2\xi_s}$	$Q_p = \dfrac{1}{2\xi_p}$

or for $\xi_s = 0$

$$i_i(t) = \frac{1}{\omega_0 L}\sin(\omega_0 t)\tag{20.8}$$

that is, the response is a damped, or for $\xi_s = 0$ undamped, sinusoidal function.

When the current changes as a step function in parallel circuit, $Ri_i(s) = 1/s$, the expression for the voltage response v_i is given by the right side of (20.7), as well, since $RY_s = Z_p/R$. Of course, now ξ_s has to be replaced by ξ_p. The time function $f(t/T)$ for various damping factors ξ are shown in Figure 20.2.

Assuming sinusoidal input variables, the frequency response for series circuit is

$$\frac{R\overline{i_i}}{\overline{v_i}} = R\overline{Y}_s(j\nu) = \frac{1}{1+jQ_s(\nu-1/\nu)} = \frac{1}{\overline{D}_s(\nu)}\tag{20.9}$$

and for parallel circuit is

$$\frac{\overline{v_i}}{R\overline{i_i}} = \frac{1}{R}\overline{Z}_p(j\nu) = \frac{1}{1+jQ_p(\nu-1/\nu)} = \frac{1}{\overline{D}_p(\nu)}\tag{20.10}$$

where $\nu = \omega/\omega_0$. Both circuits are pure resistive at resonance: $\overline{v}_i = R\overline{i}_i$ when $\nu = 1$.

The plot of the amplitude and phase of the right side of (20.9) and (20.10) as a function of ν are shown in Figure 20.3. The voltage across R and its power can be changed by varying ν. When Q is high, a small change in ν can produce a large variation in the output.

The voltage across the energy storage components, for instance, across L in series circuit, is

$$\frac{\overline{v}_L}{\overline{v}_i} = \frac{j\nu Q_s}{\overline{N}_s(\nu)}\tag{20.11}$$

and the currents in the energy storage components, for instance, in L in parallel circuit, is

$$\frac{\overline{i}_L}{\overline{i}_i} = \frac{Q_p}{j\nu\overline{N}_p(\nu)}\tag{20.12}$$

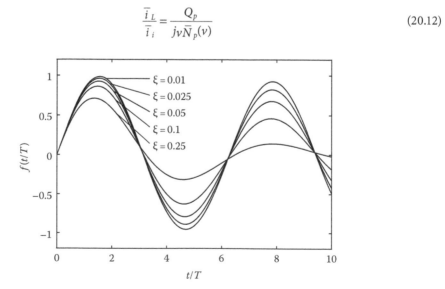

FIGURE 20.2 Time response of $f(t/T)$. ($T = 1$).

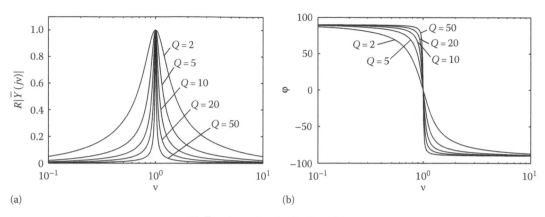

FIGURE 20.3 Frequency response of $[R\bar{Y}(j\nu)]$: amplitude (a), phase (b).

TABLE 20.2 Resonance, $\omega = \omega_0$

Series	Parallel
$\dfrac{\bar{v}_C}{\bar{v}_i} = -jQ_s$	$\dfrac{\bar{i}_C}{\bar{i}_i} = jQ_p$
$\dfrac{\bar{v}_L}{\bar{v}_i} = jQ_s$	$\dfrac{\bar{i}_L}{\bar{i}_i} = -jQ_p$
$Q_s = \dfrac{2\pi\left(\dfrac{1}{2}LI_s^2\right)}{\left(\dfrac{1}{2}RI_s^2\right)\dfrac{1}{f_0}}$	$Q_p = \dfrac{2\pi\left(\dfrac{1}{2}CV_p^2\right)}{\left(\dfrac{1}{2}\dfrac{V_p^2}{R}\right)\dfrac{1}{f_0}}$

The voltages (currents) of the energy storage components in series (parallel) resonant circuit at $\nu = 1$ is Q times as high as the input voltage (current) (Table 20.2). If $Q = 10$, the capacitor or inductor voltage (current) is 10 times the source voltage (current).

The value of L and C and their power rating is tied to the quality factor. The higher the value Q, the better the filter action, that is, the attenuation of the harmonics is better and it is easier to control the output voltage and power by a small change in the switching frequency. The definition of Q is

$$Q = \frac{2\pi \cdot \text{Peak stored energy}}{\text{Energy dissipated per cycle}} \tag{20.13}$$

Using this definition, the expressions for Q are given in Table 20.2, where I_p and V_p are the peak current in the inductor and peak voltage across the capacitor, respectively. For a given output power, the energy dissipated per cycle is specified. The only way to obtain a higher Q is to increase the peak stored energy. The price paid for higher Q is the high peak energy storage requirements in both the inductor and capacitor.

20.3 Load-Resonant Converters

In these converters, the resonant $L–C$ circuit is connected in the load. The currents in the switching semiconductors decay to zero due to the oscillation in the load circuit. Four typical converters are discussed:

1. Voltage-source series resonant converters (SRCs)
2. Current-source parallel resonant converters (PRCs)
3. Class E resonant converters
4. Series- and parallel-loaded resonant DC–DC converters

20.3.1 Input Time Functions

As a result of the on–off actions of the switching devices, the frequently produced time functions of the input variable at the terminals of the ringing load circuit are shown in Figure 20.4. The input variable x_i can be either voltage in SRCs or current in PRCs, and it can be unidirectional (Figure 20.4a) or

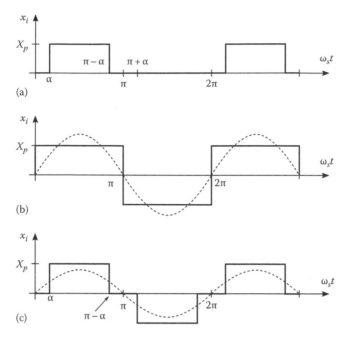

FIGURE 20.4 Frequent input time functions in case of (a) unidirectional and (b,c) bidirectional switches.

bidirectional (Figure 20.4b and c). The ringing load is excited by a variable (Figure 20.4a) that is constant in the interval $\alpha \leq \omega_s t \leq \pi - \alpha$ and short-circuited in the interval $\pi + \alpha \leq \omega_s t < 2\pi - \alpha$, where ω_s is the switching angular frequency. The circuit is interrupted during the rest of the period. The interruption interval shrinks to zero when $\omega_s \geq \omega_d$. The input variable is square-wave and quasi-square-wave in Figure 20.4b and c, respectively. The RMS value of the fundamental component is

$$X_{irms} = \frac{4}{\pi\sqrt{2}} X_p \cos\alpha \tag{20.14}$$

The output variable changes in proportion to the input. Varying angle α provides another means of controlling the output besides the switching frequency f_s.

20.3.2 Series Resonant Converters

SRCs can be implemented by employing either unidirectional (Figure 20.5) or bidirectional (Figure 20.6) switches. The unidirectional switch can be a thyristor, gate turn-off (GTO) thyristor, bipolar transistor, insulated-gate bipolar transistor (IGBT), etc., while these devices with an antiparallel diode or reverse conducting thyristor (RCT) can be used as a bidirectional switch.

Depending on the switching frequency f_s, the wave shape of the output voltage v_o can take any one of the forms shown in Figure 20.7 using the circuit of Figure 20.5. The damped resonant frequency f_d is greater than f_s in Figure 20.7a, $f_s < f_d$; equal to f_s in Figure 20.7b, $f_s = f_d$; and smaller than f_s in Figure 20.7c, $f_s > f_d$. S_1 and S_2 are alternately turned on. The terminals of the series resonant circuit are connected to the source voltage V_{DC} by S_1 or short-circuited by S_2. When neither of the switches are on, the circuit is interrupted. The voltage across the terminals of the series resonant circuit follows the time function shown in Figure 20.4a for $f_d > f_s$, and in Figure 20.4b for $f_d \leq f_s$, respectively. By turning on one of the switches, the other one will be force commutated by the close coupling of the two inductances.

The configuration shown in Figure 20.6 can be operated below resonance, $f_s < f_d$ (Figure 20.8a); at resonance, $f_s = f_d$ (Figure 20.8b); and above resonance, $f_s > f_d$ (Figure 20.8c). The voltage v_i across the terminals

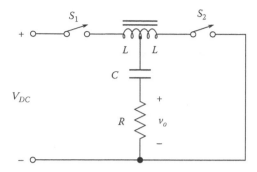

FIGURE 20.5 SRC with unidirectional switches.

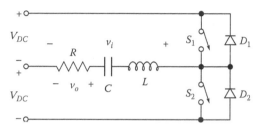

FIGURE 20.6 SRC with bidirectional switches.

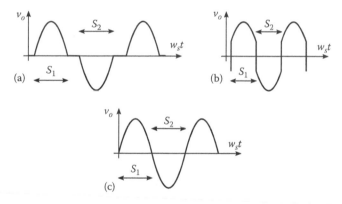

FIGURE 20.7 Output voltage waveforms for Figure 20.5: (a) $f_s < f_d$; (b) $f_s = f_d$; (c) $f_s > f_d$.

of the series resonant circuit is square wave. The harmonics of the load current can be neglected for high Q value. The output voltage v_o equals its fundamental component v_{o1}. The L–C network can be replaced by an equivalent capacitor (inductor) below (above) resonance and by a short circuit at resonance. The circuit is capacitive (inductive) below (above) resonance and purely resistive at resonance (Figure 20.8). The output voltage $v_o \cong v_{o1}$ is leading (lagging) the fundamental component v_{i1} of the input voltage below (above) resonance and in phase at resonance. Negative voltage develops across switches S_1 and S_2 during diode conduction and can be utilized to assist the turn-off processes of switches S_1 and S_2.

No switching loss develops in the switches at $f_s = f_d$ (Figure 20.8b) since the load current will be passing through zero exactly at the time when the switches change state (ZCS). However, when $f_s < f_d$ or $f_s > f_d$, the switches are subjected to lossy transitions. For instance, if $f_s < f_d$ the load current will flow through the switch at the beginning of each half-cycle and then commutate to the diode when the current changes polarity (Figure 20.8a). These transitions are lossless. However, when the switch turns on or when the diode turns off, they are subjected to simultaneous step changes in voltage and current. These transitions therefore are lossy ones. As a result, each of the four devices is subjected to only one lossy transition per cycle.

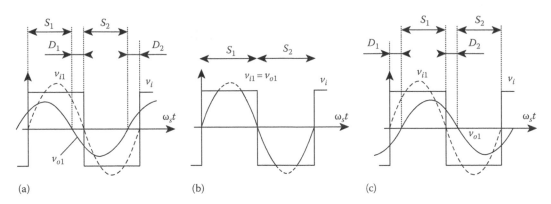

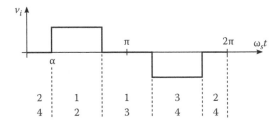

FIGURE 20.8   Output voltage waveforms for Figure 20.6: (a) capacitive, $f_s < f_d$; (b) resistive, $f_s = f_d$; (c) inductive $f_s > f_d$.

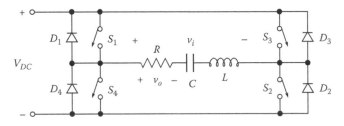

FIGURE 20.9   SRC in bridge topology.

FIGURE 20.10   Quasi-square-wave voltage for output control.

The bridge topology (Figure 20.9) extends the output power to higher range and provides an other control mode for changing the output power and voltage (Figure 20.10).

20.3.3 Discontinuous Mode

Converters with either unidirectional or bidirectional switches can be controlled in discontinuous mode as well. In this mode, the resonant current is interrupted in every half-cycle when using unidirectional switches (Figure 20.7a) and in every cycle when using bidirectional switches (Figure 20.11). The power is controlled by varying the duration of the current break as it is done in duty ratio control of DC–DC converters. Note that this control mode theoretically avoids switching losses because whenever a switch turns on or off its current is zero and no step change can occur in its current as a result of the inductance L. The shortcoming of this control mode is the distorted current waveform. In some applications, such as induction heatings and ballasts for fluorescent lamps, the sinusoidal waveform is not necessary.

20.3.4 Parallel Resonant Converters

The PRCs are the dual of the SRCs (Figure 20.12). The bidirectional switches must block both positive and negative voltages rather than conduct bidirectional current. They are supplied by a current source and the

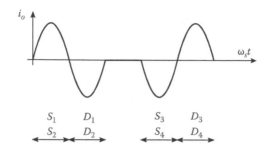

FIGURE 20.11 Discontinuous mode for bridge topology.

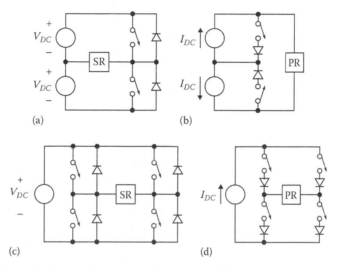

FIGURE 20.12 SRC and PRC are duals. Four-quadrant SRC and PRC topologies with (a,b) two symmetrical sources and with (c,d) four bidirectional switches in bridge connection.

converters generate a square-wave input current i_i that flows through the parallel resonant circuit (Figure 20.13). They offer better short-circuit protection under fault conditions than the SRCs with a voltage source.

When the quality factor Q is high and f_s is near resonance, the harmonics in the R–L–C circuit can be neglected. For $f_s < f_d$, the parallel L–C network is, in effect, inductive. The effective inductance shunts some of the fundamental component i_{i1} of the input current, and a reduced leading current i_{i1} flows in the load resistance (Figure 20.13a). For $f_s = f_d$, the parallel L–C filter looks like an infinitely large impedance. The total current i_{i1} passes through R and the output voltage v_{o1} is in phase with i_{i1} (Figure 20.13b). Being $v_{o1} = 0$ at switching instants, no switching loss develops in the switching devices. For $f_s > f_d$, the L–C network is an equivalent capacitor at the fundamental component i_{i1}. A part of the input current flows

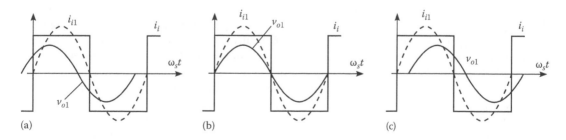

FIGURE 20.13 Waveforms for PRC: (a) inductive, $f_s < f_d$; (b) resistive, $f_s = f_d$; and (c) capacitive, $f_s > f_d$.

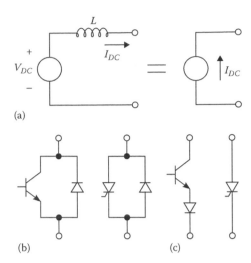

FIGURE 20.14 Implementation of current source (a), implementation of bidirectional switch for SRC (b), and for PRC (c).

through the equivalent capacitor and only the remaining portion passes through the resistor R developing the lagging voltage v_{o1} (Figure 20.13c). As a result of the current shunting through the equivalent L_e and C_e, the voltage v_{o1} is smaller in Figure 20.13a and c than in Figure 20.13b although i_{i1} is the same in all three cases. The current source is usually implemented by the series connection of a DC voltage source and a large inductor (Figure 20.14a). The bidirectional switch is implemented in practice for SRCs with the antiparallel connection of a transistor–diode or thyristor–diode pair (Figure 20.14b) and for PRCs with the series connection of a transistor-diode pair or thyristor. The condition $f_s > f_d$ must be met for PRCs in order for the thyristor to be commutated. By turning on one of the thyristors, a negative voltage is imposed across the previously conducting one, forcing it to turn off (Figure 20.12b and 20.13c). If $f_s > f_d$ and a series transistor–diode pair is used, the diode will experience switching losses at turn-off and the transistor will experience losses at turn-on (Figure 20.13c).

20.3.5 Class E Converter

The class E converter is supplied by a DC current source (Figure 20.14a) and its load R is fed through a sharply tuned series resonant circuit ($Q \geq 7$) (Figure 20.15a). The output current i_o is practically sinusoidal. It uses a single switch (transistor), which is turned on and off at zero voltage. The converter has low—theoretically zero—switching losses and a high efficiency of more than 95% at an operating frequency of several 10 kHz. Its output power is usually low, less than 100 W, and it is used mostly in high-frequency electronic lamp ballasts.

The converter can be operated in optimum and in suboptimum modes. The first mode is explained in Figure 20.15. When the switch is on (off) the equivalent circuit is shown in Figure 20.15b and c. In the optimum mode of operation, the switch (capacitor) voltage, $v_T = v_{C1}$, decays to zero with a zero slope: $I_{DC} + i_o = i_{C1} = 0$. Turning on the switch at t_0, a current pulse $i_T = I_{DC} + i_o$ will flow through the switch with a high peak value: $\hat{I}_T \cong 3I_{DC}$ (Figure 20.15d). Turning off the switch at $t = t_1$, the capacitor voltage builds up reaching a rather high value, $\hat{V}_C = 3.5\ V_{DC}$, and eventually falls back to zero at $t = t_0 + T$ (Figure 20.15e and d). The average value of v_T, and that of the capacitor voltage v_C, is V_{DC}. The average value of i_T is I_{DC}, while there is no DC current component in i_o. In nonoptimum mode of operation, $i_{C1} < 0$, when v_T reaches zero value and the diode D is needed.

The advantage of the class E converter is the simple configuration, the sinusoidal output current, the high efficiency, the high output frequency, and the low EMI. Its shortcomings are the high peak voltage and current of the switch and the large voltages across the resonant L–C components.

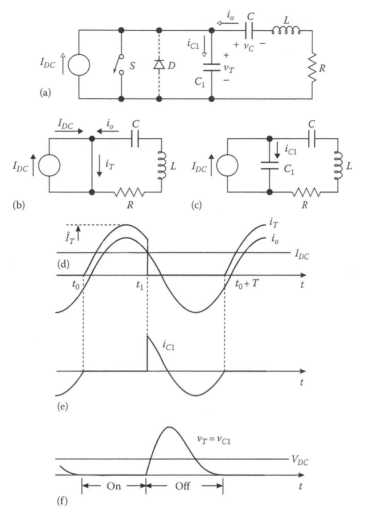

FIGURE 20.15 (a) Class E resonant converter, (b,c) equivalent circuits in case of the two states of the switch S, and (d,e,f) time functions of the converter in optimum mode.

20.3.6 Series- and Parallel-Loaded Resonant DC–DC Converters

The load R can be connected in series with L–C or in parallel with C in SRCs. The first case is called series-loaded resonant (SLR) converter, while the second one is called parallel-loaded resonant (PLR) converter. When the converter is used as a DC–DC converter, the load circuit is built up by a transformer followed by a diode rectifier, a low-pass filter, and finally the actual load resistance. The resonant circuit makes possible the use of a high-frequency transformer reducing its size and the size of the filter components in the low-pass filter.

The properties of the SLR and PLR converters are quite different in some respects. Without the transformer action, the SLR converter can only step down the voltage (20.9), while the PLR converter can both step up and step down (in discontinuous mode of operation) the voltage. The step-up action can be understood by noting that the voltage across the capacitor is Q times higher than that across R in the SRC. The PLR converter has an inherent short-circuit protection when the capacitor is shorted due to a fault in the load. The current is limited by the inductor L.

20.4 Resonant-Switch Converters

The trend to push the switching frequency to higher values, to reduce size and weight, and to suppress EMI led to the development of switch configurations providing ZCS or ZVS. As a result of zero current (voltage) during turn-on and turn-off in ZCS (ZVS), the switching power loss is greatly reduced. $L–C$ resonant circuit is built around the semiconductor switch to ensure ZCS or ZVS. Sometimes, the undesirable parasitic components, such as the leakage inductance of the transformer and the capacitance of the semiconductor switch, are utilized as components of the resonant circuit. Two ZCS and one ZVS configurations are shown in Figure 20.16. The switch S can be implemented for unidirectional and bidirectional current (Figure 20.17). Converters using ZCS or ZVS topology are termed resonant switch converters or quasi-resonant converters.

20.4.1 ZCS Resonant Converters

A step-down DC–DC converter using ZCS configuration shown in Figure 20.16a is presented in Figure 20.18a. Switch S is implemented as shown in Figure 20.17a. The $L_f – C_f$ are sufficiently large to filter the harmonic current components. Current I_o can be assumed to be constant in one switching cycle. Four equivalent circuits associated with the four intervals of each cycle of operation are shown in Figure 20.18b and c together with the waveforms.

Interval 1 ($0 \le t \le t_1$): Both the current i_L in L and the voltage v_C across C are zero prior to turning the switch on at $t = 0$. The output current flows through the freewheeling diode D. After turning the switch on, the total input voltage develops across L and i_L rises linearly ensuring ZCS and soft current change. The interval 1 ends when i_s reaches I_o and the current conduction stops in D at t_1.

Interval 2 ($t_1 \le t \le t_2$): The $L–C$ resonant circuit starts resonating and the change in i_L and v_C will be sinusoidal (Figure 20.18b and c). Interval 2 has two subintervals. The capacitor current $i_C = i_L - I_o$ is positive in $t_1 \le t \le t_2'$ and v_C rises; while it is negative in $t_2' \le t \le t_2$, v_C falls. The peak current is $\hat{I}_L = I_o + V_{DC}/Z_0$ at $t = t_m$ and peak voltage is $\hat{V}_C = 2V_{DC}$ at $t = t_2'$. V_{DC}/Z_0 must be larger than I_o, otherwise i_L will not swing back to zero.

Interval 3 ($t_2 \le t \le t_3$): Current i_L reaches zero at t_2 and the switch is turned off by ZCS. The capacitor supplies the load current and its voltage falls linearly.

Interval 4 ($t_3 \le t \le t_4$): The output current freewheels through D. The switch is turned on at t_4 again and the cycle is repeated.

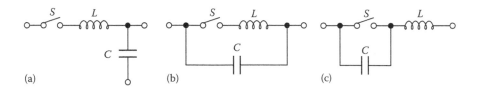

(a)   (b)   (c)

FIGURE 20.16   ZCS (a, b) and ZVS (c) configurations.

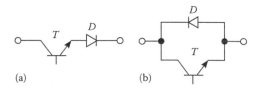

(a)   (b)

FIGURE 20.17   Switch for unidirectional (a) and for bidirectional (b) current.

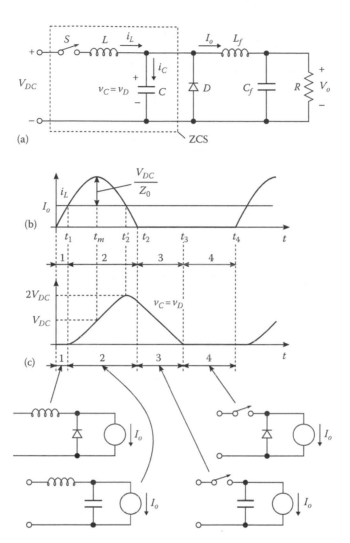

FIGURE 20.18 (a) ZCS resonant converter, (b,c) time functions of the resonant components and the equivalent circuits associated to the four operating intervals.

The output voltage V_o will equal the average value of voltage v_C. V_o can be varied by changing the interval $t_4 - t_3$, that is, the switching frequency.

Applying the ZCS configuration shown in Figure 20.16b, rather than that shown in Figure 20.16a, the operation of the converter remains basically the same. The time function of the switch current and the D diode voltage will be unchanged. The C capacitor voltage will be $v_C = V_{DC} - v_D$.

20.4.2 ZVS Resonant Converter

A ZVS resonant and step-down DC–DC converter is shown in Figure 20.19a and is obtained from Figure 20.18a by replacing the ZCS configuration with the ZVS configuration shown in Figure 20.16c. Note, that the bidirectional current switch is used. This converter's operation is very similar to that of the ZCS converter. The waveform of v_C is the same as the one for i_L in Figure 20.18b and the waveform of i_L is the same as the one for v_C when the ZCS configuration shown in Figure 20.16b is used. $I_o = $ const. in one cycle can be assumed again.

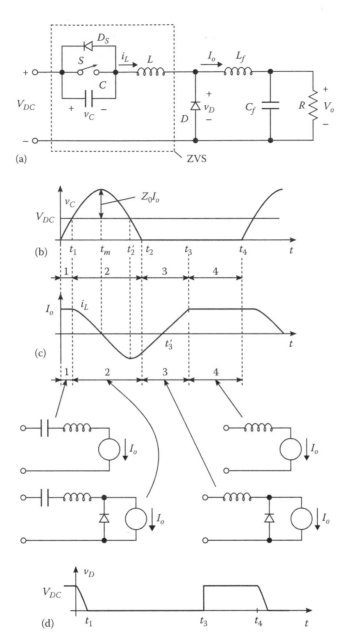

FIGURE 20.19 (a) ZVS resonant converter, (b,c) waveforms are duals of ZCS waveforms shown in Figure 20.18b and c, and (d) time function of the diode voltage v_D.

Interval 1 ($0 \leq t \leq t_1$): S is turned off at $t = 0$. The constant $i_L = I_o$ current starts passing through the capacitor C. Its voltage v_C rises linearly from zero to V_{DC}. ZVS occurs.

Interval 2 ($t_1 \leq t \leq t_2$): Diode D turns on at t_1. The L–C circuit starts resonating through D and the source. Both v_C and i_L are changing sinusoidally. When i_L drops at zero, v_C reaches its peak value: $\hat{V}_C = V_{DC} + Z_0 I_o$. The voltage v_C reaches zero at t_2. The load current must be high enough so that $Z_0 I_o > V_{DC}$; otherwise v_C will not reach zero and the switch will have to be turned on at nonzero voltage.

Interval 3 ($t_2 \leq t \leq t_3$): The diode D_S of the bidirectional switch turns on. It clamps v_C to zero and conducts i_L. The gate signal is reapplied to the switch. V_{DC} develops across L and i_L increases linearly up to I_o, which is reached at t_3. Prior to that, the current i_L changes its polarity at t_3' and S begins to conduct it.

Interval 4 ($t_3 \leq t \leq t_4$): Freewheeling diode D turns off at t_3. It is a soft transition because of the small negative slope of the current i_D. Current I_o flows through S at t_4 when S is turned off and the next cycle begins.

Diode voltage v_D develops across D only in intervals 1 and 4 (Figure 20.19d). Its average value is equal to V_o, which can be varied by interval 4, or in other words, by the switching frequency.

20.4.3 Summary and Comparison of ZCS and ZVS Converters

The main properties of ZCS and ZVS are highlighted as follows:

- The switch turn-on and turn-off occurs at zero current or at zero voltage, which significantly reduces the switching losses.
- Sudden current and voltage changes in the switch are avoided in ZCS and in ZVS, respectively. The di/dt and dv/dt values are rather small. EMI is reduced.
- In the ZCS, the peak current $I_o + V_{DC}/Z_o$ conducted by S must be more than twice as high as the maximum of the load current I_o.
- In the ZVS, the switch must withstand the forward voltage $V_{DC} + Z_0 I_o$, and $Z_0 I_o$ must exceed V_{DC}.
- The output voltage can be varied by the switching frequency.
- The internal capacitances of the switch are discharged during turn-on in ZCS, which can produce significant switching loss at high switching frequency. No such loss occurs in ZVS.

20.4.4 Two-Quadrant ZVS Resonant Converters

One drawback in the ZVS converter, shown in Figure 20.19, is that the switch peak forward voltage is significantly higher than the supply voltage. This drawback does not appear in the two-quadrant ZVS resonant converter where the switch voltage is clamped at the input voltage. In addition, this technique can be extended to the single phase and the three-phase DC-to-AC converter to supply an inductive load.

The basic principle will be presented by means of the DC–DC step-down converter shown in Figure 20.20a. Two switches, two diodes, and two resonant capacitors, $C_1 = C_2 = C$, are used. The voltage V_o can be assumed to be constant in one switching period because C_f is large. The current i_L must fluctuate in large scale and must take both positive and negative values in one switching cycle. To achieve this operation L must be rather small. One cycle consists of six intervals.

Interval 1: S_1 is on. The inductor voltage is $v_L = V_{DC} - V_o$. i_L rises linearly from zero.

Interval 2: S_1 is turned off at t_1. None of the four semiconductors conducts. The resonant circuit consisting of L and the two capacitors connected in parallel is ringing through the source and the load. Now, the impedance $Z_0 = \sqrt{2L/C}$ is high (C is small) and the peak current will be small. The voltage across C_2 approximately changes linearly and reaches zero at t_2. As a result of C_1, the voltage across S_1 changes slowly from zero.

Interval 3: D_2 conducts i_L. The inductor voltage v_L is $-V_o$. i_L is reduced linearly to zero at t_3. S_2 is turned on in this interval when its voltage is zero.

Interval 4: S_2 begins to conduct, v_L is still $-V_o$ and i_L increases linearly in the negative direction.

Interval 5: S_2 is turned off at t_4. None of the four semiconductors conducts. A similar resonant process occurs as in interval 2. As a result of C_2, the voltage across S_2 rises slowly from zero to V_{DC}.

Interval 6: v_C reaches V_{DC} at t_5. D_1 begins to conduct i_L. The inductor voltage $v_L = V_{DC} - V_o$ and i_L rises linearly with the same positive slope as in interval 1 and reaches zero at t_6. The cycle is completed.

The output voltage can be controlled by pulse width modulation (PWM) at a constant switching frequency. Assuming that the intervals of the two resonant processes, that is, interval T_2 and T_5, are small compared to the period T, the wave shape of v_C is of a rectangular form. V_o is the average value of v_C and, therefore, $V_o = DV_{DC}$, where D is the duty ratio: $D = (T_1 + T_6)/T$. Here T is the period: $T \cong T_1 + T_3 + T_4 + T_6$. During the time DT either S_1 or D_1 is on. Similarly, the output current is equal to the average value of i_L.

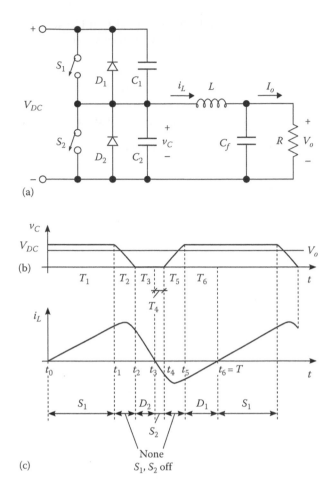

(a)

(b)

(c)

FIGURE 20.20 (a) Two-quadrant ZVS resonant converter, (b,c) time functions on the resonant components and the operating intervals.

20.5 Resonant DC Link Converters with ZVS

To avoid the switching losses in the converter, a resonant circuit is connected between the DC source and the PWM inverter. The basic principle is illustrated by the simple circuit shown in Figure 20.21a. The resonant circuit consist of the L–C–R components. The load of the inverter is modeled by the I_o current source. I_o is assumed to be constant in one cycle of the resonant circuit.

Switch S is turned off at $t = 0$ when $i_L = I_{L0} > I_o$. First, assuming a lossless circuit ($R = 0$), the equations for the resonant circuit are as follows:

$$i_L = I_o + \frac{V_{DC}}{Z_0}\sin\omega_0 t + (I_{L0} - I_o)\cos\omega_0 t \tag{20.15}$$

$$v_C = V_{DC}(1 - \cos\omega_0 t) + Z_0(I_{L0} - I_o)\sin\omega_0 t \tag{20.16}$$

where

$$\omega_0 = \frac{1}{\sqrt{LC}} \quad \text{and} \quad Z_0 = \sqrt{\frac{L}{C}}.$$

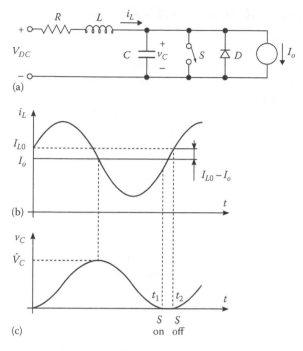

(a)

(b)

(c)

FIGURE 20.21 (a) Resonant DC link converter, (b,c) time functions on the resonant components and the operating intervals.

To turn on and off the switch at zero voltage, the capacitor voltage v_C must start from zero at the beginning and must return to zero at the end of each cycle (Figure 20.21c). Without losses and when $I_{L0} = I_o$, the voltage swing just starts off and returns to zero peaking at $2V_{DC}$. However, when $R \neq 0$, which represents the losses, the voltage swing is damped and v_C would never return to zero under the condition $I_{L0} = I_o$. To force v_C back to zero, a value of $I_{L0} > I_o$ must be chosen (Figure 20.21b). This condition adds the term $Z_0(I_{L0} - I_o)\sin \omega_0 t$ into the right side of Equation 20.16, and thus v_C can reach zero again. By controlling the time interval $t_2 - t_1$, in other words, the on-time of switch S, both $I_{L0} - I_o$ and the peak voltage $\hat{V}_C$ are regulated (Figure 20.21c).

This principle can be extended to the three-phase PWM voltage source inverter (VSI) shown in Figure 20.22. The three cross lines indicate that the configuration has three legs. Any of the two switches and two diodes in one leg can perform the same function that is done by the antiparallel connected S–D circuit in Figure 20.21a. All of the six switches can be turned on and off at zero voltage in Figure 20.22.

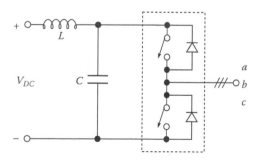

FIGURE 20.22 Resonant DC link converter for three-phase PWM-VSI.

20.6 Dual-Channel Resonant DC–DC Converter Family

20.6.1 Basic Configurations

The converters can be built up of two basic building blocks B_{to} (Figure 20.23a) and B_{off} (Figure 20.23b). Both include two controlled switches S_1 and S_2 and one inductance L. The controlled switches can conduct current flowing to point P in B_{to} and flowing off point P in B_{off} (see the arrows of the switches).

The general configuration of the converters is shown in Figure 20.24, where two switched capacitances C and βC are used beside the building blocks. There are two channels, the upper or p positive channel with block B_p and the lower or n negative channel with block B_n. The two input voltages of the converters v_{ip} and v_{in} can be supplied either by two independent voltage sources or one source by a capacitive divider. The capacitances across the input and output terminals for short-circuiting the high-frequency components of the input and output currents are not shown.

Table 20.3 summarizes the setup of the three basic configurations—the buck, the boost, and the buck and boost (B&B)—by the two building blocks and their connections to terminals x, y, and z. Terminal x and y in the positive and in the negative channel is different one, respectively (Figure 20.24). Suffix i and o refer to input and output, while suffix p and n refer to positive and negative, respectively.

The basic buck, B&B, and the boost configurations derived from the building blocks are presented in Figure 20.25. Note the letters x, y, z and a, b, c in Figure 20.25. They explain the derivation of the three configurations in Figure 20.25 from Figures 20.23 and 20.24 and Table 20.3, resulting in somewhat simpler configurations in which the capacitance βC is replaced by short circuit in the buck, and in the B&B converters, and by an interruption as well as by a short circuit of terminal 0 and 0′ in the boost converter. Further simplification can be accomplished by connecting two clamping diodes in place of the two clamping switches S_{cp} and S_{cn}. Note, that the polarity of the output voltages is reversed in the B&B configuration (Figure 20.25b). The comments made imply the feasibility to build up altogether 12

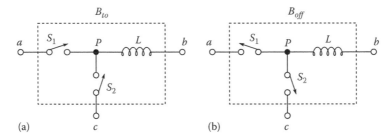

FIGURE 20.23 Basic building blocks. The controlled switches can conduct current (a) flowing to point P in B_{to} and (b) flowing off point P in B_{off}. (From Hamar, J., *EPE J.*, 17(3), 5, 2007. With permission.)

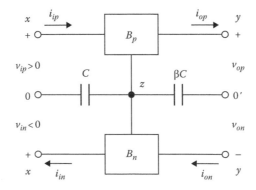

FIGURE 20.24 General configuration of the converters. (From Hamar, J., *EPE J.*, 17(3), 5, 2007. With permission.)

basic configurations. From now on, we consider only those configurations where capacitance βC is removed.

20.6.2 Steady-State Operation

First, for the sake of simplicity, the two so-called clamping switches S_{cp} and S_{cn} are replaced by diodes D_{cp} and D_{cn}. Discontinuous current conduction mode (DCM) in inductance L, and lossless symmetrical operation ($v_{ip} = -v_{in} = V_i =$ const.; $v_{op} = v_{on} = V_o =$ const.; and $R_p = R_n = R$), is assumed. The output V_{op} and V_{on} and input V_{ip} and V_{in} voltages are supposed to be constant and ripple-free due to the large capacitances (the input capacitances are not shown). The operation of the

TABLE 20.3 Set Up of the Converters

	x	y	z	B_p	B_n
Buck	a	b	c	B_{to}	B_{off}
B&B	c	a	b	B_{to}	B_{off}
Boost	b	c	a	B_{off}	B_{to}

Source: Hamar, J., *EPE J.*, 17(3), 5, 2007. With permission.

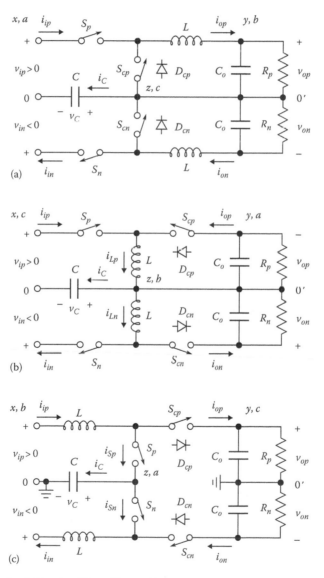

FIGURE 20.25 Basic buck (a), B&B (b), and boost (c) configurations. (From Hamar, J., *EPE J.*, 17(3), 5, 2007. With permission.)

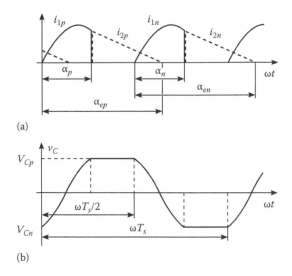

(a)

(b)

FIGURE 20.26 Time functions of (a) input and output currents and (b) capacitor voltage v_C (discontinuous operation). (From Hamar, J., *EPE J.*, 17(3), 5, 2007. With permission.)

TABLE 20.4 Composition of Currents ($V_iI_i = V_oI_o$)

	Subcircuit 1	Subcircuit 2	V_{Cp}	i_i	i_o	i_C	
Buck	$S_p, L, V_{op}, C, V_{ip}$	D_{cp}, L, V_{op}	V_{ip}	i_1	$i_1 \quad i_2$	i_1	$I_o/I_i \geq 1$ $V_o/V_i \leq 1$
B&B	S_p, L, C, V_{ip}	D_{cp}, L, V_{op}	$V_{ip}+V_{op}$	i_1	i_2	i_1	$I_o/I_i \lessgtr 1$ $V_o/V_i \lessgtr 1$
Boost	L, S_p, C, V_{ip}	$V_{ip}, L, D_{cp}, V_{op}$	V_{op}	$i_1 \quad i_2$	i_2	i_1	$I_o/I_i \leq 1$ $V_o/V_i \geq 1$

Source: Hamar, J., *EPE J.*, 17(3), 5, 2007. With permission.

three configurations drawn in Figure 20.25 can jointly be described by using Figures 20.25 and 20.26 and Table 20.4.

By turning on switch S_p, in any of subcircuits 1 (definitions are in Table 20.4), a sinusoidal current pulse $i_1 = i_{sp}$ is developed from $\omega t = 0$ to $\alpha_p = \alpha$ (Figure 20.26a, $\omega = 2\pi f_r = 1/\sqrt{LC}$). It makes the capacitor voltage v_c swing from V_{Cn} to V_{Cp} ($V_{Cn} < 0$) (Figure 20.26b and Table 20.4). The diode D_{cp} is reverse biased from $\omega t = 0$ to $\omega t = \alpha$. Reaching $v_C = V_{Cp}$ at $\omega t = \alpha$, the clamping diode D_{cp} turns on and it clamps v_C on the value V_{Cp} (Table 20.4) and the choke current commutates from S_p to D_{cp}.

In Subcircuit 2 (Table 20.4), the choke current $i_2 = i_{S_{cp}}$ is reduced to zero like a ramp from angle α to the extinction angle $\alpha_{ep} = \alpha_e$. (In continuous current conduction mode [CCM], the choke current does not reach zero before the initialization of the next period.) i_i input and i_o output currents and condenser current i_C can be composed from i_1 and i_2 for all three converters (Table 20.4).

Similar process takes place in the next half period from $T_s/2$ in the negative channel of the converter in the three configurations.

Neglecting the losses, the power balance is $V_iI_i = V_oI_o$, where I_i and I_o are average values. i_i and i_o overlap each other in the buck ($I_o \geq I_i$) and in the boost ($I_o \leq I_i$) converter, while there is no overlapping in the B&B converter (Table 20.4). Taking into account $V_o/V_i = I_i/I_o$, the voltage ratio is $V_o/V_i \leq 1$ for the buck converter, $V_o/V_i \geq 1$ for the boost converter as $I_o \geq I_i$ in the buck and $I_o \leq I_i$ in the boost converter, respectively. V_o/V_i can be higher or lower than unity for the B&B converter.

The peak value of the capacitor voltage is clamped in the positive channel on V_{ip} for buck converter, on $(V_{ip} + V_{op})$ for B & B converter, and on V_{op} for boost converter, respectively (Table 20.4). Similar statement holds for the negative channel.

An important advantage is the soft switching of S and D in DCM. The sinusoidal current in S starts from zero (zero current turn on). Both in CCM and in DCM the current commutates from S to D practically under zero voltage across S and D.

20.6.3 Configurations with Four Controlled Switches

Applying diodes as clamping switches (D_{cp}, D_{cn}), the control variables for changing the output voltage are the switching frequency f_s and the input voltage V_i. However, replacing D_{cp} and D_{cn} by controlled switches S_{cp} and S_{cn}, the current commutation between S and S_c is determined by the timing of turning on S_c independently of the value of capacitor voltage, ensuring two other control variables (α_p and α_n) in asymmetrical operation and only one additional control variable (α) in symmetrical operation. In other words, the peak value of the capacitor voltage $V_C = V_{Cp} = V_{Cn}$ can be the third control variable beside f_s and V_i. In general, the ZVS of S and S_c are lost, but the ZCS remains in DCM operation.

The nonconducting controlled switch has to be turned on by a short time before the other conducting switch is turned off. The conduction of the two switches S and S_c has to overlap each other to ensure a freewheeling path for the inductor current and to suppress voltage spikes.

One possible way for realizing the controlled switches is to use a MOSFET and a diode in series connection allowing current conduction only in one direction.

Figure 20.27 presents the voltage and current waveform of the two switches S_p and S_{cp} in DCM when the waveforms are a little bit more complex.

Figure 20.27a and b shows the waveforms for buck and for boost converter, respectively. The voltage waveforms are drawn by bold line, and the current waveforms are drawn by thin line.

It was assumed that the controlled switches are realized by the series connection of an ideal MOSFET and diode. Current can flow only in one direction through them. The two controlled switches are in complementary state, when one is on the other one is off and vice versa. The only exception is the very short overlap interval in commutation.

The voltage time functions v_{S_p} and $v_{S_{cp}}$ in buck converter are as follows: When S_p is turned on $v_{S_p} = 0$ (Figure 20.27a). During the current conduction of S_{cp}, $v_{S_p} = v_{ip} - v_C$ (Figure 20.25a) and it starts increasing when v_C decreases by current i_{1n}. After the current i_{2p} drops to zero, the voltage v_{S_p} suddenly changes to $v_{S_p} = v_{ip} - v_{op} - v_C$. On the other hand, $v_{S_{cp}} = v_C - v_{ip}$ is increased with v_C due to current i_{1p} when S_p is on (Figure 20.25a). After turning on S_{cp} voltage $v_{S_{cp}} = 0$ till α_{ep}. From $\omega t = \alpha_{ep}$, the total output voltage develops across S_{cp}; $v_{S_{cp}} = -v_{op}$.

Turning to the boost converter, similarly to the buck converter, both v_{S_p} and $v_{S_{cp}}$ have three distinct intervals. In $0 \leq \omega t \leq \alpha_p$ switch S_p is turned on, $v_{S_p} = 0$ (Figure 20.25c). Voltage $v_{S_{cp}} = v_C - v_{op}$ increases with v_C. In the second interval $\alpha_p \leq \omega t \leq \alpha_{ep}$, S_{cp} is turned on, $v_{S_{cp}} = 0$ and $v_{S_p} = v_{op} - v_C$. Finally in the third interval $\alpha_{ep} \leq \omega t \leq \omega T_s = 2\pi$, there is no current condition. At start both voltages are changed suddenly, $v_{S_p} = v_{ip} - v_C$ and $v_{S_{cp}} = v_{ip} - v_{op}$.

As a conclusion, it can be stated that the peak value of the voltage across the controlled switches mainly depends on the peak capacitor voltage V_{Cp} and in boost converter on the maximum value of the output voltage v_{op}.

The waveform in Figure 20.27 depicts the time functions of the variables in channel p. Similar waveforms hold true for channel n in symmetrical operation.

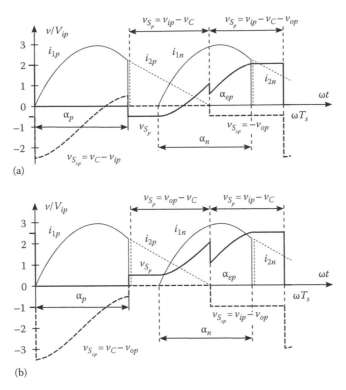

FIGURE 20.27 Voltage and current waveforms of the switches in DCM. Buck converter, $V_{op} = 0.5V_{ip}$; $V_{Cp} = |V_{Cn}| = 1.5V_{ip}$ (a). Boost converter, $V_{op} = 2V_{ip}$; $V_{Cp} = |V_{Cn}| = 1.5V_{ip}$ (b). (From Hamar, J., *EPE J.*, 17(3), 5, 2007. With permission.)

20.6.4 Control Characteristics

Unified mathematical treatment is applied. The three basic converters are described by the same unified equations by introducing the binary variables u_d for step-down (buck) and u_u for step-up (boost) converter. $u_d = 1$ for buck converter and $u_d = 0$ for the other converters. Similar statement holds for u_u. Lossless components are assumed.

The derivation of the voltage ratio V_0/V_i is based on the energy balance. The input energy pulse delivered by one input current pulse during one switching period in either the positive or in the negative channel is (V_i = const.; V_o = const.)

$$w_i = w_{ip} = w_{in} = V_i \int_0^{\alpha/\omega} i_C \, dt + V_i \int_{\alpha/\omega}^{T_s} i_o \, dt = 2CV_iV_C + u_u \frac{V_iV_o}{Rf_s} \tag{20.17}$$

The output energy in one period is

$$w_o = w_i = \frac{V_o^2}{Rf_s} = 2CV_iV_C + u_u \frac{V_iV_o}{Rf_s} \tag{20.18}$$

Dividing each term by V_i^2 and expressing V_o/V_i in (20.18), the output voltage ratio as a function of V_C/V_i and f_s is

$$\frac{V_o}{V_i}\left(\frac{V_C}{V_i}, f_s\right) = \frac{u_u}{2} + \sqrt{\frac{u_u}{4} + 2RCf_s \frac{V_C}{V_i}} \tag{20.19}$$

TABLE 20.5 Voltage V_o/V_i and Current I_i/I_o Ratio ($V_i = 1$)

$V_o / V_i(f_s^*, V_C) = I_i / I_o$	
CCM and DCM Clamping Switch	CCM and DCM Clamping Diode
$\dfrac{u_u}{2} + \sqrt{\dfrac{u_u}{4} + 2R^* f_s^* V_C}$ (see Equation 20.19)	
Buck $u_u = 0$	$\sqrt{2R^* f_s^*}$
B&B $u_u = 0$ — $2^* \sqrt{2R^* f_s^* V_C}$	$R^* f_s^* \left[1 + \sqrt{1 + 2/(R^* f_s^*)} \right]$
Boost $u_u = 1$ — $\dfrac{1}{2} + \sqrt{\dfrac{1}{4} + 2R^* f_s^* V_C}$	$1 + 2R^* f_s^*$

Control variables: $f_s/f_r = f_s^*, V_C$.

Note that (20.19) holds for all three configurations both for CCM and for DCM. By knowing V_o/V_i, the current ratio is known from the power balance $V_i I_i = V_o I_o$ as well. Assuming constant input voltage $V_i = 1$, there are two control variables: the switching frequency f_s and the peak condenser voltage V_C (or the commutation angle α).

Table 20.5 is composed on the basis of Equation 20.19. The last column in Table 20.5 refers to the cases, when clamping switches S_{cp} and S_{cn} are replaced by diodes D_{cp} and D_{cn}.

The derivation of the individual relations from the general equation (Equation 20.19) for the three converters is straightforward by substituting the binary variable u_u. Per unit quantities are used:

$$R^* = RCf_r; \quad f_s^* = \frac{f_s}{f_r}; \quad f_r = \frac{1}{2\pi\sqrt{LC}}$$

Beside the control variables f_s^* and V_C, the parameter is $R^* = RCf_r = R/(2\pi Z)$, where $Z = \sqrt{L/C}$ is the characteristic impedance.

Acknowledgments

This chapter was supported by the János Bolyai Research Scholarship of the Hungarian Academy of Sciences (HAS), by the Hungarian Research Fund (OTKA K72338), and the Control Research Group of the HAS, furthermore, the Hungarian Science and Technology Foundation in framework of JP-25/2006 project and IT-20/2007 project for the financial support. J. Hamar's and B. Buti's collaboration in Section 20.6 is gratefully acknowledged. This work is connected to the scientific program of the "Development of quality-oriented and cooperative R+D+I strategy and functional model at BME" project. This project is supported by the New Hungary Development Plan (Project ID: TÁMOP-4.2.1/B-09/1/KMR-2010-0002).

References

1. J. Hamar, B. Buti, and I. Nagy. Dual channel resonant DC–DC converter family. *EPE* (*European Power Electronics and Drives*) *Journal*, 17(3):5–15, Sept. 2007.
2. J. G. Kassakian, M. F. Schlecht, and G. C. Verghese. *Principles of Power Electronics*. Addison-Wesley, Reading, MA, 1992.
3. N. Mohan, T. M. Undeland, and W. P. Robbins. *Power Electronics: Converters, Applications and Design*, 3rd edn. John Wiley & Sons, New York, 2003.
4. E. Ohno. *Introduction to Power Electronics*. Clarendon Press, Oxford, U.K., 1988.
5. M. H. Rashid. *Power Electronics: Circuits, Devices and Applications*, 3rd edn. Prentice Hall, Englewood Cliffs, NJ, 2003.

IV

Motor Drives

21

Control of Converter-Fed Induction Motor Drives

Marian
P. Kazmierkowski
*Warsaw University
of Technology*

21.1 Introduction

Induction motors (IMs) are widely applied in many different types of industrial electric vehicles and domestic drives owing to their well-known advantages like simple construction (squirrel cage rotor), operation reliability, ruggedness, and low cost. Furthermore, in contrast to the DC brush motor with mechanical commutation, these can be used in aggressive or volatile environments, since there are no problems with spark and corrosion. These advantages, however, are superseded by control problems when using an IM in adjustable speed drives (ASDs). This is because an IM as a plant of feedback control system has a coupled and nonlinear structure. There are different methods to control IM torque and speed, varying in performance, complexity, and costs.

The most economical method is based on stator frequency and voltage control, allowing stepless speed adjustment over a wide range including field weakening (at rated voltage), where rotor speed can be achieved 2–4 times of its rated value. However, it requires the use of a power electronic frequency converter [17]. Therefore, ASD consists of a power electronic converter interfacing a power source and an IM.

The power electronic converter can be fed from an AC or a DC source. For AC power–fed drives, the DC link frequency converter consists of two stages: rectifier (AC/DC) and voltage source inverter (DC/AC). In the AC line side stage, mostly a diode rectifier with a DC link braking resistor is used. However, in cases when regenerative braking is required, an active insulated gate bipolar transistor IGBT transistor rectifier is necessary, because it allows for bidirectional energy flow (motor/generator operation). The voltage source inverter (DC/AC converter) converts the DC link voltage into variable frequency and variable magnitude AC voltage source using either sinusoidal pulse width modulation (PWM) or space vector modulation (SVM). Owing to the switch mode (ON/OF) operation of semiconductor power switches, the PWM inverters are characterized by very high efficiency and very fast operation, creating a high-quality power amplifier. In the DC line or battery-fed drives, only an inverter stage is needed.

The cost relation "DC link frequency converter/IM" is in the range of 2–5; however, in most applications, the energy saving pays off an additional investment in a time period of 4–8 years. Also, the continuous development of power semiconductor devices and cheap, powerful digital processing circuits (digital signal processors—DSP, Application Specific Integrated Circuits—ASIC, and field programmable gate arrays—FPGA) are reducing costs and improving the functionality of modern frequency converters. As a result, an annual extension rate of 7%–8% was observed on the worldwide market over the last decade, with similar prognosis for the future.

In this chapter, the main control strategies applied for voltage source inverter-fed cage-rotor IMs of low and medium power are presented in a systematic way. The chapter starts with a brief review of IM theory including space vector–based equations, which provide a basis for understanding scalar and vector control methods discussed further. The main focus is given on high-performance vector control methods: field-oriented control (FOC), direct torque control (DTC), and DTC with SVM (DTC-SVM), because these are widely offered in the market and their importance and application areas continuously expand. Finally, a conclusion and an overview of typical parameters of these control schemes are briefly discussed.

21.2 Symbols Used in the Analysis of Converter-Fed Induction Motors

$\mathbf{a} = e^{j2\pi/3} = -(1/2) + j(\sqrt{3}/2)$	complex unit vector
f_s	stator frequency
I_A, I_B, I_C	instantaneous values of stator phase currents
$\mathbf{I_r}$	rotor current space vector
I_s	stator current space vector
$I_{s\alpha}, I_{s\beta}$	stator current vector components in stationary $\alpha - \beta$ coordinates
I_{sd}, I_{sq}	stator current vector components in rotating $d - q$ coordinates
$I_{r\alpha}, I_{r\beta}$	rotor current vector components in stationary $\alpha - \beta$ coordinates
J	moment of inertia
L_M	main magnetizing inductance
L_s	stator winding self-inductance
L_r	rotor winding self-inductance
M_e	electromagnetic torque
M_L	load torque
m_s	number of phase windings
p_b	number of pole pairs
S_A, S_B, S_C	switching states for the voltage source inverter
R_r	rotor phase windings resistance
R_s	stator phase windings resistance
$T_r = \dfrac{L_r}{R_r}$	rotor time constant

T_s	sampling time
U_A, U_B, U_C	instantaneous values of stator phase voltages
$\mathbf{U_s}$	stator voltage space vector
$\mathbf{U}_v$	inverter output voltage space vectors, $v = 0, ..., 7$
$U_{s\alpha}, U_{s\beta}$	stator voltage vector components in stationary $\alpha - \beta$ coordinates
U_{sd}, U_{sq}	stator voltage vector components in rotating $d - q$ coordinates
U_{dc}	inverter DC link voltage
Ψ_A, Ψ_B, Ψ_C	flux linkages of stator phase windings
$\boldsymbol{\Psi}_\mathbf{s}$	space vector of the stator flux linkage
$\boldsymbol{\Psi}_\mathbf{r}$	space vector of the rotor flux linkage
Ψ_s	stator flux magnitude
Ψ_r	rotor flux magnitude
$\Psi_{s\alpha}, \Psi_{s\beta}$	stator flux vector components in stationary $\alpha - \beta$ coordinates
$\Psi_{r\beta}, \Psi_{r\beta}$	rotor flux vector components in stationary $\alpha - \beta$ coordinates
γ_m	motor shaft position angle in stationary $\alpha - \beta$ coordinates
γ_{sr}	rotor flux vector angle in stationary $\alpha - \beta$ coordinates
γ_{ss}	stator flux vector angle in stationary $\alpha - \beta$ coordinates
Ω_K	angular speed of the coordinate system
Ω_m	angular speed of the motor shaft, $\Omega_m = d\gamma_m/dt$
Ω_{sr}	angular speed of the rotor flux vector, $\Omega_{sr} = d\gamma_{sr}/dt$
Ω_{ss}	angular speed of the stator flux vector, $\Omega_{ss} = d\gamma_{ss}/dt$
Ω_{sl}	slip frequency
σ	total leakage factor, $\sigma = 1 - \left(L_M^2 / L_s L_r \right)$

Rectangular coordinate systems:

$\alpha - \beta$	stator-oriented stationary coordinates
$d - q$	rotor-flux-oriented rotated coordinates

21.3 Fundamentals of Induction Motor Theory

21.3.1 Space Vector–Based Equations

Modeling of the IM is based on complex space vectors, which are defined in a coordinate system, K, rotating with an angular speed, Ω_K. In absolute units and real-time representation, the following equations describe the behavior of an ideal cage-rotor IM [1,10,12,26]:

$$\mathbf{U}_{sK} = R_s \mathbf{I}_{sK} + \frac{d\boldsymbol{\Psi}_{sK}}{dt} + j\Omega_K \boldsymbol{\Psi}_{sK} \tag{21.1}$$

$$0 = R_r \mathbf{I}_{rK} + \frac{d\boldsymbol{\Psi}_{rK}}{dt} + j\left(\Omega_K - p_b\Omega_m\right)\boldsymbol{\Psi}_{rK} \tag{21.2}$$

$$\boldsymbol{\Psi}_{sK} = L_s \mathbf{I}_{sK} + L_M \mathbf{I}_{rK} \tag{21.3}$$

$$\boldsymbol{\Psi}_{rK} = L_r \mathbf{I}_{rK} + L_M \mathbf{I}_{sK} \tag{21.4}$$

$$\frac{d\Omega_m}{dt} = \frac{1}{J}\left[M_e - M_L \right] \tag{21.5}$$

The electromagnetic torque, M_e, can be expressed by the following formula:

$$M_e = p_b \frac{m_s}{2} \mathrm{Im}\left(\mathbf{\Psi}^*_{sK}\mathbf{I}_{sK}\right) \tag{21.6}$$

Remarks

- The stator and rotor quantities appearing in Equations 21.1 through 21.4 are complex space vectors represented in the common reference frame, rotating with an angular speed Ω_K (and hence the indices K in these quantities). The way they are related to the natural components of a three-phase IM can be represented (e.g., for currents) by

$$\mathbf{I}_{sK} = \frac{2}{3}\left[\mathbf{1}I_A(t) + \mathbf{a}I_B(t) + \mathbf{a}^2I_C(t) \right] \cdot e^{-j\Omega_K t} \tag{21.7}$$

$$\mathbf{I}_{rK} = \frac{2}{3}\left[\mathbf{1}I_a(t) + \mathbf{a}I_b(t) + \mathbf{a}^2I_c(t) \right] \cdot e^{-j(\Omega_K - \Omega_m)t} \tag{21.8}$$

where
  I_A, I_B, and I_C are instantaneous values of the stator winding currents
  I_a, I_b, and I_c are the instantaneous values of rotor winding currents referred to the stator circuit

Similar formulae hold for voltages $\mathbf{U}_{sK}$ and for the flux linkages $\mathbf{\Psi}_{sK}$ and $\mathbf{\Psi}_{rK}$.
- The motion Equation 21.5 is a real equation.
- Owing to the transformation of the equations to a common reference frame, the IM parameters can be regarded as independent of the rotor position.
- The electromagnetic torque formula Equation 21.6 is independent of the choice of the coordinate system, in which the space vectors are represented. This is because for any coordinate system,

$$\mathbf{\Psi}_{sK} = \mathbf{\Psi}_s e^{-j\Omega_K t}, \quad \mathbf{I}_{sK} = \mathbf{I}_s e^{-j\Omega_K t} \tag{21.9}$$

- Including Equation 21.9 in the electromagnetic torque formula (Equation 21.6), one obtains

$$M_e = \mathrm{Im}\left(\mathbf{\Psi}^*_{sK}\mathbf{I}_{sK}\right) = \mathrm{Im}\left(\mathbf{\Psi}^*_s e^{j\Omega_K t} \cdot \mathbf{I}_s e^{-j\Omega_K t}\right) = \mathrm{Im}\left(\mathbf{\Psi}^*_s \mathbf{I}_s\right) \tag{21.10}$$

- Owing to the use of complex space vectors, and assuming that symmetric sine waves are involved, it is possible to employ the symbolic method going over to the steady state, and thus to obtain a convenient bridge to the classical 50/60 Hz supply theory of IM.

21.3.2 Block Schemes

The relations described in Equations 21.1 through 21.6 can be illustrated as block schemes in terms of space vectors in the complex form [10,26], or, following resolution into two-axis components, in the real form [12,16]. When resolving vector equations, one may, in view of the motor symmetry, adopt an arbitrary coordinate. Moreover, taking advantage of the linear dependency between flux linkages and currents, the torque expression can also be written in a number of ways. It follows that there is not just one block diagram of an IM, but instead, on the basis of the set of vector Equations 21.1 through 21.6, one may construct various versions of such a scheme [12]. In going over to the two-axis model, essential differences between various models depend on speed and position of reference coordinates, input signals, and output signals. Based on the method of illustration, we consider two cases.

Case 1: Voltage-Controlled IM Represented in a Stator-Fixed System of Coordinates (α, β)
The popular cage-rotor IM description is presented in a stator-fixed coordinate system ($\Omega_K = 0$), in which the complex space vectors can be resolved into components α and β:

$$\mathbf{U}_{sK} = U_{s\alpha} + jU_{s\beta} \tag{21.11}$$

$$\mathbf{I}_{sK} = I_{s\alpha} + jI_{s\beta} \tag{21.12a}$$

$$\mathbf{I}_{rK} = I_{r\alpha} + jI_{r\beta} \tag{21.12b}$$

$$\mathbf{\Psi}_{sK} = \Psi_{s\alpha} + j\Psi_{s\beta} \tag{21.13a}$$

$$\mathbf{\Psi}_{rK} = \Psi_{r\beta} + j\Psi_{r\beta} \tag{21.13b}$$

Taking Equations 21.11 through 21.13 into account, the set of machine Equations 21.1 through 21.5, after rearranging, can be written as

$$\frac{d\Psi_{s\alpha}}{dt} = U_{s\alpha} - R_s I_{s\alpha} \tag{21.14a}$$

$$\frac{d\Psi_{s\beta}}{dt} = U_{s\beta} - R_s I_{s\beta} \tag{21.14b}$$

$$\frac{d\Psi_{r\alpha}}{dt} = -R_r I_{r\alpha} - p_b \Omega_m \Psi_{r\beta} \tag{21.15a}$$

$$\frac{d\Psi_{r\beta}}{dt} = -R_r I_{r\beta} + p_b \Omega_m \Psi_{r\alpha} \tag{21.15b}$$

$$I_{s\alpha} = \frac{1}{\sigma L_s} \Psi_{s\alpha} - \frac{L_M}{\sigma L_s L_r} \Psi_{r\alpha} \tag{21.16a}$$

$$I_{s\beta} = \frac{1}{\sigma L_r} \Psi_{s\beta} - \frac{L_M}{\sigma L_s L_r} \Psi_{r\beta} \tag{21.16b}$$

$$I_{r\alpha} = \frac{1}{\sigma L_r} \Psi_{r\alpha} - \frac{L_M}{\sigma L_s L_r} \Psi_{s\alpha} \tag{21.17a}$$

$$I_{r\beta} = \frac{1}{\sigma L_r} \Psi_{r\beta} - \frac{L_M}{\sigma L_s L_r} \Psi_{s\beta} \tag{21.17b}$$

$$\frac{d\Omega_m}{dt} = \frac{1}{J} \left[p_b \frac{m_s}{2} \left(\Psi_{s\alpha} I_{s\beta} - \Psi_{s\beta} I_{s\alpha} \right) - M_L \right] \tag{21.18}$$

where σ is the total leakage factor. Equations 21.14 through 21.18 constitute the block scheme of an IM, as depicted in Figure 21.1.

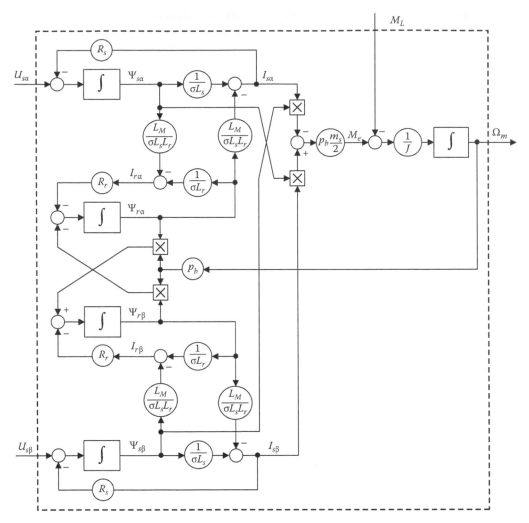

FIGURE 21.1 Block scheme of a voltage controlled cage-rotor IM in the system of α, β coordinates corresponding to Equations 21.14 through 21.18, where σ is total leakage factor.

The model thus obtained (Equations 21.14 through 21.18) corresponds directly to the two-phase motor description and can be used to build a simulation model of the IM. It can be seen from Figure 21.1 that the cage-rotor IM, as a control plant, has a coupled nonlinear dynamic structure, and two of the state variables (rotor currents and fluxes) are not measurable. Moreover, IM resistances and inductances vary considerably with a significant impact on both steady-state and dynamic performances.

Case 2: Current-Controlled IM Represented in Synchronous Coordinates (*d, q*)
Let us adopt a coordinate system $d - q$ rotating with the angular speed equal to the angular speed of the rotor flux vector ($\Omega_K = \Omega_{sr}$), which is defined as follows:

$$\Omega_{sr} = \frac{d\gamma_{sr}}{dt} \tag{21.19}$$

Let us also assume that this system of coordinates rotates concurrently with the rotor flux linkage vector, $\mathbf{\Psi_r}$, where the component $\Psi_{rq} = 0$ (Figure 21.2), and that IM is current controlled. The current control

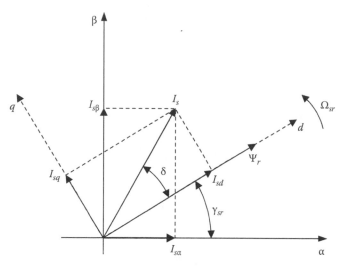

FIGURE 21.2 Vector diagram of IM in stationary $\alpha - \beta$ and rotating $d - q$ coordinates.

of supply occurs quite frequently in practical individual drive systems when an IM is fed by a current source inverter (CSI), and current controlled (CC), CC-PWM-transistor inverter [12,18,26]. When constructing a block diagram of the machine with such an assumption, a simplification can be made by omitting the stator circuit voltage Equation 21.1.

Under these assumptions, the current and flux complex space vectors can be resolved into components d and q:

$$\mathbf{I_{sK}} = I_{sd} + jI_{sq} \tag{21.20a}$$

$$\mathbf{I_{rK}} = I_{rd} + jI_{rq} \tag{21.20b}$$

$$\mathbf{\Psi_{sK}} = \Psi_{sd} + j\Psi_{sq} \tag{21.20c}$$

$$\mathbf{\Psi_{rK}} = \Psi_{rd} = \Psi_r \tag{21.20d}$$

In the $d - q$ coordinate system, the IM model Equations 21.2 through 21.5 can be written as follows:

$$0 = R_r I_{rd} + \frac{d\Psi_r}{dt} \tag{21.21a}$$

$$0 = R_r I_{rq} + \Psi_r \left(\Omega_{sr} - p_b \Omega_m\right) \tag{21.21b}$$

$$\Psi_{sd} = L_s I_{sd} + L_M I_{rd} \tag{21.22a}$$

$$\Psi_{sq} = L_s I_{sq} + L_M I_{rq} \tag{21.22b}$$

$$\Psi_r = L_r I_{rd} + L_M I_{sd} \tag{21.22c}$$

$$0 = L_r I_{rq} + L_M I_{sq} \tag{21.22d}$$

$$\frac{d\Omega_m}{dt} = \frac{1}{J} \left[p_b \frac{m_s}{2} \frac{L_M}{L_r} \Psi_r I_{sq} - M_L \right] \tag{21.23}$$

Equations 21.21b and 21.22c can be easily transformed to

$$\frac{d\Psi_r}{dt} = \frac{L_M R_r}{L_r} I_{sd} - \frac{R_r}{L_r} \Psi_r \tag{21.24}$$

The motor torque can be expressed by the rotor flux magnitude, Ψ_r, and the stator current component, I_{sq}, as follows:

$$M_e = p_b \frac{m_s}{2} \frac{L_M}{L_r} \Psi_r I_{sq} \tag{21.25}$$

Equations 21.24 and 21.25 are used to construct a block scheme of the cage-rotor IM in the $d - q$ coordinate system, which is presented in Figure 21.3.

The input quantities in this diagram are components I_{sd} and I_{sq} of the stator current vector. The output quantities are the angular shaft speed, Ω_m, and the electromagnetic torque, M_e, while the disturbance is the load torque, M_L.

21.3.3 Steady-State Characteristics

It follows from the IM vector equations in the synchronous coordinates, that is, $\Omega_K = \Omega_s$, that under steady-state conditions, all vector quantities remain constant. For this reason, the time-related

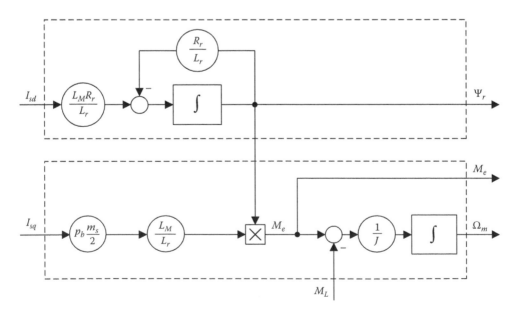

FIGURE 21.3   Block scheme of current controlled IM in $d - q$ field coordinates corresponding to Equations 21.24 through 21.25

derivatives in the voltage Equations 21.1 and 21.2 and in the equation of motion (Equation 21.5) must be neglected. Thus, one obtains a set of algebraic equations that describe the steady-state motor operation. Further, from the algebraic form of voltage and flux current equations, one can find the torque developed by a machine by representing it as a function only of the stator voltage amplitude:

$$M_e(R_s = 0) = \frac{L_M^2 R_r \Omega_{sl}}{(\Omega_{sl}\sigma L_s L_r)^2 + (R_r L_s)^2} \left(\frac{U_s}{\Omega_s}\right)^2 \tag{21.26}$$

From (21.26) we obtain, by comparing $dM_e/d\Omega_{sl}$ to zero, the breakdown slip frequency:

$$\Omega_{slk}(R_s = 0) = \pm \frac{R_r}{\sigma L_r} \tag{21.27}$$

With (21.27), Equation 21.26 can be written in the following form known as the simplified Kloss formula:

$$M_e = M_{ek} \frac{2}{\left(\Omega_{sl}/\Omega_{slk}\right)^2 + \left(\Omega_{slk}/\Omega_{sl}\right)^2} \tag{21.28}$$

where the breakdown torque is

$$M_{ek} = p_b \frac{m_s}{2} \frac{1-\sigma}{2\sigma} \frac{1}{L_s} \left(\frac{U_s}{\Omega_s}\right)^2 \tag{21.29}$$

the following properties arise from the above equations:

- The breakdown torque is independent of the rotor resistance (21.29).
- The breakdown slip frequency is proportional to the rotor resistance (21.27).
- Under constant U_s/f_s mode, the breakdown torque remains constant (21.29).

The torque curve obtained from the Kloss formula (21.28) is shown in Figure 21.4.

In many applications, the IM operates not only below but also above the rated speed. This is possible because most of the IMs can be driven up to twice the rated speed without any mechanical problems.

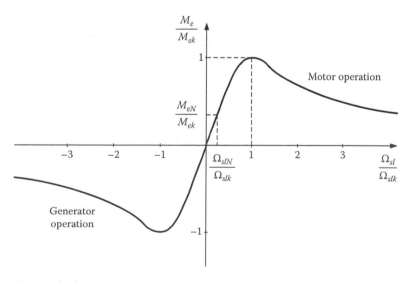

FIGURE 21.4 Torque-slip frequency characteristic obtained from the Kloss formula (21.28).

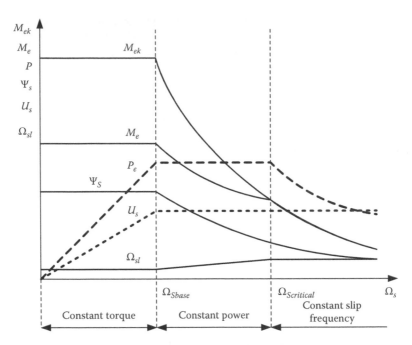

FIGURE 21.5   Control characteristics of IM in constant and weakened flux regions.

The typical characteristics are plotted in Figure 21.5. Below the rated speed, the flux amplitude is kept constant, and at the rated slip frequency, the motor can develop rated torque. Hence, this region is called the constant torque region. Increasing the stator frequency, Ω_s, above its base (rated) value, $\Omega_s > \Omega_{s(base)}$, at constant rated voltage, U_{sN}, it is possible to increase the motor speed beyond the rated speed. However, the motor flux, proportional to U_s/f_s, will be weakened. Therefore, when the slip frequency increase is proportional to the stator frequency, $\Omega_{sl} \sim \Omega_s$, the electromagnetic power

$$P_e = \Omega_m \cdot M_e \approx \Omega_s \left(\frac{\Omega_{sl}}{R_r} \right) \left(\frac{\Psi_r}{\Omega_s} \right)^2 \tag{21.30}$$

can be held constant, giving the name "constant power" to this region (Figure 21.5).

With constant stator voltage and increased stator frequency, the motor speed reaches the high-speed region, where the flux is reduced so much that the IM approaches its breakdown torque and the slip frequency cannot be increased any longer. Consequently, the torque capability is reduced according to the breakdown torque characteristic $M_e \sim M_{ek} \sim (U_s/\Omega_s)^2$. This high-speed region is called the constant slip frequency region (Figure 21.5).

21.4 Classification of IM Control Methods

Based on the space vector description, the IM control methods are divided into a scalar and a vector control. The general classification of the frequency controllers is presented in Figure 21.6. In scalar control—which is based on the relation valid for steady states—only the magnitude and frequency (angular speed) of voltage, current, and flux linkage space vectors are controlled. Thus, the scalar control system does not act on space vectors' position during transients and belongs to low-performance control implemented in an open-loop fashion. Contrary to this, in vector control—which is based on the relation valid for dynamic states—not only magnitude and frequency (angular speed), but also instantaneous positions of voltage, current, and flux space vectors are controlled. Thus, the vector control system acts on the

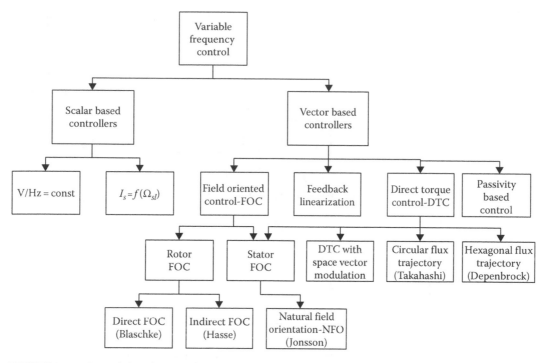

FIGURE 21.6   General classification of IM control methods.

positions of the space vectors and provides their correct orientation for both steady states and transients. This guarantee dynamically decouples fast flux and torque control and belongs to high-performance control implemented in a closed-loop fashion.

According to the above definition, the vector control can be implemented in many different ways. However, there are only several basic schemes that are offered in the market. Among these, the most popular strategies are FOC, DTC, DTC-SVM, and their variants. Another group of modern nonlinear control strategies, which includes feedback linearization control [12,16,21] and passivity-based control [20] schemes, is not discussed here because from the present industrial point of view, these represent only an alternative solution to existing FOC and DTC schemes.

21.5 Scalar Control

21.5.1 Open-Loop Constant Volts/Hz Control

In numerous industrial applications, the requirements related to the dynamic properties of drive control are of secondary importance. This is especially the case where no rapid motor speed change is required and where there are no sudden load torque changes. In such cases, one may just as well make use of open-loop constant volts/Hz (V/Hz) control systems (Figure 21.7). This method is based on the assumption that the flux amplitude is constant in a steady-state operation, and from Equation 21.1, for $\Omega_K = \Omega_s$ and $d\Psi_s/dt = 0$, one obtains the stator voltage vector equation:

$$\mathbf{U_s} = R_s \mathbf{I_s} + j2\pi f_s \Psi_s \tag{21.31}$$

where $f_s = \Omega_s/2\pi$. Thus, the stator vector magnitude can be calculated from Equation 21.31 as

$$U_s = \sqrt{\left(R_s I_s\right)^2 + \left(2\pi f_s \Psi_s\right)^2} \tag{21.32}$$

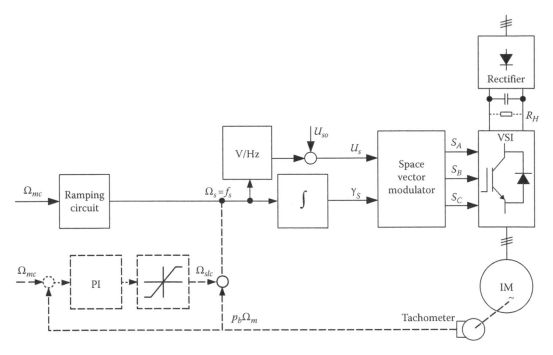

FIGURE 21.7 Constant V/Hz control scheme (dashed lines show version with limited slip frequency Ω_{slc} and speed control).

For $R_s = 0$, the relationship between stator voltage magnitude and frequency is linear and Equation 21.32 takes the form

$$\frac{U_s}{f_s} = 2\pi\Psi_s = \text{const.} \tag{21.33}$$

giving the name "constant V/Hz" (in Europe, it is called constant U/f) to this method.

For practical implementation, however, the relation of Equation 21.32 can be expressed as

$$U_s = U_{so} + 2\pi f_s \Psi_s \tag{21.34}$$

where $U_{so} = I_s R_s$, which is the offset (boost) voltage to compensate for the stator resistive drop.

The block diagram of an open-loop constant V/Hz control implemented according to Equation 21.34 for a PWM inverter–fed IM drive is shown in Figure 21.7. The control algorithm calculates the voltage magnitude, proportional to the command speed value, and the angle γ_s is obtained by the integration of this speed. The voltage vector in polar coordinates is the reference value for the space vector modulator (SVM), which delivers switching signals to the PWM inverter. The speed command signal, Ω_{mc}, determines the inverter frequency, $f_s \approx \Omega_s$, which simultaneously defines the stator voltage command according to constant V/Hz.

However, the mechanical speed, Ω_m, and hence the slip frequency, $\Omega_{sl} = \Omega_s - p_b\Omega_m$, are not precisely controlled. This can lead to motor operation in the instable region of torque–slip frequency curves (Figure 21.4), resulting in overcurrent problems. Therefore, to avoid high slip frequency values during transients, a ramping circuit is added to the stator frequency control path. The scheme is basically speed sensorless; however, when speed stabilization is necessary, speed control may be applied with slip regulation (dashed lines in Figure 21.7). The slip frequency command, Ω_{slc}, is generated by the speed proportional-integral (PI) controller. This signal is added to the tachometer signal and determines the inverter frequency command $\Omega_s = 2\pi f_s$. Owing to the limitation of slip frequency command, Ω_{slc}, the motor does not pull out either under rapid speed command changes or under load torque changes. Rapid speed reduction results in a negative slip command, and the motor goes into the generator breaking range (Figure 21.4). The regenerated energy must then either be returned to the line by the feedback converter or dissipated in the DC link dynamic breaking resistor, R_H.

21.6 Field-Oriented Control

21.6.1 Introduction

The principle of the FOC is based on an analogy to the mechanically commutated DC brush motor. In this motor, owing to separate exciting and armature winding, flux is controlled by exciting current and torque is controlled independently by adjusting the armature current. So, the flux and torque currents are electrically and magnetically separated. Contrarily, the cage-rotor IM has only a three-phase winding in the stator, and the stator current vector, $\mathbf{I_s}$, is used for both flux and torque control. So, exciting and armature current are coupled (not separated) in the stator current vector and cannot be controlled separately. The decoupling can be achieved by the decomposition of the instantaneous stator current vector, $\mathbf{I_s}$, into two components: flux current, I_{sd}, and torque-producing current, I_{sq}, in the rotor-flux-oriented coordinates (R-FOC) d–q (Figure 21.2). In this way, the control of the IM becomes identical with a separately excited DC brush motor and can be implemented using a cascaded structure with linear PI controllers [1,12,15,26].

21.6.2 Current-Controlled R-FOC Schemes

The simplest implementation of the R-FOC scheme is achieved in conjunction with a current-controlled PWM inverter. The choice of a suitable current control method affects both the parameters obtained and the final configuration of the entire system. In the standard version, the PWM current control loop operates in synchronous field-oriented coordinates d–q, as shown in Figure 21.8. The feedback stator currents, I_{sd} and I_{sq}, are obtained from the measured values I_A and I_B after phase conversion from three phase to two phase:

$$I_{s\alpha} = I_A \tag{21.35a}$$

$$I_{s\beta} = (1/\sqrt{3})(I_A + 2I_B) \tag{21.35b}$$

followed by coordinate transformation α–β/d–q:

$$I_{sd} = I_{s\alpha} \cos\gamma_{sr} + I_{s\beta} \sin\gamma_{sr} \tag{21.36a}$$

$$I_{sq} = -I_{s\alpha} \sin\gamma_{sr} + I_{s\beta} \cos\gamma_{sr} \tag{21.36b}$$

The PI current controllers generate voltage vector commands U_{sdc} and U_{sqc}, which, after coordinate transformation x–y/d–q,

$$U_{s\alpha c} = U_{sdc} \cos\gamma_{sr} - U_{sqc} \sin\gamma_{sr} \tag{21.37a}$$

$$U_{s\beta c} = U_{sdc} \sin\gamma_{sr} + U_{sqc} \cos\gamma_{sr} \tag{21.37b}$$

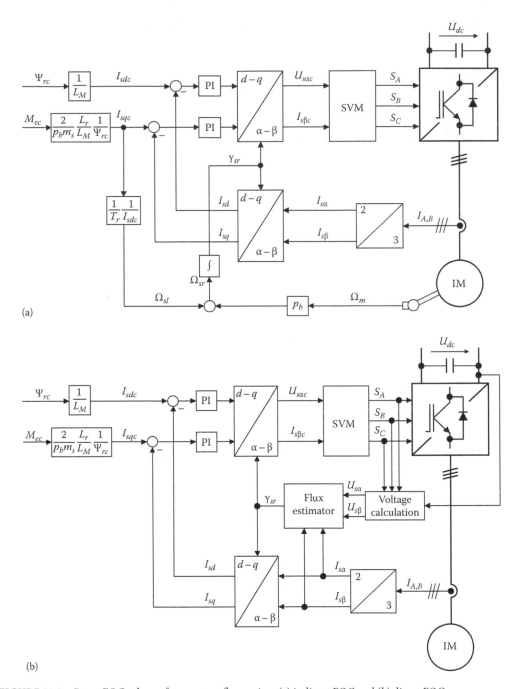

FIGURE 21.8   Rotor FOC scheme for constant flux region: (a) indirect FOC and (b) direct FOC.

are delivered to the SVM. Finally, the SVM calculates the switching signals S_A, S_B, and S_C for the power transistors of the PWM inverter.

The main information of the FOC scheme, namely, the flux vector position, γ_{sr}, necessary for coordinate transformation can be delivered in two different ways, giving generally two types of FOC (schemes) called indirect and direct FOC. Indirect FOC refers to an implementation where the flux vector position, γ_{sr}, is calculated from the reference values (feed-forward control) and the mechanical speed (position) measurement (Figure 21.8a), while direct FOC refers to the case where the flux vector position, γ_{sr}, is measured or estimated (Figure 21.8b) [1,3,13, 25,26].

21.6.2.1 Indirect R-FOC Scheme

For the indirect FOC scheme, proposed by Hasse [8] (Figure 21.8a), the rotor flux angle, γ_{sr}, is obtained from commanded currents I_{sdc} and I_{sqc}. The angular speed of the rotor flux vector can be calculated as

$$\Omega_{rs} = \Omega_{sl} + p_b \Omega_m \tag{21.38}$$

where
  Ω_{sl} is the slip angular speed
  Ω_m is the angular speed of the motor shaft (measured by a motion sensor or estimated from the measured currents and voltages [22])
  p_b is the number of pole pairs

The slip angular speed can be calculated from (21.21a) and (21.21b) as

$$\Omega_{sl} = \frac{1}{I_{sdc}} \frac{1}{T_r} I_{sqc} \tag{21.39}$$

where $T_r = L_r/R_r$ is the rotor time constant. The flux vector position angle, γ_{sr}, with respect to the stator is obtained by the integration of Equation 21.38:

$$\gamma_{sr} = \int_0^t (p_b \Omega_m + \Omega_{sl}) dt = \int_0^t \Omega_s dt \tag{21.40}$$

The commanded currents in a rotating coordinate system, I_{sdc} and I_{sqc}, are calculated from the commanded flux and torque values. Taking into consideration the equations describing IM in a field-oriented coordinate system, (21.24) and (21.25), the formulas for reference currents can be written as follows:

$$I_{sdc} = \frac{1}{L_M} \left(\Psi_{rc} + \frac{1}{T_r} \frac{d\Psi_{rc}}{dt} \right) \tag{21.41}$$

$$I_{sqc} = \frac{2}{p_b m_s} \frac{L_r}{L_M} \frac{1}{\Psi_{rc}} M_{ec} \tag{21.42}$$

Equations 21.39, 21.41, and 21.42 constitute the basis for control in both constant and weakened field regions (Figure 21.9). For constant flux operation, Equation 21.41 is simplified to

$$I_{sdc} = \frac{\Psi_{rc}}{L_M} \tag{21.43}$$

which corresponds to the situation presented in Figure 21.8a and b.

21.6.2.1.1 Parameter Sensitivity

The indirect R-FOC scheme is effective only as long as the set values of the motor parameters in the vector controller are equal to the actual motor parameter values. For the constant rotor flux operation region,

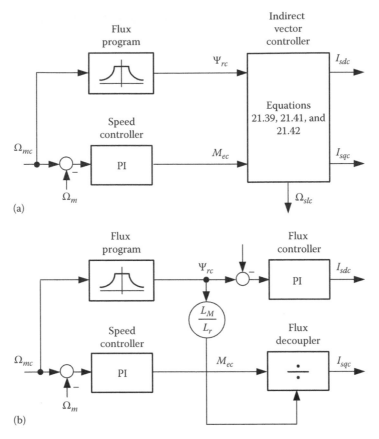

FIGURE 21.9 Variants of FOC control schemes for field-weakened operation: (a) indirect FOC and (b) direct FOC.

change of the rotor time constant, T_r, results in deviation in the slip frequency value, Ω_{sl}, calculated from Equation 21.39. The predicted rotor flux position, $\gamma_{src} = \int(p_b\Omega_m + \Omega_{slc})dt$, deviates from the actual position angle, $\gamma_{sr} = \int(p_b\Omega_m + \Omega_r)dt$, which produces a torque angle (see Figure 21.2) deviation, $\Delta\delta = \gamma_{src} - \gamma_{sr}$, and, consequently, leads to an incorrect subdivision of the stator current vector, $\mathbf{I_s}$, into two components, I_{sd} and I_{sq}. The decoupling condition of flux and torque control cannot be achieved. This leads to

- Incorrect rotor flux, Ψ_r, and torque current component, I_{sq}, values in the steady-state operating points (for $M_{ec} = $ const.)
- Second-order (nonlinear) system transient response to changes of torque command, M_{ec}

For a predetermined point of operation defined by the torque and flux current command values, I_{sqc} and I_{sdc}, it is possible to determine the effect of T_r changes on the real torque and rotor flux of the motor. These relations, derived from Equations 21.42 and 21.43 for steady states, can be conveniently presented in the form [12]

$$\frac{M_e}{M_{ec}} = \frac{T_r}{T_{rc}}\frac{1+(I_{sqc}/I_{sdc})^2}{1+\left[(T_r/T_{rc})(I_{sqc}/I_{sdc})\right]^2} \tag{21.44}$$

$$\frac{\Psi_r}{\Psi_{rc}} = \sqrt{\frac{1+(I_{sqc}/I_{sdc})^2}{1+\left[(T_r/T_{rc})(I_{sqc}/I_{sdc})\right]^2}} \tag{21.45}$$

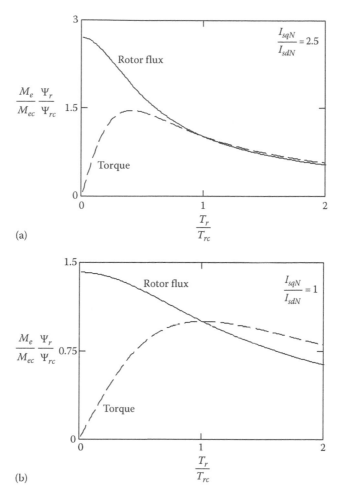

FIGURE 21.10 Detuning effect of rotor time constant T_r on steady-state characteristic for rated flux and torque current commands: (a) high-power motors and (b) low-power motors.

The normalized torque and rotor flux values are nonlinear functions of the ratio of the actual/predicted rotor time constants (T_r/T_{rc}) and the motor point of operation given by I_{sqc}/I_{sdc}. For the rated values of the field-oriented current commands, $I_{sqc} = I_{sqN}$ and $I_{sdc} = I_{sdN}$, we obtain from (21.44) and (21.45) the curves plotted in Figure 21.10 (where the saturation effect is omitted). Note that since high-power motors have a small magnetizing current (in steady state $I_{sd} = I_{Mr}$) relative to the rated current, I_{sN}, they are characterized by large values of $I_{sqN}/I_{sdN} = 2 - 3$. For low-power motors, on the other hand, we have the ratio $I_{sqN}/I_{sdN} = 1 - 2$.

Note that high-power motors are much more sensitive to the detuning of the time constant (T_r/T_{rc}) than are low-power ones.

In a similar way, one can take into account the effect of changes in the magnetizing inductance, L_M, induced by magnetic circuit saturation [1,15].

21.6.2.1.2 Parameter Adaptation

The critical parameter to the decoupling conditions of an indirect FOC scheme is the rotor time constant, T_r. It changes primarily under the influence of temperature changes of rotor resistance (R_r) and changes brought about by the saturation effect because of rotor inductance (L_r). While the temperature changes of R_r are very slow, the changes of L_r can be very fast, for example, in the case of speed reversal when the motor

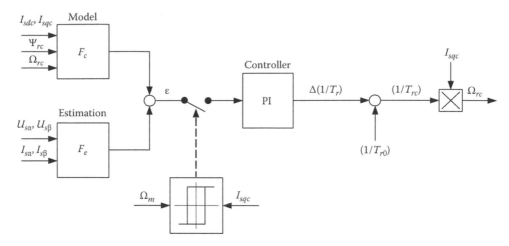

FIGURE 21.11 Basic block scheme of T_r—Adaption based on model reference adaptive system (MRAS).

changes quickly between its rated speed and the field-weakening region. It is assumed that T_r changes in the $0.75T_{r0} < T_r < 1.5T_{r0}$ range, where T_{r0} is the value at the rated load and a temperature of 75°C.

Parameter correction is effected by online adaptation. It follows from the graphs of Figure 21.10 that the correction signal for time constant changes $(1/\Delta T_r)$ may be found from the measured actual torque or flux values, or from such familiar quantities as torque current or flux current. However, these quantities are very difficult to measure or calculate over the entire range of speed control, the difficulty being comparable to that involved in flux vector estimation (see Section 21.6.2.3) in direct FOC systems.

Figure 21.11 shows the basic idea of the T_r adaptation scheme [7], which corresponds to the structure of the model reference adaptive systems (MRAS).

The reference function, F_c, is calculated from command quantities (indices c) in the field coordinates d–q. The estimated function, F_e, is calculated from measured quantities, which are usually expressed in the stator-oriented coordinates α–β. The error signal $\varepsilon = F_c - F_e$ is delivered to the PI controller, which generates the correction signal $(1/\Delta T_r)$. This correction signal is added to an initial value $(1/T_{r0})$ giving the updated time constant $(1/T_{rc})$, which finally is used for the calculation of the slip frequency, Ω_{slc}. In the steady state, when $\varepsilon \to 0$, then $T_{rc} \to T_r$. A variety of criterion functions (F) have been suggested for the identification of T_r changes (see Table 21.1). Most of them work neither for the no-load condition nor for zero speed. Therefore, in the near-zero speed region and no-load operation, the output signal of the error calculator, ε, must be blocked. The last value of $\Delta(1/T_r)$ is stored in the PI T_r controller.

Methods of online parameter identification based on the observer technique are also proposed [1,19].

TABLE 21.1 Variants of T_r—Adaption Algorithms (Figure 21.11)

	F_c	F_e	Parameter Sensitivity	Remarks
1	$-\left(\dfrac{L_M}{L_r}\right)\Omega_{slc}\Omega_{sc}I_{sdc}$	$(U_{s\alpha}I_{s\beta} - U_{s\beta}I_{s\alpha}) - \sigma L_s(pI_{s\alpha}I_{s\beta} - pI_{s\beta}I_{s\alpha})$	σL_s	No pure integration problem
2	$-\left(\dfrac{L_M}{L_r}\right)\Omega_{slc}\Omega_{sc}I_{sdc}$	$(U_{sd}I_{sq} - U_{sq}I_{sd}) - \sigma L_s\Omega_s(I_{sd}^2 - I_{sq}^2)$	σL_s	U_{sd} and U_{sq} are outputs of current controllers
3	$\left(\dfrac{L_M}{L_r}\right)\Psi_{rc}I_{sqc}$	$\Psi_{s\alpha}I_{s\beta} - \Psi_{s\beta}I_{s\alpha}$	R_s	Initial condition and drift problem (pure integration)
4	I_{sqc}	$\dfrac{\Psi_{s\alpha}I_{s\beta} - \Psi_{s\beta}I_{s\alpha}}{\left(L_r/L_M\right)\sqrt{\Psi_{s\alpha}^2 + \Psi_{s\beta}^2}}$	R_s	
5	0	$U_{sd} - R_sI_{sd} + \Omega_s\sigma L_sI_{sq}$	$R_s, \sigma L_s$	Simple, good convergence

21.6.2.2 Direct R-FOC Scheme

The main block in this scheme (proposed by Blaschke [2] and used by Siemens Company) is the flux vector estimator, which generates position γ_s and magnitude Ψ_r of the rotor flux vector, Ψ_r. The flux magnitude, Ψ_r, is controlled by a closed loop, and the flux controller generates the flux current command, I_{sdc}. Above the rated speed, field weakening is implemented by making the flux command, Ψ_{rc}, speed dependent, using a flux program generator, as shown in Figure 21.9b. In the field-weakening region, the torque current command, I_{sqc}, is calculated in the flux decoupler from the torque and flux commands, M_{ec} and Ψ_{rc}, according to Equation 21.42. If the estimated torque signal, M_e, is available, the flux decoupler can be replaced by the PI torque controller, which generates the torque current command, I_{sqc}. In both cases, the influence of variable flux on torque control is compensated. However, the stator current vector magnitude has to be limited as

$$\sqrt{I_{sdc}^2 + I_{sqc}^2} \leq I_{s\max} \tag{21.46}$$

21.6.2.3 Flux Vector Estimation

To avoid the use of additional sensors or measuring coils in the IM, methods of indirect flux vector generation have been developed, known as flux models or flux estimators. These are models of motor equations that are excited by appropriate easily measurable quantities, such as stator voltages and/or currents (U_s, I_s), angular shaft speed (Ω_m), or position angle (γ_s). There are many types of flux vector estimators, which usually are classified in terms of the input signals used [1,12,26]. Recently, only estimators based on stator currents and voltages are used, because they avoid the need for mechanical motion sensors.

21.6.2.3.1 Stator Flux Vector Estimators

Integrating the stator voltage equations represented in stationary coordinates α–β (21.14a,b), one obtains the stator flux vector components as

$$\Psi_{s\alpha} = \int_0^t (U_{s\alpha} - R_s I_{s\alpha})dt \tag{21.47a}$$

$$\Psi_{s\beta} = \int_0^t (U_{s\beta} - R_s I_{s\beta})dt \tag{21.47b}$$

The block diagram of the stator flux estimator according to Equation 21.47a and b is shown in Figure 21.12a. The stator flux can also be calculated in the scheme of Figure 21.12b operated with polar coordinates. In this scheme, coordinate transformation α–β/x–y (Equations 21.13a and b) and voltage equations in field coordinates are used.

To avoid the DC-offset problem of the open-loop integration, the pure integrator ($y = (1/s)x$) can be rewritten as

$$y = \frac{1}{s + \omega_c} x + \frac{\omega_c}{s + \omega_c} y \tag{21.48}$$

where
   x and y are the system input and output signals
   ω_c is the cutoff frequency

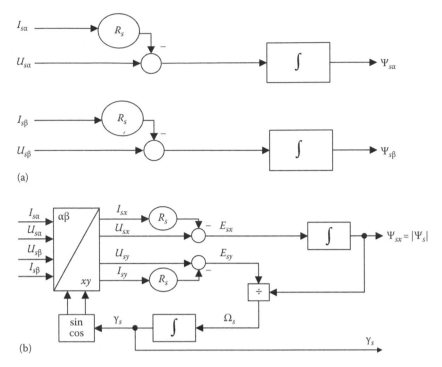

(a)

(b)

FIGURE 21.12   Stator flux vector estimators: (a) in Cartesian coordinates and (b) in polar coordinates.

The first part of Equation 21.48 represents an LP (low-pass) filter, whereas the second part implements a feedback used to compensate for the error in the output. The block diagram of the improved integrator according to Equation 21.48 is shown in Figure 21.13. It includes a saturation block that stops the integration when the output signal exceeds the reference stator flux magnitude.

In a DSP-based implementation, the voltage vector components are not measured but calculated from the inverter switching signals, S_A, S_B, and S_C, and the measured DC link voltage, U_{dc}, as follows:

$$U_{s\alpha} = \frac{2}{3} U_{dc} \left(S_A - \frac{1}{2} \left(S_B + S_C \right) \right) \tag{21.49a}$$

$$U_{s\beta} = \frac{\sqrt{3}}{3} U_{dc} \left(S_B - S_C \right) \tag{21.49b}$$

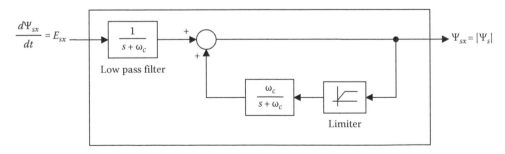

FIGURE 21.13   Block scheme of improved amplitude estimation in Figure 21.12b.

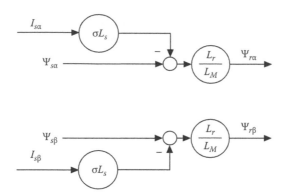

FIGURE 21.14 Rotor flux estimator based on stator flux according to Equations 21.50a and b.

However, in a very-low-speed operation, the effect of inverter nonlinearities (dead time, DC link voltage pulsations, and power semiconductor's voltage drop) has to be compensated.

21.6.2.3.2 Rotor Flux Vector Estimator

When the stator flux vector, Ψ_s, is known, the rotor flux vector can be easily calculated as

$$\Psi_{r\alpha} = \frac{L_r}{L_m}\left(\Psi_{s\alpha} - \sigma L_s I_{s\alpha}\right) \tag{21.50a}$$

$$\Psi_{r\beta} = \frac{L_r}{L_m}\left(\Psi_{s\beta} - \sigma L_s I_{s\beta}\right) \tag{21.50b}$$

The above equations are represented in Figure 21.14 as block diagrams in stationary α–β coordinates.

There are many other methods for rotor flux estimation based on speed or position measurement. Also, the observer technique is widely applied (see [1,26]).

21.6.3 Voltage-Controlled Stator-Flux-Oriented Control Scheme: Natural Field Orientation

The implementation of stator-flux-oriented coordinates (S-FOC) is much simpler for a voltage-controlled as for a current-controlled PWM inverter. Further simplifications can be achieved when instead of stator flux, the stator EMF will be used as the basis for the currents and/or voltage orientation (Figure 21.15). This avoids the integration necessary for flux calculation. Such a control scheme, known as natural field orientation (NFO), is commercially available as an ASIC [11]. Note that the NFO scheme is developed from the stator flux model of Figure 21.12b for $E_{sd} = 0$. The lack of current control loops and only R_s-dependent stator EMF estimation make the NFO scheme attractive for low-cost speed-sensorless applications. However, as shown in the oscillograms of Figure 21.16, the torque control dynamics is limited by the natural behavior of the IM (mainly by the rotor time constant, which for medium- and high-power motors can be in the range of 200 ms–1 s). Therefore, NFO can be feasible for low-power motors (up to 10 kW) or for low dynamic performance applications (like open-loop constant V/Hz control). An improvement can be achieved with an additional torque control loop (Figure 21.16), which requires online torque estimation. So, the final control scheme configuration becomes like that of DTC-SVM (see Section 21.8).

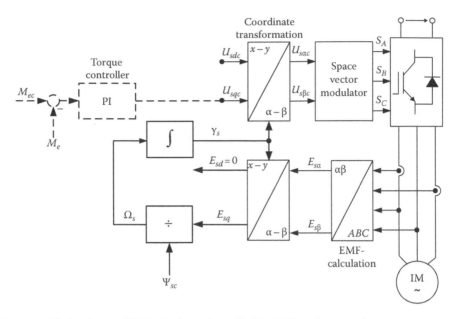

FIGURE 21.15 Block scheme of NFO (Voltage Controlled S-FOC) with optional outer torque control loop (dashed lines).

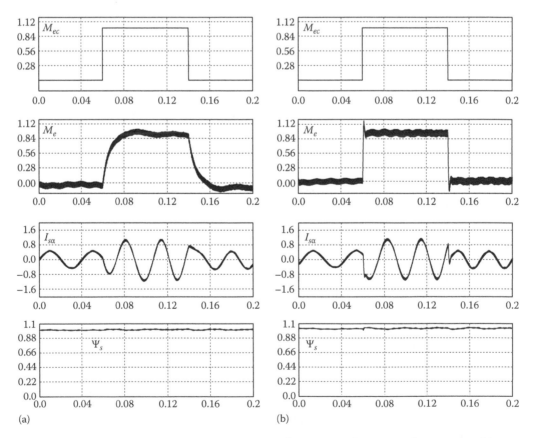

FIGURE 21.16 Torque transients in NFO control scheme of Figure 21.15 for constant flux operation: (a) conventional and (b) with outer torque control loop.

21.7 Direct Torque Control

21.7.1 Basic Principles

In the FOC strategy, the torque is controlled by the stator current component, I_{sq}, in accordance with Equation 21.25. This equation can also be written as

$$M_e = p_b \frac{m_s}{2} \frac{L_M}{L_r} \Psi_r I_s \sin\delta \tag{21.51}$$

where δ is the torque angle between the rotor flux vector and the stator current vector.

This makes the current-controlled PWM inverter very convenient for the implementation of the R-FOC scheme (Figure 21.8) and torque is controlled by adjusting the stator current vector. In the case of voltage source PWM inverter–fed IM drives, however, not only the stator current but also the stator flux vector may be used as the torque control quantity:

$$M_e = p_b \frac{m_s}{2} \frac{L_M}{L_r L_s - L_M^2} \Psi_s \Psi_r \sin\delta_\psi \tag{21.52}$$

where δ_ψ is the torque angle between rotor and stator flux vectors.

From (21.52), it can be seen that the torque depends on the stator and rotor flux magnitudes as well as the sine of angle δ_ψ. The two torque angles, δ and δ_ψ, are shown in the vector diagram of Figure 21.17. The angle δ is the torque angle in FOC algorithms, whereas δ_ψ is used in DTC techniques.

The motor voltage Equation 21.1, in stator-fixed coordinates, $\Omega_K = 0$, and for the omitted stator resistance, $R_s = 0$, reduces to

$$\frac{d\Psi_s}{dt} = \mathbf{U_s} \tag{21.53}$$

Taking into consideration the output voltage of the inverter in the above equation, the stator flux vector can be expressed as

$$\Psi_s = \int_0^t \mathbf{U}_v dt \tag{21.54}$$

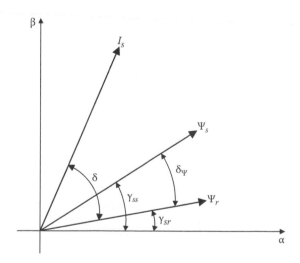

FIGURE 21.17 Vector diagram of induction motor in stator-fixed coordinates α–β.

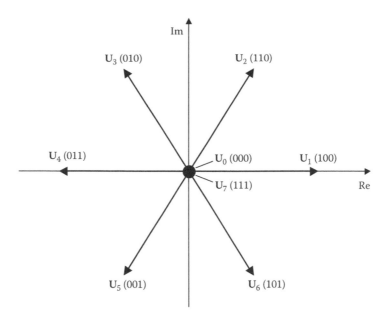

FIGURE 21.18   Inverter output voltage represented as space vectors.

where

$$\mathbf{U}_\nu = \begin{cases} \dfrac{2}{3}U_{dc}e^{j(\nu-1)\pi/3} & \nu = 1\ldots 6 \\[2mm] 0 & \nu = 0,7 \end{cases} \tag{21.55}$$

Equation 21.55 describes eight voltage vectors, which correspond to possible inverter states. These vectors are shown in Figure 21.18. There are six active vectors, $\mathbf{U}_1$–$\mathbf{U}_6$, and two zero vectors, $\mathbf{U}_0$ and $\mathbf{U}_7$.

It can be seen from (21.54), that the stator flux vector can directly be adjusted by the inverter voltage vector (21.55).

For a six-step operation, the inverter output voltage constitutes a cyclic and symmetric sequence of active vectors, so that, in accordance with (21.54), the stator flux moves with constant speed along a hexagonal path (Figure 21.19b). The introduction of zero vectors stops the flux, but does not change its path. This differs from the sinusoidal PWM operation, where the inverter output voltage constitutes a suitable sequence of two active and zero vectors and the stator flux moves along a track resembling a circle (Figure 21.20b). A magnified part of the flux vector trajectory is shown in Figure 21.21.

In any case, the rotor flux rotates continuously at the actual synchronous speed along a near-circular path, since it is smoothed by the rotor circuit filtering action.

From the point of view of torque production, it is the relative motion of the two vectors that is important, for they form the torque angle δ_ψ (Figure 21.17) that determines the instantaneous motor torque according to (21.52). By the cyclic switching of active and zero vectors, the motor torque is controlled. In the field-weakening region, zero vectors cannot be employed. Torque control is then achieved via a fast change of the torque angle, δ_ψ, by advancing (to increase the torque) or retarding (to reduce it) the phase of the stator flux vector [6,12].

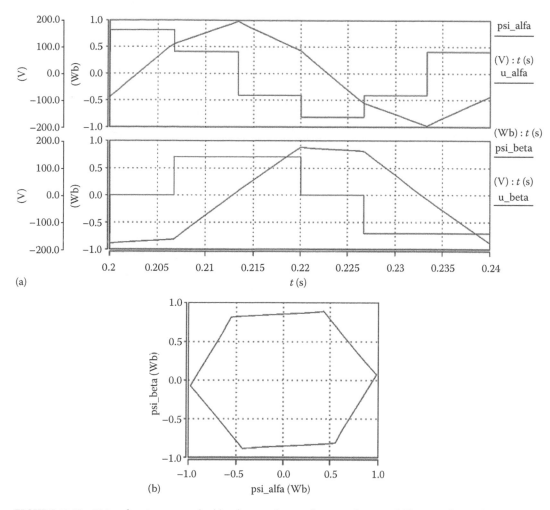

FIGURE 21.19 IM under six-step mode: (a) voltage and stator flux waveforms and (b) stator flux path.

21.7.2 Generic DTC Scheme

The generic DTC scheme consists of two hysteresis controllers (Figure 21.22). The stator flux controller imposes the time duration of the active voltage vectors, which move the stator flux along the commanded trajectory, and the torque controller determines the time duration of the zero voltage vectors, which keep the motor torque in the defined-by-hysteresis tolerance band.

At every sampling time, the voltage vector selection block chooses the inverter switching state (S_A, S_B, S_C), which reduces the instantaneous flux and torque errors. Compared to the conventional FOC scheme (Figure 21.8b), the DTC scheme has the following features:

- Simple structure.
- There are no current control loops; hence, the current is not regulated directly.
- Coordinate transformation is not required.
- Speed sensor is not required.
- There is no separate voltage pulse width modulator (PWM).
- Stator flux vector and torque estimation are required.

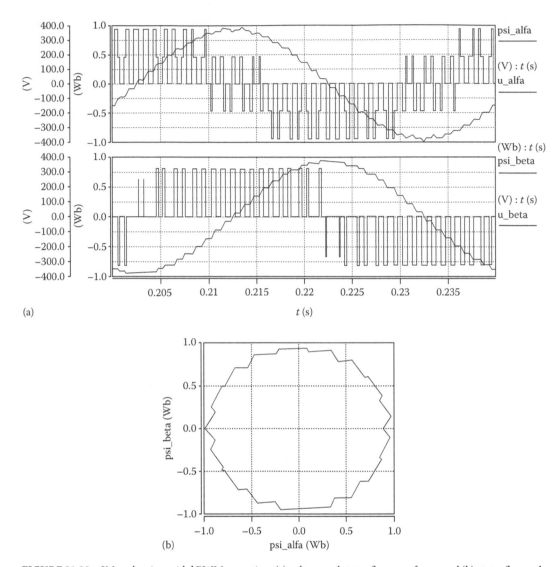

(a)

(b)

FIGURE 21.20 IM under sinusoidal PWM operation: (a) voltage and stator flux waveforms and (b) stator flux path.

Depending on how the switching sectors are selected, two different DTC schemes are possible. One, proposed by Takahashi and Noguchi [23], operates with a circular stator flux vector path, and the second one, proposed by Depenbrock [6], operates with a hexagonal stator flux vector path.

21.7.3 Switching Table-Based DTC: Circular Stator Flux Path

21.7.3.1 Basic Scheme

A block scheme of classical DTC (used by ABB Company [24]) is presented in Figure 21.23.

The stator flux magnitude, Ψ_{sc}, and the motor torque, M_c, are the command signals, which are compared with the estimated $\hat{\Psi}_s$ and $\hat{M}_e$ values, respectively. The flux, e_Ψ, and torque, e_M, errors are delivered to the hysteresis controllers. The digitized output variables, d_Ψ and d_M, and the stator flux vector position sector, $N(\gamma_s)$, obtained from the angular position $\gamma_{ss} = \mathrm{arctg}(\Psi_{s\beta}/\Psi_{s\alpha})$, select the appropriate voltage vector from the switching table. Thus, pulses S_A, S_B, and S_C are generated from the selection table to control the power switches in the inverter.

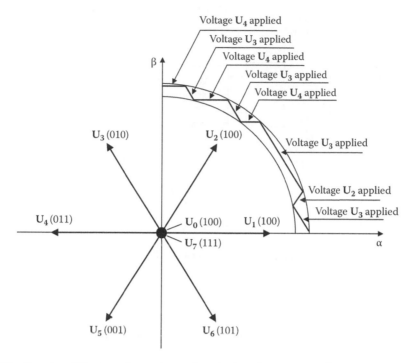

FIGURE 21.21 Forming of the stator flux trajectory by selection of appropriate voltage vectors sequence.

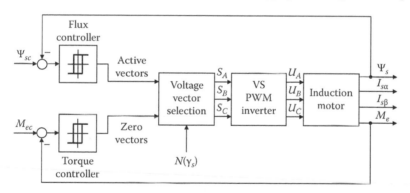

FIGURE 21.22 Generic block scheme of direct torque control (DTC).

The output signals of hysteresis controllers, d_Ψ and d_M, are defined as

$$d_\Psi = 1 \quad \text{for } e_\Psi > H_\Psi \tag{21.56a}$$

$$d_\Psi = 0 \quad \text{for } e_\Psi < -H_\Psi \tag{21.56b}$$

$$d_M = 1 \quad \text{for } e_M > H_M \tag{21.57a}$$

$$d_M = 0 \quad \text{for } e_M = 0 \tag{21.57b}$$

$$d_M = -1 \quad \text{for } e_M < -H_M \tag{21.57c}$$

where $2H_\Psi$ and $2H_M$ are flux- and torque-controller tolerance bands, respectively.

In the classical ST-DTC (switching table-based DTC) method, the plane is divided for the six sectors, as shown in Figure 21.24.

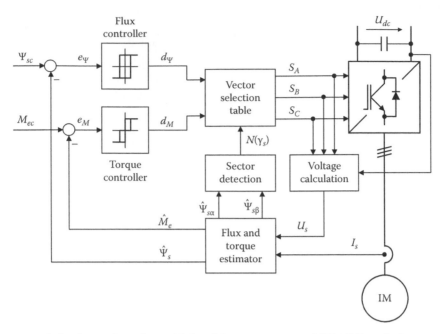

FIGURE 21.23 Block scheme of switching table based direct torque control (ST-DTC) method.

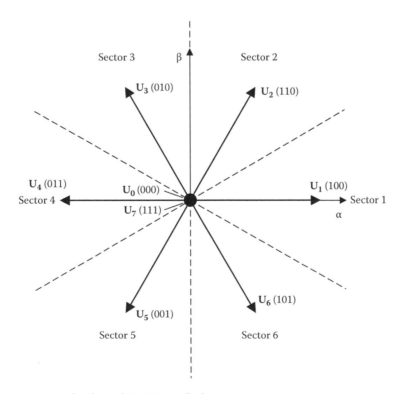

FIGURE 21.24 Sectors in the classical ST-DTC method.

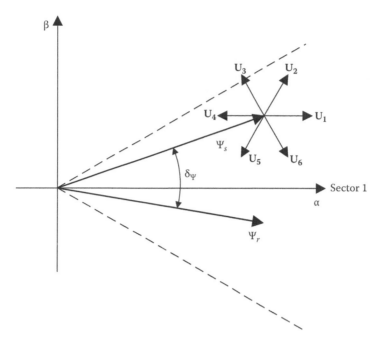

FIGURE 21.25 Selection of the optimum voltage vectors for the stator flux vector located in sector 1.

In order to increase the magnitude of the stator flux vector lying in sector 1 (Figure 21.25), the voltage vectors U_1, U_2, and U_6 can be selected. Conversely, a decrease can be obtained by selecting U_3, U_4, and U_5. By applying one of the zero vectors, U_0 or U_7, the integration in Equation 21.54 is stopped and stator flux vector also stops.

For torque control, the torque angle, δ_ψ, is used according to Equation 21.52. Therefore, to increase the motor torque, the voltage vectors U_2, U_3, and U_4 can be selected, and to decrease the motor torque, U_1, U_5, and U_6 can be selected.

The above considerations allow the construction of the selection rules, as presented in Table 21.2.

The typical signal waveforms for the steady-state operation of the classical ST-DTC method are shown in Figure 21.26. The characteristic features of the ST-DTC scheme of Figure 21.23 include the following:

- Nearly sinusoidal stator flux and current waveforms; the harmonic content is determined by the flux- and torque-controller hysteresis bands, H_ψ and H_M.
- Excellent torque dynamics.
- Flux and torque hysteresis bands determine the inverter switching frequency, which varies with the synchronous speed and load conditions.

TABLE 21.2 Optimum Switching Table of Classical DTC

d_ψ	d_M	Sector 1	Sector 2	Sector 3	Sector 4	Sector 5	Sector 6
1	1	U_2	U_3	U_4	U_5	U_6	U_1
	0	U_7	U_0	U_7	U_0	U_7	U_0
	−1	U_6	U_1	U_2	U_3	U_4	U_5
0	1	U_3	U_4	U_5	U_6	U_1	U_2
	0	U_0	U_7	U_0	U_7	U_0	U_7
	−1	U_5	U_6	U_1	U_2	U_3	U_4

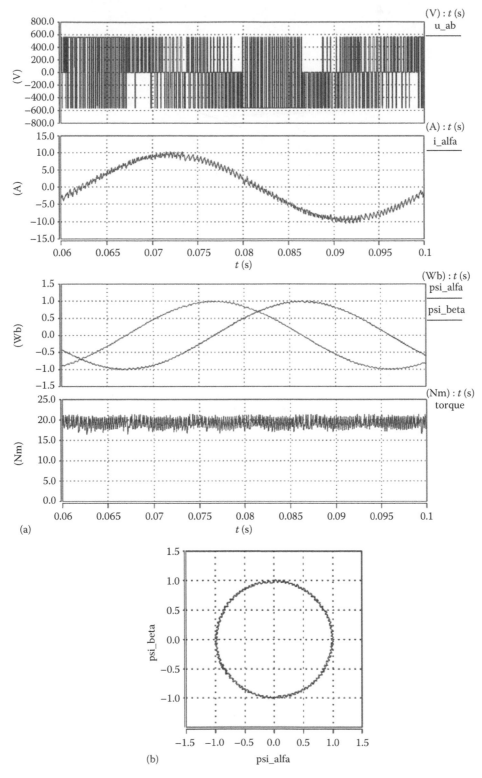

FIGURE 21.26 Steady state operation for the classical ST-DTC method ($f_s = 40$ kHz). (a) signals in time domain and (b) stator flux path.

21.7.3.2 Modified ST-DTC

Many modifications of the basic ST-DTC scheme aimed at improving starting, overload conditions, very-low-speed operation, torque ripple reduction, variable switching frequency functioning, and noise-level attenuation have been proposed during the last decade.

During starting and very-low-speed operation, the basic ST-DTC scheme selects many times the zero voltage vectors resulting in flux-level reduction, owing to the stator resistance drop. This drawback can be avoided by using either a dither signal or a modified switching table [4,26].

Torque ripple reduction can be achieved by a subdivision of the sampling period in two or three [5] equal time intervals. This creates 12 or 56 voltage vectors, respectively. The increased number of available voltage vectors allows both to subdivide the hysteresis of torque and flux controllers into more levels and to create a more accurate switching table that takes into account also the speed value.

In order to increase the torque overload capability of an ST-DTC scheme, the rotor flux instead of the stator flux magnitude should be regulated. For given commands of the rotor flux (Ψ_{rc}) and the torque (m_c), the stator flux command needed by an ST-DTC scheme can be calculated as

$$\Psi_{sc} = \sqrt{\left(\frac{L_s}{L_M}\Psi_{rc}\right)^2 + (\sigma L_s)^2 \left(\frac{L_r}{L_M}\frac{M_{ec}}{\Psi_{rc}}\right)^2} \tag{21.58}$$

However, the price for better overload capabilities is a higher parameter sensitivity of the rotor flux magnitude control.

21.7.4 Direct Self-Control: Hexagonal Stator Flux Path

21.7.4.1 Basic Direct Self-Control Scheme

The block diagram of the DSC (direct self-control) method is shown in Figure 21.27. Based on the command stator flux, Ψ_{sc}, and the actual phase components, Ψ_{sA}, Ψ_{sB}, and Ψ_{sC}, the flux comparators generate digital variables, d_A, d_B, and d_C, which correspond to active voltage vectors ($\mathbf{U}_1$–$\mathbf{U}_6$).

The hysteresis torque controller generates signal d_m, which determines zero states. For the constant flux region, the control algorithm is as follows:

$$S_A = d_C,\; S_B = d_A,\; S_C = d_B \quad \text{for } d_m = 1 \tag{21.59a}$$

$$S_A = 0,\; S_B = 0,\; S_C = 0 \quad \text{for } d_m = 0 \tag{21.59b}$$

Typical signal waveforms for the steady-state operation of the DSC method are shown in Figure 21.28. It can be seen that the flux trajectory is identical with that for the six-step mode (Figure 21.19). This follows from the fact that the zero voltage vectors stop the flux vector, but do not affect its trajectory. The dynamic performances of torque control for the DSC are similar as those for the ST-DTC.

The characteristic features of the DSC scheme of Figure 21.27 are as follows:

- Non-sinusoidal stator flux and current waveforms that, with the exception of the harmonics, are identical for both PWM and the six-step operation.
- The stator flux vector moves along a hexagonal path also under the PWM operation.
- No voltage supply reserve is necessary and the inverter capability is fully utilized.
- The inverter switching frequency is lower than in the ST-DTC scheme of Figure 21.23a, because PWM is not of sinusoidal type as it turns out by comparing the voltage patterns in Figures 21.26b and 21.28b.
- Excellent torque dynamics in constant and weakening field regions.

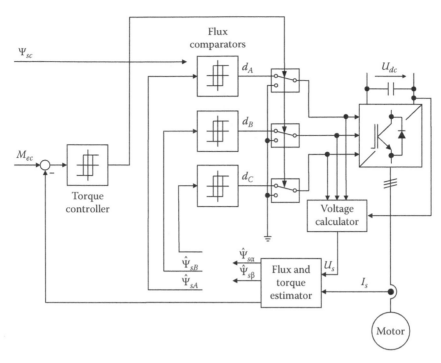

FIGURE 21.27   Block diagram of direct self control (DSC) method.

Low switching frequency and fast torque control even in the field-weakening region are the main reasons why the DSC scheme is convenient for high-power traction drives.

21.7.4.2 Indirect Self-Control

To improve the DSC performance in a low-speed region, the method called indirect self-control (ISC) has been proposed [9]. In the first stage of development, this method was used in DSC drives only for starting and for operation up to 20%–30% of the rated speed. Later, it was expanded as a new control strategy offered for inverters operated at high switching frequencies (>2 kHz). The ISC scheme, however, produces a circular stator flux path in association with a voltage PWM and, therefore, belongs to DTC-SVM schemes presented in Section 21.8.3.

21.8 DTC with Space Vector Modulation

21.8.1 Critical Evaluation of Hysteresis-Based DTC Schemes

The disadvantages of the hysteresis-based DTC schemes are variable switching frequency, violence of polarity consistency rules (avoidance ±1 switching over DC link voltage), current and torque distortion caused by sector changes, starting and low-speed operation problems, as well as high sampling frequency needed for the digital implementation of hysteresis controllers.

When a hysteresis controller is implemented in a digital signal processor (DSP), its operation is quite different from that of the analog scheme. Figure 21.29 illustrates a typical switching sequence in (a) analog and (b) discrete (also called sampled hysteresis) implementations. In the analog implementation, the torque ripple is kept exactly within the hysteresis band and the switching instants are not equally spaced. In contrast, the discrete system operates at a fixed sampling time, T_s, and if

$$2H_m \gg \frac{d_{m\,\text{max}}}{dt} \cdot T_s \qquad (21.60)$$

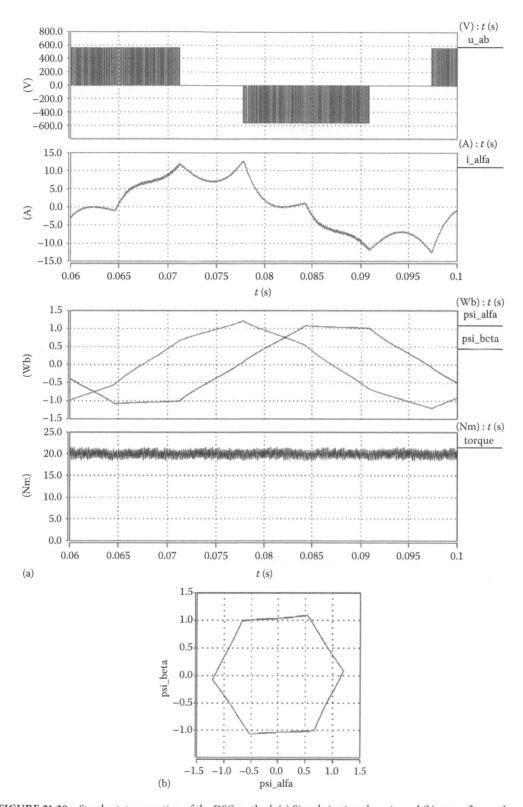

FIGURE 21.28   Steady state operation of the DSC method. (a) Signals in time domain and (b) stator flux path.

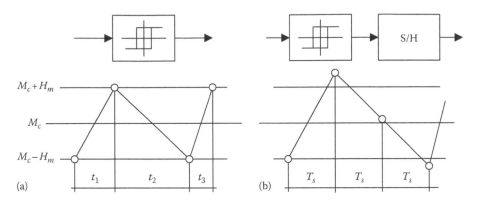

FIGURE 21.29 Operating principle of the torque hysteresis controller. (a) Analog and (b) digital.

the discrete controller operates like the analog one. However, it requires fast sampling. For lower sampling frequency, the switching instants do not occur when the estimated torque crosses the hysteresis band, but at the sampling time (see Figure 21.29b).

All the above difficulties can be eliminated when, instead of the switching table, a voltage PWM is used. Basically, the DTC strategies operating at constant switching frequency can be implemented by means of closed-loop schemes with PI, predictive/deadbeat, or neuro-fuzzy controllers. The controllers calculate the required stator voltage vector, averaged over a sampling period. The voltage vector is finally synthesized by a PWM technique, which, in most cases, is the SVM. So, differently from the conventional DTC, where hysteresis controllers operate on instantaneous values, in a DTC-SVM scheme, the linear controllers operate on values averaged over the sampling period. Therefore, the sampling frequency can be reduced from about 40 kHz in ST-DTC to 2–5 kHz in the DTC-SVM scheme.

21.8.2 DTC-SVM Scheme with Closed-Loop Torque Control

A block scheme of DTC-SVM with closed-loop torque control is presented in Figure 21.30.

For torque regulation, a PI controller is applied. The output of this PI controller produces an increment in the torque angle, $\Delta\delta_\psi$ (Figure 21.31). Assuming that rotor and flux magnitudes are approximately equal, the torque is controlled only by changing the torque angle, δ_ψ.

The reference stator flux vector is calculated as follows:

$$\mathbf{\Psi_{sc}} = \Psi_{sc}e^{j(\hat{\gamma}_{ss}+\Delta\delta_\psi)} \tag{21.61}$$

Next, the reference stator flux vector is compared with the estimated value and the stator flux vector error, $\Delta_{\psi s}$, is used for the command voltage vector calculation:

$$\mathbf{U}_{sc} = \frac{\Delta\mathbf{\Psi_s}}{T_s} + R_s\mathbf{I_s} \tag{21.62}$$

where
  T_s is the sampling time
  R_s is the stator resistance

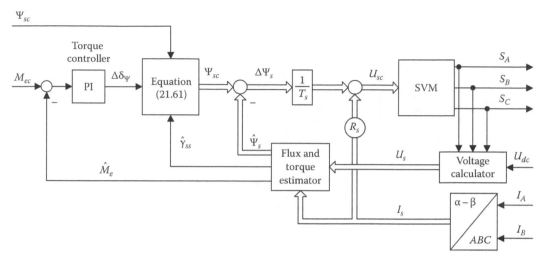

FIGURE 21.30   DTC-SVM scheme with closed-loop torque control.

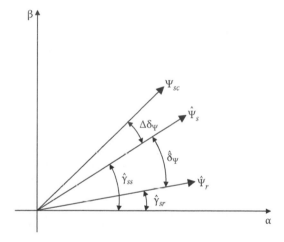

FIGURE 21.31   Vector diagram for control scheme of Figure 21.30.

The presented method has a very simple structure and only one PI torque controller. It makes the tuning procedure easier. Also, it is universal and can be applied for the control of permanent magnet synchronous motors (PMSMs) [27].

21.8.3 DTC-SVM Scheme with Closed-Loop Torque and Flux Control

A block scheme of DTC-SVM with closed-loop torque and flux control operating in Cartesian stator flux coordinates [1,4] is presented in Figure 21.32. The output of the PI flux and torque controllers is interpreted as the reference stator voltage components, U_{sdc} and U_{sqc}, in S-FOC ($d - q$).

These DC voltage commands are then transformed into stationary coordinates ($\alpha - \beta$), and the commanded values, $U_{s\alpha c}$ and $U_{s\beta c}$, are delivered to the SVM block. Note that because the commanded voltage vector is generated by flux and torque controllers, the scheme of Figure 21.32 is less sensitive to noisy feedback signals as the scheme of Figure 21.30, where the commanded voltage is calculated by flux error differentiation (21.62). Typical waveforms during speed reversal in constant and weakened flux regions are shown in Figure 21.33.

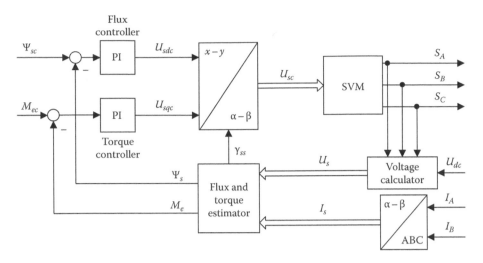

FIGURE 21.32   DTC-SVM scheme operated in stator flux Cartesian coordinates *d–q*.

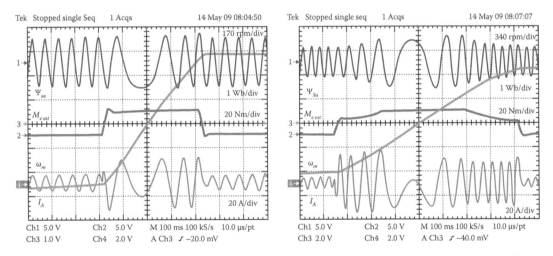

FIGURE 21.33   Speed reversal in the DTC-SVM scheme of Figure 21.32, left: constant flux operation, right: in flux weakening region.

21.9 Summary and Conclusions

Today, a number of different control schemes for accurate flux and torque control of the IM are developed. This chapter has reviewed the basic control strategies for low and medium power drives of PWM inverter–fed IMs. Starting from the space vector description of the IM, the control strategies are generally divided into scalar and vector methods.

- Scalar control is based on the IM equations at steady-state operating points and is typically implemented in open-loop schemes keeping constant V/Hz. However, such a scheme applied to a multivariable, coupled system like the IM cannot perform decoupling between inputs and outputs, resulting in problems of independent control of outputs, for example, torque and flux.
- To achieve decoupling in high-performance IM drives, vector control, also known as field-oriented control as well as direct torque control, has been developed. The FOC and DTC are now de facto standard, in highly dynamic IM industrial drives.

TABLE 21.3 Overview of Main IM Control Strategies in Low and Medium Power

	Parameters Control Strategy	Speed Control Range	Static Speed Accuracy	Torque Rise Time	Starting Torque	Cost	Typical Applications
1	Scalar Control (Constant V/Hz)	1:10 (open loop)	5%–10%	Not available	Low	Very low	Low performance: pumps, fans, compressors, HVAC, etc.
2	Scalar control with slip compensation	1:25 (open loop)	2%	Not available	Medium	Low	Low performance: conveyors, mixers, etc.
3	Natural field orientation (NFO)	1:50 (open loop)	1%	>10 ms	High	Medium	Medium performance: packing, crane, etc.
4	Field oriented control (FOC)	>1:200 (closed loop)	0%	<1–2 ms	High	High	High performance: crane, lifts, transportation, etc.
5	Direct torque control (DTC)	>1:200 (closed loop)	0%	<1–2 ms	High	High	High performance: crane, lifts, transportation, etc.
6	DTC with space vector modulation (DTC-SVM)	>1:200 (closed loop)	0%	<1–2 ms	High	High	High performance: crane, lifts, transportation, etc.
7	Servo drives	1:10,000 (closed loop)	0%	<1 ms	High	High	High performance, speed rise time <10 ms: robots, manipulators, ind. automation, etc.

- The R FOC is easily implemented in combination with a current-controlled PWM inverter.
- For a good low-speed operation performance, indirect R-FOC with a speed/position sensor is recommended. This scheme, however, is sensitive to changes of the rotor time constant, which has to be adapted online.
- DTC has a very fast torque response, a very simple structure, does not require a shaft motion sensor, and is less sensitive to IM parameter changes as in FOC.
- For a speed-sensorless operation, the DTC or the direct R-FOC scheme can be advised.
- To reduce torque ripple and fix the inverter switching frequency, the SVM has been introduced into the DTC structure, resulting in a new scheme known as DTC-SVM. Basically, this is S-FOC without current control loops. However, the DTC-SVM scheme combines advantages and eliminates disadvantages of classical DTC and FOC schemes. Therefore, it is an excellent solution for general-purpose IM (also PMSM) drives.

An overview of basic parameters and application areas of discussed control strategies is presented in Table 21.3. It can be concluded that the group of high-performance control methods achieves similar parameters and areas of applications. In FOC and DTC-SVM schemes, the control action is usually synchronized with PWM generation and is executed with the sampling time equal to the switching time ranging from 50 to 500 μs. The typical torque rise time is about 4–6 times that of the sampling time and is limited by the switching frequency.

The current trend in the control of the IM is to incorporate such techniques like model predictive control (MPC) [14] and neuro-fuzzy schemes to achieve robustness to parameter variations and increase functionality, self-commissioning, and fault-monitoring abilities.

References

1. I. Boldea and S. A. Nasar, *Electric Drives*, 2nd edn., CRC Press, Boca Raton, FL/Ann Arbor, London, U. K./Tokyo, Japan, 2006.
2. F. Blaschke, The principle of field-orientation as applied to the Transvector closed-loop control system for rotating-field machines, *Siemens Review*, 34, 217–220, 1972.
3. B. K. Bose, *Modern Power Electronics and AC Drives*, Prentice-Hall, Englewood Cliffs, NJ, 2001.
4. G. S. Buja and M. P. Kazmierkowski, Direct torque control of PWM inverter-fed ac motors—A survey, *IEEE Transactions on Industrial Electronics*, 51(4), 744–757, Aug. 2004.
5. D. Casadei, F. Profumo, G. Serra, and A. Tani, FOC and DTC: Two viable schemes for induction motors torque control, *IEEE Transactions on Power Electronics*, 17(5), 779–787, 2002.
6. M. Depenbrock, Direct self control of inverter-fed induction machines, *IEEE Transactions on Power Electronics*, 3(4), 420–429, Oct. 1988.
7. L. Garces, Parameter adaptation for the speed controlled static ac drive with a squirrel cage induction motor, *IEEE Transactions Industrial Applications*, 16, 173–178, 1980.
8. K. Hasse, Drehzahlgelverfahren fur schnelle Umkehrantriebe mit stromrichtergespeisten Asynchron—Kurzchlusslaufermotoren, *Reglungstechnik*, 20, 60–66, 1972.
9. F. Hoffman and M. Janecke, Fast torque control of an IGBT-inverter-fed three-phase ac drive in the whole speed range—Experimental Result, in *Proceedings of the EPE Conference*, Sevilla, Spain, 1995, pp. 3.399–3.404.
10. J. Holtz, The representation of ac machines dynamic by complex signal flow graphs, *IEEE Transactions on Industrial Electronics*, 42(3), 263–271, 1995.
11. R. Jönsson and W. Leonhard, Control of an induction motor without a mechanical sensor, based on the principle of natural field orientation (NFO), in *Proceedings of the IPEC'95*, Yokohama, Japan, 1995.
12. M. P. Kazmierkowski and H. Tunia, *Automatic Control of Converter Fed Drives*, Elsevier, Amsterdam, the Netherlands, 1994.
13. M. P. Kazmierkowski, R. Krishnan, and F. Blaabjerg, *Control in Power Electronics*, Academic Press, San Diego, CA, 2002.
14. M. P. Kazmierkowski, R. M. Kennel, and J. Rodrigue, Special section on predictive control in power electronics and drives, *IEEE Transactions on Industrial Electronics*, Part I, 55(12), 4309–4429, Dec. 2008; Part II, 56(6), 1823–1963, June 2009.
15. R. Krishnan, *Electric Motor Drives*, Prentice Hall, NJ, 2001.
16. Z. Krzeminski, Nonlinear control of induction motors, in *Proceedings of the 10th IFAC World Congress*, Munich, Germany, 1987, pp. 349–54.
17. N. Mohan, T. M. Undeland, and B. Robbins, *Power Electronics*, 3rd edn., John Wiley & Sons, New York, 2003.
18. D. W. Novotny and T. A. Lipo, *Vector Control and Dynamics of AC Machines*, Clarendon Press, Oxford, U.K., 1996.
19. T. Orłowska-Kowalska, Application of extended Luenberger observer for flux and rotor time-constant estimation in induction motor drives, *IEE Proceedings*, 136(Pt. D), 6, 324–330, 1989.
20. R. Ortega and A. Loria, P. J. Nicklasson, and H. Sira-Ramirez, *Passivity-Based Control of Euler-Lagrange Systems*, Springer Verlag, London, U.K., 1998.
21. M. Pietrzak-David and B. de Fornel, Non-linear control with adaptive observer for sensorless induction motor speed drives, *EPE Journal*, 11(4), 7–13, 2001.
22. K. Rajashekara, A. Kawamura, and K. Matsue, *Sensorless Control of AC Motor Drives*, IEEE Press, New York, 1996.
23. I. Takahashi and T. Noguchi, A new quick-response and high efficiency control strategy of an induction machine, *IEEE Transactions on Industrial Application*, 22(5), 820–827, Sept./Oct. 1986.

24. P. Tiiten, P. Pohjalainen, and J. Lalu, Next generation motion control method: Direct torque control (DTC), *EPE Journal*, 5(1), Mar. 1995.
25. A. M. Trzynadlowski, *Control of Induction Motors*, Academic Press, San Diego, CA, 2000.
26. P. Vas, *Sensorless Vector and Direct Torque Control*, Clarendon Press, Oxford, U.K., 1998.
27. L. Xu and M. Fu, A sensorless direct torque control technique for permanent magnet synchronous motors, in *IEEE Industry Applications Conference*, Vol.1, Phoenix, AZ, 1999, pp. 159–164.

22

Double-Fed Induction Machine Drives

Elżbieta Bogalecka
Gdańsk University
of Technology

Zbigniew
Krzemiński
Gdańsk University
of Technology

22.1 Introduction

A slip-ring induction machine with stator windings connected to the power system directly and rotor windings connected through the power converter is called double-fed machine (DFM). If the converter enables a bidirectional power flow, a DFM will become a universal electromechanical converter. Such a machine can work as a motor and as a generator in both above and below the synchronous speed. The speed range depends on the converter and machine rotor winding ratings. Usually the speed range is limited (about 1:2). This limit causes that DFM is not often used as a motor. The machine is mainly used as a variable-speed generator.

Synchronous machine operating as a constant-speed generator is the main source of the electric power. In some applications, the need of optimal conversion of mechanical to electrical power leads to a variable-speed generator with power converters and sophisticated control systems. Nowadays, the variable speed generation technologies are often used, e.g., in wind power plants, in distributed power generation systems with high-speed gas turbines, in UPS with flywheel, and in shipboard shaft generators. The main reasons for variable speed generation are more efficient energy conversion and better prime mover energy extraction.

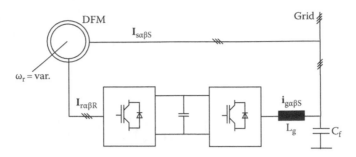

FIGURE 22.1   Scheme of the DFM.

Variable-speed generators can operate in different power systems:

1. Autonomous power system with only one source of the electric power
2. Small power systems with flexible voltage and frequency (with two or more generators of the rated power similar values)
3. Large power systems with constant grid voltage and frequency

In each system, the requirements for the generator control system are different. In the autonomous power system, the generator works individually and controls the grid voltage and frequency. The machine is a source of the active and reactive powers required by the load. In the flexible power systems, the generator influences the system voltage and frequency and vice versa. Stable and correct work of the variable-speed generator with, e.g., synchronous generator of the similar power, is required. In the large power systems, the grid voltage and frequency are forced by the power system. Nowadays in most of the applications, the variable-speed generator is connected to the grid and works only as a source of the power. It does not take part in the direct voltage and frequency control.

The DFM is a good solution for variable speed generation with limited speed range. The main advantage of the DFM, comparing to the synchronous and the induction machine, is the limited size of the converter. It depends on the machine slip; and because the DFM can work as a generator in both subsynchronous and oversynchronous area, the power of the converter usually does not exceed 25%–30% rating, while the speed range is 1:2. To provide machine operation below and above the synchronous speed, a four-quadrant converter in the rotor is required (often a bidirectional voltage-source inverter (VSI) with dc link), shown in Figure 22.1.

The machine is controlled from the rotor side. The main task of the control system is the mechanical power conversion and transmission to the grid with required power factor. There are two VSIs in the rotor circuit called the machine-side inverter and the grid-side inverter. The machine-side inverter is responsible for the machine state control and the grid-side inverter is responsible for the dc-link voltage control. The DFM working as a generator parallel to the grid is a controlled source of active and reactive power. High power quality requires independent control of both powers. A shaft sensorless operation is desirable.

22.2 Machine Model

Differential equations of a DFM obtained from the space vector theory are as follows:

$$\mathbf{u}_s = R_s \mathbf{i}_s + \frac{d\boldsymbol{\psi}_s}{d\tau} + j\omega_x \boldsymbol{\psi}_s, \tag{22.1}$$

$$\mathbf{u}_r = R_r \mathbf{i}_r + \frac{d\boldsymbol{\psi}_r}{d\tau} + j(\omega_x - \omega_r)\boldsymbol{\psi}_r, \tag{22.2}$$

$$J\frac{d\omega_r}{d\tau} = \text{Im}\left|\boldsymbol{\psi}_s^*\mathbf{i}_s\right| - m_0,$$ (22.3)

where

$\boldsymbol{\psi}_s$, $\boldsymbol{\psi}_r$ are the stator and rotor flux vectors
$\mathbf{i}_s$, $\mathbf{i}_r$ are the stator and rotor current vectors
$\mathbf{u}_s$, $\mathbf{u}_r$ are the stator and rotor voltage vectors
R_s, R_r are the stator and rotor resistances
m_0 is the load torque
J is the moment of inertia
ω_r is the rotor angular velocity
ω_x is the angular velocity of the frame of references
τ is the relative time

All variables are expressed in p.u. system.

The space vectors may be expressed in different coordinates. The angular position of the coordinates rotating with the angular speed ω_x is defined by the angle θ as shown in Figure 22.2. In the rotating coordinates vectors are expressed as follows:

$$\mathbf{u}_{sxy} = \mathbf{u}_{s\alpha\beta}e^{j\theta}, \quad \mathbf{i}_{sxy} = \mathbf{i}_{s\alpha\beta}e^{j\theta}, \quad \boldsymbol{\psi}_{sxy} = \boldsymbol{\psi}_{s\alpha\beta}e^{j\theta},$$ (22.4)

where

index xy denotes vector in the coordinates rotating with an arbitrary chosen angular velocity
index $\alpha\beta$ denotes unmoving coordinates

For example, the Equation 22.1 in the coordinates $\alpha\beta$ unmoving in respect to the stator takes the following form ($\omega_x = 0$ in (Equation 22.1)):

$$\mathbf{u}_{s\alpha\beta} = R_s\mathbf{i}_{s\alpha\beta} + \frac{d\boldsymbol{\psi}_{s\alpha\beta}}{d\tau}.$$ (22.5)

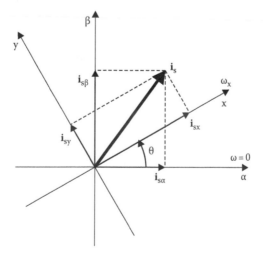

FIGURE 22.2 Vectors in different coordinate systems.

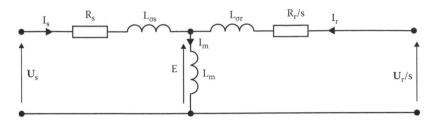

FIGURE 22.3   Equivalent circuit of the DFM.

Four variables $\boldsymbol{\psi}_s$, $\boldsymbol{\psi}_r$, $\mathbf{i}_s$, $\mathbf{i}_r$ appear in Equations 22.1 through 22.3. The dependences for stator and rotor fluxes make it possible to eliminate two variables from Equations 22.1 through 22.3:

$$\boldsymbol{\psi}_s = L_s \mathbf{i}_s + L_m \mathbf{i}_r, \tag{22.6}$$

$$\boldsymbol{\psi}_r = L_m \mathbf{i}_s + L_r \mathbf{i}_r, \tag{22.7}$$

where L_s, L_r, L_m are stator, rotor, and magnetizing inductances, respectively.

The Equations 22.6 and 22.7 result from arrows shown in Figure 22.3, where the notation for complex numbers is used. According to this notation, the active power coming to the stator or rotor is positive. If the DFM generates the active power this has a negative value.

If the stator flux and rotor current vectors are chosen as state variables, equations for vector components in the coordinate system rotating with speed ω_x, take the following form:

$$\frac{d\psi_{sx}}{d\tau} = -\frac{R_s}{L_s}\psi_{sx} + R_s\frac{L_m}{L_s}i_{rx} + \omega_x\psi_{sy} + u_{sx}, \tag{22.8}$$

$$\frac{d\psi_{sy}}{d\tau} = -\frac{R_s}{L_s}\psi_{sy} + R_s\frac{L_m}{L_s}i_{ry} - \omega_x\psi_{sx} + u_{sy}, \tag{22.9}$$

$$\frac{di_{rx}}{d\tau} = -\frac{1}{T_d}i_{rx} + \frac{R_sL_m}{L_sW_\delta}\psi_{sx} + (\omega_x - \omega_r)i_{ry} - \frac{L_m}{W_\delta}\omega_r\psi_{sy} + \frac{L_s}{W_\delta}u_{rx} - \frac{L_m}{W_\delta}u_{sx}, \tag{22.10}$$

$$\frac{di_{ry}}{d\tau} = -\frac{1}{T_d}i_{ry} + \frac{R_sL_m}{L_sW_\delta}\psi_{sy} - (\omega_x - \omega_r)i_{rx} + \frac{L_m}{W_\delta}\omega_r\psi_{sx} + \frac{L_s}{W_\delta}u_{ry} - \frac{L_m}{W_\delta}u_{sy}, \tag{22.11}$$

$$J\frac{d\omega_r}{dt} = -\frac{L_m}{L_s}(\psi_{sx}i_{ry} - \psi_{sy}i_{rx}) - m_0, \tag{22.12}$$

where

$$\frac{1}{T_d} = \frac{R_s}{L_s} + \frac{L_s^2R_r + L_m^2R_s}{L_sW_\delta}, \tag{22.13}$$

$$W_\delta = L_sL_r - L_m^2. \tag{22.14}$$

The considered machine is used as a generator connected to the grid. The instantaneous active and reactive powers of the stator windings are defined as follows:

$$p = u_{sx}i_{sx} + u_{sy}i_{sy}, \tag{22.15}$$

$$q = -u_{sx}i_{sy} + u_{sy}i_{sx}. \tag{22.16}$$

The Equations 22.8 through 22.12 and 22.15, 22.16 form a mathematical model of the DFM in the coordinates rotating with the angular velocity ω_x.

The above equations are very convenient for simulations because all variables are defined in the same coordinate system. In the real DFM, the rotor variables are defined and measured in the frame of references tied to the rotor, and the stator variables are defined and measured in the frame of references tied to the stator. The transformation from stator to rotor coordinates is defined as

$$i_{xR} = i_{xS}\cos\varphi_{RS} + i_{yS}\sin\varphi_{RS}, \tag{22.17}$$

$$i_{yR} = -i_{xS}\sin\varphi_{RS} + i_{yS}\cos\varphi_{RS}, \tag{22.18}$$

where φ_{RS} is the angle between the rotor (R) and the stator (S) coordinates. The variables measured in the rotor are transformed to the frame of references tied to the stator in the similar way.

22.3 Properties of the DFM

To analyze the machine properties the coordinate system should be selected. For the frame of references connected with the stator voltage vector

$$\mathbf{u}_s = u_{sx} + ju_{sy} = u_{sx}, \tag{22.19}$$

taking into account (from (22.6))

$$\mathbf{i}_s = \frac{1}{L_s}\boldsymbol{\psi}_s - \frac{L_m}{L_s}\mathbf{i}_r \tag{22.20}$$

and in the steady state, with additional assumption that $R_s = 0$, the following simplified expressions for the active and reactive powers are obtained:

$$p = -\frac{L_m}{L_s}u_{sx}i_{rx}, \tag{22.21}$$

$$q = \frac{u_{sx}^2}{L_s\omega_s} + \frac{L_m}{L_s}u_{sx}i_{ry}, \tag{22.22}$$

where ω_s is angular speed of the stator voltage vector.

The stator active and reactive powers depend on the rotor current vector components $\mathbf{i}_{rx}$ and $\mathbf{i}_{ry}$, respectively. From the above equations, results that decoupled control of p and q power is possible only in steady state. To control the currents (power) in the stator windings, currents in the machine rotor windings are forced. Rotor current vector is rotating in relation to the stator with the speed equal to the sum of the rotor speed ω_r and its own speed ω_{ir}. The position of the stator current vector is therefore defined by the position of the rotor current vector. Stable machine operation requires a constant angle between the stator current and the voltage vector or synchronization of both vectors. The condition of stable machine operation in steady state is

$$\omega_r + \omega_{ir} = \omega_s \tag{22.23}$$

where ω_s is the stator voltage vector rotational speed.

The best understanding of DFM properties takes place if the rotor current vector components are assumed as input variables. This is physically possible if the DFM is supplied from the current-controlled VSI. With assumption that rotor current vector components i_{rx}, i_{ry} are forced by the control system, equations for the rotor circuit dynamics (22.10) and (22.11) can be omitted for analysis purposes. The machine is now described by the following equations resulting from (22.8) and (22.9):

$$\frac{d\psi_{sx}}{d\tau} = -\frac{R_s}{L_s}\psi_{sx} + R_s\frac{L_m}{L_s}i_{rx} + \omega_s\psi_{sy} + u_{sx}, \tag{22.24}$$

$$\frac{d\psi_{sy}}{d\tau} = -\frac{R_s}{L_s}\psi_{sy} + R_s\frac{L_m}{L_s}i_{ry} - \omega_s\psi_{sx}. \tag{22.25}$$

On the basis of Equations 22.24 and 22.25 the simplified model of the DFM can be constructed, where rotor current vector components i_{rx}, i_{ry} are the inputs; the p, q powers are the outputs; and the grid voltage is uncontrolled input (Figure 22.4).

If the input variables i_{rx}, i_{ry} change fast (see Figure 22.5), the system (22.24) and (22.25) answers with oscillations damped with time constant equal to

$$T_s = \frac{L_s}{R_s}. \tag{22.26}$$

The oscillation frequency is determined, as can be seen from (22.24) and (22.25), by ω_s and can be observed on the amplitude of the stator flux vector and the angle between the stator flux vector and the stator voltage vector. These oscillations are the inner property of the DFM. Similar transients are observed in both simulated and real systems; especially for very fast systems.

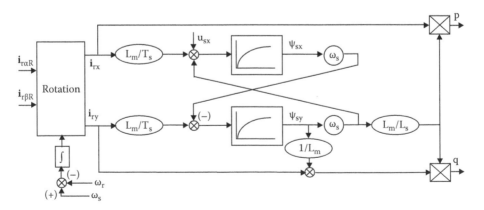

FIGURE 22.4 Model of a current-regulated DFM in the voltage-oriented frame of references.

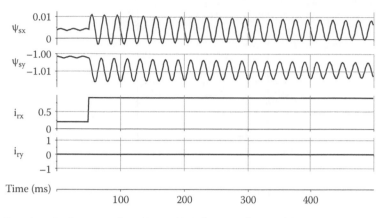

FIGURE 22.5 Transients in the system from Figure 22.4 after step change of the rotor current vector component.

Oscillations appear, as mentioned before, in the uncontrolled part of the machine and result from the direct connection of the stator to the grid. The machine control system cannot be adjusted to be very fast without damping the inner oscillations. The amplitude of oscillations is small and in many control systems it is impossible to observe them. Anyway, the existence of the stator flux oscillations is very disadvantageous if the decoupled and sensorless control is considered.

22.4 Steady-State Machine Operation

Steady state is investigated with the assumption that only the fundamental harmonic is taken into account. DFM steady-state equations can be obtained from Equations 22.8 through 22.11 in the reference frame connected with stator voltage vector. In these coordinates vectors are stationary. After transformations,

$$\mathbf{E} = j\omega_s L_m (\mathbf{I}_s + \mathbf{I}_r) \tag{22.27}$$

$$\mathbf{U}_s = R_s \mathbf{I}_s + j\omega_s L_{\sigma s} \mathbf{I}_s + j\omega_s L_m (\mathbf{I}_s + \mathbf{I}_r), \tag{22.28}$$

$$\mathbf{U}_r = R_r \mathbf{I}_r + j s \omega_s L_{\sigma r} \mathbf{I}_r + j s \omega_s L_m (\mathbf{I}_s + \mathbf{I}_r), \tag{22.29}$$

where capital letter denotes steady-state values, s denotes the slip $s = (\omega_s - \omega_r)/\omega_s$. The equivalent circuit is presented in Figure 22.3.

DFM can operate

1. Below synchronous speed: $\omega_r < \omega_s$, $0 < s < 1$
2. Above synchronous speed: $\omega_r > \omega_s$, $s < 0$
3. With synchronous speed: $\omega_r = \omega_s$, $s = 0$
4. With the slip $s > 1$, where the direction of the rotor speed is opposite to the stator flux vector speed direction

If P_s, P_r, P_m, and ΔP denote the stator, the rotor, the mechanical power and the power losses, the power balance equations are

$$P_s + P_r = P_m + \Delta P, \tag{22.30}$$

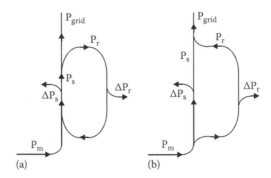

FIGURE 22.6   Power flow diagram (a) below synchronous speed, s > 0 and (b) above synchronous speed s < 0.

$$P_r = -sP_s. \tag{22.31}$$

In Figure 22.6 simplified power flow diagrams are presented during

- Generator operation below the synchronous speed: the mechanical power taken from the shaft ($P_m < 0$) and the electrical power ($P_r > 0$) taken from the rotor windings are transmitted to the grid through the stator windings ($P_s < 0$).
- Generator operation above the synchronous speed: the mechanical power taken from the shaft ($P_m < 0$) is transmitted to the power system through both stator and rotor windings ($P_s < 0$, $P_r < 0$). In this case, delivered power can exceed the stator nominal power.

In Figure 22.7, vector diagrams of the DFM during selected generator states are presented. If DFM works as a generator connected to the grid, the machine stator voltage is constant and the main flux $\Psi_m = L_m I_m$ is almost constant too. In steady state, the dependence between stator and rotor current vector magnitude is as follows:

$$\mathbf{I}_m = \mathbf{I}_s + \mathbf{I}_r = \text{const.}, \tag{22.32}$$

where $\mathbf{I}_m$ is the magnetizing current. In Figure 22.8, the locus of stator and rotor currents for different stator power factor are shown. The motor and generator mode are marked. It is visible that the machine may operate as a motor and generator with different stator power coefficients. It depends on the rotor current vector phase with respect to the stator voltage or flux vector. The machine may be excited from the stator or/and rotor side. It means that, for a defined stator active power, the active component of the

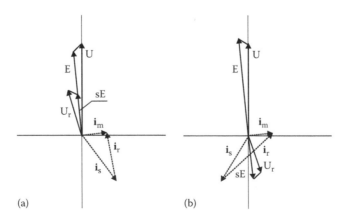

FIGURE 22.7   Vector diagram of the DFM with (a) s > 0, P < 0, Q > 0 and (b) s < 0, P < 0, Q < 0.

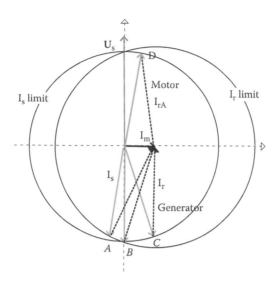

FIGURE 22.8 Locus of the stator and rotor current vectors and limitations.

stator current is also defined, but the reactive component depends on control rules and requirements about stator power coefficient. For example, if stator $\cos \varphi = 1$ is required, the stator current reactive component is equal to zero and the rotor current vector is defined by point B. The point with minimal copper losses is between B and C, and depends on stator and rotor resistances.

The same value of the active power can be delivered with different power factors and different values of reactive power. There are limitations in the reactive power production by the DFM. The rated value of the rotor current amplitude limits the range of the φ angle between the stator voltage and the stator current that can be reached. Finally, the possibility of reactive power production depends on the actual active power production. Enhancement of the reactive power production (or power factor) value brings about undesirable effects: increase of the machine rotor current and converter ratings.

22.5 Control Rules and Decoupled Control

Different schemes of control systems for the DFM were proposed [BK02,QDL05,MDD02,BDO06,P05, ESPF05]. Presented stable systems are based on forcing the currents in the rotor windings. To design the structure of the control system the space vector theory is used.

The main task of the control system is stable, independent control of the machine active and reactive power. From (22.21), (22.22) results that power depends on the rotor current vector components. Synthesis of the control system needs a choice of the reference frame. In each system, a structure of the control system and machine dynamic performances may be different.

Variables used in the control system are measured on the machine stator and rotor side, but the control is applied on the rotor side only. The rotor angular position is used to transform the variables from one to another frame of references and vice versa (22.17), (22.18). The control system has to be equipped with rotor position sensor or algorithm of the rotor angle estimation. Sensorless system is preferred.

The simplest way to force the currents in the rotor windings is through a hysteresis current controller applied to the VSI. But in such a case, the switching frequency of the VSI is not constant. Requirements for the energy quality can be fulfilled if the switching frequency is constant. A predictive current controller or standard proportional-integral (PI) controllers for the current vector components and a voltage pulse-width modulation (PWM) algorithm for the inverter can be used instead. From Equations 22.10 and 22.11 results that decoupling network is required (described in Section 22.5.3).

Because the system is nonlinear, multidimensional, and inclined to oscillations, some kind of decoupling is necessary and some kind of damping structure is suggested [KS01].

22.5.1 Decoupling Based on MM Machine Model

Most of the known up to now DFM control systems are based on the vector model of the machine. Simplifications are assumed for the control system design. The DFM is strongly nonlinear, and for precise control a decoupling is necessary. Because modern signal processors (DSP) give possibilities to implement sophisticated control algorithms, new control methods based on nonlinear control theory are applied [GK05,G07,QDL05]. In the chapter 27, *Modern Nonlinear Control*, a new model of the induction machine, called the "multiscalar model" (MM), applied in [K90] for the DFM, is presented. The following state variables are defined to obtain the MM model:

$$z_{11} = \omega_r, \quad z_{12} = \psi_{sx} i_{ry} - \psi_{sy} i_{rx}, \quad z_{21} = \psi_s^2, \quad z_{22} = \psi_{sx} i_{rx} + \psi_{sy} i_{ry}. \tag{22.33}$$

The variables z_{12} and z_{22} are the scalar and vector products of the stator flux vector and the rotor current vector. The variables defined by (22.33) do not depend on the frame of references.

Application of new $\mathbf{z} = [z_{11}, z_{12}, z_{21}, z_{22}]^T$ variables to the machine model (22.8) through (22.12) and a nonlinear feedback of the form

$$u_{r1} = \frac{w_\delta}{L_s}\left(-z_{11}\left(z_{22} + \frac{L_m}{w_\delta} z_{21}\right) + \frac{L_m}{w_\delta} u_{sf1} - u_{si1} + \frac{1}{T_v} m_1\right), \tag{22.34}$$

$$u_{r2} = \frac{w_\delta}{L_s}\left(-\frac{R_s L_m}{L_s w_\delta} z_{21} - \frac{R_s L_m}{L_s} i_r^2 + z_{11} z_{12} + \frac{L_m}{w_\delta} u_{sf2} - u_{si2} + \frac{1}{T_v} m_2\right), \tag{22.35}$$

where

$$\begin{aligned} u_{r1} = u_{ry}\psi_{sx} - u_{rx}\psi_{sy}, \quad u_{sf1} = u_{sy}\psi_{sx} - u_{sx}\psi_{sy}, \quad u_{si1} = u_{sx} i_{ry} - u_{sy} i_{rx} \\ u_{r2} = u_{rx}\psi_{sx} + u_{ry}\psi_{sy}, \quad u_{sf2} = u_{sx}\psi_{sx} + u_{sy}\psi_{sy}, \quad u_{si2} = u_{sx} i_{rx} + u_{sy} i_{ry} \end{aligned} \tag{22.36}$$

transforms the DFM model into two linear, independent subsystems (Figure 22.9):

* Mechanical subsystem

$$\frac{dz_{11}}{d\tau} = \frac{L_m}{JL_s} z_{12} - \frac{1}{J} m_0, \tag{22.37}$$

$$\frac{dz_{12}}{d\tau} = \frac{1}{T_v}\left(-z_{12} + m_1\right), \tag{22.38}$$

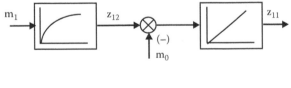

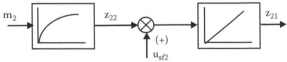

FIGURE 22.9 Model of the DFM after nonlinear decoupling.

• Electromagnetic subsystem

$$\frac{dz_{21}}{d\tau} = -2\frac{R_s}{L_s}z_{21} + 2\frac{R_sL_m}{L_s}z_{22} + 2u_{sf2},\tag{22.39}$$

$$\frac{dz_{22}}{d\tau} = \frac{1}{T_v}(-z_{22} + m_2).\tag{22.40}$$

where

m_1, m_2 are new inputs

u_{sf2} is the distortion

$T_v = W_\delta/(L_sR_r + R_sL_r)$ is time constant

Equations 22.37 through 22.40 describe a model of the DFM based on new variables together with nonlinear feedback. The input m_1 controls the machine torque z_{12} and the input m_2 controls the variable z_{22}. From the m_1 and m_2 point of view, the system is linear and decoupled. However, in the electromagnetic subsystem (22.39), a disturbance u_{sf2}

$$u_{sf2} = u_{sx}\psi_{sx} + u_{sy}\psi_{sy}\tag{22.41}$$

appears. The influence of this disturbance is mainly visible during transients and causes the weak damped oscillations of the machine stator flux.

The expressions for instantaneous active p and reactive q powers of the stator windings in steady state (for new **z** variables) take the form

$$p = -\frac{L_m}{L_s}\omega_s z_{12}, \quad q = \frac{\omega_s}{L_s}z_{21} - \frac{L_m}{L_s}\omega_s z_{22}.\tag{22.42}$$

This means that the active power depends mainly on the variable z_{12} and the reactive power depends on the variable z_{22}. During a transient the expressions (22.42) are more complicated.

22.5.2 Decoupling Based on Vector Model

The machine equation (22.2) described in the frame of references connected with the stator voltage vector takes the form

$$\frac{d\mathbf{\psi}_r}{d\tau} = -R_r\mathbf{i}_r - j(\omega_s - \omega_r)L_r\mathbf{i}_r - j(\omega_s - \omega_r)L_m\mathbf{i}_s + \mathbf{u}_r.\tag{22.43}$$

Decoupling feedback may be calculated assuming desired differential equations for the rotor flux vector components [BDO06]. Canceling first three terms in this equation by feedback of the form

$$u_{rx} = -L_mi_{sy}(\omega_s - \omega_r) - L_ri_{ry}(\omega_s - \omega_r) + R_ri_{rx} + v_x,\tag{22.44}$$

$$u_{ry} = L_mi_{sx}(\omega_s - \omega_r) + L_ri_{rx}(\omega_s - \omega_r) + R_ri_{ry} + v_y,\tag{22.45}$$

the rotor equations are transformed into

$$\frac{d\psi_{rx}}{d\tau} = v_x,\tag{22.46}$$

$$\frac{d\psi_{ry}}{d\tau} = v_y,\tag{22.47}$$

where v_x, v_y are the control variables that are defined by PI action:

$$v_x = k_P \varepsilon_y + k_I \int \varepsilon_y \, d\tau, \tag{22.48}$$

$$v_y = -k_P \varepsilon_x - k_I \int \varepsilon_x \, d\tau, \tag{22.49}$$

where $\varepsilon_x = i_{sx} - i_{sx-set}$, $\varepsilon_y = i_{sy} - i_{sy-set}$.

Such (or similar) decoupling makes it possible to directly control the stator current vector components (or stator active and reactive power) without inner control loops. The decoupling may also be calculated from other differential equations. Anyway, a part of the designed system may remain out of control, and as result the system may be inclined to oscillations.

22.5.3 Decoupling Based on Rotor Current Equation

In order to independently control the rotor current vector components, coupling between x and y axes should be reduced. Compensation of the part of the following differential equations for the rotor current vector components in the frame of references fixed to the stator flux vector (like (22.10), (22.11))

$$\frac{di_{rx}}{d\tau} = -\frac{1}{T_d} i_{rx} + \frac{R_s L_m}{L_s W_\delta} \psi_{sx} + (\omega_s - \omega_r) i_{ry} - \frac{L_m}{W_\delta} \omega_r \psi_{sy} + \frac{L_s}{W_\delta} u_{rx} - \frac{L_m}{W_\delta} u_{sx}, \tag{22.50}$$

$$\frac{di_{ry}}{d\tau} = -\frac{1}{T_d} i_{ry} + \frac{R_s L_m}{L_s W_\delta} \psi_{sy} - (\omega_s - \omega_r) i_{rx} + \frac{L_m}{W_\delta} \omega_r \psi_{sx} + \frac{L_s}{W_\delta} u_{ry} - \frac{L_m}{W_\delta} u_{sy}, \tag{22.51}$$

of the form [TTO03]

$$u_{rx} = \frac{W_\delta}{L_s} (\omega_s - \omega_r) i_{ry} + v_x \tag{22.52}$$

$$u_{ry} = -\frac{W_\delta}{L_s} (\omega_s - \omega_r) i_{rx} + \frac{L_m}{L_s} \omega_r (L_s i_{sx} + L_m i_{rx}) + v_y \tag{22.53}$$

can be used. Variables v_x, v_y are the new control signals. Such a decoupling network compensates only the main couplings, but improves the quality of the rotor current vector component dynamics.

22.6 Overall Control System

The variable speed generation system consists of the DFM with stator connected directly to the grid and a rotor connected through the inverter. The DFM with an active and reactive power control system is coupled through the LC filter (see Figure 22.1) with the grid that can be treated as a voltage source. In such a system, weak damped oscillations depending on both system parameters and transmitted through the filter power (direction and value) may appear. Machine converter consists of the grid-side inverter and the rotor-side inverter connected together through the dc-link capacitor. Both the inverters are equipped with PWM algorithms and control systems. The grid-side inverter controls the dc-link voltage and the reactive current component. The machine-side inverter controls the active power of the whole system (stator and rotor windings and filter) and controls the stator reactive power. In the DFM control system, other important variables can be controlled too, e.g., stator and/or rotor current components.

There is no single method to control the double-fed induction generator, and many control structures can be proposed. The main task of the control system is the active and reactive power decoupled control, but in many applications additional requirements are defined:

- Speed and position sensorless control
- Smooth active power production in spite of prime mover torque disturbances like, e.g., wind gusts in wind power plants
- Stable operation with relatively large sampling time of microprocessor control board

Dependencies (22.21), (22.22) define the way of controlling the active and reactive power of only stator windings. In the real system, the machine consumes/delivers the power through both stator and rotor windings (see Figure 22.6), and the power of the whole system has to be controlled. Machine stator and rotor reactive powers are controlled independently, while the stator active power depends on the rotor active power according to (22.31). The reactive power at the stator terminals is controlled by the appropriate component of the rotor current vector, but the rotor-side reactive power is controlled by the grid-side inverter. In many applications, the unity power factor of the whole system is required. The rotor-side power coefficient is therefore set to one, the rotor-side reactive power reference value is set to zero, and the stator reactive power reference value is set to zero too. Because the active power of the stator and the rotor depend on each other, it is enough to control only the stator active power while the rotor active power may be treated as a disturbance.

The structure of the DFM control system is presented in Figure 22.10. The rotor-side inverter control system is responsible for the full system active and reactive power control. The powers are controlled in separate loops. Outputs of the active and reactive power controllers act as set values for variables that are controlled in the inner loops. These variables result from Equations 22.42, 22.21, 22.22. Proper choosing of inner variables and the coordinate system determine the control quality. The machine active and reactive powers may be also controlled directly without any inner loops

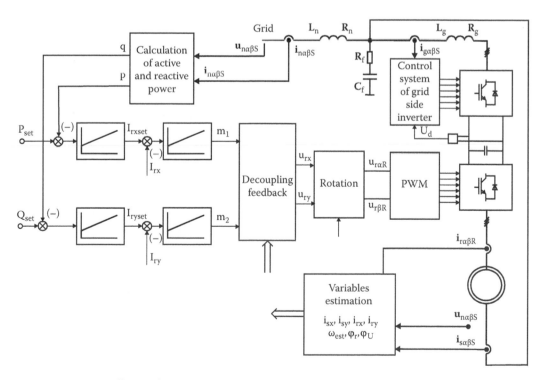

FIGURE 22.10   Overall control system structure.

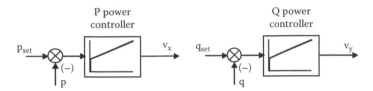

FIGURE 22.11   Structure of the control system with direct power control.

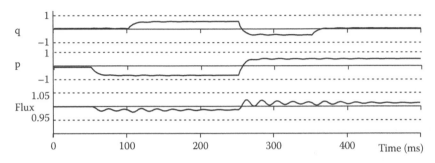

FIGURE 22.12   Transients in the system from Figure 22.10 during power control.

[KG05] (see Figure 22.11). Additional decoupling (see Section 22.5) is necessary in each structure. After appropriate transformations of control variables the rotor voltage vector components $u_{r\alpha R}$, $u_{r\beta R}$ are used in PWM algorithm. There are additional blocks in the control system: the variables transformations, calculations, and estimation. As a result, stable, fast, and decoupled control of both powers is obtained (see Figure 22.12).

Structure of the grid-side inverter control system is simpler. The outer loop controls the dc-link voltage u_d. The inner loops control the grid-side inverter current vector components in coordinates connected usually with grid voltage.

22.6.1 Control System Based on MM Model

The MM model, described in Section 5.1, is used for the control system synthesis. Nonlinear feedback based on this model enables decoupled control of the z_{12} and z_{22} variables (see Equation 22.42) and also the stator windings active and reactive power decoupled control (in steady state). Power controller outputs act as set values for z_{12} and z_{22} variables that are controlled in inner loops. The m_1, m_2 variables are the inputs for the decoupling feedback defined by (22.34), (22.35). Simulation-obtained transients in this system are presented in Figure 22.12. DFM may be controlled also in the system with only one controller in every channel (see Figure 22.11). In this case, the m_1, m_2 variables exist at the outputs of the power controllers. During transients presented in Figure 22.13, a small coupling appears. Presented structures work correctly if the rotor position is estimated.

It should be noted that the part of the system is uncontrolled. As it was explained earlier, during transients weak damped oscillations of the stator flux vector, z_{21} (22.39), (22.41) appear. The amplitude is small, but causes small active and reactive power oscillations. Oscillations of the z_{21} variable come into being after each change of the power reference value, and they disappear slowly. The presented simulation results have been made for PI-type controllers, and it is not possible to eliminate the oscillations by the simple change of the controller settings. Undesirable effects can be reduced by limitation of the reference values derivatives (ramp function), but then the system is slow. The other method is the use of the controller structure different from PI (e.g., neural controller) or the use of additional damping feedback based on the stator flux magnitude derivative [SV06].

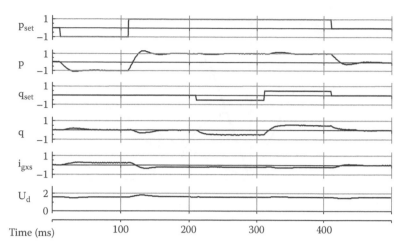

FIGURE 22.13 Transients in the system with one controller in every channel.

22.6.2 Control System Based on Vector Model

The structure of the control system based on vector model is presented in Figure 22.10, but the meaning of the variables is different. Rotor current vector components i_{rx}, i_{ry} in the frame of references connected with stator flux or stator voltage vector exist at the power controller outputs. Decoupling described in Section 22.4.2 or similar may be used. Because there are few possibilities to choose decoupling algorithm and frame of references [MDD02, BDO06], the features of the system may differ from each other. The results do not differ strongly from those presented in Figures 22.12 and 22.13.

22.7 Estimation of Variables

22.7.1 Calculation of the Angle between the Stator and the Rotor

For the correct operation of the control system, the angle between the stator and the rotor and the rotor angular velocity are necessary. The simplest way is the application of the encoder to measure the rotor angular position, but it is inconvenient in many applications, e.g., for the high-power wind power generators. To avoid a rotor position encoder, a sensorless system is preferred, where the angle of rotor position is estimated.

To perform speed sensorless control of the squirrel cage induction machine, the observer technology is used. In DFM, speed observer is not necessary because the stator windings are accessible through the slip rings and measurements can be made.

There are a few possibilities to estimate the rotor speed and the angular position of the double-fed induction machine. One of the sensorless control methods of the DFM is based on the determination (measurement, calculation) of the same current (e.g., rotor current) in different frames of references. The first one is connected with unmoving stator and the other one is connected with the rotor. The selected vector in different frames of references has the same amplitude but different angles expressed as follows (see Figure 22.14):

$$\varphi_i^s = \varphi_i^r + \varphi_{rs}, \tag{22.54}$$

where φ_i^s, φ_i^r are the angles of the current vector in the stator and rotor frame of references. The angle φ_{rs} between coordinate systems is equal to the angle between the stator and the rotor.

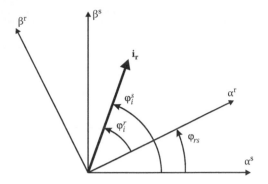

FIGURE 22.14 Rotor current vector in different coordinate systems.

Trigonometric functions of the angle between two vectors may be calculated as follows:

$$\cos(\varphi_{rs}) = \frac{i_{r\alpha}^r i_{r\alpha}^s + i_{r\beta}^r i_{r\beta}^s}{i_r^2}, \tag{22.55}$$

$$\sin(\varphi_{rs}) = \frac{i_{r\alpha}^r i_{r\beta}^s - i_{r\beta}^r i_{r\alpha}^s}{i_r^2}, \tag{22.56}$$

where superscript r, s denote the current vector components determined respectively in the frame of references connected to the rotor and the stator.

Components of the rotor current vector $i_{r\alpha}^r$, $i_{r\beta}^r$ can be measured directly, but $i_{r\alpha}^s$, $i_{r\beta}^s$ have to be estimated. The main dependence of any induction machine is as follows (Figure 22.3):

$$\mathbf{i}_m = \mathbf{i}_s + \mathbf{i}_r, \tag{22.57}$$

where $\mathbf{i}_m$, $\mathbf{i}_s$, $\mathbf{i}_r$ are the magnetizing, stator and rotor current vectors. The magnetizing current is defined as follows:

$$\mathbf{i}_m = \frac{\boldsymbol{\Psi}_m}{L_m}, \tag{22.58}$$

where
$\boldsymbol{\psi}_m$ is the main flux vector
L_m is the mutual inductance

The main flux is equal to

$$\boldsymbol{\Psi}_m = \boldsymbol{\Psi}_s - L_{\sigma s}\mathbf{i}_s, \tag{22.59}$$

where
$\boldsymbol{\psi}_s$ is the stator flux
$L_{\sigma s}$ is the stator leakage inductance

The stator flux may be calculated from the following differential equation:

$$\frac{d\boldsymbol{\psi}_s}{dt} = -R_s\mathbf{i}_s + \mathbf{u}_s, \tag{22.60}$$

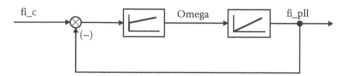

FIGURE 22.15 Estimation of rotor position with PLL.

where $\mathbf{u}_s$ is the stator flux vector. From (22.57) through (22.60), the rotor current vector components in stator α, β coordinates can be calculated on the basis of the stator current and voltage measurements. The stator current vector in (22.57) is directly measured. Stator flux vector may be simply estimated from (22.60) or on the basis of the steady-state equations received by assumption that $R_s = 0$.

The rotor current vector components in steady state, in the frame of references connected to the stator are

$$i^s_{r\alpha} = -\frac{1}{\omega_s L_m}u^s_{s\beta} - \frac{L_s}{L_m}i^s_{s\alpha}, \quad i^s_{r\beta} = \frac{1}{\omega_s L_m}u^s_{s\alpha} - \frac{L_s}{L_m}i^s_{s\beta}. \tag{22.61}$$

If the current vector is calculated on the basis of simplified dependencies, the angle between the stator and rotor is calculated with small error.

Dependencies similar to (22.55), (22.56) may be written for stator current. In this case, the stator current has to be calculated in the frame of references connected with the rotor.

22.7.2 Application of Phase-Locked Loop for the Estimation of Rotor Speed and Position

The rotor position angle calculated from equations derived in section 7 cannot be used in transformations because of disturbances that may destroy the stability of the control system. Smooth rotor position angle may be received together with rotor angular velocity from the system with phase-locked loop (PLL) presented in Figure 22.15. The PI controller have to be applied in this structure. The rotor angular velocity is received on input to the integrator. The rotor angular velocity is needed in decoupling feedback. Other schemes are possible too.

22.8 Remarks about Digital Realization of the Control System

22.8.1 Compensation of the Delay Time Caused by Sampling

In the high-power converters, the carrier frequency of pulse with modulation is in the range between 2 and 3 kHz because of switching losses limitation. A high sampling period appears therefore in the control system. It was investigated by simulations that the delay caused by the sampling period does not influence the nonlinear feedback. The variables appearing in the nonlinear feedback are constant in steady states, change slowly in transients, and controllers compensate for the delay.

The estimated stator flux vector have to be rotated by the angle resulting from its rotation in sampling period to compensate the delay.

22.8.2 Measurements of Currents and Voltages

The stator and rotor currents have to be sampled at the same time resulting from application of a PWM strategy to the generation of rotor voltage. The point of sampling instant have to be chosen exactly in the middle of time when zero voltage vector appears on the inverter output as shown in Figure 22.16. The reason is that the simplified dependencies derived in Section 22.7 are valid for fundamentals only.

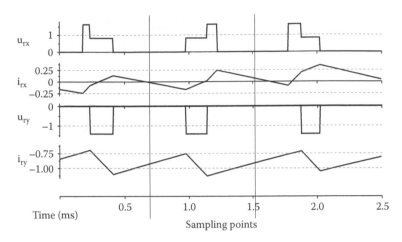

FIGURE 22.16 The choice of the sampling point.

It can be seen from Figure 22.16 that the instant value of rotor current is equal to its fundamental in the middle of zero voltage vector.

The method presented in Section 22.7 makes it possible to determine the angle between the rotor and stator in instant k. The value of this angle in instant k + 1 needed in the control system may be predicted from the following equation:

$$\varphi_{rs}(k+1) = 2\varphi_{rs}(k) - \varphi_{rs}(k-1). \tag{22.62}$$

References

[BDO06] C. Battle, A. Doria-Cereza, R. Ortega, A robustly adaptive PI controller for the double fed induction machine, *Proceedings of the 32nd Annual Conference IECON*, Paris, France, pp. 5113–5118, 2006.

[BK02] E. Bogalecka, Z. Krzeminski, Sensorless control of double fed machine for wind power generators, *Proceedings of the EPE-PEMC Conference*, Croatia, 2002.

[ESPF05] S. El Khil, I. Slama-Belkhodja, M. Pietrzak-David, B. de Fornel, Rotor flux oriented control of double fed machine, *11th European Conference on Power Electronics and Applications (EPE'2005)*, Dresden, Germany, 2005, pp. A73713.

[G07] A. Geniusz, Power control of an induction machine, U.S. Patent Application Publication, No. US2007/0052394/A1, 2007.

[GK05] A. Geniusz, Z. Krzeminski, Control system based on the modified MM model for the double fed machine, *Conference PCIM'05*, Nurenberg, Germany, 2005.

[K90] Z. Krzeminski, Control system of doubly fed induction machine based on multiscalar model, *IFAC 11th World Congress on Automatic Control*, Tallinn, Estonia, vol. 8, 1990.

[KS01] C. Kelber, W. Schumacher, Active damping of flux oscillations in doubly fed AC machines using dynamic variations of the systems structure, *9th European Conference on Power Electronics and Applications (EPE'2001)*, Graz, Austria, 2001.

[MDD02] S. Muller, M. Deicke, R. W. De Doncker, Double fed induction generator systems for wind turbines, *IEEE Industry Applications Magazine*, 3, 26–33, May/June 2002.

[P05] A. Peterssonn, Analysis, modelling and control of double fed induction generators for wind turbines, PhD thesis, Chalmers University of Technology, Goteborg, Sweden, 2005.

[QDL05] N. P. Quang, A. Dittrich, P. N. Lan, Doubly fed induction generator in wind power plant: Nonlinear control algorithms with direct decoupling, *Proceedings of the EPE Conference*, Dresden, Germany, 2005.

[SV06] I. Schmidt, K. Veszpremi, *Field oriented current vector control of double fed induction wind generator*, 1-4244-0136-4/06, 2006 IEEE.

[TTO03] A. Tapia, G. Tapia, J. X. Ostolaza, J. R. Saenz, Modelling and control of a wind turbine driven doubly fed induction generator, *IEEE Transactions on Energy Conversion*, 18(2), 194–204, June 2003.

23

Standalone Double-Fed Induction Generator

Grzegorz Iwański
Warsaw University of Technology

Włodzimierz Koczara
Warsaw University of Technology

23.1 Introduction

Standalone, AC voltage, power generation systems with conversion of mechanical energy mainly use wound rotor synchronous generators (WRSGs) operated with fixed speed, related to the reference frequency, e.g., 50 or 60 Hz. Power systems, like wind turbines or water plants, in which fixed speed is difficult to obtain, can be adopted to standalone variable-speed operation and provide standard fixed frequency AC voltage. Normalized voltage can be obtained by the use of full-range power electronics converter, as a coupler interface between variable-speed generator and an isolated load (Figure 23.1). WRSG or permanent magnet synchronous generator (PMSG) based systems can be equipped with an AC/DC diode rectifier with an optional DC/DC converter and a DC/AC converter (Figure 23.1a). In case of cage induction generator (CIG) (Figure 23.1b), back-to-back converter is necessary (controlled AC/DC and DC/AC). Power inverter, responsible for generation of standard AC voltage, in standalone mode requires an output L_f–C_f filter, to obtain high-quality generated voltage [1–3].

Other variable-speed power generation system, which is recently often applied in grid-connected wind turbines, consists of doubly fed induction generator (DFIG) and rotor-connected power electronics converter (Figure 23.2) [4–6]. The typical speed range of DFIG generation system equals ±33% around synchronous speed. For that speed range, the power electronics converter is limited to 33% of DFIG rated power.

In comparison to the total system power, the DFIG corresponds to 75% and converter corresponds to 25% of maximum produced power, due to the fact, that during over-synchronous speed operation, power is delivered via stator side as well as rotor and power electronics converter. The variable-speed systems with DFIG, driven by wind turbines, are dedicated only to grid-connected systems, if not supported by energy storage or other power source.

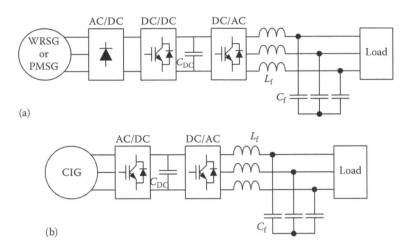

(a)

(b)

FIGURE 23.1   Standalone variable-speed power generation systems with full-range converter and (a) synchronous generator and (b) cage induction generator.

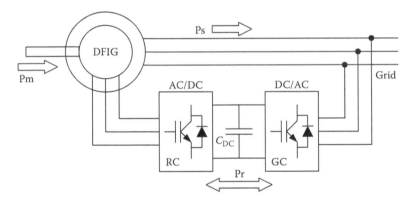

FIGURE 23.2   Typical power topology of grid-connected power generation system based on DFIG.

23.2 Standalone DFIG Topology

In opposite to the grid-operated DFIG systems, where the stator and the power electronics converter have to be connected to the power network of imposed voltage parameters, the standalone DFIG system supplies an isolated load (Figure 23.3). The stator of the slip-ring induction machine, excited by the rotor current, produces normalized voltage. The rotor current is controlled by an AC/DC rotor converter RC, whose DC side is connected to a DC/AC grid converter GC. Independent of the load power and actual speed, the rotor converter RC has to maintain the fixed-voltage amplitude and frequency on the stator side, whereas similarly to the grid-connected systems, the grid converter GC has to maintain the DC link voltage on the reference level. The C_f capacitor provides filtration of the output stator AC voltage.

23.2.1 Model of Standalone DFIG

Fundamental electrical equations of the DFIG, equipped with stator-connected filtering capacitance C_f, in the frame connected with stator voltage vector, are as follows:

$$u_s = R_s i_s + \frac{d\psi_s}{dt} + j\omega_s\psi_s \tag{23.1}$$

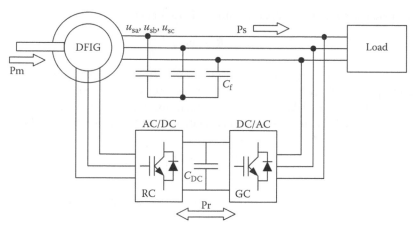

FIGURE 23.3 Topology of standalone DFIG.

$$u_r = R_r i_r + \frac{d\psi_r}{dt} + j\left(\omega_s - p_b \omega_m\right)\psi_r \qquad (23.2)$$

$$\psi_s = L_s i_s + L_m i_r \qquad (23.3)$$

$$\psi_r = L_m i_s + L_r i_r \qquad (23.4)$$

$$i_s = -C_f \frac{du_s}{dt} + i_{ld} - j\omega_s C_f u_s \qquad (23.5)$$

where
  u_s and u_r are the stator and rotor voltages
  ψ_s and ψ_r are the stator and rotor flux
  i_s and i_r are the stator and rotor currents
  R_s, R_r are the stator and rotor resistances
  L_s, L_r, L_m are the stator, rotor, and magnetizing inductances
  p_b represents the number of poles pairs
  ω_s is the synchronous speed
  ω_m is the mechanical speed
  C_f is the filtering capacitance
  i_{ld} is the load current

Considering that the rotor is supplied from the current-controlled voltage source inverter RC, which can be treated as current source, (23.2) can be neglected and then the relations between stator voltage and rotor current can be described.

  Standalone DFIG systems described in some publications are not equipped with filtering capacitors [7–9]. For filtering capacitance C_f equal zero and resistive load, model of standalone DFIG supplied from current controlled voltage source inverter (VSI), with neglected stator resistance, is

$$u_s = \frac{R_o L_m}{Z_s}\frac{di_r}{dt} + j\left(\frac{\omega_s R_o L_m}{Z_s}\right)i_r - \frac{L_s}{Z_s}\frac{du_s}{dt} \qquad (23.6)$$

where R_o is the load resistance and Z_s are the stator and load impedances:

$$Z_s = R_o + j\omega_s L_s \tag{23.7}$$

The differential of the rotor current existing in (23.6) indicates that the generated voltage u_s is distorted by rotor current ripples produced by PWM converter of the rotor side. The negative sign by stator voltage derivative is responsible for partial damping of the voltage distortions. However, for low power, the voltage distortions are significant. In limited case, no-load operation, R_o and Z_s are infinite and (23.6) is reduced to

$$u_s = L_m \frac{di_r}{dt} + j\omega_s L_m i_r \tag{23.8}$$

which indicates that the stator voltage is distorted by rotor current ripples of PWM frequency.

High-power-grid-connected DFIG systems are equipped with rotor-connected series inductances, which reduce rotor current ripples. Inductance filters are not enough to obtain high-quality stator voltage in standalone operated systems, especially during low-load operation, when the system damping ratio is low and PWM frequency distortions are transformed into the stator side. Additional inductance L_{radd} adds to rotor leakage inductance in mathematical model, as these inductances are connected in series. Considering (23.2) and (23.4), the rotor current ripples can be determined.

$$\frac{di_r}{dt} = \frac{1}{L_m + L_{r\sigma} + L_{radd}} \left(u_r - R_r i_r - L_m \frac{di_s}{dt} - j(\omega_s - p_b \omega_m) L_m i_s \right) - j(\omega_s - p_b \omega_m) i_r \tag{23.9}$$

To obtain effective filtration, the rotor-connected additional inductance has to be comparable with magnetizing inductance, which makes the system more expensive and heavier. However, the voltage of unloaded generator still will be distorted, as the rotor current is forced by PWM rotor converter and (23.8) remains the same.

The system with filtering capacitors on the stator provides a high-quality stator voltage without PWM frequency distortions. The worst-case scenario for the system stability is no-load operation, as of the lowest damping ratio of the system. For high-power DFIG, the stator resistance can be neglected and the generated voltage with good approximation can be described by

$$u_s = \frac{L_m}{W} \frac{di_r}{dt} + j\frac{\omega_s L_m}{W} i_r - j\frac{2\omega_s L_s C_f}{W} \frac{du_s}{dt} - \frac{L_s C_f}{W} \frac{d^2 u_s}{dt^2} \tag{23.10}$$

where

$$W = 1 - \omega_s^2 L_s C_f \tag{23.11}$$

The negative sign by second-order derivative of the stator voltage (23.10) is the component responsible for effective damping of the voltage distortions.

23.2.2 Selection of the Filtering Capacitors

For high-frequency harmonics, the model of unloaded DFIG can be simplified to LC filter, which consists of equivalent generator leakage inductance $L_{rs\sigma}$ and C_f. Selection of the filtering capacitor is needed to obtain high quality and stability of the generated voltage. The first criteria is a resonant frequency of LC filter. On the logarithmic scale, the resonant frequency has to be obtained in the middle between operational frequency

(50 or 60 Hz) and switching frequency. The second criteria is that the capacitor C_f must not fully compensate the generator magnetizing reactive power. In calculation of the reactive power, not the operating frequency (50 Hz or 60 Hz), but the frequency corresponding to maximum possible mechanical speed has to be taken into consideration. In some cases, full compensation of induction machine results in self-excitation. In case of DFIG, the remanence flux rotates with mechanical speed, while the flux originated from rotor current rotates with synchronous speed. Normally the rotor flux, originated from the rotor current, is much higher than remanence flux, but in self-excitation conditions (full compensation or overcompensation) these two fluxes are comparable and cannot be synchronized, and the induced stator voltage is unstable.

For typical induction machine, the leakage factor is close to 0.04–0.06. The stator and rotor leakage inductances can be represented for high-frequency by equivalent leakage inductance $L_{rs\sigma}$:

$$L_{r\sigma} + L_{s\sigma} = L_{rs\sigma} \approx \sigma L_m \qquad (23.12)$$

where σ is the leakage factor equal to

$$\sigma = 1 - \frac{L_m^2}{L_r L_s} \qquad (23.13)$$

The approximation error in (23.12) is negligible.

The filtering capacitor C_f, which does not overcompensate the reactive power of the DFIG, at the frequency f_m, related to the maximum possible mechanical speed, has to meet the requirement

$$C_f < \frac{1}{4\pi^2 f_m^2 L_m} \qquad (23.14)$$

At the same time, to obtain given resonant frequency f_r of the output filter $L_{rs\sigma} C_f$, which has to be significantly smaller than the switching frequency, the capacitance must be equal to

$$C_f = \frac{1}{4\pi^2 f_r^2 \sigma L_m} \qquad (23.15)$$

23.2.3 Initial Excitation of Standalone DFIG

In the grid-connected systems, the problem of generating system startup does not exist, as the power network provides energy for magnetization of the stator and for the preliminary charge of DC link capacitors in power electronics converter. Standalone power systems, based on the induction generators, require some initial energy to obtain excitation, needed in the stator voltage generation. It can be obtained using many ideas.

The example way to obtain initial excitation is to connect additional capacitors (Figure 23.4), which fully compensate the reactive power of induction machine. The self-excitation is known from fixed-speed standalone CIGs. However, in case of standalone DFIG, additional capacitors can be connected by stator switch (SS), only for initial self-excitation and preliminary charging of the DC link; as for steady-state operation the self-excitation phenomena can disturb the voltage, controlled by power electronics converter. Additional capacitors can be used when load power is very high. The self-excitation will be dumped by the load, whereas the capacitors can improve the voltage quality during nonlinear load supply. During low-load operation, the influence of remanence flux and the flux produced by the rotor current may be similar, but those fluxes cannot be synchronized, as they rotate with different speeds. Additional capacitors can provide high voltage on the stator side, similarly to the case with CIG. However, it may require short circuit on the rotor side. It can be made by rotor switch RS or by thyristor crowbar used for protection of power electronics converter against the effects of the grid short circuit.

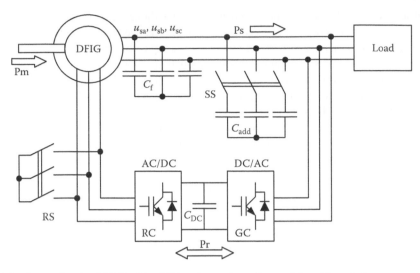

FIGURE 23.4   Power circuit of DFIG with additional capacitors C_{add} used for self-excitation.

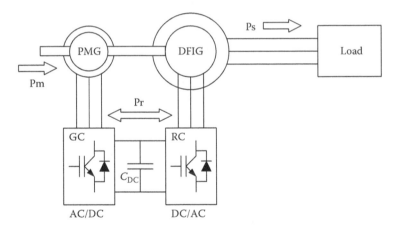

FIGURE 23.5   Power circuit of DFIG with additional PMG.

Systems, with permanent magnet generators (PMG), which are mechanically coupled on the same shaft as DFIG, and the electric coupling of PMG stator with DFIG rotor is made by power converter are already known and described [10,11] (Figure 23.5). The problem of initial excitation is eliminated, but the possible delivered power does not exceed the power of DFIG stator. For over-synchronous speed the electrical power is provided to the PMG, which operates as a motor, and only the DFIG stator power is available.

Combination of classical DFIG topology, and fractional power PMG, coupled on the same shaft, used only to charge the DC link in power converter, keeps the advantages of DFIG with back-to-back converter connected to the rotor and grid, and eliminates the problem with initial excitation (Figure 23.6). PMG can be replaced with small capacity energy storage (battery) used for DC link charging. This battery can be recharged, when the power generation system operates, and reused by the next system startup.

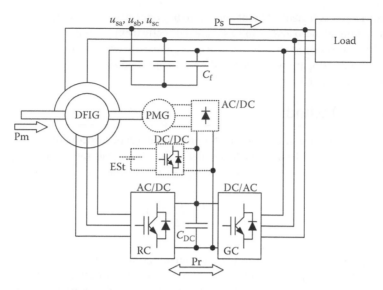

FIGURE 23.6 Power circuit of DFIG with additional low-power permanent magnet exciter or battery.

23.2.4 Stator Configurations

Two main configurations of stator windings connection can be used in standalone DFIG. The first one is three-wire system on both rotor and stator side (Figure 23.7a); the second configuration is the three-wire system on the rotor side and four-wire system on the stator side (Figure 23.7b). Both systems can be used for balanced and unbalanced standalone operation. In the three-wire system on the stator side, the

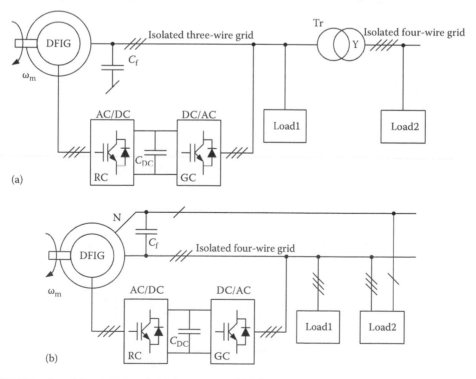

FIGURE 23.7 Standalone DFIG with (a) three-wire and (b) four-wire established isolated grid.

load, which requires access to neutral point and has to be connected by three- to four-wire transformer. Other balanced and unbalanced loads can be supplied directly from the stator [12]. For low-power DFIG systems, when the load is placed close to the generator, four-wire system can be used [13] on the stator side, as the DFIG is a rotating transformer and the star-connected stator can provide symmetrical voltage, even if three-wire system is on the rotor side.

23.3 Control Method

It can be seen in (23.6) that in the steady state, there is proportional dependence of the output voltage from the rotor current in standalone DFIG. For the system loaded with resistive load, the output voltage is also proportional to the rotor current and to obtain reference voltage vector, the rotor current has to meet the equation:

$$i_r = \left(\frac{1}{R_0} - j\frac{1 - \omega_s^2 L_s C_f}{\omega_s L_m} \right) u_{sref} \tag{23.16}$$

The real component is adequate to the resistive load current, whereas the imaginary part represents the excitation current that is also partially compensated by the stator-connected capacitors. For other loads (e.g., *RL*), similar equations can be derived and the stator voltage dependence on the rotor current is always proportional. Thus, the simple PI controller of the stator voltage amplitude can be implemented with the output signal responsible for reference amplitude of the rotor current vector $|i_r|^*$. The linear dependence of the output voltage amplitude from the rotor current amplitude is known from classic synchronous generator. In fact, the WRSG is a special case of doubly fed machine [14].

Independent of the stator voltage amplitude control, the fixed frequency has to be obtained. The stator voltage is produced by the excitation current, injected to the rotor. The obtained frequency f_s corresponds to the field angular speed related to the stator and is a result of mechanical rotation and the rotation provided by the rotor current frequency (23.17).

$$f_s = \frac{1}{2\pi}(p\omega_m + \omega_{ir}) \tag{23.17}$$

where
  p is the number of poles pairs
  ω_m is the rotor angular speed
  ω_{ir} is the angular speed of the rotor current vector

Using (23.17), simple control system, based on the mechanical speed sensor, can be applied and the stator voltage stabilized (Figure 23.8). However, this simple control does not allow to control the stator

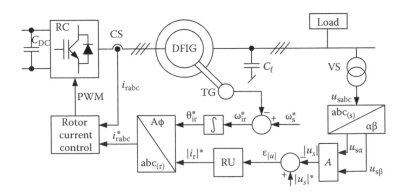

FIGURE 23.8 Simple voltage control with rotor speed sensor.

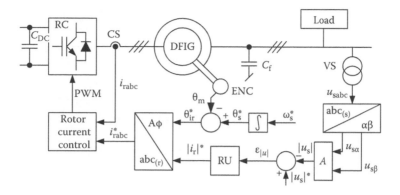

FIGURE 23.9   Simple voltage control with rotor position encoder.

voltage phase due to the fact, that the speed sensor does not indicate the rotor position and every error in the speed measurement result in the error of the rotor current angular speed ω_{ir}^* and phase angle θ_{ir}^*, therefore consequently in the stator voltage frequency f_s and phase.

More precise control of the stator voltage frequency can be obtained by replacing the speed sensor with rotor position encoder. Reference rotor current angle θ_{ir}^* is obtained from the equation

$$\theta_{ir}^* = \theta_s^* - p\theta_m \tag{23.18}$$

where

θ_s^* is the reference angle, obtained with the integration of the reference synchronous angular speed ω_s^*
θ_m is the rotor position angle

The angle θ_s^* is not a reference phase of the voltage, as there is a phase shift between rotor current and stator voltage. However, this phase shift is fixed for given load and can be neglected. In this control, there is only an error of the phase, as for different load there are different phase shifts between rotor current and stator voltage. However, the error of the frequency does not occur. The control method based on the rotor position encoder is shown in Figure 23.9.

Modern power generation systems have to be designed as sensorless, what significantly increases their reliability. High precision of the rotor position required in some drives application is not necessary in power generation system. Moreover, the variable-speed standalone DFIG system does not require any determination of the rotor position. For given mechanical speed, the relation of output stator frequency f_s with the rotor current frequency f_{ir} is linear with constant component corresponding to mechanical speed (Figure 23.10).

Negative f_{ir} frequency represents negative rotation of the field in relation to the rotor: over-synchronous speed operation; zero means synchronous operation with DC current injected in each phase of the rotor: synchronous speed operation. Linear dependence of the stator voltage amplitude and frequency from the rotor current amplitude and frequency can be used in the primitive scalar voltage control methods [11]. The example of scalar voltage control method is presented in Figure 23.11. For the given, unknown mechanical speed, PI frequency controller RF adjusts the rotor current frequency to obtain stator frequency on the fixed reference level. However, calculation of the frequency twice per period based on the zero crossing detection makes the frequency loop inaccurate, even if the frequency change is slow.

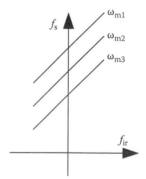

FIGURE 23.10 Dependence of the output frequency f_s of the rotor current frequency f_{ir} for different mechanical speed $\omega_{m1} - \omega_{m3}$.

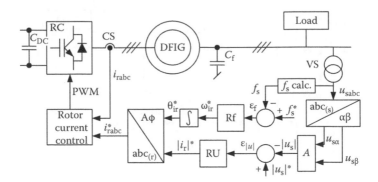

FIGURE 23.11 Simple sensorless scalar control of standalone DFIG stator voltage.

23.3.1 Sensorless Control of the Stator Voltage Vector

Sensorless control of the stator voltage vector u_s requires two coordinates. Necessity of the control of the voltage vector magnitude $|u_s|$, allows to use polar coordinates system, in which the second coordinate is the vector angle α_{us}, related to one of the axes of reference frame. Control of the voltage vector magnitude uses orthogonal components, calculated in every switching period, and this magnitude is the same in both stationary and rotating frame. Control of voltage vector magnitude $|u_s|$, with reference magnitude of the rotor current vector $|i_r|^*$, as an output signal from RU controller is linear. The remaining problem is the control of voltage vector angle. It requires determination of adequate reference angular speed ω_{ir}^* of the rotor current vector i_r and its phase θ_{ir}^*.

Sensorless control method, which allows to determine the adequate angle θ_{ir}^* of the rotor current vector i_r, to provide fixed frequency f_s and reference phase α_{us}^* of the stator voltage vector u_s, is presented in Figure 23.12. Frequency control loop operates similarly to the PLL structure. However, in classical PLL, two signals have the same frequency, and only the phase of one signal is adopted to obtain synchronization with the second signal. In DFIG, the rotor and stator frequencies are different and it is necessary to adopt both frequency and phase of the rotor current to obtain synchronization of the actual and reference stator voltage vectors.

In the dq system, rotating with reference angular speed ω_s^*, corresponding to the reference frequency f_s^*, the reference voltage vector u_s^* is overlapped with d-axis (reference angle α_{us}^* is equal to zero). The actual angle α_{us} of the voltage vector u_s is calculated using dq components:

$$\alpha_{us} = \arctan\frac{u_{sq}}{u_{sd}} \tag{23.19}$$

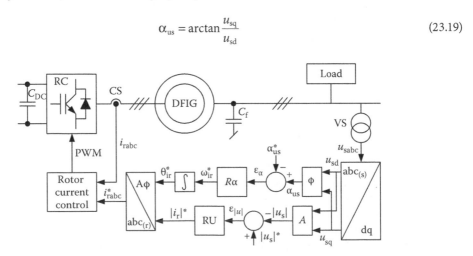

FIGURE 23.12 Method of synchronization of the actual and reference stator voltage vector in standalone operated DFIG.

The angular displacement requires adequate movement of the rotor current vector i_r, and it is obtained by the change of the reference angular speed ω_{ir}^* of the rotor current vector i_r. Rotor current angular speed ω_{ir}^* is changed by the PI type, $R\alpha$ regulator, then the speed is integrated to obtain reference absolute position θ_{ir}^* of the rotor current vector i_r. Therefore, simultaneously adequate frequency and phase of the rotor current is determined. Reference amplitude $|i_r|^*$ and phase θ_{ir}^* of the rotor current vector i_r in the polar frame $A\phi$ can be used for transformation to other coordinate system. Obtained reference signals of the rotor current components can be further used for inner rotor current control loop, necessary to obtain fast response and stability of the system. In practice, no voltage control method of the rotor converter in doubly fed induction machine systems is applied. Rotor current can be controlled by P or PI controllers in rotor-connected three phase $abc_{(r)}$ coordinates as well as in $\alpha\beta_{(r)}$ frame—stationary in relation to the rotor. However, in these frames, the rotor current reference signals are not fixed. Therefore, to eliminate the steady-state error, more advanced controllers than PI have to be used. Simple PI controller can be applied in the structure presented in Figure 23.13, where reference rotor current vector components are transformed to the frame connected with the rotor current vector i_r itself.

The transformation of the polar components $|i_r|^*$, θ_{ir}^* of the reference rotor current vector i_r^* to the xy frame, connected with the rotor current vector is described by (23.20) and (23.21).

$$\begin{bmatrix} i_{r\alpha}^* \\ i_{r\beta}^* \end{bmatrix} = |i_r^*| \begin{bmatrix} \cos\theta_{ir}^* \\ \sin\theta_{ir}^* \end{bmatrix} \tag{23.20}$$

$$\begin{bmatrix} i_{rx}^* \\ i_{ry}^* \end{bmatrix} = i_{r\alpha}^* \begin{bmatrix} \cos\theta_{ir}^* \\ \sin\theta_{ir}^* \end{bmatrix} + i_{r\beta}^* \begin{bmatrix} \sin\theta_{ir}^* \\ -\cos\theta_{ir}^* \end{bmatrix} = \begin{bmatrix} |i_r|^* \\ 0 \end{bmatrix}, \tag{23.21}$$

The characteristic feature of the selected xy frame is, that the reference component i_{rx}^* is equal to the reference amplitude $|i_r|^*$, whereas reference i_{ry}^* component equals zero. Thus, the calculation of (23.20)

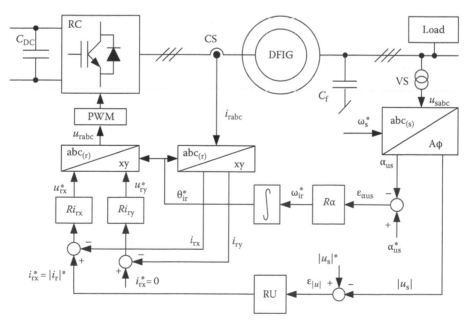

FIGURE 23.13 Control of the stator voltage and rotor current in the polar frame.

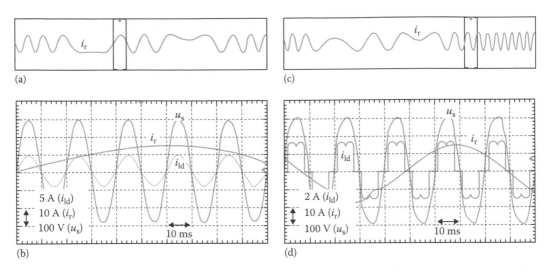

FIGURE 23.14 Oscillograms presenting phase rotor current i_r, stator voltage u_s, and load current i_{ld} during variable-speed operation of standalone DFIG with (a, b) linear and (c, d) nonlinear load.

and (23.21) can be neglected and adequate signals used directly as the references for the rotor current controllers Ri_{rx}, Ri_{ry}. The measured signals of the rotor current i_{ra}, i_{rb}, i_{rc}, are transformed to i_{rx}, i_{ry}, by well-known equations, and used as a feedback signals for the rotor current controllers Ri_{rx}, Ri_{ry}. The structure, with stator voltage and rotor current represented in polar frame, shown in the Figure 23.13, allows for independent control of amplitude and angle of the stator voltage vector. That also means independent control of the amplitude and frequency of the stator voltage, which is the goal in every standalone power generation system, as well as the voltage phase, what is needed for synchronization and soft connection of standalone operated system with power grid [15]. Adjusting the rotor current frequency and phase to the mechanical speed (Figure 23.14a and c provides fixed amplitude and frequency of the stator voltage (Figure 23.14b and d)).

Stator-connected capacitors C_f provide high quality of the stator voltage, despite the nonlinear character of the supplied load (Figure 23.14d). Control of the amplitude $|u_s|$ in polar frame allows to obtain fast response of the system during step loading (Figure 23.15a through c) and step unloading (Figure 23.15d through f) of the standalone DFIG.

Sensorless control of the stator voltage u_s in polar reference frame is robust. Nevertheless, the frequency control is not linear, and what is more important the dependence of the stator voltage phase α_{us} from the rotor current frequency f_{ir} is not precise. For the correct rotor current frequency, the phase α_{us} of the stator voltage u_s, depends only on the phase of the rotor current. However, incorrect rotor current frequency provides permanent increasing or decreasing of the phase angle α_{us} of the stator voltage vector u_s represented in rotating dq frame. The voltage vector phase in dq frame is calculated with the function, which returns the values in the range of $-\pi$ to π. As a result, for incorrect rotor current frequency, the voltage vector angle α_{us} has periodical dependence from the absolute phase angle α_{abs} of the vector (Figure 23.16a).

The structure with series PI controller and integration block is generally able to handle this problem, as for variable-speed operation, including start up of loaded system; the determination of adequate rotor current angular speed ω_{ir} can be done during first period of the α_{us} signal. However, start up of the system has to be obtained with unloaded DFIG, as the load should not be supplied by the voltage with incorrect initial parameters. It is a problem for linear controller to determine the adequate angular speed of the rotor current vector, as after α_{us} overflow the sign of the error $\varepsilon_{\alpha us}$ changes. The error $\varepsilon_{\alpha us}$ has

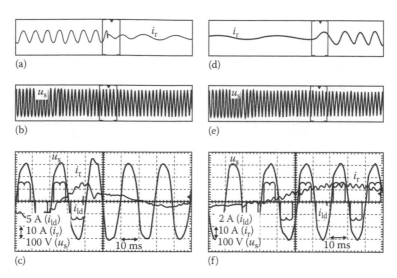

FIGURE 23.15 Oscillograms presenting phase rotor current i_r, stator voltage u_s, and nonlinear load current i_{ld} during (a, b, c) step loading and (d, e, f) step unloading of the standalone DFIG.

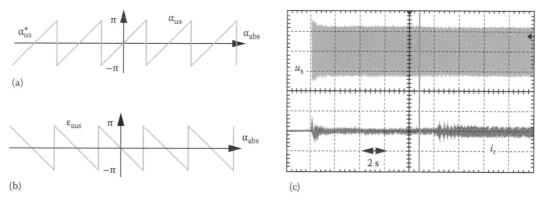

FIGURE 23.16 (a) Reference α_{us}^* and actual α_{us} angle with overflow, (b) periodical error $\varepsilon_{\alpha us}$ of angle controller, and (c) rotor phase current i_r and stator phase voltage u_s during startup of unloaded DFIG.

periodical character with average value equal to zero (Figure 23.16b). It is observed, that during startup of the unloaded DFIG power generation system the determination of adequate rotor current frequency takes more than 10 s (Figure 23.16c).

The frequency control, based on the voltage vector phase in dq frame, can be modified. The use of overflow detection (phase change from $-\pi$ to π or opposite) eliminates the problem with periodical character of the error $\varepsilon_{\alpha us}$. Following the overflow detection, the reference angle α_{us}^* is increased by 2π, or decreased by -2π depending on the sign change of angle α_{us}, and simultaneously saturated at values -2π and 2π if exceeds this range (Figure 23.17a). The phase error $\varepsilon_{\alpha us}$ is also saturated with values $-\pi$, π (Figure 23.17b). This modification eliminates the periodical component (sawtooth wave) of the angle error $\varepsilon_{\alpha us}$, and introduces constant component in two ranges, and linear dependence of the angle error $\varepsilon_{\alpha us}$ from the absolute value of the voltage vector angle α_{us} in dq frame. Modification of reference angle α_{us}^* and the error $\varepsilon_{\alpha us}$ provides shorter transient states (Figure 23.17c), due to the nonperiodical character of the angle controller error $\varepsilon_{\alpha us}$.

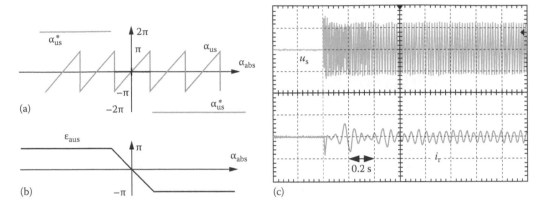

FIGURE 23.17 (a) Reference α_{us}^* and actual α_{us} angle with overflow detection, (b) nonperiodical error $\varepsilon_{\alpha us}$ of angle controller, and (c) rotor phase current i_r and stator phase voltage u_s during start up of unloaded DFIG with modified angle control loop.

References

1. W. Koczara and N. Al Khayat, Variable speed integrated generator VSIG as a modern controlled and decoupled generation system of electrical power, *European Conference on Power Electronics and Applications*), Aalborg, Denmark, September 11–14, 2007, 10 pp.
2. A. Roshan, R. Burgos, A.C. Baisden, F. Wang, and D. Boroyevich, A D-Q frame controller for a full-bridge single phase inverter used in small distributed power generation systems, in *Proceedings of the IEEE APEC*, Anaheim, CA, February 2007, pp. 641–647.
3. A. Kulka, T. Undeland, S. Vazquez, and L.G. Franquelo, Stationary frame voltage harmonic controller for standalone power generation, *European Conference on Power Electronics and Applications (EPE' 07)*, Aalborg, Denmark, September 2–5, 2007, pp. 1–10.
4. R. Datta and V.T. Ranganathan, Direct power control of grid connected wound rotor induction machine without rotor position sensors, *IEEE Transactions on Power Electronics*, 16(3), 390–399, May 2001.
5. T.K.A. Brekken and N. Mohan, Control of a doubly fed induction wind generator under unbalanced grid voltage conditions, *IEEE Transaction on Energy Conversion*, 22(1), 129–135, March 2007.
6. G. Abad, M.A. Rodriguez, and J. Poza, Two-level VSC-based predictive direct power control of the doubly fed induction machine with reduced power ripple at low constant switching frequency, *IEEE Transactions on Energy Conversion*, 23(2), 570–580, June 2008.
7. R. Cardenas, R. Pena, J. Proboste, G. Asher, and J. Clare, MRAS observer for sensorless control of standalone doubly fed induction generators, *IEEE Transactions on Energy Conversion*, 20, 710–718, December 2005.
8. A.K. Jain and V.T. Ranganathan, Wound rotor induction generator with sensorless control and integrated active filter for feeding nonlinear loads in a stand-alone grid, *IEEE Transactions on Industrial Electronics*, 55(1), 218–228, January 2008.
9. Y. Kawabata, T. Oka, E. Ejiogu, and T. Kawabata, Variable speed constant frequency stand-alone power generator using wound-rotor induction machine, *Fourth International Power Electronics and Motion Control Conference (IPEMC 2004)*, Xian, China, 2004, Vol. 3, pp. 1778–1784.
10. S. Breban et al., Variable speed small hydro power plant connected to AC grid or isolated loads, *European Power Electronics and Drives Association Journal*, 17(4), 29–36, January 2008.
11. C. Mi, M. Filippa, J. Shen, and N. Natarajan, Modeling and control of a variable-speed constant-frequency synchronous generator with brushless exciter, *IEEE Transactions on Industrial Applications*, 40(2), 565–573, March 2004.

12. M. Chomat, L. Schreier, and J. Bendl, Control method for doubly fed machine supplying unbalanced load, *9th European Conference on Power Electronics and Applications (EPE)*, Toulouse, France, 2003 (CD Proceedings).

13. G. Iwanski and W. Koczara, Sensorless direct voltage control of the stand-alone slip-ring induction generator, *IEEE Transactions on Industrial Electronics*, 54(2), 1237–1239, April 2007.

14. L. Jiao, B.T. Ooi, G. Joos, and F. Zhou, Doubly-fed induction generator (DFIG) as a hybrid of asynchronous and synchronous machines, *Electric Power System Research*, 76(1–3), 33–37, September 2005.

15. G. Iwanski and W. Koczara, DFIG based power generation system with UPS function for variable speed applications, *IEEE Transactions on Industrial Electronics*, 55(8), 3047–3054, August 2008.

24
FOC: Field-Oriented Control

Emil Levi
*Liverpool John Moores
University*

24.1 Introductory Considerations

Variable speed electric drives are nowadays utilized in almost every walk of life, from the most basic devices, such as hand-held tools and other home appliances, to the most sophisticated ones, such as electric propulsion systems in cruise ships and high-precision manufacturing technologies. Depending on the application, the control variable may be the motor's torque, speed, or position of the rotor shaft. In the most demanding applications, the requirement is to be able to control the electric machine's electromagnetic torque in order to be able to provide a controlled transition from one operating speed (position) to another speed (position). This means that the control of the drive must be able to achieve desired dynamic response of the controlled variable in a minimum time interval. This can only be achieved if the motor's electromagnetic torque can be practically instantaneously stepped from the previous steady-state value to the maximum allowed value, which is in turn governed by the allowed maximum current. Variable speed electric drives that are capable of achieving such a performance are usually called high-performance drives, since the control is effective not only in steady state but in transient as well. Common features of all high-performance drives are that they require information on instantaneous rotor position (speed), operation is with closed-loop control, and the machine is supplied from a power electronic converter. Applications that necessitate use of a high-performance drive are numerous and include robotics, machine tools, elevators, rolling mills, paper mills, spindles, mine winders, electric traction, electric and hybrid electric vehicles, and the like.

A principal schematic outlay of a high-performance electric drive is shown in Figure 24.1 and it applies equally to all types of electric machinery. Electromagnetic torque of an electric machine can be expressed as a product of the flux-producing current and torque-producing current, so that the control system in Figure 24.1 has two parallel paths. Flux-producing current reference is shown as a constant; however, this may or may not be the case, as discussed later. Torque-producing current is in principle the output of the torque controller. However, torque controller of Figure 24.1 is usually not present in high-performance drives, since the torque-producing current reference can be obtained directly from

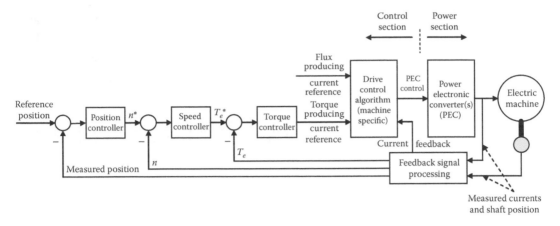

FIGURE 24.1 Schematic outlay of a high-performance variable speed electric drive ($n = (60/2\pi)\omega$).

the reference torque by means of a simple scaling (or the output of the speed controller can be made to be directly the torque-producing current reference). This is so since the torque and the torque-producing current are, when a high-performance control algorithm is applied, related through a constant. The control structure in Figure 24.1 is composed of cascaded controllers (typically of proportional plus integral [PI] type). An asterisk stands for reference quantities, while θ, ω, and T_e designate further on instantaneous values of electrical rotor position, electrical rotor angular speed (speed is shown in figures as n in rpm; this is not to be confused with phase number n) and electromagnetic torque developed by the motor, respectively. The cascaded structure is based on the fundamental equations that govern rotor rotation, which are for a machine with P pole pairs given with (T_L stands for load torque, k is the friction coefficient, and J is the inertia of rotating masses)

$$T_e - T_L = \frac{J}{P}\frac{d\omega}{dt} + \frac{1}{P}k\omega \tag{24.1a}$$

$$\theta = \int \omega\, dt \tag{24.1b}$$

High-performance drives typically involve measurement of the rotor position (speed) and motor supply currents, as indicated in Figure 24.1. Since the machine's torque is governed by currents rather than voltages, measured currents are used in the block "Drive control algorithm" to incorporate the closed-loop current control (CC) algorithm. What this means is that the power electronic converter is current-controlled, so that applied voltages are such as to minimize the errors in the current tracking.

Until the early 1980s of the last century, the separately excited dc motor was the only available electric machine that could be used in a high-performance drive. A dc motor is by virtue of its construction ideally suited to meeting control specifications for high performance. However, due to numerous shortcomings, dc motor drives are nowadays replaced with ac drives wherever possible. To explain the requirements on high-performance control, consider a separately excited dc motor. Stator of such a machine can be equipped with either a winding (excitation winding) or with permanent magnets. The role of the stator is to provide excitation flux in the machine, which is in the case of permanent magnets constant, while it is controllable if there is an excitation winding. For the sake of explanation, it is assumed that the stator carries permanent magnets, which provide constant flux, ψ_m, so that the upper input into the "Drive control algorithm" block in Figure 24.1 does not exist. The permanent magnet flux is stationary in space and it acts along a magnetic axis, as schematically illustrated in Figure 24.2, where the cross section of the machine is shown. Rotor of the machine carries a winding (armature winding)

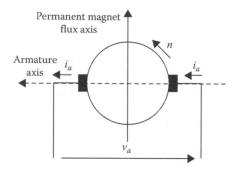

FIGURE 24.2 Cross-section of a two-pole permanent magnet excited dc machine, illustrating mutual position of the flux axis and armature axis.

access to which is provided by means of stationary brushes and an assembly on the rotor, called commutator. The supply is from a dc source (in principle, a power electronic converter of dc–dc or ac–dc type, depending on the application), which provides dc armature current as the input into the rotor winding. The brushes are placed in an axis orthogonal to the permanent magnet flux axis (Figure 24.2). Since the brushes are stationary, flux and the armature terminal current are at all times at 90°. It is this orthogonal position of the torque-producing current (armature current i_a) and the permanent magnet flux ψ_m that enables instantaneous torque control of the machine by means of instantaneous change of the armature current. This follows from the electromagnetic torque equation of the machine, which is given by (K is a constructional constant)

$$T_e = K\psi_m i_a \tag{24.2}$$

It also follows that since the torque-producing (armature) current and the torque are related through a constant, armature current reference in Figure 24.1 can be obtained by scaling the torque reference with the constant (which is normally embedded in the speed controller PI gains), so that the torque controller is not required. On the basis of these explanations and (24.2) it is obvious that the machine's torque can be stepped if armature current can be stepped. This of course requires current-controlled operation of the armature dc supply, so that the armature voltage is varied in accordance with the armature current requirements.

It is important to remark here that, inside the rotor winding, the current is actually ac. It has a frequency equal to the frequency of rotor rotation, since the commutator converts dc input into ac output current and therefore performs, together with fixed stationary brushes, the role of a mechanical inverter (in motoring operation; in generation it is the other way round, so that the commutator acts as a rectifier). As the rotor winding is rotating in the stationary permanent magnet flux, a rotational electromotive force (emf) is induced in the rotor winding according to the basic law of electromagnetic induction, $e = K\psi_m\omega$.

The machine in Figure 24.2, with constant permanent magnet excitation, can operate with variable speed in the base speed region only (i.e., up to the rated speed), since operation above base speed (field weakening region) requires the means for reduction of the flux in the machine. This is so since the armature voltage cannot exceed the rated voltage of the machine, which corresponds to rated speed, rated torque operation. To operate at a speed higher than rated, one has to keep the induced emf as for rated speed operation. Since speed goes up flux must come down, something that is not possible if permanent magnets are used but is achievable if there is an excitation winding. In such a case "flux-producing current reference" of Figure 24.1 has a constant rated value up to the rated speed and is further gradually reduced to achieve operation with speeds higher than rated (hence the name, field weakening region). However, due to the orthogonal position of the flux and armature axes, flux and torque control do not mutually impact on each other as long as the flux-producing current is kept constant. It is hence said that torque and flux control are decoupled (or independent) and this is the normal mode of operation in the base speed region. Once when field weakening region is entered, dynamic decoupled flux and torque control is not possible any more since reduction of the flux impacts on torque production.

The preceding discussion can be summarized as follows: high-performance operation requires that torque of a motor is controllable in real time; instantaneous torque of a separately excited dc motor is directly controllable by armature current as flux and torque control are inherently decoupled; independent flux and torque control are possible in a dc machine due to its specific construction that involves commutator with brushes whose position is fixed in space and perpendicular to the flux position; instantaneous flux and torque control require use of current controlled dc source(s); current and position (speed) sensing is necessary in order to obtain feedback signals for real-time control.

Substitution of dc drives with ac drives in high-performance applications has become possible only relatively recently. From the control point of view, it is necessary to convert an ac machine into its equivalent dc counterpart so that independent control of two currents yields decoupled flux and torque control. The set of control schemes that enable achievement of this goal is usually termed "field-oriented control (FOC)" or "vector control" methods. The principal difficulty that arises in all multiphase machines (with a phase number $n \geq 3$) is that the operating principles are based on the rotating field (flux) in the machine (note that the machines customarily called two-phase machines are in essence four-phase machines, since spatial displacement of phases is 90°; in two-phase machines phase pairs in spatial opposition are connected into one phase). As a consequence, the flux that was stationary in a separately excited dc machine is now rotating in the cross section of the machine at a synchronous speed, determined with the stator winding supply frequency. Thus, the stationary flux axis of Figure 24.2 now becomes an axis that rotates at synchronous speed. Since decoupled flux and torque control require that flux-producing current is aligned with the flux axis, while the torque-producing current is in an axis perpendicular to the flux axis, the control of a multiphase machine has to be done using a set of orthogonal coordinates that rotates at the synchronous speed (speed of rotation of the flux in the machine). The situation is further complicated by the fact that, in a multiphase machine, there are in principle three different fluxes (or flux linkages, as they will be called further on), stator, air-gap, and rotor flux linkage. While in steady-state operation they all have synchronous speed of rotation, the instantaneous speeds during transients differ. Hence a decision has to be made with regard to which flux the control should be performed. Basic outlay of the drive remains as in Figure 24.1. However, while in the case of a dc drive the block "drive control algorithm" in essence contains only current controllers, in the case of a multiphase ac machine this block becomes more complicated. The reason is that using design of the drive control as for a dc machine, where there exist flux and torque-producing dc current references, means that the control will operate in a rotating set of coordinates (rotating reference frame). In other words, current components used in the control (flux- and torque-producing currents) are not currents that physically exist in the machine. Instead, these are the fictitious current components that are related to physically existing ac phase currents through a coordinate transformation. This coordinate transformation produces, from dc current references, ac current references for the supply of the stator winding of a multiphase machine. Thus, what commutator with brushes does in a dc machine (dc–ac conversion) has to be done in ac machines using a mathematical transformation in real time.

Fundamental principles of FOC (vector control), which enable mathematical conversion of an ac multiphase machine into an equivalent dc machine, were laid down in the early 1970s of the last century for both induction and synchronous machines [1–5]. What is common for both dc and ac high-performance drives is that the supply sources are current-controlled power electronic converters, current feedback and position (speed) feedback are required, and torque is controlled in real time. However, stator winding of multiphase ac machines is supplied with ac currents, which are characterized with amplitude, frequency, and phase rather than just with amplitude as in dc case. Thus, an ac machine has to be fed from a source of variable output voltage, variable output frequency type. Power electronic converters of dc–ac type (inverters) are the most frequent source of power in high-performance ac drives.

Application of vector-controlled ac machines in high-performance drives became a reality in the early 1980s and has been enabled by developments in the areas of power electronics and microprocessors. Control systems that enable realization of decoupled flux and torque control in ac motor drives are relatively complex, since they involve a coordinate transformation that has to be executed in real time. Application of microprocessors or digital signal processors is therefore mandatory.

In what follows the basic principles of FOC are summarized. The discussion is restricted to the multiphase machines with sinusoidal magnetomotive force distribution. The range of available multiphase ac machine types is huge and includes both singly-fed and doubly-fed (with or without slip rings) machines. The coverage is here restricted to singly-fed machines, with supply provided at the stator side. The considered machine types are induction machines with a squirrel-cage rotor winding, permanent magnet synchronous machines (PMSMs) (with surface mounted and interior permanent magnets and without rotor cage, i.e., damper winding), and synchronous reluctance (Syn-Rel) motors (without damper winding). This basically encompasses the most important types of ac machines as far as the servo (high performance) drives are concerned. FOC of synchronous motors with excitation and damper windings (used in the high-power applications) and of slip ring (wound rotor) induction machines (used as generators in wind electricity generation) is thus not covered and the reader is referred to the literature referenced shortly for more information. Considerations here cover the general case of a multiphase machine with three or more phases on stator ($n \geq 3$) since the basic field–oriented control principles are valid in the same manner regardless of the actual number of phases. It has to be noted that the complete theory of vector control has been developed under the assumption of an ideal variable voltage, variable frequency, symmetrical and balanced sinusoidal stator winding multiphase supply. Hence, the fact that such a supply does not exist and a nonideal (power electronic) supply has to be used instead is just a nuisance, which has no impact on the control principles (this being in huge contrast with another group of high-performance control schemes for multiphase electric drives, direct torque control (DTC) schemes, where the whole idea of the control is based around the utilization of the nonideal power electronic converter as the supply source; DTC is beyond the scope of this chapter).

Since the 1980s of the last century, FOC has been extensively researched and has by now reached a mature stage, so that it is widely applied in industry when high performance is required. It has also been treated in a number of textbooks [6–25] at varying levels of complexity and detail. Assuming that the machine is operated as a speed-controlled drive, a generic schematic block diagram of a field-oriented multiphase singly-fed machine in closed-loop speed control mode can be represented, as shown in Figure 24.3. Since the machine is supplied from stator side only, flux- and torque-producing current references refer now to stator current components and are designated with indices d and q. Here d applies to the flux axis and q to the axis perpendicular to the d-axis, while index s stands for stator. This scheme is valid for

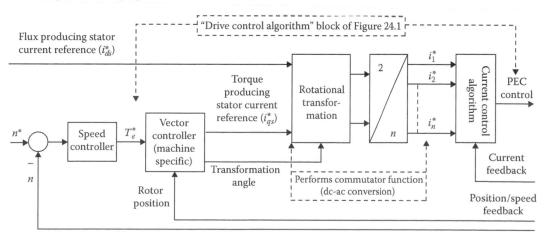

FIGURE 24.3 Basic vector control scheme for a multiphase machine with CC in the stationary reference frame.

both synchronous and induction machines and the type of the machine impacts on the setting of the flux-producing current reference and on the structure of the "vector controller" block. It is assumed in Figure 24.3 that CC algorithm is applied to the machine's stator phase currents (so-called current control in the stationary reference frame; phases are labeled with numerical indices 1 to *n*). As indicated in Figure 24.3, blocks "CC algorithm," "vector controller," "Rotational transformation" and "2/*n*" are now constituent parts of the block "Drive control algorithm" of Figure 24.1. Blocks "Rotational transformation" and "2/*n*" take up the role of the commutator with brushes in dc machines, by doing the dc–ac conversion (inversion) of control signals (flux- and torque-producing stator current references).

Vector control schemes for synchronous machines are, in principle, simpler than the equivalent ones for an induction machine. This is so since the frequency of the stator-winding supply uniquely determines the speed of rotation of a synchronous machine. If there is excitation, it is provided by permanent magnets (or dc excitation current in the rotor winding). Rotor carries with it the excitation flux as it rotates and the instantaneous spatial position of the rotor flux is always fixed to the rotor. Hence, if rotor position is measured, position of the excitation flux is known. Such a situation leads to relatively simple vector control algorithms for PMSMs, which are therefore considered first. The situation is somewhat more involved in Syn-Rel machines. Rotor is of salient pole structure but without either magnets or excitation winding, so that excitation flux stems from the ac supply of the multiphase stator winding. By far the most complex situation results in induction machines where not only that the excitation flux stems from stator winding supply, but the rotor rotates asynchronously with the rotating field. This means that, even if the rotor position is measured, position of the rotating field in the machine remains unknown. Vector control of induction machines is thus the most complicated case and is considered last.

The starting point for derivation of an FOC scheme is, regardless of the type of the multiphase machine, a mathematical model obtained using transformations of the general theory of electrical machines. For all synchronous machine types, such a model is always developed in the common reference frame firmly fixed to the rotor, while for induction machines the speed of the common reference frame is arbitrarily selectable. All the standard assumptions of the general theory apply: those that are the most relevant further on are the assumption of sinusoidal field (flux) spatial distribution and constancy of all the parameters of the machine, including magnetizing inductance(s) where applicable (meaning that the nonlinearity of the ferromagnetic material is neglected).

As noted already, the FOC schemes are developed assuming ideal sinusoidal supply of the machine. If the control scheme is of the form illustrated in Figure 24.3, where CC is performed using stator phase currents, then the current-controlled voltage source (say, an inverter) is treated as an ideal current source and the machine is said to be current fed. In simple words, it is assumed that the multiphase power supply can deliver any required stator voltage, such that the actual stator currents perfectly track the reference currents of Figure 24.3. This greatly simplifies the overall vector control schemes, since dynamics of the stator (stator voltage equations) can be omitted from consideration. Note that for an *n*-phase machine with a single neutral point, the control scheme of Figure 24.3 implies existence of (*n*–1) current controllers. These are typically of hysteresis or ramp-comparison type and are the same regardless of the ac machine type. CC of the supply is not considered here, nor are the PWM control schemes that are relevant when CC is not in the stationary reference frame. It is therefore assumed further on that whatever the machine type and the actual FOC scheme used, the source is capable of delivering ideal sinusoidal stator currents (or voltages, as discussed shortly).

24.2 Field-Oriented Control of Multiphase Permanent Magnet Synchronous Machines

Consider a multiphase star-connected PMSM, with spatial shift between any two consecutive phases of $2\pi/n$, and let the phase number *n* be an odd number without any loss of generality. The neutral point of the stator winding is isolated. Permanent magnets are on the rotor and they can be surface mounted (surface-mounted permanent magnet synchronous machine [SPMSM]) or embedded in the rotor (interior permanent magnet synchronous machine [IPMSM]). In the former case the air-gap of the machine can be considered as uniform, while in the latter case the air-gap length is variable, since permanent magnets have a permeability that is practically the same as for the air. Thus SPMSMs are characterized with a rather large air gap (which will make operation in the field weakening region difficult, as discussed later), while the air gap of the IPMSMs is small, but the magnetic reluctance is variable, due to the saliency effect produced by the embedded magnets. Rotor of the machine does not carry any windings, regardless of the way in which the magnets are placed.

Mathematical model of an IPMSM can be given in the common reference frame firmly attached to the rotor with the following equations:

$$v_{ds} = R_s i_{ds} + \frac{d\psi_{ds}}{dt} - \omega\psi_{qs}$$

$$v_{qs} = R_s i_{qs} + \frac{d\psi_{qs}}{dt} + \omega\psi_{ds}$$

(24.3)

$$v_{xis} = R_s i_{xis} + \frac{d\psi_{xis}}{dt} \quad i = 1\ldots(n-3)/2$$

$$v_{yis} = R_s i_{yis} + \frac{d\psi_{yis}}{dt} \quad i = 1\ldots(n-3)/2$$

$$v_{0s} = R_s i_{0s} + \frac{d\psi_{0s}}{dt}$$

(24.4)

$$\psi_{ds} = L_d i_{ds} + \psi_m$$

$$\psi_{qs} = L_q i_{qs}$$

(24.5)

$$\psi_{xis} = L_{ls} i_{xis} \quad i = 1\ldots(n-3)/2$$

$$\psi_{yis} = L_{ls} i_{yis} \quad i = 1\ldots(n-3)/2$$

$$\psi_{0s} = L_{ls} i_{0s}$$

(24.6)

$$T_e = P\left[\psi_m i_{qs} + (L_d - L_q) i_{ds} i_{qs}\right]$$

(24.7)

where index l stands for leakage inductance, v, i, and ψ denote voltage, current, and flux linkage, respectively, d and q stand for the components along permanent magnet flux axis (d) and the axis perpendicular to it (q), and s denotes stator. Inductances L_d and L_q are stator winding self-inductances along d- and q-axis.

Voltage and flux linkage equations (24.3) through (24.6) represent an n-phase machine in terms of sets of new n variables, obtained after transforming the original machine model in phase-variable domain by means of a power invariant transformation matrix that relates original phase variables and new variables through

$$[f_{dq}] = [D][C][f_{1,2,\ldots,n}] = [T][f_{1,2,\ldots,n}]$$

$$[T] = [D][C]$$

(24.8)

where f stands for voltage, current, or flux linkage and $[D]$ and $[C]$ are the rotational transformation matrix and decoupling transformation matrix (block "2/n" in Figure 24.3) for stator variables, respectively. For an n-phase machine with an odd number of phases, these matrices are

$$[D] = \begin{array}{c} ds \\ qs \\ x_{1s} \\ y_{1s} \\ \cdots \\ 0_s \end{array} \left[\begin{array}{cccccc} \cos\theta_s & \sin\theta_s & 0 & 0 & \cdots & 0 \\ -\sin\theta_s & \cos\theta_s & 0 & 0 & \cdots & 0 \\ 0 & 0 & 1 & 0 & \cdots & 0 \\ 0 & 0 & 0 & 1 & \cdots & 0 \\ \cdots & \cdots & \cdots & \cdots & \cdots & 0 \\ 0 & 0 & 0 & 0 & \cdots & 1 \end{array} \right] \tag{24.9}$$

$$[C] = \sqrt{\frac{2}{n}} \begin{array}{c} \alpha \\ \beta \\ x_1 \\ y_1 \\ x_2 \\ y_2 \\ \cdots \\ x_{(n-3)/2} \\ y_{(n-3)/2} \\ 0 \end{array} \left[\begin{array}{cccccc} 1 & \cos\alpha & \cos 2\alpha & \cos 3\alpha & \cdots & \cos(n-1)\alpha \\ 0 & \sin\alpha & \sin 2\alpha & \sin 3\alpha & \cdots & \sin(n-1)\alpha \\ 1 & \cos 2\alpha & \cos 4\alpha & \cos 6\alpha & \cdots & \cos 2(n-1)\alpha \\ 0 & \sin 2\alpha & \sin 4\alpha & \sin 6\alpha & \cdots & \sin 2(n-1)\alpha \\ 1 & \cos 3\alpha & \cos 6\alpha & \cos 9\alpha & \cdots & \cos 3(n-1)\alpha \\ 0 & \sin 3\alpha & \sin 6\alpha & \sin 9\alpha & \cdots & \sin 3(n-1)\alpha \\ \cdots & \cdots & \cdots & \cdots & \cdots & \cdots \\ 1 & \cos\left[(n-1)/2\right]\alpha & \cos 2\left[(n-1)/2\right]\alpha & \cos 3\left[(n-1)/2\right]\alpha & \cdots & \cos\left[(n-1)^2/2\right]\alpha \\ 0 & \sin\left[(n-1)/2\right]\alpha & \sin 2\left[(n-1)/2\right]\alpha & \sin 3\left[(n-1)/2\right]\alpha & \cdots & \sin\left[(n-1)^2/2\right]\alpha \\ 1/\sqrt{2} & 1/\sqrt{2} & 1/\sqrt{2} & 1/\sqrt{2} & \cdots & 1/\sqrt{2} \end{array} \right]$$

$$\tag{24.10}$$

Due to the selected power-invariant form of the transformation matrices, the inverse transformations are governed with $[T]^{-1} = [T]^t$, $[D]^{-1} = [D]^t$, $[C]^{-1} = [C]^t$. Angle of transformation θ_s in (24.9) is identically equal to the rotor electrical position, so that

$$\theta_s = \theta = \int \omega \, dt \tag{24.11}$$

As the d-axis of the common reference frame then coincides with the instantaneous position of the permanent magnet flux, this means that the given model is already expressed in the common reference frame firmly attached to the permanent magnet flux.

The pairs of d-q equations (24.3) and (24.5) constitute the flux/torque-producing part of the model, as is evident from torque equation (24.7). Since in a star-connected winding, with isolated neutral, zero-sequence current cannot flow, the last equation of (24.4) and (24.6) can be omitted. The model then contains, in addition to the d-q equations, $(n-3)/2$ pairs of x-y component equations in (24.4) and (24.6), which do not contribute to the torque production and are therefore not transformed with rotational transformation (24.9) (i.e., their form is the one obtained after application of decoupling transformation (24.10) only). It has to be noted however, that the reference value of zero for all of these components (which will exist in the model for $n \geq 5$) is implicitly included in the control scheme of Figure 24.3, since reference phase currents are built from d-q current references only. Equations 24.4 and 24.6 are of the same form for all the multiphase ac machines considered here (all types of synchronous and induction machines).

For a SPMSM machine, the set of equations (24.3), (24.5), and (24.7) further simplifies since the air-gap is regarded as uniform and hence $L_s = L_d = L_q$. Thus (24.3) and (24.5) reduce to

$$v_{ds} = R_s i_{ds} + L_s \frac{di_{ds}}{dt} - \omega L_s i_{qs}$$

$$v_{qs} = R_s i_{qs} + L_s \frac{di_{qs}}{dt} + \omega(L_s i_{ds} + \psi_m)$$

(24.12)

while the torque equation takes the form

$$T_e = P\psi_m i_{qs}$$

(24.13)

By comparing (24.13) with (24.2), it is obvious that the form of the torque equation is identical as for a separately excited dc motor. The only but important difference is that the role of the armature current is now taken by the q-axis stator current component. Assuming that the machine is current-fed (i.e., CC is executed in the stationary reference frame), stator current dynamics of (24.12) are taken care of by the fast CC loops and the global control scheme of Figure 24.3 becomes as in Figure 24.4. Since the machine has permanent magnets that provide excitation flux, there is no need to provide flux from the stator side and the stator current reference along d-axis is set to zero. According to (24.11), the measured rotor electrical position is the transformation angle of (24.9).

The control scheme of Figure 24.4 is a direct analog of the corresponding control scheme of permanent magnet excited dc motors, where the role of the commutator with brushes is now replaced with the mathematical transformation $[T]^{-1}$. A few remarks are due. Figure 24.4 includes a limiter after the speed controller. This block is always present in high-performance drives (although it was not included in Figures 24.1 and 24.3, for simplicity) and limiting ensures that the maximum allowed stator current (normally governed by the power electronic converter) is not exceeded. Next, as already noted, a constant that relates torque and stator q-axis current reference according to (24.13) and which is shown in Figure 24.4 will normally be incorporated into speed controller gains, so that the limited output of the speed controller will actually directly be the stator q-axis current reference.

The control scheme of Figure 24.4 satisfies for control in the base speed region. If it is required to operate the machine at speeds higher than rated, it is necessary to weaken the flux so that the voltage applied to the machine does not exceed the rated value. However, permanent magnet flux cannot be changed and the only way to achieve operation at speeds higher than rated is to keep the term $\omega(L_s i_{ds} + \psi_m)$ of (24.12)

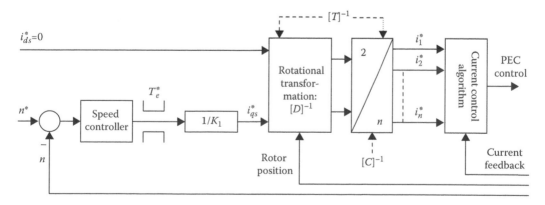

FIGURE 24.4 Vector control of a PMSM with surface-mounted magnets in the base speed region ($K_1 = P\psi_m$).

constant and equal to its value at rated speed (index n stands further on for rated values), $\omega_n \psi_m$. Hence, at any speed higher than rated, stator d-axis current reference must take a negative value, such that

$$i_{ds}^* = -\frac{|\omega_n - \omega|}{\omega}\frac{\psi_m}{L_s} \quad \omega > \omega_n \tag{24.14}$$

Achievable field weakening region with SPMSMs is rather limited since the required d-axis current reference of (24.14) becomes significant at rather low values of speed above base speed, due to the large effective air-gap (and hence small inductance value). What this means is that the available current limit gets quickly fully utilized by the d-axis current, thus leaving no margin for the q-axis current (and hence torque production).

An illustration of the machine windings and position of the axes used in the FOC scheme of Figure 24.4 is shown in Figure 24.5. A three-phase machine is assumed and the stator phases are labeled as a, b, c (rather than 1,2,3); permanent magnets are represented with a fictitious field (f) winding along d-axis and stator current space vector, defined as

$$\underline{i}_s = i_{ds} + ji_{qs} = \sqrt{i_{ds}^2 + i_{qs}^2}\,\exp(j\delta)$$
$$i_s = \sqrt{i_{ds}^2 + i_{qs}^2} \tag{24.15}$$

is shown in an arbitrary position, as though it has positive both d- and q-axis components. As noted, in the base speed region stator d-axis current component is zero, meaning that the complete stator current space vector of (24.15) is aligned with the q-axis. Stator current is thus at 90° ($\delta = 90°$) with respect to the flux axis in motoring, while the angle is −90° ($\delta = -90°$) during braking. In the field weakening d-axis current is negative to provide an artificial effect of the reduction in the flux linkage of the stator winding, so that $\delta > 90°$ in motoring. If the machine operates in field weakening region, simple q-axis current limiting of Figure 24.4 is not sufficient any more, since the total stator current of (24.15) must not exceed the prescribed limit, while d-axis current is now not zero any more. Hence, the q-axis current must have a variable limit, governed by the maximum allowed stator current i_{smax} and the value of the d-axis current command of (24.14). A more detailed discussion is available in [19].

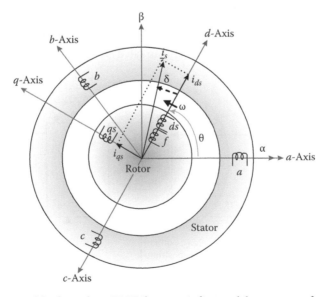

FIGURE 24.5　Illustration of the three-phase SPMSM's stator windings and the common reference frame used in FOC.

In PMSMs, since there is no rotor winding, flux linkage in the air-gap and rotor is taken as being the same and this is the flux linkage with which the reference frame has been aligned for FOC purposes in Figure 24.4. Schematic representation of Figure 24.5 is the same regardless of the number of stator phases as long as the CC is implemented, as shown in Figure 24.4. The only thing that changes is the number of stator winding phases and their spatial shift.

An illustration of a three-phase SPMSM performance, obtained from an experimental rig, is shown next. PI speed control algorithm is implemented in a PC and operation in the base speed region is studied. Stator d-axis current reference is thus set to zero at all times, so that the drive operates in the base speed region only (rated speed of the motor is 3000 rpm). The output of the speed controller, stator q-axis current command, is after D/A conversion supplied to an application-specific integrated circuit that performs the coordinate transformation $[T]^{-1}$ of Figure 24.4. Outputs of the coordinate transformation chip, stator phase current references, are taken to the hysteresis current controllers that are used to control a 10 kHz switching frequency IGBT voltage source inverter. Stator currents are measured using Hall-effect probes. Position is measured using a resolver, whose output is supplied to the resolver to digital converter (another integrated circuit). One of the outputs of the R/D converter is the speed signal (in analog form) that is taken to the PC (after A/D conversion) as the speed feedback signal for the speed control loop. Speed reference is applied in a stepwise manner. Speed PI controller is designed to give an aperiodic speed response to application of the rated speed reference (3000 rpm) under no-load conditions, using the inertia of the SPMSM alone. Figure 24.6 presents recorded speed responses to step speed references equal to 3000 and 2000 rpm. Speed command is always applied at 0.25 s. As can be seen from Figure 24.6, speed response is extremely fast and the set speed is reached in around 0.25–0.3 s without any overshoot.

SPMSM is next mechanically coupled to a permanent magnet dc generator (load), whose armature terminals are left open. An effective increase in inertia is therefore achieved, of the order of 3 to 1. As the dc motor rated speed is 2000 rpm, testing is performed with this speed reference, Figure 24.7. Operation in the current limit now takes place for a prolonged period of time, as can be seen in the accompanying q-axis current reference and phase a current reference traces included in Figure 24.7 for the 2000 rpm reference speed. Due to the increased inertia, duration of the acceleration transient is now considerably longer, as is obvious from the general equation of rotor motion (24.1a). In final steady state, stator q-axis current reference is of constant nonzero value, since the motor must develop some torque (consume some real power) to overcome the mechanical losses according to (24.1a), as well as the core losses in the ferromagnetic material of the stator.

If a machine's electromagnetic torque can be instantaneously stepped from a constant value to the maximum allowed value, then the speed response will be practically linear, as follows from (24.1a). Stepping of torque requires stepping of the q-axis current in the machine. Due to the very small time constant of the stator winding (very small inductance) in a SPMSM, stator q-axis current component

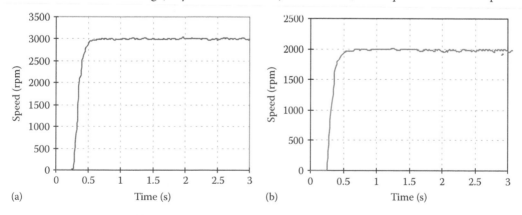

FIGURE 24.6 Experimentally recorded SPMSM's speed response to step speed reference application under no-load conditions: (a) 3000 rpm and (b) 2000 rpm. (From Ibrahim, Z. and Levi, E., *EPE J.*, 12(2), 37, 2002. With permission.)

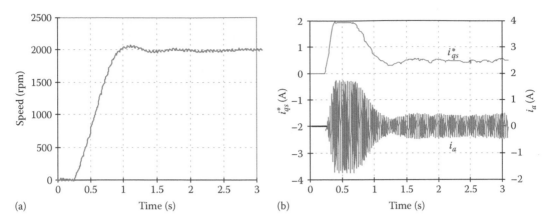

(a)   Time (s)   (b)   Time (s)

FIGURE 24.7 Response of the SPMSM to 2000 rpm speed reference with a substantially increased inertia: (a) speed and (b) stator *q*-axis current reference, and phase *a* current reference. (From Ibrahim, Z. and Levi, E., *EPE J.*, 12(2), 37, 2002. With permission.)

changes extremely quickly (although not instantaneously) and, as a consequence, speed response to step change of the speed reference is practically linear during operation in the torque (stator *q*-axis current) limit. This is evident in Figures 24.6 and 24.7.

An important property of any high-performance drive is its load rejection behavior (i.e., response to step loading/unloading). For this purpose, during operation of the SPMSM with constant speed reference of 1500 rpm the armature terminals of the dc machine, used as the load, are suddenly connected to a resistance in the armature circuit, thus creating an effect of step load torque application. Speed response, recorded during the sudden load application at 1500 rpm speed reference, is shown in Figure 24.8. Since load torque application is a disturbance, the speed inevitably drops during the transient. How much the speed will dip from the reference value depends on the design parameters of the speed controller and on the maximum allowed stator current value, since this is directly proportional to the maximum electromagnetic torque value.

Control scheme of Figure 24.4, which in turns corresponds to the one of Figure 24.1, assumes that the CC is in the stationary reference frame, exercised upon machine's phase currents. This was the preferred solution in the 1980s and early 1990s of the last century, which was based on utilization of digital electronics for the control part, up to the creation of stator phase current references. The CC algorithm for power electronic converter (PEC) control was typically implemented using analog electronics. Due to the rapid developments in the speed of modern microprocessors and DSPs and reduction in their cost, a completely digital solution

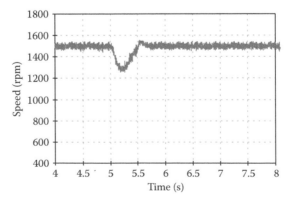

FIGURE 24.8 Speed response at constant speed reference of 1500 rpm to step loading of the SPMSM. (From Ibrahim, Z. and Levi, E., *EPE J.*, 12(2), 37, 2002. With permission.)

is predominantly utilized nowadays. This means that the outputs of the DSP (or a microprocessor) are basically firing signals for the PEC semiconductor switches. Such a solution normally involves a different CC scheme, which is now realized in the rotating coordinates. In simple words, rather than controlling ac phase currents, one now controls their d–q components. This requires that the stator voltage equations (24.12) are now included in the consideration, since the ultimate output of the control system are basically semiconductor switch control signals. Thus, the vector control system generates at first d–q axis stator current references, in the same manner as in Figure 24.4, but this is followed now by stator d–q axis CC in the rotating reference frame. Inspection of (24.12) shows that there is a coupling between stator d–q current and voltage components. The outputs of stator d–q current controllers are therefore defined, using (24.12), as

$$v'_{ds} = R_s i_{ds} + L_s \frac{di_{ds}}{dt}$$

$$v'_{qs} = R_s i_{qs} + L_s \frac{di_{qs}}{dt}$$

(24.16)

and the total stator voltage d–q references are created by summing the outputs of the PI current controllers with decoupling voltages, according to

$$v^*_{ds} = v'_{ds} + e_d$$

$$v^*_{qs} = v'_{qs} + e_q$$

(24.17)

Comparison of (24.16) and (24.17) with (24.12) shows that the decoupling voltages e are in the general case given by

$$e_d = -\omega L_s i_{qs}$$

$$e_q = \omega \left(L_s i_{ds} + \psi_m \right)$$

(24.18)

If the machine operates in the base speed region only, with zero stator d-axis current reference setting, decoupling voltage along q-axis contains only the rotational-induced emf due to the permanent magnet flux. Calculations according to (24.18) require information on the speed of rotation (which is available), knowledge of stator inductance, and stator d–q axis current components. Either values obtained from measured phase currents or reference values can be used as d–q current components in (24.18). In this context, it is important to note that, since CC is now based on d–q stator current components, it is necessary to convert measured stator phase currents into rotating reference frame using coordinate transformation $[T]$. What this means is that the control system requires two coordinate transformations, rather than one as the case was in Figure 24.4. Phase currents are transformed into d–q components, while reference d–q axis stator voltages are transformed into phase voltage references (i.e., one transformation is required in each direction).

Another important remark is that, at least theoretically, it appears that when CC is implemented in the rotating reference frame, it is necessary to use only two current controllers (d–q pair) regardless of the number of phases of the machine. This is, however, in practice not sufficient. In an n-phase machine, there are (n–1) independent currents, and hence any nonideal behavior of the machine/PEC will lead to poor control if there are only two current controllers. This issue will be addressed in detail in the section on induction motor FOC (the problems are the same, regardless of the machine type).

Illustration of the vector control scheme when CC is realized in the rotating reference frame is shown in Figure 24.9. Decoupling voltage terms are often omitted and this is a satisfactory solution if the drive

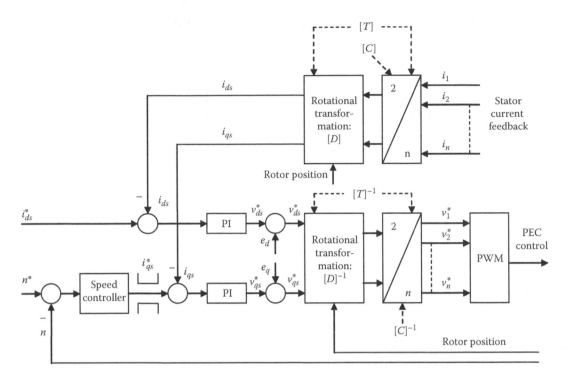

FIGURE 24.9 Fully digital control of a multiphase SPMSM drive with CC in rotational reference frame.

supply is operated at a high switching frequency and fast current controllers are used. As noted, this CC scheme suffices in practice only if the machine is three-phase, although the general *n*-phase case is illustrated again.

FOC of an IPMSM can be in essence the same as for a SPMSM, where now one has to account for the different inductances of the machine, according to (24.5). If the stator *d*-axis current reference is set to zero, then the control schemes remain the same as in Figures 24.4 and 24.9 for CC in the stationary reference frame and CC in the rotating reference frame, respectively (the only difference is in expressions for decoupling voltages (24.18) where there are now different inductances along the two axes). However, operating an IPMSM with zero stator *d*-axis current reference setting is not optimal, since the second component of the torque in (24.7), reluctance torque, is zero and is not utilized. Hence, it is customary to operate IPMSMs with nonzero stator *d*-axis current reference setting in the base speed region as well. The value of the reference is determined by observing that for any required torque value there will be an optimal setting of the *d*-axis current that minimizes the total stator current. In other words, control is done in such a way that the operation with maximum torque per ampere of stator current is obtained (the strategy is usually called just MTPA). To explain the idea, consider Figure 24.5 and (24.15). Stator current *d*–*q* axis components can be given as functions of the total stator current and the angle δ as

$$i_{ds} = i_s \cos \delta$$
$$i_{qs} = i_s \sin \delta$$

$$(24.19)$$

Electromagnetic torque of the machine (24.7) can then be given as

$$T_e = P\left[\psi_m i_{qs} + (L_d - L_q)i_{ds}i_{qs} \right]$$
$$T_e = Pi_s\left[\psi_m \sin \delta + \Delta L i_s 0.5 \sin 2\delta \right]$$

$$(24.20)$$

where $\Delta L = L_d - L_q$. Operation with MTPA will be achieved at a certain value of the angle δ, which is found by differentiating (24.20) with respect to δ and equating the first derivative to zero. Hence

$$\psi_m \cos\delta + \Delta L i_s \cos 2\delta = 0$$

$$\cos^2\delta + \left[\frac{\psi_m}{(2\Delta L i_s)}\right]\cos\delta - 0.5 = 0$$

(24.21)

An important remark is due here. In contrast to synchronous machines of salient pole rotor structure with excitation winding and Syn-Rel machines, where $L_d > L_q$ and hence $\Delta L > 0$, in IPMSMs the opposite holds true. This is so since the permanent magnets are in the d-axis and due to their low permeability they present high magnetic reluctance, so that the d-axis inductance is small. Hence, for IPMSMs, $L_d < L_q$ and hence $\Delta L < 0$. The net consequence of this is that reluctance torque component in (24.20) gives a positive contribution to the torque only if stator d-axis current reference is negative. Hence the solution of the quadratic equation (24.21) that will lead to MTPA operation is the one with the negative sign of the stator d-axis current.

24.3 Field-Oriented Control of Multiphase Synchronous Reluctance Machines

Syn-Rel machines for high-performance variable speed drives have a salient pole rotor structure without any excitation and without the cage winding. The model of such a machine is obtainable directly from (24.3) through (24.7) by setting the permanent magnet flux to zero. If there are more than three phases, then stator equations (24.4) and (24.6) also exist in the model but remain the same and are hence not repeated. Thus, from (24.3), (24.5), and (24.7), one has the model of the Syn-Rel machine, which is again given in the reference frame firmly attached to the rotor d-axis (axis of the minimum magnetic reluctance or maximum inductance):

$$v_{ds} = R_s i_{ds} + L_d \frac{di_{ds}}{dt} - \omega L_q i_{qs}$$

$$v_{qs} = R_s i_{qs} + L_q \frac{di_{qs}}{dt} + \omega L_d i_{ds}$$

(24.22)

$$T_e = P\left(L_d - L_q\right)i_{ds}i_{qs}$$

(24.23)

It follows from (24.23) that the torque developed by the machine is entirely dependent on the difference of the inductances along d- and q-axis. Hence constructional maximization of this difference, by making L_d/L_q ratio as high as possible, is absolutely necessary in order to make the Syn-Rel a viable candidate for real-world applications. For this purpose, it has been shown that, by using an axially laminated rotor rather than a radially laminated rotor structure, this ratio can be significantly increased. From FOC point of view, it is however irrelevant what the actual rotor construction is (for more details see [13]).

As the machine's model is again given in the reference frame firmly attached to the rotor and the real axis of the reference frame again coincides with the rotor magnetic d-axis, transformation expressions that relate the actual phase variables with the stator d–q variables (24.9) through (24.11) are the same as for PMSMs. Rotor position, being measured once more, is the angle required in the transformation matrix (24.9). Thus one concludes that FOC schemes for a Syn-Rel will inevitably be very similar to those of an IPMSM.

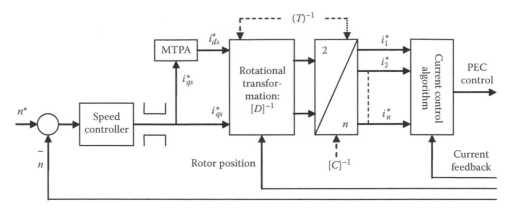

FIGURE 24.10   FOC of a multiphase Syn-Rel using CC in the stationary reference frame.

Since in a Syn-Rel there is no excitation on rotor, excitation flux must be provided from the stator side and this is the principal difference, when compared to the PMSM drives. Here again a question arises as to how to subdivide the available stator current into corresponding d–q axis current references. The same idea of MTPA control is used as with IPMSMs. Using (24.19), electromagnetic torque (24.23) can be written as

$$T_e = 0.5P\left(L_d - L_q\right)i_s^2 \sin 2\delta \tag{24.24}$$

By differentiating (24.24) with respect to angle δ, one gets this time a straightforward solution $\delta = 45°$ as the MTPA condition. This means that the MTPA results if at all times stator d-axis and q-axis current references are kept equal. FOC scheme of Figure 24.4 therefore only changes with respect to the stator d-axis current reference setting and becomes as illustrated in Figure 24.10. The q-axis current limit is now set as $\pm i_{s\,\max}/\sqrt{2}$, since the MTPA algorithm sets the d- and q-axis current references to the same values.

The same modifications are required in Figure 24.9, where additionally now the permanent magnet flux needs to be set to zero in the decoupling voltage calculation (24.18). Otherwise the FOC scheme is identical as in Figure 24.9 and is therefore not repeated.

It should be noted that the simple MTPA solution, obtained above, is only valid as long as the saturation of the machine's ferromagnetic material is ignored. In reality, however, control is greatly improved (and also made more complicated) by using an appropriate modified Syn-Rel model, which accounts for the nonlinear magnetizing characteristics of the machine in the two axes.

As an illustration, some responses collected from a five-phase Syn-Rel experimental rig are given in what follows. To enable sufficient fluxing of the machine at low load torque values, the MTPA is modified and is implemented according to Figure 24.11, with a constant d-axis reference in the initial part. The upper limit on the d-axis current reference is implemented in order to avoid heavy saturation of the magnetic circuit. Phase currents are measured using LEM sensors and a DSP performs closed-loop inverter phase CC in the stationary reference frame, using digital form of the ramp-comparison method. Inverter switching frequency is 10 kHz. The five-phase Syn-Rel is 4-pole, 60 Hz with 40 slots on stator. It was obtained from a 7.5 HP, 460 V three-phase induction machine by designing new stator laminations, a five-phase stator winding, and by cutting out the original rotor (unskewed, with 28 slots), giving a ratio of the magnetizing d-axis to q-axis inductances of approximately 2.85. The machine is equipped with a resolver and control operates in the speed-sensored mode at all times.

Response of the drive during reversing transient with step speed reference change from 800 to −800 rpm under no-load conditions is illustrated in Figure 24.12, where the traces of measured speed, stator q-axis current reference (which in turn determines the stator current d-axis reference, according to Figure 24.11), and reference and measured phase current are shown. It can be seen that the quality of

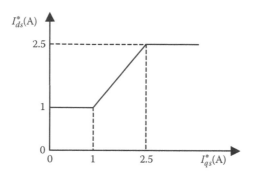

FIGURE 24.11 Variation of the stator d-axis current reference against the q-axis current reference (rms values) for the five-phase Syn-Rel in the experimental setup. (From Levi, E. et al., *IEEE Trans. Energ. Convers.*, 22(2), 281, 2007. With permission.)

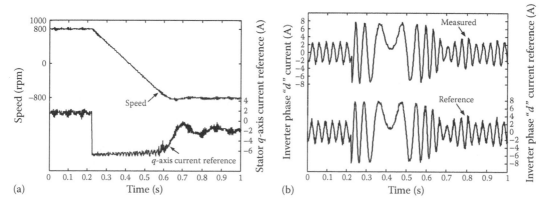

FIGURE 24.12 Reversing transient of the five-phase Syn-Rel drive from 800 to −800 rpm: (a) speed response and stator q-axis current reference (peak value, $\sqrt{2}I_{qs}^{*}$) and (b) reference and measured phase current. (From Levi, E. et al., *IEEE Trans. Energ. Convers.*, 22(2), 281, 2007. With permission.)

the transient speed response is practically the same as with a SPMSM (Figure 24.6 and 24.7), since the same linearity of the speed change profile is observable again. In final steady-state operation at −800 rpm the machine operates with q-axis current reference of more than 1 A rms, although there is no load. This is again the consequence of the mechanical and iron core losses that exist in the machine but are not accounted for in the vector control scheme (mechanical loss appears, according to (24.1a), as a certain nonzero load torque). Measured and reference phase current are in an excellent agreement, indicating that the CC of the inverter operates very well.

24.4 Field-Oriented Control of Multiphase Induction Machines

Similar to synchronous machines, FOC schemes for induction machines are also developed using mathematical models obtained by means of general theory of ac machines. An n-phase squirrel cage induction motor can be described in a common reference frame that rotates at an arbitrary speed of rotation ω_a with the flux–torque-producing part of the model

$$v_{ds} = R_s i_{ds} + \frac{d\psi_{ds}}{dt} - \omega_a \psi_{qs}$$

$$v_{qs} = R_s i_{qs} + \frac{d\psi_{qs}}{dt} + \omega_a \psi_{ds}$$

(24.25a)

$$v_{dr} = 0 = R_r i_{dr} + \frac{d\psi_{dr}}{dt} - (\omega_a - \omega)\psi_{qr}$$

$$v_{qr} = 0 = R_r i_{qr} + \frac{d\psi_{qr}}{dt} + (\omega_a - \omega)\psi_{dr} \tag{24.25b}$$

$$\psi_{ds} = (L_{ls} + L_m)i_{ds} + L_m i_{dr}$$

$$\psi_{qs} = (L_{ls} + L_m)i_{qs} + L_m i_{qr} \tag{24.26a}$$

$$\psi_{dr} = (L_{lr} + L_m)i_{dr} + L_m i_{ds}$$

$$\psi_{qr} = (L_{lr} + L_m)i_{qr} + L_m i_{qs} \tag{24.26b}$$

$$T_e = P\left(\psi_{ds}i_{qs} - \psi_{qs}i_{ds}\right) = P\frac{L_m}{L_r}\left(\psi_{dr}i_{qs} - \psi_{qr}i_{ds}\right) \tag{24.27}$$

This is at the same time the complete model of a three-phase squirrel cage induction machine. If stator has more than three phases, the model also includes the non-flux/torque-producing equations (24.4) and (24.6), which are of the same form for all n-phase machines with sinusoidal magnetomotive force distribution. As the rotor is short-circuited, no x-y voltages of nonzero value can appear in the rotor (since there is not any coupling between stator and rotor x-y equations, [26]), so that x-y (as well as zero-sequence) equations of the rotor are always redundant and can be omitted. Index l again stands for leakage inductances, indices s and r denote stator and rotor, and L_m is the magnetizing inductance.

Relationship between phase variables and variables in the common reference frame is once more governed with (24.9) and (24.10) for stator quantities. What is however very different is that the setting of the stator transformation angle according to (24.11) would be of little use, since rotor speed is different from the synchronous speed. In simple terms, rotor rotates asynchronously with the rotating field, meaning that rotor position does not coincide with the position of a rotating flux in the machine. The other difference, compared to a PMSM, is that the rotor does not carry any means for producing the excitation flux. Hence the flux in the machine has to be produced from the stator supply side, this being similar to a Syn-Rel.

Torque equation can be given in different ways, including the two that are the most relevant for FOC, (24.27), in terms of stator flux and rotor flux linkage d–q axis components. It is obvious from (24.27) that the torque equation of an induction machine will become identical in form to a dc machine's torque equation (24.2) if q-component of either stator flux or rotor flux is forced to be zero. Thus, to convert an induction machine into its dc equivalent, it is necessary to select a reference frame in which the q-component of either the stator or rotor flux linkage will be kept at zero value (the third possibility, of very low practical value, is to choose air-gap [magnetizing] flux instead of stator or rotor flux, and keep its q-component at zero). Thus, FOC scheme for an induction machine can be developed by aligning the reference frame with the d-axis component of the chosen flux linkage. While selection of the stator flux linkage for this purpose does have certain applications, it results in a more complicated FOC scheme and is therefore not considered here. By far the most frequent selection, widely utilized in industrial drives, is the FOC scheme that aligns the d-axis of the common reference frame with the rotor flux linkage.

As with synchronous motor drives, CC of the power supply can be implemented using CC in stationary or in rotating reference frame. Since with CC in the stationary reference frame one may assume that the supply is an ideal current source, so that again

$$i_1^* = i_1, \;\; i_2^* = i_2 \;\ldots\; i_n^* = i_n \tag{24.28}$$

stator voltage equations (24.25a) can be omitted from consideration. The common reference frame in which control is now executed is the rotor flux reference frame, so that the FOC is usually termed rotor flux-oriented control (RFOC). The reference frame is characterized with

$$\theta_s = \phi_r \quad \omega_a = \omega_r \quad \omega_r = \frac{d\phi_r}{dt} \tag{24.29}$$

where ω_r and ϕ_r stand for instantaneous speed and position of the rotating rotor flux in the cross section of the machine. Thus, the angle of transformation in (24.9) becomes instantaneous rotor flux position.

As the d-axis of the common reference frame coincides with d-axis component of the rotor flux linkage, while q-axis component of rotor flux linkage is kept at zero, then in this specific reference frame, one has

$$\psi_{dr} = \psi_r \quad \psi_{qr} = 0 \quad \frac{d\psi_{qr}}{dt} = 0 \tag{24.30}$$

Rotor voltage equations (24.25b) are in this reference frame given by

$$0 = R_r i_{dr} + \frac{d\psi_{dr}}{dt}$$
$$0 = R_r i_{qr} + (\omega_r - \omega)\psi_{dr} \tag{24.31}$$

Rotor current d–q axis components can be expressed from rotor flux linkage equations (24.26b)

$$\psi_r = (L_{lr} + L_m)i_{dr} + L_m i_{ds} \implies i_{dr} = \frac{(\psi_r - L_m i_{ds})}{L_r}$$
$$0 = (L_{lr} + L_m)i_{qr} + L_m i_{qs} \implies i_{qr} = -\left(\frac{L_m}{L_r}\right)i_{qs} \tag{24.32}$$

Substitution of (24.32) into (24.31) and (24.30) into (24.27) leads to the complete model of a current-fed rotor flux oriented induction machine in the form:

$$\psi_r + T_r \frac{d\psi_r}{dt} = L_m i_{ds} \tag{24.33}$$

$$(\omega_r - \omega)\psi_r T_r = L_m i_{qs} \tag{24.34}$$

$$T_e = P\left(\frac{L_m}{L_r}\right)\psi_r i_{qs} \tag{24.35}$$

where $T_r = L_r/R_r$ is the rotor time constant.

It follows from (24.35) that the electromagnetic torque can be changed instantaneously by stepping the stator q-axis current reference, provided that the rotor flux is kept constant. Inspection of (24.33) reveals that rotor flux is independent of the torque-producing q-axis current and that its value is uniquely determined with the stator d-axis current setting. The response of rotor flux to stator d-axis current application is exponential and, after approximately three rotor time constants, rotor flux achieves steady-state constant value. Hence in all industrial drives stator d-axis current is applied immediately at the drive power-up, so that at the time of speed reference application the machine is already fully fluxed (i.e., rotor flux has already stabilized at the rated value).

The third equation of (24.33) through (24.35) is the equation that relates machine's slip speed $\omega_{sl} = \omega_r - \omega$ with the stator q-axis current component. By expressing i_{qs} from (24.34) and substituting it into (24.35), correlation between torque and slip speed is obtained in the form

$$T_e = P\left(\frac{\psi_r^2}{R_r}\right)\omega_{sl} \tag{24.36}$$

It follows from (24.36) that relationship between torque and slip speed is the linear one, so that there is, theoretically, no pull-out (maximum) torque. In practice, maximum achievable torque is governed by the maximum allowed stator current.

An illustration of the rotor flux oriented reference frame and the stator current d–q axis components in such a reference frame is shown in Figure 24.13. Stator current and its components are still governed with (24.15) and (24.19), this being the same as for a PMSM. However, stator d-axis current component is now always of nonzero value, as the case was with a Syn-Rel.

In any steady-state operation rotor flux of (24.33) is governed with $\psi_r = L_m i_{ds}$. This means that, according to (24.32) steady-state rotor d-axis current component is zero. This expression also gives a clue as to how stator d-axis current reference should be set in the base speed region, where rotor flux is kept constant. In essence, stator d-axis current reference is set as equal to the machine's no-load (magnetizing) current with rated voltage supply (obtainable from no-load test), since under no-load conditions rotor current is practically zero and magnetizing flux and rotor flux are equal.

Basic form of RFOC scheme for a current-fed multiphase induction machine, assuming operation in the base speed region only, is illustrated in Figure 24.14. What remains to be explained is the way in which the instantaneous rotor flux position angle ϕ_r is acquired. This is in essence the only but important difference between the FOC of a Syn-Rel (Figure 24.10) and the RFOC of an induction machine.

Rotor flux spatial position cannot be easily measured. It therefore has to be somehow estimated on the basis of the measurable signals and an appropriate model of the machine. In practice, the most important method of rotor flux position calculation is based on utilization of the Equation 24.34 in a feed-forward

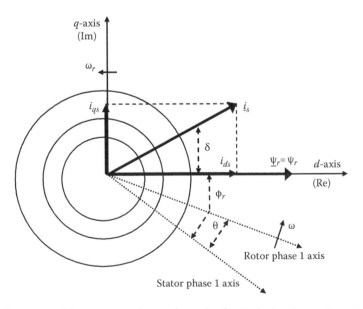

FIGURE 24.13 Illustration of the common reference frame firmly attached to the rotating rotor flux linkage of a multiphase induction machine.

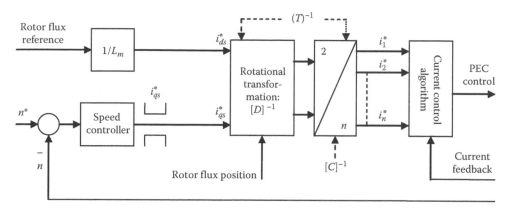

FIGURE 24.14 Basic form of an RFOC scheme for a multiphase induction machine, with CC in the stationary reference frame (base speed region only).

manner, using reference value of the stator q-axis current component. Since $\omega_{sl} = \omega_r - \omega$, then rotor flux speed of rotation can be calculated as $\omega_r = \omega + \omega_{sl}^*$. Then from (24.34) and (24.29) it follows that

$$\omega_{sl}^* = \frac{L_m i_{qs}^*}{T_r \psi_r^*} \tag{24.37}$$

$$\phi_r = \int \omega_r dt = \int \left(\omega + \omega_{sl}^* \right) dt = \theta + \int \omega_{sl}^* dt \tag{24.38}$$

Thus, calculation of the rotor flux position requires only measurement of the rotor position. However, stator current measurement is still necessary for the CC of the power supply.

RFOC scheme in which the rotor flux position is obtained by means of (24.37) and (24.38) is known as indirect rotor flux oriented control (IRFOC) scheme. Since in the base speed region stator d-axis current reference is constant, then $\psi_r = L_m i_{ds}$ and relationship between slip speed and stator q-axis current reference (24.37) reduces to

$$\omega_{sl}^* = \frac{i_{qs}^*}{T_r i_{ds}^*} = SG i_{qs}^* \tag{24.39}$$

where SG stands for "slip gain" constant $1/(T_r i_{ds}^*)$. IRFOC scheme is shown, for operation in the base speed region, in Figure 24.15. The angle required for rotational transformation is calculated according to (24.38) and (24.39).

An acceleration transient of a five-phase induction machine with IRFOC scheme and CC in the stationary reference frame is illustrated in Figure 24.16. Closed-loop inverter phase CC is of the digital ramp-comparison type and the inverter switching frequency is 10 kHz. The five-phase machine is 4-pole, 60 Hz with 40 slots on stator. It was obtained from a 7.5 HP, 460 V three-phase induction machine by designing new stator laminations and a five-phase stator winding (the rotor is the original one, unskewed, with 28 slots). Speed response and stator q-axis current reference are shown, together with the reference and actual phase current of one of the inverter (motor) phases. By comparing the results of Figure 24.16 with the corresponding ones for SPMSM and Syn-Rel (Figures 24.6, 24.7, and 24.12), it is evident that the same quality of the transient response has been achieved.

Similar results (speed response and stator phase current) are shown in Figure 24.17 as well, this time for the IRFOC of a three-phase 2.3 kW, 380 V, 4-pole, 50 Hz induction machine. Ramp-comparison CC is used again, with 10 kHz inverter switching frequency and acceleration and deceleration transients

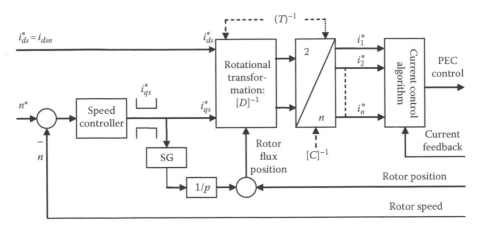

FIGURE 24.15 Indirect RFOC scheme for operation of an induction machine in the base speed region (p = Laplace operator; $1/p$ = integrator).

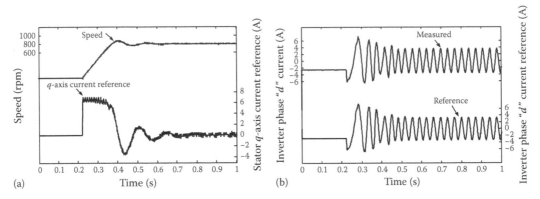

FIGURE 24.16 Acceleration of a five-phase induction motor drive with IRFOC from 0 to 800 rpm: (a) speed and q-axis current reference and (b) reference and measured phase current. (From Levi, E. et al., *IEEE Trans. Energ. Convers.*, 22(2), 281, 2007. With permission.)

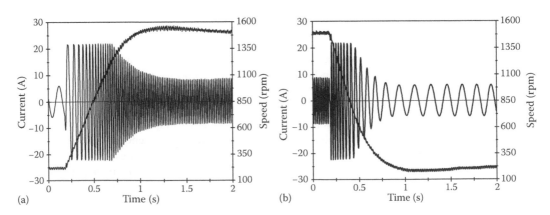

FIGURE 24.17 Acceleration from 200 to 1500 rpm (a) and deceleration from 1500 to 200 rpm (b) of a three-phase induction machine under no-load conditions with IRFOC scheme.

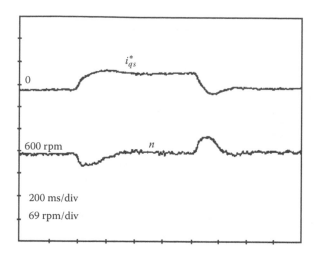

FIGURE 24.18 Experimentally recorded response to step application and removal of the rated load torque of a 0.75 kW three-phase induction machine with IRFOC. (From Levi, E. et al., Saturation compensation schemes for vector controlled induction motor drives, in *IEEE Power Electronics Specialists Conference PESC*, San Antonio, TX, pp. 591–598, 1990. With permission.)

under no-load conditions are shown. Comparison of Figures 24.16 and 24.17 shows that the same quality of dynamic response is achievable regardless of the number of phases on the stator of the machine.

Load rejection properties of a three-phase 0.75 kW, 380 V, 4-pole, 50 Hz induction motor drive with IRFOC are illustrated in Figure 24.18, where at constant speed reference of 600 rpm rated load torque is at first applied and then removed. The response of the stator q-axis current reference and rotor speed are shown. Once more, speed variation during sudden loading/unloading is inevitable, as already discussed in conjunction with Figure 24.8.

IRFOC scheme discussed so far suffices for operation in the base speed region, where rotor flux (stator d-axis current) reference is kept constant. If the drive is to operate above base speed, it is necessary to weaken the field. Since flux is produced from stator side, this now comes to a simple reduction of the stator d-axis current reference for speeds higher than rated. The necessary reduction of the rotor flux reference is, in the simplest case, determined in very much the same way as for a PMSM. Since supply voltage of the machine must not exceed the rated value, then at any speed higher then rated product of rotor flux and speed should stay the same as at rated speed. Hence

$$\omega_n \psi_m = \omega \psi_r \quad \omega > \omega_n$$

$$\psi_r^* = \psi_m \left(\frac{\omega_n}{\omega} \right)$$

(24.40)

Since change of rotor speed takes place at a much slower rate than the change of rotor flux (i.e., mechanical time constant is considerably larger than the electromagnetic time constant), industrial drives normally base stator current d-axis setting in the field weakening region on the steady-state rotor flux relationship, $i_{ds}^* = \psi_r^*/L_m$. However, since modern induction machines are designed to operate around the knee of the magnetizing characteristic of the machine (i.e., in saturated region), while during operation in the field weakening region flux reduces and operating point moves toward the linear part of the magnetizing characteristic, it is necessary to account in the design of the IRFOC aimed at wide-speed operation for the nonlinearity of the magnetizing curve (i.e., variation of the parameter L_m). One rather simple and widely used solution is illustrated in Figure 24.19, where only creation of stator d–q axis current references and the reference slip speed is shown. The rest of the control scheme is the same as in Figure 24.15.

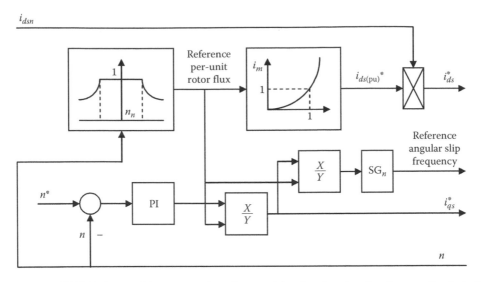

FIGURE 24.19 IRFOC scheme with compensation of magnetizing flux de-saturation for operation in both base speed and field weakening region. Inverse magnetizing curve of the machine is embedded in the controller as an analytical function in per unit form.

The scheme of Figure 24.19 sets the rotor flux reference in per unit (normalized with respect to the rated rotor flux value) according to (24.40). Stator d-axis current setting in per unit is further obtained by passing the rotor flux through the nonlinear magnetizing characteristic of the machine (which has to be determined experimentally). A simple two-parameter analytical approximation of this curve suffices for industrial drives [18]. Slip gain (SG) in Figure 24.19 is the one governed by the rated rotor flux (i.e., rated stator d-axis current), which in turn corresponds to the rated rotor time constant value, $SG_n = 1/(T_{rn}i_{dsn})$. Note that, since rotor flux reference is now a variable quantity, both stator q-axis current reference calculation and slip speed reference calculation involve divisions. Further, since rotor time constant is $T_r = L_r/R_r$ and $L_r = L_{lr} + L_m$, variation of the magnetizing inductance causes variation of the rotor time constant. Since magnetizing inductance is typically 10 times or more the rotor leakage inductance, then the approximation $L_m/L_{mn} \approx L_r/L_{rn}$ holds true (here index n once more refers to rated operating conditions). Using this approximation and the constant slip gain value of $SG_n = 1/(T_{rn}i_{dsn})$, reference slip speed in the field weakening region can be determined according to

$$\omega_{sl}^* = \frac{L_m i_{qs}^*}{T_r \psi_r^*} = \frac{SG_n i_{qs}^*}{\psi_{r(pu)}^*} \tag{24.41}$$

Similarly, stator q-axis current reference is calculated from the torque reference (the output of the speed controller) as

$$i_{qs}^* = \frac{T_e^*}{\left(P\psi_r^* L_m/L_r\right)} \approx \frac{T_e^*}{\left(P(\psi_{r(pu)}^* \psi_{rn})L_{mn}/L_{rn}\right)} = \frac{KT_e^*}{\psi_{r(pu)}^*} \tag{24.42}$$

where constant K is the same as for operation in the base speed region ($K = (P\psi_{rn}L_{mn}/L_{rn})^{-1}$) and is regarded in Figure 24.19 as having been already incorporated in the PI speed controller gains. Current limiting is for simplicity not shown in Figure 24.19 but is, as always, necessary and present. Using IRFOC of Figure 24.19 means that the decrease of the rotor flux reference in the field-weakening region is automatically followed by proper stator d–axis current adjustment, since the nonlinearity of the magnetization characteristic is taken into account in the stator d-axis current reference calculation.

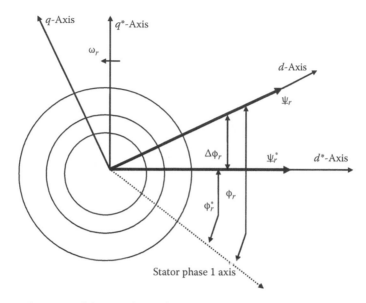

FIGURE 24.20 Misalignment of the actual rotor flux–oriented reference frame and the reference frame determined by IRFOC due to rotor time constant detuning.

As is already obvious from previous considerations, accurate IRFOC requires correct setting of the rotor time constant in the controller. This is a machine-specific parameter, which is in the control scheme regarded as a constant. Unfortunately, however, this is a parameter that can undergo substantial variations during operation of the machine. Since $T_r = L_r/R_r$, variation can occur due to both variation of the rotor inductance and rotor resistance. Variation of the rotor inductance is predominantly related to the variable stator d-axis current setting (as in Figure 24.19) and compensation is relatively simple, as shown above for operation in the field-weakening region. Variation of rotor resistance is however a much more difficult problem, since this is a parameter determined by thermal conditions of the rotor. In drives that operate intermittently and, when operated, are subjected to temporary overloads, variation of rotor resistance from cold to hot condition can easily be up to 60% (or ±30% with respect to the average value). The net consequence of the difference between the rotor time constant value in the controller and actual value in the machine is that rotor flux position is wrongly calculated, so that the control system operates in a misaligned reference frame, as illustrated in Figure 24.20. Basically, there is a detuning between true rotor time constant value and the value used in the controller. Hence, the q-axis component of the rotor flux is not zero, as "thought" by the controller, and the torque equation is of the form given in (24.27) rather than as for true rotor flux orientation, (24.35).

How severe the consequences of detuning are depends on the rated power and on the operating mode of the machine. The most pronounced effects are in drives operated in torque control mode, while in speed- or position-controlled drives, the consequences are much less severe due to the filtering effect of the drive's inertia. As an illustration, Figure 24.21 shows speed response of an IRFOC scheme, operated in open-loop torque control mode, with correct and incorrect setting of the rotor time constant. The machine operates at a certain speed with rated stator d-axis current setting and zero load torque. Torque reference (i.e., stator q-axis current reference) is an alternating square wave of plus/minus rated value. If the rotor time constant in the controller is at correct value, torque is of the form given in (24.35). Hence, the torque developed by the motor follows the alternating square-wave reference and the speed response is, according to (24.1a), a triangular function. However, if wrong value of the rotor time constant is used, the torque contains the second component associated with rotor flux q-component, as in (24.27), so that actual torque does not follow the reference. As a consequence, speed response deviates from the triangular waveform. The deviations become more and more severe as the value in the controller differs more and more from the correct value.

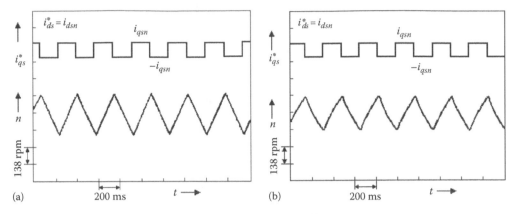

FIGURE 24.21 Experimentally recorded speed response to the alternating square wave *q*-axis stator current command for the rated *d*-axis current and open speed loop: (a) correctly and (b) incorrectly (1.7 times the correct value) set rotor time constant in the IRFOC scheme. (From Toliyat, H.A. et al., *IEEE Trans. Energ. Convers.*, 18(2), 271, 2003. With permission.)

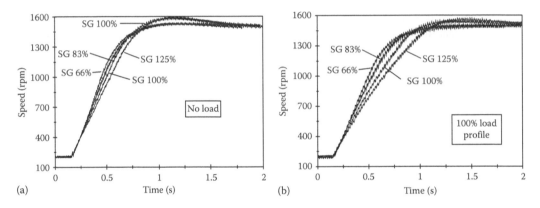

FIGURE 24.22 Speed response during acceleration transient with various slip gains (SG = 66%, 83%, 100%, and 125%) (a) under no-load conditions and (b) with 100% load profile.

In speed (or position)-controlled drives, the impact of the incorrect setting of the rotor time constant (slip gain) in the controller is suppressed to some extent by the filtering effect of the drive's inertia. This is illustrated in Figure 24.22, where an acceleration transient (from 200 to 1500 rpm) is shown for the 2.3 kW three-phase machine already considered in conjunction with Figure 24.17. The same transient is recorded for a number of settings of the slip gain (SG) of Figure 24.15. Two different loading conditions are considered: no-load acceleration and acceleration with a variable load that requires rated motor torque at 1500 rpm (100% load profile). Since the load is a dc generator, then the load torque in the latter case continuously increases (approximately linearly) as the speed increases. Various speed responses are identified with the percentage value of the slip gain setting with respect to the rated value, $SG_n = 1/(T_{rn}i_{dsn})$. It can be seen from Figure 24.22 that the impact of incorrect setting of the slip gain is more pronounced for higher load torques. As the total stator current is limited, a higher transient torque can be developed with a lower slip gain setting. Therefore, acceleration is more rapid and speed responses are faster for lower slip gain settings. Although use of a smaller slip gain value yields faster response than the correct slip gain value, subsequent steady-state operation is characterized with higher stator current across the entire base speed region for light loads and across most of the base speed region for heavy loads. It can also be seen that deviation of the speed response from the desired linear one during operation in the current limit (see Figure 24.17 where response for SG = 100% under no-load conditions has already been shown and the phase current trace has been included) is relatively small, even for significant differences between the actual slip gain value and the one used in the controller. The time interval required to reach the steady-state operating conditions can be significantly different, depending on the slip gain setting. However, the settling time also depends on the PI speed controller design.

The only method of determining rotor flux position, discussed here, is the one based on utilization of the measured rotor shaft position and reference stator d–q axis current components in a feed-forward manner. Although this is the dominant solution in industrial drives, it has to be noted that there are numerous other ways of calculating this angle. For this purpose, one may use some or all of the easily measurable signals, such as rotor position, stator currents, and stator voltages (they are usually not measured directly; instead, they are reconstructed using measured dc link voltage in inverter fed drives and the knowledge of the semiconductor switching signals). More detailed discussion of these methods is beyond the scope of this chapter.

Similarly, as for synchronous motor vector controlled drives, CC in RFOC induction motor drives can be implemented using CC in the rotating reference frame. This requires that, once more, stator voltage equations (24.25a) are taken into consideration. By expressing stator flux d–q axis components as functions of stator current and rotor flux d–q components, using (24.26), and then substituting into (24.25a) and applying the rotor flux orientation conditions (24.29) and (24.30), stator d–q axis voltage equations take the form

$$v_{ds} = R_s i_{ds} + \sigma L_s \frac{di_{ds}}{dt} + \frac{L_m}{L_r}\frac{d\psi_r}{dt} - \omega_r \sigma L_s i_{qs}$$

$$v_{qs} = R_s i_{qs} + \sigma L_s \frac{di_{qs}}{dt} + \omega_r \frac{L_m}{L_r}\psi_r + \omega_r \sigma L_s i_{ds} \tag{24.43}$$

where $\sigma = 1 - L_m^2/(L_s L_r)$ is the total leakage coefficient of the machine. The parameter $L_s' = \sigma L_s$ is the transient stator inductance. Equations 24.43 show that the d- and q-axis components of stator voltage and stator current are not decoupled. In order words, each of the two voltage components is a function of both stator current components, as the case was with synchronous machines as well. If the decoupled control of stator d- and q-axis currents is to be achieved, it is necessary to introduce appropriate decoupling circuit in the control system. If the output variables of current controllers are defined again as

$$v_{ds}' = R_s i_{ds} + L_s'\frac{di_{ds}}{dt}$$

$$v_{qs}' = R_s i_{qs} + L_s'\frac{di_{qs}}{dt} \tag{24.44}$$

The required reference values of axis voltages v_{ds}^* and v_{qs}^* are obtained as

$$v_{ds}^* = v_{ds}' + e_d \quad v_{qs}^* = v_{qs}' + e_q \tag{24.45}$$

where auxiliary variables e_d and e_q are calculated as

$$e_d = \frac{L_m}{L_r}\frac{d\psi_r}{dt} - \omega_r L_s' i_{qs}$$

$$e_q = \omega_r \frac{L_m}{L_r}\psi_r + \omega_r L_s' i_{ds} \tag{24.46}$$

Equations 24.44 and 24.45 are the same as for a PMSM, (24.16) and (24.17), and the only difference is in the expressions for decoupling voltages, (24.46). If the machine is operated in the base speed region, derivative of rotor flux in the first of (24.46) is zero. Further, $\psi_r = L_m i_{ds}$, so that (24.46) reduce to a simple form

$$e_d = -\omega_r \sigma L_s i_{qs}$$

$$e_q = \omega_r \frac{L_s}{L_m} \psi_r = \omega_r L_s i_{ds} \tag{24.47}$$

Here again, one can use either stator current d–q reference currents or d–q axis current components calculated from the measured phase currents. The principal RFOC scheme, assuming that rotor flux position is again determined according to the indirect field orientation principle, is shown in Figure 24.23 (current limiting block is not shown for simplicity).

Operation in the field weakening region can again be realized by using the stator d-axis current and slip speed reference setting as in Figure 24.19. Since the rotor flux reference will change slowly, the rate of change of rotor flux in (24.46) is normally neglected in the decoupling voltage calculation, so that e_d calculation remains as in (24.47). However, since rotor flux reference reduces with the increase in speed, e_q calculation has to account for the rotor flux (stator d-axis current) variation.

As noted in the section on RFOC of PMSMs, vector control with only two current controllers, as in Figure 24.23, suffices for three-phase machines. While, in theory, this should also be perfectly sufficient for machines with more than three phases, in practice various nonideal characteristics of the PEC supply (for example, inverter dead time) and the machine (any asymmetries in the stator winding) lead to the situation where the performance with only two current controllers is not satisfactory [26]. To illustrate this statement, an experimental result is shown in Figure 24.24 for a five-phase induction machine (which has already been described in conjunction with Figure 24.16). Control scheme of Figure 24.23

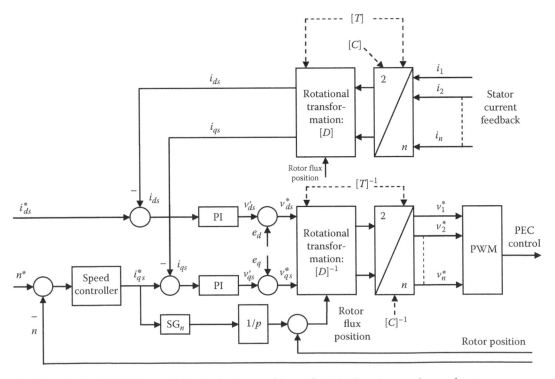

FIGURE 24.23 IRFOC of a multiphase induction machine with CC in the rotating reference frame.

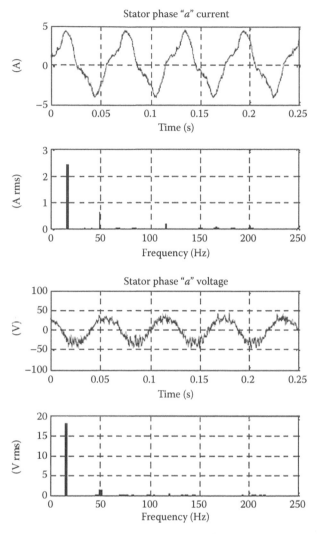

FIGURE 24.24 Stator phase *a* current and voltage (time-domain waveforms and spectra) for operation of a five-phase induction machine with IRFOC according to Figure 24.23 in steady state without load at 500 rpm (16.67 Hz). (From Jones, M. et al., *IEEE Trans. Energ. Convers.*, 24(4), 860, 2009. With permission.)

is applied, sinusoidal PWM is utilized with a triangular carrier wave at 10 kHz, stator *d*-axis current reference setting is 2.6 A rms, and the machine runs under no-load conditions at 500 rpm (16.67 Hz stator frequency) in steady state. Stator current and stator phase voltage have been measured and low-pass filtered with a filter cut-off frequency of 1.6 kHz.

One expects to see a sinusoidal stator phase current. However, the waveform of the phase current is heavily distorted (Figure 24.24) and the spectrum contains significant low-order harmonics, in particular the third and the seventh (around 20% and 10% of the fundamental, respectively). Although the corresponding voltage harmonics are much smaller (9% and 3%, respectively), the current harmonics are significant due to the very small impedance presented to these harmonics. These harmonics are caused by the inverter dead time and they in essence map into the *x-y* stator voltage components of (24.4) [26]. As can be seen from (24.4), the impedance for these harmonics is the stator leakage impedance and it is small, meaning that even a relatively small amount of these voltage harmonics causes substantial stator current harmonics. In order to suppress these unwanted harmonics one has to utilize a CC scheme with,

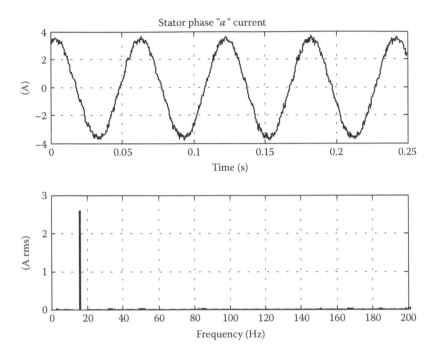

FIGURE 24.25 Stator phase *a* current for operation of a five-phase induction machine with IRFOC according to Figure 24.23 but with an added (second) pair of current controllers (with zero reference current setting) and operating conditions the same as in Figure 24.24. (From Jones, M. et al., *IEEE Trans. Energ. Convers.*, 24(4), 860, 2009. With permission.)

in general, $(n-1)$ current controllers (as the case was with CC in the stationary reference frame, Figures 24.15 and 24.16). In principle, two additional current controllers for the x-y stator current component pair are required in the case of a five-phase machine [26]. Adding the second pair of current controllers gives for the same operating conditions as in Figure 24.24 stator phase current shown in Figure 24.25, which is now without practically any low-order harmonics.

24.5 Concluding Remarks

FOC of ac machines is a vast area, to which numerous books have been devoted in entirety in recent times. An attempt has been made in this chapter to introduce the idea of vector control by explaining the physical background of the FOC and to present the basic control schemes for synchronous and induction machines with supply from stator side only. Numerous important aspects have been either only briefly mentioned or not addressed at all. For example, it has been assumed at all times that the machine is equipped with a position sensor. While this is still the case in the most demanding applications, in many other applications position sensor has been replaced with a rotor position (speed) estimator, leading to so-called sensorless FOC (for more details, see for example [17]). This is so since the position sensor is costly, it requires space for mounting and cabling for power supply and position signal transmission, and it reduces reliability of the drive. Similarly, it has been assumed at all times that the FOC schemes are based on constant parameter models of the machines and the problem of parameter variations has been only briefly addressed. Numerous more sophisticated machine models exist nowadays, which are predominantly aimed at providing modified vector control schemes with an automatic compensation of some of the parasitic phenomena that are neglected in the constant parameter models (for example, main flux saturation and ferromagnetic core losses). Further, a whole range of online identification methods has been developed over the years to provide accurate information about the rotor resistance (rotor time constant) value during operation of a vector-controlled induction machine. It has also been assumed here that the stator d-axis current reference setting is in essence constant for all considered machines, except in the field weakening region. However, vector-controlled machines may be operated with a variable flux (variable stator d-axis current setting) even in the base speed region, for example, for optimum efficiency control. Last but not least, various much more sophisticated approaches exist for estimation of the instantaneous rotor flux position in induction motor drives using various modern control theory approaches (observers, model reference adaptive control, extended Kalman filters, etc; see [7] for example for more details).

References

1. K. Hasse, *Zur Dynamik drehzahlgeregelter Antriebe mit stromrichtergespeisten Asynchron-Kurzschluβläufermaschinen*, PhD thesis, TH Darmstadt, Darmstadt, West Germany, 1969.
2. F. Blaschke, Das Prinzip der Feldorientierung, die Grundlage für die TRANSVECTOR-Regelung von Drehfeldmaschinen, *Siemens-Zeitschrift*, 45(10), 757–760, 1971.
3. K.H. Bayer, H. Waldmann, and M. Weibelzahl, Die TRANSVECTOR-Regelung für den feldorientierten Betrieb einer Synchronmaschine, *Siemens-Zeitschrift*, 45(10), 765–768, 1971.
4. K. Hasse, Drezhalregelverfahren für schnelle Umkehrantriebe mit stromrichtergespeisten Asynchron-Kurzschluβläufermotoren, *Regelungstechnik*, 20, 60–66, 1972.
5. F. Blaschke, *Das Verfahren der Feldorientierung zur Regelung der Drehfeldmaschine*, PhD thesis, TU Braunschweig, Braunschweig, West Germany, 1974.
6. H. Späth, *Steurverfahren für Drehstrommaschinen*, Springer-Verlag, Berlin, West Germany, 1985.
7. P. Vas, *Vector Control of AC Machines*, Clarendon Press, Oxford, U.K., 1990.
8. I. Boldea and S.A. Nasar, *Vector Control of AC Drives*, CRC Press, Boca Raton, FL, 1992.
9. S.A. Nasar and I. Boldea, *Electric Machines: Dynamics and Control*, CRC Press, Boca Raton, FL, 1993.
10. S.A. Nasar, I. Boldea, and L.E. Unnewehr, *Permanent Magnet, Reluctance and Self-synchronous Motors*, CRC Press, Boca Raton, FL, 1993.
11. A.M. Trzynadlowski, *The Field Orientation Principle in Control of Induction Motors*, Kluwer Academic Publishers, Norwell, MA, 1994.
12. M.P. Kazmierkowski and H. Tunia, *Automatic Control of Converter-Fed Drives*, Elsevier, Amsterdam, the Netherlands, 1994.
13. I. Boldea, *Reluctance Synchronous Machines and Drives*, Clarendon Press, Oxford, U.K., 1996.
14. D.W. Novotny and T.A. Lipo, *Vector Control and Dynamics of AC Drives*, Clarendon Press, Oxford, U.K., 1996.
15. W. Leonhard, *Control of Electrical Drives*, 2nd edn., Springer-Verlag, Berlin, Germany, 1996.
16. B.K. Bose (ed.), *Power Electronics and Variable Frequency Drives: Technology and Applications*, IEEE Press, Piscataway, NJ, 1997.
17. P. Vas, *Sensorless Vector and Direct Torque Control*, Oxford University Press, New York, 1998.
18. E. Levi, Magnetic variables control, in *Encyclopaedia of Electrical and Electronics Engineering*, Vol. 12, J.G.Webster(ed.), John Wiley & Sons, New York, 1999, pp. 242–260.
19. R. Krishnan, *Electric Motor Drives: Modeling, Analysis and Control*, Prentice Hall, Upper Saddle River, NJ, 2001.
20. A.M. Trzynadlowski, *Control of Induction Motors*, Academic Press, San Diego, CA, 2001.
21. B.K. Bose, *Modern Power Electronics and AC Drives*, Prentice Hall, Upper Saddle River, NJ, 2002.
22. J. Chiasson, *Modeling and High-Performance Control of Electric Machines*, John Wiley & Sons, Hoboken, NJ, 2005.
23. S.A. Nasar and I. Boldea, *Electric Drives*, CRC Press, Boca Raton, FL, 2006.
24. S.N. Vukosavic, *Digital Control of Electrical Drives*, Springer, New York, 2007.
25. N.P. Quang and J.A. Andreas, *Vector Control of Three-Phase AC Machines*, Springer, New York, 2008.
26. E. Levi, R. Bojoi, F. Profumo, H.A. Toliyat, and S.Williamson, Multiphase induction motor drives—A technology status review, *IET—Electric Power Applications*, 1(4), 489–516, 2007.

27. Z. Ibrahim, and E. Levi, An experimental investigation of fuzzy logic speed control in permanent magnet synchronous motor drives, *European Power Electronics and Drives Journal*, 12(2), 37–42, 2002.

28. E. Levi, M. Jones, A. Iqbal, S.N. Vukosavic, and H.A.Toliyat, An induction machine/Syn-Rel two-motor five-phase series-connected drive, *IEEE Transactions on Energy Conversion*, 22(2), 281–289, 2007.

29. E. Levi, S. Vukosavic, and V. Vuckovic, Saturation compensation schemes for vector controlled induction motor drives, *IEEE Power Electronics Specialists Conference PESC*, San Antonio, TX, 1990, pp. 591–598.

30. H.A. Toliyat, E. Levi, and M. Raina, A review of RFO induction motor parameter estimation techniques, *IEEE Transactions on Energy Conversion*, 18(2), 271–283, 2003.

31. M. Jones, S. Vukosavic, D. Dujic, and E. Levi; A synchronous current control scheme for multiphase induction motor drives, *IEEE Transactions on Energy Conversion*, 24(4), 860–868, 2009.

25

Adaptive Control of Electrical Drives

Teresa
Orłowska-Kowalska
Wroclaw University
of Technology

Krzysztof Szabat
Wroclaw University
of Technology

25.1 Introduction

The parameters and also the nature of the controlled drive systems can change with the operating conditions. If classical controllers with fixed parameters are used, then such change can introduce a difference in dynamic behavior of the drive. It can result in weaker or stronger damping, and thus in a tendency to instability or in an increase of system response rising time. If the specifications of the drive system do not admit such behavior, then adaptive controllers must be used [L85,BSS90,KT94]. It is demanded of an adaptive controller that the control system shall satisfy a defined control index independently of parameter changes in the controlled system. A number of drive systems incorporate controlled elements that are subject to parameter variations. In most cases, they are of limited extent (e.g., changes of the converter amplification with variation in supply voltage), or no very precise demands are made as regards dynamic behavior. But in some cases, operation conditions cause even significant parameter changes in the drive system, due to temperature, saturation, wear and tear of the drive system elements, etc. In the further part of this chapter, only those cases will be investigated in which marked parameter variations appear and for which, under certain circumstances, application of an adaptive controller is necessary.

In the controlled electrical drive systems, the following possible parameter variations can occur [L85,BSS90,KT94]:

1. Change of winding electromagnetic time constant due to the temperature rise or material deterioration
2. Change of the mechanical time constant due to moment of inertia changes of the drive
3. Change of the flux value, in drive with the field weakening operation
4. Change of the drive system structure (e.g., due to the transition from continuous to discontinuous armature current in a rectifier-fed DC motor drive)

These cases will be discussed in some detail in this chapter.

25.2 Adaptive Control Structure: Basis

According to the control theory, the adaptive control systems can be divided into three classes [ÄW95,SB89]:

- Gain scheduling systems (GS)
- Self-tuning regulators (STR)
- Model reference adaptive systems (MRAS)

Gain scheduling is one of the earliest and most intuitive approaches to adaptive control, introduced in 1950s and 1960s. The idea consists in finding auxiliary process variables (other than the plant outputs used for feedbacks) that correlate well with the changes in process dynamics. If these variables can be measured, they can be used to change the regulator parameters and thus compensate for parameter variations. A block diagram of a system with such control concept is presented in Figure 25.1. So the gain scheduling is open-loop compensation and can be viewed as a system with feedback control in which the feedback gains are adjusted by feedforward compensation [ÄB95]. There is no feedback from the performance of the closed-loop system, which compensates for an incorrect schedule.

This approach was called *gain scheduling* because the scheme was originally used to accommodate changes in process gain. With regard to nomenclature, it is controversial whether gain scheduling should be considered as an adaptive system or not, because the parameters are changed in open loop, with no real "learning" or intelligence [SB89]. Nevertheless, gain scheduling is very poplar in practice and is a very useful technique for reducing the effects of parameter variations in the case, when in the controlled system there are auxiliary variables that relate well to the characteristics of the process dynamics.

A different scheme is obtained if the process parameters are updated and the regulator parameters are obtained from the solution of a design problem. A block diagram of such system is presented in Figure 25.2.

The adaptive regulator can be thought of as composed of two loops. The inner loop consists of the plant and an ordinary linear feedback regulator. The parameters of the regulator are adjusted by the outer loop, which is composed of a specific algorithm for parameter estimation (recursive identification algorithm, observer, Kalman filter, neural network (NN)) and a design calculation. It should be noticed that the system can be viewed as an automation of process modeling and design, in which the process model and the control design are updated at each sampling period. A controller of this construction

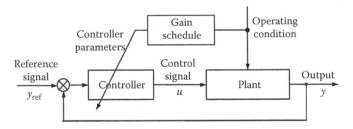

FIGURE 25.1   Block diagram of a system with gain scheduling.

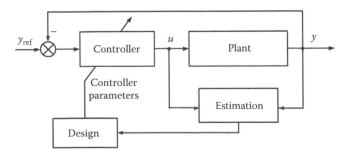

FIGURE 25.2   Block diagram of an STR.

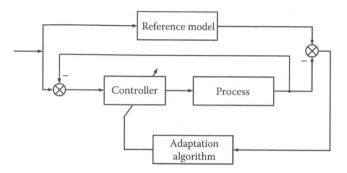

FIGURE 25.3 Block diagram of a system with an MRAS.

is called a *self-tuning regulator* to emphasize that the controller automatically tunes its parameters to obtain the desired properties of the closed-loop system [ÄW95, SB89]. The STR scheme is very flexible with respect to the choice of the underlying design ad estimation methods. Many different methods have been explored. The regulator parameters are updated indirectly via the design calculations in the self-tuner shown in Figure 25.2.

The third adaptive control concept, called *model reference adaptive system* (MRAS), was originally proposed to solve a problem in which the specifications are given in terms of a reference model that tells how the process output ideally should respond to the command signal [ÄW95]. A block diagram of such system is presented in Figure 25.3.

In this case, the reference model is in parallel with the system and the regulator can be thought of as consisting of two loops. The inner loop, which is an ordinary feedback loop, is composed of the plant and the regulator. The parameters of the regulator are adjusted by the outer (adaptation) loop in such a way that the error e between the process output y and the model output y_m becomes small. The outer loop is thus also a regulator loop.

The key problem is to determine the adjustment algorithm so that a stable system, which brings the error e to zero, is obtained. In the earliest applications of this scheme, the following update, called *gradient update*, was used [ÄW95,SB98]:

$$\frac{d\theta}{dt} = -\gamma e(\theta)\frac{d}{d\theta}\left(e^2(\theta)\right) = -2\gamma e(\theta)\frac{d}{d\theta}\left(e(\theta)\right) = -2\gamma e(\theta)\frac{d}{d\theta}\left(y(\theta)\right) \tag{25.1}$$

where
  e denotes the model error
  θ are the adjustable parameters of the controller
  $\partial e/\partial\theta$ are the sensitivity derivatives of the error with respect to the adjustable parameters θ
  γ is the positive constant called the adaptation rate

The gradient of e with respect to θ is equal to the gradient of the process output y with respect to θ, since the model output y_m is independent of θ and represents the sensitivity of the output error to variations in the controller parameter θ. This rule can be explained as follows: if we assume that the parameters θ change much slower than the other system variables, to make the square of the error small, it seems reasonable to change the parameters in the direction of the negative gradient of e^2. Unfortunately, the usage of this gradient update (25.1) encountered several problems as the sensitivity function $\partial y/\partial\theta$ usually depends on the unknown plant parameters, and thus is unavailable. At this point, the so-called *MIT rule* (because the algorithm was done at the Massachusetts Institute of Technology), which replaced the unknown parameters by their estimates at time t, was proposed [ÄW95,SB98]. The approximation of the sensitivity derivatives can be generated as outputs of a linear system driven by process inputs and outputs.

The MRAS schemes are called *direct* methods, because the adjustment rules tell directly how the regulator parameters should be updated. On the contrary, the STR are called *indirect* methods, as they first identify the plane parameters and then use these estimates to update the controller parameter through some fixed transformation (resulting from the controller design rules). The MRAS schemes update the controller parameters directly (no explicit estimate or identification of the plant parameters is made). It is easy to see that the inner control loop of STR could be the same as the inner loop of a MRAS design. In other words, the MRAS schemes can be seen as a special case of the STR schemes, with an identity transformation between updated parameters and controller parameters. So, sometimes it is reasonable to distinguish between direct and indirect schemes rather than between model-reference and self-tuning algorithms.

25.3 Gain Scheduling in the Drive Systems

The gain scheduling concept was mainly used in the converter-fed DC drive systems, where the armature current control loop was changed according to the operation conditions [BSS90,KT94]. The commonly applied control structure of the DC drive system is composed of a power converter-fed, separately excited (or permanent magnet) DC motor coupled to a mechanical system, a microprocessor-based speed and current controllers, current, speed and/or positions sensors used for feedback signals. Usually, cascade control structure containing two major control loops is used. The block diagram of such system is presented in Figure 25.4, where i_a, u_a are the armature current and voltage; e_m is the electromotive force; m_e, m_L are the electromagnetic and load torques; Ψ_f is the exciting flux; ω_m, ω_{ref} are the motor and reference speeds; K_a, K_p, K_i, K_T are the gain coefficients of the armature, static converter, current, and speed sensors; T_a, T_M, T_o are the electromagnetic and mechanical time constants of the motor and converter delay time constant; K_{Ri}, $K_{R\omega}$ are the gain factors of the current and speed controller; and T_{Ri}, $T_{R\omega}$ are the time constants of the current and speed controller.

The inner control loop performs a motor current (torque) regulation and consists of the power converter, electromagnetic part of the motor, current sensor, and respective current (torque) controller. The outer speed control loop consists of the mechanical part of the drive, speed sensor, speed controller, and is cascaded to the inner current control loop. It provides speed control according to its reference value. In Figure 25.4, two different current controller structures PI/I are shown, what results from different performance of the current control loop under continuous and discontinuous current modes of the power rectifier supplying the DC motor [KT94]. These specific dynamical performances of the current control loop depending on the operation mode are presented in detail in Figure 25.5.

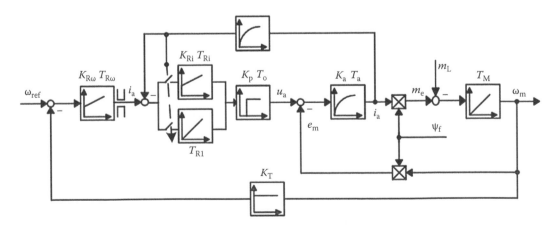

FIGURE 25.4   The control structure of the drive system with the adaptive (GS) current controller.

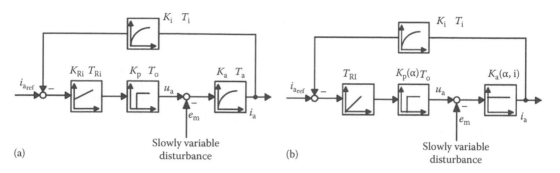

FIGURE 25.5 The current control loop for (a) the continuous and (b) discontinuous current mode.

When the controlled rectifier operates in the continuous current mode, the PI current controller is applied, tuned according to the modulus criterion, for the control loop presented in Figure 25.5a, where the controller time constant and gain factor are, respectively, adjusted as [BSS90,KT94]

$$T_{Ri} = T_a, \quad K_{Ri} = \frac{T_a}{2T_\sigma K_O}, \tag{25.2}$$

where

$K_O = K_p K_a K_i$
$T_\sigma = T_o + T_i$

The current control system with discontinuous current differs from that with continuous current in two respects.

1. Absence of the armature time constant T_a: Under discontinuous mode, if the delay angle of the converter α is changed at time t_1, the new average value of the current is reached after one pulse period has elapsed (Figure 25.6a); that is, the armature time constant no longer has any effect.
2. Change of amplification: The amplification $K_p K_a$ of the rectifier and the armature circuit, which is nearly constant with continuous armature current, changes very substantially on transition to discontinuous current (Figure 25.6b). Amplification is obtained from the slope of the characteristics. It decreases rapidly after the transition to discontinuous current. In the discontinuous current range, this amplification changes in function of the delay angle α and will be noted as $k_a(i_a, \alpha)$.

It can be proved that the armature current loop under discontinuous current mode can be shown as in Figure 25.6b, taking into account remark 1, given above. The PI controller designed for the previous

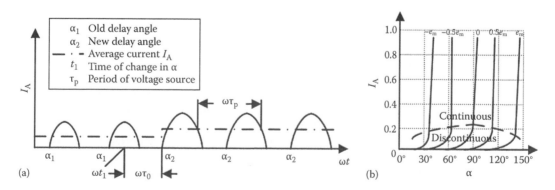

FIGURE 25.6 Current variation in the discontinuous mode under (a) the delay angle change and (b) armature current as a function of the delay angle with motor EMF as parameter.

case is not suitable now, since no inertia element with electromagnetic time constant T_a is available to compensate the numerator term in PI controller transfer function. Because $k_a(i_a, \alpha) \ll K_p K_a$, a control designed for continuous current has a much longer rise time with discontinuous current; indeed at small current value, the rise time can be of the order of seconds. If the current control is designed for a short rise time at low discontinuous current, then it becomes unstable on transition to continuous current. To avoid this instability, the current controller should be changed, to present the same dynamics of the armature current control loop whether that current is continuous or discontinuous.

Thus, an I controller must be employed, with the following integration time constant [BSS90], [KT94]:

$$T_{RI} = 2K_O T_\sigma \tag{25.3}$$

Accordingly, the adaptive current controller must satisfy the following conditions

PI—behavior with continuous current mode
I—behavior with discontinuous current mode

and the variation of the integration time constant in such manner that (25.2) and (25.3) are fulfilled in the suitable operation range.

In the speed control loop, the Kessler symmetrical optimum is used for the controller adjustment, according to [KT94]

$$K_{R\omega} = \frac{T_M}{2K_\omega T_{\sigma i}}, \quad T_{R\omega} = 4T_{\sigma i}, \tag{25.4}$$

where
$K_\omega = K_T K_{zi} \psi_f$
$T_{\sigma i} = 2T_\sigma$

In the case, if mechanical time constant T_M of the drive changes, for example, due to gear or inertia load changes, etc., then the P-amplification of the speed controller must be varied in proportion to T_M as indicated by Equation 25.4. If we know exactly the way T_M changes, we can simply modify online this gain coefficient and thus the speed controller will be adaptive in the sense of gain scheduling method.

25.4 Self-Tuning Speed Regulator for the Drive System

In industrial application only, the mechanical parameters of the motor calculated according to the nominal data are known. The parameters of the shaft and load machine are uncertain or even unknown in many cases. If these changes are not known a priori, in contrast to previous case of armature current mode changes, the difficulty that arises here is that the parameter change is not dependent upon a measurable variable (i.e., armature current or speed). So the principal task of the adaptive controller is to detect the changing value of mechanical parameters of the drive system, especially time constant T_M, which is changing in many cases. It could be done using the parameter estimation based on different online methods, like Luenberger observer, Kalman filter, or neural estimators [OJ03,OS07, OS03,SOD06]. Such estimators can calculate mechanical time constant in the real time, based on the speed and/or current measurements, and using this information the self-tuning speed controller can be designed, what will be presented in the next part of this chapter.

The general structure of the adaptive speed control loop in the speed sensorless version (without speed sensor) is presented in Figure 25.7a.

The drive system with fuzzy-logic (FL) speed controller is equipped with speed and load torque observer as well as with mechanical time constant estimator based on neural modeling approach [OJ03]. In the proposed system, simple FL controller with nine rule base is applied, whose output factor k_{du} is online modified due to the changes of mechanical time constant of the drive system.

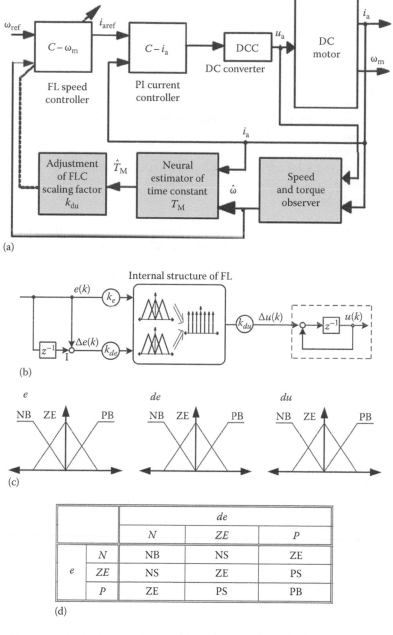

FIGURE 25.7 Structure of adaptive speed control loop for sensorless DC drive with (a) the FL controller, (b) the internal structure of the FL controller, (c) its membership functions, and (d) the rule base.

The NN can be applied to the mechanical time constant estimation of the drive system based on neural modeling concept [KO02]. It can be done if the mathematical description of the dynamical system is transformed to the form used in the description of a simple feedforward NN with single linear neuron.

In the case of mechanical equation of the drive system [p.u]

$$T_M \frac{d\omega_m}{dt} = \Psi_f i_a - m_L \tag{25.5}$$

the mentioned transformation to NN form in the discrete form is following:

$$\omega_m(k) = \omega_m(k-1) + \frac{\Delta T_s}{T_M}(i_a(k-1) - m_L(k-1)), \qquad (25.6)$$

where

$T_M = J\Omega_{oN}/M_N$

J is the inertia of the drive system

Ω_{oN}, M_N are the nominal idle-running speed and nominal torque of the motor, respectively

m_L is the load torque

i_a is the armature current

Ψ_f is the exciting flux (constant nominal value, equal 1 in per unit system)

ω_m is the motor speed

ΔT_s is the sampling time

Equation 25.6 can be viewed as the mathematical description of a simple NN with single linear neuron, three inputs and one output. Thus, it can be written in the following form:

$$\omega_m(k) = W_1\omega_m(k-1) + W_2\, i_a(k-1) - W_3\, m_L(k-1), \qquad (25.7)$$

where

$$W_1 = 1, \quad W_2 = W_3 = \frac{\Delta T_s}{T_M}. \qquad (25.8)$$

The coefficients $W_1 \div W_3$ represent the adjustable weights of this NN. For their modification, the back-propagation algorithm can be used. The estimate of the mechanical time constant is thus

$$T_M = \frac{\Delta T_s}{W_2}. \qquad (25.9)$$

From Equation 25.4 results, that proposed neural estimator of the mechanical time constant requires the input information about the motor speed and actual load torque. In the proposed sensorless drive, the motor speed and load torque are estimated using the state extended Luenberger observer (ELO), where the state vector of the DC motor drive system is extended by the new variable—the load torque:

$$\mathbf{x}_E = col(i_a, \omega_m, m_L). \qquad (25.10)$$

The full-order ELO of the general form

$$\dot{\mathbf{x}}_E = A\hat{\mathbf{x}}_E + \mathbf{B}u + \mathbf{G}(\mathbf{y} - \hat{\mathbf{y}}) \qquad (25.11)$$

takes the following form in the case of DC motor drive:

$$\frac{d\hat{i}_a}{dt} = \frac{1}{T_e}\left(-\hat{i}_a + K_t\left(u_a - \psi_f\hat{\omega}_m\right)\right) + g_1\left(i_a - \hat{i}_a\right)$$

$$\frac{d\hat{\omega}_m}{dt} = \frac{1}{T_M}\left(\psi_f\hat{i}_a - \hat{m}_L\right) + g_2\left(i_a - \hat{i}_a\right) \qquad (25.12)$$

$$\frac{d\hat{m}_L}{dt} = g_3\left(i_a - \hat{i}_a\right)$$

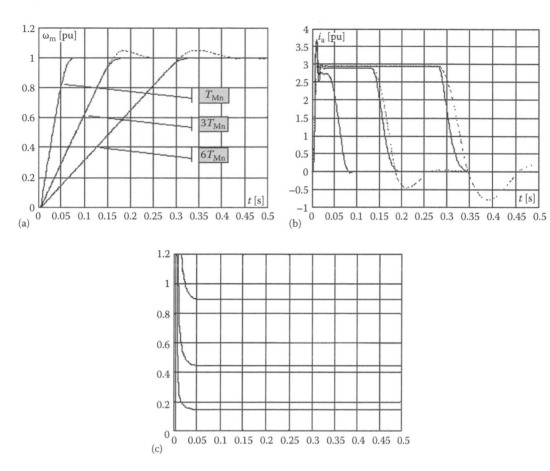

FIGURE 25.8 Step response of DC drive system with self-tuning FL controller for $\omega_{ref} = 1$: (a) transients of motor speed ω_m (b) transients of armature current i_a (doted line—response of the system without adaptation), and (c) results of T_M online estimation by NN.

where

T_e, K_t are the electromagnetic time constant and the gain factor of the motor armature winding, respectively

g_1, g_2, g_3 are the elements of the gain matrix **G** of the observer, chosen using pole placement method or genetic algorithm [OS03]

This state observer enables the estimation of rotor speed and load torque values based on the armature current measurement. For suitable choice of gain coefficients g_i, the state estimation is stable and very fast.

The proposed adaptive tuning of FL controller's scaling factor $k_{du} = f(T_M)$ ensures the assumed dynamics of the drive system: overshoot of the motor speed response equals zero in the whole range of changes of mechanical time constant, as it was assumed in the designing process of FL controller for nominal drive inertia. On the contrary, the application of similar adaptive PI controller has resulted in greater overshoots (marked by doted line in Figure 25.8), which can be explained by using the Kessler symmetrical optimum under PI speed controller design procedure. Both types of speed controllers without adaptive parameter tuning gave much worse results, with greater speed response overshoots, for time constant T_M different than nominal one.

In Figure 25.9, the transients of sensorless operation of the drive system during start-up, loading with nominal torque and speed reverse modes are demonstrated. The neural inertia estimator reconstructs

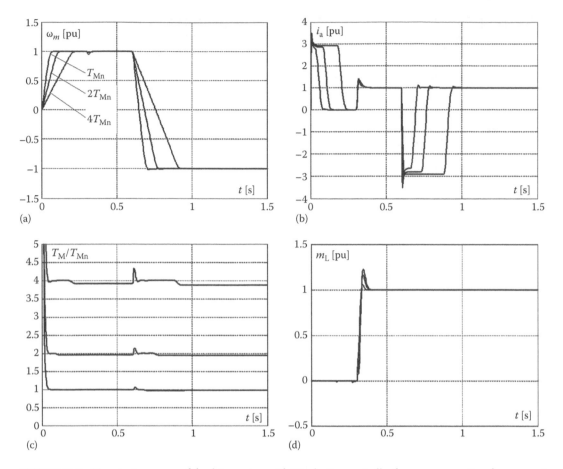

FIGURE 25.9 Transient response of the drive system with FL adaptive controller for reverse operation from $\omega_{ref} = 1$ to $\omega_{ref} = -1$: (a) transients of motor speed ω, (b) transients of motor current i_a, (c) time constant T_M, and (d) load torque m_L estimation process.

the mechanical time constant during transients. The speed estimation is performed online, but the load torque estimation is limited to the drive operation modes described by constant rotor speed. Such operation ensures the smooth transients of neural inertia identifier, speed observer, and proper work of adaptive sensorless structure, with speed controller parameters adapted according to actual value of the mechanical time constant of the drive. The similar control concept, but with the application of the Kalman filter for state variables and mechanical time constant estimation, was used in the case of indirect adaptive control methods of the drive system with elastic couplings and very good performance of the drive was obtained [SO08].

25.5 Model Reference Adaptive Structure

For the systems with changing parameters, the MRAS method gives a general approach for adjusting controller parameters so that the closed-loop transfer function will be close to a prescribed model. This is called the model-following problem [KT94,ÄW95]. One important question is how small we can make the error *e*. This depends both on the model, the system, and the type of the command signal. Usually, it is difficult to make the error equal to zero for all command signals, which will be illustrated in the following.

In the case of many drive systems, the inertia moment is changing depending on the operation conditions. So the speed controller must be updated online. Below two examples of the MRAS systems based on the same adaptation algorithm are presented, for the induction and DC motor drives [JO04,ODS06,OS07].

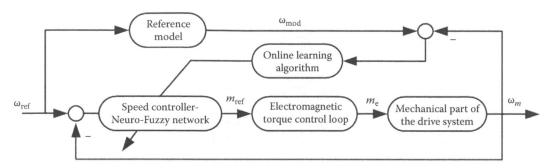

FIGURE 25.10 Structure of the adaptive control system for the electrical drive system.

The MRAS structure with the online tuned speed controller is demonstrated in Figure 25.10.

This control structure is general for the electrical drive system and can be adopted for AC or DC motor drive, under the condition, that the torque control loop is designed to provide sufficiently fast torque control, so it can be approximated by an equivalent first-order term. If this control is ensured, the driven machine could be AC or DC motor, with no difference in the outer speed control loop [ODS06,OS07]. Recently, instead of a classical PI speed controller, its neuro-fuzzy or sliding-mode neuro-fuzzy versions are proposed in many applications [JO04,ODS06,OS07,CT98,LWC98,LFW98, LLS01,OS08].

Below, the adaptive fuzzy speed controller with automatically adjusted rules is described. Control rules are tuned so that the actual output can follow the output of the reference model. The tracking error signal e_m between the desired output ω_{mod} and the actual output ω_m is used as the tuning signal.

In the present study, the PI-type neuro-fuzzy controller is used [JO04]. It describes the relationship between speed error $e(k)$, its change $\Delta e(k)$, and change of the control signal $\Delta u(k)$. The rule base of the controller is composed of a collection of *IF-THEN* rules in the following forms:

$$R_j : \text{IF } x_1 \text{ is } A_1^j \quad \text{and} \quad x_2 \text{ is } A_2^j \quad \text{THEN } y = w_i, \tag{25.13}$$

where
    x_i is the input variable of the system
    A_i^j is the specific membership function
    w_i is the consequent function

This controller can be realized as a general structure of neuro-fuzzy system shown in Figure 25.11 in the case of the nine-rules controller.

The functions of each layer are presented as follows:

Layer 1. Each input node in this layer corresponds to the specific input variable ($x_1 = e(k)$; $x_2 = \Delta e(k)$). These nodes only pass input signals to the second layer.

Layer 2. Each node performs a membership function A_i^j that can be referred to as the fuzzification procedure.

Layer 3. Each node in this layer represents the precondition part of fuzzy rule and is denoted by Π that multiplies the incoming signals and sends the results out.

Layer 4. This layer acts as a defuzzifier. The single node is denoted by Σ and sums of all incoming signals. A defuzzification process in the neuro-fuzzy network, known as the singleton defuzzification method, is described by

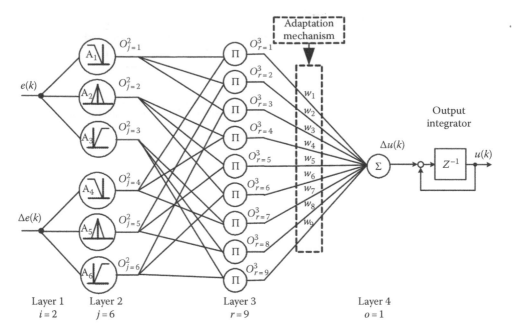

FIGURE 25.11   General structure of a neuro-fuzzy controller.

$$\Delta u = \frac{\displaystyle\sum_{j=1}^{M} w_j u_j}{\displaystyle\sum_{j=1}^{M} u_j}.$$

(25.14)

The fuzzy-neuro controller has a simple rule base with nine elements. The input membership functions have commonly used triangular shapes (see Figure 25.11). The fuzzy controller is tuned so that the actual drive output can follow the output of the reference model. The tracking error is used as the tuning signal.

The reference model is usually chosen as a standard second-order term:

$$G_{\text{mod}}(s) = \frac{\omega_n^2}{s^2 + 2\zeta\omega_n s + \omega_n^2},$$

(25.15)

where
   ζ is a damping ratio
   ω_n is a resonant frequency

The supervised gradient descent algorithm is used to tune the parameters $w_1, \ldots, w_M$ in the direction of minimizing the cost function like

$$J(k) = \frac{1}{2}\left(\omega_{\text{mod}} - \omega_m\right)^2 = \frac{1}{2}e_m^2.$$

(25.16)

The parameters adaptation is obtained using the following expression:

$$w_r(k+1) = w_r(k) - \gamma \frac{\partial J(k)}{\partial w_r(k)}.$$

(25.17)

The chain rule is used then:

$$\frac{\partial J(k)}{\partial w_r(k)} = \frac{\partial J(k)}{\partial \omega_m} \frac{\partial \omega_m}{\partial \Delta u} \frac{\partial \Delta u}{\partial w_r}, \tag{25.18}$$

where

$$\frac{\partial J(k)}{\partial \omega_m} = -(\omega_{\text{mod}} - \omega_m) = -e_m, \tag{25.19}$$

$$\frac{\partial \Delta u(k)}{\partial w_r} = O_{Nr}^3, \tag{25.20}$$

with O_n^3, the normalized firing strength of each rule.

Expression (25.18) involves the computation of the gradient of ω_m with respect to the Δu output of the controller, which is the change of reference electromagnetic torque $\Delta m_{1\,\text{ref}}$. The exact calculation of this gradient cannot be determined due to the nonlinearities and parameter uncertainty of the drive system. However, it can be assumed that the change of the drive speed with respect to the torque or armature current is a monotonic increasing process. Thus, this gradient can be approximated by some positive constant values. Owning to the nature of gradient descent search, only the sign of the gradient is critical to the iterative algorithm convergence. So, the adaptation law of controller parameters can be written as

$$w_r(k+1) = w_r(k) + \gamma e_m O_{Nr}^3 \tag{25.21}$$

where

e_m is the error between model response ω_{mod} and actual speed of the drive ω_m
O_{Nr} is the firing strength of rth rule
γ is the learning rate

However, the learning speed of the above algorithm is not satisfactory due to the slow convergence. To overcome this weakness, a modified algorithm based on local gradient PD control is used [LWC98]:

$$w_r(k+1) = w_r(k) + O_{Nr}^3\left(k_p e_m(k) + k_d \Delta e_m(k)\right) \tag{25.22}$$

Comparing (25.21) to (25.20), one can see that the coefficient k_p is equivalent to the learning rate γ. The derivative term with k_d is used to suppress a large gradient rate. The quality of reference speed tuning depends on k_d and k_p parameters of the adaptive law (25.21). The bigger values of these parameters cause the faster decrease of the system tracking error. However, too large values of adaptation coefficients introduce the high-frequency oscillations into the system state variables. Coefficients k_p and k_d can be tuned by "trial-and-error" approach or by using artificial intelligence methods, as genetic algorithms [OS07,OS04].

In Figure 25.12, the MRAS structures for the induction motor drive and DC motor drive, with the neuro-fuzzy speed controllers of PI-type and above adaptation algorithm are demonstrated.

In the case of IM drive, the control structure is based on the direct field-oriented control (DFOC) concept, with decoupling circuits for linearization of the voltage-fed induction motor [KT94]. For obtaining the sensorless drive system, suitable rotor flux and speed estimators are used [ODS06]. Examples of drive system transients, for rectangular and sinusoidal reference speed traces are demonstrated in Figure 25.13 through Figure 25.15, for the IM [ODS06] and DC [JO04] drive systems, respectively. It is worth saying that in both cases, the initial values of ANF speed controller are set to zeros (see Figures 25.13c, f, 25.14c and 25.15c). During the adaptive process, their values tend to optimal ones, which ensure drive system

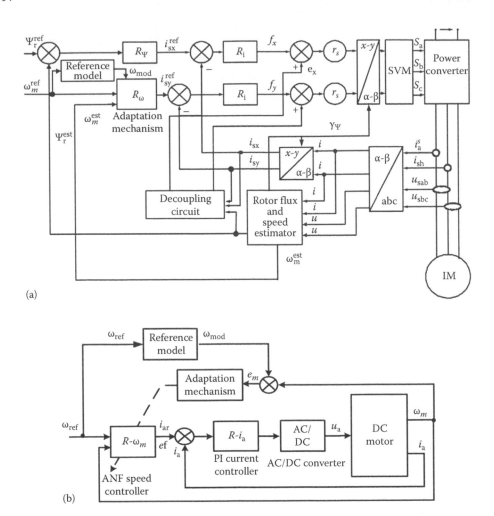

(a)

(b)

FIGURE 25.12 Structures of the MRAS speed control for the field-oriented control of (a) the IM drive and (b) the DC motor drive and with online tuned adaptive neuro-fuzzy speed controllers.

dynamics given by the reference model. Practically, after 1–2 operation periods, the system speed tracks the reference speed almost perfectly. It can be seen, especially in the case of the sinusoidal speed reference transients (Figures 25.14a and 25.15a). It proves that the algorithm of controller parameter adaptation works very well and system transients are optimal in the sense of reference model tracking performance, even for changing mechanical time constant of the drive system.

A faster dumping of the drive speed transients, resulting from the initial error between the reference model speed and the motor speed (due too initial weight factors of the NF controller equal zero), can be obtained using the sliding-mode (or PD) neuro-fuzzy speed controller [OS08], which differs from the described above in the lack of the output integrator of Δu signal in Figure 25.11. More results concerning these adaptation methods can be seen in [JO04,ODS06,OS07,CC98,LWC98,LFW98,LLS01,OS08].

25.6 Neurocontrol of Electrical Drives as Special Case of Adaptive Regulators

Recently, NN are widely used in the control of processes dynamics, resulting in the new field called *neurocontrol* that can be considered as a unconventional branch of the adaptive control theory. Similarly, as in the classical theory, neural controllers can be used in the indirect (Figure 25.2) and direct

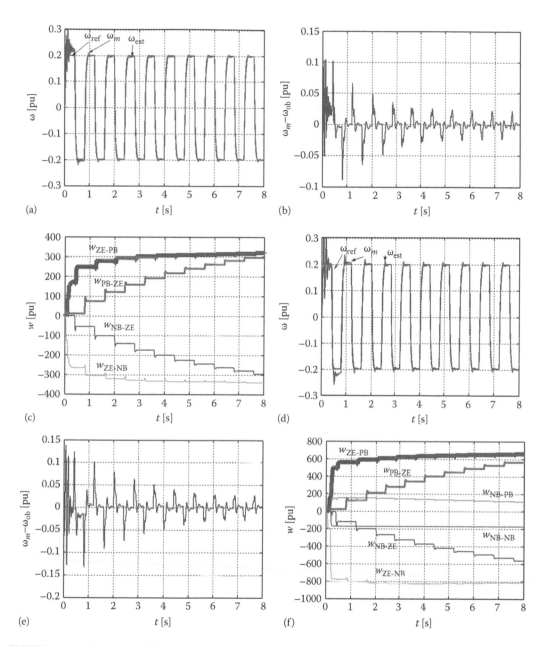

FIGURE 25.13 Transients of the IM drive with ANF controller for step changes of speed reference and $T_M = T_{MN}$ (a–c) and $T_M = 3T_{MN}$ (d–f): reference and rotor speeds (a, d), speed error (b, d), chosen controller output factors w_i (c, f).

(Figure 25.3) adaptive structures. Even in the case of the linear model of the plant, the adaptive control system is nonlinear one. So, the synthesis of the control strategy using analytical methods generates many problems and NN propose very attractive solution because of their well-known adaptation features [NP90,FS92,HIW95,NRPH00].

In Figure 25.16, general structures of the direct and indirect adaptive control systems with neural controllers for the electrical drives are demonstrated [NP90]. There are blocks with delay lines in these structures, which enable to memorize the suitable signals used with some delay by neural identifiers and controllers. In these structures, NN are trained online.

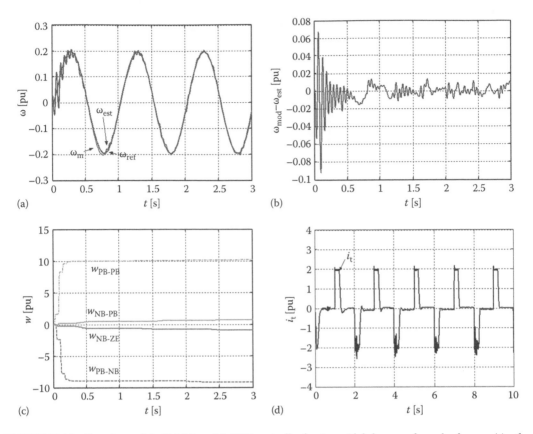

FIGURE 25.14 Transients of the IM drive with ANF controller for sinusoidal changes of speed reference: (a) reference and rotor speeds, (b) speed error, and (c) chosen controller output factors w_i.

In the direct structure (Figure 25.16a), some difficulties occur with the direct adjustment of the controller parameter based on the error $e_c(k)$, as this error is not directly accessible at the NN output and cannot be used for the corrections calculation of NN weight factors. It is especially difficult in the case of nonlinear plants, as their description (transfer function) is not known or changing with respect to the operation point. Nevertheless, with the assumption, that the plant is described by certain function $f_p(u(k))$, the backpropagation method can be only used, if the derivative $f_p'(u(k))$ is calculated. Generally, for nonlinear plants, this derivative is not known (and accessible), so its approximation must be used. It can be calculated according to [NRP00]

$$f_P'\big(u(k)\big) \approx \frac{\Delta f_P\big(u(k)\big)}{\Delta u(k)}, \tag{25.23}$$

where
$$\Delta f_p(u(k)) = f_p(u(k-1)) - f_p(u(k-2))$$
$$\Delta u(k) = u(k-1) - u(k-2)$$

The other solution consists in the application of the additional NN for modeling the plant. Such neural emulator of the plant enables the application of backpropagation method for the generation of the cluster of training samples for the neural controller based on the estimated output of the neural model $\hat{y}(k)$. It results in the natural transition from the direct to indirect control structure (Figure 25.16b).

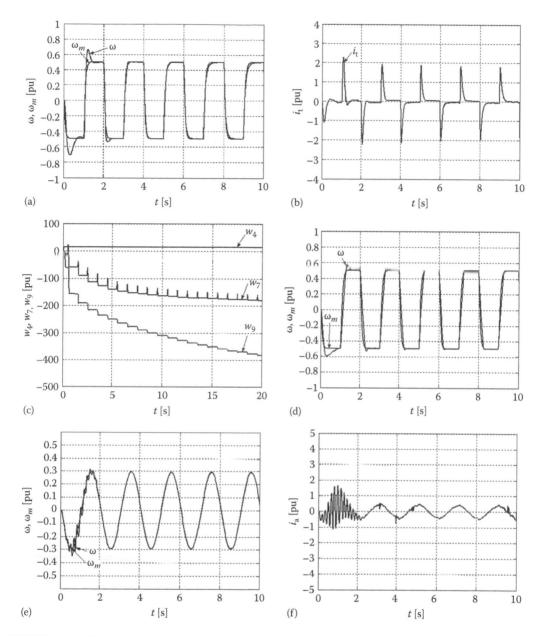

FIGURE 25.15 Transients of the DC drive with ANF controller for step (a–e) and sinusoidal (f, g) changes of speed reference and $T_M = T_{MN}$ (a–c) and $T_M = 2T_{MN}$ (d–g): reference and rotor speeds (a, d, f), armature current (b, e, g), and chosen controller output factors w_i (c).

In the indirect case, the nonlinear plant is parameterized assuming the suitable neural model structure, which parameters are adapted online on the base of identification error e_i. Then the controller parameters are adjusted by the identified model using the backpropagation method of the error between identified and reference model. Such backpropagation of the error is possible due to neural realization of the identifier [NP90,FS92].

The identification algorithms as well as control algorithms can be performed in each sampling interval or after the conversion of the data contained in certain limited period. The synchronized adaptation

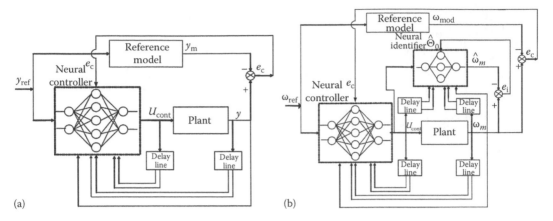

(a)   (b)

FIGURE 25.16   General structures of drive systems with neural controllers: (a) direct and (b) indirect.

of the controller and identifier is recommended in the case of lack of the external disturbances. In the other cases, the identification algorithm should be realized in every sampling interval but controller parameters can be adapted in the slower time scale. Such procedure enables robustness of the system to the external disturbances or noises and protects the proper operation of the system [HIW95,NRP00]. Some interesting application of NN to electrical drives control can be found in [WE93,BBT94,FS97,S99, GW04].

Below, chosen results obtained for the induction motor drive with simplified field-oriented control method (NFO method [JL95]) are demonstrated [GW04]. In Figure 25.17, the induction motor drive with neural speed controller in the indirect adaptive structure is shown. The neural model of

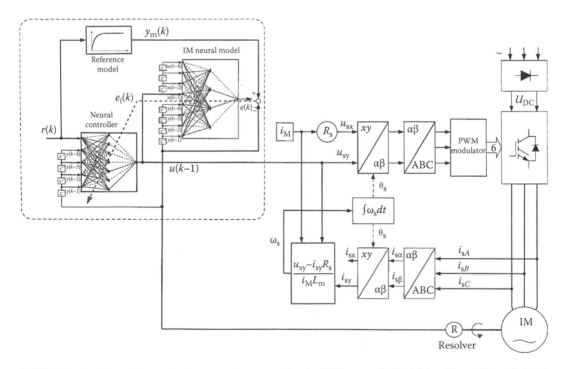

FIGURE 25.17   The indirect adaptive control structure for the NFO-controlled IM drive. (From Grzesiak, L. M. and Wyszomierski, D., *Elect. Rev.*, 53(1), 11, 2004. With permission.)

the converter-fed induction motor drive is based on the perceptron network with one hidden layer of nonlinear neurons, as shown in the figure. This network was trained *off-line* using input vector consisting actual and three delayed values of the rotor speed and voltage signal u_{sy}.

Neural controller has to minimize the control error

$$e(k) = y(k) - r(k) \tag{25.24}$$

and forms the control signal for the drive system:

$$u(k) = u_{sy}(k) = f_R\left(x_R(k)\right), \tag{25.25}$$

where

$$x_R(k) = \left[\omega_m^{ref}(k),\ \omega_m(k-1),\ \omega_m(k-2),\ \omega_m(k-3),\ \omega_m(k-4)\right]^T \tag{25.26}$$

while ω_m^{ref} is the reference speed of the drive system.

The error $e(k)$ is backpropagated through the neural model of the plant and thus the virtual error $e_i(k)$ is calculated, which next is used for weights modification of neural speed. controller. Thus, the modification of weights between the output and the hidden layer of the controller is made as follows:

$$\Delta w_j^{(R)}(k) = \eta e_i(k) y_j^{(R)}(k) \tag{25.27}$$

where

$\Delta w_j^{(R)}(k)$ are the corrections for weights between the hidden and output layer of the regulator

$y_j^{(R)}(k)$ is the output of the hidden layer

And for the connections between the hidden and the input layer of the controller are made as follows:

$$\Delta w_{ji}^{(R)}(k) = \eta \delta_j^{(R)}(k) y_i^{(R)}(k);$$
$$\delta_j^{(R)}(k) = \left(1 - (y_i^{(R)}(k))^2\right) w_{ji}^{(R)}(k) e_i(k) \tag{25.28}$$

where

$\delta_j^{(R)}$ is the error of the hidden layer

$y_i^{(R)}$ is the output signal of the input layer

$w_{ji}^{(R)}$ are the weights of the hidden and input layer of the regulator

In Figure 25.18, an example of the drive speed transient under reverse operation of the drive system is shown. It illustrates the effectiveness of the speed adaptation process.

The initial weights of neural controller are randomly chosen and as the neural controller is not trained, the initial transient error occurs in the drive system output speed. Next, weights are adapted online during the system operation and the error between the output of reference model and the drive system is minimized *online* and after few seconds, no speed overshoot is visible. It is seen that simple feedforward NN can perform well as controllers of highly nonlinear induction

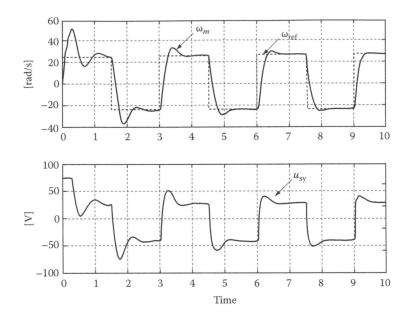

FIGURE 25.18 Transients of the reference and motor speed during reverse operation in the drive system with adaptive neural controller. (From Grzesiak, L. M. and Wyszomierski, D., *Elect. Rev.*, 53(1), 11, 2004. With permission.)

motor drive system. Although the additional NN used as process identifier requires additional training procedure performed off-line, it gives the possibility of easy transportation of the control error back and application of well-known backpropagation method to the online training of the second NN used as the neural controller.

25.7 Summary

In this chapter, the overview of adaptive control strategies used in electrical drives is presented. Starting from the control theory, main concepts of the adaptive control are shortly described and evaluated. Next, the application of gain scheduling scheme, STR, and model reference control schemes to electrical drive systems is shown. For each adaptive control strategy, the examples of DC or AC drive system are presented and applied algorithms are discussed. In the final part of the chapter, NN as a special case of adaptive controllers are introduced. The dynamical performances of all discussed systems have been evaluated and illustrated by the experimental transients obtained in the laboratory drive systems. It was shown that adaptive control concepts are very effective in electrical drives with changeable parameters.

References

[ÄW95] K.J. Äström and B. Wittenmark, *Adaptive Control*, 2nd edn., Addison-Wesley, New York, 1995.

[BBT94] M. Bertoluzzo, G. Buja, and F. Todesco, Neural network adaptive control of DC drive, *Proceedings of IECON*, Bologna, Italy, pp. 1232–1236, 1994.

[BSS90] A. Buxbaum, K. Schierau, and A. Straughen, *Design of Control Systems for DC Drives*, Springer-Veralg, Berlin, Germany, 1990.

[CT98] Y.C. Chen and C.C. Teng, A model reference control structure using a fuzzy neural network, *Fuzzy Sets and Systems*, 73, 291–312, 1995.

[FS92] F. Fukuda and T. Shibata, Theory and applications of neural networks for industrial control systems, *IEEE Transactions on Industrial Electronics*, 39(6), 472–489, 1992.

[FS97] K. Fischle and D. Schröder, Stable model reference neurocontrol for electric drive systems, *Proceedings of 7th European Conference on Power Electronics and Applications (EPE'97)*, Trondheim, Norway, vol. 2, pp. 2.432–2.437, 1997.

[GW04] L.M. Grzesiak and D. Wyszomierski, Adaptive control of AC drive based on reference model structure with neural speed controller, *Electrical Review*, 53(1), 11–16, 2004.

[HIW95] K.J. Hunt, G.R. Irvin, and K. Warwick (eds.), *Neural Network Engineering in Dynamic Control Systems*, Springer-Verlag, Berlin, Germany, 1995.

[JL95] R. Jonsson and W. Leonhard, Control of induction motor without mechanical sensor, based on the principle of natural field orientation, *Proceedings of the IPEC'95*, Yokohama, Japan, 1995.

[JO04] K. Jaszczak and T. Orlowska-Kowalska, Adaptive fuzzy-neuro control of DC drive system, *Proceedings of the 8th International Conference on Optimization of Electrical and Electronic Equipment (OPTIM'04)*, Romania, vol. 3, pp. 55–62, 2004.

[KO02] M.P. Kazmierkowski and T. Orlowska-Kowalska, NN state estimation and control in converter-fed induction motor drives, Chapter 2. In: *Soft Computing in Industrial Electronics*, Physica-Verlag, Springer, New York, pp. 45–94, 2002.

[KT94] M.P. Kazmierkowski and H. Tunia, *Automatic Control of Converter-Fed Drives*, Elsevier/PWN, Amsterdam, the Netherlands, 1994.

[L85] W. Leonhard, *Control of Electric Drives*, Springer-Verlag, Berlin, Germany, 1985.

[LFW98] F.J. Lin, R.F. Fung, and R.J. Wai, Comparison of sliding-mode and fuzzy neural network control for motor-toggle servomechanism, *IEEE Transactions on Mechatronics*, 3(4), 302–318, 1998.

[LLS01] F.J. Lin, C.H. Lin, and P.H. Shen, Self-constructing fuzzy neural network speed controller for permanent-magnet synchronous motor drive, *IEEE Transactions on Fuzzy Systems*, 9(5), 751–759, 2001.

[LWC98] F.J. Lin, R.J. Wai, and H.P. Chen, A PM synchronous servo motor drive with an on-line trained fuzzy neural network controller, *IEEE Transactions on Energy Conversion*, 13(4), 319–325, 1998.

[NP90] K.S. Narendra and K. Parthasarathy, Identification and control of dynamical systems using neural networks, *IEEE Transactions on Neural Networks*, 1(1), 4–27, 1990.

[NRPH00] M. Norgaard, O. Ravn, N.K. Poulsen, and L.K. Hansen, *Neural Networks for Modeling and Control of Dynamic Systems,* Springer, London, U.K., 2000.

[ODS06] T. Orlowska-Kowalska, M. Dybkowski, and K. Szabat, Adaptive neuro-fuzzy control of the sensorless induction motor drive system, *Proceeding of 12th International Power Electronics and Motion Control Conference (PEMC'2006)*, Portoroz, Slovenia, pp. 1836–1841, 2006.

[OJ03] T. Orlowska-Kowalska and K. Jaszczak, Sensorless adaptive fuzzy-logic control of DC drive with neural inertia estimator, *Journal of Electrical Engineering*, 3, 39–44, 2003.

[OS03] T. Orlowska-Kowalska and K. Szabat, Sensitivity analysis of state variable estimators for two-mass drive system, *Proceedings of the 10th European Power Electronics Conference (EPE'03)*, CD, Toulouse, France, 2003.

[OS04] T. Orlowska-Kowalska and K. Szabat, Optimisation of fuzzy logic speed controller for DC drive system with elastic joints, *IEEE Transactions on Industrial Applications*, 40(4), 1138–144, 2004.

[OS07] T. Orlowska-Kowalska and K. Szabat, Neural networks application for mechanical variables estimation of two-mass drive system, *Transactions on Industrial Electronics*, 54(3), 1352–1364, 2007.

[OS07] T. Orlowska-Kowalska and K. Szabat, Control of the drive system with stiff and elastic couplings using adaptive neuro-fuzzy approach, *IEEE Transactions on Industrial Electronics*, 54(1), 228–240, 2007.

[OS08] T. Orlowska-Kowalska and K. Szabat, Damping of torsional vibrations in two-mass system using adaptive sliding neuro-fuzzy approach, *IEEE Transactions on Industrial Informatics*, 4(1), 47–57, 2008.

[S99] D.L. Sobczuk, Application of ANN for control of PWM inverter fed induction motor drives, PhD thesis, Warsaw University of Technology, Warsaw, Poland, 1999.

[SB89] S. Sastry and M. Bodson, *Adaptive Control: Stability, Convergence and Robustness*, Prentice-Hall Inc., Englewood Cliffs, NJ, 1989.

[SB98] R. Sutton and A. Barto, *Reinforcement Learning: An Introduction*, MIT Press, Cambridge, MA, 1998.

[SO08] K. Szabat and T. Orlowska-Kowalska, Performance improvement of industrial drives with mechanical elasticity using nonlinear adaptive Kalman filter, *IEEE Transactions on Industrial Electronics*, 55(3), 1075–1084, 2008.

[SOD06] K. Szabat, T. Orlowska-Kowalska, and K. Dyrcz, Application of extended Kalman Filters in the control structure of two-mass system, *Bulletin of Polish Academy of Sciences*, 54(3), 315–325, 2006.

[WE93] S. Weerasooriya and M. El-Sharkawi, Laboratory implementation of neural network trajectory controller for a DC motor, *IEEE Transactions on Energy Conversion*, 8(1), 107–113, March 1993.

26

Drive Systems with Resilient Coupling

Teresa
Orłowska-Kowalska
Wroclaw University
of Technology

Krzysztof Szabat
Wroclaw University
of Technology

26.1 Introduction

The chapter presents a variety of commonly used control methods for vibration damping in two-mass drive systems. The finite stiffness of the shaft causes torsional vibrations that affect the drive system performance significantly. Torsional vibrations limit the performance of many industrial drives. They decrease the system reliability, product quality, and in some specific cases, they can even lead to instability of the whole control structure. The problem of damping of torsional vibrations originates from the rolling-mill drive, where large inertias of the motor and load parts with a long shaft create an elastic system [HSC99,SO07,PS08,DKT93]. Similar problems exist in paper and textile industry, where the electromagnetic torque goes through complex mechanical parts of the drive [VBL05,PAE00]. The damping ability of the system is also a critical issue in conveyer and cage-host drives [HJS05,HJS06]. Originally, the elastic system has been recognized in high-power applications; however, due to the progress in power electronic and microprocessor systems, which allow the electromagnetic torque control almost without delay, the torsional vibrations appear in many medium- and small-power applications. Today, they are acknowledged in servo drives, throttle drives, robot arm drives including space applications, and others [VS98,EL00,VBPP07,OBS06,FMRVR05].

26.2 Mathematical Model of the Drive

In the analysis of the drive system with a flexible coupling, the following models can be used [HSC99, SO07,PS08,DKT93,VBL05]:

- The model with distributed parameters
- The Rayleigh model
- The inertia-shaft-free model

The selection of the suitable model is a compromise between the obtained modeling accuracy and calculation complexity. In the model with distributed parameters, it is assumed that the inertias of the motor, shaft, and load machine are split through the axis of movement. This model can ensure the best accuracy of results; it is characterized by an infinite degree of freedom. However, the equations describing the system are the partial differential equations with the form inconvenient for control structure analysis. Therefore, in the analysis of the system with flexible joints, the different models, which reduce the three-dimensional phenomena, are utilized commonly.

The Rayleigh model takes into consideration continuous distribution of inertia, but it also assumes linear distribution of the mechanical stress along the mechanical system. This model is used when the inertia of the shaft is comparable to the inertia of the motor and the load machine.

When the moment of the shaft inertia J_s is small in comparison to the moments of inertia concentrated on its ends, the inertia-shaft-free model should be used. The moment of the shaft inertia should be divided by two and added to the inertia of the motor J_e and the load machine J_o, according to the following equations:

$$J_1 = J_e + \frac{J_s}{2} \tag{26.1}$$

$$J_2 = J_o + \frac{J_s}{2} \tag{26.2}$$

Such model is widely used in the analysis of the system with flexible connection (in more than 99% of published papers). The drive system is very often composed of the motor connected to the load machine through the shafts and mechanical gearboxes, which are applied in order to reduce the speed from the motor to the load side. The moment of inertia of these gearboxes is far bigger than the shaft inertias. This inertia should be taken into consideration in the mathematical model of the system. It leads to formulating three- or multimass inertia-shaft-free models. A more complicated model allows obtaining better accuracy of the calculation, but it increases computational complexity. The right choice of the model order is especially important in sensorless drives, where special estimation methods have to be applied so as to reconstruct the nonmeasurable state variables that are necessary to ensure effective damping of torsional vibrations. Nevertheless, the influence of the additional degree of freedom on the drive system dynamics is usually neglected and the simplest two-mass system model is considered.

The two-mass inertia-shaft-free model is described by the following equations:

$$T_1 \frac{d\omega_1}{dt} = m_e - m_S \tag{26.3}$$

$$T_2 \frac{d\omega_2}{dt} = m_S - m_L \tag{26.4}$$

$$T_c \frac{dm_S}{dt} = \omega_1 - \omega_2 \tag{26.5}$$

where
  ω_1 is the motor speed
  ω_2 is the load speed
  m_e is the electromagnetic torque
  m_S is the shaft torque
  m_L is the load torque
  T_1 is the mechanical time constant of the motor
  T_2 is the mechanical time constant of the load machine
  T_c is the stiffness time constant

The schematic diagram of the two-mass system is presented in Figure 26.1.

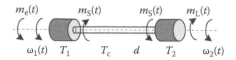

FIGURE 26.1   The schematic diagram of the two-mass system.

26.3 Methods of Torsional Vibration Damping

Torsional vibrations can appear in a drive system due to the following reasons [HSC99,SO07,PS08, DKT93,VBL05,PAE00,HJS05,VS98]:

- Changeability of the reference speed
- Changeability of the load torque
- Fluctuation of the electromagnetic torque
- Limitation of the electromagnetic torque
- Mechanical misalignment between the electrical motor and load machine
- Variations of load inertia
- Unbalance of the mechanical masses
- System nonlinearities, such as friction torque and backlash (especially, in the low-speed operation)

In order to suppress torsional oscillations, different control methods have been developed, from classical to the advanced ones. The simplest method to avoid the system state variable oscillations relies on decreasing the dynamics of the control structure, yet this method neglects the performance of the drive and is hardly ever utilized. The commonly utilized approaches can be divided into two major groups:

1. Passive methods, i.e., utilization of the mechanical dampers [ZFS00] and application of the digital filters [PAE00,VS98,EL00,DM08]
2. Active methods that include the application of the control structures based on the modern control theory [HSC99,SO07,HJS06,SO08]

Both mentioned methods allow damping the torsional vibration effectively. The passive methods though possess some series drawbacks. Mechanical dampers have to be fixed to the driving system and hence additional space is required. These elements influence the system reliability in a negative way; besides, they increase the total cost of the system. Moreover, mechanical dampers do not allow shaping the responses of the two-mass system in wide ranges. In the system with a high value of the resonant frequency, the application of the digital filters is an industrial standard. Although they can damp the vibrations effectively, the dynamics of the system can be affected. Additionally, changes of the system parameters can affect the properties of the drive significantly. The active methods allow damping the vibrations successfully and at the same time, shaping the responses of the system, which is one of their major advantages. Because modern industrial drives use microprocessor systems, the application of the advanced control structures does not increase the cost of the drive but the computational complexity of these systems is considered sometimes a serious drawback.

26.4 Passive Methods

The application of the mechanical dampers for torsional vibration dumping of the two-mass system brings about the complexity of the mechanical part of the drive and is not widely used. Such solution is discussed in [ZFS00] and used for the suppression of torsional vibrations of the robot arm. From the point of view of the control theory, the utilization of the mechanical damper causes the increase of the

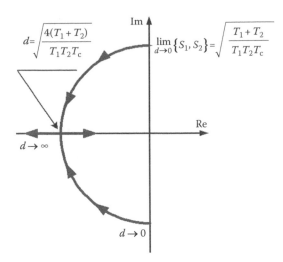

FIGURE 26.2 The effect of the increased value of the internal damping coefficient d of the shaft to the system poles location.

value of the internal damping coefficient d. This shifts the poles of the two-mass system from the imaginary axis to the real axis according to the following equation:

$$s_{1,2} = \frac{-d(T_1 + T_2) \pm (T_1 + T_2)\sqrt{d^2 - \dfrac{4T_1T_2}{(T_1 + T_2)T_c}}}{2T_1T_2} \tag{26.6}$$

The effect of the increasing value of the coefficient d (from zero to infinity) to the location of the two system poles is presented in Figure 26.2.

The next passive method utilizes digital filters for the suppression of torsional vibrations [PAE00, VS98,EL00,DM08]. The filter is located between speed and torque controllers as shown in Figure 26.3.

Usually, the Notch filter is mentioned as a tool ensuring damping of the torsional oscillations. However, as shown in [VS98], the exact cancellation of resonance modes is possible only if all the parameters of the system are precisely known. So, it is claimed that the identification of the plants is a serious problem in tuning the Notch compensator. Additionally, even small variations of drive system parameters may lower the restricted value of system damping coefficient. Consequently, the Notch series compensator can reduce the resonance modes but it cannot eliminate them completely.

The other filter that is widely used in the drive system with resonant modes is the FIR filter. The idea of the FIR filter application relies on the fact that the periodical oscillations in a signal with period T can be eliminated by adding an identical signal delayed by $T/2$ to the original signal. In the case of parameter changes, the FIR filter can damp the oscillations more effectively than the Notch filter [VS98]. Contrary

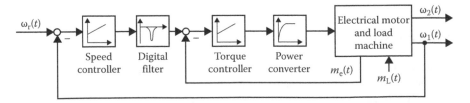

FIGURE 26.3 Classical control structure with additional digital filter.

to [VS98], other authors [PAE00] claim that the implementation of Notch filter requires no knowledge of the plant and the filter parameters can be easily set experimentally. They suggest that Notch filter could be used in the system with middle value of resonant frequency and the FIR filter in the system with a high value of resonant frequency.

A comparison between dynamical properties of the low-pass filter, Notch filter, and Bi-filter is presented in [EL00]. Authors suggest the application of the low-pass filter in the case of the resonant frequency higher than 1 kHz. Still, it can cause the loss of the system dynamics when the value of resonant frequency is small. Bi-filter can ensure the smoother motor speed transient because its application ensures the compensation of antiresonant and resonant frequencies of the system. However, Bi-filter is very sensitive to system parameter changes. The Notch filter is found to be the optimal solution from all analyzed filters, for compensation of low and middle values of resonant frequencies.

26.5 Modification of the Classical Control Structure

The classical cascade control structure of the two-mass drive system is presented in Figure 26.4. It consists of the optimized inner torque control loop. The mechanical time constants of the mechanical part of the drive are much bigger than the time constant of the inner torque loop. Therefore, for the speed controller synthesis, the delay of this optimized inner loop is usually neglected [SO07].

The closed-loop system with the PI controller is of the fourth order. Because there are only two parameters of the PI controller, it is not possible to locate all poles of the control structure independently. Usually, the double location of the poles is selected. In this case, parameters of the controller are set using following equations:

$$K_P = 2\sqrt{\frac{T_1}{T_c}}, \quad K_I = \frac{T_1}{T_2 T_c} \tag{26.7}$$

The dynamical characteristics of the drive system depend on the inertia ratio of the load and motor sides, defined as $R = T_2/T_1$. The decrease of the R value causes the bigger and slowly damped oscillations in the system step response. The oscillations are eliminated in the system with bigger value of R, yet at the same time, the dynamic is lost. In Figure 26.5, the transients of the investigated control structure for different value of R are presented (all presented results were obtained for the following mechanical parameters drive system: $T_1 = T_2 = 203\,\text{ms}$, $T_c = 2.6\,\text{ms}$).

To improve the performances of the classical control structure with the PI controller, the additional feedback loop from one selected state variable can be used. The additional feedback allows setting the desired value of the damping coefficient, yet the free value of the resonant frequency cannot be achieved simultaneously. The additional feedback can be inserted to the electromagnetic torque control loop or to the speed control loop.

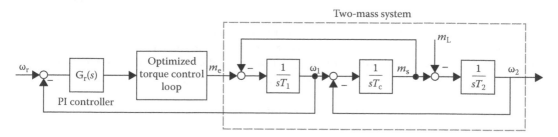

FIGURE 26.4 Classical control structure with basic feedback from the motor speed.

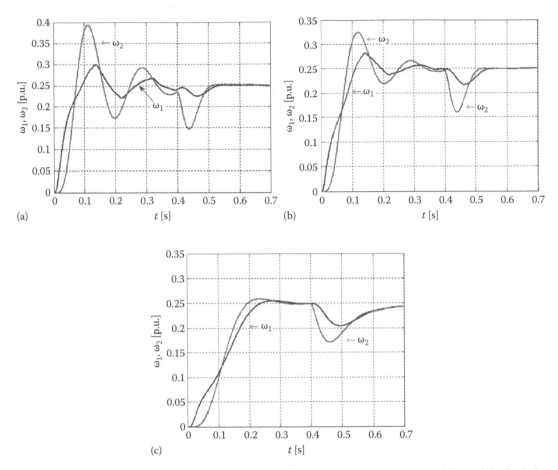

FIGURE 26.5 Transients of motor and load speeds of the two-mass system without additional feedback for (a) $R = 0.5$, (b) $R = 1$, and (c) $R = 2$.

In [SH96], the additional feedback from derivative of the shaft torque inserted to the electromagnetic torque node was presented. The authors investigated the proposed method and applied it to the two- and three-mass system. Nevertheless, the proposed estimator of the shaft torque is quite sensitive to measurement noises, so the suppression of high-frequency vibrations is difficult and additionally the system becomes less dynamic. Another modification of the control structure results from inserting the additional feedback from the shaft torque. This type of feedback is utilized in many works, i.e., [PAE00, OBS06]. The damping of the torsional vibrations is reported to be successful.

In the paper [PAE00], the feedback from the difference between motor and load speed is utilized. Although the oscillations were successfully suppressed, the authors claim the loss of response dynamics and large load impact effect. The additional feedback from the derivative of the load speed is proposed in [Z99], resulting in the same dynamic performance as for the previous control structure. Another possible modification of the classical structure is based on the insertion of an additional feedback to the speed control loop, e.g., [PAE00]. The authors argue that this feedback can ensure good dynamic characteristics and is able to damp the vibrations effectively. The same results can be obtained by applying the feedback from difference between motor and load speed.

In the paper [SO07], nine different control structures with one additional feedback are analyzed. It is shown that all these systems can be divided into three different groups according to their dynamical

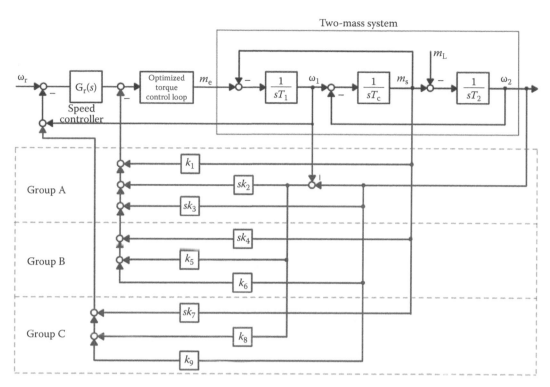

FIGURE 26.6 The control structure with different additional feedbacks.

characteristics. The analyzed structures are presented in Figure 26.6. The equations allowing setting the parameters of the control structure are presented in [SO07].

In Figure 26.7, the closed-loop poles loci of all considered control systems belonging to groups A–C and suitable load speed responses are presented. These systems are of the fourth order and the presented poles are double (Figure 26.7a). The location of the closed-loop poles of the system without the additional feedback depends only on the mechanical parameters of the drive. The system poles are situated relatively close to the imaginary axis. The response of the drive system has quite a large overshoot and settling time. The closed-loop poles location of the system with one additional feedback depends on the assumed damping coefficient, which in each case was set to $\xi_r = 0.7$.

The closed-loop poles of the system from group B (in this case B_1) have the highest value of the resonant frequency. The rising time of the speed response of the mentioned drive is approximately twice as short as that of the remaining systems. The next fastest system is the control structure belonging to group A. The dynamic characteristics of the remaining structures (group C and group B_2) are quite similar. The shape of the load speed transients of all considered systems (Figure 26.7b) confirms the closed-loop poles location analysis. It should be stressed that the above comparison has been provided for selected parameters of the two-mass system and is not universal.

To obtain a free design of the control structure parameters, i.e., the resonant frequency and the damping coefficient, the application of two feedbacks from different groups is necessary. The type of the selected feedback from a particular group is not significant, because, as was said before, feedbacks belonging to the prescribed group give the same results. In Figure 26.8, the system speed transients for two required values of the resonant frequency and the damping coefficient $\xi_r = 0.7$ are presented. It is clear that the system dynamics can be programmed freely in the linear range of the work [SO07].

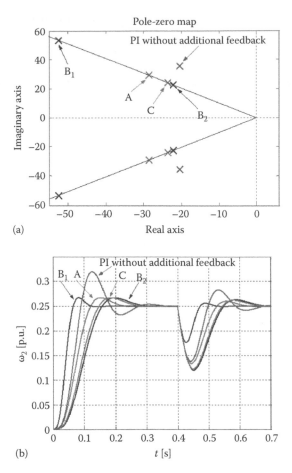

FIGURE 26.7   (a) The closed-loop poles location and (b) the load speed transient of all considered system.

26.6 Resonance Ratio Control

One of the popular control structures of the two-mass system is Resonance Ratio Control (RRC) [HSC99,LH07,OBS06,KO05]. It is commonly applied to the linear or nonlinear (with backlash) systems. The basic idea of the system is to estimate and feed back to the control structure the estimated value of the shaft torque. From this point of view, it is very similar to the structure presented in the previous section. The main difference is the design methodology. In the previous structure, the value of the feedback is determined using poles-placement method. In the RRC structure, though, the feedback gain is calculated on the basis of the frequency characteristics of the system (original and desired). The system is said to have good damping ability when the ratio of the resonant to antiresonant frequency H has a relatively big value (about 2). The block diagram of the control structure is presented in Figure 26.9.

The control structure consists of the PI speed controller, normalization factor ($k_{RC}(T_1 + T_2)$), additional feedback from the estimated shaft torque with a suitable gain, shaft torque estimator, and an optimized torque control loop. The shaft torque estimator includes the high-pass filter with a suitable value of T_q. This value is a compromise between the noise level in the system and the delay time in the estimated transients [HSC99].

The parameters of the PI controller are set using one of the popular methods. Hypothetical transients of the two-mass system for different value H are presented in Figure 26.10.

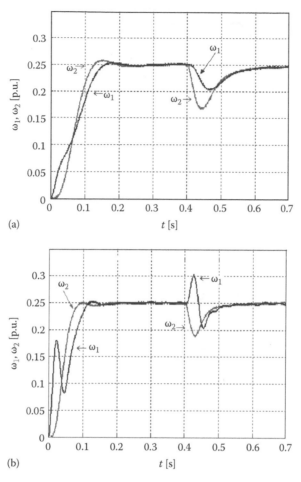

(a)

(b)

FIGURE 26.8 The transients of the two-mass system with two additional feedbacks (k_2, k_8) for $\xi_r = 0.7$ and two different values of the resonant frequency: (a) $\omega_r = 40\,\text{s}^{-1}$ and (b) $\omega_r = 60\,\text{s}^{-1}$.

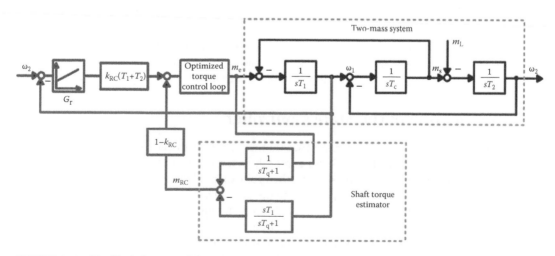

FIGURE 26.9 The block diagram of the resonance ratio control structure.

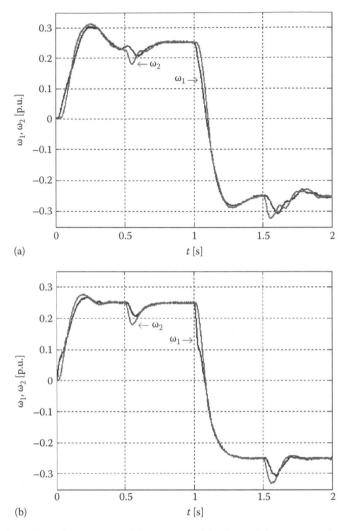

(a)

(b)

FIGURE 26.10 A hypothetical transients of the motor and load speed for setting value of the resonance to antiresonance frequency (a) $H = 1.2$ and (b) $H = 2$.

As can be concluded from Figure 26.10, the increasing value H allows to damp the system oscillation better. The overshoots and the system oscillations are reduced. It should be stated that due to the noises in the estimated transient of the shaft torque, it is impossible to set a relatively big value of H.

26.7 Application of the State Controller

The control structures presented so far are based on the classical cascade compensation schemes. Since the early 1960s, a completely different approach to the analysis of the system dynamics has been developed—the state space methodology. The application of the state space controller allows placing the system poles in an arbitrary position; so theoretically, it is possible to obtain any dynamic response of the system. The suitable location of the closed-loop system poles becomes one of the basic problems of the state space controller application. In [JS95], the selection of the system poles is realized through LQ approach. The authors emphasize the difficulty of the matrices selection in the case of the system parameter variation. The influence of the closed-loop location on the dynamic

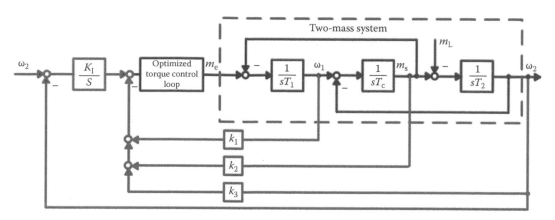

FIGURE 26.11 The two-mass drive system with the state space controller.

characteristics of the two-mass system is analyzed in [QZLW02,SHPLL01] In [SHPLL01], it is stated that the location of the system poles in the real axes improves the performance of the drive system and makes it more robust against the parameter changing. The control structure with the state controller is presented in Figure 26.11.

Because there are four parameters of the control structure, the independent location of all closed-loop poles is possible. It means that in the linear range of the work, the shape of the load speed transient can be set freely. In Figure 26.12, transients of the motor and load speeds as well as electromagnetic and load torques are presented.

Figure 26.12a and b show transients of the system for the assumed value of the resonant frequency $\omega_r = 30\ s^{-1}$ (a) and $\omega_r = 50\ s^{-1}$ (c) and the damping coefficient $\xi_r = 0.7$. The raising time of the load speed is about 120 ms (a) and 80 ms (b). Therefore, the shape of the load speed response can be set freely within some range of the system parameters. However, it should be emphasized that dynamics can be set freely only in the linear range of the work (below the maximal limit of the electromagnetic torque).

26.8 Model Predictive Control

In the industrial applications, the fulfillment of the process limitations plays a very important role. Usually, the limitation of the control signal due to the actuator saturation is taken into account. For the systems with PI/PID controllers, different antiwindup structures have been developed and presented in the [GS03]. In many control problems, there are also constraints on other variables. For instance, in the two-mass drive system, the limitation of the shaft torque has to be taken into account due to the following reasons. First, for the safety operation, the load machine can accept only some maximal value of the incoming torque and exceeding this value can damage the load machine. Second, the shaft can undertake a specific value of the torque resulting from its geometry and used material. The bigger value can also damage it and lead to the failure of the entire drive system. In spite of the problem significance, the effects of output constraints are often omitted in the control structure design. Usually, the output constraints are fulfilled by decreasing the gain of the control structure. Nevertheless, it leads to the loss dynamic of the control structure (also in the regions where constrains are not exceeded) and cannot be accepted in high-performance application.

The model predictive control (MPC) is one of the few techniques (apart from PI/PID techniques) that have a significant number of industry applications. It is generally used in chemical and process industry. The MPC algorithm adapts to the current operation point of the process generating optimal control signal. It is able to take the input and output constraints of the system directly into the controller design procedure, which is not easy in the control structure with the PI controller. So far, the real-time implementations of the MPC have been limited to objects with relatively large time constants. It results from

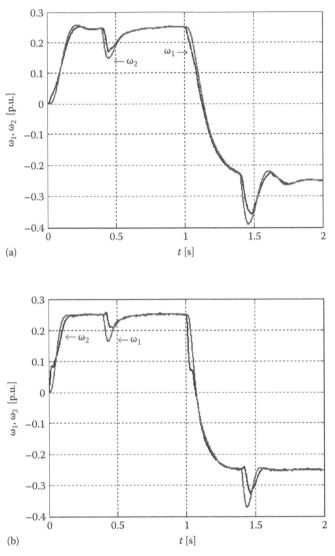

(a)

(b)

FIGURE 26.12 The transients of the two-mass system in space control structure for $\xi_r = 0.7$ and two different values of the resonant frequency: (a) $\omega_r = 30\,s^{-1}$ and (b) $\omega_r = 50\,s^{-1}$.

the fact that in every calculating step, there is a need to solve a complex optimization problem. Still, due to the progress of the microprocessor technique, the MPC strategy can be implemented in the time demanding system today, e.g., the two-mass system [CSO09]. The block diagram of the MPC control structure is shown in Figure 26.13 (where second subscript "e" appoints estimated values).

The MPC controller consists of the predictor and the optimizer (online or its explicit version [CSO09]). In every calculation step, the controller predicts the behavior of the system (in desired time) for the assumed number of the control signal. The bigger the prediction horizon, the better the achieved performance; yet, with the increased number of the prediction samples, the computational complexity increases drastically.

As has been stated earlier, the constraints of the system can be incorporated into the control algorithm. In order to illustrate these properties, the system with a different value of the shaft torque constraints is considered ($|m_s| < 1.25$ (Figure 26.14a and b); $|m_s| < 2$ (Figure 26.14c and d)).

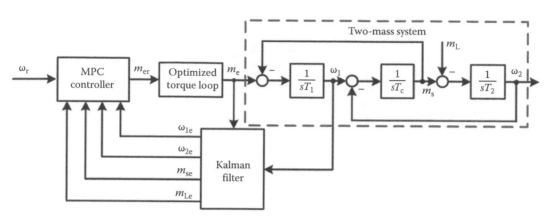

FIGURE 26.13 The block diagram of the MPC-based control structure.

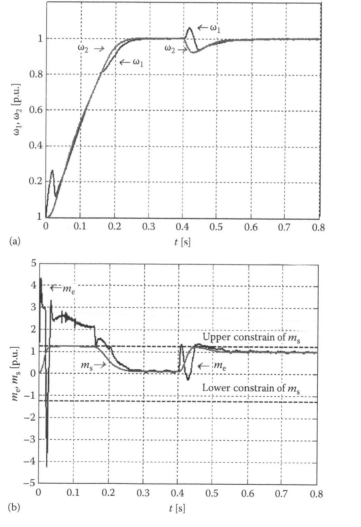

FIGURE 26.14 The transients of the MPC-based control structure for the system constrains $|m_s| < 1.25$ (a, b) and $|m_s| < 2$ (c, d).

(*continued*)

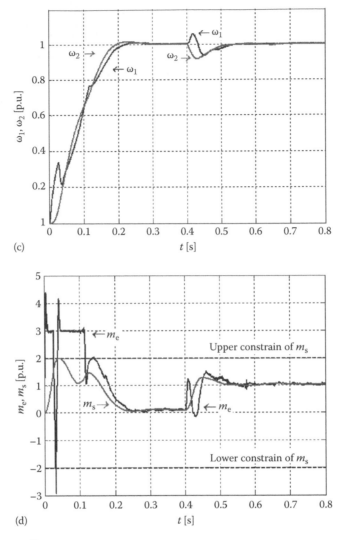

FIGURE 26.14 (continued)

The control algorithm keeps the shaft torque in the safety region. It is clearly visible in Figure 26.14b and d, that during the start up, the electromagnetic torque decreases to avoid the violation of the upper bound of shaft torque constraint. Setting of the different values of the electromagnetic and shaft torque constraints allows also minimizing the oscillations between the motor and load speed during the start up of the drive system.

26.9 Adaptive Control

For the system with changeable parameters, more advanced control concepts have been developed. In [IIM04], the applications of the robust control theory based on the H_∞ are presented. The genetic algorithm is applied to setting of the control structure parameters. The author reports good performance of the system despite the variation of the inertia of the load machine. The next approach consists in the application of the sliding-mode controller. For example, in paper [EKS99], this method is applied to controlling the SCARA robot. A design of the control structure is based on the Lyapunov function. A similar approach is used in [HJS05], where the conveyor drive is modeled as the two-mass system. The authors claim that the design structure is robust to parameter changes of the drive and external disturbances. Other application examples of the sliding-mode control can be found in [E08].

The next two frameworks of control approach rely on the use of the adaptive control theory. In the first framework, changeable parameters of the plant are identified and then the controller is retuned in accordance with the currently identified parameters (indirect adaptive control). The Kalman filter is applied in order to identify the changeable value of the inertia of the load machine [SO08]. This value is used to correct the parameters of the PI controller and two additional feedbacks. The similar adaptive strategy is presented in [HPH06].

In the other framework, the controller parameters are adjusted online on the basis of the comparison between the outputs of the reference model and the object. In [OS08], two adaptive neuro-fuzzy structures working in the MRAS structure are compared. The experimental results show the robustness of the proposed concept (direct adaptive control) against plant parameter variations.

The block diagram of the indirect adaptive control structure is shown in Figure 26.15. The main part of the control structure is composed of the linear PI controller supported by additional feedbacks from the shaft torque and the difference between the motor and the load speed. The parameters of this part are designed using the pole-placement method in order to meet the design specifications. The identification part of the control structure shown in Figure 26.15 is based on the Nonlinear Extended Kalman Filter (NEKF).

It provides information about the values of the mechanical time constant of the load machine T_2. Besides, it estimates the additional state variables, such as shaft torque, load speed, and the load torque, which are utilized in the control structure. The mathematical model of the used Kalman filter is presented in [SO08]. Figure 26.16 demonstrates transients of the motor and load speeds (a) as well as real and estimated load speed (b) electromagnetic, shaft and load torques (c), time constant of the load machine (d), and control structure parameters (e, f) for the adaptive control structure working the with NEKF. The shaft torque and the load speed, provided by the adaptive NEKF, are inserted to the feedback loops of the control structure. The estimated value of the time constant of the motor is used to change the parameters of the speed controller and gains of suitable feedbacks (K_I, K_P, k_1, k_2).

Because the load torque and the time constant of the motor are coupled, they are not estimated at the same time. The estimation of the time constant of the load machine is only activated when the control error is bigger than 0.25 (this value depends on a particular application). Simultaneously, the estimation

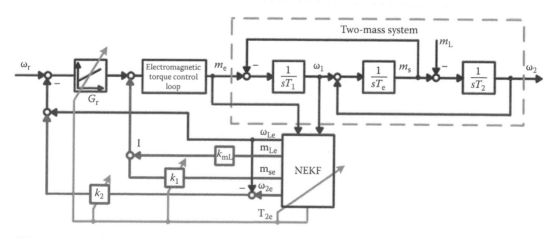

FIGURE 26.15 The block diagram of the control structure based on the indirect adaptive concept.

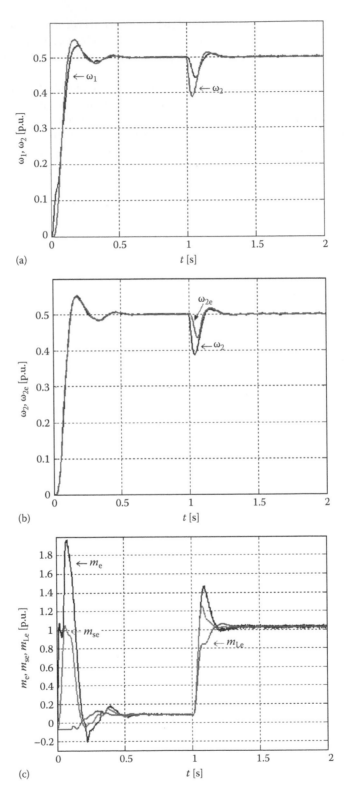

FIGURE 26.16 Transients of the real and estimated state variables: (a) motor and load speeds, (b) load speed and its estimate, (c) electromagnetic and estimated shaft and load torques,

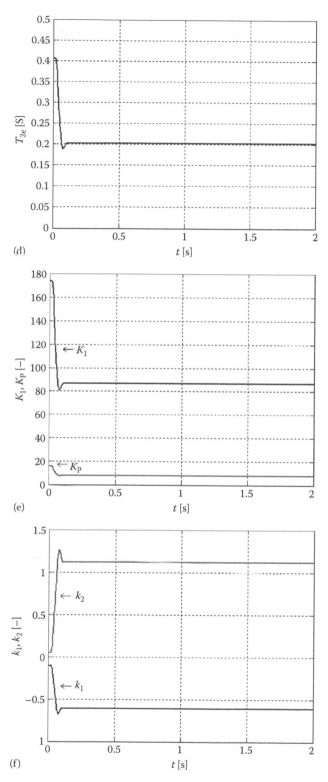

(d)

(e)

(f)

FIGURE 26.16 (continued) (d) estimated time constant of the load machine, and (e, f) adaptive control structure parameters.

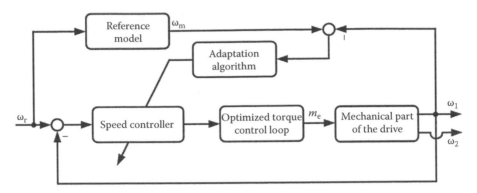

FIGURE 26.17 Structure of the direct adaptive control system.

of the load torque is terminated. The last estimated value of m_{Le} is given to the NEKF. During the drive system startup, the value of the estimated load torque is assumed to be zero. The estimation of T_2 is stopped, when the value of the control error drops to 0.01 and then the estimation of the load torque m_{Le} is activated. The value of the time constant T_{2e} utilized in the NEKF algorithm is set to its previously estimated value [SO08].

The direct adaptive control structure does not have the identification part, in contrary to the indirect adaptive concept. The controller parameters are adjusted according to the adaptation rule, depending on the currently measured model and system output variables. In Figure 26.17, the model reference adaptive control structure with the online tuning speed controller for the drive system with the elastic joint is presented [OS08]. An interesting feature of this control structure is the fact that it relies only on the motor speed.

The speed controller is tuned so that the actual drive output could follow the output of the reference model. The tracking error is used as the tuning signal. The system transients for two different ratios between the load and motor time constants, namely $R = 1$ (a, b, c) and $R = 0.25$ (d, e, f), are presented in Figure 26.18. In this case, the Sliding Neuro-Fuzzy Controller (SNFC) with retunable output weights is used as a speed controller [OS08].

In Figure 26.18a, the reference, motor, and load speeds are presented for 10 s of the system work. The system starts with the controller parameters set to zero. It means that the parameters of the drive are unknown. Despite this, the initial tracking error is small for all time of the work. The biggest difference between the load and the reference speed exists when the load torque changes rapidly (Figure 26.18b and e).

As can be concluded from the presented transients, the direct adaptive system damps torsional oscillations successfully for different values of the inertia ratio R, with the dynamics given by the reference model. This structure is very interesting due to its simplicity and requires only the motor torque (current) and speed measurements. Thus, no additional state variable observers are necessary. It should be emphasized that it cannot work properly for a very fast reference model.

26.10 Summary

The main goal of this chapter has been a presentation of different control structures that are used in order to damp torsional vibrations of the two-mass system. The different mathematical models that can be used for the analysis of a system with flexible connection are discussed. The application of the passive method is briefly described. The commonly used control structures have been shortly described. The dynamical performances of all discussed systems have been evaluated and illustrated by the experimental transients obtained in the laboratory two-mass drive system.

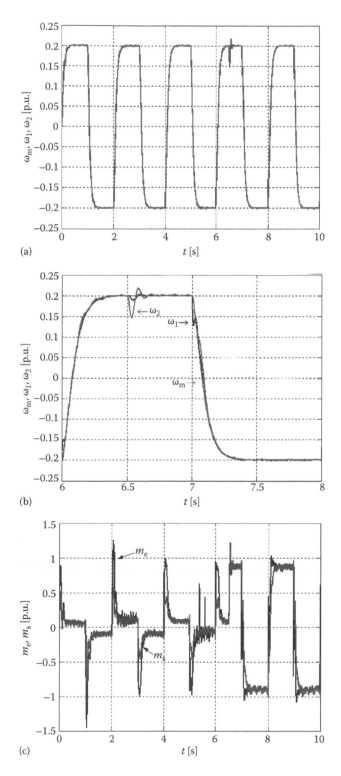

FIGURE 26.18 Transients of the two-mass system: the reference motor and load speeds (a, b, d, e) and electromagnetic and shaft torques (c, f), in the MRAS structure with SNFC speed controller for $R = 1$ (a, b, c) and $R = 0.25$ (d, e, f).

(*continued*)

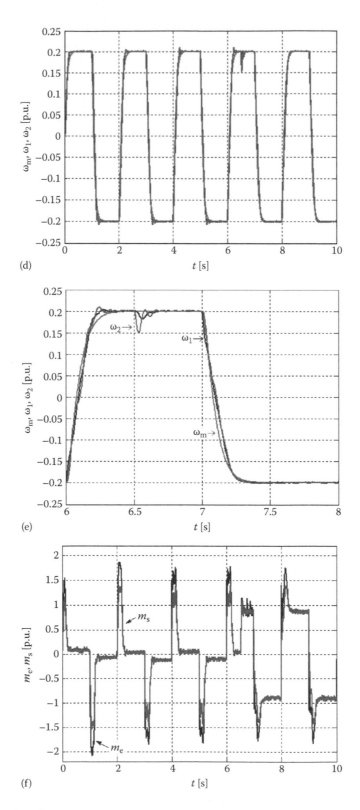

(d)

(e)

(f)

FIGURE 26.18 (continued)

References

[CSO09] M. Cychowski, K. Szabat, and T. Orlowska-Kowalska, Constrained model predictive control of the drive system with mechanical elasticity, *IEEE Transactions on Industrial Electronics*, 56(6), 1963–1973, 2009.

[DKT93] R. Dhaouadi, K. Kubo, and M. Tobise, Two-degree-of-freedom robust speed controller for high-performance rolling mill drivers, *IEEE Transactions on Industry Application*, 27(5), 919–925, 1993.

[DM08] J. Deskur and R. Muszynski, The problems of high dynamic drive control under circumstances of elastic transmission, *Proceedings of the 13th Power Electronics and Motion Control Conference EPE-PEMC 2008*, Poznan, Poland, 2008, pp. 2227–2234.

[E08] K. Erenturk, Nonlinear two-mass system control with sliding-mode and optimised proportional and integral derivative controller combined with a grey estimator, *Control Theory & Applications, IET*, 2(7), 635–642, 2008.

[EKS99] K. Erbatur, O. Kaynak, and A. Sabanovic, A study on robustness property of sliding mode controllers: A novel design and experimental investigations, *IEEE Transactions on Industrial Electronics*, 46(5), 1012–1018, 1999.

[EL00] G. Ellis and R. D. Lorenz, Resonant load control methods for industrial servo drives, *Proceedings of the IEEE Industry Application Society Annual Meeting*, Rome, Italy, 2000, vol. 3, pp. 1438–1445.

[FMRVR05] G. Ferretti, G. A. Magnani, P. Rocco, L. Vigano, and A. Rusconi, On the use of torque sensors in a space robotics application, *Proceedings on the IEEE/RSJ International Conference on Intelligent Robots and Systems (IROS'2005)*, Edmonton, Canada, 2005, pp. 1947–1952.

[GS03] A. H. Glattfelder and W. Schaufelberger, *Control Systems with Input and Output Constraints*, Springer, London, U.K., 2003.

[HJS05] A. Hace, K. Jezernik, and A. Sabanovic, Improved design of VSS controller for a linear belt-driven servomechanism, *IEEE/ASME Transactions on Mechatronics*, 10(4), 385–390, 2005.

[HJS06] A. Hace, K. Jezernik, and A. Sabanovic, SMC with disturbance observer for a linear belt drive, *IEEE Transactions on Industrial Electronics*, 53(6), 3402–3412, 2006.

[HPH06] M. Hirovonen, O. Pyrhonen, and H. Handroos, Adaptive nonlinear velocity controller for a flexible mechanism of a linear motor, *Mechatronics, Elsevier*, 16(5), 279–290, 2006.

[HSC99] Y. Hori, H. Sawada, and Y. Chun, Slow resonance ratio control for vibration suppression and disturbance rejection in torsional system, *IEEE Transactions on Industrial Electronics*, 46(1), 162–168, 1999.

[IIM04] D. Itoh, M. Iwasaki, and N. Matsui, Optimal design of robust vibration suppression controller using genetic algorithms, *IEEE Transactions on Industrial Electronics*, 51(5), 947–953, 2004.

[JS95] J. K. Ji and S. K. Sul, Kalman filter and LQ based speed controller for torsional vibration suppression in a 2-mass motor drive system, *IEEE Transactions on Industrial Electronics*, 42(6), 564–571, 1995.

[KO05] S. Katsura and K. Ohnishi, Force servoing by flexible manipulator based on resonance ratio control, *Proceedings of the IEEE International Symposium on Industrial Electronics ISIE'2005*, Dubrovnik, Croatia, 2005, pp. 1343–1348.

[LH07] W. Li and Y. Hori, Vibration suppression using single neuron-based PI fuzzy controller and fractional-order disturbance observer, *IEEE Transactions on Industrial Electronic*, 54(1), 117–126, 2007.

[OBS06] T. O'Sullivan, C. C. Bingham, and N. Schofield, High-performance control of dual-inertia servo-drive systems using low-cost integrated SAW torque transducers, *IEEE Transactions on Industrial Electronics*, 53(4), 1226–1237, 2006.

[OS08] T. Orlowska-Kowalska and K. Szabat, Damping of torsional vibrations in two-mass system using adaptive sliding neuro-fuzzy approach, *IEEE Transactions on Industrial Informatics*, 4(1), 47–57, 2008.

[PAE00] J. M. Pacas, J. Armin, and T. Eutebach, Automatic identification and damping of torsional vibrations in high-dynamic-drives, *Proceedings of the International Symposium on Industrial Electronics (ISIE'2000)*, Cholula-Puebla, Mexico, 2000, pp. 201–206.

[PS08] J. Pittner and M. A. Simaan, Control of a continuous tandem cold metal rolling process, *Control Engineering Practice*, 16(11), 1379–1390, 2008.

[QZLW02] R. Qiao, Q. M. Zhu, S. Y. Li, and A. Winfield, Torsional vibration suppression of a 2-mass main drive system of rolling mill with KF enhanced pole placement, *Proceedings of the Fourth World Congress on Intelligent Control and Automation*, Chongqing, China, 2002, pp. 206–210.

[SH96] K. Sugiura and Y. Hori, Vibration suppression in 2- and 3-mass system based on the feedback of imperfect derivative of the estimated torsional torque, *IEEE Transactions on Industrial Electronics*, 43(2), 56–64, 1996.

[SHPLL01] G. Suh, D. S. Hyun, J. I. Park, K. D. Lee, and S. G. Lee, Design of a pole placement controller for reducing oscillation and settling time in a two-inertia system, *Proceedings 24th Annual Conference of the IEEE Industrial Electronics Society IECON'01*, Denver, CO, 2001, pp. 1439–1444.

[SO07] K. Szabat and T. Orlowska-Kowalska, Vibration suppression in two-mass drive system using PI speed controller and additional feedbacks—Comparative study, *IEEE Transactions on Industrial Electronics*, 54(2), 1352–1364, 2007.

[SO08] K. Szabat and T. Orlowska-Kowalska, Performance improvement of industrial drives with mechanical elasticity using nonlinear adaptive Kalman filter, *IEEE Transactions on Industrial Electronics*, 55(3), 1075–1084, 2008.

[VBL05] M. A. Valenzuela, J. M. Bentley, and R. D. Lorenz, Evaluation of torsional oscillations in paper machine sections, *IEEE Transactions on Industry Application*, 41(2), 493–501, 2005.

[VBPP07] M. Vasak, M. Baotic, I. Petrovic, and N. Peric, Hybrid theory-based time-optimal control of an electronic Throttle, *IEEE Transactions on Industrial Electronic*, 436(3), 1483–1494, 2007.

[VS98] S. N. Vukosovic and M. R. Stojic, Suppression of torsional oscillations in a high-performance speed servo drive, *IEEE Transactions on Industrial Electronics*, 45(1), 108–117, 1998.

[Z99] G. Zhang, Comparison of control schemes for two-inertia system, *Proceedings of the International Conference on Power Electronics and Drive Systems PEDS'99*, Hong-Kong, China, 1999, pp. 573–578.

[ZFS00] G. Zhang, J. Furusho, and M. Sakaguchi, Vibration suppression control of robot arms using a homogeneous-type electrorheological fluid, *IEEE/ASE Transaction on Mechatronics*, 5(3), 302–309, 2000.

27

Multiscalar Model–Based Control Systems for AC Machines

Zbigniew
Krzemiński
*Gdańsk University
of Technology*

27.1 Introduction

Drive systems with AC machines are designed for applications with different requirements for a rotor speed control. In the closed-loop control systems, usually a torque has to be controlled and machine currents have to be limited. A general idea of designing the controlled drives is based on the application of a controller, which ensures required dynamical properties for the torque. In case of AC machines, the torque is generated as the result of mutual interaction between currents and fluxes formed by three-phase, or more general, multiphase systems. Models of AC machines derived as differential equations for phase variables are complicated and not appropriate for the synthesis of control systems. The application of known methods of controllers synthesis and tuning requires, in case of AC machines, linear and nonlinear transformations of variables and derivation of models in form of differential equations. At first, a linear transformation of phase variables to orthogonal coordinates is applied. A space vector method is used to receive new variables expressed as components of vectors. As after such transformation, the variables in orthogonal coordinates are harmonic, and the next transformation is applied to receive variables that do not contain periodical components. In a classical control theory of electrical machines, a simple rotation of frame of references is used, which, if properly orientated, results in machine model in form of differential equation with the constant values of variables in steady states. Unfortunately, after transformations based on space vector method, the differential equations of AC machines remain nonlinear although in a few cases one of the equations takes the linear form. A rule is that differential equation for the rotor angular velocity remains nonlinear because the machine torque is expressed as nonlinear dependence of the components of current and flux vector.

From the above considerations results that control of AC electrical machines requires linear and nonlinear transformations of variables. It is convenient to use space vector method, especially with regard to simple interpretation. Anyway, if new variables resulting from linear and nonlinear transformations are selected in another way, the resulting machine model may have different properties than the vector model of machine. Although the number of sets of new variables is infinite, there are simple hints for selecting a proper one. At first, the torque should be one of the new variables. The remaining variables should be selected on a basis of the analysis of the differential equations and simplicity of expression used in the synthesis of control system may be achieved. If after transformations, nonlinear differential equations remain, the application of a nonlinear feedback transforms the control system into linear form. This is known as feedback linearization.

27.2 Nonlinear Transformations and Feedback Linearization

Models of AC machines for orthogonal variables received using space vector methods and linear transformation has following form:

$$\dot{\mathbf{x}} = \mathbf{f}(\mathbf{x}) + \mathbf{B}\mathbf{u} + \mathbf{E}\mathbf{z} \qquad (27.1)$$

$$\mathbf{y} = \mathbf{C}\mathbf{x}, \qquad (27.2)$$

where
  $\mathbf{x}$ is a vector of state variables
  $\dot{\mathbf{x}}$ is derivative of $\mathbf{x}$
  $\mathbf{f}(\mathbf{x})$ is a vector of nonlinear functions
  $\mathbf{B}, \mathbf{C}$ are matrixes with constant parameters
  $\mathbf{E}$ is column matrix with one coefficient equal to 1
  $\mathbf{u}$ is a vector of controls
  $\mathbf{z}$ is a vector of disturbances
  $\mathbf{y}$ is a vector of output variables

Usually, as control variables $\mathbf{v}$ in (27.1), components of voltage vectors appear and the disturbance variable z is the load torque.

The form of the system (27.1) and (27.2) is selected from a general form of the nonlinear differential equations. The wide analysis of linearizing control of the nonlinear systems may be found in (Yurkevich, 2004).

The nonlinear transformation of variables

$$\mathbf{q} = \mathbf{h}(\mathbf{x}) \qquad (27.3)$$

should result in the following form of the machine model:

$$\dot{q}_1 = g_1 q_2 + hz, \qquad (27.4)$$

$$\dot{\mathbf{q}}_{2n} = \mathbf{g}_{2n}(\mathbf{q}_{2n}) + \mathbf{D}\mathbf{v}, \qquad (27.5)$$

where

g_1 is the angular velocity of rotor, the first state variable
$^T[g_n \dots g_2] = g_{2n}$ is the vector of remaining variables
$\mathbf{v}$ is the vector of control variables
$\mathbf{g}$ is the vector of nonlinear functions
$\mathbf{D}$ is the matrix of constant components

For the special selection of variables, at least one of the functions in the vector $\mathbf{g}_{2n}$ is linear.

After analyzing and rearranging the differential equations of machine model, the following n − k + 1 equations are received taking only nonlinear equations into consideration:

$$\dot{\mathbf{q}}_{kn} = \mathbf{g}_{kn}(\mathbf{q}_{kn}) + \mathbf{D}_{kn}\mathbf{v}, \tag{27.6}$$

where matrix $\mathbf{D}_{kn}$ is nonsingular.

Now the control variables $\mathbf{v}$ will be found to transform the controlled system into the following desired form:

$$\dot{\mathbf{q}}_{kn} = \mathbf{A}_{kN}\mathbf{q}_{kn} + \mathbf{B}_{kN}\mathbf{m}_{kn}, \tag{27.7}$$

where

$\dot{\mathbf{q}}_{kn}$ is the vector of new variables
$\mathbf{m}_{kn}$ is the vector of new variables
$\mathbf{A}_{kN}, \mathbf{B}_{kN}$ are matrixes of constant coefficients

The derivatives in (27.5) and (27.7) have to be equal to ensure the same dynamical properties of the original system and linearized by feedback. From this it follows that

$$\mathbf{v} = \mathbf{D}^{-1}\left(\mathbf{A}_{kN}\mathbf{q}_{kn} + \mathbf{B}_{kN}\mathbf{m}_{kn} - \mathbf{g}_{kn}(\mathbf{q}_{kn})\right). \tag{27.8}$$

Transformation of variables (27.3) and application of control (27.8) results in linearization of the system (27.1), and the following form is received:

$$\dot{\mathbf{q}} = \mathbf{A}_N\mathbf{q} \mid \mathbf{B}_N\mathbf{m} + \mathbf{E}z, \tag{27.9}$$

where $\mathbf{A}_N$ and $\mathbf{B}_N$ are constant matrixes.

Schemes of original system and linearized by feedback are presented in Figure 27.1.

The above algorithm is general but its application to the machine models requires special analysis in each case.

Different multiscalar models may be derived for each type of AC machine depending on the nonlinear transformation of variables, which is not unique. Moreover, the number of transformations is infinite. Usually, special requirements appear for each control system, and the new variables should fulfill specific demands. Simplicity of nonlinear feedback, limitation of selected state variables and all control variables, and dependences between variables are required in drives. Very important in the selection of nonlinear transformation is a simple and understandable physical interpretation of the new variables and relation between them.

Examples of transformation of variables and feedback linearization for widely analyzed applications of machines are further presented.

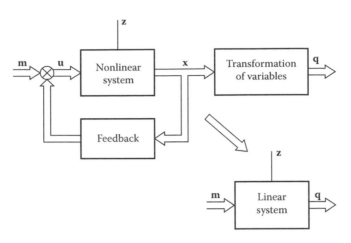

FIGURE 27.1   Scheme of original and linearized systems.

27.3 Models of the Squirrel Cage Induction Machine

27.3.1 Vector Model of the Squirrel Cage Induction Machine

A vector model of the squirrel cage induction machine may be, after linear transformation of phase variables, expressed as differential equations for the stator current vector and the rotor flux vector in unmoving frame of references in the following form:

$$\mathbf{i}_s = a_1\mathbf{i}_s + a_2\boldsymbol{\psi}_r + ja_3\omega_r\boldsymbol{\psi}_r + a_4\mathbf{u}_s, \tag{27.10}$$

$$\dot{\boldsymbol{\psi}}_r = a_5\mathbf{i}_s + a_6\boldsymbol{\psi}_r + j\omega_r\boldsymbol{\psi}_r, \tag{27.11}$$

$$\dot{\omega}_r = \frac{1}{J}(T_e - m_0), \tag{27.12}$$

where
   $\mathbf{u}_s$, $\mathbf{i}_s$, $\boldsymbol{\psi}_r$ are the stator voltage, stator current, and rotor flux vectors, respectively
   T_e is the motor torque
   m_0 is the motor load
   τ is the relative time
   ω_r is the angular rotor velocity
   $j = \sqrt{-1}$
   J is the moment of inertia
   $a_1, \dots, a_6$ are coefficients depending on machine parameters:

$$a_1 = -\frac{R_sL_r^2 + R_rL_m^2}{wL_r}; \quad a_2 = \frac{R_rL_m}{wL_r}; \quad a_3 = \frac{L_m}{w};$$

$$a_4 = \frac{L_r}{w}; \quad a_5 = \frac{R_rL_m}{L_r}; \quad a_6 = -\frac{R_r}{L_r}; \quad w = L_sL_r - L_m^2;$$

where
   R_s, R_r are the stator and rotor resistances, respectively
   L_s, L_r, L_m are the stator, rotor, and mutual inductances, respectively

All variables used in the machine models are expressed in p.u. system.

The machine torque is expressed as

$$T_e = \frac{L_m}{L_r} \mathrm{Im} \left| \boldsymbol{\psi}_r^* \mathbf{i}_s \right|. \tag{27.13}$$

After linear transformation of vectors $\mathbf{i}_s$, $\boldsymbol{\psi}_r$, the other pair of vectors may be obtained and used for the derivation of machine model.

The application of transformation of vector components to coordinates rotating with angular velocity of the rotor flux vector results in a widely used field-oriented form of Equations 27.10 through 27.12.

27.3.2 Multiscalar Models of the Squirrel Cage Induction Machine

The model received after nonlinear transformation is called a multiscalar model of the induction motor because the state variables are scalars. The multiscalar model has been received on a basis of the analysis of differential equations. It was presented in (Krzeminski, 1987) and generalized in (Krzeminski et al., 2006) together with the application of nonlinear control based on the linearization of differential equations for the highest derivative in which control variables appear. Similar results may be received by the application of methods of differential geometry as has been shown in (Marino et al., 1993). The main benefit of the application of nonlinear control is splitting of the controlled system with induction motor into two decoupled linear subsystems. The summary of multiscalar model–based control method of the induction machine has been presented in (Kazmierkowski et al., 2002).

Similar variables as used in the multiscalar model of the induction motor were proposed in (Dong and Ojo, 2006) and (Balogun and Ojo, 2009).

The simplest possibility to receive decoupled subsystems is defining a machine torque as the state variable instead of current vector component in q axis as presented in (Krzeminski, 1987) and in (Mohanty and De, 2000). The other variables of vector model in field-oriented coordinates remain untransformed.

More complex variables, being the linear combination of simple multiscalar variables, are proposed in (Lee et al., 2000). Simulation results are similar to obtained using multiscalar model.

Nonlinear feedback may be applied to the system of first-order equations as in (Krzeminski et al., 2006) or to the first- and second-order equations as presented in (Zaidi et al., 2007) and in (Salima et al., 2008). In the last case, no interior variable does not appear directly and limitation of variables in control system is difficult.

Nonlinear decoupling control requires the exact estimation of machine parameters. Methods of parameters estimation are included into nonlinear adaptive control of induction machine (Jeon et al., 2006). The other method of parameter estimation is based on the application of model reference adaptive scheme to nonlinear control system as presented in (Kaddouri et al., 2008).

There are two types of the multiscalar models of the squirrel cage induction machine (Krzeminski et al., 2006). The general form of multiscalar model of type 1 is as follows:

$$\dot{\mathbf{x}} = \mathbf{A}_x \mathbf{x} + \mathbf{g}_x(\mathbf{x}) + \mathbf{B}_x \mathbf{u}_x + \mathbf{m}, \tag{27.14}$$

where
$\dot{\mathbf{x}}$ is derivative of $\mathbf{x}$
$\mathbf{x} = [x_{11}, x_{12}, x_{21}, x_{22}]^T$
$\mathbf{g}_x(\mathbf{x}) = [0, g_{x12}(\mathbf{x}), 0, g_{x22}(\mathbf{x})]^T$

$$\mathbf{B}_x = \begin{bmatrix} 0 & 0 & 0 & b_{x42} \\ 0 & b_{x21} & 0 & 0 \end{bmatrix}^T$$

$\mathbf{u}_x = [u_{x1}, u_{x2}]^T$
$\mathbf{m} = [0\ 0\ 0\ m_0]$
$\mathbf{A}_x$ is matrix of coefficients
m_0 is the load torque

The variables **x** are defined as the rotor speed, scalar and vector products of two vectors from the vector model of the induction motor, and square of the flux vector.

The multiscalar model of type 2 may be based on the stator flux vector and takes the following form:

$$\dot{z} = A_z z + g_z(z) + B_z u_z + m, \tag{27.15}$$

where

$\mathbf{z} = [z_{11}, z_{12}, z_{21}, z_{22}]^T$

$\mathbf{g}_z(\mathbf{z}) = [0, g_{z12}(z), 0, g_{z22}(z)]^T$

$$\mathbf{B}_z = \begin{bmatrix} 0 & b_{z22} & b_{z32} & b_{z42} \\ 0 & b_{z21} & b_{z31} & b_{z41} \end{bmatrix}^T$$

$\mathbf{u}_z = [u_{z1}, u_{z2}]^T$

A_z is the matrix of coefficients and one of the elements b_{z31} and b_{z32} may be equal to zero

The state variables **z** are defined similarly as variables **x** for the induction motor model of type 1.

The main difference between models of type 1 and 2 lies in form of matrixes $\mathbf{B}_x$ and $\mathbf{B}_z$. There are only two coefficient of matrix $\mathbf{B}_x$ different from zero, and in matrix $\mathbf{B}_z$ nonzero coefficients appear in all rows except the first one.

Forms of the multiscalar models do not depend on the frame of references in which the original vectors are defined.

The model of type 1 appears if square of the rotor flux vector is chosen as the variable in the multiscalar model. All variables of type 1 model may be as follows:

$$x_{11} = \omega_r, \tag{27.16}$$

$$x_{12} = \psi_{r\alpha} i_{s\beta} - \psi_{r\beta} i_{s\alpha}, \tag{27.17}$$

$$x_{21} = \psi_{r\alpha}^2 + \psi_{r\beta}^2, \tag{27.18}$$

$$x_{22} = \psi_{r\alpha} i_{s\alpha} + \psi_{r\beta} i_{s\beta}, \tag{27.19}$$

where $i_{s\alpha}$, $i_{s\beta}$, $\psi_{r\alpha}$, $\psi_{r\beta}$ are the stator current and rotor flux vector components.

The differential equations for the variables (27.16) through (27.19) are as follows:

$$\dot{x}_{11} = \frac{L_m}{JL_r} x_{12} - \frac{1}{J} m_0, \tag{27.20}$$

$$\dot{x}_{12} = a_{1m} x_{12} - x_{11}(x_{22} + a_3 x_{21}) + a_{42} u_{x1}, \tag{27.21}$$

$$\dot{x}_{21} = 2a_5 x_{22} - 2a_6 x_{21}, \tag{27.22}$$

$$\dot{x}_{22} = a_{1m} x_{22} + x_{11} x_{12} + a_2 x_{21} + a_5 i_s^2 + a_4 u_{x2}, \tag{27.23}$$

where

$$a_{1m} = -\frac{R_r L_s + R_s L_r}{w}, \tag{27.24}$$

$$i_s^2 = \frac{x_{12}^2 + x_{22}^2}{x_{21}}, \tag{27.25}$$

$$u_{x1} = \psi_{r\alpha} u_{s\beta} - \psi_{r\beta} u_{s\alpha}, \tag{27.26}$$

$$u_{x2} = \psi_{r\alpha} u_{s\alpha} + \psi_{r\beta} u_{s\beta}, \tag{27.27}$$

where $u_{s\alpha}$, $u_{s\beta}$ are the stator voltage vector components.

The variables u_{x1} and u_{x2} are the new control variables. The stator voltage vector components are calculated from the following expressions:

$$u_{s\alpha} = \frac{\psi_{r\alpha} u_{x2} - \psi_{r\beta} u_{x1}}{\psi_r^2}, \tag{27.28}$$

$$u_{s\beta} = \frac{\psi_{r\alpha} u_{x1} + \psi_{r\beta} u_{x2}}{\psi_r^2}. \tag{27.29}$$

The derivative of the variable x_{21} does not depend directly on any control variable. This variable has to be controlled because the value of the square of the rotor flux vector decides on efficiency of the energy conversion in the machine and on value of the stator voltage.

The multiscalar variables of the model of type 2 for the stator current and stator flux vectors chosen as original variables are as follows:

$$z_{11} = \omega_r, \tag{27.30}$$

$$z_{12} = \psi_{s\alpha} i_{s\beta} - \psi_{s\beta} i_{s\alpha}, \tag{27.31}$$

$$z_{21} = \psi_{s\alpha}^2 + \psi_{s\beta}^2, \tag{27.32}$$

$$z_{22} = \psi_{s\alpha} i_{s\alpha} + \psi_{s\beta} i_{s\beta}, \tag{27.33}$$

where $\psi_{s\alpha}$, $\psi_{s\beta}$ are the stator flux vector components.

The differential equations for the variables (27.30) through (27.33) are as follows:

$$\dot{z}_{11} = \frac{1}{J} z_{12} - \frac{m_0}{J}, \tag{27.34}$$

$$\dot{z}_{12} = a_{1m} z_{12} + z_{11}(z_{22} - a_4 z_{21}) + a_3 u_{x1}, \tag{27.35}$$

$$\dot{z}_{21} = -2R_s z_{22} + 2u_{z2}, \tag{27.36}$$

$$\dot{z}_{22} = a_{1m} z_{22} - R_s i_s^2 + \frac{R_r}{w} z_{21} - z_{11} z_{12} + a_3 u_{x2}, \tag{27.37}$$

where

$$i_s^2 = \frac{z_{12}^2 + z_{22}^2}{z_{21}}, \tag{27.38}$$

$$u_{z2} = \psi_{s\alpha} u_{s\alpha} + \psi_{s\beta} u_{s\beta}. \tag{27.39}$$

The variables u_{x1}, u_{x2} are defined by (27.26) and (27.27). On the other hand, the control variable u_{x2} depends on the variable u_{z2} as follows:

$$u_{x2} = \frac{\left(\psi_{r\alpha}^2 + \psi_{r\beta}^2\right) u_{z2}}{\psi_{s\alpha} \psi_{r\alpha} + \psi_{s\beta} \psi_{r\beta}}. \tag{27.40}$$

The variables u_{x1} and u_{z2} are new control variables. The stator voltage vector components are calculated from the following expressions:

$$u_{s\alpha} = \frac{\psi_{r\alpha} u_{z2} - \psi_{s\beta} u_{x1}}{\psi_{s\alpha} \psi_{r\alpha} + \psi_{s\beta} \psi_{r\beta}}, \tag{27.41}$$

$$u_{s\beta} = \frac{\psi_{r\beta} u_{z2} + \psi_{s\alpha} u_{x1}}{\psi_{s\alpha} \psi_{r\alpha} + \psi_{s\beta} \psi_{r\beta}}. \tag{27.42}$$

It is convenient to remain the rotor flux vector components for in (27.41) and (27.42).

The differential equation for the variable z_{21} depends on the control variable u_{z2}. This variable has to be controlled because the value of the square of the stator flux vector decides on the efficiency of the energy conversion in the machine. On the other hand, the control variables appear in three equations. This means that the variable z_{22} remains uncontrolled directly and equation (27.37) describes an inner dynamics of the control system. The state variables z_{21} and z_{12} are stabilized in steady states and form constant coefficients in (27.37) of the values, which ensures stability of inner dynamics.

27.3.3 Feedback Linearization of Multiscalar Models of the Induction Motor

In accordance with the procedure described in Section 27.2, the application of the nonlinear controls of the form

$$u_{x1} = \frac{1}{a_4}\left(x_{11}\left(x_{22} + a_3 x_{21}\right) - a_{1m} m_{x1}\right), \tag{27.43}$$

$$u_{x2} = \frac{1}{a_4}\left(-x_{11} x_{12} - a_2 x_{21} - a_5 i_s^2 - a_{1m} m_{x2}\right) \tag{27.44}$$

transforms the system (27.20) through (27.23) of type 1 into two independent linear subsystems:

1. Mechanical subsystem

$$\dot{x}_{11} = \frac{L_m}{JL_r} x_{12} - \frac{1}{J} m_0, \tag{27.45}$$

$$\dot{x}_{12} = a_{1m}\left(x_{12} - m_{x1}\right). \tag{27.46}$$

2. Electromagnetic subsystem

$$\dot{x}_{21} = 2a_6x_{21} + 2a_5x_{22},$$

(27.47)

$$\dot{x}_{22} = a_{1m}\left(x_{22} - m_{x2}\right).$$

(27.48)

In similar way, the nonlinear controls of the form

$$u_{z1} = \frac{1}{a_3}\left(m_{z1} - z_{11}\left(z_{22} - a_4z_{21}\right)\right),$$

(27.49)

$$u_{z2} = \frac{1}{2T}\left(m_{z2} - z_{21}\right) + R_sz_{22}$$

(27.50)

transform the system (27.34) through (27.37) into two subsystems:

1. Mechanical subsystem

$$z_{11} = \frac{1}{J}z_{12} - \frac{m_0}{J},$$

(27.51)

$$\dot{z}_{12} = a_{1m}\left(z_{12} - m_{z1}\right).$$

(27.52)

2. Electromagnetic subsystem

$$\dot{z}_{21} = \frac{1}{T}\left(-z_{21} + m_{z2}\right),$$

(27.53)

$$\dot{z}_{22} = a_{1m}z_{22} - R_si_s^2 + \frac{R_r}{w}z_{21} - z_{11}z_{12} + a_3\frac{\left(\psi_{r\alpha}^2 + \psi_{r\beta}^2\right)}{\psi_{s\alpha}\psi_{r\alpha} + \psi_{s\beta}\psi_{r\beta}}\left(\frac{1}{2T}\left(-z_{21} + m_{z2}\right) + R_sz_{22}\right).$$

(27.54)

The Equation 27.54 remains nonlinear but, as previously mentioned, the variable z_{22} remains uncontrolled.

The variable z_{21} is directly controlled as results from (27.53).

27.4 Models of the Double-Fed Induction Machine

27.4.1 Vector Model of the Double-Fed Induction Machine

The double-fed induction machine (DFM) exploited recently as a generator is connected directly to the grid from the stator side and fed by an inverter from the rotor side. The same scheme of feeding is used if the machine is applied in motor mode. The reason of designing such systems is cost effectiveness, specially for high-power drives because only part of machine power is converted by power electronics. The rotor current may be measured in the control system and should be used in the vector model of the DFM. The vector model of the DFM takes the form proper to the synthesis of the control system if the rotor current and stator flux vectors are selected as state variables and the frame of references oriented with the rotor is used. The differential equation of the DFM takes in such a case the following form:

$$\mathbf{i}_r = b_1 \mathbf{i}_r + b_2 \mathbf{\psi}_s + jb_3 \omega_r \mathbf{\psi}_s - b_3 \mathbf{u}_s + b_4 \mathbf{u}_r, \tag{27.55}$$

$$\dot{\mathbf{\psi}}_s = b_5 \mathbf{i}_r + b_6 \mathbf{\psi}_s + j\omega_r \mathbf{\psi}_s + \mathbf{u}_s, \tag{27.56}$$

$$\dot{\omega}_r = \frac{1}{J}\left(T_e - m_0\right), \tag{27.57}$$

where $\mathbf{u}_r$, $\mathbf{i}_r$, $\mathbf{\psi}_s$ are the stator voltage, stator current, and rotor flux vectors, and $b_1, \ldots, b_6$ are coefficients depending on machine parameters:

$$b_1 = -\frac{R_r L_s^2 + R_s L_m^2}{wL_s}; \quad b_2 = \frac{R_s L_m}{wL_s}; \quad b_3 = \frac{L_m}{w};$$

$$b_4 = \frac{L_s}{w}; \quad b_5 = \frac{R_s L_m}{L_r}; \quad b_6 = -\frac{R_s}{L_s};$$

The DFM torque is expressed as follows:

$$T_e = \mathrm{Im}\left|\mathbf{\psi}_s^* \mathbf{i}_s\right|. \tag{27.58}$$

Similar to the squirrel cage machine after linear transformation of vectors $\mathbf{i}_r$, $\mathbf{\psi}_s$, the other pair of vectors may be obtained and used for the derivation of DFM model.

27.4.2 Multiscalar Model of the DFM

The multiscalar model of the DFM was analyzed in Krzeminski (2002).
    The nonlinear transformation of variables of the form

$$z_{11} = \omega_r, \tag{27.59}$$

$$z_{12} = \psi_{sx} i_{ry} - \psi_{sy} i_{rx}, \tag{27.60}$$

$$z_{21} = \psi_s^2, \tag{27.61}$$

$$z_{22} = \psi_{sx} i_{rx} + \psi_{sy} i_{ry} \tag{27.62}$$

leads to the multiscalar model of the DFM. Differential equations for the variables (27.59) through (27.62) are as follows:

$$\dot{z}_{11} = \frac{L_m}{JL_s} z_{12} - \frac{1}{J} m_0, \tag{27.63}$$

$$\dot{z}_{12} = b_{1m}z_{12} + z_{11}z_{22} + b_3z_{11}z_{21} - b_3u_{sf1} + b_4u_{r1} + u_{si1}, \tag{27.64}$$

$$\dot{z}_{21} = -2b_6z_{21} + 2b_5z_{22} + 2u_{sf2}, \tag{27.65}$$

$$\dot{z}_{22} = b_{1m}z_{22} + b_2z_{21} + b_5i_r^2 - z_{11}z_{12} - b_3u_{sf2} + b_4u_{r2} + u_{si2}, \tag{27.66}$$

where

$$u_{r1} = u_{ry}\Psi_{sx} - u_{rx}\Psi_{sy} \tag{27.67}$$

$$u_{r2} = u_{rx}\Psi_{sx} + u_{ry}\Psi_{sy} \tag{27.68}$$

$$u_{sf1} = u_{sy}\Psi_{sx} - u_{sx}\Psi_{sy} \tag{27.69}$$

$$u_{sf2} = u_{sx}\Psi_{sx} + u_{sy}\Psi_{sy} \tag{27.70}$$

$$u_{si1} = u_{sx}i_{ry} - u_{sy}i_{rx} \tag{27.71}$$

$$u_{si2} = u_{sx}i_{rx} + u_{sy}i_{ry} \tag{27.72}$$

27.4.3 Feedback Linearization of DFM

In accordance with the procedure described in Section 27.2, application of the nonlinear controls of the form

$$u_{r1} = \frac{1}{b_4}\left(-z_{11}\left(z_{22} + b_3\,z_{21}\right) + b_3u_{sf1} - u_{si1} - b_{1m}m_1\right), \tag{27.73}$$

$$u_{r2} = \frac{1}{b_4}\left(-b_2z_{21} - b_5i_r^2 + z_{11}z_{12} + b_3u_{sf2} - u_{si2} - b_{1m}m_2\right) \tag{27.74}$$

transforms the system (27.63) through (27.66) into two linear susbsystems:

1. Mechanical subsystem

$$\dot{z}_{11} = \frac{L_m}{JL_s}z_{12} - \frac{1}{J}m_0, \tag{27.75}$$

$$\dot{z}_{12} = b_{1m}\left(z_{12} - m_1\right). \tag{27.76}$$

2. Electromagnetic subsystem

$$\dot{z}_{21} = b_6z_{21} + 2b_5z_{22} + 2u_{sf2}, \tag{27.77}$$

$$\dot{z}_{22} = b_{1m}\left(z_{22} - m_2\right). \tag{27.78}$$

Components of the rotor voltage vector are calculated from the following expressions:

$$u_{rx} = \frac{u_{r1}\psi_{sy} + u_{r2}\psi_{sx}}{z_{21}},\tag{27.79}$$

$$u_{ry} = \frac{u_{r2}\psi_{sy} - u_{r1}\psi_{sx}}{z_{21}}.\tag{27.80}$$

The variable u_{sf2} is the scalar product of the stator flux vector and the voltage flux vector. From general point of view, this is control variable as the stator voltage vector appears in differential equations as control. If the stator is connected to the grid, then constant amplitude and frequency of grid voltage may be treated as parameters. The components of the voltage vector are parameters depending on time. In such case, u_{sf2} is the variable resulting from transformation of state variables.

27.5 Models of the Interior Permanent Magnet Synchronous Machine

27.5.1 Vector Model of the Interior Permanent Magnet Synchronous Machine

An interior permanent magnet synchronous machine (IPMSM) is a synchronous machine with construction of magnetic circuit allowing field weakening and utilizing the reluctance torque. The differential equations for stator current vector components in the frame of references stationary in relation to the stator are complicated and usually the rotor-oriented frame of references is used. The differential equations for the state variables of IPMSM are as follows:

$$\dot{i}_d = \frac{1}{L_d}\left(-Ri_d + \omega_r\psi_q + u_d\right),\tag{27.81}$$

$$\dot{i}_q = \frac{1}{L_q}\left(-Ri_q - \omega_r\psi_d + u_q\right),\tag{27.82}$$

$$\dot{\omega}_r = \frac{1}{J}\left(T_e - m_0\right),\tag{27.83}$$

$$\psi_d = \psi_f + L_d i_d,\tag{27.84}$$

$$\psi_q = L_q i_q,\tag{27.85}$$

where
    $i_d, i_q, \psi_d, \psi_q, u_d, u_q$ are the stator current, stator flux and voltage vector components
    ψ_f is the exciting flux
    R is the stator resistance
    L_d, L_q are the direct and quadrate inductances, respectively
    J is moment of inertia

The torque is expressed as

$$T_e = \psi_f i_q + \left(L_d - L_q\right)i_d i_q.\tag{27.86}$$

The main drawback of vector model of the IPMSM is the form of expression for the torque. If the machine torque is controlled rapidly, the reference values of d and q components of the stator current have to be calculated from complicated dependences ensuring maximum efficiency. The current vector components should be controlled simultaneously and very quickly to receive high performances of the drive. Usually, d component is controlled as function of q component, which results in compromise between rapidity and efficiency.

27.5.2 Multiscalar Model of the IPMSM

In a case of IPMSM, the following transformation of variables gives benefits of fast control and simplicity of auxiliary expression:

$$w_{11} = \omega_r, \tag{27.87}$$

$$w_{12} = \psi_f i_q + \left(L_d - L_q\right) i_d i_q, \tag{27.88}$$

$$w_{22} = \psi_f i_d + \left(L_d - L_q\right) i_d^2, \tag{27.89}$$

$$w_{21} = \left(\psi_f + \left(L_d - L_q\right) i_d\right)^2. \tag{27.90}$$

In contrary to the other types of AC machines, only three multiscalar variables are needed to form the model of the IPMSM and the variable (27.89) is used for the simplification of notation. Additionally, the following dependence appears for multiscalar variables:

$$i_s^2 = \frac{w_{12}^2 + w_{22}^2}{w_{21}}, \tag{27.91}$$

where i_s is the amplitude of the stator current.

The rotor position is required to transform the stator current components from stationary frame of references to the frame of references connected to the rotor.

The differential equations for the variables (27.87) through (27.89) are as follows:

$$\dot{w}_{11} = \frac{1}{J} x_{12} - \frac{1}{J} m_0, \tag{27.92}$$

$$\dot{w}_{12} = -\frac{R}{L_q} w_{12} - \left(1 - \frac{L_q}{L_d}\right) R i_d i_q - w_{11}\left(\frac{1}{L_q} w_{21} + w_{22}\right) + w_{11}\left(1 - \frac{L_q}{L_d}\right) L_q i_q^2 + u_1, \tag{27.93}$$

$$\dot{w}_{21} = 2\frac{L_d - L_q}{L_d}\left(-R w_{22} + L_q w_{11} w_{12} + u_2\right), \tag{27.94}$$

where

$$u_1 = \left(\psi_f + \left(L_d - L_q\right) i_d\right)\frac{1}{L_q} u_q + \left(1 - \frac{L_q}{L_d}\right) i_q u_d, \tag{27.95}$$

$$u_2 = \left(\psi_f + \left(L_d - L_q\right)i_d\right)u_d. \tag{27.96}$$

The d, q components of the stator current appear in (27.93) and (27.94) to simplify the notation of the equations.

27.5.3 Feedback Linearization of IPMSM

For the IPMSM, the linearizing feedback is as follows:

$$u_1 = -L_q\left(\frac{1}{L_d} - \frac{1}{L_q}\right)Ri_di_q + w_{11}\left(\frac{1}{L_q}w_{21} + w_{22}\right) + w_{11}\left(\frac{1}{L_d} - \frac{1}{L_q}\right)L_q^2i_q^2 + \frac{R}{L_q}m_1, \tag{27.97}$$

$$u_2 = Rw_{22} - L_qw_{11}w_{12} + \frac{L_d}{2T\left(L_d - L_q\right)}\left(-w_{21} + m_2\right), \tag{27.98}$$

where T is a time constant.

The stator voltage vector components are calculated from the following expressions:

$$u_d = \frac{1}{x_{22}}i_du_2, \tag{27.99}$$

$$u_q = \frac{L_q}{x_{22}}i_du_1 - \frac{L_d - L_q}{L_dx_{21}}L_qi_qu_2. \tag{27.100}$$

Application of (27.97) and (27.98) to (27.93) and (27.94) results in the following linearized subsystems:

1. Mechanical subsystem

$$\dot{w}_{11} = \frac{1}{J}w_{12} - \frac{1}{J}m_0, \tag{27.101}$$

$$\dot{w}_{12} = \frac{R}{L_q}\left(-w_{12} + m_1\right). \tag{27.102}$$

2. Electromagnetic subsystem

$$\dot{w}_{21} = \frac{1}{T}\left(-w_{21} + m_2\right). \tag{27.103}$$

27.5.4 Efficient Control of IPMSM

The motor torque is generated with higher efficiency if the stator flux is reduced. The function for the d component of the stator current vector in dependence on q component for maximum torque may find after simple calculations in the form

$$i_{dM} = \frac{\Psi_f}{2(L_q - L_d)} - \sqrt{\frac{\Psi_f^2}{4(L_d - L_q)^2} + i_{qM}^2},$$

(27.104)

where subscript M denotes values for maximum torque.

If the machine torque expressed by (27.86) is controlled rapidly, the reference values of d and q components of the stator current have to be calculated from dependence (27.104) ensuring maximum efficiency. The current vector components should be controlled simultaneously and very quickly to receive high performances of the drive. Usually, d component is controlled as function of q component in a way ensuring compromise between rapidity and efficiency.

More convenient formula results after replacing variables appearing in (27.104) by expressions depending on the multiscalar model variables:

$$w_{21M} = 0.5\left(\Psi_f^2 + \sqrt{\Psi_f^4 + 16w_{12M}^2(L_d - L_q)^2}\right).$$

(27.105)

The motor torque is controlled in accordance to (27.102) independently on the flux. Application of (27.105) makes it possible to control the variable x_{21} in a simple way ensuring high machine efficiency.

27.6 Structures of Control Systems for AC Machines Linearized by Feedback

The mechanical subsystems for each machine presented in Sections 27.3.3, 27.4.3, and 27.5.3 has similar structure resulting from choosing the rotor speed and the machine torque as controlled variables. Only different time constants appear in the models in dependence of machine type and rating. As the machine torque has to be limited in the drives systems, the cascaded controllers may be applied to control the rotor speed and the torque. A simple system with PI controllers for mechanical subsystem is presented in Figure 27.2. The controllers may be tuned with the application of known methods for linear systems.

For electromagnetic subsystems, two basic structures of linearized subsystems may be pointed out. The first structure consists of two inertial elements connected in series. Two controllers connected in cascade are the simplest solution for control system presented in Figure 27.3. The second structure

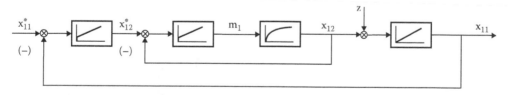

FIGURE 27.2 Controllers for mechanical subsystem.

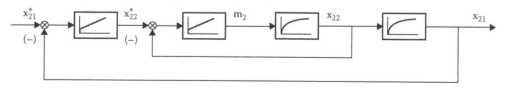

FIGURE 27.3 Controllers for electromagnetic subsystem.

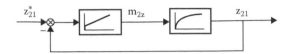

FIGURE 27.4   Controller for structure with one internal element.

consists of two inertial elements from which only one can be directly controlled. The remaining element is usually stable if all controlled variables are stabilized and do not require additional control or correction. One controller is sufficient for this structure as presented in Figure 27.4.

The control variables acting in the subsystems have to be limited because of limited power of the supplying inverters. There are two ways of limiting the control variables. The simple way is the limitation of controller outputs on constant levels resulting from inverter voltages in steady states and margin resulting from required dynamics of controlled variables. Sharing of inverter power margin between the subsystems is the problem that has to be individually solved. The other way is dynamical calculation of controller limits from available inverter output voltage vector and actual voltage vector for steady state. The dynamics of machine variables differs in such a case in dependence on working point.

Inverter output current has to be limited in the drive system to avoid damage of power electronics devices. Usually, the machine torque is limited and limitation of the second variable is calculated from expression on the current. The dependences may be complicated, especially for field weakening region of operation. For example, in case of the IPMSM working in maximum efficiency mode, the variable w_{21} is the function of the machine torque and it is sufficient to limit one variable only.

The basic structure of control system for the induction motor is presented in Figure 27.5. The rotor speed and remaining variables are estimated in a speed observer. The variables estimated in the speed observer denoted by ∧ are used to calculate the estimated multiscalar variables. To control the drive in a field weakening region, additional limiting functions are added. Transients during speed reversal of drive with induction motor controlled on a basis of the multiscalar model is presented in Figure 27.6. Good dynamical properties and limitation of the stator current are observed.

The basic structure of control system for the IPMSM is presented in Figure 27.7. Similarly to the induction motor, the rotor speed and remaining variables are estimated in a speed observer. The variables estimated in the speed observer denoted by ∧ are used to calculate the estimated multiscalar variables. Nonlinear function (NF) is applied to calculate the set value for the variable w_{21}. Transients during starting up are presented in Figure 27.8. Limitation of stator current may be observed.

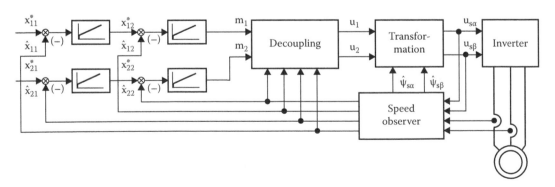

FIGURE 27.5   The basic structure of control system for the induction motor.

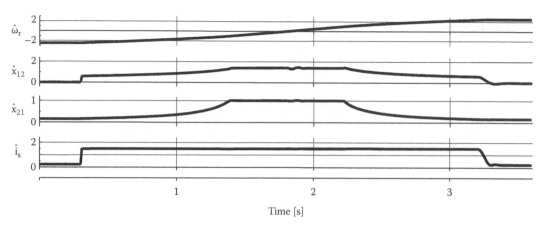

FIGURE 27.6 Transients during speed reversal of drive with induction motor controlled on a basis of the multiscalar model.

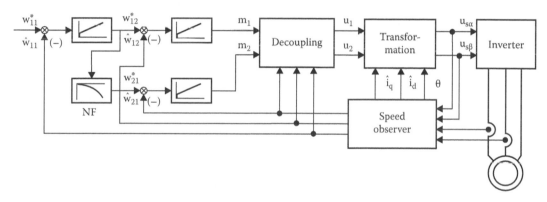

FIGURE 27.7 The basic structure of control system for the IPMSM.

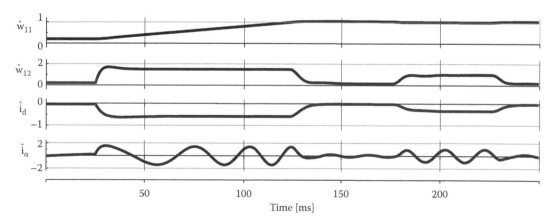

FIGURE 27.8 Transients during starting up of the IPMSM.

References

Balogun, A. and Ojo, O. Natural variable Simulation of induction machines. *IEEE AFRICON 2009*, Nairobi, Kenya, September 23–25, 2009.

Dong, G. and Ojo, O. Efficiency optimization control of induction motor using natural variables. *IEEE Transactions on Industrial Electronics*, 53(6), 1791–1798, December 2006.

Jeon, S. H., Baang, D., and Choi, J. Y. Adaptive feedback linearization control based on airgap flux model for induction motors. *International Journal of Control, Automation, and Systems*, 4(4), 414–427, August 2006.

Kaddouri, A., Akhrif, O., Ghribi, M., and Le-Huy, H. Adaptive nonlinear control of an electric motor. *Applied Mathematical Sciences*, 2(52), 2557–2568, 2008.

Kazmierkowski, M. P., Krishnan, R., and Blaabjerg, F. *Control in Power Electronics: Selected Problems.* Academic Press, London, U.K., 2002.

Krzeminski, Z. Nonlinear control of induction motor, *Proceedings of the 10th IFAC World Congress*, Munich, Germany, 1987, pp. 349–354.

Krzeminski, Z. Sensorless multiscalar control of double fed machine for wind power generators, *Proceedings of the Power Conversion Conference, 2002, (PCC Osaka 2002).* Vol. 1, Osaka, Japan, April 2–5, 2002, pp. 334–339.

Krzeminski, Z., Lewicki, A., and Włas, M. Properties of control systems based on nonlinear models of the induction motor. *COMPEL*, 25(1), 195–206, 2006. Special issue of *COMPEL*, Selected papers from the *18th Symposium on Electromagnetic Phenomena in Nonlinear Circuits*.

Lee, H. T., Chang, J. S., and Fu, L. C. Exponentially stable non-linear control for speed regulation of induction motor with field-oriented PI-controller. *International Journal of Adaptive Control and Signal Processing*, 14(2–3), 297–312, 2000.

Marino, R., Peresada, S., and Valigi, P. Adaptive input-output linearizing control of induction motors. *IEEE Transactions on Automatic Control*, 38, 208–221, 1993.

Mohanty, K. B. and De, N. K. Application of differential geometry for a high performance induction motor drive. *International Conference on Recent Advances in Mathematical Sciences*, Kharagpur, India, December 2000, pp. 225–234.

Salima, M., Riad, T., and Hocine, B. Applied input-output linearizing control or high-performance induction motor. *Journal of Theoretical and Applied Information Technology*, 4(1), 6–14, 2008.

Yurkevich, V. D. *Design of Nonlinear Control Systems with the Highest Derivative in Feedback.* World Scientific, Singapore, 2004.

Zaidi, S., Naceri, F., and Abdessamed, R. Non linear control of an induction motor. *Asian Journal of Information Technology*, 6(4), 468–473, 2007.

V

Power Electronic Applications

28

Sustainable Lighting Technology

Henry Chung
City University of Hong Kong

Shu-Yuen (Ron) Hui
City University of Hong Kong

and

Imperial College London

28.1 Introduction

For decades, improvement in energy efficiency has been the main focus in many energy-conversion applications including lighting. The continuous increase in luminous efficacy of discharge lamps such as the introduction of T5 fluorescent lamps and the arrival of new high-brightness light emitting diodes (LEDs) marked significant progresses in light sources. The introduction of electronic ballasts and new low-loss magnetic ballasts has resulted in the progressive elimination of poor-quality ballasts. The first decade of the twenty-first century is the period during which the electronic ballast technology reached maturity [1]. The proposal of replacing incandescent lamps with energy-saving lamps by many governments has created new opportunities to new lighting technology and market.

The increasing awareness of climate change and electronic waste issues has prompted the need to reexamine existing lighting technology [2]. Unlike the traditional approach of using energy saving as the only criterion, this chapter aims at describing a new "sustainable lighting technology" concept that includes three essential features as the criteria for modern lighting products. These features are

1. Energy saving
2. Long product lifetime
3. Recyclability

The principle behind the sustainable lighting technology is to use lighting energy when and where it is necessary and to the appropriate lighting level.

The concepts of "energy saving" and "environmental protection" can be easily mixed up. In fact, an "energy-saving" technology is not necessarily an "environmentally friendly" one. For genuine environmental protection, one must

1. Reduce greenhouse gas emission to protect the atmosphere
2. Reduce waste/pollution to protect soil and water

These two requirements must go hand in hand. Energy saving is a means of reducing greenhouse gas emission. If an energy-saving lighting product creates a lot of toxic chemicals and electronic waste due to short lifetime, it is not environmentally friendly. Since the dawning of the electronic age cannot be reversed, the three simultaneous objectives of the sustainable lighting technology will not "eliminate" all electronic and chemical wastes. Instead, they aim at "reducing" both global energy consumption and electronic and chemical wastes.

In lighting technology, the electrolytic capacitors are the bottleneck of electronic ballast technology. The progress in improving the lifetime of electrolytic capacitors has been slow. Figure 28.1 shows the project of four grades of electrolytic capacitors with typical lifetimes rated at 10,000, 8,000, and 5,000 h at 105°C, and 2,000 h at 85°C. As electrolytic capacitors contain electrolytes in liquid form, they are sensitive to operating temperature. The lifetime of electrolytic capacitors is halved when the operating temperature is increased by 10°C. This means that electrolytic capacitors working at a temperature 20°C above the rated temperature will have 25% of the rated lifetime. This problem is especially serious for compact fluorescent lamps (CFLs), which have the electronic ballasts totally housed inside the plastic covers with limited space and extremely limited cooling effects (Figure 28.2). The operating temperature of the electrolytic capacitor inside the CFL will even be higher if the CFL is housed in lighting fixture with no ventilation. Consequently, the short lifetime of CFLs has been a common consumer compliant. Over 2.5 billion units of CFLs were made in 2007 (China Source Report-Compact Fluorescent Lamps, 2007 Bharat Book Bureau). Considering their short lifetime, it is not difficult to imagine how fast the global electronic waste problem could deteriorate as more and more governments are trying to replace incandescent lamps with CFLs.

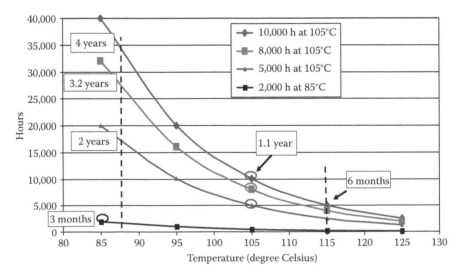

FIGURE 28.1 Projected lifetime of electrolytic capacitors. (Modified from Hui, S.Y.R. and Yan, W., Re-examination on energy saving & environmental issues in lighting applications, *Proceedings of the 11th International Symposium on Science 7 Technology of Light Sources*, Shanghai, China, May 2007 (Invited Landmark Presentation), pp. 373–374.)

FIGURE 28.2 Photograph of CFL with electronic ballasts and mercury-based fluorescent lamps.

In this chapter, commonly used lighting technologies (except LED, which is covered in another chapter) are first reviewed. In particular, ballast technology with dimming capability for discharge lamps is stressed because dimming capability provides an effective means to control lighting energy. New lighting concepts with energy saving, long lifetime, and recyclability are introduced.

28.2 Dimming Technologies

In order to use lighting energy when and where it is necessary, and to the appropriate level, it is essential to understand various dimming methods. A wide range of different types of lamps and lighting systems are used in various applications. These include incandescent lamps, fluorescent lamps, high- and low-pressure discharge lamps. For both aesthetic and energy-saving reasons, various attempts have been made in the prior art to provide such lamps with a dimming control so that the brightness of the lamps can be adjusted.

Dimming function is particularly useful for high-intensity discharge (HID) lamps, which are widely used in public lighting systems due to the HID lamps' manifold advantages such as longevity and high luminous efficacy. Unlike incandescent lamps, HID lamps generally require a long warm-up time to reach full brightness. After being shut off, they need a cooling down period before they can be restarted again unless very high striking voltage (>15 kV) is used to restart the lamp arc at high temperature. It is the complication of this "re-strike" characteristic that makes dimming a very attractive alternative to simply turning off some of the lights for energy saving, because dimming can avoid considerable warm-up time of the lamps and the use of a high-voltage ignitor. Although numerous attempts have been made in the prior art to develop dimmable electronic ballasts for individual lamps, conventional magnetic ballasts are still the most reliable, robust, cost effective, environmentally friendly and dominant choice for high-wattage discharge lamps and large-scale lighting systems, such as street lighting systems. Dimming also has other advantages such as reduction of peak power demand, increase of flexibility for multiuse spaces, safer driving in light traffic conditions, and avoidance of light pollution.

Existing dimming methods for existing lighting systems include triode for alternating current (triac)-based dimmers for incandescent lamps and gaseous discharge lamps compatible with triac dimmers, dimmable electronic ballasts for gaseous lamps, and a range of disparate techniques for dimming lamps driven by magnetic ballasts.

28.2.1 Dimming of Incandescent Lamps

Triac-based dimmers have been popularly used as the dimming devices for Edison-type incandescent lamps and some triac-dimmable fluorescent lamps [3]. The circuit connection is illustrated in Figure 28.3a. A triac dimmer consists of a triac and also a triggering circuit that controls the phase angle of turning the triac on over a cycle of the mains voltage. As shown in Figure 28.3b, by controlling the delay angle (α), the output root-mean-square voltage, and thus the power to the lamp, can be controlled. This control of AC voltage results in the ability to adjust the brightness of the lamp.

However, the wave shape of the mains input current through the triac dimmer is dependent on the delay angle. When the delay angle is nonzero, the input current will deviate from the sinusoidal shape of the mains voltage. When the delay angle is increased, the conduction time of the triac is diminished. The input current will then consist of high harmonic components and thus generates undesirable harmonics into the power system. In addition, as the input power factor is the product of the displacement factor and the distortion factor [4], the input power factor becomes small when the delay angle is large. It is because the displacement factor is equal to the cosine of the delay angle (if the delay angle is large, the displacement factor will become small) and the distortion factor deteriorates as the current harmonic content increases. The ultimate effect of this low input power factor is the presence of reactive power flow between the AC mains and the lighting system. This reactive power could cause serious defects over the

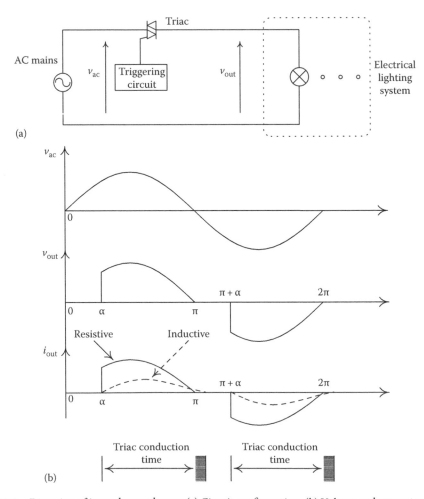

FIGURE 28.3 Dimming of incandescent lamps. (a) Circuit configuration. (b) Voltage and current waveforms.

power system. The lower the power factor, the larger the rating of the transformers and the larger the size of the conductors of transmission must be. In other words, the greater the cost of generation and transmission will be. This is the reason why supply undertakings always stress upon the consumers to increase the power factor [5].

28.2.2 Dimming of Low-Pressure Discharge Lamps with Frequency-Control Electronic Ballasts

Recently, there has been an increasing trend of using dimmable electronic ballasts for discharge lamps such as fluorescent lamps and HID lamps. A dimmable electronic ballast usually has a four-wired connection arrangement on the input side. Two connections are for the "live" and "neutral" of the AC mains, the other two are for the dimming level control signal, which is typically a DC signal within 1–10 V. A general structure of the dimmable electronic ballast is illustrated in Figure 28.4a. It consists of an active or a passive power factor correction circuit, a high-frequency DC/AC converter, and a resonant tank circuit. The power factor correction circuit and the DC/AC converter are interconnected through a DC link of high voltage. The DC/AC converter is used to drive the lamp through the resonant tank circuit. It is usually switched at a frequency slightly higher than the resonant frequency of the resonant tank circuit. The resonant tank is used to preheat the electrodes, provide a high voltage to ignite the lamp and ballast the lamp current. Dimming function is achieved by controlling the DC link voltage and/or the switching frequency of DC/AC converter. The input power factor can be kept high at any power level. As illustrated in Figure 28.4b, the waveform of the input current i_{ac} is sinusoidal and in phase with the AC mains.

A typical circuit of electronic ballast for fluorescent lamps is illustrated in Figure 28.4c, in which the AC mains is rectified by a rectifier, the power factor correction circuit is realized by a boost DC/DC converter, the DC/AC converter is realized by a half-bridge inverter circuit, and the resonant tank circuit is formed by inductors and capacitors. The DC link voltage is regulated at a level slightly higher than the peak value of the AC mains voltage. The typical value of the DC link voltage is 400 V.

Figures 28.4d through f show the three most common types of resonant tank circuits. They are the series-loaded resonant circuit (SLR) (Figure 28.4d), parallel-loaded resonant circuit (PLR) and series-parallel-loaded resonant circuit (SPLR). Based on the fundamental frequency approximation, the transfer functions of the three circuits are given as follows:

For SLR (Figure 28.4d),

$$\left| \frac{v_o(j\omega)}{v_i(j\omega)} \right| = \frac{1}{\sqrt{1 + Q^2 \left(\dfrac{\omega}{\omega_o} - \dfrac{\omega_o}{\omega} \right)^2}} \tag{28.1}$$

where $\omega_o = 1/\sqrt{LC}$ and $Q = \omega_o L / R$.

For PLR (Figure 28.4e),

$$\left| \frac{v_o(j\omega)}{v_i(j\omega)} \right| = \frac{1}{\sqrt{\left(1 - \left(\dfrac{\omega}{\omega_o} \right)^2 \right)^2 + \left(\dfrac{\omega}{\omega_o Q} \right)^2}} \tag{28.2}$$

where $\omega_o = 1/\sqrt{LC}$ and $Q = R/\omega_o L$.

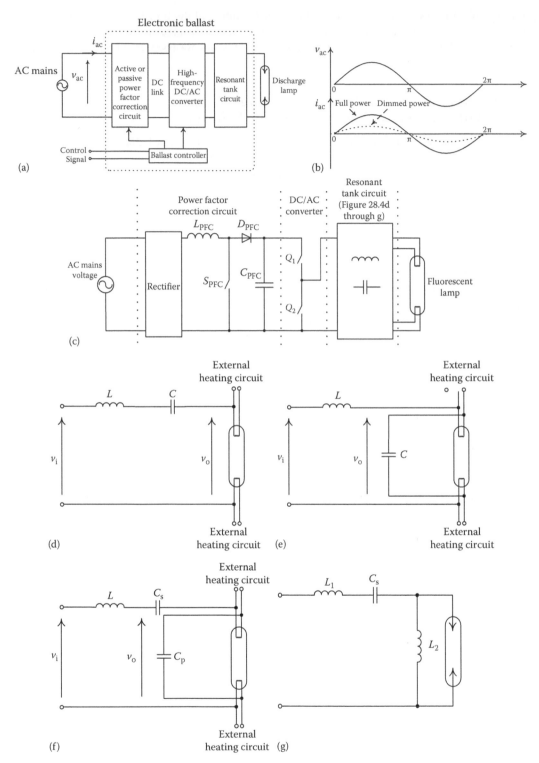

FIGURE 28.4 Electronic ballasts for discharge lamps. (a) General structure. (b) Key voltage and current waveforms. (c) Circuit schematics. (d) Series loaded resonant circuit. (e) Parallel-loaded resonant circuit. (f) Series-parallel-loaded resonant circuit. (g) Resonant tank circuit for HID lamps.

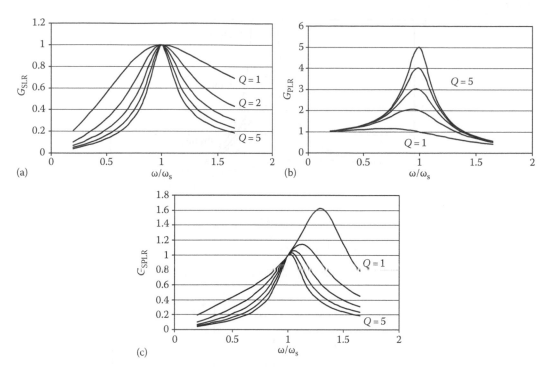

FIGURE 28.5 Frequency characteristics of different resonant circuits. (a) Parallel-loaded resonant circuit. (b) Series-loaded resonant circuit. (c) Series-parallel-loaded resonant circuit.

For SPLR (Figure 28.4f),

$$\left| \frac{v_o(j\omega)}{v_i(j\omega)} \right| = \frac{1}{\sqrt{\left(2 - \left(\frac{\omega}{\omega_s}\right)^2\right)^2 + Q_s^2 \left(\frac{\omega}{\omega_s} - \frac{\omega_s}{\omega}\right)^2}} \tag{28.3}$$

where $\omega_s = 1/\sqrt{LC_s}$ and $Q = \omega_s L/R$.

The frequency characteristics of the three resonant circuits are given in Figure 28.5. Among the three circuits, PLR is the most popular choice. The stages of operations with the PLR from preheat to dimming is illustrated in Figure 28.6.

Preheat stage: The lamp is nonconducting and its equivalent resistance is very high. Thus, the value of Q is very high. The switching frequency is held constant and is much higher than the resonant frequency for a fixed time to preheat the electrodes with a predetermined electrode current.

Ignition stage: The switching frequency is decreased toward resonance to generate a high voltage across the lamp.

Dimming: The switching frequency is further decreased to the frequency at which the lamp power is at the rated value (100%). The lamp power can be reduced by increasing the switching frequency.

Electronic ballasts for fluorescent lamps (low-pressure discharge lamps) have been widely used and have been shown that their use has an overall economic benefit [6]. Operating at high frequency (typically above 20 kHz), electronic ballasts can eliminate the flickering effects of the fluorescent lamps and

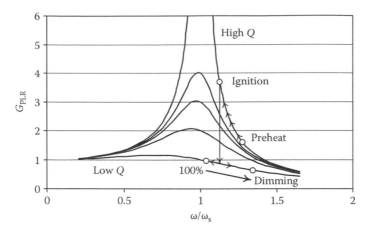

FIGURE 28.6   Illustration of the sequence of operation.

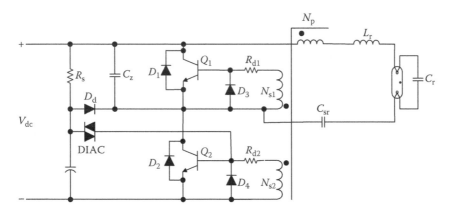

FIGURE 28.7   DC/AC inverter with self-excited gate drive.

achieve a higher efficacy than mains-frequency (50 Hz or 60 Hz)-operated magnetic ballasts. Therefore, fluorescent lamps driven by electronic ballasts consume less energy for the same light output when compared with lamps driven by magnetic ballasts.

Electronic ballasts are typically driven by integrated circuits (ICs). However, due to the cost pressure, there are many IC-less electronic ballasts. Driving of the switches Q_1 and Q_2 in Figure 28.4c can be accomplished by two possible methods. The first method is to use a self-oscillating circuit (Figure 28.7). Q_1 and Q_2 are bipolar transistors (BJTs) or MOSFETs, however, BJTs are the most dominant and feasible choice. The base driving currents or gate voltages are derived from the resonant inductor through a saturable or non-saturable transformer. The second method is to use a ballast IC. Q_1 and Q_2 are usually MOSFETs. Nevertheless, the self-oscillating inverter is the dominant solution, because its circuit is simple, robust, and cost effective.

28.2.3 Dimming of Low-Pressure Discharge Lamps with DC-Link Voltage-Control Electronic Ballasts

Frequency-control dimming method requires a fairly wide inverter frequency (typically from 45 to 110 kHz) operation for lamp power control. Typical dimming range can be from 100% to a few percent of the lamp power. However, input electromagnetic induction (EMI) filter design of wide bandwidth is needed, and switching losses and core losses will increase as the lamp power is reduced. For applications where high energy efficiency and stringent thermal requirements are needed, DC-link voltage control can be used. DC-link voltage control uses an AC-DC front-stage power circuit with output voltage step-down capability (such as flyback, SEPIC converters) to vary the DC-link voltage in Figure 28.4a as the dimming control. One application example that prefers voltage control for dimming is desk lamp in which the ballast is usually totally enclosed inside a small fixture without air ventilation (for safety reasons) and forced cooling.

The use of variable inverter DC link voltage can provide a smooth and desirable dimmer control for fluorescent lamp systems. This patented scheme [7] controls the output DC voltage V_{dc} of the front-end converter in order to control the lamp power; use constant duty cycle (near 0.5) for the switching of the half-bridge inverter in order to ensure a wide power range of continuous inductor current operation for soft-switching operation. Switching control and EMI filter design can be made easy because the inverter can be operated at constant switching frequency (or at the loaded-resonant frequency if self-excited gate/base drive is used). The Lr–Cr tank can be optimized for a given type of lamp. The standard half bridge L-C resonant converter for driving fluorescent lamps can be easily designed to operate with zero-voltage switching (ZVS) under fixed frequency. High efficiency may be obtained because the switching frequency can be chosen in the 20–30 kHz range without getting close to the infrared band around 34 kHz. The ZVS condition can be easily maintained over a wide dimming range (5%–100% of lamp power).

For the PLR circuit in Figure 28.4e, the lamp current is roughly proportional to the magnitude of the high-frequency AC voltage, V_{ac}, the magnitude of which is determined by the controllable inverter DC-link voltage V_{dc}. Thus

$$I_{lamp} \propto V_{dc} \tag{28.4}$$

The fundamental difference of voltage control and frequency control can be seen from a practical comparison of two dimmable ballasts for a 220 V, 2×36 W T8 lamp system. The ballast losses of a voltage control (Product-V) and frequency control (Product-F) are shown in Figure 28.8. Both products have a front-end AC-DC power factor correction stage and a power inverter stage, and each ballast is used to drive two T8 36 W lamps. As expected, the ballast loss of the frequency-control ballast increases as the total system power is reduced because the switching losses and magnetic core losses increase as the inverter frequency is increased for dimming purpose. On the contrary, the ballast loss of the voltage-control

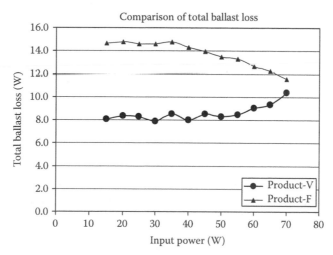

FIGURE 28.8 Total ballast losses of voltage-control (Product-V) and frequency-control (Product-F) dimmable electronic ballasts for a 220 V, 2×36 W T8 lamp system. (From Hui, S.Y.F. et al., *IEEE Trans. Power Electron.*, 21(6), 1769, November 2006. With permission.)

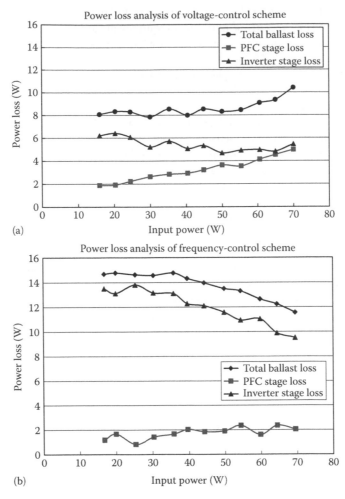

FIGURE 28.9 (a) Power loss components in a 220 V, 2 × 36 W T8 lamp system under voltage dimming control. (b) Power loss components in a 220 V, 2 × 36 W T8 lamp system under frequency dimming control.

ballast decreases because a reduced DC-link voltage will lead to a corresponding reduction in switching losses and core losses. Detailed power losses analyses of these ballasts with voltage control and frequency control are included in Figure 28.9a and b, respectively.

28.2.4 Dimming of High-Intensity Discharge Lamps with Electronic Ballasts

Among various light sources, HID lamps exhibit the best combination of the high luminous efficacy and good color rendition with the high power compact source characteristics. Through appropriate choice of dose, full spectrum (white light) sources with excellent color-rendering properties can be produced with good efficacy and compact size. HID lamps have been used in many applications, such as wide area floodlighting, stage, studio, and entertainment fighting to UV lamps.

Use of high-frequency electronic ballast can reduce the size and the weight of the ballast and improve the system efficacy. This feature is especially attractive for low-wattage HID lamps because the overall lighting system is expected to be of small size. Moreover, as the operating frequency increases, the reignition and extinction peaks disappear, resulting in a longer lamp lifetime [11]. The load characteristic of an HID lamp can be approximated as a resistor and the lamp (power) factor approaches unity. There is no flickering effect and the stroboscopic effect in the light output and the light lumen can be improved. However, the operation of high-pressure HID lamps with high-frequency current waveforms is offset by the occurrence of standing pressure waves (acoustic resonance). This acoustic resonance can lead to changes in arc position and light color or to unstable arcs. Instability in the arcs could sometimes cause the arcs to extinguish.

The common explanation for acoustic resonance is that the periodic power input from the modulated discharge current causes pressure fluctuations in the gas volume of the lamp. If the power frequency is at or close to an eigenfrequency of the lamp, traveling pressure waves will appear. These waves travel toward and reflect on the discharge tube wall. The result is standing waves with large amplitudes. The strong oscillations in the gas density can distort the discharge path, which in turn distorts the heat input that drives the pressure wave. The lamp eigenfrequencies depend on arc vessel geometry, gas filling, and gas thermodynamic state variables (such as pressure, temperature, and gas density).

Many articles on ballast circuit topologies or control methods have been proposed to avoid instability caused by acoustic resonance. Typical circuit arrangement is similar to the one shown in Figure 28.4a. There are two basic approaches of tackling acoustic resonance:

1. The output inverter is operated at a frequency well away from frequencies in the acoustic resonance range of the lamp. These ballasts can be further categorized into
   a. DC-type ballast
   b. Tuned high-frequency ballast
   c. Very high-frequency ballast
2. The switching frequency of the output inverter is modulated with fixed or random frequency. The input energy spreads over a wide spectrum so as to minimize the magnitude of the input energy in a certain frequency.

Figure 28.10 shows the circuit schematic of dimmable electronic ballast for HID lamps [8]. It controls the current flowing through the lamp and has two operating modes. In the first mode, S_3 is turned on and S_2 is turned off. Thus, the current through the lamp is regulated by sensing the voltage across the sensing resistor R_{sense2} (i.e., the lamp current) and then controlling the duty cycle of S_1 and S_4. In the second mode, S_4 is turned on and S_1 is turned off. The current through the lamp is regulated by sensing the voltage across the sensing resistor R_{sense1} and then controlling the duty cycle of S_2 and S_3. The fundamental frequency of the lamp current is low, typically 200–400 Hz, which can avoid acoustic resonance.

An HID lamp goes through several stages during the ignition process. The transitions are depicted as follows: Its lamp arc resistance is extremely large (like an open circuit) in the beginning, becomes nearly zero (short-circuit transition) for a short period, and then increases again until it reaches a steady state. Sufficient energy and a low impedance discharge path must be available for fast discharge. The authors in [9] use a series inductor–capacitor circuit and a parallel inductor—an $L_1 – C_1 – L_2$ circuit as shown in Figure 28.4f. The ballast is operated at a very high frequency. The inductance of the parallel inductor L_2

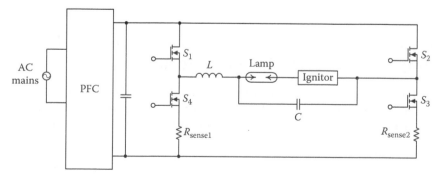

FIGURE 28.10 Circuit diagram of a dimmable electronic ballast for HID lamps.

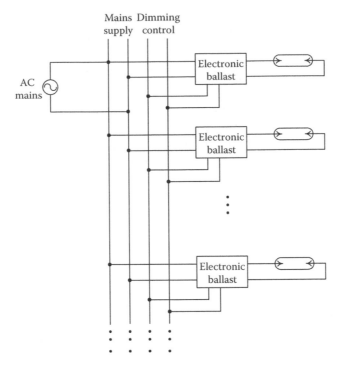

FIGURE 28.11   Schematic of large-scale dimmable electronic ballast system.

is much higher than the series inductor L_1. Multiple frequency shifting is used in starting and operating the lamp. The major limitation of this circuit is the use of a large L_2. While the resonant voltage of L_2 is large for igniting the lamp, the large impedance of L_2 ($= 2\pi f L_2$, where f is the operating frequency of the inductor) limits the rate of change di/dt of the startup discharge current in the lamp arc. Consequently, the lamp arc may have to keep on striking for many times before it can be established.

28.2.5 Dimming of Large Lighting Systems with Electronic Ballasts

Dimming large lighting systems can be performed with the use of dimmable electronic ballasts as shown in Figure 28.11. An extra 1–10 V dimming control signal has to be provided to all dimmable electronic ballasts. More sophisticated systems such as DALI have also been proposed. However, large-scale dimmable electronic ballast systems are only suitable for special lighting applications in which control of lighting is of paramount importance. The high costs of dimmable electronic ballasts and the associated control systems remain the major obstacle to their widespread use.

28.3 Sustainable Dimming Systems—Dimming of Discharge Lamps with Recyclable Magnetic Ballasts

To qualify for being a sustainable technology, a dimming system should satisfy the three criteria explained in Section 28.1. As electronic ballasts for discharge lamps are not recyclable and will end up as electronic waste, it is imperative to examine dimmable magnetic ballasts that follow the sustainable principle.

Unlike electronic ballasts, magnetic ballasts have the advantages of extremely high reliability and long lifetime, and robustness against transient voltage surge (e.g., due to lightning) and hostile working environment (e.g., high humidity and wide variation of temperature). Particularly, they offer superior

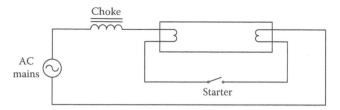

FIGURE 28.12 Typical structure of lamp circuit with magnetic ballast.

lamp-arc stability performance in HID lamps. Also, the inductor core materials and winding materials are recyclable, while electronic ballasts have more toxic and nonrecyclable materials.

The most common form of lamp circuit with starter is shown in Figure 28.12. The common starter is the glow starter consisting of a small bimetallic electrode and a fixed electrode. The two electrodes are initially separated. When the circuit is activated, the full line voltage is applied across the starter, causing a discharge between the electrodes. Current will then flow through the choke and electrodes. The heat generated from the discharge arc will cause the bimetallic electrode to bend and the two electrodes will separate. Due to a sudden change of the current through the choke, a high voltage will be created across the lamp for ignition. There are other types of starters, including thermal starters and electronic starters. Non-radiative starters are also commercially available.

In the past, the major limitation of magnetic ballasts was their lack of flexibility in achieving dimming control. Apart from the technical issues, it is also not economical to use a dimming device for each individual lamp in a lighting system formed of a large group or network of lamps. The arrangement is particularly a concern for converting a non-dimmable lighting system into a dimmable one by replacing all dimmable control gears with dimmable devices. As illustrated in Figure 28.11, the wiring and electrical installation will be complicated, since it is necessary to redesign the electrical networks for both power lines and control signals. The situation will be even more complicated in systems having multiple zones. Therefore, the installation of individual dimmable electronic ballasts in all lamp posts in a road lighting system, for example, will involve high installation cost and will also be a maintenance nightmare for the road lighting management companies, in view of the relatively poor immunity of electronic ballasts against extreme weather conditions.

Therefore, if magnetic ballasts can be made dimmable, the combined features of their long lifetime, high reliability and energy saving can make such "dimmable magnetic ballasts" an attractive solution for both indoor and outdoor applications. Moreover, it would be useful to have a technology that can dim a plurality of lamps with magnetic ballasts. Figure 28.13a shows the general non-dimmable lamp system configuration with magnetic ballasts, in which the input of the ballasts is directly connected to the AC mains through a switch gear. The switch gear is used to turn on the lamps and is controlled by various means, for example, manual control, automatic timer control, and photo sensor. To date, several dimming methods for lamps with magnetic ballasts have been reported. The ultimate purpose is to control the lamp current, and hence the lamp power, so that the lamps' brightness can be varied. The strategies are mainly acted on the input side of the ballast or at the lamp side. As depicted in Figure 28.13b, they can be categorized into several methods.

28.3.1 Method I: Control of the Supply Voltage or Current to the Lamp

Reducing the voltage supplying to the ballast is a direct way of dimming. As illustrated in Figure 28.13b, when the supply voltage v_L is reduced, the supply current i_L, the lamp current i_{lamp}, and hence, the lamp power will be decreased. This method can be realized by various voltage transformation means, such as low-frequency transformers or high-frequency switching converters.

One of the most obvious methods of altering the voltage conditions on the ballast input is to provide means whereby the voltage ratio of the supply transformers in the system may be varied. As the voltage

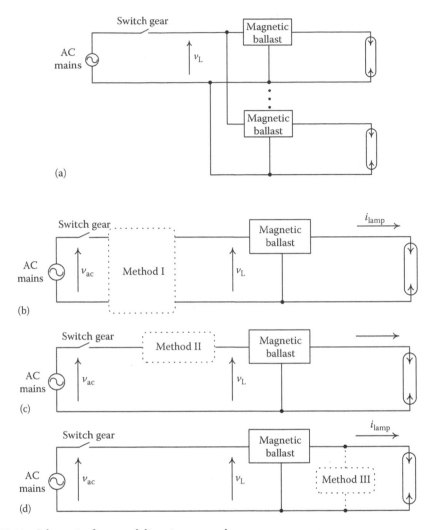

FIGURE 28.13 Schematics for several dimming approaches.

ratio is dependent on the turns ratio, it follows that if the turns ratio can be altered then the voltage ratio will be changed by the same amount [10]. Various methods have been adopted for effecting this desired change in the transformation ratio, the simplest one involving the use of a tapped winding on one side of the transformer, so that the effective turns ratio can be altered. Another one is the use of autotransformer that the turns ratio can be continuously varied. In [12], a two-winding autotransformer is used to provide two voltage levels for implementing a two-level dimming system. In [13], a multilevel dimming system that uses a more complicated transformer is proposed. All these methods involve the use of mechanical devices, such as contactors for changing the turns ratio and motors for continuously adjusting the turns ratio.

Another method is the use of high-frequency switching converters. The AC mains voltage is converted into a DC voltage by an AC/DC converter and the DC voltage is converted into an AC voltage by a DC/AC converter [4]. Thus, the overall system can flexibly provide a high-quality variable voltage and variable frequency output at v_L. However, as the input energy from the AC mains is processed twice, the overall efficiency is low. For example, if the efficiencies of the AC/DC and the DC/AC converters are 0.95, the overall efficiency is $0.95 \times 0.95 = 0.90$.

Apart from using the AC/DC/AC conversion approach, an AC/DC converter such as cycloconverter can be used to provide a controllable voltage at mains frequency. In [14], a power converter is used to chop the AC sinusoidal voltage into voltage pulses with the sinusoidal envelope. Similar approach is used in [15,16]. However, considerable current harmonics will be generated in the process, leading to harmonic pollution problems in the power system. It is unsuitable for lamps, such as HID lamps, which are sensitive to the excitation voltage. It may cause undesirable acoustic resonance and flickering effect.

Another approach [17] is the use of an external current-control power circuit to control the magnitude of the input current at mains frequency. Instead of transforming the voltage magnitude, this method adjusts the input current taken from the AC mains. As the active power taken from the AC mains is proportional to the product of the AC mains voltage and current, this can thus control the overall power delivered to the lamps.

28.3.2 Method II: Control of the Ballast-Lamp Impedance Path

Instead of transforming the AC mains voltage directly, Figure 28.13c illustrates another dimming method that the apparatus is connected in series with the lamp system. The connected apparatus is a variable reactance that it does not dissipate any active power ideally. As the overall impedance of the lamp system is adjustable, the magnitude of v_L and input current becomes adjustable.

As discussed in [19], a two-step inductor consisting of two series inductors is used for the choke in the ballast. With a switch that can bypass one of the two inductors, the overall inductance can be altered in a discrete manner.

In [20], a saturable reactor is used in the ballast that can dim the lamps continuously within a limited range. By adding an extra winding to the reactor and injecting a DC into this extra winding, the reactor core can be saturated, so that the impedance of the inductor in the ballast can be changed. The resulting effect is to adjust the current flowing to the lamps. A variant in [21] uses a variable reactance with the current to the control winding being provided by a multi-tapped autotransformer, so that different combinations of equivalent series impedance can be realized.

Instead of using passive elements, another approach [21,22] is based on creating a voltage source connecting in series with the lamp path. In [22], a DC/AC converter is connected in series with the lamp system. The DC side of the converter is connected to another AC/DC converter, which is supplied from the AC mains. Both converters have to handle active and reactive power. In other words, there is a circulating energy between the two converters. Similar idea is used in [21] that the implementation is based on using transformer coupling. Nevertheless, this circulating energy will introduce energy loss in the system. Apart from lowering the efficiency, it is also necessary to handle the thermal issue.

28.3.3 Method III: Control of the Lamp Terminal Impedance

As illustrated in Figure 28.13d, the third approach is to use an apparatus that can divert the current from the ballast. The overall effect is to reduce the lamp current i_{lamp}. In [23], a switchable capacitor is connected across a lamp. If dimming is required, the capacitor is switched on, so that part of the current from the ballast will be diverted away from the lamp into the capacitor. In this way, the lamp current and hence the lamp power can be controlled in a discrete manner.

Comparing the above methods, Methods I and II are suitable for dimming a plurality of lamps, particularly for existing installation. Method III requires the modification or installation of the dimming apparatus on each individual lamp. Although all of the above methods can dim the lamps with magnetic ballasts, they have their respective limitations of

1. Requiring expensive and bulky mechanical construction [12,13]
2. Introducing undesirable harmonic pollution to the power system [14–18]
3. Inapplicable for dimming a plurality of lamps [19,20,23]

4. Handling the total active and reactive power of the load [12–18]
5. Providing discrete dimming only [12,13,19–21,23]
6. Being practically difficult for central or automatic control [12,13,19,20,22]
7. Reducing the input power factor of the entire lighting system when the lamps are dimmed [19–21]
8. Handling dissipative circulating energy [21,22]

28.3.4 Practical Examples of Sustainable Lighting Technology

One area where sustainable lighting technology can be applied is large public lighting networks such as those used in streets, multistoried car parks, and corridors and hallways of buildings. In these public lighting systems, flickering effects are not a serious concern and so high-frequency lamp operation is not essential. Therefore, recyclable magnetic ballasts already in place can be retained. In order to minimize the amount of electronic waste, central dimming with one single controller for a large group of ballast-lamp sets can be considered. For example, if one central dimming system can control over 100 magnetic ballast–driven lamps, the amount of electronic waste can be significantly reduced and energy saving can be achieved simultaneously. Bearing in mind that the magnetic core losses can be reduced as the AC supply voltage is reduced. One ideal choice of achieving central dimming for large lighting systems is to design an AC–AC power converter without power loss. Among the technologies available, transform technology and power electronics provide two possibilities.

As discussed in Section 28.3, Methods I and II are suitable for dimming a plurality of lamps. However, it is essential to consider the energy efficiency in the actual implementation. If the traditional AC–AC power converter is used in Method I, this power converter has to handle both active and reactive power. Typical energy efficiency of power supplies is around 90%. Thus, about 10% of power will be lost in the process. This power loss is quite significant because typical power ratings of road lighting system could be in the order of several tens of kilowatts. Similar problem exists if a power transformer is used to provide discrete output-voltage levels for the large lighting systems. Discrete voltage outputs are not suitable for HID lamps because sudden step change of AC supply voltage could affect the stability of HID lamps.

One solution that has been tested successfully is the use of reactive power controller as a central dimming system [24]. Figure 28.14 shows the schematic of this concept. A power inverter is employed as a reactive power controller, which in turn provides a controllable auxiliary voltage V_a. Consequently, the output voltage V_o, which is equal to the vectorial difference of V_s and V_a, becomes a controllable AC

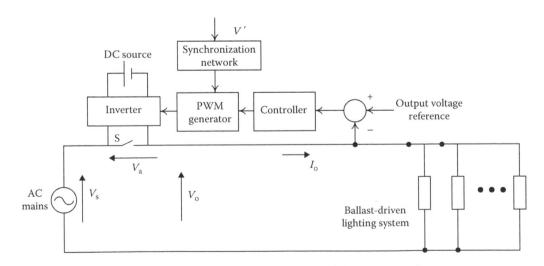

FIGURE 28.14 Schematic of a central dimming system for large-scale lighting networks.

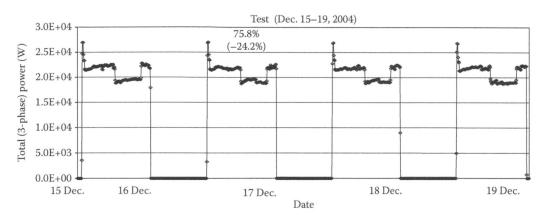

FIGURE 28.15 Power measurement of a 28 kW large-scale magnetic-ballast-driven road lighting system under central dimming control.

voltage for the lighting system. Since reactive power control does not involve active power consumption, this concept is theoretically lossless. Essentially, this approach allows the active power to go directly from the AC mains to the load. The reactive power controller only processes a portion of the system reactive power. With the help of soft switching and improvements of power devices characteristics, the energy efficiency of this central dimming system can be close to 99%, as the only power losses mainly come from the non-ideal features of the power electronic devices and passive circuit components (i.e., conduction and core losses). Figure 28.15 shows a 4 day measurement of a 28 kW road lighting system. The lighting system is activated at around 6 p.m. After a short warm-up time of 20 min, the lighting system is programmed to about 80% of the full power. From midnight to around 5 a.m., the system power is further reduced to 70% of the full power. Afterward, it goes back to 80% power until the system is turned off in the morning. An average energy saving of 24.2 has been recorded. This application example illustrates the principle of sustainable lighting technology, which allows lighting energy to be used when and where it is necessary and to the appropriate lighting level. Since a single central dimming system can control a large number of lamps (typical over 100), this type of technology can save energy effectively and retain the magnetic ballasts that have long lifetime and are recyclable. As a result, lots of electronic waste can be avoided.

28.4 Future Sustainable Lighting Technology—Ultralow-Loss Passive Ballasts for T5 Fluorescent Lamps

One effective way to save lighting energy is to replace T8 fluorescent lamps with T5 lamps. In the last two decades, it was believed that electronic ballasts were more energy efficient than magnetic ballasts. However, this understanding may not be valid for T5 lamps. T5 lamps were originally designed to be driven by electronic ballasts that can use the resonant tanks to generate high ignition voltage. For T5 28 and 35 W, the on-state lamp voltages at high-frequency operation are 167 and 209 V, respectively. These high voltage levels are close to the mains voltage of 220–240 V. Traditionally, magnetic ballasts were thought to be not suitable for driving high-voltage lamps such as T5 lamps. The technical challenges for developing magnetic ballasts that can outperform electronic ballasts for T5 lamps are

1. Sufficient ignition voltage
2. End of life detection for aged or faulty lamps
3. Provision of high lamp voltage to sustain the lamp arc after lamp ignition
4. Less ballast loss than the electronic counterparts

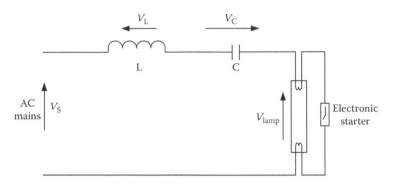

FIGURE 28.16   Circuit diagram of the ULL magnetic ballast (LC ballast) for T5 28 W lamps.

Requirements 1 and 2 can be met by using electronic starters. Electronic starters developed in the late 1990s for T8 lamps can be applied to T5 lamps in general. End-of-life detection has been a common feature among some electronic starters [25]. Since T5 lamps have on-state voltage close to the mains voltage, series inductive–capacitive (LC) ballasts (Figure 28.16) previously suggested for high-voltage lamps can be used [25,26]. As the voltage vector of the capacitor is opposite to that of an inductor, the voltage drop across the inductor can be partially or totally canceled by the voltage vector of the capacitor. Thus, requirement (3) can be met with an LC ballast.

Since T8 36 W lamps are being replaced by T5 28 W lamps, it is meaningful to use them for comparison. Table 28.1 contains a comparison of typical manufacturers' data for T5 and T8 lamps. It can be seen that high-voltage T5 lamps have high on-state voltage and low on-state current when compared with T8 lamps of similar power. For magnetic ballasts, the power losses include the conduction loss and core loss. Since conduction loss is proportional to the square of the current, the low-current feature of T5 lamps enables huge reduction of the conduction loss.

In Table 28.1, the conduction loss of a T8 magnetic ballast is used as a reference (100%). Assuming that the winding resistance of the magnetic ballasts for T5 and T8 lamps are identical, the conduction loss of the T5 magnetic ballast is only 16% that of T8 magnetic ballast. This is an 84% reduction in conduction loss. The core loss is proportional to the magnetic flux, which in turn is proportional to the current in the magnetic ballast. In this regard, the core loss of a T5 28 W magnetic ballast is only 40% that of a T8 36 W ballast, resulting in 60% reduction in core loss. Based on this theoretical assessment, significant reduction in both conduction and core losses can be achieved in magnetic ballasts for T5 lamps. Therefore, it is worthwhile to practically evaluate the energy-saving potential of magnetic ballast for T5 lamps, particularly knowing that magnetic ballasts can last for tens of years and can be recycled without creating toxic and nonbiodegradable electronic waste. The information in Table 28.1 provides the ground for developing magnetic ballasts that are more efficient than electronic ones in order to meet requirement 4.

Reference [27] describes a computer-aided analysis and practical implementation of a patented ultralow-loss (ULL) magnetic technology. Table 28.2 shows a practical comparison of such ULL magnetic ballast with electronic ballasts for a T5 high-efficient 28 W lamp at 230 V. The magnetic ballast loss is found to be less than 2.5 W. The low-current feature of high-voltage lamps such as T5 lamps is a major reason for

TABLE 28.1   Theoretical Assessment of the Power-Loss Components

Lamp Type	T8 6 W	T5 28 W
Rated voltage (V_{rms})	103	167
Rated current (A_{rms})	0.44	0.175
Conduction loss (i^2R)	100%	16%
Core loss ($\propto I$ or ϕ)	100%	40%

TABLE 28.2 Comparison of Electrical and Luminous Performance Based on the Use of the Same Philips TL5 28 W/865 Lamp

Model	Input Power (W)	Lamp Power (W)	Ballast Loss (W)	Luminous Flux (lm)	Energy Efficiency (%)	System Luminous Efficacy (lm/W)
ULL LC ballast	31.01	28.59	2.42	2318.3	92.20	74.76
Philips EB-S128 TL5 230	30.95	26.30	4.65	2188.1	84.98	70.70
Osram QT-FH 1X14-35 230240 CW	30.90	27.62	3.28	2263.8	89.39	73.26

Source: Hui, S.Y.R. et al., A 'class-A2' magnetic ballast for T5 fluorescent lamps, *IEEE Applied Power Electronics Conference (APEC)*, Technical Session: General Lighting, Palm Springs, CA, February 25, 2010.

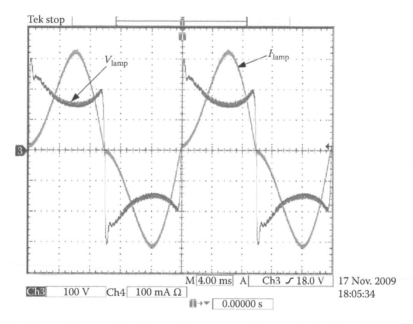

FIGURE 28.17 Measured lamp voltage V_{lamp} and lamp current I_{lamp} operated with an ULL ballast under full power operation at $V_s = 230$ V.

achieving low ballast loss in the magnetic ballast. Since the ULL ballasts of Figure 28.16 contain no electrolytic capacitor, active electronic parts, and electronic control, they offer a highly efficient, reliable, and Eco-friendly solution to T5 lamps. Typical lamp voltage and current waveforms of a T5 28 W lamp operated by a ULL ballast are included in Figure 28.17.

28.5 Conclusions

Many efforts have been made in the last two decades to improve the energy efficiency of lighting systems. It is high time to widen the scope of research to cover not only energy saving, but also lifetime and recyclability of lighting products. From the basic principle of magnetic and electronic ballasts for discharge lamps, this chapter introduces the "sustainable lighting technology" concept, which emphasizes not only energy saving, but also product lifetime and recyclability. It is important that future R & D in lighting technology will take a more holistic view on environmental protection. It is envisaged that more ballast designs without using electrolytic capacitors [28–30] and perhaps even without electronic switches and control [27,31] will be used in future lighting control. The latest emergence of recyclable ULL passive ballasts may impact the existing trend of using electronic ballasts to some extent, particularly in the public lighting systems. The passive ballast concept can in principle be applied to LED technology in order that the lifetime of the LED driver can match that of the LED devices.

References

1. J. Marcos Alonso, Chapter 22 Electronic ballasts, *Power Electronics Handbook*, Academic Press, Burlington, MA, 2007, pp. 565–591.
2. S.Y.R. Hui and W. Yan, Re-examination on energy saving & environmental issues in lighting applications, *Proceedings of the 11th International Symposium on Science 7 Technology of Light Sources*, Shanghai, China, May 2007 (Invited Landmark Presentation), pp. 373–374.
3. J. Janczak et al., Triac dimmable integrated compact fluorescent lamp, *Journal of the Illuminating Engineering Society*, 144–151, Winter 1998.
4. N. Mohan, T. Undeland, and W. Robbins, *Power Electronics: Converters, Applications, and Design*, John Wiley & Sons, Inc., New York, 2003.
5. B. M. Weedy, *Electric Power Systems*, John Wiley & Sons, Inc., New York, 1988.
6. Darnell Group, *Global Electronic Ballast Markets Technologies, Applications, Trends and Competitive Environment*, Darnell Group Inc., Corona, CA, December 2005.
7. S.Y.R. Hui, W. Yan, H. Chung, and P.W. Tam, Practical evaluation of dimming control methods for electronic ballasts, *IEEE Transactions on Power Electronics*, 21(6), 1769–1775, November 2006.
8. M. Shen, Z. Qian, and F. Z. Peng, Design of a two-stage low-frequency square-wave electronic ballast for HID lamps, *IEEE Transactions on Industry Applications*, 39(2), 424–430, March/April 2003.
9. R. Redl and J. D. Paul, A new high-frequency and high-efficiency electronic ballast for HID lamps: Topology, analysis, design, and experimental results, *Proceedings of the IEEE APEC*, Dallas, TX, March 1999, vol. 1, pp. 486–492 (also US Patent 5,677,602, 1997).
10. R. Simpson, *Lighting Control: Technology and Applications*, Focal Press, Burlington, MA, 2003.
11. W. Yan, Y. Ho, and S. Hui, Stability study and control methods for small wattage high-intensity-discharge (HID) lamps, *IEEE Transactions on Industrial Applications*, 37(5), 1522–1530, September 2001.
12. E. Daniel, Dimming system and method for magnetically ballasted gaseous discharge lamps, US Patent 6,271,635, August 7, 2001.
13. E. Persson and D. Kuusito, A performance comparison of electronic vs. magnetic ballast for power gas-discharge UV lamps, *Rad Tech'98*, Chicago, IL, 1998, pp. 1–9.
14. J.S. Spira et al., Gas discharge lamp control, US Patent 4,350,935, September 21, 1982.
15. L. Lindauer et al., Power regulator, US Patent 5,714,847, February 3, 1998.
16. J. Hesler et al., Power regulator employing a sinusoidal reference, US Patent 6,407,515, June 18, 2002.
17. B. Szabados, Apparatus for dimming a fluorescent lamp with a magnetic ballast, US Patent 6,121,734, September 19, 2000.
18. B. Szabados, Apparatus for dimming a fluorescent lamp with a magnetic ballast, US Patent 6,538,395, March 25, 2003.
19. L. Abbott et al., Magnetic ballast for fluorescent lamps, US Patent 5,389,857, February 14, 1995.
20. D. Brook, Wide range load current regulation in saturable reactor ballast, US Patent 5,432,406, July 11, 1995.
21. R. Scoggins et al., Power regulation of electrical loads to provide reduction in power consumption, US Patent 6,486,641, November 26, 2002.

22. E. Olcina, Static energy regulator for lighting networks with control of the quantity of the intensity and/or voltage, harmonic content and reactive energy supplied to the load, US Patent 5,450,311, September 1995.

23. R. Lesea et al., Method and system for switchable light levels in operating gas discharge lamps with an inexpensive single ballast, US Patent 5,949,196, July 11, 1999.

24. H.S.-H Chung, N.-M. Ho, W. Yan, P.W. Tam, and S.Y. Hui, Comparison of dimmable electromagnetic and electronic ballast systems—An assessment on energy efficiency and lifetime, *IEEE Transactions on Industrial Electronics*, 54(6), 3145–3154, December 2007.

25. D.E. Rothenbuhler, S.A. Johnson, G.A. Noble, and J.P. Seubert, Preheating and starting circuit and method for a fluorescent lamp, US Patent 5,736,817, April 7, 1998.

26. D.E. Rothenbuhler and S.A. Johnson, Resonant voltage multiplication, current-regulating and ignition circuit for a fluorescent lamps, US Patent 5,708,330, January 13, 1998.

27. S.Y.R. Hui, D.Y. Lin, W.M. Ng, and W. Yan, A 'class-A2' magnetic ballast for T5 fluorescent lamps, *IEEE Applied Power Electronics Conference (APEC)*, Technical Session: General Lighting, Palm Springs, CA, February 25, 2010.

28. Y.X. Qin, H.S.H. Chung, D.Y. Lin, and S.Y.R. Hui, Current source ballast for high power lighting emitting diodes without electrolytic capacitor, *34th Annual Conference of the IEEE Industrial Electronics, 2008 (IECON 2008)*, Orlando, FL, 2008, pp. 1968–1973.

29. P.T. Krein and R.S. Balog, Cost-effective hundred-year life for single-phase inverters and rectifiers in solar and LED lighting applications based on minimum capacitance requirements and a ripple power port, *IEEE Applied Power Electronics Conference and Exposition, 2009 (APEC 2009)*, Washington, DC, February 15–19, 2009, pp. 620–625.

30. G. Linlin, R. Xinbo, X. Ming, and Y. Kai, Means of eliminating electrolytic capacitor in AC/DC power supplies for LED lightings, *IEEE Transactions on Power Electronics*, 24, 1399–1408, 2009.

31. S.Y.R. Hui, W. Chen, S. Li, X.H. Tao, and W.M. Ng, A novel passive lighting-emitting diode (LED) driver with long lifetime, *IEEE Applied Power Electronics Conference*, Technical Session: LED Lighting I, Palm Springs, CA, February 24, 2010.

29

General Photo-Electro-Thermal Theory and Its Implications for Light-Emitting Diode Systems

Shu-Yuen (Ron) Hui
City University
of Hong Kong

and

Imperial College London

29.1 Introduction

Given that electric lighting is central to modern life and consumes about 20% of global electrical power, it is remarkable that the relatively primitive incandescent (1870s) and fluorescent (1940s) technologies are still predominant today. Light emitting diodes (LEDs) have emerged as promising lighting devices for the future. However, LEDs are still primarily restricted to decorative, display, signage, and signaling applications so far and have not reached the stage of massively entering the general and public illumination markets. While there has been increasing hope that LED technology may replace energy-inefficient incandescent lamps and mercury-based and highly toxic linear and compact fluorescent lamps (CFL) in the future, it is imperative for scientists and engineers to examine the technology in an objective way.

Among various limitations of LED, the heat dissipation and thermal degradation of luminous efficacy (i.e., reduction of lm/W due to increasing junction temperature) are probably the two most important ones. These critical issues have been the focal point in [1–6]. Despite the claims of high efficacy, such high-efficacy figures are only true at low junction temperature and are not sustainable at high temperature, which is the normal operating condition for LED applications unless expensive heatsinks and/or forced cooling can be used to keep the junction temperature at a low level. Figure 29.1 shows a typical relationship of the luminous output of LED and the junction temperature for a constant LED current (i.e., almost constant power,

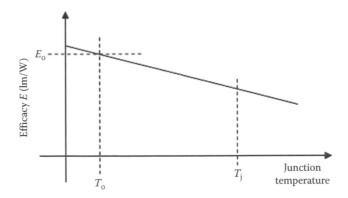

FIGURE 29.1   A typical relationship of the luminous output of LED and the junction temperature for a constant LED current.

assuming that the LED voltage does not change significantly) [7,8]. For this reason, several studies have been reported in the thermal design and management of LEDs [9–13]. The thermal modeling and measurements of the thermal resistance of LEDs have also been investigated [14–18].

In photometry, one important factor commonly used for comparing different lighting devices is the luminous efficacy (lm/W) [19]. The major hindrance to the widespread usage of LED applications in general and public illumination is the degradation of luminous flux of LEDs with the junction temperature of the LEDs [7,9,11]. This phenomenon is reflected in many LED system designs in which the maximum luminous output of LEDs does not occur at the rated power of the LEDs. In practice, the recombination of a hole and an electron results in the emission of either a photon (light) or phonon (atom vibration or heat) [20]. The drop of efficacy is caused by a non-radiative carrier-loss mechanism that becomes dominant as the current increases. Suggested reasons for such reduction include electronic leakage, lack of hole injection, carrier delocalization, Auger recombination, defects, and junction heating [30].

It is rightly pointed out that the quantum efficiency and junction thermal resistance of LED are the two limiting factors in LED technology [21]. The luminous efficacy of various LEDs typically decreases by approximately 0.2%–1% per degree Celsius rise in temperature [7]. Due to the aging effect, the actual degradation of luminous efficacy could be higher than the quoted figures. Recent research reports have highlighted the relationship of efficacy degradation and the junction temperature of the LEDs. Accelerated age tests carried out in [22] show that the light output can drop by a further 45%. For aged LEDs, the efficacy degradation rate could be up to 1% per degree Celsius. In some applications such as automobile headlights and compact lamps, the ambient temperature could be very high and the size of the heatsink is limited. This serious thermal problem has been addressed in [12,23]. The drop in luminous efficacy due to thermal problem would be serious, resulting in the reduction of luminous output [13].

Photometric parameters such as luminous flux and luminous efficacy, electrical parameters such as electric power, current, and voltage of LED, and thermal parameters such as junction and heatsink temperature and thermal resistance are closely linked together. In [7,8], the relationship between the luminous output (photometric variables) and thermal behavior has been reported. Reference [13] highlights the highly nonlinear thermal behavior of the junction-to-case thermal resistance of LED with electric power consumption of LED. The junction-to-case thermal resistance is affected by many factors such as the mounting and cooling methods [14,15], the size of the heatsink and even the orientation of the heatsink [13]. Thus, analysis on the junction thermal resistance [13,16,17] and thermal management [18,19] have been major LED research topics. To deal with various factors that affect the luminous output, control methods have been proposed to control the luminous output of LED systems [20,21]. An LED device model has been proposed to model the thermal junction resistance and the light output [22]. But this model is for an LED device and not for an LED system, including the thermal design of the heatsink and the electric power control.

In this chapter, a thermal and luminous comparison between LED and fluorescent lamps are first summarized [24]. Then a general theory that links the photometric, electrical, and thermal aspects of LED system [25] is presented. This theory is based on a simple thermal model of the LED and the heat-sink, and can be used to predict the optimal operating point (i.e., maximizing the luminous output) and provide design parameters for optimal thermal design. Tests have been carried out to verify the general theory. The examination of the theory also provides clear explanation on why the optimal operating points of some LED systems occur in an operating power less than the rated power of the LED. Practical results obtained in the experiments also highlight the major limitations of existing LEDs. Both theory and practical results provide useful insights for LED system designers and allow users to determine the advantages and disadvantages of using LED in different applications [26].

29.2 Thermal and Luminous Comparison of White High-Brightness LED and Fluorescent Lamps

29.2.1 Comparison of Heat Dissipation

Heat dissipation of a lighting device can be obtained by a simple method described in [24]. By immersing the lighting devices in silicone oil inside an insulated container with a transparent lid for the light to escape; the lighting devices can be operated without extra electricity. The heat dissipation can be absorbed by the silicone oil, the temperature of which can be used to quantify the heat dissipation.

With the heat dissipation measurement obtained, one can define a heat-dissipation factor $k_h(P_{lamp})$ for a lighting device

$$k_h\left(P_{lamp}\right) = \frac{P_{heat}}{P_{lamp}} \tag{29.1}$$

where
P_{heat} is the heat dissipation from the lamp in Watts
P_{lamp} is the total electrical input power of the lamp in Watts

This $k_h(P_{lamp})$ factor is an indicator of the amount of heat energy emitted from a lighting device for a given electrical input power of the lamp. Therefore, by comparing this $k_h(P_{lamp})$ factor, one can determine which lighting devices will generate more heat than the others.

Table 29.1 shows a comparison of some fluorescent lamps and LED devices. It is important to note that

TABLE 29.1   Comparison of Luminous Efficacy and Heat Dissipation of LEDs and Fluorescent Lamps

At Full Power	18 W T8 Fluorescent Lamp (Osram)	14 W T5 Fluorescent Lamp (Philips)	1 W LED (Philips Luxeon LXHL-PW01)	3 W LED (Philips Luxoen LXK2-PW14-V00)	LED (CREE) WREWHT-L1-0000-00D01
Rated efficacy (lm/W)	61	96	45 at 25°C junction temperature	40 at 25°C junction temperature	107 at 25° junction temperature
Measured efficacy (lm/W)	60.3	96.7	31 at 1 W (heatsink temperature of 70°C)	30 at 3 W (heatsink temperature of 80°C)	78.5 at 3 W (heatsink temperature of 76°C)
Heat dissipation factor	0.77	0.73	0.9	0.89	0.87

Source: Qin, Y.X. et al., *IEEE Trans. Power Electron.*, 24(7), 1811, July 2009. With permission.

1. The luminous efficacy of fluorescent lamps does not change noticeably with lamp temperature, while that of LED decreases significantly with increasing operating temperature.
2. Fluorescent lamps dissipate about 73%–77% of the input power as heat, but LEDs dissipate almost 90% of input power as heat.
3. Lighting devices of the same type tend to have a lower heat dissipation factor if their luminous efficacy is higher.
4. In air-conditioned buildings, the heat dissipation factor must be considered because the energy consumption of the air-conditioners affects the overall energy consumption of the building.

29.2.2 Comparison of Heat Loss Mechanism

Heat loss from light devices is achieved through radiation, convection, and conduction. Heat loss by radiation and convection is essentially free of charge from a product design point of view. However, heat loss by conduction has cost implication because it means that large heatsinks and/or fan cooling are needed. Reference [13] presents a comparison on how different light sources lose their heat. Over 90% of heat in LED has to be removed by conduction. Therefore, thermal design and management are important issues in LED technology (Table 29.2).

Figure 29.2 shows the changes of typical junction-case thermal resistance of LED devices over the last decade. The junction thermal resistance is an important factor that limits the luminous performance of the LED. For example, if the thermal resistance of a high-brightness LED is 10°C/W, there is

TABLE 29.2   Comparison of Heat Loss Mechanism among Various Lighting Devices

Light Source	Heat Lost by Radiation (%)	Heat Lost by Convection (%)	Heat Lost by Conduction (%)
Incandescent	>90	<5	<5
Fluorescent	40	40	20
High intensity discharge	>90	<5	<5
LED	<5	<5	>90

Source: Petroski, J., Spacing of high-brightness LEDs on metal substrate PCB's for proper thermal performance, *in Proceedings of the Ninth Intersociety Conference on Thermal and Thermomechanical Phenomena in Electronic Systems (ITHERM'04)*, Las Vegas, NV, June 2004, pp. 507–514. With permission.

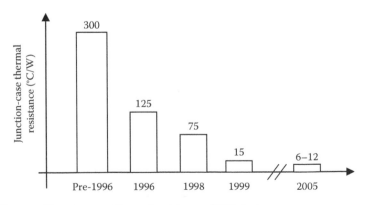

FIGURE 29.2   Changes of junction-case thermal resistance of LED devices.

a temperature difference of 50°C between the semiconductor wafer and the case if the LED is operated at 5 W. Therefore, the structural design of the LED device can affect the overall performance.

29.3 General Photo-Electro-Thermal Theory for LED Systems

This section summarizes the general photo-electro-thermal theory [25] for LED systems. This theory links the interactions of light, heat, and power together.

29.3.1 General Analysis

Let ϕ_v be the total luminous flux of an LED system consisting of N LED devices:

$$\phi_v = N \times E \times P_d \tag{29.2}$$

where
  E is the luminous efficacy (lm/W)
  P_d is the real power of one LED (W)

Figure 29.1 shows typical relationship from LED manufacturers. This curve follows an exponential decay [20], but since the operating temperature is seldom higher than 120°C, it can be approximated as

$$E = E_o \left[1 + k_e \left(T_j - T_o \right) \right] \quad \text{for } T_j \geq T_o \quad \text{and} \quad E \geq 0 \tag{29.3}$$

where
  E_o is the rated efficacy at the rated temperature T_o (typically 25°C in some LED data sheets)
  k_e is the relative rate of reduction of efficacy with increasing temperature

For example, if E reduces by 40% over a temperature increase of 100°C, then $k_e = -0.004$.
  In general, the LED power can be defined as $P_d = V_d \times I_d$. But only part of the power will be dissipated as heat. Thus, the heat generated in one LED is defined as

$$P_{\text{heat}} = k_h P_d = k_h V_d I_d \tag{29.4}$$

where k_h is a constant less than 1 and it represents the portion of LED power than turns into heat. For example, if 85% of the LED power is dissipated as heat, then $k_h = 0.85$. The measurement of k_h can be found in [24] (Figure 29.3).

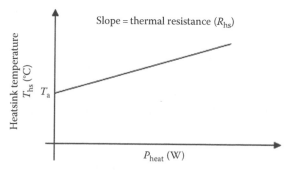

FIGURE 29.3 Typical relationship of heatsink temperature and power dissipation.

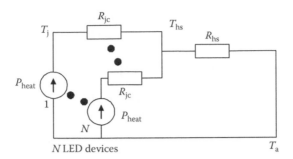

FIGURE 29.4 Simplified steady-state thermal equivalent circuit with N LEDs mounted on the same heatsink.

A simplified steady-state thermal equivalent circuit of an LED system is shown in Figure 29.4 [27,28], assuming that (1) the N LEDs are placed on the same heatsink with a thermal resistance of R_{hs}, (2) the LEDs have their junction-to-case thermal resistance R_{jc}, and (3) the thermal resistance of the insulating material (such as heatsink compound) used to isolate the LEDs from the heatsink is negligible.

Based on the model in Figure 29.4, the luminous efficacy of the LED system can be expressed as

$$E = E_o\left[1 + k_e\left(T_a - T_o\right) + k_e k_h\left(R_{jc} + NR_{hs}\right)P_d\right] \tag{29.5}$$

So, the total luminous flux ϕ_v is

$$\phi_v = NE_o\left\{\left[1 + k_e\left(T_a - T_o\right)\right]P_d + k_e k_h\left(R_{jc} + NR_{hs}\right)P_d^2\right\} \tag{29.6}$$

Because k_e is negative and less than 1, (29.6) is in the form of $\phi_v = \alpha_1 P_d - \alpha_2 P_d^2$ where a_1 and a_2 are two positive coefficients. As P_d is increased from zero, ϕ_v increases almost linearly because the second term is negligible when P_d is small. As P_d increases, the second negative term, which is proportional to the square of P_d, will reduce ϕ_v significantly. After reaching the maximum point, the ϕ_v will drop faster as P_d and R_{jc} increase (due to the increasing significance of the negative terms in (29.6)). This means that the parabola of ϕ_v is not symmetrical. Since the luminous flux function is a parabola and therefore has a maximum value, this maximum point can be obtained from $d\phi_v/dP_d = 0$.

29.3.2 Simplified Equations

From LED manufacturer data sheets [5], the degradation of the efficacy with junction temperature is usually assumed to be linear and thus k_e is assumed to be constant. This first approximation is acceptable for k_e and k_h, and will be relaxed to accommodate the changing nature of R_{jc} in the analysis later. Based on this assumption, the maximum-ϕ_v point can be obtained by putting $d\phi_v/dP_d = 0$ and

$$P_d^* = -\frac{\left[1 + k_e\left(T_a - T_o\right)\right]}{2k_e k_h\left(R_{jc} + NR_{hs}\right)} \tag{29.7}$$

where P_d^* is the LED power at which maximum ϕ_v occurs. (Note that k_e is a negative value.)

From (29.4), the corresponding LED current at which maximum ϕ_v occurs can be obtained as

$$I_d^* = -\frac{\left[1 + k_e\left(T_a - T_o\right)\right]}{2k_e k_h\left(R_{jc} + NR_{hs}\right)V_d} \tag{29.8}$$

29.3.3 Effects of Junction-to-Case Thermal Resistance R_{jc} of LED

The general theory can accommodate nonlinear junction-to-case thermal resistance R_{jc} in principle. Since R_{jc} is a complex and nonlinear function of the lamp's heat dissipation P_{heat} (which is equal to $k_h P_d$) and the thermal design of the mounting structure, the theoretical prediction is based on a simplified linear function as follows:

$$R_{jc} = R_{jco}\left(1 + k_{jc}P_d\right) \tag{29.9}$$

where

R_{jco} is the rated junction-to-case thermal resistance at 25°C
k_{jc} is a positive coefficient

A typical linear approximation of R_{jc} is shown in Figure 29.5. The luminous flux equation that incorporates a temperature-dependent R_{jc} becomes

$$\phi_v = NE_o\left\{\left[1 + k_e\left(T_a - T_o\right)\right]P_d + \left[k_e k_h\left(R_{jco} + NR_{hs}\right)\right]P_d^2 + \left[k_e k_h k_{jc} R_{jco}\right]P_d^3\right\} \tag{29.10}$$

29.3.4 Use of the General Theory for LED System Design

The use of this general theory is illustrated with an example of mounting eight Luexon K2 Cool-White 3 W LEDs on a heatsink with thermal resistance of 6.3°C/W. Parameter k_h is shown to be 0.85 at dimmed situation and 0.9 at the rated power of LED [24]. Because of its relatively small variation, it is kept constant at 0.85, although a more accurate result can be obtained if k_h is expressed as a function of P_d. The parameters required for Equation 29.6 are $k_e = -0.005$, $k_h = 0.85$, $T_a = 28°C$, $T_0 = 25°C$, $E_0 = 41\,\text{lm/W}$, $N = 8$, $R_{hs} = 6.3°C/W$, $R_{jco} = 10°C/W$ and $k_{jc} = 0.1°C/W^2$.

The measured total luminous flux for the eight LEDs is used for comparison with calculated values in Figure 29.6a. The measured and calculated total luminous flux values are plotted, not against the total power sum of the eight LEDs but against one LED power. (Note: the eight LEDs are identical and are connected in series). Using the power of one LED in the x-axis allows one to check easily if the P_d^* operating point is at the rated LED power or not. The measured efficacy values and the calculated values are

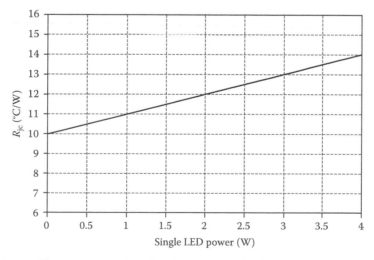

FIGURE 29.5 Assumed linear function of junction-to-case thermal resistance R_{jc}.

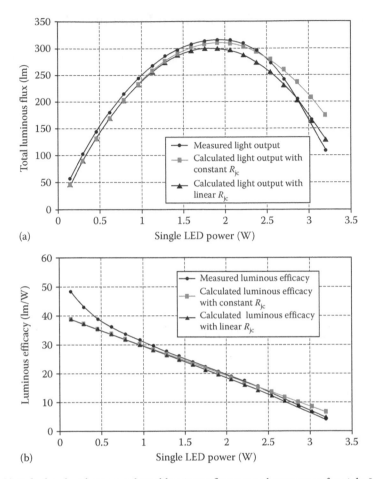

FIGURE 29.6 (a) Calculated and measured total luminous flux versus lamp power for eight Luxeon 3 W LEDs mounted on a heatsink with thermal resistance of 6.3°C/W. (b) Calculated and measured luminous efficacy versus lamp power for eight Luxeon 3 W LEDs mounted on a heatsink with thermal resistance of 6.3°C/W. (Reproduced from Hui, S.Y.R. et al., *IEEE Trans. Power Electron.*, 24(8), 1967, August 2009. With permission.)

displayed in Figure 29.6b. It is noted that the calculated values are generally consistent with measurements, except at very low power where the light output is low and the relative measurement error is large. It is important to note that $P_d^* = 1.9$ W at which the efficacy is only 20 lm/W. If these LEDs are operated at rated power (3 W), the efficacy will even drop to 8 lm/W, which is worse than that of incandescent lamps of 10–15 lm/W. The prediction from (29.10) is more accurate than that from (29.6).

29.4 Implications of the General Theory

29.4.1 Increasing Cooling Effect Can Increase Luminous Output

The maximum luminous flux will occur approximately at a lamp power P_d^* specified in (29.7). This P_d^* will shift to a lower value if $(R_{jc} + NR_{hs})$ is increased. This leads to the possibility that the P_d^* may occur at a power level that is less than the rated power $P_{d(rated)}$ of the LED. One should expect that the P_d^* could be shifted to higher power level if a larger heatsink with lower R_{hs} is used. For many applications such as vehicle headlamps and compact LED lamps (for replacement of incandescent lamps), the size of the heatsink is highly restricted and the ambient temperature is high. In these cases, there is a high possibility that P_d^* will occur at a power level less than the rated power unless proper design is adopted. In order to

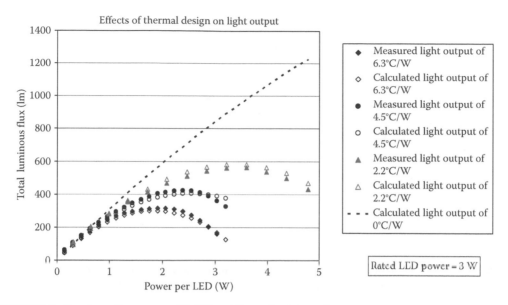

FIGURE 29.7 Luminous flux curves with different thermal resistance R_{hs}.

boost the luminous output, a better heatsink can be used. Figure 29.7 shows the predicted and measured luminous output of the eight LEDs on three different sizes of heatsinks. It can be seen that better cooling effect can increase the luminous output. If there is a perfect cooling system, the dotted line in Figure 29.7 shows the ideal limit of the luminous output in this LED system.

29.4.2 Multi-Chip versus Single-Chip LED Devices [25,29]

To reduce $(R_{\text{jc}} + NR_{\text{hs}})$ in the denominator of (29.7), one can use multi-chip LED structure in order to increase the contact area and thus reduce the effective junction-to-case thermal resistance R_{jc} of the LED package. For example, the SHARP GW5 C15L00 LED package consists of 30 units of 0.1 W LED chips (with a thermal resistance of 6°C/W), while one CREE X lamp XR-E LED (with thermal resistance of 8°C/W) consists of one high-power LED chip. The multi-chip structure of the SHARP LED has contact area larger than the single-chip structure of the CREE LED. Consequently, this SHARP LED sample has a lower thermal resistance (Figure 29.8).

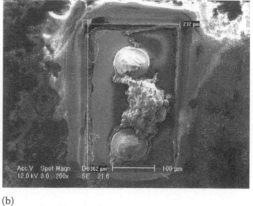

(a)   (b)

FIGURE 29.8 (a) Contact area of the single-chip CREE LED. (b) Contact area of one of the 30 chips in a SHARP LED.

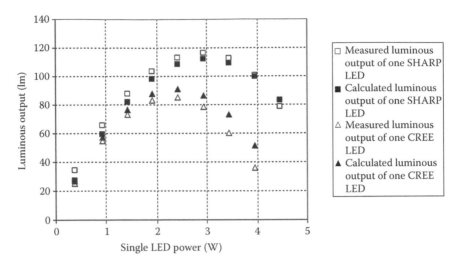

FIGURE 29.9   Measured and calculated luminous output of comparison between one SHARP and one CREE LED with the same total power on the separate heatsinks with the same thermal resistance 30°C/W.

Figure 29.9 shows the luminous performance of these two LEDs mounted separately on the same type of heatsinks with thermal resistance of 30°C/W. The results suggest that multi-chip LED could be better in terms of luminous performance than single-chip LED because of its lower junction-to-case thermal resistance.

29.4.3 Use of Multiple Low-Power LED versus Use of Single High-Power LED

Based on the same argument of using larger contact areas for heat transfer, the theory also favors the use of multiple low-power LEDs over single high-power LEDs. A comparison of using one 5 W LED and five 1 W LEDs on the identical heatsinks with thermal resistance of 8.5°C/W can illustrate this point. The results are recorded in Figure 29.10. It can be seen that, at the same power, the multiple low-power LED system generates more luminous flux than the single high-power LED system.

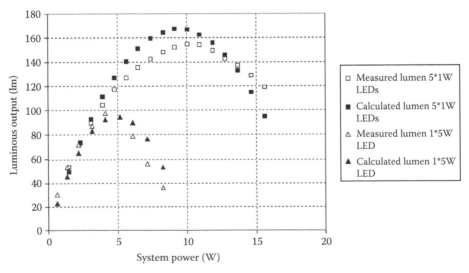

FIGURE 29.10   Measured and calculated luminous output of comparison between one 5 W LED and five 1 W LED with the same total power on the separate heatsinks with the same thermal resistance 8.5°C/W.

29.5 Conclusions

This chapter provides a theory that links the photometric, electric, and thermal aspects of LED systems together. A brief comparison of fluorescent lamps and white LEDs is given and the thermal issues of the LED technology highlighted. The general photo-electro-thermal theory and its implications are explained with the use of both theoretical and experimental data. It is hoped that this theory can assist LED manufacturers, scientists, engineers, and designers to optimize the LED technology in the future. A new method for improving the luminous efficacy is being investigated [30]. It is envisaged that LED will play a more important role in general illumination in the near future.

References

1. B. Ackermann, V. Schulz, C. Martiny, A. Hilgers, X. Zhu, Control of LEDs, in *Proceedings of the IEEE IAS'06*, Tampa, FL, October 2006, pp. 2608–2615.
2. J. M. Zhou, W. Yan, Experimental investigation on the performance characteristics of white LEDs used in illumination application, in *Proceedings of the PESC'07*, Orlando, FL, June 2007, pp. 1436–1440.
3. B. Cathy, LED light emission as a function of thermal conditions, in *IEEE Semiconductor Thermal Measurement and Management Symposium*, San Jose, CA, March 16–20, 2008, pp. 180–184.
4. P. Baureis, Compact modeling of electrical, thermal and optical LED behavior, in *Proceedings of 35th European Solid-State Device Research Conference (ESSDERC'05)*, Grenoble, France, September 2005, pp. 145–148.
5. Z. T. Ma, X. J. Wang, D. Q. Zhu, S. Liu, Thermal analysis and modeling of LED arrays integrated with an innovative liquid-cooling module, in *6th International Conference on Electronic Packaging Technology, 2005 (ICEPT'05)*, Shenzhen, China, August 2005, pp. 542–545.
6. J. Petroski, Thermal challenges facing new generation LEDs for lighting applications, *Proceedings of SPIE*, 4776, 215–222, 2003.
7. Datasheet of Luxeon Emitter, DS51, LUEXON POWER LEDS. http://www.lumileds.com/pdfs/DS51.pdf
8. J. Garcia, D. G. Lamar, M. A. Costa, J. M. Alonso, M. R. Secades, An estimator of luminous flux for enhanced control of high brightness LEDs, in *Proceedings of the IEEE PESC'08*, Rhodes, Greece, June 2008, pp. 1852–1856.
9. J. H. Cheng, C. K. Liu, Y. L. Chao, R. M. Tain, Cooling performance of silicon-based thermoelectric device on high power LED, in *Proceedings of the 24th International Conference on Thermoelectrics (ICT'05)*, Clemson, SC, June 2005, pp. 53–56.
10. X. B. Luo, S. Liu, A microjet array cooling system for thermal management of high-brightness LEDs, *IEEE Transactions on Advanced Packaging*, 30(3), 475–484, August 2007.
11. T. Zahner, Thermal management and thermal resistance of high power LEDs, in *13th International Workshop on Thermal Investigation of ICs and Systems (THERMINIC'07)*, Budapest, Hungary, September 2007, pp. 195–195.
12. J. Bielecki, A. S. Jwania, E. Khatib, T. Poorman, Thermal considerations for LED components in an automotive lamp, in *Proceedings of the Twenty Third Annual IEEE Semiconductor Thermal Measurement and Management Symposium (SEMI-THERM'07)*, San Jose, CA, March 2007, pp. 37–43.
13. J. Petroski, Spacing of high-brightness LEDs on metal substrate PCB's for proper thermal performance, in *Proceedings of the Ninth Intersociety Conference on Thermal and Thermomechanical Phenomena in Electronic Systems (ITHERM'04)*, Las Vegas, NV, June 2004, pp. 507–514.
14. K. C. Chen, R. W. Chuang, Y. K. Su, C. L. Lin, C. H. Hsiao, J. Q. Huang, K. F. Yang, High thermal dissipation of ultra high power light-emitting diodes by copper electroplating, in *Proceedings of the Electronic Components and Technology Conference, 2007 (ECTC'07)*, Reno, NV, May 2007, pp. 734–736.

15. C. Yin, Y. Lee, C. Bailey, S. Riches, C. Cartwnght, R. Sharpe, H. Ott, Thermal analysis of LEDs for liquid crystal display's backlighting, in *Proceedings of the 8th International Conference on Electronic Packaging Technology (ICEPT'07)*, Shanghai, China, August 2007, pp. 1–5.

16. Z. L. Ma, X. R. Zheng, W. J. Liu, X. W. Lin, W. L. Deng, Fast thermal resistance measurement of high brightness LED, in *Proceedings of the 6th International Conference on Electronic Packaging Technology, 2005 (ICEPT'05)*, Shenzhen, China, August 2005, pp. 614–616.

17. Q. Cheng, Thermal management of high-power white LED package, in *Proceedings of the 8th International Conference on Electronic Packaging Technology (ICEPT'07)*, Shanghai, China, August 2007, pp. 1–5.

18. J. Lalith, Y. M. Gu, N. Nadarajah, Characterization of thermal resistance coefficient of high-power LEDs, in *Sixth International Conference on Solid State Lighting*, San Diego, CA, August 2006, pp. 63370–63377.

19. R. Simpson, *Lighting Control: Technology and Applications*, Focal Press, Boston, MA, 2003.

20. E. F. Schubert, *Light-Emitting Diodes*, 2nd edn., Cambridge University Press, Cambridge, U.K., 2006.

21. S. Buso, G. Spiazzi, M. Meneghini, G. Meneghesso, Performance degradation of high-brightness light emitting diodes under DC and pulsed bias, *IEEE Transactions on Device and Materials Reliability*, 8(2), 312–322, June 2008.

22. L. Trevisanello, M. Meneghini, G. Mura, M. Vanzi, M. Pavesi, G. Meneghesso, E. Zanoni, Accelerated life test of high brightness light emitting diodes, *IEEE Transactions on Device and Materials Reliability*, 8(2), 304–311, June 2008.

23. J. F. Van, D. Michele, M. Colgan, White LED sources for vehicle forward lighting, *Proceedings of SPIE*, 4776, 195–205, 2002.

24. Y. X. Qin, D. Y. Lin, S. Y. R. Hui, A simple method for comparative study on the thermal performance of light emitting diodes (LED) and fluorescent lamps, *IEEE Transactions on Power Electronics*, 24(7), 1811–1818, July 2009.

25. S. Y. R. Hui, Y. X. Qin, General photo-electro-thermal theory for light emitting diode (LED) systems, *IEEE Transactions on Power Electronics*, 24(8), 1967–1976, August 2009.

26. Y. X. Qin, S. Y. R. Hui, Analysis of structural designs of LED devices and systems based on the general photo-electro-thermal theory, in *IEEE Energy Conversion Congress and Exposition, 2009 (ECCE)*, San Jose, CA, September 20–24, 2009, pp. 2833–2839.

27. C. J. Adkins, *An Introduction to Thermal Physics*, Cambridge University Press, Cambridge, U.K., 1987.

28. M. Arik, J. Petroski, S. Weaver, Thermal challenges in future generation solid state lighting application: Light emitting diodes, in *2002 International Society Conference on Thermal Phenomena*, Orlando, FL, 2002, pp. 113–120.

29. S. Y. R. Hui, Y. X. Qin, Analysis of the structural designs of LED devices and systems based on the general photo-electro-thermal theory, *IEEE ECCE*, City University of Hong Kong, Hong Kong, China, 2009.

30. J. Xu, M. F. Schubert, A. N. Noemaun, D. Zhu, J. K. Kim, Reduction in efficiency droop, forward voltage, ideality factor, and wavelength shift in polarization-matched GalnN/GalnN multi-quantum-well light-emitting diodes, *American Physics Letters*, 94, 01113-1–01113-3, 2009.

30

Solar Power Conversion

Giovanni Petrone
Università di Salerno

Giovanni Spagnuolo
Università di Salerno

30.1 Introduction

The amount of energy coming from the Sun and hitting the Earth surface in one day might be enough to fuel human activities all around the world for almost 1 year. The solar energy available on Earth changes with latitude, but it has been estimated as $7\,kWh/m^2/day$ in the Arizona desert area [1] in the United States, and as $2\,kWh/m^2/day$ in Russia [2]. Up to now, this natural and free resource of energy has been largely underutilized due to the full availability of fossil fuels, e.g., oil, carbon, and natural gas, and a sleeping environmental awareness. Further motivations for such vacancy are the high cost of the electrical and thermal energy produced by employing solar radiation, especially by means of photovoltaic (PV), wind (such energy can be traced back to temperature differences due to solar radiation), or thermal solar systems, if compared with the energy drawn from the grid. Such unbalance, which has been worsened by an inherent unreliability of such systems, has been due to the high research and development costs that any new technology requires, to the low demand from the market and, at least for PV systems, determined by the high costs of the raw materials because of silicon shortage. As concerns the electrical energy production, the world market received a significant encouragement by the feed in tariffs, launched by German and Japanese governments and adopted by many other countries; such incentives reward the energy produced by a power plant in place of supporting its installation without security, thus indirectly supporting the development of energy conversion systems characterized by higher efficiency. This policy has also induced a beneficial effect on the research activities of the groups working in the areas of physics and engineering for solar power conversion, so that the number of papers that have been appearing in prestigious journals and presented in international conferences all around the world has increased at an exponential rate. New technologies for employing solar energy and innovative ideas in terms of circuit topologies and control solutions aimed at improving performances, efficiency and reliability over the others, have been proposed in a large number of international patents submitted in recent years. This incredible boost of activities has led to the founding of many new companies both in the field of solar power conversion and in power electronic systems for such application, with an increased competition and a record high of the world solar PV market installations of 2826 MW in 2007, with a growth of 62% over the previous year [3–7]. Unfortunately, this scenario has not yet led to a significant reduction of the prices of the system components, probably due to the presence of the incentives. A new revolution is expected in the near future, because the reached maturity of this market will lead to the cancellation of the feed in tariffs for new plants. This political decision will change the equilibrium among the competitors because the leading companies will be those proposing products exhibiting top performances at the best prices as in any mature market, thus allowing to reach the so-called "grid parity," i.e., producing electrical energy at the same price of the energy taken from the grid.

The aim of this chapter is to give to the reader an overview of current technologies for converting solar power into electrical power by means of PV systems. Special attention has been initially dedicated to a review of the different types of cells that share the largest portions of the actual world market. Technologies that at present do not guarantee significant efficiencies and/or exhibiting low reliability and short lifetimes, but having a high potential for the future applications are also mentioned. Issues concerning the balance of system, especially the optimal control of the PV field, and grid interfacing are treated.

30.2 Solar Cells: Present and Future

About 87% of the solar cells produced all around the world is made of crystalline silicon. In 2007, more than half of the worldwide production of electronic-grade silicon was used to produce solar cells [7]. Nevertheless, until then, the silicon industry produced electronic-grade silicon exclusively for the semiconductor industry, with only a small fraction delivered to the PV industry. Such a situation, which has changed in recent years, in coincidence with the significant growth of the PV industry, led to a silicon shortage for PV industry that lifted up the solar power cost from about $4 a watt in the early 2000s to more than $4.80 per watt after 2005 [7]. According to some studies, the 15,000 tons of silicon that were available for use in solar cells in 2005 will become 123,000 tons in 2010, with a consequent reduction of PV module prices. Silicon shortages and the need of reducing the PV system prices, for the largest part due to semiconductor-grade silicon cost, led to the development of different technologies aimed at reaching the maximum efficiency but, at the same time, with the minimum amount of silicon or without the purest silicon.

A monocrystalline silicon cell needs absolutely pure semiconducting material to be produced. Monocrystalline rods are extracted from melted silicon and then sawed into thin plates in order to guarantee a level of efficiency that is considerably higher than that one characterizing polycrystalline cells. In this context, efficiency is given by the ratio between the electric power produced by the cell and the solar power incident on the cell area. Polycrystalline cells are produced by means of a more cost-efficient process: liquid silicon is poured into blocks that are subsequently sawed into plates. The lower efficiency of this type of solar cells is due to crystal defects emerging at the borders during solidification of the material, when crystal structures of varying sizes are formed. On average, in 2003, the thickness of wafers used for producing such cells was 0.32 mm, it decreased to 0.17 mm in 2008, while efficiency increased from 14% to 16% over the same period. By 2010, the aim is to reach 0.15 mm and 16.5% of wafer thickness and efficiency [8].

Amorphous or thin film modules are obtained by depositing extremely thin layers of silicon onto another material that has the role of substrate, e.g., glass, stainless steel, or plastic, and that can be flexible, thus giving them wider applications. The use of 1 μm layer thickness results in lower production costs compared to the more material-intensive crystalline technology, a price advantage that is currently counterbalanced by significantly lower efficiencies. Three types of thin film modules are commercially available: these are manufactured from amorphous silicon (a-Si), copper indium diselenide, copper indium gallium diselenide (CIS, CIGS), and cadmium telluride (CdTe). The typical a-Si efficiency approaches 8%, while CIS/CIGS and CdTe reach 11%; the former is the most important in terms of production and installation (5.2% of the total market in 2007), but EPIA expects a growth in the thin film market share to reach about 20% of the total production of PV modules by 2010. While the generation of 1 W of power currently costs approximately $1.75 to $5, thin film substrates will help bring down the prices to a more acceptable $1.3/W by 2012 [9].

The goal of reducing the amount of silicon can be also obtained by using concentrator cells: they work by focusing solar light onto a small area using an optic concentrator, with a concentrating ratio of up to 1000. The small area collecting the concentrated beam is equipped with a material made from

III–V compound semiconductors, namely multijunction gallium arsenide type, which have efficiencies of 30% and in laboratories of up to 40%. Due to the fact that such systems use the direct, not the diffuse, component of the sunlight, they need a mechanical tracking system in order to direct the system toward the Sun. Single-axis as well as two-axes tracking equipments are available on the market.

Carbon-based plastics, dyes, and nanostructures represent the new frontier of PV generation: they are much cheaper with respect to traditional semiconductors such as silicon, since they can be manufactured by means of a process that does not need the high-temperature vacuum processing used for inorganic materials. Organic PV cells also are much more flexible and lighter in weight than crystalline ones, thus suggesting further range of uses. The main limitation of such technology is in the low efficiency, which makes it suitable for those markets needing low performance levels and low manufacturing cost.

The greatest interest from investors and researchers is for bulk-heterojunction cells, invented in the early 1990s and composed of conducting polymers and carbon nanostructures called buckyballs that, in the right combinations, mimic the light-absorbing p-n junction of crystalline cells, and for dye-sensitized or Grätzel cells, characterized by liquid components.

A complete glance of the different technologies is reported in Figure 30.1. Analytical and circuit models of PV cells are useful in order to reproduce their behavior, especially for the development of circuit and systems aimed at exploiting the energy produced by the PV generator.

As for crystalline cells, two circuit models can be adopted in order to reproduce their behavior: they are referred to as single-diode and two-diode models [10]. A single-diode model, as depicted in Figure 30.2a, includes a photo current source, a diode in parallel with the source, a series resistor, and a shunt resistor. The double-diode model, depicted in Figure 30.2b, includes an additional diode for better curve fitting. Due to the exponential equation of a p-n diode junction, the equivalent analytical model consists of a system of nonlinear equations. The single-diode model allows a good reproduction of the real characteristics of a solar cell by using parameters taken from data sheets [11].

An example is reported in Figures 30.3 and 30.4, where the current–voltage and the power–voltage characteristics of a Kyocera KC120 module [11] are shown. The model also allows to reproduce the dependency

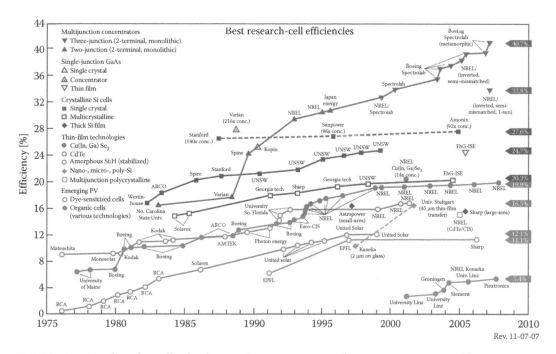

FIGURE 30.1 Timeline of PV cell technologies and energy conversion efficiencies: past, present, and future at a glance (National Renewable Energy Laboratory—USA www.nrel.gov/ncpv/thin_film/docs/kaz_best_research_cells.ppt).

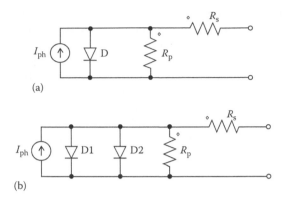

FIGURE 30.2 (a) Single-diode and (b) two-diode circuit models of a crystalline solar cell.

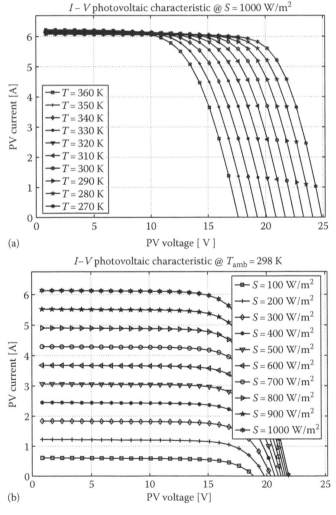

FIGURE 30.3 Current vs. voltage characteristics of a Kyocera KC120 module: (a) varying temperature at fixed irradiation level and (b) varying irradiation level at fixed temperature.

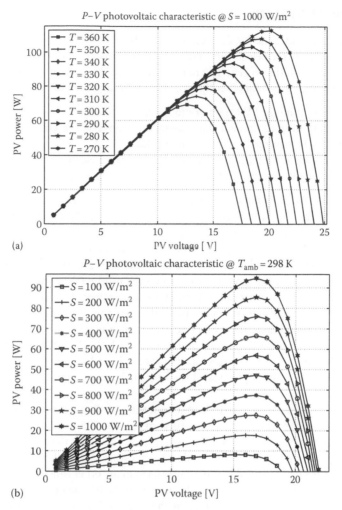

FIGURE 30.4 Power vs. voltage characteristics of a Kyocera KC120 module: (a) varying temperature at fixed irradiation level and (b) varying irradiation level at fixed temperature.

of the electrical characteristics of the cell on the environmental parameters, namely temperature and irradiation (T and S in Figure 30.3, respectively). The voltage at which the cell/module current is zero is named open-circuit voltage (V_{oc}) and the current given by the PV cell/module when its terminals are short-circuited is named short-circuit current (I_{sc}). Figure 30.3 shows that an increase in irradiation lifts I_{sc} up and has a weak effect on V_{oc}. On the contrary, a temperature increase reduces V_{oc}, but slightly affects I_{sc}. Such variations also affect the maximum of the power–voltage characteristic, namely the maximum power point (MPP), in the sense that higher irradiation levels and lower temperatures improve the power production of the PV cell/module because of the increase of the current (I_{MPP}) and of the voltage (V_{MPP}) values, respectively, when the MPP occurs. The fill-factor gives a measure of quality of the solar cell/module: it is the ratio between the actual maximum power ($P_{MPP} = V_{MPP} \cdot I_{MPP}$) and the theoretical one ($P_T = V_{oc} \cdot I_{sc}$). The more rectangle-like is the current–voltage characteristic, the higher is the fill-factor: typical values are up to 0.82. The two resistances R_s and R_p affect the fill-factor: high R_s and low R_p values worsen the fill-factor and are indices of a low-quality PV cell. Standard test conditions (STC) are used to compare the performances of different cell/modules: they refer to an irradiance value of 1000 W/m², cell junction temperature of 25°C, and a solar reference spectrum equal to AM (air mass) 1.5.

30.3 Balance of System

A complete PV system includes the PV devices, namely cells, modules, arrays, and so on, that convert sunlight into direct-current (DC) electricity. Such energy is transferred to the load or to the electrical grid by means of a subsystem that is generally referred to as the "balance of system" or BOS. It encompasses all components of a PV system other than the PV panels and includes the following:

- Structures for mounting the PV arrays or modules.
- The power-conditioning equipment that adjusts and converts the DC electricity to the proper form and magnitude required by an AC load or grid.
- Storage devices, such as batteries, for storing PV-generated electricity to be used on site during cloudy days or at night, especially whenever the access to the energy grid is not available.

As for the structure, it is usually bulky enough to be weatherproof and it is placed so that the modules orientation, both in terms of tilt and azimuth angles, ensures the highest energy production. The best orientation depends on the latitude of the site [1,2] and it is usually determined on the basis of the average energy production. Nevertheless, especially in stand-alone applications, for which it is preferable to use the PV power when it is produced, so that a double flow to and from the energy storage unit is avoided, the best orientation of the modules depends on the profile of the load power demand during the day [12]. Especially when concentrating PV modules are involved, but their adoption improves the power production even if flat-plate systems are used, the structure must be able to move the module in order to ensure that solar rays hit the module surface perpendicularly all day. One-axis or two-axis trackers can be used; the former are typically designed to track the Sun from east to west on its daily route. The latter is also able to track the seasonal course of the Sun between the northern and southern hemispheres. Naturally, the more sophisticated the system the more expensive and the more maintenance it may require.

As far as flat-plate systems are involved, the maximization of the power produced must be pursued even if the modules are mounted on a fixed structure, so that the PV array voltage/current is settled at the value ensuring that the maximum power is drawn from the array for the current weather conditions. To this aim, power-conditioning circuits are needed: such feature is described in detail in Section 30.4. In addition, power electronics is needed in order to perform many other functions, such as conversion of DC power into AC power, in order to supply AC loads or feed the AC energy grid, ensure galvanic isolation between the PV field and the AC grid, comply with standards [13] concerning the quality of the current injected into the grid, manage the state of charge of an energy store system like a battery in off-grid applications, manage the energy flux in order to avoid powering a location even though power from the electric utility is no longer present, or, if the islanding operating mode is desired, to supply a local load after disconnecting the PV array from the grid or during a power blackout. The islanding operation mode detection is an important function of grid-connected PV inverters: it is the ability of detecting that the grid has been removed on purpose, due to an accident or by damage, but the inverter continues its operation by supplying local loads. In such conditions, appropriate measures in order to protect persons and equipments must be taken [14]: they are classified into active and passive detection schemes. The former provides for injecting a disturbance into the grid and monitoring its effect, while the latter just monitors grid parameters, without affecting the power quality and avoiding interactions among multiple inverters in parallel with the grid.

As a consequence of this, the power conditioning stage is made of two stages: a DC/DC devoted to the PV array control and a DC/AC accomplishing to the inversion task. Details about both of them will be given in Sections 30.3 and 30.4.

Storage systems are also usually included into the BOS. They are often adopted in off-grid plants whenever the energy produced by the PV array must be used at night or on cloudy days. Batteries are the most common devices used to store energy, but they have a nonnegligible environmental impact since they use metals and their lifetime is considerably shorter than that of the PV array. Battery lifetime, usually ranging from 5 to 10 years, depends on charge/discharge cycle rates numbers: the deeper the battery is discharged the shorter the lifetime. This always occurs in PV applications, so that suitable technologies for long lifetime batteries for PV applications have been developed recently [15–18]. Suitable power electronic converters have been developing in order to ensure the best employment of the PV generator and management of the battery pack, e.g., in terms of charge current profile.

Since batteries decrease the efficiency of the PV system, because only about 80% of the energy injected into them can be reclaimed, a growing number of applications include supercapacitors that are used when some energy must be stored with a higher efficiency, but lower capacity, with respect to that ensured by batteries. Supercapacitors are often used in portable and automotive applications.

30.4 Maximum Power Point Tracking Function

A PV array subjected to uniform weather conditions, in terms of temperature and irradiance, exhibits a nonlinear current–voltage characteristic. The array power performance is usually expressed by means of the power–voltage curve that allows putting into evidence that there is a unique point, called the MPP, where the array produces maximum output power. In Figure 30.1, an example of the two electrical characteristics, namely current vs. voltage and output power vs. voltage, is reported. Such figures, referring to a Kyocera KC120 module [11], put into evidence the sensitivity of the module characteristics with respect to weather conditions. It must be pointed out that the open-circuit voltage, exhibited by the module in no-load conditions, is strongly affected by temperature variations and varies quite slightly with the irradiation level. The short-circuit current, instead, increases with the irradiation level, but is weakly dependent on temperature. Such changes due to the variation of the irradiance level and of the cells' working temperature, with the latter depending, in turn, on the irradiance level, on the ambient temperature, on the efficiency of the heat exchange mechanism, and on the operating point of the cells, also affect the MPP. It has been also evidenced in both figures that, due to the effects of such time-varying operating conditions, it occurs at different voltage–current couples. As a consequence, if the PV array works at a fixed voltage all day long, it cannot deliver the maximum power at every instant, so that a considerable amount of energy is certainly wasted. In literature, the possibility of fixing the PV array operating voltage at about 80% of its open-circuit voltage is suggested quite frequently, especially if a cheap and reliable PV-powered system must be realized. This is often the case of simple stand-alone systems, e.g., oriented to boat applications or campers, obtained by connecting the PV array straight to the terminals of a battery, supplying DC loads in parallel. Such solution, although simple, does not maximize the power output from the PV array because it does not track continuously the MPP according to the time-varying operating conditions it is subjected to. A typical example concerns the rough connection between a lead acid battery working at V_{bat} voltage, and a PV panel having the "nominal" MPP at the STC conditions reported on the data sheet at the same voltage, $V_{MPP} \approx V_{bat}$. When weather conditions change so that $V_{MPP} > V_{bat}$, the PV module delivers a power lower than the maximum it could give. Worst conditions occur if $V_{MPP} < V_{bat}$, because, due to the high slope of the power–voltage characteristic of any PV panel on the right side of the MPP (see Figure 30.4), a dramatic power drop can occur and, for $V_{bat} > V_{oc} > V_{MPP}$, the output power is even zero.

Due to such limitation in power production and thanks to the fact that the cost of a power electronic device that is able to accomplish the task of maximum power point tracking (MPPT) has decreased significantly in recent years, almost any PV system includes such feature operated by means of a switching converter as in Figure 30.5.

The MPPT issue has been addressed by means of fuzzy logic–based controllers, neural networks, and evolutionary approaches: many examples, even too much involved, can be found in literature [19–23]. Nevertheless, in many applications, especially those requiring low costs so that a digital signal processor (DSP) or a field programmable gate array (FPGA) device cannot be used in order to implement

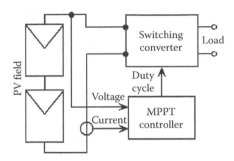

FIGURE 30.5　Schematic of MPPT operation by means of a switching converter.

sophisticated strategies, the use of simple and robust algorithms is needed. In such cases, the perturb and observe (P&O) and incremental conductance (INC) [19] techniques are widely used.

The P&O MPPT algorithm is based on the following criterion: if the operating voltage of the PV array is perturbed and if the power drawn from the PV array increases, this means that the operating point has moved toward the MPP, and therefore the operating voltage must be further perturbed in the same direction. Otherwise, if the perturbation of the operating voltage leads to a decrease of the power drawn from the PV array, the operating point has moved away from the MPP, therefore the direction of the operating voltage perturbation must be reversed. Such strategy can be even implemented on a low-cost digital controller and it requires to sense the PV array current and voltage, or even simply its actual power, by means of suitable, but even cheap, sensors. The perturbation of the PV array voltage can be operated by adopting a DC/DC converter interfacing the PV array with a load, a DC-link capacitor, or a battery bank. The P&O controller modifies the PV array voltage, namely the voltage at its input terminals, by perturbing the converter duty cycle directly or through the reference voltage at the comprator terminal of the pulse width modulator (PWM).

A drawback of P&O MPPT technique is that, at steady state, the operating point oscillates around the MPP giving rise to the waste of some amount of available energy. Several improvements of the P&O algorithm have been proposed in order to reduce the number of oscillations around the MPP in steady state as well as their amplitude. Such strategies work on the two operating parameters that affect static and dynamic performances of the P&O technique: the frequency and the amplitude of the perturbations imposed to the system. If the duty cycle d is assumed as the perturbed variable, the amplitude of its perturbations can be indicated as $\Delta d = |d(kTa) - d(k-1)Ta)| > 0$, where the time interval between the consecutive perturbations is usually named Ta. In the P&O algorithm, the sign of the duty-cycle perturbation at the $(k+1)$th sample is decided on the basis of the sign of the difference between the power $p((k+1)Ta)$ and the power $p(kTa)$ according to the rules discussed above:

$$d((k+1)Ta) - d(kTa) + (d(kTa) - d((k-1)Ta)) \times \text{sign}(p((k+1)Ta) - p(kTa)) \qquad (30.1)$$

By lowering the amplitude of the duty-cycle perturbation, Δd, the steady-state losses caused by the oscillation of the array operating point around the MPP are reduced. Unfortunately, this makes the algorithm less efficient in case of rapidly changing atmospheric conditions because a small Δd leads to a long transient until the new steady-state condition is reached. The choice of the value Ta also affects the P&O MPPT performances, both in presence of quickly varying MPP and whenever the MPP moves very slowly. The sampling interval Ta should be set higher than a proper threshold in order to avoid instability of the MPPT algorithm and to reduce the number of oscillations around the MPP in steady state. In fact, considering a fixed PV array MPP, if the algorithm samples the array voltage and current too slowly, the ability of the algorithm in tracking rapidly varying irradiation conditions is compromised. On the other side, if the PV array power is sampled too quickly, the P&O controller is subjected to possible mistakes caused by the transient behavior of the whole system, consisting of both PV array and

DC/DC converter, thus missing, even if temporarily, the current MPP of the PV array, even if it is in steady-state operation. As a consequence, the MPPT efficiency decays as the algorithm can be confused and the operating point can become unstable, entering disordered and/or chaotic behaviors. To avoid such mistake, it must be ensured that, after each duty-cycle perturbation, the system reaches the steady state before the next measurement of array voltage and current is done. In [24], the problem of choosing *Ta* is analyzed and an optimized solution, based on the tuning of the P&O algorithm according to converter's dynamics, is proposed. The optimization procedure is also illustrated for rapidly varying irradiance conditions. Better performances can be obtained if an adaptive choice of the parameters affecting the effectiveness of the MPPT algorithm is done, even if at the price of hardware complexity. In [25], a possible solution for a time-varying setting of the P&O parameters is presented, but similar approaches are also suggested [26] for the other widely used MPPT algorithm, the INC algorithm [27].

The INC algorithm is based on the observation that, at the MPP, it is

$$\frac{dP}{dv} = \frac{d(v \cdot i)}{dv} = 0 \tag{30.2}$$

consequently,

$$i + v \cdot \frac{di}{dv} = 0 \tag{30.3}$$

so that

$$\frac{di}{dv} + \frac{i}{v} = 0 \tag{30.4}$$

where *i* and *v* are the PV array current and voltage, respectively. When the operating point in the voltage vs. power plane is on the right of the MPP, then it is

$$\frac{di}{dv} + \frac{i}{v} < 0 \tag{30.5}$$

while, if the operating point is on the left of the MPP, it is

$$\frac{di}{dv} + \frac{i}{v} > 0 \tag{30.6}$$

It results that the sign of the quantity

$$\frac{di}{dv} + \frac{i}{v} \tag{30.7}$$

indicates the correct direction of perturbation leading to the MPP. At least theoretically, the INC algorithm gives a condition letting the controller to know when the MPP has been reached, so that the perturbation can be stopped and all the steady-state losses due to oscillations of the operating point oscillating around the MPP typical of the P&O implementation were overcome. Unfortunately, as discussed in [27], noises, measurement uncertainties, and quantization errors affecting the electrical variables at the PV array terminals make the condition (30.4) never exactly satisfied in practice, so that it is usually required that such condition is approximately satisfied within a given accuracy. As a consequence, even if the INC algorithm is used, the oscillations of the operating voltage cannot be ever suspended because the PV array voltage will never coincide with the MPP exactly. A disadvantage of the INC algorithm, with respect to P&O, is in the increased hardware and software complexity, also leading to increased computation times and to the consequent slowing down of the possible sampling rate of array voltage and current.

Both P&O and INC methods can be confused during those time intervals characterized by changing atmospheric conditions, because during such time intervals the operating point can move away from the MPP instead of keeping close to it [27].

There is no general agreement in the literature on which of the two methods is the better one, even if it is often said that the efficiency—expressed as the ratio between the actual array output energy and the maximum energy the array could produce under the same temperature and irradiance level—of the INC algorithm is higher than that one of the P&O algorithm. To this regard, it is worth saying that the comparisons presented in the literature are carried out without a proper optimization of P&O parameters. In [19] it is shown that the P&O method, when properly optimized, leads to an efficiency that is equal to that obtainable by the INC method. The key idea underlying the proposed optimization approach lies in the customization of the P&O MPPT parameters to the dynamic behavior of the whole system composed by the specific converter and PV array adopted. Results obtained by means of such approach clearly show that, in the design of efficient MPPT regulators, the easiness and flexibility of the P&O MPPT control technique can be exploited by optimizing it according to the specific system's dynamic.

Any optimization of the parameters influencing the effectiveness of P&O and INC algorithms in the presence of uniform working conditions of the PV array becomes useless whenever such conditions run out. In fact, such methods, but this is a limitation of almost all the MPPT techniques presented in literature, properly work when all the modules of the array operate in the same operating conditions, since they are able, although by means of different processes, to detect the unique peak of the power vs. voltage characteristic of the PV array. Unfortunately, in many real cases, the PV field does not receive a uniform irradiation and/or not all its parts (panels as well as single cells) work at the same temperature, so that mismatches among different parts of the array may arise. Such diverse conditions might be due to the weather typical of the latitude at which the PV plant is installed, to structural and architectural elements in the neighborhoods, to the morphology of the place, but even to soiling. As for the shadowing due to clouds, it has a sensible mismatching effect on large PV fields when windy days with clouds running on the field often occur. The effects of structural and architectonical elements near the field often influence the generator's performances, especially in building-integrated photovoltaic (BIPV) applications. For example, PV fields placed on roofs of buildings may suffer from shadows deriving from architectural elements: they might shade the field with almost the same law during each day of the year. This possibility is critical at high latitudes and may be due to structures that are built up after the installation of the PV plant. The position of the installation site with respect to the context has a great impact on the instantaneous PV field electrical characteristics. Especially for large PV plants, it may be useful to choose a different orientation for distinct field subsections, e.g., depending on the ground structure. This trick is often used, regardless of the site characteristics, to ensure energy peaks at fixed Sun positions during daylight. In this case, each PV field subsection would be able to guarantee the maximum energy production during a precise time interval only, which is different from those during which the other sections ensure their maxima. Nevertheless, the different orientation of separate subsections of the field may also arise from architectural constraints and the need to integrate panels into building surfaces [6]. Working conditions described above are often referred as "mismatching" and give rise to significant limitations on power production if "mismatched" PV sections are electrically connected in series. Such critical conditions may also occur if PV field subsections, panels or cells of panels, differ in rated power, area, technology used for producing different cells, or simply due to production tolerances and different drifts ascribable to aging.

The PV plant power reduction due to mismatching can be mitigated by using bypass diodes: in a series-connected string of PV elements, bypass diodes shunt those that are not able to support the high current produced by some others in the same string (e.g., the ones receiving a higher irradiance). The high string current imposed by the PV elements producing the highest power may be higher than the short-circuit current of some others. Consequently, the latter reverse their terminal voltage and absorb a part of the power produced by the elements of the string. The bypass diode limits the PV element reverse voltage and represents an alternative way for the string current. Bypass diodes may be placed in parallel to each module, to a series of modules, or to a group of cells, thus avoiding "hot spots," namely the cell burnout. Such diodes must not be confused with the blocking diodes, which are placed in series with a string in order to avoid current backflow toward cells.

The presence of bypass diodes smoothes the effects of mismatching because they provide an alternative way for the high current produced by fully isolated cells, thus making the narrow neck of the shaded cells ineffective. The other side of the coin is in the fact that bypass diodes greatly affect both power–voltage and current–voltage characteristics of the field. A matched PV field exhibits a nonlinear current–voltage characteristic and a single-peak power–voltage characteristic, but for a mismatched PV field this is not true.

Figure 30.6 shows a comparison between uniform and mismatched working conditions of a PV string composed of two Kyocera KC120 modules connected in series, each one of them equipped with one bypass diode. In literature [28], it has been demonstrated that the height of the peaks characterizing the power–voltage curve depends on the type of mismatching conditions: Figure 30.6b shows the worst case, with the absolute MPP occurring at a voltage level that is quite different with respect to the one at which the MPP occur in uniform working conditions. Consequently, the mismatched PV field might

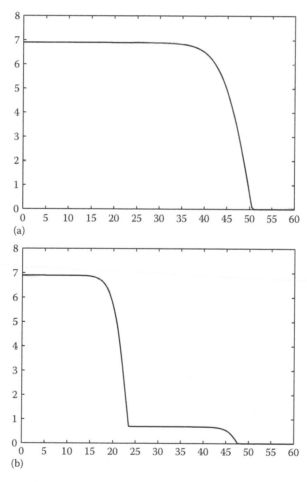

FIGURE 30.6 Current vs. voltage characteristics of a PV string made of two Kyocera KC120 modules: (a) both modules receive $S = 1\,\text{kW/m}^2$; (b) one module receives $S = 1\,\text{kW/m}^2$ and the other receives $S = 100\,\text{W/m}^2$. Horizontal axis PV voltage [V], vertical axis PV current [A].

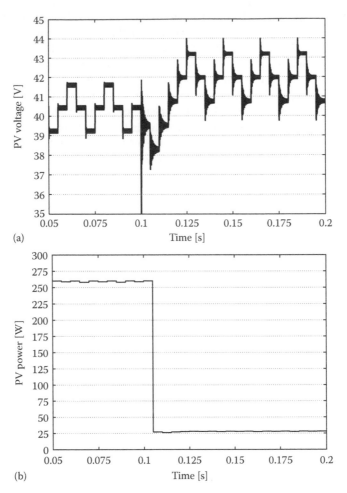

FIGURE 30.7   (a) String voltage and (b) string power before and after the irradiation drop from 1 kW/m^2 to 100 W/m^2 on one of the two models composing the string.

exhibit, as in Figure 30.6b, a multimodal power–voltage characteristic that makes standard MPPT techniques ineffective, since they are likely deceived and consequently they track a point where $dP/dv = 0$, but that is not the absolute MPP.

Figure 30.7 shows the wrong behavior of a standard P&O MPPT technique controlling the two-modules PV array mentioned above. The simulation has been done by assuming that the array initially works under uniform irradiation and temperature but, at 0.1 s, a sudden shading affects one of the two modules, reducing the irradiation it receives to one- tenth of that received by the other module. Due to the fact that the P&O strategy is essentially a local optimization technique based on the hill-climbing principle, the MPPT controller remains trapped in the voltage region where the MPP in uniform conditions were placed, so that the local, but not absolute, maximum is tracked. Unfortunately, there is no way for the controller to escape from this condition, since the array behavior does not exhibit any peculiarity. In fact, as shown in Figure 30.8, the P&O algorithm works correctly with a three-point behavior, but allows to draw from the PV array almost 25 W instead of 120 W that should be obtained if it were able to track the absolute MPP placed close to 20 V.

In literature, some techniques ensuring a reliable MPPT of the absolute MPP under mismatching conditions are presented, but they are often quite involved and not general. Such drawback can be overcome by means of a distributed MPPT approach described in Section 30.5.

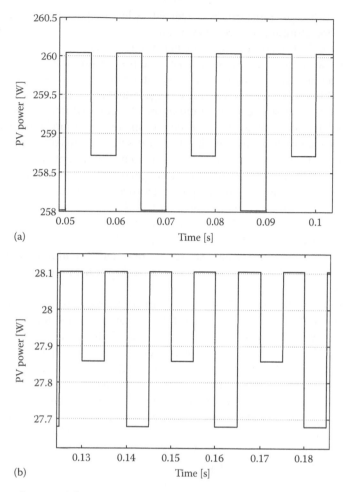

(a)

(b)

FIGURE 30.8 Magnification of the power vs. voltage characteristic reported in Figure 30.7b: (a) before the irradiation drop and (b) after the irradiation drop.

30.5 Single-Stage and Multiple-Stage Photovoltaic Inverters

The term "photovoltaic inverter" usually refers to the whole device converting the DC power drawn from the PV string into the AC power injected into the grid or delivered to a load. In the first case, the inverter is a "grid-connected" device, thus subjected to standards given by the utility companies [29–32], while in the second case the PV system is off-grid and it is usually classified as a "stand-alone" device. Such distinction is only one of many possible classifications that can be made and that are used in some overview papers published in literature. In fact, many solutions have been proposed in order to reach the highest possible efficiency even if, in many cases, different targets, such as low cost, high reliability, or high quality of the current injected into the grid, are considered as the main features.

It is the authors' opinion that a first important distinction is based on the presence of a galvanic isolation between the PV source and the grid. In fact, as confirmed by the large number of topologies described in literature [33,34], in many cases a transformer is included in the inverter. A high-frequency transformer included into the DC/DC converter does not allow limiting the amplitude of the DC current into the grid below a maximum allowable amount fixed by international standards. Such goal can be achieved if a line-frequency transformer is placed at the inverter output, so that the saturation of the distribution transformers is avoided. A low-frequency transformer placed at the inverter output and operating at grid frequency would be the worst solution, since it would be bulky and would lower the DC/AC chain conversion efficiency. Consequently, isolated topologies involving high-frequency transformers are numerous because they ensure a good tradeoff between efficiency and weight/volume, even allowing the implementation of soft-switching techniques based on parasitic parameters of the transformer (e.g., the leakage inductance). The galvanic isolation between source and grid ensures the best safety conditions at the PV source while, on the contrary, as clearly pointed out in [14,35], transformer-less inverters do not ensure separation between grid and module, so that, depending on the inverter topology, this determines fluctuations of the potential between the PV module and ground. Consequently, not only capacitive currents might flow in a person connected to ground and touching the module, but also electromagnetic interference might appear around the PV module. The need for an isolated topology is a controversial point that at present is regulated differently from country to country. Nevertheless, the use of sophisticated inverter controls and reliable isolation of the modules seem to be enough to avoid the drawbacks mentioned above, so that the trend seems to be encouraging for transformer-less architectures. In [14], some details about control systems replacing the transformer features are reported. The number of topologies that do not involve transformers at all is increasing dramatically: the more interesting and innovative ones are described in [35].

While galvanic isolation often occurs in string or multi-string inverters, namely to devices dedicated to interface one or more PV strings to the AC grid, the use of an isolated topology very often occurs in module-dedicated inverters, also known as "AC-modules." This is a new frontier for PV inverters, because it makes the PV module an autonomous entity, almost a "plug-and-play" device, capable of producing electrical energy and injecting it into the grid straightforwardly, without the need of any other component. This philosophy also gives the advantage of modularity, since any PV power plant can be enlarged simply, according to further power needs arising or funds availability. Moreover, AC-modules also allow to overcome problems related to mismatching working conditions, dramatically affecting the power production of classical PV plants using series-connected PV module strings, thanks to a distributed MPPT with this feature included in each AC-module. The differences between central, string, and module inverters are clearly evidenced in Figure 30.9.

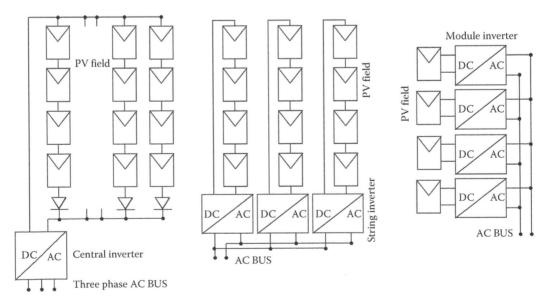

FIGURE 30.9 Different architectures of PV inverters. From left to right: central inverter, string inverter, and module inverter.

Central inverters exhibit high efficiency and have a low cost per kWp, but the power they are able to draw from the PV field drops significantly if a mismatching factor occurs (e.g., partial shadowing of the PV field). String inverters are based on a philosophy that lies between central and module inverters. In fact, if compared with the central inverters, they reduce the detrimental effect of possible causes of mismatching affecting the PV field, reduce the DC cabling, and avoid the need of inserting a block diode in series to each PV string in order to avoid current backflows. Drawbacks of string inverters are in their higher cost and increased complexity with respect to central inverters.

Multi-string inverters have a lower cost with respect to string inverters and ensure an MPPT dedicated to each string. This feature guarantees the maximum power production to each string even if they were differently oriented with respect to the Sun and/or if only some of them were shadowed. Consequently, this architecture is useful in BIPV applications, but its use is also profitable if the PV power production profile that is almost flat along the day must be obtained by fractioning the PV field in subsections, each with a different orientation angle (see Figure 30.10).

The idea underlying multi-string inverters is the basis of a new concept named distributed maximum power point tracking (DMPPT) approach [36]. This accomplishes the MPPT goal by means of a DC/DC converter dedicated to each module, which might be either a custom product, dedicated to the specific PV module, with an optimal behavior in the voltage and current range of that module, or a general-purpose product, sold independently from the module and having significant performances in a given range of voltage and currents. The PV module-dedicated DC/DC converter must exhibit a high efficiency, because a high-losses device would make the DMPPT fruitless, and a high reliability since, especially if it is built in the PV module, a lifetime comparable with that of the PV module would be required. Both efficiency and reliability are hard challenges for the design of such DC/DC converters. This is due to the relatively low power (<400 W) and high current (5–9 A) characterizing PV modules actually available in the market and to the fact that the converters' outputs need to be connected in parallel at the DC/AC stage input, so that they have to step up the PV module voltage to some hundreds of volts. In order to improve efficiency and lifetime of DC/DC converters for DMPPT, in [36] a series connection of their output ports has been proposed. This solution gives rise to

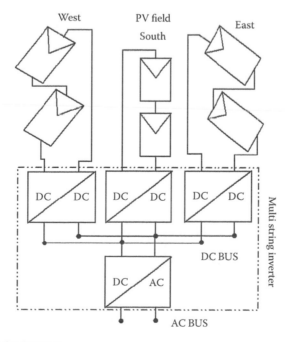

FIGURE 30.10   Multi-string inverter.

further control issues, but reveals attractive even whenever mismatched conditions occur with a low frequency and opens new prospects of spreading PV technology by self-consistent devices that give the maximum power for the actual weather conditions. This concept of PV generator reaches its apex with the AC-module, or module inverter, which is aimed at equipping the PV module with a DC/AC converter, thus simplifying the module interfacing with the AC grid and/or an AC load.

In literature, almost all the topologies proposed for AC-module implementation include a transformer, since, besides the advantages mentioned above, the high voltage gain ensured by the transformer is employed in order to lift the voltage at the module's terminals (usually <50 V) up to the grid voltage. Even in this case, the employment of a high-frequency transformer ensures compactness but lower efficiency, so that some tricks, essentially based on soft-switching techniques, are presented in literature in order to gain some efficiency points. In the AC-module application, the DC/AC converter is a high-frequency PWM full (half) bridge stage if it manages a high DC input voltage and a sinusoidal shaping of the current injected into the grid is needed. The other possibility is that the inverter receives a full-wave rectified current, obtained by means of a suitably controlled DC/DC converter, so that a line-frequency commutated unfolding bridge is needed. The former case does not require a DC/DC converter preceding the inverter, and this improves compactness, but the high-switching frequency of the bridge counterbalances the absence of the DC/DC stage in terms of efficiency. The large number of examples of the two approaches presented in literature confirms the need of reaching a tradeoff between them.

The use of a cascade connection of a DC/DC stage exploiting the MPPT function and of a DC/AC converter that injects a sinusoidal current into the grid or supplying an AC load is an almost common practice. Nevertheless, both functions might be concentrated in a unique DC/AC converter, thus leading to a single-stage inverter; or a multiple-stage device might be employed in order to obtain some additional feature. Differences between single-stage and two-stage power-processing systems, representing the largest part of architectural solutions available on the market and proposed in literature, are put into evidence in Figure 30.11.

One of the main differences between the two architectures schematized in Figure 30.11 is related to the presence and position of the bulk capacitor managing the fluctuating energy injected into the grid and the constant energy coming from the PV source. The role of such component is crucial in any power electronics interface between a PV source and the grid, because it works as a buffer component between the DC power extracted from the PV source and the pulsating power injected

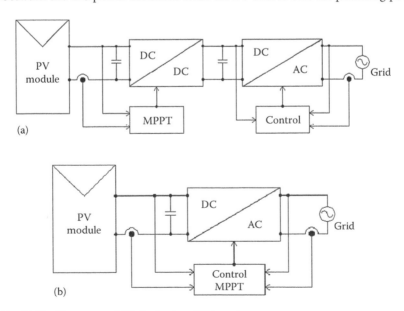

FIGURE 30.11   (a) Double-stage and (b) single-stage PV inverters.

into the grid. Consequently, its capacitance depends on the nominal power of the PV source P_{PV}, the grid frequency ω_{grid}, and the mean value and ripple amplitude of the voltage across it, V_c and ΔV_c, respectively [35]:

$$C = \frac{P_{PV}}{2 \cdot \omega_{grid} \cdot V_c \cdot \Delta V_c} \qquad (30.8)$$

Such formula is based on the assumption that the current from PV source is purely DC and that the inverter output current is in phase with the grid voltage, and both are purely sinusoidal, so that the power injected into the grid has a constant component and a sinusoidal component at twice the grid frequency. Equation 30.8 reveals one of the main drawbacks of single-stage topologies interfacing PV sources with the AC grid. Such kind of converters use a decoupling capacitor in parallel with the PV source so that, if the single-stage converter is an AC-module, the capacitor works at a low voltage level in the range [18] V, that is the typical MPP voltage of the PV modules actually available on the market. As revealed by (30.8), a low value of the capacitor voltage and, more in general, a high value of the ratio P_{PV}/V_c lifts the capacitance value up, so that the use of an electrolytic capacitor becomes mandatory. This is a weakness point of single-stage topology-based AC-modules, since it is well established that electrolytic capacitors have a short lifetime [32]. On the other side, if a single-stage inverter is used for interfacing a large PV field to the AC grid, a large capacitance might be even required, depending on the P_{PV}/V_c value, but also in order to ensure a small ΔV_c value that, when the capacitor is placed in parallel with the PV source, represents the amplitude of a PV voltage oscillation. This superimposes on the PV voltage variation/perturbation used by the MPPT algorithm, so that it must be reduced as much as possible because it might be the source of the MPPT deception and/or of a periodic deviation from the MPP causing power losses [37].

Problems related to the design of the bulk capacitor in multiple-stage inverter are not less challenging: the lower capacitance value, the higher the amplitude of the voltage oscillation across its terminals. This increases the current amplitude in the capacitor, thus potentially increasing the ohmic losses in its parasitic series resistance, but also requires a higher average voltage value on the capacitor if a buck-derived DC/AC stage follows the bulk capacitance and injects power into the AC grid. Drawbacks due to the dc-link capacitor have been avoided by using topologies that cannot be traced back to standard architectures shown in Figure 30.11. A clear classification of such alternative converters is reported in [14].

The overall performance of a PV inverter is usually evaluated in a specific operating point, but an average value computed at six different operating points, according to the definition of the European efficiency, gives useful information on the behavior of the power stage, but also of the control strategies, at different irradiation levels. The weighted sum adopted for such computation is the following [35]:

$$\eta_{european} = 0.03\eta_{5\%} + 0.06\eta_{10\%} + 0.13\eta_{20\%} + 0.10\eta_{30\%} + 0.48\eta_{50\%} + 0.20\eta_{100\%} \qquad (30.9)$$

where the percent value refers to the rated power.

30.6 Conclusions

Energy coming from the Sun is enough to support future human development, but it must be exploited by means of efficient and reliable systems, so that the price of the electrical energy produced in this way becomes equal to that produced by fossil fuels and available by means of a simple connection to the energy grid. This paper gives an overview of the main PV technologies for the conversion of solar power into electrical power. Sections 30.2 through 30.5 have been dedicated to compare the actual and rising technologies and materials realizing the energy conversion and to illustrate the features of the components included in the BOS, with a special emphasis to the power electronic systems devoted to the control of the PV array and to its interfacing toward the energy grid and loads. Features and drawbacks of different topologies and power-processing solutions as well as of various control strategies are illustrated. Tradeoffs between the two main features of PV power-processing systems, namely efficiency and lifetime, are put into evidence in order to highlight the design guidelines of future technology.

References

1. NREL: Dynamic maps, GIS data, and analysis tools-solar maps (available online: http://www.nrel. gov/gis/solar.html)
2. Photovoltaic Geographical Information System (PVGIS) (available online: http://sunbird.jrc.it/pvgis/)
3. Technology Review: A Price Drop for Solar Panels (available online http://www.technologyreview. com/Biztech/20702/?nlid = 1041)
4. A strategic research agenda for photovoltaic solar energy technology, EU PV Technology Platform, ISBN 978-92-79-05523-2, June 2007 (available online: http://www.eupvplatform.org/fileadmin/ Documents/PVPT_SRA_Complete_070604.pdf).
5. Greenpeace-EPIA, Solar Generation V 2008 (available online: http://www.epia.org/fileadmin/ EPIA_docs/documents/EPIA_SG_V_ENGLISH_FULL_Sept2008.pdf).
6. IEA Photovoltaic Power Systems Programme (www.iea-pvps.org)
7. Portal to the World of Solar Energy (www.solarbuzz.com)
8. P. Fairley, Solar-cell squabble, *IEEE Spectrum*, 45(4), 36–40, April 2008.
9. S. Upson, How free is solar energy?, *IEEE Spectrum*, 45(2), 72, February 2008.
10. S. Liu and R. A. Dougal, Dynamic multiphysics model for solar array, *IEEE Transactions Energy Conversion*, 17(2), 285–294, June 2002.
11. Kyocera KC120 Data Sheet (available online: http://www.kyocerasolar.com/pdf/specsheets/kc120_1.pdf)
12. N. Femia, G. Petrone, G. Spagnuolo, and M. Vitelli, Load matching of photovoltaic field orientation in stand-alone distributed power systems, *IEEE International Symposium on Industrial Electronics (ISIE04)*, Vol. 1, Ajaccio, France, May 4–7, 2004, pp. 1011–1016.
13. IEEE Std. 1547, IEEE Standard for Interconnecting Distributed Resources with Electric Power Systems, IEEE, New York, 2003.
14. F. Blaabjerg, R. Teodorescu, M. Liserre, and A. V. Timbus, Overview of control and grid synchronization for distributed power generation systems, *IEEE Transactions on Industrial Electronics*, 53(5), 1398–1409, Oct. 2006.
15. Battery Testing for Photovoltaic Applications (available online: http://photovoltaics.sandia.gov/ docs/battery1.htm)
16. B. Hariprakash, S. K. Martha, S. Ambalavanan, S. A. Gaffoor, and A. K. Shukla, Comparative study of lead-acid batteries for photovoltaic stand-alone lighting systems, *Journal of Applied Electrochemistry*, 38(1), 77–82, Jan. 2008.
17. IEEE 1013-2007, Recommended practice for sizing lead-acid batteries for photovoltaic (PV) systems, Institute of Electrical and Electronics Engineers, New York, 2007.
18. EUROBAT, The Association of European Automotive and Industrial Battery Manufacturers (http:// www.eurobat.org/)
19. V. Salas, E. Olías, A. Barrado, and A. Lázaro, Review of the maximum power point tracking algorithms for stand-alone photovoltaic systems, *Solar Energy Materials and Solar Cells*, 90(11), 1555–1578, July 6, 2006.
20. C. Hua, J. Lin, and C. Shen, Implementation of a DSP-controlled photovoltaic system with peak power tracking, *IEEE Transactions on Industrial Electronics*, 45(1), 99–107, Feb. 1998.
21. N. Khaehintung, T. Wiangtong, and P. Sirisuk, FPGA implementation of MPPT using variable step-size P&O algorithm for PV Applications, *International Symposium on Communications and Information Technologies (ISCIT'06)*, Bangkok, Thailand, Oct. 18–Sept. 20, 2006, pp. 212–215.

22. B. M. Wilamowski and X. Li, Fuzzy system based maximum power point tracking for PV system, *IEEE 2002 28th Annual Conference of the Industrial Electronics Society*, Vol. 4, Seville, Spain, Nov. 5–8, 2002, pp. 3280–3284.

23. N. Patcharaprakiti and S. Premrudeepreechacharn, Maximum power point tracking using adaptive fuzzy logic control for grid-connected photovoltaic system, *IEEE Power Engineering Society Winter Meeting 2002*, Vol. 1, Chicago, IL, Jan. 27–31, 2002, pp. 372–377.

24. N. Femia, G. Petrone, G. Spagnuolo, and M. Vitelli, Optimization of perturb and observe maximum power point tracking method, *IEEE Transactions on Power Electronics*, 20(4), 963–973, July 2005.

25. N. Femia, D. Granozio, G. Petrone, G. Spagnuolo, and M. Vitelli, Predictive & adaptive MPPT perturb and observe method, *IEEE Transactions on Aerospace and Electronic Systems*, 43(3), 934–950, July 2007.

26. F. Liu, S. Duan, F. Liu, B. Liu, and Y. Kang, A variable step size INC MPPT method for PV systems, *IEEE Transactions on Industrial Electronics*, 55(7), 2622–2628, July 2008.

27. K. H. Hussein, I. Muta, T. Hoshino, and M. Osakada, Maximum photovoltaic power tracking: An algorithm for rapidly changing atmospheric conditions, *IEE Proceedings on Generation, Transmission and Distribution*, 142(1), 59–64, Jan. 1995.

28. H. Patel and V. Agarwal, MATLAB-based modeling to study the effects of partial shading on PV array characteristics, *IEEE Transaction on Energy Conversion*, 23(1), 302–310, Mar. 2008.

29. NREL, A Review of PV Inverter Technology Cost and Performance Projections, NREL Subcontract Report NREL/SR-620–38771, Navigant Consulting Inc., Burlington, MA, Jan. 2006.

30. NFPA, *National Electrical Code*, National Fire Protection Association, Inc., Quincy, MA, 2008.

31. J. Wiles and W. Bower, Changes in the National Electrical Code for PV installations, *Conference Record of the 2006 IEEE Fourth World Conference on Photovoltaic Energy Conversion*, Vol. 2, Waikoloa, HI, May 2006, pp. 2331–2334.

32. G. Petrone, G. Spagnuolo, R. Teodorescu, M. Veerachary, and M. Vitelli, Reliability issues in photovoltaic power processing systems, *IEEE Transactions on Industrial Electronics*, 55(7), 2569–2580, July 2008.

33. Q. Li, and P. Wolfs, A review of the single phase photovoltaic module integrated converter topologies with three different DC link configurations, *IEEE Transactions on Power Electronics*, 23(3), 1320–1333, May 2008.

34. J. M. Carrasco, L. G. Franquelo, J. T. Bialasiewicz, E. Galvan, R. C. Portillo Guisado, M. A. M. Prats, J. I. Leon, and N. Moreno-Alfonso, Power-electronic systems for the grid integration of renewable energy sources: A survey, *IEEE Transactions on Industrial Electronics*, 53(4), 1002–1016, June 2006.

35. S. B. Kjaer, J. K. Pedersen, and F. Blaabjerg, A review of single-phase grid-connected inverters for photovoltaic modules, *IEEE Transactions on Industry Applications*, 41(5), 1292–1306, Sept.–Oct. 2005.

36. N. Femia, G. Lisi, G. Petrone, G. Spagnuolo, and M. Vitelli, Distributed maximum power point tracking of photovoltaic arrays. Novel approach and system analysis, *IEEE Transactions on Industrial Electronics*, 55(7), 2610–2621, July 2008.

37. N. Femia, G. Petrone, G. Spagnuolo, and M. Vitelli, A technique for improving P&O MPPT performances of double stage grid-connected photovoltaic systems, *IEEE Transactions on Industrial Electronics*, 56(11), 4473–4482, Nov. 2009.

31

Battery Management Systems for Hybrid Electric Vehicles and Electric Vehicles

Jian Cao
*Illinois Institute
of Technology*

Mahesh
Krishnamurthy
*Illinois Institute
of Technology*

Ali Emadi
*Illinois Institute
of Technology*

31.1 Introduction

With an increasing demand for higher fuel economy in vehicles and environment-friendly solutions, the last decade has seen a growing emphasis on electric vehicles (EVs), hybrid electric vehicles (HEVs), and plug-in hybrid electric vehicles (PHEVs). A hybrid vehicle is defined as a vehicle that uses two or more on-board energy to power it. These sources of power could be an internal combustion engine (ICE), fuel cells, ultracapacitors, battery, flywheel, etc. Hybrid vehicles have proven to save energy and reduce pollution by combining an electric motor and an ICE in such a way that the most desirable characteristics of each can be utilized.

This chapter is divided into two parts: (a) classification of hybrid vehicles and (b) battery technology for EVs and HEVs.

31.2 HEV Classification

The challenge in the successful implementation of HEVs is to enhance efficiency, improve ruggedness, reduce size, reduce cost of associated power electronics, and optimize design of electric machines and control electronics. The following section presents a broad classification of HEVs, as shown in Figure 31.1.

31.2.1 Micro-Hybrid

In a micro-hybrid vehicle, the electric motor does not add to the propulsion of the vehicle. Instead, it acts as a starter/generator and allows the ICE to stop and restart instantly and avoid idling. It is also used to charge the batteries during regenerative braking. A micro-hybrid drive train is not known to improve fuel economy as much as a mild- or full-hybrid drive train.

31.2.2 Mild Hybrid

Mild hybrids are the more popular configuration today. The motors used in these vehicles are typically not capable of providing propulsion power to the vehicle with the electric motor alone. Instead, the electric motor is used as an assistance to the ICE. In addition to the electric assistance that they provide to the transmission, a mild hybrid also has an engine that automatically shuts off when idling, and restarts when the driver presses the gas pedal. Since idling wastes a lot of fuel, this is incredibly beneficial.

31.2.3 Full Hybrid

Full hybrids are vehicles that can drive a certain distance at a low speed without the assistance of the ICE. These types of vehicles find use in city driving, since they save a significant amount of energy by using the electric motor (which has an efficiency much higher than the ICE) for propulsion over a short period of time.

31.2.4 Muscle Hybrid

A muscle hybrid is a vehicle that uses an electric motor to enhance performance and increase power. Fuel efficiency in such vehicles is marginally higher and they are not very cost-effective.

31.2.5 Plug-In Hybrid

The plug-in hybrid is the most diversed hybrid vehicle in the market. This type of vehicle uses a high-power battery (typically lithium-ion or lithium-polymer type) for operation. This battery configuration is designed to accept charge from an electrical outlet. When the battery is charged, the vehicle is capable

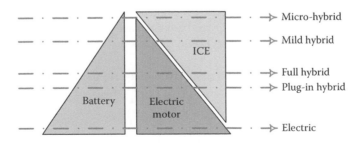

FIGURE 31.1 Classification of hybrid vehicle based on degree of hybridization.

of running strictly on electric power. If the charge on the battery gets depleted during the drive, it is then driven by the ICE. This hybrid configuration has all the advantages of a hybrid vehicle while sharing some of the benefits of an EV (ability to charge off the wall).

31.2.6 Electric Vehicle

An EV is a model that runs only on electrical input. There is no ICE on-board and this model uses a plug-in connection to charge the high-power battery. A reasonably sized electric motor is used to provide propulsion power to the motor.

There are several challenges in the use of EVs:

- Driving range: Distance driven by an EV before it needs recharging is typically lower than the distance a gasoline-powered vehicle can travel before needing a refill.
- Refueling time: Amount of time taken by batteries to fully charge is typically much higher than refilling a gas tank.
- Cost of battery: Battery packs are not yet cost-efficient.
- Bulk: Considerable space in the vehicle is allocated to battery packs. However, several chemistries are being considered for batteries and is an active area of research (Table 31.1).

31.2.7 Extended-Range Electric Vehicle

An extended-range EV (E-REV) is essentially a series plug-in hybrid vehicle with an ICE. In this vehicle, propulsion power is provided by an electric motor. The primary source of power is the battery that can be charged using an off-the-wall charger. Once the charge on the battery is depleted, the ICE turns an on-board generator to provide additional power to the battery and, therefore, extend its driving range.

31.2.8 Fuel Cell Vehicle

Fuel cell vehicles (FCV) are propelled by an electric motor. They are comprised of a high-power battery, an additional on-board power generation unit—the fuel cell. Fuel cells create electric power through a chemical process using hydrogen fuel and oxygen from the air. Hydrogen required for this kind of vehicle can be obtained from on-board high-pressure hydrogen tanks or from hydrocarbon fuels through the process of electrolysis.

TABLE 31.1 Comparison of Different Hybrid Vehicle Configurations Based on Their Drive Capabilities

Feature	HEV Type						
	Micro Hybrid	Mild Hybrid	Muscle Hybrid	Full Hybrid	Plug-in Hybrid	Electric Vehicle	Extended Range Electric Vehicle
Idle-off capability	◉	◉	◉	◉	◉	◉	◉
Regenerative braking	◉	◉	◉	◉	◉	◉	◉
Electric assist to conventional ICE		◉	◉	◉	◉		
All electric drive				◉	◉	◉	◉
Charging capability from power grid					◉	◉	◉

31.3 Hybrid Drive Train Configurations

HEVs can broadly be classified into two categories: (a) series hybrid vehicles and (b) parallel hybrid vehicles. In a series hybrid vehicle, the engine drives a generator, which in turn, powers an electric motor. In a parallel hybrid vehicle, the engine and the electric motor are coupled to drive the vehicle. A series hybrid vehicle typically offers lower fuel consumption in a city driving cycle by making the ICE consistently operate at the highest efficiency point during frequent stops/starts. A parallel hybrid vehicle can have lower fuel consumption in the highway driving cycle, in which the ICE is at the highest efficient point while the vehicle is running at constant speed.

31.3.1 Series Hybrid Vehicles

A typical configuration of a series hybrid propulsion system is shown in Figure 31.2. A series hybrid vehicle is essentially an EV with an on-board source of power for charging the batteries. The engine is coupled to a generator to provide power for charging the batteries. It is also possible to design the system in such a way that the generator acts as a load-leveling device that provides propulsion power. In this case, the size of the batteries could be reduced, but the sizes of the generator and the engine need to be increased. The power electronic components for a typical series hybrid vehicle system are (1) a converter for converting the alternator output to dc for charging the batteries and (2) an inverter for converting the dc to ac to power the propulsion motor. A dc–dc converter is required to charge the 12 V battery in the vehicle as well. In addition, an electric air-conditioning unit needs an inverter and associated control systems.

31.3.2 Parallel Hybrid Vehicles

Parallel hybrids can offer the lowest cost and the option of using the existing manufacturing capability for engines, batteries, and motors. However, a parallel hybrid vehicle needs a complex control system. There are various configurations of parallel hybrid vehicles, depending on the roles of the electric motor/generator and the engine. In a parallel hybrid vehicle, the engine and the electric motor can be used separately or together to propel a vehicle. The Toyota Prius and the Honda Insight are some examples of

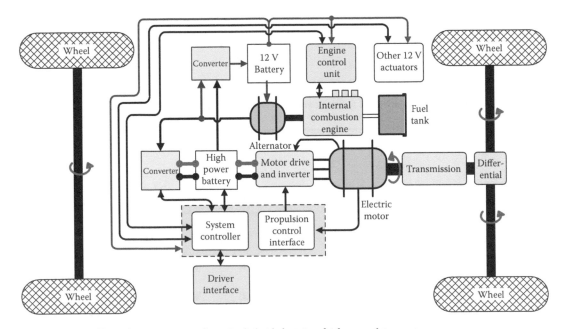

FIGURE 31.2 Typical arrangement of a series hybrid electric vehicle propulsion system.

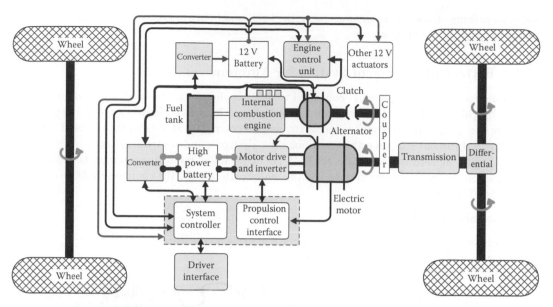

FIGURE 31.3   Typical arrangement of a parallel hybrid electric vehicle propulsion system.

parallel hybrid systems, which are commercially available. A typical configuration of a parallel hybrid propulsion system is illustrated in Figure 31.3.

31.4 Battery Electronics for EVs and HEVs

The battery system used on modern HEVs and EVs is both complex and elaborate. It is composed of several cells connected in series, thus, providing high-power levels needed by the car's propulsion system. In conventional cars, the single battery is used to power low-power electrical components such as the starter, head lights, and windshield wipers. In EVs and HEVs the systems differ, as the battery not only provides power for the low-power electrical components, but also powers the main traction system. This warrants a complex battery system for the electrical propulsion system to work.

Although the series connected cells in the EV or HEV systems are manufactured to be the same, they have some differences in volume, internal resistance, and self-discharge rate. When they are combined together for increasing the capacity, cell balancing becomes an issue. Also with large amounts of energy stored inside of the battery, the safety of driver and passengers becomes a issue. Finally, the operating condition and load profile of the battery varies significantly due to the start–stop nature of a vehicle. All of the problems addressed above create a demand for a complex battery management system (BMS) for the vehicle.

Generally, a BMS varies according to applications, ranging from basic to highly complex. For batteries used in cell phones, the BMS simply monitors the key operational parameters during charging and discharging, such as voltages, currents, and the battery internal temperature. The monitoring circuit gives a command signal to the protection circuit to disconnect the battery from the load or charger if any of the monitored parameters exceed the preset limits. BMS is also used in uninterruptible power supplies (UPS) in power plants. Since a UPS is the last defense against a power blackout, the BMS of such a system is not only used to monitor and protect the battery but it also includes battery electronics to make sure that the battery is able to deliver full power and prolong the life of the power system.

BMS for EVs and HEVs is more demanding, since the load profile varies rapidly according to the road conditions and the driver's behaviors. Therefore, the vehicle requires a fast-acting power management system, and that system must be able to interface with other control systems such as an engine power management system. Furthermore, since the operation environment of the battery system is harsher than those of other systems, which sit still in constant temperature rooms, protection of the battery is also a challenge. Under these circumstances, the issue is not only to protect the battery, but also to protect the passengers in the car.

31.4.1 Battery Management System for Automotives

An automotive BMS is more critical compared to other applications. The battery has to coordinate with the hybrid controller or the power-train controller to perform its duties while providing power or absorbing energy rapidly as the vehicle accelerates and brakes. Figure 31.4 shows the layout of a typical automotive BMS and its relation with other automotive electronic and electrical systems.

According to the hardware structure, the BMS can be separated into three major parts: monitoring and protection system, balancing system, and intelligent battery unit. The common functions of each part are listed in the following:

Monitoring and Protection System

- Monitors voltage of each cell or each module that make up the battery.
- Protects cell voltages from exceeding limits.

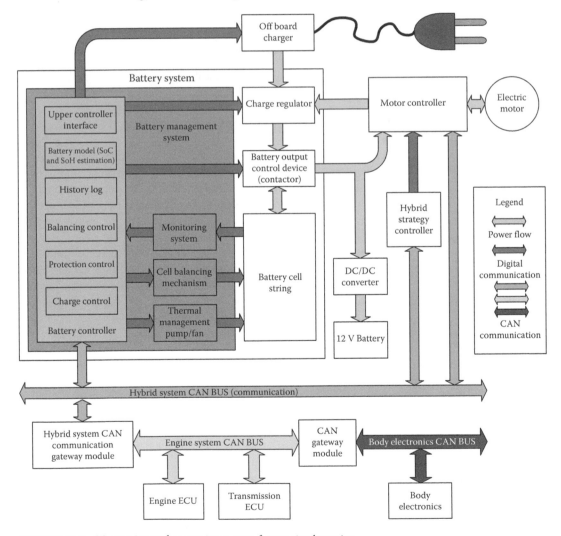

FIGURE 31.4 Monitoring and protection system for traction batteries.

- Monitors current and temperature of the battery system.
- Protects system from exceeding current and temperature limits.
- Provides mechanisms to protect the battery from uncontrolled conditions.

Balancing System

- Provides balancing mechanism to equalize the voltage differences of the cells.

Intelligent Battery Unit

- Maintains the state of charge (SoC) of the battery to allow the regenerative energy to be absorbed without the battery pack being overcharged. Also disables the regenerative braking when the battery is fully charged.
- Logs the charge and discharge of the battery to estimate the SoC of the battery and provide SoC information to the upper controller.
- Tracks performance of the battery and estimates the state of health (SoH) of the battery and report the SoH information to the upper controller.
- Provides state of the battery information to the upper controller to be displayed by the vehicle.
- Provides error code of the battery to the upper controller during diagnostic.
- Predicts range of the vehicle for all electric drive according to the SoC.
- Adjusts mode of the battery according to the command from the charger.
- Adjusts mode of the battery according to the state of the car control system.

A traction battery on an HEV or EV stores a huge amount of energy inside of it. Protecting the battery from extending beyond its operating limits is not only needed to avoid early failure of the battery, but is also the basis of protecting the safety of the driver and passengers. Financially speaking, battery components account for about half of the cost of some EVs. Consequentially, not only is early failure of the battery unsafe, but it could mean repair costs in excess of the vehicle's worth. These potential hazards along with operation at such high-voltage and high-power in harsh automotive conditions make protection of the battery system from overextending a very important issue.

Protection of batteries can be classified into two categories; one is active and the other is passive. Usually, active protection methods are done by the BMS. The BMS coordinates with the power controller of the vehicle so that the power command from the power controller does not exceed the capability of the battery. This protects the battery from overcurrents. The BMS coordinates with the thermal management system so that the battery does not operate in overtemperature conditions. The BMS also coordinates with the regenerative braking controller and the off-board charger to protect the battery from overvoltage conditions.

Passive protection involves the isolation of the battery system from other systems of the vehicle when the battery is about to exceed its limit. Reset of the battery controller or other device from the driver or a technician is needed to make the battery system reinstate. Passive battery protection relies on the monitoring of the battery parameters. The monitoring system senses cell/module voltages, battery current, and cell temperature. Any of these values exceeding the limit will result in the battery being protected.

31.4.2 Overcurrent Protection

Overcurrent protection is very important to both the life of the battery and safety of the vehicle. Failure to protect the battery from overcurrent conditions, in the best case, will destroy the power electronics devices in the motor controller. In the worst case, it will put the car in a dangerous state. The monitoring system senses the current via a current transducer, and the BMS disconnects the battery from the vehicle by controlling the output device (usually a contactor), if the continuous and instant current exceed their preset value.

Other than the intelligent controller protection, fuse protection is used to protect the battery from extreme conditions. Compared to intelligent protection, fuse protection is more reliable and straightforward because the fuse protection does not require the sensing circuit under normal operation in order to make the mechanism work. Usually, fuses are in series in the middle of the battery string. This configuration is better than having the fuse in series with the output terminal of the total battery pack. This is because when extreme conditions occur, the fuse breaks the total string of battery into two lower-voltage strings, making the overall system safer.

31.4.3 Overvoltage and Undervoltage Protection

Usually, overvoltage and undervoltage protection is done by the intelligent controller, which controls the output contactor of the battery. For a traction battery, since it is composed of hundreds of single cells connected together, it is too costly and complex to measure the voltage of each cell to make sure that they are all in a safe voltage range. The most common way is to assemble the cells in modules and measure the voltage of each or several modules. For example, the 2004–2009 Toyota Prius has individual modules composed of six cells with a nominal voltage of 7.2 V. The battery protection/management system measures the voltage of every two modules (i.e., 14.4 V, nominal voltage). The 2003–2005 Honda Civic Hybrid also has the same voltage sensing strategy. The battery protection system monitors the voltages of every two modules to see if their voltages are within the normal operating range. If an unusual voltage is sensed that could potentially harm the battery, the protection system turns off the output contactor, so that the battery pack is protected from further damage.

31.4.4 Overtemperature and Undertemperature Protection

Multiple thermistors or other temperature sensors are mounted on the battery body to sense the temperature of the cells. If the temperature of any sensing spot exceeds the preset temperature limits, the protection system disconnects any load applied to the battery by controlling the output devices.

31.4.4.1 Other Protections

A manual switch is usually in series in the middle of the battery string to serve as a service switch. The service switch and the fuse are usually placed together in an easily accessible location in the vehicle so the technician can open the service switch and replace the fuse, if necessary. By opening the service switch, the battery is broken into two lower-voltage strings.

31.4.5 Examples of Traction Battery Sensing System in HEVs

Toyota Prius and Honda Civic Hybrid are the two best-selling hybrid vehicles in the market. Figure 31.5 shows the configuration of the 2003–2005 Honda Civic Hybrid battery pack. In this pack, D size nickel-metal hydride (Ni-MH) cells are used. The nominal voltage of the D size Ni-MH cells is 1.2 V and the nominal capacity is 6.0 Ah. Six of these D size Ni-MH cells are connected in series to form a 7.2 V nominal module. Twenty of these 7.2 V modules are connected in series to form the total battery system with a nominal voltage of 144 V and nominal capacity of 6.0 Ah. The battery management and protection system monitors the voltages of every two 7.2 V modules. Three thermistors are distributed equally among the 20 modules to monitor the cell temperatures. A circuit breaker, which can work as a manual switch and a fuse, is connected between modules 8 and 9. If the fuse or the circuit breaker works, the 144 V battery string is broken into two lower-voltage strings. A current transducer is mounted near the battery positive before the relays. Two relays are used at the output of the battery pack—the main output relay, and a lower duty relay. The low duty relay is in series with a 12 Ω cement resistor. The relay with the resistor is used to precharge the capacitors in the motor drive.

Figure 31.6 shows the internal structure of the battery pack on a 2004–2009 Toyota Prius Hybrid Vehicle. The Prius has a higher battery voltage than Civic Hybrid: 28 of the 7.2 V battery modules are connected in series to form a battery system with a nominal voltage of 201.6 V. For protection, the Prius

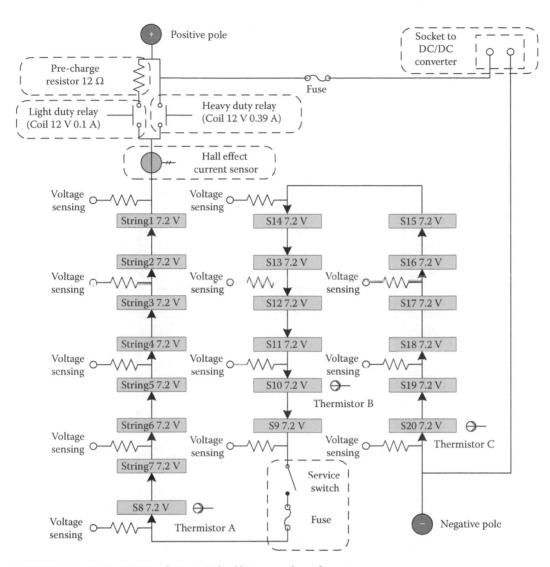

FIGURE 31.5 2003–2005 Honda Civic Hybrid battery pack configuration.

battery has a very similar strategy to the Civic Hybrid battery. Table 31.2 is a list of similarities of the two battery systems.

From the table we can see that neither of the batteries has a balancing system. Actually, it is reasonable, because in an HEV, the SoC of the battery is always maintained at a certain value (usually 60%). In other words, the battery will never be fully charged. So balancing is not an issue. Furthermore, Ni-MH batteries can be balanced by using passive methods.

31.4.6 Cell Balancing Methods for Traction Batteries

As mentioned in the protection section of this chapter, the SoC of the battery in HEVs is always maintained at a certain value (usually 60%). And all production HEVs through 2008 used Ni-MH batteries. Therefore, balancing is not an issue. The battery system on an EV is a totally different case. Figure 31.7 shows the typical operation ranges of batteries used in HEVs and EVs.

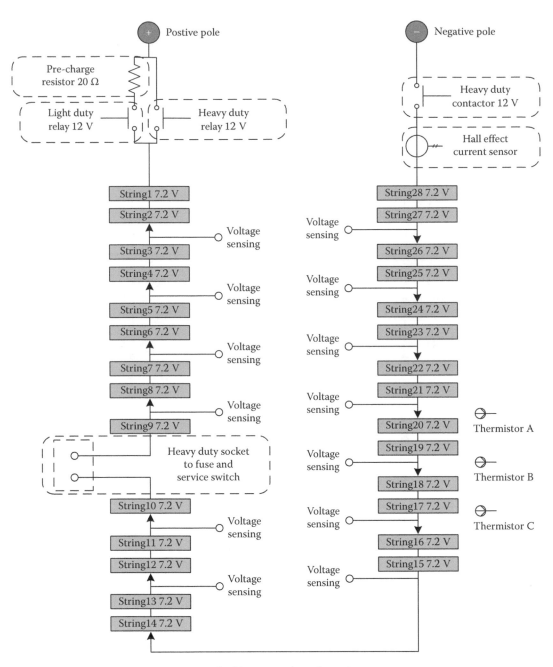

FIGURE 31.6 2004–2009 Toyota Prius Hybrid battery pack configuration.

In an EV, aside from providing power for the secondary electrical system, the battery's main task is in providing automotive power. The battery alone powers the car's propulsion system. While an HEV can move without a well-maintained battery, this is not possible in an EV. Therefore, the battery on an EV is vital and is always fully charged when plugged-in. When the battery pack needs to be fully charged, the balancing problem becomes evident immediately, especially when lithium-ion batteries are used. So far, there is no mass produced HEV or EV equipped with battery balancing mechanism. In this section, possible balancing methods are discussed and a comparative analysis is given to find their suitability to EV applications.

TABLE 31.2   Similarities between Honda Civic Hybrid and Toyota Prius Battery Packs

Battery	Honda Civic Hybrid Battery Pack (144 V)	Toyota Prius Hybrid Battery Pack (201.6 V)
Cells	Ni-MH D Size 1.2 V	Ni-MH Panasonic Prismatic 1.2 V
Module	6 cells in series	6 cells in series
Structure	120s (20 modules in series)	168s (28 modules in series)
Thermistors	3	3
Current sensors	1 current transducer at positive	1 current transducer at negative
Voltage sensing	Every 2 modules	Every 2 modules
Output relays	1 (Positive)	2 (Positive and Negative)
Pre-charge relay	Yes	Yes
Pre-charge resistor	Cement, 12 Ω	Cement, 20 Ω
Segmentation protection	Yes	Yes
Balancing	No	No

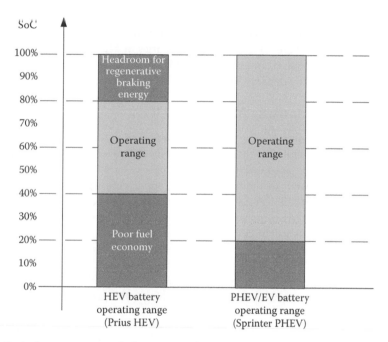

FIGURE 31.7   Typical operation range for batteries used in EVs and HEVs.

31.4.6.1 Why Batteries Need to Be Balanced

Balancing is the most important concept concerning the life of the battery system because without the balancing system, the individual cell voltages drift apart over time. The capacity of the total pack also decreases more quickly during operation, which results in the failure of the total battery system. This condition is especially severe when the battery has a long string of cells (high-voltage battery systems) and frequent regenerative braking (charging) is done via the battery pack.

Imbalance of cells in battery systems is very common and is the result of many sources. The sources fall into two major categories—internal and external. The internal sources include manufacturing variance in physical volume, variations in internal impedance, and differences in self-discharge rate. The external source is mainly caused by some multirank pack protection ICs draining unequally from the different series ranks in the pack. Thermal difference across the pack is another external source because it results in different self-discharge rates of the cells.

Balancing methods can be either passive or active. Passive balancing method can only be used for lead-acid and nickel-based batteries because lead-acid and nickel-based batteries can be brought into over-charge conditions without permanent cell damage. When the overcharge is not very severe, the excess energy is released by an increased cell body temperature. When the overcharge is too much, the energy will be released by gassing via the gassing valve equipped on the cells. This is the natural method of balancing a series string of such cells. However, overcharge balancing is only effective on a small number of series cells because balancing problems grow exponentially with the number of series cells. Generally, this method is a cost-effective solution for low-voltage lead-acid and nickel-based battery systems.

The basic idea of an active balancing is to use external circuits to actively transport energy among cells so as to balance the cells. The active balancing method can be used for most modern battery systems because they do not rely on the characteristic of cells for balancing. This method is the only applicable balancing method for lithium-based batteries, since the temperature of the lithium-based batteries must be rigorously controlled in the safety operation range. Generally, active balancing method should be used for a lithium-ion battery pack, which has three cells or more connected in series.

This section discusses various active balancing methods. Sorted by energy flow, active balancing methods can be grouped into four categories. They are dissipative method, single cell-to-pack method, pack-to-single cell method, and single cell-to-single cell method.

Sorted by circuit topology, there are three major categories of active balancing methods. They are shunting method, shuttling method, and energy converter method. In the following discussions, balancing methods will be sorted by their circuit topology.

31.4.7 Shunt Active Balancing Methods

Shunt active balancing method is the most straightforward concept of cell balancing. This method removes the excess energy from higher voltage cell(s) to let them wait for the lower-voltage cell(s) to catch up. The shunting method can be separated into two subcategories based on whether or not the method is dissipative. In using the dissipative methods, according to application, a good balance between heat dissipation and effectiveness of balancing must be made, since excessive heat dissipation increases the difficulty of thermal management. Also, uneven temperature of the cells aggravates the imbalance between the cells. Five shunting methods will be covered in this section.

31.4.7.1 Dissipative Shunting Resistor

Dissipative resistor is a special balancing method because of its reliability and simplicity. Figure 31.8 shows the basic dissipative resistor balancing circuit topology. The same topology could work in two

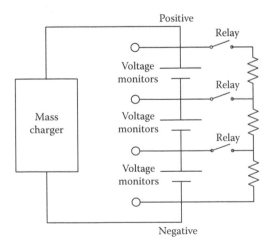

FIGURE 31.8 Basic dissipative resistor schematic.

modes—continuous mode and detecting mode. In continuous mode, all of the relays are controlled by the same signal, i.e., ON or OFF at the same time. They are turned ON during charging. The cell with a higher voltage has less charging current in order for other cells to be charged. This is effective during the entire charging process and could be quite effective if the resistor value is properly selected. The advantage of this mode is that there is no need for complex control.

While in the detecting mode, voltage monitors are added to each cell, and an intelligent controller senses the imbalance conditions and determines if the dissipative resistor needs to be connected to remove the excessive energy from the cell.

In either mode, the value of the resistor should be determined according to application. If the resistor value is chosen so that the dissipating current is smaller than 10 mA/Ah, the physical size of the resistor can be small. A 10 mA/Ah resistor could balance cells at a rate of 1% per hour. So, this circuit could drain the battery pack in a few days. When the battery is in stand-alone mode, the dissipative resistors should be removed from the battery pack by controlling the relays. Although this topology is not a very effective active balancing method, it could be used in many applications for low-cost solutions.

31.4.7.2 Analog Dissipative Shunting

Analog dissipative shunting shares the same idea as the resistor shunting. The only difference is that instead of using resistors, it uses transistors as the dissipation component. Figure 31.9 shows the typical analog shunting circuit.

In an analog shunting circuit, when a cell reaches the maximum charge voltage (set by the voltage reference and the voltage divider), the current is proportionally shunted around the cell and that cell is charged at constant voltage. This charge continues until the last cell in the string reaches the maximum charge voltage. In this method, the current is only shunted at the end of charging, so compared to dissipative resistor in continuous mode, it has less energy loss. Compared to dissipative resistor in detecting mode, this method costs less because it does not need intelligent control. In addition, this circuit approach could be expanded to larger battery packs with more battery cells.

31.4.7.3 Pulse Width Modulation–Controlled Shunting

Pulse width modulation (PWM)-controlled shunting is a type of nondissipative shunting method. In this method, the BMS senses the voltage difference of the two neighboring cells. By applying a PWM square wave on the gating of the pair of metal oxide semiconductor field effect transistors (MOSFET) the BMS controls the current difference of the two neighboring cells. As a result, the average current flows

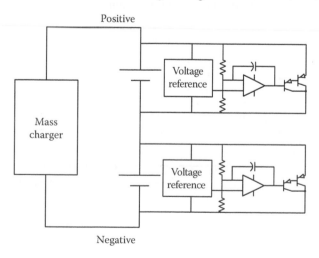

FIGURE 31.9 Conceptual layout for analog shunting circuit.

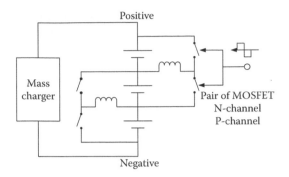

FIGURE 31.10 Conceptual layout for PWM-controlled shunting technique.

through the higher voltage cell will be lower than the normal cell. Figure 31.10 shows the PWM shunting circuit. Disadvantages of this circuit are that it needs accurate voltage sensing and it is relatively complex. It requires $2(n-1)$ switches and $n-1$ inductors for n cells.

31.4.7.4 Resonant Converter

The resonant converter is another version of the previous PWM shunting method. Instead of using intelligent control to sense and generate a PWM gating signal, resonant circuits are used to both transfer energy and drive the MOSFETs.

Figure 31.11 shows the resonant converter balancing circuit. The inductor L1 and capacitor C1 are used to both transfer energy and drive the MOSFET. This circuit needs a start-up circuit to start the resonance. When the voltage across L1 is positive, Q2 is ON; with the decrease of inductor voltage, Q2 turns OFF. When voltage increases in the other direction, Q1 turns ON, and L1 and C1 resonate with the first cell. The resonance causes a reverse current in L1, turns Q2 OFF and turns Q1 ON, which is the starting of another cycle of the resonance. If cell1 has a higher voltage than cell2, the average current flowing through inductor L2 is positive to balance the two cells. One set of resonant circuits are needed for every pair of neighboring cells. The circuit is complex and requires a resonance start-up circuit.

31.4.7.5 Boost Shunting

In the boost shunting method individual cell voltages are measured and the switch for the cell with higher voltage will be activated by the main controller. The switch is controlled by PWM signal. Figure 31.12 shows the schematic layout for the boost shunting circuit.

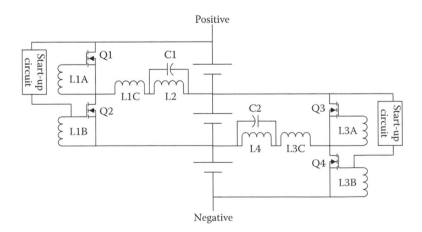

FIGURE 31.11 Resonant converter.

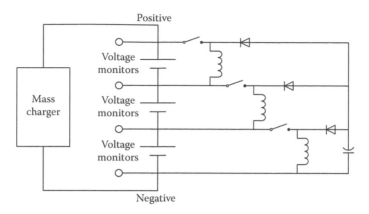

FIGURE 31.12 Boost shunting circuit schematic.

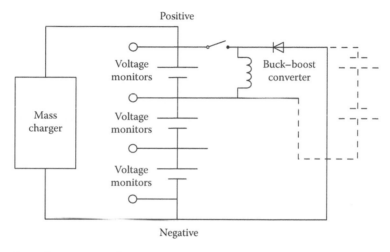

FIGURE 31.13 Boost shunting equivalent circuit.

When it is operational, the circuit acts as a boost converter. The boost converter diverts the extra energy to the other cells in the string. The equivalent circuit is shown in Figure 31.13.

The circuit is relatively simple and fewer components are used as compared to other advanced balancing methods.

31.4.7.6 Complete Shunting

To get the best results in expensive UPS, cells inside of the battery system are individually charged. However, this requires an expensive parallel charger. The complete shunting method can be a substitute for these systems. Complete shunting is shown in Figure 31.14.

In this circuit, only one mass charger is needed. The mass charger is a current-controlled converter. When one cell reaches its max voltage, the cell is completely shunted using two switches. The charge finishes until the last cell in the string is fully charged. This method seems to be quite straightforward. However, when a string is long, it may need a cascaded buck converter for which the output voltage range is very wide.

31.4.8 Shuttling Active Balancing Methods

The shuttling active balancing method utilizes external energy storage devices (usually capacitors) to shuttle the energy among cells so as to balance the cells. There are two shuttling topologies—switched

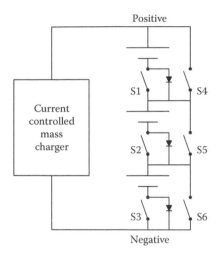

FIGURE 31.14   Basic dissipative resistor.

capacitor topology and single switched capacitor topology. Switched capacitor requires $(n - 1)$ capacitors to balance n cells whereas single switched capacitor needs only one capacitor to balance n cells. The single switched capacitor is a derivation of the switched capacitor. Shuttling balancing method utilizes external energy storage devices (usually a capacitor) to shuttle the energy between adjacent cells.

31.4.8.1 Switched Capacitor

Switched capacitor circuit is shown in Figure 31.15. In this topology, to balance n cells $2n$ switches and $(n - 1)$ capacitors are required. The control strategy is very simple because there are only two states. In the first state, C1 is in parallel with B1. C1 is charged or discharged to obtain the same voltage as B1. After this process, the system turns to the other state. In this state, C1 is in parallel with B2. After cycles of this process, B1 and B2 should be balanced. The same thing happens to C2 and the total battery pack can be balanced.

The advantages of the switched capacitor topology are that it does not need intelligent control and it can work in both recharging and discharging operation. This is very important for HEVs whose batteries do not have an end-of-charge state.

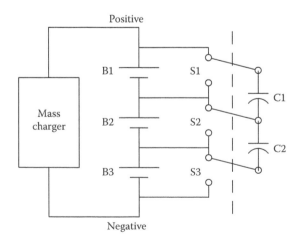

FIGURE 31.15   Conceptual arrangement for switched capacitor balancing.

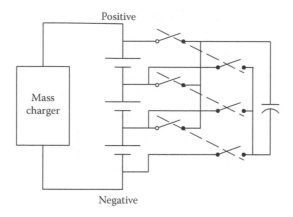

FIGURE 31.16   Single switched capacitor.

31.4.8.2 Single Switched Capacitor

Single switched capacitor topology is shown in Figure 31.16. Single switched capacitor circuit is a derivation of the switched capacitor. The difference is that this method uses only one capacitor to shuttle the energy.

The control strategy is simple. The speed of balancing is only $1/n$ of regular switched capacitor method. (n is the number of cells). However, for this topology, more advanced control strategies can be used to switch between the highest and the lowest voltage cell, which is called the cell-to-cell method. The balancing speeds are also much higher. For this topology, n switches and one capacitor are needed to balance n cells.

31.4.9 Energy Converter Active Balancing Methods

The energy converters here are defined as isolated converters. In these converters, the input and output side of the converters have isolated grounds.

31.4.9.1 Step-Up Converter

A step-up converter balancing circuit is shown in Figure 31.17. This method uses isolated boost converters to remove the excess energy from single cell to the total pack. The inputs of the converters are

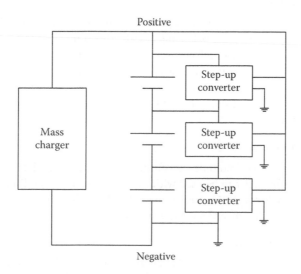

FIGURE 31.17   Step-up converter configuration for active balancing.

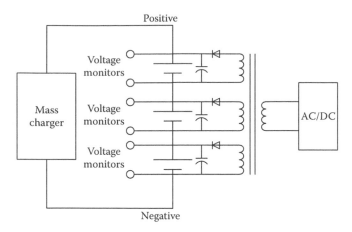

FIGURE 31.18 Multi-winding transformer arrangement.

connected to each cell to be balanced. The outputs of the boost converters are connected together and connected to the total battery pack. By sensing the voltage of the cells, the intelligent controller commands the operation of the converters to balance the cells.

This method is relatively expensive, but suitable for modular design. From energy flow point of view, this is the single cell-to-pack method. Special consideration is needed if the battery pack has very long string cells because the step-up converter needs to boost a single cell voltage to the total pack voltage.

31.4.9.2 Multi-Winding Transformer

In the multi-winding transformer topology, a shared transformer has a single magnetic core with secondary taps for each cell. Current from the cell stack is switched into the transformer primary and induces currents in each of the secondary. The secondary with the least reactance has the most induced current.

Figure 31.18 shows the multi-winding transformer balancing circuit topology. The major part of this circuit is a multi-winding transformer, which must be customized according to the number of cells. This trait restricts it from being modularized. The circuit is complex, the cost is high, and the multi-winding transformer is still being researched.

31.4.9.3 Ramp Converter

Ramp converter topology shares the same idea as multi-winding transformers. It is also an improvement on the multi-winding transformer balancing circuit. This ramp converter topology only requires one secondary winding for each pair of cells instead of one per cell.

Figure 31.19 shows the arrangement for the ramp converter technology for balancing. During operation, on one half-cycle, most of the current is used to charge the odd number of lowest voltage cells; while on the other half-cycle, most of the current is used to charge the even number lowest voltage cells via the so-called ramp.

31.4.9.4 Multiple Transformers

Figure 31.20 shows the topology of multiple transformer balancing method. In the multiple transformers topology, several transformers can be used with the same result by coupling the primary windings instead of coupling via a single magnetic core. Compared to the multi-winding transformer scheme, this method is better for modular design, although it is still expensive.

31.4.9.5 Switched Transformer

A switched transformer is actually a selectable energy converter. The input of the converter is the total battery pack, whereas the output of the converter is connected to a series of switches, which are

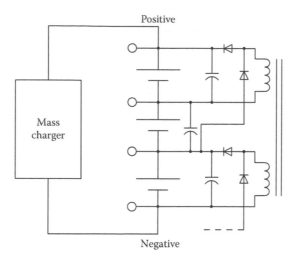

FIGURE 31.19 Ramp converter setup.

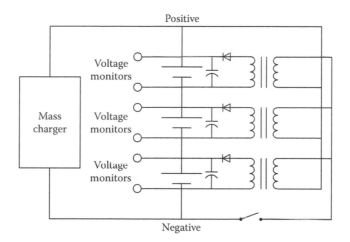

FIGURE 31.20 Multiple transformer configuration.

used to select which cell the output connects to. Figure 31.21 shows the topology of single transformer balancing method.

This topology is actually a pack-to-cell topology. The controller detects the unbalanced (lower voltage) cell and then controls the switches to connect the transformer (isolated converter) to it.

31.4.10 Comparative Analysis

A comparison chart of the balancing methods was shown in Table 31.3. Among all these methods, dissipative resistor, boost shunting, and switched capacitors are three good methods for different applications. Dissipative resistor in continuous mode is good for low-power applications, because the resistors are operating in continuous mode, they can be small and they do not need much thermal management. Another advantage of this method is that it is inexpensive. Boost shunting is good for either high- or low-power application. And its relatively low cost and simple control make it a good candidate for many applications. Switched capacitor is good for HEV applications, not only because it could be operated in both charging and discharging, but also because it has very simple control.

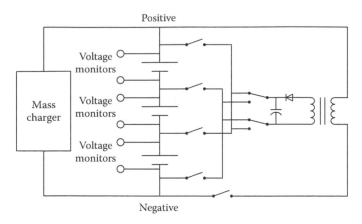

FIGURE 31.21   Switched transformer.

TABLE 31.3   Comparison between Balancing Methods

Balancing Methods	Balancing Nature	Major Components Needed to Balance an *n* Cell String	Best Effective Period	Modular Design Capability
Dissipative shunt resistor	Shunting	*n* Switches, *n* Resistors	Recharging	Easy
Analog shunting	Shunting	*n* Transistors	Recharging	Very easy
PWM-controlled shunting	Shunting	$2(n-1)$ Switches, $n-1$ Inductors	Recharging	Moderate
Boost shunting	Shunting	*n* Switches, *n* Inductors	Recharging	Moderate
Complete shunting	Shunting	$2n$ Switches, *n* Diodes	Recharging	Moderate
Switched capacitor	Shuttling	$2n$ Switches, $n-1$ Capacitors	Recharging and Discharging	Easy
Single switched capacitor	Shuttling	$2n$ Switches, 1 Capacitor	Recharging and discharging	Poor
Step-up converter	Energy converter	*n* Isolated Boost Converters	Recharging	Easy
Multi-winding transformer	Energy converter	$1n$ Winding Transformer	Recharging	Very poor
Ramp converter	Energy converter	$1\ n/2$ Winding transformer	Recharging	Very poor
Multiple transformer	Energy converter	*n* Transformers	Recharging	Easy
Switched transformer	Energy converter	$n+3$ Switches 1 Transformer	Recharging	Moderate
Resonant converter	Energy converter	$2(n-1)$ Switches, $2n$ inductors	Recharging	Easy

31.4.11 Intelligent Battery Unit

The intelligent battery unit is the brain of the BMS. It uses its "intelligence" to determine battery operating conditions and give commands to the protection, balancing, and communication systems. In an automotive battery design, the intelligent battery unit usually includes the following important parts or functions: battery model, SoC determination and control, SoH determination, history log function, and communication.

31.4.12 Battery Model

The battery model is embedded inside of the battery controller used to describe the characteristic of the battery in all of its possible working conditions such as temperature, voltage, discharge rate, and SoH. Figure 31.22 shows the general discharge rate characteristics of batteries. Figure 31.23 shows how the battery discharge time is influenced by ambient temperature. These two figures show only the big picture of battery characteristics. In a practical battery model, since the target battery is known and the characteristics are obtained, the specific characteristics are covered by the battery model. The model can be used to predict the reaction of the battery at any external and internal conditions.

The most beneficial function of embedding the battery model is to help in estimating the SoC of the battery system. When the voltage-based SoC determination method is used, the characteristic curve of the battery is a function of discharge rate, SoH of battery, voltage, temperature, and even environment humidity. With the help of a battery model, a more accurate SoC can be obtained because the battery model can be used to calibrate the errors according to the operating condition of the battery. This is explained further in the SoC determination subsection of this chapter. When coulomb-counting SoC determination method is used, the battery model describes the charging efficiency, self-discharge rate,

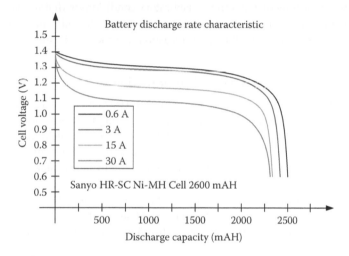

FIGURE 31.22   Battery discharge rate characteristics for Ni-MH cell.

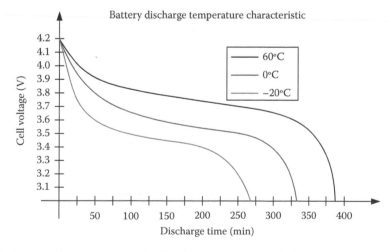

FIGURE 31.23   Battery discharge temperature characteristics for Ni-MH battery.

and the influences of other error sources. Therefore, the SoC determination method can use the information provided by the battery model to calibrate the results.

Together with the history log function, the battery model can be used in estimating the SoH of the battery as well. The SoH determination is discussed later in this chapter.

31.4.13 State of Charge Determination

SoC is a term used to describe the energy left in a battery pack, usually expressed by percentage. When a new battery is fully charged the SoC is considered 100% of its nominal capacity. On the other hand, when it is completely drained, the SoC is considered 0%. For an EV driver, the SoC is similar to the fuel gauge on a gasoline vehicle, which is used to tell the driver the range left before recharging is needed. The SoC information is also very important for other function blocks of the vehicle, such as hybrid control strategy for an HEV and charge control for an EV. There are many methods used to determine the SoC of a battery pack. Some widely used methods are explained and discussed below.

31.4.13.1 Direct Measurement

The SoC of a battery can be directly measured in laboratory conditions with nominal current discharge of the battery. This method is very accurate but not practical because the whole battery must be discharged before the SoC is revealed. So the direct measurement method is only used by manufactures for battery testing.

31.4.13.2 Voltage-Based State of Charge Determination

Voltage-based SoC determination method is a very straightforward method. It is based on the principle that the voltage of the battery drops while it discharges. Although the SoC does not reduce linearly with a decrease in voltage, several points can be selected and piecewise linearization can be employed to approximate the SoC. This method is used in many low-cost battery indicators: several segments of LEDs are used to display the SoC of the battery. Several threshold voltage values are used to determine the SoC so as to turn on the corresponding LED to indicate the SoC.

One problem with this method is that for some battery chemistries, the voltage versus SoC curve is so flat that they cannot be easily linearized. Figure 31.24 shows the typical discharge characteristic curve of Ni-MH, Li-ion, and lead-acid batteries. From this figure, it can be seen that the discharge curve

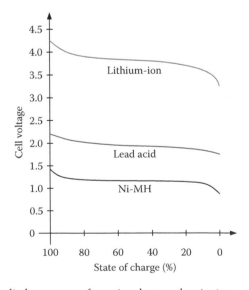

FIGURE 31.24 Comparative discharge curves for various battery chemistries.

of Ni-MH battery is much flatter than that of Li-ion and lead-acid batteries. Thus, the use of voltage-based SoC determination poses more problems for use on Ni-MH battery than on Li-ion and lead-acid batteries.

Discharge characteristics of a battery highly depend on battery temperature, discharge rate, and SoH of the battery. Therefore, only if the battery is well maintained so that the characteristic curve does not drift away from the preset curve, the load drains a relatively constant current, and if the operating temperature of the battery is relatively constant, the SoC estimation method can be informative. Otherwise, a battery model is needed in order to estimate the SoC.

31.4.13.3 Coulomb-Counting State of Charge Determination Method

This method measures the inflow and outflow current in order to determine the remaining capacity in the battery pack. The implementation of this method is to do a current integration over time. When microcontrollers with built-in analog-to-digital converters (ADC) are used for coulomb-counting, sampling method is used. The measurement of current can be done by using shunt resistors or hall sensors. Hall sensors are more widely used in automotive battery packs because they will not result in power loss compared to shunt resistors. Figure 31.25 shows the implementation of the sampling method based coulomb-counting method.

Coulomb-counting method can be accurate in several charge and discharge cycles. With the increased cycles of charge and discharge, errors will be accumulated because this method cannot count the self-discharge portion of energy.

Another effect, which must be taken into consideration, is the recharging efficiency. When using the coulomb-counting method to measure the incoming charge energy, errors occur because the charging efficiency depends on the condition of the battery other than 100%. Several methods can be used to eliminate the effect of self-discharge and other sources of errors. Two commonly used methods are

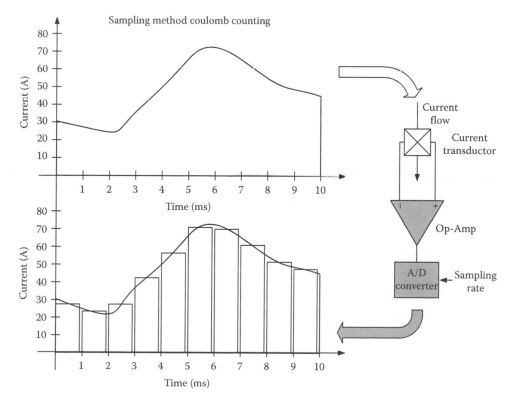

FIGURE 31.25 Implementation of sampling method based on Coulomb counting.

the battery-model method and the reset method. The battery-model method has been discussed in the battery model part of this chapter. The battery model basically describes the behavior of the battery in different working conditions and the performance of the battery over time. Reset calibration is a practical way for coulomb-counting method and it is especially beneficial for EV batteries since the battery of an EV has the chance to be fully charged. When the battery is fully charged, the reset calibrate method can reset the coulomb counting and tell the system that the battery has a 100% SoC.

31.4.13.4 Other SoC Determination Methods

For lead-acid batteries, a method to measure the changes in the weight of the active chemicals can be effective to determine the SoC of the battery. However this method can only be used in lead-acid batteries.

31.4.14 State of Health Determination

SoH is a term used to describe the capability of delivering energy of a battery to the load. It is also an indication of if the battery needs to be replaced or how long the battery can last before needing to be replaced. When concerned about the capability of delivering energy, it is assumed that a brand new battery, which can deliver 100% of its nominal capacity is in good SoH. In the battery industry, it was defined that a battery, which can deliver only 70% of its nominal capacity is a battery in poor SoH. Actually, battery manufactures claim that the cycle life of a battery is also based on the 70% of capacity value. It means the cycle life of the battery before it still has 70% of its nominal capacity.

Measurement of SoH can be done by comparing the current capacity of the battery pack with the capacity of a new battery. This requires the SoH determination system to get access to the history information of the capacity, or some preset capacity value should be provided. Other methods to determine the SoH are the measurement of internal resistance changes, self-discharge rate changes, and so on, because these parameters also change with the battery SoH.

31.4.15 History Log Function

The history log function is one possible and useful function for BMS in automotive application. In the BMS, a dedicated memory chip or memory area is used to do the history log. The history information can be easily obtained via external diagnostic tools. The history log function records the following information:

- Charge and discharge cycles
- Discharge capacity of each cycle
- Maximum and minimum voltage and temperature
- Maximum discharge current of the battery

By evaluating the history log, it can be determined whether or not the battery had ever been abused, which would be helpful in a warranty claim or dispute.

31.4.16 Charge Regulation

Since the intelligent unit on the BMS not only has all the information (voltage, current, temperature, and history) about the battery pack, but also takes control of the battery, the charge regulation should be done by this unit as well. The intelligent unit on the battery does all the sensing, control, and charge termination, which results in a greatly simplified off-board charger. The off-board charger only needs to respond to the command from the battery pack during the charging process.

In automotive application, it is better if the battery can be charged faster. For a controlled fast charge, the termination is important. An "out of control" charging process is dangerous. Thus, determination of

TABLE 31.4   List of Charging Methods and Termination Methods

	Ni-Cd	Ni-MH	Lead-Acid	Li-Ion
Slow charge	CC + Trickle/Timer	CC + Timer	CC-CV + Trickle/ Timer	CC + Voltage limit
Fast charge 1	CC + NDV	CC + ZDV	CC-CV + I_{min}	CC-CV + I_{min}
Fast charge 2	CC + dT/dt	CC + dT/dt		
Backup termination 1	Temp. cut off	Temp. cut off	Timer	Temp. cut off
Backup termination 2	Timer	Timer		Timer

Note: CC, Constant Current; CC CV, Constant Current Constant Voltage; NDV, Negative Delta Voltage; ZDV, Zero Delta Voltage.

charge termination is an important function in the charge regulation. Table 31.4 is a list of termination methods for different chemistry batteries.

31.4.16.1 Charging Control of Nickel-Based Batteries

Any battery, including nickel-based batteries, can be charged using constant current with a proper termination method. Termination methods for nickel-based batteries are negative delta voltage (NDV) termination and zero delta voltage (ZDV) termination, which are based on the charging characteristics of the batteries.

31.4.16.1.1 Negative Delta Voltage Termination Method

The NDV termination method is applicable for Ni–Cd batteries because only a Ni–Cd battery presents a negative dV/dt, when it is fully charged. In other words, when the battery is not fully charged, the battery voltage increases with the elapse of charging time. However, when the battery is fully charged, the voltage begins to drop with the continuing of charging. The NDV method detects the dV/dt of the battery and if the dV/dt reaches the preset negative value, the charging process is terminated. Figure 31.26 shows the idea of NDV and ZDV method.

31.4.16.1.2 Zero Delta Voltage Termination Method

Unlike Ni–Cd batteries, Ni–MH batteries do not present such a noticeable negative dV/dt, when they are fully charged. Usually, the dV/dt is very small or even zero. This is a helpful characteristic in determining

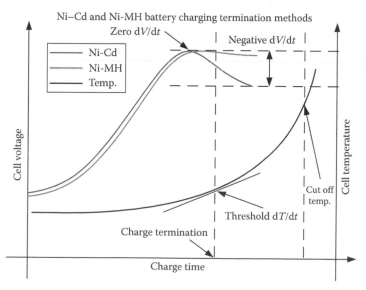

FIGURE 31.26   Charging termination methods for Ni–Cd and Ni-MH battery types.

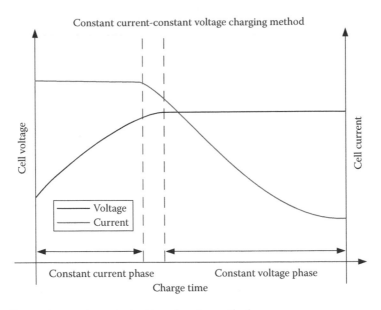

FIGURE 31.27   Constant current-constant voltage charging method.

the termination as well. Therefore, the ZDV termination methods detect the dV/dt, whenever the dV/dt is equal to or less than zero in a period of time, then the charging process can be considered finished.

31.4.16.2 Charging Control of Lithium-Ion and Lithium-Polymer Batteries

Charging a lithium-ion or lithium-polymer battery without voltage limiting can be dangerous. This is because when their voltage exceeds the maximum value, the extra charging energy could damage the cell or cause fire. Constant current constant voltage (CC-CV) charging method is usually used for lithium batteries. The CC-CV charging method has two phases, the constant current phase and the constant voltage phase. When the battery voltage is low, the battery is charged with constant current. Before the voltage reaches its maximum value, the controlled charger switches the charging phase to a constant voltage state in which the current reduces to a trickle current. When the current drops below a preset minimal value (I_{min}), which means the battery is fully charged, the charging process will be terminated. Figure 31.27 shows the concept of CC-CV charging method. The CC-CV + I_{min} method is the recommended method for fast charging of lead-acid batteries as well.

31.4.17 Communication

The standard for vehicle communication is the controller area network (CAN) bus. It was originally developed in 1983 by Bosch in Germany with the purpose of solving the communication problem between multiple microcontrollers on vehicles. An automotive BMS, therefore, communicates with other controllers in the hybrid system on the vehicle via hybrid system CAN bus. Usually hybrid system CAN bus shares information with the power-train controller, display modules and safety controls via CAN communication gateways.

32

Electrical Loads in Automotive Systems

Mahesh
Krishnamurthy
*Illinois Institute
of Technology*

Jian Cao
*Illinois Institute
of Technology*

Ali Emadi
*Illinois Institute
of Technology*

32.1 Introduction

Owing to major advancements in power electronics and improvements in the efficiency of motor drives, vehicular technology standards have been raised significantly. Automotive electronics have been adapted to cater to the consumers' needs. Figure 32.1 shows some of the more common loads that are seen in automotives today. The following sections briefly introduce the role played by power electronics and motor drives in these applications.

An additional torque called assist torque is applied to the steering column from an attached electric motor. Using feedback of torque and vehicle velocity for on-board sensors, the control unit calculates relevant torque and current commands for the motor. This calculated amount of torque is then applied to steering column by motor through a reduction gearbox. Since the electric motor only provides assist torque to the driver during turns and is unused otherwise, it exhibits better efficiency than its hydraulic counterpart. The configuration of a typical power steering system is shown in Figure 32.2.

32.2 Electric Power Steering System

32.2.1 Conventional Power Steering System

Electrification of the power steering system in automotives has several advantages. It inherently offers light steering at lower speeds and higher stability at high speeds. This system typically comprises of a steering wheel, a steering column, an intermediate shaft, rack and pinion, and steering linkages. A torque sensor is placed between the steering wheel and the steering column. The purpose of this sensor is to measure the torque applied by the driver as a feedback to the control unit.

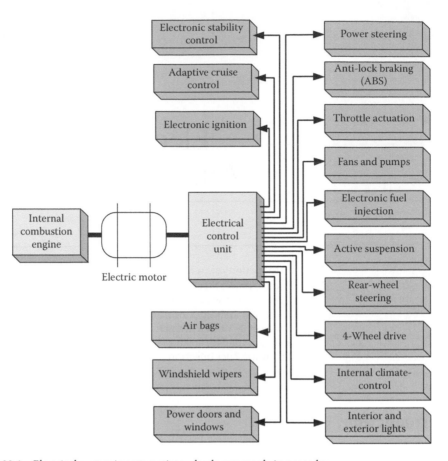

FIGURE 32.1 Electrical system in automotives—loads commonly in use today.

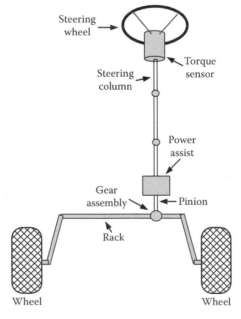

FIGURE 32.2 Typical power steering system in automotives.

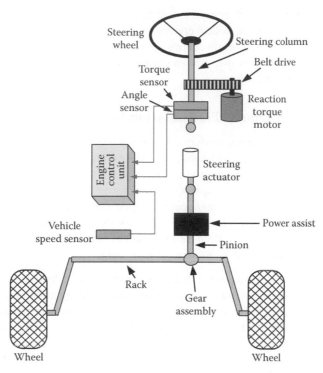

FIGURE 32.3 Block diagram representation of a steer-by-wire concept in automotives.

32.2.2 Steer-by-Wire System

Based on construction, there are four different types of electric power steering systems. They are the column-assist, the pinion-assist, the rack-assist, or the fully steer-by-wire. Figure 32.3 shows the layout and placement of a typical steer-by-wire system. The steer-by-wire has three basic subsystems in this arrangement:

- The steering wheel subsystem contains the torque sensor, steering angle sensor, and steering wheel motor. The torque and angle sensors are used to signal the commanded turn angles and the rate of change of angular position from the driver to the control unit. The electric motor is used to provide a counter-torque to the driver so that the reaction of the road and vehicle velocity can be felt. This reaction is helpful in adjusting the commanded angle and sharpness of the turn.
- The controller subsystem includes the vehicle speed sensor and an electronic control unit (ECU). The ECU controls the steering wheel motor and the front wheel motor for the driver's steering feel and for improving vehicle maneuverability and stability.
- The front wheel motor subsystem includes the position sensor; rack pinion gear; other reduction gear arrangements; and the front wheel motor, which positions the tire according to the data provided by the driver via the steering wheel subsystem. The steering wheel subsystem and the front wheel subsystem transmit the information obtained from sensors to the ECU to calculate the desired reactive torque and front wheel angle. Figure 32.3 illustrates the placement and layout of a typical steer-by-wire system for a vehicle.

32.3 Electronic Stability Control System

Electronic stability control (ESC) system is a safety feature that is used to detect and avoid the car from skidding. This is a software-based control that is mainly used to protect the passengers in the vehicle. This system uses information of torque angle, vehicle speed, yaw angle, etc., to determine the direction that the vehicle is being commanded to turn. It compares this data to the direction that the vehicle is moving to make an informed decision. If the ESC system perceives the possibility of a skid, it applies brakes to individual wheels to enable the system to regain control. This system is typically designed to be able to apply brakes to each wheel individually. It is an improvement over manual control mainly because the drive has a collective control over all four wheels.

32.3.1 Continuously Variable Transmission System

In a conventional transmission system, mechanical power is transferred from the engine to the wheels using gear ratios to regulate speed of the wheels. Since the number of combinations available is finite, the engine is often forced to modulate its speed, which affects its power output and efficiency. A continuously variable transmission (CVT) is capable of allowing an infinite number of ratios within a finite range. This allows the engine to continuously operate at speeds that offer highest efficiency or best performance. In its transition from a mechanical to an electric system, the CVT has undergone several implementations. The following classification demonstrates the approach used in purely electrical CVT systems along with a brief overview of its mechanical equivalent:

- *Hydraulic CVT*: These transmission systems are comprised of a hydraulic pump and a motor. Typically, the engine drives the pumps and transmits power using high-pressure oil. The hydraulic motor converts this power into mechanical power, which is transferred to the load. Since the pressure from the pump can be regulated, this transmission system has an indefinite transmission ratio available. This type of CVT is typically used in agricultural machines.
- *Belt-driven CVT*: This system is most commonly used in automobiles. As the name suggests, it includes a pulley and belt assembly. While one pulley is connected to the engine shaft, the other is connected to the output shaft. A belt is used to link the two pulleys. Variation in the distance between the two half pulleys allows a variation in effective diameter for the belt. The transmission ratio is therefore a function of the two effective diameters.
- *Electrical CVT*: The purely electric CVT is similar in operation to the hydraulic CVT. It is primarily comprised of two electric machines. The machine that is connected to the engine operates as a as generator, while the machine connected to the wheels is used as a motor. Electric power is passed through the control. One of the requirements for this system is that the sizes (maximum continuous power) for both machines have to be equal to the maximum power of the engine. Also there is no rigid connection between the motor and generator, which gives it an infinitely varying gear ratio.

32.3.2 Ignition Systems

The ignition system in an automotive is to ignite the fuel–air mixture in the cylinder of the engine. Intermediate power electronic circuits are responsible to create between 20 and 50 kV across an air gap at the tip of the spark plug. This is done up to several thousand times per minute. These sparks have to be timed very accurately to ignite the mixture in the right cylinder. If the timing of this system is off, it could result in a drastic reduction in the performance of the engine. There are two basic ignition techniques: mechanical and electronic. They are explained in brief in the section below.

- *Mechanical ignition system*: This ignition type uses a mechanical system to time the high-voltage sparks generated to ignite the fuel. It can be broadly divided into two sections: the low-voltage circuit (operates at battery-level), which generates the signal to fire the spark plug; and the high-voltage ignition coil, which is responsible for stepping battery voltage to the required range. The ignition coil used in this system operates similar to a step-up transformer. Current is applied from the battery through a current-limiting resistor or resistance wire to the primary coil, and is grounded through ignition points in the distributor. This creates current through the

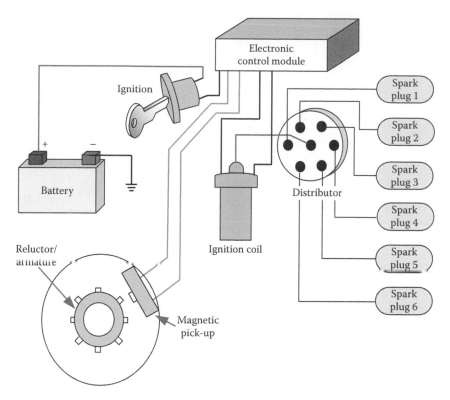

FIGURE 32.4 Conceptual layout of an electronic ignition system.

secondary coils, resulting in the buildup of a magnetic field. As the distributor cam rotates along with the engine, contact points are forced apart, breaking the circuit and stopping the flow of current. This interruption causes the magnetic circuit to collapse and produce a high-voltage spark. The spark is sent to the right spark plug through the distributor.

- *Electronic ignition system*: Electronic ignition systems (Figure 32.4) are different from the mechanical system in the fact that they do not use a distributor. Instead, they have an armature (also called a reluctor), a pickup coil, and an electronic control module. The setup of the electronic ignition is similar to mechanical unit. Current from the battery flows through the ignition switch to the primary coil windings. This causes the buildup of a strong magnetic field. When the rotating armature nears a pick-up coil, the electronic module is commanded to turn OFF the primary current, which causes the field to collapse. This causes a high-voltage spark in the secondary winding, which is transferred to the spark plug through the distributor.
- *Distributor-Less ignition system*: As the name suggests, this ignition technique does not use a distributor to transmit the high voltage to the spark plug and has no moving parts. Instead, the spark plugs are fired directly from the coils. Timing of the sparks is controlled by on-board computers: ignition control unit (ICU) and the engine control module.

For example, in a four-stroke engine with one ignition coil connected to two cylinders, each cylinder is paired with another that has the opposite firing sequence. The basic arrangement is as shown in Figure 32.5. For this engine configuration, when one cylinder (called the event cylinder) is in the compression stroke another one (also called the waste cylinder) is in the exhaust stroke. When the coil discharges, both spark plugs fire at the same time to complete the circuit. Since the polarity of the primary and the secondary windings are fixed, the firing sequence of these two plugs is always

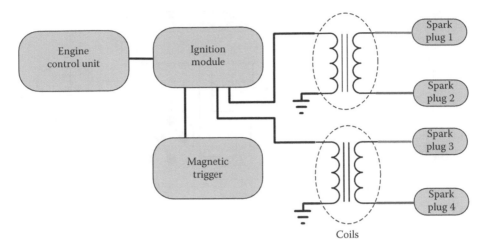

FIGURE 32.5   Conceptual arrangement for a distributor-less ignition system.

opposite. Since there are no moving parts to this system, there is no issue with wear. In addition, there is no drag on the engine since no distributor is used.

32.3.3 Antilock Braking

Owing to the rapid advancement in signal processing capabilities and power electronics, several additional features have been added to improve safety and reliability of the vehicle. Antilock braking system (ABS) is one such feature that is extremely useful in avoiding skidding of tires due to sudden deceleration for specific road conditions. When the vehicle decelerates too fast, it is possible for one or more tires to lose traction, causing the wheels to "lock." This causes the driver to lose the ability to steer the vehicle while stopping, which can be dangerous. The ABS senses any such rapid deceleration and applies pulses of braking force to the locked wheel while applying normal braking power to the remaining wheels. This "pulsing" is felt by the driver on the brake pedals of a vehicle (with ABS) when the brake is pressed hard. This action allows the driver to retain control while braking during unusual driving conditions. This system is even more effective in electric and hybrid electric vehicles with hub-motors built into each wheel. The ABS system has four main components that effectively describe its operation.

- *Speed sensors*: The vehicle speed sensor estimates speed of the vehicle and detect any sudden decelerations in motion initiated by the driver. This allows the control unit to anticipate any possibility of locking of wheels. These sensors are typically placed on the wheels or the differential of the vehicle.
- *Valves*: The valves used in the ABS system are used for three purposes:
  - To pass pressure from the master cylinder to the brakes.
  - To close the line and prevent brake pressure from rising further in case the driver applies additional pressure on the pedals.
  - To release pressure from the brakes.
- *Pump*: The pump is used in the ABS system to reestablish the pressure in the line after it has been released from the valves.
- *Electrical control unit*: This unit is the "brain" of the ABS. It is responsible for monitoring the speed sensor, commands from the driver, and pressure in the line to control the pump and the valves.

32.4 Electronic Fuel Injection

An electronic fuel injector replaces the conventional carburetor in older ICE-driven vehicles. Simply stated, an electronic fuel injector is an electromechanical valve that is controlled based on information obtained from various on-board sensors. It ensures an appropriate combination of fuel-to-air ratio for the ICE. There are three main parameters that are governed by the fuel injector:

1. Quantity of fuel injected—based on input from electronic signals from the mass air-flow sensor, oxygen sensor, throttle position sensor, coolant temperature sensor, and engine speed sensor for the vehicle.
2. The fuel delivery system is responsible to maintain the pressure on the injector. This is useful in maintaining the appropriate amount of fuel during operation of the engine.
3. The air induction system is responsible for delivering air to the engine based on driver demand to form the air–fuel mixture.

32.4.1 Automotive Pumps

In a vehicle, the gas (petrol/ diesel) tank and the engine are typically located at opposite ends. As a result, fuel needs to be drawn toward the engine. This task is performed by a fuel pump. They are broadly classified into two categories: mechanical and electrical. A mechanical fuel pump is typically located close to the engine. This pump uses power from the engine to create a negative pressure. This vacuum is then used to draw the gas toward it.

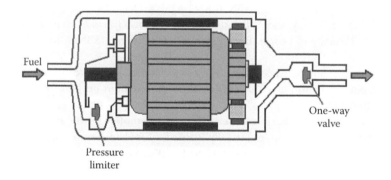

FIGURE 32.6 Conceptual placement of an electric motor in a fuel pump system for automotives.

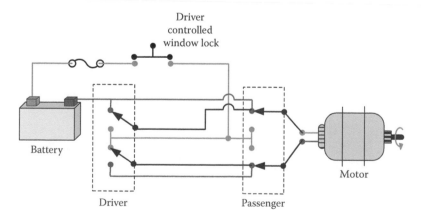

FIGURE 32.7 Conceptual schematic for flow of control in a basic power window system.

An electric fuel pump is controlled by a computer and pumps fuel in accordance with throttle pressure. It is used to provide a constant flow of fuel that is sprayed as a fine mist into the compression chamber. This pump is typically submerged in the gasoline in the tank to keep it cool and create a positive pressure. Since gasoline in its liquid form does not explode, this placement is also done to reduce risk. Apart from this application, some of the other applications that use pumps include cooling system, power steering, etc. Figure 32.6 shows a diagram that shows the position of the electric motor and the approximate placement of inlet and outlet valves.

32.4.2 Power Doors and Windows

There are different ways to wire power windows and doors. A basic control system is shown in Figure 32.7. In this system, power is fed to the central locking unit on the driver's door through the circuit breaker. When the switch is disengaged (turned OFF), power is distributed to a contact at the center of each of the four window switches. At this point, individual control is allowed to passengers at each door. The button pressed by the person decides the polarity in which the motor rotates and can be used to roll the window up or down.

Bibliography

1. Emadi, A., *Handbook of Automotive Power Electronics and Motor Drives*, Boca Raton, FL: CRC Press, ISBN: 0-8247-2361-9, May 2005.
2. Emadi, A., M. Ehsani, and J. M. Miller, *Vehicular Electric Power Systems: Land, Sea, Air, and Space Vehicles*, New York: Marcel Dekker, ISBN: 0-8247-4751-8, December 2003.
3. Ehsani, M., Y. Gao, S. E. Gay, and A. Emadi, *Modern Electric, Hybrid Electric, and Fuel Cell Vehicles: Fundamentals, Theory, and Design*, Boca Raton, FL: CRC Press, ISBN: 0-8493-3154-4, December 2004.
4. Emadi, A., Y.-J. Lee, and K. Rajashekara, Power electronics and motor drives in electric, hybrid electric, and plug-in hybrid electric vehicles, *IEEE Transactions on Industrial Electronics*, 55(6): 2237–2245, 2008.
5. Lukic, S. M., and A. Emadi, Performance analysis of automotive power systems: Effects of power electronic intensive loads and electrically-assisted propulsion systems, *56th IEEE Vehicular Technology Conference (VTC)*, Vancouver, Canada, 2002, Vol. 3, pp. 1835–1839.

33

Plug-In Hybrid Electric Vehicles

Sheldon S.
Williamson
Concordia University

Xin Li
Concordia University

33.1 Introduction

In the recent past, hybrid electric and plug-in hybrid electric vehicles (HEVs/PHEVs) have been widely accepted as viable alternatives to conventional vehicles, due to their environmentally friendly and energy-wise features. In the form of a modified HEV, PHEVs are equipped with sufficient on-board electric power, to support daily driving (on an average of 40 miles/day in North America) in all-electric mode, using only the energy stored in batteries, without consuming a drop of fuel. This, in turn, causes the embedded internal combustion engine (ICE) to merely use a minimal amount of fossil fuel, to support further driving beyond 40 miles. Eventually, a large reduction in green house gas (GHG) emissions is experienced. PHEVs can reduce fuel consumption by charging its battery from the grid or, possibly in the near future, from various forms of green and renewable energy resources [1,2].

From a structural point of view, a PHEV is an HEV with sufficiently sized rechargeable on-board battery pack, which is allowed to employ either a series topology, parallel topology, or a series–parallel combined topology, as shown in Figures 33.1 and 33.2.

Essentially, PHEVs could drive in an all-electric mode until the on-board battery is completely drained. Fuel energy is used for further traveling or whenever needed, such as in the case of aggressive driving conditions. Thus, PHEVs are considered highly electric-intensive vehicles, which may have excellent performance over a regular series HEV.

33.2 PHEV Technology

A PHEV drivetrain involves several stages of energy-consuming components. The drivetrain mainly includes the high-energy-density energy storage system, overall controller, power electronic converters, and the electric motor. From the analytical point of view, Figure 33.3 shows a typical layout of a parallel PHEV drivetrain topology, as an example.

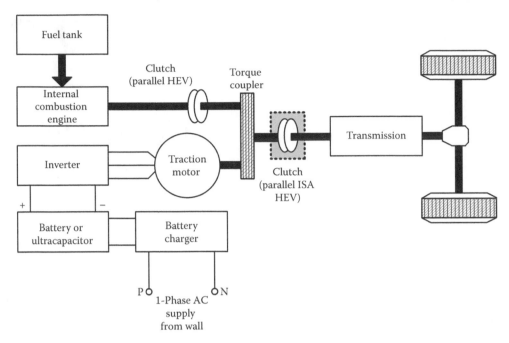

FIGURE 33.1   Typical layout of a parallel PHEV drive train topology.

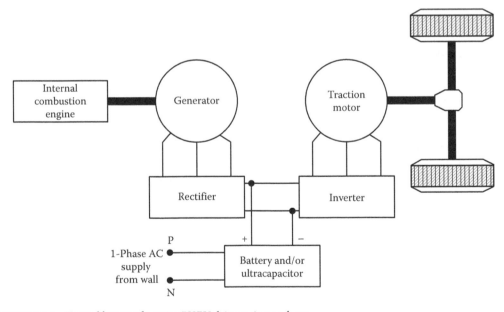

FIGURE 33.2   Typical layout of a series PHEV drive train topology.

In the case of a series PHEV, as shown in Figure 33.1, two different energy sources are combined in succession. It is important to note here that the electric motor offers the only traction, making it an electric-intensive vehicle, more suitable for city driving. The ICE works at its optimal operating point as an on-board generator, maintaining battery charge, by meeting the predetermined state of charge (SOC) requirements.

In case of a parallel PHEV, as is clear from Figure 33.2, the vehicle has two traction sources, both electric as well as mechanical. This type of configuration offers great freedom for choosing suitable

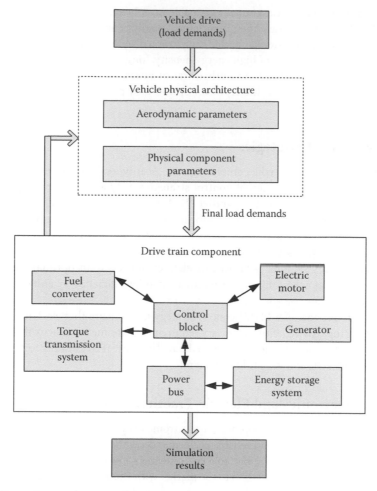

FIGURE 33.3 Power flows within a parallel PHEV drive train system.

combinations of traction sources. By combining the two different traction sources, a smaller engine can be used. In addition, a parallel PHEV arrangement requires a relatively smaller battery capacity compared to that in case of series PHEV, which results in the drivetrain mass to be lighter.

33.2.1 PHEV Energy Storage System

PHEVs require reliable and constant electric energy to support running in all-electric mode. Therefore, the on-board batteries should possess higher energy densities compared to those used in regular HEVs, in order to store enough energy within a small volume, as well as to depict long life cycles, so that they can be charged and discharged frequently. Conventionally, the lead-acid (PbA) battery has been the favored variety of energy storage, but its low energy density means that it cannot be considered as the best candidate for PHEV propulsion.

There exist three types of advanced batteries that are popularly considered as viable candidates for PHEV energy storage. The nickel–metal hydride (Ni-MH) battery is one of the options, because of its high energy density, shorter charging time, and long life cycle. At the same time, it presents an immature recycling system. The lithium-ion (Li-ion) battery chemistry is considered as a definite future trend but, compared to Ni-MH, it depicts much lower durability, which is an issue that is being focussed upon by auto battery manufacturers. The third candidate is an ultracapacitor (UC)/flywheel energy storage system. Ultracapacitors, also known as "double-layer" capacitors, have typically small values of resistance and high power densities. Their capacitance usually ranges between 600 and 2500 F. On the other hand, the flywheel energy storage option, constructed by environmentally friendly materials, offers great characteristics in terms of high energy density, long cycle-life, and high reliability, which are well suited for heavy-duty vehicles and urban transit buses. In recent years, due to considerable improvement in volumetric density, the flywheel option also presents a promising alternative for passenger cars [3].

33.2.2 PHEV Control Strategies

The PHEV control strategy is indeed a complex system. A universally optimized control strategy is nearly impossible to formulate. Different control strategies only achieve certain goals by sacrificing certain features. For example, in the parallel PHEV topology, the parallel electric assist control strategy generally aims at using the electric motor for additional power when needed by the vehicle and maintains the battery SOC at a certain predetermined level. In other words, the electric motor is used when driving torque is below the set value, or it is used to help shift the ICE operating points to higher efficiency regions [4,5]. In this chapter, however, an optimal control strategy is introduced, which aims at achieving higher motor-controller efficiency by enhancing regenerative braking efficiency.

In summary, the main goal of a PHEV control strategy is to ensure that electric energy is the first energy to be used during driving, if the necessary requirements are met satisfactorily. Fuel is only used for further traveling, after electric energy is drained out. The critical constraints to be satisfied usually comprise of the battery SOC and vehicle load demand.

33.2.3 Power Electronics and Electric Traction Motor

The biggest challenge for PHEV development, apart from bringing existing and future technologies together, is to have the most stable and reliable performance, while at the same time offering the most comfortable driving experience at a reasonable cost. Advanced power electronic devices and electric motors play major roles in bringing PHEVs to the market with the aforementioned excellence, reliability, and affordability [6].

DC/DC and DC/AC converters are strategically designed for PHEV applications. Looking ahead to the inevitable future, when PHEVs develop toward pure electric propulsion systems (EVs), power electronics will obviously play an integral role in the drivetrain system. Therefore, issues such as efficiency, reliability, and cost-effective and compact designing of power electronic devices become major challenges. Advanced power electronic converter designs and control techniques are already implemented into prototype PHEVs. For example, soft-switching techniques are used to reduce the switching stresses and to lower the overall losses [7]. Furthermore, advanced universal converters have been recently proposed in literature, to ensure small PHEV charger sizes, which facilitate efficient charging under various possible scenarios.

In terms of electric motor options, in general, there exist four major types in the market. These include permanent magnet (PM) DC motors, AC induction motors (IM), PM brushless motors, and future switched reluctance motors (SRM). Recent studies point out that the IM and PM motors are the two popularly adopted candidates [8]. The IM is an obvious choice, because of its reliability, low maintenance, low cost, and operating capabilities in aggressive environments. The PM brushless motor, however, is popularly utilized in modern PHEV designs, due to its light weight, smaller volume, higher efficiency, and rapid heat dissipation. However, SRMs are believed to be potential candidates in the future, because of their higher efficiency and superior torque-speed characteristics. At the same time, the major challenge for SRM development includes minimizing the inherent developed torque ripples and related control issues.

33.3 PHEV Charging Infrastructures

Essentially, an HEV that can be pre-charged from the grid is broadly termed as a PHEV. As a modified HEV model, an advanced attribute of the PHEV is that the control strategy guarantees the usage of electric energy, and treats it as the first priority. As aforementioned, fuel is only used for further traveling, after electric energy is drained out.

33.3.1 Charging from Grid

For charging purposes, PHEVs can be directly connected to any residential or commercial building outlets, to charge the on-board battery. Complete charging of the battery takes between 6–8 h, if charged from a conventional residential power outlet, provided the vehicle supports daily all-electric driving autonomy. For typical PHEV driving characteristics, the upper limit of battery SOC is usually set to as high a value as 95%, while the lower limit of SOC is set to as low as 20%. In other words, about 75% of the total battery energy is typically used, keeping in mind that it cannot be literally fully charged or completely discharged. Thus, the actual capacity of battery pack can be simply calculated by Equation 33.1, where SOC_{hi} and SOC_{low} are the upper and lower limits of SOC, respectively:

$$E_{real} = \frac{E_{req}}{SOC_{hi} - SOC_{low}}$$

(33.1)

However, from the environmental point of view, GHG emissions cannot be completely eliminated, especially if utility companies use thermal generation sources. In such cases, charging PHEVs for long hours from a utility grid eventually contributes to GHG emissions, from a well-to-wheels (WTW) perspective.

33.3.2 Charging from Renewable Energy Sources

Assuming PHEVs can be charged solely from conventional utility sources, it is thus a valid assumption, that moving into the future, a large number of PHEV users will most definitely exist. Hence, the overall influence of charging on-board PHEV energy storage systems (ESS) cannot be neglected. As a supplement to traditional charging infrastructures, and keeping in mind their impact on overall GHG emissions, charging PHEVs from renewable energy could be a exceedingly practical option in the near future.

Consider charging from solar energy (PV), as an example. Based on the PHEV energy requirement, an appropriately sized PV panel can be designed, considering charging on the worst day (under minimum solar radiation received during a day). This can be calculated based on the variation of mean solar radiant energy (per month), at the specific charging location. Considering a typical PV array efficiency ($\eta_{PV} = 15\%$) and DC/DC converter efficiency ($\eta_{DC/DC} = 95\%$), the area of the PV array can be calculated by Equation 33.2:

$$A = \frac{E_{req}}{R_{day}\eta_{PV}\eta_{DC/DC}}$$

(33.2)

33.3.3 Power Flow Control Strategies

PHEVs are capable of not only absorbing energy from the grid, but considering it as a mobile energy storage device; it is also capable of injecting energy back to the grid, when appropriate. Furthermore, using the PHEV as an uninterruptible power supply (UPS), during power outages, will also help boost the reliability of a single house, an intelligent building, or the grid. It is easy to implement this strategy in a single house or an intelligent building, but it is tough to implement it at the grid level, because a highly complex power flow strategy is needed, in order to measure the amount of power required, and when it is required. Recent literature shows exhaustive studies being conducted on designing intelligent power systems, in order to facilitate the usage of PHEVs for simultaneous grid-to-vehicle (G2V) and vehicle-to-grid (V2G) power flows, as well as usage in UPS applications.

33.4 PHEV Efficiency Considerations

The overall efficiency analysis of PHEVs can be categorized into two parts: well-to-tank (WTT) efficiency and tank-to-wheels (TTW) efficiency (also called drivetrain efficiency). Collectively, the WTT and TTW efficiencies represent the overall fuel-cycle WTW efficiency. More specifically, the calculation of PHEV efficiency involves the investigation of losses incurred during raw material extraction and transportation, during the WTT stage, and calculation of charging/discharging losses, control strategy losses, and regenerative braking losses, during the TTW stage.

33.4.1 PHEV Well-to-Tank Efficiency

The WTT analysis of PHEV is a complex procedure, which involves numerous processes. For instance, at the raw material/fossil fuel extraction stage, the WTT efficiency analysis includes extraction, transportation, and storage losses. At the same time, utility generating efficiency needs to be considered, based on the type of generating fuel being used. Generally, however, the WTT processes typically depict efficiencies in the range of 88%–92%, which is much higher than the TTW energy efficiency of the vehicle.

33.4.2 PHEV Tank-to-Wheels Efficiency

A detailed TTW efficiency (drivetrain efficiency) analysis must be conducted to determine the effectiveness of a specific PHEV drivetrain. Generally, the drivetrain efficiency can be simply yielded out by calculating the losses at each power stage in either a series or parallel PHEV drivetrain structure. However, in order to have a fair efficiency comparison, some parameters that directly affect fuel consumption, such as charging and discharging, regenerative braking efficiency, as well as control strategy design, should be taken into consideration.

33.4.2.1 Charging and Discharging Efficiency

Figure 33.4 demonstrates a typical power system layout of a PHEV. The charger plugs into the grid, to charge the high-energy on-board battery unit. A bidirectional DC/DC converter connects the battery to the high-voltage DC bus and is also used to deliver the energy back to the battery, during regenerative breaking events.

The charging and discharging efficiency in a drivetrain reflects the efficiency of energy exchange in the energy storage system. Therefore, battery efficiency and the efficiencies of power electronics converters are included in the overall calculation.

33.4.2.2 PHEV Control Strategies

The effect of the designed control strategy on the drivetrain efficiency is practically nonnegligible. In case of PHEVs, the control strategy acts more like a charge depleting (CD) strategy in regular HEVs. In CD mode, electric energy is not derived from fuel usage, and hence calculating drivetrain losses in CD mode becomes extremely straightforward. For this purpose, the output energy from the battery needs to be monitored and compared with the energy demands from a practically tested driving pattern.

33.4.2.3 Regenerative Braking Efficiency

Regenerative braking efficiency is introduced to assess the performance of the PHEV drivetrain system in using the available mechanical traction energy, to in turn, produce electric energy [9]. Regenerative

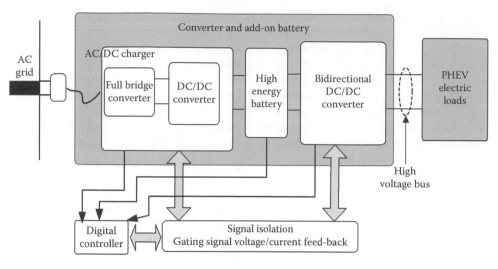

FIGURE 33.4 Power system schematic of a PHEV.

braking converts vehicle momentum into electricity and stores it into the on-board energy storage device, thereby playing an important role in improving overall drivetrain efficiency. Regenerative braking efficiency can be broadly defined as the ratio of regenerative braking energy recovery to the total energy used for braking, as expressed in Equation 33.3. In order to compute regenerative braking efficiency, however, the total energy used for braking events has to be characterized first, which is essentially the total negative traction energy, as defined in Equation 33.4. Second, the energy recovery over regenerative braking events can be found by retrieving the regenerated current, multiplied by the bus voltage (V_{bus}), as described in Equation 33.5:

$$\eta_{\text{REGEN}} = \frac{E_{\text{regen}}}{E_{\text{neg.trac}}} \cdot 100\% \tag{33.3}$$

$$E_{\text{neg.trac}} = \sum_{i=1}^{n} \left(|P_i| \cdot \Delta t \right), \quad P_i \prec 0 \tag{33.4}$$

$$E_{\text{regem}} = I_{\text{regen}} \cdot V_{\text{bus}} = \left(I_{\text{battery.regen}} + I_{\text{acces}} \right) \cdot V_{\text{bus}} \tag{33.5}$$

where
    η_{REGEN} is the regenerative braking efficiency
    $E_{\text{neg.trac}}$ is the negative traction energy
    E_{regem} is the regenerative braking energy recovery
    $I_{\text{battery.regen}}$ is the electric current flowing to the battery pack due to regenerative braking
    I_{acces} is the electric current used by vehicle accessory loads

33.5 Conclusions

As highlighted in this chapter, a PHEV system involves numerous mechanical as well as electric components and concepts. As introduced in the first section, two popular topologies for PHEVs exist. Although the series PHEV topology has been recently targeted as the prime choice for PHEV applications, it is unclear as to whether or not it is indeed the most efficient option. Hence, a thorough efficiency study of PHEV drivetrains becomes an interesting avenue of research.

Currently, keeping a more environmentally friendly and a highly efficient PHEV design in mind, any striking progress in the auto industry is possible. From typical PHEV drivetrain configurations to in-wheel motors, fuel economy improvements and GHG emission reductions are easily achievable [10]. For example, in order to improve drivetrain efficiency, modified planetary gear systems have been successfully designed and tested, whereby the shaft is placed inside the motor, in order to avoid losses occurring during mechanical power transmission. Although such a trivial change might improve the drivetrain efficiency with lesser cost implications, research is constantly being carried out to discover innovative drivetrain configurations, which can integrate additional usage of electric propulsion and reduce mechanical losses.

As a fast-developing industry, it is hard to give an explicit conclusion to the PHEV technology, in general. However, huge potentials across the world certainly represent a promising future for advanced electric-propulsion-based vehicular technologies. It is predicted that future vehicular technologies will most definitely incorporate hybrid/plug-in hybrid propulsion systems. It is hard to say that, maybe, the pure electric propulsion (EV) system is the ultimate vehicular system of the future, but currently, PHEVs have their own mission, presenting the potential of a greener, more energy-efficient driving experience to future customers. The automotive future is hard to predict, but it is indeed promising for the power electronics and motor drives industry.

References

1. X. Li, L. A. C. Lopes, and S. S. Williamson, Charging plug-in electric vehicles with solar energy, in *Proceedings of the 33rd Annual Conference of Solar Energy Society of Canada & the 3rd Canadian Solar Building Research Network Conference*, New Brunswick, Canada, August 2008.
2. X. Li and S. S. Williamson, Efficiency and suitability analyses of varied drive train architectures for plug-in hybrid electric vehicle (PHEV) applications, in *Proceedings of the IEEE Vehicular Power and Propulsion Conference*, Harbin, China, September 2008, pp. 1–6.
3. J. Moreno, M. E. Ortuzar, and J. W. Dixon, Energy-management system for a hybrid electric vehicle, using ultracapacitors and neutral networks, *IEEE Transactions on Industrial Electronics*, 53(2), 614–623, April 2006.
4. S. Wang, K. Huang, Z. Jin, and Y. Peng, Parameter optimization of control strategy for parallel hybrid electric vehicle, in *Proceedings of the 2nd IEEE Conference on Industrial Electronics and Applications*, Harbin, China, May 2007, pp. 2010–2012.
5. X. Li and S. S. Williamson, Efficiency analysis of hybrid electric vehicle (HEV) traction motor-inverter drive for varied driving load demands, in *Proceedings of the IEEE Applied Power Electronics Conference*, Austin, TX, February 2008, pp. 280–285.
6. K. Rajashekara, Power electronics applications in electric/hybrid electric vehicles, in *Proceedings of the 29th Annual Conference of the IEEE Industrial Electronics Society*, Roanoke, VA, November 2003, vol. 3, pp. 3029–3030.
7. M. Ehsani, K. M. Rahman, M. D. Bellar, and A. J. Severinsky, Evaluation of soft switching for EV and HEV motor drives, *IEEE Transactions on Industrial Electronics*, 48(1), 82–89, February 2001.
8. X. Li and S. S. Williamson, Assessment of efficiency improvement techniques for future power electronics intensive hybrid electric vehicle drive trains, in *Proceedings of the IEEE Electrical Power Conference*, Montreal, Canada, October 2007, pp. 268–273.
9. M. Ehsani, Y. Gao, S. E. Gay, and A. Emadi, *Modern Electric, Hybrid Electric, and Fuel Cell Vehicles: Fundamentals, Theory, and Design*, Boca Raton, FL: CRC Press, December 2004.
10. X. Li and S. S. Williamson, Comparative investigation of series and parallel hybrid electric vehicle (HEV) efficiencies based on comprehensive parametric analysis, in *Proceedings of the IEEE Vehicle Power and Propulsion Conference*, Arlington, TX, September 2007, pp. 499–505.

VI

Power Systems

IV

Power Systems

34

Three-Phase Electric Power Systems

Charles A. Gross
Auburn University

34.1 Case for Balanced Polyphase Power Systems

Electric power is transmitted from source to load in the circuit shown in Figure 34.1a, and is proportional to the root mean square (RMS) voltage between conductors (V) and the RMS current (I). There are certain voltages determined by IEEE/ANSI standards from which one may choose. The selection is made for a given application based on several factors, including availability, cost, power level, and safety. Given that the voltage has been selected, the transmitted power is proportional to current; hence, the more power to be delivered to the load, the greater the current. The upper limit on current-carrying capacity (i.e., ampacity) for a conductor is proportional to the cross-sectional area (A), as is its cost. So, for a given load, we need "2A worth" of conductors, where A is the requisite cross-sectional area required to carry the current to and from the load.

Now suppose the load triples. We modify the system by using three sources, serving one-third the total load from three identical circuits, as shown in Figure 34.1b. Even though we have tripled the capacity, we have also tripled the conductor cost, since we now need a total of "6A" conductors. We further modify the system by making a common return path, as shown in Figure 34.1c. Apparently, this accomplishes little since the return conductor must have size 3A since it must have three times the ampacity, and we still need a total "6A" worth of conductors.

But suppose we modified the source so that three voltages, equal in magnitude, and 120° phase displaced, were used. Analysis shows that the currents in conductors a, b, and c are equal, and the return current (n) is zero!. Hence, we need only "3A worth" of conductor to serve the same load as in Figure 34.1b, a 50% savings!

These dramatic savings are only fully realized for the perfectly balanced case, and most systems install a neutral to accommodate load imbalance. However, experience shows that even for worse-case imbalance, n usually carries no more current than a, b, or c, and hence need be no larger than A. Even so, this represents a significant 33% savings and, frequently, conductor n is sized even smaller. These same benefits accrue to any polyphase system; the reason "three" is usually chosen is because it results in economical apparatus (transformer, generator, motor, transmission line) design.

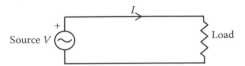

(a) Serving a load from a single-phase source

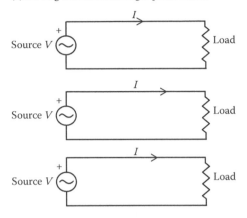

(b) Serving three times a load

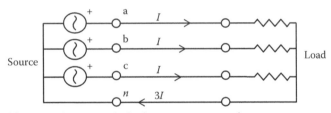

(c) Serving three times the load in an interconnected system

FIGURE 34.1 Basic single-phase power delivery schemes. (a) An electric power delivery circuit. (b) A modified system with triple capacity. (c) System (b) with a common return conductor.

We pause to define some terms. Let the conductors a, b, and c be referred to as the "phases" and conductor *n* as the neutral. The arrangement of Figure 34.1c, shall be called "three-phase." It is fundamental to three-phase systems that the voltages be equal in magnitude, and 120° phase displaced, or nearly so.

34.2 Balanced Three-Phase Circuit Analysis

Consider Figure 34.2, which is fundamental to our three-phase discussion.

For balanced operation the voltages are defined as

$$v_{an}(t) = V_{max} \cdot \cos(\omega t) = \sqrt{2} \cdot V \cdot \cos(\omega t)$$

$$v_{bn}(t) = V_{max} \cdot \cos(\omega t - 120°) = \sqrt{2} \cdot V \cdot \cos(\omega t + 120°) \tag{34.1}$$

$$v_{cn}(t) = V_{max} \cdot \cos(\omega t + 120°) = \sqrt{2} \cdot V \cdot \cos(\omega t + 120°)$$

and are shown in Figure 34.3. All are SI units unless otherwise noted.

The voltages convert to the following phasors:

$$\bar{V}_{an} = V\underline{/0°} \quad \bar{V}_{bn} = V\underline{/-120°} \quad \bar{V}_{cn} = V\underline{/+120°} \tag{34.2}$$

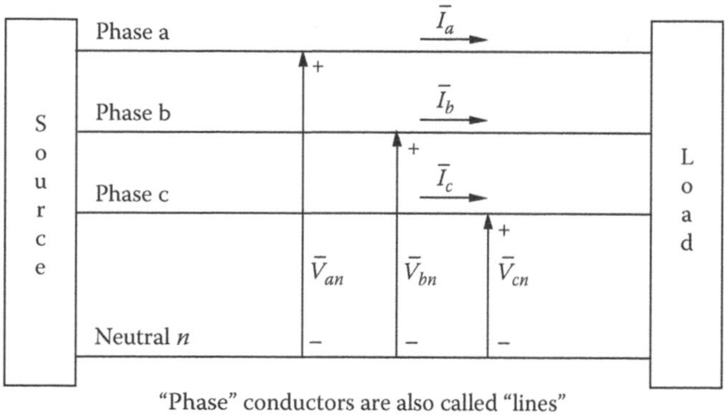

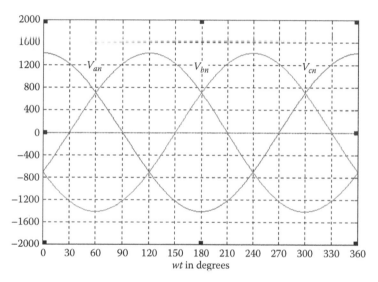

FIGURE 34.2 The basic three-phase situation.

FIGURE 34.3 The instantaneous phase-to-neutral voltages (Equation 34.1).

The notation and terminology is

$$v(t) = \sqrt{2} \cdot V \cdot \cos(\omega t + \alpha) = \text{instantaneous voltage}$$

$$\bar{V} = V \underline{/\alpha} = \text{phasor voltage} \tag{34.3}$$

$$V = |\bar{V}| = \text{magnitude of } \bar{V} \text{ (RMS value)}$$

$$V_{\text{PHASE}} = V_P = V_{an} = V_{bn} = V_{cn} \tag{34.4}$$

Balanced operation is central to the three-phase system performance, because to the extent the system is unbalanced, the advantages of three phases are diminished. Fortunately, most practical systems are reasonably balanced, and extreme unbalance is exceptional (Figure 34.3).

For balanced operation, there are two possibilities, called "phase sequences":

- "Phase b" voltage lags "phase a" voltage by 120° (sequence abc)
- "Phase b" voltage leads "phase a" voltage by 120° (sequence acb)

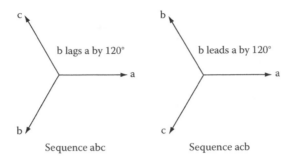

FIGURE 34.4 Phase sequence.

Throughout this discussion, as in most others, a phase sequence of abc is assumed. If this is not the case for a given application, reverse the labels of b and c phases (Figure 34.4).

In addition to the phase voltages, there are three other voltages that are important in three-phase systems. By Kirchhoff's voltage law (KVL),

$$\overline{V}_{ab} = \overline{V}_{an} - \overline{V}_{bn} = V\underline{/0°} - V\underline{/-120°} = \sqrt{3} \cdot V\underline{/30°}$$

$$\overline{V}_{bc} = \overline{V}_{bn} - \overline{V}_{cn} = \sqrt{3} \cdot V\underline{/-90°} \qquad (34.5)$$

$$\overline{V}_{cb} = \overline{V}_{cn} - \overline{V}_{bn} = \sqrt{3} \cdot V\underline{/150°}$$

Collectively, we call these the "line" voltages, such that

$$V_{\text{LINE}} = V_L = V_{ab} = V_{bc} = V_{ca} \qquad (34.6)$$

Observe that

$$V_L = \sqrt{3} \cdot V_P = \sqrt{3}V \qquad (34.7)$$

which is one of the most important results in balanced three-phase analysis. The complete voltage phasor diagram is shown in Figure 34.5. Plots of all six voltages are provided in Figure 34.6. For a three-phase system, it is customary to provide the line voltage as the nominal voltage.

Example 34.1

Given a 1732 V three-phase system, write the phase and line voltages.

$$\overline{V}_{an} = 1000\underline{/0°}\,V \quad \overline{V}_{bn} = 1000\underline{/-120°}\,V \quad \overline{V}_{cn} = 1000\underline{/120°}\,V$$

$$\overline{V}_{ab} = 1732\underline{/30°}\,V \quad \overline{V}_{bc} = 1732\underline{/-90°}\,V \quad \overline{V}_{cb} = 1732\underline{/150°}\,V \qquad (34.8)$$

$$V_p = 1000\,V \quad V_l = 1732V$$

Note that balanced operation, sequence abc, and the use of the phase a voltage as phase reference is assumed, unless otherwise specified.

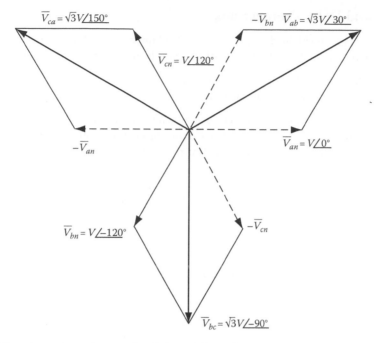

FIGURE 34.5 The voltage phasor diagram for a balanced three-phase system.

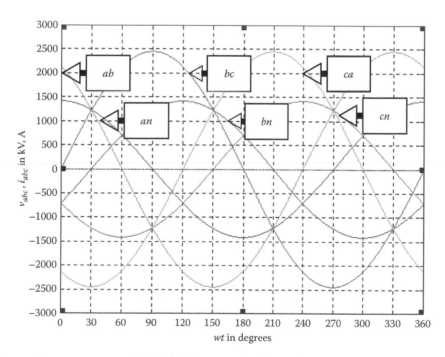

FIGURE 34.6 The instantaneous voltages in a balanced three-phase system.

34.2.1 Wye and Delta Connections

The currents in the system will be determined by the load impedances. If the currents (a, b, c) are to be balanced (i.e., equal in magnitude; 120° phase displaced), the impedances must meet two conditions:

- The three-phase connection must be symmetrical.
- The three phase impedances must be equal.

The two connections that meet the first criteria are shown in Figure 34.7.

Note that in the wye connection, the neutral can be "brought out" and connected into a four-wire system, whereas in the delta connection, the neutral always leads to an open circuit.

The currents in the wye case are

$$\overline{I}_a = \frac{\overline{V}_{an}}{\overline{Z}_{an}} \quad \overline{I}_b = \frac{\overline{V}_{bn}}{\overline{Z}_{bn}} \quad \overline{I}_c = \frac{\overline{V}_{cn}}{\overline{Z}_{cn}} \tag{34.9}$$

In the balanced case, the impedances are equal:

$$\overline{Z}_Y = \overline{Z}_{an} = \overline{Z}_{bn} = \overline{Z}_{cn} = Z_Y \underline{/\theta_Y} \tag{34.10}$$

So that

$$\overline{I}_a = \frac{\overline{V}_{an}}{\overline{Z}_Y} = \frac{V\underline{/0°}}{Z_Y\underline{/\theta_Y}} = \frac{V}{Z_Y}\underline{/-\theta_Y} \tag{34.11}$$

Given balanced three-phase voltage is applied to a balanced three-phase wye-connected load, it is straightforward to show that the a, b, and c phase currents are balanced.

The currents in the delta case are

$$\overline{I}_{ab} = \frac{\overline{V}_{ab}}{\overline{Z}_{ab}} \quad \overline{I}_{bc} = \frac{\overline{V}_{bc}}{\overline{Z}_{bc}} \quad \overline{I}_{ca} = \frac{\overline{V}_{ca}}{\overline{Z}_{ca}}$$

$$\overline{I}_a = \overline{I}_{ab} - \overline{I}_{ca} \quad \overline{I}_b = \overline{I}_{bc} - \overline{I}_{ab} \quad \overline{I}_c = \overline{I}_{ca} - \overline{I}_{bc} \tag{34.12}$$

In the balanced case, the impedances are equal:

$$\overline{Z}_\Delta = \overline{Z}_{ab} = \overline{Z}_{bc} = \overline{Z}_{ca} = Z_\Delta \underline{/\theta_\Delta} \tag{34.13}$$

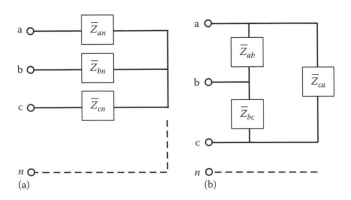

FIGURE 34.7   Balanced three-phase connections. (a) The wye (Y) connection and (b) the delta (Δ) connection.

So that

$$\bar{I}_{ab} = \frac{\bar{V}_{ab}}{\bar{Z}_\Delta} = \frac{V\sqrt{3}\angle 30°}{Z_\Delta\angle\theta_\Delta} = \frac{V\sqrt{3}}{Z_\Delta}\angle 30° - \theta_\Delta \tag{34.14}$$

Given balanced three-phase voltage is applied to a balanced three-phase delta-connected load, it is straightforward to show that the a, b, and c phase currents are balanced.

$$\bar{I}_a = \bar{I}_{ab} - \bar{I}_{ca}$$

$$= \left(\frac{V\sqrt{3}}{Z_\Delta}\angle 30° - \theta_\Delta\right) - \left(\frac{V\sqrt{3}}{Z_\Delta}\angle 30° - \theta_\Delta + 150°\right)$$

$$= \sqrt{3} \cdot \frac{V\sqrt{3}}{Z_\Delta}\angle -\theta_\Delta = \frac{3V}{Z_\Delta}\angle -\theta_\Delta \tag{34.15}$$

Now for wye–delta loads to be equivalent, it is necessary that the two loads draw the same currents:

$$\bar{I}_a(Y) = \bar{I}_a(\Delta)$$

$$\frac{V}{Z_Y}\angle -\theta_Y = \frac{3V}{Z_\Delta}\angle -\theta_\Delta \tag{34.16}$$

$$\bar{Z}_\Delta = 3\bar{Z}_Y$$

Example 34.2

Convert the given wye load into an equivalent delta.

$$\bar{Z}_\Delta = 3 \cdot \bar{Z}_Y = 3(8 + j6) = 24 + j18\,\Omega \tag{34.17}$$

It is possible to convert unbalanced delta-connected impedances to wye, and vice versa, in three-wire (a, b, c) situations. To convert delta elements to wye

$$\overline{Z}_{an} = \frac{\overline{Z}_{ab} \cdot \overline{Z}_{ca}}{\overline{Z}_{ab} + \overline{Z}_{bc} + \overline{Z}_{ca}}$$

$$\overline{Z}_{bn} = \frac{\overline{Z}_{bc} \cdot \overline{Z}_{ab}}{\overline{Z}_{ab} + \overline{Z}_{bc} + \overline{Z}_{ca}} \qquad (34.18)$$

$$\overline{Z}_{cn} = \frac{\overline{Z}_{ca} \cdot \overline{Z}_{bc}}{\overline{Z}_{ab} + \overline{Z}_{bc} + \overline{Z}_{ca}}$$

To convert wye elements to delta

$$\overline{Z}_{ab} = \frac{\overline{Z}_{an} \cdot \overline{Z}_{bn} + \overline{Z}_{bn} \cdot \overline{Z}_{cn} + \overline{Z}_{cn} \cdot \overline{Z}_{an}}{\overline{Z}_{cn}}$$

$$\overline{Z}_{bc} = \frac{\overline{Z}_{an} \cdot \overline{Z}_{bn} + \overline{Z}_{bn} \cdot \overline{Z}_{cn} + \overline{Z}_{cn} \cdot \overline{Z}_{an}}{\overline{Z}_{an}} \qquad (34.19)$$

$$\overline{Z}_{ca} = \frac{\overline{Z}_{an} \cdot \overline{Z}_{bn} + \overline{Z}_{bn} \cdot \overline{Z}_{cn} + \overline{Z}_{cn} \cdot \overline{Z}_{an}}{\overline{Z}_{bn}}$$

Unbalanced three-phase systems are usually analyzed using a technique called symmetrical components, a topic beyond the scope of this section.

Example 34.3

Given that the voltages of Example 34.1 are applied to loads of Example 34.2, compute all currents.

The wye case

$$\overline{I}_a = \frac{\overline{V}_{an}}{\overline{Z}_{an}} = \frac{1000\underline{/0°}}{8+j6} = 100\underline{/36.9°}\,A$$

$$\overline{I}_b = \frac{\overline{V}_{bn}}{\overline{Z}_{bn}} = \frac{1000\underline{/-120°}}{8+j6} = 100\underline{/-156.9°}\,A \qquad (34.20)$$

$$\overline{I}_c = \frac{\overline{V}_{cn}}{\overline{Z}_{cn}} = \frac{1000\underline{/120°}}{8+j6} = 100\underline{/83.1°}\,A$$

The delta case

$$\overline{I}_{ab} = \frac{\overline{V}_{ab}}{\overline{Z}_{ab}} = \frac{1732\underline{/30°}}{24+j18} = 57.73\underline{/-6.9°}\,A$$

$$\overline{I}_{bc} = \frac{\overline{V}_{bc}}{\overline{Z}_{bc}} = \frac{1732\underline{/-90°}}{24+j18} = 57.73\underline{/-126.9°}\,A$$

$$\overline{I}_{ca} = \frac{\overline{V}_{ca}}{\overline{Z}_{ca}} = \frac{1732\underline{/-150°}}{24+j18} = 57.73\underline{/113.1°}\,A \qquad (34.21)$$

$$\overline{I}_a = \overline{I}_{ab} - \overline{I}_{ca} = 100\underline{/-36.9°}\,A$$

$$\overline{I}_b = \overline{I}_{bc} - \overline{I}_{ab} = 100\underline{/-156.9°}\,A$$

$$\overline{I}_c = \overline{I}_{ca} - \overline{I}_{bc} = 100\underline{/83.1°}\,A$$

Note that balanced operation, sequence abc, and use of the phase a voltage as phase reference is assumed, unless otherwise specified.

34.3 Power Considerations

There are essentially five types of power in ac circuit analysis. In a single-phase situation,

$$p(t) = v(t) \cdot i(t)$$

$$= \left(V\sqrt{2}\cos(\omega t + \alpha) \right) \cdot \left(I\sqrt{2}\cos(\omega t + \beta) \right)$$

$$= V \cdot I \cdot \cos(\alpha - \beta) + V \cdot I \cdot \cos(2\omega t + \alpha + \beta)$$

$$= \text{instantaneous power}$$

$$\overline{S} = P + jQ = \overline{V} \cdot \overline{I}^* = \text{complex power} \qquad (34.22)$$

$$S = |\overline{S}| = V \cdot I = \text{apparent power}$$

$$P = \operatorname{Re}[\overline{S}] = \text{average ("real") power}$$

$$Q = \operatorname{Im}[\overline{S}] = \text{reactive ("imaginary") power}$$

Given that voltages and current are in volts and amperes, respectively, all five powers must be in watts. However, by tradition, "watts" is used only for $p(t)$ and P, using "VA" (volt-ampere) and "var" (volt-ampere reactive) for S and Q, respectively.

We define

$$\theta = \alpha - \beta \quad \text{Then...}$$

$$p(t) = V \cdot I \cdot \cos(\theta) + V \cdot I \cdot \cos(2\omega t + \alpha + \beta)$$

$$\overline{S} = \overline{V} \cdot \overline{I}^* = V \cdot I \angle \theta \qquad (34.23)$$

$$= V \cdot I \cdot \cos(\theta) + jV \cdot I \cdot \sin(\theta)$$

$$S = V \cdot I \quad P = V \cdot I \cdot \cos(\theta) \quad Q = V \cdot I \cdot \sin(\theta) \ \overline{S} = VI\angle\theta$$

There is a concept called "power factor," which is defined as

$$\text{Power factor} = pf = \frac{P}{S} \qquad (34.24)$$

For the sinusoidal case,

$$pf = \frac{P}{S} = \frac{V \cdot I \cdot \cos(\theta)}{V \cdot I} = \cos(\theta) \qquad (34.25)$$

Power factors are said to be "lagging" or "leading," depending on the phase position of the current relative to the voltage. See Figure 34.8.

Note that R-C loads will have a leading pf; R loads, unity pf; and R-L loads, lagging pf.

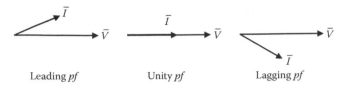

Leading *pf*   Unity *pf*   Lagging *pf*

FIGURE 34.8   Leading and lagging power factor.

Example 34.4

Given the system of Examples 34.1 through 34.3, find the a-phase powers and *pf*.

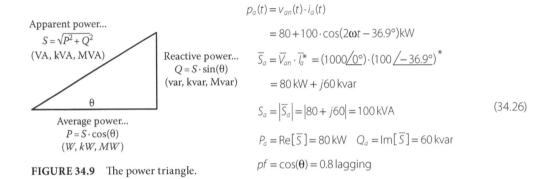

$$p_a(t) = v_{an}(t) \cdot i_a(t)$$

Apparent power...
$$S = \sqrt{P^2 + Q^2}$$
(VA, kVA, MVA)

$$= 80 + 100 \cdot \cos(2\omega t - 36.9°)\,\text{kW}$$

Reactive power...
$$Q = S \cdot \sin(\theta)$$
(var, kvar, Mvar)

$$\overline{S}_a = \overline{V}_{an} \cdot \overline{I_a}^* = (1000\underline{/0°}) \cdot (100\underline{/-36.9°})^*$$

$$= 80\,\text{kW} + j60\,\text{kvar}$$

$$S_a = \left| \overline{S}_a \right| = |80 + j60| = 100\,\text{kVA} \qquad (34.26)$$

$$\theta$$

Average power...
$$P = S \cdot \cos(\theta)$$
(W, kW, MW)

$$P_a = \text{Re}[\overline{S}] = 80\,\text{kW} \quad Q_a = \text{Im}[\overline{S}] = 60\,\text{kvar}$$

FIGURE 34.9   The power triangle.

$$pf = \cos(\theta) = 0.8\,\text{lagging}$$

Suppose we plot complex power in the complex plane (i.e., make an argand diagram). We get a diagram as shown in Figure 34.9, called the "power triangle."

Example 34.5

Draw the power triangle for Example 34.4

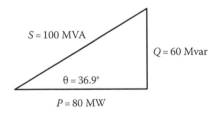

$S = 100$ MVA

$Q = 60$ Mvar

$\theta = 36.9°$

$P = 80$ MW

In a general three-phase four-wire (abcn) situation,

$$p_{3\phi}(t) = p_{an}(t) + p_{bn}(t) + p_{cn}(t)$$

$$= v_{an}(t) \cdot i_a(t) + v_{bn}(t) \cdot i_b(t) + v_{cn}(t) \cdot i_c(t) \qquad (34.27)$$

$$\overline{S}_{3\phi} = \overline{S}_a + \overline{S}_b + \overline{S}_c$$

$$= \overline{V}_{an} \cdot \overline{I_a}^* + \overline{V}_{bn} \cdot \overline{I_b}^* + \overline{V}_{cn} \cdot \overline{I_c}^*$$

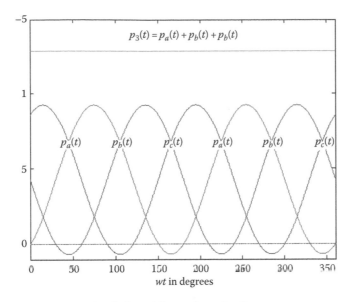

FIGURE 34.10 Instantaneous power in a balanced three–phase situation.

If the system is balanced, it is straightforward to show that (Figure 34.10)

$$
\begin{aligned}
p_{3\phi}(t) &= p_{an}(t) + p_{bn}(t) + p_{cn}(t) \\
&= V \cdot I \cdot \cos(\theta) + V \cdot I \cdot \cos(2\omega t + \alpha + \beta) \\
&\quad + V \cdot I \cdot \cos(\theta) + V \cdot I \cdot \cos(2\omega t + \alpha + 240°) \\
&\quad + V \cdot I \cdot \cos(\theta) + V \cdot I \cdot \cos(2\omega t + \alpha - 240°) \\
&= 3V \cdot I \cdot \cos(\theta)
\end{aligned}
\tag{34.28}
$$

Hence, we see the second advantage to balanced three-phase systems. **The three-phase instantaneous power is constant!**

Contrast this with the single-phase instantaneous power, which is oscillatory with radian frequency 2ω. Continuing, it is straightforward to show that

$$
\begin{aligned}
\overline{S}_{3\phi} &= 3 \cdot \overline{V}_{an} \cdot \overline{I}_a^* \\
\overline{S}_{3\phi} &= 3 \cdot V \cdot I_L = 3 \cdot \frac{V_L}{\sqrt{3}} \cdot I_L \\
&= \sqrt{3} \cdot V_L \cdot I_L \\
P_{3\phi} &= \sqrt{3} \cdot V_L \cdot I_L \cdot \cos(\theta) \\
Q_{3\phi} &= \sqrt{3} \cdot V_L \cdot I_L \cdot \sin(\theta)
\end{aligned}
\tag{34.29}
$$

Example 34.6

Given that Examples 34.1 through 34.3 represent one phase of a balanced three-phase situation, find the three-phase powers.

$$S_{3\phi} = \sqrt{3} \cdot V_L \cdot I_L = \sqrt{3} \cdot (1732) \cdot (100) = 300\,\text{kVA}$$

$$P_{3\phi} = \sqrt{3} \cdot V_L \cdot I_L = \cos(\theta) = (300)(0.8) = 300\,\text{kW}$$

$$Q_{3\phi} = \sqrt{3} \cdot V_L \cdot I_L = \sin(\theta) = (300)\,(0.6) = 180\,\text{kvar} \qquad (34.30)$$

$$P_{3\phi}(t) = 240\,\text{kW}$$

$$\overline{S}_{3\phi} = 240 + j180$$

The power triangle for Example 34.6 is

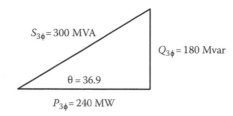

The power factor for Example 34.4 is

$$pf = \cos(\theta) = \cos(36.9°) = 0.8\,\text{lagging} \qquad (34.31)$$

Utilities prefer that their customers operate at unity power factor (pf) because this corresponds to maximum (real) delivered power at minimum current. To encourage maximum pf operation, the utility rate structure imposes a penalty for low pf operation. Hence, it is to a customer's advantage to "correct" their pf to near unity. This is called "the pf correction problem", and is discussed here.

Typically the load pf is lagging (inductive), which requires that the correcting elements must be capacitive; hence the term "pf correcting capacitors". The capacitor bank is installed in parallel with the load, and on the load side of the metering point as shown Figure 34.11a.

Example 34.7

A utility supplies a customer's three-phase load of 1000 kVA @ 12.47 kV; pf = 0.6 lagging. The customer wishes to install pf correcting capacitors to correct the pf to 0.92 lagging. Size the capacitors.

$$\overline{S}_L = \text{Load power} = P_L + jQ_L = 600 + j\,800$$

$$\overline{S}_C = \text{Compensating (capacitive) power} = 0 - jQ_C$$

$$\overline{S}_S = \text{Source power} = P_S + jQ_S$$

$$\overline{S}_S = \overline{S}_L + \overline{S}_C = 600 + j\,800 + 0 - jQ_C$$

$$= 600 + j\,(800 - Q_C) \qquad (34.32)$$

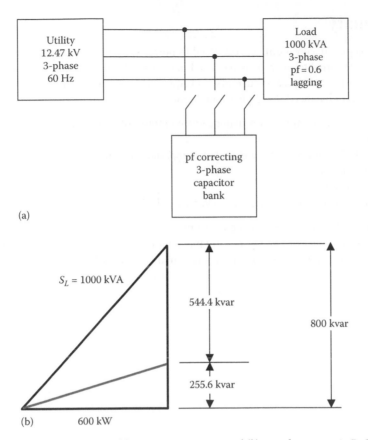

FIGURE 34.11 Power factor correction. (a) System components and (b) complex powers in P–Q coordinates.

Since the requested $pf = 0.92$ lagging, $\theta = 23.07°$. Then,

$$\bar{S}_S = 600 + j(800 - Q_C) = S_S \underline{/23.07°}$$

$$= 600 + j255.6 \tag{34.33}$$

Therefore $Q_C = 800 - 255.6 = 544.4$ kvar

Output data: capacitive compensation required

	Wye	Delta
Reactor ratings		
Line volt = 12,470 V; total $Q = 544.401$ kvar		
Single-phase element ratings	Wye	Delta
Reactive power rating (kvar)	181.467	181.467
Voltage rating (V)	7,199.6	12,470.0
Current rating (A)	25.2	14.6
Reactor impedance (Ω)	285.64	856.91
Reactor capacitance (μF)	9.287	3.096

34.4 Summary

Three-phase concepts are commonly encountered in power systems, where power levels exceed 10 kW. Such systems are designed to operate as balanced, meaning equal in magnitude, and 120° phase displaced, when applied to voltages and currents, and equal, when applied to impedance and power.

Some advantages of balanced three-phase operation are

1. Transmission of twice the power using a given amount of conductor, compared with the single-phase case.
2. DC transmission of instantaneous power, compared with the oscillatory nature of single-phase power transmission.
3. Choice of two voltage levels.
4. Some transmission capability remains, even with the loss of two phases.

For analysis of balanced three-phase systems, symmetry can be used to great advantage, eliminating two-thirds of the work, making calculations not much more complicated that those for the single-phase case. Unbalanced operation is considered an abnormality, and is best analyzed using symmetrical components.

35

Contactless Energy Transfer

Marian P.
Kazmierkowski
*Warsaw University
of Technology*

Artur Moradewicz
Electrotechnical Institute

Jorge Duarte
*Eindhoven University
of Technology*

Elena Lomonowa
*Eindhoven University
of Technology*

Christoph Sonntag
*Eindhoven University
of Technology*

35.1 Introduction

Recently, contactless energy transfer (CET) systems have been developed and investigated widely [see list of References]. This innovative technology creates new possibilities to supply mobile devices with electrical energy because elimination of cables, connectors, and/or slip rings increases reliability and maintenance-free operation of such critical systems as in aerospace, biomedical, and robotics applications. Figure 35.1 shows a classification of the CET systems.

As "medium" for CET could be used electromagnetic waves, including light, acoustic waves (sound), as well as electric fields. In the most popular applications, the core of CET system is inductive or capacitive coupling between power source and load, and high switching frequency converter.

The capacitive coupling (Figure 35.2b) is used in low power range (e.g., supply systems for sensors) whereas inductive coupling (Figure 35.2a) allows transferring power from a few milliwatts up to hundred kilowatts [20]. It should be noted that there is no commonly accepted nomenclature in CET systems. Some authors use the term "wireless" [1,5,16,21,32–34] instead of "contactless" energy transfer or power supply. However, the term "wireless" energy transfer (or power supply) is used mostly to describe systems where energy is transferred on longer distances (several meters), like for cellular phones or wireless sensor technologies [5,20,34].

In this chapter, only *inductive coupled* CET systems are discussed. The potential applications for such a technology are practically endless and can range from the transfer of energy between low-power home

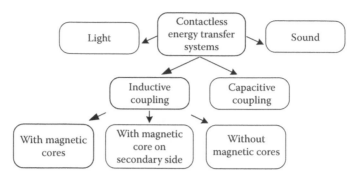

FIGURE 35.1 Classification of contactless/wireless energy transfer systems. (Reproduced from Sonntag, C. L. W. et al., Load position detection and validation on variable-phase contactless energy transfer desktops <http://repository.tue.nl/661798>. In *Proceedings of the IEEE Energy Conversion Congress and Exposition, ECCE 2009*, San Jose, 20–24 September, 2009, pp. 1818–1825. With permission.)

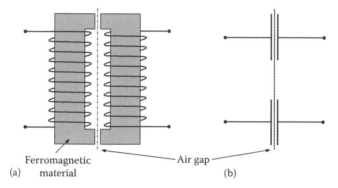

FIGURE 35.2 Inductive (a) and capacitive (b) coupling used in CET systems. (Reproduced from Sonntag, C. L. W. et al., Load position detection and validation on variable-phase contactless energy transfer desktops <http://repository.tue.nl/661798>. In *Proceedings of the IEEE Energy Conversion Congress and Exposition, ECCE 2009*, San Jose, 20–24 September, 2009, pp. 1818–1825. With permission.)

and office devices to high-power industrial applications. Medical, marine, and other applications where physical electrical contact might be dangerous (battery chargers), impossible, or very problematic, are all prospective candidates for the use of CET systems.

Because of many parameters used in specification of a CET system, it has to be designed and adapted to individual conditions and there is no one universal solution. In spite of many works presenting individual solutions of inductive coupled CET systems [3,7–14,18,19,22,23,29–31,38], there is no commonly accepted control and design methodology. Because of high switching frequency ($f_{sw} \geq 20\,\mathrm{kHz}$) used in CET converters, most of the reported systems have been built in hardware technology [22,23,29,30] and implemented control and protection methods were characteristic for hardware-based approach. However, in the last time more sophisticated methods are development and implemented in digital signal processors or programmable logical controllers, like field-programmable gate array (FPGA) circuits, which has been reported in Ref. [25,26].

35.2 Basic Principles of Operation

Figure 35.3 shows the block diagram of typical inductive coupled CET systems. It consists of primary side DC/AC resonant converter, which converts DC into high-frequency AC energy. Next the AC energy via transformer with inductive coupling factor k is transmitted to the secondary side. The secondary side is not connected electrically with primary and, therefore, can be movable (linearly or/and rotating),

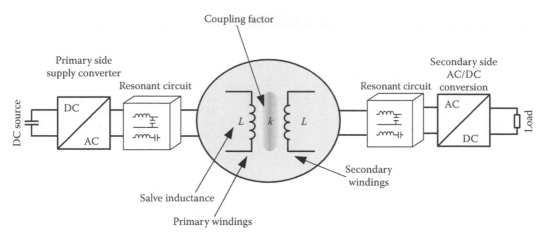

FIGURE 35.3 Block scheme of CET system. (Reproduced from Sonntag, C. L. W. et al., Load position detection and validation on variable-phase contactless energy transfer desktops <http://repository.tue.nl/661798>, In *Proceedings of the IEEE Energy Conversion Congress and Exposition, ECCE 2009*, San Jose, 20–24 September, 2009, pp. 1818–1825. With permission.)

giving flexibility, mobility, and safety for supplied loads. In the secondary side, the high-frequency AC energy is converted safely by AC/DC converter to meet the requirements specified by the load parameters. In most cases, a diode rectifier with capacitive filter is used as AC/DC converter.

However, in some applications an active rectifier or inverter (for stabilized DC or AC loads) is required [12–14]. Hence, the inductive coupled CET system consists mainly of a large air gap transformer and resonant converter.

35.2.1 Compensation Topologies

In conventional applications, a transformer is used for galvanic insulation between source and load, and its operation is based on high magnetic coupling factor k between primary and secondary windings. Because of used two-halves cores and/or air gap, CET transformers usually operate under much lower magnetic coupling factor. As a result, the main inductance L_{12} (see Table 35.1) is very small whereas leakage inductances (L_{11}, L_{22}) are large as compared with conventional transformers. Consequently, increase in magnetizing current causes higher conducting losses. Also, winding losses increase because of large leakage inductances. Another disadvantage of transformers with relatively large gap is electromagnetic compatibility (EMC) problem (strong radiation). To minimize the above disadvantages of CET transformers, several power conversion topologies have been proposed, which can be classified into the following categories: the flyback, resonant, quasi-resonant, and self-resonant. The common thing for all these topologies is that they all utilize the energy stored in the transformer. In most applications, a resonant soft switching technique is used because it allows both compensation of the transformer leakage inductance and reduction of power converter switching losses. To form resonant circuits and compensate for the large CET transformer leakage inductances, two methods can be applied [16,17]: *S*-series or *P*-parallel, giving four basic topologies: *SS*, *SP*, *PS*, and *PP* (first letter denotes *primary* and second a *secondary* compensation—see Table 35.1).

The *PS* and *PP* topologies require an additional series inductor to regulate the inverter current flowing into the parallel resonant tank. This additional inductor increases EMC distortion, converter size, and the total cost of CET system.

Assuming the same numbers of primary and secondary winding, $N_1 = N_2$, the basic parameters of *SS* and *SP* topologies have been summarized in Table 35.2, where $G_V = u_2/u_1$—voltage output-input transfer function of CET system; $\omega = \omega_s/\omega_0$—normalized frequency; k—magnetic coupling factor; and R_0—load resistance on the secondary side.

TABLE 35.1   Variants of Leakage Inductance Compensation Circuits

Simplified Circuits	Abbreviation	Comments	Sensitivity for Coupling and Load Changes
1	Series–series SS	Systems with intermediate DC voltage bus	Sensitive for load changes
2	Series–parallel SP	(Current source output) battery charging	Sensitive for coupling changes
3	Parallel–series PS	Systems with intermediate DC voltage bus	Sensitive for coupling changes
4	Parallel–parallel PP	(Current source output) battery charging	Sensitive for coupling changes

Comparing *SS* with *SP* parameters, it can be seen that selected topology influences strongly the correct choice of the primary capacitance. An important advantage of *SS* topology is that primary capacitance is independent of either magnetic coupling factor or the load. Contrary to this, the *SP* topology depends on coupling factor and requires higher value of capacitance for stronger magnetic coupling.

35.2.2 Resonant Power Converters

Resonant power converters contain *resonant L–C networks*, also called *resonant circuit* (*RC*) or *resonant tank network*, whose voltage and current waveforms vary sinusoidally during one or more subintervals of each switching period. These converters contain low total harmonic distortion because switching frequency is equal to first harmonic frequency. Basic power converter topologies used in CET systems are presented in Figure 35.4. The full-bridge (Figure 35.4c) inverter, composed of four switches and an RC, is commonly used in high power applications. The half-bridge inverter (Figure 35.4a) has only two switches and two others can be replaced by capacitors (Figure 35.4b). The output voltage μ_1 of the full-bridge converter is doubled when compared with half-bridge topology.

TABLE 35.2 Basic Parameters of Compensation Circuits

Circuit/parameter	SS Compensation	SP Compensation
Voltage transfer G_V function	$G_{Vss} = \sqrt{\left(1+\frac{1-k}{k}\left(1-\frac{1}{\omega^2}\right)\right)^2 + \left(Q_{ss}\left(\omega-\frac{1}{\omega}\right)\left(1+\frac{1-k}{2k}\left(1-\frac{1}{\omega^2}\right)\right)\right)^2}$	$G_{Vsp} = \dfrac{1}{\sqrt{\left[1+(1-k^2)\cdot(1-(\omega/\omega_0)^2)\right]^2 + Q_{ss}^2\,((\omega/\omega_0)\cdot(\omega_0/\omega))^2}}$
Resonance angular frequency	$\omega_0 \cong 1/\sqrt{L_r C_r}$	$\omega_0 \cong 1/\sqrt{L_r C_r}$
Resonant capacitor	$C_{r1} = C_{r2}$	$C_{r1} = \dfrac{1}{1-k^2}C_{r2}$
Circuit quality factor	$Q_{ss} = \dfrac{\omega(L_{11}+L_{22})}{R_{es}} = \dfrac{\omega L_r}{R_{es}}$	$Q_{sp} = \dfrac{R_{ep}}{\omega(L_{11}+L_{22})} = \dfrac{R_{ep}}{\omega L_r}$
Equivalent load resistance	$R_{es} = \dfrac{8}{\pi^2}R_0$	$R_{ep} = \dfrac{\pi^2}{8}R_0$

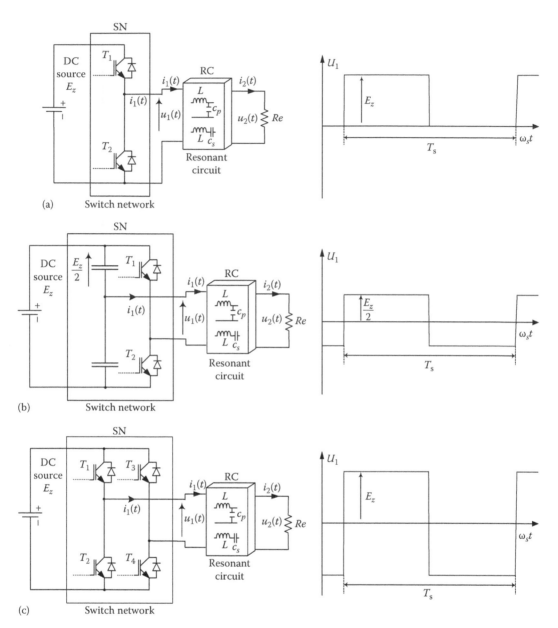

FIGURE 35.4 Basic topologies of series resonant converter and resonant circuit voltage $u_1(t)$ waveforms. (a) Half-bridge unipolar converter, (b) half-bridge bipolar converter, and (c) full-bridge converter.

The main advantage of resonant technique is reduction of switching losses via mechanisms known as zero current switching (ZCS) and zero voltage switching (ZVS). The *switch-on* and/or *switch-off* converter semiconductor components can occur only at zero crossing of the resonant quasi-sinusoidal waveforms. This eliminates some of the switching loss mechanisms. Hence, switching losses are reduced, and resonant converters can operate at switching frequencies that are considerably higher than in comparable pulse width modulation (PWM) hard switching converters. ZVS can also eliminate or reduce some of the electromagnetic emission sources, also known as electromagnetic interference. Another advantage is that both ZVS and ZCS converters can utilize transformer leakage inductance and diode junction capacitors as well as the output parasitic capacitor of the power switch.

However, resonant converters exhibit several disadvantages. Although, the components of an RC can be chosen such that good performance with high efficiency is obtained at a single operating point, typically it is difficult to optimize the resonant components in such a way that good performance is obtained over a wide range of load currents and input voltages variations. Significant currents may circulate through the tank components, even when the load is removed, leading to poor efficiency at light loads. Therefore, the converter used in CET system has to be carefully designed [9,11,23].

Typical steady-state waveforms of the voltages u_1, u_2; currents i_1, i_2; and primary side power P_1 in an insulated gate bipolar transistor-based CET resonant full-bridge converter with *series-series* compensation for operation at resonance frequency are presented in Figure 35.5. The rotatable transformer air gap is 25.5 mm, the load resistance 10 Ω, and transferred power is 2.5 kW [25,26]. It can be seen that resonant converter operates with zero primary current switching.

35.3 Review of CET Systems

Depending on the power range and air gap length, different transformer cores can be used. A general overview representing a typical construction of inductive coupling used in CET systems is shown in Figure 35.6. It can be seen that for high power and low air gap, transformers with magnetic cores in primary and secondary side are applied. Contrary to this, for large air gap and low power, air transformers (coreless) are preferred. A special case is a *sliding transformer* that can have construction for linear or circular movement (see Section 35.6). The final configuration of CET systems also depends strongly on the number of loads to be supplied. In such cases, transformers with multi-winding secondary or primary side are used. In the next subsection, some selected examples of inductive coupled CET systems are presented.

35.4 CET Systems with Multiple Secondary Winding

The CET system of Figure 35.3 can be equipped with multiple secondary winding, as shown in Figure 35.7. This is a very flexible solution in which several isolated and/or moving loads can be supplied. In situations when stabilized AC or DC loads are required, an additional active DC/AC or DC/DC converter has to be added (Figure 35.7). Of course, it results in additional losses and efficiency reduction. Based on this idea, in [12,13], a CET system has been proposed which can be compared to a plug-and-socket extension cable. Instead of inserting a plug into a socket, a connection between supply line (cable) and loads (clamps) is established using CET. Also, ABB Corporate Research, Ladenburg, Germany, has developed a factory communication and wireless power supply system for sensors and actuators called WISA [1,27,33,34]. In this solution, a coreless single winding primary side (constructed in form of a frame) is coupled with distributed multiple secondary windings to supply sensors and actuators with 10 mW output power each.

The transformers used in the system of Figure 35.7 can have different construction: stationary, rotating, rotatable, with magnetic core, or coreless. As an example a rotating transformer with double parallel connected secondary windings is used in CET systems for the power supply of airborne radar systems [29].

35.5 CET Systems with Cascaded Transformers

In Figure 35.8, a CET system used in power supply for robots and manipulators [10,11] is shown. The indirect DC link AC/DC/AC power converter generates a square wave voltage of 200–600 V and 20–60 kHz frequency. This voltage is fed to the primary winding of first rotatable transformer located on the first axis of the robot. The transformer secondary side is connected to the next DC link AC/DC/AC power converter, which using PWM technique generates variable frequency AC voltage to supply first three-phase motor. The transformer secondary side is also connected to the primary side of the next rotatable

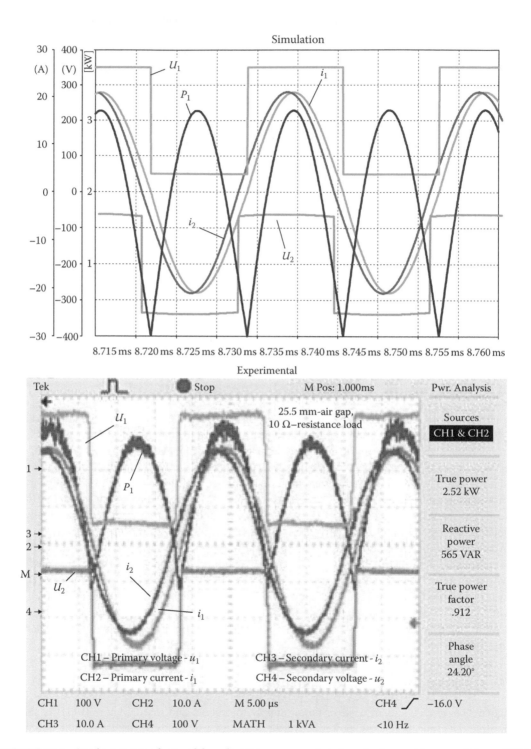

FIGURE 35.5 Steady-state waveforms of the voltages u_1, u_2; currents i_1, i_2; and primary side power P_1 in an IGBT transistor-based CET resonant full-bridge converter with *series-series* compensation (operation at resonance frequency).

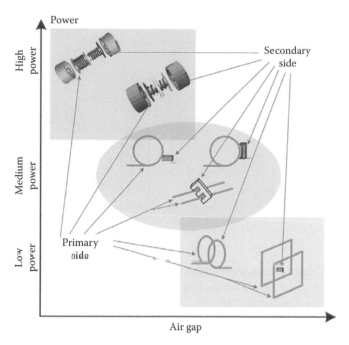

FIGURE 35.6 Power range of inductive coupling-based CET systems versus air gap wide.

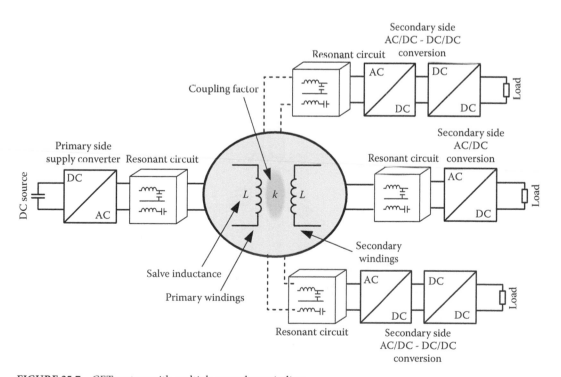

FIGURE 35.7 CET system with multiple secondary winding.

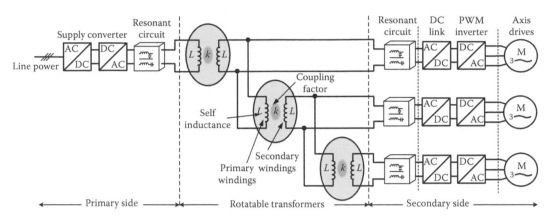

FIGURE 35.8   CET system with cascaded transformers.

transformer, which is located on the second joint of the robot. The transformer feeds the second axis drive in similar way as described above for the first machine. More transformers may be added to create arrangement of an AC bus throughout the robot. Similar system is applied for multilayer optical disc used in data storage systems [14]. However, the output power in optical disc is in the range of 20–30 mW, whereas in robots supply is 10–20 kW.

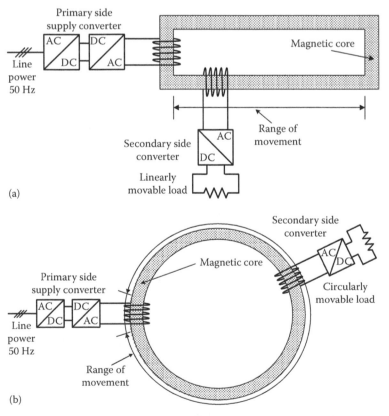

FIGURE 35.9   Basic configuration of CET system with sliding transformer; (a) for linear movement, (b) for circular movement.

35.6 CET Systems with Sliding Transformers

The contactless electrical energy delivery systems used in long distance are based on *sliding transformers* with long primary windings [4,22,24]. Basically, two configurations are applied: primary winding forming elongated loop as long as range of receiver movement is required (Figure 35.9a) or circular form for circular movement (Figure 35.9b). The output converter(s) and load(s) are directly connected to secondary winding placed on movable magnetic core.

The sliding magnetic core constructions enable movement of secondary winding along the primary winding loop (Figures 35.9 and 35.10). The sliding transformer gives possibility to construct long contactless, electrical energy delivery systems for mobile receivers. These transformer cores are composed of many strips of magnetic materials. Regarding magnetic and mechanical properties, the amorphous or nanocrystalline magnetic materials are preferable. However, when high dynamic properties of mobile receiver are required, some problem may appear because of core inertia. Heavy magnetic core is fixed with the energy receiver (Figure 35.10); therefore, it increases mass of the secondary side. The length of primary winding is in the range of 1–70 m and output power 1–200 kW [24].

35.7 CET Systems with Multiple Primary Winding

35.7.1 Introduction

Electronic devices like mobile phones, multimedia- and music players, laptops, and many more are used daily by countless people all around the world. Many of these devices are fitted with a battery, which allows them to operate independently and without drawing power continuously from the power utility network. However, these devices need to be periodically recharged, since their batteries can only store a finite amount of power. These devices operate with relatively low DC voltage levels (typically 5–12 V) compared to the high utility voltage of 240 V AC (120 V AC in the United States), and thus almost always require an AC-to-DC converter (also called a charger) to accomplish this. With a multitude of different devices available around the world, it makes for quite a lot of chargers. Also, most devices come with their own unique chargers. Using various different chargers with unique specifications and plugs can be bothersome and irritating. From a consumer point of view, charging these devices using only one universal charger would be great; it would be even better if this charger could charge multiple devices at a time without even plugging them into a socket, but by simply placing them close to the charger itself.

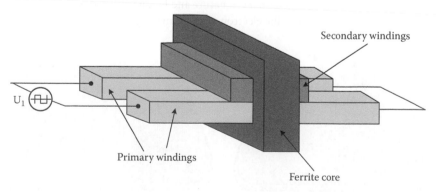

FIGURE 35.10 Example of sliding transformer construction for linearly moving secondary. (Sonntag, C. L. W. et al., Load position detection and validation on variable-phase contactless energy transfer desktops <http://repository.tue.nl/661798>. In *Proceedings of the IEEE Energy Conversion Congress and Exposition, ECCE 2009*, San Jose, 20–24 September, 2009, pp. 1818–1825. With permission.)

One increasingly popular technology for powering these devices without using adaptors is through the use of CET system. CET is the process in which electrical energy is transferred among two or more electrical devices through inductive coupling as opposed to energy transfer through conventional "plug and socket" connectors.

Different CET charging platforms for these applications have been proposed. One approach is based on a CET charging platform with a single spiral inductor. Here the inductive coupling between the primary inductor and a similar inductor installed into a mobile phone is used to transfer power. With only a single primary and secondary winding however, the phone needs to be placed in a very specific position so that the windings overlap each other exactly. The phone is thus restricted to a certain area wherein it can be charged.

In CET applications where a high degree of freedom regarding the placement of CET devices is required, the use of multiple primary inductors is a popular choice [1,35–37]. One such CET charging platform is presented in Figure 35.11a. Here multiple planar inductors arranged to form a matrix are embedded in a CET charging platform or into a section of an office table (Figure 35.11b). When small consumer electronic devices, like mobile phones, PDAs, multimedia- and music players, and even laptops fitted with similar inductors are placed on the platform, power is transferred from the CET platform (transmitter) to the CET devices (receivers) through inductive coupling.

35.7.2 Planar Inductor Windings

At the heart of any CET system lies the primary and secondary inductors that form the inductive link and allow power to be transferred between the transmitter and receiver. Their geometries play a vital role in determining the power transfer capability and efficiency of the system. In applications like these, where the size of the inductors, especially the secondary, is very limited, spiral planar winding inductors are often used. Hexagon spiral windings, in particular, use the available surface area very effectively and can be placed in a two-dimensional hexagonal lattice or matrix without any openings between the windings. Furthermore, the distribution of the magnetic field produced by these windings is unique as they produce a strong z-component, which make them especially suited for applications where the primary and secondary inductor placements are parallel to each other. Produced as copper tracks on (flexible) PCB, they can also be easily and cheaply manufactured.

Figure 35.12a shows an actual hexagon spiral winding produced on PCB; Figure 35.12b shows its graphical representation; and Figure 35.12c shows a matrix of hexagon spiral windings.

35.7.3 Electromagnetic Design

The design of a CET system is multidisciplinary in nature and is concentrated in various research fields. First, and perhaps most importantly, is the electromagnetic investigation. Here, methods for modeling

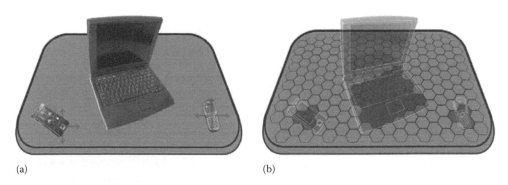

(a)   (b)

FIGURE 35.11 (a) CET receiver objects randomly placed on a CET-enabled platform for charging. (b) The CET platform showing multiple inductors underneath the CET platform.

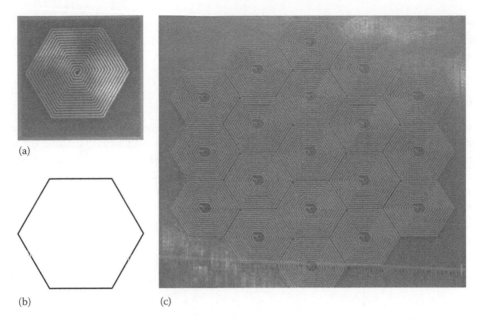

FIGURE 35.12 (a) An actual hexagon spiral winding produced as copper tracks on a PCB, and (b) its graphical representation used in this work. (c) A matrix of hexagon spiral windings. (Sonntag, C. L. W. et al., Load position detection and validation on variable-phase contactless energy transfer desktops <http://repository.tue.nl/661798>. In *Proceedings of the IEEE Energy Conversion Congress and Exposition, ECCE 2009*, San Jose, 20–24 September, 2009, pp. 1818–1825. With permission.)

the various important parameters of the CET inductors are created. These include methods for estimating the distribution of the magnetic field intensity, the self- and mutual winding inductances, and their AC resistances. For CET systems void of any soft magnetic materials, as the variable-phase CET desktop presented in [35], the magnetic vector potential and Biot–Savart methods can be used.

In CET systems where soft magnetic materials are used for shielding purposes, these methods can no longer be directly used. Here, the finite element method is a popular choice.

Figure 35.13a and b shows the distribution of the magnetic field as calculated by the developed model. Here Figure 35.13a shows the distribution of the magnetic field in a *xy*-plane parallel to the winding at a height of 1 mm above the winding, and Figure 35.13b shows the magnetic field at 5 mm.

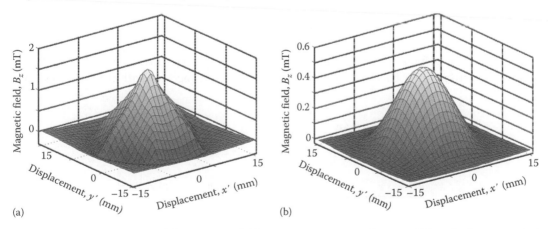

FIGURE 35.13 Distribution of the magnetic field *z*-component in a *xy*-plane parallel to the hexagon spiral inductor calculated at (a) 1 mm above the inductor, and (b) 5 mm above the inductor.

In general, CET platforms for these applications use planar inductors with radii between 10 and 30 mm, switching frequencies between 500 kHz and several megahertz. The air gaps are usually limited between 1 and 10 mm.

35.7.4 Power Electronics Implementation

The power electronics investigation focuses on the design and implementation of the power electronic systems needed to transfer and control the power delivered over the inductive link. Using the lumped parameters obtained from the electromagnetic investigation, the CET link is modeled as a lossy transformer. The coupling between the primary and secondary windings is often much weaker compared to traditional iron-cored transformers, and to increase the overall power transfer efficiency, resonance is often used.

The circuit equations that govern the transfer of power from the primary winding to the secondary load can be written in phasor notation:

$$V_A = j\omega L_A I_A + \frac{I_A}{j\omega C_A} + R_A I_A - j\omega M_{AB} I_B, \tag{35.1}$$

$$j\omega M_{AB} I_A = j\omega L_B I_B + \frac{I_B}{j\omega C_B} + R_B I_B + Z_L I_B. \tag{35.2}$$

where, as shown in Figure 35.14,

L_A and L_B are the inductances
R_A and R_B are the resistances
C_A and C_B are the series resonant capacitors
I_A and I_B are the winding currents of the primary and secondary circuits, respectively
V_A is the fundamental primary switching voltage
M_{AB} is the mutual inductance between the primary and secondary windings
Z_L is the secondary load, with V_L the voltage over it

If the series-series capacitor compensation is chosen to operate in resonance with their respective windings, Equations 35.1 and 35.2 can then be further rewritten as

$$V_A \approx \left(R_A + \frac{\omega^2 M_{AB}^2}{R_B + Z_L} \right) I_A. \tag{35.3}$$

The switching voltage is usually generated using a half-bridge inverter (see Figure 35.4b), a MOSFET driver, and a buck converter to power them. This, together with the series resonant

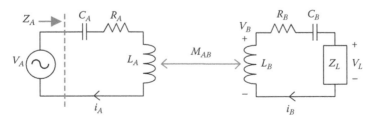

FIGURE 35.14   A simplified schematic diagram of power transfer to load Z_L.

capacitor, acts as a band-pass filter, allowing only the fundamental switching current to flow through in the primary winding.

To control the primary current, and to keep it constant during loaded and non-loaded operations, a PI or hysteresis current controller can also be implemented. This can be programmed in a micro-controller.

Coil commutation circuits are also used in certain CET platforms for switching the current into the different primary windings.

35.7.5 Operational Features

The CET platform operational features refer to certain technical and non-technical attributes and characteristics that need to be implemented into the system in order to make it user friendly, safe, and operate logically, within its working environment.

One of these important operational features is the location and authentication of valid CET devices placed on the CET platform. From a practical point of view, a CET platform used in an office environment might also contain objects that are not CET-enabled and should not be charged. Some of these objects can be metallic in nature, like a bunch of keys, a soft-drink can, pens, coins, etc. Exciting the primary windings close to these conductive objects could create eddy currents and result in undesired heating of the objects. Other objects, like magnets and ferrites, with high permeability can also interfere with the normal operation of a CET desktop or platform, and should thus be avoided. In [37] a method is devised to locate the position and distinguish between three possible object types placed on the CET platform. These are: metallic objects, magnetic objects, and valid CET devices. Figure 35.15a shows a CET platform filled with various CET and non-CET devices. Figure 35.15b shows an image of an actual metallic and magnetic materials placed on an implemented CET platform for testing. Here object *A* is a 60 mm aluminum office key, object *B* is a toroidal ferrite core, object *C* is two ferrite E-cores, and object *D* is a piece of copper plate. Due to their unique influences on the primary winding impedances, they can all be located and distinguished.

Other operational features which could be implemented in a CET platform include periodic scanning of the CET platform to locate newly placed CET devices. Also, audible user feedback tones can be implemented to notify the user when a new CET device is located and powered or when a device is fully charged or removed from the platform. From a safety perspective, the constant monitoring of the current and voltage levels of excited primary windings could indicate potential problems if the values fall outside certain allowed range.

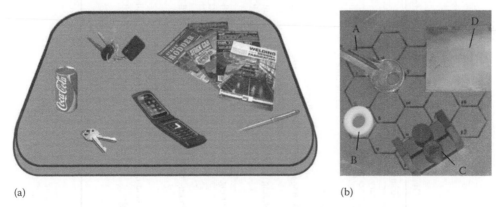

(a)   (b)

FIGURE 35.15 (a) A CET platform with various CET and non-CET objects randomly placed on its surface, and (b) image from an actual CET platform showing a few different metallic and ferrite materials that the CET platform can distinguish.

TABLE 35.3 Overview of Inductive Coupled CET Systems

| | Transformer Construction | | DC/AC Converter | | | | | | |
	Primary Side	Secondary Side	Topology	Freq. (kHz)	Output Power (W)	Output Voltage (V)	Air Gap Length (mm)	Max. Efficiency (%)	Application (–)
1	Single winding ferrite core	Single winding ferrite core	Full-bridge MOSFET/IGBT	20–100	1–150 kW	15–350	0.2–1; 1–300	≥ 90 ≥ 80	Battery chargers [11,15,17–20,38,39]
2	Single coreless	Triply ferrite core moving	Flyback MOSFET	125	0.1	3.0 DC		—	Biomedical [28,32]
3	Single winding ferrite	Double ferrite rotating	Full-bridge MOSFET	100	1000	54 DC	0.25–2	≥90	Biomedical [28,32]
4	Single winding coreless	Multi-winding ferrite core movable	Full-bridge MOSFET	80	2 × 240	240 AC 50 Hz	2–5	≈90	Multiple users; mobile devices [5,12,13]
5	Single winding coreless	Multi-winding coreless movable	Full-/half-bridge MOSFET	120	Each load 0.01	5–15	1000–7000	—	Industrial sensors and actuators, ABB [1,16–17,27,33–34]
6	Single winding ferrite core	Single winding ferrite core rotatable/linear	Full-bridge IGBT	20–40	10–60 kW	3 × 230 V AC	0.2–2	≥92	Robots and manipulators [10,14,25,26]
7	Multiple winding coreless (desktop)	Single winding coreless movable	Half-bridge MOSFET	100–400	30–300	12	2–5	≈90	Stationary (laptops, phone) or mobile (actuators) [3,6,7–9,35–37]

35.8 Summary and Conclusion

A brief review of basic CET systems, with special focus on inductively coupled solution, is given in this chapter. Several groups of application with typical specification are summarized in Table 35.3.

Key conclusions include the following:

- The CET systems are used in power range from milliwatts (biomedicine, sensors, actuators, etc.) till several hundred kilowatts (cranes, fast battery charging).
- The final efficiency achieved by inductively coupled CET systems is in the range of 60%–90% for low and high power applications, respectively.
- In high power (>1 kW), transformers with core winding are applied.
- In low power (<100 mW), air gap coupling and very high transmission frequency from 100 kHz till several (MHz) is preferred.
- For long-distance mobile loads, CET systems with sliding transformers are used.
- There is no one standard solution of CET system; every design has to take into account several specific parameters and user conditions.

References

1. Ch. Apneseth, D. Dzung, S. Kjesbu, G. Scheible, and W. Zimmermann. Introduction wireless proximity switches, *ABB Rev.*, 4, 2002, 42–49.
2. T. Bieler, M. Perrottek, V. Nguyer, and Y. Perriard. Contactless power and information transmission, *IEEE Trans. Ind. Appl.*, 38(5), 2002, 1266–1272.
3. J. de Boeij, E. Lomonova, J. Duarte, and A. Vandenput. Contactless energy transfer to a moving actuator, in *Proceedings of IEEE-ISIE 2006*, Montreal, Canada, 2006 (CD).
4. J. T. Boys and G. A. J. Elliot. An appropriate magnetic coupling co-efficient for the design and comparison of ICPT pickups, *IEEE Trans. Power Electron.*, 22(1), January 2007, 333–335.
5. D. Bess. Fuel cells and wireless power transfer at the 2008 International Consumer Electronics Show, *Bodo's Power, Electronics in Motion and Conversion*, Laboe, Germany, February 2008, pp. 16–17.
6. K. W. E. Cheng and Y. Lu. Development of a contactless power converter, in *Proceedings of IEEE ICIT'02*, Bangkok, Thailand, 2002 (CD).
7. G. A. Covic, G. Elliot, O. H. Stielau, R. M. Green, and J. T. Boys. The design of a contact-less energy transfer system for a people mover system, in *Proceedings of the International Conference on Power System Technology (PowerCon)*, Perth, Australia, Vol. 2, December 4–7, 2000, pp. 79–84.
8. G. A. Covic, J. T. Boys, M. L. G. Kisin, and H. G. Lu. A three-phase inductive power transfer system for roadway-powered vehicles, *IEEE Trans. Ind. Electron.*, 54(6), 2007, 3370–3378.
9. A. Ecklebe and A. Lindemann, Analysis and design of a contactless energy transmission system with flexible inductor positioning for automated guided vehicles, in *Proceedings of the 32nd Annual Conference of the IEEE Industrial Electronics Society, IECON 2006—*, Paris, France, November 7–10, 2006, pp. 1721–1726.
10. A. Esser and H. Ch. Skudelny. A new approach to power supply for robots, *IEEE Trans. Ind. Appl.*, 27(5), 1991, 872–875.
11. A. Esser. Contactless charging and communication for electric vehicles, *IEEE Ind. Appl. Mag.*, November/December, 1995, 4–11.
12. F. F. A. Van der Pijl, P. Bauer, J. A. Ferreira, and H. Polinder, Design of an inductive contactless power system for multiple users, in *Proceedings of IEEE IAS Annual Meeting*, Tampa, Florida, October 8–12, 2006, pp. 343–349.
13. F. F. A. Van der Pijl, P. Bauer, J. A. Ferreira, and H. Polinder. Quantum control for an experimental contactless energy transfer system for multiple users, in *Proceedings of IEEE IAS Annual Meeting*, New Orleans, Louisiana, September 23–27, 2007, pp. 1876–1883.

14. Y. Fujita, A. Hirotsune, and Y. Amano. Contactless power supply for layer-selection type record-able multi-layer optical disk, in *Proceedings of IEEE Optical Data Storage Topical Meeting*, Montreal, Quebec City, Canada, 23–26 April, 2006 (on CD).

15. J. G. Hayes, M. G. Egan, J. M. Murphy, S. E. Schulz, and J. T. Hall. Wide-load-range resonant converter supplying the SAE J-1773 electric vehicle inductive charging interface, *IEEE Trans. Ind. Appl.*, 35(4), 884–895, 1999.

16. J. Hirai, T. W. Kim, and A. Kawamura. Wireless transmission of power and information and information for cable less linear motor drive, *IEEE Trans. Power Electron.*, 15(1), 2000, 21–27.

17. J. Hirai, T. W. Kim, and A. Kawamura. Study on intelligent battery charging using inductive transmission of power and information. *IEEE Trans. Power Electron.*, 15(2), 2000, 335–344.

18. Y. Jang and M. M. Jovanovic. A contactless electrical energy transmission system for portable-telephone battery chargers, *IEEE Tran. Ind. Electron.*, 50(3), 520–527, 2003.

19. C.-G. Kim, D.-H. Seo, J.-S. You, J.-H. Park, and B. H. Cho, Design of a contactless battery charger for cellular phone, *IEEE Trans. Ind. Electron.*, 48(6), 1238–1247, 2001.

20. K. W. Klontz et al. An electric vehicle charging system with universal inductive interface, in *Proceedings of the PCC-Yokohama*, Yokohama, Japan, 19–21 April, 1993, pp. 227–232.

21. A. Kurs et al. Wireless power transfer via strongly couplet magnetic resonances, *Sciencexpress*, www.sciencexpress.org, Published online 7 June 2007; 10.1126/science.114354.

22. J. Lastowiecki and P. Staszewski. Sliding transformer with long magnetic circuit for contactless electrical energy delivery to mobile receivers, *IEEE Trans. Ind. Electron.*, 53(6), 2006, 1943–1948.

23. R. Mecke and C. Rathage. High frequency resonant converter for contactless energy transmission over large air gap, in *Proceedings of IEEE-PESC*, Aachen, Germany, 20–25 June, 2004, pp. 1737–1743.

24. J. Meins, R. Czainski, and F. Turki. Phase characteristics of resonant contactless high power supplies, *Przeglad Elektrotechniczny*, 11, 2007, 10–13.

25. A. Moradewicz and M. P. Kazmierkowski. Resonant converter based contactless power supply for robots and manipulators, *J. Autom. Mobile Robot. Intell. Syst.*, 2(3), 20–25, 2008.

26. A. Moradewicz and M. P. Kazmierkowski. FPGA based control of series resonant converter for contactless power supply, in *Proceedings of IEEE-ISIE Conference*, Cambridge, U.K., 2008 (on CD).

27. K. O'Brien, G. Scheible, and H. Gueldner. Analysis of wireless power supplies for industrial automation systems, in *Proceedings of IEEE-IECON'03*, Roanoke, VA (CD).

28. K. Onizuka et al. Chip-to-chip inductive wireless power transmission system for SiP applications, in *Proceedings of IEEE-CICC*, San Jose, California, 10–13 September, 2006, pp. 15-1-1–15-1-4.

29. K. D. Papastergiou and D. E. Macpherson. An airborne radar power supply with contactless transfer of energy—Part I: Rotating transformer; Part II: Converter design, *IEEE Trans. Ind. Electron.*, 54(5), October 2007, 2874–2893.

30. A. G. Pedder, A. D. Brown, and J. A. Skinner. A contactless electrical energy transmission system, *IEEE Trans. Ind. Electron.*, 46(1), 1999, 23–30.

31. M. Ryu, Y. Park, J. Baek, and H. Cha. Comparison and analysis of the contactless power transfer systems using the parameters of the contactless transformer, in *Proceedings of IEEE Power Electronics Specialists Conference*, Jeju, South Korea, 18–22 June, 2006 (CD).

32. N. Samad et al. Design of a wireless power supply receiver for biomedical applications, in *Proceedings of IEEE-APCCAS*, 2006, pp. 674–677.

33. G. Scheible, J. Endersen, D. Dzung, and J. E. Frey. Unplugged but connected: Design and implementation of a truly-wireless real-time sensor/actuator interface, *ABB Rev.*, 3 and 4, 2005, 70–73; 65–68.

34. G. Scheible, J. Schutz, and C. Apneseth. Novel wireless power supply system for wireless communication devices in industrial automation systems, in *Proceedings of IEEE-IECON*, Seville, Spain, 5–8 November, 2002.

35. C. L. W. Sonntag, E. A. Lomonova, and J. L. Duarte, Variable-phase contactless energy transfer desktop. Part I: Design, in *Proceedings of The International Conference on Electrical Machines and Systems, ICEMS 2008*, Wuhan, China, October 2008, pp. 1–6 (on CD).

36. C. L. W. Sonntag, E. A. Lomonova, and J. L. Duarte, Implementation of the Neumann formula for calculating the mutual inductance between planar PCB inductors, in *Proceedings of the 18th International Conference on Electrical Machines, ICEM 2008*, Vilamoura, Portugal, September 2008, pp. 1–6 (on CD).

37. C. L. W. Sonntag, J. L. Duarte, and A. J. M. Pemen, Load position detection and validation on variable-phase contactless energy transfer desktops <http://repository.tue.nl/661798>. In *Proceedings of the IEEE Energy Conversion Congress and Exposition, ECCE 2009*, San Jose, 20–24 September, 2009, pp. 1818–1825.

38. Ch-S. Wang, O. H. Stielau, and G. A. Covic. Design considerations for contactless electric vehicle battery charger, *IEEE Trans. Ind. Electron.*, 52(5), 2005, 1308–1313.

39. W. Lim, J. Nho, B. Choi, and T. Ahn. Low-profile contactless battery charger using planar printed circuit board. Windings as energy transfer device, *IEEE Trans. Ind. Electron.*, 2002, 579–584.

36

Smart Energy Distribution

Friederich Kupzog
Vienna University
of Technology

Peter Palensky
Austrian Institute
of Technology

Energy distribution has mainly been a domain of power electronics, but not of industrial communication systems in the past. However, in the light of climate change and strong efforts to reduce CO_2 emissions, the concept of "smart" energy distribution and "smart power grids" has emerged. Such smart grids are defined through the combination of power and information as well as communication technology, whereas the latter allow smart control algorithms to be applied. This chapter describes the motivations, drivers, developments, and implications of smart grids.

36.1 Evolution of Smart Energy Distribution

The electrical power systems as they exist today are the result of more than 100 years of technical development. Since Edison's first installations, the guiding design principle was that of large central power plants and a large number of small distributed consumers, which are connected with the generation sites by means of a power grid. For availability reasons, isolated systems got connected and large national and transnational grids evolved.

The electric power grid is a classical distributed system [TS06, p. 2], even if it is not consisting of computers but of electrical generators and loads. Seen from an abstract viewpoint, it consists of a large number of interacting entities (nodes or energy resources) that are connected with communication channels, the power lines. Communication is realized by influencing and observing ubiquitous physical parameters like power flows or the grid frequency. The growing power grid was, together with the telephony system that had its growing period nearly at the same time [CD04,C04], one of the first and largest distributed systems that have been designed by electrical engineers. This duality of key infrastructure systems, one for electrical energy and the other for communication, which began in the 1880s and 1890s, still exists today. However, while the telephone system has merged into the Internet and thus made through a number of revolutionary technological changes, the power system has remained with comparably little changes. The primary reason for this is that changes in the power grid are associated with extraordinary high investment costs (compared to those of smaller components in the telecom sector), resulting in long investment cycles of several decades.

The electric power grid cannot store electrical energy in large amounts for long times. The generated power has to match the consumed power at all times. In some way, the current demand has to be communicated to the generation sites, so that they can adjust to it (theoretically, also the generation amount could be communicated to the loads so that they would adjust to it). Further, a suitable way of load-sharing between the generators is needed, which can be seen as a protocol that determines in detail which generator reacts when and how much. All this has to work over hundreds or even thousands of kilometers. This technical challenge was brilliantly met without the use of any explicit data communication by the introduction of power-frequency control [K04]. Today, however, information and communication technology (ICT) is seen as one of the key concepts to maintain efficient and secure provision of electrical energy. The reason for this is a paradigm shift: In addition to the centralized generation, smaller generation units in larger numbers are more and more integrated into the grid, the so-called dispersed or distributed generation [JAC00]. Electricity generation from renewable sources is often only realizable as distributed generation. This is because compared to the traditional generation from fossil resources, the energy density of renewable energy sources is low. The number of generation units is comparably high, but they have a rather low individual power output compared to large centralized power plants. Units are set up at locations where the availability of the energy source is good (e.g., strong winds, flowing water). This results in generation units being scattered over the power grid infrastructure in a spatially distributed manner. It can be argued that a more distributed organization of the power infrastructure is also going along with an increase in reliability. Bottlenecks can be avoided and regional electricity supply can be maintained using local sources even if the backbone grid fails. This, however, can only be achieved with more complex system operation basing on an extensive use of automation infrastructures.

The integration of a high density of distributed generation into the existing power grids leads to a number of different issues. The two most prominent ones are the fact that generators connect to the grid at positions that have initially been designed for loads only and the volatile nature of generation from renewable sources. The strong growth of electricity generation in the medium voltage grid, where most of the installed distributed generation injects its power, leads to grid voltage problems [JAC00]. In times of low demand, the grid voltages at the feeding points reach the limits set by grid operators and regulation authorities, so that no more units could be installed without significant grid investments [KBP07]. The second issue is that due to the volatile nature of generation from renewable energy resources, it becomes more and more difficult to predict the amount of electricity generation. As a result, in future there will be a stronger need for balance energy [S02]. Especially the rising amount of distributed generation will significantly increase the uncertainty in balance prediction and therefore increase the need for more balance energy provision [SIF07].

Therefore, the rising energy demand and the necessity to increase energy efficiency and generation from renewable resources imply that power quality and security of supply can only be maintained (before even improved) if the basic management mechanisms of the power grid are adapted to the changing situation. This results in the need for large investments and a number of innovative technical solutions. The "smart power grid" serves as the umbrella for a harmonized and coordinated application of such new technical solutions, which heavily rely on ICT (Figure 36.1).

Smart power grids can be defined as in [LFP09]: "Smart grids are power grids, with a coordinated management based on bi-directional communication between grid components, generators, energy storages and consumers to enable an energy-efficient and cost-effective system operation that is ready for future challenges of the energy system."

There are two key drivers for the development of smart grids. The first is the integration of renewable energy resources into the power grids, as mentioned before. The second is the advance in ICTs. Innovations in communication systems, especially in the areas of signal processing and in production technologies, have resulted in the deployment of communication systems that enable comparably high data throughput for low costs. Wireless transmission of data is state of the art at this stage for remote control in medium voltage grids, a fact that shows that this technology has reached an adequate level of maturity and is accepted by the grid operators, who traditionally are very concerned about the reliability of information technology in the grid. The technological advance on the side of information technology on one hand and the beginning shortage in energy supply (including the need for CO_2 reduction) on the other hand also result in an economic paradigm shift: As costs for energy rise and costs for communication fall, the relation between both begin to change and the unit cost for energy is ultimately becoming higher than the unit cost for communication in many application areas (Figure 36.2).

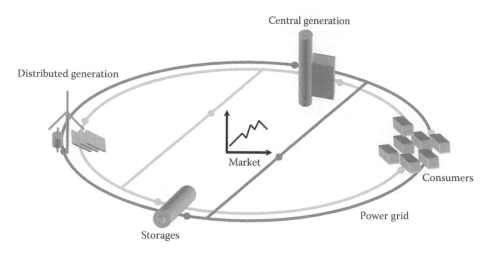

FIGURE 36.1 Definition picture of a smart grid. (National Technology Platform—Smart Grids Austria. See http://smartgrids at, visited 02/2009.)

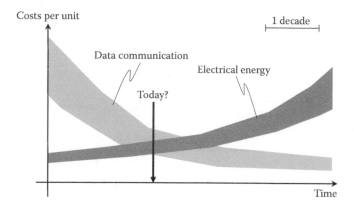

FIGURE 36.2 Estimated development of costs for communication and energy.

36.2 Key Concepts of Smart Grids

Smart grids are made out by the application of a number of key concepts that are discussed in the text following. All of them need an underlying industrial communication infrastructure, which can be affordable when deployed for multiple applications.

Supervisory control and data acquisition (SCADA) and substation automation are the traditional domain of narrowband automation infrastructure in power grids, especially in medium and high voltage grids. The SCADA infrastructure is used to connect network components in the field with control centers, so that they can be supervised and remote-controlled. The media used for SCADA can be very diverse and span from glass fiber over wireless solutions to distribution line carrier, a special form of power line communication for medium voltage grids.

Substations are currently the endpoint of the utilities' automation infrastructure. From this point on, there is usually no more online data coming from the grid. This situation is going to be changed by AMI (automated metering infrastructure) or similar initiatives. The problem, however, with existing substation automation and communication is that many are based on proprietary technology. This is now changed by IEC 61850 [IEC05]. Several existing efforts (EPRI UCA 2.0, IEC 60870) are converging to one unified standard for telemetry and remote control. It is specifically designed for local area networks like Ethernet and therefore more than an encapsulation of control commands. Using a substation bus (10 Mb/s to 1 Gb/s) and a process bus (100 Mb/s to 10 Gb/s), it connects meters, protocol relays and converters, human machine interfaces, and substation equipment. It uses a strictly object-oriented approach to model the application and has types and formats for all necessary data. Peers are either in a client-server or multicast relation and exchange information via messages. The protocol stack offers slim and real-time-capable transports as well as interoperable, IP-based services. The fast services have direct access to the data link layer, while all others use a sophisticated protocol stack that ensures easy management and commissioning.

Active distribution grids allow the integration of a high density of distributed generation in existing medium voltage infrastructure by an active control of generation power on the basis of voltage or power flow measurements at critical points in the grid. As shown in Figure 36.3, the main barrier for connecting new generators to the grid is that power feed-in increases the grid voltage at the feed-in point. The voltage has to be kept in an allowed band (e.g., ±10% of nominal value) by the grid operator in any case. The worst case occurs when there is no load but strong energy generation on the feeder, as shown in Figure 36.3, Example B. In an active distribution grid, the generation of the distributed generators is managed according to the voltage at critical points. If the voltage rises too high, reactive power management is performed. If this is not effective enough, even the active power can be curtailed [KBP07].

Such an active management of generated power in a medium voltage feeder is basically a form of multi-objective control. The challenge here is that sensors, controllers, and actuators are very far from each other. Voltage and power information has to be communicated over dozens of miles once every

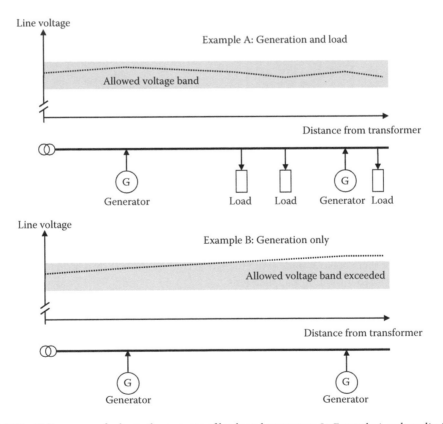

FIGURE 36.3 Voltage over a feeder in the presence of loads and generators. In Example A, voltage limits are not exceeded. In Example B, no load reduces the voltage, but generation is still active (e.g., in a night with strong winds). Without active generation management, the generators would be disconnected from the line by voltage protection switches.

6 s or so. The automation infrastructure used has to be highly reliable. Often, the protocols are transported over a variety of different media, depending on the available communication links.

Smart meters can be part of a smart grid, but they are not the same as smart grids. Although the origins of smart metering technology lies in remote meter reading, other aspects also play a role in smart meters than consumed kilowatt-hours. These are consumption profiles, power quality monitoring, and remote switching of loads.

It can be expected that power grids will in future be operated closer to their limits as it is currently the case. One of the reasons for that is that the pattern and kind of investments into the grid infrastructure will change due to the liberalization of power markets. For maintaining the high standards in power quality, today it is already considered to be necessary to monitor power quality variables such as voltage, flicker, and harmonics using online measurements in the grid. This is another driving factor for an increasing flood of online measurement data from the grid.

Smart metering systems can generate snapshots of the consumption state of the whole grid so that grid operators can examine in detail how much power was flowing to where in the moment of the snapshot. Therefore, smart meters are interconnected by means of communication links, usually narrowband power line communication to data concentrators at the transformer stations. From here, backbone networks (e.g., glass fiber) bring the data to control centers. Communication infrastructures are essential features of smart metering systems, but in many grids they are still nonexistent.

While in some countries smart meters are area-wide deployed, in other countries the debate about their benefits is still underway. On the positive side, these systems simplify the accounting and consumers can be promptly informed about their energy consumption. More data are available from the grid, and network development planning can be done on the basis of real data instead of worst-case models. Failure detection becomes easier and voltage bands can be used more efficiently. On the negative side, the costs are very high and it is basically assumed that the consumer will pay the price. Further, there is a severe lack of standards. Long-term reliability and data security questions are not yet completely answered.

One of the largest projects about smart metering in the United States is the AMI initiative; similar projects exist around the world [AMI08]. AMI is seen as the next step after remote metering: bidirectional communication between utilities, customers, and grid operators. The main concerns of AMI are affordable and secure acquisition and management of billing-relevant data. Although usually considered as non-critical transport, there are high requirements in accuracy and reliability. The expectations toward AMI are

- Reduced management costs for billing
- Increased insight into consumption patterns
- Reduced energy consumption because of immediate feedback information to the customer
- Identification and correction of leakages and losses
- Improved grid stability due to integration with demand response programs

The communication technology for AMI is typically separated into two parts. The wide area link is classically based on Internet technology, transported over UMTS, DSL, and other available Internet connections. Inside the customers' facilities, the choice often lies on wireless home networks, for instance based on ZigBee or Z-wave.

Automated demand response (DR), that is, remote switching of customers' appliances, plays a key role in most smart grid conceptions. A number of different terms are used in this context, such as demand side management, DR, or load shifting. The general idea is to gain influence on the load side of the power grid and make use of flexibilities in the timing of energy consumption. Such measures are seen as a supporting tool to match supply and demand under the condition of supply from fluctuating renewable energy resources, whose generation patterns do not match the demand curves. In countries with a high blackout frequency, DR can be a key concept to better distribute the available generation and transmission capacities.

Load shifting, in particular, does not aim to reduce energy consumption in long term, but to reduce peak loads by shifting consumption to off-peak times. As a short-term method, it allows improving the balance of supply and demand without the decrease of functionality for end users. Modern load shifting happens hidden and unrealized by the energy subscriber. The control of load shifts, distributed storages, and curtailment of interruptible loads are the major tools for this strategy. Based on their specific processing, properties, and energy storage functionality, there is the possibility to reschedule energy consumption of certain loads. Energy can either be stored in real energy storages, such as thermal storages, or as conceptual energy storages that can be exploited by rescheduling a process to a later point in time (load shift) [KR07]. Load shifting can be performed in various processes, for example, washing, cleaning, heating, chilling, and pumping. These electricity-consuming processes have, depending on the application, certain degrees of freedom in their time schedule. Many representatives of these classes of potentially shiftable loads can be found within buildings, especially large functional buildings.

DR is taking influence on loads, for the benefit of the grid or the customer's energy bill. While demand response can refer to either incentive-driven (e.g., by time-of-use tariffs) or automated means to change the behavior of electrical loads, automated demand response implies communication from "the grid" to energy-consuming appliances connected to the grid. The granularity of this communication could be reduced by addressing the whole building instead of every single part of equipment. Functional buildings account for a significant share of energy consumption and at the same time have usually large load shift potentials. Further, they are often equipped with building automation and control systems, which can interpret the demand response command from the grid and translate them into dedicated actions for the electrical consumers within a building. Therefore, the "building-to-grid" approach for demand response (Figure 36.4) is very promising.

Automated DR is mostly load shedding, setpoint adjustments, duty-cycling, and load shifting, done in an automated fashion. So-called "aggregators"—service providers, mediating between a utility and customers—typically install such systems to manage their customers' facilities. The open auto-DR specification is the first attempt to standardize these systems.* Its core component is the demand response automation server (DRAS, see Figure 36.5) and a set of standardized messages (events).

A utility or grid operator can issue a "demand response event" (e.g., a grid emergency) and depending on who subscribed to which demand response program, the DRAS distributes the required information to the clients (energy management control systems, aggregators, or directly the loads) in a secure and reliable manner. This concept can and will be extended to also serve developments like "Building2Grid" or "Plugin Hybrid Vehicles."

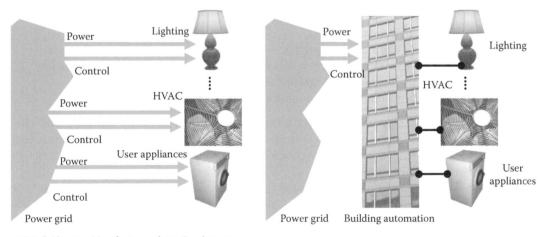

HVAC: Heating, Ventilation and Air Conditioning

FIGURE 36.4 "Appliance-to-grid" versus "building-to-grid" approach. The granularity of this communication could be reduced by addressing the whole building instead of every single part of equipment.

---

* http://drrc.lbl.gov

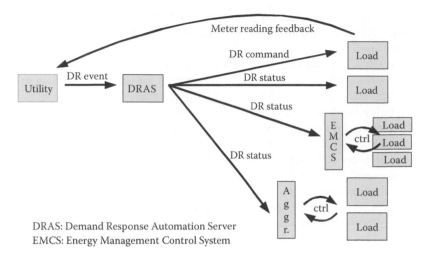

DRAS: Demand Response Automation Server
EMCS: Energy Management Control System

FIGURE 36.5 Open auto demand response architecture.

36.3 Smart Grid Vision

The future smart grid will be characterized by an intensified flow of information compared to the state-of-the-art power grid, where the dominating flow of energy is only accompanied by sporadic (monthly or yearly) meter readings. The communication system will be used for many different applications (Figure 36.6), which altogether justify the large investments needed to build the infrastructure. The challenges for this development are not only of technical and economic, but also of organizational nature. It is hoped that by means of a smart infrastructure, an efficient and cost-effective power grid operation is achievable that is ready for future challenges of the energy system.

The actual realization of smart grids, however, is currently hindered by a kind of hen-egg-problem. This is at least the case in central Europe. On one side, the common communication infrastructure is the defining element of a smart grid and serves for many smart applications. It is one of the key investments to be done. However, this investment is delayed because it seems to have no direct profit seen on its own. Also, the investor, who could be the grid operator, will only invest in an infrastructure that serves his or her own concerns (such as smart metering). Further services that he or she could provide for other stakeholders in the liberalized electricity market, such as plant operators, suppliers, or energy consumers, mostly cannot be paid regard due to the lack of standards of how such services should exactly look like.

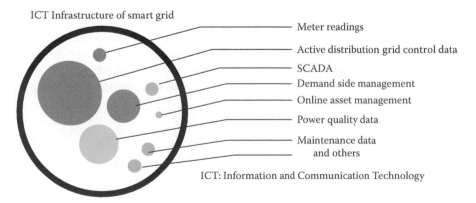

FIGURE 36.6 Anticipated application data flowing over smart grid communication infrastructures.

On the other side, the potential smart applications cannot or not efficiently be implemented without the basic communication infrastructure. Also, here, investments are delayed due to missing communication infrastructure.

A way out of this deadlock situation could be that a modular step-by-step strategy is developed that defines building blocks for basic and upgradable ICT services for smart grids, which can be used by all potential applications. Depending on the communication requirements of the applications, some applications can be used with the basic version and some only with upgraded versions of the ICT service set. For smart metering, for example, only small bandwidth connections with best-effort service are needed, while for active voltage control higher bandwidth and real-time service is required. Then, incentives for strategic investments by grid operators in the basic ICT services could be set. This approach would enable those applications with moderate service requirements to be realized and further lower the threshold for the implementation of applications with higher ICT service demand. The approach requires standardization of smart services and an upgradability of the ICT infrastructure to ensure that previous investments are still of use when the infrastructure is extended. It further requires actual communication components for smart grids that enable a step-by-step extension of the ICT infrastructure.

References

[AMI08] D.G. Hart, Using AMI to realize the smart grid. Conversion and delivery of electrical energy in the 21st century, *2008 IEEE Power and Energy Society General Meeting*, Pittsburgh, PA, 2008.

[C04] H.N. Casson, *The History of the Telephone*, Kessinger Publishing, Fairfield, IA, 2004, pp. 1–5, ISBN 1419166.

[CD04] J. Casazza and F. Delea, *Understanding Electric Power Systems*, IEEE Press, Piscataway, NJ, 2004, pp. 1–10, ISBN 0471446521.

[IEC05] R.E. Mackiewicz, Overview of IEC 61850 and Benefits, IEEE PES TD, 2005.

[JAC00] N. Jenkins, R. Allan, P. Crossley, D. Kirschen, and G. Strbac, *Embedded Generation*, The Institution of Electrical Engineers, London, U.K., 2000, ISBN 0 85296 774 8.

[K04] P. Kundur, *Power System Stability and Control*, McGrawHill, New York, 1994, pp. 581–592.

[KBP07] F. Kupzog, H. Brunner, W. Prüggler, T. Pfajfar, and A. Lugmaier, DG DemoNet-concept—A new algorithm for active distribution grid operation facilitating high DG penetration, *5th International IEEE Conference on Industrial Informatics (INDIN 2007)*, Vienna, Austria, July 2007.

[KR07] F. Kupzog and C. Roesener, A closer look on load management, *Fifth International IEEE Conference on Industrial Informatics (INDIN 2007)*, Vienna, Austria, July 2007.

[LFP09] A. Lugmaier, H. Fechner, W. Prueggler, and F. Kupzog, National technology platform—Smart grids Austria, *20th International Conference on Electricity Distribution*, Prague, Czech Republic, June 8–11, 2009 (to be published).

[S02] G. Strbac, *Quantifying the System Costs of Additional Renewables in 2020*, Manchester Centre of Electrical Energy, UMIST, Manchester, U.K., Technical Report, October 2002, Report to the U.K. Department of Trade and Industry (available online: www.berr.gov.uk/files/file21352.pdf).

[SIF07] J.A. Short, D.G. Infield, and L.L. Freris, Stabilization of grid frequency through dynamic demand control, *IEEE Transactions on Power Systems*, 22(3), 1284–1293, August 2007.

[TS06] A.S. Tanenbaum and M. van Steen, *Distributed Systems: Principles and Paradigms*, Pearson Prentice Hall, 2006, Upper Saddle River, NJ, ISBN 0132392275.

37

Flexible AC Transmission Systems

Jovica V. Milanović
*The University
of Manchester*

Igor Papič
University of Ljubljana

Ayman A.
Alabduljabbar
*King Abdul Aziz City
for Science and Technology*

Yan Zhang
ABB Corporate Research

37.1 Introduction

The electric power system is one of the largest and most complex man-made systems, encompassing billions of components, tens of millions of kilometers of transmission lines, and thousands of generators serving a diverse, huge number of consumers. The function of a power system is to generate electric energy economically and with the minimum ecological disturbance and to transfer this energy over transmission lines and distribution networks with the maximum efficiency and reliability for delivery to customers at virtually fixed voltage and frequency. The conventional power system structure is highly hierarchical where power flows are typically unidirectional (i.e., from generating plant to end user). Most of the electricity is still generated in large, centrally managed power plants, due to economies of scale and location of resources (e.g., coal, water), and transported in bulk to the areas through a meshed transmission grid (with built-in redundancy to increase security and availability) and finally delivered to the consumers through passive typically radial distribution systems. The end users are typically non-responsive consumers and do not participate in system operation.

The constant growth of demand for electricity over the years resulted in the appearance of "bottle necks" in electrical power transmission corridors (i.e., the amount of power that could be transferred from one point in the network to the other started to become more and more limited due to physical capacities of transmission lines) and in increasing difficulty to ensure appropriate regulation and control of key attributes of the electrical power transfer that would meet increasing customer demands for high quality of electricity supply (i.e., efficiency of power transfer, reliability of supply, and delivery to customers at almost fixed voltage and frequency). It was soon realized that relying solely on the construction of new primary plants to solve the problem is not a viable option for economic, environmental, and public acceptance reasons. The investments in new generating plants and transmission lines are extremely expensive and take years to complete. The transmission lines, generating plants, and large storage facilities have a visual and environmental impact that will limit their acceptance by the public.

The task of delivering the electricity in a reliable, secure, and controllable manner relied in the past, and still does to a large extent, on the supervisory control over the transmission system. This has been achieved largely by means of control equipment such as the tap-changing transformers, shunt and series reactors, the capacitor banks, the protection apparatus and systems, etc. With the growth of the transmission system, due to increased interconnections among different (geographical) regions and demand for operating the system under more stressed conditions in order to satisfy the growing and more versatile demand in more and more environmentally and economically aware surrounding, the ability of the conventional equipment to control the system became limited and the need for fast and frequent self-operating equipment that introduces additional degrees of freedom in system operation appeared.

The flexible alternating current transmission system (FACTS) devices, often referred to as flexible alternating current transmission systems (FACTS) represent a dependable solution to the task of advanced control of transmission systems in a new operating environment. The terms "FACTS" and "FACTS device" will be used interchangeably in the rest of this chapter as this is often the case in published literature. The Electric Power Research Institute (EPRI) introduced this technology originally during the 1980s, since then, it has been constantly evolving [1]. The FACTS technology is largely based on application of the high-voltage power electronic switches enabling, through fast and sophisticated control, modulation of key parameters that govern the operation of transmission systems including series and shunt impedances, currents, voltages, phase angles, and real and reactive power flows. In addition to their advanced control capabilities they are also environmentally friendly. They are built from safe materials and do not produce any kind of emissions or waste during the operation that may pollute the environment [2].

37.2 Basic FACTS Technology

Two different technical approaches influenced the development of FACTS devices. The first group of devices employs reactive elements or a tap-changing transformer with thyristor switches as controllable elements. The second group uses self-commutated static converters as controlled voltage sources.

Ever since the first thyristor has been developed, the main design objectives for power semiconductors were low switching losses, high switching rates, and minimal conduction losses. Subsequent innovations in FACTS technology were mainly driven by those objectives [1,3,4].

37.2.1 · Power Semiconductors

The most widely used power semiconductors in FACTS technology are a conventional thyristor, a gate turn-off (GTO) thyristor, and an insulated-gate bipolar transistor (IGBT).

The conventional thyristor is a device that can be triggered (turned on) with a pulse at the gate and afterward remains in conducting mode (turned on) until the next current zero-crossing. Therefore, only one switching per half cycle is possible. This property limits the controllability of the device. Conventional thyristors have the highest current and blocking voltage among conventionally used power semiconductors, therefore, fewer semiconductors are required for an application. They are used as switches for capacitors or inductors and are still the preferred devices for applications with the highest voltage and power levels. Thyristors are an essential part of the most frequently used FACTS devices including the biggest high-voltage DC (HVDC) transmission systems with voltage levels exceeding 500 kV and power ratings of several thousands MVA [4].

GTO thyristors are devices that can be switched off with a current pulse at the gate. They were developed to increase the controllability of conventional thyristors. This technology has grown very rapidly, and high-power GTOs are now available. The latter, are nowadays replaced by insulated-gate commutated thyristors (IGCT), which combine the advantage of a conventional thyristor, i.e., low conducting losses, with a low switching losses [4].

The IGBT can be switched on with a positive voltage signal and switched off by removing the voltage signal. Therefore, a very simple gate drive unit can be used to control the IGBT. It is becoming more and more important for FACTS technology. The voltage and power level of applications are being increased to 300 kV and 1000 MVA, respectively, for an HVDC transmission with voltage-sourced converters (VSCs) [4]. The capabilities of modern IGBTs make them applicable in the wide range of power system applications.

37.2.2 Thyristor-Based FACTS Devices

In high-power applications semiconductor elements are used primarily as switches. To accommodate switching in an AC system, two unidirectional conducting devices are connected in an antiparallel configuration. This group of FACTS controllers employs conventional thyristors. Most of them have a common characteristic that the necessary reactive power, required for compensation, is generated or absorbed by a traditional capacitor or reactor banks. The thyristor switches are used only to control the combined reactive impedance that these banks present to an AC system, as shown in Figure 37.1.

37.2.3 Converter-Based FACTS Devices

The second group of FACTS devices employs self-commutated VSCs. A VSC basically represents rapidly controllable static synchronous AC voltage source. Compared to the first group of FACTS devices, the VSC-based devices generally have superior performance characteristics. Figure 37.2 illustrates the basic scheme of a two-level three-phase VSC consisting of six power transistors with a parallel power diode connected in reverse and a capacitor on the DC side. A suitable switching pattern must be defined for the switch on and switch off capability. The simplest solution is the combination of a triangular voltage with a reference voltage as control variables, i.e., pulse width modulation (PWM) [1,4].

Three stages of an output voltage (plus, minus, and zero) can be achieved with a three-level converter. Whereas, increasing frequency of switching not only reduces harmonics injected into the network, but also increases the switching losses. A compromise between harmonic injection (and consequent

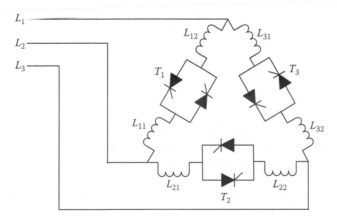

FIGURE 37.1   Schematic representation of a three-phase thyristor controlled reactor.

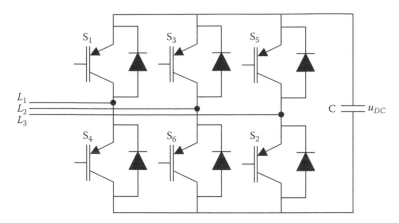

FIGURE 37.2 Basic scheme of a two-level voltage sourced converter.

requirement for output harmonic filters) and switching losses must be found for practical applications. In high-power applications more complex (multi-pulse) converters are used. In these converters, the number of used semiconductor elements increases the cost of devices more than reduction in switching losses, or harmonic injection, would justify.

37.3 Types and Modeling of FACTS

As already mentioned, FACTS can be generally classified into two main groups based on the physical nature of the control action that they provide. One group of devices acts on reactance (reactive imped-ances), i.e., changing the power flow through the control of impedance, while the other uses static con-verters as voltage sources to inject or absorb power in the power system as appropriate.

The first group includes devices such as the Static VAr Compensator (SVC), the thyristor-controller series capacitor (TCSC), and the thyristor-controlled phase-shifting transformer (TCPST). SVC acts on the voltage magnitude, TCSC acts on the transmission line impedance, and TCPST acts on the transmission angle. It can be seen that each device controls one of the three parameters governing power transmission.

The static synchronous compensator (STATCOM), the static synchronous series compensator, (SSSC, sometimes it is referred to as solid-state series controller), the unified power flow controller (UPFC), and the interline power flow controller (IPFC) make up the second group of FACTS devices. These are converter-based FACTS controllers, with a synchronous voltage source, VSC, capable of generating internal reactive power, as well as exchanging real power with the network. Similarly to the SVC, the STATCOM acts on the voltage and the SSSC acts effectively on the transmission reactance. The UPFC can influence any of the three parameters, while the IPFC is able to provide real power transfer as well as reactive series compensation.

Major types of FACTS used in transmission systems around the world are listed in the following and described briefly. The most widely used, or distinctive in the way of design and/or operation, are discussed in separate subsections.

- SVC is a shunt-connected device consisting of a combination of power electronics controlled reac-tor and capacitor whose major role is to regulate the voltage at the point of connection by varying injected reactive power through modulation of susceptance.
- STATCOM is a shunt-connected solid-state synchronous condenser that controls either bus volt-age magnitude or injected reactive power at the bus by varying its output current.
- SSSC is a series-connected solid-state synchronous condenser that controls either bus voltage magnitude or injected reactive power at one of the terminals of the series-connected transformer by varying its output current. (Similar to STATCOM but series connected.)

- TCSC is a series-connected device consisting of a series capacitor (which may also be thyristor controlled) paralleled by a thyristor-controlled reactor (TCR) whose major role is to ensure smooth variable series compensation through modulation of reactance and thus controls power transfer through the line. (Similar to SVC but series connected.)
- TCR is a shunt-connected thyristor-controlled reactor whose effective reactance is varied in a continuous manner by partial conduction of thyristor valve in order to regulate bus voltage magnitude.
- TCVR is a series-connected TCR whose effective reactance is varied in a continuous manner by partial conduction of thyristor valve in order to regulate voltage magnitude at one of the terminals of the series-connected transformer. (Similar to TCR but series connected.)
- TCPST is a series-connected TCR whose effective reactance is varied in a continuous manner by partial conduction of thyristor valve in order to regulate voltage phase angle at one of the terminals of the series-connected transformer (the main difference with respect to TCVR is in the way how the required voltage component is injected, i.e., in phase or at an angle with respect to line voltage).
- UPFC is a combination of a STATCOM and SSSC connected in a way that they share a common DC capacitor. It is able to control, simultaneously or selectively, the transmission line impedance, the bus voltage magnitude, and the real and reactive power flow through the line. Additionally, it can also provide independently controllable shunt reactive compensation.

37.3.1 Static VAr Compensators

The SVC is a shunt-connected device based on application of conventional thyristors. It is in principle a shunt-connected variable reactor that has the ability to exchange reactive power with an AC power system in a smoothly controlled manner in order to regulate the voltage at the point of connection. Assuming that the SVC is placed in the middle (typically) of the transmission line connecting two buses, the primary objective of the SVC is to maintain the voltage magnitude at the regulated point at a predetermined value. Voltage regulation is achieved by injecting the required amount of reactive power at the point of connection. In doing so, it indirectly increases the power transmission capability of the line. Therefore, the relationship between the maximum power transmitted and the SVC action is indirect since the actual control over the transmitted power can be achieved mainly by the series line impedance and the angle difference between the two buses. Figure 37.3 shows the basic structure of the SVC [3,5].

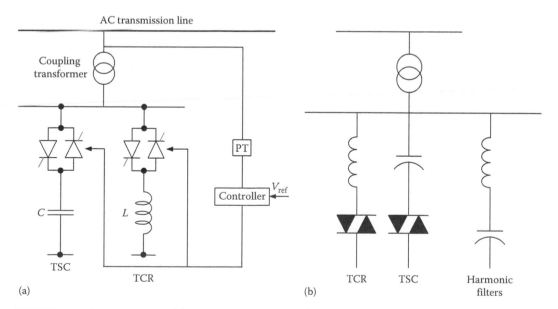

FIGURE 37.3   Basic structures of the SVC: (a) design without harmonic filter and (b) design with harmonic filter.

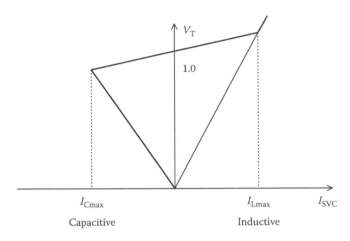

FIGURE 37.4 The SVC *V–I* characteristic.

It can be seen from Figure 37.3 that the SVC consists of two major parts that effectively control the injected (positive or negative) reactive power (*Q*). They are the thyristor-switched capacitor (TSC) and the TCR. Additionally, an integral part of an SVC may be a transformer if it is connected to a high-voltage bus. The SVC can be connected directly to the medium-voltage bus without a transformer. Finally, an SVC may, typically, contain a harmonic filter to ensure that there is no injection of unacceptably high harmonics in the network resulting from switching operation of thyristors. The coordinated switching between TSC and TCR controls the VAr output, which in turn maintains the bus voltage at the specified value. The basic SVC structure contains normally a number of TSCs and TCRs. A fixed-capacitor (FC) bank may also be included as a part of the SVC. There are more than 750 SVCs installed in the power networks around the world, and their ratings are generally problem dependant [3]. SVC ratings from +45/−30 MVAr to +425/−125 MVAr have been reported in the literature [3,6].

37.3.1.1 SVC *V–I* Characteristic

Figure 37.4 shows the SVC voltage–current characteristic at the regulated bus. It shows that the SVC regulates the bus voltage V_T by either injecting reactive current in case of voltage drop or absorbing reactive current in case of voltage increase. The SVC *V–I* characteristic is limited in both the capacitive and the inductive region. In the capacitive region, if the capacitive current reaches the limit, the SVC behaves like an FC, i.e., it is no longer controllable. Thus, a further drop in the voltage causes a significant reduction in the generated reactive power as it is proportional to the voltage squared. This behavior is one of the major drawbacks of an SVC, since it cannot support the voltage adequately when it is strongly required. On the other hand, if the inductive current reaches the limit, the SVC becomes a fixed reactor.

37.3.1.2 Modeling of SVC for Steady-State Studies

Since the function of the SVC is to exchange reactive power with the AC power system at the point of connection, it will appear from the AC power system perspective as an equivalent to a parallel connection of shunt capacitor and a shunt reactor as shown in Figure 37.5 [5].

Based on the previous description, the SVC can be modeled in steady state as a variable-shunt susceptance (β_{SVC}) as shown in Figure 37.6 [7].

The general form of power flow equations for the bus *k* is

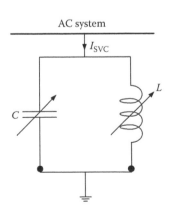

FIGURE 37.5 The SVC in steady state.

$$P_k = \sum_{m=1}^{N} V_k V_m \left[G_{km} \cos\theta_{km} + B_{km} \sin\theta_{km} \right] \qquad (37.1)$$

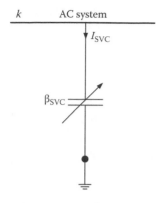

$$Q_k = \sum_{m=1}^{N} V_k V_m \left[G_{km} \sin\theta_{km} - B_{km} \cos\theta_{km} \right] \tag{37.2}$$

In (37.1) and (37.2), the calculation of the real and reactive power involves the consideration of all the branches connected to bus k. In practice, because of the presence of the SVC variable shunt susceptance at bus k only, the change in the power flow equations for bus k including the SVC appears only when $m = k$ as follows:

$$P = V_k^2 \left[G_{kk} \cos\theta_{kk} + (B_{kk} + \beta_{SVC}) \sin\theta_{kk} \right] \tag{37.3}$$

$$Q = V_k^2 \left[G_{kk} \sin\theta_{kk} - (B_{kk} + \beta_{SVC}) \cos\theta_{kk} \right] s \tag{37.4}$$

FIGURE 37.6 The model of SVC in steady state.

where G and B are the corresponding conductance and susceptance, of the network Y_{bus} matrix, respectively.

37.3.1.3 Modeling of SVC for Transient Studies

A simplified mathematical model of the SVC for transient studies is briefly discussed below. The equivalent circuit of the SVC is shown in Figure 37.7. The SVC is connected to the network bus u_i through the coupling impedance (R_p, L_p). It comprises a parallel combination of a TCR and a TSC. It is assumed that the capacitor is switched on. The model does not take into consideration dynamics related to capacitor switching. Under dynamic conditions the reactance of the TCR changes, which is modeled by varying the factor b_{TCR} that can take values between zero and one. The circuit also consists of resistance in series with the reactance in order to represent the losses of the TCR.

For the purposes of derivation of the mathematical model, per-unit system is adopted as specified by (37.5). i_B and u_B are the base current and voltage values, respectively, and ω_B is the synchronous angular speed of the fundamental network voltage component. Mathematical descriptions are given in the rotating d–q reference frame [8,9]. Under steady-state conditions, all the quantities in the model are constant values, which is suitable for the derivation of control algorithms:

$$i_p' = \frac{i_p}{i_B}; \quad i_{TCR}' = \frac{i_{TCR}}{i_B}; \quad u_{SVC}' = \frac{u_{SVC}}{u_B}$$

$$z_B = \frac{u_B}{i_B}; \quad L_p' = \frac{\omega_B L_p}{z_B}; \quad R_p' = \frac{R_p}{z_B}; \quad C_{TSC}' = \frac{1}{\omega_B C_{TCR} z_B}; \quad L_{TCR}' = \frac{\omega_B L_{TCR}}{z_B}; \quad R_{TCR}' = \frac{R_{TCR}}{z_B} \tag{37.5}$$

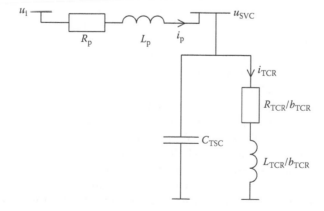

FIGURE 37.7 Equivalent circuit of the SVC for transient studies.

Considering the above assumptions and instantaneous values of variables shown in Figure 37.7, the state equations of the SVC in matrix format and using d–q coordinate system are given by

$$
\frac{d}{dt}
\begin{bmatrix}
i'_{pd} \\
i'_{pq} \\
i'_{TCRd} \\
i'_{TCRq} \\
u'_{SVCd} \\
u'_{SVCq}
\end{bmatrix}
=
\begin{bmatrix}
\dfrac{-R'_p\omega_B}{L'_p} & \omega & 0 & 0 & \dfrac{-\omega_B}{L'_p} & 0 \\[2ex]
-\omega & \dfrac{-R'_p\omega_B}{L'_p} & 0 & 0 & 0 & \dfrac{-\omega_B}{L'_p} \\[2ex]
0 & 0 & \dfrac{-R'_{TCR}\omega_B}{L'_{TCR}} & \omega & b_{TCR}\dfrac{\omega_B}{L'_{TCR}} & 0 \\[2ex]
0 & 0 & -\omega & \dfrac{-R'_{TCR}\omega_B}{L'_{TCR}} & 0 & b_{TCR}\dfrac{\omega_B}{L'_{TCR}} \\[2ex]
\omega_B C'_{TSC} & 0 & -\omega_B C'_{TSC} & 0 & 0 & \omega \\[2ex]
0 & \omega_B C'_{TSC} & 0 & -\omega_B C'_{TSC} & -\omega & 0
\end{bmatrix}
\begin{bmatrix}
i'_{pd} \\
i'_{pq} \\
i'_{TCRd} \\
i'_{TCRq} \\
u'_{SVCd} \\
u'_{SVCq}
\end{bmatrix}
+
\begin{bmatrix}
\dfrac{\omega_B}{L'_p}u'_{id} \\[2ex]
\dfrac{\omega_B}{L'_p}u'_{iq} \\[2ex]
0 \\[1ex]
0 \\[1ex]
0 \\[1ex]
0
\end{bmatrix}
$$

$$(37.6)$$

The response time of TCR power electronics is modeled by a first-order block (37.7) with the time constant T_{SVC}:

$$
\frac{d}{dt}b_{TCR} = -\frac{1}{T_{SVC}}b_{TCR} + \frac{1}{T_{SVC}}b_{TCRref}
\tag{37.7}
$$

It should be noted that (37.6) represents the simplified mathematical model of the SVC, whereas the overall dynamic behavior of the device primarily depends on the applied control system. The SVC can operate in susceptance control mode, but the application of voltage control is more common [1,6]. The output from the applied controller represents the value b_{TCRref}. Modeling of an SVC for electromagnetic transient studies requires detailed representation of all SVC nonlinearities (semiconductor elements), as well as different control and protection functions.

37.3.2 Static Compensator

The STATCOM is also a shunt-connected device as SVC; however, it is a VSC-based device that maintains the bus voltage by injecting a variable AC current through a transformer and generates required reactive power at its terminal. Its operational principle allows the exchange of both real and reactive power with the power system if it is equipped with a DC energy storage [3,6]. A STATCOM is in principle a static equivalent of the rotating synchronous condenser, which exchanges the reactive power with the system at much faster rate since there are no rotating parts (and such inertia) involved. The function of the STATCOM is the same as that of the SVC. It enables much more robust voltage support than the SVC. This increase in robustness comes with a higher price tag compared with the SVC of a similar size [6]. Further comparison with the SVC shows that the STATCOM is smaller in physical size (about 30%–40% reduction in overall size of the SVC [10]). The attainable response time and the bandwidth of the closed-voltage regulation loop of the STATCOM are also significantly better than those of the SVC. STATCOM can also incorporate suitable energy storage and thus facilitate real power exchange with the host AC system. This potential real power exchange capability provides a new tool for enhancing dynamic compensation, improving power system efficiency and, potentially, preventing power outages [1].

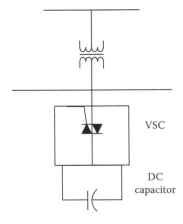

FIGURE 37.8 Basic structures of STATCOM.

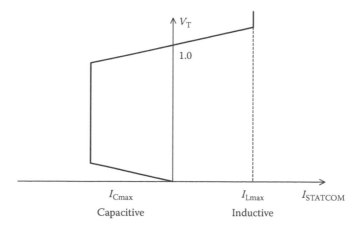

FIGURE 37.9 The STATCOM *V–I* characteristic.

The basic structure of STATCOM is shown in Figure 37.8. It consists of three-phase VSC, DC capacitor and transformer. The VSC uses self-commutated power electronic devices, GTO thyristors, or IGBTs to synthesize a voltage from a DC voltage source. Generally, a GTO thyristor is used for higher voltage applications and IGBT is for lower voltages. The capacitor on the DC side acts as a DC voltage source [7].

(*Note*: An SSSC is a series-connected FACTS device very similar to STATCOM, a part from its series connection. The SSSC is in fact a solid-state synchronous condenser that controls either bus voltage magnitude or injected reactive power at one of the terminals of the series-connected transformer. The injected voltage is perpendicular to the line current. The SSSC acts as a controllable voltage source whose voltage magnitude is controlled independently of the line current. By exchanging only reactive power with the system, the SSSC affects primarily the real power flow through a transmission line [11–14].)

37.3.2.1 STATCOM *V–I* Characteristic

Figure 37.9 shows the *V–I* characteristic of STATCOM. It can be seen from the figure that even at very low voltage, unlike the SVC, the STATCOM can continue to operate with rated leading (or lagging) current and inject/absorb required reactive power. In contrast, the current injection of an SVC is proportional to terminal voltage and it reduces at lower voltages with voltage squared. STATCOM is therefore able to provide better voltage support than the SVC when the voltage becomes severely depressed.

37.3.2.2 Modeling of STATCOM for Steady-State Studies

Referring to its equivalence with a rotating synchronous condenser, STATCOM can be modeled as a conventional synchronous generator (see Figure 37.10) with zero real power output in series with the impedance of the connecting transformer ($\underline{Z}_T$). If higher flexibility in modeling is required, then it can be represented as a variable voltage source ($\underline{E} = E\angle\theta = E\cos\theta + jE\sin\theta$) whose magnitude ($E$) and phase angle ($\theta$) can be adjusted in each phase separately. In such a case, the limits should be set for voltage magnitude in accordance with the size of the DC capacitor, while the voltage phase angle can take any value between 0° and 360° [7].

Alternatively, STATCOM can be also represented as a variable current source for steady-state short-circuit calculations [15]. Since its contribution to the system varies with its size and the connected bus voltage, the injected current has to be calculated carefully beforehand [15].

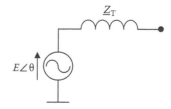

FIGURE 37.10 The model of STATCOM for steady state studies.

37.3.2.3 Modeling of STATCOM for Transient Studies

The equivalent circuit of the STATCOM model for transient studies is shown in Figure 37.11. Sinusoidal voltage sources are connected to the

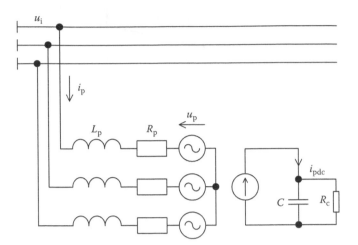

FIGURE 37.11 Equivalent circuit of STATCOM.

network through the reactance of the coupling transformer. The circuit also consists of resistance in series with the reactance in order to represent the losses of the transformer. The current magnitude of the shunt-connected device depends on the difference between the system voltage and the adjustable output voltage of the converter. The DC circuit is represented by a current source connected to capacitor C. The shunt connection of resistance R_c enables the representation of losses in the DC circuit.

Similarly as before, a per-unit system is adopted as specified by (37.8), i_B and u_B are the base values of current and voltage, respectively, and ω_B is the synchronous angular speed of the fundamental network voltage component. Mathematical descriptions are also given in the rotating d–q reference frame [8,9]:

$$i'_p = \frac{i_p}{i_B}; \quad u'_p = \frac{u_p}{u_B}; \quad u'_i = \frac{u_i}{u_B}; \quad u'_{dc} = \frac{u_{dc}}{u_B}$$

$$z_B = \frac{u_B}{i_B}; \quad L'_p = \frac{\omega_B L_p}{z_B}; \quad R'_p = \frac{R_p}{z_B}; \quad C' = \frac{1}{\omega_B C z_B}; \quad R'_c = \frac{R_c}{z_B}$$

(37.8)

Both components (d and q) of the converter output voltage depend on the DC voltage. A set of equations that defines these voltage components is given by (37.9), where k_p is a coefficient that includes the transformer ratio, relates the DC and AC voltage, and takes into account converter type. Angle δ_p represents the phase shift of the converter output voltage from the reference position, and the control parameter (converter factor) m_p can take any value between zero and one.

$$u'_{pd} = u'_{dc} k_p m_p \cos\delta_p = u'_{dc} k_p d_{pd}$$

$$u'_{pq} = u'_{dc} k_p m_p \sin\delta_p = u'_{dc} k_p d_{pq}$$

(37.9)

As can be observed from (37.10), both adjustable parameters of the converter output voltage, m_p and δ_p, are used to determine the average switching function d_{pd} in the direction of the d-axis and d_{pq} in the direction of the q-axis (37.3):

$$d_{pd} = m_p \cos\delta_p$$

$$d_{pq} = m_p \sin\delta_p$$

(37.10)

DC circuit dynamics is described by (37.11). The initial value of the DC voltage depends on the structure of the converter. It is determined in such a way that the device with $m_p = 1$ operates in a capacitive area—the converter output voltage is higher than the network voltage.

$$\frac{d}{dt}u'_{dc} = -\omega_B C'\left(i'_{pdc} + \frac{u'_{dc}}{R'_C}\right) \tag{37.11}$$

The balance equation for the real power is given by

$$u'_{dc}i'_{pdc} = \frac{3}{2}\left(u'_{pd}i'_{pd} + u'_{pq}i'_{pq}\right) \tag{37.12}$$

The influence of the converter operation on the small DC capacitor is described using the DC current source as shown by

$$i'_{pdc} - \frac{3}{2}\left(k_p d_{pd} i'_{pd} + k_p d_{pq} i'_{pq}\right) \tag{37.13}$$

Finally, the state equations of the STATCOM in the matrix format and using the d–q coordinate system are given by

$$\frac{d}{dt}\begin{bmatrix} i'_{pd} \\ i'_{pq} \\ u'_{dc} \end{bmatrix} = \begin{bmatrix} \dfrac{-R'_p \omega_B}{L'_p} & \omega & \dfrac{-k_p \omega_B}{L'_p}d_{pd} \\ -\omega & \dfrac{-R'_p \omega_B}{L'_p} & \dfrac{-k_p \omega_B}{L'_p}d_{pq} \\ \dfrac{3k_p \omega_B C'}{2}d_{pd} & \dfrac{3k_p \omega_B C'}{2}d_{pq} & \dfrac{-C'\omega_B}{R'_c} \end{bmatrix}\begin{bmatrix} i'_{pd} \\ i'_{pq} \\ u'_{dc} \end{bmatrix} + \begin{bmatrix} \dfrac{\omega_B}{L'_p}u'_{id} \\ \dfrac{\omega_B}{L'_p}u'_{iq} \\ 0 \end{bmatrix} \tag{37.14}$$

As it can be observed, there are two adjustable parameters (m_p, δ_p) and three state variables. Only two variables can be controlled independently. STATCOM does not have large energy storage capacity, thus only reactive power can be exchanged with the system in a steady state. The reactive current component can be controlled independently and the other free parameter is used for maintaining constant DC voltage across the DC capacitor.

An even more characteristic, or classical, mode of operation of STATCOM is when only one controllable parameter is used. The control factor m_p is set to 1 and with the time-limited phase shift, δ_p, a certain amount of the energy can be absorbed or sent to the network. In this way, the magnitude of the DC voltage across the capacitor can be controlled, and consequently the magnitude of the converter output AC voltage or the reactive current component.

The mathematical model of the STATCOM with sinusoidal sources can be used to derive an appropriate control system [8,16–18]. The outputs from the applied controller represent the values of the switching functions d_{pd} and d_{pq}. Mathematical modeling of STATCOM for electromagnetic transient studies requires detailed representation of the VSC and a relevant control system, where limits of controllable parameters would also need to be considered.

37.3.3 Thyristor-Controlled Series Capacitor

The TCSC is one of the series-connected FACTS devices based on thyristor valves. The primary function of a TCSC is to vary the line impedance through smooth modulation of reactance by appropriately switching off the thyristor valves. This ability to alter the series reactance of the transmission line offers

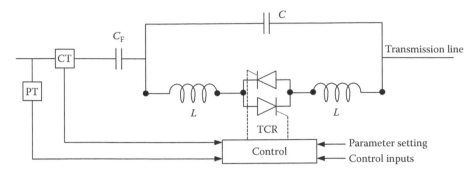

FIGURE 37.12 The TCSC basic scheme.

a direct control over the transmitted power across the line. The TCSC therefore represents an excellent series compensator that is particularly useful for long transmission lines [1].

Figure 37.12 shows one of the basic TCSC designs [3]. It consists of a series capacitor C in parallel with a TCR. The degree of series compensation provided to the line is controlled by the thyristor conduction period. The practical applications of TCSC may involve several cascading modules of this type [6].

The installation of the TCSCs started in early 1990s with three devices installed in the USA [3]. The rating of the TCSC depends on the total power transfer across the transmission line that it is installed in. Recent examples of TCSC installations include a 107.5 MVAr device in Brazil in a network with total generation capacity of 62 GW and a 123 MVAr device in Sweden [6].

The ability of the TCSC to change the transmission line impedance can be used to achieve several tasks. In order to control the targeted parameters in the transmission line, such as the real power flow, the control law changes the reference signal of the TCSC controller in order to generate the desired value of the series compensation. Constant power (CP) and constant angle (CA) controls are two principle features of TCSC control [6]. An example of TCSC application is shown in Figure 37.13.

In case of CP control, the objective is to maintain the desired level of the real power flow in the TCSC compensated line (P_{23k}) by changing the TCSC variable reactance. The desired real power flow level (P_{23ko}) in line 2–3k is normally selected as the reference signal.

The CA type of control is applied when there are predefined transmission paths along the TCSC compensated line, e.g., as shown in Figure 37.13 (line 2–3m). The control law in this case is to keep the total power transmitted across the parallel circuit (line 2–3m) constant. This is achieved by changing the TCSC series compensation such that any real power change in line 1–2 is absorbed. The reference signal in this case is $P_{12o} + P_{23ko}$. Assuming that the voltage magnitudes at bus 2 and bus 3 are regulated, that the transmission line resistance is negligible, and that the impedance of line 2–3m is fixed, the real power flow constancy in line 2–3m entails that the difference in voltage angles of bus 2 and bus 3 is constant [6].

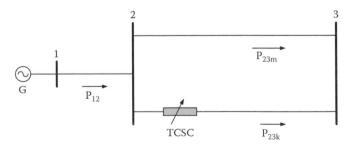

FIGURE 37.13 The example of TCSC application.

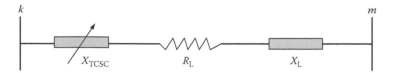

FIGURE 37.14 The model of TCSC in steady state.

37.3.3.1 Modeling of TCSC for Steady-State Studies

The TCSC is modeled in the steady-state studies as a variable capacitive reactance in series with the impedance of the compensated transmission line [1,3,6,7] as shown in Figure 37.14.

Usually, the value of the X_{TCSC} is only a fraction of the transmission line reactance X_L. After adding the TCSC to a transmission line, the effective impedance of the line (neglecting the resistance) becomes

$$X_{eff} = X_L - X_C = (1-k)X_L \tag{37.15}$$

where k is the degree of series compensation given by

$$k = \frac{X_C}{X_L} \quad 0 \le k < 1 \tag{37.16}$$

As far as the power flow equations are concerned, the inclusion of TCSC in the network also imposes some changes in power flow equations. Figure 37.15 shows the lumped π-equivalent model of the transmission line used to illustrate the required changes.

The presence of the TCSC between bus k and bus m will introduce changes in the original Y_{bus} matrix of the system, and these changes will appear in the power flow equations. Assuming that there are only two connections at each bus (bus k and bus m) as shown in Figure 37.15, the injected powers at bus k and bus m become

$$P_k = G_{kk}V_k^2 + V_kV_m\left(G_{km}\cos\theta_{km} + B_{km}\sin\theta_{km}\right) \tag{37.17}$$

$$Q_k = -B_{kk}V_k^2 + V_kV_m\left(G_{km}\sin\theta_{km} - B_{km}\cos\theta_{km}\right) \tag{37.18}$$

$$P_m = G_{mm}V_m^2 + V_mV_k\left(G_{mk}\cos\theta_{mk} + B_{mk}\sin\theta_{mk}\right) \tag{37.19}$$

$$Q_m = -B_{mm}V_m^2 + V_mV_k\left(G_{mk}\sin\theta_{km} - B_{mk}\cos\theta_{mk}\right) \tag{37.20}$$

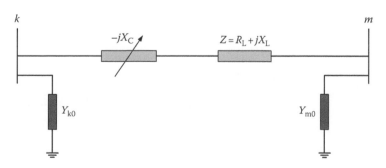

FIGURE 37.15 The transmission line model including TCSC ($-jX_C$).

where

$$Y_{kk} = Y_{k0} + Y_{km} \tag{37.21}$$

$$Y_{mm} = Y_{m0} + Y_{mk} \tag{37.22}$$

$$Y_{k0} = G_{k0} + jB_{k0} \tag{37.23}$$

$$Y_{m0} = G_{m0} + jB_{m0} \tag{37.24}$$

$$Y_{km} = Y_{mk} = G_{km} + jB_{km} \tag{37.25}$$

$$G_{km} = \frac{R_{\mathrm{L}}}{R_{\mathrm{L}}^2 + \left(X_{\mathrm{L}} - X_{\mathrm{C}}\right)^2} \tag{37.26}$$

$$B_{km} = -\frac{X_{\mathrm{L}} - X_{\mathrm{C}}}{R_{\mathrm{L}}^2 + \left(X_{\mathrm{L}} - X_{\mathrm{C}}\right)^2} \tag{37.27}$$

37.3.3.2 Modeling of TCSC for Transient Studies

The equivalent circuit of the TCSC for development of mathematical model for transient studies is shown in Figure 37.16. The line inductance is represented by L_{s}. The circuit also consists of series resistance R_{s} in order to represent line losses. The TCSC is a capacitive reactance compensator consisting of a capacitor bank (C_{FC}) in parallel with a TCR in order to provide a smoothly variable series capacitive reactance. The values for C_{FC} and L_{TCR} are chosen such that the factor b_{TCR} can take any value between 0 (inductive TCSC reactance) and 1 (capacitive TCSC reactance), whereby values around the resonance point should be avoided. The voltage drop across the TCSC reactance due to line current i_{s} is denoted by u_{TCSC}.

The per-unit system (37.28) was adopted for the variables in the model. Again i_{B} and u_{B} are the base values for current and voltage, respectively, and ω_{B} is the synchronous angular speed of the fundamental network voltage component [6].

$$i_{\mathrm{s}}' = \frac{i_{\mathrm{s}}}{i_{\mathrm{B}}}; \quad i_{\mathrm{TCR}}' = \frac{i_{\mathrm{TCR}}}{i_{\mathrm{B}}}; \quad u_{\mathrm{TCSC}}' = \frac{u_{\mathrm{TCSC}}}{u_{\mathrm{B}}}; \quad u_{\mathrm{i}}' = \frac{u_{\mathrm{i}}}{u_{\mathrm{B}}}; \quad u_{\mathrm{o}}' = \frac{u_{\mathrm{o}}}{u_{\mathrm{B}}}$$

$$z_{\mathrm{B}} = \frac{u_{\mathrm{B}}}{i_{\mathrm{B}}}; \quad R_{\mathrm{TCR}}' = \frac{R_{\mathrm{TCR}}}{z_{\mathrm{B}}}; \quad L_{\mathrm{TCR}}' = \frac{\omega_{\mathrm{B}} L_{\mathrm{TCR}}}{z_{\mathrm{B}}}; \quad C_{\mathrm{FC}}' = \frac{1}{z_{\mathrm{B}} \omega_{\mathrm{B}} C_{\mathrm{FC}}}; \quad L_{\mathrm{s}}' = \frac{\omega_{\mathrm{B}} L_{\mathrm{s}}}{z_{\mathrm{B}}}; \quad R_{\mathrm{s}}' = \frac{R_{\mathrm{s}}}{z_{\mathrm{B}}} \tag{37.28}$$

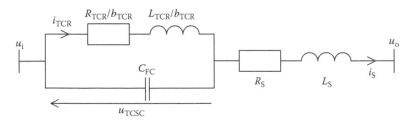

FIGURE 37.16 Equivalent circuit of TCSC.

The mathematical model is developed in the rotating d–q reference frame as before. Considering the instantaneous variables shown in Figure 37.16, the state equations of the TCSC are given by

$$
\frac{d}{dt}
\begin{bmatrix}
i'_{sd} \\
i'_{sq} \\
i'_{TCRd} \\
i'_{TCRq} \\
u'_{TCSCd} \\
u'_{TCSCq}
\end{bmatrix}
=
\begin{bmatrix}
\frac{-R'_s\omega_B}{L'_s} & \omega & 0 & 0 & \frac{-\omega_B}{L'_s} & 0 \\
-\omega & \frac{-R'_s\omega_B}{L'_s} & 0 & 0 & 0 & \frac{-\omega_B}{L'_s} \\
0 & 0 & \frac{-R'_{TCR}\omega_B}{L'_{TCR}} & \omega & b_{TCR}\frac{\omega_B}{L'_{TCR}} & 0 \\
0 & 0 & -\omega & \frac{-R'_{TCR}\omega_B}{L'_{TCR}} & 0 & b_{TCR}\frac{\omega_B}{L'_{TCR}} \\
\omega_B C'_{FC} & 0 & -\omega_B C'_{FC} & 0 & 0 & \omega \\
0 & \omega_B C'_{FC} & 0 & -\omega_B C'_{FC} & -\omega & 0
\end{bmatrix}
\begin{bmatrix}
i'_{sd} \\
i'_{sq} \\
i'_{TCRd} \\
i'_{TCRq} \\
u'_{TCSCd} \\
u'_{TCSCq}
\end{bmatrix}
+
\begin{bmatrix}
\frac{\omega_B}{L'_s}\left(u'_{id}-u'_{od}\right) \\
\frac{\omega_B}{L'_s}\left(u'_{iq}-u'_{oq}\right) \\
0 \\
0 \\
0 \\
0
\end{bmatrix}
$$

$$(37.29)$$

The response time of the TCSC power electronics is modeled using relationship (37.30) with the time constant T_{TCSC}:

$$
\frac{d}{dt}b_{TCR} = -\frac{1}{T_{TCSC}}b_{TCR} + \frac{1}{T_{TCSC}}b_{TCRref}
$$

$$(37.30)$$

As in the case of the SVC modeling, the overall dynamic behavior of the TCSC primarily depends on the applied control system. As mentioned before, the TCSC can operate in a constant-current control mode or, in the case of parallel transmission paths, in a constant-angle control mode [1,6]. The output from the controller is the reference value for the factor b_{TCR}, i.e., b_{TCRref}.

37.3.4 Thyristor-Controlled Voltage Regulator and Thyristor-Controlled Phase Shifting Transformer

The TCVR and TCPST are the other two types of FACTS devices that use thyristor valves as basic building components to perform fast switching, and therefore ensure smooth and continuous control of desired variable. They are very similar in design and therefore will be discussed together.

A TCVR is series-connected TCR whose effective reactance is varied in continuous manner by partial conduction of thyristor valve in order to regulate voltage magnitude at one of the terminals of the series-connected transformer. (Its function is very similar to that of TCR except that it is connected in series.) The role of TCPST on the other hand is to regulate the voltage phase angle at one of the terminals of the series-connected transformer. The main difference between TCVR and TCPST is in the way how the required voltage component is injected, i.e., in phase (TCVR) or at an angle (TCPST) with respect to line voltage [1], as illustrated in Figure 37.17. The principle of operation of both, the TCVR and the TCPST, is based largely on the principles of operation of the classical on-load tap-changing transformer [1].

A significant disadvantage of these two devices is that they cannot generate nor absorb reactive power. After they introduce desired changes in the bus voltage magnitudes and phase angles, it is left largely to the power system to handle the resulting change in the reactive power demand, so, a lack of reactive power support in the system in such cases might lead to a system security problem [1,3].

37.3.4.1 Modeling of TCVR and TCPST for Steady-State Studies

Figure 37.18 shows a model of TCVR and TCPST in a transmission line where a represents the turns ratio and α represents the phase shift. The original line is between k_1 and m; however, after inserting the TCVR (with $a \neq 0$, $\alpha = 0$) or TCPST (with $a = 0$, $\alpha \neq 0$), a new bus k is added to facilitate the demonstration of the change in the power flow equations.

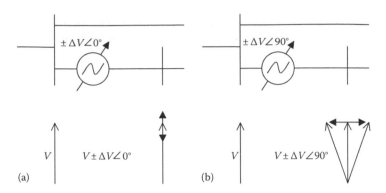

FIGURE 37.17 (a) Voltage magnitude and (b) phase angle regulation by the TCVR and TCPST, respectively.

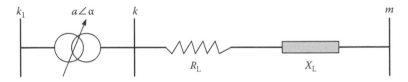

FIGURE 37.18 A general model of TCVR and TCPST.

Following the connection of TCVR and TCPST the injected real and reactive power equations for bus k and bus m become [7,19]

$$P_k = G_{kk}a^2V_k^2 + aV_kV_m\left(G_{km}\cos(\theta_{km}+\alpha)+B_{km}\sin(\theta_{km}+\alpha)\right) \qquad (37.31)$$

$$Q_k = -B_{kk}a^2V_k^2 + aV_kV_m\left(G_{km}\sin(\theta_{km}+\alpha)-B_{km}\cos(\theta_{km}+\alpha)\right) \qquad (37.32)$$

$$P_m = G_{mm}V_m^2 + V_maV_k\left(G_{mk}\cos(\theta_{mk}+\alpha)+B_{mk}\sin(\theta_{mk}+\alpha)\right) \qquad (37.33)$$

$$Q_m = -B_{mm}V_m^2 + V_maV_k\left(G_{mk}\sin(\theta_{km}+\alpha)-B_{mk}\cos(\theta_{mk}+\alpha)\right) \qquad (37.34)$$

37.3.4.2 Modeling of TCVR and TCPST for Transient Studies

The equivalent circuit of the TCVR/TCPST for transient studies is shown in Figure 37.19. The transformer series branch inductance together with the line inductance is represented by L_s. The device is

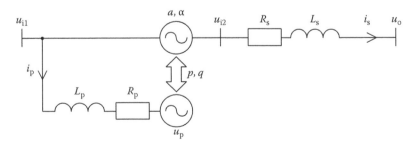

FIGURE 37.19 Equivalent circuit of TCVR/TCPST.

represented in the model by two controllable sinusoidal voltage sources with balanced real and reactive power exchange. Dynamics in the circuit between the two branches are neglected. The controllable parameter a or α of the series voltage source enables the main function of the device—voltage or power flow control. The zero exchange of the real and reactive power with the network is maintained by shunt-connected voltage source (i_p, u_p). The circuit also contains a series resistance in order to account for the line losses. The inductance and the losses of the shunt branch are represented by L_p and R_p, respectively.

The adopted per-unit system for model development is given by

$$i_p' = \frac{i_p}{i_B}; \quad i_s' = \frac{i_s}{i_B}; \quad u_{i1}' = \frac{u_{i1}}{u_B}; \quad u_{i2}' = \frac{u_{i2}}{u_B}; \quad u_o' = \frac{u_o}{u_B}$$

$$z_B = \frac{u_B}{i_B}; \quad L_p' = \frac{\omega_B L_p}{z_B}; \quad R_p' = \frac{R_p}{z_B}; \quad L_s' = \frac{\omega_B L_s}{z_B}; \quad R_s' = \frac{R_s}{z_B}$$

(37.35)

Voltage components, with consideration of controllable parameters, in the rotating d-q reference frame are described by

$$u_{i1d}' = m_i \cos \delta_i$$

$$u_{i1q}' = m_i \sin \delta_i$$

$$u_{i2d}' = m_i (1+a)\cos(\delta_i + \alpha)$$

$$u_{i2q}' = m_i (1+a)\sin(\delta_i + \alpha)$$

$$u_{od}' = m_o \cos \delta_o$$

$$u_{oq}' = m_o \sin \delta_o$$

(37.36)

where m_i and m_o are per-unit magnitudes of the input and output voltage, respectively, and the δ_i and δ_o are the respective phase angles. Considering the instantaneous variables shown in Figure 37.19, the state equations of the TCVR/TCPST are given by

$$\frac{d}{dt}\begin{bmatrix} i_{pd}' \\ i_{pq}' \\ i_{sd}' \\ i_{sq}' \end{bmatrix} = \begin{bmatrix} \frac{-R_p'\omega_B}{L_p'} & \omega & 0 & 0 \\ -\omega & \frac{-R_p'\omega_B}{L_p'} & 0 & 0 \\ 0 & 0 & \frac{-R_s'\omega_B}{L_s'} & \omega \\ 0 & 0 & -\omega & \frac{-R_s'\omega_B}{L_s'} \end{bmatrix} \begin{bmatrix} i_{pd}' \\ i_{pq}' \\ i_{sd}' \\ i_{sq}' \end{bmatrix} + \begin{bmatrix} \frac{\omega_B}{L_p'}\left(u_{i1d}' - u_{pd}'\right) \\ \frac{\omega_B}{L_p'}u_{iq}'\left(u_{i1q}' - u_{pq}'\right) \\ \frac{\omega_B}{L_s'}\left(u_{i2d}' - u_{od}'\right) \\ \frac{\omega_B}{L_s'}\left(u_{i2q}' - u_{oq}'\right) \end{bmatrix}$$

(37.37)

$$0 = \begin{bmatrix} u_{pd}' & u_{pq}' & \left(u_{i1d}' - u_{i2d}'\right) & \left(u_{i1q}' - u_{i2q}'\right) \\ -u_{pq}' & u_{pd}' & -\left(u_{i1q}' - u_{i2q}'\right) & \left(u_{i1d}' - u_{i2d}'\right) \end{bmatrix} \begin{bmatrix} i_{pd}' \\ i_{pq}' \\ i_{sd}' \\ i_{sq}' \end{bmatrix}$$

The response times of the TCVR and TCPST power electronics are modeled using first-order blocks (37.38) with the time constant T_{TCVR} and T_{TCPST}, respectively.

$$\frac{d}{dt}a = -\frac{1}{T_{TCVR}}a + \frac{1}{T_{TCVR}}a_{ref}$$

$$\frac{d}{dt}\alpha = -\frac{1}{T_{TCPST}}\alpha + \frac{1}{T_{TCPST}}\alpha_{ref}$$

(37.38)

The overall dynamic behavior of the TCVR or TCPST primarily depends on the applied control system as in the case of previously discussed FACTS devices. The output from the TCVR controller is the reference value a_{ref} and the output from the TCPST controller is the reference value α_{ref}.

37.3.5 Unified Power Flow Controller

The UPFC consists of two VSCs using GTO thyristors that operate from a common DC circuit consisting of a DC storage capacitor as illustrated in Figure 37.20 [1]. It could be described as a device consisting of a parallel and a series branch. Each converter can independently generate or absorb reactive power. This arrangement enables free flow of real power in either direction between the AC terminals of the two converters.

The function of the parallel converter is to supply or absorb the real power demanded by the series branch. This converter is connected to the AC terminal through a parallel-connected transformer. If required, it may also generate or absorb reactive power, which can provide independent parallel reactive compensation for the line. The second series-connected converter provides the main function of the UPFC by injecting an AC voltage with controllable magnitude and phase angle. The transmission line current flows through this voltage source resulting in a real and reactive power exchange with the AC system. A parallel branch provides the real power exchange at the AC terminal, while the reactive power exchange is generated internally by the converter.

37.3.5.1 Modeling of UPFC for Steady-State Studies

The UPFC equivalent circuit for steady-state modeling is shown in Figure 37.21. The equivalent circuit consists of two ideal voltage sources:

$$\underline{V}_p = V_p\left(\cos\theta_p + j\sin\theta_p\right)$$

$$\underline{V}_s = V_s\left(\cos\theta_s + j\sin\theta_s\right)$$

(37.39)

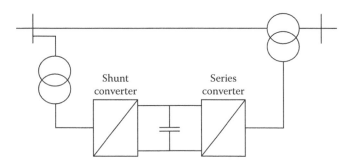

FIGURE 37.20   Basic scheme of a UPFC.

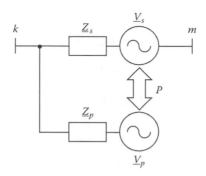

FIGURE 37.21   UPFC steady state equivalent circuit.

Based on the equivalent circuit shown in Figure 37.21, the real and reactive power equations at node k are given by [4,7,20]

$$
\begin{aligned}
P_k &= V_k^2 G_{kk} + V_k V_m \left(G_{km} \cos(\theta_k - \theta_m) + B_{km} \sin(\theta_k - \theta_m) \right) \\
&\quad + V_k V_s \left(G_{km} \cos(\theta_k - \theta_s) + B_{km} \sin(\theta_k - \theta_s) \right) \\
&\quad + V_k V_p \left(G_p \cos(\theta_k - \theta_p) + B_p \sin(\theta_k - \theta_p) \right) \\
Q_k &= -V_k^2 B_{kk} + V_k V_m \left(G_{km} \sin(\theta_k - \theta_m) - B_{km} \cos(\theta_k - \theta_m) \right) \\
&\quad + V_k V_s \left(G_{km} \sin(\theta_k - \theta_s) - B_{km} \cos(\theta_k - \theta_s) \right) \\
&\quad + V_k V_p \left(G_p \sin(\theta_k - \theta_p) - B_p \cos(\theta_k - \theta_p) \right)
\end{aligned}
\tag{37.40}
$$

The real and reactive power equations at node m are given by

$$
\begin{aligned}
P_m &= V_m^2 G_{mm} + V_m V_k \left(G_{mk} \cos(\theta_m - \theta_k) + B_{mk} \sin(\theta_m - \theta_k) \right) \\
&\quad + V_m V_s \left(G_{mm} \cos(\theta_m - \theta_s) + B_{mm} \sin(\theta_m - \theta_s) \right) \\
Q_m &= -V_m^2 B_{mm} + V_m V_k \left(G_{mk} \sin(\theta_m - \theta_k) - B_{mk} \cos(\theta_m - \theta_k) \right) \\
&\quad + V_m V_s \left(G_{mm} \sin(\theta_m - \theta_s) - B_{mm} \cos(\theta_m - \theta_s) \right)
\end{aligned}
\tag{37.41}
$$

The real and reactive power equations for the series converter are given by

$$
\begin{aligned}
P_s &= V_s^2 G_{mm} + V_s V_k \left(G_{km} \cos(\theta_s - \theta_k) + B_{km} \sin(\theta_s - \theta_k) \right) \\
&\quad + V_s V_m \left(G_{mm} \cos(\theta_s - \theta_m) + B_{mm} \sin(\theta_s - \theta_m) \right) \\
Q_s &= -V_s^2 B_{mm} + V_s V_k \left(G_{km} \sin(\theta_s - \theta_k) - B_{km} \cos(\theta_s - \theta_k) \right) \\
&\quad + V_s V_m \left(G_{mm} \sin(\theta_s - \theta_m) - B_{mm} \cos(\theta_s - \theta_m) \right)
\end{aligned}
\tag{37.42}
$$

and for the shunt converter by

$$
\begin{aligned}
P_p &= -V_p^2 G_p + V_p V_k \left(G_p \cos(\theta_p - \theta_k) + B_p \sin(\theta_p - \theta_k) \right) \\
Q_p &= V_p^2 B_p + V_p V_k \left(G_p \sin(\theta_p - \theta_k) - B_p \cos(\theta_p - \theta_k) \right)
\end{aligned}
\tag{37.43}
$$

where

$$\underline{Y}_{kk} = G_{kk} + jB_{kk} = \underline{Z}_s^{-1} + \underline{Z}_p^{-1}$$

$$\underline{Y}_{mm} = G_{mm} + jB_{mm} = \underline{Z}_s^{-1}$$

$$\underline{Y}_{km} = G_{km} + jB_{km} = -\underline{Z}_s^{-1}$$ (37.44)

$$\underline{Y}_p = G_p + jB_p = -\underline{Z}_p^{-1}$$

Assuming a lossless converter operation, the UPFC neither absorbs nor injects real power from/to the AC system. The DC voltage remains constant. Hence, the real power supplied to the shunt converter must satisfy the real power demanded by the series converter.

$$P_p + P_s = 0 \tag{37.45}$$

37.3.5.2 Modeling of UPFC for Transient Studies

As mentioned above, the UPFC consists of a shunt- and series-connected VSC operating with a common DC circuit. Each of the two converters can independently generate or absorb reactive power. This structure enables free flow of the real power in either direction between the AC terminals of both converters [9,11,21]. Figure 37.22 shows an equivalent circuit of the UPFC.

On the AC side, the UPFC can be represented by sinusoidal voltage sources with a controllable magnitude and phase angle. The circuit also consists of series and shunt impedance representing coupling transformers. The influence of both converters connected in series and shunt to the DC system can be represented by two current sources in the common DC circuit connected to the capacitor C. Shunt connection of resistance R_c enables representation of losses in the DC circuit.

The derivations and mathematical descriptions are based on the transformation of a balanced three-phase system into an orthogonal synchronously rotating coordinate system (d–q). The adopted per-unit system is given by

$$i_p' = \frac{i_p}{i_B}; \quad i_s' = \frac{i_s}{i_B}; \quad i_{dc}' = \frac{i_{dc}}{i_B}; \quad u_p' = \frac{u_p}{u_B}; \quad u_s' = \frac{u_s}{u_B}; \quad u_i' = \frac{u_i}{u_B}; \quad u_o' = \frac{u_o}{u_B}; \quad u_{dc}' = \frac{u_{dc}}{u_B}$$

$$z_B = \frac{u_B}{i_B}; \quad L_p' = \frac{\omega_B L_p}{z_B}; \quad R_p' = \frac{R_p}{z_B}; \quad L_s' = \frac{\omega_B L_s}{z_B}; \quad R_s' = \frac{R_s}{z_B}; \quad C' = \frac{1}{\omega_B C z_B}; \quad R_c' = \frac{R_c}{z_B}$$

(37.46)

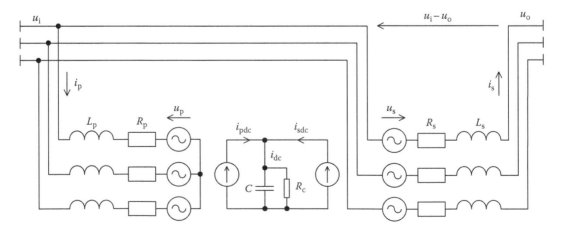

FIGURE 37.22 Equivalent circuit of the UPFC using sinusoidal voltage sources.

Components of both converter output voltages depend on the DC voltage. Equations (37.47) and (37.48) can be written for the respective voltages where k_p and k_s are factors including a shunt and a series transformer ratio and relating the DC and AC voltage of each converter while taking into account converter type. The angles δ_p and δ_s represent the phase shift of each converter output voltage with respect to the reference position. The control factors m_p for shunt converter and m_s for series converter can take any value between 0 and 1.

$$u'_{pd} = u'_{dc}k_p m_p \cos\delta_p = u'_{dc}k_p d_{pd}$$
$$u'_{pq} = u'_{dc}k_p m_p \sin\delta_p = u'_{dc}k_p d_{pq}$$
(37.47)

$$u'_{sd} = u'_{dc}k_s m_s \cos\delta_s = u'_{dc}k_s d_{sd}$$
$$u'_{sq} = u'_{dc}k_s m_s \sin\delta_s = u'_{dc}k_s d_{sq}$$
(37.48)

Adjustable parameters of the converter output voltages m_p, δ_p, m_n, and δ_n can be used to determine the average switching functions d_{pd} in the direction of the d-axis and d_{pq} in the direction of the q-axis for the shunt converter, and similarly average switching functions d_{sd} and d_{sq} for the series converter.

The common balance equation for the real power using d–q components is given by

$$u'_{dc}i'_{dc} = \frac{3}{2}\left(u'_{pd}i'_{pd} + u'_{pq}i'_{pq} + u'_{sd}i'_{sd} + u'_{sq}i'_{sq}\right)$$
(37.49)

In the rotating reference frame, $(d$–$q)$, the influence of both converters on the DC capacitor can be represented by a common DC current source:

$$i'_{dc} = i'_{pdc} + i'_{sdc} = \frac{3}{2}\left(k_p d_{pd}i'_{pd} + k_p d_{pq}i'_{pq} + k_s d_{sd}i'_{sd} + k_s d_{sq}i'_{sq}\right)$$
(37.50)

The state-space mathematical model of the UPFC in the matrix format and using reference frame, $(d$–$q)$, is given by

$$\frac{d}{dt}\begin{bmatrix} i'_{pd} \\ i'_{pq} \\ i'_{sd} \\ i'_{sq} \\ u'_{dc} \end{bmatrix} = \begin{bmatrix} \dfrac{-R'_p\omega_B}{L'_p} & \omega & 0 & 0 & \dfrac{-k_p\omega_B}{L'_p}d_{pd} \\[2mm] -\omega & \dfrac{-R'_p\omega_B}{L'_p} & 0 & 0 & \dfrac{-k_p\omega_B}{L'_p}d_{pq} \\[2mm] 0 & 0 & \dfrac{-R'_s\omega_B}{L'_s} & \omega & \dfrac{-k_s\omega_B}{L'_s}d_{sd} \\[2mm] 0 & 0 & -\omega & \dfrac{-R'_s\omega_B}{L'_s} & \dfrac{-k_s\omega_B}{L'_s}d_{sq} \\[2mm] \dfrac{3k_p\omega_B C'}{2}d_{pd} & \dfrac{3k_p\omega_B C'}{2}d_{pq} & \dfrac{3k_s\omega_B C'}{2}d_{sd} & \dfrac{3k_s\omega_B C'}{2}d_{sq} & \dfrac{-C'\omega_B}{R'_c} \end{bmatrix}\begin{bmatrix} i'_{pd} \\ i'_{pq} \\ i'_{sd} \\ i'_{sq} \\ u'_{dc} \end{bmatrix} + \begin{bmatrix} \dfrac{\omega_B}{L'_p}u'_{id} \\[2mm] \dfrac{\omega_B}{L'_p}u'_{iq} \\[2mm] \dfrac{\omega_B}{L'_s}\left(u'_{id}-u'_{od}\right) \\[2mm] \dfrac{\omega_B}{L'_s}\left(u'_{iq}-u'_{oq}\right) \\[2mm] 0 \end{bmatrix}$$
(37.51)

It can be seen from (37.51) that there are five state variables and only four controllable parameters. The two current components of the series branch and the reactive current component of the shunt branch can be controlled independently. Indirectly, with the second free parameter of the shunt converter (the current d-component), the constant DC voltage across the common DC capacitor is maintained.

The UPFC can also operate in a variable DC voltage mode where the series branch still enables the exchange of the real power with the system, and the influence of the DC voltage variations can be compensated for with the proper setting of the series converter control factor m_s.

The mathematical model of the UPFC with sinusoidal voltage sources may be used for the derivation of the appropriate control system [22–24]. The outputs from the applied controller represent the values of average switching functions d_{pd} and d_{pq} for the shunt converter, and d_{sd} and d_{sq} for the series converter.

37.4 Areas of Applications of FACTS Devices in Power Systems

FACTS were originally developed to facilitate better control and operation of power transmission networks (and ultimately power systems as a whole) that started to face more and more constrains in their daily operation. Those constraints were further aggravated over the years by expansion restrictions imposed by environmental and social factors and by electricity market rules.

Modern FACTS devices are capable of performing a range of different tasks and functions that contribute to safe, secure, and economic operation of power systems. The contribution of FACTS devices to the enhancement of various attributes of power system can be broadly classified into four categories:

- Voltage stability and reactive power compensation
- Transfer capability and power flow control (or congestion management)
- Transient and small disturbance stability
- Reliability

In the sequel, the contributions that FACTS devices make within each of those categories will be discussed separately. This does not mean, however, that the influence of any particular FACTS device is restricted to a single area while others remain unaffected. In reality, these devices usually affect several areas of power system operation simultaneously. Their "non-intended" contribution to system operation, i.e., the one that they were not originally designed for, may be either positive or negative, as reported in the literature.

37.4.1 Voltage Stability and Reactive Power Compensation

The TCSC and SVC were used in [25] for the advanced VAr planning in a power system network in order to prevent voltage collapse following a contingency, while at the same time ensuring that the bus voltages remain within the statutory limits. The same devices were employed in [26], to enhance voltage stability of the network. It was found that in addition to enhancing voltage stability, these devices greatly enlarged the region of small disturbance voltage stability. Enhanced voltage support by an SVC was also reported in [27]. Comparative study between a TCSC and SVC [28] found that although both of them can support network voltages very well, the TCSC contributes to better power system voltage stability for a wide range of loading conditions. A UPFC and a STATCOM were used in [29,30], respectively, to improve the voltage stability of the network. In addition to voltage stability improvement, the UPFC demonstrated excellent voltage control and series reactive power control capabilities [29].

In summary, series and shunt FACTS can be used effectively for voltage stability and reactive power compensation. They can greatly enlarge the region of voltage stability for a wide range of loading conditions and play an essential role in improving the voltage profile of the system.

37.4.2 Available Transfer Capability and Power Flow Control (Congestion Management)

The UPFC was used in [31–37] for the enhancement of available transfer capability (ATC) and power flow control. It was found that the UPFC is a cost-effective solution for congestion management in a number of cases [33] and that its advanced power flow control contributes to minimization of electricity generation cost [36]. The issue of minimization of generation costs by optimal settings of parameters of various FACTS devices (TCSC, TCPST, UPFC, and SVC) was also addressed in [38].

The enhancement of ATC by a TCSC [39,40], a TCPST [40], or a combination of FACTS devices (SVCs and TCSCs) [41–43] was also reported. It was found that the rating of FACTS devices plays an essential role in determining their contribution to congestion management [41] and that the location of devices in the network [43] should be carefully chosen as, depending on their location, they could have both positive and negative influence on system stability.

To summarize, a wide range of FACTS device can be effectively used for the enhancement of ATC and reduction in cost of electricity generation. Their rating and location in the power system though have to be carefully determined as they play important roles in their total impact on chosen network attributes.

37.4.3 Transient and Small Disturbance Stability

Apart from voltage regulation and stability improvement and congestion management, the FACTS devices have been also extensively used for the improvement of power system angular stability. The UPFC alone was used for the enhancement of system angular stability in [44–46] and for damping of torsional oscillations in [46]. For the similar purpose, damping of torsional oscillations, STATCOM was used in [47]. The SVC, TCSC, STATCOM, and thyristor-controlled phase angle regulator (TCPAR) [41,48–51] were also used for the improvement of transient stability. It was demonstrated in [48] that the SVC's contribution to transient stability is bigger than that of the TCSC's, though it depended more on its location than that was the case with a TCSC. Interestingly, a reasonably small degree of compensation (4%–5%) by TCSC is found to be sufficient for adequate damping of the inter-area power system oscillations. Many other derivatives of these basic types of FACTS devices have also been used for damping of power system oscillations [52–55]. In most cases, conventional power system damping controllers (e.g., power system stabilizer) have been added in auxiliary control loops of FACTS to facilitate damping of electromechanical oscillations [55–57].

In all studies, without exception, it has been found that FACTS devices can greatly contribute to both small disturbance and transient power system angular stability. The contribution is particularly significant in damping of inter-area electromechanical modes that are very difficult to control by conventional, generator installed, damping controllers.

37.4.4 Reliability

In addition to the above more intuitive areas of FACTS' impact on system performance, a contribution of FACTS devices to power system reliability has been also verified. The impact of SVC, TCPAR, and UPFC on system reliability was addressed in [58–60]. It was found that this contribution is largely indirect through enhancing the system's transfer capability. Therefore, the contribution to system reliability can be considered as an additional benefit resulting from application of FACTS device to address some of the previously mentioned concerns (i.e., congestion management, voltage stability and control, angular stability.)

37.5 Rating of FACTS Devices

The rating of FACTS is strongly influenced by their intended function and by the way in which they are connected to the network, i.e., in series or in shunt.

The SVC, for example, is a shunt-connected device with two operating regions, inductive and capacitive. For the inductive mode, the rating should be chosen in order to protect the bus voltage from temporary overvoltages. On the other hand, in the capacitive mode of operation, it should be able to inject the required reactive power in order to maintain the bus voltage within statutory limits during the rated power flow over the transmission line [6].

The TCSC is installed in series with the transmission line of interest. It entails that the TCSC, in terms of insulation, must be able to withstand the full line current over the transmission line and the line voltage at its terminals. This, of course, affects its cost. The TCSC rating is generally determined as a percentage of the total MVAr losses over the compensated transmission line at the full line current [1,6]. For instance, if the full line current is 2000 A and the voltage drop is 60 kV, the total MVAr losses will be 120 MVAr. Assuming that the TCSC should compensate up to 25% of the total transmission line reactance, the TCSC rating will be 120 × 0.25 = 30 MVAr. The usual range of TCSC reactance (X_{TCSC}) variation with respect to compensated transmission line reactance (X_L) is from $0.8X_L$ to $0.2X_L$ [61,62].

The TCVR and TCPST are also installed in series with the transmission line; however, unlike the TCSC, they exchange the full real and reactive power flows with the transmission line. This makes their required rating high compared to other devices [3,63,64]. The TCVR turns ratio is typically chosen to be between 0.9 and 1.1 p.u., while the phase shift provided by the TCPST is usually in the range of ±5° [61,62]. A conventional phase shifter can vary angle approximately within ±30° in discrete steps of about 1° or 2° [3].

37.6 Cost of FACTS Devices

The previous sections described FACTS devices at some detail and showed that these versatile and technically advanced systems can offer huge advantages to the operation and control of electric power systems. In spite of their unquestionably high technical performance and control abilities, the number of real-life applications is still not as high as one might expect based on potential advantages that they can offer to power systems. One of the major drawbacks of this rather slow uptake is their very high price. The following subsections address the cost of FACTS device in a bit more detail.

37.6.1 Cost Structure

The cost of FACTS devices has two main components: initial installation cost and operation and maintenance cost. The initial installation cost includes the cost of device plus delivery and installation charges, professional fees, and sales tax. The total installation cost is usually expressed as a function of rated electrical capacity of FACTS device. The other cost component, operation, and maintenance cost is incurred over the lifetime of the system. Operating costs include maintenance and service, insurance, and any applicable taxes. A rule of thumb estimate for annual operating expenses is 5%–10% of the initial system cost. A typical initial installation cost structure for FACTS device can be laid out as follows [65]: hardware, 55%; engineering and project management, 15%; civil works, 12%; installation, 10%; freight and insurance, 4%; and commissioning, 4%.

37.6.2 Price Guideline

The cost of a FACTS installation depends on many factors, such as power rating, type of device, system voltage, system requirement, environmental conditions, regulatory requirement, etc. Due to the variety of options available for optimum design, it is impossible to give a generic cost figure for a FACTS installation. The approximate prices of SVC and a range of VSC-based devices are shown in Table 37.1.

Similarly rough guidelines are provided in Table 37.2 [66] for the purchase and installation cost of a medium-size (between 100 and 500 kVA) devices.

TABLE 37.1 Cost of FACTS

Type	Cost ($/kVAr)
STATCOM	40
SVC	35
UPFC	40
TCSC	50

Source: Grunbaum, R. et al., Improving the efficiency and quality of AC transmission systems, in Joint World Bank/ABB Power Systems Paper, San Mateo, CA, March 24, 2000.

TABLE 37.2 Price of FACTS

Size (kVA)	Cost (Euro/kVA)
>500	<150
100–500	150–250
<100	>250

Source: Didden, D.M., Voltage disturbances—Considerations for choosing the appropriate sag mitigation device, in *Power Quality Application Guide*, Copper Development Association, Brussels, Belgium, 2005.

TABLE 37.3 Price of FACTS

Type	Cost ($/kVAr)				Installation Costs Included
	100 MVAr	200 MVAr	300 MVAr	400 MVAr	
SVC	60	50	45	40	NO
	100	80	70	60	YES
STATCOM	90	75	68	60	NO
	130	115	110	100	YES

Source: Habur, K. and O'Leary, D., FACTS—For cost effective and reliable transmission of electrical energy, 2004, available: http://www.worldbank.org/html/fpd/em/transmission/facts_siemens.pdf.

In [10], the costs of SVC and STATCOM are shown as exponentially decreasing functions of the rating of device with and without installation costs. The extrapolated values from those curves are summarized in Table 37.3.

Based on the above figures one can only conclude that the price of FACTS devices varies within a wide range and that it is very difficult to give a generic cost for any of them.

37.7 Conclusion

The FACTS devices are undoubtedly versatile and technically advanced systems that can offer huge advantages to operation and control of electric power systems. Their ability to control quickly and efficiently one or more power system parameters will particularly be required in the future power networks characterized by proliferation of renewable, often stochastic or intermittent, generation built (typically) in the areas with limited power transfer infrastructure (leading to creation of power transfer bottlenecks), by market-driven demand for increased cross-border power transfers (beyond existing typical maximum of 10%) and ever-increasing demands for higher quality and security of electricity supply and participation of distributed demand and storage in peak shaving.

The major issue that hampered their proliferation in the past was their high cost. In the foreseeable future it is expected that the cost of FACTS devices will continue to fall, chiefly due to reduction in cost of power electronic components, while their effectiveness will further improve. Envisaged reduction in cost will certainly further contribute to their competitiveness. More importantly, since FACTS devices generally contribute to the enhancement of several electrical power network functions simultaneously, the benefits resulting from their installation will also start to be assessed more comprehensively, i.e., taking into account several contributions at the same time, and it will be easy to prove that they are cost-effective options for facilitating flexible electric power networks of the future.

References

1. N. G. Hingorani and L. Gyugyi, *Understanding FACTS: Concepts and Technology of Flexible AC Transmission Systems*, vol. 1, IEEE, Piscataway, NJ, 2000.
2. K. Habur and D. O'leary, FACTS for cost effective and reliable transmission of electrical energy (Siemens 2004, [online] available http://www.worldbank.org/html/fpd/em/transmission/facts_siemens.pdf).
3. Y. H. Song and A. T. Johns, *Flexible AC Transmission Systems (FACTS)*, The Institute of Electrical Engineering, London, U.K., 1999.
4. X. P. Zhang, C. Rehtanz, and B. Pal, *Flexible AC Transmission Systems: Modelling and Control*, Springer, Berlin, Germany, 2006.
5. P. Kundur, *Power System Stability and Control*, McGraw-Hill, New York, 1994.

6. R. M. Mathur and R. K. Varma, *Thyristor-Based FACTS Controllers for Electrical Transmission Systems*, John Wiley & Sons, Inc., New York, 2002.

7. E. Acha, C. R. Fuerte-Esquivel, H. Ambriz-Pérez, and C. Angeles-Camacho, *FACTS: Modelling and Simulation in Power Networks*, John Wiley & Sons, Chichester, U.K., 2004.

8. C. D. Schauder and H. Mehta, Vector analysis and control of advanced static VAr compensators, Conference Publication no. 345 of *the IEE Fifth International Conference on AC and DC Power Transmission*, London, U.K., pp. 266–272, September 1991.

9. I. Papič, Mathematical analysis of FACTS devices based on a voltage source converter. Part 1, Mathematical models, *Electr. Power Syst. Res.*, 56, 139–148, 2000.

10. K. Habur and D. O'Leary, FACTS—For cost effective and reliable transmission of electrical energy, 2004. Available: http://www.worldbank.org/html/fpd/em/transmission/facts_siemens.pdf.

11. I. Papič, Mathematical analysis of FACTS devices based on a voltage source converter. Part 2, Steady state operational characteristics, *Electr. Power Syst. Res.*, 56, 149–157, 2000.

12. R. Mihalič and I. Papič, Mathematical models and simulation of a static synchronous series compensator, *Proceedings of the International Conference on Electric Power Engineering PowerTech '99*, Budapest, Hungary, August 29–September 2, 1999. Budapest, Hungary: IEEE Hungary Section, cop., pp. 1–6, 1999.

13. R. Mihalič and I. Papič, Static synchronous series compensator—A mean for dynamic power flow control in electric power systems, *Electr. Power Syst. Res.*, 45, 65–72, 1998.

14. I. Papič and A. M. Gole, Enhanced control system for a static synchronous series compensator with energy storage, *Proceedings of the Seventh International Conference on AC-DC Power Transmission*, London, U.K., November 28–30, 2001 (Conference publication, no. 485). London, U.K.: Institution of Electrical Engineers, cop., pp. 327–332, 2001.

15. J. V. Milanović and Y. Zhang, Modelling of FACTS devices for voltage sag mitigation studies in large power systems, *IEEE Transactions on Power Delivery*, 25(3), 2010.

16. B. Blažič and I. Papič, STATCOM control for operation with unbalanced voltages, *EPE-PEMC 2006: Conference Proceedings*, Portoroz, Itlay: IEEE, cop., pp. 1454–1459, 2006.

17. B. Blažič and I. Papič, A new mathematical model and control of D-StatCom for operation under unbalanced conditions, *Electr. Power Syst. Res.*, 72(3), 279–287, 2004.

18. B. Blažič and I. Papič, Improved D-StatCom control for operation with unbalanced currents and voltages, *IEEE Trans. Power Deliv.*, 21(1), 225–233, 2006.

19. P. Preedavichit and S. C. Srivastava, Optimal reactive power dispatch considering FACTS devices, *Electr. Power Syst. Res.*, 46, 251–257, 1998.

20. C. R. Fuerte-Esquivel and E. Acha, Unified power flow controller: A critical comparison of Newton-Raphson UPFC algorithms in power flow studies, *IEE Proc. Gen. Trans. Distrib.*, 144(5), 437–444, 1997.

21. A. Nabavi-Niaki and M. R. Iravani, Steady-state and dynamic models of unified power flow controller (UPFC) for power system studies, *IEEE/PES Winter Meeting, 96 WM, 257-6 PWRS*, Baltimore, MD, January 1996.

22. X. Lombard and P. G. Therond, Control of unified power flow controller: Comparison of methods on the basis of a detailed numerical model, *IEEE/PES Summer Meeting, 96 SM 511-6 PWRS*, Denver, CO, July 1996.

23. I. Papič, P. Žunko, D. Povh, and M. Weinhold, Basic control of unified power flow controller, *IEEE Trans. Power Syst.*, 12(4), 1734–1739, 1997.

24. I. Papič and P. Žunko, UPFC converter-level control system using internally calculated system quantities for decoupling, *Electr. Power Energy Syst.*, 25(8), 667–675, 2003.

25. N. Yorino, E. E. El-Araby, H. Sasaki, and S. Harada, A new formulation for FACTS allocation for security enhancement against voltage collapse, *IEEE Trans. Power Syst.*, 18, 3–10, 2003.

26. X. Li, L. Bao, X. Duan, Y. He, and M. Gao, Effects of FACTS controllers on small-signal voltage stability, presented at *2000 IEEE Power Engineering Society Winter Meeting. Conference Proceedings*, Singapore, January 23–27, 2000.

27. M. H. Haque, Determination of steady state voltage stability limit of a power system in the presence of SVC, presented at *Proceedings of Power Tech*, Porto, Portugal, pp. 10–13, September 2001.

28. F. A. El-Sheikhi, Y. M. Saad, S. O. Osman, and K. M. El-Arroudi, Voltage stability assessment using modal analysis of power systems including flexible AC transmission system (FACTS), presented at *Large Engineering Systems Conference on Power Engineering (LESCOPE)*, Montreal, Que., Canada, May 7–9, 2003.

29. N. Dizdarevic and M. Majstrovic, FACTS-based reactive power compensation of wind energy conversion system, presented at *IEEE PowerTech*, Bologna, Italy, June 23–26, 2003.

30. R. A. Mukhedkar, T. S. Davies, and H. Nouri, Influence of FACTS on power system voltage stability, presented at *Proceedings of AC and DC Transmission*, London, U.K., November 28–30, 2001.

31. Y. Xiao, Y. H. Song, and Y. Z. Sun, Application of stochastic programming for available transfer capability enhancement using FACTS devices, presented at *Power Engineering Society Summer Meeting*, Seattle, WA, July 16–20, 2000.

32. K. Belacheheb and S. Saadate, Compensation of the electrical mains by means of unified power flow controller (UPFC)-comparison of three control methods, presented at *Proceedings of International Conference on Harmonics and Quality of Power*, Orlando, FL, October 1–4, 2000.

33. J. Brosda and E. Handschin, Congestion management methods with a special consideration of FACTS-devices, presented at *Proceedings of Power Tech*, Porto, Portugal, September 10–13, 2001.

34. C. Bulac, M. Eremia, R. Balaurescu, and V. Stefanescu, Load flow management in the interconnected power systems using UPFC devices, presented at *IEEE Bologna PowerTech*, Bologna, Italy, June 23–26, 2003.

35. S. Bruno and M. La Scala, Unified power flow controllers for security-constrained transmission management, *IEEE Trans. Power Syst.*, 19, 418–26, 2004.

36. K. Belacheheb and S. Saadate, UPFC control for line power flow regulation, presented at *Proceedings of International Conference on Harmonics and Quality of Power*, Athens, Greece, October 14–16, 1998.

37. J. Brosda, E. Handschin, A. L'Abbate, C. Leder, and M. Trovato, Visualization for a corrective congestion management based on FACTS devices, presented at *IEEE Bologna PowerTech*, Bologna, Italy, June 23–26, 2003.

38. P. Bhasaputra and W. Ongsakul, Optimal power flow with multi-type of FACTS devices by hybrid TS/SA approach, presented at *International Conference on Industrial Technology on 'Productivity Reincarnation through Robotics and Automation,'* Bankok, Thailand, December 11–14, 2002.

39. C. Schaffner and G. Andersson, Value of controllable devices in a liberalized electricity market, presented at *Proceedings of AC and DC Transmission*, London, U.K., November 28–30, 2001.

40. A. Oudalov, R. Cherkaoui, A. J. Germond, and M. Emery, Coordinated power flow control by multiple FACTS devices, presented at *IEEE Bologna PowerTech*, Bologna, Italy, June 23–26, 2003.

41. C. Praing, T. Tran-Quoc, R. Feuillet, J. C. Sabonnadiere, J. Nicolas, K. Nguyen-Boi, and L. Nguyen-Van, Impact of FACTS devices on voltage and transient stability of a power system including long transmission lines, presented at *Power Engineering Society Summer Meeting*, Seattle, WA, July 16–20, 2000.

42. S. C. Srivastava and P. Kumar, Optimal power dispatch in deregulated market considering congestion management, presented at *Proceedings of International Conference on Electric Utility Deregulation and Restructuring, and Power Technologies 2000*, London, U.K., April 4–7, 2000.

43. X. Yu, C. Singh, S. Jakovljevic, D. Ristanovic, and G. Huang, Total transfer capability considering FACTS and security constraints, presented at *IEEE PES Transmission and Distribution Conference and Exposition*, Dallas, TX, September 7–12, 2003.

44. K. M. Son and R. H. Lasseter, A Newton-type current injection model of UPFC for studying low-frequency oscillations, *IEEE Trans. Power Deliv.*, 19, 694–701, 2004.

45. R. Caldon, A. Mari, A. Scala, and R. Turri, Application of modal analysis for the enhancement of the performances of a UPFC controller in power oscillation damping, presented at *Proceedings of 10th Mediterranean Electrotechnical Conference—MELECON*, Lemesos, Cyprus, May 29–31, 2000.

46. W. Bo and Z. Yan, Damping subsynchronous oscillation using UPFC-a FACTS device, presented at *PowerCon*, Kunming, China, October 13–17, 2002.

47. K. V. Patil, J. Senthil, J. Jiang, and R. M. Mathur, Application of STATCOM for damping torsional oscillations in series compensated AC systems, *IEEE Trans. Energy Conversion*, 13, 237–43, 1998.

48. Y. L. Tan and Y. Wang, Effects of FACTS controller line compensation on power system stability, *IEEE Power Eng. Rev.*, 18, 45–7, 1998.

49. D. D. Rasolomampionona, AGC and FACTS stabilization device coordination in interconnected power system control, presented at *IEEE Bologna PowerTech*, Bologna, Italy, June 23–26, 2003.

50. G. Chunlin, T. Luyuan, and W. Zhonghong, Stability control of TCSC between interconnected power networks, presented at *PowerCon*, Kunming, China, October 13–17, 2002.

51. M. H. Haque, Improvement of first swing stability limit by utilizing full benefit of shunt FACTS devices, *IEEE Trans. Power Syst.*, 19, 1894–902, 2004.

52. S. H. Hosseini and P. D. Azar, Damping of large signal electromechanical oscillations of power systems using multi-functional power transfer controller, presented at *IEEE TENCOM'02. IEEE Region 10 Conference on Computer, Communications, Control and Power Engineering*, Beijing, China, October 28–31, 2002.

53. K. Kobayashi, M. Goto, K. Wu, Y. Yokomizu, and T. Matsumura, Power system stability improvement by energy storage type STATCOM, presented at *IEEE Bologna PowerTech*, Bologna, Italy, June 23–26, 2003.

54. Y. J. Fang and D. C. Macdonald, Dynamic quadrature booster as an aid to system stability, *IEE Proc. Gen. Transm. Distrib.*, 145, 41–7, 1998.

55. S. Abazari, J. Mahdavi, M. Ehsan, and M. Zolghadri, Transient stability improvement by using advanced static VAr compensator, presented at *IEEE Bologna PowerTech*, Bologna, Italy, June 23–26, 2003.

56. A. M. Hemeida and G. El-Saady, Damping power systems oscillations using FACTS combinations, presented at *39th International Universities Power Engineering Conference (UPEC)*, Bristol, U.K., September 6–8, 2004.

57. S. Arabi, P. Kundur, P. Hassink, and D. Matthews, Small signal stability of a large power system as affected by new generation additions, presented at *Power Engineering Society Summer Meeting*, Seattle, WA, July 16–20, 2000.

58. G. M. Huang and Y. Li, Composite power system reliability evaluation for systems with SVC and TCPAR, presented at *IEEE Power Engineering Society General Meeting*, Toronto, Ont., Canada, July 13–17, 2003.

59. M. Fotuhi-Fikruzabad, R. Billinton, S. O. Faried, and S. Aboreshaid, Power system reliability enhancement using unified power flow controllers, presented at *Proceedings of International Conference on Power System Technology (POWERCON 2000)*, Perth, Australia, December 4–7, 2000.

60. S. N. Singh and A. K. David, A new approach for placement of FACTS devices in open power markets, *IEEE Power Eng. Rev.*, 21, 58–60, 2001.

61. S. Gerbex, R. Cherkaoui, and A. J. Germond, Optimal location of multi-type FACTS devices in a power system by means of genetic algorithms, *IEEE Trans. Power Syst.*, 16, 537–44, 2001.

62. L. Cai and I. Erlich, Optimal choice and allocation of FACTS devices using genetic algorithms, *CD Rom of the 12th Intelligent System Application to Power Systems Conference*, Lemnos, Greece, August 2003.

63. P. Paterni, S. Vitet, M. Bena, and A. Yokoyama, Optimal location of phase shifters in the French network by genetic algorithm, *IEEE Trans. Power Syst.*, 14, 37–42, 1999.

64. L. Ippolito and P. Siano, Selection of optimal number and location of thyristor-controlled phase shifters using genetic based algorithms, *IEE Proc. Gen. Transm. Distrib.*, 151, 630–637, 2004.
65. R. Grunbaum, R. Sharma, and J.-P. Charpentier, Improving the efficiency and quality of AC transmission systems, Joint World Bank/ABB Power Systems Paper, San Mateo, CA, March 24, 2000.
66. D. M. Didden, Voltage disturbances—Considerations for choosing the appropriate sag mitigation device, in *Power Quality Application Guide*, Copper Development Association, Brussels, Belgium, 2005.

38

Filtering Techniques for Power Quality Improvement

Salem Rahmani
École de Technologie
Supérieure

Kamal Al-Haddad
École de Technologie
Supérieure

38.1 Introduction

Power quality is generally evaluated in terms of harmonic content in both supply voltage and current. For an ideal system, harmonics are typically caused by the use of nonlinear loads found in domestic equipment such as switch-mode power electronics converters, ballast for fluorescent lamps, computers, televisions, and other nonlinear loads used in tertiary and industrial applications such as power electronics operated adjustable speed drives, arc furnaces, and welding equipment [1]. The nonlinear loads draw nonsinusoidal current from the network, an important harmonic content of the supply voltage with regards to the fundamental. The presence of such harmonics in the system can cause a number of unwanted effects for sensitive electronic loads such as industrial process controllers, hospital monitoring equipments, and laboratory measurement devices, and computers malfunction or fail to operate when connected to an ac line that has high harmonic voltage content. Also, electric utility transmission and distribution equipment may be susceptible to ac line harmonics. Furthermore, transmission lines, motors, and transformers could have higher operating losses; capacitor banks may fail due to over current, protective relays may not operate properly.

Generally, at the point of common coupling (PCC), the impact of the loads on the supply voltage can easily be measured and identified. Two types of harmonic-producing loads can be characterized at the bus bar that connects the supply voltage to different loads [2,3]:

- Current type harmonics-producing loads; these loads are found as diode rectifiers and phase-controlled thyristor rectifiers feeding sufficient inductance connected to the dc side.
- Voltage harmonics-producing loads, such as diode rectifiers feeding sufficient filtering dc capacitors.

These two types of harmonic sources have completely distinctive dual properties and characteristics. Based on their natural distinctive properties, both current and voltage type of harmonic-producing loads have their own suitable filter configurations.

Various mitigation techniques for reducing harmonics in the power system have been developed with time. Traditionally, passive filters such as low pass, high pass, band pass, and tuned filters have been used to eliminate low-order and high-order harmonics, and sometimes tuned filters are used to attenuate specific harmonics. Moreover, these filters contribute to the improvement of the power factor (PF), but their bulky size, limited compensation ability, and susceptibility to resonance with the source impedance constitute the major drawbacks of the technology [1–4].

Other industrial applications use power factor correction (PFC) devices for reactive power and current harmonics compensation. In these circuits, switched capacitor banks are typically connected in parallel to current-source-type loads. Seen from the load side, the capacitance of the PFC and the source inductor create a parallel resonant circuit. Looking from the source side, the PFC capacitors and the line inductor represent a series resonant circuit. To overcome the drawbacks of passive filters integrated with the PFC equipment, typical active power filter (APF) topologies may be used [5–6]. They are preferred over the passive filters because of their filtering characteristics and their capability of improving the system stability by avoiding possible resonance between the filter components and the mains impedance [7–14].

APFs have been known as an effective tool for harmonic mitigation as well as reactive power compensation, voltage regulation, load balancing, and voltage flicker compensation. They can be classified according to the converter type used (voltage source or current source); the number of phases (single-phase, three-phase application to three or four wires); and their topologies, which include shunt [7–40], series [41–51], hybrid [52–78], and unified-power quality conditioner (UPQC), which is a combination of series and shunt active filters [79–84]. Shunt active filters are connected in parallel to electrical systems and can substantially improve current distortions, reactive power, load unbalance, and neutral current. It operates by injecting harmonic current into the utility system with the same magnitudes as the harmonic generated by a given nonlinear load, but with opposite phases. Unfortunately, it cannot compensate voltage-source type of nonlinear loads. In fact, lots of electronic appliances used in power system, such as frequency converters, switch-mode power supplies, and uninterruptible power supplies (UPSs) as well as electronic ballasts, etc., have a large filter capacitor on the dc side of the rectifier circuit. They intrinsically belong to voltage-source nonlinear type of loads. The harmonics generated by such voltage-source nonlinear load can effectively be suppressed by using a series APF. Indeed, series active filters suppress and isolate voltage-based distortions such as voltage harmonics, voltage unbalance, voltage flickers, and voltage sags and swells. APFs have the capability of damping the harmonic resonance between an existing passive filter and the supply impedance, but they suffer from high kVA ratings. The boost converter constituting the shunt active filter requires a high dc-link voltage in order to compensate effectively higher order harmonics. On the other hand, a series active filter needs a transformer capable to withstand full load current in order to compensate for voltage distortion.

Combining the advantages of both passive and active filters, hybrid filter topologies are appealing. They have been developed achieving the desired damping performance with a significant reduction of KVA effort required by the power active filter [52–78]. They are cost-effective solutions to controlling voltage variations and distortions as well as suppressing harmonics. Passive filters are also used in this topology to carry the fundamental current component in a series active filter and the fundamental voltage component in a shunt active filter. UPQCs are the most effective devices to improve power quality. Its configuration consists of a series and a shunt active filter that usually share the energy source. The series active filter cancels voltage harmonics and the shunt active filter cancels current harmonics. Active filter systems have been also developed for dc/dc converters. These configurations of active filters are used for two purposes. The first purpose is to remove high-frequency electromagnetic interference (EMI) from input current of converter. The second reason is to remove the voltage ripple from the output voltage of the converter. Among all the configurations of active filters, UPQC is known as the best tool for power quality improvement. The latter is used to cancel both current-based and voltage-based distortions.

This chapter describes harmonic-producing loads, effects of harmonics on utility line, and harmonic mitigation methods, especially, passive active and also hybrid filters. Different topologies of active filters, their applications, configurations, control methods, modeling and analysis, and stability issues are detailed; moreover, simulation results are given to show the performance of every topology studied.

38.2 Harmonic Production and Characteristics

Harmonics are periodic voltages and currents signals having frequencies that are integral multiples of the fundamental frequency. In single-phase, 60 Hz power systems, odd harmonics such as 3rd, 5th, 7th, … are present on the ac side with the third harmonic being dominant, whereas even harmonics are found on the dc side. In three-phase three-wire, 60 Hz power systems, only non-triplen odd harmonics such as 5th, 7th, 11th, 13th, … are present. Harmonic distortion of supply voltage is caused because of the supply impedance and the presence of rich harmonic currents drawn by residential, commercial, and industrial loads such as switch-mode power converters, adjustable speed drives, elevators, electronic ballasts, air conditioners, arc welders, battery chargers, copy machines/printers, personal or mainframe computers, UPSs, silicon-controlled rectifier (SCR) drives, and x-ray equipment. The term total harmonic distortion (THD) gives the measure of harmonics content in a signal and is generally used to denote the level of harmonics present in the voltage or current signals. The quality of the energy became a major concern because of the recommended international standards such as IEEE-519, "IEEE Recommended Practices and Requirements for Harmonic Control in Electrical power Systems" and IEC-6002-3. The IEEE standard 519-1992 establishes the recommended guidelines for harmonic currents and voltage control in utility distribution systems. The standard specifies harmonics current and voltage limits at the PCC. The European harmonic standard, IEC-555, proposes absolute harmonic limits for individual equipment loads.

38.3 Characterization of the Disturbances

Several parameters are used in power systems to characterize distortion and harmonic content of a waveform and their effects: the PF, the THD, the distortion factor (DF), and the crest factor (CF).

The nonlinear load current i_L is generally expressed by

$$i_L(\theta_s) = \sum_{h=1}^{\infty} I_{Lh}\sqrt{2}\sin(h\theta_s - \phi_h) = I_{L1}\sqrt{2}\sin(\theta_s - \phi_1) + \sum_{h=2}^{\infty} I_{Lh}\sqrt{2}\sin(h\theta_s - \phi_h) \qquad (38.1)$$

where
  ϕ_1 is the phase angle of the fundamental load current
  $\theta_s = \omega t$, ω is the frequency of the network
  I_{L1} is the rms amplitude of the fundamental load current
  I_{Lh} is the rms amplitude of the hth harmonic load current
  ϕ_h is the phase angle of the hth harmonic load current

The total rms load current is

$$I_{rms} = \sqrt{\sum_{h=1}^{\infty} I_{Lh}^2}$$

(38.2)

38.3.1 Power Factor

The apparent power, (volt-amperes), of a power system is given as

$$S = V_{rms} I_{rms} = V_{rms} \sqrt{\sum_{h=1}^{\infty} I_{Lh}^2}$$

(38.3)

The real power, also known as active transmitted power (watts), is expressed as

$$P = V_{rms} I_{L1} \cos\phi_1$$

(38.4)

The expression of the PF includes the harmonic content of the current is

$$PF = \frac{P}{S} = \frac{I_{L1}}{\sqrt{\sum_{n=1}^{\infty} I_{Ln}^2}} \cos\phi_1$$

(38.5)

The reactive power (volt-ampere-reactive, var) is

$$Q = V_{rms} I_{L1} \sin\phi_1$$

(38.6)

In order to estimate the participation of the harmonics in the apparent power, one uses the concept of distortion power D defined as

$$D = V_{rms} \sqrt{\sum_{h=2}^{\infty} I_{Lh}^2}$$

(38.7)

The apparent power can thus be put in the form

$$S = \sqrt{P^2 + Q^2 + D^2}$$

(38.8)

The PF can therefore be expressed as

$$PF = \frac{P}{\sqrt{P^2 + Q^2 + D^2}}$$

(38.9)

Note that the PF is degraded by the presence of harmonics and the exchange of reactive power between the source and the load.

38.3.2 Total Harmonic Distortion

The THD makes it possible to evaluate the difference between the real waveform and the sinusoidal waveform for the current or the voltage. It is used to quantify the levels of the current flowing in the distribution system or the voltage level at the PCC where the utility can supply other customers. It is defined as the ratio of the rms amplitude of harmonics to the rms amplitude of the fundamental component of the voltage or current as given in the following equation:

$$\text{THD\%} = 100 \frac{\sqrt{\sum_{h=2}^{\infty} I_{Lh}^2}}{I_{L1}} \tag{38.10}$$

By using the THD applies to current, the PF becomes

$$\text{PF} = \frac{\cos \phi_1}{\sqrt{1 + \text{THD}^2}} \tag{38.11}$$

38.3.3 Distortion Factor

The DF is defined as the ratio between the rms value of the fundamental current and the rms value of the same current:

$$\text{DF} = \frac{I_{L1}}{I_{rms}} \tag{38.12}$$

When the current is perfectly sinusoidal DF = 1, the latter decreases when the current is distorted.

38.3.4 Crest Factor

Another important quantity that characterizes the quality of the source current is the CF. The CF is the ratio between peak values to the total rms value of the same current:

$$\text{CF} = \frac{I_{peak}}{I_{rms}} \tag{38.13}$$

For a sinusoidal waveform, CF is equal to 1.41. The peak factor can reach values higher than 4 and 5 for much distorted waves, especially for diode rectifiers feeding capacitive loads.

38.4 Types of Harmonic Sources

Nonlinear loads can generally be classified into two types, namely, voltage-source type of nonlinear loads (or voltage-fed type of harmonic-producing load) and current-source type nonlinear loads (or current-fed type of harmonic-producing load). These two types of harmonic sources have completely distinctive dual properties and characteristics. A voltage-source type nonlinear load may consist of loads like a diode or thyristor rectifier with a large smoothing capacitor at the load end. It is used in electronic equipments, household appliances, ac drives, and in power converters such as switch-mode power converters, UPSs, variable frequency drives (VFD), etc. Harmonics generated by these loads have

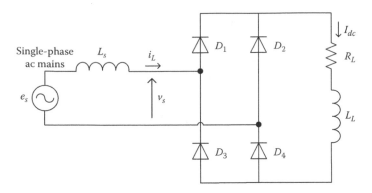

FIGURE 38.1   Single-phase current-source type nonlinear load.

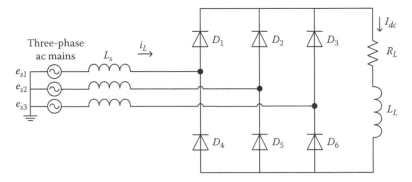

FIGURE 38.2   Three-phase current-source nonlinear load.

become a major issue. The recommended types of compensating filters for this type of harmonic source are series passive, active, and hybrid filters (a combination of passive and active filters) [41–57].

A current-source type of nonlinear load may consist of a diode or thyristor rectifier with sufficient inductance on the dc side such that it produces a constant direct current. It is used in applications such as dc drives, battery chargers, etc. The current at the input of the rectifier contains a large amount of harmonics due to the switching operation of the rectifier. It is recommended in [1] that for optimum harmonics compensation with these types of loads, parallel passive filters, active filters, and hybrid filters (a combination of the shunt or series passive and the shunt or series active filter) should be used due to the high dc-side impedance of the load that will force the compensating current to flow into the source side instead of the load side [16–40, 58–78]. Figures 38.1 and 38.2 show typical single-phase and three-phase current-source nonlinear loads. These bridge rectifiers feeding on the dc side an inductor $L_L = 10\,mH$ in series with a resistor $R_L = 12\,\Omega$. The single-phase supply voltage (v_s), the load current (i_L), and its spectrum analysis are shown in Figure 38.3. The measured THD of the load current is 11.31%.

The simulation results of the three-phase current-source nonlinear load are presented in Figure 38.4. The supply voltage (v_{s1}), the load current (i_{L1}), and the spectrum of the load current in phase 1 are depicted in the same figure. The measured THD of the current generated by the nonlinear load is approximately 27.06%. The results show that the current contains a large number of odd harmonics. This distorted current causes a distorted voltage drop on the supply conductors, leads to voltage distortion in the supply systems, and results in poor power quality.

Figures 38.5 and 38.6 show, respectively, single-phase and three-phase voltage-source type of nonlinear loads. These rectifiers are feeding a dc load constituted by a dc capacitor $C_L = 1000\,\mu F$ connected in parallel with a resistor $R_L = 12\,\Omega$. Figure 38.7 illustrates the supply voltages (v_s), the load current (i_L), and the spectral analysis of the supply voltage and the load current. The THD of the supply voltage and

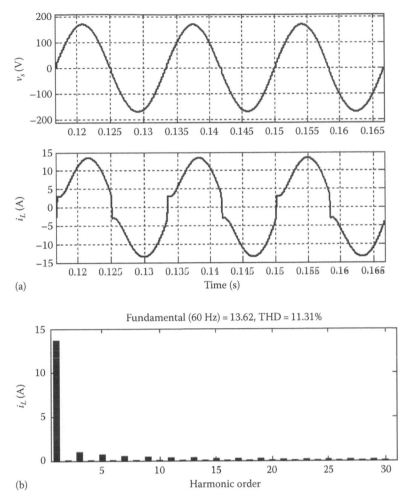

(a)

(b)

FIGURE 38.3 Steady-state response of single-phase current-source nonlinear load: (a) voltage and current waveforms and (b) spectrum of load current.

load current are 9.45% and 115.98%, respectively. It is interesting to note that the load current contains a large amount of odd harmonics, with the third harmonic being dominant.

In Figure 38.8, the supply voltage (v_{s1}), the load current (i_{L1}), and the spectrum of the supply voltage and load current in phase 1 are presented. One can notice that the THD of the supply voltage and the load current in phase 1 are 7.74% and 72.37%, respectively. It can be seen that the current and voltage waveforms of voltage-source nonlinear loads are much more distorted than those of the current-source nonlinear loads. This important distortion of the voltage is created due to discontinuity in the supply current.

38.5 Filters Used to Enhance Power Quality

Harmonic reduction is becoming more and more relevant due to the limitations required by interactional standards such as the IEC 1000-3-2 or EN61000-3-2 and IEEE-519. Several mitigation methods are available that permit substantial reduction of harmonics components. Different power filters have been installed in power systems to keep the harmonic distortion within acceptable limits. Conventionally, passive filters alone have been broadly used for harmonic mitigation; these devices have the advantages

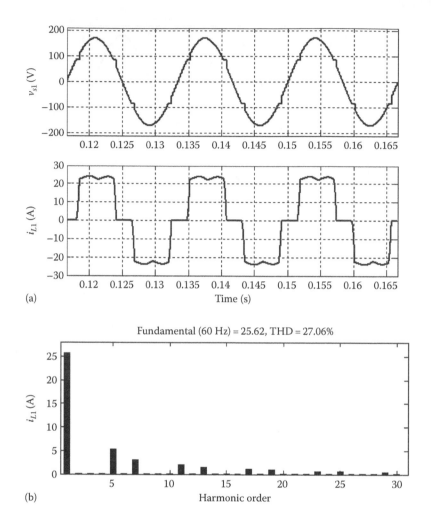

(a)

Fundamental (60 Hz) = 25.62, THD = 27.06%

(b)

FIGURE 38.4 Steady-state response of three-phase current-source nonlinear load: (a) voltage and current waveforms in phase 1 and (b) spectrum of load current.

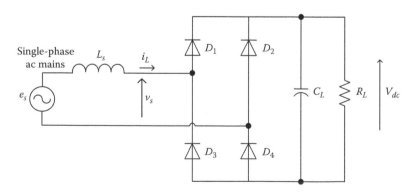

FIGURE 38.5 Single-phase voltage-source nonlinear load.

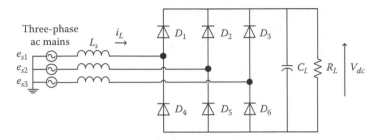

FIGURE 38.6 Three-phase voltage-source nonlinear load.

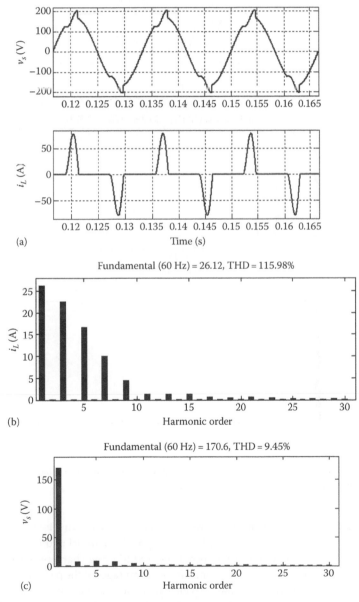

FIGURE 38.7 Steady-state response of single-phase voltage-source nonlinear load: (a) voltage and current waveforms, (b) spectrum of load current, and (c) spectrum of source voltage.

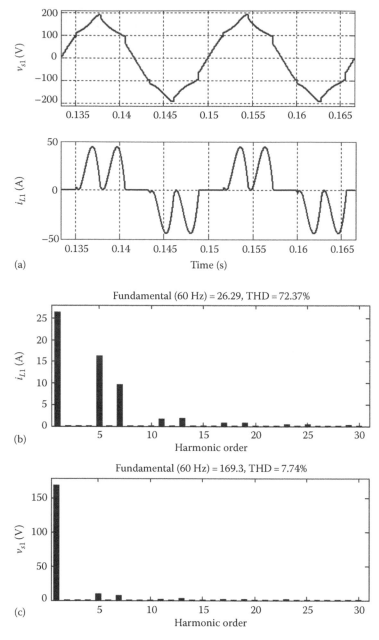

FIGURE 38.8 Steady-state waveforms of three-phase voltage-source nonlinear load: (a) supply voltage and current in phase 1, (b) load-current spectrum, and (c) source-voltage spectrum.

of being simple to design, not expensive to install, reliable, and require low maintenance efforts. However, they have several drawbacks, such as large size, possible parallel and/or series resonance that could be created with both load and utility impedances, and filtering characteristics strongly affected by source and load impedances [1–4]. To overcome the disadvantages of the passive filters, various types of APF have been developed to improve power performance. But, APF topologies suffer from high cost due to high KVA rating of the converter, and are less reliable. Hybrid active filters (HAFs) provide improved performance and have become a cost-effective solution to harmonic elimination, particularly for high-power nonlinear loads. The other alternative is the use of a UPQC to compensate voltage and current problem simultaneously. However, the use of UPQC is an expensive solution.

38.5.1 Passive Filters

Passive filters are combinations of inductors, capacitors, and damping resistors connected in series or in parallel to present the appropriate high or low impedance to the current or voltage harmonics. Various topologies of passive filter are available and have different compensation characteristics and applications. They are generally used as shunt passive filters or series passive filters.

38.5.1.1 Shunt Passive Filters

The shunt passive filter is a series-tuned resonant circuit having low impedance at the tuned frequencies. It can also provide limited reactive power compensation and voltage regulation. Single-tuned, first-order, second-order, and third-order high-pass passive filters are commonly used configurations. Generally, one or more passive filter branches are designed for low-order harmonics and then one high-pass filter is designed for the rest of the higher order harmonics. The shunt passive filter is very effective for compensating current-source nonlinear loads type of generated harmonics. These filters though quite useful pose various practical problems. The filter may create a series or parallel resonance with the source impedance resulting in the amplification of the harmonics with negative consequences. The frequency variation of the power system and tolerances in filter components affect its compensation characteristics. As a result, the size of the components in each tuned branch becomes impractical if the frequency variation is large. Overload occurs when the load harmonics level increases and consequently high current and voltages circulate in the passive branches; therefore, protection circuits are generally added to prevent such cases. Moreover, the supply impedance strongly influences the performance of the shunt passive filter, and since the source impedance may not be easily determined, the performance of the shunt passive filter becomes difficult to predict. Figures 38.9 through 38.12 show the most used passive filter configurations.

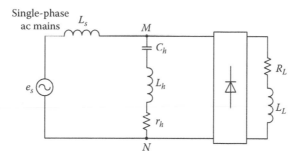

FIGURE 38.9 Series-tuned second-order resonant branch (inductance, capacitance, and resistance in series).

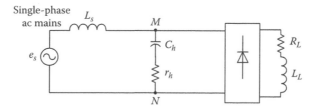

FIGURE 38.10 First-order high-pass passive filter.

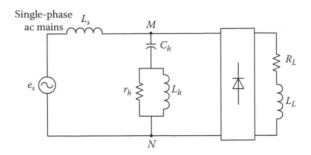

FIGURE 38.11   Second-order high-pass passive filter.

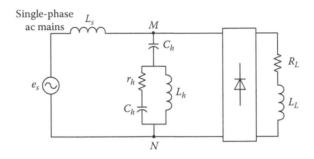

FIGURE 38.12   Third-order high-pass passive filter.

The equivalent dynamic circuit of a series-tuned, second-order resonant branch at harmonic scale as seen between points M and N is shown in Figure 38.13.

The source impedance is given by the following equation:

$$Z_{s1} = jL_s h\omega \tag{38.14}$$

where
L_s is the source inductance
$\omega = 2\pi f$ is the fundamental source frequency
j is the imaginary unit with the property $j^2 = -1$

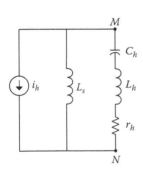

FIGURE 38.13   Equivalent harmonic diagram seen as of points M and N.

After connecting the filter, the impedance for the harmonic h seen between points M and N becomes

$$Z_{s2} = \frac{jL_s h\omega\left(1 - L_h C_h (h\omega)^2 + jr_h C_h(h\omega)\right)}{1 - (L_s + L_h)C_h(h\omega)^2 + jr_h C_h(h\omega)} \tag{38.15}$$

If the inductance resistance is neglected ($r_h \approx 0$), then Z_{s2} becomes

$$Z_{s2} = \frac{jL_s h\omega\left(1 - L_h C_h (h\omega)^2\right)}{1 - (L_s + L_h)C_h(h\omega)^2} \tag{38.16}$$

The quality factor (Q) of the passive filter, which is defined as the ratio of capacitive reactance (X_c) or inductive reactance (X_L) to the resistance (r_h) at tuned frequency, becomes infinite. Therefore, Q can be expressed as

$$Q = \frac{L_h \omega}{r_h} = \frac{1}{C_h \omega r_h}$$

(38.17)

Consequently, the tuned resonant frequency of the filter at the hth harmonic rank is given as

$$f_h = \frac{1}{2\pi h \sqrt{L_h C_h}}$$

(38.18)

where
f_h is the tuned frequency
C is the filter capacitance
L is the filter inductance

On the other hand, the parallel impedance, also known as anti-resonance impedance, which can involve amplification and overvoltage at the frequency f_{rh}, can be expressed as in Equation 38.18a. The amplification factor depends on the quality factor of the filter:

$$f_{rh} = \frac{1}{2\pi h \sqrt{(L_h + L_s)C_h}}$$

(38.18a)

Indeed, the decrease of the quality factor of the filter inductance reduces the overvoltage at the resonance frequency. Also, at the resonance frequency, the impedance is not null and the specific hth harmonic is not completely deviated. The ratio of the impedance after filtering to the impedance before filtering as a function of frequency when the resonant filter is tuned to fifth harmonic is given in Figure 38.14.

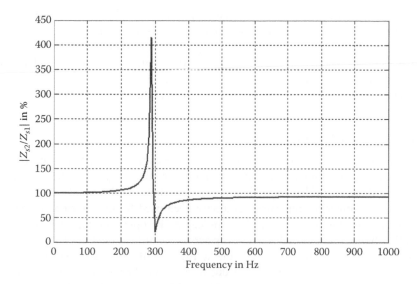

FIGURE 38.14 Compensation characteristic of series-tuned resonant circuit.

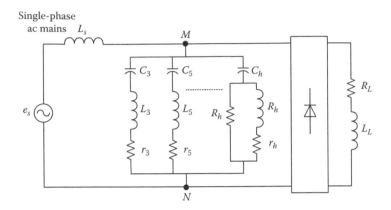

FIGURE 38.15 Single-phase tuned shunt passive filter.

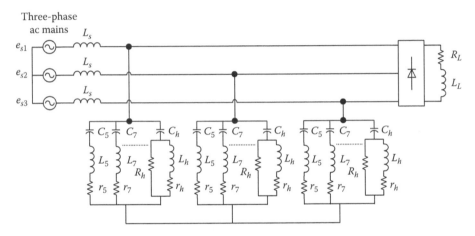

FIGURE 38.16 Three-phase three-wire tuned shunt passive filter.

To eliminate several harmonics, the idea consists in placing a tuned filter by harmonic. The elimination of k harmonics requires the parallel connection of k-tuned filters. In practice, each passive filter element employs three tuned filters, the first two being for the lowest dominant harmonics followed by high-pass filter elements. Figures 38.15 and 38.16 show the single-phase and three-phase tuned shunt passive filters. The ratio of the impedance after filtering with the impedance before filtering as a function of frequency when the resonant shunt passive filter is tuned at the fifth and seventh harmonic is given by Figure 38.17.

38.5.1.2 Series Passive Filters

Series passive filters are constituted of parallel resonant branches connected in series with the nonlinear loads. They provide high impedance to the harmonic currents and prevent them from flowing into the power system. These filters also help reduce the current ripple on the dc side of the rectifier circuit. They are of low cost, simple to implement, and have been used to limit harmonics caused by large loads. The series passive filters suffer heavily from lagging PF operation for a whole range of operation. On the other hand, a finite small voltage drop across the finite inductive reactance and resistance of the coil occurs at fundamental frequencies due to the difficulty in designing sharply tuned filters, and large drop occurs at harmonic frequencies due to current at harmonic frequencies escaping from the block that has been created by these filters. The series passive filter has been found suitable for

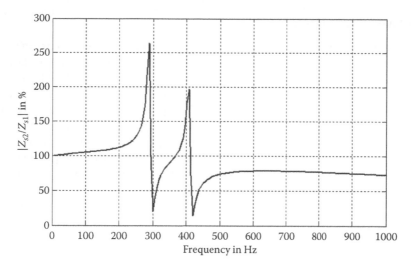

FIGURE 38.17 Compensation characteristic of series tuned shunt passive filter that is tuned at the fifth and seventh harmonic.

voltage-fed type of harmonic-producing loads. Generally, each series passive filter element employs three tuned filters, the first two being for the lowest dominant harmonics followed by high-pass filter elements. In each series passive filter element, two lossless LC components are connected in parallel for creating a harmonic dam to block harmonic currents. All the three components of the series passive filter are connected in series. Figures 38.18 and 38.19 show the general schemes of single-phase and three-phase series passive filters.

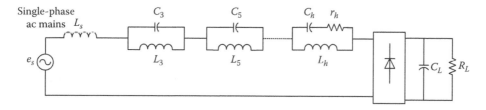

FIGURE 38.18 Single-phase series passive filter.

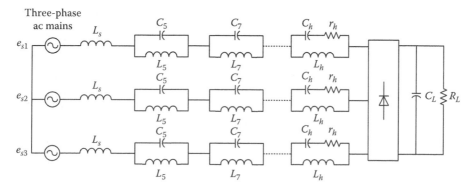

FIGURE 38.19 Three-phase three-wire series passive filter.

38.5.2 Active Power Filter

To overcome the limitation of passive filters, APFs were developed to provide better dynamic control of current harmonics and voltage distortion control. This is achievable thanks to the developments in solid-state switching devices and control technology in recent years. APFs can be classified based by a number of elements in topology, supply system, and the types of converter used in their circuits. They are single-phase (two-wire), three-phase three-wire, and three-phase four-wire voltage- or current-source inverters used to generate the compensating voltage or current that is injected into the line. Current source active filters employ an inductor as the dc energy storage device. In voltage-source active filters, a capacitor acts as the energy storage element. Voltage source active filters are cheaper, lighter, and easier to control compared to current-source active filters. Several APF design topologies as illustrated in the block diagram shown in Figure 38.20 have been proposed. They can be classified as follows: shunt active power filter (SAPF), series APF, hybrid shunt active filter, hybrid series active filter, and UPQC. They use PWM-controlled current-fed or voltage-fed converters with inductive and capacitive energy storage elements, respectively.

38.5.2.1 Shunt Active Power Filter

The shunt active filter operates by injecting harmonic current into the utility system with the same magnitudes as the harmonic currents generated by a given nonlinear load, but with opposite phases to maintain a sinusoidal current at the PCC. The major aim of the shunt active filter is to compensate harmonic currents yielding an improvement of the PF. It can also be used as a static var compensator in power system networks for compensating for other disturbances such as voltage flicker and imbalance. The SAPF offers some advantages such as the following: source-side inductance does not affect the harmonic compensation capability of the SAPF system, cost-effective for low to medium KVA industrial loads, can damp harmonic propagation in a distribution feeder, do not create displacement PF problems, and utility loadings. On the other hand, the APF topologies suffer from high KVA rating of the power electronic inverter for high-power industrial loads. This is due to the fact that the converter must withstand the line frequency, utility voltage, and supply harmonic current. In addition, it does not compensate for the harmonic in the load voltage. Figures 38.21 and 38.22 show a single-phase and a three-phase voltage-fed shunt active filter. A large capacitor connected to the dc bus of the converter behaves as a voltage source. A single-phase and a three-phase current-fed shunt active filter are shown in Figures 38.23 and 38.24. They use an inductive element for energy storage.

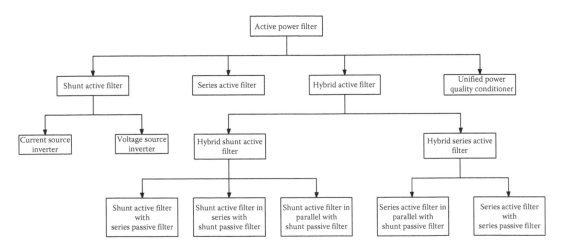

FIGURE 38.20   Configurations of APF for power quality improvement.

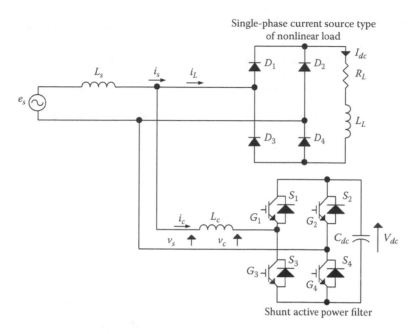

FIGURE 38.21   Single-phase voltage-fed shunt active filter.

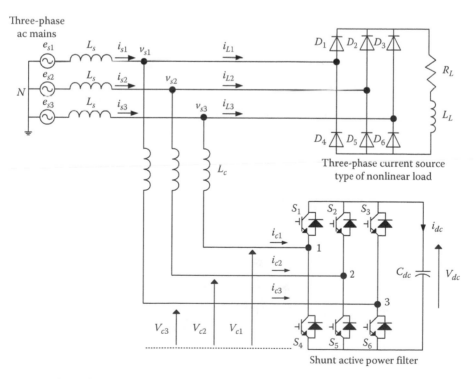

FIGURE 38.22   Three-phase voltage-fed shunt active filter.

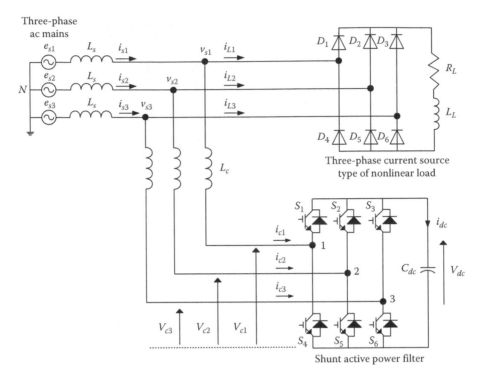

FIGURE 38.23   Single-phase current-fed shunt active filter.

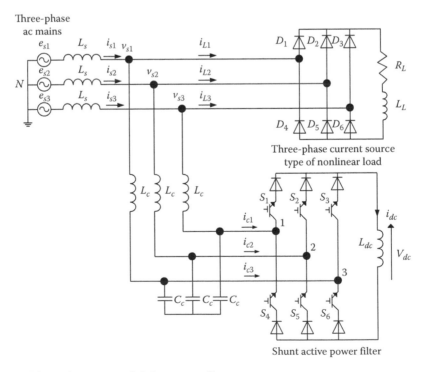

FIGURE 38.24   Three-phase current-fed shunt active filter.

The inductor behaves as a controllable nonsinusoidal current source to compensate for the harmonic current requirement of nonlinear loads. A diode is used in series with the self-commutating device for reverse voltage blocking.

A four-pole switch type, a capacitor midpoint type, and a three single-phase bridge configuration of a four-wire SAPF are shown in Figures 38.25 through 38.27.

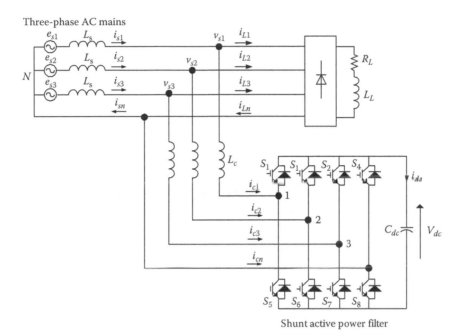

FIGURE 38.25 Four-pole four-wire shunt active filter.

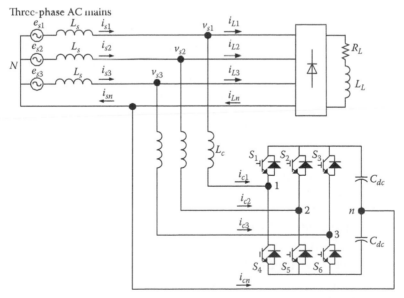

FIGURE 38.26 Capacitor midpoint four-wire shunt active filter.

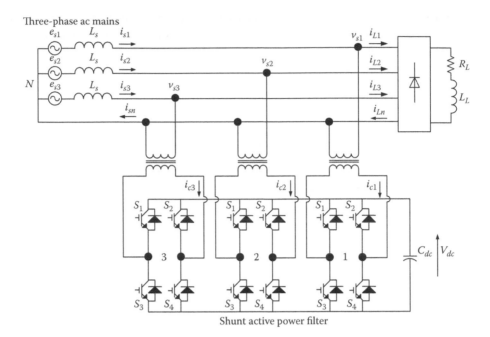

FIGURE 38.27 Three single-phase bridge four-wire shunt active filter.

38.5.2.2 Series Active Power Filter

The series APF is connected in series with the utility system through a matching transformer so that it prevents harmonic currents from reaching the supply system or compensates the distortion in the load voltage. It is controlled in such a way that it can present zero impedance, at the PCC, to the fundamental frequency and high impedance to harmonic frequencies to prevent harmonic currents from flowing into the system. It injects the necessary voltage needed for compensation of voltage harmonics, voltage sags, and swells in dynamic voltage restoration, voltage flicker, and other voltage disturbances that distort the desired sinusoidal waveform at the PCC. It is also used to damp out harmonic propagation caused by resonance with line impedance and shunt passive filters. The series active filter is effective for compensating such voltage-source nonlinear loads. The function of the series active filter is not to directly compensate for the current harmonics of the load, but to isolate the current harmonics between the load and the source. A drawback to the series active compensator is its inability to directly compensate for current harmonics, balance the load current, suppress neutral currents, and compensate the reactive power. In addition, it carries full load current and must withstand large power ratings. In the event of a failure of the filter's transformer, the load will lose the power supply. Series active filters are designed either as controllable voltage sources (voltage-fed converter type) or as controllable current sources (current-fed converter type). A single-phase and a three-phase voltage-fed series active filter are shown in Figures 38.28 and 38.29. Figures 38.30 and 38.31 show a single-phase and a three-phase current-fed series active filter. Figure 38.32 shows a three-phase four-wire voltage-fed series active filter.

38.5.2.3 Unified Power Quality Conditioner

The UPQC is a combination of series and SAPF, which are connected back to back and sharing a common self-supporting dc link. The series filter is controlled as a voltage source; hence, it is used for voltage compensation while the shunt filter compensates for harmonic currents. Hence, UPQC has the advantages of both the series and shunt filter, simultaneously. Although its main drawback is its

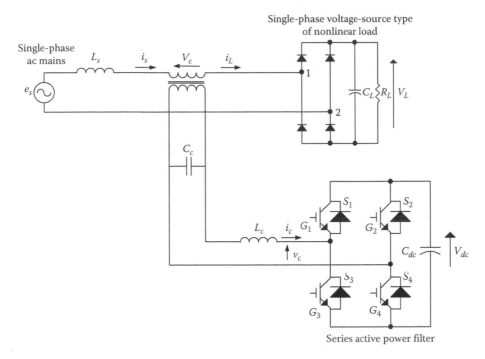

FIGURE 38.28 Single-phase voltage-fed series active filter.

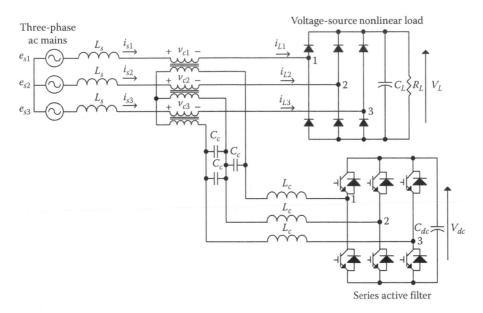

FIGURE 38.29 Three-phase voltage-fed series active filter.

high cost and complexity of control, interest in UPQC is growing due to its superior performance. It can compensate significant power quality issues, such as, voltage harmonics, voltage sag, voltage swell, voltage unbalance, voltage flicker, current harmonics, load reactive power, current unbalance, and neutral current. A single-phase voltage-fed and current-fed UPQC are shown in Figures 38.33 and 38.34. Figures 38.35 and 38.36 show the schematic of a three-phase three-wire and a three-phase four-wire UPQC.

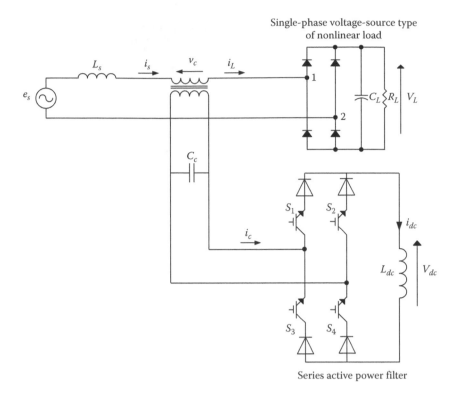

FIGURE 38.30   Single-phase current-fed series active filter.

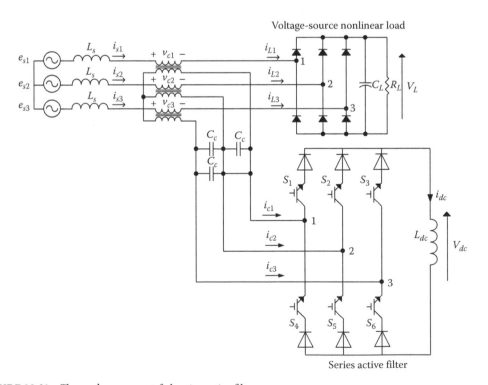

FIGURE 38.31   Three-phase current-fed series active filter.

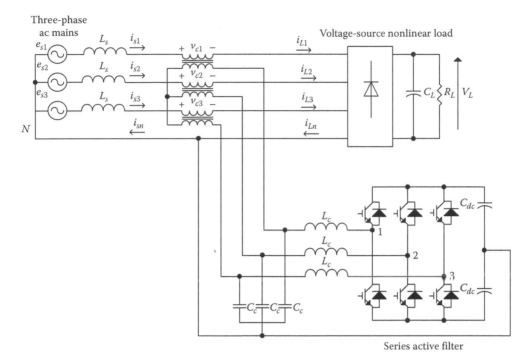

FIGURE 38.32 Three-phase four-wire voltage-fed series active filter.

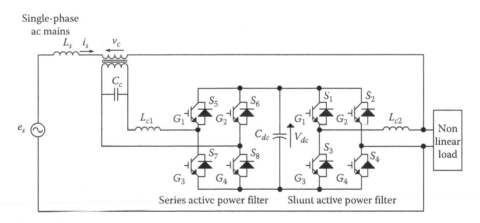

FIGURE 38.33 Single-phase voltage-fed UPQC.

38.5.2.4 Hybrid Filters

APFs have the capability of damping harmonic resonance between an existing passive filter and the supply impedance, but they require a large current rating with low efficiency and harmful disturbance to neighborhood appliances. HAF topologies that combine the advantages of both active and passive filters are more appealing in terms of cost and performance. They are cost-effective by reducing the KVA rating of the active filter as much as possible while offering harmonic isolation and voltage regulation. Two types of HAFs have been developed: a shunt HAF and a series HAF. The shunt hybrid filters consisting of shunt active filter and shunt or series passive filters connected in series or in parallel with each other combine the advantages of both filters. This is an attempt to reduce the high KVA rating of the shunt active filter without compromising its functions. A single-phase and a three-phase shunt HAF are shown

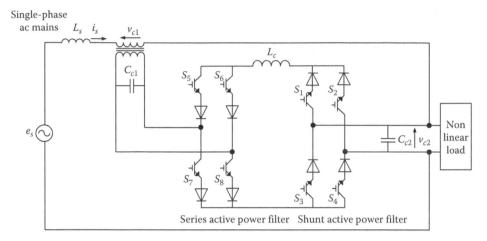

FIGURE 38.34   Single-phase current-fed UPQC.

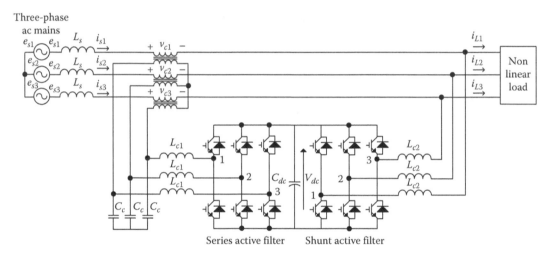

FIGURE 38.35   Circuit configuration of a three-phase three-wire voltage-fed UPQC.

in Figures 38.37 through 38.42. The reduced switch hybrid filters along with load arrangement are shown in Figures 38.38 and 38.40. Since one leg of the active filter is eliminated by a center tap capacitor, the gate drive circuit requirement and associated electronic circuit including the number of current sensors are eliminated. Thus, these topologies reduce the overall cost of the system. Figures 38.43 and 38.44 show the schematic of a three-phase series hybrid filter. The series hybrid filter is a series or parallel combination of a series active filter and series or shunt passive filter. The series active filter acts as a harmonic "isolator." It is to improve the filtering characteristics and to solve the problems of the passive filter. Hence, the rating of the series active filter is much smaller than that of a conventional parallel active filter.

38.6 Control of Active Filters

The quality and performance of the APF depends partly on the modulation and control method used to implement the compensation scheme. Several control strategies can be used to regulate the current produced by the filter: variable switching frequency, such as hysteresis and sliding-mode controls allow direct control of the current, but make the design of the output filter quite difficult as well as the reduction

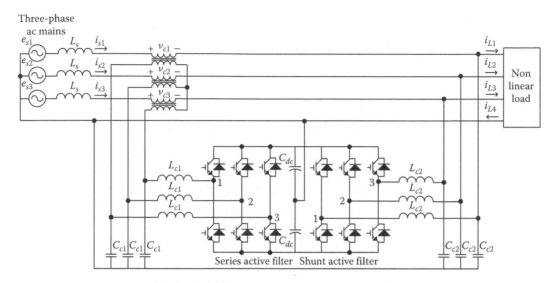

FIGURE 38.36 Circuit configuration of a three-phase four-wire voltage-fed UPQC.

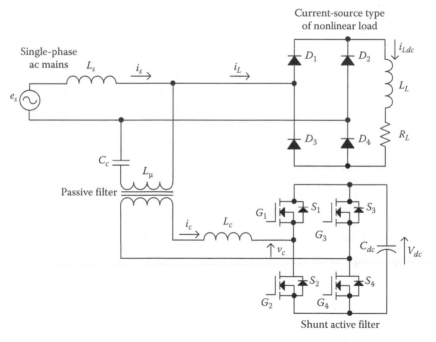

FIGURE 38.37 Single-phase voltage-fed shunt hybrid filter.

of the noise level. PWM control [12,85–88] eliminates these problems, but the dynamic response of the current feedback loop reduces the ability of the filter to compensate for fast current transitions. Many algorithms in time and frequency domains are proposed to extract or estimate compensating harmonic references for controlling the APFs [89–102]. The most popular are the time-domain methods such as the notch filter, the instantaneous reactive power theory (IRPT), the synchronous reference frame (SRF) theory, high-pass filter method, low-pass filter method, unity PF method, sliding-mode of control, passivity-based control, proportional integral (PI) controller, flux-based controller, and sine multiplication method. The main advantage of these time-domain control methods compared to the frequency-domain

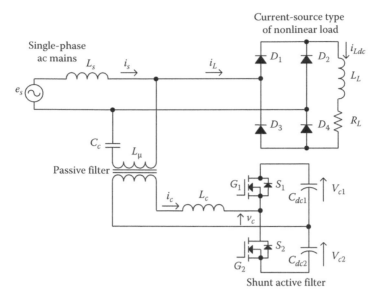

FIGURE 38.38   Single-phase voltage-fed reduced switch shunt hybrid filter.

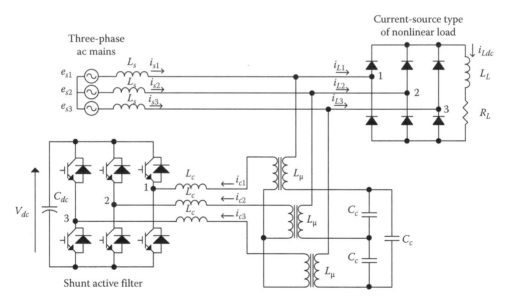

FIGURE 38.39   Three-phase three-wire voltage-fed shunt hybrid filter.

methods based on the fast Fourier transformation (FFT) is the fast response obtained. On the other side, frequency domain methods provide accurate individual and multiple harmonic load-current detection. The discrete Fourier transform, the Kalman filter, and the artificial neural networks are the most harmonic estimation techniques. There are two control techniques used to generate switching signals for APFs, namely, the direct and the indirect control techniques [12–15]. The direct method detects harmonics in the loads and injects current through an active filter to cancel the harmonics. On the other hand, the indirect method senses the harmonics in an ac network, and injects harmonic currents using feedback control to reduce the harmonics. It has been demonstrated that the direct method is not robust enough when the time delay in the control circuits is considered, and the indirect method is more reliable. Indeed, in the literature, it was reported that the current-type nonlinear load exhibits a step

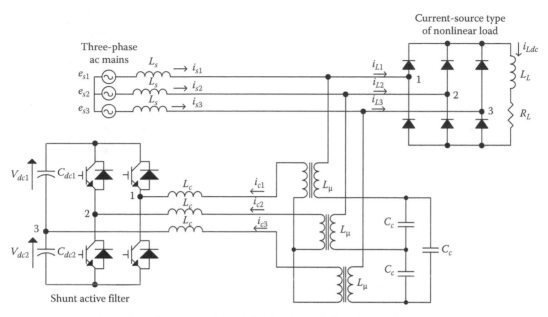

FIGURE 38.40 Three-phase three-wire voltage-fed reduced switch shunt hybrid filter.

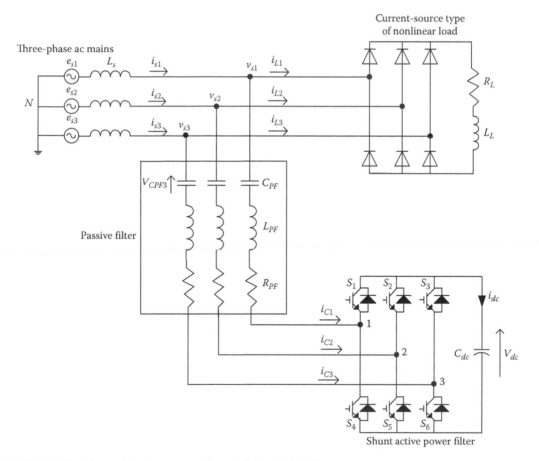

FIGURE 38.41 Three-phase three-wire voltage-fed shunt hybrid filter.

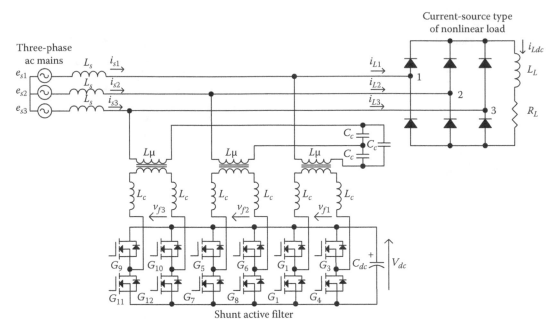

FIGURE 38.42   Three single-phase bridge three-wire voltage-fed shunt hybrid filter.

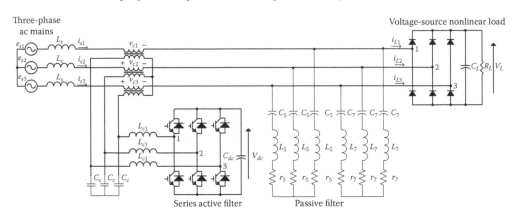

FIGURE 38.43   Three-phase three-wire voltage-fed series hybrid filter.

wave shape, and there is an instantaneous change from one step to another. This requires instantaneous compensation, but the inherent delay in the compensation using direct current control scheme results in switching ripples in the supply current. It is also essential to find why the direct control algorithm of APF suffers from this problem of switching ripples. The reference APF current are fast varying nonsinusoidal signals and the direct current control algorithm works on the principle of feed-forward control, where, the reference current of the APF is compared with its sensed current. Therefore, at a point in the ac cycle, the direct current controller does not have accurate information about the shape of the actual (sensed) supply current. Therefore, even if there are switching ripples in the supply current, the direct current controller does not compensate the ripples due to lack of exact information.

Section 38.7 addresses the analysis of the single-phase shunt active power filtering scheme with two control techniques. Direct and indirect current control techniques are presented with the use of a unipolar PWM (U-PWM) and a bipolar PWM (B-PWM) applied to the single-phase shunt active power filter (SPSAPF) to compensate the current harmonics and the reactive power [12–15]. It is demonstrated that by

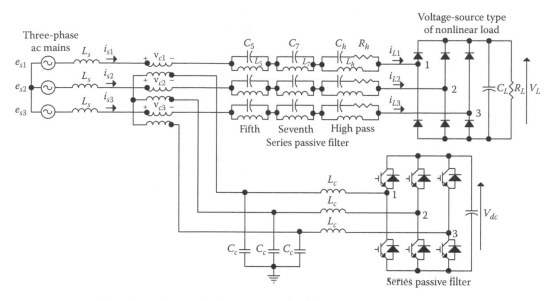

FIGURE 38.44 Three-phase three-wire voltage-fed series hybrid filter.

using the averaging technique, the direct consequence of using the U-PWM is that the transfer function of the SPSAPF becomes a pure gain, which simplifies the tuning of the regulator parameters. Also, the U-PWM pushes back the first significant harmonic rays toward twice the switching frequency $2f_{sw}$. Furthermore, it eliminates the rays groups that are centered on the odd multiples of the switching frequency. In addition to the current compensation loop, a voltage loop is also designed in order to regulate the dc bus voltage and to stabilize it at a designed value. The current and voltage regulators are designed by applying the linear control theory on a small-signal frequency domain model of the filter. This mathematical model is derived by using the state-space average modeling technique and, then, applying the small-signal linearization process. The SPSAPF is analyzed based on effective THD levels and response to changing dynamics. The results concerning the two PWM control techniques are verified by simulation, and experimental results on a 1 kVA prototype obtained using the direct and indirect current control strategies confirm the predicted performance and the superiority of the indirect current control technique with the U-PWM.

38.7 Single-Phase Shunt Active Power Filter Topology

Figure 38.45 shows the complete system where the SPSAPF is connected in parallel with a nonlinear load consisting of a single-phase diode rectifier. The SPSAPF consists of a single-phase PWM full-bridge voltage-source inverter, a dc bus capacitor C_{dc}, and an inductor L_c, which is required to attenuate the high-frequency ripples generated by the voltage-source inverter (VSI). A single-phase diode bridge rectifier feeding a series RL circuit represents the nonlinear load. The converter losses are represented by shunt resistance R_{dc} connected in parallel to a dc bus capacitor C_{dc}.

38.7.1 Extraction of Reference Signals

The performance of an active filter is greatly influenced by the method used for extracting the current reference. There are two control techniques for line current wave shaping in an active filter: the direct current control and the indirect current control. In the direct current control technique, the closed-loop current error is the difference between the desired current i_c^* and the real current i_c at the ac input of the shunt active filter. Whereas, in the indirect current control strategy, the current error is the gap between the source current reference i_s^* and the sensed source current i_s.

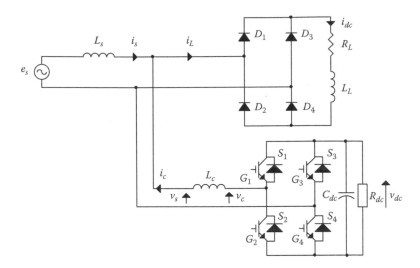

FIGURE 38.45   Configuration of the studied system.

38.7.1.1 The Indirect Current Control Technique of SPSHPF

The generation of the reference current is based on the determination of the amplitude of the fundamental active current i_{Lf}, which is done with the help of the classic demodulation technique. The nonlinear load current i_L is decomposed of the fundamental component i_{Lf} and the harmonic components i_{Lh} as follows:

$$i_L(\theta_s) = \hat{i}_{L1}\sin(\theta_s - \phi_1) + \sum_{h=2}^{\infty} \hat{i}_{Lh}\sin\left(h\theta_s - \phi_h\right) = i_{Lf} + i_{Lh} \tag{38.19}$$

where
    ϕ_1 is the phase angle of the fundamental load current
    $\theta_s = wt$, w being the mains frequency
    $\hat{i}_{L1}$ is the peak value of the fundamental current
    $\hat{i}_{Lh}$ is the peak value of the hth harmonic load current
    ϕ_h is the angle of the hth harmonic load current

$$i_{Lf} = I_{L1}\sin(\theta_s - \phi_1) \tag{38.20}$$

$$i_{Lh} = \sum_{h=2}^{\infty} \hat{i}_{Lh}\sin(h\theta_s - \phi_h) \tag{38.21}$$

On the other hand, the fundamental current can be divided into two components, namely, fundamental active $i_{Lfa} = \hat{i}_{L1}\cos\phi_1\sin\theta_s$ and fundamental reactive $i_{Lfr} = \hat{i}_{L1}\sin\phi_1\cos\theta_s$.

This technique allows the compensation of harmonics and reactive power at the same time. The objective is to cancel the harmonics and to compensate for the reactive power; therefore, the reference current for the active filter i_s^* is equal to the load fundamental active current i_{Lfa}:

$$i_s^* = i_{Lfa} = i_L - (i_{Lh} + i_{Lfr}) \tag{38.22}$$

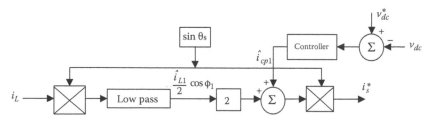

FIGURE 38.46 Indirect current control algorithm of SPSAPF system.

In order to simplify the filtering of the load current i_{Lh}, the fundamental component i_{Lfa} is transformed into the dc component. Multiplying both sides of Equation 38.19 by $\sin \theta_s$,

$$i_L(\theta_s)\sin\theta_s = \frac{\hat{i}_{L1}}{2}\cos\phi_1 - \frac{\hat{i}_{L1}}{2}\cos(2\theta_s - \phi_1) + \sin\theta_s \sum_{h=2}^{\infty} \hat{i}_{Lh}\sin(h\theta_s - \phi_h) \tag{38.23}$$

This equation shows the presence of a dc component and the ac components of which minimal frequency is equal to twice the frequency network (120 Hz). A low-pass filter, with a relatively low cut-off frequency is used to prevent the high-frequency component. However, it is indispensable to respect a good compromise between the effective filtering of frequencies parasites and the fast dynamics of the extraction algorithm.

The compensator active current $\hat{i}_{cp1}$ is obtained from the bus voltage regulation loop. The error between the reference value V_{dc}^* and the sensed feedback value V_{dc} is processed toward a PI controller giving $\hat{i}_{cp1}$ signal. This current is added to $2^* i_{Lfiltered}$, leading the peak value of the reference current. In order to reconstitute the fundamental active reference current, the peak value is multiplied by $\sin \theta_s$. The block diagram that generates i_s^* is shown in Figure 38.46.

38.7.1.2 The Direct Current Control Technique of SPSHPF

The proposed block diagram of the control algorithm of an active filter with direct current control is shown in Figure 38.47.

38.7.2 Principle of the Unipolar PWM Control

The generation of the U-PWM control pattern is illustrated in Figure 38.48. It is based on two comparisons: (1) between a control signal β and a triangular high-frequency carrier, and (2) between the opposite of the control signal ($-\beta$) and the same carrier. A typical waveform of the control signal β is given in Figure 38.49. Such a signal, which is slow time-varying in each half-period-wide interval, is generally delivered by a closed-loop PI controller that adjusts continuously the current i_c at the input of the active filter to track a current reference i_c^* such that

$$i_L - i_c^* = i_s^* = \hat{i}_s \sin\omega_s t \tag{38.24}$$

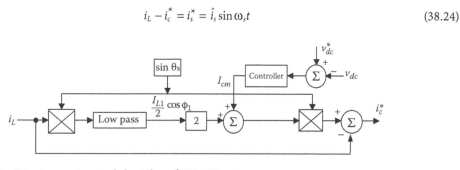

FIGURE 38.47 Direct current control algorithm of SPSAPF system.

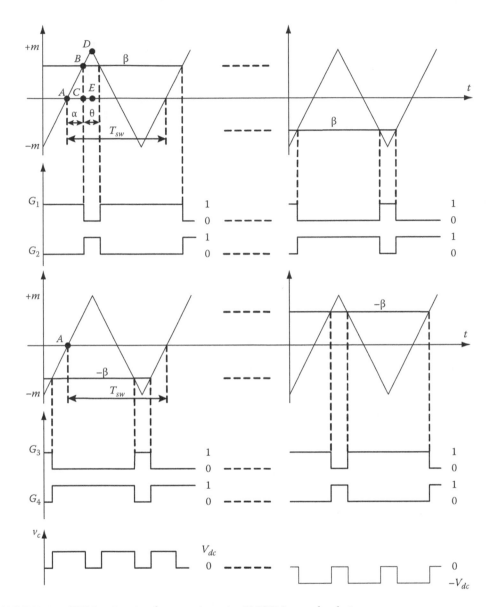

FIGURE 38.48 PWM gating signals generation using U-PWM control technique.

where ω_s denotes the angular frequency of the mains. The waveform given in Figure 38.49 allows compensating both the undesirable harmonics in the source current and the reactive power required by the nonlinear load.

Considering the frequency range of the control signal β and the triangular carrier, one may consider that the modulating signal is practically constant during a switching period T_{sw}. In this case, referring to Figure 38.48, we have

$$d_1 = 1 - \frac{\theta}{T_{sw}} = \frac{1}{2}\left(1 + \frac{\beta}{m}\right) \quad \text{and} \quad d_2 = \frac{\theta}{T_{sw}} = \frac{1}{2}\left(1 - \frac{\beta}{m}\right) \tag{38.25}$$

where

  d_1 and d_2 are, respectively, the duty cycles of switches T_1 and T_3 over one sampling period T_{sw}
  m is the peak value of the triangular carrier

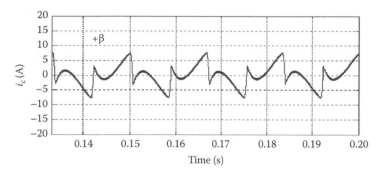

FIGURE 38.49 Modulating signal β.

One can, thus, deduce the average value of the voltage v_c at the input of the inverter over a switching period T_{sw}:

$$\langle v_c \rangle_{T_{sw}} = (d_1 - d_2)V_{dc} = \frac{V_{dc}}{m}\beta = G\beta \tag{38.26}$$

It should be noted that the average voltage $\langle v_c \rangle_{T_{sw}}$ is proportional to the control variable β. For operation below the switching frequency, one can assume that the PWM inverter transfer function is equivalent to the constant G.

38.7.2.1 Harmonic Analysis of Active Power Filter Inverter Input Voltage v_c

According to Figure 38.48, the voltage v_c is an alternating modulated square wave signal. Its frequency is twice the control signal of S_1 and S_3, and its magnitude is lower than the peak of its fundamental. In the following, let us define $v_c = v_{c1}$ when the standard PWM is used and $v_c = v_{c2}$ when the U-PWM is used.

38.7.2.2 Harmonic Analysis with Standard PWM ($v_c = v_{c1}$)

Under the assumption that the voltage v_{c1} remains periodic over a few successive switching cycles, i.e., $d_1 T_{sw}$ = constant, by applying the Fourier transform to v_c waveform shown in Figure 38.50, a first local decomposition is applied to the voltage v_{c1}:

$$v_{c1}(t) = a_0 + \sum_{k=1}^{\infty} a_k \cos\left(k\omega_{sw}t\right) \tag{38.27}$$

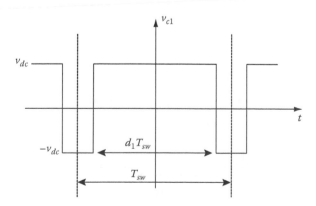

FIGURE 38.50 Voltage waveform of v_{c1} obtained by B-PWM.

where

$$a_0 = \langle v_{c1} \rangle = \frac{1}{T_{sw}} \int_0^{T_{sw}} v_{c1}(t)\,dt \quad \text{and} \quad a_k = \frac{2}{T_{sw}} \int_0^{T_{sw}} v_{c1}(t)\cos(k\omega_{sw}t)\,dt$$

Computing the coefficient a_0, which represents the local mean value of v_{c1}, leads to

$$a_0 = \langle v_{c1} \rangle = \frac{1}{T_{sw}}\left[V_{dc}d_1 T_{sw} - V_{dc}\left(T_{sw} - d_1 T_{sw}\right)\right] = V_{dc}\left(2d_1 - 1\right) \tag{38.28}$$

The calculation of coefficient a_k gives the following expression:

$$a_k = \frac{4}{T_{sw}} \int_0^{T_{sw}/2} v_{c1}(t)\cos(k\omega_{sw}t)\,dt = \frac{4V_{dc}}{k\pi}\sin(kd_1\pi) \tag{38.29}$$

Using Equations 38.27 through 38.29, we obtain finally

$$v_{c1} = V_{dc}(2d_1 - 1) + \sum_{k=1} \frac{4V_{dc}}{k\pi}\sin\left(kd_1\pi\right)\cos\left(k2\pi T_{sw}t\right) \tag{38.30}$$

38.7.2.3 Harmonic Analysis with U-PWM ($v_c = v_{c2}$)

We now seek the expression of v_{c2} with the U-PWM. This voltage can be constituted by subtracting terminal voltages of switches S_2 and S_4, respectively. These voltages v_{s2} and v_{s4} are illustrated in Figure 38.51.

The voltages v_{s2} and v_{s4} are obtained by calculating the coefficients a_0 and a_k of the relations (38.28) and (38.29). From which we can obtain

$$\begin{cases} v_{s2} = V_{dc}d_1 + \displaystyle\sum_{k=1} \frac{2V_{dc}}{k\pi}\sin\left(kd_1\pi\right)\cos\left(k\omega_{sw}t\right) \\[3mm] v_{s4} = V_{dc}d_2 + \displaystyle\sum_{k=1} \frac{2V_{dc}}{k\pi}\sin\left(kd_2\pi\right)\cos(k\omega_{sw}t) \end{cases} \tag{38.31}$$

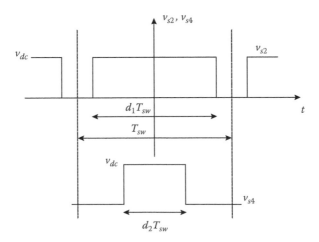

FIGURE 38.51 Voltage waveform of v_{s2} and v_{s4} obtained with U-PWM.

The relation (38.31) gives us the expression for voltage v_{c2}:

$$v_{c2} = v_{S2} - v_{S4} = V(d_1 - d_2) + \sum_{k=1} \frac{4V_{dc}}{k\pi} \left(\sin\left(\frac{k\pi}{2}(d_1 - d_2)\right) \cos\left(\frac{k\pi}{2}(d_1 + d_2)\right) \right) \cos k\omega_{sw}t \qquad (38.32)$$

38.7.2.4 Relation between v_{c1} and v_{c2}

In the following, let us find the mathematical relation between the voltage signal for both B-PWM v_{c1} and U-PWM v_{c2}.

The modulating signal β must contain two components: the fundamental active current ($\hat{\beta}_{1a} \sin \omega_s t$), which is necessary to maintain a constant terminal voltage across the C_{dc} capacitor; and all the reactive ($\hat{\beta}_1 \cos \omega_s$) and harmonic $\left(\sum_{h=2}^{\infty} \hat{\beta}_h \sin(h\omega_s t) \right)$ current components to compensate the nonlinear load. For simplification purposes, one can suppose that the modulating signal contains only the third harmonic components. Using Equation 38.26, one can obtain

$$\begin{cases} d_1 = \frac{1}{2}\left(1 + \frac{\beta}{m}\right) \\ d_2 = \frac{1}{2}\left(1 - \frac{\beta}{m}\right) \end{cases} \quad \text{with } \beta = \hat{\beta}_3 \sin(3\omega_s t) = \hat{\beta}_3 \sin(6\pi f_s t) \qquad (38.33)$$

where $\hat{\beta}_3$ is the amplitude of the modulating signal, which is the third harmonic current.

Hence, by replacing d_1 and d_2 in Equations 38.30 and 38.32, we get

$$\begin{cases} v_{c1} = V_{dc}\frac{\hat{\beta}_3}{m}\sin 3\omega_s t + \sum_{k=1} \frac{4V_{dc}}{k\pi}\sin\left(\frac{k\pi}{2}\left(1 + \frac{\hat{\beta}_3}{m}\sin 3\omega_s t\right)\right) \cos k\omega_{sw}t \\ v_{c2} = V_{dc}\frac{\hat{\beta}_3}{m}\sin 3\omega_s t + \sum_{k=1} \frac{4V_{dc}}{k\pi}\sin\left(\frac{k\pi}{2}\frac{\hat{\beta}_3}{m}\sin 3\omega_s t\right) \cos\left(\frac{k\pi}{2}\right) \cos k\omega_{sw}t \end{cases} \qquad (38.34)$$

Let $z = ((k\pi/2)(\hat{\beta}_3/m))\sin 3\omega_s t$, Equation 38.34 becomes

$$\begin{cases} v_{c1} = V_{dc}\frac{\hat{\beta}_3}{m}\sin 3\omega_s t + \sum_{k=1} \frac{4V_{dc}}{k\pi}\left(\sin\frac{k\pi}{2}\cos z + \cos\frac{k\pi}{2}\sin z\right) \cos k\omega_{sw}t \\ v_{c2} = V_{dc}\frac{\hat{\beta}_3}{m}\sin 3\omega_s t + \sum_{k=1} \frac{4V_{dc}}{k\pi}\sin z \cos\frac{k\pi}{2} \cos k\omega_{sw}t \end{cases} \qquad (38.35)$$

Functions z, $\cos z$, and $\sin z$ are represented by Figure 38.52. $\omega_3 = 3\omega_s$ is the third harmonic frequency.

Taking into account the periodicity and the observable symmetry of the functions of Figure 38.52, we can write the expressions of $\cos z$ and $\sin z$ as

$$\begin{cases} \cos z = \lambda_0 + \lambda_2 \cos 2\omega_3 t + \lambda_4 \cos 4\omega_3 t + \cdots \\ \sin z = \delta_1 \sin \omega_3 t + \delta_3 \sin 3\omega_3 t + \delta_5 \sin 5\omega_3 t + \cdots \end{cases} \qquad (38.36)$$

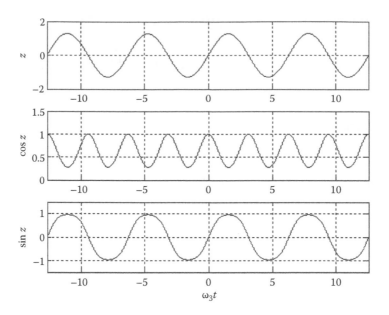

FIGURE 38.52 Waveform of functions z, $\cos z$, and $\sin z$.

By replacing $\cos z$ and $\sin z$ in the relation (38.35), one obtains

$$
\begin{cases}
v_{c1} = V_{dc}\dfrac{\hat{\beta}_3}{m}\sin\omega_3 t + \displaystyle\sum_{k=1}\dfrac{4V_{dc}}{k\pi}\Bigg[\left(\sin\dfrac{k\pi}{2}\right)\left(\displaystyle\sum_{j=0}\lambda_{2j}\cos 2j\omega_3 t\right)\cos k\omega_{sw}t \\[4mm]
\hspace{4cm} +\left(\cos\dfrac{k\pi}{2}\right)\left(\displaystyle\sum_{j=0}\delta_{(2j+1)}\sin(2j+1)\omega_3 t\right)\cos k\omega_{sw}t\Bigg] \\[6mm]
v_{c2} = V_{dc}\dfrac{\hat{\beta}_3}{m}\sin\omega_3 t + \displaystyle\sum_{k=1}\dfrac{4V_{dc}}{k\pi}\Bigg[\left(\cos\dfrac{k\pi}{2}\right)\left(\displaystyle\sum_{j=0}\delta_{(2j+1)}\sin(2j+1)\omega_3 t\right)\cos k\omega_{sw}t\Bigg]
\end{cases}
\tag{38.37}
$$

Taking into account the parity of k ($k = 2i$ or $k = 2i + 1$), relation (38.37) becomes

$$
\begin{cases}
v_{c1} = V_{dc}\dfrac{\hat{\beta}_3}{m}\sin\omega_3 t + \displaystyle\sum_{i=0}\dfrac{4V_{dc}}{(2i+1)\pi}\Bigg[(-1)^i\left(\displaystyle\sum_{j=0}\lambda_{2j}\cos 2j\omega_3 t\right)\cos(2i+1)\omega_{sw}t\Bigg] \\[4mm]
\hspace{2cm} +\displaystyle\sum_{i=1}\dfrac{2V_{dc}}{i\pi}\Bigg[(-1)^i\left(\displaystyle\sum_{j=0}\delta_{(2j+1)}\sin(2j+1)\omega_3 t\right)\cos(2i\omega_{sw}t)\Bigg] \\[6mm]
v_{c2} = V_{dc}\dfrac{\hat{\beta}_3}{m}\sin\omega_3 t + \displaystyle\sum_{i=1}\dfrac{2V_{dc}}{i\pi}\Bigg[(-1)^i\left(\displaystyle\sum_{j=0}\delta_{(2j+1)}\sin(2j+1)\omega_3 t\right)\cos(2i\omega_{sw}t)\Bigg]
\end{cases}
\tag{38.38}
$$

Finally, the spectral of v_{c1} and v_{c2} are given in the following relation:

$$
\begin{cases}
v_{c1} = V_{dc}\dfrac{\hat{\beta}_3}{m}\sin\omega_3 t + \sum_{i=0}\dfrac{4V_{dc}}{(2i+1)\pi}\left\{(-1)^i\left[\sum_{j=0}\dfrac{\lambda_{2j}}{2}\begin{bmatrix}\cos((2i+1)\omega_{sw}+2j\omega_3)t\\+\cos((2i+1)\omega_{sw}-2j\omega_3)t\end{bmatrix}\right]\right\}\\[2em]
\quad + \sum_{i=1}\dfrac{2V_{dc}}{i\pi}\left\{(-1)^i\left[\sum_{j=0}\dfrac{\delta_{(2j+1)}}{2}\begin{bmatrix}\sin(2i\omega_{sw}+(2j+1)\omega_3)t\\-\sin(2i\omega_{sw}-(2j+1)\omega_3)t\end{bmatrix}\right]\right\}\\[2em]
v_{c2} = V_{dc}\dfrac{\hat{\beta}_3}{m}\sin\omega_3 t + \sum_{i=1}\dfrac{2V_{dc}}{i\pi}\left\{(-1)^i\left[\sum_{j=0}\dfrac{\delta_{(2j+1)}}{2}\begin{bmatrix}\sin(2i\omega_{sw}+(2j+1)\omega_3)t\\-\sin(2i\omega_{sw}-(2j+1)\omega_3)t\end{bmatrix}\right]\right\}
\end{cases}
\tag{38.39}
$$

The relation (38.39) enables us to express v_{c2} according to v_{c1}:

$$
v_{c2} = v_{c1} - \sum_{i=0}\dfrac{4V_{dc}}{(2i+1)\pi}\left\{(-1)^i\left[\sum_{j=0}\dfrac{\lambda_{2j}}{2}\begin{bmatrix}\cos((2i+1)\omega_{sw}+2j\omega_3)t\\+\cos((2i+1)\omega_{sw}-2j\omega_3)t\end{bmatrix}\right]\right\}
\tag{38.40}
$$

Taking account of all the harmonics, the active component, and the reactive component, the modulating frequency is equal to the network frequency ω_s, from where the following relation is deduced:

$$
v_{c2} = v_{c1} - \sum_{i=0}\dfrac{4V_{dc}}{(2i+1)\pi}\left\{(-1)^i\left[\sum_{j=0}\dfrac{\lambda_{2j}}{2}\Big[\cos\big((2i+1)\omega_{sw}+2j\omega_s\big)t+\cos\big((2i+1)\omega_{sw}-2j\omega_s\big)t\Big]\right]\right\}
\tag{38.41}
$$

From Equation 38.41, one can deduce that, for the same $(2h+1)$th harmonic frequency, denoted by f_{2h+1}, switching frequency f_{sw}, and modulation depth $\hat{\beta}_{2h+1}/m$, the spectrum of v_{c2} is identical to the spectrum of v_{c1}, except that the groups of rays centered on the odd multiples of the switching frequency have disappeared. Figure 38.53a shows the spectra of v_{c1} and Figure 38.53b shows the spectra of v_{c2}.

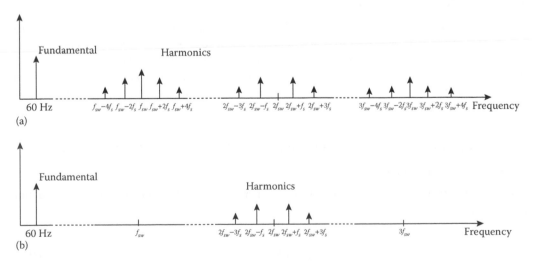

FIGURE 38.53   Comparison between the spectra of the voltage v_c obtained by (a) the B-PWM and (b) the U-PWM.

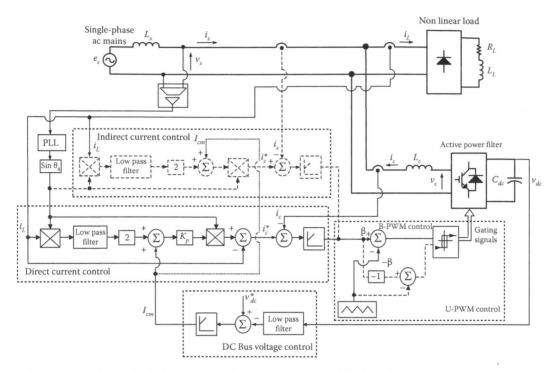

FIGURE 38.54 System block diagram with direct or indirect control blocks and B-PWM or U-PWM techniques in dashed line.

38.7.3 PWM's Principle of Gating Signal Generation

The reference supply (harmonic) current obtained from the control algorithm is compared with sensed supply (filter) current. As above, the error signal is fed to a controller having a limiter at its output. Consequently, the controlling signal β and its opposite −β are compared with a triangular carrier wave resulting in the switching signals to the gates of solid-state switching devices of VSI as shown is Figure 38.54.

38.7.4 Control of Active Power Filter

Figure 38.54 shows the complete diagram used for the identification of the active fundamental load current. This diagram includes the direct and indirect current control algorithm implemented with the B-PWM and U-PWM controllers. Besides the estimation of active fundamental load current, a real current reference component is derived from a PI regulator that controls the dc bus voltage of the inverter as shown in Figure 38.54. The component is summed to create the demanded reference supply current for the inner current regulator loop.

38.7.5 Small-Signal Modeling of the Single-Phase Active Power Filter

38.7.5.1 Averaged Model

By applying the average modeling technique to the filter in Figure 38.45, the following state equations are obtained:

$$L_c \frac{d\overline{i_c}}{dt} = \left(d_1 - d_2\right)\overline{v_{dc}} - \overline{v_s} \tag{38.42}$$

$$C_{dc} \frac{d\overline{v_{dc}}}{dt} + \frac{\overline{v_{dc}}}{R_{dc}} = -\left(d_1 - d_2\right) \overline{i_c} \tag{38.43}$$

where $\overline{i_c}$, $\overline{v_s}$, and $\overline{v_{dc}}$ denote, respectively, the averaged values, computed on the basis of a switching period, of the filter input current i_c, filter input voltage v_s, and the voltage v_{dc} at the dc side of the filter. d_1 and d_2 are, respectively, the duty cycles of the upper switches in the first and second leg of the filter.

Taking into account the following relation between d_1 and d_2,

$$d_1 + d_2 = 1 \tag{38.44}$$

Equations 38.43 and 38.44 can be rearranged to obtain a system having two outputs (which are also the state variables) $\overline{i_c}$ and $\overline{v_{dc}}$, and a single control input d_1:

$$L_c \frac{d\overline{i_c}}{dt} = \left(2d_1 - 1\right) \overline{v_{dc}} - \overline{v_s} \tag{38.45}$$

$$C_{dc} \frac{d\overline{v_{dc}}}{dt} + \frac{\overline{v_{dc}}}{R_{dc}} = -\left(2d_1 - 1\right) \overline{i_c} \tag{38.46}$$

38.7.5.2 Linear Control System

The control system is illustrated in Figure 38.54. A successive two-loops strategy is employed. The inner loop embeds an indirect current control technique. Here, and contrarily to the direct control method where the controlled current is the one at the filter input, it is the source current that is sensed and injected in an inner feedback system in order to shape it. This approach allows simplifying considerably the generation of the current reference and offers better dynamics due to the absence of discontinuities in the current reference waveform. The outer loop, which is designed to be enough slower than the inner one for stability considerations, ensures voltage regulations at the dc side of the filter by compensating the power losses in the semiconductors and the reactive elements of the filter (recall that these losses are represented by the fictive resistance R_{dc} in Figure 38.45).

In order to choose adequately the inner and outer regulators, the knowledge of the filter's small-signal transfer functions is required. As noticed in Figure 38.54, the determination of the transfer functions, on the basis of which the regulators are calculated, follows two steps.

38.7.5.2.1 Inner Subsystem Transfer Function

First, the inner control loop is considered. Here, a transfer function relating the inner output variable $\overline{i_s}$ to the control input d_1 has to be established. This is easily done by applying first small-signal linearization to Equation 38.45. We get

$$L_c \frac{d\left(\delta \overline{i_c}\right)}{dt} = \left(2D_1 - 1\right)V_{dc} + \left(2D_1 - 1\right)\delta\overline{v_{dc}} + 2\left(\delta \, d_1\right)V_{dc} - \left(V_s + \delta\overline{v_s}\right) \tag{38.47}$$

where δx denotes the small-variation of a time variable x around its static value X. Furthermore, the static regime is obtained by setting all the time derivatives and the small variations to zero. It yields

$$0 = \left(2D_1 - 1\right)V_{dc} - V_s \tag{38.48}$$

which gives

$$D_1 = \frac{V_s}{2V_{dc}} + \frac{1}{2} = \frac{1}{2} \tag{38.49}$$

knowing that the static value V_s of the voltage at the ac side of the filter is zero.

By replacing expression (38.49) into Equation 38.47 yields

$$L_c \frac{d\left(\delta \bar{i}_c\right)}{dt} = 2\left(\delta\ d_1\right)V_{dc} - \delta \bar{v}_s \tag{38.50}$$

The computation of the required transfer function of the inner subsystem becomes obvious. It yields from (38.50)

$$G_{d_1 i_c}\left(s\right) \equiv \left. \frac{\bar{i}_c\left(s\right)}{d_1\left(s\right)} \right|_{\bar{v}_s(s)=0} = \frac{2V_{dc}}{L_c s} \tag{38.51}$$

where
   $x(s)$ represents the Laplace transform of the time variable δx
   s being the Laplace variable

Now, considering that

$$\bar{i}_s = \bar{i}_L - \bar{i}_c \tag{38.52}$$

$\bar{i}_L$ being the current injected into the nonlinear load, allows us to express directly the transfer function between the source current $\bar{i}_s$ and the duty cycle d_1:

$$G_{d_1 i_s}\left(s\right) \equiv \left. \frac{\bar{i}_s(s)}{d_1(s)} \right|_{\substack{\bar{v}_s(s)=0 \\ \bar{i}_L(s)=0}} = -\frac{2V_{dc}}{L_c s} \tag{38.53}$$

Note that the input voltage $\bar{v}_s$ and the load current $\bar{i}_L$ are regarded as disturbance signals in the control design process. Thus, it appears clearly that the inner subsystem exhibits an integrator-like behavior, which makes quite easy the determination of the inner regulator transfer function that would ensure optimal dynamic characteristics to the inner current loop. A typical regulator is a first-order low-pass filter, as it will be demonstrated subsequently. The regulator's parameters are chosen in order to ensure, first, a maximized open-loop gain in the bandwidth and, thus, a minimized current error at the input of the regulator and, second, the attenuation of the current harmonics at multiple switching frequency.

38.7.5.2.2 Outer Subsystem Transfer Function

The development of the system transfer function for the design of the outer control loop is carried out by taking into account the presence of the inner one. Furthermore, it will be assumed that the inner regulator is suitably chosen, so that the inner controlled variable $\bar{i}_s$ follows perfectly the current reference i_s^*:

$$\bar{i}_s \cong i_s^* = \hat{i}_s \sin\left(\omega_s t\right) \tag{38.54}$$

where
   $\hat{i}_s$ represents the peak value of the source current
   ω_s is the mains angular frequency

As described in Figure 38.54, the peak value $\hat{i}_s$ is delivered by the outer regulator and is therefore considered as the input signal of the outer subsystem. The next step, then, is to develop the transfer function that relates the outer controlled variable $\overline{v}_{dc}$ to the input control variable $\hat{i}_s$, on the basis of which the design of the linear outer regulator will be carried out. Using expressions (38.52) and (38.54) into Equation 38.55 yields the steady-state time expression of the duty cycle

$$d_1^*\left(t\right)=\frac{1}{2}+\frac{1}{2\overline{v}_{dc}}\left[\overline{v}_s+L_c\frac{d\left(\overline{i}_L-i_s^*\right)}{dt}\right] \tag{38.55}$$

with

$$\overline{v}_s=\hat{v}_s\sin\left(\omega_s t\right) \tag{38.56}$$

and

$$\overline{i}_L=\sum_{k=1,3,5,\ldots}^{\infty}\hat{i}_{Lpk}\sin\left(k\omega_s t\right)+\hat{i}_{Lqk}\cos\left(k\omega_s t\right) \tag{38.57}$$

In (38.57), $\hat{i}_{Lpk}$ and $\hat{i}_{Lqk}$ designate the coefficients of the kth harmonic obtained by Fourier series decomposition of the signal $\overline{i}_L$. By replacing expressions (38.56) and (38.57) into (38.55), we obtain

$$d_1^*\left(t\right)=\frac{1}{2}+\frac{1}{2\overline{v}_{dc}}\left[\hat{v}_s\sin\left(\omega_s t\right)-L_c\omega_s\,\hat{i}_s\cos\left(\omega_s t\right)+L_c\omega_s\sum_{k=1,3,5,\ldots}^{\infty}k\left(\hat{i}_{Lpk}\cos\left(k\omega_s t\right)-\hat{i}_{Lqk}\sin\left(k\omega_s t\right)\right)\right] \tag{38.58}$$

Practically, the adopted value of the filter inductor L_c is too small, and can be further decreased if the switching frequency is increased. Consequently, we can write approximately

$$d_1^*\left(t\right)\cong\frac{1}{2}+\frac{\hat{v}_s}{2\overline{v}_{dc}}\sin\left(\omega_s t\right) \tag{38.59}$$

Recalling expression (38.46), and using Equations 38.52, 38.54, 38.57, and 38.59, yields

$$C_{dc}\frac{d\overline{v}_{dc}}{dt}+\frac{\overline{v}_{dc}}{R_{dc}}=\frac{\hat{v}_s}{\overline{v}_{dc}}\sin\left(\omega_s t\right)\left[\hat{i}_s\sin\left(\omega_s t\right)-\sum_{k=1,3,5,\ldots}^{\infty}\hat{i}_{Lpk}\sin\left(k\omega_s t\right)+\hat{i}_{Lqk}\cos\left(k\omega_s t\right)\right] \tag{38.60}$$

which can be rewritten using trigonometric properties as follows:

$$C_{dc}\frac{d\overline{v}_{dc}}{dt}+\frac{\overline{v}_{dc}}{R_{dc}}=\frac{\hat{v}_s}{2\overline{v}_{dc}}\left[\left(\hat{i}_s-\hat{i}_{Lp1}\right)\left[1-\cos\left(2\omega_s t\right)\right]-\hat{i}_{Lq1}\sin\left(2\omega_s t\right)\right.$$

$$-\sum_{k=1,3,5,\ldots}^{\infty}\hat{i}_{Lpk}\left\{\cos\left[(k-1)\omega_s t\right]-\cos\left[(k+1)\omega_s t\right]\right\}$$

$$\left.+\hat{i}_{Lqk}\left\{\sin\left[(k+1)\omega_s t\right]+\sin\left[(k-1)\omega_s t\right]\right\}\right] \tag{38.61}$$

The outer loop is designed to be much slower than the inner one for stability considerations. In addition, in order to avoid the distortion of the current reference i_s^* and to eliminate, consequently, the possibility of creation of additional harmonics into the mains, the outer-loop control signal delivered by the outer regulator must be free of harmonics, especially, the one at twice the mains frequency, which is omnipresent in the voltage v_{dc} at the dc side of the filter. This can be accomplished by limiting the open-loop bandwidth of the outer loop at frequencies enough lower than twice the mains frequency. Considering this assumption, all the harmonics that appear in expression (38.61) will have a negligible influence on the dc voltage v_{dc} and can, thus, be disregarded. Therefore, expression (38.61) can be expressed approximately by

$$C_{dc}\frac{d\overline{v_{dc}}}{dt} + \frac{\overline{v_{dc}}}{R_{dc}} = \frac{\hat{v}_s}{2\overline{v_{dc}}}\left(\hat{i}_s - \hat{i}_{Lp1}\right) = -\frac{\hat{v}_s}{2\overline{v_{dc}}}\hat{i}_{cp1}$$ (38.62)

where $\hat{i}_{cp1} \equiv \hat{i}_{Lp1} - \hat{i}_s$ represents the peak value of the active fundamental current absorbed by the filter, which is in phase with the input voltage v_s, and which is used to compensate the power losses in the filter and to maintain the dc voltage level at a desired value.

Equation 38.62 can be also written as follows:

$$C_{dc}\frac{d\left(\overline{v_{dc}}^2\right)}{dt} + 2\frac{\overline{v_{dc}}^2}{R_{dc}} = \hat{v}_s\left(\hat{i}_s - \hat{i}_{Lp1}\right)$$ (38.63)

By applying the small-signal linearization to Equation 38.63, and assuming $\hat{i}_{Lp1}$ a disturbance signal for the outer loop design, it yields

$$C_{dc}\frac{d\left(\delta\overline{v_{dc}}\right)}{dt} + 2\frac{\delta\overline{v_{dc}}}{R_{dc}} = \frac{\hat{v}_s}{2V_{dc}}\delta\,\hat{i}_s$$ (38.64)

The transfer function that relates the dc voltage $\overline{v_{dc}}$ to the outer control input $\hat{i}_s$ is then derived forwardly as follows:

$$G_{i_s v_{dc}}(s) \equiv \left.\frac{\overline{v_{dc}}(s)}{\hat{i}_s(s)}\right|_{\hat{i}_{Lp1}(s)\equiv 0} = \frac{R_{dc}\hat{v}_s}{4V_{dc}}\frac{1}{1+\dfrac{R_{dc}C_{dc}}{2}s}$$ (38.65)

By considering the power losses in the filter negligible (i.e., R_{dc} tends to infinity), expression (38.65) is reduced to

$$G_{i_s v_{dc}}(s) = \frac{\hat{v}_s}{2V_{dc}C_{dc}s}$$ (38.66)

This is also an integrator-like behavior, which makes quite easy the calculation of the outer regulator transfer function that would ensure optimal dynamic characteristics to the outer voltage loop. A typical regulator is a first-order low-pass filter, as it will be demonstrated subsequently. The regulator's parameters are chosen in order to ensure, first, an infinite open-loop gain at the zero frequency and, thus, a zero steady-state voltage error at the input of the regulator and, second, the attenuation of the voltage harmonic at twice the mains frequency.

38.7.5.2.3 Design of the Regulators

The block diagram of the implementation circuit is shown in Figure 38.54. The reference current i_s^* obtained from the control algorithm is compared with the sensed current i_s. As above, the error signal is fed to the current controller having a limiter at its output. Consequently, the controlling signal is compared with a triangular carrier resulting in the gating signals.

38.7.5.2.3.1 DC Bus Voltage Controller

If the regulation of the dc bus voltage works properly, the terminal voltage of the capacitor is equal to its reference V_{dc}^*, from which the following relation can be drawn:

$$G_{i_s v_{dc}}(s) = \frac{\hat{v}_s}{2 V_{dc}^* C_{dc} s} \tag{38.67}$$

As the open-loop transfer function given by Equation 38.67 presents an integral action, proportional action of the controller can be satisfactory. Therefore, the controller transfer function can be written as $K_v/(1 + \tau_v s)$, where K_v represents a gain with a first-order low-pass filtering.

The transfer function in closed loop of the voltage control is written as

$$\frac{V_{dc}}{V_{dc}^*} = \frac{K_v \hat{v}_s}{2\tau_v C_{dc} V_{dc}^* s^2 + 2 C_{dc} V_{dc}^* s + K_v \hat{v}_s} \tag{38.68}$$

It is a second-order system of the form

$$\frac{V_{dc}}{V_{dc}^*} = \frac{w_v^2}{s^2 + 2\zeta_v w_v s + w_v^2} \tag{38.69}$$

where

$$\begin{cases} w_v = \sqrt{\dfrac{K_v \hat{v}_s}{2\tau_v C_{dc} V_{dc}^*}} \\[4mm] \zeta_v = \sqrt{\dfrac{C_{dc} V_{dc}^*}{2\tau_v K_v \hat{v}_s}} \end{cases} \tag{38.70}$$

where
  w_v is the natural frequency
  ζ_v is the damping factor

The parameters of the regulator (K_v and τ_v) are calculated so that the system ensures a desired response, i.e., the damping coefficient and pulsation are selected in an optimal way. The values are given in Table 38.1.

38.7.5.2.3.2 Supply Current Regulator

The supply current controller transfer function can be written as $K_i/(1 + \tau_i s)$, where K_i represents a gain with a first-order low-pass filtering. The transfer function in closed loop of the current control can be written as

$$\frac{i_s}{i_s^*} = \frac{2 K_i V_{dc}^*}{\tau_i L_c s^2 + L_c s + 2 K_i V_{dc}^*} \tag{38.71}$$

TABLE 38.1   System Parameters Used for Simulation

Line voltage, and frequency	$V_s = 120\,\mathrm{V}$ (rms), $f_s = 60\,\mathrm{Hz}$
Line impedance	$L_s = 0.3\,\mathrm{mH}$, $R_s = 0.1\,\Omega$
Load impedance	$L_L = 20\,\mathrm{mH}$, $R_L = 20\,\Omega$
Active filter parameters	$L_c = 5\,\mathrm{mH}$, $R_{dc} = 5000\,\Omega$, $C_{dc} = 2000\,\mu\mathrm{F}$
Filter dc bus voltage	$V_{dc} = 350\,\mathrm{V}$
Switching frequency	$f_{sw} = 5\,\mathrm{kHz}$
Parameters of the current regulator	$K_i = 0.14$, $\tau_i = 10\,\mu\mathrm{s}$,
Parameters of the voltage regulator	$K_v = 206$, $\tau_v = 10\,\mu\mathrm{s}$

It is a second-order system of the form

$$\frac{i_s}{i_s^*} = \frac{w_i^2}{s^2 + 2\zeta_i w_i s + w_i^2} \tag{38.72}$$

with

$$\begin{cases} w_i = \sqrt{\dfrac{2K_i V_{dc}^*}{\tau_i\, L_c}} \\[3mm] \zeta_i = \dfrac{1}{2}\sqrt{\dfrac{L_c}{2\tau_i K_i V_{dc}^*}} \end{cases} \tag{38.73}$$

The parameters of the regulator (K_i and τ_i) are calculated so that the system ensures a fast dynamic response, and the current i_s of the supply takes a sinusoidal form, and in phase with the voltage. The values of K_i and τ_i are given in Table 38.1.

38.7.6 Simulation Results

In order to validate the accuracy of these controllers, the system is simulated using the parameters as given in Table 38.1. The circuit of Figure 38.54 was implemented in the Simulink® toolbox of MATLAB® under various conditions.

38.7.6.1 Compensation for Direct and Indirect Current Control Techniques Implemented with the Bipolar PWM Controller

The simulation results of the SPSAPF with direct and indirect current control algorithms implemented with the B-PWM controller are presented in Figure 38.55a and b, respectively, where load current (i_L), the SPSAPF current (i_c), the supply voltage (v_s), the supply current (i_s), and the dc bus voltage of the SPSAPF (v_{dc}) are depicted. The harmonic spectra of the supply current before and after compensation with direct and indirect current control schemes implemented with the B-PWM controller are shown in Figure 38.56a through c, respectively. The THD of the supply current is reduced from 28.5% before compensation to 10.4% after compensation with the direct current control technique and to 6.3% with the indirect current control technique. These results show that when using indirect current control technique, the current ripple of the supply current has eliminated. Moreover, it is observed that the B-PWM controller suffers from the problem of high-frequency harmonics in the supply current.

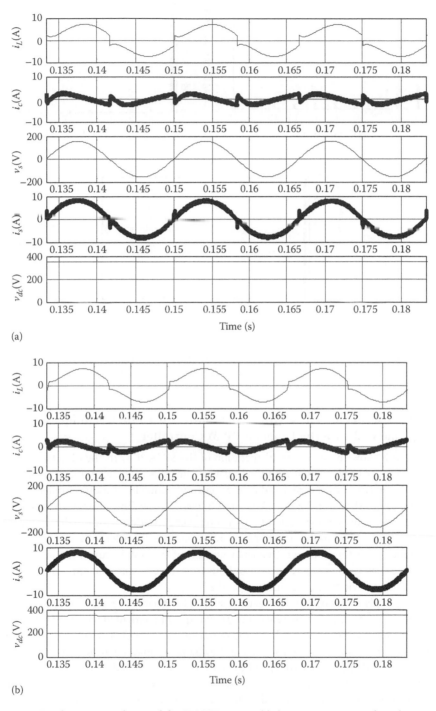

(a)

(b)

FIGURE 38.55 Steady-state waveforms of the SPSAPF system: (a) direct current control implemented with the B-PWM, (b) indirect current control implemented with the B-PWM.

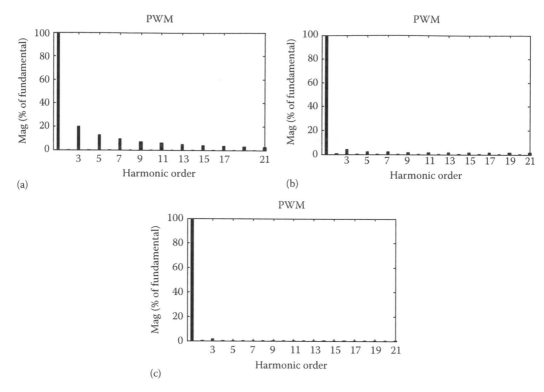

FIGURE 38.56 Spectrum of the source current: (a) before compensation and after compensation with, (b) direct current control implemented with the B-PWM, and (c) indirect current control implemented with the B-PWM.

38.7.6.2 Compensation for These Direct and Indirect Current Techniques Implemented with the Unipolar PWM Controller

Figure 38.57a and b show the steady-state operation of the SPSAPF with the direct and indirect current control techniques implemented with the U-PWM controller. The harmonic spectra of the supply current after compensation with direct and indirect current control techniques implemented with the U-PWM controller are shown in Figure 38.58a and b, respectively. The THD of the supply current is reduced from 28.5% before compensation to 4.4% after compensation with the direct current control technique and to 1% with the indirect current control technique. In order to show the efficiency of the U-PWM controller regarding the high-frequency content of the supply current, one can compare, as shown in Figure 38.59, the harmonic spectra of both techniques. Note that this comparison is made within the vicinities of switching frequencies f_{sw} and $2f_{sw}$.

In practice, the loads power demand are usually subject to variations. Hence, it is necessary to examine the performance of the indirect current control technique implemented with the U-PWM controller under such disturbances. Figure 38.60 shows the response of the SPSAPF system for a step increase of 100% of the load current at t = 166.7 ms. During the change of load condition, the system maintains unity PF operation and the dc bus voltage of SPSAPF is also regulated to the reference value. These results confirm that by using indirect current control technique with the U-PWM control strategy, substantial improvements have been observed in harmonic content of the supply current as well as on dynamic response of the SPSAPF.

It is observed that the indirect current control algorithm of SPSAPF implemented with the U-PWM controller is free from switching ripples and high-frequency harmonics in the supply current. The reference supply current is a slowly varying signal (60 Hz), therefore at any instant on the ac cycle, the

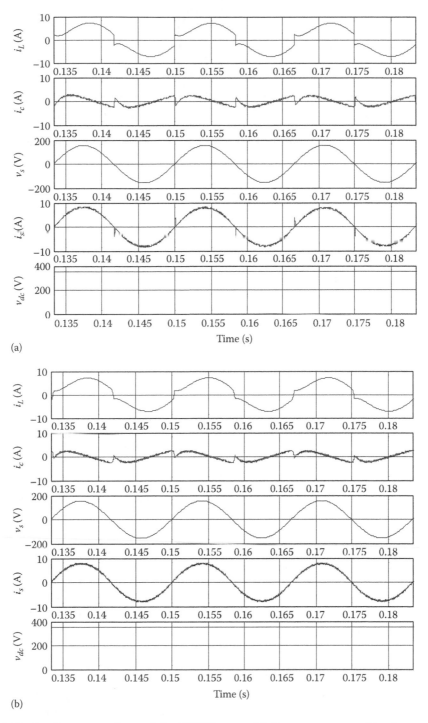

FIGURE 38.57 Steady-state waveforms of the SPSAPF system: (a) direct current control implemented with the U-PWM, (b) indirect current control implemented with the U-PWM.

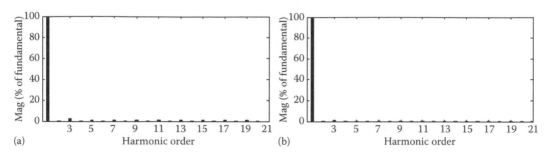

FIGURE 38.58 Spectrum of source current after compensation with (a) direct current control implemented with the U-PWM and (b) indirect current control implemented with the U-PWM.

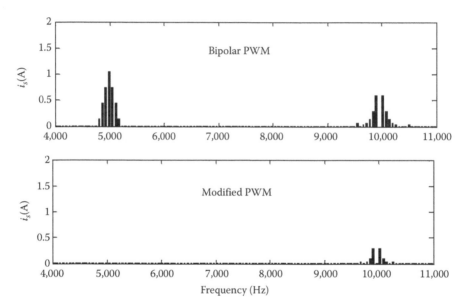

FIGURE 38.59 Spectrum analysis of source currents obtained using indirect current control with B-PWM and U-PWM.

indirect current controller has exact information about the shape of the supply current and, hence, it takes a desired corrective action to fully compensate the switching ripples in the supply current. The supply current is observed very close to sinusoidal wave and it remains in phase with the supply voltage, therefore unity PF is maintained at ac mains. The SPSAPF supplies the reactive power demand of the load locally and it also compensates its harmonics. The THD of supply current for different control techniques is summarized in Table 38.2.

38.7.7 Experimental Validation

To experimentally validate the developed model of SPSAPF, various tests are conducted. Both direct and indirect current control techniques with both PWM controllers have been implemented. The experimental setup parameters as a 574 VA diode rectifier is taken as the nonlinear load; the supply voltage is a 110 V, 60 Hz. The SPSAPF is made of 4-IGBT modules IXGH 40 N60 of Ixys. The dc voltage is set at 350 V. The filter inductor is selected to be 5 mH and the dc bus capacitor is 2000 μF. The switching frequency of the IGBT devices is taken as 5 kHz.

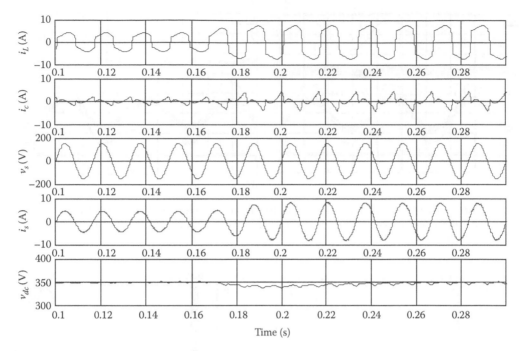

FIGURE 38.60 System response to a 100% step of load increase.

TABLE 38.2 THD Value of the Compensated Source Current

Before compensation	28.5%
After compensation using current direct control implemented with the B-PWM	10.4%
After compensation using indirect current control implemented with the B-PWM	6.3%
After compensation using direct current control implemented with the U-PWM	4.4%
After compensation using indirect current control implemented with the U-PWM	1%

38.7.7.1 Compensation for These Direct and Indirect Current Control Techniques Implemented with the Bipolar PWM Controller

The steady-state results of the direct current control technique implemented with the B-PWM controller are shown in Figure 38.61a, and the waveforms of the indirect current control technique implemented with the B-PWM controller are shown in Figure 38.61b. These results demonstrate the capability of the indirect current control technique to eliminate the ripples. The low-frequency analysis of the load and supply currents is shown in Figure 38.62a through c. The THD of this supply current is decreased from 28.83% to 10.9% with the direct current control technique implemented with the B-PWM controller and from 28.83% to 6.7% with the indirect current control technique implemented with the B-PWM controller.

38.7.7.2 Compensation for These Direct and Indirect Current Control Techniques Implemented with the Unipolar PWM Controller

Figure 38.63a and b show the steady-state results of the direct current control technique implemented with the U-PWM controller and the waveforms of the indirect current control technique implemented with the U-PWM controller. These results demonstrate the capability of the U-PWM

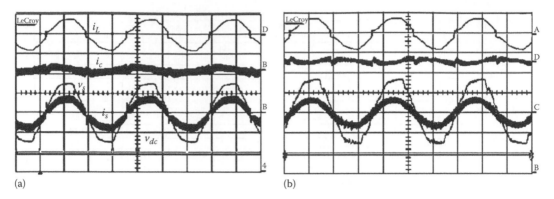

(a)   (b)

FIGURE 38.61 Steady-state waveforms: (a) direct current control implemented with the B-PWM and (b) indirect current control implemented with the B-PWM, v_s, [100 V/div], v_{dc} [400 V/div], i_s [10 A/div], i_L [10 A/div] and i_c [10 A/div], time (5 mS/div).

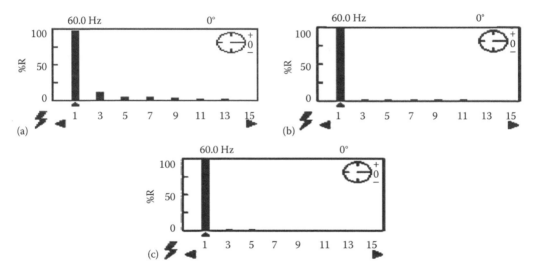

FIGURE 38.62 Spectrum of the source current: (a) before compensation and after compensation with (b) current control implemented with the B-PWM, and (c) indirect current control implemented with the B-PWM.

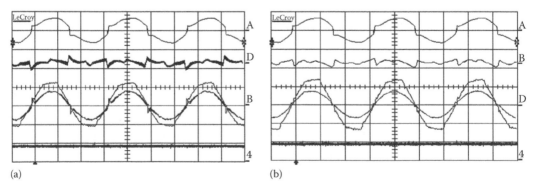

(a)   (b)

FIGURE 38.63 Steady-state waveforms: (a) direct current control implemented with the U-PWM and (b) indirect current control implemented with the U-PWM, v_s, [100 V/div], v_{dc} [400 V/div], i_s [10 A/div], i_L [10 A/div] and i_c [10 A/div], Time (5 mS/div).

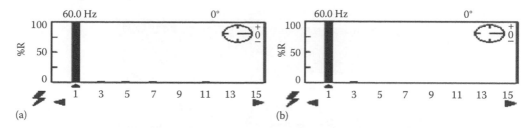

(a) (b)

FIGURE 38.64 Spectrum of the source current after compensation with (a) direct current control implemented with the U-PWM and (b) indirect current control implemented with the U-PWM.

controller to better compensate the low-frequency harmonics. The low-frequency analysis of supply current is shown in Figure 38.64a and b. The THD of this supply current is decreased from 28.83% to 4.9% with the direct current control technique implemented with the U-PWM controller and from 28.83% to 2.2% with the indirect current control technique implemented with the U-PWM controller.

In order to evaluate the impact of U-PWM controller, one has to perform a spectrum analysis of the same steady-state waveforms at the switching frequency. The results of the analysis are shown in Figure 38.65a through c.

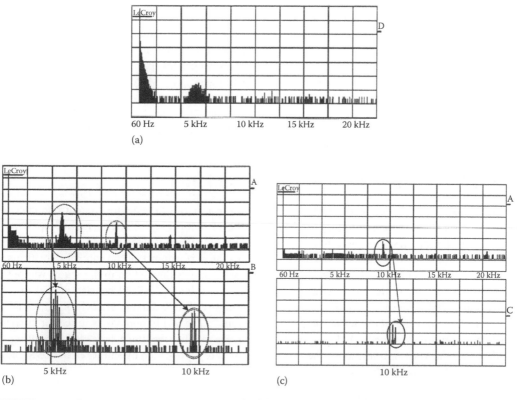

FIGURE 38.65 Source current spectrum covering both low- and high-frequency components: (a) before compensation and after compensation with, (b) indirect current control implemented with the B-PWM (top caption amplitude 9.2 dB/div, bottom caption zoom view 5.5 dB/div) and (c) indirect current control implemented with the U-PWM (top caption amplitude 9.2 dB/div, bottom caption zoom view 9.2 dB/div).

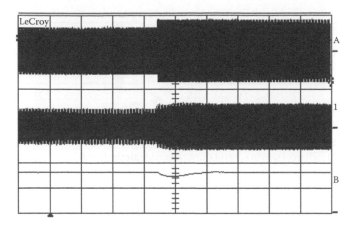

FIGURE 38.66   DC source control behavior in case of nonlinear load transient regime i_L [10 A/div], i_s [10 A/div], v_{dc} [200 V/div].

In order to test the performance of the SPSAPF system, a step change in the load is applied by an increase of 70% of the load current. Figure 38.66 shows the response of the dc bus voltage controller. The voltage fluctuation on the dc bus depends on the compensation speed of the outer loop that regulates the dc bus voltage. A sudden increase in the load power of the rectifier load results in a decrease of the dc side voltage of the active filter, which recovers within few cycles.

From the experimental results, it is observed that harmonic currents and reactive power generated by the nonlinear load could be effectively compensated with these two methods. Furthermore, the indirect current control technique of SPSAPF is free from switching ripples and the U-PWM controller has the advantage of pushing the harmonics toward high-frequency range. The first significant spectrum bar of v_c is located at the neighborhoods of twice the switching frequency $2f_{sw}$. Moreover, the U-PWM controller eliminates the groups from lines that are centered on the odd multiples of the switching frequency, and it attenuates the lines that are located around the frequency $2f_{sw}$. Thereby, through adopting indirect current control technique with the U-PWM controller, the performance of the SPSAPF is improved significantly. The supply current after compensation is close to a sinusoidal wave.

38.8 Three-Phase Shunt Active Power Filter

In this section, both indirect and direct current control techniques to generate current reference for APFs are presented. These techniques are based on the instantaneous active current component i_d method, extracting. These control techniques are based on synchronous rotating frame transformation derived from the mains voltages. Simulation results show the performance of the control algorithms to compensate for harmonics and reactive power. Figure 38.67 shows the system under study, where the SAPF is connected in parallel between the line and the nonlinear load. The SAPF consists of a full-bridge voltage-source PWM inverter, a dc-side capacitor C_{dc}, and three line inductors, namely, L_c. These latter are required to limit the ripple of the compensator current i_c. The SAPF topology is suited for current-source type of nonlinear loads. A three-phase diode bridge rectifier feeding a series R-L circuit on its dc side represents the nonlinear load. The converter losses are represented by shunt resistance R_{dc} connected in parallel with the dc bus capacitor.

38.8.1 Current References Extraction

Generally, the control of a shunt APF is composed of two interconnected loops. The first inner loop controls the currents in order to achieve a desired filter current that is usually based on the sensed

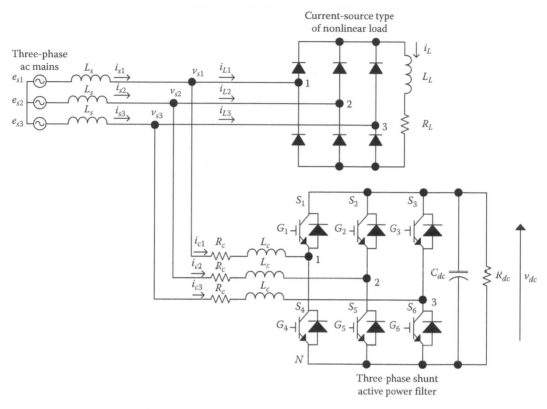

FIGURE 38.67 System configuration.

load currents, actual filter currents, supply voltages, etc. The second outer loop regulates the dc bus voltage. This voltage may vary during the transient regime depending on the control algorithm used and the losses of the inverter. The overall filter control loops should insure the following criteria:

- The generation of a suitable current reference set point that is determined from harmonic load current extraction procedure
- Sending the suitable switching signals pattern to the controllable semiconductor device's gates so that the filter current tracks its reference
- Achieve good regulation of the dc bus voltage

To determine the set of the load harmonics content, the SRF harmonic method is used [19,20]. This presents the advantage of being robust against the fluctuations of supply frequency and guarantees the conservation of electrical amplitude values (currents and voltages); moreover, a phase-locked loop (PLL) allows synchronizing the SRF frequency with that of the network. Furthermore, this method possesses good qualities in term of stability and of transient (speed and quality of the response during a step of load). Finally, it directly delivers Park current references components.

38.8.1.1 Phase-Locked Loop Circuit

A numerical PLL (Figure 38.68) determines the phase angle θ of the source voltage, which is required for the dq transformation [19]. Therefore, the three-phase voltages v_{s1}, v_{s2}, and v_{s3} are measured and the reactive or q-component v_{sq} of these voltages is calculated. A PI controller is used to control the angular frequency. So, if the q-component is zero the phase voltage and the angle θ are in phase.

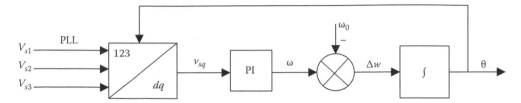

FIGURE 38.68   Phase-locked loop.

38.8.1.2 The Direct Current Control Technique of SAPF

In the direct current control technique the switching signals for SAPF devices are obtained by comparison of reference (i_{c1}^*, i_{c2}^*, and i_{c3}^*) and sensed (i_{c1}, i_{c2}, and i_{c3}) currents of the SAPF [19]. This technique allows the compensation of either harmonics, reactive power, or both. This is due to the fact that the reactive energy draws a nonzero dc component ($\bar{i}_q$) along the q axis. The sensed three-phase load currents i_{L1}, i_{L2}, and i_{L3} in the Park reference frame are transformed into the rotating reference frame dq by using the following matrix transformation:

$$\begin{pmatrix} i_d \\ i_q \end{pmatrix} = C \begin{pmatrix} i_{L1} \\ i_{L2} \\ i_{L3} \end{pmatrix} \tag{38.74}$$

and the transformation matrix C is given by

$$C = \frac{2}{3} \begin{pmatrix} \cos\theta & \cos\left(\theta - \dfrac{2\pi}{3}\right) & \cos\left(\theta + \dfrac{2\pi}{3}\right) \\ \sin\theta & \sin\left(\theta - \dfrac{2\pi}{3}\right) & \sin\left(\theta + \dfrac{2\pi}{3}\right) \end{pmatrix} \tag{38.75}$$

where θ represents the actual phase angle of the line voltage space vector. i_d and i_q are the components of the resulting source current space vector in the fundamental mains frequency rotating coordinate system. The obtained instantaneous currents i_d and i_q are divided into dc and ac component as

$$\begin{cases} i_d = \bar{i}_d + \tilde{i}_d \\ i_q = \bar{i}_q + \tilde{i}_q \end{cases} \tag{38.76}$$

The active dc component $\bar{i}_d$ represents the positive sequence at the fundamental frequency of the sensed load current. The reactive dc component $\bar{i}_q$ represents the positive sequence at fundamental of the reactive power. The ac components $\tilde{i}_d$ and $\tilde{i}_q$ represent the total harmonic content of the load current. These dq components are obtained at the output of a high-pass filter having i_d and i_q as inputs. A low-pass filter with a subtracted forward action synthesizes the high-pass filter. The dc components are then eliminated, and only the ac components remains in the output signals $\tilde{i}_d$ and $\tilde{i}_q$. The error signal between the reference value v_{dc}^* and the sensed feedback value v_{dc} is processed through a PI controller to obtain I_{cm}, which is added to the oscillating harmonic current component $\tilde{i}_d$ yielding the reference current i_d^*. The SAPF reference currents used for compensation are defined as follows:

- Harmonics compensation only

$$\begin{pmatrix} i_{c1}^* \\ i_{c2}^* \\ i_{c3}^* \end{pmatrix} = C^{-1} \begin{bmatrix} I_{cm} - \tilde{i}_{Ld} \\ -\tilde{i}_{Lq} \end{bmatrix} \tag{38.77}$$

- Reactive power compensation only

$$\begin{pmatrix} i_{c1}^* \\ i_{c2}^* \\ i_{c3}^* \end{pmatrix} = C^{-1} \begin{bmatrix} 0 \\ -\overline{i}_{Lq} \end{bmatrix} \tag{38.78}$$

- Harmonics and reactive power compensation

$$\begin{pmatrix} i_{c1}^* \\ i_{c2}^* \\ i_{c3}^* \end{pmatrix} = C^{-1} \begin{bmatrix} I_{cm} - \tilde{i}_{Ld} \\ -\tilde{i}_{Lq} - \overline{i}_{Lq} \end{bmatrix} \tag{38.79}$$

With C^{-1}: inverse Park transforms

$$C^{-1} = \begin{pmatrix} \cos\theta & \sin\theta \\ \cos\left(\theta - \dfrac{2\pi}{3}\right) & \sin\left(\theta - \dfrac{2\pi}{3}\right) \\ \cos\left(\theta + \dfrac{2\pi}{3}\right) & \sin\left(\theta + \dfrac{2\pi}{3}\right) \end{pmatrix} \tag{38.80}$$

Note that the signal I_{cm} is the peak value of the fundamental current I_c used to compensate APF losses. In this case, the active filter needs to compensate for the harmonics and the reactive power. The block diagram of the synchronous rotating dq reference frame of the SAPF direct current control algorithm is shown in Figure 38.69.

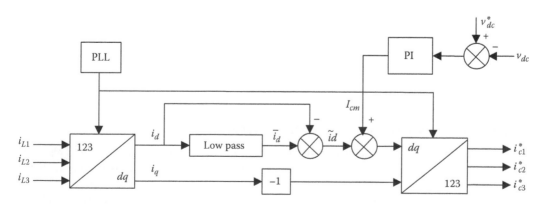

FIGURE 38.69 Direct current control algorithm of SAPF system.

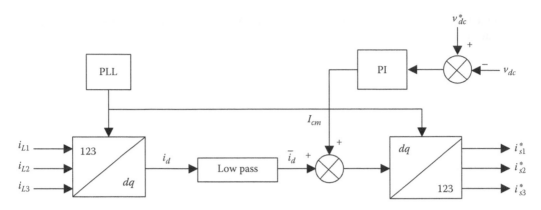

FIGURE 38.70   Indirect current control algorithm of SAPF system.

38.8.1.3 The Indirect Current Control Technique of SAPF

In the indirect current control technique, the switching signals for SAPF devices are obtained by comparison of reference (i_{s1}^*, i_{s2}^* and i_{s3}^*) and sensed (i_{s1}, i_{s2} and i_{s3}) source currents [19,20]. The d and q axes reference supply currents are expressed as follows:

$$i_{sd}^* = \overline{i_d} + I_{cm}$$

$$i_{sq}^* = 0 \tag{38.81}$$

The block diagram of the indirect current control of active filter is shown in Figure 38.70.

38.8.2 Control Technique Principle

The inverter control technique uses the B-PWM principle. The control signals (β_i, $i = 1,2,3$ and $-1 < \beta_i < 1$, see Figure 38.73) are compared to a triangular carrier V_{PWM}. Each of the three legs has an independent functioning, and the switches of the same leg work in a complementary fashion. The inverter output voltage evaluated with respect to the point N takes two values, namely, v_{dc} and zero depending on the sign of $\beta_i - V_{PWM}$. If the PWM frequency is sufficiently high, we can consider that the output mean value of the instantaneous PWM voltage over a switching period is very close to $\beta_i v_{dc}$ [25].

Moreover, if the sum $\beta_1 + \beta_2 + \beta_3$ is zero, the following current equations hold:

$$\begin{cases} L_c \dfrac{di_{c1}}{dt} = v_{s1} - R_c i_{c1} - \beta_1 v_{dc} \\[2mm] L_c \dfrac{di_{c2}}{dt} = v_{s2} - R_c i_{c2} - \beta_2 v_{dc} \\[2mm] L_c \dfrac{di_{c3}}{dt} = v_{s3} - R_c i_{c3} - \beta_3 v_{dc} \end{cases} \tag{38.82}$$

38.8.2.1 Signal Generation for the Inverter Switches

In the direct current control, the reference currents (i_{c1}^*, i_{c2}^*, and i_{c3}^*) designed from the control algorithm are compared with the sensed currents (i_{c1}, i_{c2}, and i_{c3}). Then the residual signal feeds a controller having a limiter at its output. The control law provides the modulation β_i considered as the direct image of the voltage mean value at the input of the inverter, which is compared with a triangular carrier resulting in the switching signal to the gate (Figure 38.71).

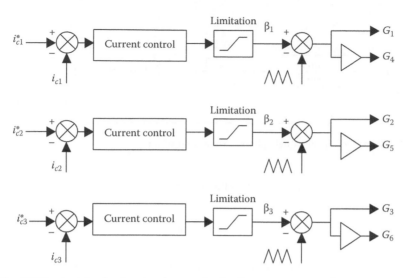

FIGURE 38.71 PWM principle of gating signal generation for direct current control.

In the indirect current control, the modulation β_i is obtained from the filter output current regulator as shown in Figure 38.72.

38.8.2.2 Current Control Loop Design

Assume that the sum $\beta_1 + \beta_2 + \beta_3$ is zero, in other words, the current of a given phase depends only on the output regulator of the same phase. Note that this assumption is valid only in a two-loop structure, but may be considered as a big simplification in a three-loop structure, especially when the system is not perfectly balanced and is not in its steady-state regime. Figure 38.73 gives a block diagram of a single-phase current control loop.

The regulator used in the different loops is of PI type. The transfer function of a typical PI regulator is given by

$$C(s) = k_p + \frac{k_i}{s} = k_p\left(1 + \frac{1}{\rho s}\right) \tag{38.83}$$

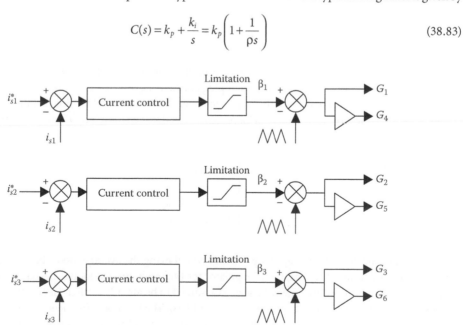

FIGURE 38.72 PWM principle of gating signal generation for indirect current control.

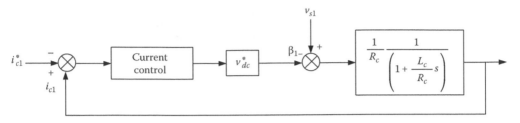

FIGURE 38.73   Block diagram of a single-phase current control loop.

The current expression for a phase may be decomposed as

$$i_c = F_1 i_c^* + F_2 v_{s1} \tag{38.84}$$

where F_1 and F_2 are the closed-loop transfer function, respectively, to the current reference and to the perturbation signal. Their exact expressions are

$$F_1(s) = \frac{1 + \rho s}{1 + \left(\rho + \dfrac{1}{k}\right)s + \dfrac{1}{k}\tau_e s^2} \tag{38.85}$$

$$F_2(s) = -\frac{1}{v_{dc}k_i}\frac{s}{1 + \left(\rho + \dfrac{1}{k}\right)s + \dfrac{1}{k}\tau_e s^2} \tag{38.86}$$

with

$$\tau_e = \frac{L_s}{R_c}, \qquad \rho = \frac{k_p}{k_i}, \qquad k = \frac{v_{dc}k_i}{R_c}$$

The coefficients k_p and k_i are chosen so that the overall closed system behaves as an optimal second order having two complex conjugate poles with a damping coefficient $\xi = 0.707$. They may be expressed in terms of the damping coefficient and the natural frequency ω_n by Equation 38.87. For instance, the system is represented by a couple of interconnected loops. An internal fast current loop i_c and an outer slow voltage v_{dc} loop.

$$\begin{cases} k_i = \dfrac{L_s \omega_n^2}{v_{dc}} \\[2ex] k_p = \dfrac{2\xi}{\omega_n}k_i - \dfrac{R_c}{v_{dc}} \end{cases} \tag{38.87}$$

The PI regulator for the outer voltage loop is designed following the average model basis by considering a perfect current tracking. Whereas, the PI regulator for the inner current loop is designed independently of the outer one, in other words, the voltage v_{dc} is assumed to be already settled. Hence, the PI regulator is synthesized regarding two transfer functions $F_1(s)$ and $F_2(s)$. The harmonic frequency range is $(1 < h < 40)$. With the given parameters: damping ratio $\xi = 0.707$, and natural frequency $\omega_n = 50{,}000\,\text{rad/s}$, one can obtain: $k_p = 0.1167$ and $k_i = 4{,}167$, the bandwidth of the regulator should be able to respond to these criteria.

TABLE 38.3 System Parameters Used for Simulation

Line voltage, and frequency	$V_s = 120\,\text{V (rms)}, f_s = 60\,\text{Hz}$
Line impedance	$L_s = 0.3\,\text{mH}$
Load impedance	$L_L = 10\,\text{mH}, R_L = 12\,\Omega$
Active filter parameters	$L_c = 5\,\text{mH}, R_c = 0.1\,\Omega, R_{dc} = 5000\,\Omega,$ $C_{dc} = 500\,\mu\text{F}$
Filter dc bus voltage	$V_{dc} = 350\,\text{V}$
Switching frequency	$f_{sw} = 5\,\text{kHz}$

38.8.3 Simulation Results

In order to validate the accuracy of the direct and indirect control algorithms, the system described previously was built and simulated. The source current waveforms of the simulation results have been analyzed to obtain their THD under varying load conditions. It is simulated using the parameters as given in Table 38.3. The goal of the simulation is to present four different aspects: (a) harmonics and reactive power compensation for the direct current control technique, (b) harmonics and reactive power compensation for the indirect current control technique, (c) response of indirect current control to load variations, (d) compensation of harmonics and reactive power for indirect current control under distorted ac source.

38.8.3.1 Harmonics and Reactive Power Compensation for Direct Current Control

The simulation results of the system with direct current control algorithm are shown in Figure 38.74, where the load current (i_{L1}), SAPF current (i_{c1}), supply voltage (v_{s1}), supply current (i_{s1}), and dc bus voltage of the SAPF (v_{dc}) are presented. The output voltage of the dc bus is stabilized at 350 volts. One can observe that the direct current control algorithm suffers from the problem of excessive switching ripples that are caused by the discontinuity abrupt in the load current that causes inadequate control response; therefore, is an instantaneous compensation of the load harmonics necessary to avoid such control loss? The harmonic

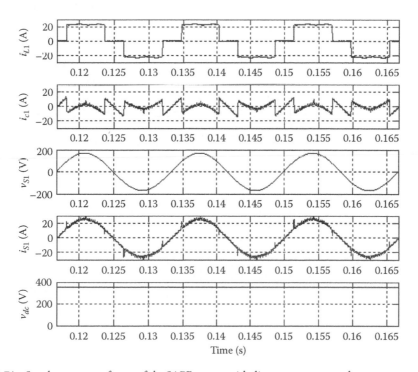

FIGURE 38.74 Steady-state waveforms of the SAPF system with direct current control.

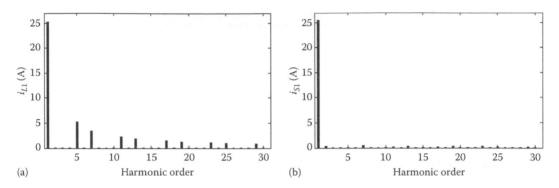

(a) Harmonic order   (b) Harmonic order

FIGURE 38.75 Spectrum of phase 1: (a) load current and (b) source current after compensation with the direct current control of SAPF.

spectrum of the current is measured 29.17%, whereas the compensated supply current THD is 3.68%. A graphical representation of both load and supply currents are depicted in Figure 38.75a and b.

38.8.3.2 Harmonics and Reactive Power Compensation for Indirect Current Control

The simulation results of the system using indirect current control algorithm are shown in Figure 38.76 where the load current (i_{L1}), SAPF current (i_{c1}), supply voltage (v_{s1}), supply current (i_{s1}), and dc bus voltage of the SAPF (v_{dc}) are depicted. It is found from this figure that the supply current exhibit a ripple-free sinusoidal shape. The harmonic spectrum of the supply current is shown in Figure 38.77. The THD of the source current is reduced from 29.17% before compensation to 1.94% after compensation.

38.8.3.3 Response of SAPF with the Indirect Current Control Load Variation

In practice, the load power demand is usually subject to variations. Hence, it is necessary to examine the performance of the system under such disturbances. Figure 38.78 shows the response of the SAPF

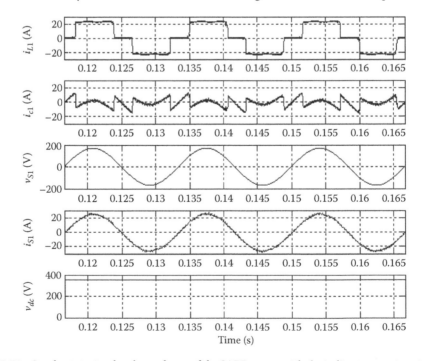

Time (s)

FIGURE 38.76 Steady-state simulated waveforms of the SAPF system with the indirect current control.

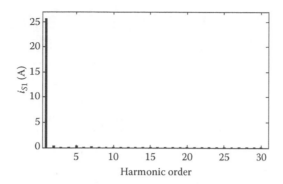

FIGURE 38.77 Spectrum of phase 1 source current after compensation with the indirect current control of SAPF.

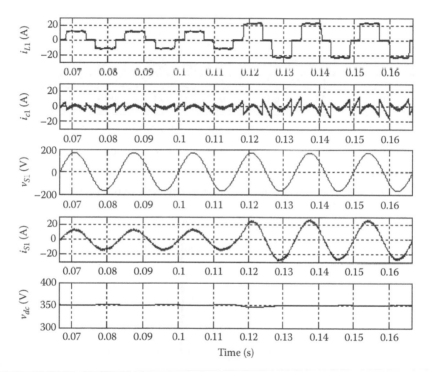

FIGURE 38.78 Indirect current control response to load variation.

system for a step increase of 100% of the load current at t = 116.7 ms. In other terms, the value of the load resistance is changed from 24 to 12 Ω, during which the system maintains full compensation by forcing unity PF operation without occurrence of switching ripples in the supply current. The results confirm the good performance of the compensator for a rapid change in the load current. Thus, the indirect current control algorithm offers better response.

38.8.3.4 Harmonics and Reactive Power Compensation for Indirect Current Control under Distorted AC Source

The purpose of the compensation is to obtain a sinusoidal current whatever the level of distortion in the source voltage. The three-phase load currents (i_{L123}), the SAPF currents (i_{c123}), the distorted supply voltages (v_{s123}), the supply currents (i_{s123}), and the dc bus voltage of the SAPF during steady-state operation are shown in Figure 38.79. These waveforms show the controller capability to compensate for nonlinear load

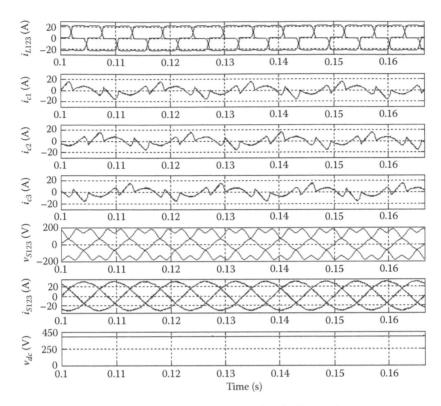

FIGURE 38.79 Steady-state response of indirect current control under distorted ac source.

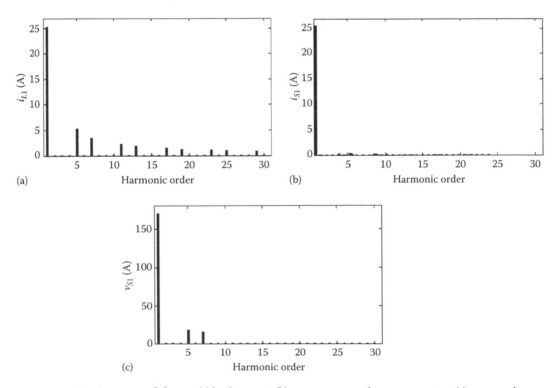

FIGURE 38.80 Spectrum of phase 1: (a) load current, (b) source current after compensation, (c) source voltage.

currents under this severe distorted three-phase supply voltage. It is important to notice that the supply currents are kept balanced and free of harmonics. Figure 38.80 illustrates the harmonic spectrum of phase 1 supply voltage, load, and supply currents. The imposed THD on the source voltage was 13.78%. The measured THD of the phase 1 source current is, therefore, reduced from 29.17% before compensation to 1.66% after compensation. These results demonstrate the robustness of the proposed controller.

38.9 Summary

This chapter presents the power-quality related problems and some mitigation techniques. Current power quality issues are discussed and the problems stated. The harmonics are defined and their causes and effects are explained. Various methods in use in mitigating the power quality problems are identified and presented. A large number of active filter configurations are available to compensate harmonic current, reactive power, neutral current, unbalance current, voltage sag, swell, and flicker. A study of two current control algorithms has been made and implemented on SPSAPF, in order to compensate current harmonics and reactive power generated by the nonlinear load. It has been shown that the indirect current control method offer ripples- and distortion-free supply current. Through its simplicity, it requires less hardware and offers improved performance. It is robust against abrupt changes in the load. In addition, the bipolar and U-PWM techniques are used to generate the gate signals for the switches. The U-PWM technique has the major advantage of eliminating the groups of harmonics that are centered on the odd multiples of the switching frequency. Furthermore, it requires a relatively reduced output filter, due to the high selected cut-off frequency that can be chosen in the neighborhood of the switching frequency. Simulation and experimental results have confirmed the superiority of the indirect current control technique performance compared to that of the direct current control technique when applied to a SPSAPF and the predicted good performance of the U-PWM controller with respect to the B-PWM controller. Two control methods for three-phase SAPF are applied to compensate harmonics, reactive power under distorted ac source has been discussed. The results of indirect current control are free from distortion in supply currents during steady-state and transient operating conditions. The indirect current control is robust against abrupt changes in the load and distorted ac source. Moreover, the later is easy to implement, and shows robustness and very good performance during both steady-state and transient operations.

References

1. B. Singh, K. Al-Haddad, and A. Chandra, A review of active filters for power quality improvement, *IEEE Transactions on Industrial Electronics*, 46(5), 960–971, Oct. 1999.
2. S. Rahmani, A. Hamadi, and K. Al-Haddad, A new three phase hybrid passive filter to dampen resonances and compensate harmonics and reactive power for any type of load under distorted source conditions, *Specialists Conference IEEE-PESC*, Orlando, FL, June 17–21, 2007.
3. Ab. Hamadi, S. Rahmani, and K. Al-Haddad, A hybrid passive filter configuration for VAR control and harmonic compensation, *IEEE Transactions on Industrial Electronics*, 57(7), 2419–2434, July 2010.
4. Y.-M. Chen, Passive filter design using genetic algorithms, *IEEE Transactions on Industrial Electronics*, 50(1), 202–207, Feb. 2003.
5. H. Akagi, Trends in active power line conditioners, *IEEE Transactions Power Electronics*, 9(3), 263–268, May 1994.
6. H. Akagi, New trends in active filters for power conditioning, *IEEE Transactions on Industry Applications*, 32(6), 1312–1322, Nov.–Dec. 1994.
7. S. Kim and P. N. Enjeti, A modular single-phase power-factor-correction scheme with a harmonic filtering function, *IEEE Transactions on Industrial Electronics*, 50(2), 328–335, Apr. 2003.

8. H. Komurcugil and O. Kukrer, A new control strategy for single-phase shunt active power filters using a Lyapunov function, *IEEE Transactions on Industrial Electronics*, 53(1), 305–312, Feb. 2006.

9. M. Cirrincione, M. Pucci, and G. Vitale, A single-phase DG generation unit with shunt active power filter capability by adaptive neural filtering, *IEEE Transactions on Industrial Electronics*, 55(5), 2093–2110, May 2008.

10. J. Miret, M. Castilla, J. Matas, J. M. Guerrero, and J. C. Vasquez, Selective harmonic-compensation control for single-phase active power filter with high harmonic rejection, *IEEE Transactions on Industrial Electronics*, 56(8), 3117–3127, Aug. 2009.

11. M. Cirrincione, M. Pucci, G. Vitale, and A. Miraoui, Current harmonic compensation by a single-phase shunt active power filter controlled by adaptive neural filtering, *IEEE Transactions on Industrial Electronics*, 56(8), 3128–3143, Aug. 2009.

12. S. Rahmani, K. Al-Haddad, and H. Y. Kanaan, A comparative study of two PWM techniques for single-phase shunt active power filters employing direct current control strategy, *Journal IET Proceedings—Electric Power Applications*, 1(3), 376–385, September 2008.

13. S. Rahmani, K. Al-Haddad, H. Y. Kanaan, and F. Fnaiech, A comparative study of two PWM techniques for single-phase shunt active power filters employing direct current control strategy, *Specialists Conference IEEE-PESC*, Recife, Brazil, June 12–16, 2005, pp. 2758–2763.

14. S. Rahmani, K. Al-Haddad, and H. Y. Kanaan, Experimental design and simulation of a modified PWM with a new indirect current control technique applied to a single-phase shunt active power filter, *International Symposium on Industrial Electronics IEEE ISIE2005*, Dubrovnik, Croatia, June 20–23, 2005, pp. 519–524.

15. S. Rahmani, K. Al-Haddad, H. Y. Kanaan, and F. Fnaiech, Implementation and simulation of a modified PWM with a two current control techniques applied to a single-phase shunt hybrid power filter, *Specialists Conference IEEE-PESC*, Recife, Brazil, June 12–16, 2005, pp. 2345–2350.

16. V. Soares, P. Vedelho, and G. D. Marques, An instantaneous active and reactive current component method for active filters, *IEEE Transactions on Power Electronics*, 15(4), 660–669, July 2000.

17. P. Mattavelli, A closed-loop selective harmonic compensation for active filters, *IEEE Transactions on Industry Applications*, 37(1), 81–89, Sept./Oct. 2001.

18. A. Chandra, B. Singh, B. N. Singh, and K. Al-Haddad, An improved control algorithm of shunt active filter for voltage regulation, harmonic elimination, power-factor correction, and balancing of nonlinear loads, *IEEE Transactions on Power Electronics*, 15(3), 495–507, May 2000.

19. S. Rahmani, K. Al-Haddad, and F. Fnaiech, A new indirect current control algorithm based on the instantaneous active current for reduced switch active filters, *Tenth European Conference on Power Electronics and Applications EPE 2003*, Toulouse, France, September 2–4, 2003.

20. S. Rahmani, K. Al-Haddad, and F. Fnaiech, A general algorithm applied to three phase shunt active power filter to compensate for source and load perturbations simultaneously, *International Symposium on Industrial Electronics IEEE (ISIE 2004)*, Ajaccio, France, May 4–7, 2004, pp. 777–782.

21. Y. G. Jung, W.-Y. Kim, Y.-Ch. Lim, S.-H. Yang, and F. Harashima, The algorithm of expanded current synchronous detection for active power filters considering three-phase unbalanced power system, *IEEE Transactions on Industrial Electronics*, 50(5), 1000–1006, Oct. 2003.

22. C. Qiao, T. Jin, and K. M. Smedley, One-cycle control of three-phase active power filter with vector operation, *IEEE Transactions on Industrial Electronics*, 51(2), 455–463, Apr. 2004.

23. P. Mattavelli and F. P. Marafao, Repetitive-based control for selective harmonic compensation in active power filters, *IEEE Transactions on Industrial Electronics*, 51(5), 1018–1024, Oct. 2004.

24. S. J. Ovaska and O. Vainio, Evolutionary-programming-based optimization of reduced-rank adaptive filters for reference generation in active power filters, *IEEE Transactions on Industrial Electronics*, 51(4), 910–916, Aug. 2004.

25. S. Rahmani, K. Al-Haddad, and F. Fnaiech, A model reference generating an optimal dc voltage for a three phase shunt active power filter, *Eleventh International Conference on Harmonics and Quality of Power (ICHQP'04)*, New York, September 12–15, 2004, pp. 22–27.

26. M. Salo and H. Tuusa, A new control system with a control delay compensation for a current-source active power filter, *IEEE Transactions on Industrial Electronics*, 52(6), 1616–1624, Dec. 2005.

27. M. Cichowlas, M. Malinowski, M. P. Kazmierkowski, D. L. Sobczuk, P. Rodriguez, and J. Pou, Active filtering function of three-phase PWM boost rectifier under different line voltage conditions, *IEEE Transactions on Industrial Electronics*, 52(2), 410–419, Apr. 2005.

28. O. Abdeslam, P. Wira, J. Merckle, D. Flieller, and Y.-A. Chapuis, A unified artificial neural network architecture for active power filters, *IEEE Transactions on Industrial Electronics*, 54(1), 61–76, Feb. 2007.

29. S.-Y. Kim and S.-Y. Park, Compensation of dead-time effects based on adaptive harmonic filtering in the vector-controlled AC motor drives, *IEEE Transactions on Industrial Electronics*, 54(3), 1768–1777, June 2007.

30. S. A. Gonzalez, R. Garcia-Retegui, and M. Benedetti, Harmonic computation technique suitable for active power filters, *IEEE Transactions on Industrial Electronics*, 54(5), 2791–2796, Oct. 2007.

31. K. Gulez, A. A. Adam, and H. Pastaci, Torque ripple and EMI noise minimization in PMSM using active filter topology and field-oriented control, *IEEE Transactions on Industrial Electronics*, 55(1), 251–257, Jan. 2008.

32. L. Asiminoaei, P. Rodriguez, F. Blaabjerg, and M. Malinowski, Reduction of switching losses in active power filters with a new generalized discontinuous-PWM strategy, *IEEE Transactions on Industrial Electronics*, 55(1), 467–471, Jan. 2008.

33. L. Asiminoaei, E. Aeloiza, P. N. Enjeti, and F. Blaabjerg, Shunt active-power-filter topology based on parallel interleaved inverters, *IEEE Transactions on Industrial Electronics*, 55(3), 1175–1189, March 2008.

34. Z. Shu, Y. Guo, and J. Lian, Steady-state and dynamic study of active power filter with efficient FPGA-based control algorithm, *IEEE Transactions on Industrial Electronics*, 55(4), 1527–1536, Apr. 2008.

35. M. Malinowski and S. Bernet, A simple voltage sensorless active damping scheme for three-phase PWM converters with an LCL filter, *IEEE Transactions on Industrial Electronics*, 55(4), 1876–1880, Apr. 2008.

36. C. Lascu, L. Asiminoaei, I. Boldea, and F. Blaabjerg, Frequency response analysis of current controllers for selective harmonic compensation in active power filters, *IEEE Transactions on Industrial Electronics*, 56(2), 337–347, Feb. 2009.

37. S. K. Jain, P. Agarwal, and H. O. Gupta, Simulation and experimental investigations on a 3-phase 4-wire shunt active power filter for power quality improvement, in *Proceedings of the ElectrIMACS'02*, Montreal, Canada, August 18–21, 2002.

38. H. Y. Kanaan, A. Hayek, S. Georges, and K. Al-Haddad, Averaged modelling, simulation and linear control design of a PWM fixed frequency three-phase four-wire shunt active power filter for a typical industrial load, in *Proceedings of the Third IEE International Conference on Power Electronics, Machines and Drives (PEMD'06)*, Dublin, Ireland, April 04–06, 2006.

39. R. Grino, R. Cardoner, R. Costa-Castello, and E. Fossas, Digital repetitive control of a three-phase four-wire shunt active filter, *IEEE Transaction on Industrial Electronics*, 54(3), 1495–1503, June 2007.

40. S. Orts-Grau, F. J. Gimeno-Sales, S. Segui-Chilet, A. Abellan-Garcia, M. Alcaniz-Fillol, and R. Masot-Peris, Selective compensation in four-wire electric systems based on a new equivalent conductance approach, *IEEE Transactions on Industrial Electronics*, 56(8), 2862–2874, Aug. 2009.

41. D. le Roux, H. du, T. Mouton, and H. Akagi, Digital control of an integrated series active filter and diode rectifier with voltage regulation, *IEEE Transactions on Industry Applications*, 39(6), 1814–1820, Nov./Dec. 2003.

42. A. Hamadi, K. Al-Haddad, and S. Rahmani, Series active filter to mitigate power quality for medium size industrial loads: Multi pulses transformers and modern AC drives, *International Symposium on Industrial Electronics IEEE ISIE 2006*, Montréal, Canada, July 9–13, 2006.

43. L. A. Morán, I. Pastorini, J. Dixon, and R. Wallace, A fault protection scheme for series active power filters, *IEEE Transactions on Power Electronics*, 14(5), 928–938, Sept. 1999.

44. H. Fujita and H. Akagi, An approach to harmonic current-free AC/DC power conversion for large industrial loads: The integration of a series active filter with a double-series diode rectifier, *IEEE Transactions on Industry Applications*, 33(5), 1233–1240, Sept./Oct. 1997.

45. Z. Pan, F. Z. Peng, and S. Wang, Power factor correction using a series active filter, *IEEE Transactions on Power Electronics*, 20(1), 148–153, Jan. 2005.

46. S. Srianthumrong, H. Fujita, and H. Akagi, Stability analysis of a series active filter integrated with a double-series diode rectifier, *IEEE Transactions on Power Electronics*, 17(1), 117–124, Jan. 2002.

47. J. W. Dixon, G. Venegas, and L. A. Moran, A series active power filter based on a sinusoidal current-controlled voltage-source inverter, *IEEE Transactions on Industrial Electronics*, 44(5), 612–620, Oct. 1997.

48. G.-M. Lee, D.-C. Lee, and Jul-Ki Seok, Control of series active power filters compensating for source voltage unbalance and current harmonics, *IEEE Transactions on Industrial Electronics*, 51(1), 132–139, Feb. 2004.

49. S. Inoue, T. Shimizu, and K. Wada, Control methods and compensation characteristics of a series active filter for a neutral conductor, *IEEE Transactions on Industrial Electronics*, 54(1), 433–440, Feb. 2007.

50. Z. Wang, Q. Wang, W. Yao, and J. Liu, A series active power filter adopting hybrid control approach, *IEEE Transactions on Power Electronics*, 16(3), 301–310, May 2001.

51. Y. Kanaan, K. Al-Haddad, M. Aoun, A. Abou Assi, J. Bou Sleiman, and C. Asmar, Averaged modeling and control of a three-phase series active power filter for voltage harmonic compensation, in *Proceedings of the IEEE IECON'03*, Vol. 1, Roanoke, VA, November 2–6, 2003, pp. 255–260.

52. W. Wu, L. Tong, M. Y. Li, Z. M. Qian, Z. Y. Lu, and F. Z. Peng, A novel series hybrid active power filter, in *35th Annual IEEE Power Electronics Specialists Conference (PESC'04)*, Aachen, Germany, 2004, pp. 3045–3049.

53. F. Z. Peng, H. Akagi, and A. Nabae, Compensation characteristics of the combined system of shunt passive and series active filters, *IEEE Transactions on Industry Applications*, 29(1), 144–152, Jan./Feb. 1993.

54. S. Rahmani, K. Al-Haddad, and F. Fnaiech, A series combination of series active and series passive filters adopting hybrid control, in *IEEE-SMC 2002*, Hammamet, Tunisia, October 6–9, 2002.

55. S. Rahmani, K. Al-Haddad, and F. Fnaiech, A series hybrid power filter to compensate harmonic currents and voltages, *IEEE Industrial Electronics Conference (IECON 2002)*, Seville, Spain, November 5–8, 2002, pp. 644–649.

56. H. Fujita and H. Akagi, A practical approach to harmonic compensation in power systems—Series connection of passive and active filters, *IEEE Transactions on Industry Applications*, 27(6), 1020–1025, Nov./Dec. 1991.

57. Ab. Hamadi, S. Rahmani, and K. Al-Haddad, A new hybrid series active filter configuration to compensate voltage sag, swell, voltage and current harmonics and reactive power, in *IEEE International Symposium on Industrial Electronics (ISIE 2009)*, Seoul, Korea, July 5–9, 2009.

58. M. Rastogi, N. Mohan, and A.-A. Edris, Hybrid-active filtering of harmonic currents in power systems, *IEEE Transactions on Power Delivery*, 10(4), 1994–2000, Oct. 1995.

59. S. Rahmani, K. Al-Haddad, and F. Fnaiech, A new control technique based on the instantaneous active current applied to shunt hybrid power filters, in *Specialists Conference IEEE-PESC 2003*, Accopulco, Mexico, June 15–19, 2003, pp. 808–813.

60. S. Rahmani, K. Al-Haddad, and F. Fnaiech, A three phase shunt hybrid power filter adopted a general algorithm to compensate harmonics, reactive power and unbalanced load under nonideal mains voltages, in *International Conference on Industrial Technology (IEEE ICIT04)*, Hammamet, Tunisia, December 8–10, 2004, pp. 651–656.

61. Ab. Hamadi, S. Rahmani, W. Santana, and K. Al-Haddad, A novel shunt hybrid power filter for the mitigation of power system harmonics, in *Proceedings of the Electrical Power Conference, 2007 (IEEE-EPC 2007)*, Montreal, Canada, October 25–26, 2007, pp. 117–122.

62. S. Rahmani, A. Hamadi, and K. Al-Haddad, A new combination of shunt hybrid power filter and thyristor controlled reactor for harmonics and reactive power compensation, in *Ninth Annual Electrical Power and Energy Conference (EPEC 2009)*, October 22–23, 2009, Montreal, Canada.

63. F. Z. Peng, H. Akagi, and A. Nabae, A new approach to harmonic compensation in power systems—A combined system of shunt passive and series active filters, *IEEE Transactions on Industry Applications*, 26(6), 983–990, Nov./Dec. 1990.

64. M. Al-Zamil and D. A. Torrey, A passive series, active shunt filter for high power applications, *IEEE Transactions on Power Electronics*, 16(1), 101–109, Jan. 2001.

65. J.-H. Sung, S. Park, and K. Nam, New hybrid parallel active filter configuration minimising active filter size, *IEE Proceedings—Electric Power Applications*, 147(2), 93–98, Mar. 2000.

66. S. Kim and P. N. Enjeti, A new hybrid active power filter (APF) topology, *IEEE Transactions on Power Electronics*, 17(1), 48–54, Jan. 2002.

67. P.-T. Cheng, S. Bhattacharya, and D. M. Divan, Control of square-wave inverters in high-power hybrid active filter systems, *IEEE Transactions on Industry Applications*, 34(3), 458–472, May/June 1998.

68. H. Fujita, T. Yamasaki, and H. Akagi, A hybrid active filter for damping of harmonic resonance in industrial power systems, *IEEE Transactions on Power Electronics*, 15(2), 215–222, Mar. 2000.

69. D. Basic, V. S. Ramsden, and P. K. Muttik, Harmonic filtering of high-power 12-pulse rectifier loads with a selective hybrid filter system, *IEEE Transactions on Industrial Electronics*, 48(6), 1118–1127, Dec. 2001.

70. D. Alexa and A. Sirbu, Optimized combined harmonic filtering system, *IEEE Transactions on Industrial Electronics*, 48(6), 1210–1218, Dec. 2001.

71. S. Senini and P. J. Wolfs, Analysis and design of a multiple-loop control system for a hybrid active filter, *IEEE Transactions on Industrial Electronics*, 49(6), 1283–1292, Dec. 2002.

72. G. van Schoor, J. D. van Wyk, and I. S. Shaw, Training and optimization of an artificial neural network controlling a hybrid power filter, *IEEE Transactions on Industrial Electronics*, 50(3), 546–553, June 2003.

73. Luo, C. Tang, Z. K. Shuai, W. Zhao, F. Rong, and K. Zhou, A novel three-phase hybrid active power filter with a series resonance circuit tuned at the fundamental freque, *IEEE Transactions on Industrial Electronics*, 56(7), 2431–2440, July 2009.

74. V. F. Corasaniti, M. B. Barbieri, P. L. Arnera, and M. I. Valla, Hybrid active filter for reactive and harmonics compensation in a distribution network, *IEEE Transactions on Industrial Electronics*, 56(3), 670–677, Mar. 2009.

75. N. He, D. Xu, and L. Huang, The application of particle swarm optimization to passive and hybrid active power filter design, *IEEE Transactions on Industrial Electronics*, 56(8), 2841–2851, Aug. 2009.

76. V. F. Corasaniti, M. B. Barbieri, P. L. Arnera, and M. I. Valla, Hybrid power filter to enhance power quality in a medium-voltage distribution network, *IEEE Transactions on Industrial Electronics*, 56(8), 2885–2893, Aug. 2009.

77. S. Senini and P. J. Wolfs, Hybrid active filter for harmonically unbalanced three phase three wire railway traction loads, *IEEE Transactions on Power Electronics*, 15(4), 702–710, July 2000.

78. D. Detjen, J. Jacobs, R. De Doncker, and H.-G. Mall, A new hybrid filter to damped resonances and compensate harmonic currents in industrial power systems with power factor correction equipment, *IEEE Transactions on Power Electronics*, 16(6), 821–827, Nov. 2001.

79. H. Fujita and H. Akagi, The unified power quality conditioner: The integration of series- and shunt-active filters, *IEEE Transactions on Power Electronics*, 13(2), 315–322, Mar. 1998.

80. A. Elnady, W. El-khattam, and M. M. A. Salama, Mitigation of AC arc furnace voltage flicker using the unified power quality conditioner, in *IEEE Power Engineering Society Meeting*, Chicago, IL, 2002, pp. 735–739.

81. M. Forghani and S. Afsharnia, Online wavelet transform-based control strategy for UPQC control system, *IEEE Transactions on Power Delivery*, 22(1), 481–491, 2007.

82. B. Han, B. Bae, H. Kim, and S. Baek, New configuration of UPQC for medium voltage application, *IEEE Transactions on Power Delivery*, 21(3), 1438–1444, 2006.

83. N. Jayanti, M. Basu, F. Conlon, and G. Kevin, Rating requirements of a unified power quality conditioner (UPQC) for voltage ride through capability enhancement, in *Third IET International Conference on Power Electronics, Machines and Drives*, Dublin, Ireland, 2006, pp. 632–636.

84. A. Jindal, A. Ghosh, and A. Joshi, Interline unified power quality conditioner, *IEEE Transactions on Power Delivery*, 22(1), 364–372, 2007.

85. J. Holtz, Pulsewidth modulation—A survey, *IEEE Transactions on Industrial Electronics*, 39, 410–420, Dec. 1992.

86. J. Holtz, Pulsewidth modulation for electronic power conversion, *Proceedings of the IEEE*, 82(8), 1194–1214, Aug. 1994.

87. G. Amler, A PWM current-source inverter for high quality drives, *EPE Journal*, 1(1), 21–32, July 1991.

88. A. Khambadkone and J. Holtz, Low switching frequency high-power inverter drive based on field-oriented pulsewidth modulation, in *EPE European Conference on Power Electronics and Applications*, Florence, Italy, 1991, pp. 4/672–677.

89. M. K. Mishra and K. Karthikeyan, An investigation on design and switching dynamics of a voltage source inverter to compensate unbalanced and nonlinear loads, *IEEE Transactions on Industrial Electronics*, 56(8), pp. 2802–2810, Aug. 2009.

90. Lavopa, P. Zanchetta, M. Sumner, and F. Cupertino, Real-time estimation of fundamental frequency and harmonics for active shunt power filters in aircraft electrical systems, *IEEE Transactions on Industrial Electronics*, 56(8), 2875–2884, Aug. 2009.

91. W. Lenwari, M. Sumner, and P. Zanchetta, The use of genetic algorithms for the design of resonant compensators for active filters, *IEEE Transactions on Industrial Electronics*, 56(8), 2852–2861, Aug. 2009.

92. M. Sani and S. Filizadeh, An optimized space vector modulation sequence for improved harmonic performance, *IEEE Transactions on Industrial Electronics*, 56(8), 2894–2903, Aug. 2009.

93. R. S. Herrera, P. Salmeron, and H. Kim, Instantaneous reactive power theory applied to active power filter compensation: Different approaches, assessment, and experimental results, *IEEE Transactions on Industrial Electronics*, 55(1), 184–196, Jan. 2008.

94. G. Escobar, P. G. Hernandez-Briones, P. R. Martinez, M. Hernandez-Gomez, and R. E. Torres-Olguin, A repetitive-based controller for the compensation of 6l ± 1 harmonic components, *IEEE Transactions on Industrial Electronics*, 55(8), 3150–3158, Aug. 2008.

95. F. Defay, A. M. Llor, and M. Fadel, A predictive control with flying capacitor balancing of a multicell active power filter, *IEEE Transactions on Industrial Electronics*, 55(9), 3212–3220, Sept. 2008.

96. K. K. Shyu, M. J. Yang, Y. M. Chen, and Y. F. Lin, Model reference adaptive control design for a shunt active-power-filter system, *IEEE Transactions on Industrial Electronics*, 55(1), 97–106, Jan. 2008.

97. K. Drobnic, M. Nemec, D. Nedeljkovic, and V. Ambrozic, Predictive direct control applied to AC drives and active power filter, *IEEE Transactions on Industrial Electronics*, 56(6), 1884–1893, June 2009.

98. B. Singh and J. Solanki, An implementation of an adaptive control algorithm for a three-phase shunt active filter, *IEEE Transactions on Industrial Electronics*, 56(8), 2811–2820, Aug. 2009.

99. F. D. Freijedo, J. Doval-Gandoy, O. Lopez, P. Fernandez-Comesana, and C. Martinez-Penalver, A signal-processing adaptive algorithm for selective current harmonic cancellation in active power filters, *IEEE Transactions on Industrial Electronics*, 56(8), 2829–2840, Aug. 2009.

100. B. Kedjar and K. Al-Haddad, DSP-based implementation of an LQR with integral action for a three-phase three-wire shunt active power filter, *IEEE Transactions on Industrial Electronics*, 56(8), 2821–2828, Aug. 2009.
101. P. Kirawanich and R. M. O'Connell, Fuzzy logic control of an active power line conditioner, *IEEE Transactions on Power Electronics*, 19(6), 1574–1585, Nov. 2004.
102. N. Mendalek, K. Al Haddad, L. A-Dessaint, and F. Fnaiech, Nonlinear control technique to enhance dynamic performance of a shunt active power filter, *IEE Proceedings—Electric Power Applications*, 150(4), 373–379, July 2003.

Index

E

N